Lecture Notes in Artificial Intelligence 12975

Subseries of Lecture Notes in Computer Science

Series Editors

Randy Goebel
University of Alberta, Edmonton, Canada
Yuzuru Tanaka
Hokkaido University, Sapporo, Japan
Wolfgang Wahlster
DFKI and Saarland University, Saarbrücken, Germany

Founding Editor

Jörg Siekmann
DFKI and Saarland University, Saarbrücken, Germany

More information about this subseries at http://www.springer.com/series/1244

Nuria Oliver · Fernando Pérez-Cruz ·
Stefan Kramer · Jesse Read ·
Jose A. Lozano (Eds.)

Machine Learning and Knowledge Discovery in Databases

Research Track

European Conference, ECML PKDD 2021
Bilbao, Spain, September 13–17, 2021
Proceedings, Part I

 Springer

Editors
Nuria Oliver (iD)
ELLIS - The European Laboratory
for Learning and Intelligent Systems
Alicante, Spain

Avda Universidad, San Vicente del Raspeig
Alicante, Spain

Vodafone Institute for Society
and Communications
Berlin, Germany

Data-Pop Alliance
New York, USA

Stefan Kramer
Johannes Gutenberg University of Mainz
Mainz, Germany

Jose A. Lozano (iD)
Basque Center for Applied Mathematics
Bilbao, Spain

Fernando Pérez-Cruz (iD)
ETHZ and EPFL
Zürich, Switzerland

Jesse Read (iD)
École Polytechnique
Palaiseau, France

ISSN 0302-9743   ISSN 1611-3349 (electronic)
Lecture Notes in Artificial Intelligence
ISBN 978-3-030-86485-9   ISBN 978-3-030-86486-6 (eBook)
https://doi.org/10.1007/978-3-030-86486-6

LNCS Sublibrary: SL7 – Artificial Intelligence

This Springer imprint is published by the registered company Springer Nature Switzerland AG
The registered company address is: Gewerbestrasse 11, 6330 Cham, Switzerland

Preface

This edition of the European Conference on Machine Learning and Principles and Practice of Knowledge Discovery in Databases (ECML PKDD 2021) has still been affected by the COVID-19 pandemic. Unfortunately it had to be held online and we could only meet each other virtually. However, the experience gained in the previous edition joined to the knowledge collected from other virtual conferences allowed us to provide an attractive and engaging agenda.

ECML PKDD is an annual conference that provides an international forum for the latest research in all areas related to machine learning and knowledge discovery in databases, including innovative applications. It is the leading European machine learning and data mining conference and builds upon a very successful series of ECML PKDD conferences. Scheduled to take place in Bilbao, Spain, ECML PKDD 2021 was held fully virtually, during September 13–17, 2021. The conference attracted over 1000 participants from all over the world. More generally, the conference received substantial attention from industry through sponsorship, participation, and also the industry track.

The main conference program consisted of presentations of 210 accepted conference papers, 40 papers accepted in the journal track and 4 keynote talks: Jie Tang (Tsinghua University), Susan Athey (Stanford University), Joaquin Quiñonero Candela (Facebook), and Marta Kwiatkowska (University of Oxford). In addition, there were 22 workshops, 8 tutorials, 2 combined workshop-tutorials, the PhD forum, and the discovery challenge. Papers presented during the three main conference days were organized in three different tracks:

- Research Track: research or methodology papers from all areas in machine learning, knowledge discovery, and data mining.
- Applied Data Science Track: papers on novel applications of machine learning, data mining, and knowledge discovery to solve real-world use cases, thereby bridging the gap between practice and current theory.
- Journal Track: papers that were published in special issues of the Springer journals Machine Learning and Data Mining and Knowledge Discovery.

We received a similar number of submissions to last year with 685 and 220 submissions for the Research and Applied Data Science Tracks respectively. We accepted 146 (21%) and 64 (29%) of these. In addition, there were 40 papers from the Journal Track. All in all, the high-quality submissions allowed us to put together an exceptionally rich and exciting program.

The Awards Committee selected research papers that were considered to be of exceptional quality and worthy of special recognition:

- Best (Student) Machine Learning Paper Award: Reparameterized Sampling for Generative Adversarial Networks, by Yifei Wang, Yisen Wang, Jiansheng Yang and Zhouchen Lin.

- First Runner-up (Student) Machine Learning Paper Award: "Continual Learning with Dual Regularizations", by Xuejun Han and Yuhong Guo.
- Best Applied Data Science Paper Award: "Open Data Science to fight COVID-19: Winning the 500k XPRIZE Pandemic Response Challenge", by Miguel Angel Lozano, Oscar Garibo, Eloy Piñol, Miguel Rebollo, Kristina Polotskaya, Miguel Angel Garcia-March, J. Alberto Conejero, Francisco Escolano and Nuria Oliver.
- Best Student Data Mining Paper Award: "Conditional Neural Relational Inference for Interacting Systems", by Joao Candido Ramos, Lionel Blondé, Stéphane Armand and Alexandros Kalousis.
- Test of Time Award for highest-impact paper from ECML PKDD 2011: "Influence and Passivity in Social Media", by Daniel M. Romero, Wojciech Galuba, Sitaram Asur and Bernardo A. Huberman.

We would like to wholeheartedly thank all participants, authors, Program Committee members, area chairs, session chairs, volunteers, co-organizers, and organizers of workshops and tutorials for their contributions that helped make ECML PKDD 2021 a great success. We would also like to thank the ECML PKDD Steering Committee and all sponsors.

September 2021

Jose A. Lozano
Nuria Oliver
Fernando Pérez-Cruz
Stefan Kramer
Jesse Read
Yuxiao Dong
Nicolas Kourtellis
Barbara Hammer

Organization

General Chair

Jose A. Lozano · Basque Center for Applied Mathematics, Spain

Research Track Program Chairs

Nuria Oliver · Vodafone Institute for Society and Communications, Germany, and Data-Pop Alliance, USA
Fernando Pérez-Cruz · Swiss Data Science Center, Switzerland
Stefan Kramer · Johannes Gutenberg Universität Mainz, Germany
Jesse Read · École Polytechnique, France

Applied Data Science Track Program Chairs

Yuxiao Dong · Facebook AI, Seattle, USA
Nicolas Kourtellis · Telefonica Research, Barcelona, Spain
Barbara Hammer · Bielefeld University, Germany

Journal Track Chairs

Sergio Escalera · Universitat de Barcelona, Spain
Heike Trautmann · University of Münster, Germany
Annalisa Appice · Università degli Studi di Bari, Italy
Jose A. Gámez · Universidad de Castilla-La Mancha, Spain

Discovery Challenge Chairs

Paula Brito · Universidade do Porto, Portugal
Dino Ienco · Université Montpellier, France

Workshop and Tutorial Chairs

Alipio Jorge · Universidade do Porto, Portugal
Yun Sing Koh · University of Auckland, New Zealand

Industrial Track Chairs

Miguel Veganzones · Sherpa.ia, Portugal
Sabri Skhiri · EURA NOVA, Belgium

Award Chairs

Myra Spiliopoulou   Otto-von-Guericke-University Magdeburg, Germany
João Gama   University of Porto, Portugal

PhD Forum Chairs

Jeronimo Hernandez   University of Barcelona, Spain
Zahra Ahmadi   Johannes Gutenberg Universität Mainz, Germany

Production, Publicity, and Public Relations Chairs

Sophie Burkhardt   Johannes Gutenberg Universität Mainz, Germany
Julia Sidorova   Universidad Complutense de Madrid, Spain

Local Chairs

Iñaki Inza   University of the Basque Country, Spain
Alexander Mendiburu   University of the Basque Country, Spain
Santiago Mazuelas   Basque Center for Applied Mathematics, Spain
Aritz Pèrez   Basque Center for Applied Mathematics, Spain
Borja Calvo   University of the Basque Country, Spain

Proceedings Chair

Tania Cerquitelli   Politecnico di Torino, Italy

Sponsorship Chair

Santiago Mazuelas   Basque Center for Applied Mathematics, Spain

Web Chairs

Olatz Hernandez Aretxabaleta   Basque Center for Applied Mathematics, Spain
Estíbaliz Gutièrrez   Basque Center for Applied Mathematics, Spain

ECML PKDD Steering Committee

Andrea Passerini   University of Trento, Italy
Francesco Bonchi   ISI Foundation, Italy
Albert Bifet   Télécom ParisTech, France
Sašo Džeroski   Jožef Stefan Institute, Slovenia
Katharina Morik   TU Dortmund, Germany
Arno Siebes   Utrecht University, The Netherlands
Siegfried Nijssen   Université Catholique de Louvain, Belgium

Luís Moreira-Matias	Finiata GmbH, Germany
Alessandra Sala	Shutterstock, Ireland
Georgiana Ifrim	University College Dublin, Ireland
Thomas Gärtner	University of Nottingham, UK
Neil Hurley	University College Dublin, Ireland
Michele Berlingerio	IBM Research, Ireland
Elisa Fromont	Université de Rennes, France
Arno Knobbe	Universiteit Leiden, The Netherlands
Ulf Brefeld	Leuphana Universität Lüneburg, Germany
Andreas Hotho	Julius-Maximilians-Universität Würzburg, Germany
Ira Assent	Aarhus University, Denmark
Kristian Kersting	TU Darmstadt University, Germany
Jefrey Lijffijt	Ghent University, Belgium
Isabel Valera	Saarland University, Germany

Program Committee

Guest Editorial Board, Journal Track

Richard Allmendinger	University of Manchester
Marie Anastacio	Leiden University
Ana Paula Appel	IBM Research Brazil
Dennis Assenmacher	University of Münster
Ira Assent	Aarhus University
Martin Atzmueller	Osnabrueck University
Jaume Bacardit	Newcastle University
Anthony Bagnall	University of East Anglia
Mitra Baratchi	University of Twente
Srikanta Bedathur	IIT Delhi
Alessio Benavoli	CSIS
Viktor Bengs	Paderborn University
Massimo Bilancia	University of Bari "Aldo Moro"
Klemens Böhm	Karlsruhe Institute of Technology
Veronica Bolon Canedo	Universidade da Coruna
Ilaria Bordino	UniCredit R&D
Jakob Bossek	University of Adelaide
Ulf Brefeld	Leuphana Universität Luneburg
Michelangelo Ceci	Universita degli Studi di Bari "Aldo Moro"
Loïc Cerf	Universidade Federal de Minas Gerais
Victor Manuel Cerqueira	University of Porto
Laetitia Chapel	IRISA
Silvia Chiusano	Politecnico di Torino
Roberto Corizzo	American University, Washington D.C.
Marco de Gemmis	Università degli Studi di Bari "Aldo Moro"
Sébastien Destercke	Università degli Studi di Bari "Aldo Moro"
Shridhar Devamane	Visvesvaraya Technological University

Carlotta Domeniconi	George Mason University
Wouter Duivesteijn	Eindhoven University of Technology
Tapio Elomaa	Tampere University of Technology
Hugo Jair Escalante	INAOE
Nicola Fanizzi	Università degli Studi di Bari "Aldo Moro"
Stefano Ferilli	Università degli Studi di Bari "Aldo Moro"
Pedro Ferreira	Universidade de Lisboa
Cesar Ferri	Valencia Polytechnic University
Julia Flores	University of Castilla-La Mancha
Germain Forestier	Université de Haute Alsace
Marco Frasca	University of Milan
Ricardo J. G. B. Campello	University of Newcastle
Esther Galbrun	University of Eastern Finland
João Gama	University of Porto
Paolo Garza	Politecnico di Torino
Pascal Germain	Université Laval
Fabian Gieseke	University of Münster
Josif Grabocka	University of Hildesheim
Gianluigi Greco	University of Calabria
Riccardo Guidotti	University of Pisa
Francesco Gullo	UniCredit
Stephan Günnemann	Technical University of Munich
Tias Guns	Vrije Universiteit Brussel
Antonella Guzzo	University of Calabria
Alexander Hagg	Hochschule Bonn-Rhein-Sieg University of Applied Sciences
Jin-Kao Hao	University of Angers
Daniel Hernández-Lobato	Universidad Autónoma de Madrid
Jose Hernández-Orallo	Universitat Politècnica de València
Martin Holena	Institute of Computer Science, Academy of Sciences of the Czech Republic
Jaakko Hollmén	Aalto University
Dino Ienco	IRSTEA
Georgiana Ifrim	University College Dublin
· Felix Iglesias	TU Wien
Angelo Impedovo	University of Bari "Aldo Moro"
Mahdi Jalili	RMIT University
Nathalie Japkowicz	University of Ottawa
Szymon Jaroszewicz	Institute of Computer Science, Polish Academy of Sciences
Michael Kamp	Monash University
Mehdi Kaytoue	Infologic
Pascal Kerschke	University of Münster
Dragi Kocev	Jozef Stefan Institute
Lars Kotthoff	University of Wyoming
Tipaluck Krityakierne	University of Bern

Peer Kröger	Ludwig Maximilian University of Munich
Meelis Kull	University of Tartu
Michel Lang	TU Dortmund University
Helge Langseth	Norwegian University of Science and Technology
Oswald Lanz	FBK
Mark Last	Ben-Gurion University of the Negev
Kangwook Lee	University of Wisconsin-Madison
Jurica Levatic	IRB Barcelona
Thomar Liebig	TU Dortmund
Hsuan-Tien Lin	National Taiwan University
Marius Lindauer	Leibniz University Hannover
Marco Lippi	University of Modena and Reggio Emilia
Corrado Loglisci	Università degli Studi di Bari
Manuel Lopez-Ibanez	University of Malaga
Nuno Lourenço	University of Coimbra
Claudio Lucchese	Ca' Foscari University of Venice
Brian Mac Namee	University College Dublin
Gjorgji Madjarov	Ss. Cyril and Methodius University
Davide Maiorca	University of Cagliari
Giuseppe Manco	ICAR-CNR
Elena Marchiori	Radboud University
Elio Masciari	Università di Napoli Federico II
Andres R. Masegosa	Norwegian University of Science and Technology
Ernestina Menasalvas	Universidad Politécnica de Madrid
Rosa Meo	University of Torino
Paolo Mignone	University of Bari "Aldo Moro"
Anna Monreale	University of Pisa
Giovanni Montana	University of Warwick
Grègoire Montavon	TU Berlin
Katharina Morik	TU Dortmund
Animesh Mukherjee	Indian Institute of Technology, Kharagpur
Amedeo Napoli	LORIA Nancy
Frank Naumann	University of Adelaide
Thomas Dyhre	Aalborg University
Bruno Ordozgoiti	Aalto University
Rita P. Ribeiro	University of Porto
Pance Panov	Jozef Stefan Institute
Apostolos Papadopoulos	Aristotle University of Thessaloniki
Panagiotis Papapetrou	Stockholm University
Andrea Passerini	University of Trento
Mykola Pechenizkiy	Eindhoven University of Technology
Charlotte Pelletier	Université Bretagne Sud
Ruggero G. Pensa	University of Torino
Nico Piatkowski	TU Dortmund
Dario Piga	IDSIA Dalle Molle Institute for Artificial Intelligence Research - USI/SUPSI

Gianvito Pio	Università degli Studi di Bari "Aldo Moro"
Marc Plantevit	LIRIS - Université Claude Bernard Lyon 1
Marius Popescu	University of Bucharest
Raphael Prager	University of Münster
Mike Preuss	Universiteit Leiden
Jose M. Puerta	Universidad de Castilla-La Mancha
Kai Puolamäki	University of Helsinki
Chedy Raïssi	Inria
Jan Ramon	Inria
Matteo Riondato	Amherst College
Thomas A. Runkler	Siemens Corporate Technology
Antonio Salmerón	University of Almería
Joerg Sander	University of Alberta
Roberto Santana	University of the Basque Country
Michael Schaub	RWTH Aachen
Lars Schmidt-Thieme	University of Hildesheim
Santiago Segui	Universitat de Barcelona
Thomas Seidl	Ludwig-Maximilians-Universitaet Muenchen
Moritz Seiler	University of Münster
Shinichi Shirakawa	Yokohama National University
Jim Smith	University of the West of England
Carlos Soares	University of Porto
Gerasimos Spanakis	Maastricht University
Giancarlo Sperlì	University of Naples Federico II
Myra Spiliopoulou	Otto-von-Guericke-University Magdeburg
Giovanni Stilo	Università degli Studi dell'Aquila
Catalin Stoean	University of Craiova
Mahito Sugiyama	National Institute of Informatics
Nikolaj Tatti	University of Helsinki
Alexandre Termier	Université de Rennes 1
Kevin Tierney	Bielefeld University
Luis Torgo	University of Porto
Roberto Trasarti	CNR Pisa
Sébastien Treguer	Inria
Leonardo Trujillo	Instituto Tecnológico de Tijuana
Ivor Tsang	University of Technology Sydney
Grigorios Tsoumakas	Aristotle University of Thessaloniki
Steffen Udluft	Siemens
Arnaud Vandaele	Université de Mons
Matthijs van Leeuwen	Leiden University
Celine Vens	KU Leuven Kulak
Herna Viktor	University of Ottawa
Marco Virgolin	Centrum Wiskunde & Informatica
Jordi Vitrià	Universitat de Barcelona
Christel Vrain	LIFO – University of Orléans
Jilles Vreeken	Helmholtz Center for Information Security

Willem Waegeman	Ghent University
David Walker	University of Plymouth
Hao Wang	Leiden University
Elizabeth F. Wanner	CEFET
Tu Wei-Wei	4paradigm
Pascal Welke	University of Bonn
Marcel Wever	Paderborn University
Man Leung Wong	Lingnan University
Stefan Wrobel	Fraunhofer IAIS, University of Bonn
Zheng Ying	Inria
Guoxian Yu	Shandong University
Xiang Zhang	Harvard University
Ye Zhu	Deakin University
Arthur Zimek	University of Southern Denmark
Albrecht Zimmermann	Université Caen Normandie
Marinka Zitnik	Harvard University

Area Chairs, Research Track

Fabrizio Angiulli	University of Calabria
Ricardo Baeza-Yates	Universitat Pompeu Fabra
Roberto Bayardo	Google
Bettina Berendt	Katholieke Universiteit Leuven
Philipp Berens	University of Tübingen
Michael Berthold	University of Konstanz
Hendrik Blockeel	Katholieke Universiteit Leuven
Juergen Branke	University of Warwick
Ulf Brefeld	Leuphana University Lüneburg
Toon Calders	Universiteit Antwerpen
Michelangelo Ccci	Università degli Studi di Bari "Aldo Moro"
Duen Horng Chau	Georgia Institute of Technology
Nicolas Courty	Université Bretagne Sud, IRISA Research Institute Computer and Systems Aléatoires
Bruno Cremilleux	Université de Caen Normandie
Philippe Cudre-Mauroux	University of Fribourg
James Cussens	University of Bristol
Jesse Davis	Katholieke Universiteit Leuven
Bob Durrant	University of Waikato
Tapio Elomaa	Tampere University
Johannes Fürnkranz	Johannes Kepler University Linz
Eibe Frank	University of Waikato
Elisa Fromont	Université de Rennes 1
Stephan Günnemann	Technical University of Munich
Patrick Gallinari	LIP6 - University of Paris
Joao Gama	University of Porto
Przemyslaw Grabowicz	University of Massachusetts, Amherst

Eyke Hüllermeier	Paderborn University
Allan Hanbury	Vienna University of Technology
Daniel Hernández-Lobato	Universidad Autónoma de Madrid
José Hernández-Orallo	Universitat Politècnica de València
Andreas Hotho	University of Wuerzburg
Inaki Inza	University of the Basque Country
Marius Kloft	TU Kaiserslautern
Arno Knobbe	Universiteit Leiden
Lars Kotthoff	University of Wyoming
Danica Kragic	KTH Royal Institute of Technology
Sébastien Lefèvre	Université Bretagne Sud
Bruno Lepri	FBK-Irst
Patrick Loiseau	Inria and Ecole Polytechnique
Jorg Lucke	University of Oldenburg
Fragkiskos Malliaros	Paris-Saclay University, CentraleSupelec, and Inria
Giuseppe Manco	ICAR-CNR
Dunja Mladenic	Jozef Stefan Institute
Katharina Morik	TU Dortmund
Sriraam Natarajan	Indiana University Bloomington
Siegfried Nijssen	Université catholique de Louvain
Andrea Passerini	University of Trento
Mykola Pechenizkiy	Eindhoven University of Technology
Jaakko Peltonen	Aalto University and University of Tampere
Marian-Andrei Rizoiu	University of Technology Sydney
Céline Robardet	INSA Lyon
Maja Rudolph	Bosch
Lars Schmidt-Thieme	University of Hildesheim
Thomas Seidl	Ludwig-Maximilians-Universität München
Arno Siebes	Utrecht University
Myra Spiliopoulou	Otto-von-Guericke-University Magdeburg
Yizhou Sun	University of California, Los Angeles
Einoshin Suzuki	Kyushu University
Jie Tang	Tsinghua University
Ke Tang	Southern University of Science and Technology
Marc Tommasi	University of Lille
Isabel Valera	Saarland University
Celine Vens	KU Leuven Kulak
Christel Vrain	LIFO - University of Orléans
Jilles Vreeken	Helmholtz Center for Information Security
Willem Waegeman	Ghent University
Stefan Wrobel	Fraunhofer IAIS, University of Bonn
Min-Ling Zhang	Southeast University

Area Chairs, Applied Data Science Track

Francesco Calabrese	Vodafone
Michelangelo Ceci	Università degli Studi di Bari "Aldo Moro"
Gianmarco De Francisci Morales	ISI Foundation
Tom Diethe	Amazon
Johannes Fründkranz	Johannes Kepler University Linz
Han Fang	Facebook
Faisal Farooq	Qatar Computing Research Institute
Rayid Ghani	Carnegie Mellon Univiersity
Francesco Gullo	UniCredit
Xiangnan He	University of Science and Technology of China
Georgiana Ifrim	University College Dublin
Thorsten Jungeblut	Bielefeld University of Applied Sciences
John A. Lee	Université catholique de Louvain
Ilias Leontiadis	Samsung AI
Viktor Losing	Honda Research Institute Europe
Yin Lou	Ant Group
Gabor Melli	Sony PlayStation
Luis Moreira-Matias	University of Porto
Nicolò Navarin	University of Padova
Benjamin Paaßen	German Research Center for Artificial Intelligence
Kitsuchart Pasupa	King Mongkut's Institute of Technology Ladkrabang
Mykola Pechenizkiy	Eindhoven University of Technology
Julien Perez	Naver Labs Europe
Fabio Pinelli	IMT Lucca
Zhaochun Ren	Shandong University
Sascha Saralajew	Porsche AG
Fabrizio Silvestri	Facebook
Sinong Wang	Facebook AI
Xing Xie	Microsoft Research Asia
Jian Xu	Citadel
Jing Zhang	Renmin University of China

Program Committee Members, Research Track

Hanno Ackermann	Leibniz University Hannover
Linara Adilova	Fraunhofer IAIS
Zahra Ahmadi	Johannes Gutenberg University
Cuneyt Gurcan Akcora	University of Manitoba
Omer Deniz Akyildiz	University of Warwick
Carlos M. Alaíz Gudín	Universidad Autónoma de Madrid
Mohamed Alami	Ecole Polytechnique
Chehbourne Abdullah Alchihabi	Carleton University
Pegah Alizadeh	University of Caen Normandy

Reem Alotaibi	King Abdulaziz University
Massih-Reza Amini	Université Grenoble Alpes
Shin Ando	Tokyo University of Science
Thiago Andrade	INESC TEC
Kimon Antonakopoulos	Inria
Alessandro Antonucci	IDSIA
Muhammad Umer Anwaar	Technical University of Munich
Eva Armengol	IIIA-SIC
Dennis Assenmacher	University of Münster
Matthias Aßenmacher	Ludwig-Maximilians-Universität München
Martin Atzmueller	Osnabrueck University
Behrouz Babaki	Polytechnique Montreal
Rohit Babbar	Aalto University
Elena Baralis	Politecnico di Torino
Mitra Baratchi	University of Twente
Christian Bauckhage	University of Bonn, Fraunhofer IAIS
Martin Becker	University of Würzburg
Jessa Bekker	Katholieke Universiteit Leuven
Colin Bellinger	National Research Council of Canada
Khalid Benabdeslem	LIRIS Laboratory, Claude Bernard University Lyon I
Diana Benavides-Prado	Auckland University of Technology
Anes Bendimerad	LIRIS
Christoph Bergmeir	University of Granada
Alexander Binder	UiO
Aleksandar Bojchevski	Technical University of Munich
Ahcène Boubekki	UiT Arctic University of Norway
Paula Branco	EECS University of Ottawa
Tanya Braun	University of Lübeck
Katharina Breininger	Friedrich-Alexander-Universität Erlangen Nürnberg
Wieland Brendel	University of Tübingen
John Burden	University of Cambridge
Sophie Burkhardt	TU Kaiserslautern
Sebastian Buschjäger	TU Dortmund
Borja Calvo	University of the Basque Country
Stephane Canu	LITIS, INSA de Rouen
Cornelia Caragea	University of Illinois at Chicago
Paula Carroll	University College Dublin
Giuseppe Casalicchio	Ludwig Maximilian University of Munich
Bogdan Cautis	Paris-Saclay University
Rémy Cazabet	Université de Lyon
Josu Ceberio	University of the Basque Country
Peggy Cellier	IRISA/INSA Rennes
Mattia Cerrato	Università degli Studi di Torino
Ricardo Cerri	Federal University of Sao Carlos
Alessandra Cervone	Amazon
Ayman Chaouki	Institut Mines-Télécom

Paco Charte	Universidad de Jaén
Rita Chattopadhyay	Intel Corporation
Vaggos Chatziafratis	Stanford University
Tianyi Chen	Zhejiang University City College
Yuzhou Chen	Southern Methodist University
Yiu-Ming Cheung	Hong Kong Baptist University
Anshuman Chhabra	University of California, Davis
Ting-Wu Chin	Carnegie Mellon University
Oana Cocarascu	King's College London
Lidia Contreras-Ochando	Universitat Politècnica de València
Roberto Corizzo	American University
Anna Helena Reali Costa	Universidade de São Paulo
Fabrizio Costa	University of Exeter
Gustavo De Assis Costa	Instituto Federal de Educação, Ciência e Tecnologia de Goiás
Bertrand Cuissart	GREYC
Thi-Bich-Hanh Dao	University of Orleans
Mayukh Das	Microsoft Research Lab
Padraig Davidson	Universität Würzburg
Paul Davidsson	Malmö University
Gwendoline De Bie	ENS
Tijl De Bie	Ghent University
Andre de Carvalho	Universidade de São Paulo
Orphée De Clercq	Ghent University
Alper Demir	İzmir University of Economics
Nicola Di Mauro	Università degli Studi di Bari "Aldo Moro"
Yao-Xiang Ding	Nanjing University
Carola Doerr	Sorbonne University
Boxiang Dong	Montclair State University
Ruihai Dong	University College Dublin
Xin Du	Eindhoven University of Technology
Stefan Duffner	LIRIS
Wouter Duivesteijn	Eindhoven University of Technology
Audrey Durand	McGill University
Inês Dutra	University of Porto
Saso Dzeroski	Jozef Stefan Institute
Hamid Eghbalzadeh	Johannes Kepler University
Dominik Endres	University of Marburg
Roberto Esposito	Università degli Studi di Torino
Samuel G. Fadel	Universidade Estadual de Campinas
Xiuyi Fan	Imperial College London
Hadi Fanaee-T.	Halmstad University
Elaine Faria	Federal University of Uberlandia
Fabio Fassetti	University of Calabria
Kilian Fatras	Inria
Ad Feelders	Utrecht University

Songhe Feng	Beijing Jiaotong University
Àngela Fernández-Pascual	Universidad Autónoma de Madrid
Daniel Fernández-Sánchez	Universidad Autónoma de Madrid
Sofia Fernandes	University of Aveiro
Cesar Ferri	Universitat Politécnica de Valéncia
Rémi Flamary	École Polytechnique
Michael Flynn	University of East Anglia
Germain Forestier	Université de Haute Alsace
Kary Främling	Umeå University
Benoît Frénay	Université de Namur
Vincent Francois	University of Amsterdam
Emilia Gómez	Joint Research Centre - European Commission
Luis Galárraga	Inria
Esther Galbrun	University of Eastern Finland
Claudio Gallicchio	University of Pisa
Jochen Garcke	University of Bonn
Clément Gautrais	KU Leuven
Yulia Gel	University of Texas at Dallas and University of Waterloo
Pierre Geurts	University of Liège
Amirata Ghorbani	Stanford University
Heitor Murilo Gomes	University of Waikato
Chen Gong	Shanghai Jiao Tong University
Bedartha Goswami	University of Tübingen
Henry Gouk	University of Edinburgh
James Goulding	University of Nottingham
Antoine Gourru	Université Lumière Lyon 2
Massimo Guarascio	ICAR-CNR
Riccardo Guidotti	University of Pisa
Ekta Gujral	University of California, Riverside
Francesco Gullo	UniCredit
Tias Guns	Vrije Universiteit Brussel
Thomas Guyet	Institut Agro, IRISA
Tom Hanika	University of Kassel
Valentin Hartmann	Ecole Polytechnique Fédérale de Lausanne
Marwan Hassani	Eindhoven University of Technology
Jukka Heikkonen	University of Turku
Fredrik Heintz	Linköping University
Sibylle Hess	TU Eindhoven
Jaakko Hollmén	Aalto University
Tamas Horvath	University of Bonn, Fraunhofer IAIS
Binbin Hu	Ant Group
Hong Huang	UGoe
Georgiana Ifrim	University College Dublin
Angelo Impedovo	Università degli studi di Bari "Aldo Moro"

Nathalie Japkowicz	American University
Szymon Jaroszewicz	Institute of Computer Science, Polish Academy of Sciences
Saumya Jetley	Inria
Binbin Jia	Southeast University
Xiuyi Jia	School of Computer Science and Technology, Nanjing University of Science and Technology
Yuheng Jia	City University of Hong Kong
Siyang Jiang	National Taiwan University
Priyadarshini Kumari	IIT Bombay
Ata Kaban	University of Birmingham
Tomasz Kajdanowicz	Wroclaw University of Technology
Vana Kalogeraki	Athens University of Economics and Business
Toshihiro Kamishima	National Institute of Advanced Industrial Science and Technology
Michael Kamp	Monash University
Bo Kang	Ghent University
Dimitrios Karapiperis	Hellenic Open University
Panagiotis Karras	Aarhus University
George Karypis	University of Minnesota
Mark Keane	University College Dublin
Kristian Kersting	TU Darmstadt
Masahiro Kimura	Ryukoku University
Jiri Klema	Czech Technical University
Dragi Kocev	Jozef Stefan Institute
Masahiro Kohjima	NTT
Lukasz Korycki	Virginia Commonwealth University
Peer Kröger	Ludwig Maximilian University of Münich
Anna Krause	University of Würzburg
Bartosz Krawczyk	Virginia Commonwealth University
Georg Krempl	Utrecht University
Meelis Kull	University of Tartu
Vladimir Kuzmanovski	Aalto University
Ariel Kwiatkowski	Ecole Polytechnique
Emanuele La Malfa	University of Oxford
Beatriz López	University of Girona
Preethi Lahoti	Aalto University
Ichraf Lahouli	Euranova
Niklas Lavesson	Jönköping University
Aonghus Lawlor	University College Dublin
Jeongmin Lee	University of Pittsburgh
Daniel Lemire	LICEF Research Center and Université du Québec
Florian Lemmerich	University of Passau
Elisabeth Lex	Graz University of Technology
Jiani Li	Vanderbilt University
Rui Li	Inspur Group
Wentong Liao	Lebniz University Hannover

Jiayin Lin	University of Wollongong
Rudolf Lioutikov	UT Austin
Marco Lippi	University of Modena and Reggio Emilia
Suzanne Little	Dublin City University
Shengcai Liu	University of Science and Technology of China
Shenghua Liu	Institute of Computing Technology, Chinese Academy of Sciences
Philipp Liznerski	Technische Universität Kaiserslautern
Corrado Loglisci	Università degli Studi di Bari "Aldo Moro"
Ting Long	Shanghai Jiaotong University
Tsai-Ching Lu	HRL Laboratories
Yunpu Ma	Siemens AG
Zichen Ma	The Chinese University of Hong Kong
Sara Madeira	Universidade de Lisboa
Simona Maggio	Dataiku
Sara Magliacane	IBM
Sebastian Mair	Leuphana University Lüneburg
Lorenzo Malandri	University of Milan Bicocca
Donato Malerba	Università degli Studi di Bari "Aldo Moro"
Pekka Malo	Aalto University
Robin Manhaeve	KU Leuven
Silviu Maniu	Université Paris-Sud
Giuseppe Marra	KU Leuven
Fernando Martínez-Plumed	Joint Research Centre - European Commission
Alexander Marx	Max Plank Institue for Informatics and Saarland University
Florent Masseglia	Inria
Tetsu Matsukawa	Kyushu University
Wolfgang Mayer	University of South Australia
Santiago Mazuelas	Basque center for Applied Mathematics
Stefano Melacci	University of Siena
Ernestina Menasalvas	Universidad Politécnica de Madrid
Rosa Meo	Università degli Studi di Torino
Alberto Maria Metelli	Politecnico di Milano
Saskia Metzler	Max Planck Institute for Informatics
Alessio Micheli	University of Pisa
Paolo Mignone	Università degli studi di Bari "Aldo Moro"
Matej Mihelčić	University of Zagreb
Decebal Constantin Mocanu	University of Twente
Nuno Moniz	INESC TEC and University of Porto
Carlos Monserrat	Universitat Politécnica de Valéncia
Corrado Monti	ISI Foundation
Jacob Montiel	University of Waikato
Ahmadreza Mosallanezhad	Arizona State University
Tanmoy Mukherjee	University of Tennessee
Martin Mundt	Goethe University

Mohamed Nadif	Université de Paris
Omer Nagar	Bar Ilan University
Felipe Kenji Nakano	Katholieke Universiteit Leuven
Mirco Nanni	KDD-Lab ISTI-CNR Pisa
Apurva Narayan	University of Waterloo
Nicolò Navarin	University of Padova
Benjamin Negrevergne	Paris Dauphine University
Hurley Neil	University College Dublin
Stefan Neumann	University of Vienna
Ngoc-Tri Ngo	The University of Danang - University of Science and Technology
Dai Nguyen	Monash University
Eirini Ntoutsi	Free University Berlin
Andrea Nuernberger	Otto-von-Guericke-Universität Magdeburg
Pablo Olmos	University Carlos III
James O'Neill	University of Liverpool
Barry O'Sullivan	University College Cork
Rita P. Ribeiro	University of Porto
Aritz Pèrez	Basque Center for Applied Mathematics
Joao Palotti	Qatar Computing Research Institute
Guansong Pang	University of Adelaide
Pance Panov	Jozef Stefan Institute
Evangelos Papalexakis	University of California, Riverside
Haekyu Park	Georgia Institute of Technology
Sudipta Paul	Umeå University
Yulong Pei	Eindhoven University of Technology
Charlotte Pelletier	Université Bretagne Sud
Ruggero G. Pensa	University of Torino
Bryan Perozzi	Google
Nathanael Perraudin	ETH Zurich
Lukas Pfahler	TU Dortmund
Bastian Pfeifer	Medical University of Graz
Nico Piatkowski	TU Dortmund
Robert Pienta	Georgia Institute of Technology
Fábio Pinto	Faculdade de Economia do Porto
Gianvito Pio	University of Bari "Aldo Moro"
Giuseppe Pirrò	Sapienza University of Rome
Claudia Plant	University of Vienna
Marc Plantevit	LIRIS - Universitè Claude Bernard Lyon 1
Amit Portnoy	Ben Gurion University
Melanie Pradier	Harvard University
Paul Prasse	University of Potsdam
Philippe Preux	Inria, LIFL, Universitè de Lille
Ricardo Prudencio	Federal University of Pernambuco
Zhou Qifei	Peking University
Erik Quaeghebeur	TU Eindhoven

Tahrima Rahman	University of Texas at Dallas
Herilalaina Rakotoarison	Inria
Alexander Rakowski	Hasso Plattner Institute
María José Ramírez	Universitat Politècnica de València
Visvanathan Ramesh	Goethe University
Jan Ramon	Inria
Huzefa Rangwala	George Mason University
Aleksandra Rashkovska	Jožef Stefan Institute
Joe Redshaw	University of Nottingham
Matthias Renz	Christian-Albrechts-Universität zu Kiel
Matteo Riondato	Amherst College
Ettore Ritacco	ICAR-CNR
Mateus Riva	Télécom ParisTech
Antonio Rivera	Universidad Politécnica de Madrid
Marko Robnik-Sikonja	University of Ljubljana
Simon Rodriguez Santana	Institute of Mathematical Sciences (ICMAT-CSIC)
Mohammad Rostami	University of Southern California
Céline Rouveirol	Laboratoire LIPN-UMR CNRS
Jože Rožanec	Jožef Stefan Institute
Peter Rubbens	Flanders Marine Institute
David Ruegamer	LMU Munich
Salvatore Ruggieri	Università di Pisa
Francisco Ruiz	DeepMind
Anne Sabourin	Télécom ParisTech
Tapio Salakoski	University of Turku
Pablo Sanchez-Martin	Max Planck Institute for Intelligent Systems
Emanuele Sansone	KU Leuven
Yucel Saygin	Sabanci University
Patrick Schäfer	Humboldt Universität zu Berlin
Pierre Schaus	UCLouvain
Ute Schmid	University of Bamberg
Sebastian Schmoll	Ludwig Maximilian University of Munich
Marc Schoenauer	Inria
Matthias Schubert	Ludwig Maximilian University of Munich
Marian Scuturici	LIRIS-INSA de Lyon
Junming Shao	University of Science and Technology of China
Manali Sharma	Samsung Semiconductor Inc.
Abdul Saboor Sheikh	Zalando Research
Jacquelyn Shelton	Hong Kong Polytechnic University
Feihong Shen	Jilin University
Gavin Smith	University of Nottingham
Kma Solaiman	Purdue University
Arnaud Soulet	Université François Rabelais Tours
Alessandro Sperduti	University of Padua
Giovanni Stilo	Università degli Studi dell'Aquila
Michiel Stock	Ghent University

Lech Szymanski	University of Otago
Shazia Tabassum	University of Porto
Andrea Tagarelli	University of Calabria
Acar Tamersoy	NortonLifeLock Research Group
Chang Wei Tan	Monash University
Sasu Tarkoma	University of Helsinki
Bouadi Tassadit	IRISA-Université Rennes 1
Nikolaj Tatti	University of Helsinki
Maryam Tavakol	Eindhoven University of Technology
Pooya Tavallali	University of California, Los Angeles
Maguelonne Teisseire	Irstea - UMR Tetis
Alexandre Termier	Université de Rennes 1
Stefano Teso	University of Trento
Janek Thomas	Fraunhofer Institute for Integrated Circuits IIS
Alessandro Tibo	Aalborg University
Sofia Triantafillou	University of Pittsburgh
Grigorios Tsoumakas	Aristotle University of Thessaloniki
Peter van der Putten	LIACS, Leiden University and Pegasystems
Elia Van Wolputte	KU Leuven
Robert A. Vandermeulen	Technische Universität Berlin
Fabio Vandin	University of Padova
Filipe Veiga	Massachusetts Institute of Technology
Bruno Veloso	Universidade Portucalense and LIAAD - INESC TEC
Sebastián Ventura	University of Cordoba
Rosana Veroneze	UNICAMP
Herna Viktor	University of Ottawa
João Vinagre	INESC TEC
Huaiyu Wan	Beijing Jiaotong University
Beilun Wang	Southeast University
Hu Wang	University of Adelaide
Lun Wang	University of California, Berkeley
Yu Wang	Peking University
Zijie J. Wang	Georgia Tech
Tong Wei	Nanjing University
Pascal Welke	University of Bonn
Joerg Wicker	University of Auckland
Moritz Wolter	University of Bonn
Ning Xu	Southeast University
Akihiro Yamaguchi	Toshiba Corporation
Haitian Yang	Institute of Information Engineering, Chinese Academy of Sciences
Yang Yang	Nanjing University
Zhuang Yang	Sun Yat-sen University
Helen Yannakoudakis	King's College London
Heng Yao	Tongji University
Han-Jia Ye	Nanjing University

Kristina Yordanova	University of Rostock
Tetsuya Yoshida	Nara Women's University
Guoxian Yu	Shandong University, China
Sha Yuan	Tsinghua University
Valentina Zantedeschi	INSA Lyon
Albin Zehe	University of Würzburg
Bob Zhang	University of Macau
Teng Zhang	Huazhong University of Science and Technology
Liang Zhao	University of São Paulo
Bingxin Zhou	University of Sydney
Kenny Zhu	Shanghai Jiao Tong University
Yanqiao Zhu	Institute of Automation, Chinese Academy of Sciences
Arthur Zimek	University of Southern Denmark
Albrecht Zimmermann	Université Caen Normandie
Indre Zliobaite	University of Helsinki
Markus Zopf	NEC Labs Europe

Program Committee Members, Applied Data Science Track

Mahdi Abolghasemi	Monash University
Evrim Acar	Simula Research Lab
Deepak Ajwani	University College Dublin
Pegah Alizadeh	University of Caen Normandy
Jean-Marc Andreoli	Naver Labs Europe
Giorgio Angelotti	ISAE Supaero
Stefanos Antaris	KTH Royal Institute of Technology
Xiang Ao	Institute of Computing Technology, Chinese Academy of Sciences
Yusuf Arslan	University of Luxembourg
Cristian Axenie	Huawei European Research Center
Hanane Azzag	Université Sorbonne Paris Nord
Pedro Baiz	Imperial College London
Idir Benouaret	CNRS, Université Grenoble Alpes
Laurent Besacier	Laboratoire d'Informatique de Grenoble
Antonio Bevilacqua	Insight Centre for Data Analytics
Adrien Bibal	University of Namur
Wu Bin	Zhengzhou University
Patrick Blöbaum	Amazon
Pavel Blinov	Sber Artificial Intelligence Laboratory
Ludovico Boratto	University of Cagliari
Stefano Bortoli	Huawei Technologies Duesseldorf
Zekun Cai	University of Tokyo
Nicolas Carrara	University of Toronto
John Cartlidge	University of Bristol
Oded Cats	Delft University of Technology
Tania Cerquitelli	Politecnico di Torino

Prithwish Chakraborty	IBM
Rita Chattopadhyay	Intel Corp.
Keru Chen	GrabTaxi Pte Ltd.
Liang Chen	Sun Yat-sen University
Zhiyong Cheng	Shandong Artificial Intelligence Institute
Silvia Chiusano	Politecnico di Torino
Minqi Chong	Citadel
Jeremie Clos	University of Nottingham
J. Albert Conejero Casares	Universitat Politécnica de Vaécia
Evan Crothers	University of Ottawa
Henggang Cui	Uber ATG
Tiago Cunha	University of Porto
Padraig Cunningham	University College Dublin
Eustache Diemert	CRITEO Research
Nat Dilokthanakul	Vidyasirimedhi Institute of Science and Technology
Daizong Ding	Fudan University
Kaize Ding	ASU
Michele Donini	Amazon
Lukas Ewecker	Porsche AG
Zipei Fan	University of Tokyo
Bojing Feng	National Laboratory of Pattern Recognition, Institute of Automation, Chinese Academy of Science
Flavio Figueiredo	Universidade Federal de Minas Gerais
Blaz Fortuna	Qlector d.o.o.
Zuohui Fu	Rutgers University
Fabio Fumarola	University of Bari "Aldo Moro"
Chen Gao	Tsinghua University
Luis Garcia	University of Brasília
Cinmayii Garillos-Manliguez	University of the Philippines Mindanao
Kiran Garimella	Aalto University
Etienne Goffinet	Laboratoire LIPN-UMR CNRS
Michael Granitzer	University of Passau
Xinyu Guan	Xi'an Jiaotong University
Thomas Guyet	Institut Agro, IRISA
Massinissa Hamidi	Laboratoire LIPN-UMR CNRS
Junheng Hao	University of California, Los Angeles
Martina Hasenjaeger	Honda Research Institute Europe GmbH
Lars Holdijk	University of Amsterdam
Chao Huang	University of Notre Dame
Guanjie Huang	Penn State University
Hong Huang	UGoe
Yiran Huang	TECO
Madiha Ijaz	IBM
Roberto Interdonato	CIRAD - UMR TETIS
Omid Isfahani Alamdari	University of Pisa

Guillaume Jacquet	JRC
Nathalie Japkowicz	American University
Shaoxiong Ji	Aalto University
Nan Jiang	Purdue University
Renhe Jiang	University of Tokyo
Song Jiang	University of California, Los Angeles
Adan Jose-Garcia	University of Exeter
Jihed Khiari	Johannes Kepler Universität
Hyunju Kim	KAIST
Tomas Kliegr	University of Economics
Yun Sing Koh	University of Auckland
Pawan Kumar	IIIT, Hyderabad
Chandresh Kumar Maurya	CSE, IIT Indore
Thach Le Nguyen	The Insight Centre for Data Analytics
Mustapha Lebbah	Université Paris 13, LIPN-CNRS
Dongman Lee	Korea Advanced Institute of Science and Technology
Rui Li	Sony
Xiaoting Li	Pennsylvania State University
Zeyu Li	University of California, Los Angeles
Defu Lian	University of Science and Technology of China
Jiayin Lin	University of Wollongong
Jason Lines	University of East Anglia
Bowen Liu	Stanford University
Pedro Henrique Luz de Araujo	University of Brasilia
Fenglong Ma	Pennsylvania State University
Brian Mac Namee	University College Dublin
Manchit Madan	Myntra
Ajay Mahimkar	AT&T Labs
Domenico Mandaglio	Università della Calabria
Koji Maruhashi	Fujitsu Laboratories Ltd.
Sarah Masud	LCS2, IIIT-D
Eric Meissner	University of Cambridge
João Mendes-Moreira	INESC TEC
Chuan Meng	Shandong University
Fabio Mercorio	University of Milano-Bicocca
Angela Meyer	Bern University of Applied Sciences
Congcong Miao	Tsinghua University
Stéphane Moreau	Université de Sherbrooke
Koyel Mukherjee	IBM Research India
Fabricio Murai	Universidade Federal de Minas Gerais
Taichi Murayama	NAIST
Philip Nadler	Imperial College London
Franco Maria Nardini	ISTI-CNR
Ngoc-Tri Ngo	The University of Danang - University of Science and Technology

Anna Nguyen	Karlsruhe Institute of Technology
Hao Niu	KDDI Research, Inc.
Inna Novalija	Jožef Stefan Institute
Tsuyosh Okita	Kyushu Institute of Technology
Aoma Osmani	LIPN-UMR CNRS 7030, Université Paris 13
Latifa Oukhellou	IFSTTAR
Andrei Paleyes	University of Cambridge
Chanyoung Park	KAIST
Juan Manuel Parrilla Gutierrez	University of Glasgow
Luca Pasa	Università degli Studi Di Padova
Pedro Pereira Rodrigues	University of Porto
Miquel Perelló-Nieto	University of Bristol
Beatrice Perez	Dartmouth College
Alan Perotti	ISI Foundation
Mirko Polato	University of Padua
Giovanni Ponti	ENEA
Nicolas Posocco	Eura Nova
Cedric Pradalier	GeorgiaTech Lorraine
Giulia Preti	ISI Foundation
A. A. A. Qahtan	Utrecht University
Chuan Qin	University of Science and Technology of China
Dimitrios Rafailidis	University of Thessaly
Cyril Ray	Arts et Metiers Institute of Technology, Ecole Navale, IRENav
Wolfgang Reif	University of Augsburg
Kit Rodolfa	Carnegie Mellon University
Christophe Rodrigues	Pôle Universitaire Léonard de Vinci
Natali Ruchansky	Netflix
Hajer Salem	AUDENSIEL
Parinya Sanguansat	Panyapiwat Institute of Management
Atul Saroop	Amazon
Alexander Schiendorfer	Technische Hochschule Ingolstadt
Peter Schlicht	Volkswagen
Jens Schreiber	University of Kassel
Alexander Schulz	Bielefeld University
Andrea Schwung	FH SWF
Edoardo Serra	Boise State University
Lorenzo Severini	UniCredit
Ammar Shaker	Paderborn University
Jiaming Shen	University of Illinois at Urbana-Champaign
Rongye Shi	Columbia University
Wang Siyu	Southwestern University of Finance and Economics
Hao Song	University of Bristol
Francesca Spezzano	Boise State University
Simon Stieber	University of Augsburg

Sponsors

Invited Talks Abstracts

WuDao: Pretrain the World

Jie Tang

Tsinghua University, Beijing, China

Abstract. Large-scale pretrained model on web texts have substantially advanced the state of the art in various AI tasks, such as natural language understanding and text generation, and image processing, multimodal modeling. The downstream task performances have also constantly increased in the past few years. In this talk, I will first go through three families: augoregressive models (e.g., GPT), autoencoding models (e.g., BERT), and encoder-decoder models. Then, I will introduce China's first homegrown super-scale intelligent model system, with the goal of building an ultra-large-scale cognitive-oriented pretraining model to focus on essential problems in general artificial intelligence from a cognitive perspective. In particular, as an example, I will elaborate a novel pretraining framework GLM (General Language Model) to address this challenge. GLM has three major benefits: (1) it performs well on classification, unconditional generation, and conditional generation tasks with one single pretrained model; (2) it outperforms BERT-like models on classification due to improved pretrain-finetune consistency; (3) it naturally handles variable-length blank filling which is crucial for many downstream tasks. Empirically, GLM substantially outperforms BERT on the SuperGLUE natural language understanding benchmark with the same amount of pre-training data.

Bio: Jie Tang is a Professor and the Associate Chair of the Department of Computer Science at Tsinghua University. He is a Fellow of the IEEE. His interests include artificial intelligence, data mining, social networks, and machine learning. He served as General Co-Chair of WWW'23, and PC Co-Chair of WWW'21, CIKM'16, WSDM'15, and EiC of IEEE T. on Big Data and AI Open J. He leads the project AMiner.org, an AI-enabled research network analysis system, which has attracted more than 20 million users from 220 countries/regions in the world. He was honored with the SIGKDD Test-of-Time Award, the UK Royal Society-Newton Advanced Fellowship Award, NSFC for Distinguished Young Scholar, and KDD'18 Service Award.

The Value of Data for Personalization

Susan Athey

Stanford Graduate School of Business, Stanford, California

Abstract. This talk will present methods for assessing the economic value of data in specific contexts, and will analyze the value of different types of data in the context of several empirical applications.

Bio: Susan Athey is the Economics of Technology Professor at Stanford Graduate School of Business. She received her bachelor's degree from Duke University and her PhD from Stanford, and she holds an honorary doctorate from Duke University. She previously taught at the economics departments at MIT, Stanford and Harvard. She is an elected member of the National Academy of Science, and is the recipient of the John Bates Clark Medal, awarded by the American Economics Association to the economist under 40 who has made the greatest contributions to thought and knowledge. Her current research focuses on the economics of digitization, marketplace design, and the intersection of econometrics and machine learning. She has worked on several application areas, including timber auctions, internet search, online advertising, the news media, and the application of digital technology to social impact applications. As one of the first "tech economists," she served as consulting chief economist for Microsoft Corporation for six years, and now serves on the boards of Expedia, Lending Club, Rover, Turo, and Ripple, as well as non-profit Innovations for Poverty Action. She also serves as a long-term advisor to the British Columbia Ministry of Forests, helping architect and implement their auction-based pricing system. She is the founding director of the Golub Capital Social Impact Lab at Stanford GSB, and associate director of the Stanford Institute for Human-Centered Artificial Intelligence.

AI Fairness in Practice

Joaquin Quiñonero Candela

Facebook

Abstract. In this talk I will share learnings from my journey from deploying ML at Facebook scale to understanding questions of fairness in AI. I will use examples to illustrate how there is not a single definition of AI fairness, but several ones that are in contradiction and that correspond to different moral interpretations of fairness. AI fairness is a process, and it's not primarily an AI issue. It therefore requires a multidisciplinary approach.

Bio: Joaquin Quiñonero Candela leads the technical strategy for Responsible AI at Facebook, including areas like fairness and inclusiveness, robustness, privacy, transparency and accountability. As part of this focus, he serves on the Board of Directors of the Partnership on AI, an organization interested in the societal consequences of artificial intelligence, and is a member of the Spanish Government's Advisory Board on Artificial Intelligence. Before this he built the AML (Applied Machine Learning) team at Facebook, driving product impact at scale through applied research in machine learning, language understanding, computer vision, computational photography, augmented reality and other AI disciplines. AML also built the unified AI platform that powers all production applications of AI across the family of Facebook products. Prior to Facebook, Joaquin built and taught a new machine learning course at the University of Cambridge, worked at Microsoft Research, and conducted postdoctoral research at three institutions in Germany, including the Max Planck Institute for Biological Cybernetics. He received his PhD from the Technical University of Denmark.

Safety and Robustness for Deep Learning with Provable Guarantees

Marta Kwiatkowska

University of Oxford, Oxford, England

Abstract. Computing systems are becoming ever more complex, with decisions increasingly often based on deep learning components. A wide variety of applications are being developed, many of them safety-critical, such as self-driving cars and medical diagnosis. Since deep learning is unstable with respect to adversarial perturbations, there is a need for rigorous software development methodologies that encompass machine learning components. This lecture will describe progress with developing automated verification and testing techniques for deep neural networks to ensure safety and robustness of their decisions with respect to input perturbations. The techniques exploit Lipschitz continuity of the networks and aim to approximate, for a given set of inputs, the reachable set of network outputs in terms of lower and upper bounds, in anytime manner, with provable guarantees. We develop novel algorithms based on feature-guided search, games, global optimisation and Bayesian methods, and evaluate them on state-of-the-art networks. The lecture will conclude with an overview of the challenges in this field.

Bio: Marta Kwiatkowska is Professor of Computing Systems and Fellow of Trinity College, University of Oxford. She is known for fundamental contributions to the theory and practice of model checking for probabilistic systems, focusing on automated techniques for verification and synthesis from quantitative specifications. She led the development of the PRISM model checker (www.prismmodelchecker.org), the leading software tool in the area and winner of the HVC Award 2016. Probabilistic model checking has been adopted in diverse fields, including distributed computing, wireless networks, security, robotics, healthcare, systems biology, DNA computing and nanotechnology, with genuine flaws found and corrected in real-world protocols. Kwiatkowska is the first female winner of the Royal Society Milner Award, winner of the BCS Lovelace Medal and was awarded an honorary doctorate from KTH Royal Institute of Technology in Stockholm. She won two ERC Advanced Grants, VERIWARE and FUN2MODEL, and is a coinvestigator of the EPSRC Programme Grant on Mobile Autonomy. Kwiatkowska is a Fellow of the Royal Society, Fellow of ACM, EATCS and BCS, and Member of Academia Europea.

Contents – Part I

Transfer and Multi-task Learning

Semi-supervised and Few-Shot Learning

Learning Algorithms and Applications

Online Learning

Routine Bandits: Minimizing Regret on Recurring Problems

Hassan Saber[1]([⊠]), Léo Saci[2]([⊠]), Odalric-Ambrym Maillard[1]([⊠]),
and Audrey Durand[3]([⊠])

[1] Université de Lille, Inria, CNRS, Centrale Lille UMR 9189 – CRIStAL,
59000 Lille, France
{hassan.saber,odalric.maillard}@inria.fr
[2] ENS Paris-Saclay, Gif-Sur-Yvette, France
leo.saci@ens-paris-saclay.fr
[3] Canada CIFAR AI Chair, Université Laval, Mila, Quebec City, Canada
audrey.durand@ift.ulaval.ca

Abstract. We study a variant of the multi-armed bandit problem in which a learner faces every day one of $\mathcal{B}$ many bandit instances, and call it a routine bandit. More specifically, at each period $h \in [\![1, H]\!]$, the same bandit $b_\star^h$ is considered during $T > 1$ consecutive time steps, but the identity $b_\star^h$ is unknown to the learner. We assume all rewards distribution are Gaussian standard. Such a situation typically occurs in recommender systems when a learner may repeatedly serve the same user whose identity is unknown due to privacy issues. By combining bandit-identification tests with a KLUCB type strategy, we introduce the KLUCB for Routine Bandits (KLUCB-RB) algorithm. While independently running KLUCB algorithm at each period leads to a cumulative expected regret of $\Omega(H \log T)$ after H many periods when $T \to \infty$, KLUCB-RB benefits from previous periods by aggregating observations from similar identified bandits, which yields a non-trivial scaling of $\Omega(\log T)$. This is achieved without knowing which bandit instance is being faced by KLUCB-RB on this period, nor knowing a priori the number of possible bandit instances. We provide numerical illustration that confirm the benefit of KLUCB-RB while using less information about the problem compared with existing strategies for similar problems.

Keywords: Multi-armed bandits · Transfer learning · KL-UCB

1 Introduction

The stochastic multi-armed bandit [5,17,19,23], is a popular framework to model a decision-making problem where a learning agent (*learner*) must repeatedly choose between several real-valued unknown sources of random observations (*arms*) to sample from in order to maximize the cumulative values (*rewards*) generated by these choices in expectation. This framework is commonly applied to recommender systems where arms correspond to items (e.g., ads, products) that can be recommended and rewards correspond to the success of the

© Springer Nature Switzerland AG 2021
N. Oliver et al. (Eds.): ECML PKDD 2021, LNAI 12975, pp. 3–18, 2021.
https://doi.org/10.1007/978-3-030-86486-6_1

recommendation (e.g., click, buy). An optimal strategy to choose actions would be to always play an arm with highest expected reward. Since the distribution of rewards and in particular their mean are unknown, in practice a learner needs to trade off *exploiting* arms that have shown good rewards until now with *exploring* arms to acquire information about the reward distributions. The stochastic multi-armed bandit framework has been well-studied in the literature and optimal algorithms have been proposed [7,14–16,24].

When a recommender system is deployed on multiple users, one does not typically assume that the best recommendation is the same for all users. The naive strategy in this situation is to consider each user as being a different bandit instance and learning from scratch for each user. When users can be recognized (e.g., characterized by features), this information can be leveraged to speed up the learning process by sharing observations across users. The resulting setting is known as contextual bandit [18,20]. In this paper, we tackle the case where users cannot be or do not want to be identified (e.g., for privacy reasons), but where we assume that there exists a (unknown) finite set of possible user profiles (bandit instances), such that information may be shared between the current user and some previously encountered users.

Outline and Contributions. To this end, we introduce the *routine bandit* problem (Sect. 2), together with lower bounds on the achievable cumulative regret that adapt the bound from [17] to the routine setting. We then extend the KLUCB [10] algorithm, known to be optimal under the classical stochastic bandit setting, into a new strategy called KLUCB-RB (Sect. 3) that leverages the information obtained on previously encountered bandits. We provide a theoretical analysis of KLUCB-RB (Sect. 4) and investigate the performance of the algorithm using extensive numerical experiments (Sect. 5). These results highlight the empirical conditions required so that past information can be efficiently leveraged to speed up the learning process. The main contributions of this work are 1) the newly proposed routine bandit setting, 2) the KLUCB-RB algorithm that solves this problem with asymptotically optimal regret minimization guarantees, and 3) an empirical illustration of the conditions for past information to be beneficial to the learning agent.

2   The Routine Bandit Setting

A routine bandit problem is specified by a time horizon $T \geqslant 1$ and a finite set of distributions $\nu = (\nu_b)_{b \in \mathcal{B}}$ with means $(\mu_{a,b})_{a \in \mathcal{A}, b \in \mathcal{B}}$, where $\mathcal{A}$ is a finite set of arms and $\mathcal{B}$ is a finite set of bandit configurations. Each $b \in \mathcal{B}$ can be seen as a classical multi-armed bandit problem defined by $\nu_b = (\nu_{a,b})_{a \in \mathcal{A}}$. At each period $h \geqslant 1$ and for all time steps $t \in [\![1, T]\!]$, the learner deals with a bandit $b_*^h \in \mathcal{B}$ and chooses an arm $a_t^h \in \mathcal{A}$, based only on the past. The learner then receives and observes a reward $X_t^h \sim \nu_{a_t^h, b_*^h}$. The goal of the learner is to maximize the expected sum of rewards received over time (up to some unknown number of periods $H \geqslant 1$). The distributions are unknown, which makes the problem non-trivial. The optimal strategy therefore consists in playing repeatedly on each period h, an optimal arm $a_\star^h \in \mathrm{argmax}_{a \in \mathcal{A}} \mu_{a,b_\star^h}$, which has mean $\mu_\star^h = \mu_{a_\star^h, b_\star^h}$. The goal of the learner is equivalent to minimizing the cumulative *regret* with respect to an optimal strategy:

$$R(\nu, H, T) = \mathbb{E}_\nu \left[\sum_{h=1}^{H} \sum_{t=1}^{T} \left(\mu_\star^h - X_t^h \right) \right]. \tag{1}$$

Related Works. One of the closest setting to routine bandits is the sequential transfer scenario [12], where the cardinality $|\mathcal{B}|$ and quantities H and T are known ahead of time, and the instances in $\mathcal{B}$ are either known perfectly or estimated with known confidence. Routine bandits also bear similarity with clustering bandits [11], a contextual bandit setting [18] where contexts can be clustered into finite (unknown) clusters. While both settings are recurring bandit problems, routine bandits assume no information on users (including their number) but users are recurring for several iterations of interaction, while clustering bandits assume that each user is seen only once, but is characterized by features such that they can be associated with previously seen users. Finally, latent bandits [21] consider the less structured situation when the learner faces a possibly different user at every time.

Assumptions and Working Conditions. The configuration ν, the set of bandits $\mathcal{B}$, and the sequence of bandits $(b_\star^h)_{h \geqslant 1}$ are *unknown* (in particular $|\mathcal{B}|$ and the identity of user $b_\star^h$ are unknown to the learner at time t). The learner only knows that $\nu \in \mathcal{D}$, where $\mathcal{D}$ is a given set of bandit configurations. In order to leverage information from the bandit instances encountered, we should consider that bandits reoccur. We denote by $\beta_b^h = \sum_{h'=1}^{h} \mathbb{I}_{\{b_\star^{h'} = b\}}/h$ the frequency of bandit $b \in \mathcal{B}$ at period $h \geqslant 1$ and assume $\beta_b^H > 0$. The next two assumptions respectively allow for two bandit instances b and b' to be distinguishable from their means when $b \neq b'$ and show consistency in their optimal strategy when $b = b'$.

Assumption 1 (Separation). *Let us consider* $\gamma_\nu := \min_{b \neq b'} \min_{a \in \mathcal{A}} \{|\mu_{a,b} - \mu_{a,b'}|, 1\}$. *We assume* $\gamma_\nu > 0$.

Assumption 2 (Unique optimal arm). *Each bandit* $b \in \mathcal{B}$ *has a unique optimal arm* a_b^*.

Assumption 2 is standard. Finally, we consider normally-distributed rewards. Although most of our analysis (e.g., concentration) would extend to exponential families of dimension 1, Assumption 3 increases readability of the statements.

Assumption 3 (Gaussian arms). *The set* $\mathcal{D}$ *is the set of bandit configurations such that for all bandit* $b \in \mathcal{B}$, *for all arm* $a \in \mathcal{A}$, $\nu_{a,b}$ *is a one-dimensional Gaussian distribution with mean* $\mu_{a,b} \in \mathbb{R}$ *and variance* $\sigma^2 = 1$.

For $\nu \in \mathcal{D}$, we define for an arm $a \in \mathcal{A}$ and a bandit $b \in \mathcal{B}$ their gap $\Delta_{a,b} = \mu_b^* - \mu_{a,b}$ and their total number of pulls over H periods $N_{a,b}(H,T) = \sum_{h=1}^{H} \sum_{t=1}^{T} \mathbb{I}_{\{a_t^h = a, b_\star^h = b\}}$. An arm is optimal for a bandit if their gap is equal to zero and sub-optimal if it is positive. Thanks to the chain rule, the regret rewrites as

$$R(\nu, H, T) = \sum_{b \in \mathcal{B}} \sum_{a \neq a_b^*} \mathbb{E}_\nu \left[N_{a,b}(H, T) \right] \Delta_{a,b}. \tag{2}$$

Remark 1 (Fixed horizon time). We assume the time horizon T to be the same for all periods $h \in [\![1, H]\!]$ out of clarity of exposure of the results and simplified definition of consistency (Definition 1). Considering a different time T_h for each h would indeed require a substantial rewriting of the statements (e.g. think of the regret lower bound), which we believe hinders readability and comparison to classical bandits.

We conclude this section by adapting for completeness the known lower bound on the regret [2,13,17] for *consistent* strategies to the routine bandit setting. We defer the proof to Appendix A [6].

Definition 1 (Consistent strategy). *A strategy is H-consistent on $\mathcal{D}$ if for all configuration $\nu \in \mathcal{D}$, for all bandit $b \in \mathcal{B}$, for all sub-optimal arm $a \neq a_\star^b$, for all $\alpha > 0$,*

$$\lim_{T \to \infty} \mathbb{E}_\nu \left[\frac{N_{a,b}(H, T)}{N_b(H, T)^\alpha} \right] = 0,$$

where $N_b(H, T) = \beta_b^H H T$ is the number of time steps the learner has dealt with bandit b.

Proposition 1 (Lower bounds on the regret). *Let us consider a consistent strategy. Then, for all configuration $\nu \in \mathcal{D}$, it must be that*

$$\liminf_{T \to \infty} \frac{R(\nu, H, T)}{\log(T)} \geqslant c_\nu^* := \sum_{b \in \mathcal{B}} \sum_{a \neq a_b^*} \frac{\Delta_{a,b}}{KL(\mu_{a,b} | \mu_b^\star)},$$

where $KL(\mu | \mu') = (\mu' - \mu)^2 / 2\sigma^2$ denotes the Kullback-Leibler divergence between one-dimensional Gaussian distributions with means $\mu, \mu' \in \mathbb{R}$ and variance $\sigma^2 = 1$.

This lower bound differs (it is larger) from structured lower bound that can exclude some set of arms, as in [2,21] using prior knowledge on $\mathcal{B}$, which here is not available. On the other hand, we remark that the right hand side of the bound does not depend on H, which suggests that one at least asymptotically, one can learn from the recurring bandits. In the classical bandit setting, lower bounds on the regret [17] have inspired the design of the well-known KLUCB [10] algorithm. In the next section, we build on this optimal strategy to propose a variant for the routine bandit.

3   The KLUCB-RB Strategy

Given the current period h, the general idea of this optimistic strategy consists in aggregating observations acquired in previous periods $1 \ldots h - 1$ where bandit instances are tested to be the same as the current bandit b_*^h. To achieve this, KLUCB-RB relies both on concentration of observations gathered in previous periods and the consistency of the allocation strategy between different periods.

Notations. The number of pulls, the sum of the rewards and the empirical mean of the rewards from the arm a in period $h \geqslant 1$ at time $t \geqslant 1$, are respectively denoted by $N_a^h(t) = \sum_{s=1}^{t} \mathbb{I}_{\{a_s^h = a\}}$, $S_a^h(t) = \sum_{s=1}^{t} \mathbb{I}_{\{a_s^h = a\}} X_s^h$ and $\widehat{\mu}_a^h(t) = S_a^h(t)/N_a^h(t)$ if $N_a^h(t) > 0$, 0 otherwise.

Strategy. For each period $h \geqslant 1$ we compute an empirical best arm for bandit b_*^h as the arm with maximum number of pulls in this period: $\overline{a}_*^h \in \underset{a \in \mathcal{A}}{\operatorname{argmax}}\, N_a^h(T)$.[1]
Similarly, in the current period $h \geqslant 1$, for each time step $t \in [\![1, T]\!]$, we consider an arm with maximum number of pulls: $\overline{a}_t^h \in \underset{a \in \mathcal{A}}{\operatorname{argmax}}\, N_a^h(t)$ (arbitrarily chosen).

At each period $h \in [\![2, H]\!]$ each arm is pulled once. Then at each time step $t \geqslant |\mathcal{A}| + 1$, in order to possibly identify the current bandit b_*^h with some bandits b_*^k from a previous period $k \in [\![1, h-1]\!]$, we introduce for all arm $a \in \mathcal{A}$, the test statistics

$$Z_a^{k,h}(t) = \infty \cdot \mathbb{I}_{\{\overline{a}_t^h \neq \overline{a}_*^k\}} + \left|\widehat{\mu}_a^h(t) - \widehat{\mu}_a^k(T)\right| - \mathrm{d}\left(N_a^h(t), \delta^h(t)\right) - \mathrm{d}\left(N_a^k(T), \delta^h(t)\right), \quad (3)$$

where the deviation for $n \geqslant 1$ pulls with probability $1 - \delta$, for $\delta > 0$, and probability $\delta^h(t)$ are, respectively,

$$\mathrm{d}(n, \delta) = \sqrt{2\left(1 + \frac{1}{n}\right) \frac{\log\left(\sqrt{n+1}/\delta\right)}{n}} \qquad \delta^h(t) = \frac{1}{4|\mathcal{A}|} \times \frac{1}{h-1} \times \frac{1}{t(t+1)}.$$

The algorithm finally computes the test

$$\mathrm{T}^{k,h}(t) := \max_{a \in \mathcal{A}} Z_a^{k,h}(t) \leqslant 0. \quad (4)$$

After t rounds in current period h, the previous bandit b_*^k is suspected of being the same as b_*^h if the test $\mathrm{T}^{k,h}(t)$ is true. From Eq. 3, we note that this requires the current mostly played arm to be the same as the arm that was mostly played in period k, which happens if there is consistency in the allocation strategy for both periods under Assumption 2. We then define aggregated numbers of pulls and averaged means: For all arm $a \in \mathcal{A}$, for all period $h \geqslant 1$, for all time step $t \geqslant 1$,

$$\overline{N}_a^h(t) := N_a^h(t) + \sum_{k=1}^{h-1} \mathbb{I}_{\{\mathrm{T}^{k,h}(t)\}} N_a^k(T), \qquad \overline{K}_t^h := \sum_{k=1}^{h-1} \mathbb{I}_{\{\mathrm{T}^{k,h}(t)\}},$$

$$\overline{S}_a^h(t) := S_a^h(t) + \sum_{k=1}^{h-1} \mathbb{I}_{\{\mathrm{T}^{k,h}(t)\}} S_a^k(T), \qquad \overline{\mu}_a^h(t) = \overline{S}_a^h(t)/\overline{N}_a^h(t).$$

and follow a KLUCB strategy by defining the index of arm $a \in \mathcal{A}$ in period $h \geqslant 1$ at time step $t \geqslant 1$ as

$$u_a^h(t) = \min\left\{U_a^h(t), \overline{U}_a^h(t)\right\}, \quad (5)$$

---

[1] Ties are broken arbitrarily.

8   H. Saber et al.

where

$$U_a^h(t) := \widehat{\mu}_a^h(t) + \sqrt{\frac{2f(t)}{N_a^h(t)}},\qquad(6)$$

$$\overline{U}_a^h(t) := \overline{\mu}_a^h(t) + \sqrt{\frac{2f\left(\overline{K}_t^h T + t\right)}{\overline{N}_a^h(t)}},\qquad(7)$$

with the function f being chosen, following [7] for classical bandits, as

$$f(x) := \log(x) + 3\log\log(\max\{e, x\}), \forall x \geqslant 1.$$

One recognizes that Eq. 6 corresponds to the typical KLUCB upper bound for Gaussian distributions. The resulting KLUCB-RB strategy is summarized in Algorithm 1.

Algorithm 1. KLUCB-RB

Initialization (period $h = 1$): follow a KLUCB strategy for bandit $b_\star^1$.
for period $h \geqslant 2$ **do**
  Pull each arm once
  for time step $t \in [\![|\mathcal{A}|, T-1]\!]$ **do**
    Compute for each previous period $k \in [\![1, h-1]\!]$ the test $\mathrm{T}^{k,h}(t) := \max_{a \in \mathcal{A}} Z_a^{k,h}(t) \leqslant 0$
    Aggregate data from periods with positive test and compute for each arm $a \in \mathcal{A}$ the index $u_a^h(t)$ according to equations (5)-(6)-(7).
    Pull an arm with maximum index $a_{t+1}^h \in \mathrm{argmax}_{a \in \mathcal{A}} u_a^h(t)$
  end for
end for

Theoretical Guarantees. The next result shows that the number of sub-optimal pulls done by KLUCB-RB is upper-bounded in a near-optimal way.

Theorem 1 (Upper bounds). *Let us consider a routine bandit problem specified by a set of Gaussian distributions $\nu \in \mathcal{D}$ and a number of periods $H \geqslant 1$. Then under KLUCB-RB strategy, for all $0 < \varepsilon < \varepsilon_\nu$, for all bandit $b \in \mathcal{B}$, for all sub-optimal arm $a \neq a_b^\star$,*

$$\mathbb{E}_\nu[N_{a,b}(H,T)] \leqslant \frac{f(\beta_b^H HT)}{KL(\mu_{a,b} + \varepsilon| \mu_b^\star)}$$

$$+ \sum_{h=1}^H \mathbb{I}_{\{b_\star^h = b\}} \left[\tau_\nu^h + 4|\mathcal{A}| \left(\frac{1}{\varepsilon^2} + 1\right)\left(5 + \frac{8h\,f(hT)}{T\,KL(\mu_{a,b} + \varepsilon| \mu_b^\star)}\right)\right],$$

where, for all period $h \geqslant 2$, $\tau_\nu^h := 2\varphi\left(8|\mathcal{A}|\left[\varepsilon_\nu^{-2} + 65\gamma_\nu^{-2}\log\left(128|\mathcal{A}|(4h)^{1/3}\gamma_\nu^{-2}\right)\right]\right)$, $\varphi : x \geqslant 1 \mapsto x\log(x)$, $\varepsilon_\nu = \min_{b \in \mathcal{B}}\min_{a \neq a_b^\star} \Delta_{a,b}/2$ and $\gamma_\nu = \min_{b \neq b'}\min_{a \in \mathcal{A}} \{|\mu_{a,b} - \mu_{a,b'}|, 1\}$.

This implies that the dependency on the time horizon T in these upper bounds is asymptotically optimal with regard to the lower bound on the regret given in Proposition 1. From Eq. 2, by considering the case when the time horizon T tends to infinity, we deduce that KLUCB-RB achieves asymptotic optimality.

Corollary 1 (Asymptotic optimality). *With the same notations and under the assumptions as in Theorem 1, KLUCB-RB achieves*

$$\limsup_{T \to \infty} \frac{R(\nu, H, T)}{\log(T)} \leqslant c_\nu^\star,$$

where $c_\nu^\star$ is defined as in Proposition 1.

For comparison, let us remark that under the strategy that runs a separate KLUCB type strategy for each period, the regret normalized by $\log(T)$ asymptotically scales as $H \sum_{b \in \mathcal{B}} \beta_b^H \sum_{a \neq a_b^\star} \Delta_{a,b} / \mathrm{KL}(\mu_{a,b} | \mu_b^\star)$. KLUCB-RB strategy then performs better than this naive strategy by a factor of the order of $H/|\mathcal{B}|$. Also, up to our knowledge, this result is the first showing provably asymptotic optimal regret guarantee in a setting when an agent attempts at transferring information from past to current bandits without contextual information. In the related but different settings considered in [11,12,21], only logarithmic regret was shown, however asymptotic optimality was not proved for the considered strategies. Also, let us remind that $|\mathcal{B}|$ does not need to be known ahead of time by the KLUCB-RB algorithm.

4   Sketch of Proof

This section contains a sketch of proof for Theorem 1. We refer to Appendix B [6] for more insights and detailed derivations. The first preoccupation is to ensure that KLUCB-RB is a consistent strategy. This is achieved by showing that KLUCB-RB aggregates observations that indeed come from the same bandits with high probability. In other words, we want to control the number of previously encountered bandits falsely identified as similar to the current one.

Definition 2 (False positive). *At period $h \geqslant 2$ and step $t \geqslant 1$, a previous period $k \in [\![1, h-1]\!]$ is called a false positive if the test $\mathrm{T}^{k,h}(t)$ is true while previous bandit $b_\star^k$ differs from current bandit $b_\star^h$.*

Combining the triangle inequality and time-uniform Gaussian concentration inequalities (see e.g., [1]), we prove necessary condition for having $Z_a^{k,h}(t) \leqslant 0$ for some arm $a \in \mathcal{A}$ at current period h and time step t, while having $b_\star^k \neq b_\star^h$.

Lemma 1 (Condition for false positives). *If there exists a false positive at period $h \geqslant 2$ and time step $t > |\mathcal{A}|$, then with probability $1 - 1/t(t+1)$, it must be that*

$$\min_{k \in [\![1, h-1]\!]: b_\star^k \neq b_\star^h} \min_{a \in \mathcal{A}} |\mu_{a,b^h} - \mu_{a,b^k}| \leqslant 4 \, \mathrm{d} \left(\frac{t}{|\mathcal{A}|}, \delta^h(t) \right).$$

The proof of this key result is provided in Appendix B.1 [6]. It relies on time-uniform concentration inequalities. We now introduce a few quantities.

Let us first consider at period $h \geqslant 2$ the time step

$$t_\nu^h := \max\left\{t \geqslant |\mathcal{A}| : \gamma_\nu \leqslant 4\,\mathrm{d}\left(\frac{t}{|\mathcal{A}|}, \delta^h(t)\right)\right\} + 1, \tag{8}$$

beyond which there is no false positives with high probability. We define for all $a \neq a_\star^h$, for all $0 < \varepsilon < \varepsilon_\nu := \min_{b \in \mathcal{B}} \min_{a \neq a_b^\star}\{\Delta_{a,b}, 1\}/2$ the subsets of times when there is a false positive

$$\mathcal{T}_a^h := \{t \geqslant t_\nu^h : a_{t+1}^h = a \text{ and } \mathcal{K}_+^h(t) \neq \mathcal{K}_\star^h(t)\} \qquad \mathcal{T}^h := \bigcup_{a \neq a_\star^h} \mathcal{T}_a^h, \tag{9}$$

where we introduced for convenience the sets $\mathcal{K}_+^h := \{k \in [\![1, h-1]\!] : T^{k,h}(t)$ is true$\}$ and $\mathcal{K}_\star^h(t) := \{k \in [\![1, h-1]\!] : b_\star^k = b_\star^h \text{ and } \overline{a}_t^k = \overline{a}_\star^k\}$. We also consider the times when the mean of the current pulled arm is poorly estimated or the best arm $a_\star^h$ is below its mean (either for the current period or by aggregation) and define

$$\mathcal{C}_{a,\varepsilon}^h := \left\{t \geqslant 1 : a_{t+1}^h = a \text{ and } \left(\left|\widehat{\mu}_a^h(t) - \mu_a^h\right| > \varepsilon \text{ or } u_{a_\star^h}^h(t) = U_{a_\star^h}^h(t) < \mu_\star^h\right)\right\}$$

$$\mathcal{C}_\varepsilon^h := \bigcup_{a \neq a_\star^h} \mathcal{C}_{a,\varepsilon}^h \tag{10}$$

$$\overline{\mathcal{C}}_{a,\varepsilon}^h := \mathcal{T}_a^h \cup \left\{t \geqslant t_\nu^h : t \notin T^h, a_{t+1}^h = a \text{ and } \left(\left|\overline{\mu}_a^h(t) - \mu_a^h\right| > \varepsilon \text{ or } u_{a_\star^h}^h(t) = \overline{U}_{a_\star^h}^h(t) < \mu_\star^h\right)\right\}$$

$$\overline{\mathcal{C}}_\varepsilon^h := \bigcup_{a \neq a_\star^h} \overline{\mathcal{C}}_{a,\varepsilon}^h. \tag{11}$$

The size of these (bad events) sets can be controlled by resorting to concentration arguments. The next lemma borrows elements of proof from [8] for the estimation of the mean of current pulled arm and [7] for the effectiveness of the upper confidence bounds on the empirical means of optimal arms. We adapt these arguments to the routine-bandit setup, and provide additional details in the appendix.

Lemma 2 (Bounded subsets of times). *For all period $h \geqslant 2$, for all arm $a \in \mathcal{A}$, for all $0 < \varepsilon < \varepsilon_\nu$,*

$$\mathbb{E}_\nu\left[|\mathcal{T}^h|\right] \leqslant 1 \qquad \mathbb{E}_\nu\left[|\mathcal{C}_{a,\varepsilon}^h|\right] \leqslant 4\varepsilon^{-2} + 2 \qquad \mathbb{E}_\nu\left[\left|\overline{\mathcal{C}}_{a,\varepsilon}^h\right|\right] \leqslant 4\varepsilon^{-2} + 3.$$

By definition of the index (Eq. 7), we have

$$\forall t > |\mathcal{A}|, \; N_a^h(t)\mathrm{KL}\!\left(\widehat{\mu}_a^h(t)\,\middle|\, U_a^h(t)\right) = f(t)$$
$$\overline{N}_a^h(t)\mathrm{KL}\!\left(\overline{\mu}_a^h(t)\,\middle|\, \overline{U}_a^h(t)\right) = f\left(\overline{K}_t^h T + t\right).$$

We then provide logarithmic upper bounds on the aggregated number of pulls $\overline{N}_a^h(t)$ to deduce the consistency of KLUCB-RB strategy. The following non-trivial result combines standard techniques with the key mechanism of the algorithm.

Lemma 3 (Consistency). *Under KLUCB-RB strategy for all period $h \geqslant 2$, for all $0 < \varepsilon < \varepsilon_\nu$, for all sub-optimal arm $a \neq a_\star^h$, for all $t > |\mathcal{A}|$ such that $a_{t+1}^h = a$,*

$$\text{if } t \notin \mathcal{C}_{a,\varepsilon}^h, \ N_a^h(t) \leqslant \frac{f(t)}{KL(\mu_a^h + \varepsilon \mid \mu_\star^h)}, \quad \text{if } t \geqslant t_\nu^h \text{ and } t \notin \overline{\mathcal{C}}_{a,\varepsilon}^h, \overline{N}_a^h(t) \leqslant \frac{f\left(\underline{K}_t^h T + t\right)}{KL(\mu_a^h + \varepsilon \mid \mu_\star^h)},$$

where $\underline{K}_t^h := \min\left\{ \overline{K}_t^h, \beta_{b^h}^{h-1}(h-1) \right\}$. In particular this implies

$$\forall t \geqslant 1, \forall a \neq a_\star^h, \quad N_a^h(t) \leqslant \frac{f(t)}{KL(\mu_a^h + \varepsilon \mid \mu_\star^h)} + \left|\mathcal{C}_{a,\varepsilon}^h\right| + N_a^h\left(|\mathcal{A}| + 1\right),$$

where $N_a^h\left(|\mathcal{A}| + 1\right) \leqslant 2$ and $\mathbb{E}_\nu\left[\left|\mathcal{C}_{a,\varepsilon}^h\right|\right] \leqslant 4\varepsilon^{-2} + 2$.

Thanks to Eq. 5 that involves the minimum of the aggregated index $\overline{U}_a^h(t)$ on past episodes and (not aggregated) indexes $U_a^h(t)$ for the current epoch, the proof proceeds by considering the appropriate sets of time, namely $t \notin \mathcal{C}_{a,\varepsilon}^h$ or $t \notin \overline{\mathcal{C}}_{a,\varepsilon}^h$ depending on the situation. In particular, we get for the considered a that the maximum index $u_a^h(t)$ is either greater than $u_{a_\star^h}^h(t) = U_{a_\star^h}^h(t)$ or $u_{a_\star^h}^h(t) = \overline{U}_{a_\star^h}^h(t)$, which in turns enable to have a control either on $N_a^h(t)$ or $\overline{N}_a^h(t)$. In order to obtain the last statement, it essentially remains to consider the maximum time $t' \in [\![|\mathcal{A}| + 1; t]\!]$ such that $a_{t'+1}^h = a$ and $t' \notin \mathcal{C}_{a,\varepsilon}^h$.

In order to be asymptotically optimal (in the sense of Corollary 1), the second preoccupation is to ensure with high probability that we aggregate all of the observations coming from current bandit $b_\star^h$ when computing the indexes. From the definition of $\mathcal{T}^h$ (Eq. 9) and Lemma 2, this amounts to ensure that the current most pulled arm and the most pulled arms of previous periods are the optimal arms of the corresponding periods with high probability. By using the consistency of KLUCB-RB, we prove necessary conditions for the most pulled arms being different from the optimal ones.

Lemma 4 (Most pulled arms). *For all period $h \geqslant 2$, for all $0 < \varepsilon < \varepsilon_\nu$, for all $t \geqslant t_\nu^h$ such that $t \notin \mathcal{T}^h$ and $\overline{a}_t^h \neq a_\star^h$,*

$$\frac{t + \left|\mathcal{K}_\star^h(t)\right| T}{2} - \left(f(t) + \left|\mathcal{K}_\star^h(t)\right| f(T)\right) \sum_{a \neq a_\star^h} \frac{1}{KL(\mu_a^h + \varepsilon \mid \mu_\star^h)} - \left(1 + \left|\mathcal{K}_\star^h(t)\right|\right) |\mathcal{A}| \leqslant \sum_{k \in \mathcal{K}_\star^h(t) \cup \{h\}} \left|\mathcal{C}_\varepsilon^k\right|.$$

Let us remind that $\mathcal{K}_\star^h(t)$, defined after Lemma 1, counts the previous phases before h facing the same bandit as the current one, and for which the most-played arm until then agree. Then, by combining Lemma 3 and Lemma 4 we obtain randomized upper bounds on the number of pulls of sub-optimal arms.

Proposition 2 (Randomized upper bounds). *Under KLUCB-RB strategy, for all bandit $b \in \mathcal{B}$, for all sub-optimal arm $a \neq a_b^\star$, for all $0 < \varepsilon < \varepsilon_\nu$,*

$$N_{a,b}(H,T) \leqslant \frac{f(\beta_b^H HT)}{KL(\mu_{a,b} + \varepsilon \mid \mu_b^\star)}$$

$$+ \sum_{h=1}^{H} \mathbb{I}_{\{b_\star^h = b\}} \left[T_{\nu,\varepsilon}^h + 4\left|\mathcal{C}_\varepsilon^h\right| + \left|\overline{\mathcal{C}}_\varepsilon^h\right| + \frac{f(hT)}{KL(\mu_{a,b} + \varepsilon \mid \mu_b^\star)} \sum_{k=1}^{h} \frac{8\left|\mathcal{C}_\varepsilon^k\right|}{T} + \mathbb{I}_{\{T \in \mathcal{T}^k\}} \right],$$

where $T_{\nu,\varepsilon}^h := \max \left\{ t \geqslant t_\nu^h : \dfrac{t}{4} - \displaystyle\sum_{a \neq a_\star^h} \dfrac{f(t)}{KL(\mu_a^h + \varepsilon \mid \mu_\star^h)} \leqslant |\mathcal{A}| \right\} + 1$ *for* $h \geqslant 2$, *with* t_ν^h *defined in Eq.* (8).

We prove Theorem 1 by averaging the randomized upper bounds from Proposition 2.

5   Numerical Experiments

We now perform experiments to illustrate the performance of the proposed KLUCB-RB under different empirical conditions. We compare KLUCB-RB with a baseline strategy which consists in using a KLUCB that restarts from scratch at every new period, that is the default strategy when no information (features) is provided to share information across periods. We also include a comparison with the sequential transfer algorithm tUCB [12] which constitutes interesting baseline to compare with, since it transfers the knowledge of past periods to minimize the regret in a very similar context. Through the periods $h \in [\![1, H]\!]$, tUCB incrementally estimates the mean vectors by the Robust Tensor Power method [3,4], then yielding a deviation of rate $\mathcal{O}(1/\sqrt{h})$ over the empirical means. Thus, it needs to know in advance the total number of instances $|\mathcal{B}|$. Besides the RTP method requires the mean vectors to be linearly independent mutually, which forces the number of arms $|\mathcal{A}|$ to be larger than $|\mathcal{B}|$, while KLUCB-RB can tackle this kind of distributions. The next comparisons between KLUCB-RB and tUCB will mainly illustrate the ability of the former to make large profits from the very first periods, while the later needs to get a sufficiently high confidence over the models estimates before beginning to use knowledge from the previous periods.

All experiments are repeated 100 times. Sequence $(b^h)_{1 \leqslant h \leqslant H}$ is chosen randomly each time. All the different strategies are compared based on their cumulative regret (Eq. 1). Additional experiments are provided in Appendix C [6].

5.1   More Arms Than Bandits: A Beneficial Case

We first investigate how Assumption 1 can be relaxed in practice. Indeed KLUCB-RB is designed such that only data from previous periods $k < h$ for which the most pulled arm $\bar{a}_\star^k$ is the same as the current most pulled arm $\bar{a}_t^h$ may be aggregated. Consequently, let us define $\gamma_\nu^\star := \min\limits_{b \neq b'} \min\limits_{a \in \mathcal{A}^\star} |\mu_{a,b} - \mu_{a,b'}|$ with $\mathcal{A}^\star$ being the set of arms optimal on at least one instance $b \in \mathcal{B}$. Assuming that KLUCB-RB converges to the optimal action in a given period, it is natural in practice to relax Assumption 1 from $\gamma_\nu > 0$ to $\gamma_\nu^\star > 0$. Let us consider a routine two-bandit setting $\mathcal{B} = \{b_1, b_2\}$ with actions $\mathcal{A}$ such that

$$b_1 : (\mu_{1,b_1}, \mu_{2,b_1}) = (\frac{\Delta}{2}, -\frac{\Delta}{2}) \qquad \text{and} \qquad \forall a \geqslant 3, \ \mu_{a,b_1} = \mu \quad (12)$$

$$b_2 : (\mu_{1,b_2}, \mu_{2,b_2}) = (\frac{\Delta}{2} - \gamma, -\frac{\Delta}{2} + \gamma) \qquad \text{and} \qquad \forall a \geqslant 3, \ \mu_{a,b_2} = \mu, \quad (13)$$

with $\mu = -\frac{\Delta}{2}$, and $\gamma = 0.85\Delta$, and where $\Delta = 10\sqrt{\frac{\log(HT)}{T}}$ is set to accomodate the convergence of KLUCB in the experiment. Note that Assumption 1 is not satisfied anymore since $\gamma_\nu = 0$, but that $\gamma_\nu^* = \gamma$. Figure 1 shows the average cumulative regret with one standard deviation after $H = 500$ periods of $T = 10^3$ rounds on settings where $|\mathcal{A}^*| = 2$ and $|\mathcal{A}| \geqslant 2$.

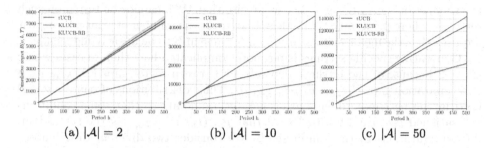

(a) $|\mathcal{A}| = 2$ (b) $|\mathcal{A}| = 10$ (c) $|\mathcal{A}| = 50$

Fig. 1. Cumulative regret of KLUCB, KLUCB-RB and tUCB along $H = 500$ periods of $T = 10^3$ **rounds**, for different action sets.

We observe that KLUCB-RB can largely benefit from relying on previous periods when the number of arms exceeds the number of optimal arms, which naturally happens when $|\mathcal{A}| > |\mathcal{B}|$. This can also happen for $|\mathcal{A}| \leqslant |\mathcal{B}|$ if several bandits $b \in \mathcal{B}$ share the same optimal arm. Besides, Fig. 2 shows a remake of the same experiment, that is $\Delta = 10\sqrt{\frac{\log(H \times 10^0)}{10^3}}$, where the number of rounds per period is decreased from 10^3 to $T = 100$. We can see that KLUCB-RB still yields good satisfying performances, although T is not large enough to enable a sure identification at each period of the current instance.

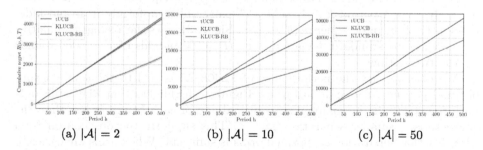

(a) $|\mathcal{A}| = 2$ (b) $|\mathcal{A}| = 10$ (c) $|\mathcal{A}| = 50$

Fig. 2. Cumulative regret of KLUCB, KLUCB-RB and tUCB along $H = 500$ periods of $T = 100$ **rounds**, for different action sets.

5.2   Increasing the Number of Bandit Instances

We now consider experiments where we switch among $|\mathcal{B}| = 5$ four-armed bandits. This highlights the kind of settings which may cause more difficulties to KLUCB-RB in distinguishing the different instances: the lesser is the number of arms $|\mathcal{A}|$ compared to the number of bandits $|\mathcal{B}|$, the harder it should be for KLUCB-RB to distinguish efficiently the different instances, in particular when the separation gaps are tight. Let us precise that tUCB cannot be tested on such settings, where the number of models $|\mathcal{B}|$ exceeds the number of arms $|\mathcal{A}|$, since it requires that the mean vectors $(\mu_{a,b})_{a\in\mathcal{A}}$ for all b in $\mathcal{B}$ to be linearly independent.

Generating specific settings is far more complicated here than in cases where $|\mathcal{B}| = 2$ because of the intrinsic dependency between regret gaps $(\Delta_{a,b})_{a\in\mathcal{A},b\in\mathcal{B}}$ and separation gaps $(|\mu_{a,b} - \mu_{a,b'}|)_{a\in\mathcal{A},b\neq b'}$. Thus, distributions of bandits $\nu \in \mathcal{D}$ used in the next experiments are generated randomly so that some conditions are satisfied (see Eq. 14, 15). Recall that $\nu : (\nu_{b_1}, \ldots, \nu_{b_{|\mathcal{B}|}})$ is the set of bandit configurations in the bandit set $\mathcal{B}$. We consider two different distributions $\nu^{(1)}$ and $\nu^{(2)}$, resulting in associated sets of bandits $\mathcal{B}_1$ and $\mathcal{B}_2$, satisfying the condition $C(\nu)$ in order to ensure the convergence of algorithms at each period:

$$C(\nu) : \forall b \in \mathcal{B}, \quad 8\sqrt{\frac{\log(HT)}{T}} \leqslant \min_{a\neq a_b^\star} \Delta_{a,b} \leqslant 12\sqrt{\frac{\log(HT)}{T}}. \tag{14}$$

Let $\gamma(\alpha) := \alpha\sqrt{\frac{\log(HT)}{T}}$. We generate two sets of bandits $\mathcal{B}_1$ and $\mathcal{B}_2$ such as to ensure that $\nu^{(1)}$ and $\nu^{(2)}$ satisfy

$$\gamma(12) \leqslant \gamma_{\nu^{(1)}}^\star \leqslant \gamma(16) \qquad \gamma(4) \leqslant \gamma_{\nu^{(2)}}^\star \leqslant \gamma(8). \tag{15}$$

Figure 7 (Appendix C.3 [6]) shows the bandit instances in the two generated bandit sets.

All experiments are conducted under the fair frequency $\beta = 1/|\mathcal{B}|$. More precisely, once a period $h \geqslant 1$ ends, $b_\star^{h+1}$ is sampled uniformly in $\mathcal{B}$ and independently of the past sequence $(b_\star^k)_{1\leqslant k\leqslant h}$. Figure 3 shows the average cumulative regret with one standard deviation after $H = 100$ periods of $T = 5000$ rounds for the two settings.

We observe that the performance of KLUCB-RB is tied to the smallest suboptimal gap for all bandit instances. Figure 3a highlights that KLUCB-RB outperforms KLUCB if the minimal sub-optimal gap of each bandit is less than the characteristic smaller separation gap $\gamma_\nu^\star$. This supports the observation from Sect. 5.1 that separation on optimal arms is sufficient. When arms are easier to separate than bandits, one might as well restart a classical KLUCB from scratch on each period (Fig. 3b). Note that situations where $0 < \gamma_\nu \ll \min_{b\in\mathcal{B}} \min_{a\neq a_b^\star} \Delta_{a,b}$ may not result in a catastrophic loss in learning performances if the arms in $\mathcal{A}^\star$ are *close enough* not to distort estimates computed on aggregated samples of from false positive models (see Appendix C [6]).

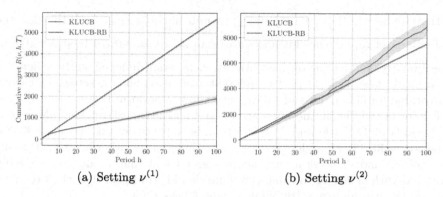

(a) Setting $\nu^{(1)}$ (b) Setting $\nu^{(2)}$

Fig. 3. Cumulative regret of KLUCB and KLUCB-RB along $H = 100$ periods of $T = 5000$ rounds over three generated settings of $|\mathcal{B}| = 5$ bandit instances with $|\mathcal{A}| = 4$ arms per instance.

5.3 Critical Settings

We saw previously that settings where bandit instances are difficult to distinguish may yield poor performance (see Sect. 5.2, Fig. 3b). Indeed, to determine if two estimated bandit models might result from the same bandit, both KLUCB-RB and tUCB rely on a compatibility over each arm, i.e. the intersection of confidence intervals. Therefore, it is generally harder to distinguish rollouts from many different distributions (that is the cardinal of $|\mathcal{B}|$ is high) when $|\mathcal{A}|$ is low and differences between arms are tight. To illustrate that, we consider an experiment on the setting described in Fig. 8 (Appendix C.3 [6]), composed of 4-armed bandits. We recall that tUCB requires in particular $|\mathcal{A}| \geqslant |\mathcal{B}|$. Thus we choose a set $|\mathcal{B}|$ of cardinal 4 in order to include a comparison of our algorithm with tUCB.

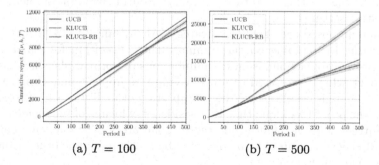

(a) $T = 100$ (b) $T = 500$

Fig. 4. Cumulative regret of KLUCB, KLUCB-RB and tUCB along $H = 500$ period for different numbers of rounds.

Here we have $|\mathcal{A}^{\star}| = \{0, 1, 3\}$ and $\gamma_{\nu}^{\star} := \min\limits_{b \neq b'} \min\limits_{a \in \mathcal{A}^{\star}} |\mu_{a,b} - \mu_{a,b'}| = 0.15$, while the minimal regret gaps of each instances are $(\min\limits_{a \neq a_b^{\star}} \Delta_{a,b})_{b \in \mathcal{B}} = (0.74, 0.80, 0.81, 0.89)$. Consequently, finding the optimal arm at each period independently is here far less difficult than separating the different instances. Such a setting is clearly unfavorable for KLUCB-RB and we expect KLUCB to perform better.

Figure 4a and Fig. 4b the cumulative regret for the three strategies, along $H = 500$ periods of $T = 100$ and $T = 500$ rounds respectively. As expected, KLUCB outperforms KLUCB-RB under this critical setting. On the other hand tUCB seems more robust and displays a cumulative regret trend that would be improving compared with KLUCB in the long run. One should still recall that tUCB requires knowing the cardinality of $|\mathcal{B}|$, while KLUCB-RB does not.

We may notice (Fig. 4a) that if the number of rounds T is sufficiently small, that is KLUCB does not have enough time to converge for each bandit, then KLUCB-RB does not perform significantly worse than KLUCB for the first periods. Then, as T rises (Fig. 4b), KLUCB begins to converge while KLUCB-RB still aggregate samples from confusing instances, which yields an explosion of the cumulative regret curve. We then expect for such setting that KLUCB will need far more longer periods ($T \to \infty$) to reach a regime in which it will discard all false positive rollouts and takes advantage over KLUCB. On the contrary, tUCB takes advantage of the knowledge of $|\mathcal{B}|$ and then waits to have enough confidence over the mean vectors of the 4 models to exploit them.

6   Conclusion

In this paper we introduced the new routine bandits framework, for which we provided lower bounds on the regret (Proposition 1). This setting applies well to problems where, for example, customers anonymously return to interact with a system. These dynamics are known to be of interest to the community, as evidenced by the existing literature [11,12,21]. Routine bandits complement well these existing settings.

We then proposed the KLUCB-RB strategy (Algorithm 1) to tackle the routine bandit setting by building on the seminal KLUCB algorithm for classical bandits. We proved upper bounds on the number of sub-optimal plays by KLUCB-RB (Theorem 1), which were used to prove asymptotic upper bounds on the regret (Corollary 1). This result shows the asymptotic optimality of the strategy and thanks to the proof technique that we considered, which is of independent interest, we further obtained finite-time regret guarantees with explicit quantities. We indeed believe the proof technique may be useful to handle other structured setups beyond routine bandits.

We finally provided extensive numerical experiments to highlight the situations where KLUCB-RB can efficiently leverage information from previously encountered bandit instances to improve over a classical KLUCB. More importantly, we highlighted the cost to pay for re-using observations from previous periods, and showed that easy tasks may be better tackled independently. This is akin to an agent that would behave badly by relying on a wrong inductive bias. Fortunately, there are many situations where one can leverage knowledge from bandit instances faced in the past. This would notably be the case if the agent has to select products to recommend from a large set ($\mathcal{A}$) and it turns out that there exists a much smaller set of products ($\mathcal{A}^\star$) that is preferred by users (Sect. 5.1).

Our results notably show that transferring information from previously encountered bandits can be highly beneficial (e.g., see Fig. 1 and 3a). However, the lack of prior knowledge about previous instances (including the cardinality of the set of instances) introduces many challenges in transfer learning. For example, attempting to leverage knowledge from previous instances could result in negative transfer if bandits cannot be distinguished properly (e.g., see Fig. 4).

Therefore, reducing the cost incurred for separating bandit instances should constitute a relevant angle to tackle as future work. Another natural line of other future work could investigate extensions of KLUCB-RB to the recurring occurrence of other bandit instances, e.g., linear bandits, contextual bandits, and others.

Acknowledgments. This work has been supported by CPER Nord-Pas-de-Calais/FEDER DATA Advanced data science and technologies 2015–2020, the French Ministry of Higher Education and Research, Inria, the French Agence Nationale de la Recherche (ANR) under grant ANR-16-CE40-0002 (the BADASS project), the MEL, the I-Site ULNE regarding project R-PILOTE-19-004-APPRENF, and CIFAR.

References

1. Abbasi-Yadkori, Y., Pál, D., Szepesvári, C.: Improved algorithms for linear stochastic bandits. In: Advances in Neural Information Processing Systems, pp. 2312–2320 (2011)
2. Agrawal, R., Teneketzis, D., Anantharam, V.: Asymptotically efficient adaptive allocation schemes for controlled IID processes: finite parameter space. IEEE Trans. Autom. Control **34**(3) (1989)
3. Anandkumar, A., Ge, R., Hsu, D., Kakade, S.: A tensor spectral approach to learning mixed membership community models. In: Conference on Learning Theory, pp. 867–881. PMLR (2013)
4. Anandkumar, A., Ge, R., Hsu, D.J., Kakade, S.M., Telgarsky, M.: Tensor decompositions for learning latent variable models. J. Mach. Learn. Res. **15**(1), 2773–2832 (2014)
5. Bubeck, S., Cesa-Bianchi, N.: Regret analysis of stochastic and nonstochastic multi-armed bandit problems. Foundations and Trends in Machine Learning abs/1204.5721 (2012). http://arxiv.org/abs/1204.5721
6. Saber, H., Saci, L., Maillard, O.A., Durand, A.: Routine bandits: minimizing regret on recurring problems. hal-03286539 (2021). https://hal.archives-ouvertes.fr/hal-03286539
7. Cappé, O., Garivier, A., Maillard, O.A., Munos, R., Stoltz, G.: Kullback- Leibler upper confidence bounds for optimal sequential allocation. Ann. Stat. **41**(3), 1516–1541 (2013)

8. Combes, R., Proutiere, A.: Unimodal bandits: regret lower bounds and optimal algorithms. In: Proceedings of the 31st International Conference on Machine Learning (ICML), pp. 521–529 (2014)
9. Garivier, A.: Informational confidence bounds for self-normalized averages and applications. In: 2013 IEEE Information Theory Workshop (ITW), pp. 1–5. IEEE (2013)
10. Garivier, A., Cappé, O.: The KL-UCB algorithm for bounded stochastic bandits and beyond. In: Proceedings of the 24th Annual Conference on Learning Theory (COLT), pp. 359–376 (2011)
11. Gentile, C., Li, S., Zappella, G.: Online clustering of bandits. In: Proceedings of the ICML, pp. 757–765 (2014)
12. Gheshlaghi Azar, M., Lazaric, A., Brunskill, E.: Sequential transfer in multi-armed bandit with finite set of models. In: Proceedings of the NIPS, pp. 2220–2228. Curran Associates, Inc. (2013)
13. Graves, T.L., Lai, T.L.: Asymptotically efficient adaptive choice of control laws in controlled Markov chains. SIAM J. Control. Optim. 35(3), 715–743 (1997)
14. Honda, J., Takemura, A.: An asymptotically optimal bandit algorithm for bounded support models. In: Proceedings of the 23rd Annual Conference on Learning Theory, Haifa, Israel (2010)
15. Korda, N., Kaufmann, E., Munos, R.: Thompson sampling for 1-dimensional exponential family bandits. In: Advances in Neural Information Processing Systems (NIPS), pp. 1448–1456 (2013)
16. Lai, T.L.: Adaptive treatment allocation and the multi-armed bandit problem. Ann. Stat. 1091–1114 (1987)
17. Lai, T.L., Robbins, H.: Asymptotically efficient adaptive allocation rules. Adv. Appl. Math. 6(1), 4–22 (1985)
18. Langford, J., Zhang, T.: The Epoch-Greedy algorithm for multi-armed bandits with side information. In: Platt, J.C., et al. (eds.) NIPS. MIT Press (2007)
19. Lattimore, T., Szepesvári, C.: Bandit Algorithms. Cambridge University Press, Cambridge (2020)
20. Lu, T., Pál, D., Pál, M.: Contextual multi-armed bandits. In: Teh, Y.W., Titterington, M. (eds.) Proceedings of the 13th International Conference on Artificial Intelligence and Statistics, vol. 9, pp. 485–492 (2010)
21. Maillard, O.A., Mannor, S.: Latent bandits. In: International Conference on Machine Learning (ICML) (2014)
22. Peña, V.H., Lai, T.L., Shao, Q.M.: Self-normalized Processes: Limit Theory and Statistical Applications. Springer, Heidelberg (2008)
23. Robbins, H.: Some aspects of the sequential design of experiments. Bull. Am. Math. Soc. 58(5), 527–535 (1952)
24. Thompson, W.: On the likelihood that one unknown probability exceeds another in view of the evidence of two samples. Biometrika 25, 285–294 (1933)

Conservative Online Convex Optimization

Martino Bernasconi de Luca, Edoardo Vittori, Francesco Trovò[⊠],
and Marcello Restelli

Dipartimento di Elettronica, Informazione e Bioingegneria, Politecnico di Milano,
Piazza Leonardo da Vinci, 32, Milan, Italy
{martino.bernasconideluca,edoardo.vittori,francesco1.trovo,
marcello.restelli}@polimi.it

Abstract. Online learning algorithms often have the issue of exhibiting poor performance during the initial stages of the optimization procedure, which in practical applications might dissuade potential users from deploying such solutions. In this paper, we study a novel setting, namely *conservative online convex optimization*, in which we are optimizing a sequence of convex loss functions under the constraint that we have to perform at least as well as a known default strategy throughout the entire learning process, a.k.a. conservativeness constraint. To address this problem we design a meta-algorithm, namely *Conservative Projection* (CP), that converts any no-regret algorithm for online convex optimization into one that, at the same time, satisfies the conservativeness constraint and maintains the same regret order. Finally, we run an extensive experimental campaign, comparing and analyzing the performance of our meta-algorithm with that of state-of-the-art algorithms.

Keyword: Online learning

1 Introduction

In the classic Empirical Risk Minimization (ERM) framework [38], the objective is to solve a stochastic optimization problem by minimizing the empirical loss function over a given set of training examples drawn from the unknown distribution. However, using the ERM approach in production exposes the learner to the issue of concept drift [37], *i.e.,* the risk that the distribution producing a training dataset may differ from the one observed during the operational life of the model. A solution to this issue is offered by techniques deriving from the Online Convex Optimization (OCO) field [32], which aim at minimizing a sequence of convex loss functions w.r.t. to the best-fixed strategy in hindsight. Nonetheless, even if they ensure convergence to the optimal solution, OCO techniques notoriously have poor empirical performance during the early stages of the learning process [29], which might dissuade potential users from deploying such solutions. To model this issue, we define a novel *Conservative Online Convex Optimization* (COCO) framework in which the learner has to perform online asymptotically as well as the best-fixed decision in hindsight while satisfying a *conservativeness*

© Springer Nature Switzerland AG 2021
N. Oliver et al. (Eds.): ECML PKDD 2021, LNAI 12975, pp. 19–34, 2021.
https://doi.org/10.1007/978-3-030-86486-6_2

constraint, *i.e.*, during the operational life of the system it has to perform no worse than a given fixed strategy. Furthermore, we propose the Conservative Projection (CP) algorithm, a newly-designed Online learning meta-algorithm, applicable to any OCO algorithms, that exploits the strengths of both ERM and OCO solutions.

Learning an optimal strategy while satisfying a conservativeness constraint during the exploration phase is of paramount importance in multiple domains. For instance, an intuitive example can be found in automatic spam filters [6]. Generally, companies optimize offline models on historical data, *e.g.*, past e-mails, until such classifiers perform satisfactorily given the collected dataset. When deploying this product, the company would like to maintain at least the above-mentioned performance while continually optimizing the model, integrating newly collected data, and possibly adapting to data distribution changes. Another field that benefits from being conservative is the financial field, *e.g.*, the asset management sector [7]. In this context, the goal of portfolio managers is to beat a specific market index (a weighted average of a set of stocks), *i.e.*, to perform better than the chosen index and, concurrently, maximize the collected wealth.

The idea of learning while guaranteeing the performance of a fixed and known policy has also been studied in the fields of Reinforcement Learning (RL) [13] and Multi-Armed Bandits (MAB) [40]. To the best of our knowledge, no work explicitly tackles the problem of conservativeness in the OCO framework as defined in this paper. To solve this problem, we extend the techniques from the OCO literature, and propose a meta-algorithm, namely CP, which extends *any* Online learning algorithm to satisfy the requirements of the COCO framework. Thanks to the use of a pseudo-loss and a projection in a so-called *conservative ball*, the proposed CP algorithm provides anytime guarantees w.r.t. a fixed default strategy. Specifically, the contributions of this work are:

- the definition of the novel COCO framework, where the objective is to obtain sub-linear regret while performing better than the default strategy during the entire learning process;
- the CP algorithm, which provides a solution to the above-mentioned problem for *any* OCO algorithm. Furthermore, we provide theoretical evidence that CP performs at least as well as the default strategy and that its regret is of the same order as that of the original OCO algorithm considered;
- an in-depth empirical evaluation of the CP algorithm in terms of regret and conservativeness on both simulated and real-world problems, comparing it with state-of-the-art algorithms from the OCO literature.

2 Background

Problems closely related to those of conservativeness have been commonly addressed by *safe* RL techniques. In [15], the authors provide a comprehensive overview of the different definitions of *safety* in RL. The most common assumption is to have access to a safe policy, and the goal is to improve that policy monotonically throughout the learning process. The seminal paper for this setting in [18], which proposes a conservative policy iteration algorithm with monotonic improvement guarantees for mixtures of greedy policies. This approach is generalized to stationary and stochastic policies in [28,31]. Building on the former, in [25–27] the authors have designed monotonically improving policy gradient algorithms for Gaussian, Lipschitz, and, recently, smoothing policies. This setting differs substantially from ours as the underlying environment is assumed to be stochastic.

In the bandit setting, the authors of [23] analyzed the same problem, characterizing the Pareto regret frontier in the stochastic case, *i.e.*, a surface determined by the admissible regret bounds for each arm. Following these seminal works, the interest of the MAB community in conservative exploration has grown in recent years, starting with the work presented in [40], where the authors modified the well-known UCB algorithm [2,3] to guarantee the safety constraint. Later, the idea was applied to contextual linear bandits in [19] and later improved in [14], as wells as to GPUCB, as presented in [34,35]. We inherit the concept of *safety as conservatism* from these works on stochastic bandit feedback and apply it to the context of adversarial full-information feedback.

In the Expert Learning literature, a work similar to ours is [30]. In this work, the authors design a strategy, named (A, B)-prod, that provides regret guarantees w.r.t. the regret of two generic strategies A, and B. However, their conservativeness definition is not comparable to ours, since it does not hold anytime. The question of bounding the regret not only to the best action but also to other strategies is addressed in [17,21], in which the authors proved, for the full information setting, that there exists an algorithm that guarantees a regret of $\mathcal{O}(\sqrt{T})$, with a specific constant for each expert. In particular, the main focus of the paper is to characterise the admissible vectors $\{r_k\}_{k \in K}$ guaranteeing a regret $R_T^k \leq r_k$ w.r.t. each expert k. Even if these works cover a more general theoretical framework than ours, *i.e.*, multi-objective regret minimization, the algorithms therein do not guarantee that their loss is strictly smaller than that of a given expert, and, therefore, their results cannot be compared with ours.

3   Problem Formulation

Let us build on the standard Online Convex Optimization framework [32] in which a learning agent, at each round t, has to select a parameter $\theta_t \in \Theta$, representing a strategy, where $\Theta \subset \mathbb{R}^d$ is a closed and convex set of a finite d dimensional Euclidean space. At each round t, the agent receives a loss $f_t(\theta_t)$ where $f_t : \Theta \rightarrow [\epsilon_l, \epsilon_u]$ is a convex and differentiable function, where ϵ_l, ϵ_u are the minimum and maximum value of the function $f_t(\cdot)$, respectively, and $0 \leq \epsilon_l < \epsilon_u$. The objective of the learning agent $\mathfrak{U}$ is to minimize the regret $R_T(\mathfrak{U})$ over a given time horizon $T \in \mathbb{N}$, *i.e.*, the difference between the loss suffered by the algorithm $\mathfrak{U}$ and the one suffered from the best fixed decision in hindsight, formally defined as:

$$R_T(\mathfrak{U}) := L_T - \bar{L}_T,$$

where $L_T := \sum\limits_{t=1}^{T} f_t(\theta_t)$ and $\bar{L}_T := \sum\limits_{t=1}^{T} f_t(\bar{\theta})$ are the loss accumulated by the running algorithm $\mathfrak{U}$ and the smallest loss obtainable by a clairvoyant selection of the parameters, i.e., $\bar{\theta} := \arg\inf\limits_{\theta \in \Theta} \sum\limits_{t=1}^{T} f_t(\theta)$, respectively.

In the COCO setting, we are interested in those algorithms $\mathfrak{U}$ for which the regret $R_T(\mathfrak{U})$ is bounded by a sub-linear function of the time horizon T, and, at the same time, perform throughout the optimization at least as well as an established default parameter $\tilde{\theta} \in \Theta$, selected at the beginning of the learning process. While the former requirement represents the so-called no-regret property of an algorithm [9], the latter one is formally defined as follows:

Definition 1. *An online algorithm $\mathfrak{U}$ is said to be* conservative *if it satisfies the following conservativeness constraint for each $t \in [T]$:*

$$L_t \le (1+\alpha)\tilde{L}_t, \tag{1}$$

where $\alpha > 0$ is the conservativeness level required by the problem, and $\tilde{L}_t := \sum\limits_{k=1}^{t} f_k(\tilde{\theta})$ is the cumulative loss of the default parameter $\tilde{\theta}$ over t rounds.[1,2]

From now on, we will refer to the quantity $Z_t(\mathfrak{U}) := (1+\alpha)\tilde{L}_t - L_t$ as the *budget* of the algorithm $\mathfrak{U}$, i.e., the advantage in terms of loss accumulated by $\mathfrak{U}$ over time w.r.t. the one provided by a constant choice of the default parameter $\tilde{\theta}$. We also assume that there exists $\mu > \epsilon_l$ s.t. $\tilde{L}_t \ge \mu t$, which imply that the fixed strategy $\tilde{\theta}$ is sub-optimal.

We remark that, in this work, we require the constraint in Eq. (1) to be satisfied at each round $t \in [T]$. Indeed, any Online learning algorithm $\mathfrak{U}$ providing a regret of $R_t(\mathfrak{U}) \le \xi\sqrt{t}$ is guaranteed to satisfy the above constraint for $t > \left(\frac{\xi}{\alpha\mu}\right)^2$, instead we require that it holds for each $t \in [T]$.[3] Therefore, satisfying the condition imposed by our constraint requires the design of ad-hoc algorithms. Conversely, the design of algorithms which have a higher grade of conservativeness, i.e., $\alpha \le 0$ is not a viable option due to the following:

Theorem 1. *In the OCO setting, there is no algorithm $\mathfrak{U}$ which obtains $L_t \le \tilde{L}_t$, unless $\theta_t = \tilde{\theta}$ for all $t \in [T]$.*

Proof. Let k be the first round in which the algorithm $\mathfrak{U}$ plays $\theta_k \ne \tilde{\theta}$. If the loss function is $f_k(x) := f_k(\tilde{\theta}) + ||\tilde{\theta} - x||_2$, then, by the convexity of the space Θ, we can find $c \in (0,1)$ and $z \in \Theta$ s.t. $\theta_k = c\tilde{\theta} + (1-c)z$. This implies that $f_k(\theta_k) = f_k(\tilde{\theta}) + (1-c)||\tilde{\theta} - z||_2 > f_k(\tilde{\theta})$, showing that $L_t > \tilde{L}_t$.

---

[1]   The conservativeness constraint in Eq. (1) is expressed in terms of losses, as commonly done in the OCO framework. Its reformulation in terms of rewards, as commonly done in the RL and MAB fields, is straightforward.

[2]   We denote with $[T]$ the set $\{1,\dots,T\}$.

[3] This comes from the fact that $L_t - \tilde{L}_t \le R_t(\mathfrak{U})$ and $\xi\sqrt{t} \le \alpha\mu t$ holds for $t > \left(\frac{\xi}{\alpha\mu}\right)^2$.

In other words, it is impossible to guarantee that an algorithm does strictly better than or equal to a given default parameter $\tilde{\theta}$ over the entire time horizon T, unless one always plays the default parameter.

4   The Conservative Projection Algorithm

We begin this section by characterizing a set of parameters in the parameter space Θ which guarantees that their choice implies the conservativeness of an algorithm at round t. Then, we select a specific parameter from this set, thus defining the CP algorithm, and, subsequently, we show it is conservative and it has sub-linear bounds for the regret.

4.1   The Conservative Ball

Let us define the following:

Definition 2. *A conservative ball $B(\tilde{\theta}, r_t) \in \mathbb{R}^d$ is a d-dimensional ball centered in $\tilde{\theta}$ with radius:*

$$r_t := \left[1 - \left(\frac{L_{t-1} - (1+\alpha)\tilde{L}_{t-1} - \alpha\epsilon_l}{DG} + 1\right)^+\right] D, \qquad (2)$$

where $D := \sup\limits_{x,y \in \Theta} ||x - y||_2$ is a bound on the diameter of the parameter space Θ, $G := \sup\limits_{x \in \Theta} ||\nabla f_t(x)||_2$ is the upper bound on the norm of the gradient of the loss $f_t(\cdot)$, $|| \cdot ||_2$ denotes the L2 norm of a vector, and $(a)^+$ denotes the maximum between the quantity a and zero.

From now on, we refer to this ball as the *conservative ball* since this choice of r_t implies that playing any of the parameters $\theta \in B(\tilde{\theta}, r_t)$ at round t guarantees that the accrued budget $Z_t(\mathfrak{U})$ does not become negative. Formally:

Theorem 2. *Let $B(\tilde{\theta}, r_t)$ be the conservative ball defined in Eq. (2) and assume that Eq. (1) is satisfied at round $t - 1$. Then, each parameter $\theta \in B(\tilde{\theta}, r_t) \cap \Theta$ satisfies Eq. (1) at round t.*

Proof. Given $\theta \in B(\tilde{\theta}, r_t) \cap \Theta$ we have:

$$f_t(\theta) - (1+\alpha)f_t(\tilde{\theta}) \leq \langle \nabla f_t(\theta), \theta - \tilde{\theta} \rangle - \alpha f_t(\tilde{\theta}) \leq Gr_t - \alpha\epsilon_l, \qquad (3)$$

where the first inequality is given by the convexity of $f_t(\cdot)$, and the second inequality is given by the Cauchy-Schwarz inequality and by the fact that $\theta \in B(\tilde{\theta}, r_t)$ implies $||\tilde{\theta} - \theta||_2 \leq r_t$. Let us consider two cases: $r_t < D$, and $r_t = D$.

Case $r_t < D$: In this case, the value of the radius is $r_t = \frac{(1+\alpha)\tilde{L}_{t-1} - L_{t-1} + \alpha\epsilon_l}{G}$. By substituting it in Eq. (3), we conclude that:

$$f_t(\theta) - (1+\alpha)f_t(\tilde{\theta}_t) \leq (1+\alpha)\tilde{L}_{t-1} + L_{t-1}. \qquad (4)$$

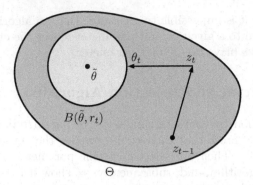

Fig. 1. Visual representation of CP. In green the conservative ball $B(\tilde{\theta}, r_t)$, and in red the parameter set Θ. The CP algorithm selects the parameter θ_t for round t by projecting the parameter z_t, selected by $\mathcal{A}$, on the conservative ball. (Color figure online)

Case $r_t = D$: From the fact that $r_t \geq 0$ and using Eq. (2), we obtain that:

$$\frac{L_{t-1} - (1+\alpha)\tilde{L}_{t-1} - \alpha\epsilon_l}{GD} + 1 \leq 0 \tag{5}$$

$$GD - \alpha\epsilon_l \leq (1+\alpha)\tilde{L}_{t-1} + L_{t-1}. \tag{6}$$

Combining the above result with the inequality in Eq. (3), provides the same result presented in Eq. (4).

The proof is concluded by rearranging the terms of Eq. (4).

Notice that the projection of a generic parameter z_t on the ball $B(\tilde{\theta}, r_t)$ can be computed analytically and efficiently. Indeed, the projection operation on the conservative ball satisfies the following:

$$\theta_t = \Pi_{B(\tilde{\theta}, r_t)}(z_t) = \beta_t\tilde{\theta} + (1 - \beta_t)z_t, \tag{7}$$

where

$$\beta_t = \begin{cases} 1 - \frac{r_t}{||z_t - \tilde{\theta}||_2} & z_t \notin B(\tilde{\theta}, r_t) \\ 0 & z_t \in B(\tilde{\theta}, r_t) \end{cases}. \tag{8}$$

In what follows, we choose z_t as the parameter provided by a generic OCO algorithm at round t.

4.2   Description of the CP Algorithm

Theorem 2 provides a way to choose a sequence of parameters over time, for which the conservativeness constraint is satisfied. The CP algorithm uses this result by choosing, at each round t, the parameter θ_t in the ball $B(\tilde{\theta}, r_t)$ as close as possible to the prediction z_t provided by the OCO algorithm fed using the pseudo-loss function $g_{t-1}(z_{t-1}) := (1 - \beta_{t-1})f_{t-1}(z_{t-1})$, $i.e.$, it selects a convex

---

Algorithm 1. CP

Require: Online learning algorithm $\mathcal{A}$, conservativeness level $\alpha > 0$, default parameter $\tilde{\theta} \in \Theta$
 1: Set $\tilde{L}_0 \leftarrow 0$, $L_0 \leftarrow 0$, and $\beta_0 \leftarrow 1$
 2: **for** $t \in [T]$ **do**
 3:     Get point z_t from $\mathcal{A}$ applied to loss $g_{t-1}(z_{t-1})$
 4:     Compute r_t as in Eq. 2
 5:     Select $\theta_t = \Pi_{B(\tilde{\theta}, r_t)}(z_t)$
 6:     Suffer loss $f_t(\theta_t)$
 7:     Observe $f_t(z_t)$ and $f_t(\tilde{\theta})$
 8:     Set $g_t(z_t) \leftarrow (1 - \beta_t) f_t(z_t)$
 9: **end for**

---

combination of the default parameter $\tilde{\theta}$ and z_t. The intuition behind this choice is that we want to choose θ_t as close as possible to the no-regret prediction z_t of the OCO algorithm, that is guaranteed to have sub-linear regret. Furthermore, we show that this algorithm increases the radius r_t over time, and therefore, in finite-time, the conservative ball includes the parameter z_t, allowing CP to have a sub-linear regret. Finally, we remark that the CP algorithm is designed so that the more the default parameter $\tilde{\theta}$ is distant from the optimal one, the more the value of the radius r_t increases, which, in its turn, decreases the cost of guaranteeing conservativeness.

The pseudo-code of the CP algorithm is presented in Algorithm 1, and its visual representation is depicted in Fig. 1. The algorithm requires as input a generic Online learning algorithm $\mathcal{A}$, which selects the parameter z_t to play at each round t, a conservativeness level $\alpha > 0$, and the default parameter $\tilde{\theta} \in \Theta$. At first, we set the initial value of the cumulative losses $L_0 = 0$, that of the default parameter $\tilde{L}_0 = 0$ (Line 1), and we set the parameter $\beta_0 = 1$. Afterwards, at each round t, z_t is chosen by the algorithm $\mathcal{A}$ by considering the pseudo-loss $g_t(x)$ (Line 3). Thanks to a projection operation (Line 5), which projects z_t into the conservative ball $B(\tilde{\theta}, r_t)$, the resulting parameter θ_t satisfies the conservativeness constraint in Eq. (1). Finally, the algorithm suffers the loss $f_t(\theta_t)$, and observes $f_t(z_t)$ and $f_t(\tilde{\theta})$, *i.e.*, the loss of the algorithm $\mathcal{A}$ and the default parameter $\tilde{\theta}$, respectively (Lines 7–8).

Notice that, from a computational point of view, the CP algorithm has a small computational overhead w.r.t. the original Online learning algorithm $\mathcal{A}$, *i.e.*, an overhead proportional to d, due to the additional projection on the conservative ball and the evaluation of the losses $f_t(\theta_t)$, and $f_t(\tilde{\theta})$.

4.3   Analysis of the CP Algorithm

In this section, we prove that CP has the desired conservativeness property and maintains the sub-linear regret of the subroutine algorithm $\mathcal{A}$. Since the CP algorithm selects a parameter θ_t inside the conservative ball $B(\tilde{\theta}, r_t)$, a straightforward corollary of Theorem 2 guarantees that the conservativeness constraint is satisfied. Formally:

Corollary 3. *The CP algorithm applied to a generic Online learning algorithm $\mathcal{A}$ is conservative.*

Once we established the conservativeness of our approach, we need to prove that the CP algorithm has sub-linear regret. Intuitively, we need to show that the radius r_t grows over time, and eventually includes the entire space Θ, so that from a specific round we are allowed to follow the no-regret choice z_t. Formally, we show the following:

Theorem 3. *Consider any OCO algorithm $\mathcal{A}$ which guarantees a regret of $R_T(\mathcal{A}) \leq \xi\sqrt{T}$. The CP algorithm using $\mathcal{A}$ as subroutine has the following regret bound:*

$$R_T(CP) \leq \xi\sqrt{T} + \tau DG, \tag{9}$$

for any $T > \tau$, where:

$$\tau := \frac{2\alpha\mu(DG + \alpha\mu) + \xi\left(\sqrt{\xi^2 + 4\alpha\mu(DG + \alpha\mu)} + \xi\right)}{2\alpha^2\mu^2}. \tag{10}$$

Proof. Using the convexity of the loss functions on the regret and the definition of θ_t in Eq. (7), we have:

$$L_T - \tilde{L}_T \leq \sum_{t=1}^{T}[\beta_t f_t(\tilde{\theta}) + (1 - \beta_t)f_t(z_t) - f_t(\tilde{\theta})]$$

$$= \sum_{t=1}^{T}(1 - \beta_t)[f_t(z_t) - f_t(\tilde{\theta})] \tag{11}$$

$$\leq \sup_{\theta \in \Theta}\left(\sum_{t=1}^{T}(1 - \beta_t)[f_t(z_t) - f_t(\theta)]\right) \leq \xi\sqrt{T}. \tag{12}$$

This shows that the CP algorithm has sub-linear regret w.r.t. an algorithm that always chooses the default parameter $\tilde{\theta}$ over the entire time horizon T.

Combining Eq. (8) and (2), we have:

$$\beta_t \leq 1 - \frac{r_t}{||z_t - \tilde{\theta}||_2} \leq 1 + \frac{L_{t-1} - (1 + \alpha)\tilde{L}_{t-1} - \alpha\epsilon_l}{DG} \tag{13}$$

$$\leq 1 + \frac{\xi\sqrt{t} - (t - 1)\mu\alpha}{DG}, \tag{14}$$

where we used the bound in Eq. (12), the fact that the space Θ has radius D, and that $\tilde{\theta}$ is not a no-regret strategy, and, hence, there exists a $\mu > \epsilon_l > 0$ s.t. $\tilde{L}_{t-1} > \mu(t - 1)$.

On the other hand, we assumed that $\mathcal{A}$ is a no-regret strategy and, therefore, the regret of the algorithm $\mathcal{A}$ is sub-linear, this means that there exists a round $\tau > 0$ s.t. Eq. (14) is negative, and, consequently, for $t > \tau$, defined in Eq. (10)

we have $\beta_t = 0$. The value of τ is provided by the solution of the following equation $1 + \frac{\xi\sqrt{\tau} - \tau\mu\alpha}{DG} = 0$.

What we showed above also proves that the CP algorithm for $t > \tau$ eventually plays the same parameter as $\mathcal{A}$ since for all $t > \tau$ the pseudo-losses $g_t(\cdot)$ and the true losses $f_t(\cdot)$ coincide. Indeed, the regret of the CP algorithm can be written as:

$$R_T(CP) \leq \sum_{t=1}^{\tau}\left[\beta_t f_t(\tilde{\theta}) + (1 - \beta_t)f_t(z_t) - f_t(\bar{\theta})\right] + \sum_{t=\tau+1}^{T}(f_t(z_t) - f_t(\bar{\theta})) \quad (15)$$

$$\leq \sum_{t=1}^{\tau}\beta_t\left[f_t(\tilde{\theta}) - f_t(z_t)\right] + \sum_{t=1}^{T}[f_t(z_t) - f_t(\bar{\theta})] \quad (16)$$

$$\leq \sum_{t=1}^{\tau}\beta_t\langle\nabla f_t(\tilde{\theta}), \tilde{\theta} - z_t\rangle + \sum_{t=1}^{T}[f_t(z_t) - f_t(\bar{\theta})] \quad (17)$$

$$\leq \tau DG + \xi\sqrt{T}, \quad (18)$$

where the inequality in Eq. (15) uses the convexity of $f_t(\cdot)$. Equation (16) comes from the extension of the time horizon from $\{\tau, \ldots, T\}$ to $\{1, \ldots, T\}$. Equation (17) follows from the convexity of $f_t(\cdot)$ and the inequality in Eq. (18) follows from the Cauchy-Schwarz inequality on the first term while the second term is the regret of the used no-regret algorithm $\mathcal{A}$.

A regret of order $\mathcal{O}(\sqrt{T})$ is tight in general OCO problems [1], but there exists specific settings in which a $\mathcal{O}(\log T)$ regret can be achieved, e.g., in the case of H-strongly convex losses or in the case of exp-concave losses [16]. In such settings, the CP algorithm guarantees $\mathcal{O}(\log T)$ regret together with the conservative constraint, formally:

Theorem 4. *Consider any OCO algorithm $\mathcal{A}$ which guarantees a regret of $R_T(\mathcal{A}) \leq \rho\log(T)$. The CP algorithm using $\mathcal{A}$ as subroutine has the following regret bound:*

$$R_T(CP) \leq \rho\log(T) + \tau DG, \quad (19)$$

for any $T > \tau$, where:

$$\tau := \frac{\alpha e^2\mu(DG + \alpha\mu) + 2\rho\left(\sqrt{\alpha e^2\mu(DG + \alpha\mu) + \rho^2} + \rho\right)}{e^2\alpha^2\mu^2}. \quad (20)$$

Proof. The proof is similar to that of Theorem 4, we only report the steps that are significantly different from it. From Eq. (12), which holds also in this setting, we obtain:

$$L_T - \tilde{L}_T \leq \rho\log(T). \quad (21)$$

This shows that the regret w.r.t. an algorithm which always chooses the default parameter $\tilde{\theta}$ is of the order $\mathcal{O}(\log(T))$. Following the same steps used to derive Eq. (14), we have that $\beta_t \leq 1 + \frac{\rho\log(t) - t\mu\alpha}{DG}$. Therefore, β_t is zero after τ rounds, where τ is defined in Eq. (20).[4] Finally, using the same argument used to derive Eq. (18), we obtain the bound present in the theorem.

Notice that for Theorem 3 and 4 we have that $\tau \propto 1/\mu$, meaning that for default parameters $\tilde{\theta}$ with smaller accrued losses w.r.t. the optimum $\bar{\theta}$ (and hence smaller μ), the CP algorithm is required to wait longer to play the action prescribed by the no-regret strategy $\mathcal{A}$. Moreover, the bound shows a dependence $\tau \propto 1/\alpha$, meaning that a tighter conservative constraint makes the problem more challenging for the CP algorithm.[5]

5   Experiments

This section provides the experimental study of the proposed algorithm for the COCO setting, where we use OGD [41] as subroutine. We evaluate the performance of the CP-OGD in three settings: a synthetically generated regression problem, and two real-world classification scenarios. We compare our performances to OGD [41], the non-conservative version of the proposed algorithm, AdaGrad [12], a state-of-the-art algorithm of online optimization which has theoretical guarantees on the regret, the Conservative Switching (CS) algorithm, a naive conservative baseline, and the Constrained Reward Doubling Guess (CRDG). CS is a budget-first algorithm we designed. This algorithm plays the fixed default action until enough budget has been accrued, then it plays the no regret strategy. We described it and provide its theoretical properties in Appendix B.1. As for CP, in CS we consider OGD as subroutine and, thus, will refer to it as CS-OGD. CRDG is a conservative baseline obtained by combining the Reward Doubling Guess algorithm [33], originally designed for unconstrained online optimization setting, with the *Constraint Set Reduction* procedure presented in [10]. We provide the its detailed pseudo-code and a discussion on its theoretical properties in Appendix B.2.

For CP-OGD, CS-OGD, and OGD we initialize the learning rate $\eta_t = \frac{K}{\sqrt{t}}$, where $K = \frac{D}{G\sqrt{2}}$ is chosen to minimize the theoretical regret bound of OGD, while for AdaGrad we initialize the parameter $\alpha_t = \frac{1}{\sqrt{t}}$, as prescribed in [20]. For the CRDG algorithm we set $\epsilon = \mu\alpha/2$ to guarantee the conservativeness constraint in Definition 1 with level α (as CP-ODG and CS-ODG do). We evaluate the algorithms in terms of regret $R_t(\mathfrak{U})$, and budget $Z_t(\mathfrak{U})$. The code to run the experiments is available at: https://github.com/martinobdl/safe_OCO.

5.1   Synthetic Regression Dataset

We analyze a synthetic online linear regression environment, where the agent is presented with a vector $x_t \sim U([0,1]^d) \subset \mathbb{R}^d$, *i.e.*, a d dimensional vector drawn

---

[4] The derivation of τ is provided in Lemma 4, reported in Appendix A for space reasons.

[5] In Appendix D.3 we performed experiments to explore the relationship between the conservativeness and performance of the CP algorithm.

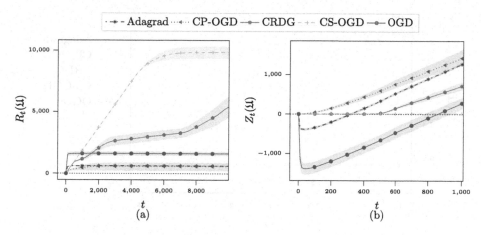

Fig. 2. Results on the synthetically generated regression dataset: (a) regret $R_T(\mathfrak{U})$; (b) magnification of the budget $Z_t(\mathfrak{U})$ over the first 10^3 samples. $\tilde{\theta}$ has been chosen so that $\tilde{D} = 0.5$.

uniformly from $[0,1]^d$, and the target value is generated as $y_t = \langle x_t, \bar{\theta} \rangle + \gamma_t$, where $\bar{\theta} \in \Theta = [-1,1]^d$ is the unknown optimal parameter, and γ_t is a noise term that we considered $i.i.d.$ with zero mean. Each algorithm provides a prediction $\hat{y}_t = \langle x_t, \theta_t \rangle$, where θ_t is the chosen parameter for round t, and suffers a loss $f_t(\theta_t) := (\langle x_t, \theta_t - \bar{\theta} \rangle - \gamma_t)^2$.

We set $d = 40$, γ_t from a truncated Gaussian distribution $\mathcal{N}(0, 0.15^2)$ with values in $[-1,1]$, $T = 10^4$, and we fix $\bar{\theta} := [0, \dots, 0]$. The conservativeness level is set to $\alpha = 0.01$. In this setting, the bound on the gradient is $G = 2(\sqrt{2d}+1)\sqrt{d}$, the minimum and maximum loss are $\epsilon_l = 0$ and $\epsilon_u = \sqrt{2d}+1$, respectively, and the bound on the diameter of the decision space is $D = \sqrt{2d}$. We ran the experiment 30 times and averaged the results. The confidence intervals on the mean, represented in the figures as semi-transparent areas, are the 95% confidence intervals computed by statistical bootstrap. Multiple default parameters $\tilde{\theta}$ have been considered in this setting so that $\tilde{D} := ||\tilde{\theta} - \bar{\theta}||_2 \in \{0.5, 1, \dots, 3, 3.5\}$.

Results. Figure 2 shows the results for experiments where the default parameter has a distance from the optimum of $\tilde{D} = 0.5$. Figure 2a shows that all the algorithms, but CRDG, on average converge to the optimal solution since the regret $R_T(\mathfrak{U})$ is asymptotically approaching a constant value. In particular, AdaGrad and CP-OGD perform comparably in terms of regret, OGD has slightly worse performance, and CRDG and CS-OGD provide a regret more than 3 times larger than the other algorithms over the analyzed time horizon T. The magnification of the budget $Z_t(\mathfrak{U})$ over the first 1,000 rounds, provided in Fig. 2b, shows that OGD and AdaGrad have a negative budget during the first ≈ 900 and ≈ 300 rounds, respectively, while CP-OGD, CS-OGD, and CRDG guarantee the conservativeness constraint at each round, *i.e.*, they have $Z_t(\mathfrak{U}) > 0$, for all $t \in [T]$.

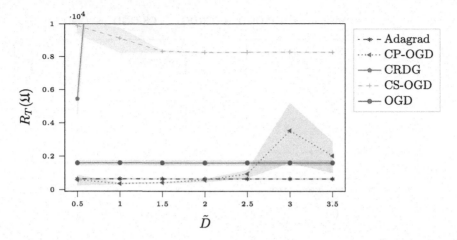

Fig. 3. Regret $R_T(\mathfrak{U})$ at the end of the time horizon T as the distance from the optimum $\tilde{D}$ varies for the synthetic dataset.

These results suggest that the proposed CP-OGD is the only algorithm, among the tested ones, capable of maintaining a small regret while, at the same time, being conservative.

Figure 3 presents the behaviour of the regret $R_T(\mathfrak{U})$ as the distance $\tilde{D}$ between the optimum and the default parameter varies. For values of the distance $\tilde{D} < 2.5$, *i.e.*, default parameters which are close to the optimum one, CP-OGD provides a smaller regret than that of all the other algorithms on average. Instead, if $\tilde{D} \geq 3$, the fact that it is constrained to maintain a positive budget penalizes CP-OGD in terms of regret. In such a situation, OGD and AdaGrad provide a smaller regret than CP-OGD. This suggests that the proposed approach might provide a large regret if the default parameter $\tilde{\theta}$ is far from the optimum one $\bar{\theta}$.

5.2   Online Classification: The IMDB Dataset

The second set of experiments has been run on the IMDB dataset [24], consisting of 50,000 reviews of movies and labels classifying the reviews as positive or negative. Data has been preprocessed as done by [20]. The general setup for the online logistic regression model is as follows: the algorithm processes a single feature vector $x_t \in \{0,1\}^d$ with $d = 10^4$, predicts the probability of belonging to the positive class as $\hat{y}_t \in [0,1]$ as $\hat{y}_t = \sigma(\langle x_t, \theta_t \rangle)$, where $\theta_t \in \Theta = [-2,2]^d$ and suffers a loss given by the binary cross entropy defined as $f_t(\theta_t) = -[y_t \log(\hat{y}_t) + (1 - y_t)\log(1 - \hat{y}_t)]$, where $y_t \in \{0,1\}$ is the true sample class.[6] In this setting the gradient is bounded by $G = \sqrt{d}$, the diameter by $D = 2\sqrt{d}$ and we set $\alpha = 0.01$. To bound the maximum and minimum loss needed by the CS algorithm, we clipped the loss between $\epsilon_l = 1e^{-4}$ and $\epsilon_u = 10$. The default parameter $\tilde{\theta}$ has been

---

[6] $\sigma(x) := 1/(1 + exp(-x))$ is the sigmoid function.

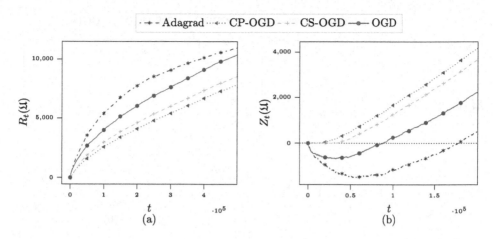

Fig. 4. Results for the IMDB movie dataset: (a) regret, (b) budget for the first 2.5×10^5 samples.

generated by training a batch logistic regression using $1,000$ samples at random from the dataset. Notice that the IMDB dataset is known to be a challenging setting for OGD [20] due to the sparse nature of its input, setting for which an adaptive step size algorithm, like AdaGrad, generally performs better in terms of regret than the single-pass ones. We could not run the CRDG algorithm on the IMDB dataset since its computational requirements in this setting were too demanding due to the large number of features.

Results. Figure 4a shows the regret $R_t(\mathfrak{U})$ for the analyzed algorithms. Both CP-OGD and CS-OGD outperform AdaGrad and OGD in terms of regret. This happens because AdaGrad and OGD surpass the performance of the default parameter only after many rounds. In fact, this specific setting is challenging for OGD [20], while CP-OGD and CS-OGD exploit successfully the information provided by $\tilde{\theta}$. The results suggest that conservative algorithms might also outperform their non-conservative counterparts in some specific challenging optimization problems. Furthermore, Fig. 4b shows that, even in this setting, the budget of the OGD and AdaGrad algorithms is negative for the first $\approx 100,000$ and $\approx 200,000$ rounds, respectively, while the budget of the CP-OGD and CS-OGD is positive for all $t \in [T]$, which is in line with the theoretical analysis we provided before.

5.3   Online Classification: The SpamBase Dataset

The SpamBase dataset is taken from the UCI repository and contains $4,601$ emails labeled as spam or ham [11]. The dataset has been normalized so that the input vector $x_t \in [0,1]^d$, with $d = 57$. The safe parameter $\tilde{\theta}$ has been generated by training a batch logistic regression on 100 samples chosen at random from the dataset. The values for the parameters not explicitly defined in this section are the same as those used in the IMDB experiment.

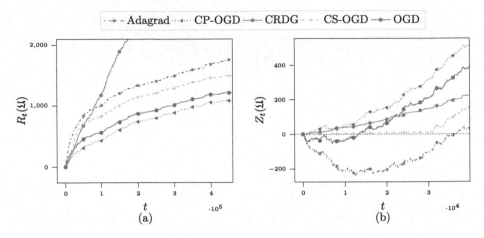

Fig. 5. Results for the SpamBase dataset: (a) regret, (b) budget for the first 4×10^4 samples.

Results. Figure 5a shows the regret suffered by the algorithms on the SpamBase dataset. Even in this case, CP-OGD outperforms all the others, and, by looking at Fig. 5b, we see that also in this experiment the budget of OGD and AdaGrad is negative for $\approx$10,000 and $\approx$30,000 rounds, respectively. Finally, the CRDG algorithm satisfies the budget constrain during the entire learning time horizon but accumulates a large regret over the time horizon.

6   Conclusions

The focus of this paper is to solve the problem of conservative optimization in an online setting with adversarial environments, in which we require an algorithm to provide sub-linear regret while performing at least as well as a given fixed strategy. To solve this problem, we proposed the CP algorithm, showed that it satisfies the conservativeness constraint, and proved that it maintains the same regret order the OCO algorithm it uses as a subroutine. Furthermore, we ran an extensive experimental campaign on synthetic and real-wolrd data, showing that the CP algorithm is competitive in terms of regret with OGD, AdaGrad, CS, and CRDG while also behaving conservatively.

An interesting direction is whether the assumption that the default strategy $\tilde{\theta}$ is fixed can be relaxed to include specific classes of time-varying strategies. Another line of research that might be promising is the use of the definition of the conservative ball to design algorithms also for the unconstrained online optimization setting.

References

1. Abernethy, J., Bartlett, P.L., Rakhlin, A., Tewari, A.: Optimal strategies and minimax lower bounds for online convex games. University of California, Berkeley, United States of America, Technical report (2008)
2. Auer, P., Cesa-Bianchi, N., Fischer, P.: Finite-time analysis of the multiarmed bandit problem. Mach. Learn. **47**(2–3), 235–256 (2002)
3. Auer, P., Ortner, R.: UCB revisited: improved regret bounds for the stochastic multi-armed bandit problem. Period. Math. Hung. **61**(1–2), 55–65 (2010)
4. Azuma, K.: Weighted sums of certain dependent random variables. Tohoku Math. J. Second Ser. **19**(3), 357–367 (1967)
5. Besson, L., Kaufmann, E.: What doubling tricks can and can't do for multi-armed bandits. arXiv preprint arXiv:1803.06971 (2018)
6. Blanzieri, E., Bryl, A.: A survey of learning-based techniques of email spam filtering. Artif. Intell. Rev. **29**(1), 63–92 (2008)
7. Browne, S.: Beating a moving target: optimal portfolio strategies for outperforming a stochastic benchmark. In: Handbook of the Fundamentals of Financial Decision Making: Part II, pp. 711–730. World Scientific (2013)
8. Cesa-Bianchi, N., Conconi, A., Gentile, C.: On the generalization ability of on-line learning algorithms. IEEE Trans. Inform. Theory **50**(9), 2050–2057 (2004)
9. Cesa-Bianchi, N., Lugosi, G.: Prediction, Learning, and Games. Cambridge University Press, Cambridge (2006)
10. Cutkosky, A., Orabona, F.: Black-box reductions for parameter-free online learning in banach spaces. In: Conference On Learning Theory (COLT), pp. 1493–1529. PMLR (2018)
11. Dua, D., Graff, C.: UCI machine learning repository (2017). http://archive.ics.uci.edu/ml
12. Duchi, J., Hazan, E., Singer, Y.: Adaptive subgradient methods for online learning and stochastic optimization. J. Mach. Learn. Res. **12**(7) (2011)
13. Garcelon, E., Ghavamzadeh, M., Lazaric, A., Pirotta, M.: Conservative exploration in reinforcement learning. In: International Conference on Artificial Intelligence and Statistics (AISTATS), pp. 1431–1441 (2020)
14. Garcelon, E., Ghavamzadeh, M., Lazaric, A., Pirotta, M.: Improved algorithms for conservative exploration in bandits. In: Conference on Artificial Intelligence (AAAI), pp. 3962–3969 (2020)
15. Garcıa, J., Fernández, F.: A comprehensive survey on safe reinforcement learning. J. Mach. Learn. Res. **16**(1), 1437–1480 (2015)
16. Hazan, E., Agarwal, A., Kale, S.: Logarithmic regret algorithms for online convex optimization. Mach. Learn. **69**(2–3), 169–192 (2007)
17. Hutter, M., Poland, J.: Adaptive online prediction by following the perturbed leader. J. Mach. Learn. Res. **6**(Apr), 639–660 (2005)
18. Kakade, S., Langford, J.: Approximately optimal approximate reinforcement learning. In: International Conference on Machine Learning (ICML), vol. 2, pp. 267–274 (2002)
19. Kazerouni, A., Ghavamzadeh, M., Yadkori, Y.A., Van Roy, B.: Conservative contextual linear bandits. In: Neural Information Processing Systems (NeurIPS), pp. 3910–3919 (2017)
20. Kingma, D.P., Ba, J.: Adam: a method for stochastic optimization. arXiv preprint arXiv:1412.6980 (2014)

21. Koolen, W.M.: The pareto regret frontier. In: Neural Information Processing Systems (NeurIPS), pp. 863–871 (2013)
22. Lacoste, A., Luccioni, A., Schmidt, V., Dandres, T.: Quantifying the carbon emissions of machine learning. arXiv preprint arXiv:1910.09700 (2019)
23. Lattimore, T.: The pareto regret frontier for bandits. In: Neural Information Processing Systems (NeurIPS), pp. 208–216 (2015)
24. Maas, A., Daly, R.E., Pham, P.T., Huang, D., Ng, A.Y., Potts, C.: Learning word vectors for sentiment analysis. In: Annual Meeting of the Association for Computational Linguistics: Human Language Technologies, pp. 142–150 (2011)
25. Papini, M., Pirotta, M., Restelli, M.: Adaptive batch size for safe policy gradients. In: Neural Information Processing Systems (NeurIPS), pp. 3591–3600 (2017)
26. Papini, M., Pirotta, M., Restelli, M.: Smoothing policies and safe policy gradients. arXiv preprint arXiv:1905.03231 (2019)
27. Pirotta, M., Restelli, M., Bascetta, L.: Policy gradient in Lipschitz Markov decision processes. Mach. Learn. **100**(2–3), 255–283 (2015)
28. Pirotta, M., Restelli, M., Pecorino, A., Calandriello, D.: Safe policy iteration. In: International Conference on Machine Learning (ICML), pp. 307–315 (2013)
29. Reddi, S.J., Kale, S., Kumar, S.: On the convergence of Adam and beyond. arXiv preprint arXiv:1904.09237 (2019)
30. Sani, A., Neu, G., Lazaric, A.: Exploiting easy data in online optimization. In: Neural Information Processing Systems (NeurIPS) (2014)
31. Schulman, J., Levine, S., Abbeel, P., Jordan, M.I., Moritz, P.: Trust region policy optimization. In: International Conference on Machine Learning (ICML), pp. 1889–1897 (2015)
32. Shalev-Shwartz, S., et al.: Online learning and online convex optimization. Found. Trends Mach. Learn. **4**(2), 107–194 (2011)
33. Streeter, M., McMahan, H.B.: No-regret algorithms for unconstrained online convex optimization. arXiv preprint arXiv:1211.2260 (2012)
34. Sui, Y., Burdick, J., Yue, Y., et al.: Stagewise safe Bayesian optimization with gaussian processes. In: International Conference on Machine Learning (ICML), pp. 4781–4789. PMLR (2018)
35. Sui, Y., Gotovos, A., Burdick, J., Krause, A.: Safe exploration for optimization with gaussian processes. In: International Conference on Machine Learning (ICML), pp. 997–1005. PMLR (2015)
36. Tange, O.: GNU parallel 2018. Lulu. com (2018)
37. Tsymbal, A.: The problem of concept drift: definitions and related work. Comput. Sci. Dept. Trinity Coll. Dublin **106**(2), 58 (2004)
38. Vapnik, V.: Principles of risk minimization for learning theory. In: Neural Information Processing Systems (NeurIPS), pp. 831–838 (1992)
39. Vittori, E., de Luca, M.B., Trovò, F., Restelli, M.: Dealing with transaction costs in portfolio optimization: online gradient descent with momentum. In: ACM International Conference on AI in Finance (ICAIF), pp. 1–8 (2020)
40. Wu, Y., Shariff, R., Lattimore, T., Szepesvári, C.: Conservative bandits. In: International Conference on Machine Learning (ICML), pp. 1254–1262 (2016)
41. Zinkevich, M.: Online convex programming and generalized infinitesimal gradient ascent. In: International Conference on Machine Learning (ICML), pp. 928–936 (2003)

Knowledge Infused Policy Gradients with Upper Confidence Bound for Relational Bandits

Kaushik Roy, Qi Zhang, Manas Gaur[✉], and Amit Sheth

Artificial Intelligence Institute, University of South Carolina, Columbia, USA
{kaushikr,mgaur}@email.sc.edu, qz5@cse.sc.edu, amit@sc.edu

Abstract. Contextual Bandits find important use cases in various real-life scenarios such as online advertising, recommendation systems, healthcare, etc. However, most of the algorithms use flat feature vectors to represent context whereas, in the real world, there is a varying number of objects and relations among them to model in the context. For example, in a music recommendation system, the user context contains what music they listen to, which artists create this music, the artist albums, etc. Adding richer relational context representations also introduces a much larger context space making exploration-exploitation harder. To improve the efficiency of exploration-exploitation knowledge about the context can be infused to guide the exploration-exploitation strategy. Relational context representations allow a natural way for humans to specify knowledge owing to their descriptive nature. We propose an adaptation of Knowledge Infused Policy Gradients to the Contextual Bandit setting and a novel Knowledge Infused Policy Gradients Upper Confidence Bound algorithm and perform an experimental analysis of a simulated music recommendation dataset and various real-life datasets where expert knowledge can drastically reduce the total regret and where it cannot.

1 Introduction

Contextual Bandits (CB) are an extension of the classical Multi-Armed-Bandits (MAB) setting where the arm choice depends also on a specific context [1]. As an example, in a music recommendation system, the choice of song recommendation (the arm choice) depends on the user context (user preferences concerning genre, artists, etc.). In the real world, the context is often multi-relational but most CB algorithms do not model multi-relational context and instead use flat feature vectors that contain attribute-value pairs [2]. While relational modeling allows us to enrich user context, it further complicates the exploration-exploitation problem due to the introduction of a much larger context space. Initially, when much of the space of context-arm configurations are unexplored, aggressive exploitation may yield sub-optimal total regret. Hence, a principled exploration-exploitation

© Springer Nature Switzerland AG 2021
N. Oliver et al. (Eds.): ECML PKDD 2021, LNAI 12975, pp. 35–50, 2021.
https://doi.org/10.1007/978-3-030-86486-6_3

strategy that encodes high uncertainty initially that tapers off with more information is required to effectively achieve near-optimal total regret. The Upper-Confidence-Bound (UCB) algorithm uses an additional term to model initial uncertainty that tapers off during each arm pull [3]. However, though the UCB provides a reasonable generalized heuristic, the exploration strategy can further be improved with more information about the reward distribution, for example, if it is known that the expected reward follows a Gaussian distribution. This is what Thompson Sampling does - incorporates a prior distribution over the expected rewards for each arm and updates a Bayesian posterior [4]. If external knowledge is available the posterior can be reshaped with knowledge infusion [5]. An example of this knowledge for the IMDB dataset described in Sect. 7 can be seen in Fig. 1 and the detailed formulation for the knowledge used is described in Sect. 4. A couple of issues arise with posterior reshaping: a) The choice of reshaping function is difficult to determine in a principled manner, and b) The form of the prior and posterior is usually chosen to exploit a likelihood-conjugate before analytically compute posterior estimates as sampling is typically inefficient. Similarly, the choice of reshaping function needs to either be amenable to efficient sampling for exploration or analytically computed. Thus, we observe that we can instead directly optimize for the optimal arm choice through policy gradient methods [6]. Using a Bayesian formulation for optimization of policy in functional space, we can see that the knowledge infused reshape function can be automatically learned by an adaption of the *Knowledge Infused Policy Gradients* (KIPG) algorithm for the Reinforcement Learning (RL) setting to the CB setting [7], which takes as input a state and knowledge, and outputs an action.

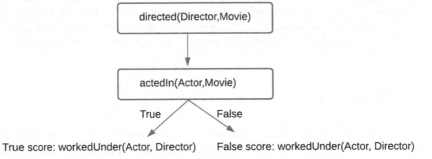

Fig. 1. Example of expert knowledge in the IMDB dataset. This says that if a director directed a movie in which an actor acted, there is a chance that the actor worked under the director.

The CB setting presents a unique challenge for knowledge infusion. Since arm pulling happens in an online fashion, the human knowledge about the user is uncertain until the human observes some arm choices. First, we adapt the KIPG algorithm from the RL to the CB setting and then we improve upon it to make it less aggressive in its knowledge infusion strategy when the human is still uncertain about the user's preferences. For this reason, we develop a UCB style uncertainty measure that considers the initial uncertainty as the human gathers more information about the user context, before providing knowledge. Thus, we develop a *Knowledge Infused Policy Gradient Upper Confidence Bound* (KIPGUCB) algorithm to incorporate human uncertainty in providing knowledge in the knowledge infusion strategy. Our methodological contributions are as follows:

- We adapt KIPG for the RL setting to the CB setting to reduce the total regret with high-quality knowledge.
- We develop a novel relational CB algorithm KIPGUCB that reduces regret through knowledge infusion with both high-quality and noisy knowledge using exploration.
- Theoretically, we observe that KIPG is fundamentally a gradient ascent method and derive a regret bound that depends on the knowledge. We also derive a confidence bound for when the knowledge is noisy.
- Empirically, through experiments on various real-life datasets, we perform analysis of settings where KIPGUCB achieves a drastic reduction in total regret. We compare KIPGUCB to KIPG without a confidence bound and compare against the Relational Boosted Bandits algorithm (RB2) [8], a state-of-the-art contextual bandit algorithm for relational domains.

2   Problem Setting

We consider the problem setting of Bernoulli Contextual Bandits with relational features. Formally, at each step k, when an arm $i \in [N] := \{1, 2, ..., N\}$ is pulled from among N arms, the reward $r_k(i) \in \{0, 1\}$ is Bernoulli. Also, pulling an arm i depends on a relational context $c_k(i)$. Since $\pi_k(i)$, which represents the probability of choosing arm i given context $c_k(i)$, is expected to be high if $P(r_k(i) = 1|c(i))$ is high, we directly maximize the total reward over K arm choices, $\sum_{k=1}^{K} \pi_k(i) r_k(i)$. Here $\pi_k(i) = \sigma(\Psi_k(i))$, and σ is the sigmoid function. $\Psi_k(i)$ is a relational function that includes the relational context $c_k(i)$.

3   Knowledge Infused Policy Gradients

In this section, we develop the formulation for the KIPG adaptation to the CB setting. We first describe policy gradients for CB, extend it to functional spaces and then use Bayes rule to derive the KIPG formulation. In next section, we show the connection of KIPG to Thompson Sampling with posterior reshaping and the Exponential Weight for Exploration and Exploitation (Exp3) algorithm [9], which is also derived from a gradient ascent procedure (mirror ascent) that can be seen as an instance of KIPG.

Policy Gradients for Contextual Bandits with Flat Feature Vectors.
In policy gradient methods the probability of picking an arm i given context
$c(i)$, is parameterized as $\pi(i) = \sigma(\theta(i)^T c(i))$. We want to maximize the expected
reward over K arm pulls $\sum_{k=1}^{K} \pi_k(i) r_k(i)$. We update the parameters for arm i,
at each $k+1$, using gradient ascent as $\theta_{k+1}(i) = \theta_k(i) + \eta \nabla_{\theta_k(i)} (\sum_k \pi_k(i) r_k(i))$.
Here we note that the gradient $\nabla_{\theta_k(i)} \pi_k(i) = \pi_k(i) \nabla_{\theta_k(i)} \log(\pi_k(i))$ and thus we
optimize:

$$\theta_{k+1}(i) = \theta_k(i) + \eta(\sum_k \pi_k(i) \nabla_{\theta_k(i)} \log(\pi_k(i)) r_k(i))$$

Policy Gradients for Contextual Bandits in Functional Space. In func-
tional space the $\theta(i)^T c(i)$ is replaced by a function $\Psi(i)$ i.e. $\pi(i) = \sigma(\Psi(i))$, where
$\Psi(i)$ is a relational function that includes context $c(i)$. Thus, the policy gradient
update becomes

$$\Psi_k(i) = \Psi_k(i) + \eta(\sum_k \pi_k(i) \nabla_{\Psi_k(i)} \log(\pi_k(i)) r_k(i)).$$

Here, $\Psi_k(i)$ at each iteration of policy gradients is grown stage wise. We start
with a $\Psi_0(i)$ and update $\Psi_K(i) = \Psi_0(i) + \sum_{k=1}^{K} \eta \delta_k(i)$, where each $\delta_k(i)$ fits a
function to $\pi_k(i) \nabla_{\Psi_k(i)} \log(\pi_k(i)) r_k(i)$ [10]. In our experiments this function is a
TILDE regression tree [11]. However, we derive a Bayesian formulation for $\pi_k(i)$
for knowledge infusion. Thus, After pulling an arm i at step k, and observing
rewards $r_k(i)$, and context $c_k(i)$, using Bayes rule we can write

$$P(\Psi_k(i)|r_k(i)) = \frac{P(r_k(i)|\Psi_k(i))P(\Psi_k(i))}{\int_{\Psi_k(i)} P(r_k(i)|\Psi_k(i))P(\Psi_k(i))}.$$

Using the sigmoid function we can set $P(r_k(i)|\Psi_k(i)) = \sigma(\Psi_k(i)) = \frac{e^{\Psi_k(i)}}{(1+e^{\Psi_k(i)})}$
and use the Bayesian posterior to obtain a prior informed policy as

$$\pi_k(i) = \frac{\sigma(\Psi_k(i))P(\Psi_k(i))}{\int_{\Psi_k(i)} \sigma(\Psi_k(i))P(\Psi_k(i))}.$$

To optimize using policy gradients, again we note that $\nabla_{\Psi_k(i)}(\pi_k(i)) = \pi_k(i) \nabla_{\Psi_k(i)} \log(\pi_k(i))$ If we use a form for $P(\Psi_k(i))$, for which the normalization doesn't depend on $\Psi_k(i)$ such as a Laplace or a Gaussian distribution, we
can take the log on both sides without loss of generality to derive the gradient
$\nabla_{\Psi_k(i)} \log(\pi_k(i))$:

$$\log(\pi_k(i)) \propto \log(\sigma(\Psi_k(i))) + \log(P(\Psi_k(i))),$$

taking the gradient gives us

$$(I_k(i) - \sigma(\Psi_k(i))) + \nabla_{\Psi_k(i)} \log(P(\Psi_k(i))),$$

where $I_k(i)$ is the indicator function representing if arm i was chosen at step k.
Now we can employ functional gradient ascent by fitting a weak learner (such as a

TILDE tree for relational context, or linear function for propositional context) to the gradient $\pi_k(i)\nabla_{\Psi_k(i)}\log(\pi_k(i))$. Note here that $\log(P(\Psi_k(i)))$ will determine the nature of knowledge infused into the policy gradient learning setup at each k. We call this approach Knowledge Infused Policy Gradients (KIPG).

4   Formulation of Knowledge Infusion

At each k, the prior over functions $\Psi_k(i)$ for each arm $P(\Psi_k(i))$ determines the knowledge infusion process. We now show the formulation for infusing arm preferences as knowledge as we use this in our experiments. Depending on the problem needs, the user may pick their choice of $P(\Psi_k(i))$ to be any distribution. Since our knowledge is given as weighted preferences over arm choices, we will cover two intuitive ways to formulate the knowledge and derive the formulation we use in our experiments.

$P(\Psi_k(i)) = Normal(\mu, \Sigma)$: Given a context included in $\Psi_k(i)$, if we want to prefer the arm choice i, we can specify this knowledge using a two step procedure. First we set $\Psi_k(i)_{knowledge} = \alpha$, where $\alpha \geq 1$. Then we set $P(\Psi_k(i)) = Normal(\mu = \Psi_k(i)_{knowledge} - \sigma(\Psi_k(i)), \Sigma = I)$. Similarly if the arm choice i is not preferred, $\Psi_k(i)_{knowledge} = -\alpha$. Here α controls how quickly knowledge infusion takes place.

$P(\Psi_k(i)) = Laplace(x, b)$: Specifying α is a tricky thing to do for a human and we would like them to able to just simply specify preference over arm choice given a context instead, if they are an expert. To model an expert

- First we set $\Psi_k(i)_{knowledge} = \text{LUB}\{\alpha\}$, where $\text{LUB}\{\alpha\}$ stands for the least upper bound from among a set of $\alpha \in \{\alpha\}$. The interpretation is that α has to be at least that high to qualify as expert knowledge. We set $\text{LUB}\{\alpha\} = K \cdot \max \pi_k(i)\nabla_{\Psi_k(i)}\log(\pi_k(i))r_k(i) = K \cdot 1 \cdot K = K^2$ as the maximum value of $\pi_k(i) = 1$ and the maximum value of $\nabla_{\Psi_k(i)}\log(\pi_k(i)) \cdot r_k(i)$ is $1 \cdot K$ as the maximum value of $\sum_{k=1}^{K} r_k(i) = K$. Thus we set $\Psi_k(i)_{knowledge} = \text{LUB}\{\alpha\} = K^2$. The interpretation is the human has to be at least as sure as the correction required to the error in arm choice i.e. the max gradient to qualify as an expert. Therefore to prefer arm i, $\alpha = K^2$ and if arm i is not preferred, $\alpha = -K^2$.
- Next, we replace the $Normal(\mu, \Sigma)$ distribution with the $Laplace(x = |\Psi_k(i)_{knowledge} - \Psi_k|, b = 1)$ distribution. Thus, we obtain that $\nabla_{\Psi_k(i)}\log(P(\Psi_k(i)) = \text{sign}(\Psi_k(i)_{knowledge} - \sigma(\Psi_k(i))) = \pm 1$. If the expert prefers the arm i, $\delta_k(i) = \delta_k(i) + 1$ and if the expert does not prefer the arm i, $\delta_k(i) = \delta_k(i) - 1$. This is very intuitive as it means that the $\Psi_k(i)$, representing chance of arm i being pulled is simply increased or decreased by an additive factor depending on expert's preference, thus preventing the need to carefully specify α.
- With this insight, it suffices for the human expert to specify knowledge as a tuple

$$\textbf{knowledge} : (\mathbf{c_k}(i), \textbf{prefer}(i) = \{0, 1\}),$$

Algorithm 1. Knowledge Infused Policy Gradients - KIPG

1: Initialize $\Psi_0(i) = 0$ $\forall$ arms i
2: **for** $k \leftarrow 1$ to K **do**
3:     set $\pi_k(i) = \sigma(\Psi_{k-1}(i))$
4:     Draw arm $i^* = \arg\max_i i \sim \pi_k(i)$ $\triangleright$ observe reward $r_k(i^*)$ and context $c_k(i^*)$
5:     Compute $\nabla_{\Psi_k(i^*)} \log(\pi_k(i^*))$ as $\triangleright \pm$ Depending on preference

$$(I_k(i^*) - \pi_k(i^*) \pm 1)$$

6:     Compute gradient as $\pi_k(i^*)\nabla_{\Psi_k(i^*)} \log(\pi_k(i^*))(r_k(i^*)+1)$ $\triangleright$ add 1 smoothing
7:     Fit $\delta_k(i^*)$ to gradient using TILDE tree
8:     Set $\Psi_k(i^*) = \Psi_{k-1}(i^*) + \eta\delta_k(i^*)$
9: return $\pi_K(i)$

---

which simply means that at step k, given the context $c_k(i)$, arm i is either preferred ($prefer(i) = 1$)) or not preferred ($prefer(i) = 0$). This is much more natural and easy for the expert human to specify. Note that if the human had a reason to specify α quantifying how quickly they want the knowledge infusion to take place depending on how sure they are (expert level), we can use the *Normal* or *Laplace* distribution form to specify without the use of LUB$\{\alpha\}$. Algorithm 1 shows the pseudocode for KIPG with expert knowledge infusion. Also, we add 1 to $r_k(i)$ so that the gradient doesn't vanish when $r(i) = 0$.

Example of Knowledge in the IMDB Dataset Using the Laplacian Formulation. At a step k, we can define knowledge over the actors set $\mathbf{A} = x\{actor1, actor2, actor3, ..\}$ with respect to a directors set $\mathbf{D} = \{director1, director2, ..\}$ and a movies set $\mathbf{M} = \{movie1, movie2, ..\}$ as,

$$(directed(\mathbf{D}, \mathbf{M}) \wedge actedIn(\mathbf{A}, \mathbf{M}), prefer(workedUnder(\mathbf{A}, \mathbf{D})) = 1).$$

This means that *The set of actors* $\mathbf{A}$, *worked under the set of directors* $\mathbf{D}$, *in the movies in the set* $\mathbf{M}$. In this example, ($directed(\mathbf{D}, \mathbf{M}) \wedge actedIn(\mathbf{A}, \mathbf{M})$ is the context $c(i)$, i is the arm label workedUnder.

Connection with Previous Work on Relational Preferences. Odom et al. [12] have previously specified relational preference knowledge in supervised learning and imitation learning settings. Using their approach, at step k, the knowledge would be incorporated by an additive term to the gradient term $(I_k(i) - \sigma(\Psi_k(i)))$. This term is $n_k(i)_t - n_k(i)_f$, where $n_k(i)_t$ is the number of knowledge sources that prefer arm i and $n_k(i)_f$ is the number of knowledge sources that do not prefer arm i, at step k. We prove in Theorem 1 that the approach of Odom et al. [12] is a specific instance of KIPG with multiple knowledge sources. For our experiments, we specify only a single source of knowledge at all steps k.

Theorem 1. *At step k, For S multiple knowledge sources, that either prefer or don't prefer arm i, $k1, k2, ..kS$, assuming independence, let $P(\Psi_k(i)) = \prod_{s=1}^{S} Laplace(|\Psi_k(i) - \Psi_k(i)_{ks}|, b = 1)$. Here $\Psi_k(i)_{ks} = \Psi_k(i)_{knowledge} \ \forall s \in S$. Then we have $\nabla_{\Psi_k} \log(\pi_k(i)) = n_k(i)_t - n_k(i)_f$.*

Proof. We know that with assuming a $Laplace(x, b)$ distribution and setting $\Psi_k(i)_{ks} = \Psi_k(i)_{knowledge} = \text{LUB}\{\alpha\} \ \forall s \in S$, we get $\nabla_{\Psi_k(i)} \log(P(\Psi_k(i))) = \sum_{s=1}^{S} \text{sign}(\text{LUB}\{\alpha\} - \sigma(\Psi_k(i)))$. We know also that $\text{sign}(\text{LUB}\{\alpha\} - \sigma(\Psi_k(i))) = \pm 1$ depending on if the expert prefers the arm i or not. Thus we get, $\sum_{s=1}^{S} \text{sign}(\text{LUB}\{\alpha\} - \sigma(\Psi_k(i))) = n_k(i)_t - n_k(i)_f$.

Connection with Thompson Sampling. We now formalize the connection between Thompson Sampling with posterior reshaping and KIPG. For arm $i \in [N]$, at every step of arm pulling $k \in [K]$, a reward $r_k(i)$ and a context $c_k(i)$ is emitted. In Thompson Sampling, the posterior $P(\Theta_k(i)|r_k(i), c_k(i))$ for parameter $\Theta_k(i)$ representing $P(r_k(i)|c_k(i))$ is updated at each step k as

$$\frac{P(r_k(i)|\Theta_k(i), c_k(i)) \Pr(\Theta_k(i)|c_k(i))}{\int_{\Theta_k(i)} P(r_k(i)|\Theta_k(i), c_k(i)) \Pr(\Theta_k(i)|c_k(i))}.$$

Finally, the optimal arm choice corresponds to the arm that has the max among the sampled $\Theta_k(i) \sim P(\Theta_k(i)|r_k(i), c_k(i))$ for each arm i. The posterior $P(\Theta_k(i)|r_k(i), c_k(i))$, can be reshaped for example by using $P(\Theta_k(i) = \mathbf{F}(\Theta_k(i)|r_k(i), c_k(i))$. The reshaping changes the sufficient statistics such as mean, variance, etc. This $\mathbf{F}$ can be informed by some knowledge of the domain. We encounter a couple of issues with Posterior Reshaping for knowledge infusion. First, that the choice of $\mathbf{F}$ is difficult to determine in a principled manner. Second, the choice of $\mathbf{F}$ must be determined such that it is amenable to sampling for exploration. Sampling itself is very inefficient for problems of appreciable size. Thus, we observe that we can instead directly optimize for the optimal arm choice through policy gradient methods. Using a Bayesian formulation for optimization of policy in functional space, we can see that the reshaped posterior after K iterations of arm pulling (where K is sufficiently high), corresponds to learning an optimal function $\Psi(i)$ since $\Psi(i)$ is high if $\mathbf{F}(\Theta_k(i)|r_k(i), c_k(i))$, representing $P(r(i) = 1|c(i))$, is high.

Connection with Exp3. Exp3 maximizes the total expected reward over K arm pulls $f = \sum_{k=1}^{K} \pi_k(i) r_k(i)$. Using the proximal definition of gradient descent and deriving the mirror descent objective after each arm pull, we have

$$\pi_k(i) = \arg\max_{\pi(i)} ((\gamma \cdot \pi(i) \cdot \nabla_{\pi_{k-1}(i)}(f)) + \mathcal{D}(\pi_{k-1}(i), \pi(i))).$$

where γ is the learning rate. Choosing $\mathcal{D}(\pi(i), \pi_{k-1}(i)) = \Phi(\pi_{k-1}(i)) - (\Phi(\pi(i)) + \nabla\Phi(\pi_{k-1}(i))(\pi_{k-1}(i) - \pi(i)))$, where Φ is a convex function, we get

$$\nabla\Phi(\pi_k(i)) = \nabla\Phi(\pi_k(i)) + \gamma \cdot \nabla_{\pi_{k-1}(i)}(f).$$

Since π is a probability we need to choose a convex Φ such that it works with probability measures. So we will choose $\Phi(\pi) = \sum_i \pi(i) \log \pi(i)$ to be negative entropy and we have

$$\log(\pi_k(i)) = \log(\pi_{k-1}(i)) + \gamma \cdot \nabla_{\pi_{k-1}(i)}(f).$$

Setting $\pi_{k-1}(i) = \sigma(\Psi_k(i))$, we get,

$$\log(\pi_k(i)) \propto \log(\sigma(\Psi_k(i))) + \log(e^{\gamma \cdot \nabla_{\pi_{k-1}(i)}(f)}),$$

where $\log P(\Psi_k(i)) = \log(e^{\gamma \cdot \nabla_{\pi_{k-1}(i)}(f)})$. Thus we see that *Exp3 can be seen as a case of applying a specific prior probability in KIPG*.

5   Regret Bound for KIPG

We now derive a bound for the total regret after K steps of KIPG to understand the convergence of KIPG towards the optimal arm choice. Since KIPG is fundamentally a gradient ascent approach, we can use analysis similar to the regret analysis for online gradient ascent to derive the regret bound [13]. Using $a^2 - (a-b)^2 = 2ab - b^2$ and letting $a = (\Psi_k(i) - \Psi^*(i))$ and $b = \nabla_{\Psi(i)_k} \sum_{k=1}^{K} \pi_k(i) r_k(i)$, We know that for a sequence over K gradient ascent iterations, $\{\Psi_k(i) | k \in [K]\}$, we have

$$(\Psi_k(i) - \Psi^*(i))^2 \leq (\Psi_{k-1}(i) - \Psi^*(i))^2 - 2\gamma(\pi_k(i) r_k(i) - \pi^*(i) r(i^*)) + \gamma^2 \mathcal{L}$$

where $\mathcal{L} \geq \nabla_{\Psi_k(i)} \sum_{k=1}^{K} \pi_k(i) r_k(i)$ is an upper bound on the gradient (Lipschitz constant) and γ is the learning rate. Using a telescoping sum over K iterations we have

$$(\Psi_K(i) - \Psi^*(i))^2 \leq (\Psi_0(i) - \Psi^*(i))^2 - 2\sum_{k=1}^{K}(\gamma(\pi_k(i) r_k(i) - \pi^*(i) r(i^*))) + \sum_{k=1}^{K} \gamma^2 \mathcal{L}$$

and therefore

$$\sum_{k=1}^{K}(\gamma(\pi_k(i) r_k(i) - \pi^*(i) r(i^*))) \leq \frac{\max_{\Psi_k(i)}(\Psi_k(i) - \Psi^*(i))^2 + \mathcal{L}^2 \sum_{k=1}^{K} \gamma^2}{2\sum_{k=1}^{K} \gamma}.$$

Solving for γ by setting $\nabla_\gamma(R.H.S) = 0$, we finally have our total regret bound over K steps as:

$$\sum_{k=1}^{K}(\gamma(\pi_k(i) r_k(i) - \pi^*(i) r(i^*))) \leq \frac{\max_{\Psi_k(i)}(\Psi_k(i) - \Psi^*(i))^2 \mathcal{L}}{\sqrt{K}}.$$

This regret bound has a very intuitive form. It shows that the regret is bounded by how far off the learned $\Psi(i)$ from the true $\Psi^*(i)$ for each arm i. Thus we expect that in the experiments, with quality knowledge infusion this gap is drastically reduced over K steps to result in a low total regret.

6   KIPG-Upper Confidence Bound

So far we have developed KIPG for the Bandit Setting and derived a regret bound. Since KIPG estimates $\pi(i)$ after each arm pull, we can sample from $\pi(i)$ and choose the max like in Thompson Sampling. However, since the arm to pull is being learned online, the uncertainty in the arm choice even with knowledge needs to be modeled. The human providing knowledge needs to observe a few user-arm pulls to gradually improve their confidence in the knowledge provided. As the data is not available offline to study by the human, it is unlikely that the knowledge provided is perfect initially. Thus, we now derive a confidence bound to quantify the uncertainty in the arm choice. At step k, let the arm choice be denoted by i^*. First we notice that $Z = |\pi_k(i^*) - \pi^*(i^*)|$, is binomial distributed at step k. Also, $\pi_k(i^*)$ is binomial distributed. However, for both we will use a Gaussian approximation and note that for this Gaussian, $\mu(Z) = 0$ and $\sigma(Z) \leq \mathbb{E}[(\pi_k(i^*) - \pi^*(i^*))^2]$, thus making this a *sub-Gaussian* [14,15]. Using Markov's inequality we have [16]:

$$P(Z > \epsilon) \leq e^{-k\epsilon}\mathbb{E}[kZ] \implies P(e^{kZ} > e^{k\epsilon}) \leq \mathbb{E}[e^{kZ}] \cdot e^{-k\epsilon}$$

where e^{kZ} is the moment-generating-function for Z. We know that e^{kZ} is convex and thus $e^{kZ} \leq \gamma(e^{kb}) + (1-\gamma)e^{ka}$ for $Z \in [a, b]$ and $\gamma \in [0, 1]$. Thus we obtain $Z \leq \gamma b + (1-\gamma)a$, which gives us $\gamma \geq \frac{Z-a}{b-a}$, therefore we know

$$e^{kZ} \leq \frac{-ae^{kb} + be^{ka}}{b-a} + \frac{Z(e^{kb} - e^{ka})}{b-a}.$$

Taking expectation on both sides we get $\mathbb{E}[e^{kZ}] \leq \frac{-ae^{kb}+be^{ka}}{b-a}$. Let $e^{g(k)} = \frac{-ae^{kb}+be^{ka}}{b-a}$, we get $g(k) = ka + \log(b - ae^{k(b-a)}) - \log(b-a)$. Using Taylor series expansion for $g(k)$ upto the second order term as $g(0) + \nabla(g(k))k + \frac{\nabla^2(g(k))k^2}{2}$, we get

$$\nabla^2(g(k)) = \frac{ab(b-a)^2(-e^{k(b-a)})}{(ae^{k(b-a)} - b)^2}.$$

We note that $ae^{t(b-a)} \geq a \implies ae^{t(b-a)} - b \geq a - b \implies (ae^{t(b-a)} - b)^{-2} \leq (b-a)^{-2}$. We know $-e^{k(b-a)} \leq -1$, therefore we obtain

$$\nabla^2(g(k)) \leq \frac{-ab(b-a)^2}{(b-a)^2} = -ab \leq \frac{(b-a)^2}{4} \implies g(k) \leq \frac{(b-a)^2}{4}\frac{k^2}{2}.$$

We know that $\mathbb{E}[e^{kZ}] \leq e^{g(k)} \implies \mathbb{E}[e^{kZ}] \leq e^{\frac{k^2(b-a)^2}{8}}$. Once again from the Markov inequality, we have

$$P(Z > \epsilon) \leq e^{-k\epsilon}\mathbb{E}[kZ] \implies P(|\pi_k(i^*) - \pi^*(i^*)| > \epsilon) \leq e^{-k\epsilon + \frac{k^2(b-a)^2}{8}}.$$

Using $k = \frac{4\epsilon}{(b-a)^2}$, by solving for the minimum of $e^{-k\epsilon + \frac{k^2(b-a)^2}{8}}$ we get $P(|\pi_k(i^*) - \pi^*(i^*)| > \epsilon) \leq e^{\frac{-2\epsilon^2}{(b-a)^2}}$. As $0 \leq (b-a) \leq 1$, we have $P(|\pi_k(i^*) - \pi^*(i^*)| > \epsilon) \leq e^{-2\epsilon^2}$ and, after K time steps,

$$P(|\pi_K(i^*) - \pi^*(i^*)| > \epsilon) \le e^{-2K\epsilon^2}.$$

Solving for ϵ we get, $\epsilon \le \frac{-\log(P(|\pi_K(i^*)-\pi^*(i^*)|>\epsilon))}{2K}$. Thus, we draw the next optimal arm choice i at $k+1$ as follows:

$$\arg\max_i \left\{ i \sim \pi_{k+1}(i) = \sigma(\Psi_k(i) + \frac{-\log(P(Z > \epsilon))}{2k}) \right\},$$

where $Z = |\pi_k(i^*) - \pi^*(i^*)|$. This confidence bound also has an intuitive form as it is reasonable that the expectation $\mathbb{E}(I(|\pi_k(i^*) - \pi^*(i^*)| > \epsilon))$ gets closer to the truth (less probable) as more arms are pulled, where I is the indicator function. Since we never actually know $\pi^*(i^*)$, we set to the current best arm choice. We expect that knowledge infusion will allow the error between the current best arm choice and $\pi^*(i^*)$ to get smaller. As P is usually initially set high and decayed as k increases causing $\log(P)$ to increase, we achieve this effect by simply using $-\log(|\pi_k(i^*) - \pi^*(i^*)|)$. Algorithm 2 shows how a simple modification to the pseudocode in Algorithm 1 can incorporate the bound derived.

Algorithm 2. KIPG Upper Confidence Bound - KIPGUCB

1: Initialize $\Psi_0(i) = 0 \; \forall$ arms i
2: **for** $k \leftarrow 1$ to K **do**
3:     set $\pi_k(i) = \sigma(\Psi_{k-1}(i))$
4:     Draw arm $i^* = \arg\max_i i \sim \pi_k(i)$ ▷ observe reward $r_k(i^*)$ and context $c_k(i^*)$
5:     Set $\pi^*(i^*) = I(\pi_k(i^*) = i^*)$
6:     Compute $\nabla_{\Psi_k(i^*)} \log(\pi_k(i^*))$ as   ▷ $\pm$ Depending on preference

$$\left(I_k(i^*) - \pi_k(i^*) \pm 1 - \frac{\log(|\pi_k(i^*) - \pi^*(i^*)|)}{2k} \right)$$

7:     Compute gradient as $\pi_k(i^*)\nabla_{\Psi_k(i^*)} \log(\pi_k(i^*))(r_k(i^*) + 1)$ ▷ add 1 smoothing
8:     Fit $\delta_k(i^*)$ to gradient using TILDE tree
9:     Set $\Psi_k(i^*) = \Psi_{k-1}(i^*) + \eta\delta_k(i^*)$
10: return $\pi_K(i)$

---

7   Experiments

The knowledge used in our experiments comes from domain experts, an example of which is seen in Sect. 4. We aim to answer the following questions:

1. How effective is the knowledge for bandit arm selection?
2. How effective is the UCB exploration strategy for bandit arm selection?

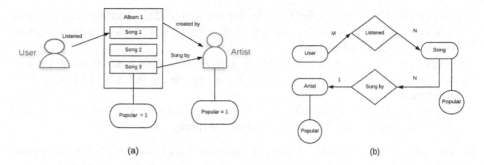

(a)   (b)

Fig. 2. Illustration of the Entity-Relationship Schema diagram for the Music Recommendation system being simulated (b) and a particular instantiation (a). Users listen to songs by artists. M Users can listen to N Songs, N Songs can be written by 1 Artist and Artists and Songs can be popular among Users.

7.1   Simulated Domains

Simulation Model: We perform experiments on a simulated music recommendation dataset. The dataset simulates songs, artists, users, and albums where there are the following user behaviors:

– Behavior A: The users are fans of one of the artists in the dataset.
– Behavior B: The users follow the most popular song.
– Behavior C: They follow the most popular artist.

We will denote the set of behaviors by **Behaviors**. Figure 2(b) shows an illustration for the Schema for the simulation model depicting that M users can listen to N songs and N songs can be sung by N artists, etc. Artists and Songs have attributes "Popular" denoting if a particular artist or a song is popular among users.

Context Induction: Once the simulation model is used to generate different users based on a predefined behavior $\in$ **Behaviors**. We need now to generate different possible user contexts from this dataset. Since the whole dataset is not available to us offline, we construct a dataset by 50 random arm choices to induce contexts. The contexts will be represented using predicate logic clauses: antecedent ($\land$ preconditions representing possible user context) $\implies$ consequent (user song choice). For this, an inductive bias needs to be provided to induce sensible clauses. Such an inductive bias is included as background knowledge to the induction program. We use the method in Hayes et al. [17] to automatically construct the inductive bias from the schema in Fig. 2(b). The clauses induced are kept if they satisfy minimum information criteria i.e. if they discriminate at least one user from another in their song choice, in the dataset. The clauses induced using the provided inductive bias and are as follows:

- sungBy(B, C) ∧ ¬ popular(C) ⟹ listens(A, B). This context says *User A listens to song B if song B is sungBy artist C. Also, C is not a popular artist,* which describes **behavior A**.
- sungBy(B, C) ∧ popular(C) ⟹ listens(A, B). This context says *User A listens to song B if song B is sungBy a popular artist C,* which describes **behavior C**.
- listened(C, B) ⟹ listens(A, B). This context says *user A listens to song B if user C listened to B,* which describes **behavior B**.

We use satisfiability of these clause antecedents as features for TILDE regression tree stumps. Figure 3 shows an example, where sigmoid of the regression values represents arm choice probability $\pi(i)$.

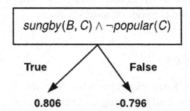

Fig. 3. Example of a TILDE regression tree stump for song choice. The tree depicts that if *if song B is sungBy artist C and also, C is not a popular artist, User A listens to B* with probability $\sigma(0.806)$. Else, *User A listens to B* with probability $\sigma(-0.796)$.

Results. We compare the RB2 algorithm with KIPG and KIPGUCB. For each type of user, at time step k, a recommendation is provided depending on the algorithm used. The regret drawn from comparison to the ground truth (GT) recommendation is recorded. The regret equation for an algorithm $\mathcal{A}$ is:

$$R_{\mathcal{A}} = \sum_{k=1}^{K} (r^{GT} - \pi_k(i^*)_{\mathcal{A}} r_k(i^*)),$$

where i^* is the optimal arm drawn from arg max over $\pi(i)$ samples at step k (See Algorithm 1, 2 - line 4). r^{GT} is the reward if the ground truth optimal arm is drawn at k.

Perfect Knowledge: The human providing knowledge may have some previous knowledge about a user in the system. In this case, it is expected that the knowledge is pretty good from the start. In this setting, we expect the regret is ordered as $R_{KIPG} < R_{KIPGUCB} < R_{RB2}$ for most $k = 1$ to K. We expected this trend since RB2 uses no knowledge and KIPGUCB moves slower towards knowledge initially. Given that the knowledge is perfect, we expect KIPG to perform the best. We set $K = 500$. Figure 4 (top row) shows that the experiments corroborate this.

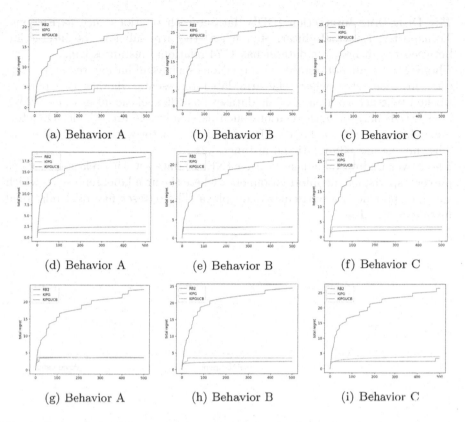

Fig. 4. Comparison of R_{RB2}, R_{KIPG}, $R_{KIPGUCB}$ for the perfect (*top*), nearly perfect (*middle*), and noisy (*bottom*) knowledge settings.

Noisy Knowledge: In this setting the human again observes some user arm interactions to improve the knowledge that they provide. In this case however, the humans observation skills are less sharp. We simulate this scenario by using noisy knowledge for $k = 1$ to 50, where perfect knowledge is provided 60% of the time instead of 80%. Here, we expect that for most $k = 1 to K$, where $K = 500$, $R_{KIPGUCB} < R_{KIPG} < R_{RB2}$. We expect this as a perfection rate of 60% means that the tempering of *Knowledge Infusion* by KIPGUCB initially leads to better total regret for KIPGUCB. Figure 4 (bottom row) shows this result.

7.2 Real-World Datasets

We also evaluate the algorithms in the following real-world datasets:

– The Movie Lens dataset with relations such as user age, movietype, movie rating, etc., where the arm label is the genre of a movie. The dataset has 166486 relational instances [18].

- The Drug-Drug Interaction (DDI) dataset with relations such as Enzyme, Transporter, EnzymeInducer, etc., where the arm label is the interaction between two drugs. The dataset has 1774 relational instances [19].
- The ICML Co-author dataset with relations such as affiliation, research interests, location, etc., where the arm label represents whether two persons worked together on a paper. The dataset has 1395 relational instances [20].
- The IMDB dataset with relations such as Gender, Genre, Movie, Director, etc., where the arm label is WorkUnder, i.e., if an actor works under a director. The dataset has 938 relational instances [21].
- The Never Ending Language Learner (NELL) data set with relations such as players, sports, league information, etc., where the arm label represents which specific sport does a particular team plays. The dataset has 7824 relational instances [22] (Fig. 5).

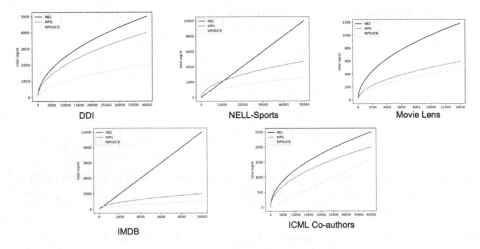

Fig. 5. Performance plots computed using total regret of RB2, KIPG, and KIPGUCB for the datasets for $k = 1$ to K. We see that KIPG and KIPG-UCB perform significantly better with expert knowledge in Movie Lens and IMDB compared to others. This is because it is relatively easier for an expert to provide knowledge in these domains. On the contrary, in the NELL-Sports, because of noisy knowledge, initially, the performance of KIPGUCB dips compared to RB2, but it increased thereafter.

We used 10 boosted trees for all the experiments and results are averaged over 5 runs. It is seen that while the total regret remains high for all the datasets over several steps of learning, both the expert knowledge and the exploration strategy using the UCB method are effective in increasing performance. The performance increase is more pronounced in the Movie Lens and IMDB datasets as the expert knowledge are relatively easier to provide for human experts. For the DDI dataset and the ICML Co-authors dataset, it is not straightforward to specify which drugs might interact or which authors may work together in a diverse academic setting. Since the knowledge comes from an expert and systematically targets faster convergence to the optimal distribution, knowledge infusion is expected to perform better. If the knowledge were noisy, the error accumulation over time may have lead to sub-optimal results. In the NELL-sports dataset, it can be seen that RB2 initially outperforms both KIPG and KIPGUCB. Policy gradient algorithms that have been studied under the contextual bandits setting were sometimes unstable [23]. In our approach, the knowledge infusion leads to the policy gradients being more stable, as seen from the curves.

8 Conclusion and Future Work

In this study, we develop a novel algorithm KIPGUCB to perform knowledge infusion in CB settings. We show that the regret bound depends on the knowledge and hence the total regret can be reduced if the right knowledge is available. Furthermore, we develop a confidence bound to account for initial uncertainty in provided knowledge in online settings. Though we have developed a general framework for knowledge infusion, we have yet to explore knowledge forms beyond preference knowledge. Furthermore, the knowledge may depend on latent behaviors that cannot be modeled such as a bias by an actor towards a particular director. Also, the actor's bias towards directors may keep changing as more data is seen. This type of non-stationarity and partial observability in context will be interesting to model. Also, if knowledge is noisy and fails to lower total regret, identifying the right descriptive question to ask the human to elicit new knowledge is an interesting future direction. Relational descriptions make tackling this issue plausible. Finally, it will be interesting to mathematically evaluate when the knowledge should be incorporated at all. We aim to tackle these issues in future work.

References

1. Langford, J., Zhang, T.: The epoch-greedy algorithm for multi-armed bandits with side information. In: Advances in Neural Information Processing Systems, vol. 20, pp. 817–824 (2007)
2. Zhou, L.: A survey on contextual multi-armed bandits. arXiv preprint arXiv:1508.03326 (2015)
3. Lai, T.L., Robbins, H.: Asymptotically efficient adaptive allocation rules. Adv. Appl. Math. 6(1), 4–22 (1985)
4. Thompson, W.R.: On the likelihood that one unknown probability exceeds another in view of the evidence of two samples. Biometrika 25(3/4), 285–294 (1933)
5. Chapelle, O., Li, L.: An empirical evaluation of Thompson sampling. In: Advances in Neural Information Processing Systems, pp. 2249–2257 (2011)
6. Peters, J., Bagnell, J.A.: Policy gradient methods. Scholarpedia 5(11), 3698 (2010)
7. Roy, K., Zhang, Q., Gaur, M., Sheth, A.: Knowledge infused policy gradients for adaptive pandemic control. arXiv preprint arXiv:2102.06245 (2021)
8. Kakadiya, A., Natarajan, S., Ravindran, B.: Relational boosted bandits (2020)

9. Auer, P., Cesa-Bianchi, N., Freund, Y., Schapire, R.E.: The nonstochastic multi-armed bandit problem. SIAM J. Comput. **32**(1), 48–77 (2002)
10. Kersting, K., Driessens, K.: Non-parametric policy gradients: a unified treatment of propositional and relational domains. In: Proceedings of the 25th International Conference on Machine Learning, pp. 456–463 (2008)
11. Blockeel, H., De Raedt, L.: Top-down induction of first-order logical decision trees. Artif. Intell. **101**(1–2), 285–297 (1998)
12. Odom, P., Khot, T., Porter, R., Natarajan, S.: Knowledge-based probabilistic logic learning. In: Twenty-Ninth AAAI Conference on Artificial Intelligence (2015)
13. Hazan, E., Rakhlin, A., Bartlett, P.L.: Adaptive online gradient descent. In: Advances in Neural Information Processing Systems, pp. 65–72 (2008)
14. Peizer, D.B., Pratt, J.W.: A normal approximation for binomial, F, beta, and other common, related tail probabilities, I. J. Am. Stat. Assoc. **63**(324), 1416–1456 (1968)
15. Buldygin, V.V., Kozachenko, Y.V.: Sub-Gaussian random variables. Ukr. Math. J. **32**(6), 483–489 (1980)
16. Cohen, J.E.: Markov's inequality and Chebyshev's inequality for tail probabilities: a sharper image. Am. Stat. **69**(1), 5–7 (2015)
17. Hayes, A.L., Das, M., Odom, P., Natarajan, S.: User friendly automatic construction of background knowledge: mode construction from ER diagrams. In: Proceedings of the Knowledge Capture Conference, pp. 1–8 (2017)
18. Motl, J., Schulte, O.: The CTU prague relational learning repository. arXiv preprint arXiv:1511.03086 (2015)
19. Dhami, D.S., Kunapuli, G., Das, M., Page, D., Natarajan, S.: Drug-drug interaction discovery: kernel learning from heterogeneous similarities. Smart Health **9**, 88–100 (2018)
20. Dhami, D.S., Yan, S., Kunapuli, G., Natarajan, S.: Non-parametric learning of Gaifman models. arXiv preprint arXiv:2001.00528 (2020)
21. Mihalkova, L., Mooney, R.J.: Bottom-up learning of Markov logic network structure. In: Proceedings of the 24th International Conference on Machine Learning, pp. 625–632 (2007)
22. Mitchell, T., et al.: Never-ending learning. Commun. ACM **61**(5), 103–115 (2018)
23. Chung, W., Thomas, V., Machado, M.C., Roux, N.L.: Beyond variance reduction: understanding the true impact of baselines on policy optimization. arXiv preprint arXiv:2008.13773 (2020)

Exploiting History Data
for Nonstationary Multi-armed Bandit

Gerlando Re, Fabio Chiusano, Francesco Trovò[(✉)], Diego Carrera,
Giacomo Boracchi, and Marcello Restelli

Dipartimento di Elettronica, Informazione e Bioingegneria, Politecnico di Milano,
Piazza Leonardo da Vinci, 32, Milan, Italy
{gerlando.re,fabio.chiusano}@mail.polimi.it,
{francesco1.trovo,diego.carrera,giacomo.boracchi,
marcello.restelli}@polimi.it

Abstract. The Multi-armed Bandit (MAB) framework has been applied successfully in many application fields. In the last years, the use of active approaches to tackle the nonstationary MAB setting, i.e., algorithms capable of detecting changes in the environment and re-configuring automatically to the change, has been widening the areas of application of MAB techniques. However, such approaches have the drawback of not reusing information in those settings where the same environment conditions recur over time. This paper presents a framework to integrate past information in the abruptly changing nonstationary setting, which allows the active MAB approaches to recover from changes quickly. The proposed framework is based on well-known *break-point prediction* methods to correctly identify the instant the environment changed in the past, and on the definition of *recurring concepts* specifically for the MAB setting to reuse information from recurring MAB states, when necessary. We show that this framework does not change the order of the regret suffered by the active approaches commonly used in the bandit field. Finally, we provide an extensive experimental analysis on both synthetic and real-world data, showing the improvement provided by our framework.

Keywords: Multi-armed bandit · Non-stationary MAB · Break-point prediction · Recurring concepts

1 Introduction

The stochastic Multi-Armed Bandit (MAB) setting has been widely used in real-world applications in sequential decision-making problems, e.g., for clinical trials [4], network routing [17], dynamic pricing [21], and internet advertising [16]. In the stochastic MAB framework, a learner selects an option – commonly referred to as *arm* – among a given finite set and observes a corresponding stochastic reward. The learning goal is to maximize the rewards collected during the entire learning process. The success of this framework is mainly due to

© Springer Nature Switzerland AG 2021
N. Oliver et al. (Eds.): ECML PKDD 2021, LNAI 12975, pp. 51–66, 2021.
https://doi.org/10.1007/978-3-030-86486-6_4

its strong theoretical properties [7], which, in practice, turns into very effective results.

Over the past few years, researchers have targeted new strategies to increase the flexibility of the MAB framework, thus foresee new applications to more complex scenarios. One of the most interesting extensions of MAB techniques consists of handling scenarios where the distribution of rewards varies over time. This is a relatively common situation in real-world dynamic pricing [21] and online advertising problems [13], where the distributions of reward for each arm can be considered stationary only over short time intervals as they might evolve due to changes of the competitors' strategies or abrupt modification of the user behaviour. While the most general situation where reward distributions are allowed to arbitrarily change over time is not tractable by this framework, it is possible to design efficient and theoretically grounded learning algorithms under some mild assumption on change type and regularity.

One of the most studied settings, which commonly occurs in practical applications, is that of the so called *abruptly changing MAB* environments, where each arm reward expected value is a piece-wise constant function of time and is allowed to change a finite number of times. MAB algorithms operating in this setting follow two mainstream approaches to cope with nonstationarity: passive [9,22], and active [8,14]. Passive methods use only the most recent rewards to define the next arm to be selected. Thus, they progressively discard rewards gathered in the far past as soon as new samples are collected. Conversely, active MAB algorithms incorporate detection procedures to spot the change and adapt the decision policy only when necessary. This approaches, from now on addressed as Change Detection MABs (CD-MABs), couple a stationary MAB procedure with a Change Detection Test (CDT) [5], as for instance in [14]. Even if from a theoretical point of view the two approaches have similar guarantees, it has been shown that the active approaches are performing generally better when their empirical performances are tested [14].

In the CD-MAB framework, a CDT is used to monitor the distribution of rewards, and as soon as this gathers enough empirical evidence to state that a change has occurred, it triggers a detection and restarts from scratch the classical MAB procedure. In practice, a change detected on a specific arm triggers a reset of both the statistics of the CDT and the corresponding arm. The major limitation of this approach is that it discards the information gathered in the past by MAB, while this could be potentially used in two situations. On the one hand, samples gathered between the occurrence and the detection of the change can be used to reconfigure the MAB over the specific arm and avoid a complete restart from scratch. On the other hand, when the process presents some regularity over time, e.g., seasonal effects, it would be ideal to identify when the arm goes back in a state that was already encountered and use the information learned about that distribution to have a fast recover after the detection.

In this paper, we present the Break-point and Recurrent MAB (BR-MAB), which extends generic CD-MABs to reuse data collected before the detection and replaces the MAB cold restart with a better initialization. Most remarkably, our neat approach still makes theoretical analysis amenable in these non-stationary settings. In particular, our novel contributions are:

- we propose a technique based on *break-point prediction* [11], to reuse the most informative samples for the current distribution gathered before the change has been detected;
- we propose a technique to identify the so-called *recurrent phases* in the MAB setting, to handle cases in which seasonality effect are present;
- we integrate these techniques in a single framework, called BR-MAB, which allow their application to a generic CD-MAB;
- we show that, BR-MAB applied to CUSUM-UCB maintains the theoretical guarantees of the original active non-stationary MAB;
- we provide extensive empirical analysis to show the improvement provided by BR-MAB, when applied to a CD-MAB, comparing its performance with the state-of-the-art techniques for non-stationary MAB settings.

2   Related Works

The algorithms designed to tackle non-stationary MAB problems with a limited number of changes are divided into passive and active approaches.

From the passive approaches, we mention the D-UCB algorithm [9], which deals with nonstationarity by giving less importance to rewards collected in the near past by weighting them by a discount factor. Conversely, the SW-UCB algorithm [9] fixes a window size and feeds a UCB-like algorithm only with the most recently collected samples. They provide guarantees on the upper-bound for the pseudo-regret of order $O(\sqrt{NB_N} \log N)$ and $O(\sqrt{NB_N \log N})$, respectively, where N is the time horizon of the learning process, and B_N is the number of changes present in the environment up to time N. Another well-analyzed passive method is the SW-TS [22], which applies the sliding window approach to the Bayesian Thompson Sampling algorithm. It provides a bound on the pseudo-regret of $O(\sqrt{N} \log N)$, if the number of changes is constant w.r.t. N. We want to remark that, in general, the passive approach does not allow for incorporating information coming from past data since their intrinsic strategy consists of systematically discarding them. Therefore, they are not appealing candidates for the approach proposed here.

For what concerns the active approaches, i.e., those algorithms using a CDT to actively detect changes in the expected values of the arms' reward distributions, the bandit literature offers a wide range of techniques [6,8,14,15]. More specifically, the CUSUM-UCB method [14] uses the CUSUM CDT to detect changes and a UCB-like approach as MAB strategy. This method provides theoretical upper bound for its regret of order $O(\sqrt{NB_N} \log(N/B_N))$. The Monitored-UCB [8] is a UCB-like policy with random exploration which uses a windowed CDT to provide a regret bound of $O(\sqrt{NB_N} \log(N))$. The GLR-klUCB [6] uses a KL-UCB algorithm in combination with a Generalized Likelihood Ratio (GLR) test as a change detection algorithm to get a regret of $O(\sqrt{NB_N} \log(N))$. Notably, the approach we propose here can be applied to any of the aforementioned active approach.

Finally, other well known and efficient methods are Adapt-EvE [10], an actively adaptive policy that uses UCB1-Tuned as a sub-algorithm and employs the Page-Hinkley test [12] to detect decreases in the mean of the optimal arm. Whenever a change-point is detected, a meta-bandit transient phase starts, whose goal is to choose between two options: reset the sub-algorithm or not. Instead, the BOCD-TS [15] uses Thompson Sampling with a Bayesian Change Point Detection algorithm. The upper-bound for these methods is unknown, hence they are accounted as heuristic algorithms.

Garivier et al. [9] showed that the problem of abruptly changing MAB has a lower bound for the expected pseudo-regret of order $\Omega(\sqrt{N})$. We recall that, in settings in which the optimal arm expected value can change without any restriction, only trivial upper bounds for the dynamic pseudo-regret $\overline{R}_N(\mathfrak{U})$ are known [2]. Conversely, if stricter assumptions holds, e.g., the occurrence of global changes, better guarantees can be derived.

3   Problem Formulation

We model our problem as a stochastic abruptly changing MAB setting, similar to what has been defined in [9], in which the arms reward distributions are constant during sequences of rounds, and they change at specific rounds unknown to the learner. Formally, at each round n over a finite time horizon N, the learner selects an arm $a_{i(n)}$ among a finite set of K arms $\mathcal{A} := \{a_1, \ldots, a_K\}$ and observes a realization of the reward $x_{i(n),n}$ from the chosen arm $a_{i(n)}$. The rewards for each arm a_i are modeled by a sequence of independent random variables $X_{i,n}$ from a distribution whose parameters are unknown to the learner. As customary in the MAB literature, here we consider Bernoulli distributed rewards, i.e., $X_{i,n} \sim Be(\mu_{i,n})$, where $\mu_{i,n}$ is the expected value of the reward for arm a_i at round n.[1] During the learning process, we denote as *breakpoints* those rounds in which the expected reward of at least one arm a_i changes. Formally, a break-point $b \in \{1, \ldots, N\}$ is a round in which for at least an arm a_i we have $\mathbb{E}[X_{i,b-1}] \neq \mathbb{E}[X_{i,b}]$. In the analysed setting, we have a set of B_N breakpoints $\mathcal{B} := \{b_1, \ldots, b_{B_N}\}$ that occur before round N (for sake of notation we define $b_0 = 1$), and whose location is unknown to the learner. The breakpoints determine a set of phases $\{\mathcal{F}_1, \ldots, \mathcal{F}_{B_N}\}$, where each phase $\mathcal{F}_\phi$ is a sequence of rounds between two consecutive breakpoints:

$$\mathcal{F}_\phi = \{n \in \{1, \ldots, N\} \mid b_{\phi-1} \leq n < b_\phi\}. \tag{1}$$

With abuse of notation, we denote with $\mu_{i,\phi} := \mathbb{E}[X_{i,n}]$, with $n \in \mathcal{F}_\phi$, the expected value of the reward of the arm a_i during the phase $\mathcal{F}_\phi$. Figure 1 illustrates an example of a specific setting with two arms a_1 and a_2 in which three phases $\mathcal{F}_1$, $\mathcal{F}_2$, and $\mathcal{F}_3$ occurs over the time horizon. Note that, differently from the classical MAB setting, a single optimal arm over the entire time horizon might not exist. Indeed, during each phase $\mathcal{F}_\phi$ we define $a_\phi^* := \arg\max_i \mu_{i,\phi}$ the arm having the largest expected reward $\mu_\phi^* := \max_i \mu_{i,\phi}$. A *policy* $\mathfrak{U}$ is a function

---

[1] The extension to other finite support distributions is straightforward and the theoretical results here provided are still valid.

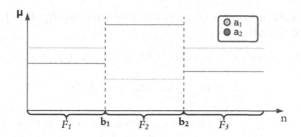

Fig. 1. Example of a nonstationary setting.

$\mathfrak{U}(h) = a_{i(n)}$ that chooses the arm $a_{i(n)}$ to play at round n according to history h, defined as the sequence of past plays and obtained rewards.

Our goal is to design a policy $\mathfrak{U}$ that minimizes the loss w.r.t. the optimal decision in terms of reward. This loss, namely the *dynamic pseudo-regret*, is:

$$\overline{R}_N(\mathfrak{U}) := \mathbb{E}\left[\sum_{n=1}^{N} \mu_n^* - \mu_{i(n),n}\right], \tag{2}$$

where $\mu_n^* := \max_{i \in \{1,\dots K\}} \mu_{i,n}$ is the optimal expected reward at round n.

In this work, we are interested in reusing the information coming from the situation in which an arm a_i has a value of the expected reward that recurs over the different phases. This models the possibility that an arm behaviour is recurring over time due to seasonality effects. Formally:

Definition 1. *A recurrent phase on arm a_i occurs when there exist two phases $\mathcal{F}_\phi, \mathcal{F}_{\phi'}$, with $\phi \neq \phi'$, s.t. $\mu_{i,\phi} = \mu_{i,\phi'}$, i.e., when the arm over the two phases has the same expected reward.*[2]

The rationale behind the above definition is that the information gathered from an arm are valid in the future, no matter how the other arms' rewards are changing, and, thus, they can be reused as long as the arm has the same reward distribution. In Fig. 1, two recurrent phases are present, i.e., as $\mathcal{F}_1$ and $\mathcal{F}_3$, since the arm a_1 has $\mu_{1,1} = \mu_{1,3}$. Notice that, if a concept recurs during phase $\mathcal{F}_2$, one might reuse the samples collected during phase $\mathcal{F}_1$ to speed up learning.

Finally, it is common in the CD-MAB literature to require two assumptions [8,14]. At first, we require a minimum magnitude for the change s.t. it is possible to detect it:

Assumption 1. $\exists\, \varepsilon \in (0,1]$, *known to the learner, such that for each arm a_i whose expected reward changes between consecutive phases ϕ and $\phi+1$, we have:*

$$|\mu_{i,\phi} - \mu_{i,\phi+1}| \geq \varepsilon. \tag{3}$$

---

[2] Since we are considering Bernoulli reward, having the same expected value also implies to have the same distribution. This definition can be easily generalized to handle other distributions, requiring that the distribution repeats over different phases.

The second assumption prevents two consecutive breakpoints from being too close in terms of rounds:

Assumption 2. *There exist a number M, known to the learner, such that:*

$$\min_{\phi \in \{1,...,B_N\}} (b_\phi - b_{\phi-1}) \geq KM. \tag{4}$$

With reference to Fig. 1, the two assumptions are stating that the two break-points b_1, and b_2 must be such that $(b_2 - b_1) > KM$, and that $|\mu_{0,\phi} - \mu_{0,\phi+1}| > \varepsilon$, and $|\mu_{1,\phi} - \mu_{1,\phi}| > \varepsilon$ for each $\phi \in \{1, 2\}$. These two assumptions are natural in MAB algorithms adopting CDT as tools to detect changes, e.g., [8,14,15] since they state that the changes are detectable by the CDT in a limited amount of rounds (Assumption 1) and allow to set the CDT at the beginning of the learning process and after each change is detected (Assumption 2). Therefore, the knowledge of ϵ and M is customary when designing algorithms following the active framework and allows them to outperform passive ones in terms of empirical performance significantly.

4   The BR-MAB Algorithm

In what follows, we present the BR-MAB algorithm, which can be seen as a generalization of the CD-MAB framework presented in [14] that learns from historical information after each detected change. The BR-MAB algorithm builds upon the definition of a concept C_i as follows:

Definition 2. *A concept $C_i = \{x_1, \ldots, x_C\}$ is a set of rewards collected over time for the arm a_i, which are deemed to belong to the same phase.*

This definition is used in BR-MAB to store information about past phases and identify recurrent phases. In this case, we refer to *recurrent concepts*.

The pseudo-code of the BR-MAB algorithm is presented in Algorithm 1, and takes as input any nonstationary active CD-MAB policy (namely both a change-detection test to be used on each arm and an arm-selection policy), a break-point prediction procedure $\mathfrak{B}$, and a test $\mathcal{E}$ to evaluate when two concepts can be conveniently aggregated. At first, the algorithm initializes all the parameters for the selected CD-MAB and, for each arm a_i, the set of tracked recurrent concepts $\mathfrak{C}_i$, the actual concept being observed C_i^{now}, and a binary variable cf_i to check if a concept had been used in the past for that arm (Line 1). Then, at each round $n \in \{1, \ldots, N\}$, the algorithm selects an arm $a_{i(n)}$ accordingly to the CD-MAB policy (Line 3), uses the reward to update the CD-MAB (Line 4), and updates the concept currently in use $C_{i(n)}^{now}$ for the selected arm $a_{i(n)}$, i.e., adds the currently collected reward $x_{i(n),n}$ to the set $C_{i(n)}^{now}$ (Line 5). Subsequently, the CDT of the CD-MAB is being executed and when this detects a change in the currently selected arm $a_{i(n)}$, the break-point procedure B, detailed in Sect. 4.1, is activated to estimate the break-point r (Line 7). As a result, the rewards collected during rounds $\{r, \ldots, n\}$ corresponding to the arm $a_{i(n)}$ are used to update the information of the arm $a_{i(n)}$ in the CD-MAB (Line 8). Moreover, the algorithm removes the rewards selected by the break-point procedure from the current concept $C_{i(n)}^{now}$, adds the current concept $C_{i(n)}^{now}$ to the set of available concepts $\mathfrak{C}_i$ (Line 10), and resets it using the reward of arm $a_{i(n)}$ collected after the break-point (Line 11). Finally, BR-MAB sets $cf_i = 0$, to state that the arm a_i is eligible of using one of the concept in $\mathfrak{C}_{i(n)}$ if it is recurring (Line 12). After the change detection phase occurred, the algorithm tries to detect if a concept in $\mathfrak{C}_{i(n)}$ is recurrent. More specifically, if no concept has been already used for the arm $a_{i(n)}$ ($cf_{i(n)} = 0$), for each concept C present in $\mathfrak{C}_{i(n)}$, it checks if it can be considered equivalent to the current concept $C_{i(n)}^{now}$ using the test $\mathcal{E}$ (Line 16), detailed in Sect. 4.2. If the test $\mathcal{E}$ passes, the current concept $C_{i(n)}^{now}$ is updated with the rewards contained into the concept C (Line 18), and C is removed from $\mathfrak{C}_{i(n)}$ (Line 10). Finally, the CD-MAB procedure is updated using the reward present in the recurrent concept C.

4.1 Break-Point Prediction Procedure

In this section, we present the break-point prediction procedure $\mathcal{B}$ that identifies the position of the break-point after the CDT provides a detection. This problem is commonly addressed in the statistical literature by the change-point formulation [11]. These tests perform a retrospective and offline analysis over a sequence of observations that presumably contains a change and determine whether there is enough statistical evidence to confirm the sequence contains a change and case its location. Change-point formulation has also been extended to detect changes in streaming data from a Bernoulli [19] or arbitrary [18] distributions. In this case, the change-point formulation provides change-detection capabilities, and the break-point estimate is automatically provided after each detection.

The CUSUM test [5] is a popular option for the CDT used for monitoring the stream of rewards in the CD-MAB is the CUSUM test. In this case, the test already provide after each detection a break-point estimate. Let t' be the time when a change has been detected on the arm a_i (or possibly $t' = 0$), and let $\{x_{i,t(1)}, \ldots x_{i,t(M)}\}$ be the sequence of last M rewards collected from arm a_i from the current phase at rounds $\{t(1), \ldots, t(M)\}$. The CUSUM test uses such rewards to estimate the expected values of the reward of a_i, namely $\hat{m}_i := \sum_{h=1}^{M} \frac{x_{i,t(h)}}{M}$. When monitoring the next rounds $h \in \{t' + M + 1, \ldots\}$, the CUSUM test computes the following statistics to detect an increase/decrease in the expected reward $\mu_{i,\phi}$:

$$g_{i,h}^{+} = \begin{cases} \max\{0, g_{i,h-1}^{+} + x_{i,h} - \hat{m}_i - \varepsilon\} & \text{if } i(h) = i \\ g_{i,h-1}^{+} & \text{otherwise} \end{cases}, \tag{5}$$

$$g_{i,h}^{-} = \begin{cases} \max\{0, g_{i,h-1}^{-} + \hat{m}_i - x_{i,h} - \varepsilon\} & \text{if } i(h) = i \\ g_{i,h-1}^{+} & \text{otherwise} \end{cases}, \tag{6}$$

where the quantities has been initialized as $g_{i,t'+M}^{+} = 0$ and $g_{i,t'+M}^{-} = 0$, and ε is defined in Assumption 1. Changes are detected as soon as one of these statistics

---

Algorithm 1. BR-MAB

Require: non-stationary algorithm CD-MAB, break-point prediction procedure $\mathcal{B}$, recurrent concept equivalence test $\mathcal{E}$

1: $\mathfrak{C}_i \leftarrow \emptyset$, $C_i^{now} \leftarrow \emptyset$, $cf_i \leftarrow 0\ \forall i \in \{1,\dots,K\}$
2: **for** $n \in \{1,\dots,N\}$ **do**
3:     Play $a_{i(n)}$ according to CD-MAB
4:     Collect reward $x_{i(n),n}$ and update the CD-MAB accordingly
5:     Update the concept $C_{i(n)}^{now} \leftarrow C_{i(n)}^{now} \cup \{x_{i(n),n}\}$
6:     **if** a change has been detected by the CD-MAB **then**   ▷ change detection
7:         Run $\mathcal{B}$ to identify the change round r ▷ break-point prediction
8:         Update arm $a_{i(n)}$ in the CD-MAB using rewards from rounds $\{r,\dots,n\}$
9:         Remove rewards collected from $a_{i(n)}$ from rounds $\{r,\dots,n\}$ from $C_{i(n)}^{now}$
10:         $\mathfrak{C}_{i(n)} \leftarrow \mathfrak{C}_{i(n)} \cup \{C_{i(n)}^{now}\}$
11:         Initialize $C_{i(n)}^{now}$ with the rewards of arm $a_{i(n)}$ collected at rounds $\{r,\dots,n\}$
12:         $cf_{i(n)} \leftarrow 0$
13:     **end if**
14:     **if** $cf_{i(n)} = 0$ **then**
15:         **for** $C \in \mathfrak{C}_{i(n)}$ **do**
16:             **if** $\mathcal{E}(C, C_{i(n)}^{now})$ **then**   ▷ recurrent concept test
17:                 $cf_{i(n)} \leftarrow 1$
18:                 $C_{i(n)}^{now} \leftarrow C \cup C_{i(n)}^{now}$ ▷ concept merge
19:                 $\mathfrak{C}_{i(n)} \leftarrow \mathfrak{C}_{i(n)} \setminus C$
20:                 Update arm $a_{i(n)}$ in the CD-MAB using the rewards in C
21:             **end if**
22:         **end for**
23:     **end if**
24: **end for**

---

exceed a suitable threshold. Let us assume that this occurs at time t'', the round corresponding to the break-point is then identified as:

$$r = \arg\min_{h\in\{t',\dots,t''\}} g_{i,h}^+, \quad \text{or} \quad r = \arg\min_{h\in\{t',\dots,t''\}} g_{i,h}^-, \tag{7}$$

depending on whether the detection comes from monitoring $g_{i,h}^+$ or $g_{i,h}^-$, respectively. If there are multiple values attaining the minimum in Eq. (7), we set r as the most recent value. Once the break-point prediction occurred, we initialize the CUSUM as described above and reset the two statistics $g_{i,h}^+$ and $g_{i,h}^-$ before restarting monitoring.

4.2 Recurrent Concepts Equivalence Test

After a change has been detected, we need to assess whether the currently expected reward of an arm a_i, represented in the concept C_i^{now}, corresponds to any of the previously encountered phases using the concepts stored in $\mathfrak{C}_i$. Inspired by [1], we solve this problem by an equivalence test $\mathcal{E}(\cdot,\cdot)$ that consists in a Two One Sided Test (TOST) [20]. More specifically, let C_i^{now} be the current concept associated to the arm a_i, and let C be any concept from the collection of previously seen concepts $\mathcal{C} \in \mathfrak{C}_i$. The TOST determines whether there is enough statistical evidence to claim that the expected rewards in the two concepts $\mathcal{C}_i^{now}$ and $\mathcal{C}$ differ less than a given threshold.

The TOST formulates the following statistical tests over the expected values μ' and μ'' of the rewards in $\mathcal{C}_i^{now}$ and $\mathcal{C}$, respectively:

$$
\begin{array}{llll}
\text{Test 1} & H_0 : \mu' - \mu'' \le -d & \text{vs.} & H_1 : \mu' - \mu'' > -d, \quad (8) \\
\text{Test 2} & H_0 : \mu' - \mu'' \ge d & \text{vs.} & H_1 : \mu' - \mu'' < d, \quad (9)
\end{array}
$$

where $d > 0$ is the equivalence bound, indicating a difference between rewards that is deemed as negligible when identifying recurrent phases. When the TOST rejects both the null hypothesis, we argue that there is enough statistical evidence that the difference $|\mu' - \mu''|$ lies within $(-d, d)$ Therefore, the test $\mathcal{E}(\mathcal{C}_i^{now}, \mathcal{C})$ asserts that the two concept are recurrent, and they are merged into a single concept in the BR-MAB algorithm.

In particular, it uses two two-sample z-test to compare proportions, formally it requires to compute the following test statistics:

$$
z_{-d} = \frac{(\hat{\mu}' - \hat{\mu}'') + d}{\sqrt{\dfrac{\hat{\mu}'(1 - \hat{\mu}')}{n'} + \dfrac{\hat{\mu}''(1 - \hat{\mu}'')}{n''}}}, \text{ and } z_d = \frac{(\hat{\mu}' - \hat{\mu}'') - d}{\sqrt{\dfrac{\hat{\mu}'(1 - \hat{\mu}')}{n'} + \dfrac{\hat{\mu}''(1 - \hat{\mu}'')}{n''}}},
$$

$$(10)$$

where $\hat{\mu}'$ and $\hat{\mu}''$ are the empirical means of the reward stored in the concepts $\mathcal{C}_i^{now}$ and $\mathcal{C}$, respectively, and $n' := |\mathcal{C}_i^{now}|$ and $n'' := |\mathcal{C}|$ are their cardinality. In this test, we fix a significance level α_z, and we reject both null hypothesis when the test statistic z_{-d} is above the $1 - \alpha_z$ quantiles of a normal distribution and z_d is below the α_z quantiles of a normal distribution.

Even though representing in each concept $\mathcal{C}$ the set of rewards is not very efficient in terms of memory requirements, in our case, a much more compact representation is possible. In fact, in the case of Bernoulli rewards, the TOST requires only the mean of the rewards collected in the concept $\mathcal{C}$ and the concept cardinality, which can be updated incrementally and stored in just two values.

4.3   Regret Analysis for Generic CD-MABs

At first we consider the CD-MAB setup, where there is no break-point prediction nor the recurrent concept identification. Assume to have a stationary stochastic MAB policy $\mathcal{P}$ ensuring an upper bound on the expected pseudo-regret of $C_1(\log N) + C_2$ over a time horizon of N for the stochastic stationary MAB problem (being $C_1, C_2 \in \mathbb{R}^+$ suitable constants), and a CDT procedure $\mathcal{D}$ ensuring an expected detection delay of $\mathbb{E}[D]$ and an expected number of false positive of $\mathbb{E}[F]$. We prove the following:

Theorem 3. *The expected pseudo-regret of a CD-MAB algorithm, where the arm selection is performed using $\mathcal{P}$ with probability $1 - \alpha$ and randomly selecting* *an arm with probability α and that uses $\mathcal{D}$ on a generic abruptly changing MAB setting, is upper bounded by:*

$$\overline{R}_N(CD\text{-}MAB) \leq (1 + B_N + \mathbb{E}[F])KM + (B_N + \mathbb{E}[F])\left(C_1 \log \frac{N}{B_N} + C_2\right)$$

$$+ \frac{KB_N\mathbb{E}[D]}{\alpha} + \alpha N, \tag{11}$$

where we assume that the CDT requires M samples for each arm to be initialized.

Proof. Due to space limitations, the proof is deferred to Appendix A.

The contribution to the regret in the right-hand side of Eq. (11) is composed by the following components (from left to right): *i*) the samples required for the initialization of the CDT at the beginning of the learning procedure and each time a change is detected, *ii*) the regret of the stationary MAB procedure repeated every time a change is detected, *iii*) the loss due to the detection delay, and *iv*) the loss due to random sampling performed over the time horizon N.

This result generalizes that in [14], in which the authors provide an upper bound to the expected pseudo-regret of the same order for an algorithm using as stationary MAB procedure the UCB1 algorithm [3]. In the same work, the authors also present theoretical results for the specific choice of UCB1 as stationary MAB and CUSUM as CDT and provide a bound of the order of $O(\sqrt{B_N N \log \frac{N}{B_N}})$, when the values of the threshold of the CUSUM h and the exploration parameter α are adequately set. Notably, Theorem 3 provides the same order of pseudo-regret of the CUSUM-UCB when substituting in Eq. (11) the guarantees provided by CUSUM and those of UCB1.

4.4   Regret Analysis for the Break-Point Prediction Procedure

Here, we analyse the theoretical guarantees provided by a specific instance of the BR-MAB algorithm, using CUSUM-UCB as CD-MAB procedure and using a generic break-point prediction procedure $\mathcal{B}$. Indeed, updating CUSUM-UCB after each detection, exploiting a bounded number of reward values recovered by the break-point prediction procedure $\mathcal{B}$, allow us to provide theoretical guarantees on the performance of BR-MAB. We show that:

Theorem 4. *Consider the BR-MAB algorithm with the CUSUM-UCB as CD-MAB procedure and a break-point procedure $\mathcal{B}$, s.t. number of rewards selected by this procedure are less than $\frac{\xi}{4} \log N_t$. Using such an algorithm on the abruptly changing MAB setting provides an upper bounded on the pseudo-regret of:*

$$R_N(\mathfrak{U}) \leq O\left(\sqrt{NB_N \log N/B_N}\right), \tag{12}$$

where ξ is the parameter used in the UCB bound for the CUSUM-UCB algorithm, $N_t := \sum_i N_{i,t}$ is the number of samples collected from the instant a change has been detected.

Proof. Due to space limitations, the proof is deferred to Appendix A.

We remark that any break-point procedure $\mathcal{B}$ can be adapted to satisfy the constraint in 4, by using $\max\{r, t - \frac{\xi}{4} \log N_t\}$, where r is the round at which $\mathcal{B}$ predicted the break-point and t is the current time instant. Notice that the limitation in terms of samples is required to avoid that the estimated expected value for an arm, used in the CUSUM-UCB to take decisions, is biased significantly by the presence of samples coming from the previous phase.

5 Experiments

In what follows, we conduct experiments to evaluate the empirical improvement provided by the proposed BR-MAB approach on generic CD-MAB algorithms. At first, we present a toy example to show the effect of using the BR-MAB approach on a CD-MAB algorithm. After that, we evaluate the proposed algorithm on synthetically generated data, and a real-world problem of online ads selection.

In the experiments, we evaluated two flavours of our BR-MAB algorithm applied to the CUSUM-UCB algorithm: the former exploiting only the break-point prediction procedure $\mathcal{B}$, denoted from now on with BR-CUSUM-UCB($\mathcal{B}$,/), and the latter using both the break-point prediction procedure $\mathcal{B}$ and the recurrent concept equivalence test $\mathcal{E}$, denoted by BR-CUSUM-UCB($\mathcal{B},\mathcal{E}$). This allows us to separately evaluate the improvements provided solely by the break-point prediction in BR-MAB. We compare our method against: *i*) the UCB1 algorithm [3], an algorithm designed for stationary stochastic bandits, *ii*) D-UCB and *iii*) SW-UCB [9], which are algorithms for non-stationary MAB adopting the passive approach to deal with changes in the environment, *iv*) CUSUM-UCB [14], the version of the CD-MAB algorithm without using our framework. We set the parameters required by each one of the tested algorithms as suggested by the corresponding papers. A summary of the parameters is provided by Table 2 provided in Appendix C. We evaluate the different algorithms in terms of empirical pseudo-regret $R_n(\mathfrak{U})$ over the time horizon. The experiments have been repeated for 200 independent simulations. The code used for the experiments is available at https://github.com/gerlaxrex/BR-MAB.

5.1 Toy Example

The aim of this experiment is to compare the behaviour over time of the upper confidence bounds of the CUSUM-UCB algorithm, BR-CUSUM-UCB($\mathcal{B}$,/), and BR-CUSUM-UCB($\mathcal{B},\mathcal{E}$). In this experiment, we model $K = 2$ arms over a time horizon of $N = 10^5$ with $B_N = 4$ break-points. We tested the three algorithms on an abruptly changing scenario where the expected rewards $\mu_{i,\phi}$ varies over time as depicted in Fig. 2a.

In Figs. 2b, 2c, and 2d we provide the estimated expected value (solid line) and the confidence bounds (shaded areas) used for the arm selection by the CUSUM-UCB, BR-CUSUM-UCB($\mathcal{B}$,/), and BR-CUSUM-UCB($\mathcal{B},\mathcal{E}$) algorithm,

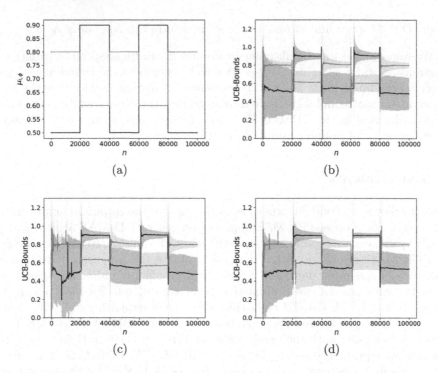

Fig. 2. Toy example: (a) expected rewards for the arms, upper confidence bounds for (b) CUSUM-UCB, (c) BR-CUSUM-UCB($\mathcal{B}$,/), (d) BR-CUSUM-UCB($\mathcal{B}$,$\mathcal{E}$).

respectively. The sole introduction of the $\mathcal{B}$ procedure improves the estimate of the mean value at the beginning of the phases, since the mean values are initialized using the samples collected before the detection of the change. This is evident at times $n = 20,000$ and $n = 40,000$ where the CUSUM-UCB algorithm features downward spikes, while ours take advantage of the samples collected before the detection to reinitialize the empirical expected value of the reward and reduce the variance in reward' estimates.

Comparing Figs. 2b and 2d in the interval $60,000 \leq t \leq 100,000$ of, we observe that the test $\mathcal{E}$ to identify recurrent concepts makes the upper confidence bounds tighter, especially those corresponding to the optimal arm in each phase. This means that the amount of exploratory pulls required by BR-CUSUM-UCB($\mathcal{B}$,$\mathcal{E}$) to identify the optimal arm are greatly reduced, which also reduces the regret suffered.

Moreover, the management of recurring concepts also mitigate the impact of false positive detection. This is evident in Figs. 2d and 2d, when two false positive detections occurring at $t \approx 7,000$ and $t \approx 12,000$ (small spikes in the figure on the orange arm statistics). While the BR-CUSUM-UCB($\mathcal{B}$,/) algorithm recovers slowly from these false detections, the BR-CUSUM-UCB($\mathcal{B}$,$\mathcal{E}$) algorithm experiences only a slight spike in the mean, while the upper confidence bounds continues to decrease monotonically. This suggests that the reuse

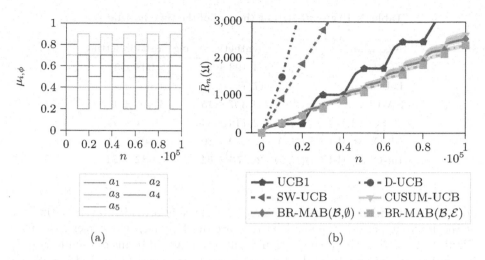

Fig. 3. Synthetic setting: (a) reward expected value, (b) empirical pseudo-regret over the learning process. The shaded areas represent the 95% confidence intervals for the mean.

of information provided by recurrent concept is also useful to recover promptly to a false positive detection of the CDT adopted in the CD-MAB.

5.2 Synthetic Setting

The first experiment was carried out in a setting with $K = 5$ arms, on a time horizon of $N = 10^5$ rounds, with $B_N = 9$ break-points, evenly distributed over time. The expected reward of the arms over time is depicted in Fig. 3a.

Results. Figure 3b shows the empirical pseudo-regret $R_n(\mathfrak{U})$ over time of the different algorithms. In this specific setting, the two passive approaches D-UCB and SW-UCB are those providing the worst performances, since the value of the regret gets larger than $3,500$ after $t \approx 30,000$. UCB1, which in principle should not be able to adapt after changes, is performing better than passive approaches. This is due to the fact that the arm originally optimal in the first phase $\mathcal{F}_1$ is also optimal in the phases $\mathcal{F}_3$, $\mathcal{F}_5$, $\mathcal{F}_7$, and $\mathcal{F}_9$, therefore, the information gathered in the past are helping in the selection performed by UCB1. Conversely, in the even index phases, where a different arm is optimal, the UCB1 algorithm experience an almost linear increase of the regret, due to the fact that it focus on the arm optimal in the initial phase, overall providing evidence that it is not suited for such a scenario. After $t = 25,000$ rounds the CUSUM-UCB keeps its regret below all the above-mentioned algorithms, showing the superiority of the active approaches. Even using this approach, we have that the increase of the regret is accentuated as soon as a change occurred. This effect is mitigated by BR-CUSUM-UCB($\mathcal{B}$,/) thanks to the samples recovered by the $\mathcal{B}$ procedure. Indeed, on average BR-CUSUM-UCB($\mathcal{B}$,/) is performing better than CUSUM-UCB but

Table 1. Regret $R_N(\mathfrak{U})$ at the end of the time horizon N.

Algorithm	Synthetic Setting	Yahoo! Setting
UCB1	$3,193 \pm 17$	908 ± 5
D-UCB	$17,758 \pm 8$	$1,653 \pm 1$
SW-UCB	$9,307 \pm 15$	$1,599 \pm 1$
CUSUM-UCB	$2,719 \pm 84$	831 ± 35
BR-CUSUM-UCB$(\mathcal{B},/)$	$2,619 \pm 80$	805 ± 34
BR-CUSUM-UCB$(\mathcal{B},\mathcal{E})$	$\mathbf{2,273 \pm 61}$	$\mathbf{682 \pm 21}$

no statistical evidence for its superior performance is provided, even at the end of the learning period (the shaded areas are overlapping). Conversely, the BR-CUSUM-UCB$(\mathcal{B}, \mathcal{E})$ is getting a significant advantage in terms of pseudo-regret, by exploiting the fact that all the even phases are recurrent, as well as all the odd ones. The proposed approach is able to incrementally gain advantage over the other algorithms as the number of recurring phases increases.

The regret at the end of the time horizon N is presented in Table 1, second column. Even if there is no significance that the BR-CUSUM-UCB$(\mathcal{B},/)$ algorithm performs better than CUSUM-UCB, on average it decreases the pseudo-regret of $\approx 4\%$ in the synthetic setting. Instead, the BR-CUSUM-UCB$(\mathcal{B}, \mathcal{E})$ provides a significant improvement of $\approx 15\%$ over CUSUM-UCB. This suggests that the information provided by previous phases, in a setting where the environment presents recurrent phases multiple times, might provide a large improvement to nonstationary MAB algorithms.

5.3 Yahoo! Setting

The second experiment used a dataset of click percentage of online articles, more specifically the ones corresponding to the first day ($T = 90,000$) of the Yahoo! Dataset [23]. In this setting the use of a CDT-MAB approach is appropriate since the user behaviour is known to vary over time, and the recommender system wants to maximize the visualization of the most interesting article at each time over the day. We selected $K = 5$ article at random from the available ones, and a phase $\mathcal{F}_\phi$ is defined computing their average click-through rate each $5,000$ seconds and keeping the arms expected reward constant over this period.

Results. The results corresponding to the empirical pseudo-regret are presented in Fig. 4. Also in this scenario, the two passive approaches, D-UCB and SW-UCB, are providing the worst performance, with a regret at the end of the time horizon of almost twice the value of the other considered algorithms. UCB1 is performing worse than CUSUM-UCB, which means that in this specific setting, the active approach is a valid solution to tackle this problem. The adoption of the break-point prediction procedure $\mathcal{B}$ used by BR-CUSUM-UCB$(\mathcal{B},/)$ is not achieving a significant improvement, even when looking at Table 1, third

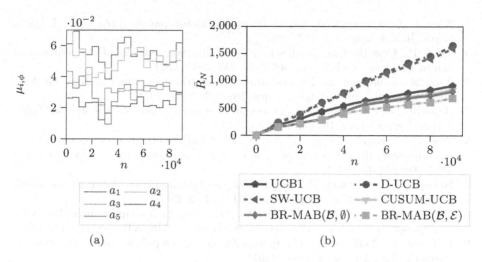

Fig. 4. Yahoo! setting: (a) reward expected value, (b) empirical pseudo-regret over the learning process. The shaded areas are the 95% confidence intervals.

column, where we have a slightly smaller regret a the end of the time horizon of ≈2.5% on average. Conversely, when adopting also a technique to integrate the samples coming from recurrent concepts, we have a significant improvement in terms of regret for $t > 40,000$ w.r.t. the one of CUSUM-UCB, which leads to a improvement of ≈15% at the end of the time horizon. This strengthens the idea that the presented BR-MAB framework outperforms standard active techniques.

6   Conclusion and Future Works

We propose BR-MAB, a general framework extending CD-MAB algorithms to better handle non-stationary MAB setting. The rationale behind BR-MAB consists in gathering, after having detected a change, all the possible information that is consistent with the current state of the arm. More specifically, BR-MAB adopts a break-point prediction technique to recover rewards acquired in between the detection and the unknown change-time instant, and a procedure to identify recurrent phases of the arm. Our analysis demonstrates that including information collected by the break-point prediction procedure preserves the guarantees on the pseudo-regret in the CUSUM-UCB case. Moreover, experiments indicate that identifying recurrent concepts is beneficial in terms of accumulated regret, also thanks to a better recovery after false positive detections. Ongoing work concerns a further investigation to achieve tighter theoretical guarantees on specific settings, like the case of changes affecting all the arms simultaneously.

References

1. Alippi, C., Boracchi, G., Roveri, M.: Just-in-time classifiers for recurrent concepts. IEEE Trans. Neural Netw. Learn. Syst. **24**(4), 620–634 (2013)

2. Auer, P.: Using confidence bounds for exploitation-exploration trade-offs. J. Mach. Learn. Res. **3**(Nov), 397–422 (2002)
3. Auer, P., Cesa-Bianchi, N., Fischer, P.: Finite-time analysis of the multiarmed bandit problem. Mach. Learn. **47**(2–3), 235–256 (2002)
4. Aziz, M., Kaufmann, E., Riviere, M.K.: On multi-armed bandit designs for dose-finding clinical trials. J. Mach. Learn. Res. **22**, 1–38 (2021)
5. Basseville, M., Nikiforov, I.V.: Detection of Abrupt Changes - Theory and Application. Prentice Hall, Hoboken (1993)
6. Besson, L., Kaufmann, E.: The generalized likelihood ratio test meets klUCB: an improved algorithm for piece-wise non-stationary bandits. arXiv preprint arXiv:1902.01575 (2019)
7. Bubeck, S., Cesa-Bianchi, N.: Regret analysis of stochastic and nonstochastic multi-armed bandit problems. CoRR abs/1204.5721 (2012)
8. Cao, Y., Wen, Z., Kveton, B., Xie, Y.: Nearly optimal adaptive procedure with change detection for piecewise-stationary bandit. In: AISTATS, pp. 418–427 (2019)
9. Garivier, A., Moulines, E.: On upper-confidence bound policies for switching bandit problems. In: ALT, pp. 174–188 (2011)
10. Hartland, C., Gelly, S., Baskiotis, N., Teytaud, O., Sebag, M.: Multi-armed bandit, dynamic environments and meta-bandits, November 2006. https://hal.archives-ouvertes.fr/hal-00113668/file/MetaEve.pdf, working paper
11. Hawkins, D.M., Qiu, P., Kang, C.W.: The changepoint model for statistical process control. J. Qual. Technol. **35**(4), 355–366 (2003)
12. Hinkley, D.: Inference about the change-point from cumulative sum tests. Biometrika **58** (1971)
13. Italia, E., Nuara, A., Trovò, F., Restelli, M., Gatti, N., Dellavalle, E.: Internet advertising for non-stationary environments. In: AMEC, pp. 1–15 (2017)
14. Liu, F., Lee, J., Shroff, N.B.: A change-detection based framework for piecewise-stationary multi-armed bandit problem. In: AAAI (2018)
15. Mellor, J.C., Shapiro, J.L.: Thompson Sampling in switching environments with Bayesian online change point detection. CoRR abs/1302.3721 (2013)
16. Nuara, A., Trovo, F., Gatti, N., Restelli, M.: A combinatorial-bandit algorithm for the online joint bid/budget optimization of pay-per-click advertising campaigns. In: AAAI, vol. 32 (2018)
17. Parvin, M., Meybodi, M.R.: MABRP: a multi-armed bandit problem-based energy-aware routing protocol for wireless sensor network. In: AISP, pp. 464–468. IEEE (2012)
18. Ross, G.J., Adams, N.M.: Two nonparametric control charts for detecting arbitrary distribution changes. J. Qual. Technol. **44**(2), 102–116 (2012)
19. Ross, G.J., Tasoulis, D.K., Adams, N.M.: Sequential monitoring of a Bernoulli sequence when the pre-change parameter is unknown. Comput. Statist. **28**(2), 463–479 (2013)
20. Schuirmann, D.J.: A comparison of the two one-sided tests procedure and the power approach for assessing the equivalence of average bioavailability. J. Pharmacokinet. Biopharm. **15**(6), 657–680 (1987)
21. Trovò, F., Paladino, S., Restelli, M., Gatti, N.: Improving multi-armed bandit algorithms in online pricing settings. Int. J. Approx. Reason. **98**, 196–235 (2018)
22. Trovò, F., Paladino, S., Restelli, M., Gatti, N.: Sliding-window Thompson Sampling for non-stationary settings. J. Artif. Intell. Res. **68**, 311–364 (2020)
23. Yahoo!: R6b - Yahoo! front page today module user click log dataset, version 2.0 (2011)

High-Probability Kernel Alignment Regret Bounds for Online Kernel Selection

Shizhong Liao(iD) and Junfan Li$^{(\boxtimes)}$(iD)

College of Intelligence and Computing, Tianjin University, Tianjin 300350, China
{szliao,junfli}@tju.edu.cn

Abstract. In this paper, we study data-dependent regret bounds for online kernel selection in the regime online classification with the hinge loss. Existing work only achieves $O(\|f\|^2_{\mathcal{H}_\kappa} T^\alpha), \frac{1}{2} \leq \alpha < 1$ regret bounds, where $\kappa \in \mathcal{K}$, a preset candidate set. The worst-case regret bounds can not reveal kernel selection improves the performance of single kernel leaning in some benign environment. We develop two adaptive online kernel selection algorithms and obtain the first high-probability regret bound depending on $\mathcal{A}(\mathcal{I}_T, \kappa)$, a variant of kernel alignment. If there is a kernel in the candidate set matching the data well, then our algorithms can improve the learning performance significantly and reduce the time complexity. Our results also justify using kernel alignment as a criterion for evaluating kernel function. The first algorithm has a $O(T/K)$ per-round time complexity and enjoys a $O(\|f\|^2_{\mathcal{H}_{i*}} \sqrt{K \mathcal{A}(\mathcal{I}_T, \kappa_{i*})})$ high-probability regret bound. The second algorithm enjoys a $\tilde{O}(\beta^{-1} \sqrt{T \mathcal{A}(\mathcal{I}_T, \kappa_{i*})})$ per-round time complexity and achieves a $\tilde{O}(\|f\|^2_{\mathcal{H}_{i*}} K^{\frac{1}{2}} \beta^{\frac{1}{2}} T^{\frac{1}{4}} \mathcal{A}(\mathcal{I}_T, \kappa_{i*})^{\frac{1}{4}})$ high-probability regret bound, where $\beta \geq 1$ is a balancing factor and $\kappa_{i*} \in \mathcal{K}$ is the kernel with minimal $\mathcal{A}(\mathcal{I}_T, \kappa)$.

Keywords: Model selection · Online learning · Kernel method

1 Introduction

Model selection aims at choosing inductive bias that matches learning tasks, and thus is central to the learning performance of algorithms. For online kernel learning, one of the model selection problems is how to choose a suitable RKHS (or kernel function), in which the data are represented with a low complexity. A simple representation of the data makes algorithms enjoy superior learning

This work was supported in part by the National Natural Science Foundation of China under grants No. 62076181.

Electronic supplementary material The online version of this chapter (https://doi.org/10.1007/978-3-030-86486-6_5) contains supplementary material, which is available to authorized users.

performance. This problem is also termed as online kernel selection, related to the more general online model selection [7,16]. An adversary sends a learner a sequence of examples $\{(\mathbf{x}_t, y_t)\}_{t=1}^T$. The learner chooses a sequence of kernels $\{\kappa_{I_t}\}_{t=1}^T$ from a preset kernel space $\mathcal{K}$, and a sequence of hypotheses $\{f_t\}_{t=1}^T$. At each round t, the loss is $\ell(f_t(\mathbf{x}_t), y_t)$. The learner should be competitive with the unknown optimal RKHS, $\mathcal{H}_{i^*}$. We use the regret to measure the performance,

$$\mathrm{Reg}_T(\mathcal{H}_{i^*}) := \sum_{t=1}^T \ell(f_t(\mathbf{x}_t), y_t) - \min_{f \in \mathcal{H}_{i^*}} \sum_{t=1}^T \ell(f(\mathbf{x}_t), y_t). \tag{1}$$

$\kappa_{i^*} \in \mathcal{K}$ is the optimal kernel for the data and induces $\mathcal{H}_{i^*}$. To this end, a stronger guarantee is to adapt to any $\mathcal{H}_\kappa, \kappa \in \mathcal{K}$ up to a small cost.

To achieve a sub-linear regret bound with respect to (w.r.t) any $\mathcal{H}_\kappa$, the main challenge is the high time complexity. The per-round time complexity of evaluating kernel functions and making prediction would be $O(KT)$, if we do not limit the model size, where K is the number of base kernels. Most of existing online kernel selection researches focus on achieving a $O(\|f\|_{\mathcal{H}_\kappa}^2 T^\alpha)$, $\alpha < 1$ regret bound, and keeping a constant per-round time complexity. One of approaches embeds implicit RKHSs to relatively low-dimensional random feature spaces [13,17,19], in which the time complexity of evaluating kernel functions and prediction is linear with D, the number of random features. The algorithm proposed in [13] has a $O(\|f\|_{\mathcal{H}_\kappa}^2 K^{\frac{1}{3}} T^{\frac{2}{3}})$ expected regret bound and suffers a $O(D)$ time complexity. Similarly, an algorithm with a $O(\|f\|_{\mathcal{H}_\kappa}^2 \sqrt{T})$ regret bound and a $O(KD)$ time complexity was proposed in [19]. The other approach maintains a fixed budget with size B [24]. An algorithm with a $O(B \ln T)$ regularized regret bound (or a $\tilde{O}(\|f\|_{\hat{\mathcal{H}}}^2 T^{\frac{2}{3}} + BT^{\frac{1}{3}})$ standard regret bound by setting λ to the optimal value $O(T^{-\frac{1}{3}})$ in [24]) and a $O(B + KB^2/T)$ time complexity was proposed, where $\hat{\mathcal{H}}$ is a surrogate hypothesis space.

However, the $O(\|f\|_{\mathcal{H}_\kappa}^2 T^\alpha)$ regret bound is pessimistic in the sense that (i) it can not distinguish the convergence rate w.r.t. T when choosing different kernel; (ii) it can not reveal kernel selection improves the learning performance in some benign environment. Recalling that the cumulative losses of algorithms are upper bounded by $\min_{f \in \mathcal{H}_\kappa} \sum_{t=1}^T \ell(f(\mathbf{x}_t), y_t) + O(\|f\|_{\mathcal{H}_\kappa}^2 T^\alpha)$. If the minimal cumulative losses in $\mathcal{H}_\kappa$ are small, then the regret bound is the dominated term and is hard to compare among different base kernels. To resolve the two issues, we should require regret bounds adapting to the data complexity in each RKHS. Thus a fundamental problem of online kernel selection is how to provide data-dependent regret bounds. It would be easy to solve the problem without considering the computational constraints. Our question is whether it is possible to achieve the two goals simultaneously. In this paper, we answer the question affirmatively.

We define a variant of kernel alignment, denoted by $\mathcal{A}(\mathcal{I}_T, \kappa)$, for measuring the complexity of data represented in $\mathcal{H}_\kappa$, which reveals the matching between the label matrix and kernel matrix. Different kernel embeds the instances into different RKHS, and thus induces different data complexity. A good kernel should be the one that represents data simply. We establish two computationally efficient algorithms achieving high-probability kernel alignment regret bounds. The first algorithm achieves a $O(\|f\|^2_{\mathcal{H}_{i*}} \sqrt{K\mathcal{A}(\mathcal{I}_T, \kappa_{i*})})$ regret and suffers a $O(T/K)$ per-round time complexity. The second algorithm enjoys a favorable regret-performance trade-off, which can provide a $\tilde{O}(\|f\|^2_{\mathcal{H}_{i*}} K^{\frac{1}{2}} \beta^{\frac{1}{2}} T^{\frac{1}{4}} \mathcal{A}(\mathcal{I}_T, \kappa_{i*})^{\frac{1}{4}})$ regret bound and suffer a $\tilde{O}(\beta^{-1} \sqrt{T\mathcal{A}(\mathcal{I}_T, \kappa_{i*})})$ time complexity, where $\beta \geq 1$ is a balancing factor. The algorithms are based on the adaptive and optimistic online mirror descent framework and two novel model evaluation strategies. We also reveal a new result: if there is a good kernel in the candidate set, then online kernel selection can improve the learning performance and computational efficiency significantly relative to online kernel learning using a bad kernel. Numerical experiments on benchmark datasets are conducted to verify our theoretical results.

1.1 Related Work

Yang et. al. [22] proposed the first online kernel selection algorithm, OKS, for alleviating the high time complexity of offline kernel selection and multi-kernel learning. For online kernel selection, OKS can provide a $O(\|f\|^2_{\mathcal{H}_\kappa} \sqrt{KT})$ regret bound and suffers a $O(T)$ per-round time complexity. Foster et al. [6] studied online model selection in Banach space, and proposed a multi-scale expert advice algorithm which achieves regret bounds scaling with the loss range of individual hypothesis space. The multi-scale algorithm can achieve data-dependent regret bounds, only if there are computationally efficient sub-algorithms. A related but different work is online multi-kernel learning [11,19], where algorithms make a prediction $f_t(\mathbf{x}_t)$ by a convex combination of K base predictions $\{f_{t,i}(\mathbf{x}_t)\}^K_{i=1}$. Existing algorithms can also not achieve data-dependent regret bounds.

Another related research is achieving data-dependent regret bounds for online kernel learning. If the loss function satisfies specific curvature property [2,10,23], such as the square loss and the logistic loss, then there exist computationally efficient online kernel learning algorithms that achieve regret bound depending on the smallest cumulative losses [23], or the effective dimension [2,10]. However, the hinge loss does not enjoy the curvature property. Thus achieving data-dependent regret bounds is more difficult. Our algorithms can be applied to online kernel learning so long as $\mathcal{K}$ only contains a single kernel.

2   Problem Setting

Let $\mathcal{I}_T = \{(\mathbf{x}_t, y_t)\}_{t \in [T]}$ be a sequence of examples, where $\mathbf{x}_t \in \mathbb{R}^d$ is an instance, $y_t \in \{-1, 1\}$ and $[T] = \{1, \ldots, T\}$. Let $\kappa(\cdot, \cdot) : \mathbb{R}^d \times \mathbb{R}^d \to \mathbb{R}$ be a positive semidefinite kernel function, and $\mathcal{K} = \{\kappa_1, \ldots, \kappa_K\}$. Assuming that $\kappa_i(\mathbf{x}, \mathbf{x}) \in [1, D_i]$ for all $i \in [K]$ and $D = \max_i D_i$. Let $\mathcal{H}_i = \{f | f : \mathbb{R}^d \to \mathbb{R}\}$ be the RKHS associated with κ_i, such that (i) $\langle f, \kappa_i(\mathbf{x}, \cdot) \rangle_{\mathcal{H}_i} = f(\mathbf{x})$; (ii) $\mathcal{H}_i = \overline{\mathrm{span}(\kappa_i(\mathbf{x}_t, \cdot) : t \in [T])}$. We define $\langle \cdot, \cdot \rangle_{\mathcal{H}_i}$ as the inner product in $\mathcal{H}_i$, which induces the norm $\|f\|_{\mathcal{H}_i} = \sqrt{\langle f, f \rangle_{\mathcal{H}_i}}$. Let $\ell(f(\mathbf{x}), y) = \max\{0, 1 - yf(\mathbf{x})\}$ be the hinge loss, and $\mathcal{D}_{\psi_{t,i}}(\cdot, \cdot) : \mathcal{H}_i \times \mathcal{H}_i \to \mathbb{R}$ be the Bregman divergence induced by a strongly convex regularizer $\psi_{t,i}(\cdot) : \mathcal{H}_i \to \mathbb{R}$.

For a sequence $\mathcal{I}_T$, if an oracle gives the optimal kernel $\kappa_{i*} \in \mathcal{K}$, then we can learn a sequence of hypotheses in $\mathcal{H}_{i*}$. Lacking such prior, the learner hopes to develop a kernel selection algorithm and generate a sequence of hypotheses $\{f_t\}_{t=1}^T$, which is competitive to that generated by the same algorithm running in $\mathcal{H}_{i*}$. The regret of the algorithm w.r.t. $\mathcal{H}_i, i \in [K]$ is defined by (1), where we replace $\mathcal{H}_{i*}$ with $\mathcal{H}_i$. A general goal is to keep $\text{Reg}_T(\mathcal{H}_i) = O(\text{Poly}(\|f_i^*\|_{\mathcal{H}_i})T^\alpha)$, $\alpha < 1$. Thus the average loss of the algorithm converges to that of the optimal hypothesis in $\mathcal{H}_{i*}$. The worst-case optimal regret bound is obtained at $\alpha = \frac{1}{2}$.

If the minimal cumulative losses in all RKHSs are smaller than $O(\sqrt{T})$, then such a worst-case regret bound is unsatisfactory, since it can not reveal kernel selection improves the learning performance. Hence, it is necessary to establish some kind of regret bound adapting to the complexity of data represented in RKHS. The representation of $(\mathbf{x}_t, y_t)$ in $\mathcal{H}_i$ is $(\phi_i(\mathbf{x}_t), y_t)$, where ϕ_i is the feature mapping induced by κ_i. We can use the variance of the examples in $\mathcal{H}_i$, i.e., $\sum_{t=1}^T \|y_t\phi_i(\mathbf{x}_t, \cdot) - \mu_{T,i}\|_{\mathcal{H}_i}^2$, where $\mu_{T,i} = \frac{1}{T}\sum_{\tau=1}^T y_\tau\phi_i(\mathbf{x}_\tau)$, to measure the data complexity [9]. Next, we present a formal definition.

Definition 1 (Alignment). *For any sequence of examples $\mathcal{I}_T$ and kernel function κ_i, the alignment is defined as follows*

$$\mathcal{A}(\mathcal{I}_T, \kappa_i) := \sum_{t=1}^T \kappa_i(\mathbf{x}_t, \mathbf{x}_t) - \frac{1}{T}\mathbf{Y}_T^\top\mathbf{K}_{\kappa_i}\mathbf{Y}_T.$$

The alignment is an extension of *kernel polarization* [1], a classical kernel selection criterion. If $\mathbf{K}_{\kappa_i}$ is the ideal kernel matrix $\mathbf{Y}_T\mathbf{Y}_T^\top$, then $\mathcal{A}(\mathcal{I}_T, \kappa_i) = 0$. Thus the alignment can be a criterion for evaluating the goodness of kernel function κ_i on $\mathcal{I}_T$. Model selection aims at adapting to $\mathcal{H}_{i*}$ induced by κ_{i*}, the unknown optimal kernel. A natural question is that does there exist some computationally efficient algorithm that achieves high-probability regret bounds depending on $\mathcal{A}(\mathcal{I}_T, \kappa_{i*})$? Our main contribution is to answer this question affirmatively.

3 A Nearly Optimal High-Probability Regret Bound

We first show a simple algorithm achieving the kernel alignment regret bound without considering the computational complexity.

3.1 Warm-Up

At a high level, our approach is based on the adaptive and optimistic online mirror descent framework (AO$_2$MD) [4,18]. We explain AO$_2$MD in a fixed RKHS $\mathcal{H}_i$. Let $\mathbb{H}_i = \{f \in \mathcal{H}_i : \|f\|_{\mathcal{H}_i} \leq U, U \geq D\}$. At any round t, let $f_{t,i}, f_{t-1,i}' \in \mathbb{H}_i$ and $\nabla_{t,i} := \nabla_{f_{t,i}}\ell(f_{t,i}(\mathbf{x}_t), y_t)$. AO$_2$MD running in $\mathbb{H}_i$ is defined as follows,

$$f_{t,i} = \arg\min_{f \in \mathbb{H}_i} \left\{ \langle f, \bar{\nabla}_{t,i} \rangle + \mathcal{D}_{\psi_{t,i}}(f, f_{t-1,i}') \right\}, \tag{2}$$

$$f_{t,i}' = \arg\min_{f \in \mathbb{H}_i} \left\{ \langle f, \nabla_{t,i} \rangle + \mathcal{D}_{\psi_{t,i}}(f, f_{t-1,i}') \right\}, \tag{3}$$

where $\{f'_{t,i}\}_{t=0}^{T-1}$ is a sequence of auxiliary hypotheses. The solutions of (2) and (3) are shown in supplementary material. The main idea of AO$_2$MD is to select an optimistic estimator of $\nabla_{t,i}$, denoted by $\bar{\nabla}_{t,i}$, and execute the first mirror updating (2). After obtaining $f_{t,i}$, we output the prediction $\hat{y}_t = \text{sign}(f_{t,i}(\mathbf{x}_t))$. When receiving the label y_t, we observe the true gradient $\nabla_{t,i}$ and execute the second mirror updating (3). If the data evolves slowly, then it is possible to find a good estimator $\bar{\nabla}_{t,i}$. The final regret bound depends on the cumulative difference $\sum_{t=1}^{T} \|\nabla_{t,i} - \bar{\nabla}_{t,i}\|^2_{\mathcal{H}_i}$, where we define $\bar{\nabla}_{1,i} = 0$.

A simple approach for obtaining a regret bound depending on the alignment is to reduce online kernel selection to a problem of prediction with expert advice. Let $\mathcal{E}(K)$ be an algorithm for prediction with expert advice. We can instantiate an AO$_2$MD algorithm in each $\mathbb{H}_i, i \in [K]$, and then aggregate the K algorithms with $\mathcal{E}(K)$. The following theorem shows the data-dependent regret bound.

Theorem 1. *Let $\bar{\nabla}_{t,i} = \nabla_{r_i(t),i}$ where $r_i(t) = \max_\tau\{\tau < t : y_\tau f_{\tau,i}(\mathbf{x}_t) < 1\}$ and $\mathcal{E}(K)$ be some algorithm for expert advice. There exists an online kernel selection algorithm such that, for all $\kappa_i \in \mathcal{K}$, with probability at least $1 - \delta$,*

$$\text{Reg}_T(\mathbb{H}_i) = O\left(\sqrt{L_T(f_i^*)\ln K \ln\frac{\ln(2T)}{\delta}} + (\|f_i^*\|^2_{\mathcal{H}_i} + 1)\sqrt{\mathcal{A}(\mathcal{I}_T, \kappa_i)}\right).$$

where $L_T(f_i^*) = \min_{f \in \mathbb{H}_i}\sum_{t=1}^{T} \ell(f(\mathbf{x}_t), y_t) \leq \mathcal{A}(\mathcal{I}_T, \kappa_i)$. The per-round time complexity of the algorithm is $O(TK)$.

To construct the algorithm, we just let $\mathcal{E}(K)$ be the weighted majority algorithm [3] that enjoys a high-probability small-loss regret bound (We give a proof in supplementary material). The algorithm description is presented in supplementary material due to the space limit. The $O(TK)$ per-round time complexity comes from the unbounded number of support vectors and running K AO$_2$MD algorithms. For a large number of base kernels, the time complexity is prohibitive. Thus such a simple algorithm is not practical. Next, we develop a more efficient algorithm enjoying a $O(T/K)$ per-round time complexity.

3.2   A More Efficient Algorithm

The simple algorithm in Theorem 1 evaluates all of the base kernels at each round. Thus the time complexity is linear with K. To resolve this issue, an intuitive approach is to reduce online kernel selection to a K-armed bandit problem [22]. However, such an approach induces two new technique challenges, i.e.,

(i) If $\bar{\nabla}_{t,i} = \nabla_{r_i(t),i}$, then we can not obtain a regret bound depending on $\sum_{t=1}^{T} \|\nabla_{t,i} - \bar{\nabla}_{t,i}\|^2_{\mathcal{H}_i}$, which has been proved in [21].
(ii) The true gradient $\nabla_{t,i}$ can not be observed unless κ_i is selected. If we use an importance-weighted estimator, such as $\nabla_{t,i}/p_{t,i}$, then the second moment is linear with $\max_{t \in [T]} \frac{1}{p_{t,i}}$, which could be much large.

To solve the first challenge, we choose the optimistic estimator $\bar{\nabla}_{t,i} := \mu_{t-1,i}$, where $\mu_{t-1,i} = \sum_{\tau=1}^{t-1} \frac{-y_\tau}{t-1} \kappa_i(\mathbf{x}_\tau, \cdot)$, $t \geq 2$. However, computing $\mu_{t-1,i}$ requires to store all of the received examples. To avoid this issue, we use the "Reservoir Sampling (RS)" technique [8,20] to construct an unbiased estimator of $\mu_{t-1,i}$, denoted by $\tilde{\mu}_{t-1,i}$. Let V be a fixed budget, which we call "Reservoir". At the beginning of round t, let $\tilde{\mu}_{t-1,i} = -\frac{1}{|V|}\sum_{(\mathbf{x},y)\in V} y\kappa_i(\mathbf{x},\cdot)$ and $\tilde{\mu}_{0,i} = 0$. Thus we define the optimistic estimator $\bar{\nabla}_{t,i} := \tilde{\mu}_{t-1,i}$. The RS technique constructs V as follows. At the end of round t, $(\mathbf{x}_t, y_t)$ is added into V with probability $\min\{1, \frac{M}{t}\}$, where $M > 1$ is the maximal size of V. If $|V| = M$ and we are to add the current example, then an old example should be removed uniformly. Note that we just need to maintain a single V, since $\bar{\nabla}_{t,i}, i = 1, \ldots, K$ can be computed by the same examples.

To solve the second challenge, assuming that we can pay additional computational cost for obtaining more information. In this way, we define a K-armed bandit problem with an additional observation. Next we propose a new decoupling exploration-exploitation scheme for obtaining more information. Let Δ_{K-1} be a $(K-1)$-dimensional probability simplex. At each round t,

- Exploitation: select a kernel κ_{I_t}, $I_t \sim \mathbf{p}_t \in \Delta_{K-1}$,
- Exploration: uniformly select a kernel $\kappa_{J_t} \in \mathcal{K}$.

Such an exploration procedure makes κ_{J_t} independent of κ_{I_t}. Based on the exploration procedure, we construct the following variance-reduced estimator,

$$\tilde{\nabla}_{t,i} = \frac{\nabla_{t,i} - \bar{\nabla}_{t,i}}{\mathbb{P}[i = J_t]}\mathbb{I}_{i=J_t} + \bar{\nabla}_{t,i}, \; \forall i \in [K].$$

In this way, the second moment of $\tilde{\nabla}_{t,i}$ is linear with $1/\mathbb{P}[i = J_t] = K$. A more intuitive estimator should incorporate the information of κ_{I_t}. However, we abandon the gradient information ∇_{t,I_t} for the goal of keeping a $O(T/K)$ per-round time complexity. AO$_2$MD is as follows,

$$f_{t,i} = \arg\min_{f \in \mathbb{H}_i} \{\langle f, \bar{\nabla}_{t,i}\rangle + \mathcal{D}_{\psi_{t,i}}(f, f'_{t-1,i})\}, \tag{4}$$

$$f'_{t,i} = \arg\min_{f \in \mathbb{H}_i} \{\langle f, \tilde{\nabla}_{t,i}\rangle + \mathcal{D}_{\psi_{t,i}}(f, f'_{t-1,i})\}. \tag{5}$$

Let $\psi_{t,i}(f) = \frac{1}{2\lambda_{t,i}}\|f\|_{\mathcal{H}_i}^2$. Then the projection of any $g \in \mathcal{H}_i$ onto $\mathbb{H}_i$ is defined by $f = \min\{1, \frac{1}{\|g\|_{\mathcal{H}_i}}U\}g$.

Let $\mathcal{M}(K)$ be some algorithm for a K-armed bandit problem, which outputs $\mathbf{p}_t$ at the beginning of round t. We first select a kernel κ_{I_t}, $I_t \sim \mathbf{p}_t$, and compute f_{t,I_t} using the first mirror updating (4). After that, we output the prediction $\hat{y}_t = \text{sign}(f_t(\mathbf{x}_t))$, where $f_t = f_{t,I_t}$. Then we explore another kernel κ_{J_t} for obtaining the gradient information ∇_{t,J_t}. The final step is to update $\mathbf{p}_t$. To this end, we define some criterion for evaluating each base kernel. Since $|f_{t,i}(\mathbf{x}_t)| = |\langle f_{t,i}, \kappa_i(\mathbf{x}_t, \cdot)\rangle| \leq U\sqrt{D}$, we have $\ell(f_{t,i}(\mathbf{x}_t), y_t) \leq 1 + U\sqrt{D}$. Let $c_{t,i} = \frac{\ell(f_{t,i}(\mathbf{x}_t), y_t)}{1 + U\sqrt{D}}$

be the criterion. The denominator scales the criterion to $[0, 1]$. At the end of round t, we send $\mathbf{c}_t = (c_{t,1}\mathbb{I}_{I_t=1}, \ldots, c_{t,K}\mathbb{I}_{I_t=K})$ to $\mathcal{M}(K)$.

We name this approach B(AO)$_2$KS (Bandit with Additional Observations for Adaptive Online Kernel Selection) and present the pseudo-code in Algorithm 1.

Algorithm 1. B(AO)$_2$KS

Input: $\lambda_i, i = 1, \ldots, K$, D, U, M.
Initialization: $\forall \kappa_i \in \mathcal{K}$, $f'_{0,i} = 0$, $V = \emptyset$.

1: **for** $t = 1, 2, \ldots, T$ **do**
2:     Select a kernel $\kappa_{I_t} \sim \mathbf{p}_t$ ($\mathbf{p}_t$ is output by $\mathcal{M}(K)$),
3:     Compute $\bar{\nabla}_{t,I_t} = \frac{-1}{|V|} \sum_{(\mathbf{x},y) \in V} y \kappa_{I_t}(\mathbf{x}, \cdot)$,
4:     Update f_{t,I_t} according to (4) and output prediction $\hat{y}_t = \text{sign}(f_{t,I_t}(\mathbf{x}_t))$,
5:     Sample a kernel $\kappa_{J_t} \in \mathcal{K}$ uniformly,
6:     **for** $\kappa_i \in \mathcal{K}$ **do**
7:         **if** $\kappa_i = \kappa_{J_t}$ **then**
8:             Compute $\bar{\nabla}_{t,J_t} = \frac{-1}{|V|} \sum_{(\mathbf{x},y) \in V} y \kappa_{J_t}(\mathbf{x}, \cdot)$,
9:             Update f_{t,J_t} according to (4) and compute ∇_{t,J_t},
10:         **end if**
11:         Compute estimator $\tilde{\nabla}_{t,i} = K(\nabla_{t,i} - \bar{\nabla}_{t,i}) \cdot \mathbb{I}_{i=J_t} + \bar{\nabla}_{t,i}$,
12:         Update $f'_{t,i}$ according to (5),
13:     **end for**
14:     Compute $c_{t,I_t} = \frac{1}{1+U\sqrt{D}} \max\{0, 1 - y_t f_{t,I_t}(\mathbf{x}_t)\}$,
15:     Send $\mathbf{c}_t = (c_{t,1}\mathbb{I}_{I_t=1}, \ldots, c_{t,K}\mathbb{I}_{I_t=K})$ to $\mathcal{M}(K)$,
16:     Sample a Bernoulli random variable $\delta_t \sim \text{Ber}(1, M/t)$,
17:     **if** $\delta_t = 1$ and $t > M$, **then** sample $(\mathbf{x}_{j_t}, y_{j_t}) \in V$ and $V = V \cup (\mathbf{x}_t, y_t) \setminus (\mathbf{x}_{j_t}, y_{j_t})$,
18:     **if** $\delta_t = 1$ and $t \leq M$, **then** $V = V \cup (\mathbf{x}_t, y_t)$,
19: **end for**

3.3 Regret Bound

We first establish an important technique lemma about the reservoir estimator.

Lemma 1. *Let $T > M$. For all $i = 1, \ldots, K$, with probability at least $1 - \delta$,*

$$\sum_{t=1}^{T} \|\tilde{\mu}_{t,i} - \mu_{t,i}\|^2_{\mathcal{H}_i} \leq \frac{\mathcal{A}(\mathcal{I}_T, \kappa_i)}{M} \ln \frac{T}{M} + \frac{8D_i}{3} \ln \frac{K}{\delta} + \sqrt{\frac{8D_i \mathcal{A}(\mathcal{I}_T, \kappa_i) \ln T \ln \frac{K}{\delta}}{M}}.$$

Lemma 1 is an extension of the statistic guarantee of reservoir sampling estimator in [8], where the expected unbiasedness of $\tilde{\mu}_{t,i}$ was proved. Next we present a sufficient condition for obtaining the data-dependent regret bound, which gives a strong constraint on the bandit algorithm $\mathcal{M}(K)$.

Assumption 1. *Let $c_t \in [0, 1]^K$ be any loss vector. For any K-armed adversarial bandit problem, with probability at least $1 - \delta$, the regret of $\mathcal{M}(K)$ satisfies*

$$\sum_{t=1}^{T} c_{t,I_t} - \min_{i \in [K]} \sum_{t=1}^{T} c_{t,i} = \tilde{O}\left(\sqrt{K\mathcal{C}_{T,*}}\right), \quad \mathcal{C}_{T,*} = \min_{i \in [K]} \sum_{t=1}^{T} c_{t,i}.$$

Assumption 1 requires $\mathcal{M}(K)$ achieving a high-probability small-loss bound. There are some superior bandit algorithms satisfying Assumption 1, such as GREEN-IX [15] and the online mirror descent based algorithm proposed in [12].

The following theorem gives the high-probability regret bound induced by the hypothesis sequence $\{f_{t,i}\}_{t=1}^{T} \subseteq \mathbb{H}_i$, $i = 1, \ldots, K$.

Theorem 2. *Let* $\psi_{t,i}(f) = \frac{1}{\lambda_i} \|f\|_{\mathcal{H}_i}^2$ *and* $\delta \in (0,1)$. *For all base kernel* $\kappa_i \in \mathcal{K}$ *and any* $f \in \mathbb{H}_i$, *with probability at least* $1 - 4\delta$, *the regret of the hypothesis sequence* $\{f_{t,i}\}_{t=1}^{T}$ *induced by* $\mathrm{B(AO)_2KS}$ *satisfies*

$$L_T(f_{1:T,i}) - L_T(f) \leq \frac{\|f\|_{\mathcal{H}_i}^2}{2\lambda_i} + 11\lambda_i \sqrt{D_i} K g_i(T,M) \mathcal{A}(\mathcal{I}_T, \kappa_i) \ln^{\frac{3}{4}} \frac{K}{\delta}$$

$$+ 20\lambda_i K^2 U D_i \ln \frac{K}{\delta} + 13K \sqrt{D_i} U \ln \frac{K}{\delta} + 7U \sqrt{K g_i(T,M) \mathcal{A}(\mathcal{I}_T, \kappa_i)} \ln^{\frac{3}{4}} \frac{K}{\delta},$$

where $L_T(f_{1:T,i}) = \sum_t \ell(f_{t,i}(\mathbf{x}_t), y_t)$, $L_T(f) = \sum_t \ell(f(\mathbf{x}_t), y_t)$ *and* $g_i(T,M) = \frac{M + D_i \ln \frac{T}{M}}{M}$. *Let* $\lambda_i = (22\sqrt{D_i} K g_i(T,M) \mathcal{A}(\mathcal{I}_T, \kappa_i))^{-\frac{1}{2}}$. *The regret is*

$$L_T(f_{1:T,i}) - L_T(f) = \tilde{O}\left((\|f\|_{\mathcal{H}_i}^2 + U)\sqrt{K \mathcal{A}(\mathcal{I}_T, \kappa_i)} \ln^{\frac{3}{4}} \frac{K}{\delta} \right).$$

Combining Assumption 1 and Theorem 2, we obtain the regret induced by the hypothesis sequence $\{f_t\}_{t=1}^{T}$ w.r.t. any $f \in \mathbb{H}_i$, $i = 1, \ldots, K$.

Theorem 3. *Under the condition of Assumption 1 and Theorem 2, for all* $\kappa_i \in \mathcal{K}$, *with probability at least* $1 - 5\delta$, *the regret of* $\mathrm{B(AO)_2KS}$ *w.r.t.* $\mathbb{H}_i$ *satisfies*

$$\mathrm{Reg}_T(\mathbb{H}_i) = \tilde{O}\left(\sqrt{L_T(f_i^*) K} + (\|f_i^*\|_{\mathcal{H}_i}^2 + U)\sqrt{K \mathcal{A}(\mathcal{I}_T, \kappa_i)} \ln^{\frac{3}{4}} \frac{K}{\delta} \right),$$

where $f_i^* = \arg\min_{f \in \mathbb{H}_i} L_T(f)$, *and* $L_T(f_i^*) \leq \mathcal{A}(\mathcal{I}_T, \kappa_i)$.

Compared with Theorem 1, the regret of $\mathrm{B(AO)_2KS}$ only increases by a factor of order $O(\sqrt{K})$, which is nearly optimal. If we just consider online kernel learning using some non-optimal kernel κ_i, then $\mathrm{B(AO)_2KS}$ obtains a regret bound of order $O(\|f_i^*\|_{\mathcal{H}_i}^2 \sqrt{\mathcal{A}(\mathcal{I}_T, \kappa_i)})$. Let κ_{i*} be the optimal kernel. After executing online kernel selection, $\mathrm{B(AO)_2KS}$ achieves a $O(\|f_{i*}^*\|_{\mathcal{H}_{i*}}^2 \sqrt{K \mathcal{A}(\mathcal{I}_T, \kappa_{i*})})$ regret. If κ_{i*} matches well with $\mathcal{I}_T$, i.e., $\mathcal{A}(\mathcal{I}_T, \kappa_{i*})$ is small, then $\mathrm{B(AO)_2KS}$ improves the learning performance significantly relative to online kernel learning using κ_i. Existing $O(\|f_i^*\|_{\mathcal{H}_i}^2 T^\alpha)$ regret bounds may not reveal that kernel selection improves the learning performance, since we can not distinguish $O(\|f_i^*\|_{\mathcal{H}_i}^2 T^\alpha)$ from $O(\|f_{i*}^*\|_{\mathcal{H}_{i*}}^2 T^\alpha)$. Besides, our result also reveals that the information-theoretic cost induced by executing online kernel selection rather than executing online kernel learning using κ_{i*} is of order $\tilde{O}(\sqrt{L_T(f_i^*) K})$, which could be very small.

3.4   Time Complexity Analysis

The computational cost of B(AO)$_2$KS is dominated by computing $\nabla_{t,i}$ and the projection operation. Let S_i be the set of support vectors used to construct $\{f_{t,i}\}_{t=1}^{T}$. The time complexity of computing $\nabla_{t,i}$ depends on $|S_i|$. The support vectors in S_i comes from (i) reservoir updating, and (ii) the second mirror updating (5). After round $T - 1$, the updating times of reservoir is of order $O(M \log T)$, which is proved by Lemma 7 in supplementary material. At any round t, $(\mathbf{x}_t, y_t)$ is added into S_i via the second mirror updating only if $\kappa_i = \kappa_{J_t}$. Let $S_{i,1} = \{(\mathbf{x}_t, y_t) \in S_i : \kappa_i = \kappa_{J_t}\}$. Since $\mathbb{P}[i = J_t] = 1/K$, it is easy to prove that $|S_{i,1}| = O(T/K)$ with a high probability. The projection operation can be executed incrementally, which only induces a $O(M)$ time complexity. The incremental computation procedure is presented in supplementary material. Thus the per-round time complexity of B(AO)$_2$KS is $O(T/K)$.

Remark 1. For online classification with the hinge loss, OKS [22] achieves a $O(\|f_i^*\|_{\mathcal{H}_i}^2 \sqrt{KT})$ expected regret bound and suffers a $O(T)$ per-round time complexity. Recently, ISKA [24] provides a $\tilde{O}(\|f_i^*\|_{\hat{\mathcal{H}}}^2 T^{\frac{2}{3}} + BT^{\frac{1}{3}})$ expected regret bound, where $\hat{\mathcal{H}} = \cup_{i=1}^{K} \mathcal{H}_i$, and enjoys a $O(B + KB^2/T)$ per-round time complexity. The online multi-kernel learning algorithm, Raker [19], uses random feature technique to approximate kernel function, which enjoys a $O(\|f_i^*\|_{\mathcal{H}_i}^2 \sqrt{T})$ regret bound and suffers a $O(KD)$ per-round time complexity where D is the number of random features. Raker can provide a $O(\sqrt{T})$ regret bound only if $D = \Omega(T)$, which yields a $O(KT)$ per-round time complexity. The same weakness of the above three algorithms is that the $O(\|f_i^*\|_{\mathcal{H}_i}^2 T^{\alpha}), \frac{1}{2} \leq \alpha < 1$ regret bound is worse than $O(\|f_i^*\|_{\mathcal{H}_i}^2 \sqrt{K \mathcal{A}(\mathcal{I}_T, \kappa_i)})$ in the case of $\mathcal{A}(\mathcal{I}_T, \kappa_i) = o(T/K)$.

In the next section, we will propose another algorithm, which relates the time complexity with the alignment and could further reduce the time complexity.

4   Regret-Performance Trade-Off

The computational cost of B(AO)$_2$KS comes from the unbounded number of support vectors. Although many effective approaches have been proposed to solve this issue, such as budgeted online kernel leaning [5,25], Nyström method [2] and random feature technique [14]. However, existing approaches can not provide regret bounds relying on the alignment. To resolve the two issues, we will propose a novel budgeted AO$_2$MD for online kernel learning. The keys include (i) how to select the optimistic estimator, and (ii) how to maintain the budget, especially construct an adaptive example adding strategy.

To solve the first challenge, we still use the reservoir sampling technique to construct the optimistic estimator $\bar{\nabla}_{t,i} := \tilde{\mu}_{t-1,i}$. The key is the second challenge. Let S_i be the budget constructing the hypothesis sequence $\{f_{t,i}\}_{t=1}^{T}$. We propose an adaptive sampling strategy. At the beginning of round t, we still execute the first mirror updating (4) for obtaining $f_{t,i}$. If $y_t f_{t,i}(\mathbf{x}_t) < 1$, then we observe the gradient $\nabla_{t,i}$ and define a Bernoulli random variable $b_{t,i}$ satisfying

$$\mathbb{P}[b_{t,i} = 1] = \frac{\|\nabla_{t,i} - \bar{\nabla}_{t,i}\|_{\mathcal{H}_i}}{Z_{t,i}}, \ Z_{t,i} = \beta_i \left(\|\nabla_{t,i} - \bar{\nabla}_{t,i}\|_{\mathcal{H}_i} + \|\bar{\nabla}_{t,i}\|_{\mathcal{H}_i} \right),$$

where $Z_{t,i}$ is a normalizing constant, and $\beta_i \geq 1$ is a *balancing factor* used to balance the regret and time complexity. If $b_{t,i} = 1$, then we add the current example into the budget, i.e. $S_i = S_i \cup \{(\mathbf{x}_t, y_t)\}$. Otherwise, S_i keeps unchanged.

Different from B(AO)$_2$KS, we reduce online kernel selection to a problem of prediction with expert advice. Let $\mathcal{E}(K)$ be the algorithm for expert advice in Theorem 1. Although evaluating all of the base kernels increases the time complexity by K times, the affection can be counteracted by tuning the balancing factor. At the beginning of round t, we first select κ_{I_t}, $I_t \sim \mathbf{p}_t$ and output $\hat{y}_t = \text{sign}(f_t(\mathbf{x}_t))$, where $f_t = f_{t,I_t}$. Then we explore all of the unselected kernels. Similarly, we update $f_{t,j}$ and compute the gradient $\nabla_{t,j}$. To update the auxiliary hypothesis, we define the variance-reduced gradient estimator $\tilde{\nabla}_{t,i}$ as follows

$$\tilde{\nabla}_{t,i} = \nabla_{t,i} \mathbb{I}_{y_t f_{t,i}(\mathbf{x}_t) \geq 1} + \left[\frac{\nabla_{t,i} - \bar{\nabla}_{t,i}}{\mathbb{P}[b_{t,i} = 1]} \mathbb{I}_{b_{t,i} = 1} + \bar{\nabla}_{t,i} \right] \mathbb{I}_{y_t f_{t,i}(\mathbf{x}_t) < 1}, \forall i = 1, \ldots, K.$$

To update the probability distribution $\mathbf{p}_t$, let $c_{t,i} = \frac{\max\{0, 1 - y_t f_{t,i}(\mathbf{x}_t)\}}{1 + U\sqrt{D}}$. At the end of round t, we send $\mathbf{c}_t = (c_{t,1}, \ldots, c_{t,K})$ to $\mathcal{E}(K)$.

We name this approach EA$_2$OKS (Expert Advice for Adaptive Online Kernel Selection) and present the pseudo-code in Algorithm 2.

Algorithm 2. EA$_2$OKS

Input: $\lambda_i, \beta_i, i = 1, \ldots, K, D, U, M$.
Initialization: $\forall \kappa_i \in \mathcal{K}, f'_{0,i} = 0, S_i = \emptyset, V = \emptyset$.
1: **for** $t = 1, 2, \ldots, T$ **do**
2:     Select a kernel $\kappa_{I_t} \sim \mathbf{p}_t$ ($\mathbf{p}_t$ is output by $\mathcal{E}(K)$),
3:     Compute $\bar{\nabla}_{t,I_t} = \frac{-1}{|V|} \sum_{(\mathbf{x},y) \in V} y\kappa_{I_t}(\mathbf{x}, \cdot)$,
4:     Update f_{t,I_t} according to (4), and output prediction $\hat{y}_t = \text{sign}(f_{t,I_t}(\mathbf{x}_t))$,
5:     **for** $\kappa_i \in \mathcal{K}$ **do**
6:         **if** $\kappa_i \neq \kappa_{I_t}$, **then** update $f_{t,i}$ according to (4),
7:         **if** $y_t f_{t,i}(\mathbf{x}_t) < 1$ **then**
8:             Compute $\mathbb{P}[b_{t,i} = 1] = \|\nabla_{t,i} - \bar{\nabla}_{t,i}\|_{\mathcal{H}_i}/Z_{t,i}$,
9:             Sample $b_{t,i} \sim \text{Ber}(\mathbb{P}[b_{t,i} = 1], 1)$,
10:             **if** $b_{t,i} = 1$, **then** $S_i = S_i \cup (\mathbf{x}_t, y_t)$,
11:             Compute $\tilde{\nabla}_{t,i} = \frac{\nabla_{t,i} \mathbb{I}_{b_{t,i} = 1}}{\mathbb{P}[b_{t,i} = 1]} + \left(1 - \frac{\mathbb{I}_{b_{t,i} = 1}}{\mathbb{P}[b_{t,i} = 1]} \right) \bar{\nabla}_{t,i}$,
12:             Updating $f'_{t,i}$ according to (5),
13:             Compute $c_{t,i} = \frac{1}{1 + U\sqrt{D}} \max\{0, 1 - y_t f_{t,i}(\mathbf{x}_t)\}$,
14:         **end if**
15:     **end for**
16:     Send $\mathbf{c}_t = (c_{t,1}, \ldots, c_{t,K})$ to $\mathcal{E}(K)$,
17:     Update Reservoir V (line 16-18 in Algorithm 1),
18: **end for**

4.1   Regret Bound

Theorem 4 gives an upper bound on the number of support vectors in each S_i, which implies the per-round time complexity of EA$_2$OKS.

Theorem 4. *For all $i = 1, \ldots, K$, with probability at least $1 - 2\delta$, EA$_2$OKS guarantees that the number of support vectors in S_i satisfies*

$$|S_i| \le 4M \ln \frac{T}{M} \sqrt{\ln \frac{K}{\delta}} + \frac{4}{3} \ln \frac{K}{\delta} + \frac{10}{\beta_i} \sqrt{\frac{M + D_i \ln \frac{T}{M}}{M}} T \mathcal{A}(\mathcal{I}_T, \kappa_i) \ln^{\frac{3}{4}} \left(\frac{K}{\delta} \right).$$

The time complexity of EA$_2$OKS depends on the alignment $\mathcal{A}(\mathcal{I}_T, \kappa_i)$ implying that selecting different kernel function not only has an impact on the learning performance of online kernel learning algorithms, but also the time complexity. More discussions are shown in Remark 2. Next we present the high-probability regret bound induced by the hypothesis sequence $\{f_{t,i}\}_{t=1}^T$, $i = 1, \ldots, K$.

Theorem 5. *Let $\psi_{t,i}(f) = \frac{1}{\lambda_i} \|f\|_{\mathcal{H}_i}^2$, $\beta_i \ge 1$ and $\delta \in (0,1)$. For all base kernel $\kappa_i \in \mathcal{K}$ and any $f \in \mathbb{H}_i$, with probability at least $1 - 4\delta$, the regret of the hypothesis sequence $\{f_{t,i}\}_{t=1}^T$ induced by EA$_2$OKS satisfies*

$$L_T(f_{1:T,i}) - L_T(f) \le \frac{\|f\|_{\mathcal{H}_i}^2}{2\lambda_i} + 18\lambda_i \beta_i \sqrt{D_i g_i(M,T) T \mathcal{A}(\mathcal{I}_T, \kappa_i) \ln \frac{K}{\delta}} +$$

$$(6\lambda_i \beta_i^2 + 7U\beta_i) D_i^\theta \ln \frac{K}{\delta} + 9(2\lambda_i \beta_i^{\frac{3}{2}} + U\beta_i^{\frac{1}{2}}) D_i^{\frac{\theta}{2}} g_i(M,T)^{\frac{1}{4}} T^{\frac{1}{4}} \mathcal{A}(\mathcal{I}_T, \kappa_i)^{\frac{1}{4}} \ln^{\frac{3}{4}} \frac{K}{\delta},$$

where $g_i(T,M) = (M + D_i \ln \frac{T}{M})/M$ and $\theta \in \{1/2, 2\}$. Let $\beta_i < \sqrt{T\mathcal{A}(\mathcal{I}_T, \kappa_i)}$ and $\lambda_i = (36\beta_i \sqrt{Tg_i(M,T)\mathcal{A}(\mathcal{I}_T, \kappa_i)})^{-\frac{1}{2}}$. The regret is thus of order

$$L_T(f_{1:T,i}) - L_T(f) = \tilde{O} \left((\|f\|_{\mathcal{H}_i}^2 + U) \sqrt{\beta_i} T^{\frac{1}{4}} \mathcal{A}(\mathcal{I}_T, \kappa_i)^{\frac{1}{4}} \ln^{\frac{3}{4}} \frac{K}{\delta} \right).$$

Now we can show the final high-probability regret bound.

Theorem 6. *Let $\mathcal{E}(K)$ be the algorithm in Theorem 1. Under the condition of Theorem 5, for all base kernel $\kappa_i \in \mathcal{K}$, with probability at least $1 - 5\delta$, the regret of EA$_2$OKS w.r.t. $\mathbb{H}_i$ satisfies*

$$\mathrm{Reg}_T(\mathbb{H}_i) = \tilde{O} \left(\sqrt{L_T(f_i^*) \ln \frac{K}{\delta}} + (\|f_i^*\|_{\mathcal{H}_i}^2 + U) \sqrt{\beta_i} T^{\frac{1}{4}} \mathcal{A}(\mathcal{I}_T, \kappa_i)^{\frac{1}{4}} \ln^{\frac{3}{4}} \frac{K}{\delta} \right).$$

The per-round time complexity is of order $\tilde{O} \left(\sum_{i=1}^K \beta_i^{-1} \sqrt{T\mathcal{A}(\mathcal{I}_T, \kappa_i)} \right)$.

Remark 2 (regret-performance trade-off). Theorem 6 reveals that the per-round time complexity depends on $1/\beta_i$, and the regret only depends on $\sqrt{\beta_i}$. Thus β_i balances the regret and time complexity. If $\beta_i = K^\epsilon, \epsilon \ge 0$, a universal value, then the time complexity is $\tilde{O}\left(K^{-\epsilon}\sum_{i=1}^{K}\sqrt{T\mathcal{A}(\mathcal{I}_T,\kappa_i)}\right)$, while the regret only increases by a factor of $K^{\frac{\epsilon}{2}}$. For $\epsilon \geq 2$, EA$_2$OKS is more efficient than B(AO)$_2$KS, but also suffers larger regret. A better approach of setting β_i is to incorporate the information of $\mathcal{A}(\mathcal{I}_T, \kappa_i)$. A more interesting result is shown in Corollary 1.

Corollary 1. *Let the optimal kernel* $\kappa_{i*} = \mathrm{argmin}_{i\in[K]}\mathcal{A}(\mathcal{I}_T, \kappa_i)$, *and the balance factor* β_i *satisfy the following condition*

$$\beta_i\sqrt{\mathcal{A}(\mathcal{I}_T, \kappa_{i*})} = K\beta\sqrt{\mathcal{A}(\mathcal{I}_T, \kappa_i)}, \quad \beta \geq 1, \quad i = 1, \dots, K.$$

The regret of EA$_2$OKS *satisfies, with probability at least* $1 - 5\delta$,

$$\mathrm{Reg}_T(\mathbb{H}_i) = \begin{cases} \tilde{O}\left(\sqrt{L_T(f_{i*}^*)K\ln\frac{K}{\delta}} + (\|f_{i*}^*\|_{\mathcal{H}_{i*}}^2 + U)\beta_K^{\frac{1}{2}}T^{\frac{1}{4}}\mathcal{A}(\mathcal{I}_T,\kappa_{i*})^{\frac{1}{4}}\right) & i = i^* \\ \tilde{O}\left(\sqrt{L_T(f_i^*)K\ln\frac{K}{\delta}} + (\|f_i^*\|_{\mathcal{H}_i}^2 + U)\beta_K^{\frac{1}{2}}T^{\frac{1}{4}}\frac{\mathcal{A}(\mathcal{I}_T,\kappa_i)^{\frac{1}{2}}}{\mathcal{A}(\mathcal{I}_T,\kappa_{i*})^{\frac{1}{4}}}\right) & i \neq i^* \end{cases}$$

where $\beta_K = K\beta$. *The per-round time complexity is of order* $O(\frac{1}{\beta}\sqrt{T\mathcal{A}(\mathcal{I}_T, \kappa_{i*})})$.

For kernel selection, it is unnecessary to compare with all of the base kernels. Any algorithm just needs to be competitive with the case in which we know the optimal kernel κ_{i*} in advance. Thus, we allow the algorithm to achieve a worse regret bound w.r.t. the non-optimal RKHS $\mathbb{H}_i, i \neq i^*$. The significance of Corollary 1 is that EA$_2$OKS can keep the same regret bound w.r.t. $\mathbb{H}_{i*}$, and reduce the per-round time complexity to $O(\beta^{-1}\sqrt{T\mathcal{A}(\mathcal{I}_T, \kappa_{i*})})$.

We analyze the time complexity. According to Theorem 4, the exact time complexity of EA$_2$OKS is of order $O\left(\sum_{i=1}^{K}|S_i|\right)$. Let $M = O(\ln T)$. Then the time complexity of EA$_2$OKS is the one claimed In Remark 2 or the comments after Corollary 1. We omit the time complexity of projection operation, since it can be executed incrementally in $O(M)$ time.

4.2   Budgeted EA$_2$OKS

Inspired by Corollary 1, we can set a same threshold for all S_i, i.e., $|S_i| \leq B$. At the end of round $t - 1$, if $|S_i| = B$, then we reset $S_i = \emptyset$ and $f_{t-1,i}' = 0$. We name the algorithm BEA$_2$OKS (Budgeted EA$_2$OKS). Due to BEA$_2$OKS is much similar with EA$_2$OKS and the space limit, the algorithm description is presented in supplementary material. Combining Theorem 4 and Corollary 1, we further obtain Corollary 2.

Corollary 2. *Let* $\kappa_{i*} = \mathrm{argmin}_{i\in[K]}\mathcal{A}(\mathcal{I}_T, \kappa_i)$. *If* B *satisfies the condition*

$$B = 4M\ln\frac{T}{M}\sqrt{\ln\frac{K}{\delta}} + \frac{4}{3}\ln\frac{K}{\delta} + \frac{10}{\beta_{i*}}\sqrt{g_{i*}(M,T)T\mathcal{A}(\mathcal{I}_T,\kappa_{i*})}\ln^{\frac{3}{4}}\left(\frac{K}{\delta}\right),$$

and $\beta_{i*} = K$, *then with probability at least* $1 - 5\delta$, S_{i*} *will not restart, and*

$$\mathrm{Reg}_T(\mathbb{H}_{i*}) = \tilde{O}\left(\sqrt{L_T(f_{i*}^*)K\ln\frac{K}{\delta}} + (\|f_{i*}^*\|_{\mathcal{H}_{i*}}^2 + U)K^{\frac{1}{2}}T^{\frac{1}{4}}\mathcal{A}(\mathcal{I}_T,\kappa_{i*})^{\frac{1}{4}}\right).$$

The per-round time complexity of $\mathrm{BEA_2OKS}$ *is of order* $O(\sqrt{T\mathcal{A}(\mathcal{I}_T, \kappa_{i*})})$.

According to Corollary 2 and 1, we claim that $\mathrm{BEA_2OKS}$ and $\mathrm{EA_2OKS}$ (let $\beta = 1$) are equivalent in the sense that they enjoy the same regret bound w.r.t. $\mathbb{H}_{i*}$ and the same per-round time complexity. The superiority of $\mathrm{BEA_2OKS}$ is that it only needs to tune B, while $\mathrm{EA_2OKS}$ needs to tune $\beta_i, i = 1, \ldots, K$.

5 Experiments

In this section, we conduct experiments to verify our theoretical results.

5.1 Experimental Setting

We only verify $\mathrm{B(AO)_2KS}$, $\mathrm{EA_2OKS}$ and $\mathrm{BEA_2OKS}$, but do not run the algorithm in Theorem 1, since the $O(KT)$ time complexity is prohibitive. For $\mathrm{B(AO)_2KS}$, let $\mathcal{M}(K)$ be GREEN-IX [15]. For $\mathrm{EA_2OKS}$ and $\mathrm{BEA_2OKS}$, let $\mathcal{E}(K)$ be the exponentially weighted average algorithm (chapter 4.2 in [3]) whose learning rate (Corollary 2.4 in [3]) is tuned by doubling trick. We use four binary classification datasets downloaded from LIBSVM website[1], including *w7a* (Num: 24692, Fea: 300), *w8a* (Num: 49749, Fea: 300), *a9a* (Num: 48842, Fea: 123) and *ijcnn1* (Num: 141691, Fea: 22). Let $\mathcal{K} = \{\sigma_i\}_{i=1}^K$ contain K Gaussian kernels, where $\kappa_i(\mathbf{u}, \mathbf{v}) = \exp(-\|\mathbf{u} - \mathbf{v}\|^2/(2\sigma_i^2))$. We implement all algorithms with R on a Windows machine with 2.5 GHz Core i7 CPU, execute each experiment 10 times with random permutation of all datasets and average all of the results[2].

We will execute three experiments. For all experiments, let $M = \lceil \ln T \rceil$ and $U = 20$. $D - 1$ for Gaussian kernel. The first experiment aims at verifying the influence of K on $\mathrm{B(AO)_2KS}$. We choose 3 groups of $\mathcal{K}$, denoted by $\mathcal{K}_{12} = \{2^{-2:0.5:3.5}\}$, $\mathcal{K}_8 = \{2^{-1:0.5:2.5}\}$ and $\mathcal{K}_4 = \{2^{-1:1:2}\}$, and name the corresponding algorithm $\mathrm{B(AO)_2KS}$-12, $\mathrm{B(AO)_2KS}$-8 and $\mathrm{B(AO)_2KS}$-4. According to Theorem 2, the optimal learning rate λ_i should be $\tilde{O}(1/\sqrt{K\mathcal{A}(\mathcal{I}_T, \kappa_i)})$. However, $\mathrm{B(AO)_2KS}$ is not parameter-free. In this paper, we set $\lambda_i = 1/\sqrt{K\mathcal{A}(\mathcal{I}_T, \kappa_{i*})}$ for all $i \in [K]$, and set $\mathcal{A}(\mathcal{I}_T, \kappa_{i*}) = \sqrt{T}$ which is an optimistic estimator.

The second experiment aims at proving the advantage of the data-dependent regret bounds and time complexity. The baseline algorithms include two online kernel selection algorithms: OKS [22] and ISKA [24], and two online kernel learning algorithms: Forgetron-σ [5] and BOGD-σ [25]. For Forgetron-σ and BOGD-σ, we set σ to the best value in hindsight. The other hyper-parameters are set to the recommended value in original papers. For $\mathrm{EA_2OKS}$ and $\mathrm{BEA_2OKS}$, we set $\beta_i = K^{\frac{3}{2}}$. Although the optimal B is unknown for $\mathrm{BEA_2OKS}$, a feasible approach is to set a slightly large value, which ensures S_{i*} will not restart and Corollary 2 holds on. We select $\mathcal{K} = \{2^{-2:1:3}\}$. For fair comparison, we set the stepsize of gradient descent (or λ_i in this paper) to $5/\sqrt{T}$ for all algorithms.

---

[1] https://www.csie.ntu.edu.tw/%7Ecjlin/libsvmtools/datasets/.

[2] The codes are available at https://github.com/JunfLi-TJU/KARegret-OKS.

The third experiment shows the influence of the balancing factor β_i and budget B on EA_2OKS and BEA_2OKS, and compares the two algorithms further. We adopt the same $\mathcal{K}$ and λ_i used in the second experiment.

5.2   Experimental Results

Table 1 shows the results of the first experiment. Overall, the experimental results coincide with our theoretical analyses (Theorem 3). If $\mathcal{K}_{12}$, $\mathcal{K}_8$ and $\mathcal{K}_4$ contain the same optimal kernel, then the smaller K is, the better learning performance and the longer running time is. The result of $B(AO)_2KS$-12 on $a9a$ does not satisfy the rule. The reason is that $\mathcal{K}_{12}$ contains many kernels performing badly on $a9a$ which leads to a large number of support vectors.

Table 1. The influence of K on $B(AO)_2KS$

Algorithm	w8a		a9a	
	Mistake (%)	Time (s)	Mistake (%)	Time (s)
$B(AO)_2KS$-4	$\mathbf{2.15} \pm 0.09$	444.88 ± 23.81	$\mathbf{12.72} \pm 0.08$	320.55 ± 5.63
$B(AO)_2KS$-8	2.53 ± 0.04	271.27 ± 13.57	15.39 ± 0.12	216.96 ± 4.53
$B(AO)_2KS$-12	2.79 ± 0.03	245.85 ± 8.49	17.47 ± 0.18	226.98 ± 6.05

Table 2 reports the results of the second experiment. In the second column, B is the budget size and $|S_{i*}| = \min_i |S_i|$ in EA_2OKS. Note that we use the kernel with minimal $|S_i|$ as a proxy for the optimal kernel κ_{i*}. $B(AO)_2KS$ enjoys the best learning performance, but also suffers a slightly larger time complexity. OKS has the longest running time on all datasets. For BOGD-σ and Forgetron-σ, we select the best σ in hindsight for constructing strong baseline algorithms, which is unprocurable in practice. Note that BEA_2OKS enjoys the same prediction performance with EA_2OKS, and has lower running time, since we set $B \approx |S_{i*}|$ in EA_2OKS. Overall, BEA_2OKS provides the best regret-time complexity trade-off except for the *ijcnn1* dataset on which ISKA performs better.

Table 3 shows the results of the third experiment. #rs is the average restart times of S_{i*} in BEA_2OKS, or the average $|S_{i*}|$ in EA_2OKS. For a same β, BEA_2OKS keeps the same prediction accuracy with EA_2OKS, and improves the efficiency significantly. The reason is that #rs is 0. Thus BEA_2OKS just needs a small budget to keep the regret w.r.t. $\mathbb{H}_{i*}$ achieved by EA_2OKS. In this case, there is no sense to increase B. For a same B, the smaller β is, the better learning performance and the longer running time is. The experimental results coincide with Corollary 2.

Table 2. Performance comparison among different online kernel selection algorithms

Algorithm	B-$|S_{i*}|$	w7a		B-$|S_{i*}|$	w8a	
		Mistake (%)	Time (s)		Mistake (%)	Time (s)
BOGD-σ	500	2.75 ± 0.03	44.66	800	3.26 ± 0.34	143.31
Forgetron-σ	500	5.29 ± 0.07	34.90	800	4.74 ± 0.06	119.92
OKS	–	2.55 ± 0.16	201.26	–	$\mathbf{2.38 \pm 0.10}$	753.08
ISKA	250	4.72 ± 2.20	53.73	400	3.16 ± 0.24	181.27
B(AO)$_2$KS	–	$\mathbf{2.56 \pm 0.06}$	100.57	–	2.48 ± 0.07	364.27
EA$_2$OKS	127	3.12 ± 0.04	112.21	201	3.03 ± 0.02	394.78
BEA$_2$OKS	200	3.17 ± 0.08	40.74	400	3.02 ± 0.02	116.56
Algorithm	B-$|S_{i*}|$	a9a		B-$|S_{i*}|$	jicnn1	
		Mistake (%)	Time (s)		Mistake (%)	Time (s)
BOGD-σ	1500	17.85 ± 0.06	114.31	3500	9.57 ± 0.00	149.93
Forgetron-σ	1700	24.38 ± 0.16	170.34	3500	13.04 ± 0.08	164.99
OKS	–	18.71 ± 0.22	715.39	–	8.90 ± 0.12	477.14
ISKA	1400	16.95 ± 0.05	296.29	1500	8.48 ± 0.05	135.79
B(AO)$_2$KS	–	$\mathbf{15.16 \pm 0.11}$	308.75	–	$\mathbf{7.58 \pm 0.17}$	391.48
EA$_2$OKS	1062	17.88 ± 0.23	267.68	1025	8.75 ± 0.11	227.43
BEA$_2$OKS	1200	17.83 ± 0.16	167.12	1500	8.79 ± 0.16	188.20

Table 3. Parameter influence on EA$_2$OKS and BEA$_2$OKS on w8a dataset

Algorithm	Mistake (%) Time (s) #rs	Mistake (%) Time (s) #rs
EA$_2$OKS	$\beta = K$	$\beta = K^2$
	2.99 ± 0.02 1060.17 ± 26.0 343	3.35 ± 0.15 215.28 ± 6.15 140
BEA$_2$OKS	$\beta = K$, $B = 400$	$\beta = K^2$, $B = 400$
	2.99 ± 0.02 132.33 ± 2.28 0	3.43 ± 0.19 123.92 ± 5.96 0
BEA$_2$OKS	$\beta = K$, $B = 600$	$\beta = K^2$, $B = 600$
	3.00 ± 0.03 174.77 ± 2.20 0	3.43 ± 0.22 144.39 ± 7.68 0

6   Conclusion

In this paper, we develop several computationally efficient online kernel selection algorithms, which achieve the first kernel alignment regret bound improving previous worst-case regret bounds. Theoretical analyses reveal that if there is a good kernel in the candidate set, then our algorithms can not only improve the learning performance relative to single kernel learning, but also suffer a low time complexity. Experimental results verify the effectiveness and efficiency of our algorithms. An important question is whether it is possible to achieve the $O(\|f\|_{\mathcal{H}_{i*}}^2 \sqrt{\mathcal{A}(\mathcal{I}_T, \kappa_{i*})})$ regret bound with a $O(\mathcal{A}(\mathcal{I}_T, \kappa_{i*}))$ time complexity.

References

1. Baram, Y.: Learning by kernel polarization. Neural Comput. **17**(6), 1264–1275 (2005)
2. Calandriello, D., Lazaric, A., Valko, M.: Efficient second-order online kernel learning with adaptive embedding. In: Advances in Neural Information Processing Systems, vol. 30, pp. 6140–6150 (2017)
3. Cesa-Bianchi, N., Lugosi, G.: Prediction, Learning, and Games. Cambridge University Press, Cambridge (2006)
4. Chiang, C., et al.: Online optimization with gradual variations. In: Proceedings of the 25th Annual Conference on Learning Theory, pp. 6.1–6.20 (2012)
5. Dekel, O., Shalev-Shwartz, S., Singer, Y.: The forgetron: a kernel-based perceptron on a budget. SIAM J. Comput. **37**(5), 1342–1372 (2008)
6. Foster, D.J., Kale, S., Mohri, M., Sridharan, K.: Parameter-free online learning via model selection. In: Advances in Neural Information Processing Systems, vol. 30, pp. 6022–6032 (2017)
7. Foster, D.J., Rakhlin, A., Sridharan, K.: Adaptive online learning. In: Advances in Neural Information Processing Systems, vol. 28, pp. 3375–3383 (2015)
8. Hazan, E., Kale, S.: Better algorithms for benign bandits. In: Proceedings of the Twentieth Annual ACM-SIAM Symposium on Discrete Algorithms, pp. 38–47 (2009)
9. Hazan, E., Kale, S.: Extracting certainty from uncertainty: regret bounded by variation in costs. Mach. Learn. **80**(2–3), 165–188 (2010)
10. Jézéquel, R., Gaillard, P., Rudi, A.: Efficient online learning with kernels for adversarial large scale problems. In: Advances in Neural Information Processing Systems, vol. 32, pp. 9427–9436 (2019)
11. Jin, R., Hoi, S.C.H., Yang, T.: Online multiple kernel learning: algorithms and mistake bounds. In: Proceedings of the 21st International Conference on Algorithmic Learning Theory, pp. 390–404 (2010)
12. Lee, C., Luo, H., Wei, C., Zhang, M.: Bias no more: high-probability data-dependent regret bounds for adversarial bandits and MDPs. In: Advances in Neural Information Processing Systems, vol. 33, pp. 15522–15533 (2020)
13. Li, J., Liao, S.: Online kernel selection with multiple bandit feedbacks in random feature space. In: Proceedings of the 11th International Conference on Knowledge Science, Engineering and Management, pp. 301–312 (2018)
14. Lu, J., Hoi, S.C.H., Wang, J., Zhao, P., Liu, Z.: Large scale online kernel learning. J. Mach. Learn. Res. **17**(47), 1–43 (2016)
15. Lykouris, T., Sridharan, K., Tardos, É.: Small-loss bounds for online learning with partial information. In: Proceedings of the 31st Conference on Learning Theory, pp. 979–986 (2018)
16. Muthukumar, V., Ray, M., Sahai, A., Bartlett, P.: Best of many worlds: robust model selection for online supervised learning. In: Proceedings of the 22nd International Conference on Artificial Intelligence and Statistics, pp. 3177–3186 (2019)
17. Nguyen, T.D., Le, T., Bui, H., Phung, D.: Large-scale online kernel learning with random feature reparameterization. In: Proceedings of the Twenty-Sixth International Joint Conference on Artificial Intelligence, pp. 2543–2549 (2017)
18. Rakhlin, A., Sridharan, K.: Online learning with predictable sequences. In: Proceedings of the 26th Annual Conference on Learning Theory, pp. 993–1019 (2013)
19. Shen, Y., Chen, T., Giannakis, G.B.: Random feature-based online multi-kernel learning in environments with unknown dynamics. J. Mach. Learn. Res. **20**(22), 1–36 (2019)

20. Vitter, J.S.: Random sampling with a reservoir. ACM Trans. Math. Softw. **11**(1), 37–57 (1985)
21. Wei, C., Luo, H.: More adaptive algorithms for adversarial bandits. In: Proceedings of the 31st Annual Conference on Learning Theory, pp. 1263–1291 (2018)
22. Yang, T., Mahdavi, M., Jin, R., Yi, J., Hoi, S.C.H.: Online kernel selection: algorithms and evaluations. In: Proceedings of the Twenty-Sixth AAAI Conference on Artificial Intelligence, pp. 1197–1202 (2012)
23. Zhang, L., Yi, J., Jin, R., Lin, M., He, X.: Online kernel learning with a near optimal sparsity bound. In: Proceedings of the 30th International Conference on Machine Learning, pp. 621–629 (2013)
24. Zhang, X., Liao, S.: Online kernel selection via incremental sketched kernel alignment. In: Proceedings of the Twenty-Seventh International Joint Conference on Artificial Intelligence, pp. 3118–3124 (2018)
25. Zhao, P., Wang, J., Wu, P., Jin, R., Hoi, S.C.H.: Fast bounded online gradient descent algorithms for scalable kernel-based online learning. In: Proceedings of the 29th International Conference on Machine Learning, pp. 1075–1082 (2012)

Reinforcement Learning

Periodic Intra-ensemble Knowledge Distillation for Reinforcement Learning

Zhang-Wei Hong[1]([✉]), Prabhat Nagarajan[2], and Guilherme Maeda[2]

[1] Massachusetts Institute of Technology, Cambridge, MA, USA
zwhong@mit.edu
[2] Preferred Networks, Inc., Tokyo, Japan
{prabhat,gjmaeda}@preferred.jp

Abstract. Off-policy ensemble reinforcement learning (RL) methods have demonstrated impressive results across a range of RL benchmark tasks. Recent works suggest that directly imitating experts' policies in a supervised manner before or during the course of training enables faster policy improvement for an RL agent. Motivated by these recent insights, we propose *Periodic Intra-Ensemble Knowledge Distillation* (PIEKD). PIEKD is a learning framework that uses an ensemble of policies to act in the environment while periodically sharing knowledge amongst policies in the ensemble through knowledge distillation. Our experiments demonstrate that PIEKD improves upon a state-of-the-art RL method in sample efficiency on several challenging MuJoCo benchmark tasks. Additionally, we perform ablation studies to better understand PIEKD.

Keywords: Ensemble learning · Reinforcement learning · Distillation

1 Introduction

In reinforcement learning (RL), the goal is to train a policy to interact with an environment, such that this policy yields the maximal expected return. While typical RL methods merely train a single parameterized policy, ensemble methods that share experiences amongst several function approximators [24,25] have been able to achieve superior performance in the context of reinforcement learning (RL). Unlike typical RL methods, Osband et al. [25] train an ensemble of neural network (NN) policies with distinct initial weights (i.e. parameters of NNs) simultaneously, by sharing experiences amongst the policies. These shared experiences are collected by first randomly selecting a policy from the ensemble to perform an episode. This episode of experiences is added to a shared experience replay buffer [20] used to train all members of the ensemble. Learning from shared experience allows for more efficient policy learning, since randomly initialized policies result in extensive exploration in the environment. Though reinforcement learning from shared experiences has shown considerable improvement over single-policy RL methods, other lines of work [13] show that directly

---

Z.-W. Hong—Work done during an internship at Preferred Networks, Inc.

N. Oliver et al. (Eds.): ECML PKDD 2021, LNAI 12975, pp. 87–103, 2021.
https://doi.org/10.1007/978-3-030-86486-6_6

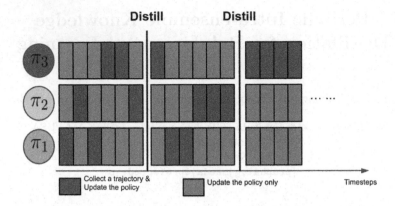

Fig. 1. An overview of Periodic Intra-Ensemble Knowledge Distillation. We select a policy from the ensemble to act in the environment, and use this experience to update all policies. Periodically, we distill the best-performing policy to the rest of the ensemble.

imitating an expert's experiences in a supervised manner can accelerate reinforcement learning.

Motivated by these results that demonstrate that direct imitation can accelerate RL, we propose Periodic Intra-Ensemble Knowledge Distillation (PIEKD), a framework that not only trains an ensemble of policies via common experiences but also shares the knowledge of the best-performing policy amongst the ensemble. Previous works on ensemble RL have shown that randomly initialized policies can result in adequate behavioral diversity [24]. Thus PIEKD first begins by initializing each policy in the ensemble with different weights to perform extensive exploration in the environment. As the behaviors of these policies are diverse in nature, at any given time during the course of training, one policy is naturally superior to the other policies. This policy is then used to improve the quality of the other policies in the ensemble, without having to improve solely through experience. To use the best policy to improve other policies, PIEKD employs knowledge distillation [14], which is effective at transferring knowledge between neural networks. By using knowledge distillation, we can encourage policies in the ensemble to act in a manner similar to the best policy, enabling them to rapidly improve and continue optimizing for the optimal policy from better starting points. Prior work [26] has shown that we can successfully distill several specialized policies into a single multitask policy, demonstrating that distillation can successfully augment behaviors into a policy without destroying existing knowledge. These results suggest that in PIEKD, despite the use of distillation between policies, their inherent knowledge is still preserved, improving individual policies without destroying the diversity amongst policies. An abstract overview of PIEKD is depicted in Fig. 1.

This paper's primary contribution is Periodic Intra-Ensemble Knowledge Distillation (PIEKD), a simple yet effective framework for off-policy RL that jointly trains an ensemble of policies while periodically performing knowledge sharing. We demonstrate empirically that PIEKD can improve the state-of-the-art soft-actor critic (SAC) [12] on a suite of challenging MuJoCo tasks, exhibiting superior sample efficiency. We further validate the effectiveness of distillation for knowledge sharing by comparing against other forms of sharing knowledge.

The remainder of this paper is organized as follows. Section 2 discusses related work in ensemble RL and knowledge distillation. Section 3 provides a brief overview of the reinforcement learning formulation. Section 4 describes PIEKD. Section 5 presents our experimental findings. Lastly, Sect. 6 summarizes our contributions and outlines potential avenues for future work.

2 Related Work

The works that are most related to PIEKD [24,25] train multiple policies via shared experience for the same task through RL, where the shared experiences are collected by all policies in the ensemble and stored in a common buffer, as our method does. Differing from those works [24,25], we additionally periodically perform knowledge distillation between policies of the ensemble. Other related methods aggregate multiple policies to select actions [10,32]. Abel et al. [1] sequentially train a series of policies, boosting the learning performance by using the errors of a prior policy. However, rather than performing decision aggregation or sequentially-boosted training, we focus on improving the performance of each individual policy via knowledge sharing amongst jointly trained policies.

Rusu et al. [26] train a single neural network to perform multiple tasks by transferring multiple pretrained policies to a single network through distillation. Hester et al. [13] and Nair et al. [22] accelerate RL agents' training progress through human experts' guidance. Rather than experts' policies, Nagabandi et al. [21], Levine and Koltun [19], and Zhang et al. [33] leverage model-based controllers' behaviors, facilitating training for RL agents. Additionally, Oh et al. [23] train RL agents to imitate past successful self-experiences or policies. Orthogonal to the aforementioned works, PIEKD periodically exploits the current best policy within the ensemble, and shares its knowledge amongst the ensemble.

In other machine learning areas, Zhang et al. [34] train multiple models that mutually imitate each other's outputs on classification tasks. Our distillation procedure is not mutual, but flows in a single direction, from a superior teacher policy to other student policies in the ensemble. Subsequent work by Lan et al. [18] trains an ensemble of models to imitate a stronger teacher model that aggregates all of the ensemble models' predictions. Our method contrasts from the above methods by periodically electing the teacher for distillation to other ensemble members. We maintain separation between ensemble members rather than aggregate them into a single policy.

Teh et al. [31] and Ghosh et al. [9] distill multiple task-specific policies to a central multi-task policy and constrain the mutual divergence between each task-specific policy and the central one. Galashov et al. [7] learn a task-specific policy while bounding the divergence between this task-specific policy and some generic policy that can perform basic task-agnostic behaviors. Czarnecki et al. [4] gradually transfer the knowledge of a simple policy to a complex policy during the course of joint training. Our work differs from the aforementioned works in several aspects. First, our method periodically elects a teacher policy for sharing knowledge rather than constraining the mutual policy divergence [7,9, 31]. Second, our method does not rely on training heterogeneous policies (e.g., a simple policy and a complex policy [4]), which makes our method more generally applicable. Finally, as opposed to Teh et al. [31] and Ghosh et al. [9], we consider single-task settings rather than multi-task settings.

Population-based methods similarly employ multiple policies in separate copies of the environment to find the optimal policy. Evolutionary Algorithms (EA) [8,17,27] randomly perturb the parameters of policies in the population, eliminate underperforming policies by evaluating the policies' performances in the environment, and produce new generations of policies from the remaining policies. Unlike EA, our method does not involve on separate copies of the environment or eliminating existing policies from the population. Instead, our method focuses on continuously improving the existing policies. In addition to EA, other work [16] done concurrently to ours adds a regularization term that forces each agent to imitate the best agent's policy when performing policy updates at each step. Differing from PIEKD, they train multiple agents in separate copies of the environment in parallel. Without relying on multiple copies of the environment, our method is more applicable in cases of costly environment interaction or costly setup of multiple environments (e.g., robot learning in the real world).

3   Background

In this section we describe the general framework of RL. RL formalizes a sequential decision-making task as a *Markov decision process* (MDP) [30]. An MDP consists of a state space $\mathcal{S}$, a set of actions $\mathcal{A}$, a (potentially stochastic) transition function $\mathcal{T} : \mathcal{S} \times \mathcal{A} \rightarrow \mathcal{S}$, a reward function $\mathcal{R} : \mathcal{S} \times \mathcal{A} \rightarrow \mathbb{R}$, and a discount factor $\gamma \in [0, 1]$. An RL agent performs episodes of a task where the agent starts in a random initial state s_0, sampled from the initial state distribution ρ_{s_0}, and performs actions, which transitions the agent to new states for which the agent receives rewards. More generally, at timestep t, an agent in state s_t performs an action a_t, receives a reward r_t, and transitions to a new state s_{t+1}, according to the transition function $\mathcal{T}$. The discount factor γ is used to indicate the agent's preference for short-term rewards over long-term rewards.

An RL agent performs actions according to its policy, a probability distribution $\pi_\phi : \mathcal{S} \times \mathcal{A} \mapsto [0, 1]$, where ϕ denotes the parameters of the policy, which may be the parameters of a neural network. RL methods iteratively update ϕ via rollouts of experience $\tau = \{(s_t, a_t, r_t, s_{t+1})\}_{t=0}^{T-1}$, seeking within the parameter space Φ the optimal ϕ^* that maximizes the expected return $\mathbb{E}_{s \sim \rho_{s_0}} \left[\sum_{t=0}^{T-1} \gamma^t r_t | s_0 = s \right]$ at each t within an episode.

4 Method

In this section, we formally present the technical details of our method, Periodic Intra-Ensemble Knowledge Distillation (PIEKD). We start by providing an overview of PIEKD and then describe its components in detail.

4.1 Overview

PIEKD maintains an ensemble of policies that collect different experiences on the same task, and then periodically shares knowledge amongst the policies in the ensemble. PIEKD is separated into three phases: ensemble initialization, joint training, and intra-ensemble knowledge distillation. First, the ensemble initialization phase randomly initializes an ensemble of policies with different parameters to achieve behavioral diversity. In the joint training stage, a policy randomly selected from the ensemble executes an episode in the environment and its experience is then stored in a shared experience replay buffer that is used to train each policy. In the last stage, we perform intra-ensemble knowledge

---

Algorithm 1. Periodic Intra-Ensemble Knowledge Distillation for Off-policy Actor Critic

---

Require: an environment $\mathcal{E}$, an off-policy actor-critic method ω, an ensemble size K, a parameter space Φ, a set of parameterized policies and critics $\{\pi_{\phi_k}\}_{k=0}^{K-1}$ and $\{Q_{\theta_k}\}_{k=0}^{K-1}$, recent episodic performance statistics $\{R_k\}_{k=0}^{K-1}$, an episode length T, a distillation interval I, an experience buffer $\mathcal{D}$

1:
2: **i. Ensemble initialization**
3:   $\phi_k \sim \text{Uniform}(\Phi), \forall k \in [0, K)$
4:   $\mathcal{D} \leftarrow \{\}$
5:   $R_k \leftarrow \{\}, \forall k \in [0, K)$
6:   $t_{acc} \leftarrow 0$
7: **while** not converged **do**
8:     **ii. Joint training**
9:     $k_e \sim \text{Uniform}([0, K))$ ▷ Policy selection
10:     $\tau \leftarrow \text{ROLLOUT}(\mathcal{E}, \pi_{\phi_{k_e}})$
11:     $\mathcal{D} \leftarrow \mathcal{D} \cup \tau$
12:     $\text{UPDATEPOLICY}(\pi_{\phi_k}, \mathcal{D}, \omega), \forall k \in [0, K)$
13:     $\text{UPDATECRITIC}(Q_{\theta_k}, \mathcal{D}, \omega), \forall k \in [0, K)$
14:     $\text{UPDATESTAT}(R_{k_e}, \tau)$ ▷ Update statistics
15:     $t_{acc} \leftarrow t_{acc} + T$
16:     **iii. Intra-Ensemble Knowledge Distillation**
17:     **if** $t_{acc} \geq I$ **then**
18:         $k_t \leftarrow \text{argmax}_k R_k$ ▷ Teacher election
19:         $\text{DISTILLPOLICY}(\phi_k, \phi_{k_t}, \mathcal{D}), \forall k \in [0, K)$ (Eq. 1)
20:         $\text{DISTILLCRITIC}(\theta_k, \theta_{k_t}, \mathcal{D}), \forall k \in [0, K)$ (Eq. 2)
21:         $t_{acc} \leftarrow 0$
22:     **end if**
23: **end while**

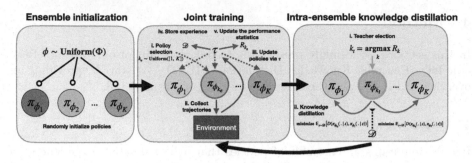

Fig. 2. An overview of the three phases of periodic intra-ensemble knowledge distillation: ensemble initialization, joint training, and intra-ensemble knowledge distillation.

distillation, where we elect a teacher policy from the ensemble used to guide the other policies towards better behaviors. To this end, we distill [14] the best-performing policy to the others. Algorithm 1 and Fig. 2 summarize our method. In this paper, we apply PIEKD to the state-of-the-art off-policy RL algorithm, soft actor-critic (SAC) [11].

4.2   Ensemble Initialization

In the ensemble initialization phase, we randomly initialize K policies in the ensemble. Each policy is instantiated with a model parameterized by ϕ_k, where k denotes the policy's index in the ensemble. ϕ_k is initialized by sampling from the uniform distribution over parameter space Φ which contains all possible values of ϕ_k: $\phi_k \sim \text{Uniform}(\Phi)$. Despite the simplicity of uniform distributions used for initialization, Osband et al. [24] show that uniformly random initialization can provide adequate behavioral diversity. In this paper, we represent each $\phi_k, \forall k \in [0, K)$ as a neural network (NN), though other parametric models can be used.

Since SAC learns both a policy and a critic function that values states or state-action pairs from past experiences stored in a replay buffer [20], we create a shared replay buffer for all policies in the ensemble and randomly initialize an NN critic function Q_{θ_k} for each policy π_{ϕ_k}. θ_k denotes the NN's weights for the critic Q_{θ_k}.

4.3   Joint Training

Each joint training phase consists of I timesteps. For each episode, we select a policy in the ensemble to act in the environment (hereinafter, we refer this process as "policy selection") The policy selection strategy is a way of selecting a policy $\pi_{\phi_{k_e}}$ from the ensemble to perform an episode τ in the environment. This episode τ is stored in a shared experience replay buffer $\mathcal{D}$, and the policy's recent episodic performance statistic R_{k_e} is updated according to the return achieved in τ, where R_{k_e} is the average episodic return in the most recent M episodes.

The episodic performance statistics $\{R_k\}_{k=0}^{K}$ and $\mathcal{D}$ will later be used in the intra-ensemble distillation phase (Sect. 4.4). In this paper, we adopt a simple uniform random policy selection strategy: $k_e \sim \text{Uniform}([0, K))$. To perform RL updates on the agent's policy,

After selecting a policy $\pi_{\phi_{k_e}}$ which performs an episode τ, we store this τ in $\mathcal{D}$ (line 11). Then, we can sample data from $\mathcal{D}$ and update all policies and critics using SAC (line 12–13). Since off-policy RL methods like SAC do not require that τ is necessarily generated by the policy that is being updated, they enable our policies to learn from the trajectories generated by other policies of the ensemble. The details of the update routine for the policy and the critic are taken from the original SAC paper [12].

4.4 Intra-ensemble Knowledge Distillation

The intra-ensemble knowledge distillation phase consists of two stages: *teacher election* and *knowledge distillation*. The teacher election stage (line 18) selects a policy from the ensemble to serve as the teacher for other policies. In our experiments, we use the natural selection criteria of the selecting the best-performing teacher. Specifically, we select the policy that has the highest average recent episodic performance recorded in the joint training phase (Sect. 4.3), namely $k_t = \arg\max_k R_k$, where k_t is the index of the teacher. Rather than use a policy's most recent episodic performance, we use its average return over its previous M episodes, to minimize the noise in our estimate of the policy's performance.

Next, the elected teacher guides the other policies in the ensemble towards better policies (line 19–20). This is done through knowledge distillation [14], which has been shown to be effective at guiding a neural network to behave similarly to another. To distill from the teacher to the students (i.e., the other policies in the ensemble), the teacher samples experiences from the buffer $\mathcal{D}$ and instructs each student to match the teacher's outputs on these samples. After distillation, the students acquire the teacher's knowledge, enabling them to correct their low-rewarding behaviors and reinforce their high-rewarding behaviors, without forgetting their previously learned behaviors [26,31]. Specifically, the policy distillation process is formalized as updating each ϕ_k in the direction of

$$\nabla_{\phi_k} \mathbb{E}_{s \sim \mathcal{D}} \left[D_{KL}(\pi_{\phi_{k_t}}(.|s) || \pi_{\phi_k}(.|s)) \right], \tag{1}$$

where Kullback–Leibler (KL) divergence (D_{KL}) is a principled way to measure the similarity between two probability distributions (i.e., policies). Several prior works [28,29] employed KL-divergence to quantify the difference between two policies. Note that when applying PIEKD to SAC, we must additionally distill the critic function from the teacher to the students. To do so, we train the students' critic functions to match the teacher's critic function, where each student's critic is updated by the gradients

$$\nabla_{\theta_k} \mathbb{E}_{(s,a) \sim \mathcal{D}} \left[(Q_{\theta_{k_t}}(s, a) - Q_{\theta_k}(s, a))^2 \right], \tag{2}$$

where θ_k and θ_{k_t} denote parameters of critic functions. $Q_{\theta_{k_t}}$ and Q_{θ_k} denote the critic function corresponding to the teacher's policy and the student's policy, respectively.

5   Experiments

Our experiments are designed to answer the following questions: (1) Can PIEKD improve upon the data efficiency of state-of-the-art RL? (2) Is knowledge distillation effective at sharing knowledge? (3) Is it necessary to choose the best-performing agent to be the teacher? Next, we show our experimental findings for each of the aforementioned questions, and discuss their implications.

5.1   Experimental Setup

Implementation. Our goal is to demonstrate how PIEKD improves the sample efficiency of an RL algorithm. Since soft actor-critic (SAC) [12] exhibits state-of-the-art performance across several continuous control tasks, we build on top of the ChainerRL implementation of SAC [6]. We directly use the hyperparameters for SAC from the original paper [12] in all of our experiments[1]. Unless stated otherwise, the hyperparameters used for PIEKD (Algorithm 1) are $I = 5000$, and $K = 3$. The value of I is tuned via grid search over $[1000, 2000, \cdots, 10000]$. We tried different ensemble size configurations ($K \in \{2, 3, 5\}$) and decided on $K = 3$, though all ensemble sizes had similar performance. For the remainder of our experiments, we term PIEKD applied to SAC as *SAC-PIEKD*.

Benchmarks.   We use OpenAI Gym [2]'s MuJoCo benchmark tasks, as used in the original SAC [12] paper. We choose most of the tasks selected in the original paper [12] to evaluate the performance of our method. The description for each task can be found in the source code for OpenAI Gym [2].

Evaluation. We adapt the evaluation approach from the original SAC paper [12]. We train each agent for 1 million timesteps, and run 20 evaluation episodes after every 10000 timesteps (i.e., number of interactions with the environment), where the performance is the mean of these 20 evaluation episodes. We repeat this entire process across 5 different runs, each with different random seeds. We plot the mean value and confidence interval of episodic return at each stage of training. The mean value and confidence interval are depicted by the solid line and shaded area, respectively. The confidence interval is estimated by the bootstrap method. At each evaluation point, we report the highest mean episodic return amongst the agents in the ensemble. In some curves, we additionally report the lowest mean episodic return amongst the agents in the ensemble.

---

[1] https://github.com/pfnet-research/piekd.

5.2   Effectiveness of PIEKD

In order to evaluate the effectiveness of intra-ensemble knowledge distillation, we compare *SAC-PIEKD*, against two baselines: *Vanilla-SAC* and *Ensemble-SAC*. *Vanilla-SAC* denotes the original SAC; *Ensemble-SAC* is the analogous variant of Osband et al. [24]'s method for ensemble Q-learning, except on SAC. At its core, Osband's method involves an ensemble of policies that act with the environment and generate experiences. These experiences are then used to train the entire ensemble using an off-policy RL algorithm, such as Q-learning or off-policy actor-critic methods. Thus, our *Ensemble-SAC* baseline denotes the training of an ensemble of policies through SAC while sharing knowledge amongst the ensemble in a shared replay buffer. Effectively, *Ensemble-SAC* is *SAC-PIEKD* without the intra-ensemble knowledge distillation phase. For both *Ensemble-SAC* and *SAC-PIEKD* we set the ensemble size K to be 3.

Our results are shown in Fig. 3. Note that we also plot the worst evaluation in the ensemble at each evaluation phase to provide insight into the general performance of the ensemble. In all tasks, we outperform all baselines, including *Vanilla-SAC* and *Ensemble-SAC*, in terms of sample efficiency. Visually we can see that throughout training, we have consistently better performance at similar amounts of experience, indicating that our method can achieve higher performance with the same number of experiences relative to our baselines.

SAC-PIEKD usually reaches the best baseline's convergent performance in half of the environment interactions. We even find that in the majority of tasks, our worst evaluation in the ensemble outperforms the baseline methods. This demonstrates that all policies of the ensemble are significantly improving, and our method's superior performance is not simply a consequence of selecting the best agent in the ensemble. In particular, *SAC-PIEKD*'s superiority over *Ensemble-SAC* highlights the effectiveness of supplementing shared experiences (*Ensemble-SAC*) with knowledge distillation. In summary, Fig. 3 demonstrates the effectiveness of PIEKD on enhancing the data efficiency of RL algorithms.

5.3   Effectiveness of Knowledge Distillation for Knowledge Sharing

In this section, we investigate the advantage of using knowledge distillation for knowledge sharing. We consider two alternative approaches towards sharing knowledge, other than distillation. First, we consider sharing knowledge by simply providing agents with additional policy updates (in lieu of distillation updates) using the shared experiences. We also consider directly copying the neural network as opposed to performing distillation. Below, we compare these two approaches against knowledge distillation.

Section 5.2 has shown that *Ensemble-SAC*, which updates all agents' policies through shared experiences fails to learn as efficiently as *SAC-PIEKD*. However, *SAC-PIEKD* uses additional gradient updates during the knowledge distillation phase, whereas *Ensemble-SAC* only performs joint training, and lacks an additional knowledge distillation phase. It is unclear whether additional policy

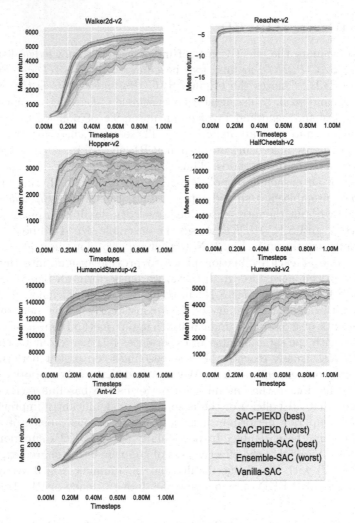

Fig. 3. Performance evaluation of PIEKD. *SAC-PIEKD* represents the implementation of our method upon SAC; *Vanilla-SAC* stands for the original SAC; *Ensemble-SAC* is an analogous variant of Osband et al. [24]'s method on *Vanilla-SAC* (effectively *SAC-PIEKD* without intra-ensemble knowledge distillation). See Sect. 5.2 for details. Notice that in most domains, *SAC-PIEKD* can reach the convergent performance of the baselines in less than half the training time.

updates in lieu of knowledge distillation can achieve the same effects. To investigate this, we compare *SAC-PIEKD* with *Vanilla-SAC (extra)* and *Ensemble-SAC (extra)*, which respectively correspond to *Vanilla-SAC* and *Ensemble-SAC* (see Sect. 5.2) that are trained with extra policy update steps with the same number of updates and minibatch sizes that *SAC-PIEKD* performs. A policy update here refers to a training step that updates the policy and value function [12], if required, by RL algorithms. Figure 4 compares the performance of our baselines to *SAC-PIEKD*. We see that *SAC-PIEKD* reaches higher performance more rapidly than the baselines. This observation shows that knowledge distillation is more effective than policy updates for knowledge sharing.

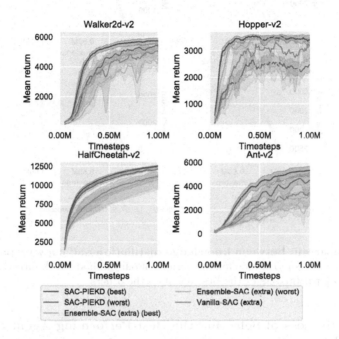

Fig. 4. Comparison between knowledge distillation and extra policy updates. *Vanilla-SAC (extra)* and *Ensemble-SAC (extra)* stand for *Vanilla-SAC* and *Ensemble-SAC* variants that use extra policy updates, respectively (see Sect. 5.3 and Sect. 3 for details).

We additionally study whether the naive method of directly copying parameters from the best-performing agent can also be an effective way to share knowledge between neural networks. We compare a variant of our method, which we denote as *SAC-PIEKD (hardcopy)*, against *SAC-PIEKD*. In *SAC-PIEKD (hardcopy)*, rather than perform intra-ensemble knowledge distillation, we simply copy the parameters of the teacher policy and critic into the student policies and critics. Figure 5 depicts the performance of this variant. We see that *SAC-PIEKD (hardcopy)* performs worse than both *Ensemble-SAC* and *SAC-PIEKD*. Thus, it is clear that knowledge distillation is superior to naively copying the best agent's parameters. In fact, it can be counterproductive to explicitly copy parameters, as *Ensemble-SAC* outperforms copying without any knowledge sharing. This experiment suggests that knowledge sharing in PIEKD does not damage ensemble diversity significantly, since the extreme case of directly copying the teacher's parameters significantly hinders performance, perhaps reducing to training a single policy as in *Vanilla-SAC*.

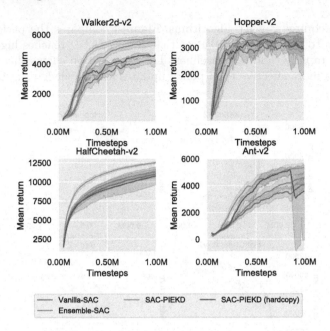

Fig. 5. Comparison between knowledge distillation and copying parameters. *SAC-PIEKD (hardcopy)* stands for the variant of our method which directly copies the neural network parameters of the best agent to the others.

5.4 Effectiveness of Selecting the Best-Performing Agent as the Teacher

During teacher election, we opted for the natural strategy of selecting the best-performing agent. However, in order to investigate its importance, we compared the performance of *SAC-PIEKD* when we select the best policy to be the teacher as opposed to selecting a random policy to be the teacher. This is depicted in Fig. 6, where *SAC-PIEKD (random teacher)* denotes the selection of a random policy to be the teacher and the standard *SAC-PIEKD* refers to the selection of the highest-performing policy to be the teacher. We see that using the highest-performing teacher for distillation appears to be slightly better than selecting a random teacher, though not significantly. Interestingly, we see that using a random teacher performs better than *Ensemble-SAC*. This result suggests that selecting the best teacher is not necessarily of high importance, as a random teacher yields benefits. While this warrants further investigation, perhaps the diverse knowledge is being shared through distillation, which may elicit the success we see in *SAC-PIEKD (random teacher)*. Another possibility is that by bringing policies closer together, the off-policy error [5] stemming from RL updates on a shared replay buffer is reduced, improving performance. However, we can conclude that selecting the highest-performing teacher, while somewhat beneficial, is nonessential, and we leave the investigation of these open questions for future work.

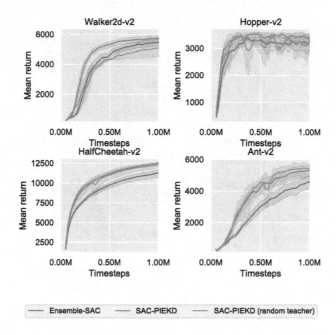

Fig. 6. Comparison between the selecting the best-performing teacher vs. a random teacher. *SAC-PIEKD (random teacher)* refers to the variant of our *SAC-PIEKD* where a randomly chosen teacher is used for knowledge distillation. This figure demonstrates that it can be more effective to select the best-performing agent as the teacher.

5.5   Ablation Study on Ensemble Size

In this subsection, we study the influence of the ensemble size K. We test our method with three ensemble sizes $K = 2$, $K = 3$, and $K = 5$. Figure 7 shows the performance of these configurations. We find that *SAC-PIEKD* performs approximately the same across all three ensemble sizes. Even with an ensemble size of 2, we see better performance than *Ensemble-SAC* (as $K = 2$ is on-par with $K = 3$, which outperforms *Ensemble-SAC*, as shown before). Thus, our method can reap benefits even from small ensembles, and is not extremely sensitive to the ensemble size.

5.6   Ablation Study on Distillation Interval

In this subsection, we investigate the distillation interval (I). We plot the performance of different configurations of I in Fig. 8. We can see that *SAC-PIEKD (I = 1000)* only achieves better performance than *Ensemble-SAC* in 1 of 4 tasks, suggesting that an interval that is too short may hinder the performance of PIEKD. Also, we can observe that *SAC-PIEKD (I = 100000)* is more sample inefficient. This observation suggests that I affects the learning speed.

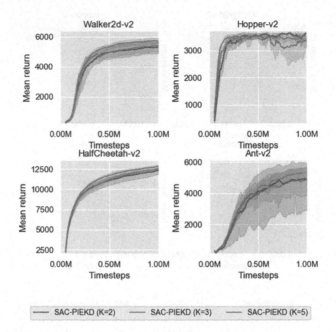

Fig. 7. Performance comparison under different ensemble sizes. Three different ensemble configurations with 2, 3, and 5 agents lead to similar performance. This result shows that PIEKD does not require a large ensemble size.

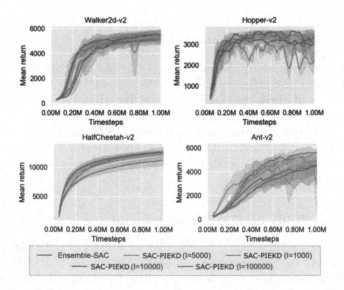

Fig. 8. Performance comparison under different distillation intervals I**.** This figure shows that a short interval can degrade performance while a long interval can impair sample efficiency.

6 Conclusion

In this paper, we introduce Periodic Intra-Ensemble Knowledge Distillation (PIEKD), a method that jointly trains an ensemble of RL agents while continually sharing information via knowledge distillation. Our experimental results demonstrate that PIEKD improves the performance and data efficiency of a state-of-the-art RL method on several challenging MuJoCo tasks. Also, we show that knowledge distillation is more effective than the other approaches for knowledge sharing. We found that selecting the best-performing agent to serve as the teacher can be somewhat beneficial for improving performance. Finally, our ablation study showed that a large ensemble is not needed for improving performance.

PIEKD opens up several avenues for future work. First, encouraging diversity within the ensemble may lead to more efficient exploration [3,15]. Additionally, while we used a simple uniform policy selection strategy, a more efficient policy selection strategy may further accelerate learning. Lastly, while our ensemble members used identical architectures, PIEKD may benefit from using heterogeneous ensembles. For example, different networks may have different architectures that are conducive to learning different skills, which can then be distilled within the ensemble.

Acknowledgements. The authors would like to thank Aaron Havens for suggestions for interesting experiments. We thank Yasuhiro Fujita for suggesting experiments and providing technical support. We thank Jean-Baptiste Mouret for useful feedback on our draft and formulation. Lastly, we thank Pieter Abbeel and Daisuke Okanohara for helpful advice on related works and the framing of the paper.

References

1. Abel, D., Agarwal, A., Diaz, F., Krishnamurthy, A., Schapire, R.E.: Exploratory gradient boosting for reinforcement learning in complex domains. In: International Conference on Machine Learning Workshop on Abstraction in Reinforcement Learning (2016)
2. Brockman, G., et al.: OpenAI Gym. arXiv preprint arXiv:1606.01540 (2016)
3. Conti, E., Madhavan, V., Such, F.P., Lehman, J., Stanley, K., Clune, J.: Improving exploration in evolution strategies for deep reinforcement learning via a population of novelty-seeking agents. In: Advances in Neural Information Processing Systems, pp. 5027–5038 (2018)
4. Czarnecki, W., et al.: Mix & match agent curricula for reinforcement learning. In: International Conference on Machine Learning, pp. 1087–1095. PMLR (2018)
5. Fujimoto, S., Meger, D., Precup, D.: Off-policy deep reinforcement learning without exploration. In: International Conference on Machine Learning, pp. 2052–2062. PMLR (2019)
6. Fujita, Y., Nagarajan, P., Kataoka, T., Ishikawa, T.: ChainerRL: a deep reinforcement learning library. J. Mach. Learn. Res. **22**(77), 1–14 (2021)
7. Galashov, A., et al.: Information asymmetry in KL-regularized RL. In: International Conference on Learning Representations (2019)
8. Gangwani, T., Peng, J.: Policy optimization by genetic distillation. In: International Conference on Learning Representations (2018)

9. Ghosh, D., Singh, A., Rajeswaran, A., Kumar, V., Levine, S.: Divide-and-conquer reinforcement learning. In: International Conference on Learning Representations (2018)
10. Gimelfarb, M., Sanner, S., Lee, C.G.: Reinforcement learning with multiple experts: a Bayesian model combination approach. In: Advances in Neural Information Processing Systems, pp. 9528–9538 (2018)
11. Haarnoja, T., Ha, S., Zhou, A., Tan, J., Tucker, G., Levine, S.: Learning to walk via deep reinforcement learning. In: Robotics: Science and Systems XV (2019)
12. Haarnoja, T., et al.: Soft actor-critic algorithms and applications. arXiv preprint arXiv:1812.05905 (2018)
13. Hester, T., et al.: Deep Q-learning from demonstrations. In: Thirty-Second AAAI Conference on Artificial Intelligence (2018)
14. Hinton, G., Vinyals, O., Dean, J.: Distilling the knowledge in a neural network. arXiv preprint arXiv:1503.02531 (2015)
15. Hong, Z.W., Shann, T.Y., Su, S.Y., Chang, Y.H., Fu, T.J., Lee, C.Y.: Diversity-driven Exploration Strategy for Deep Reinforcement learning. In: Advances in Neural Information Processing Systems, pp. 10489–10500 (2018)
16. Jung, W., Park, G., Sung, Y.: Population-guided parallel policy search for reinforcement learning. In: International Conference on Learning Representations (2020)
17. Khadka, S., Tumer, K.: Evolution-guided policy gradient in reinforcement learning. In: Advances in Neural Information Processing Systems, pp. 1188–1200 (2018)
18. Lan, X., Zhu, X., Gong, S.: Knowledge distillation by on-the-fly native ensemble. In: Advances in Neural Information Processing Systems (2018)
19. Levine, S., Koltun, V.: Guided policy search. In: International Conference on Machine Learning, pp. 1–9 (2013)
20. Mnih, V., et al.: Human-level control through deep reinforcement learning. Nature **518**(7540), 529 (2015)
21. Nagabandi, A., Kahn, G., Fearing, R.S., Levine, S.: Neural network dynamics for model-based deep reinforcement learning with model-free fine-tuning. In: 2018 IEEE International Conference on Robotics and Automation, pp. 7559–7566. IEEE (2018)
22. Nair, A., McGrew, B., Andrychowicz, M., Zaremba, W., Abbeel, P.: Overcoming exploration in reinforcement learning with demonstrations. In: 2018 IEEE International Conference on Robotics and Automation, pp. 6292–6299. IEEE (2018)
23. Oh, J., Guo, Y., Singh, S., Lee, H.: Self-imitation learning. In: International Conference on Machine Learning, pp. 3878–3887. PMLR (2018)
24. Osband, I., Blundell, C., Pritzel, A., Van Roy, B.: Deep exploration via bootstrapped DQN. In: Advances in Neural Information Processing Systems, pp. 4026–4034 (2016)
25. Osband, I., Roy, B.V., Russo, D.J., Wen, Z.: Deep exploration via randomized value functions. J. Mach. Learn. Res. **20**(124), 1–62 (2019)
26. Rusu, A.A., et al.: Policy distillation. In: International Conference on Learning Representations (2016)
27. Salimans, T., Ho, J., Chen, X., Sidor, S., Sutskever, I.: Evolution strategies as a scalable alternative to reinforcement learning. arXiv preprint arXiv:1703.03864 (2017)
28. Schulman, J., Levine, S., Abbeel, P., Jordan, M., Moritz, P.: Trust region policy optimization. In: International Conference on Machine Learning, pp. 1889–1897 (2015)
29. Schulman, J., Wolski, F., Dhariwal, P., Radford, A., Klimov, O.: Proximal policy optimization algorithms. arXiv preprint arXiv:1707.06347 (2017)

30. Sutton, R.S., Barto, A.G., et al.: Introduction to Reinforcement Learning, vol. 135. MIT Press, Cambridge (1998)
31. Teh, Y., et al.: Distral: robust multitask reinforcement learning. In: Advances in Neural Information Processing Systems, pp. 4496–4506 (2017)
32. Tham, C.K.: Reinforcement learning of multiple tasks using a hierarchical CMAC architecture. Robot. Auton. Syst. **15**(4), 247–274 (1995)
33. Zhang, T., Kahn, G., Levine, S., Abbeel, P.: Learning deep control policies for autonomous aerial vehicles with MPC-guided policy search. In: 2016 IEEE International Conference on Robotics and Automation, pp. 528–535. IEEE (2016)
34. Zhang, Y., Xiang, T., Hospedales, T.M., Lu, H.: Deep mutual learning. In: Proceedings of the IEEE Conference on Computer Vision and Pattern Recognition, pp. 4320–4328 (2018)

Learning to Build High-Fidelity and Robust Environment Models

Weinan Zhang[1]([✉]), Zhengyu Yang[1], Jian Shen[1], Minghuan Liu[1],
Yimin Huang[2], Xing Zhang[2], Ruiming Tang[2], and Zhenguo Li[2]

[1] Shanghai Jiao Tong University, Shanghai, China
{wnzhang,yzydestiny}@sjtu.edu.cn
[2] Huawei Noah's Ark Lab, Beijing, China
{yimin.huang,li.zhenguo}@huawei.com

Abstract. This paper is concerned with *robust learning to simulate* (RL2S), a new problem of reinforcement learning (RL) that focuses on learning a high-fidelity environment model (i.e., simulator) for serving diverse downstream tasks. Different from the environment learning in model-based RL, where the learned dynamics model is only appropriate to provide simulated data for the specific policy, the goal of RL2S is to build a simulator that is of high fidelity when interacting with *various* policies. Thus the robustness (i.e., the ability to provide accurate simulations to various policies) of the simulator over diverse corner cases (policies) is the key challenge to address. Via formulating the policy-environment as a dual Markov decision process, we transform RL2S as a novel robust imitation learning problem and propose efficient algorithms to solve it. Experiments on continuous control scenarios demonstrate that the RL2S enabled methods outperform the others on learning high-fidelity simulators for evaluating, ranking and training various policies.

Keywords: Simulator · Imitation learning · Robust learning

1 Introduction

Due to the powerful function approximation and representation learning properties, deep reinforcement learning (DRL) has achieved remarkable success in domains where the environment is a known simulator, such as Go [26] and Atari games [12]. However, such success highly relies on the online training paradigm where the agents must interact with the environment to collect massive data, and thus are still limited to simulated games where the interactions are of low cost. In many high-stakes real-world tasks like autonomous driving, online education and healthcare, directly training a policy in an online manner is always expensive and impractical. To overcome this problem, building high-fidelity simulators, where various agents with different policies can evaluate their performance and improve their policies without any real-world sampling cost, becomes a promising solution.

© Springer Nature Switzerland AG 2021
N. Oliver et al. (Eds.): ECML PKDD 2021, LNAI 12975, pp. 104–121, 2021.
https://doi.org/10.1007/978-3-030-86486-6_7

Recently, great efforts have been devoted to building high-fidelity simulators. Existing work on simulator building can be roughly categorized into two groups: rule-based and learning-based methods. In rule-based methods, complex physical models or hand-craft rules are used to build high-fidelity simulators. For example, Zhang et al. [34] designed a scalable vehicle simulator based on car-following models in traffic signal control. And Zhao et al. [35] used the information in an offline dataset to decide the dynamics of the simulator for recommender system tasks. On the other hand, learning-based methods, by leveraging machine learning to model from real-world transitions, tend to be more potential given sufficient data. For instance, Zheng et al. [36] imitated the behaviors of social vehicles and gained better performance than the existing rule-based simulator. And Shi et al. [25] formulated the interaction between customers and an online retail platform as a multi-agent system, which enables to learn the behavior of customers and the platform simultaneously to get a simulated retail platform for training the policy of personalized recommendation.

A good simulator provides high-quality interactions with multiple policies for evaluation and improvement, which requires the simulator to be sufficiently robust to provide high-fidelity simulation for various policies, including various corner cases. Specifically, in our paper, corner cases refer to the policies in which we need more data to optimize the simulator for stable simulation. To realize robustness which is a big challenge in simulator building, we must consider corner cases. However, to our knowledge, existing methods fail to solve the corner cases [25,36]. For rule-based methods, limited rules are impossible to cover various corner cases in complex environments. On the other hand, learning-based simulators are prone to provide inaccurate simulations in data areas with low frequency.

Based on these considerations, we propose RL2S (Robust Learning to Simulate) to build a high-fidelity and robust simulator that can provide stable simulations to *various* policies, including the corner cases. The general framework of RL2S is illustrated in Fig. 1. By formulating the simulator learning as a dual Markov decision process (DMDP), RL2S utilizes imitation learning algorithms to learn the simulator based on the data sampled from the real environment. In order to learn a robust simulator to handle possible corner cases, RL2S optimizes a Conditional Value at Risk (CVaR) objective [28] to make sure the learned simulator can serve the simulations of various policies.

In a nutshell, the main contributions of this paper are threefold.

- To the best of our knowledge, this paper is the first work that explicitly introduces the robustness objective in simulator building.
- To achieve robustness and fidelity simultaneously, we formulate the problem of simulator learning as a DMDP for the first time and propose a simple yet effective method named RL2S to solve it.
- Demonstrated by the experiments on several continuous control benchmarks, our proposed RL2S yields better worst-case performance among a set of test policies without sacrificing the average performance over all tasks compared to the existing methods.

2   Related Work

Our work considers simulator building. Beyond that, three important RL topics are highly related, i.e., model-based RL, offline policy evaluation and robust RL.

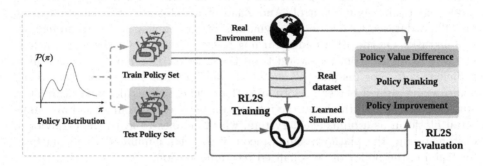

Fig. 1. RL2S framework and experiment settings.

2.1   Simulator Building

There is a rich literature on how to build a good simulator to support RL or decision-making tasks, ranging from widely used rule-based methods to the recent learning-based methods.

Most simulators can be easily built from existing rules. For instance, based on car-following model which can decide the speed of each vehicle based on the information like leading vehicles and traffic signal, Zhang et al. [34] built the CityFlow simulator and Zhou et al. [37] built SMARTS simulator to facilitate the research of reinforcement learning on traffic signal control and autonomous driving. However, the rules used to build the simulator are impossible to fully recover the dynamics of the real environment, especially in complex scenes like autonomous driving and healthcare. Therefore, the learned policy in these simulators may fail in the real environment because of the incomplete or even incorrect rules.

Recently, in order to build high-fidelity simulators, learning-based methods have been paid more attention. Xu et al. [31] utilized generative adversarial imitation learning (GAIL) [4] to build the physical model in MuJoCo. Zheng et al. [36] considered the vehicle in real traffic conditions as an agent and applied GAIL to recover the agent's policies. To avoid the physical costs of online experiments of recommending policy, Shi et al. [25] modeled the interaction in the online retail platform as a multi-agent system and used multi-agent adversarial imitation learning to learn the behavior of customers and the behavior of the online retail platform at the same time. Considering the unobserved latent variables lying behind the data, Shang et al. [24] introduced a hidden policy to model these hidden elements. However, a simulator learned by these methods cannot promise stable performance across different policies without taking corner cases into account. Therefore, it is still impractical to apply such a simulator in high-stakes scenes. In this paper, RL2S designs a novel training paradigm to consider corner cases and realizes the robustness.

2.2 Model-Based Reinforcement Learning

Model-based reinforcement learning (MBRL) learns a dynamics model which serves as an environment with data sampled from the real world. And the dynamics model is leveraged to generate simulated data which will be used for agent learning along with real data. To learn the dynamics model, researchers adopt various function approximators and objectives. For function approximation, time-varying linear models [7] are effective shallow models, while pure neural networks [14] or Gaussian processes with neural networks [1] are the mainstream choices for model learning in deep RL. For objective design, effective solutions include multi-step L2-norm [10], log-likelihood [1], adversarial loss [30], etc.

Although both MBRL and our RL2S require learning the dynamics model, there are mainly two differences between MBRL and RL2S in dynamics model learning. On the one hand, MBRL is usually online and its final goal is to optimize the policy while RL2S is offline and its goal is to build a robust simulator. On the other hand, the learned dynamics model in MBRL is only appropriate to provide simulated data for the current policy to learn. By contrast, RL2S is committed to building a robust simulator that can provide guaranteed simulation to types of policies, including various corner cases.

2.3 Offline Policy Evaluation

Offline policy evaluation, which aims at evaluating the performance of the given policy based on a pre-collected offline dataset, is a special case of off-policy policy evaluation. Current work on offline policy evaluation can be roughly divided into the Direct Method (DM), Importance Sampling (IS) and Hybrid Method (HM). DM directly learns the dynamic model and then uses it to estimate the performance of the given policy [17]. IS computes the importance weights to correct the mismatch between the given policy and the behavior policy which generates the offline datasets [13]. And HM is a combination of DM and IS [29].

Although RL2S is related to the DM in offline policy evaluation, RL2S focuses on building a robust simulator that can serve various downstream tasks with guaranteed performance instead of evaluating the performance of a particular policy.

2.4 Robust Reinforcement Learning

Current work on robust reinforcement learning focuses on optimizing the worst case of the algorithms, which can be roughly divided into two classes, policy-based methods and environment-based methods. Policy-based methods apply the idea of adversarial attack and introduce an adversarial policy to minimize the cumulative rewards that the original policy can get. In detail, Pinto et al. [19] formulated the problem as a two-player Markov zero-sum game, while Zhang et al. [33] proposed state-adversarial MDP which used the adversarial policy to add noise to the state for minimizing the original policy's performance. In all, these algorithms realize the target of optimizing the worst case via a learnable adversarial policy. Different from policy-based methods, environment-based methods choose the environment in which the policy leads to the worst performance and optimize the policy in the chosen environment. Nilim et al. [16] and Rajeswaran et al. [21] optimized the worst performance among a pre-defined finite set of environments, while Lin et al. [9] sought the worst environment in a distribution through adversarial learning. Considering that the environment-based methods fit well in our setting, we adopt environment-based methods in RL2S. Specifically, since only finite pre-defined environments are accessible, we take CVaR as the objective, which is also applied in [21], to realize robustness in RL2S.

3 Preliminaries

3.1 Markov Decision Process

An RL task can be formulated as a Markov decision process (MDP), represented as a tuple $\mathcal{M} = \langle \mathcal{S}, \mathcal{A}, p, p_0, r, \gamma \rangle$. $\mathcal{S} = \{s\}$ is the space of the environment state. $\mathcal{A} = \{a\}$ is the action space of the agent. $p(s'|s,a) : \mathcal{S} \times \mathcal{A} \mapsto \Omega(\mathcal{S})$ is the dynamics model, also called the state transition probability of the environment and $\Omega(\mathcal{S})$ is the set of distributions over $\mathcal{S}$. $p_0 : S \mapsto \mathbb{R}$ is the distribution of the initial state s_0. $r(s,a) : \mathcal{S} \times \mathcal{A} \mapsto \mathbb{R}$ is the reward function. And $\gamma \in [0,1]$ is the discounted factor for future rewards.

When the agent interacts with the environment with a policy π, the occupancy measure $\rho^{p,\pi}(s,a)$ is defined as the unnormalized cumulative discounted probability of occurrence of the state-action pair (s,a) under policy π and transition p:

$$\rho^{p,\pi}(s,a) = \sum_{t=0}^{\infty} \gamma^t P(s_t = s, a_t = s | p, \pi)$$

$$= \pi(a|s) \sum_{t=0}^{\infty} \gamma^t P(s_t = s | p, \pi) = \pi(a|s) \rho^{p,\pi}(s). \tag{1}$$

As such, the agent interacts with the environment to optimize its policy for maximizing the policy value function V, defined as the expectation of cumulative discounted reward, with environment transition p as

$$\max_{\pi} V(p,\pi) = \mathbb{E}_{(s,a) \sim \rho^{p,\pi}(s,a)}[r(s,a)] = \sum_{(s,a)} \rho^{p,\pi}(s,a) r(s,a). \tag{2}$$

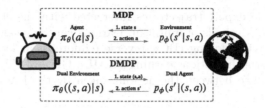

Fig. 2. Dual Markov decision process (DMDP) regards the original environment as the agent while the original agent as the new environment. State and action in DMDP are defined as (s, a) and s' in the original MDP respectively.

3.2   Dual Markov Decision Process

Dual Markov decision process (DMDP) is first introduced by Zhang et al. [32], which provides a new perspective on the environment and the agent in the opposite way. Specifically, we can regard the original environment as an agent while the original agent as the new environment. Thus we construct a DMDP $\mathcal{M}^E$ based on the original MDP $\mathcal{M}^A$ by letting state s_t^E as the state-action pair $\langle s_t^A, a_t^A \rangle$ and action a^E as the next state s_{t+1}^A. In such a way, learning a policy in $\mathcal{M}^E$ is equal to learning a dynamics model in $\mathcal{M}^A$. The formal definition of DMDP is provided in Definition 1 and an overview of the DMDP is given in Fig. 2.

Definition 1 (Dual Markov Decision Process). *For any MDP-policy pair $\langle \mathcal{M}^A, \pi^A \rangle$, where $\mathcal{M}^A = \langle \mathcal{S}^A, \mathcal{A}^A, p^A, p_0^A, r^A, \gamma^A \rangle$, a pair $\langle \mathcal{M}^E, \pi^E \rangle$, where $\mathcal{M}^E = \langle \mathcal{S}^E, \mathcal{A}^E, p^E, p_0^E, r^E, \gamma^E \rangle$ is called DMDP-policy pair if it satisfies:*

- $\mathcal{S}^E = \mathcal{S}^A \times \mathcal{A}^A = \{\langle s^A, a^A \rangle | s^A \in \mathcal{S}^A, a^A \in \mathcal{A}^A\}$, *a state in $\mathcal{M}^E$ corresponds to a state-action pair in $\mathcal{M}^A$;*
- $\mathcal{A}^E = \mathcal{S}^A = \{s^A | s^A \in \mathcal{S}^A\}$, *an action in $\mathcal{M}^E$ corresponds to a state in $\mathcal{M}^A$;*
- $p^E(s_i^E, a_k^E, s_j^E) = p^E(\langle s_i^A, a_i^A \rangle, s_k^A, \langle s_j^A, a_j^A \rangle) = \begin{cases} \pi^A(a_j^A | s_k^A) & s_k^A = s_j^A \\ 0 & s_k^A \neq s_j^A \end{cases}$, *the transition in $\mathcal{M}^E$ depends on the policy in $\mathcal{M}^A$;*
- $r^E(s_i^E, a^E) = r^E(\langle s_i^A, a_i^A \rangle, s^A) = r^A(s_i^A, a_i^A, s^A) = r^A(s_i^A, a_i^A)$, *the rewards in $\mathcal{M}^E$ are the same as in $\mathcal{M}^A$;*
- $\gamma^E = \gamma^A$, *the discounted factors are the same;*
- $p_0^E(s^E) = p_0^E(\langle s^A, a^A \rangle) = p_0^A(s^A)\pi^A(a^A | s^A)$, *the initial state distribution in $\mathcal{M}^E$ depends on the initial state distribution and the action distribution in $\mathcal{M}^A$;*
- $\pi^E(a^E | s^E) = \pi^E(s_{i'}^A | \langle s_i^A, a^A \rangle) = p^A(s_i^A, a^A, s_{i'}^A)$, *a policy in $\mathcal{M}^E$ corresponds to the dynamics in $\mathcal{M}^A$.*

3.3   Imitation Learning

Imitation learning (IL) [5] studies the task of Learning from Demonstrations (LfD), which aims to learn a policy from expert demonstrations that typically consists of the expert trajectories interacted with the environment without reward signals. Methods of imitation learning can be generally divided into three classes: behavior cloning (BC), inverse reinforcement learning (IRL) and generative adversarial imitation learning (GAIL). General IL objective tries to minimize the distance between the actions taken by the learned policy π and expert policy π_E via

$$\min_{\pi} \mathbb{E}_{s \sim \rho^{P,\pi}(s)} [\|\pi(\cdot|s) - \pi_E(\cdot|s)\|]. \tag{3}$$

However, it is always difficult to optimize Eq. (3) since only the expert trajectories are accessible instead of the expert policy itself. Thus, behavior cloning (BC) [20] provides a straightforward method by maximizing the likelihood of expert trajectories via

$$\min_{\pi} \mathbb{E}_{(s,a) \sim \rho^{P,\pi_E}(s,a)} [-\log \pi(a|s)], \tag{4}$$

which suffers from the covariate shift problem when sampling trajectories [22].

Another intriguing IL method is inverse reinforcement learning (IRL) [15], which tries to recover the reward function in the environment based on expert demonstrations. IRL normally suffers from a high complexity for its bi-level optimization (outer loop for reward learning, and inner loop for policy training).

Inspired by IRL, generative adversarial imitation learning (GAIL) [4] shows that the objective of MaxEntIRL [38] is a dual problem of occupancy measure matching, thus can be solved through generative models such as GAN [2]. GAIL builds a surrogate reward function by learning a parameterized discriminator D to classify the experience data from expert π_E and the imitating policy π_θ, while learning the imitating policy π_θ is guided by the reward computed by D_ψ via policy gradient methods like TRPO [23].

In this paper, we apply GAIL to learn the simulator in the RL2S training part as shown in Fig. 1 for its high effectiveness and scalability with good theoretical guarantee but no compounding error concern. Specifically, we consider the Wasserstein distance version of GAIL, as presented in [8], which alleviates the instability of GAIL training by minimizing the Wasserstein distance between the occupancy measure of state-action pairs collected by π and π_E instead of the Jensen-Shannon divergence. The overall objective is written as

$$\min_{\theta} \max_{\psi} \mathbb{E}_{(s,a) \sim \rho^{P,\pi_E}(s,a)} [D_\psi(s,a)] - \mathbb{E}_{(s,a) \sim \rho^{P,\pi_\theta}(s,a)} [D_\psi(s,a)]. \tag{5}$$

4   Robust Learning to Simulate

In this section, we give a detailed introduction of our robust learning to simulate (RL2S) framework. As shown in Fig. 1, we sample sets of policies from a policy distribution $\mathcal{P}(\pi)$ in the beginning and use them to interact with the real environment for data collection. Specifically, we assume the reward function $r(s,a)$ is well defined and known. This can be easily achieved since reward function is commonly defined by humans in various scenarios. Then the goal collapses into learning the state transition of the environment p^* via building a high-fidelity dynamics model p_ϕ parameterized by ϕ. Since we have formulated a DMDP which regards the transition p_ϕ as the dual policy and the policy π_θ as the dynamics, we are able to solve the problem with imitation learning algorithms, based on the pre-collected data. Section 4.2 presents the process described above comprehensively when the sample space of $\mathcal{P}(\pi)$ only contains one policy. To realize the robustness, we propose a novel framework named RL2S and apply CVaR as the objective function, which is introduced in Sect. 4.3 in detail.

4.1 Problem Definition

In our task, we seek to recover the real environment as a simulator that can accurately simulate the interactions between the environment and various policies.

Consider a policy distribution as $\mathcal{P}(\pi)$. When interacting a dynamics p, Theorem 2 of [27] shows the one-to-one correspondence between π and $\rho^{p,\pi}$. Thus for each policy $\pi \sim \mathcal{P}(\pi)$, its interactions with the real environment p^* can be measured by the occupancy $\rho^{p^*,\pi}$ with a set of sampled trajectories as

$$\tau^{p^*,\pi} = \{[s_0, a_0, s_1, a_1, \ldots, s_T]\} = \{(s, a, s') \sim \pi\}, \tag{6}$$

where T is the episode length. As such, the dataset of interaction experiences generated by the policy in $\mathcal{P}(\pi)$ with the real environment p^* can be denoted as

$$\tau^{p^*,\mathcal{P}} = \{(s, a, s') \sim \pi\}_{\pi \sim \mathcal{P}(\pi)}. \tag{7}$$

In our paper, we define the corner cases as the worst ϵ-percentile policies in $\mathcal{P}(\pi)$ under the metric of *value difference* (VD) which measures the absolute difference of the policy's value when interacting with the real environment p^* and the simulator p_ϕ. The definition of the *value difference* is

$$\begin{aligned} \mathrm{VD}(p^*, p_\phi, \pi) &= |V(p^*, \pi) - V(p_\phi, \pi)| \\ &= |\mathbb{E}_{(s,a,s')\sim\rho^*}[r(s,a)] - \mathbb{E}_{(s,a,s')\sim\rho_\phi}[r(s,a)]|, \end{aligned} \tag{8}$$

where $\rho_\phi(s, a, s')$ and $\rho^*(s, a, s')$ stands for the policy's occupancy measure in p_ϕ and p^*, respectively. Then the set of corner cases Π_c is

$$\Pi_c = \{\pi | \pi \sim \mathcal{P}(\pi), \mathrm{VD}(p^*, p_\phi, \pi) \geq \delta_\epsilon\}, \tag{9}$$

where δ_ϵ is the threshold for the ϵ-percentile corner cases: $\mathbb{P}(\mathrm{VD}(p^*, p_\phi, \pi) \geq \delta_\epsilon) = \epsilon$. Upon such a definition, we wish to build a robust simulator p_ϕ from $\tau^{p^*,\mathcal{P}}$. By robust, we mean that the learned p_ϕ can provide stable simulation for various policies in $\mathcal{P}(\pi)$, including corner cases in Π_c. To explicitly seek a robust simulator, we optimize for the CVaR objective which aims at minimizing the expectation of VD for the worst ϵ-percentile of policies in $\mathcal{P}(\pi)$ (i.e., policies in Π_c) and the definition of CVaR is

$$\min_{\phi} \int_{\pi} \mathbb{I}(\pi \in \varPi_c) \mathrm{VD}(p^*, p_\phi, \pi) \mathcal{P}(\pi) d\pi, \tag{10}$$

where $\mathbb{I}(\cdot)$ is the indicator function.

4.2   Single Behavior Policy Setting

Let us begin with a single policy π_θ. Under the perspective of DMDP, we can regard π_θ as a dual environment. Then, there is an expert dual policy p^* sampling a set of experience data τ^{p^*, π_θ} via interacting with the dual environment π_θ. Hence, it is natural to perform imitation learning from such a demonstration dataset τ^{p^*, π_θ} to obtain an imitating policy p_ϕ.

Specifically, in this paper, we choose GAIL due to its high flexibility and low compounding error [31]. Formally, the generator, discriminator and the environment are denoted as $p_\phi(s'|s, a)$, $D_\psi(s, a, s')$, $\pi_\theta((s', a')|s')$. The min-max optimization objective can be written as

$$\min_{\phi} \max_{\psi} \mathbb{E}_{(s,a,s')\sim(\pi_\theta, p^*)}[D_\psi(s, a, s')] - \mathbb{E}_{(s,a,s')\sim(\pi_\theta, p_\phi)}[D_\psi(s, a, s')]. \tag{11}$$

4.3   Robust Policy Setting

Making high-quality simulations under a single behavior policy is simple. However, providing stable simulation to various policies is non-trivial, which is the ultimate goal of RL2S. Suppose that we aim to provide stable simulation to policies in $\mathcal{P}(\pi)$, including corner cases. After that, we assume a set of collected experience datasets $\tau^{p^*, \mathcal{P}}$. Similar to the single policy setting, we can apply GAIL for modeling the dynamics as

$$\min_{\phi} \mathbb{E}_{\pi \sim \mathcal{P}(\pi)} \left[\max_{\psi_\pi} \mathbb{E}_{(s,a,s')\sim(\pi, p^*)}[D_{\psi_\pi}(s, a, s')] - \mathbb{E}_{(s,a,s')\sim(\pi, p_\phi)}[D_{\psi_\pi}(s, a, s')] \right], \tag{12}$$

where D_{ψ_π} stands for the discriminator corresponding to each dual environment $\pi \sim \mathcal{P}(\pi)$. However, Eq. (12) can hurt the learning of p_ϕ for the following reasons.

1) $D_{\psi_{\pi_1}}$ and $D_{\psi_{\pi_2}}$ may output totally different values for the same transition (s, a, s') due to the difference of π_1 and π_2, which can make the training of p_ϕ unstable.
2) The data for individual policy π is not sufficient to train a good discriminator.

Considering that the discriminator only cares about the fidelity of the transition (s, a, s'), i.e., the transition to s' conditioned on (s, a), which is only related to an environment (simulator) instead of any policy, we can just build one discriminator for the overall occupancy as

$$\min_{\phi} \max_{\psi} \mathbb{E}_{\pi \sim \mathcal{P}(\pi)} \left[\mathbb{E}_{(s,a,s')\sim(\pi, p^*)}[D_\psi(s, a, s')] - \mathbb{E}_{(s,a,s')\sim(\pi, p_\phi)}[D_\psi(s, a, s')] \right]. \tag{13}$$

In practice, given a finite set of policies $\Pi = \{\pi_m\}$ sampled from $\mathcal{P}(\pi)$, then the expectation over $\mathcal{P}(\pi)$ in Eq. (13) becomes the empirical mean over Π as

$$\min_{\phi} \max_{\psi} \sum_{m=1}^{|\Pi|} \left\{ \mathbb{E}_{(s,a,s')\sim\pi_m,p^*}[D_\psi(s,a,s')] - \mathbb{E}_{(s,a,s')\sim\pi_m,p_\phi}[D_\psi(s,a,s')] \right\},$$
(14)

where the normalization term $1/|\Pi|$ is omitted for simplicity.

Although the simulator learned by Eq. (13) has the best performance in expectation, it can still have extremely poor performance in some corner cases without considering the variability in performance for different policies from the distribution $\mathcal{P}(\pi)$. Thus such a learning objective still cannot provide stable simulation for various policies in $\mathcal{P}(\pi)$. To this end, RL2S optimizes the CVaR objective and focuses on minimizing VD on the worst ϵ-percentile policies (i.e., Π_c defined in Eq. (9)), via

$$\min_{\phi} \max_{\psi} \int_\pi \mathbb{I}(\pi \in \Pi_c) \Big[\mathbb{E}_{(s,a,s')\sim(\pi,p^*)}[D_\psi(s,a,s')]$$
$$- \mathbb{E}_{(s,a,s')\sim(\pi,p_\phi)}[D_\psi(s,a,s')] \Big] \mathcal{P}(\pi)d\pi.$$
(15)

In practice, RL2S tests the performance of the policies every K steps to update Π_c.

When optimizing Eq. (15), we only use the worst ϵ-percentile dual environments to train p_ϕ, so the lower bound of the performance of the learned simulator in $\mathcal{P}(\pi)$ can be improved and p_ϕ can provide stable simulation to a broader range of policies in $\mathcal{P}(\pi)$. Moreover, the training procedure of RL2S focuses more on the mispredicted transitions so that the corner cases can be alleviated.

5   Experiments

5.1   Experimental Protocol

Current work on simulator building compares the statistics in simulation and real world [36] or uses the learned simulator to support the policies' training [25] to present the performance of the learned simulator. In our experiments, we apply both paradigms to show the efficacy of RL2S. Specifically, we conduct three experiments: policy value difference evaluation, policy ranking and policy improvement.

- For the tasks of policy value difference evaluation [31] and policy ranking [18], we aim at evaluating the reward under the occupancy measure (i.e., $V(p_\phi, \pi)$) where policy value difference evaluation cares about the absolute difference between the $V(p^*, \pi)$ and $V(p_\phi, \pi)$ while policy ranking focuses on the relative order of the policies in terms of their values.
- For the task of policy improvement, we fine-tune the policies in $\mathcal{P}(\pi)$ through simulation data sampled from p_ϕ.

In our experiments, following the previous work [21], we sample two finite sets of policies, i.e., Π and Π', from $\mathcal{P}(\pi)$ for training and test, respectively. For Hopper, Walker2d and HalfCheetah, $|\Pi|$ and $|\Pi'|$ are set to 17 and 8. For Ant, $|\Pi|$ and $|\Pi'|$ are set to 32 and 14^1. For each policy, we sample 1000 transitions.

In the implementation, we adopt spectral norm [11] in realizing the Lipschitz constraint of discriminator D_ψ in RL2S. And we normalize the state based on mean and variance computed on the collected dataset $\tau^{p^*,\Pi}$.

Policy Value Difference Evaluation. At the beginning of the training stage, each sampled policy $\pi \in \Pi$ interacts with the real environment p^* to collect a set of transitions $\{(s,a,s')\}_\pi$. Thus the overall real data of state transitions is $B = \cup_{\pi \in \Pi}\{(s,a,s')\}_\pi$. With B as the training data, RL2S optimizes a simulator p_ϕ via Eq. (15).

In this task, we compute the VD as defined in Eq. (8) for policies in $\mathcal{P}(\pi)$. Different from the setting in Xu et al. [31], which only focuses on learning a simulator for a particular policy, RL2S aims at achieving robustness through optimizing the worst cases in $\mathcal{P}(\pi)$. Thus, to take the robustness and worse-case performance into consideration, we define *maximum value difference* (MVD). Moreover, we define *average value difference* (AVD) to test whether RL2S degrades the average performance. The definitions of AVD and MVD are written as

$$
\begin{aligned}
\text{AVD}(p^*, p_\phi, \mathcal{P}) &= \mathbb{E}_{\pi \sim \mathcal{P}}[\text{VD}(p^*, p_\phi, \pi)] \simeq \frac{1}{|F|} \sum_{\pi \in F} \text{VD}(p^*, p_\phi, \pi), \\
\text{MVD}(p^*, p_\phi, \mathcal{P}) &= \max_{\pi \sim \mathcal{P}} \text{VD}(p^*, p_\phi, \pi) \simeq \max_{\pi \in F} \text{VD}(p^*, p_\phi, \pi),
\end{aligned}
\tag{16}
$$

where F is a finite set of policies sampled from $\mathcal{P}(\pi)$. Since the target of RL2S is to build a robust simulator that can provide stable simulation to various policies in $\mathcal{P}(\pi)$, especially to unseen policies, we depict the curves of AVD and MVD on the test policy set Π' during training to explore the robustness of p_ϕ on unseen policies.

Policy Ranking. For policy ranking, we use the learned p_ϕ to rank the policies in Π'. Suppose A is list of $V(p_\phi, \pi)$ for $\pi \in \Pi'$ in descending order, B is the corresponding $V(p^*, \pi)$ normalized to [0,1] and C is obtained by sorting B in descending order. We apply *Kendall rank correlation coefficient* (τ) and *normalized discounted cumulative gain* (nDCG), which are well-recognized ranking performance metrics. The definition of Kendall rank correlation coefficient is

$$
\tau = \frac{1}{|\Pi'| \times (|\Pi'| - 1)} \sum_{i \neq j} \text{sgn}(A[i] - A[j]) \times \text{sgn}(B[i] - B[j]),
\tag{17}
$$

---

[1] Considering Ant has an especially larger state and action dimension than other environments, we sample more policies for training and test.

where sgn(x) indicates the sign of x. For nDCG, we take the normalized performance as the relevance of each policy, which is defined as

$$\text{nDCG@}k = \sum_{i=1}^{k} \frac{2^{B[i]} - 1}{\log_2(i+1)} \Big/ \sum_{i=1}^{k} \frac{2^{C[i]} - 1}{\log_2(i+1)}. \tag{18}$$

Policy Improvement. To present the performance of the learned simulator p_ϕ, we fine-tune the policies in Π' based on p_ϕ. In detail, policy $\pi \in \Pi'$ interacts with p_ϕ to collect data and uses them to improve itself for limited iterations based on SAC [3]. We measure the performance improvement $I_{\phi,\pi}$ as $I(p_\phi, \pi) = (C_\pi^1 - C_\pi^0)/C_\pi^0$ where C_π^1 is the π's performance after fine-tuning and C_π^0 is the original performance. To measure the robustness and indicate whether RL2S harms the performance in expectation, we introduce two metrics for policy improvement, i.e., *minimum performance improvement* (MPI) and *average performance improvement* (API), as

$$\text{MPI} = \min_{\pi \sim \mathcal{P}} I(p_\phi, \pi) \simeq \min_{\pi \in \Pi'} I(p_\phi, \pi),$$

$$\text{API} = \mathbb{E}_{\pi \sim \mathcal{P}}[I(p_\phi, \pi)] \simeq \frac{1}{|\Pi'|} \sum_{\pi \in \Pi'} I(p_\phi, \pi). \tag{19}$$

5.2 Studied Environments and Baselines

Following the previous work [31], we compare the results of RL2S against the baseline methods on simple-to-complex continuous control benchmarking environments, including Hopper, Walker2d, HalfCheetah and Ant from MuJoCo.

In this paper, we focus on learning-based methods, thus the rule-based methods are not included for comparison. Since almost all of the existing learning-based methods apply imitation learning to learn the simulator where GAIL is utilized in [24,25,31,36] and BC is adopted in most MBRL methods [1,6], we take GAIL and BC as baselines. Furthermore, considering that imitation learning is an important module in RL2S, we want to explore the effect of the robustness objective on different methods of imitation learning, so we take RL2S-BC, which replaces the GAIL in RL2S with BC, as the baseline method. Moreover, although some experiments in our paper are similar to offline policy evaluation, RL2S focuses on building a robust simulator that can serve various downstream tasks instead of evaluating the performance of the given policy. So the methods in offline policy evaluation are not included in our experiments.

5.3 Performance on Policy Value Difference Evaluation

The learning curves of each method on four environments are presented in Fig. 3, from which we have the following observations.

1) RL2S achieves the best performance in four environments on MVD, which validates the robustness and effectiveness of RL2S on corner cases.

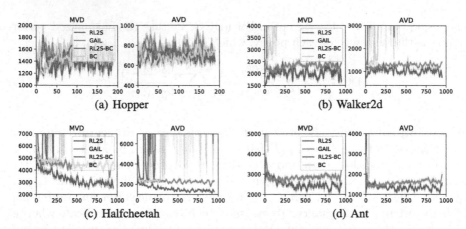

Fig. 3. Learning curves of RL2S on MVD and AVD over four environments. The x-axis is the number of training epochs, and the y-axis is the performance on MVD or AVD.

2) RL2S yields a larger improvement over GAIL in complex environments (HalfCheetah and Ant) than that in simple environments (Hopper and Walker2d), suggesting that robustness objective has better performance in complex environments.
3) Although the objective function of RL2S only aims at optimizing the lower bound of the performance in the distribution, we find that RL2S achieves the best performance on AVD in all environments, which illustrates that our robustness objective can improve the robustness of the learned simulator without harming the average performance. The reason would be that optimizing the matching of occupancy measure on corner cases in GAIL could further improve the matching on the overall occupancy measure.
4) BC-based methods (i.e., RL2S-BC, BC) have much low (or even diverged) performance in all environments except the simplest Hopper. And RL2S-BC may not achieve better performance than BC, which can be attributed to the large compounding error of BC as claimed in Lemma 3 and Theorem 3 of a theoretic analysis from Xu et al. [31].

5.4   Performance on Policy Ranking

In this task, we use the learned simulator to rank the policies in the test policy set Π'. Due to the low computational cost of this task, we expand the size of Π' to 20 for all the environments to make the results more reasonable. In policy ranking, a higher value of nDCG or τ stands for better performance. Table 1 reports nDCG@1, nDCG@3, nDCG@5, nDCG@10 and τ, from which we can get the following observations.

1) The simulator learned by RL2S consistently achieves the best performance in terms of τ and nDCG over all environments, suggesting that RL2S not only

Table 1. Results of policy ranking.

Env	Metric	RL2S	GAIL	RL2S-BC	BC
Hopper	τ	**0.3474**	−0.1895	−0.2842	−0.0632
	nDCG@1	**0.9464**	0.0177	0.0177	0.5868
	nDCG@3	**0.8326**	0.2448	0.1507	0.3816
	nDCG@5	**0.6843**	0.3971	0.2310	0.4501
	nDCG@10	**0.7326**	0.5111	0.3515	0.5130
Walker	τ	**0.2316**	−0.0737	−0.5474	0.0632
	nDCG@1	**0.7954**	0.6591	0.6184	0.6672
	nDCG@3	**0.9069**	0.7311	0.5614	0.5497
	nDCG@5	**0.8988**	0.7474	0.5971	0.6357
	nDCG@10	**0.8859**	0.7622	0.7030	0.7348
HalfCheetah	τ	**0.8737**	0.8211	0.4947	0.6000
	nDCG@1	**1.0000**	0.9878	0.6191	0.9878
	nDCG@3	**0.9691**	0.9385	0.7839	0.9652
	nDCG@5	**0.9922**	0.9667	0.8013	0.9430
	nDCG@10	**0.9944**	0.9882	0.8690	0.9190
Ant	τ	**0.8000**	0.6211	0.2632	0.4105
	nDCG@1	**1.0000**	0.8601	0.6650	**1.0000**
	nDCG@3	**1.0000**	0.9186	0.6888	0.9106
	nDCG@5	**0.9872**	0.9609	0.6988	0.9587
	nDCG@10	**0.9943**	0.9358	0.8093	0.9174

brings robustness to the fidelity of the learned simulator, but also provides a good policy selection solution (via ranking).

2) Although BC-based methods are inherently unsuitable for VD, they may have good performance in policy ranking as the ranking task cares more about the relative superiority instead of the absolute value difference.

5.5   Performance on Policy Improvement

In order to show the effectiveness of the learned simulator on improving a given policy, we use the learned simulator to fine-tune the policies in the test policy set Π'. We show the performance improvement of RL2S on Hopper and HalfCheetah in Fig. 4. And in Table 2, the results of the MPI and API achieved by different simulators learned via individual methods are listed, from which we can obtain the following observations.

1) RL2S consistently achieves the best performance on both MPI and API, which shows that RL2S can get an effective and robust simulator for policy improvement.

Table 2. Performance improvement after fine-tuning the policy with the learned simulator.

Env	Metric	RL2S	GAIL	RL2S-BC	BC
Hopper	MPI	**0.0293**	0.0039	0.0124	0.0090
	API	**2.9400**	0.9758	0.9587	0.9235
Walker	MPI	**0.0067**	0.0065	0.0032	−0.0045
	API	**0.2306**	0.1458	0.2178	0.1915
HalfCheetah	MPI	**0.0143**	0.0058	0.0129	0.0099
	API	**0.0717**	0.0529	0.0598	0.0630
Ant	MPI	**−0.021**	−0.027	−0.0232	−0.0352
	API	**0.0379**	0.0325	0.0295	0.0308

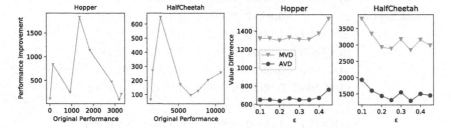

Fig. 4. Improvement w.r.t. original value.

Fig. 5. Analysis on ϵ.

2) BC-based methods have good performance because of the alleviation of the compounding error due to short rollout length [6] while utilizing p_ϕ to collect data.
3) RL2S-BC gets better performance than BC in terms of MPI over all environments, suggesting that our robustness objective can empower BC with robustness in policy improvement. However, when improving the robustness, RL2S-BC may deteriorate the performance of BC in terms of API.
4) In Ant, the most complex environment studied in our experiments, all methods obtain negative MPI, which means it is still challenging to improve the performance of corner case policies based on the learned simulator.

5.6   Analysis on Hyperparameter ϵ

Here, we analyze the influence of the hyperparameter ϵ on the robustness of the learned simulator p_ϕ. In Hopper and HalfCheetah, the results of different ϵ on policy value difference are shown in Fig. 5. We can see that too small or big ϵ can lead to poor performance, while other values stably achieve satisfactory performance. As a result, we set ϵ to 0.25.

6 Conclusion

In this paper, we propose RL2S to learn a robust simulator (i.e., environment model), which to the best of our knowledge is the first work to handle corner case simulation in simulator building. To achieve robustness, we utilize a special training procedure that only samples the worst ϵ-percentile data to train the environment model. Results of extensive experiments demonstrate the potential of RL2S achieving superior performance in robustness without harming the average performance, compared to the widely used baselines. Moreover, the simulator learned by RL2S can also improve the performance of policy training, which sheds some light on further research of model-based reinforcement learning. For future work, we plan to investigate RL2S solutions in discrete-state environments or the continuous-state ones with saltation transitions. We will also investigate more sophisticated methods under the problem of RL2S.

Acknowledgements. The authors from Shanghai Jiao Tong University are supported by "New Generation of AI 2030" Major Project (2018AAA0100900), Shanghai Municipal Science and Technology Major Project (2021SHZDZX0102) and National Natural Science Foundation of China (62076161, 81771937). The work is also sponsored by Huawei Innovation Research Program.

References

1. Chua, K., Calandra, R., McAllister, R., Levine, S.: Deep reinforcement learning in a handful of trials using probabilistic dynamics models. In: NeurIPS, pp. 4759–4770 (2018)
2. Goodfellow, I.J., et al.: Generative adversarial nets. In: NIPS (2014)
3. Haarnoja, T., Zhou, A., Abbeel, P., Levine, S.: Soft actor-critic: off-policy maximum entropy deep reinforcement learning with a stochastic actor. In: ICML, pp. 1861–1870. PMLR (2018)
4. Ho, J., Ermon, S.: Generative adversarial imitation learning. In: Advances in Neural Information Processing Systems, pp. 4565–4573 (2016)
5. Hussein, A., Gaber, M.M., Elyan, E., Jayne, C.: Imitation learning: a survey of learning methods. ACM Comput. Surv. (CSUR) **50**(2), 21 (2017)
6. Janner, M., Fu, J., Zhang, M., Levine, S.: When to trust your model: model-based policy optimization. In: NeurIPS, pp. 12519–12530 (2019)
7. Levine, S., Finn, C., Darrell, T., Abbeel, P.: End-to-end training of deep visuomotor policies. J. Mach. Learn. Res. **17**(1), 1334–1373 (2016)
8. Li, Y., Song, J., Ermon, S.: InfoGAIL: interpretable imitation learning from visual demonstrations. In: Advances in Neural Information Processing Systems, pp. 3812–3822 (2017)
9. Lin, Z., Thomas, G., Yang, G., Ma, T.: Model-based adversarial meta-reinforcement learning. In: Advances in Neural Information Processing Systems, vol. 33 (2020)
10. Luo, Y., Xu, H., Li, Y., Tian, Y., Darrell, T., Ma, T.: Algorithmic framework for model-based deep reinforcement learning with theoretical guarantees. In: ICLR (Poster) (2019)

11. Miyato, T., Kataoka, T., Koyama, M., Yoshida, Y.: Spectral normalization for generative adversarial networks. In: International Conference on Learning Representations (2018)
12. Mnih, V., et al.: Human-level control through deep reinforcement learning. Nature **518**(7540), 529–533 (2015)
13. Nachum, O., Chow, Y., Dai, B., Li, L.: DualDICE: behavior-agnostic estimation of discounted stationary distribution corrections. arXiv preprint arXiv:1906.04733 (2019)
14. Nagabandi, A., Kahn, G., Fearing, R.S., Levine, S.: Neural network dynamics for model-based deep reinforcement learning with model-free fine-tuning. In: ICRA (2018)
15. Ng, A.Y., Russell, S.: Algorithms for inverse reinforcement learning. In: ICML (2000)
16. Nilim, A., El Ghaoui, L.: Robustness in Markov decision problems with uncertain transition matrices. In: NIPS, pp. 839–846. Citeseer (2003)
17. Paduraru, C.: Off-policy evaluation in Markov decision processes. Ph.D. thesis, Ph.D. dissertation. McGill University (2012)
18. Paine, T.L., et al.: Hyperparameter selection for offline reinforcement learning. arXiv preprint arXiv:2007.09055 (2020)
19. Pinto, L., Davidson, J., Sukthankar, R., Gupta, A.: Robust adversarial reinforcement learning. In: International Conference on Machine Learning, pp. 2817–2826. PMLR (2017)
20. Pomerleau, D.A.: Efficient training of artificial neural networks for autonomous navigation. Neural Comput. **3**(1), 88–97 (1991)
21. Rajeswaran, A., Ghotra, S., Ravindran, B., Levine, S.: EPOpt: learning robust neural network policies using model ensembles. In: ICLR (2016)
22. Ross, S., Bagnell, D.: Efficient reductions for imitation learning. In: AISTATS, pp. 661–668. JMLR Workshop and Conference Proceedings (2010)
23. Schulman, J., Levine, S., Abbeel, P., Jordan, M., Moritz, P.: Trust region policy optimization. In: International Conference on Machine Learning, pp. 1889–1897 (2015)
24. Shang, W., Yu, Y., Li, Q., Qin, Z., Meng, Y., Ye, J.: Environment reconstruction with hidden confounders for reinforcement learning based recommendation. In: KDD (2019)
25. Shi, J.C., Yu, Y., Da, Q., Chen, S.Y., Zeng, A.X.: Virtual-Taobao: virtualizing real-world online retail environment for reinforcement learning. In: AAAI, vol. 33, pp. 4902–4909 (2019)
26. Silver, D., et al.: Mastering the game of go with deep neural networks and tree search. Nature **529**(7587), 484–489 (2016)
27. Syed, U., Bowling, M., Schapire, R.E.: Apprenticeship learning using linear programming. In: ICML, pp. 1032–1039. ACM (2008)
28. Tamar, A., Glassner, Y., Mannor, S.: Optimizing the CVaR via sampling. In: Proceedings of the AAAI Conference on Artificial Intelligence, vol. 29 (2015)
29. Thomas, P., Brunskill, E.: Data-efficient off-policy policy evaluation for reinforcement learning. In: International Conference on Machine Learning, pp. 2139–2148. PMLR (2016)
30. Wu, Y.H., Fan, T.H., Ramadge, P.J., Su, H.: Model imitation for model-based reinforcement learning. arXiv preprint arXiv:1909.11821 (2019)
31. Xu, T., Li, Z., Yu, Y.: Error bounds of imitating policies and environments. In: Advances in Neural Information Processing Systems, vol. 33 (2020)

32. Zhang, H., et al.: Learning to design games: Strategic environments in reinforcement learning. IJCAI (2018)
33. Zhang, H., Chen, H., Xiao, C., Li, B., Boning, D., Hsieh, C.J.: Robust deep reinforcement learning against adversarial perturbations on observations. arXiv:2003.08938 (2020)
34. Zhang, H., et al.: CityFlow: a multi-agent reinforcement learning environment for large scale city traffic scenario. In: The World Wide Web Conference, pp. 3620–3624 (2019)
35. Zhao, X., Xia, L., Zhang, L., Ding, Z., Yin, D., Tang, J.: Deep reinforcement learning for page-wise recommendations. In: RecSys, pp. 95–103 (2018)
36. Zheng, G., Liu, H., Xu, K., Li, Z.: Learning to simulate vehicle trajectories from demonstrations. In: ICDE, pp. 1822–1825. IEEE (2020)
37. Zhou, M., et al.: Smarts: scalable multi-agent reinforcement learning training school for autonomous driving. In: Conference on Robot Learning (2020)
38. Ziebart, B.D., Maas, A.L., Bagnell, J.A., Dey, A.K.: Maximum entropy inverse reinforcement learning. In: AAAI, Chicago, IL, USA, vol. 8, pp. 1433–1438 (2008)

Ensemble and Auxiliary Tasks for Data-Efficient Deep Reinforcement Learning

Muhammad Rizki Maulana$^{(\boxtimes)}$ (ID) and Wee Sun Lee (ID)

School of Computing, National University of Singapore, Singapore, Singapore
`rizki@u.nus.edu, leews@comp.nus.edu.sg`

Abstract. Ensemble and auxiliary tasks are both well known to improve the performance of machine learning models when data is limited. However, the interaction between these two methods is not well studied, particularly in the context of deep reinforcement learning. In this paper, we study the effects of ensemble and auxiliary tasks when combined with the deep Q-learning algorithm. We perform a case study on ATARI games under limited data constraint. Moreover, we derive a refined bias-variance-covariance decomposition to analyze the different ways of learning ensembles and using auxiliary tasks, and use the analysis to help provide some understanding of the case study. Our code is open source and available at https://github.com/NUS-LID/RENAULT.

Keywords: Deep reinforcement learning · Ensemble learning · Multi-task learning

1   Introduction

Ensemble learning is a powerful technique to improve the performance of machine learning models on a diverse set of problems. In reinforcement learning (RL), ensembles are mostly used to stabilize learning and reduce variability [1,5,17], and in few cases, to enable exploration [4,22]. Orthogonal to the utilization of ensembles, auxiliary tasks also enjoy widespread use in RL to aid learning [13,15,18,20].

The interplay between these two methods has been studied within a limited capacity in the context of simple neural networks [29] and decision trees [27]. In reinforcement learning, this interaction – to the best of our knowledge – has not been studied at all.

Our principal aim in this work is to study ensembles and auxiliary tasks in the context of deep Q-learning algorithm. Specifically, we apply ensemble learning on the well-established Rainbow agent [9,10], and additionally augment

Electronic supplementary material The online version of this chapter (https://doi.org/10.1007/978-3-030-86486-6_8) contains supplementary material, which is available to authorized users.

it with auxiliary tasks. We study the problem theoretically through the use of bias-variance-covariance analysis and by performing an empirical case study on a popular reinforcement learning benchmark, ATARI games, under the constraint of low number of interactions [14]. ATARI games offer a suite of diverse problems, improving the generality of the results, and data scarcity makes the effect of ensembles more pronounced. Moreover, under the constraint of low data, it is naturally preferable to trade-off the extra computational requirement of using an ensemble for the performance gain that it provides.

We derive a more refined analysis of the bias-variance-covariance decomposition for ensembles. The usual analysis assumes that each member of the ensemble is trained on a different dataset. Instead, we focus our analysis on a single dataset, used with multiple instantiations of a randomized learning algorithm. This is commonly how ensembles are actually used in practice; in fact, the multiple datasets that are used for training members of the ensemble are often constructed from a single dataset through the use of randomization. Additionally, we introduce some new "weak" auxiliary tasks that provide small improvements, based on model learning and learning properties of objects and events. We show how ensembles can be used for combining multiple "weak" auxiliary tasks to provide stronger improvements.

Our case study and analysis shows that,

- Independent training of ensemble members works well. Joint training of the entire ensemble reduces Q-learning error but, surprisingly did not perform as well as an independently trained ensemble.
- The new auxiliary tasks are "weakly" helpful. Combining them together using an ensemble can provide a significant performance boost. We observe reduction in variance and covariance with the use of auxiliary tasks in the ensemble.
- Despite their benefits, using all auxiliary tasks on each predictor in ensemble may results in poorer performance. Analysis indicates that this could cause higher bias and covariance due to loss of diversity.

It is interesting to note that our ensemble, despite its simplicity, achieves better performance on 13 out of 26 games compared to recent previous works. Moreover, our ensemble with auxiliary tasks achieves significantly better human mean and median normalized performance; 1.6× and 1.55× better than data-efficient Rainbow [9], respectively.

2   Related Works

Reinforcement Learning and Auxiliary Tasks. Rainbow DQN [10] combines multiple important advances in DQN [21] such as learning value distribution [3] and prioritizing experience [23] to improve performance. Other works try to augment RL by devising useful auxiliary tasks. [13] proposed auxiliary tasks in the form of reward prediction as a classification of positive, negative, or neutral reward. [13] also proposed to predict changing pixels for the downsampled image. Recently, [18] proposed the use of contrastive loss to learn better representation. Other auxiliary tasks have been explored as well, such as depth prediction [20] and terminal prediction [15]. These auxiliary tasks are less general; they require domain with 3D inputs and problem with episodic nature, respectively. Although much research has been done with auxiliary tasks in RL, to the best of our knowledge, none of them investigated the use of auxiliary tasks in the context of ensemble RL.

Ensemble in Reinforcement Learning. Ensemble methods have been explored in RL for various purposes [1,5,17,22]. [1] investigated the effect of ensemble in RL, especially pertaining to the reduction of target approximation error. In the model-based RL, [5] used ensemble to reduce modelling errors, and [17] accelerated policy learning by generating experiences through ensemble of dynamic models. In the context of policy gradients, [7] utilized ensemble value function as a critique to reduce function approximation error. [22] proposed the use of ensemble for exploration by training an ensemble based on bootstrap with random initialization and randomly sampled policy from the ensemble. [4] extended the idea and replaced the policy sampling with UCB. Finally, [19] proposed to combine ensemble bootstrap with random initialization [22], weighted Bellman backup, and UCB [4]. While they also studied the ensemble in the similar context, they did not attempt to explain the gain afforded by the ensemble, nor did they studied the effect of combining ensemble with auxiliary tasks.

3   Background

3.1   Markov Decision Process and RL

A sequential decision problem is often modeled as a Markov Decision Process (MDP). An MDP is defined with a 5-tuple $< \mathbb{S}, \mathbb{A}, R, T, \gamma >$ where $\mathbb{S}$ and $\mathbb{A}$ denote the set of states and actions, R and T represent the reward and transition functions, and $\gamma \in [0, 1)$ is a discount factor of the MDP. Reinforcement Learning aims to find an optimal solution of a decision problem of unknown MDP. One of the well known model-free RL algorithms is Deep Q Learning (DQN) [21], which learns a state-action (s,a) value function $Q(s, a; \theta)$ with neural networks parameterized by θ. The Q-function is used to select action when interacting with environment; of which the experience is accumulated in the replay buffer $\mathbb{D}$ for learning. We refer the reader to Appendix A for more details about MDP and RL.

3.2   Rainbow Agent

The Rainbow agent [10] extends the DQN by introducing various advances in reinforcement learning. It uses Double-DQN [26] to minimize the overestimation error. Instead of sampling uniformly from the replay buffer $\mathbb{D}$, it assigns priority to each instance based on the temporal difference (TD) error [23]. Its architecture decomposes advantage function from value function $Q(s, a) = V(s) + A(s, a)$ and learn them in an end-to-end manner [28]. Moreover, categorical value distribution is learned in place of the expected state-action value function [3]. Thus, the loss function is given as follows. We denote the scalar-valued Q-function corresponding to the the distributional Q-function as $\hat{Q}$ for simplicity.

$$\mathcal{L}(\theta) = \mathbb{E}\left[D_{\mathrm{KL}}[g_{s,a,r,s'}||Q(s,a;\theta)]\right] \tag{1}$$

$$g_{s,a,r,s'} = \Phi_{Q(s',\hat{a}';\theta')}\left(r + \gamma\mathcal{S}\right) \tag{2}$$

$$\hat{a}' = \arg\max_{a'} \hat{Q}(s',a';\theta) \tag{3}$$

where $\Phi_{Q(s',\hat{a}';\theta')}$ denotes a distributional projection [3] based on categorical atom probabilities given by $Q(s',\hat{a}';\theta')$ for support $\mathcal{S}$. Q returns a column vector Softmax output with $|\mathcal{S}|$ rows instead of a scalar value and $\mathcal{S}$ is a column vector support of the categorical distribution. The scalar-valued Q function is computed by $\hat{Q}(s,a) = \mathcal{S}^T Q(s,a)$. We refer the reader to the original paper [3] for a detailed explanation.

Multi-step returns is also employed to achieve faster convergence [24]. To aid exploration, NoisyNets [6] is utilized; it works by perturbing the parameter space of the Q-function by injecting learnable Gaussian noise.

Recently, [9] proposes a set of hyperparameters for Rainbow that works well on ATARI games under 100K interactions [14].

4   Rainbow Ensemble

Several forms of ensemble agents have been proposed in the literature [1,19,22] with different variations of complexities. For ease of analysis, we propose to use a simple ensemble similar to Ensemble DQN [1]. The original Ensemble DQN was not combined with the recent advances in DQN such as distributional value function [3], prioritized experience replay [23], and NoisyNets [6]. Here, we describe our ensemble, that we call *REN* (Rainbow ENsemble), which combines a simple ensemble approach with modern DQN advances in Rainbow.

REN is based on the following simple ensemble estimator:

$$Q_{ens}^{(M)}(s,a) = \sum_{m=1}^{M} \frac{1}{M} Q_m(s,a;\theta_m) \tag{4}$$

where θ_m is the parameter for the m-th Q function. It outputs a distributional estimate of the Q-function. The scalar-valued ensemble Q-function is given by $\hat{Q}_{ens}^{(M)}(s,a) = \mathcal{S}^T Q_{ens}^{(M)}(s,a)$. We use this Q-function as our policy by taking the action that maximizes this state-action function,

$$a = \arg\max_{a} \hat{Q}_{ens}^{(M)}(s,a) \tag{5}$$

The ensemble distributional Q-function is used to compute the TD target $g_{s,a,r,s'}$ following Eq. 2, thus allowing reduction in the target approximation error (TAE) due to averaging [1]. The agent learns the estimator by optimizing the following loss,

$$\mathcal{L}(\theta_m) = \mathbb{E}_{D_m}\left[D_{\mathrm{KL}}[g_{s,a,r,s'}||Q(s,a;\theta_m)]\right] \tag{6}$$

where m is the index of the member of the ensemble and $\mathbb{D}_m$ is the dataset (buffer) for the m-th member. The loss is computed for each member m and optimized independently.

Rainbow uses prioritized experience replay which assigns priority to each transition. This allows for important transitions – transitions with higher error – to be sampled more frequently. Since REN consists of multiple members, we adopt the use of multiple prioritized experience replay for each member; i.e., for each member m, we have a prioritized replay $\mathbb{D}_m$ with priority updated following the loss $\mathcal{L}(\theta_m)$. This allows each member to individually adjust their priority based on their individual errors, thus potentially enabling the ensemble to perform better.

Finally, NoisyNets is used for exploration in Rainbow. It takes the form of stochastic layers in the Q-function; each stochastic layer samples its parameters from a distribution of parameters modeled as a learnable Gaussian distribution. In our case, we have M Q-functions, with each containing the stochastic layers of NoisyNets.

5   Auxiliary Tasks for Ensemble RL

Auxiliary tasks have often been shown to improve performance in learning problems. However, the combination of auxiliary tasks and ensembles has not been extensively studied, particularly in the context of reinforcement learning.

We use the following framework, where each member of the ensemble can be trained together with a different set of auxiliary tasks, for combining ensembles with auxiliary tasks. Let $\mathcal{T}_m = \{t_{m,1}, ..., t_{m,N_m}\}$ be the set of auxiliary tasks for member m, we seek to optimize the following loss,

$$\mathcal{L}_A(\theta_m) = \mathcal{L}(\theta_m) + \mathbb{E}\Big[\sum_{n=1}^{N_m} \alpha_{m,n}\mathcal{L}_{t_{m,n}}(\theta_m)\Big] \tag{7}$$

where $\alpha_{m,n}$ is the strength parameter and $\mathcal{L}_{t_{m,n}}(\theta_m)$ is the auxiliary task's specific loss for task $t_{m,n}$. Each task may additionally includes a set of parameters that will be optimized jointly with the parameters of the member.

Some questions immediately arise. Should every auxiliary tasks be used with every member of the ensemble? The other extreme would be using a single distinct auxiliary task with each member of the ensemble. If each auxiliary task is *weak* in the sense of only providing a small improvement, can they be combined in the ensemble to provide much stronger improvements? We examine some of these questions in our analysis and case study.

The framework can be viewed as the generalization of MTLE [29] and MTForest [27], where each member of the ensemble is trained with the auxiliary task of predicting the value of a distinct component of the input vector. An instantiation of this framework with REN as the ensemble is denoted as *RENAULT* (Rainbow ENsemble with AUxiLiary Tasks).

We propose to use model learning (i.e., learning transition function and reward function) and learning to predict properties related to objects and events as our auxiliary tasks. Model learning has already been used in model-based RL [14], whereas predicting properties related to objects and events appears to be quite natural – rewards and hence the returns are often associated with events in the environment.

We only consider tasks that can easily be integrated with the ensemble. Some methods such as CURL [18] requires substantial changes to the base algorithm such as requiring momentum target network and data augmentation input, making their use in *RENAULT* difficult.

5.1 Network Architecture

Before delving into each auxiliary task, we will describe our network architecture in detail. Our network consists of two main components: a feature/latent state extraction function $h(s) = z$ and a latent Q function $q(z)$. The feature function h is a two layer convolution neural networks. Due to the use of dueling architecture, $q(z) = \frac{1}{|A|} \sum_{a=1}^{|A|} \text{adv}(z, a) + v(z)$, where $|A|$ is the action space of the problem, adv is a latent advantage function, and v is a latent value function. Both adv and v are two layer fully-connected networks with ReLU as a first layer activation function. We use adv_1 (v_1) to denote the first layer of the adv (v) function.

5.2 Model Learning as Auxiliary Tasks

Our first auxiliary tasks are based on model learning. Model learning is widely used in the context of model-based RL; but here we are using them as auxiliary tasks for DQN. They are easy to use with DQN; each task operates independently and requires no additional changes to the base algorithm. The detail on each model learning task is provided below.

Latent State Transition. We learn a deterministic latent transition function which maps a latent state $z = h(s)$ and action a to its next latent state through a parameterized function $T(z'|z, a; \theta)$. Given the actual next state s', we seek to minimize the loss between the predicted latent state z' and $h(s')$. We use smooth L1 loss [11] as our objective function.

Inverse Dynamic. Inverse dynamic [2] is a function that learns to predict the action that causes a transition from a certain state s to another state s'. Given $z = h(s)$ and $z' = h(s')$, we seek to learn a parameterized function $T^{-1}(\hat{a}|z, z'; \theta)$ by minimizing the loss of predicted action $\hat{a}$ with the real action a via cross entropy loss.

Reward Function. Let $w_1 = \text{adv}_1(z, \cdot)$ and $w_2 = v_1(z)$ be hidden representations corresponding to the output of the first layer of latent advantage function and latent value function for a latent state $z = h(s)$. Given an action a, and let $w = [w_1, w_2]$ be the concatenation of hidden representations w_1 and w_2, we seek to learn a reward function $r(w, a; \theta)$ by minimizing the distributional histogram loss [12] with the real reward $r(s, a)$. The use of reward function as an auxiliary prediction is not new [13]. However, we specify the task as a distributional prediction instead of classification, generalizing their formulation.

5.3   Object and Event Based Auxiliary Tasks

Our second set of auxiliary tasks aim to learn features that are useful for object and event based prediction. We propose two novel auxiliary tasks: change of moment and total change of intensity, to encourage learning features related to objects and events, respectively. The proposed tasks are simple, self-contained, and fairly general when objects and events are present, making them ideal for RENAULT. The detail of these two tasks are given as follows.

Change of Moment. We adopt the concept of moment in physics; a way to account for the distribution of physical quantities based on the product of distance and the quantities. In our case, we use pixels in place of the physical quantities. Thus, the moment corresponds to the distribution of the pixels, which roughly characterizes the distribution of the objects in the screen. For a given image state $s \in \mathbb{R}^{C \times W \times H}$ with channel C, width W, and height H, the moment is computed by $\mu(s) = \frac{1}{C} \sum_c^C \sum_{x,y}^{W,H} d(x,y) \times s_{c,x,y}$ where d is a distance function to some reference point. We use coordinate $(0,0)$ as a reference point and euclidean distance as a distance function. We learn a function that captures the change of moment between a state s and its corresponding next state s' given an action a: $\delta_\mu(z, a; \theta) \approx \mu(s') - \mu(s)$, where $z = h(s)$ is the latent state of s. For stability, we normalize the change of moment by a squared total distance given by d. The function is optimized with smooth L1 loss.

Total Change of Intensity. An event is often characterized by the change of total pixel intensity. For example, objects disappearing due to destruction results in the loss of total pixel intensity, spawning of enemies increases the total pixel intensity, and an explosion triggers dramatic total change of intensity. As such, learning total change of intensity can be a sufficiently strong signal to learn to associate rewards with events. Similar idea regarding learning changes of pixels intensity has been explored by [13], however, they propose to predict changes of intensity in the downsampled image patches using architecture similar to autoencoder. In contrast, we opt for a simpler objective of predicting the total change of intensity instead.

Given an image state s and its corresponding next state s' given an action a, we denote the channel-mean of the state as $\hat{s}$ and the next state as $\hat{s}'$. In addition, we denote the latent state of a state s as $z = h(s)$. We seek to learn a total change of intensity function $\delta_i(z, a; \theta) \approx ||\hat{s} - \hat{s}'||_2$. Since the value is bounded, we adopt the Histogram distributional loss similar to our reward function prediction.

6   Theoretical Analysis

In this section, we perform analysis to help understand the possible gains afforded by REN and RENAULT. We analyze the generalization error through bias-variance-covariance decomposition. Such an analysis is obviously inadequate for reinforcement learning, but we use it to potentially uncover good ways to use ensembles. We then run experiments to see which of the methods actually help for the case study.

We seek to decompose the generalization error of ensembles into bias, variance, and covariance in the form similar to one proposed by [25]. Our analysis differ in that our decomposition focuses on ensembles learned through the use of a single dataset with a randomized learning algorithm. In contrast, their decomposition assumes that each member is trained on a different dataset.

We begin with the case of a single estimator. For the purpose of analysis, we assume our targets are generated by a fixed target function corresponding to the optimal Q-function $f^*(x)$, possibly corrupted by noise, and that our inputs are generated by a fixed policy. We want to learn a function $f(x; \theta)$ to approximate the unknown target function by using a set of N training samples $\{(x_i, y_i)\}_{i=1}^N$. For convenience, we denote $z^N = \{z_i\}_{i=1}^N$ to be a realization of a random set $Z^N = \{Z_i\}_{i=1}^N$, where $z_i = (x_i, y_i)$ and $Z_i = (X_i, Y_i)$. The parameter θ of $f(x; \theta)$ is learnt by a randomized algorithm $\mathcal{A}(r, z^N)$, where r is a random number drawn independently from a random set $\mathcal{R}$. We will use $f(x; r, z^N)$ to refer to this parameterized function.

Given a separate test vector $Z_0 = (X_0, Y_0)$, the generalization error of the function f is $GE(f) = E_{Z^N, \mathcal{R}}[E_{Z_0}[(Y_0 - f(X_0; \mathcal{R}, Z^N))^2]]$. Let

$$\text{Var}(f|X_0) = \mathbb{E}\left[\left(f(X_0; \mathcal{R}, Z^N) - \mathbb{E}[f(X_0; \mathcal{R}, Z^N)]\right)^2\right]$$
$$\text{Bias}(f|X_0) = \mathbb{E}[f(X_0; \mathcal{R}, Z^N)] - f^*(X_0)$$

where $\mathbb{E}$ denotes the expectation $\mathbb{E}_{\mathcal{R}, Z^N}$, and let $\sigma^2 = \mathbb{E}_{X_0, Y_0}[(f^*(X_0) - Y_0)^2]$ be an irreducible error.

Theorem 1 (Generalization error of random algorithm). *The generalization error of the estimator f can be decomposed as follows.*

$$GE(f) = E_{X_0}[\text{Var}(f|X_0) + \text{Bias}(f|X_0)^2] + \sigma^2. \tag{8}$$

All proofs are provided in Appendix C.

Now, we will consider the case of ensemble estimators. Let there be M estimators $\{f_m\}_{m=1}^M$; each estimator f_m is trained using algorithm $\mathcal{A}$ with an individual random number $r^{(m)}$. Note that $r^{(m)}$ is a realization of $\mathcal{R}^{(m)}$. Given an input x, the output of the ensemble is:

$$f_{ens}^{(M)}(x) = \frac{1}{M} \sum_{m=1}^M f(x; r^{(m)}, z^N) \tag{9}$$

Following Eq. 8, the generalization error is given by

$$GE(f_{ens}^{(M)}) = \mathbb{E}_{X_0}[\text{Var}(f_{ens}^{(M)}|X_0) + \text{Bias}(f_{ens}^{(M)}|X_0)^2] + \sigma^2. \tag{10}$$

Let

$$\text{Cov}(f_m, f_{m'}|X_0) = \mathbb{E}\Big[\left(f(X_0; \mathcal{R}^{(m)}, Z^N) - \mathbb{E}[f(X_0; \mathcal{R}^{(m)}, Z^N)]\right)$$
$$\left(f(X_0; \mathcal{R}^{(m')}, Z^N) - \mathbb{E}[f(X_0; \mathcal{R}^{(m')}, Z^N)]\right)\Big].$$

$$\overline{\text{Bias}}(X_0) = \frac{1}{M} \sum_{m=1}^{M} \text{Bias}(f_m|X_0)$$

$$\overline{\text{Var}}(X_0) = \frac{1}{M} \sum_{m=1}^{M} \text{Var}(f_m|X_0)$$

$$\overline{\text{Cov}}(X_0) = \frac{1}{M(M-1)} \sum_{m} \sum_{m' \neq m} \text{Cov}(f_m, f_{m'}|X_0).$$

Theorem 2 (Generalization error of ensemble with random algorithm). *The generalization error of the ensemble estimator $f_{ens}^{(M)}$ can be decomposed as:*

$$GE(f_{ens}^{(M)}) = \mathbb{E}_{X_0}\left[\overline{\text{Bias}}(X_0)^2 + \frac{1}{M}\overline{\text{Var}}(X_0) + \left(1 - \frac{1}{M}\right)\overline{\text{Cov}}(X_0)\right] + \sigma^2 \quad (11)$$

Theorem 1 and 2 follow the proof from [8] and [25], respectively. Although the results look similar, there are subtle differences in terms of the assumption with respect to the availability of multiple datasets and the use of randomness.

By analysing, the relationship between $\overline{\text{Var}}(X_0)$ and $\overline{\text{Cov}}(X_0)$, we obtain the following results about REN.

Theorem 3. $\overline{\text{Var}}(X_0) \leq \overline{\text{Cov}}(X_0)$. *Hence, if ensemble estimator $f_{ens}^{(M)}$ consists of M identical estimators f that differ only in the random numbers used, then $GE(f_{ens}^{(M)}) \leq GE(f)$.*

This result states that ensembles with the members trained the same way cannot hurt performance. The equal case can happen, e.g. when algorithm $\mathcal{A}$ performs convex optimization, it will converge to the same minima regardless of random number used, resulting in all members of the ensemble being the same. In contrast, non-convex optimization algorithms such as SGD converges to a minima that depends on the randomness, thus will likely result in lower error due to reduction in the covariance term. Hence, REN will achieve *at least* the same performance as Rainbow, and possibly better, under the idealized assumptions.

Instead of training each member of the ensemble $f(x; r^{(m)}, z^N)$ separately, training the entire ensemble $f_{ens}^{(M)}(x)$ directly on the training set would result in lower training set error and possibly better generalization. We experiment with this as well in the case study.

For a single network, auxiliary tasks usually reduce the variance as they provide additional information that help constrain the network. However, this may come at the cost of additional bias as the network needs to optimize for multiple objectives. To further understand the effects of auxiliary losses, we decompose the ensemble squared bias.

Proposition 1. *The* $\overline{\text{Bias}}(X_0)^2$ *of ensemble estimator* $f_{ens}^{(M)}$ *can be decomposed as follows (X_0 is omitted for readability),*

$$\frac{1}{M^2}\Big[\sum_m^M \text{Bias}(f_m)^2 + \sum_m \sum_{m'\neq m} \text{Cob}(f_m, f_m')\Big] \tag{12}$$

where $\text{Cob}(f_m, f_m'|X_0) = \text{Bias}(f_m|X_0)\text{Bias}(f_{m'}|X_0)$ *denote the product of bias; we refer to this as co-bias.*

This result suggests that lower ensemble generalization error can be obtained by increasing the number of negative co-bias by having a diverse set of positively and negatively biased members. This is more likely to be achieved in RENAULT if each member is assigned a unique set of auxiliary tasks. In contrast, assigning the same set of auxiliary tasks to each member results in $\overline{\text{Bias}}(X_0) = \text{Bias}(f|X_0)$ because $\forall_m \text{Bias}(f_m|X_0) = \text{Bias}(f|X_0)$.

Proposition 2. *Let* $\bar{f}_{m,Z} = \mathbb{E}_{\mathcal{R}^{(m)}}[f_m|Z^N]$ *be a conditional expectation of* f_m *over random number* $\mathcal{R}^{(m)}$ *conditioned on* Z^N. *If the estimators are trained independently, then,* $Cov(f_m, f_m') = E_{Z^N}[(\bar{f}_{m,Z^N} - E[f_m])(\bar{f}_{m',Z^N} - E[f_{m'}])]$.

This result suggests that RENAULT may also reduce covariance if appropriate auxiliary tasks are assigned to each member. Otherwise, if all members are of the same model f, then $Cov(f_m, f_m') = E_{Z^N}[(\bar{f}_{Z^N} - E[f])^2]$ which is the variance of the averaged estimator.

Limitations. Our analysis assumes a fixed target; this is not available in RL. Instead, we have an estimate of the target (optimal Q value) based on TD return. The dataset in RL is also generated by a non-stationary policy, thus the distribution of the dataset keeps on changing during learning. Additionally, exploration also plays an important role in the learning of Q-function. Thus, it is important to note that our analysis will only provide partial insights regarding the methods; it serves to suggest possible ways to improve the algorithms, but the suggestions may not always help.

7   Experiments

In this section, we perform a set of experiments on REN and RENAULT. We compare them to prior methods. We examine whether joint training is better than independent training. We examine whether the auxiliary tasks help the ensembles and how to best use the auxiliary tasks. Before delving into each experiment, we will explain the problem domain of our case study, our architecture and hyperparameters, and our methods in detail.

Problem Domain. We evaluate REN and RENAULT on a suite of Atari games from Atari Learning Environment (ALE) benchmark. We follow the evaluation procedure of [14]; particularly, we limit the environment interaction to 100K interactions (400K frames with action repeated 4 frames) and evaluate on a sub-set of 26 games. We measure the raw performance score and human-normalized score, calculated as $100 \times$ (Method score − Random score)/(Human score − Random score).

Architecture and Hyperparameters. We follow the data-efficient Rainbow (DE-Rainbow) architecture and hyperparameters proposed by [9] and made no change to them. REN introduces a hyperparameter M which controls the number of members of the ensemble. RENAULT introduces task and member specific hyperparameters $\alpha_{m,n}$ that control the strength of each auxiliary task. Additionally, each auxiliary task adopts different architecture; we give their description in Appendix E.1.

In our preliminary experiment, we found that $M < 5$ degrades performance and higher M does not increase performance significantly while requiring more resources. Thus we fix $M = 5$ throughout the experiment.

Our Methods. REN has two variants; one that is canonical according to our description in Sect. 4, and one that optimizes all members jointly, which we refer to as *REN-J*. RENAULT also has two variants based on how we distribute the auxiliary tasks. The first variant, which we simply refer to as RENAULT, follows the suggestion of the preceding section to distribute the auxiliary tasks. As the number of auxiliary tasks equals to the number of members, we simply assign one *unique* task for each member. In contrast, the second variant assigns *all* tasks to each member, thus we call this variant *RENAULT-all*. For simplicity, RENAULT uses $\alpha_{m,n} = 1$ for all member m and task n. For RENAULT-all, we set $\alpha_{m,n} = \frac{1}{N_m}$, where $N_m = 5$ is the number of auxiliary tasks for member m. This is to ensure that the auxiliary tasks do not overwhelm the main task.

Further experimental details can be found in Appendix E.

7.1   Comparison to Prior Works

We compare the performance of REN and RENAULT to SimPLe [14], data-efficient Rainbow (DE-Rainbow) [9], and Overtrained Rainbow (OT-Rainbow) [16]. Two other recent works, CURL [18] and SUNRISE [19] use game-dependent hyperparameters instead of using the same hyperparameters for all games, making their results not directly comparable to ours. The results are given in Table 1. We report the mean of three independent runs for our methods. We take the highest reported scores for SimPLe and human baselines, as they are reported differently in prior work [16,26].

REN improves the performance of its baseline, data-efficient Rainbow on 20 out of 26 games and achieves better performance on 13 games. It also improves the mean and median human normalized performance $1.45\times$ and $1.26\times$, respectively. RENAULT further enhances the performance of REN, gaining $1.6\times$ and $1.55\times$ mean and median human normalized performance improvements. Additionally, it won on 21 games when compared to data-efficient Rainbow and exceeds REN's performance on 17 games.

Table 1. Performance on ATARI games on 100K interactions. Human Mean and Human Median indicate the mean and the median of the human-normalized score. The last two rows show the number of games won against DE-Rainbow and REN, respectively.

	SimPLe	OT-Rnbw	DE-Rnbw	REN	RNLT	REN-J	RNLT-all
alien	405.2	824.7	739.9	828.7	883.7	800.3	**890.0**
amidar	88.0	82.8	188.6	195.4	**224.4**	120.2	137.2
assault	369.3	351.9	431.2	608.5	**651.4**	504.0	524.9
asterix	**1089.5**	628.5	470.8	578.3	631.7	645.0	520.0
bank_heist	8.2	**182.1**	51.0	63.3	125.0	64.7	92.3
battle_zone	5184.4	4060.6	10124.6	**17500.0**	14233.3	12666.7	9000.0
boxing	9.1	2.5	0.2	**10.9**	5.1	5.2	4.9
breakout	**12.7**	9.8	1.9	3.7	3.4	2.7	3.0
chopper_command	**1246.9**	1033.3	861.8	713.3	896.7	980.0	563.3
crazy_climber	**39827.8**	21327.8	16185.3	16523.3	39460.0	23613.3	22123.3
demon_attack	169.5	711.8	508.0	759.3	693.0	665.5	**822.7**
freeway	20.3	25.0	27.9	28.9	29.3	24.5	**29.4**
frostbite	254.7	231.6	866.8	**2507.7**	1210.3	2284.7	1167.0
gopher	771.0	**778.0**	349.5	246.7	542.7	521.3	323.3
hero	1295.1	6458.8	6857.0	3817.2	6568.8	6499.3	**7260.5**
jamesbond	125.3	112.3	301.6	518.3	**628.3**	276.7	420.0
kangaroo	323.1	605.4	779.3	753.3	540.0	**893.3**	840.0
krull	**4539.9**	3277.9	2851.5	3105.1	2831.3	2667.2	3827.0
kung_fu_master	**17257.2**	5722.2	14346.1	12576.7	15703.3	9616.7	13423.3
ms_pacman	762.8	941.9	1204.1	1496.0	**2002.7**	1240.7	1705.0
pong	**5.2**	1.3	−19.3	−16.8	−12.0	−18.7	−10.8
private_eye	58.3	**100.0**	97.8	66.7	66.7	−35.2	**100.0**
qbert	559.8	509.3	1152.9	1428.3	583.3	**2416.7**	1014.2
road_runner	5160.4	2090.7	9600.0	11446.7	**13280.0**	5676.7	7550.0
seaquest	370.9	286.9	354.1	622.7	**671.3**	555.3	387.3
up_n_down	2152.6	2847.6	2877.4	3568.0	**4235.7**	3388.0	3459.0
Human Mean	36.45%	26.41%	28.54%	41.36%	**45.64%**	30.78%	38.32%
Human Median	9.85%	20.37%	16.14%	20.41%	**25.08%**	21.97%	23.42%
vs DE-Rnbw	10 (-3)	12 (-1)	-	20 (+7)	**21 (+8)**	18 (+5)	19 (+6)
vs REN	8 (-5)	10 (-3)	6 (-7)	-	**17 (+4)**	8 (-5)	13 (0)

7.2 Bias-Variance-Covariance Measurements

To gain additional insights into our methods, we perform an empirical analysis by measuring their bias, variance, and covariance. Measuring bias requires the optimal Q-function which is unknown in RL. We measure the approximation to ensemble bias $\widehat{\text{Bias}}(\theta)$ based on TD return in place of the real bias. We denote the ensemble bias based on this approximation as $\overline{\widehat{\text{Bias}}}$. The detail of the measurements is given in Appendix D.

Table 2. Measurement of bias approximation, variance, covariance, irreducible error σ^2, and an approximation of generalization error ($\widehat{\text{GE}}$) of all methods. For Rainbow, $\overline{\text{Bias}}$, $\overline{\text{Var}}$, and $\overline{\text{Cov}}$ denotes the estimator bias, variance, and covariance, respectively.

	$\widehat{\overline{\text{Bias}}}^2$	$\overline{\text{Var}}$	$\overline{\text{Cov}}$	σ^2	$\widehat{\text{GE}}$
REN	0.08	1.09	0.99	1.02	2.28
RENAULT	0.08	0.82	0.66	0.63	1.58
RENAULT-all	0.09	0.81	0.71	0.74	1.72
REN-J	0.07	1.07	0.51	0.52	1.41
Rainbow	0.08	*0.84*	–	0.70	1.81

The result of the measurements is given in Table 2.

We can see from Table 2 that $\overline{\text{Cov}} < \overline{\text{Var}}$ in REN as expected from Proposition 2. If the datasets used in the ensembles had been independent as well, we would have $\overline{\text{Cov}} = 0$, so the effects of independent randomization is more limited. RENAULT reduced the variance of REN as expected from the use of auxiliary tasks and running different tasks on different members of the ensemble appears to further reduce the covariance of RENAULT.

Comparison of REN and Rainbow also shows that our bias-variance-covariance measurements are not adequate for perfectly understanding the performance of the different algorithms. In particular, the generalization error of Rainbow is smaller than REN but REN had better performance. It is possible that the bias estimate using TD return is not a good proxy for the real bias; the TD return may be arbitrarily far from the optimal Q. Another possible reason could be that RL is much more than generalization error, which does not capture other aspects of RL such as exploration.

7.3 On Independent Training of Ensemble

Jointly optimizing all members of the ensemble would give better training error and possibly better generalization error. We compare the performance of REN with its variant, REN-J, that directly optimize the following loss:

$$\mathcal{L}(\theta_{ens}) = \mathbb{E}\big[D_{\text{KL}}[g_{s,a,r,s'}||Q_{ens}^{(M)}(s,a;\theta_{ens})]\big] \tag{13}$$

where $\theta_{ens} = \{\theta_m\}_{m=1}^M$. Since REN-J is essentially one single big neural network, it uses a single prioritized experience replay $\mathbb{D}$ which is updated based on $\mathcal{L}(\theta_{ens})$.

Table 2 shows that REN-J indeed generalized better than REN. In particular, joint optimization substantially reduced $\overline{\text{Cov}}$. However, REN surprisingly gives better overall performance compared REN-J. REN improves upon REN-J on 18 out of 26 games. It also improves the mean human normalized performance 1.34×, although with a slight reduction of median performance of 0.93×. When compared to data-efficient Rainbow, REN gains on two more games than REN-J.

Contrary to expectation, in this case study, it is preferable to train an ensemble by optimizing each member independently, rather than treating the ensemble as a single monolithic neural network and optimize all members jointly to reduce its generalization error.

7.4   The Importance of Auxiliary Tasks

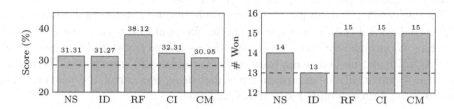

Fig. 1. Human normalized mean score (Left) and the number of games won (Right) of each member of the ensemble with (NS) latent next state prediction, (ID) inverse dynamic, (RF) reward function, (CI) total change of intensity, (CM) change of moment. As a reference, the performance of data-efficient Rainbow is indicated by a dotted line.

Table 1 shows that RENAULT improves REN on all counts. It wins on 17 games, gained 1.1× and 1.23× human mean and median normalized performance, as well as increasing the win count against Rainbow to 21 games. This demonstrates the significant benefit of augmenting ensembles with auxiliary tasks, at least in this case study. Moreover, this is achieved without any tuning to the auxiliary task hyperparameter $\alpha_{m,n}$; we simply set it to 1 for all member m and task n. We also simply distribute the auxiliary tasks as such that each member is augmented with one unique task. Careful tuning of the hyperparameter and task distribution may yield even better performance improvements.

To understand the role of each auxiliary task, we analyze each of their contribution. Figure 1 shows the contribution of each member of the ensemble that is endowed with a particular auxiliary task. It is interesting to see that although each task is weakly helpful (only offers modest performance improvement), they offer significant performance boost when combined with ensembles. The best performing auxiliary tasks in terms of games won are reward function, total change of intensity (CI), and change of moment (CM) prediction. This demonstrate the usefulness of our novel auxiliary tasks; we discuss this more in Appendix F.1.

In the opposite extreme, inverse dynamic (ID) seems to be less useful among the auxiliary tasks. Surprisingly, retraining RENAULT without ID reduces its performance substantially (see Appendix F.2). This suggests that ensemble improvements are not merely from individual gain, but also from diversity, through improved co-bias and covariance.

7.5   On Distributing the Auxiliary Tasks

Our theoretical result suggests that distributing the auxiliary tasks may be better than assigning all tasks on each member of the ensemble. To confirm this, we compare RENAULT with its variant which assigns *all* auxiliary tasks to each member, RENAULT-all.

Table 1 shows that RENAULT-all performs worse than RENAULT, achieving lower mean and median human normalized score; this is in line with our expectation. While it may also be the case that suboptimal hyperparameters plays some roles in causing the performance degradation, this comparison is fair as we also did not perform tuning for RENAULT.

Finally, RENAULT-all has larger ensemble bias and covariance compared to RENAULT in Table 2. The larger ensemble bias could be because each network now has to optimize for more objectives. Propositions 1 and 2 also suggest that RENAULT could be benefiting from reduced co-bias and covariance. The reduction could potentially be due to each member being less correlated when trained on the same dataset compared to RENAULT-all.

8   Conclusions

In this work, we study ensembles and auxiliary tasks in the context of deep Q-learning. We proposed a simple agent that creates an ensemble of Q-functions based on Rainbow, and additionally augments it with auxiliary tasks. We provide theoretical analysis and an experimental case study. Our methods improve significantly upon data-efficient Rainbow. We show that, although each auxiliary task only improves performance slightly, they significantly boost performance when combined using an ensemble.

Our study focuses on the interaction between ensembles, auxiliary tasks, and DQN on learning. However, RL is a multi-faceted problem with many important components including exploration. Future work includes studying their interaction with exploration, which may provide important insights and answers to some of the questions which eludes our understanding in this work.

Acknowledgements. We thank Lung Sin Kwee for useful discussions. This research is supported in part by the National Research Foundation, Singapore under its AI Singapore Program (AISG Award No: AISG2-RP-2020-016).

References

1. Anschel, O., Baram, N., Shimkin, N.: Averaged-DQN: variance reduction and stabilization for deep reinforcement learning. In: International Conference on Machine Learning, pp. 176–185. PMLR (2017)
2. Badia, A.P., et al.: Never give up: learning directed exploration strategies. In: International Conference on Learning Representations (2019)
3. Bellemare, M.G., Dabney, W., Munos, R.: A distributional perspective on reinforcement learning. In: International Conference on Machine Learning, pp. 449–458 (2017)
4. Chen, R.Y., Sidor, S., Abbeel, P., Schulman, J.: UCB exploration via Q-ensembles. arXiv preprint arXiv:1706.01502 (2017)

5. Chua, K., Calandra, R., McAllister, R., Levine, S.: Deep reinforcement learning in a handful of trials using probabilistic dynamics models. In: Advances in Neural Information Processing Systems, pp. 4754–4765 (2018)
6. Fortunato, M., et al.: Noisy networks for exploration. In: International Conference on Learning Representations (2018)
7. Fujimoto, S., Hoof, H., Meger, D.: Addressing function approximation error in actor-critic methods. In: International Conference on Machine Learning, pp. 1587–1596 (2018)
8. Geman, S., Bienenstock, E., Doursat, R.: Neural networks and the bias/variance dilemma. Neural Comput. **4**(1), 1–58 (1992)
9. van Hasselt, H.P., Hessel, M., Aslanides, J.: When to use parametric models in reinforcement learning? In: Advances in Neural Information Processing Systems, pp. 14322–14333 (2019)
10. Hessel, M., et al.: Rainbow: combining improvements in deep reinforcement learning. In: Proceedings of the AAAI Conference on Artificial Intelligence, vol. 32 (2018)
11. Huber, P.J.: Robust estimation of a location parameter. In: Kotz, S., Johnson, N.L. (eds.) Breakthroughs in Statistics, pp. 492–518. Springer, Heidelberg (1992). https://doi.org/10.1007/978-1-4612-4380-9_35
12. Imani, E., White, M.: Improving regression performance with distributional losses. In: International Conference on Machine Learning, pp. 2157–2166 (2018)
13. Jaderberg, M., et al.: Reinforcement learning with unsupervised auxiliary tasks. arXiv preprint arXiv:1611.05397 (2016)
14. Kaiser, L., et al.: Model based reinforcement learning for Atari. In: International Conference on Learning Representations (2019)
15. Kartal, B., Hernandez-Leal, P., Taylor, M.E.: Terminal prediction as an auxiliary task for deep reinforcement learning. In: Proceedings of the AAAI Conference on Artificial Intelligence and Interactive Digital Entertainment, vol. 15, pp. 38–44 (2019)
16. Kielak, K.: Do recent advancements in model-based deep reinforcement learning really improve data efficiency? arXiv preprint arXiv:2003.10181 (2020)
17. Kurutach, T., Clavera, I., Duan, Y., Tamar, A., Abbeel, P.: Model-ensemble trust-region policy optimization. In: International Conference on Learning Representations (2018)
18. Laskin, M., Srinivas, A., Abbeel, P.: Curl: contrastive unsupervised representations for reinforcement learning. In: International Conference on Machine Learning, pp. 5639–5650. PMLR (2020)
19. Lee, K., Laskin, M., Srinivas, A., Abbeel, P.: Sunrise: a simple unified framework for ensemble learning in deep reinforcement learning. arXiv preprint arXiv:2007.04938 (2020)
20. Mirowski, P., et al.: Learning to navigate in complex environments. arXiv preprint arXiv:1611.03673 (2016)
21. Mnih, V., et al.: Human-level control through deep reinforcement learning. Nature **518**(7540), 529–533 (2015)
22. Osband, I., Blundell, C., Pritzel, A., Van Roy, B.: Deep exploration via boot-strapped DQN. In: Advances in Neural Information Processing Systems, pp. 4026–4034 (2016)
23. Schaul, T., Quan, J., Antonoglou, I., Silver, D.: Prioritized experience replay. arXiv preprint arXiv:1511.05952 (2015)
24. Sutton, R.S., Barto, A.G.: Reinforcement Learning: An Introduction (2011)

25. Ueda, N., Nakano, R.: Generalization error of ensemble estimators. In: Proceedings of International Conference on Neural Networks (ICNN'96), vol. 1, pp. 90–95. IEEE (1996)
26. Van Hasselt, H., Guez, A., Silver, D.: Deep reinforcement learning with double Q-learning. In: Proceedings of the AAAI Conference on Artificial Intelligence, vol. 30 (2016)
27. Wang, Q., Zhang, L., Chi, M., Guo, J.: MTForest: ensemble decision trees based on multi-task learning (2008)
28. Wang, Z., Schaul, T., Hessel, M., Hasselt, H., Lanctot, M., Freitas, N.: Dueling network architectures for deep reinforcement learning. In: International Conference on Machine Learning, pp. 1995–2003. PMLR (2016)
29. Ye, Q., Munro, P.W.: Improving a neural network classifier ensemble with multi-task learning. In: The 2006 IEEE International Joint Conference on Neural Network Proceedings, pp. 5164–5170. IEEE (2006)

Multi-agent Imitation Learning with Copulas

Hongwei Wang$^{(\boxtimes)}$, Lantao Yu, Zhangjie Cao, and Stefano Ermon

Computer Science Department, Stanford University, Stanford, CA 94305, USA
{hongweiw,lantaoyu,caozj,ermon}@cs.stanford.edu

Abstract. Multi-agent imitation learning aims to train multiple agents to perform tasks from demonstrations by learning a mapping between observations and actions, which is essential for understanding physical, social, and team-play systems. However, most existing works on modeling multi-agent interactions typically assume that agents make independent decisions based on their observations, ignoring the complex dependence among agents. In this paper, we propose to use copula, a powerful statistical tool for capturing dependence among random variables, to explicitly model the correlation and coordination in multi-agent systems. Our proposed model is able to separately learn marginals that capture the local behavioral patterns of each individual agent, as well as a copula function that solely and fully captures the dependence structure among agents. Extensive experiments on synthetic and real-world datasets show that our model outperforms state-of-the-art baselines across various scenarios in the action prediction task, and is able to generate new trajectories close to expert demonstrations.

Keywords: Multi-agent systems · Imitation learning · Copulas

1 Introduction

Recent years have witnessed great success of reinforcement learning (RL) for single-agent sequential decision making tasks. As many real-world applications (e.g., multi-player games [6,27] and traffic light control [7]) involve the participation of multiple agents, multi-agent reinforcement learning (MARL) has gained more and more attention. However, a key limitation of RL and MARL is the difficulty of designing suitable reward functions for complex tasks with implicit goals (e.g., dialogue systems) [10,22,26,30]. Indeed, hand-tuning reward functions to induce desired behaviors becomes especially challenging in multi-agent systems, since different agents may have completely different goals and state-action representations [35].

Imitation learning [11,24] provides an alternative approach to directly programming agents by taking advantage of expert demonstrations on how a task should be solved. Although appealing, most prior works on multi-agent imitation learning typically assume agents make independent decisions after observing a

© Springer Nature Switzerland AG 2021
N. Oliver et al. (Eds.): ECML PKDD 2021, LNAI 12975, pp. 139–156, 2021.
https://doi.org/10.1007/978-3-030-86486-6_9

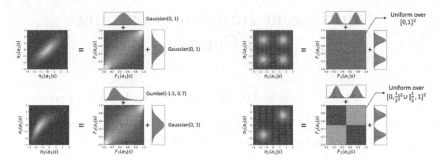

(a) Same copula but different marginals (b) Same marginals but different copulas

Fig. 1. In each subfigure, the left part visualizes the joint policy $\pi(a_1, a_2|s)$ on the joint action space $[-3,3]^2$ and the right part shows the corresponding marginal policies (e.g., $\pi_1(a_1|s) = \int_{a_2} \pi(a_1, a_2|s)\mathrm{d}a_2$) as well as the copula $c(F_1(a_1|s), F_2(a_2|s))$ on the unit cube. Here F_i is the cumulative distribution function of the marginal $\pi_i(a_i|s)$ and $u_i := F_i(a_i|s)$ is the uniformly distributed random variable obtained by probability integral transform with F_i. More details and definitions can be found in Sect. 3.1.

state (i.e., mean-field factorization of the joint policy) [16,30,35,36], ignoring the potentially complex dependencies that exist among agents. Recently, [33] and [19] proposed to implement correlated policies with opponent modeling, which incurs unnecessary modeling cost and redundancy, while still lacking coordination during execution.

Compared to the single-agent setting, one major and fundamental challenge in multi-agent learning is how to model the dependence among multiple agents in an effective and scalable way. Inspired by probability theory and statistical dependence modeling, in this work, we propose to use copulas [14,21,29] to model multi-agent behavioral patterns. Copulas are powerful statistical tools to describe the dependence among random variables, which have been widely used in quantitative finance for risk measurement and portfolio optimization [5]. Using a copulas-based multi-agent policy enables us to separately learn marginals that capture the local behavioral patterns of each individual agent and a copula function that only and fully captures the dependence structure among the agents. Such a factorization is capable of modeling arbitrarily complex joint policy and leads to *interpretable, efficient and scalable* multi-agent imitation learning. As a motivating example (see Fig. 1), suppose there are two agents, each with one-dimensional action space. In Fig. 1a, although two joint policies are quite different, they actually share the same copula (dependence structure) and one marginal. Our proposed copula-based policy is capable of capturing such information and more importantly, we may leverage such information to develop efficient algorithms for such transfer learning scenarios. For example, when we want to model team-play in a soccer game and one player is replaced by his/her substitute while the dependence among different roles are basically the same regardless of players, we can immediately obtain a new joint policy by switching in the new player's marginal while keeping the copula and other marginals unchanged. On the other hand, as shown in Fig. 1b, two different joint policies may share the same marginals while having different copulas, which implies that the mean-field policy in previous works (only modeling marginal policies and making independent decisions) cannot differentiate these two scenarios to achieve coordination correctly.

Towards this end, in this paper, we propose a copula-based multi-agent imitation learning algorithm, which is interpretable, efficient and scalable for modeling complex multi-agent interactions. Extensive experimental results on synthetic and real-world datasets show that our proposed method outperforms state-of-the-art multi-agent imitation learning methods in various scenarios and generates multi-agent trajectories close to expert demonstrations.

2   Preliminaries

We consider the problem of multi-agent imitation learning under the framework of Markov games [18], which generalize Markov Decision Processes to multi-agent settings, where N agents are interacting with each other. Specifically, in a Markov game, $\mathcal{S}$ is the common state space, $\mathcal{A}_i$ is the action space for agent $i \in \{1, \ldots, N\}$, $\eta \in \mathcal{P}(\mathcal{S})$ is the initial state distribution and $P : \mathcal{S} \times \mathcal{A}_1 \times \ldots \times \mathcal{A}_N \rightarrow \mathcal{P}(\mathcal{S})$ is the state transition distribution of the environment that the agents are interacting with. Here $\mathcal{P}(\mathcal{S})$ denotes the set of probability distributions over state space $\mathcal{S}$. Suppose at time t, agents observe $s[t] \in \mathcal{S}$ and take actions $\boldsymbol{a}[t] := (a_1[t], \ldots, a_N[t]) \in \mathcal{A}_1 \times \ldots \times \mathcal{A}_N$, the agents will observe state $s[t+1] \in \mathcal{S}$ at time $t + 1$ with probability $P(s[t + 1] \mid s[t], a_1[t], \ldots, a_N[t])$. In this process, the agents select the joint action $\boldsymbol{a}[t]$ by sampling from a stochastic joint policy $\boldsymbol{\pi} : \mathcal{S} \rightarrow \mathcal{P}(\mathcal{A}_1 \times \ldots \times \mathcal{A}_N)$. In the following, we will use subscript $-i$ to denote all agents except for agent i. For example, $(a_i, \boldsymbol{a}_{-i})$ represents the actions of all agents; $\pi_i(a_i|s)$ and $\pi_i(a_i|s, \boldsymbol{a}_{-i})$ represent the marginal and conditional policy of agent i induced by the joint policy $\boldsymbol{\pi}(\boldsymbol{a}|s)$ (through marginalization and Bayes' rule, respectively).

Suppose we have access to a set of demonstrations $\mathcal{D} = \{\tau^j\}_{j=1}^M$ provided by some expert policy $\boldsymbol{\pi}^E(\boldsymbol{a}|s)$, where each expert trajectory $\tau^j = \{(s^j[t], \boldsymbol{a}^j[t])\}_{t=1}^T$ is collected by the following sampling process:

$$s^1 \sim \eta(s), \boldsymbol{a}[t] \sim \boldsymbol{\pi}^E(\boldsymbol{a}|s[t]), s[t + 1] \sim P(s|s[t], \boldsymbol{a}[t]), \text{for } t \geq 1.$$

The goal is to learn a parametric joint policy $\boldsymbol{\pi}^\theta$ to approximate the expert policy $\boldsymbol{\pi}^E$ such that we can do downstream inferences (e.g., action prediction and trajectory generation). The learning problem is off-line as we cannot ask for additional interactions with the expert policy or the environment during training, and the reward is also unknown.

3   Modeling Multi-agent Interaction with Copulas

Many modeling methods for multi-agent learning tasks employ a simplifying mean-field assumption that the agents make independent action choices after observing a state [2,30,35], which means the joint policy can be factorized as follows:

$$\boldsymbol{\pi}(a_1,\ldots,a_N|s) = \prod_{i=1}^{N} \pi_i(a_i|s). \tag{1}$$

Such a mean-field assumption essentially allows for independent construction of each agent's policy. For example, multi-agent behavior cloning by maximum likelihood estimation is now equivalent to performing N single-agent behavior cloning tasks:

$$\max_{\boldsymbol{\pi}} \mathbb{E}_{(s,\boldsymbol{a})\sim\rho_{\boldsymbol{\pi}_E}}[\log \boldsymbol{\pi}(\boldsymbol{a}|s)] = \sum_{i=1}^{N} \max_{\pi_i} \mathbb{E}_{(s,a_i)\sim\rho_{\boldsymbol{\pi}_E,i}}[\log \pi_i(a_i|s)], \tag{2}$$

where the occupancy measure $\rho_{\boldsymbol{\pi}} : \mathcal{S} \times \mathcal{A}_1 \times \ldots \times \mathcal{A}_N \to \mathbb{R}$ denotes the state action distribution encountered when navigating the environment using the joint policy $\boldsymbol{\pi}$ [25,32] and $\rho_{\boldsymbol{\pi},i}$ is the corresponding marginal occupancy measure.

However, when the expert agents are making correlated action choices (e.g., due to joint plan and communication in a soccer game), such a simplifying modeling choice is not able to capture the rich dependency structure and coordination among agent actions. To address this issue, recent works [19,33] propose to use a different factorization of the joint policy such that the dependency among N agents can be preserved:

$$\boldsymbol{\pi}(a_i, \boldsymbol{a}_{-i}|s) = \pi_i(a_i|s, \boldsymbol{a}_{-i})\boldsymbol{\pi}_{-i}(\boldsymbol{a}_{-i}|s), for\ i \geq 1. \tag{3}$$

Although such a factorization is general and captures the dependency among multi-agent interactions, several issues still remain. First, the modeling cost is increased significantly, because now we need to learn N different and complicated opponent policies $\boldsymbol{\pi}_{-i}(\boldsymbol{a}_{-i}|s)$ as well as N different marginal conditional policies $\pi_i(a_i|s, \boldsymbol{a}_{-i})$, each with a deep neural network. It should be noted that there are many redundancies in such a modeling choice. Specifically, suppose there are N agents and $N > 3$, for agent 1 and N, we need to learn opponent policies $\boldsymbol{\pi}_{-1}(a_2,\ldots,a_N|s)$ and $\boldsymbol{\pi}_{-N}(a_1,\ldots,a_{N-1}|s)$ respectively. These are potentially high dimensional and might require flexible function approximations. However, the dependency structure among agent 2 to agent $N - 1$ are modeled in both $\boldsymbol{\pi}_{-1}$ and $\boldsymbol{\pi}_{-N}$, which incurs unnecessary modeling cost. Second, when executing the policy, each agent i makes decisions through its marginal policy $\pi_i(a_i|s) = \mathbb{E}_{\boldsymbol{\pi}_{-i}(\boldsymbol{a}_{-i}|s)}(a_i|s, \boldsymbol{a}_{-i})$ by first sampling $\boldsymbol{a}_{-i}$ from its opponent policy $\boldsymbol{\pi}_{-i}$ then sampling its action a_i from $\pi_i(\cdot|s, \boldsymbol{a}_{-i})$. Since each agent is performing such decision process independently, coordination among agents are still impossible due to sampling randomness. Moreover, a set of independently learned conditional distributions are not necessarily consistent with each other (i.e., induced by the same joint policy) [35].

In this work, to address above challenges, we draw inspiration from probability theory and propose to use copulas, a statistical tool for describing the dependency structure between random variables, to model the complicated multi-agent interactions in a scalable and efficient way.

3.1   Copulas

When the components of a multivariate random variable $x = (x_1, \ldots, x_N)$ are jointly independent, the density of x can be written as:

$$p(x) = \prod_{i=1}^{N} p(x_i). \tag{4}$$

When the components are not independent, this equality does not hold any more as the dependencies among $x_1, \ldots, x_N$ can not be captured by the marginals $p(x_i)$. However, the differences can be corrected by multiplying the right hand side of Eq. (4) with a function that *only and fully* describes the dependency. Such a function is called a copula [21], a multivariate distribution function on the unit hyper-cube with uniform marginals.

Intuitively, consider a random variable x_i with continuous cumulative distribution function F_i. Applying *probability integral transform* gives us a random variable $u_i = F_i(x_i)$, which has standard uniform distribution. Thus one can use this property to separate the information in marginals from the dependency structures among $x_1, \ldots, x_N$ by first projecting each marginal onto one axis of the hyper-cube and then capture the pure dependency with a distribution on the unit hyper-cube.

Formally, a copula is the joint distribution of random variables $u_1, \ldots, u_N$, each of which is marginally uniformly distributed on the interval $[0, 1]$. Furthermore, we introduce the following theorem that provides the theoretical foundations of copulas:

Theorem 1 (Sklar's Theorem [28]). *Suppose the multivariate random variable $(x_1, \ldots, x_N)$ has marginal cumulative distribution functions $F_1, \ldots, F_N$ and joint cumulative distribution function F, then there exists a unique copula $C : [0, 1]^N \to [0, 1]$ such that:*

$$F(x_1, \ldots, x_N) = C\big(F_1(x_1), \ldots, F_N(x_N)\big). \tag{5}$$

When the multivariate distribution has a joint density f and marginal densities $f_1, \ldots, f_N$, we have:

$$f(x_1, \ldots, x_N) = \prod_{i=1}^{N} f_i(x_i) \cdot c\big(F_1(x_1), \ldots, F_N(x_N)\big), \tag{6}$$

where c is the probability density function of the copula. The converse is also true. Given a copula C and marginals $F_i(x_i)$, then $C\big(F_1(x_1), \ldots, F_N(x_N)\big) = F(x_1, \ldots, x_N)$ is a N-dimensional cumulative distribution function with marginal distributions $F_i(x_i)$.

Theorem 1 states that every multivariate cumulative distribution function $F(x_1, \ldots, x_N)$ can be expressed in terms of its marginals $F_i(x_i)$ and a copula $C\big(F_1(x_1), \ldots, F_N(x_N)\big)$. Comparing Eq. (4) with Eq. (6), we can see that a copula function encoding correlations between random variables can be used to correct the mean-field approximation for arbitrarily complex distribution.

3.2   Multi-agent Imitation Learning with Copulas

A central question in multi-agent imitation learning is how to model the dependency structure among agent decisions properly. As discussed above, the framework of copulas provides a mechanism to decouple the marginal policies (individual behavioral patterns) from the dependency left in the joint policy after removing the information in marginals. In this work, we advocate copula-based policy for multi-agent learning because copulas offer unique and desirable properties in multi-agent scenarios. For example, suppose we want to model the interactions among players in a soccer game. By using copulas, we will obtain marginal policies for each individual player as well as dependencies among different roles (e.g., forwards and midfielders). Such a multi-agent learning framework has the following advantages:

- **Interpretable.** The learned copula density can be easily visualized to intuitively analyze the correlation among agent actions.
- **Scalable.** When the marginal policy of agents changes but the dependency among different agents remain the same (e.g., in a soccer game, one player is replaced by his/her substitute, but the dependence among different roles are basically the same regardless of players), we can obtain a new joint policy efficiently by switching in the new agent's marginal while keeping the copula and other marginals unchanged.
- **Succinct.** The copula-based factorization of the joint policy avoids the redundancy in previous opponent modeling approaches [19,33] by separately learning marginals and a copula.

Learning. We first discuss how to learn a copula-based policy from a set of expert demonstrations. Under the framework of Markov games and copulas, we factorize the parametric joint policy as:

$$\pi(a_1,\ldots,a_N|s;\boldsymbol{\theta}) = \prod_{i=1}^{N} \pi_i(a_i|s;\theta_i) \cdot c\big(F_1(a_1|s;\theta_1),\ldots,F_N(a_N|s;\theta_N)|s;\theta_c\big), \quad (7)$$

where $\pi_i(a_i|s;\theta_i)$ is the marginal policy of agent i with parameters θ_i and F_i is the corresponding cumulative distribution function; the function c (parameterized by θ_c) is the density of the copula on the transformed actions $u_i = F_i(a_i|s;\theta_i)$ obtained by processing original actions with probability integral transform.

The training algorithm of our proposed method is presented as Algorithm 1. Given a set of expert demonstrations $\mathcal{D}$, our goal is to learn marginal actions of agents and their copula function. Our approach consists of two steps.[1] We first learn marginal action distributions of each agent given their current state (lines 1–6). This is achieved by training $MLP_{marginal}$ that takes as input a state s and output the parameters of marginal action distributions of N agents

---

[1] An alternate approach is to combine the two steps together and use end-to-end training, but this does not perform well in practice because the copula term is unlikely to converge before marginals are well-trained.

Algorithm 1: Training procedure

Input: The number of trajectories M, the length of trajectory T, the number
of agents N, demonstrations $\mathcal{D} = \{\tau^i\}_{i=1}^{M}$, where each trajectory
$\tau^i = \{(s^i[t], a^i[t])\}_{t=1}^{T}$

Output: Marginal action distribution MLP $MLP_{marginal}$, state-dependent
copula MLP MLP_{copula} or state-independent copula density $c(\cdot)$

   // **Learning marginals**

1 **while** $MLP_{marginal}$ *not converge* **do**

2   **for** *each state-action pair* $(s, (a_1, \cdots, a_N))$ **do**

3   Calculate the conditional marginal action distributions for all agents:
        $\{f_j(\cdot|s)\}_{j=1}^{N} \leftarrow MLP_{marginal}(s)$;

4   **for** *agent* $j = 1, \cdots, N$ **do**

5   Calculate the likelihood of the observed action a_j: $f_j(a_j|s)$;

6   Maximize $f_j(a_j|s)$ by optimizing $MLP_{marginal}$ using SGD;

   // **Learning copula**

7 **while** MLP_{copula} *or* $c(\cdot)$ *not converge* **do**

8   **for** *each state-action pair* $(s, (a_1, \cdots, a_N))$ **do**

9         $\{f_j(\cdot|s)\}_{j=1}^{N} \leftarrow MLP_{marginal}(s)$;

10   **for** *agent* $j = 1, \cdots, N$ **do**

11          $F_j(\cdot|s) \leftarrow$ the CDF of $f_j(\cdot|s)$;

12   Transform a_j to uniformly distributed value: $u_j \leftarrow F_j(a_j|s)$;

13   Obtain $u = (u_1, \cdots, u_N) \in [0,1]^N$;

14   **if** *copula is set as state-dependent* **then**

15   Calculate the copula density $c(\cdot|s) \leftarrow MLP_{copula}(s)$;

16   Calculate the likelihood of u: $c(u|s)$;

17   Optimize MLP_{copula} by maximizing $\log c(u|s)$ using SGD;

18   **else**

19   Calculate the likelihood of u: $c(u)$;

20   Optimize parameters of $c(\cdot)$ using maximum likelihood or
         non-parametric methods;

21 **return** $MLP_{marginal}$, MLP_{copula} *or* $c(\cdot)$

---

given the input state (line 3).[2] In our implementation, we use mixture of Gaussians to realize each marginal policy $\pi_i(a_i|s; \theta_i)$ such that we can model complex multi-modal marginals while having a tractable form of the marginal cumulative distribution functions. Therefore, the output of $MLP_{marginal}$ consists of the means, covariance, and weights of components for the N agents' Gaussian mixtures. We then calculate the likelihood of each observed action a_j based on agent j's marginal action distribution (line 5), and maximize the likelihood by optimizing the parameters of $MLP_{marginal}$ (line 6).

---

[2] Here we assume that each agent is aware of the whole system state. But our model can be easily generalized to the case where agents are only aware of partial system state by feeding the corresponding state to their MLPs.

Algorithm 2: Inference procedure

Input: Marginal action distribution MLP $MLP_{marginal}$, state-dependent
copula MLP MLP_{copula} or state-independent copula density $c(\cdot)$,
current state s

Output: Predicted action $\hat{a}$

// Sample from copula

1 **if** *copula is set as state-dependent* **then**

2 $\quad$ Calculate (parameters of) state-dependent copula density
   $c(\cdot|s) \leftarrow MLP_{copula}(s)$;

3 $\quad$ Sample a copula value $u = (u_1, \cdots, u_N)$ from $c(\cdot|s)$;

4 **else**

5 $\quad$ Sample a copula value $u = (u_1, \cdots, u_N)$ from $c(\cdot)$;

// Transform copula value to action space

6 Calculate (parameters of) the conditional marginal action distributions for all
  agents: $\{f_j(\cdot|s)\}_{j=1}^N \leftarrow MLP_{marginal}(s)$;

7 **for** *agent* $j = 1, \cdots, N$ **do**

8 $\quad$ $F_j(\cdot|s) \leftarrow$ CDF of $f_j(\cdot|s)$;

9 $\quad$ $\hat{a}_j \leftarrow F_j^{-1}(u_j|s)$;

10 $\hat{a} \leftarrow (\hat{a}_1, \cdots, \hat{a}_j)$;

11 **return** $\hat{a}$

After learning marginals, we fix the parameters of marginal MLPs and start learning the copula (lines 7–20). We first process the original demonstrations using probability integral transform and obtain a set of new demonstrations with uniform marginals (lines 8–13). Then we learn the density of copula (lines 14–20). Notice that the copula can be implemented as either *state-dependent* (lines 14–17) or *state-independent* (lines 18–20): For state-dependent copula, we use MLP_{copula} to take as input the current state s and outputs the parameters of copula density $c(\cdot|s)$ (line 15). Then we calculate the likelihood of copula value u (line 16) and maximize the likelihood by updating MLP_{copula} (line 17). For state-independent copula, we directly calculate the likelihood of copula value u under $c(\cdot)$ (line 19) and learn parameters of $c(\cdot)$ by maximizing the likelihood of copula value (line 20).

The copula density ($c(\cdot)$ or $c(\cdot|s)$) can be implemented using parametric methods such as Gaussian or mixture of Gaussians. It is worth noticing that if copula is state-independent, it can also be implemented using non-parametric methods such as kernel density estimation [8,23]. In this way, we no longer learn parameters of copula by maximizing likelihood as in lines 19–20, but simply store all copula values u for density estimation and sampling in inference stage. We will visualize the learned copula in experiments.

Inference. In inference stage, the goal is to predict the joint actions of all agents given their current state s. The inference algorithm is presented as Algorithm 2, where we first sample a copula value $u = (u_1, \cdots, u_N)$ from the learned copula, either state-dependent or state-independent (lines 1–5), then apply inverse prob-

---

Algorithm 3: Generation procedure

Input: Inference module (Algorithm 2), initial state $s[0]$, required length L,
environment $\mathcal{E}$
Output: Generated trajectory $\hat{\tau}$
1 **for** $l = 0, \cdots, L$ **do**
2   Feed state $s[l]$ to the inference module and get the predicted action $\hat{a}[l]$;
3   Execute $\hat{a}[l]$ in environment $\mathcal{E}$ and get a new state $s[l+1]$;
4 $\hat{\tau} = \{(s[l], \hat{a}[l])\}_{l=0}^{L}$;
5 **return** $\hat{\tau}$;

---

ability transform to transform them to the original action space: $\hat{a}_j = F_j^{-1}(u_j|s)$ (lines 7–10). Note that an analytical form of the inverse cumulative distribution function may not always be available. In our implementation, we use binary search to approximately solve this problem since F_j is a strictly increasing function, which is shown to be highly efficient in practice. In addition, we can also sample multiple i.i.d. copula values from $c(\cdot|s)$ or $c(\cdot)$ (line 3 or 5), transform them into the original action space, and take their average as the predicted action. This strategy is shown to be able to improve the accuracy of action prediction (in terms of MSE loss), but requires more running time as a trade-off.

Generation. The generation algorithm is presented as Algorithm 3. To generate new trajectories, we repeatedly predict agent actions given the current state (line 2), then execute the generated action and obtain an updated state from the environment (line 3).

Complexity Analysis. The computational complexity of the training and the inference algorithms is analyzed as follows. The complexity of each round in Algorithm 1 is $O(MTN)$, where M is the number of trajectories in the training set, T is the length of each trajectory, and N is the number of agents. The complexity of Algorithm 2 is $O(N)$. The training and the inference algorithms scales linearly with the size of input dataset.

4 Related Work

The key problem in multi-agent imitation learning is how to model the dependence structure among multiple agents. [16] learn a latent coordination model for players in a cooperative game, where different players occupy different roles. However, there are many other multi-agent scenarios where agents do not cooperate for a same goal or they do not have specific roles (e.g., self-driving). [4] adopt parameter sharing trick to extend generative adversarial imitation learning to handle multi-agent problems, but it does not model the interaction of agents. Interaction Network [3] learns a physical simulation of objects with binary relations, and CommNet [31] learns dynamic communication among agents. But they fail to characterize the dependence among agent actions explicitly.

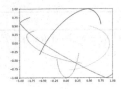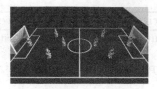

Fig. 2. Experimental environments: PhySim, Driving, and RoboCup.

Researchers also propose to infer multi-agent relationship using graph techniques or attention mechanism. For example, [15] propose to use graph neural networks (GNN) to infer the type of relationship among agents. [12] introduces attention mechanism into multi-agent predictive modeling. [17] combine generative models and attention mechanism to capture behavior generating process of multi-agent systems. These works address the problem of reasoning relationship among agents rather than capturing their dependence when agents are making decisions.

Another line of related work is deep generative models in multi-agent systems. For example, [36] propose a hierarchical framework with programmatically produced weak labels to generate realistic multi-agent trajectories of basketball game. [34] use GNN and variational recurrent neural networks (VRNN) to design a permutation equivariant multi-agent trajectory generation model for sports games. [13] combine conditional variational autoencoder (CVAE) and long-short term memory networks (LSTM) to generate behavior of basketball players. Most of the existing works focus on agent behavior forecasting but provide limited information regarding the dependence among agent behaviors.

5 Experiments

5.1 Experimental Setup

Datasets. We evaluate our method in three settings. **PhySim** is a synthetic physical environment where 5 particles are connected by springs. **Driving** is a synthetic driving environment where one vehicle follows another along a single lane. **RoboCup** is collected from an international scientific robot competition where two robot teams (including 22 robots) compete against each other. Experimental environments are shown in Fig. 2. The detailed dataset description is provided in Appendix A.

Baselines. We compare our method with the following baselines: **LR** is a logistic regression model that predicts actions of agents using all of their states. **SocialL-STM** [1] predicts agent trajectory using RNNs with a social pooling layer in the hidden state of nearby agents. **IN** [3] predicts agent states and their interactions using deep neural networks. **CommNet** [31] simulates the inter-agent communication by broadcasting the hidden states of all agents and then predicts their actions. **VAIN** [12] uses neural networks with attention mechanism for multi-agents modeling. **NRI** [15] designs a graph neural network based model to learn

Table 1. Root mean squared error (RMSE) between predicted actions and ground-truth actions for all methods.

Methods	PhySim	Driving	RoboCup
LR	0.064 ± 0.002	0.335 ± 0.007	0.478 ± 0.009
SocialLSTM	0.186 ± 0.032	0.283 ± 0.024	0.335 ± 0.051
IN	0.087 ± 0.013	0.247 ± 0.033	0.320 ± 0.024
CommNet	0.089 ± 0.007	0.258 ± 0.028	0.311 ± 0.042
VAIN	0.082 ± 0.010	0.242 ± 0.031	0.315 ± 0.028
NRI	0.055 ± 0.011	0.296 ± 0.018	0.401 ± 0.042
Copula	$\mathbf{0.037} \pm 0.005$	$\mathbf{0.158} \pm 0.019$	$\mathbf{0.221} \pm 0.024$

Table 2. Negative log-likelihood (NLL) of test trajectories evaluated by different types of copula. Uniform copula assumes no dependence among agent actions. KDE copula uses kernel density estimation to model the copula, which is state-independent. Gaussian mixtures copula uses Gaussian mixture model to characterize the copula, which is state-dependent.

Copula type	PhySim	Driving	RoboCup
Uniform	8.994 ± 0.001	-0.571 ± 0.024	3.243 ± 0.049
KDE	$\mathbf{1.256} \pm 0.006$	$\mathbf{-0.916} \pm 0.017$	$\mathbf{0.068} \pm 0.052$
Gaussian mixture	2.893 ± 0.012	-0.621 ± 0.028	3.124 ± 0.061

the interaction type among multiple agents. Since most of the baselines are used for predicting future states given historical state, we change the implementation of their objective functions and use them to predict the current action of agents given historical states. Each experiment is repeated 3 times, and we report the mean and standard deviation. Hyper-parameter settings of baselines as well as our method are presented in Appendix B.

5.2 Results

We compare our method with baselines in the task of action prediction. The results of root mean squared error (RMSE) between predicted actions and ground-truth actions are presented in Table 1. The number of Gaussian mixture components in our method is set to 2 for all datasets. The results demonstrate that all methods performs the best on PhySim dataset, since agents in PhySim follow simple physical rules and the relationships among them are linear thus easy to infer. However, the interactions of agents in Driving and RoboCup datasets are more complicated, which causes LR and NRI to underperform other baselines. The performance of IN, CommNet, and VAIN are similar, which is in accordance with the result reported in [12]. Our method is shown to outperform all baselines significantly on all three datasets, which demonstrates that explicitly characterizing dependence of agent actions could greatly improve the performance of multi-agent behavior modeling.

Table 3. Negative log-likelihood (NLL) of new test trajectories in which the action distribution of one agent is changed. We evaluate the new test trajectories based on whether to use the old marginal action distributions or copula, which results in four combinations.

Combinations	PhySim	Driving	RoboCup
Old marginals + old copula	10.231 ± 0.562	15.184 ± 1.527	4.278 ± 0.452
Old marginals + new copula	8.775 ± 0.497	13.662 ± 0.945	4.121 ± 0.658
New marginals + old copula	1.301 ± 0.016	0.447 ± 0.085	0.114 ± 0.020
New marginals + new copula	1.259 ± 0.065	-0.953 ± 0.024	0.077 ± 0.044

To investigate the efficacy of copula, we implement three types of copula function: Uniform copula means we do not model dependence among agent actions. KDE copula uses kernel density estimation to model the copula function, which is state-independent. Gaussian mixtures copula uses Gaussian mixture model to characterize the copula function, of which the parameters are output by an MLP taking as input the current state. We train the three models on training trajectories, then calculate negative log-likelihood (NLL) of test trajectories using the three trained models. A lower NLL score means that the model assigns high likelihood to given trajectories, showing that it is better at characterizing the dataset. The NLL scores of the three models on the three datasets are reported in Table 2. The performance of KDE copula and Gaussian copula both surpasses uniform copula, which demonstrates that modeling dependence among agent actions is essential for improving model expressiveness. However, Gaussian copula performs worse than KDE copula, because Gaussian copula is state-dependent thus increases the risk of overfitting. Notice that the performance gap between KDE and Gaussian copula is less on PhySim, since PhySim dataset is much larger so the Gaussian copula can be trained more effectively.

5.3   Generalization of Copula

One benefit of copulas is that copula captures the pure dependence among agents, regardless of their own marginal action distributions. To demonstrate the generalization capabilities of copulas, we design the following experiment. We first train our model on the original dataset, and learn marginal action distributions and copula function (which is called *old marginals* and *old copula*). Then we substitute one of the agents with a new agent and use the simulator to generate a new set of trajectories. Specifically, this is achieved by doubling the action value of one agent (for example, this can be seen as substituting an existing particle with a lighter one in PhySim). We retrain our model on new trajectories and learn *new marginals* and *new copula*. We evaluate the likelihood of new trajectories based on whether to use the old marginals or old copula, which, accordingly, results in four combinations. The NLL scores of four combinations are presented in Table 3. It is clear, by comparing the first and the last row, that "new marginals + new copula" significantly outperform "old marginals + old copula", since new marginals and new copula are trained on new trajectories and therefore characterize the new joint distribution exactly. To see the influence of marginals and copula more clearly, we further compare the results in row 2 and 3, where we use new copula or new marginals separately. It is clear that the model performance does not drop significantly if we use the old copula and new marginals (by comparing row 3 and 4), which demonstrates that the copula function basically stays the same even if marginals are changed. The result supports our claim that the learned copula is generalizable in the case where marginal action distributions of agents change but the internal inter-agent relationship stays the same.

5.4 Copula Visualization

Another benefit of copulas is that it is able to intuitively demonstrate the correlation among agent actions. We choose the RoboCup dataset to visualize the learned copula. As shown in Fig. 3a, we first randomly select a game (the 6th game) between cyrus2017 and helios2017 and draw trajectories of 10 players in the left team (L2 ∼ L11, except the goalkeeper). It is clear that the 10 players fulfill specific roles: L2 ∼ L4 are defenders, L5 ∼ L8 are midfielders, and L9 ∼ L11 are forwards. Then we plot the copula density between the x-axis (the hor-

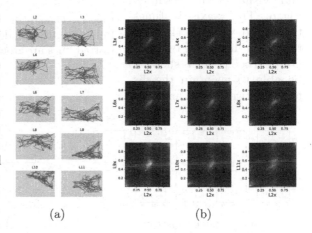

(a)   (b)

Fig. 3. (a) Trajectories of 10 players (except the goal-keeper) of the left team in one RoboCup game; (b) Copula density between x-axis of L2 and x-axis of another player (L3 ∼ L11).

izontal direction) of L2 and the x-axis of L3 ∼ L11, respectively, as shown in Fig. 3b. These figures illustrate linear correlation between their moving direction along x-axis, that is, when L2 moves forward other players are also likely to move forward. However, the correlation strength differs with respect to different players: L2 exhibits high correlation with L3 and L4, but low correlation with L9 ∼ L11. This is because L2 ∼ L4 are all defenders so they collaborate more closely with each other, but L9 ∼ L11 are forwards thus far from L2 in the field.

5.5 Trajectory Generation

The learned copula can also be used to generate new trajectories. We visualize the result of trajectory generation on RobuCup dataset. As shown in Fig. 4, the

dotted lines denote the ground-truth trajectories of the 10 player in an attack
from midfield to the penalty area. The trajectories generated by our copula model
(Fig. 4b) are quite similar to the demonstration as they exhibit high consistency.
It is clear that midfielders and forwards (No. 5 ~ No. 11) are basically moving
to the same direction, and they all make a left turn on their way to penalty
area. However, the generated trajectories by independent modeling show little
correlation since the players are all making independent decisions.

We also present the result
of trajectory generation on
Driving dataset. We ran-
domly select 10 original trajectories and 10 trajectories
generated by our method,
and visualize the result in
Fig. 5. The x-axis is timestamp and y-axis is the location (coordinate) of two cars.
Our learned policy is shown to
be able to maintain the distance between two cars.

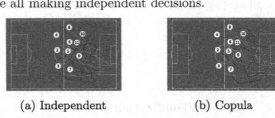

(a) Independent   (b) Copula

Fig. 4. Generated trajectories (solid lines) on
RoboCup using independent modeling or copula.
Dotted lines are ground-truth trajectories.

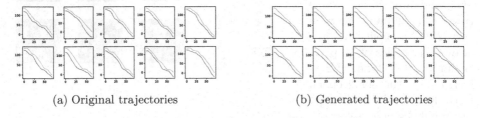

(a) Original trajectories   (b) Generated trajectories

Fig. 5. Original and generated trajectories on Driving dataset. The x-axis is timestamp
and y-axis is the location (1D coordinate) of two cars.

6   Conclusion and Future Work

In this paper, we propose a copula-based multi-agent imitation learning algorithm that is interpretable, efficient and scalable to model complex multi-agent
interactions. Sklar's theorem allows us to separately learn marginal policies that
capture the local behavioral patterns of each individual agent and a copula function that only and fully captures the dependence structure among the agents.
Compared to previous multi-agent imitation learning methods based on independent policies (mean-field factorization of the joint policy) or opponent modeling, our method is capable of modeling complex dependence among agents and
achieving coordination without any modeling redundancy.

We point out two directions of future work. First, the copula function is generalizable only if the dependence structure of agents (i.e., their role assignment) is unchanged. Therefore, it is interesting to study how to efficiently apply the learned copula to the scenario with evolving dependence structure. Another practical question is that whether our proposed method can be extended to the setting of decentralized execution, since the step of copula sampling (line 3 or 5 in Algorithm 2) is shared by all agents. A straightforward solution is to set a fixed sequence of random seeds for all agents in advance, so that the copula samples obtained by all agents are the same at each timestamp, but how to design a more robust and elegant mechanism is still a promising direction.

Acknowledgements. This research was supported by TRI, NSF (1651565, 1522054, 1733686), ONR (N00014-19-1-2145), AFOSR (FA9550-19-1-0024), ARO (W911NF21 10125), and FLI.

Appendix

A   Dataset Details

PhySim is collected from a physical simulation environment where 5 particles move in a unit 2D box. The state is locations of all particles and the action is their acceleration (there is no need to include their velocities in state because accelerations are completely determined by particle locations). We add Gaussian noise to the observed values of actions. Particles may be pairwise connected by springs, which can be represented as a binary adjacency matrix $\mathbf{A} \in \{0, 1\}^{N \times N}$. The elasticity between two particles scales linearly with their distance. At each timestamp, we randomly sample an adjacency matrix from $\{\mathbf{A}_1, \mathbf{A}_2\}$ to connect all particles, where $\mathbf{A}_1$ and $\mathbf{A}_2$ are set as complimentary (i.e. $\mathbf{A}_1 + \mathbf{A}_2 + \mathbf{I} = 1$) to ensure that they are as different as possible. Therefore, the marginal action distribution of each particle given a system state is Gaussian mixtures with two components. Here the coordination signal for particles can be seen as the hidden variable determining which set of springs ($\mathbf{A}_1$ or $\mathbf{A}_2$) is used at current time. We generate $10,000$ training trajectories, $2,000$ validation trajectories, and $2,000$ test trajectories for experiments, where the length of each trajectory is 500.

Driving is generated by CARLA[3] [9], an open-source simulator for autonomous driving research that provides realistic urban environments for training and validation of autonomous driving systems. To generate the driving data, we design a car following scenario, where a leader car and a follower car drive in the same lane. We make the leader car alternatively accelerate to a speed upper bound and slow down to stopping. The leader car does not care about the follower and drives following its own policy. The follower car tries to follow closely the leader car while keeping a safe distance. Here the state is the locations and velocities of the two cars, and the action is their accelerations. We generate $1,009$ trajectories in total, and split the whole data into training,

---

[3] https://carla.org/.

validation, and test set with ratio of 6 : 2 : 2. The average length of trajectories is 85.5 in Driving dataset.

RoboCup [20] is collected from an international scientific robot football competition in which teams of multiple robots compete against each other. The original dataset contains all pairings of 10 teams with 25 repetitions of each game (1, 125 games in total). The state of a game (locations and velocities of 22 robots) is recorded every 100 ms, resulting in a trajectory of length 6, 000 for each game (10 min). We select the 25 games between two teams, cyrus2017 and helios2017, as the data used in this paper. The state is locations of 10 robots (except the goalkeeper) in the left team, and the action is their velocities. The dataset is split into training, validation, and test set with ratio of 6 : 2 : 2.

B   Implementation Details

For our proposed method, to learn the marginal action distribution of each agent (i.e. Gaussian mixtures), we use an MLP with one hidden layer to take as input a state and output the centers of their Gaussian mixtures. To prevent overfitting, the variance of these Gaussian mixtures is parameterized by a free variable for each particle, and the weights of mixtures are set as uniform. Each dimension of states and actions in the original datasets are normalized to range $[-1, 1]$. For PhySim, the number of particles are set to 5. Learning rate is set to 0.01, and the weight of L2 regularizer is set to 10^{-5}. For Driving, learning rate is 0.005 and L2 regularizer weight is 10^{-5}. For RoboCup, learning rate is 0.001 and L2 regularizer weight is 10^{-6}.

For LR, we use the default implementation in Python sklearn package. For SocialLSTM [1], the dimension of input is set as the dimension of states in each dataset. The spatial pooling size is 32, and we use an 8×8 sum pooling window size without overlaps. The hidden state dimension in LSTM is 128. The learning rate is 0.001. For IN [3], all MLPs are with one hidden layer of 32 units. The learning rate is 0.005. For CommNet [31], all MLPs are with one hidden layer of 32 units. The dimension of hidden states is set to 64, and the number of communication round is set to 2. The learning rate is 0.001. For VAIN [12], the encoder and decoder functions are implemented as a fully connected neural network with one hidden layer of 32 units. The dimension of hidden states is 64, and the dimension of attention vectors is 10. The learning rate is 0.0005. For NRI [15], we use an MLP encoder and an MLP decoder, with one hidden layer of 32 units. The learning rate is 0.001.

References

1. Alahi, A., Goel, K., Ramanathan, V., Robicquet, A., Fei-Fei, L., Savarese, S.: Social lstm: Human trajectory prediction in crowded spaces. In: Proceedings of the IEEE Conference on Computer Vision and Pattern Recognition, pp. 961–971 (2016)
2. Albrecht, S.V., Stone, P.: Autonomous agents modelling other agents: a comprehensive survey and open problems. Artif. Intell. **258**, 66–95 (2018)

3. Battaglia, P., Pascanu, R., Lai, M., Rezende, D.J., et al.: Interaction networks for learning about objects, relations and physics. In: Advances in Neural Information Processing Systems, pp. 4502–4510 (2016)
4. Bhattacharyya, R.P., Phillips, D.J., Wulfe, B., Morton, J., Kuefler, A., Kochender-fer, M.J.: Multi-agent imitation learning for driving simulation. In: 2018 IEEE/RSJ International Conference on Intelligent Robots and Systems (IROS), pp. 1534–1539. IEEE (2018)
5. Bouyé, E., Durrleman, V., Nikeghbali, A., Riboulet, G., Roncalli, T.: Copulas for finance-a reading guide and some applications. SSRN 1032533 (2000)
6. Brown, N., Sandholm, T.: Superhuman AI for multiplayer poker. Science **365**(6456), 885–890 (2019)
7. Chu, T., Wang, J., Codecà, L., Li, Z.: Multi-agent deep reinforcement learning for large-scale traffic signal control. IEEE Trans. Intell. Transp. Syst. **21**(3), 1086–1095 (2019)
8. Davis, R.A., Lii, K.-S., Politis, D.N.: Remarks on some nonparametric estimates of a density function. In: Selected Works of Murray Rosenblatt. SWPS, pp. 95–100. Springer, New York (2011). https://doi.org/10.1007/978-1-4419-8339-8_13
9. Dosovitskiy, A., Ros, G., Codevilla, F., Lopez, A., Koltun, V.: CARLA: an open urban driving simulator. In: Proceedings of the 1st Annual Conference on Robot Learning, pp. 1–16 (2017)
10. Fu, J., Luo, K., Levine, S.: Learning robust rewards with adversarial inverse rein-forcement learning. arXiv preprint arXiv:1710.11248 (2017)
11. Ho, J., Ermon, S.: Generative adversarial imitation learning. In: Advances in Neural Information Processing Systems, pp. 4565–4573 (2016)
12. Hoshen, Y.: Vain: attentional multi-agent predictive modeling. In: Advances in Neural Information Processing Systems, pp. 2701–2711 (2017)
13. Ivanovic, B., Schmerling, E., Leung, K., Pavone, M.: Generative modeling of mul-timodal multi-human behavior. In: 2018 IEEE/RSJ International Conference on Intelligent Robots and Systems (IROS), pp. 3088–3095. IEEE (2018)
14. Joe, H.: Dependence modeling with copulas. Chapman and Hall/CRC (2014)
15. Kipf, T.N., Fetaya, E., Wang, K.-C., Welling, M., Zemel, R.S.: Neural relational inference for interacting systems. In: International Conference on Machine Learning (2018)
16. Le, H.M., Yue, Y., Carr, P., Lucey, P.: Coordinated multi-agent imitation learning. In: Proceedings of the 34th International Conference on Machine Learning, pp. 1995–2003 (2017)
17. Li, M.G., Jiang, B., Zhu, H., Che, Z., Liu, Y.: Generative attention networks for multi-agent behavioral modeling. In: AAAI, pp. 7195–7202 (2020)
18. Littman, M.L.: Markov games as a framework for multi-agent reinforcement learn-ing. In: Machine learning proceedings 1994, pp. 157–163. Elsevier (1994)
19. Liu, M.: Multi-agent interactions modeling with correlated policies. arXiv preprint arXiv:2001.03415 (2020)
20. Michael, O., Obst, O., Schmidsberger, F., Stolzenburg, F.: Robocupsimdata: a robocup soccer research dataset. arXiv preprint arXiv:1711.01703 (2017)
21. Nelsen, R.B.: An Introduction to Copulas. Springer Science & Business Media (2007)
22. Ng, A.Y., Russell, S.J., et al.: Algorithms for inverse reinforcement learning. In: ICML, vol. 1, p. 2 (2000)
23. Parzen, E.: On estimation of a probability density function and mode. Ann. Math. Stat. **33**(3), 1065–1076 (1962)

24. Pomerleau, D.A.: Efficient training of artificial neural networks for autonomous navigation. Neural Comput. **3**(1), 88–97 (1991)
25. Puterman, M.L.: Markov Decision Processes: Discrete Stochastic Dynamic Programming. Wiley, New York (2014)
26. Russell, S.: Learning agents for uncertain environments. In: Proceedings of the Eleventh Annual Conference on Computational Learning Theory, pp. 101–103 (1998)
27. Silver, D., et al.: Mastering the game of go without human knowledge. Nature **550**(7676), 354–359 (2017)
28. Sklar, A.: Fonctions de Répartition à n dimensions et leurs marges. Publications de L'Institut de Statistique de L'Université de Paris **8**, 229–231 (1959)
29. Sklar, M.: Fonctions de repartition an dimensions et leurs marges. Publ. inst. statist. univ. Paris **8**, 229–231 (1959)
30. Song, J., Ren, H., Sadigh, D., Ermon, S.: Multi-agent generative adversarial imitation learning. In: Advances in Neural Information Processing Systems, pp. 7461–7472 (2018)
31. Sukhbaatar, S., Fergus, R., et al.: Learning multiagent communication with backpropagation. In: Advances in Neural Information Processing Systems, pp. 2244–2252 (2016)
32. Syed, U., Bowling, M., Schapire, R.E.: Apprenticeship learning using linear programming. In: Proceedings of the 25th International Conference on Machine Learning, pp. 1032–1039 (2008)
33. Tian, Z., Wen, Y., Gong, Z., Punakkath, F., Zou, S., Wang, J.: A regularized opponent model with maximum entropy objective. arXiv preprint arXiv:1905.08087 (2019)
34. Yeh, R.A., Schwing, A.G., Huang, J., Murphy, K.: Diverse generation for multi-agent sports games. In: Proceedings of the IEEE Conference on Computer Vision and Pattern Recognition, pp. 4610–4619 (2019)
35. Yu, L., Song, J., Ermon, S.: Multi-agent adversarial inverse reinforcement learning. In: International Conference on Machine Learning (2019)
36. Zhan, E., Zheng, S., Yue, Y., Sha, L., Lucey, P.: Generating multi-agent trajectories using programmatic weak supervision. arXiv preprint arXiv:1803.07612 (2018)

CMIX: Deep Multi-agent Reinforcement Learning with Peak and Average Constraints

Chenyi Liu[1,4,5]([⊠]), Nan Geng[1,4,5], Vaneet Aggarwal[2], Tian Lan[3],
Yuan Yang[1,4,5], and Mingwei Xu[1,4,5]

[1] Department of Computer Science and Technology,
Tsinghua University, Beijing, China
{liucheny19,gn16,}@mails.tsinghua.edu.cn,
yangyuan_thu@mail.tsinghua.edu.cn,
xumw@tsinghua.edu.cn
[2] School of Industrial Engineering and School of Electrical and Computer
Engineering, Purdue University, West Lafayette, USA
vaneet@purdue.edu
[3] School of Engineering and Applied Science, George Washington University,
Washington, D.C., USA
tlan@gwu.edu
[4] Beijing National Research Center for Information Science and Technology,
Beijing, China
[5] Peng Cheng Laboratory (PCL), Shenzhen, China

Abstract. In many real-world tasks, a team of learning agents must ensure that their optimized policies collectively satisfy required peak and average constraints, while acting in a decentralized manner. In this paper, we consider the problem of multi-agent reinforcement learning for a constrained, partially observable Markov decision process – where the agents need to maximize a global reward function subject to both peak and average constraints. We propose a novel algorithm, CMIX, to enable centralized training and decentralized execution (CTDE) under those constraints. In particular, CMIX amends the reward function to take peak constraint violations into account and then transforms the resulting problem under average constraints to a max-min optimization problem. We leverage the value function factorization method to develop a CTDE algorithm for solving the max-min optimization problem, and two gap loss functions are proposed to eliminate the bias of learned solutions. We evaluate our CMIX algorithm on a blocker game with travel cost and a large-scale vehicular network routing problem. The results show that CMIX outperforms existing algorithms including IQL, VDN, and QMIX, in that it optimizes the global reward objective while satisfying both peak and average constraints. To the best of our knowledge, this is the first proposal of a CTDE learning algorithm subject to both peak and average constraints.

Keywords: Multi-agent reinforcement learning · Peak constraint · Average constraint

© Springer Nature Switzerland AG 2021
N. Oliver et al. (Eds.): ECML PKDD 2021, LNAI 12975, pp. 157–173, 2021.
https://doi.org/10.1007/978-3-030-86486-6_10

1 Introduction

Multi-agent reinforcement learning (MARL) has shown great promise in many cooperative tasks such as intelligent transportation system [2,16], network optimization [28], and robot swarms [13]. Existing MARL algorithms – e.g., IQL [27], VDN [26], and QMIX [24] – often leverage centralized training with decentralized execution (CTDE). However, in many real-world decision making problems, a team of distributed learning agents must also ensure that their optimized policies collectively satisfy required instantaneous peak constraints [3] and long-term average constraints [10]. For instance, optimizing a vehicular network must meet the requirements of both peak network performance (e.g., the maximum latency constraint [25]) and average network performance (the average transmission rate [15] or average latency [1] constraint). We note that since the exact action space satisfying these constraints cannot be determined prior to training, new CTDE algorithms are needed to simultaneously address both peak and average constraints.

In this paper, we propose a new MARL algorithm with CTDE, called CMIX, for maximizing the discounted cumulative reward subject to both peak and average constraints. We note that while the problem of constrained reinforcement learning (RL) has been studied under either peak constraints [3,9,11,12] or average constraints [4,6,10,22], most of the existing algorithms tackle only one type of the two constraints – but not both – and focus on single-agent tasks with centralized execution. To the best of our knowledge, there are no MARL learning algorithms that can address both peak and average constraints especially in the realm of CTDE algorithms for multiple agents. In contrast, CMIX is able to optimize distributed agents' policies under both peak and average constraints, while leveraging value function factorization to allow distributed executions.

More precisely, CMIX casts a Markov decision process (MDP) under both peak and average constraints into an equivalent max-min optimization problem. First, by setting a lower bound on the MDP's objective value, we can replace the objective with a new average constraint. Second, we introduce a penalty into the reward functions of the average constraints so that the peak constraints can be absorbed into the set of average constraints. These allow us to obtain a multi-objective constrained problem with only average constraints, whose solution can be solved through an equivalent max-min optimization problem. Finally, a search algorithm is designed to find an appropriate lower bound, such that an optimal solution of the original constrained problem can be obtained by solving the max-min optimization problem.

We further leverage value function factorization and propose a CTDE algorithm to solve the max-min optimization problem, thus providing a solution to the MDP under both peak and average constraints. To this end, individual agent networks are used to estimate the per-agent action-values with respect to each average constraint and conditioned on only local observations, while a neural mixing module consisting of multiple mixing networks combines the outputs of the agent networks to produce an estimate of the joint action-values with respect to the average constraints. Two structures of the mixing module are developed, i.e., a module with mixing networks having independent parameters (CMIX-M) and a module with all mixing networks sharing the same parameters (CMIX-S). The neural parameters can be updated through an end-to-end method similar to QMIX [24]. However, directly applying the TD-error loss function is likely resulting in a biased solution due to the fact that CMIX solves the original max-min optimization problem approximately in a CTDE paradigm. To address the issue, we propose two gap loss functions for CMIX-M and CMIX-S, respectively, to minimize the gap between the original max-min optimization problem and the approximated one solved by CMIX. The neural parameters are updated by minimizing a linear combination of TD-error loss and gap loss. To the best of our knowledge, this is the first MARL algorithm that enables CTDE under both peak and average constraints.

We conduct extensive evaluations on a blocker game with travel cost and on a large-scale vehicular network routing problem. Evaluation results compares CMIX with state-of-the-art CTDE learning algorithms – including IQL [27], VDN [26], and QMIX [24] – as well as algorithms for constrained MDP such as C-IQL, which is implemented by extending IQL in a similar way to CMIX. The results show that CMIX outperforms state-of-the-art CTDE learning algorithms in the sense that it can optimize the global objective while satisfying both peak and average constraints. Further evaluations validate that gap loss improves the objective value of CMIX while enabling CMIX to comply with both peak and average constraints.

2   Background

A cooperative multi-agent sequential decision-making task in a stochastic, partially observable environment can be modeled as a *decentralized partially observable Markov decision process (Dec-POMDP)* [21], denoted by a tuple $G = \langle \mathcal{S}, \mathcal{N}, \{\mathcal{A}_i\}_{i \in \mathcal{N}}, P, r, \{\mathcal{Z}_i\}_{i \in \mathcal{N}}, \gamma \rangle$. $s \in \mathcal{S}$ denotes the global state of the environment. At each time step t, each agent $i \in \mathcal{N} \equiv \{1, \ldots, N\}$ gets a local observation $z_i \in \mathcal{Z}_i$ and chooses a local action $a_i \in \mathcal{A}_i$, forming a joint action $\boldsymbol{a} \in \mathcal{A} \equiv \Pi_{i=1}^{N} \mathcal{A}_i$. Then the environment evolves transforms from the current state to a new state according to the state transition function $P(s'|s, \boldsymbol{a}) : \mathcal{S} \times \mathcal{A} \times \mathcal{S} \to [0, 1]$ and returns a global reward $r(s, \boldsymbol{a})$ to the agents. Given a joint policy $\boldsymbol{\pi} := (\pi_i)_{i \in \mathcal{N}}$, the joint action-value function at time step t is defined as $Q^{\pi}(s^t, \boldsymbol{a}^t) := \mathbb{E}_{\pi}[R^t|s^t, \boldsymbol{a}^t]$, where $R^t = \sum_{\tau=0}^{\infty} \gamma^{\tau} r^{t+\tau}$ is the discounted cumulative reward. The goal is to find an optimal policy $\boldsymbol{\pi}^*$ which results in the optimal value function $Q^* = \max_{\pi} Q^{\pi}(s^t, \boldsymbol{a}^t)$.

2.1   QMIX

CTDE paradigm [23] is promising in solving the optimization problems of Dec-POMDP. During training, the learning algorithm of CTDE trains agents centrally and gets access to the global state s. During execution, each agent i can only make decisions according to its local observation z_i. There have recently been many MARL algorithms proposed in the CTDE paradigm, among which value-decomposition methods attract much attention recently.

QMIX [23,24] is one of the representative MARL algorithms belonging to value-decomposition methods, which factorizes joint action-value function Q_{tot} into a combination of local action-value functions. QMIX combines the per-agent action-value function via a state-dependent, differentiable monotonic function: $Q_{tot}(s, \boldsymbol{a}) = f(Q_1(z_i, a_1), \dots, Q_N(z_N, a_N); s)$, where $\frac{\partial f}{\partial Q_i} \geq 0, \forall i \in \mathcal{N}$. Since the mixer $f(\cdot)$ is a monotonic function to per-agent action-value input, we can maximize the joint action value by maximizing per-agent action values locally.

QMIX is trained much like deep Q-network (DQN) [19]. A buffer replay mechanism and a TD-error loss function are taken for training agents. Particularly, for a mini-batch of B samples $(s_b, \boldsymbol{a}_b, r_b, s_b')$ ($b = 1, \dots, B$), QMIX learns parameters θ by minimizing $\mathcal{L}_{TD-error} = \frac{1}{B} \sum_{b=1}^{B} (Q_{tot}(s_b, \boldsymbol{a}_b; \theta) - y_b)^2$, where $y_b = r_b + \gamma \max_{\boldsymbol{a}'} Q_{tot}(s_b', \boldsymbol{a}'; \theta^-)$ is the target value of the b-th sample and θ^- denotes the parameters of a target network which are periodically copied from θ. The monotonic mixing function f is parameterized as a feed-forward network, whose non-negative weights are generated by hypernetworks [14] with the global state as input.

2.2 Constrained Reinforcement Learning

Constrained reinforcement learning tries to find a policy π^* to maximize the global discounted cumulative reward subject to some constraints such as *peak* constraints [3,9,11,12] and *average* constraints [4,6,10,22]. Formally, the objective function of these constrained problems is $\max_\pi \mathbb{E}_\pi [\sum_{t=0}^{\infty} \gamma^t r(s^t, \boldsymbol{a}^t)]$. Without loss of generality, we assume that there exist J peak constraints and K average constraints. Peak constraints limit instantaneous returns, which can be formulated as $c_j(s^t, \boldsymbol{a}^t) \geq 0, \quad \forall t, j = 1, \dots, J$. While average constraints focus on long-term limitations, which can be stated as $\mathbb{E}_\pi [\sum_{t=0}^{\infty} \gamma^t r_k(s^t, \boldsymbol{a}^t)] \geq 0, k = 1, \dots, K$. $r, \{c_j\}_{j=1}^J$, and $\{r_k\}_{k=1}^K$ are unknown functions, but their returned values can be acquired at each time step.

Existing researchers usually develop constrained RL algorithms for dealing with the two kinds of constraints separately. For the problems with only peak constraints, many approaches [3,9] choose to introduce a penalty into the original reward function, which has been validated to be effective.

There are also some researches focusing on the problems with only average constraints. [10] converts the optimization problem with average constraints to the multi-objective constrained problem by adding a lower bound δ to the objective function. The multi-objective constrained problem aims to find a feasible policy while satisfying the original average constraints as well as the new average constraint on the objective function. Such a constraint satisfaction problem can be solved through a max-min optimization problem which maximizes the minimum margin of average constraints.[1] Therefore, with an appropriate δ-

---

[1] The margin of an average constraint represents the left hand of the average constraint.

search algorithm, we can approximate the optimal solution that maximizes the discounted cumulative reward under the original average constraints.

However, most of the constrained RL algorithms are for single-agent settings, and the algorithms considering both peak and average constraints are missing especially in the paradigm of MARL.

3   Problem Formulation

In this paper, we consider the problem of the Dec-POMDP under both *peak* and *average* constraints. The problem can be formulated as

$$\max_{\boldsymbol{\pi}} \quad \mathbb{E}_{\boldsymbol{\pi}} \left[\sum_{t=0}^{\infty} \gamma^t r(s^t, \boldsymbol{a}^t) \right] \tag{1}$$

$$\text{s.t.} \quad c_j(s^t, \boldsymbol{a}^t) \geq 0, \quad \forall t, j = 1, \ldots, J \tag{2}$$

$$\mathbb{E}_{\boldsymbol{\pi}} \left[\sum_{t=0}^{\infty} \gamma^t r_k(s^t, \boldsymbol{a}^t) \right] \geq 0, \quad k = 1, \ldots, K \tag{3}$$

Next, we need to find an optimal joint policy $\boldsymbol{\pi}^*$ which optimizes the discounted cumulative reward in Eq.(1) and satisfies both peak constraints on instantaneous returns in Eq.(2) and average constraints on long-term returns in Eq.(3).

4   CMIX

We propose CMIX to solve the Dec-POMDP problem under both peak and average constraints. First, we convert the original MDP to a multi-objective constrained problem, whose solution can be obtained by solving an equivalent max-min optimization problem. Second, we leverage value function factorization to develop a CTDE algorithm for solving the max-min optimization problem approximately. We further analyze the gap between the original max-min optimization problem and the approximated one solved by CMIX and propose two gap loss functions to eliminate the bias of learned solutions.

4.1   Multi-objective Constrained Problem

The constrained Dec-POMDP problem can be converted into a multi-objective constrained problem by setting a lower bound δ for the objective function [10]. The bounded objective function becomes a new average constraint, which is equivalent to

$$\mathbb{E}_{\boldsymbol{\pi}} \left[\sum_{t=0}^{\infty} \gamma^t (r(s^t, \boldsymbol{a}^t) - \delta(1 - \gamma)) \right] \geq 0. \tag{4}$$

Then, we obtain a multi-objective constrained problem which aims to find a feasible policy while satisfying the peak constraints of Eq. (2) as well as the average constraints of Eq. (3) and Eq. (4).

Consider that peak and average constraints have very different forms. So directly dealing with the two kinds of constraints together is much challenging. To address the issue, we design that the peak constraints of Eq. (2) can be absorbed into the average constraints of Eq. (3) and Eq. (4) by importing a penalty to reward functions. In particular, we define a penalty function as $p(s^t, \boldsymbol{a}^t) = \sum_{j=1}^{J} \min_{\lambda_j \geq 0} \lambda_j c_j(s^t, \boldsymbol{a}^t)$. When computing rewards, a penalty $p(s^t, \boldsymbol{a}^t)$ will be added to the global reward r and the rewards for average constraints $\{r_k\}_{k=1}^{K}$. The penalty $p(s^t, \boldsymbol{a}^t)$ equals zero when all the peak constraints are satisfied; otherwise, a large minus value will be added to the computed reward values.

Now, we obtain a multi-objective constrained problem with only average constraints, i.e.,

Find π

$$\text{s.t.}\quad \mathbb{E}_{\pi}\left[\sum_{t=0}^{\infty} \gamma^t (r(s^t, \boldsymbol{a}^t) - \delta(1-\gamma) + p(s^t, \boldsymbol{a}^t))\right] \geq 0,$$

$$\mathbb{E}_{\pi}\left[\sum_{t=0}^{\infty} \gamma^t (r_k(s^t, \boldsymbol{a}^t) + p(s^t, \boldsymbol{a}^t))\right] \geq 0, k = 1, \ldots, K$$

Since a violation of any peak constraints will result in very small rewards, the above problem with only average constraints is equivalent to the multi-objective constrained problem with both peak and average constraints. Also note that the penalty $p(s^t, \boldsymbol{a}^t)$ can not be realized directly but there exists a large body of literature on reward engineering for the penalty design [8,9]. In practice, we use a simple approximator to $p(s^t, \boldsymbol{a}^t)$ as $\hat{p}(s^t, \boldsymbol{a}^t) = -C\sum_{j=1}^{J} \mathbf{1}_{c^j(s^t, \boldsymbol{a}^t)<0}$, where C is a positive constant bounding the variance of global reward function r and the reward functions for average constraints $\{r_k\}_{k=1}^{K}$, and $\mathbf{1}_{c^j(s^t, \boldsymbol{a}^t)<0}$ equals 1 when $c^j(s^t, \boldsymbol{a}^t) < 0$ otherwise 0. We can find that the approximator equals zero when all the peak constraints are satisfied and a large minus value otherwise, which is similar to the original penalty $p(s^t, \boldsymbol{a}^t)$.

The above multi-objective constrained problem can be solved through an equivalent max-min optimization problem, noted as a zero-sum Markov-Bandit game in [10]. Particularly, the max-min optimization problem focuses on finding a feasible policy to maximize the minimum margin of the average constraints. It is easy to find that if the original multi-objective constrained problem is feasible, the policy, that maximizes the minimum margin of the average constraints, should be a feasible solution to the original problem. We define action-value functions for representing the margins of the average constraints. Formally, let o indicates the index of average constraints. Then, we define

$$\text{if } o = 0: \ Q(s, \boldsymbol{a}, o) = \mathbb{E}_{\pi}\left[\sum_{t=0}^{\infty} \gamma^t (r(s^t, \boldsymbol{a}^t) - \delta(1-\gamma) + p(s^t, \boldsymbol{a}^t))\right], \text{ and}$$

$$\text{for } o = 1, \ldots, K: \ Q(s, \boldsymbol{a}, o) := Q(s, \boldsymbol{a}, k) = \mathbb{E}_{\pi}\left[\sum_{t=0}^{\infty} \gamma^t \left(r_k(s^t, \boldsymbol{a}^t) + p(s^t, \boldsymbol{a}^t)\right)\right].$$

The Bellman equation of the max-min optimization problem is given as follows

$$
\begin{aligned}
Q(s, \boldsymbol{a}, o) &= \quad r(s, \boldsymbol{a}, o) + \gamma \cdot \mathbb{E}_{\boldsymbol{\pi}(s')}\left[Q(s', \boldsymbol{\pi}(s'), o)\right], \\
\boldsymbol{\pi}(s) &= \quad \underset{\boldsymbol{\pi}(s)}{\arg\max}\,\underset{o \in O}{\min}\, \mathbb{E}_{\boldsymbol{\pi}(s)}\left[Q(s, \boldsymbol{\pi}(s), o)\right],
\end{aligned}
\tag{5}
$$

where $r(s, \boldsymbol{a}, o) := r(s^t, \boldsymbol{a}^t) - \delta(1 - \gamma) + p(s^t, \boldsymbol{a}^t)$ for $o = 0$ and $r(s, \boldsymbol{a}, o) := r(s, \boldsymbol{a}, k) = r_k(s^t, \boldsymbol{a}^t) + p(s^t, \boldsymbol{a}^t)$ for $o = k = 1, \cdots, K$. Note that $\boldsymbol{\pi}(s)$ is a distribution over the joint action space and can be obtained by solving a max-min optimization problem. By solving Eq. (5), we can maximize the minimum margin of the average constraints and find the feasible policy for the original multi-objective constrained problem.

This max-min optimization problem aims to provide a feasible solution to the multi-objective constrained problem for a given δ. It is easy to prove that if δ equals the optimal objective value of the original constrained Dec-POMDP problem, the obtained feasible solution is also an optimal solution to the original constrained Dec-POMDP problem. To find the optimal δ, we can set $\delta = \mathbb{E}\left[r(s, \boldsymbol{a})\right]/(1 - \gamma)$ in the max-min optimization problem learning process. In this way, the learning algorithm can learn the optimal δ and the corresponding policy $\boldsymbol{\pi}^*$ self-adaptively. In our proposed algorithm, we take an adaptive δ-search algorithm by setting $\delta_t = \varepsilon \cdot \delta_{t-1} + (1 - \varepsilon) \cdot r(s^t, \boldsymbol{a}^t)/(1 - \gamma)$ at each time step, where $\varepsilon \in [0, 1]$ is an adjustable parameter.

4.2 CMIX Architecture

To efficiently solve the constrained Dec-POMDP problem, we propose a neural architecture called CMIX by extending the original QMIX.

When applying QMIX in solving the Bellman equation in Eq.(5), we need to solve a max-min problem

$$
\boldsymbol{\pi}^*(s) = \underset{\boldsymbol{\pi}(s)}{\arg\max}\,\underset{o \in U}{\min}\, \mathbb{E}_{\boldsymbol{\pi}(s)}[Q_{tot}(s, \boldsymbol{\pi}(s), o)],
\tag{6}
$$

where $O = \{0, 1, \ldots, K\}$ is the set of constraints and $Q_{tot}(s, \boldsymbol{\pi}(s), o)$ is the joint action value calculated by the monotonic function f_o: $Q_{tot}(s, \boldsymbol{\pi}(s), o) = f_o(Q_1^o, Q_2^o, ..., Q_N^o)$. So, there are total of $1 + K$ mixing functions. For brevity, we omit the notations of each agent's local observations and actions, and let Q_i^o denote the i-th agent's action-value function with respect to the o-th average constraint.

Recall that each agent takes actions independently in the original QMIX. So naturally we need to make each agent solve a local max-min problem for taking an action. Formally, the local max-min problem of agent i can be stated as

$$
\hat{\pi}_i(z_i) = \underset{\pi_i(z_i)}{\arg\max}\,\underset{o \in O}{\min}\, \mathbb{E}_{\pi_i(z_i)}[Q_i^o].
\tag{7}
$$

It is easy to prove that by solving Eq.(7) we optimize a similar but not same problem to Eq.(6). We define function g as

$$
g(x_1, x_2, \ldots, x_n) := \underset{o \in O}{\min}\, f_o(x_1, x_2, \ldots, x_n).
$$

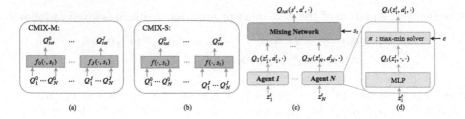

Fig. 1. (a) The mixing network structure of CMIX-M. (b) The mixing network structure of CMIX-S. (c) The overall CMIX architecture. (d) The agent network structure.

Since $\{f_o\}$ are monotonic functions, g is also monotonic. Then the actual max-min problem solved by Eq.(7) is

$$\hat{\pi}(s) = \arg\max_{\pi(s)} \mathbb{E}_{\pi(s)}[g(\min_{o \in O} Q_1^o, \ldots, \min_{o \in O} Q_N^o)]. \tag{8}$$

According to Eq. (8), we propose a CMIX architecture with two kinds of mixing module structures. In particular, we propose CMIX-M, the CMIX architecture with multiple mixing networks. That is to say, the mixing functions f_o ($\forall o \in O$) have independent parameters. Besides, we propose CMIX-S, the CMIX architecture with a single mixing network. The mixing functions in CMIX-S share parameters, i.e., $f_o = f$ ($\forall o \in O$). Figure 1 shows the overall CMIX architecture with the two kinds of mixing module structures. Agent network i adopts DQN [18] and stores Q_i^o ($\forall o \in O$). Each agent takes local state z_i^t as input and outputs the corresponding Q values $Q_i^o(s_i^t, a_i^t)$ ($\forall o \in O$) after taking a local action a_i^t. The mixing network module combines the outputs of the agents monotonically, producing the values of $Q_{tot}^o(\forall o \in O)$. To guarantee the monotonicity, hypernetworks with absolute activation functions in output layers are used to generate the weights of mixing networks f_o, which follows [24][2].

4.3   Gap Loss Function

Given the CMIX architecture, the neural parameters θ can be learned in an end-to-end fashion. For a transition $(s, \boldsymbol{a}, r, \{r_k\}_{k=1,\ldots,K}, s')$, the TD-error loss can be computed by

$$\mathcal{L}_{TD-error} = \sum_o (Q_{tot}(s, \boldsymbol{a}, o; \theta) - y^o)^2, \tag{9}$$

where $y^o = r(s, \boldsymbol{a}, o) + \gamma \cdot \mathbb{E}_{\hat{\pi}(s')}[Q(s', \hat{\pi}(s'), o; \theta^-)]$ is the target value and $\hat{\pi}(s')$ can be obtained with Eq.(7).

However, applying the TD-error loss function may be likely to result in a biased solution. As analyzed previously, the problem solved by CMIX is actually

---

[2] For brevity, we do not show the hypernetwork part of the mixing module in Fig. 1.

the max-min problem of Eq. (8). Since $\{f_o\}$ are monotonic functions, and g is also monotonic. We can get

$$\min_{o \in O} f_o(Q_1^o, \ldots, Q_N^o) \geq \min_{o \in O} g(Q_1^o, \ldots, Q_N^o) \geq g(\min_{o \in O} Q_1^o, \ldots, \min_{o \in O} Q_N^o), \ \forall \boldsymbol{\pi}(s) \tag{10}$$

The left hand of Eq. (10) is the true objective function to be maximized in Eq. (6) while the right hand is the objective function we actually maximized in CMIX. We can see that CMIX may lead to a biased solution $\hat{\boldsymbol{\pi}}(s)$ to the original global max-min problem of Eq. (6) due to the gap in objective functions.

To address the issue, we design two loss functions for CMIX-M and CMIX-S, respectively.

Loss Function for CMIX-M: To eliminate the bias caused by the gap between true objective function $\min_{o \in O} f_o(Q_1^o, \ldots, Q_N^o)$ and the one actually used in CMIX-M $g(\min_{o \in O} Q_1^o, \ldots, \min_{o \in O} Q_N^o)$, we propose a loss function called gap loss $\mathcal{L}_{gap}$ to minimize the gap, i.e.,

$$\mathcal{L}_{gap} = \left(\min_{o \in O} f_o(Q_1^o, \ldots, Q_N^o) - g(\min_{o \in O} Q_1^o, \ldots, \min_{o \in O} Q_N^o) \right)^2. \tag{11}$$

In the training phase, we combine the TD-error loss of Eq. (9) with the gap loss to update parameters. The final loss is computed by

$$\mathcal{L}_{final} = \mathcal{L}_{TD-error} + \beta \mathcal{L}_{gap}, \tag{12}$$

where β is a coefficient for adjusting the weight of the gap loss.

Loss Function for CMIX-S: In CMIX-S we assume that all the mixing functions f_o share parameters, i.e., $g = f_o = f$ ($\forall o \in O$). Then the original max-min problem becomes $\arg\max_{\boldsymbol{\pi}(s)} \min_{o \in O} \mathbb{E}_{\boldsymbol{\pi}(s)}[f(Q_1^o, \ldots, Q_N^o)]$, and the actual max-min problem solved by CMIX is $\hat{\boldsymbol{\pi}}(s) = \arg\max_{\boldsymbol{\pi}(s)} \mathbb{E}[f(\min_{o \in O} Q_1^o, \ldots, \min_{o \in O} Q_N^o)]$. According to the above equations, we design the gap loss $\mathcal{L}_{gap}$ for CMIX-S as

$$\mathcal{L}_{gap} = \left(\min_{o \in O} f(Q_1^o, \ldots, Q_N^o) - f(\min_{o \in O} Q_1^o, \ldots, \min_{o \in O} Q_N^o) \right)^2. \tag{13}$$

The final loss can be computed in the way same as Eq. (12).

Note that the gap loss functions in Eq.(11) and (13) are fully differentiable and can be easily optimized by existing gradient descent methods.

4.4   CMIX Algorithm

In Algorithm 1, we outline the pseudocode of CMIX. In the beginning, we initialize the neural parameters of θ and θ^-, and the replay buffer $\mathcal{D}$. From line 2 to line 16, we train the agents for a number of epoch$_{max}$ epochs, and each epoch contains a total of step$_{max}$ update steps. The state will be initialized

---

Algorithm 1: CMIX Algorithm

1: Initialize parameters θ, target $\theta^- = \theta$, and the replay buffer $\mathcal{D} = \emptyset$
2: **for** epoch $= 1, \cdots,$ epoch$_{max}$ **do**
3:    Initialize state s^0
4:    **for** $t = 0, \cdots,$ step$_{max}$ **do**
5:       Collect observations $\{z_i^t\}_{i \in \mathcal{N}}$ for all agents
6:       **for** each agent i **do**
7:          $\hat{\pi}_i(z_i^t) = \arg \max\limits_{\pi_i \in \Pi_i} \min\limits_{o \in O} \mathbb{E}[Q_i(z_i^t, \pi_i(z_i^t), o)]$

8:          $a_i^t = \begin{cases} sample(\hat{\pi}_i(z_i^t)) & \text{with prob. } 1 - \epsilon \\ randint(1, |\mathcal{A}_i|) & \text{with prob. } \epsilon \end{cases}$

9:       **end for**
10:       Execute the joint action $\boldsymbol{a}^t$ and store the transition
         $(s^t, \boldsymbol{a}^t, r, \{r_k^t\}_{k=1,...,K}, s^{t+1})$ in $\mathcal{D}$
11:       Sample a mini-batch B from $\mathcal{D}$
12:       Compute $\mathcal{L}_{final}$ of Eq. (12) for each sample in B
13:       Update θ by minimizing the average loss of $\mathcal{L}_{final}$ with respect to B
14:       Update target $\theta^- = \theta$ periodically
15:    **end for**
16: **end for**

---

at the beginning of each epoch (line 3). In line 5, the agents' local observations are collected. From line 6 to line 9, each agent takes an action through an ϵ−greedy method. In line 10, the joint action $\boldsymbol{a}^t$ is executed, and the transition $(s^t, \boldsymbol{a}^t, r, \{r_k^t\}_{k=1,...,K}, s^{t+1})$ is stored in the buffer $\mathcal{D}$. In line 11, a mini-batch B of samples are selected from $\mathcal{D}$ randomly. Then, the loss $\mathcal{L}_{final}$ of Eq. (12) is computed with respect to each sample in line 12. In line 13, the computed loss guides the update of θ. Finally, the target θ^- will be updated from θ after a specified interval.

5   Experiments

We evaluate CMIX on two different tasks: a blocker game with travel cost and a cooperative routing optimization task in large-scale vehicular networks. In our numerical results, CMIX is compared with the state-of-the-art CTDE learning algorithms, including IQL [27], VDN [26], and QMIX [23]. Since these algorithms do not consider constraints during training, we also compare performance with C-IQL where each agent independently optimizes the global objective with the consideration of peak and average constraints. C-IQL is implemented by extending IQL in a way similar to CMIX.

5.1   Blocker Game with Travel Cost

We consider a non-trivial blocker game with travel cost, which is an extension of the blocker game in [29]. The agents should cooperate to reach the bottom row

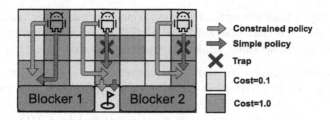

Fig. 2. Blocker game with travel cost.

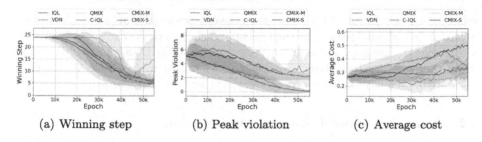

(a) Winning step   (b) Peak violation   (c) Average cost

Fig. 3. Convergence results over training epoch in the blocker game of Fig. 2. Each curve with shadow shows the mean values and the variants of the observed metric. The average cost no larger than 0.3 represents that the average constraint is fulfilled.

of the map as quickly as possible, while the blockers move left or right to block the agents. Each cell on the map is assigned a cost, and an agent will take the cost when it moves into the cell. We also place some traps on the map, which the agents are not allowed to move into.

Figure 2 shows the blocker game with travel cost in our evaluation. In the game, it costs -1 reward per time-step before the agents reach the destination, and an extra small reward will be returned to the agents when they are getting closer to the bottom. The agents try to minimize the winning step, i.e., the movement steps needed by the agents for reaching the bottom, subject to a peak constraint and an average constraint. The agents should not move into traps (i.e., peak constraint). Besides, the average cost taken by these agents in one game should be bounded. The upper bound of average cost is set to 0.3 in our evaluation. With the optimal strategy labeled by green arrows, the agents can win the game in 5 steps, satisfying both peak and average constraints. The blocker game is challenging for the agents in the sense of cooperation with only decentralized policy and local observations, and the peak and average constraints make the task even more difficult to complete. The convergence results of different algorithms in the blocker game with travel cost are presented in Fig. 3.

Winning Step: Figure 3 (a) shows the convergence results of winning step. We can see that the winning step decreases with the increment of the epoch. QMIX gets the smallest winning step finally without considering constraints. CMIX-M, CMIX-S, and IQL get similar performance on winning step and outperform

VDN and C-IQL which either have larger variance or take more training epochs to converge. We note that CMIX-M and CMIX-S optimize winning step under peak and average constraints, while IQL does not consider constraints.

Peak Violation: Figure 3 (b) shows the convergence results of peak violation, i.e., the number of violated peak constraints in each epoch. We can see that CMIX-M, CMIX-S, and C-IQL get results very close to zero after convergence. That is to say, the peak constraint is mostly satisfied by CMIX-M, CMIX-S, and C-IQL. However, the other algorithms, i.e., QMIX, IQL, and VDN, receive significant peak violations due to the ignorance of peak constraints.

Average Cost: Figure 3 (c) shows the convergence results of average cost, i.e., the average cost with respect to each agent and each step. We can see only CMIX-M satisfies the average cost after convergence, while CMIX-S and C-IQL violate the average constraints slightly. Note that compared with CMIX-M, CMIX-S using a single mixing network has a relatively limited representation ability.

5.2   Vehicular Network Routing Optimization

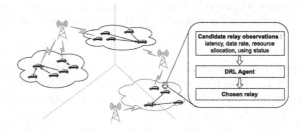

Fig. 4. The cooperative routing optimization problem in a vehicular network.

We apply the MARL algorithms in a cooperative routing optimization problem in vehicular networks shown in Fig. 4. In the scenario with three cells, each vehicle can establish a V2I (Vehicle-to-Infrastructure) link with the base station in the local cell or V2V (Vehicle-to-Vehicle) links with neighboring vehicles. These V2I and V2V links have different transmission data rates and link latencies. We attach an RL agent on each vehicle and consider downlink data transmission, i.e., data needs to be delivered from base stations to destination vehicles. For a destination vehicle, its downlink data can be delivered through i) the path of the direct V2I link, or ii) the path consisting of a V2V link with a neighboring vehicle (i.e., a relay) and a V2I link between the relay and the corresponding base station. The routing decision, that an agent on the vehicle needs to make, is choosing a proper path for downlink data transmission according to local observations of the candidate relays. All the vehicles need to be coordinated to maximize the total transmission rate with the consideration of proportional fairness, while satisfying some peak and average latency constraints. Particularly, the path latency of each vehicle should not exceed a threshold, i.e., peak

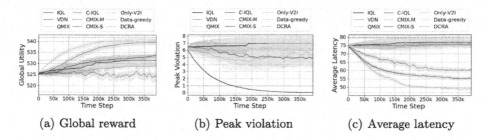

(a) Global reward   (b) Peak violation   (c) Average latency

Fig. 5. Convergence results over training step in the vehicular routing problem. The average latency no larger than 60 represents that the average constraint is fulfilled.

constraints, and the average path latency of all the vehicles in the network is bounded by a soft upper bound.

In our evaluation, we consider a vehicular network with three cells (also three base stations) with tens of randomly generated vehicles (30–60 vehicles). Link latencies and link transmission rates are also set randomly.

Besides the baselines in the blocker game, we also consider three other baselines. i) Only-V2I: Downlink data is delivered to each vehicle through the directly connected V2I link. ii) Data-greedy: Each vehicle chooses the path with the largest data rate. iii) DCRA [15]: An iterative vehicular routing optimization algorithm that can improve the transmission rate with the consideration of proportional fairness.

Global Utility: Figure 5 (a) shows the convergence of global utility. We can see that QMIX and VDN outperform others because they do not consider constraints. Then, both CMIX-S and CMIX-M outperform Only-V2I significantly, and CMIX-S provides even larger global utility than Data-greedy, which validates the effectiveness of CMIX. Note that both IQL and C-IQL suffer performance degradation since it is difficult to learn a consistent strategy for the agents without a mixing architecture. Also, note that CMIX-S shows a better performance than CMIX-M. This is because CMIX-M having more parameters is more challenging to converge than CMIX-S in such a large-scale and complicated task.

Peak Violation: Figure 5 (b) shows the convergence of peak violation. Note that, the curves of C-IQL, CMIX-M, and CMIX-S are overlapped. We can see the three algorithms can fulfill the peak constraint after convergence. Note that Only-V2I using the one-hop path of the directly connected V2I link still violates the peak constraint sometimes.

Average Latency: Figure 5 (c) shows the convergence of average latency. We can see only CMIX-M and CMIX-S satisfy the average constraint. The other baselines do not meet the average constraint.

5.3 Gap Loss Coefficient

We evaluate the performance of CMIX under different settings of the gap loss coefficient weight β in the blocker game with travel cost. $\beta = 0$ means not using

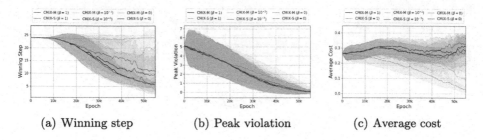

(a) Winning step   (b) Peak violation   (c) Average cost

Fig. 6. Convergence results of CMIX with different coefficient β over training epoch in the blocker game of Fig. 2. The average cost bound is set to 0.3.

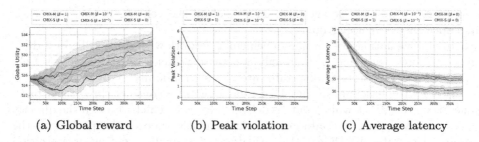

(a) Global reward   (b) Peak violation   (c) Average latency

Fig. 7. Convergence results of CMIX with different coefficient β over training step in the vehicular routing problem. The average latency bound is set to 60.

the gap loss function. Figure 6 shows the convergence result over training epoch. We find that the CMIX algorithm converges more stably and to a better result when β is large enough, which illustrates the positive effect of gap loss in bias elimination. Moreover, the results indicate that gap loss plays a more important role in the CMIX-M algorithm.

We also evaluate the effect of β in the vehicular network routing optimization task. Figure 7 shows the convergence result over training step. We can find that CMIX-S with gap loss gives better results, which illustrates the effectiveness of gap loss for CMIX in large-scale tasks. Also, note that CMIX-M performances worse with larger β since CMIX-M has more parameters and the gap loss possibly makes it more difficult to converge in large-scale tasks.

6   Related Work

Reinforcement Learning Under Peak or Average Constraints. There are lots of approaches [3,4,6,9–12,22] focusing on RL under either peak or average constraints. However, most of the constrained RL algorithms tackle only one type of the two constraints – but not both – and focus on single-agent setting.

Multi-agent Reinforcement Learning Under Constraints. Existing MARL algorithms are often developed in the paradigm of CTDE, i.e., centralized training with decentralized execution. Independent learning algorithms like IQL [27] treat other agents as part of the environment, which usually do not converge well. Another kind of approaches called centralized learning [7,13] learn a fully centralized state-action value function and then use it to guide the optimization of policies for decentralized agents. However, centralized learning suffers bad scalability due to combinatorial complexity, especially in large-scale cooperative tasks. Recently, another paradigm lying between independent learning and centralized learning has attracted much attention. A typical method called QMIX [24] as well as its extensions [17,23,29] combines the Q-values of agents through a mixing module, which can coordinate agents efficiently and result in good scalability. MARL under constraints has already attracted some attentions [5,20]. However, to the best of our knowledge, there are no MARL learning algorithms that can address both peak and average constraints especially in the realm of CTDE algorithms.

7   Conclusion

In this paper, we propose the CMIX algorithm for Dec-POMDP problem under both peak and average constraints. To this end, the original problem is converted into a multi-objective constrained problem, which can be solved through an equivalent max-min optimization problem. We leverage the value function factorization to develop a novel neural architecture in the CTDE paradigm for solving the max-min optimization problem approximately. We further analyze the gap between the original max min optimization problem and the approximated ones solved by CMIX and propose two gap loss functions to eliminate the bias of learned solutions. Evaluations on a blocker game with travel cost and a large-scale vehicular network routing problem validate the effectiveness of CMIX. We note that the proposed CMIX approach can be integrated with other algorithms using value function factorization, e.g., [17,29], which will be left to our future work.

Acknowledgment. This work is supported by the National Key R&D Program of China under Grant No. 2018YFB1800302, the National Natural Science Foundation of China under Grant (61625203, 61832013, 61872209, and 92038302), and the Independent Scientific Research Project of NUDT (ZZKYZX-03-02-02).

References

1. Abedi, A., Ghaderi, M., Williamson, C.: Distributed routing for vehicular ad hoc networks: throughput-delay tradeoff. In: IEEE MASCOTS (2010)
2. Al-Abbasi, A.O., Ghosh, A., Aggarwal, V.: Deeppool: distributed model-free algorithm for ride-sharing using deep reinforcement learning. IEEE TITS **20**, 4714–4727 (2019)
3. Bai, Q., Gattami, A., Aggarwal, V.: Provably efficient model-free algorithm for MDPs with peak constraints. arXiv preprint arXiv:2003.05555 (2021)

4. Chow, Y., Nachum, O., Duenez-Guzman, E., Ghavamzadeh, M.: A Lyapunov-based approach to safe reinforcement learning. In: NeurIPS (2018)
5. Diddigi, R.B., Danda, S.K.R., Bhatnagar, S., et al.: Actor-critic algorithms for constrained multi-agent reinforcement learning. arXiv preprint arXiv:1905.02907 (2019)
6. Ding, D., Wei, X., Yang, Z., Wang, Z., Jovanović, M.R.: Provably efficient safe exploration via primal-dual policy optimization. In: AISTATS (2021)
7. Foerster, J., Farquhar, G., Afouras, T., Nardelli, N., Whiteson, S.: Counterfactual multi-agent policy gradients. In: AAAI (2018)
8. Garcıa, J., Fernández, F.: A comprehensive survey on safe reinforcement learning. JMLR **16**, 1437–1480 (2015)
9. Gattami, A.: Reinforcement learning of Markov decision processes with peak constraints. arXiv preprint arXiv:1901.07839 (2019)
10. Gattami, A., Bai, Q., Aggarwal, V.: Reinforcement learning for multi-objective and constrained Markov decision processes. In: AISTATS (2021)
11. Geibel, P.: Reinforcement learning for MDPs with constraints. In: Fürnkranz, J., Scheffer, T., Spiliopoulou, M. (eds.) ECML 2006. LNCS (LNAI), vol. 4212, pp. 646–653. Springer, Heidelberg (2006). https://doi.org/10.1007/11871842_63
12. Geibel, P., Wysotzki, F.: Risk-sensitive reinforcement learning applied to control under constraints. JAIR **24**, 81–108 (2005)
13. Gupta, J.K., Egorov, M., Kochenderfer, M.: Cooperative multi-agent control using deep reinforcement learning. In: Sukthankar, G., Rodriguez-Aguilar, J.A. (eds.) AAMAS 2017. LNCS (LNAI), vol. 10642, pp. 66–83. Springer, Cham (2017). https://doi.org/10.1007/978-3-319-71682-4_5
14. Ha, D., Dai, A., Le, Q.V.: Hypernetworks. arXiv preprint arXiv:1609.09106 (2016)
15. Kassir, S., de Veciana, G., Wang, N., Wang, X., Palacharla, P.: Enhancing cellular performance via vehicular-based opportunistic relaying and load balancing. In: IEEE INFOCOM (2019)
16. Li, Z., Wang, C., Jiang, C.J.: User association for load balancing in vehicular networks: an online reinforcement learning approach. IEEE TITS **18**, 2217–2228 (2017)
17. Mahajan, A., Rashid, T., Samvelyan, M., Whiteson, S.: Maven: multi-agent variational exploration. arXiv preprint arXiv:1910.07483 (2019)
18. Mnih, V., et al.: Playing Atari with deep reinforcement learning. arXiv preprint arXiv:1312.5602 (2013)
19. Mnih, V., et al.: Human-level control through deep reinforcement learning. Nature **518**, 529–533 (2015)
20. Nguyen, D.T., Yeoh, W., Lau, H.C., Zilberstein, S., Zhang, C.: Decentralized multi-agent reinforcement learning in average-reward dynamic DCOPs. In: AAAI (2014)
21. Oliehoek, F.A., Amato, C., et al.: A Concise Introduction to Decentralized POMDPs, vol. 1. Springer, Heidelberg (2016). https://doi.org/10.1007/978-3-319-28929-8
22. Prashanth, L., Ghavamzadeh, M.: Variance-constrained actor-critic algorithms for discounted and average reward MDPs. Mach. Learn. **105**, 367–417 (2016). https://doi.org/10.1007/s10994-016-5569-5
23. Rashid, T., Farquhar, G., Peng, B., Whiteson, S.: Weighted QMIX: expanding monotonic value function factorisation for deep multi-agent reinforcement learning. In: NeurIPS (2020)
24. Rashid, T., Samvelyan, M., Schroeder, C., Farquhar, G., Foerster, J., Whiteson, S.: QMIX: monotonic value function factorisation for deep multi-agent reinforcement learning. In: ICML (2018)

25. Saleet, H., Langar, R., Naik, K., Boutaba, R., Nayak, A., Goel, N.: Intersection-based geographical routing protocol for VANETs: a proposal and analysis. IEEE TVT **60**, 4560–4574 (2011)
26. Sunehag, P., et al.: Value-decomposition networks for cooperative multi-agent learning based on team reward. In: Springer AAMAS (2018)
27. Tan, M.: Multi-agent reinforcement learning: independent vs. cooperative agents. In: ICML (1993)
28. Wang, F., Wang, F., Liu, J., Shea, R., Sun, L.: Intelligent video caching at network edge: a multi-agent deep reinforcement learning approach. In: IEEE INFOCOM (2020)
29. Yang, Y., et al.: Multi-agent determinantal Q-learning. In: ICML (2020)

Model-Based Offline Policy Optimization with Distribution Correcting Regularization

Jian Shen, Mingcheng Chen, Zhicheng Zhang, Zhengyu Yang,
Weinan Zhang[✉], and Yong Yu

Shanghai Jiao Tong University, Shanghai, China
{rockyshen,mcchen,zyyang,yyu}@apex.sjtu.edu.cn,
{zhangzhicheng1,wnzhang}@sjtu.edu.cn

Abstract. Offline Reinforcement Learning (RL) aims at learning effective policies by leveraging previously collected datasets without further exploration in environments. Model-based algorithms, which first learn a dynamics model using the offline dataset and then conservatively learn a policy under the model, have demonstrated great potential in offline RL. Previous model-based algorithms typically penalize the rewards with the uncertainty of the dynamics model, which, however, is not necessarily consistent with the model error. Inspired by the lower bound on the return in the real dynamics, in this paper we present a model-based alternative called DROP for offline RL. In particular, DROP estimates the density ratio between model-rollouts distribution and offline data distribution via the DICE framework [45], and then regularizes the model-predicted rewards with the ratio for pessimistic policy learning. Extensive experiments show our DROP can achieve comparable or better performance compared to baselines on widely studied offline RL benchmarks.

Keywords: Offline Reinforcement Learning · Model-based
Reinforcement Learning · Occupancy measure

1 Introduction

Reinforcement learning (RL) has achieved great success in various simulated domains, such as Go [33] and video games [11]. However, these promising results mostly rely on numerous online trial-and-error learning. Unfortunately, such an online learning manner is typically costly, even dangerous, and thus cannot be directly applied to complex real-world problems, such as recommender systems [2]. Instead, in real-world applications, there usually exist large and diverse datasets that are previously collected by one or multiple logging policies. These scenarios motivate the study of offline RL [21], also known as batch RL [20], which provides a promising alternative for widespread use of RL. At a colloquial

---

J. Shen and M. Chen—Equal contribution.

© Springer Nature Switzerland AG 2021
N. Oliver et al. (Eds.): ECML PKDD 2021, LNAI 12975, pp. 174–189, 2021.
https://doi.org/10.1007/978-3-030-86486-6_11

level, offline RL algorithms aim to learn highly rewarding policies by leveraging the static offline datasets without further interactions with environments.

Although it seems to be promising, offline RL faces enormous challenges. Previous works have observed that directly utilizing existing off-policy RL algorithms (*e.g.*, DDPG [22]) in an offline setting without online data collection performs poorly [7]. These failures are mainly caused by evaluating the target Q-function on out-of-training-distribution state-actions in the bootstrapping process [18]. Such distribution shift issue may introduce errors when updating the Q-function, making policy optimization unstable and potentially diverging.

In literature, many efforts have been devoted to mitigating the state-action distribution shift challenge in offline RL. One fruitful line of prior offline RL works consists of model-free algorithms [1,7,14,18,19,26,28,32,40], which constrain the learned policy to be close to the data collecting policy or incorporate conservatism into the Q-function training. However, such model-free algorithms can only learn on the states in the offline dataset, making it overly conservative to learn an effective policy. Therefore, it is necessary to leave the offline data support for better policy optimization. To that end, model-based counterparts have been studied to offer such a possibility by using a dynamics model.

However, directly applying model-based techniques to offline settings also suffers from the state-action distribution shift between model learning and model using. Therefore previous model-based offline RL algorithms [16,42] learn a pessimistic model by penalizing the model predicted rewards according to the estimated model uncertainty and then interact with the model to sample transitions for policy training. Due to the capability of sampling states that are not contained in the offline dataset, model-based algorithms have shown to generalize better. However, the existing model-based offline RL methods rely on heuristic uncertainty quantification techniques, and there is no guarantee that the estimated uncertainty is in proportion to the groundtruth model prediction error.

Based on this consideration, in this paper we present density ratio regularized offline policy learning (DROP), a simple yet effective model-based algorithm for offline RL. DROP directly builds upon a theoretical lower bound of the return in the real dynamics, providing a sound theoretical guarantee for our algorithm. To be more specific, DROP leverages the GradientDICE technique [45] to estimate the density ratio between the model rollout distribution and the offline data distribution, and then regularizes the model predicted rewards according to the density ratio. As an intuitive example, the reward will be severely penalized if the state-action pair is much more likely to sampled by the model than in the offline dataset, in which case the penalization could be viewed as inducing a conservative estimate of the value for out-of-support state-action pairs. We show through extensive experiments that our proposed DROP can achieve comparable or better performance compared with prior model-free and model-based offline RL methods on the offline RL benchmark D4RL [6].

2   Preliminary

We first introduce the notations used throughout the paper, and briefly discuss the problem setup of offline RL and the basic idea of model-based RL.

2.1 Markov Decision Processes

A Markov decision process (MDP) is defined by the tuple $\mathcal{M} = (\mathcal{S}, \mathcal{A}, T, r, \gamma, \mu_0)$, where $\mathcal{S}$ and $\mathcal{A}$ are the state and action spaces, respectively. $\mu_0(s)$ denotes the initial state distribution, and $\gamma \in (0, 1)$ is the discount factor. $T(s' \mid s, a)$ is the transition density of state s' given action a made under state s, and the reward function is denoted as $r(s, a)$. In (online) reinforcement learning, an agent interacts with the MDP, which is also called its environment, and aims to learn a policy $\pi(a|s)$ to maximize the expectation of the return (the sum of discounted rewards) with the collected transitions (s, a, s', r):

$$\max_{\pi} \mathcal{J}(\pi, \mathcal{M}) := \mathbb{E}_{s_0 \sim \mu_0, a_t \sim \pi(\cdot|s_t), s_{t+1} \sim T(\cdot|s_t, a_t)} \left[\sum_{t=0}^{\infty} \gamma^t r(s_t, a_t) \right]. \tag{1}$$

For a policy π, we define its discounted state visitation distribution on the real dynamics as $\mu_T^\pi(s) := (1 - \gamma) \sum_{t=0}^{\infty} \gamma^t P_{T,t}^\pi(s)$, where $P_{T,t}^\pi(s)$ is the probability of visiting state s at time t. Similarly, we also define the normalized occupancy measure [12] of policy π on dynamics T: $\rho_T^\pi(s, a) := (1 - \gamma) \cdot \pi(a \mid s) \sum_{t=0}^{\infty} \gamma^t P_{T,t}^\pi(s)$, and then we have $\rho_T^\pi(s, a) = \pi(a|s)\mu_T^\pi(s)$. Using this definition, we can equivalently express the RL objective as follows:

$$\mathcal{J}(\pi, \mathcal{M}) = \mathbb{E}_{\rho_T^\pi(s,a)}[r(s, a)] = \int \rho_T^\pi(s, a) r(s, a) \, ds \, da. \tag{2}$$

2.2 Offline RL

In offline RL (also known as batch RL [20]), the agent cannot interact with the environment to collect additional transitions. Instead, the agent is provided with a static dataset of transitions $\mathcal{D}_b = \{(s_i, a_i, s_i', r_i)\}_{i=1}^N$, which is collected by one or a mixture of behavior policies, also called logging policies, denoted by π_b. In this case, the state-action pairs in dataset $\mathcal{D}_b$ can be viewed as sampling from the distribution $\rho_T^{\pi_b}$. Typically, we do not assume the behavior policy is known in the formulation of offline RL, but we can approximate it via imitation learning over the dataset $\mathcal{D}_b$ if needed. The goal of offline RL is still to search a policy that maximizes the objective function $\mathcal{J}(\pi, \mathcal{M})$ in Eq. 1 or 2.

2.3 Model-Based RL

In practice, the groundtruth state transition T is unknown and model-based RL methods learn a dynamics model $\hat{T}$ to mimic the real one via maximum likelihood estimation, using the data $\mathcal{D}_b$ collected in the real environment. If the reward function r is unknown, a reward function $\hat{r}$ can also be learned in the dynamics model. Similarly, we define $\rho_{\hat{T}}^\pi(s, a)$ to represent the discounted occupancy measure visited by π under the model $\hat{T}$. Once the model is learned, we can construct a model MDP $\hat{\mathcal{M}} = (\mathcal{S}, \mathcal{A}, \hat{T}, \hat{r}, \gamma, \mu_0)$. Then subsequently, any off-the-shelf model-free algorithms can be used to interact with the model MDP to find the optimal policy: $\pi^* = \arg\max_\pi \mathcal{J}(\pi, \hat{\mathcal{M}})$. Besides, one can also use planning algorithms [4, 27] in the model to derive a high-performance agent.

3 A Lower Bound of the True Expected Return

As discussed above, model-based offline RL methods aim to optimize $\mathcal{J}(\pi, \mathcal{M})$ by optimizing $\mathcal{J}(\pi, \hat{\mathcal{M}})$ instead. However, it is not guaranteed that the policy optimized under the model MDP $\hat{\mathcal{M}}$ can achieve good performance in the real environment $\mathcal{M}$ due to the potential model bias. Unfortunately, the model bias is inevitable due to the state-action distribution shift between the offline data and the model rollout trajectories, and in the offline settings this phenomenon is more severe since the error can not be corrected by collecting new data. Therefore, it is important to control the trade-off between the potential gain in performance by leaving the offline data support and the consequent increased model bias.

To overcome this challenge, previous works like MOPO [42] construct a lower bound of the true expected return $\mathcal{J}(\pi, \mathcal{M})$ in the following form:

$$\mathcal{J}(\pi, \mathcal{M}) \geq \mathcal{J}(\pi, \hat{\mathcal{M}}) - C. \tag{3}$$

Once the lower bound is constructed, one can naturally design algorithms to optimize the RL objective by maximizing the lower bound. Following the theoretical analysis in previous works, we first present the following lemma.

Lemma 1. *([23], Lemma 4.3; [42], Lemma 4.1; [31], Lemma E.1) Let two MDPs $\mathcal{M}$ and $\hat{\mathcal{M}}$ share the same reward function $r(s, a)$, but have two different dynamics transition functions $T(\cdot|s, a)$ and $\hat{T}(\cdot|s, a)$, respectively. Define $G_{\hat{T}}^{\pi}(s, a) := \mathbb{E}_{s' \sim \hat{T}(\cdot|s,a)}[V_T^{\pi}(s')] - \mathbb{E}_{s' \sim T(\cdot|s,a)}[V_T^{\pi}(s')]$. For any policy π, we have*

$$\mathcal{J}(\pi, \hat{\mathcal{M}}) - \mathcal{J}(\pi, \mathcal{M}) = \kappa \cdot \mathbb{E}_{(s,a) \sim \rho_{\hat{T}}^{\pi}}[G_{\hat{T}}^{\pi}(s, a)], \tag{4}$$

where $\kappa = \gamma(1 - \gamma)^{-1}$.

MOPO then further bounds the value discrepancy $G_{\hat{T}}^{\pi}(s, a)$ by integral probability metric (IPM) [24]. According to the definition of IPM, let $\mathcal{F}$ be a collection of functions from $\mathcal{S}$ to $\mathbb{R}$, under the assumption that $V_T^{\pi} \in \mathcal{F}$, we have

$$G_{\hat{T}}^{\pi}(s, a) \leq \sup_{f \in \mathcal{F}} \left| \mathbb{E}_{s' \sim \hat{T}(\cdot|s,a)}[f(s')] - \mathbb{E}_{s' \sim T(\cdot|s,a)}[f(s')] \right| =: d_{\mathcal{F}}(\hat{T}(\cdot|s, a), T(\cdot|s, a)), \tag{5}$$

where $d_{\mathcal{F}}$ is the IPM defined by the function class $\mathcal{F}$. By choosing different $\mathcal{F}$, IPM reduces to many well-known distance metrics between probability distributions, such as Wasserstein distance [37] and maximum mean discrepancy [9]. Now it becomes the following lower bound

$$\mathcal{J}(\pi, \mathcal{M}) \geq \mathbb{E}_{s' \sim \hat{T}(\cdot|s,a)}[r(s, a) - \kappa \cdot d_{\mathcal{F}}(\hat{T}(\cdot|s, a), T(\cdot|s, a))]. \tag{6}$$

According to this lower bound, MOPO estimates the uncertainty $u(s, a)$ in model prediction and then reshapes the reward by $\tilde{r}(s, a) = \hat{r}(s, a) - \eta u(s, a)$. However, there exists a neglected gap between the practical algorithm (estimated uncertainty $u(s, a)$) and the theoretical result (real model prediction error $d_{\mathcal{F}}$). To bridge this gap, we now present the main theoretical result in our paper.

Theorem 1. *Let two MDPs $\mathcal{M}$ and $\hat{\mathcal{M}}$ share the same reward function $r(s,a)$, but with different dynamics transition $T(\cdot|s,a)$ and $\hat{T}(\cdot|s,a)$, respectively. Assume there exists a function collection $\mathcal{F}_1 = \{f : \mathcal{S} \times \mathcal{A} \to \mathbb{R} \mid \|f\|_\infty \leq \alpha\}$ such that $G_{\hat{T}}^\pi(s,a) \in \mathcal{F}_1$ and $\mathcal{F}_2 = \{f : \mathcal{S} \to \mathbb{R} \mid \|f\|_\infty \leq \beta\}$ such that $V_{\hat{T}}^\pi(s) \in \mathcal{F}_2$, we have that*

$$\mathcal{J}(\pi, \mathcal{M}) \geq \mathcal{J}(\pi, \hat{\mathcal{M}}) - \frac{\kappa\alpha}{\sqrt{2}}\sqrt{D_{KL}(\rho_{\hat{T}}^\pi \| \rho_T^{\pi_b})} - \frac{\kappa\beta}{\sqrt{2}}\mathbb{E}_{(s,a)\sim\rho_T^{\pi_b}}\left[\sqrt{D_{KL}(T\|\hat{T})}\right]. \quad (7)$$

Proof. According to Lemma 1, we have

$$
\begin{aligned}
&\mathcal{J}(\pi, \hat{\mathcal{M}}) - \mathcal{J}(\pi, \mathcal{M}) \\
=&\kappa \cdot \mathbb{E}_{(s,a)\sim\rho_{\hat{T}}^\pi}[G_{\hat{T}}^\pi(s,a)] \\
=&\kappa \cdot \mathbb{E}_{(s,a)\sim\rho_{\hat{T}}^\pi}[G_{\hat{T}}^\pi(s,a)] - \kappa \cdot \mathbb{E}_{(s,a)\sim\rho_T^{\pi_b}}[G_{\hat{T}}^\pi(s,a)] + \kappa \cdot \mathbb{E}_{(s,a)\sim\rho_T^{\pi_b}}[G_{\hat{T}}^\pi(s,a)] \\
\leq&\kappa \cdot \sup_{f\in\mathcal{F}_1}\left|\mathbb{E}_{(s,a)\sim\rho_T^{\pi_b}}[f(s,a)] - \mathbb{E}_{(s,a)\sim\rho_{\hat{T}}^\pi}[f(s,a)]\right| \qquad (8) \\
&+\kappa \cdot \mathbb{E}_{(s,a)\sim\rho_T^{\pi_b}}\left[\sup_{g\in\mathcal{F}_2}\left|\mathbb{E}_{s'\sim\hat{T}(\cdot|s,a)}[g(s')] - \mathbb{E}_{s'\sim T(\cdot|s,a)}[g(s')]\right|\right] \\
=&\kappa\alpha \cdot d_{\mathrm{TV}}(\rho_T^{\pi_b}, \rho_{\hat{T}}^\pi) + \kappa\beta \cdot \mathbb{E}_{(s,a)\sim\rho_T^{\pi_b}}[d_{\mathrm{TV}}(\hat{T}(\cdot|s,a), T(\cdot|s,a))],
\end{aligned}
$$

where the last inequality holds due to the IPM form of total variation distance. To be more specific, when using the witness function class $\mathcal{F} = \{f : \|f\|_\infty \leq 1\}$, IPM $d_{\mathcal{F}}(\mathbb{P}, \mathbb{Q})$ reduces to the total variation $d_{\mathrm{TV}}(\mathbb{P}, \mathbb{Q})$ where $\mathcal{P}$ and $\mathcal{Q}$ are two probability distributions. Then by applying Pinsker's inequality, we have

$$\mathcal{J}(\pi, \hat{\mathcal{M}}) - \mathcal{J}(\pi, \mathcal{M}) \leq \frac{\kappa\alpha}{\sqrt{2}}\sqrt{D_{\mathrm{KL}}(\rho_{\hat{T}}^\pi \| \rho_T^{\pi_b})} + \frac{\kappa\beta}{\sqrt{2}}\mathbb{E}_{(s,a)\sim\rho_T^{\pi_b}}\left[\sqrt{D_{\mathrm{KL}}(T\|\hat{T})}\right],$$

$$(9)$$

which completes the proof.

$\square$

Theorem 1 gives a lower bound of the true expected return in the true environment. In this bound, the last term corresponds to the model training error on offline data which is sampled from $\rho_T^{\pi_b}$, since training the model via maximum likelihood is an empirical approximation of minimizing the Kullback–Leibler divergence between the model and the true environment. To optimize the first term and third term in this lower bound, we are supposed to train a dynamics model on offline data and then optimize a policy on the model, which forms the naive version of model-based offline RL algorithm. Then the key is how to optimize the second term, which corresponds to the KL divergence between the model rollout data distribution and the offline data distribution. It is intuitively reasonable since this state-action distribution shift problem will lead to large bias in model predicted rollouts and result in a poor policy. Therefore, a principled technique to minimize this distribution shift problem is needed.

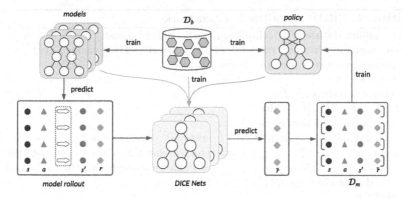

Fig. 1. Illustration of our DROP framework, which uses the distribution ratio correction (DICE) framework to penalize the rewards predicted by the model ensemble.

4   Method

In this section, we give a detailed introduction of our DROP method, which is able to optimize the lower bound in Theorem 1. The overall framework is illustrated in Fig. 1, and the detailed training procedure is shown in Algorithm 1.

4.1   Overall Framework

The primary contribution of DROP is a principled way to minimize the KL divergence between the offline data distribution $\rho_T^{\pi_b}$ and the model rollout data distribution $\rho_{\hat{T}}^{\pi}$. By the definition of KL divergence, the second term can be written as (for clarity, the constant coefficient is neglected):

$$D_{\mathrm{KL}}(\rho_{\hat{T}}^{\pi} \| \rho_T^{\pi_b}) = \mathbb{E}_{(s,a) \sim \rho_{\hat{T}}^{\pi}} \left[\log(\rho_{\hat{T}}^{\pi}(s,a)/\rho_T^{\pi_b}(s,a)) \right]. \tag{10}$$

If we can obtain the real occupancy measure ratio $\tau^*(s,a) = \rho_{\hat{T}}^{\pi}(s,a)/\rho_T^{\pi_b}(s,a)$, we can use the negative log-ratio $-\log \tau^*(s,a)$ as an additional reward when training the policy. In this way, the second term of the lower bound in Theorem 1 can be optimized.

Then it comes to the question of how to obtain the occupancy measure ratio. One direct way to estimate this ratio is constructing a binary classification problem where the model rollout data from $\rho_{\hat{T}}^{\pi}(s,a)$ is labeled positive and the offline data in $\rho_T^{\pi_b}(s,a)$ is labeled negative [43]. In practice, however, the amount of model rollout data and the offline data is not balanced since in model-based methods we hope to use the model to generate much more data to help improve the policy when offline data is limited. This imbalanced classification problem may cause the learning process to be biased [17], which could lead to poor estimation. To sidestep the above obstacle, we will turn to the DICE (DIstribution Correction Estimation) framework to estimate the ratio by exploiting the Markov property of the environments.

Algorithm 1. DROP Algorithmic Framework
***

Require: Offline dataset D_b, rollout horizon h, reward penalty coefficient η.
1: Learn ensemble dynamics models $\{\hat{T}_\theta^i\}_{i=1}^B : S \times A \mapsto \Pi(S)$ using $D_b(s)$.
2: Randomly initialize π_ϕ, empty buffer $\mathcal{D}_m$.
3: **for** G epochs **do**
4:    sample a state $s_0 \sim D_b$.
5:    **for** $j = 0, 1, \ldots, h - 1$ **do**
6:       sample an action $a_j \sim \pi(s_j)$.
7:       randomly choose a model $\hat{T}^k$ from $\{\hat{T}_\theta\}_{i=1}^B$ and sample $s_{j+1}, r_j \sim \hat{T}^k(s_j, a_j)$.
8:       $u_j = \tau^k(s_j, a_j)$.   ▷ Use corresponding DICE net to estimate the ratio
9:       $\tilde{r}_j \leftarrow \hat{r}_j - \eta \cdot \log u_j$.   ▷ Penalize reward with the calculated ratio
10:      add sample $(s_j, a_j, s_{j+1}, \tilde{r}_j)$ to buffer $\mathcal{D}_m$.
11:   **end for**
12:   Train π_ϕ several times using SAC with mini-batches sampled from $\mathcal{D}_b \cup \mathcal{D}_m$.
13:   Train the DICE networks $\{\tau, f, \nu\}$ using $\mathcal{D}_b$, $\{\hat{T}_\theta\}_{i=1}^B$ and π_ϕ.
14: **end for**
***

By incorporating the ratio estimation and policy regularization into an effective model-based method MBPO [13], we obtain our algorithm DROP. To be more specific, DROP first uses the offline data $\mathcal{D}_b$ to train an ensemble of dynamics models $\{\hat{T}_\theta\}_{i=1}^B$ parameterized by θ, where B is the ensemble size. Given a state-action pair (s, a), the model ensemble can predict the reward $\hat{r}$ and next state $\hat{s}'$. The dynamics model ensemble will be fixed after sufficient training. Then a policy π_ϕ parameterized by ϕ collects samples $\{(s, a, \hat{s}', \hat{r})\}$ by interacting with the dynamics models. To alleviate the distribution shift problem, we use DICE networks to predict the distribution density ratio $u = \tau(s, a)$ and then penalize the predicted reward with the ratio $\tilde{r} = \hat{r} - \eta \cdot \log u$ according to Eq. (10) where η is the coefficient to control the degree of penalty. The modified samples $\{(s, a, \hat{s}', \tilde{r})\}$ are added into the model buffer $\mathcal{D}_m$. Then the policy π_ϕ is trained using data from both $\mathcal{D}_b$ and $\mathcal{D}_m$. After several iterations of policy optimization, we update the DICE networks using data constructed by the offline data, dynamics models and the current policy since the ratio $\rho_{\hat{T}}^\pi(s, a)/\rho_T^{\pi_b}(s, a)$ is related to these three parts. Below we elaborate on several modeling details of model learning, policy optimization and density ratio estimation.

Dynamics Model Learning. We use a bootstrapped ensemble of probabilistic dynamics models $\{\hat{T}_\theta(s'|s, a)\}_{i=1}^B$ to capture both epistemic and aleatoric uncertainty [4]. Each individual dynamics model is a probabilistic neural network which outputs a Gaussian distribution with diagonal covariance conditioned on the state s_n and the action a_n: $\hat{T}_\theta^i(s_{n+1} \mid s_n, a_n) = \mathcal{N}(\mu_\theta^i(s_n, a_n), \Sigma_\theta^i(s_n, a_n))$ where μ and Σ are the mean and covariance, respectively. Each model is trained using the offline data buffer $\mathcal{D}_b$ with a negative log-likelihood loss:

$$\mathcal{L}_{\hat{T}}^i(\theta) = \sum_{n=1}^N \left[\mu_\theta^i(s_n, a_n) - s_{n+1}\right]^\top \Sigma_\theta^{i-1}(s_n, a_n)$$
$$\left[\mu_\theta^i(s_n, a_n) - s_{n+1}\right] + \log \det \Sigma_\theta^i(s_n, a_n). \tag{11}$$

Policy Optimization. Following previous works [4,13], we randomly choose a state from offline data $\mathcal{D}_b$ and use the current policy π_ϕ to perform h-step rollouts on the model ensemble. In detail, at each step, a probabilistic model from the ensemble is selected at random to predict the reward and the next state. The policy is trained on both offline data and model generated data using soft actor-critic (SAC) [10] by minimizing the expected KL-divergence: $\mathcal{L}_\pi(\phi) = \mathbb{E}_s[D_{\mathrm{KL}}(\pi_\phi(\cdot|s) \parallel \exp(Q(s_t,\cdot) - V(s)))]$.

4.2   Ratio Estimation via DICE

In an off-policy evaluation where we want to estimate the performance of a target policy using data generated by a behavioral policy, one promising solution is to re-weight the reward by the occupancy measure density ratio. Recently several works have proposed to estimate this ratio such as DualDICE [25], GenDICE [44] and GradientDICE [45]. In this paper, we adapt the GradientDICE architecture to estimate the density ratio in Eq. (10). We will briefly introduce the general framework and refer the readers to [45] for more detail. To begin with, the occupancy measure satisfies the following equation:

$$\rho_{\hat{T}}^{\pi}(s', a') = (1 - \gamma)\mu_0(s')\pi(a'|s') + \gamma \int \rho_{\hat{T}}^{\pi}(s, a)\hat{T}(s'|s, a)\pi(a'|s')\,\mathrm{d}s\,\mathrm{d}a, \quad (12)$$

where μ_0 is the initial state distribution. Then using $\rho_T^{\pi_D}(s, a)\tau(s, a)$ to replace $\rho_{\hat{T}}^{\pi}(s, a)$ and minimizing some divergence between the LHS and RHS of Eq. 12 over the function τ with an additional constraint can finally estimate the ratio $r^*(s, a)$. Denoting $\delta(s', a') = \gamma \int \rho_T^{\pi_D}(s, a)\tau(s, a)\hat{T}(s'|s, a)\pi(a'|s')\,\mathrm{d}s\,\mathrm{d}a + (1 - \gamma)\mu_0(s')\pi(a'|s') - \rho_T^{\pi_D}(s', a')\tau(s', a')$, we have the following objective function

$$\mathcal{L}(\tau) = \frac{1}{2}\mathbb{E}_{\rho_T^{\pi_D}}\left[\left(\frac{\delta(s, a)}{\rho_T^{\pi_D}(s, a)}\right)^2\right] + \frac{\lambda}{2}(\mathbb{E}_{\rho_T^{\pi_D}}[\tau(s, a)] - 1)^2, \quad (13)$$

where $\lambda > 0$ is a constant coefficient. By further applying Fenchel conjugate [29] and the interchangeability principle as in [44], optimizing the final objective of GradientDICE is a minimax problem as

$$\min_\tau \max_{f,\nu} \mathcal{L}_{\mathrm{DICE}}(\tau, \nu, f)$$

$$= (1 - \gamma)\mathbb{E}_{s\sim\mu_0, a\sim\pi(\cdot|s)}[f(s, a)] + \gamma\mathbb{E}_{(s,a)\sim\rho_T^{\pi_D}, s'\sim\hat{T}(\cdot|s,a), a'\sim\pi(\cdot|s')}[\tau(s, a)f(s', a')]$$

$$- \mathbb{E}_{(s,a)\sim\rho_T^{\pi_D}}[\tau(s, a)f(s, a) + \frac{1}{2}f(s, a)^2] + \lambda(\mathbb{E}_{(s,a)\sim\rho_T^{\pi_D}}[\nu\tau(s, a) - \nu] - \frac{1}{2}\nu^2),$$

where we have $\tau : \mathcal{S} \times \mathcal{A} \rightarrow \mathbb{R}$, $f : \mathcal{S} \times \mathcal{A} \rightarrow \mathbb{R}$ and $\nu \in \mathbb{R}$. In practical implementation, we use feed-forward neural networks to model the function f and τ. Moreover, since we use an ensemble of models and the ratio is related to one specific model, we need to construct a separate DICE network for each dynamics model. Finally, we could use the offline states to approximate the initial state distribution $\mu_0(s)$ since the simulated rollouts are generated starting from some state in $\mathcal{D}_b$.

5 Experiment

Through our experiments, we aim to answer the following questions: i) How does DROP perform compared to prior model-free and model-based offline RL algorithms? ii) How does the quality of uncertainty quantification used in DROP compare against prior offline model-based methods?

5.1 Comparative Evaluation

Compared Methods. We compare DROP with several representative baselines. Firstly, BC (behavior cloning) and SAC-off, directly learning a policy using offline data by supervised learning and soft actor-critic algorithm, serves as basic baselines. For model-free baselines, we compare to BEAR [18], BRAC-v [40], AWR [28] and CQL [19], which are all effective offline RL methods (as discussed in Sect. 6) with CQL achieving SoTA results. For model-based baselines, MBPO [13] and MOPO [42] are taken into comparison. By comparing to MBPO, the effectiveness of reward penalty can be investigated while the comparison with MOPO helps reveal the potential of density ratio as the reward penalty. Note that we don't compare to MOReL [16] to avoid unfair comparison since there are other different factors (*e.g.*policy training algorithm and known reward assumption) besides the reward penalty form. We leave incorporating our density ratio regularization into MOReL framework for future work.

Offline Datasets. All the compared methods are evaluated over 12 offline datasets provided in a large-scale open-source benchmark D4RL [6]. The 12 datasets have different experimental settings with three Gym-MuJoCo tasks (Hopper, Walker2d, and Halfcheetah), and four types of offline datasets. More specifically, the four types of dataset are defined as follows:

(1) The "random" dataset consists of data generated by executing a randomly initialized SAC policy for 1M steps.
(2) The "medium" dataset is generated by executing a suboptimal SAC policy (achieves "medium" level that depends on specific environment) for 1M steps.
(3) The "medium-replay" dataset collects all the samples observed in the replay buffer during the process of training a SAC policy to "medium" level.
(4) The "medium-expert" dataset mixes equal amount of samples generated by executing an "expert" level and a "medium" level SAC policies.

Implementation Details. We implement DROP using TensorFlow[1]. We train the GraidientDICE networks with fixed learning rate $1e-4$ and mini-batch size of 256. The hyperparameter λ in DICE objective is searched in range of $[0.1, 1.0]$, and the reward penalty coefficient is searched in $\{0.1, 0.5, 1.0, 2.0, 5.0\}$. We pretrain the DICE net with offline dataset and initialized SAC policy for

---

[1] Experiment code can be found at https://github.com/cmciris/DROP.

Table 1. Evaluation results on D4RL benchmark. The values reported are normalized scores roughly to the range between 0 and 100, which are calculated with $normalized\ score = 100 \times \frac{score - random\ score}{expert\ score - random\ score}$ according to [6]. Results of model-free offline methods and BC are taken from the original paper of D4RL [6]. MOPO and DROP are evaluated over six random seeds. We bold the best, and underline the second results across all methods.

Dataset	Env.	DROP	MOPO	MBPO	BC	SAC-off	BEAR	BRAC-v	AWR	CQL
random	hopp.	**18.2**	11.2	4.5	9.8	11.3	11.4	<u>12.2</u>	10.2	10.8
medium	hopp.	<u>52.1</u>	26.3	4.9	29.0	0.8	<u>52.1</u>	31.1	35.9	**58.0**
med-replay	hopp.	**88.2**	<u>78.6</u>	49.8	11.8	3.5	33.7	0.6	28.4	48.6
med-expert	hopp.	54.7	33.9	56.0	**111.9**	1.6	96.3	0.8	27.1	<u>98.7</u>
random	walk.	<u>9.2</u>	**12.1**	8.6	1.6	4.1	7.3	1.9	1.5	7.0
medium	walk.	73.4	11.9	12.7	6.6	0.9	59.1	**81.1**	17.4	<u>79.2</u>
med-replay	walk.	**39.2**	<u>39.0</u>	22.2	11.3	1.9	19.2	0.9	15.5	26.7
med-expert	walk.	65.3	59.2	7.6	6.4	−0.1	40.1	<u>81.6</u>	53.8	**111.0**
random	halfch.	**38.0**	33.8	30.7	2.1	30.5	25.1	31.2	2.5	<u>35.4</u>
medium	halfch.	**50.2**	42.3	28.3	36.1	−4.3	41.7	46.3	37.4	<u>44.4</u>
med-replay	halfch.	**58.6**	<u>54.4</u>	47.3	38.4	−2.4	38.6	<u>47.7</u>	40.3	46.2
med-expert	halfch.	**65.7**	<u>64.2</u>	9.7	35.8	1.8	53.4	41.9	52.7	62.4

adequate steps in the beginning of training. Since log function has extremely negative value when the input is near 0, we clip the ratio into $[\sigma_1, \sigma_2]$, and we search σ_1 from $\{0.01, 0.1\}$ and search σ_2 from $\{10, 20, 50\}$. Besides, we find that other monotonically increasing functions such as $\tanh(\cdot)$ can also be used to replace $\log(\cdot)$ in practice. The model rollout length h used in DROP is set the same as in MOPO for fair comparison.

Results. The comparative results are presented in Table 1. From the comparison we observe that: (i) Our proposed DROP algorithm outperforms the SoTA results in 7 out of the 12 dataset settings, and achieves comparable results (the second best) in other 2 out of 5 dataset. Especially in the halfcheetah environment, DROP shows its superior performance. This verifies the effectiveness of DROP. (ii) DROP performs better than the model-based baseline MOPO in 11 out of the 12, which demonstrates the strength of using density ratio as the reward penalty. (iii) Compared to the SoTA model-free baseline CQL, DROP outperforms it in 8 out of 12 datasets, although it achieves extremely high score in some settings like hopper medium-expert and walker2d medium-expert. By comparison, DROP achieves relatively stable performance across all the settings with hyperparameters searched in a small range while the performance of model-free offline baselines is closely related to the careful tuning in the specific environment and dataset type as mentioned in [40].

5.2 Empirical Analysis

To answer the question ii), we empirically analyze the behaviour of reward penalty used in the two model-based offline methods DROP and MOPO [42]. By

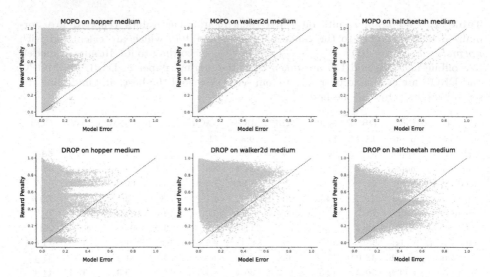

Fig. 2. Visualization of correlation between groundtruth model error and different reward penalty forms. In each Fig. 10000 state-action tuples randomly sampled from the offline dataset are plotted. The top row shows the results of MOPO while the bottom row shows the results of DROP and columns from left to right are the environments of hopper, walker2d, and halfcheetah.

rolling out trajectories on the dynamics model, the agent is allowed to explore around the support of offline data and thus is able to seek possible improvements to behavioral policies without any further interaction with real environment. In this way, how to control the trade-off between informative exploration around the support of offline data and avoiding going far away from the support without exploiting the model deficiency becomes a key problem in model-based offline RL algorithms. MOPO solves this problem by estimating the model uncertainty as the maximum standard deviation of the learned models in the ensemble $u_{\mathrm{MOPO}}(s,a) = \max_i ||\Sigma_\phi^i(s,a)||_{\mathrm{F}}, i \in \{1,2,\dots,B\}$ and then penalizing the predicted rewards with the uncertainty. However, this uncertainty quantification used in MOPO is not guaranteed to consistently reflect the error between $\hat{T}(\cdot|s,a)$ and $T(\cdot|s,a)$. Differently, instead of using heuristic uncertainty quantification, DROP uses the density ratio $u_{\mathrm{DROP}}(s,a) = \frac{\rho_{\hat{T}}^\pi(s,a)}{\rho_T^{\pi^b}(s,a)}$ as reward penalty in a principled manner.

   To better compare the behavior of different reward penalty forms, we visualize the correlation between the groundtruth model error and the two reward penalty where the model error is computed as $||\hat{T}(s,a) - T(s,a)||_2$, as shown in Fig. 2. We present the comparative results in all three environments, hopper, walker2d, and halfcheetah, on medium dataset, while the results of the other dataset types are similar. We normalize the reward penalty and model errors to $[0,1]$ interval, and thus scattered points should lie along the diagonal line $y = x$ in an ideal situation. As shown in Fig. 2, the scattered points in the fig-

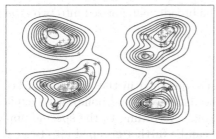 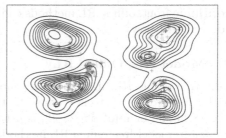

(a) MOPO behaves conservatively.   (b) DROP behaves optimistically.

Fig. 3. Visualization of offline data support and different penalized reward. Black lines show the distribution of offline data. We randomly choose 500 state-action pairs from model rollouts, and plot them with colors range form yellow to red according to the reward penalty. Points with high reward penalty are drawn in red while points with low reward penalty are drawn in yellow. (Color figure online)

ures of MOPO tend to be located at the left top of the figures, which indicates that the uncertainty quantification strategy in MOPO is overly conservative. In contrast, in the figures of DROP the scattered points are closer to the diagonal lines compared to MOPO, which means the reward penalty in DROP is less conservative. This may be the reason why DROP performs better than MOPO on D4RL benchmark since MOPO can not effectively utilize the data with low true model error.

To verify the above findings, we further visualize model rollout samples with computed reward penalty and the offline samples with corresponding data support in halfcheetah medium. To be more specific, we embed offline state-action pairs into a 2-dimensional space with the Uniform Manifold Approximation and Projection (UMAP) technique, and then project the model rollout samples using the same mapping. The visualization is shown in Fig. 3. The black lines represent the distribution of offline samples, which is estimated by Kernel Density Estimation (KDE). Thus it can briefly indicates the offline data support on the 2-dimensional space. Again we find that MOPO is quite conservative in assigning the reward penalty. For example, the penalty for the data in the minor mode at the right bottom is quite large. In contrast, DROP is able to give appropriate reward penalty.

6 Related Work

Offline RL [21], also known as batch RL [20], studies the problem of how to train an RL agent only using static offline dataset without any further interactions with the true environment. It means no exploration is allowed and the agent should focus on making full exploitation of existing knowledge. This property makes offline RL technique naturally suitable for practical application scenarios requiring high exploration cost, like healthcare [8,38,41], recommender systems [3,5,34,35], dialogue systems [14,15,46] and autonomous driving [30]. On the algorithmic front, offline RL methods can be broadly categorized into two groups as described below.

6.1 Model-Free Offline RL

Model-free offline RL algorithms are mostly designed in the principle of keeping the learned policy to stay close to the data collecting policy with explicit [7,14,18,26,40] or implicit [28,32] constraints. According to the specific implementation, model-free offline RL methods can be further grouped into two categories. The first type is Q-value based methods. For example, BCQ [7] optimizes over a subset of actions generated by a trained generative model instead of optimizing over all actions, and thus are less susceptible to over-estimation of Q-values. BEAR [18] employs maximum mean discrepancy (MMD) [9] constraint to make the policy stay close to the behavior policy approximated by a generative model, which guides the target policy to choose those actions that lie in the support of offline distribution. CQL [19] adds regularization in the original Bellman error objective to penalize the Q value of out-of-distribution data, which enables a conservative value function. BRAC [40] penalizes the value function with policy (BRAC-p) or value (BRAC-v) regularization by discrepancy measure (e.g., MMD or KL-divergence) between the learned policy and the behaviour policy. On the other hand, imitation learning based methods like MARWIL [39] and AWR [28] directly train a policy with a reweighted imitation objective. The reweighting coefficient advantage value is estimated by regression and thus the value overestimation due to the Bellman update will not occur.

6.2 Model-Based Offline RL

Model-free offline RL methods can only train the policy with offline data, which may limit the ability to learn a better policy. In contrast, by introducing a dynamics model, model-based offline RL algorithms [16,36,42], is able to provide pseudo exploration around the offline data support for the agent, and thus has potential to learn a better policy with sub-optimal offline dataset. Typically, previous model-based methods rely on uncertainty quantification. MOReL [16] constructs a pessimistic MDP and employs discrepancy of the model prediction over the ensemble as uncertainty to penalize the rewards. A concurrent work MOPO [42] uses maximum learned variance over the ensemble as the uncertainty and performs a similar reward shaping. MOOSE [36] trains a dynamics model as well as a VAE, and uses the reconstruction error as the uncertainty. Similarly, the estimated uncertainty is then used to penalize the policy training. In contrast, our proposed DROP do not rely on the uncertainty quantification. Instead, DROP estimates the density ratio and uses the ratio as the reward penalty to tackle the distribution shift problem.

7 Conclusion

In this paper, we propose a simple yet effective model-based method called DROP for offline reinforcement learning. DROP adds a penalty on rewards to discourage the policy from visiting out-of-distribution state-action tuples like previous work MOPO [42], but uses a novel reward penalty, *i.e.*, the density ratio between the model generated data distribution and the offline data distribution. This form of reward penalty is directly inspired by the lower bound of the true expected return derived in this paper, and thus has strong theoretical guarantee for policy improvement compared to heuristic uncertainty. We validate the performance of our proposed DROP in widely used benchmark D4RL, and the results show DROP achieves promising performance. One major limitation of DROP is the higher computational complexity introduced by the DICE network training. For future work, we plan to investigate more accurate and quick density ratio estimation strategy to boost the performance and computational efficiency of DROP. Also, incorporating some model-free techniques into DROP may be a promising way to improve performance in some specific datasets, such as walker2d medium-expert.

Acknowledgements. The corresponding author Weinan Zhang is supported by "New Generation of AI 2030" Major Project (2018AAA0100900), Shanghai Municipal Science and Technology Major Project (2021SHZDZX0102) and National Natural Science Foundation of China (62076161, 61772333).

References

1. Agarwal, R., Schuurmans, D., Norouzi, M.: Striving for simplicity in off-policy deep reinforcement learning (2019)
2. Chen, H., et al.: Large-scale interactive recommendation with tree-structured policy gradient. In: Proceedings of the AAAI Conference on Artificial Intelligence, vol. 33, pp. 3312–3320 (2019)
3. Chen, M., Beutel, A., Covington, P., Jain, S., Belletti, F., Chi, E.H.: Top-k off-policy correction for a reinforce recommender system. In: Proceedings of the Twelfth ACM International Conference on Web Search and Data Mining, pp. 456–464 (2019)
4. Chua, K., Calandra, R., McAllister, R., Levine, S.: Deep reinforcement learning in a handful of trials using probabilistic dynamics models. In: Advances in Neural Information Processing Systems, pp. 4754–4765 (2018)
5. Covington, P., Adams, J., Sargin, E.: Deep neural networks for Youtube recommendations. In: Proceedings of the 10th ACM Conference on Recommender Systems, pp. 191–198 (2016)
6. Fu, J., Kumar, A., Nachum, O., Tucker, G., Levine, S.: D4rl: datasets for deep data-driven reinforcement learning. arXiv preprint arXiv:2004.07219 (2020)
7. Fujimoto, S., Meger, D., Precup, D.: Off-policy deep reinforcement learning without exploration. In: International Conference on Machine Learning, pp. 2052–2062. PMLR (2019)
8. Gottesman, O., et al.: Evaluating reinforcement learning algorithms in observational health settings. Computing Research Repository (CoRR) (2018)
9. Gretton, A., Borgwardt, K.M., Rasch, M.J., Schölkopf, B., Smola, A.: A kernel two-sample test. J. Mach. Learn. Res. **13**(Mar), 723–773 (2012)
10. Haarnoja, T., Zhou, A., Abbeel, P., Levine, S.: Soft actor-critic: off-policy maximum entropy deep reinforcement learning with a stochastic actor. arXiv preprint arXiv:1801.01290 (2018)

11. Hessel, M., et al.: Rainbow: combining improvements in deep reinforcement learning. In: Proceedings of the AAAI Conference on Artificial Intelligence, vol. 32 (2018)
12. Ho, J., Ermon, S.: Generative adversarial imitation learning. In: Advances in Neural Information Processing Systems, pp. 4565–4573 (2016)
13. Janner, M., Fu, J., Zhang, M., Levine, S.: When to trust your model: model-based policy optimization. In: Advances in Neural Information Processing Systems, pp. 12498–12509 (2019)
14. Jaques, N., et al.: Way off-policy batch deep reinforcement learning of implicit human preferences in dialog. arXiv preprint arXiv:1907.00456 (2019)
15. Karampatziakis, N., Kochman, S., Huang, J., Mineiro, P., Osborne, K., Chen, W.: Lessons from real-world reinforcement learning in a customer support bot. Computing Research Repository (CoRR) (2019)
16. Kidambi, R., Rajeswaran, A., Netrapalli, P., Joachims, T.: Morel: model-based offline reinforcement learning. arXiv preprint arXiv:2005.05951 (2020)
17. Krawczyk, B.: Learning from imbalanced data: open challenges and future directions. Progr, Artif. Intell. **5**(4), 221–232 (2016). https://doi.org/10.1007/s13748-016-0094-0
18. Kumar, A., Fu, J., Tucker, G., Levine, S.: Stabilizing off-policy Q-learning via bootstrapping error reduction. arXiv preprint arXiv:1906.00949 (2019)
19. Kumar, A., Zhou, A., Tucker, G., Levine, S.: Conservative Q-learning for offline reinforcement learning. arXiv preprint arXiv:2006.04779 (2020)
20. Lange, S., Gabel, T., Riedmiller, M.: Batch reinforcement learning. In: Wiering, M., van Otterlo, M. (eds.) Reinforcement Learning, vol. 12, pp. 45–73. Springer, Heidelberg (2012). https://doi.org/10.1007/978-3-642-27645-3_2
21. Levine, S., Kumar, A., Tucker, G., Fu, J.: Offline reinforcement learning: tutorial, review, and perspectives on open problems. arXiv preprint arXiv:2005.01643 (2020)
22. Lillicrap, T.P., et al.: Continuous control with deep reinforcement learning. arXiv preprint arXiv:1509.02971 (2015)
23. Luo, Y., Xu, H., Li, Y., Tian, Y., Darrell, T., Ma, T.: Algorithmic framework for model-based deep reinforcement learning with theoretical guarantees. arXiv preprint arXiv:1807.03858 (2018)
24. Müller, A.: Integral probability metrics and their generating classes of functions. Adv. Appl. Probab. **29**(2), 429–443 (1997)
25. Nachum, O., Chow, Y., Dai, B., Li, L.: DualDICE: behavior-agnostic estimation of discounted stationary distribution corrections. arXiv preprint arXiv:1906.04733 (2019)
26. Nachum, O., Dai, B., Kostrikov, I., Chow, Y., Li, L., Schuurmans, D.: AlgaeDICE: policy gradient from arbitrary experience. arXiv preprint arXiv:1912.02074 (2019)
27. Nagabandi, A., Kahn, G., Fearing, R.S., Levine, S.: Neural network dynamics for model-based deep reinforcement learning with model-free fine-tuning. In: 2018 IEEE International Conference on Robotics and Automation (ICRA), pp. 7559–7566. IEEE (2018)
28. Peng, X.B., Kumar, A., Zhang, G., Levine, S.: Advantage-weighted regression: simple and scalable off-policy reinforcement learning. arXiv preprint arXiv:1910.00177 (2019)
29. Rockafellar, R.T.: Convex Analysis, vol. 36. Princeton University Press, Princeton (1970)
30. Sallab, A.E., Abdou, M., Perot, E., Yogamani, S.: Deep reinforcement learning framework for autonomous driving. Electron. Imaging **2017**(19), 70–76 (2017)

31. Shen, J., Zhao, H., Zhang, W., Yu, Y.: Model-based policy optimization with unsupervised model adaptation. In: Advances in Neural Information Processing Systems, vol. 33 (2020)
32. Siegel, N.Y., et al.: Keep doing what worked: behavioral modelling priors for offline reinforcement learning. arXiv preprint arXiv:2002.08396 (2020)
33. Silver, D., et al.: Mastering the game of go with deep neural networks and tree search. Nature **529**(7587), 484–489 (2016)
34. Strehl, A., Langford, J., Kakade, S., Li, L.: Learning from logged implicit exploration data. Computing Research Repository (CoRR) (2010)
35. Swaminathan, A., Joachims, T.: Batch learning from logged bandit feedback through counterfactual risk minimization. J. Mach. Learn. Res. **16**(1), 1731–1755 (2015)
36. Swazinna, P., Udluft, S., Runkler, T.: Overcoming model bias for robust offline deep reinforcement learning. arXiv preprint arXiv:2008.05533 (2020)
37. Villani, C.: Optimal Transport: Old and New, vol. 338. Springer, Heidelberg (2008). https://doi.org/10.1007/978-3-540-71050-9
38. Wang, L., Zhang, W., He, X., Zha, H.: Supervised reinforcement learning with recurrent neural network for dynamic treatment recommendation. In: Proceedings of the 24th ACM SIGKDD International Conference on Knowledge Discovery & Data Mining, pp. 2447–2456 (2018)
39. Wang, Q., Xiong, J., Han, L., Sun, P., Liu, H., Zhang, T.: Exponentially weighted imitation learning for batched historical data. In: NeurIPS, pp. 6291–6300 (2018)
40. Wu, Y., Tucker, G., Nachum, O.: Behavior regularized offline reinforcement learning. arXiv preprint arXiv:1911.11361 (2019)
41. Yu, C., Ren, G., Liu, J.: Deep inverse reinforcement learning for sepsis treatment. In: 2019 IEEE International Conference on Healthcare Informatics (ICHI), pp. 1–3. IEEE (2019)
42. Yu, T., et al.: MOPO: model-based offline policy optimization. arXiv preprint arXiv:2005.13239 (2020)
43. Zadrozny, B.: Learning and evaluating classifiers under sample selection bias. In: Proceedings of the Twenty-First International Conference on Machine Learning, p. 114 (2004)
44. Zhang, R., Dai, B., Li, L., Schuurmans, D.: GenDICE: generalized offline estimation of stationary values. arXiv preprint arXiv:2002.09072 (2020)
45. Zhang, S., Liu, B., Whiteson, S.: GradientDICE: rethinking generalized offline estimation of stationary values. In: International Conference on Machine Learning, pp. 11194–11203. PMLR (2020)
46. Zhou, L., Small, K., Rokhlenko, O., Elkan, C.: End-to-end offline goal-oriented dialog policy learning via policy gradient. Computing Research Repository (CoRR) (2017)

Disagreement Options: Task Adaptation Through Temporally Extended Actions

Matthias Hutsebaut-Buysse[(✉)], Tom De Schepper, Kevin Mets,
and Steven Latré

Department of Computer Science, University of Antwerp – imec, Antwerp, Belgium
{matthias.hutsebaut-buysse,tom.deschepper,kevin.mets,
steven.latre}@uantwerpen.be

Abstract. Embodied AI, learning through interaction with a physical environment, typically requires large amounts of interaction with the environment in order to learn how to solve new tasks. Training can be done in parallel, using simulated environments. However, once deployed in e.g., a real-world setting, it is not yet clear how an agent can quickly adapt its knowledge to solve new tasks.

In this paper, we propose a novel Hierarchical Reinforcement Learning (HRL) method that allows an agent, when confronted with a novel task, to switch between exploiting prior knowledge through temporally extended actions, and environment exploration. We solve this trade-off by utilizing the *disagreement* between action distributions of selected previously acquired policies. Selection of relevant prior tasks is done by measuring the cosine similarity of their attached natural language goals in a pre-trained word-embedding.

We analyze the resulting temporal abstractions, and we experimentally demonstrate the effectiveness of them in different environments. We show that our method is capable of solving new tasks using only a fraction of the environment interactions required when learning the task from scratch.

Keywords: Hierarchical Reinforcement Learning · Task adaptation

1 Introduction

Humans acquire a wide range of different skills over a lifetime. We are capable of solving complex new problems by quickly adapting, and combining these skills. For example, when learning how to ride a motorbike, balancing skills learned from riding a bicycle might be re-utilized.

But how do we know which skills can be useful when confronted with a new task? We could use *trial-and-error* learning, and test which of our prior skills works best in a new situation. This approach is commonly used in Hierarchical Reinforcement Learning (HRL) approaches [2,19].

However, when able to communicate, language is a much more efficient instrument to communicate how different skills can be transferred, in order to solve

© Springer Nature Switzerland AG 2021
N. Oliver et al. (Eds.): ECML PKDD 2021, LNAI 12975, pp. 190–205, 2021.
https://doi.org/10.1007/978-3-030-86486-6_12

new tasks. For example, one could say to someone who is learning how to ride a motorbike that: *riding a motorbike is just like riding a bicycle*. Or, in order to find a new object, one typically can explain how to find it in terms of the relation with other objects we already are able to localize: e.g., *the microwave is on top of the fridge*.

Embodied AI is a sub-field of AI interested in acquiring intelligent behavior through physical interaction with the environment. Various tasks have been proposed [1] such as *PointGoal* (navigating to specific points in the environment), *ObjectGoal* (navigating to an instance of an object category), and *AreaGoal* (navigating to a specific type of room).

Deep Reinforcement Learning (DRL) methods [16, 18, 27, 28] have been proposed to utilize high-dimensional visual sensor data in order to tackle these problems. The most successful attempts utilize intermediate models capable of building internal (semantic) maps in order to perform efficient exploration [5, 10].

However, these approaches typically start training from scratch, and offer no solution on how to efficiently extend the capabilities of an agent over its lifetime [23]. This is especially an important problem in real-world embodied systems (e.g., a collaborative robot). In this setting, an agent typically has no access to large amounts of compute, and needs to come up with new solutions in a reasonable timeframe.

In order to work towards real-world embodied systems, capable of quickly adapting their knowledge to novel tasks, inspired by the way humans learn through communication, we introduce a novel HRL [24] method. Our method formulates an answer to two important questions: a) which prior skills are useful when learning how to solve a new task? b) how can we solve the trade-off between utilizing prior knowledge, and acquiring new skills by exploring the environment?

We answer these questions by utilizing pre-trained word-embeddings to select source tasks based on their goal descriptions in natural language. We utilize the disagreement between prior policy action distributions in order to decide when to exploit the priors, and when to explore novel paths.

Our answers to these two questions allow an agent to use prior knowledge efficiently as temporally extended actions.

2   Preliminaries

We consider a goal-conditional Reinforcement Learning (RL) setting, and model the problem as a Semi-Markov Decision Process (SMDP), defined by the tuple $\langle S, A, \mathcal{P}, r, \gamma \rangle$. In each episode the agent is tasked with reaching a goal $g_t \in G$. On each time step t the environment produces a state $s_t \in S$ according to an to the agent unknown transition function $\mathcal{P}(s_{t+1}|s_t, a_t)$. The agent consists of a two-level hierarchy [24]. The top level samples an option $\omega_i \sim \pi(s_t, g_t)$ from its policy-over-options. Only options for which the current state is part of the option its initiation set $s_t \in \mathcal{I}_{w_i}$ are considered. The policy-over-options either invokes a single primitive action a_t (point option), or follows a temporally extended action through the intra-option policy of the option $\pi_{\omega_i}(s_t)$, which produces a sequence of primitive actions until the termination condition $\beta(s_t)$ of the active option is triggered. After utilizing a primitive action, the agent receives a reward scalar $r_t(s_t, g_t, a_t, s_{t+1})$.

The goal of the RL problem consists of maximizing the sum of rewards, discounted by a factor $\gamma \in [0, 1]$:

$$\mathop{\mathbb{E}}_{\pi, \mathcal{P}} \left[\sum_{t=0}^{\infty} \gamma^t r_t (s_t, g_t, a_t, s_{t+1}) \right] \tag{1}$$

In order to maximize this return, we opted to use Sample Efficient Actor-Critic with Experience Replay (ACER) [25] as it utilizes recent variance reduction techniques, parallel training, and off-policy updates using an experience replay buffer. More specifically we choose ACER because of the following properties:

- Focus on sample efficiency through the usage of an experience replay buffer, which allows usage of environment experiences multiple times.
- Off-policy updates through importance sampling allows for our adaptation method to utilize actions sampled from a different distribution (the prior policies).
- The policy directly outputs a distribution over actions which we can compare with other policies.

In ACER on each training iteration there is an on-policy update after taking n rollout steps. Afterwards there are also one or multiple off-policy updates by taking samples from a replay buffer.

We make use of a word-embedding in order to transform the goal object g_t, described using a word in natural language, to a continuous numerical vector $z_t \in \mathbb{R}^d$ [4]. Essential is that d is much smaller than the size of the entire vocabulary.

3   Disagreement Options

Our method is concerned with utilizing prior knowledge as temporally extended actions (options) in order to increase the sample efficiency, the required interactions with the environment, when learning new tasks.

The approach can be divided into two distinct sub-systems, which each address an important question. The *task similarity* system (Sect. 3.1) is concerned with selecting useful prior knowledge which will be best suited in order to solve the novel task. For example: would a *bicycle riding* skill be more useful than a *car driving* skill when learning how to ride a motorbike? Once we have selected which priors we would like to use, the *task adaptation* phase (Sect. 3.2) is initiated in order to train a new policy by intelligently reasoning when to utilize prior knowledge as temporally extended actions, and when to explore the environment. The agent assumes the presence of a set of prior policies, we discuss some possibilities on how to acquire such priors in Sect. 3.3.

The pseudocode of the entire approach is presented in Algorithm 1.

Algorithm 1. Disagreement Options

$\mathcal{M}(\cdot)$: Pre-trained word-embedding
$\mathcal{B}$: disagreement score buffer with max size α
$\pi(s_t, g_t)$: new policy under training

1: **while** agent rollout in progress **do**
2:　　Observe state s_t and goal g_t
3:　　$x \sim \mathcal{U}(0,1)$
4:　　**if** $x < \mathcal{H}(\pi(s_t, g_t)) - 0.1$ **then**
5:　　　　Find 2 closest prior policies (π_{z1}, π_{z2}) according to:
　　　　　$z_i = argmax_{z_i}(cos(\mathcal{M}(g_t), \mathcal{M}(z_i)))$
6:　　　　Calculate disagreement score:
　　　　　$d1 = D_{KL}(\pi_{z_1}(s_t, z_1) || \pi_{z_2}(s_t, z_2))$
　　　　　$d2 = D_{KL}(\pi_{z_2}(s_t, z_2) || \pi_{z_1}(s_t, z_1))$
　　　　　$d = min(d1, d2)$
7:　　　　Add disagreement score to buffer $\mathcal{B}$
8:　　　　**if** $\sum_i^\alpha \mathcal{B}_i / \alpha > \beta$ **then**
9:　　　　　$x \sim \mathcal{U}(0,1)$
10:　　　　**if** $x < 0.5$ **then**
11:　　　　　　Perform action $a_t \sim \pi_{z_1}(s_t, z_1)$
12:　　　　**else**
13:　　　　　　Perform action $a_t \sim \pi_{z_2}(s_t, z_2)$
14:　　　　**end if**
15:　　**end if**
16:　　**else**
17:　　　Perform action $a_t \sim \pi(s_t, g_t)$
18:　　**end if**
19:　　store $\langle s_t, a_t, s_{t+1}, g_t, r_{t+1} \rangle$ in ACER experience replay buffer
20: **end while**
21: Perform ACER on-policy update
22: Perform n ACER off-policy updates

3.1　Task Similarity: How to Select Relevant Priors?

The agent is provided with a library of different prior policies $\{\pi_{g_1}, ..., \pi_{g_i}\}$, all capable of reliably performing one or multiple different tasks $\{g_1, ..., g_i\}$. In order to decide which prior policies are useful as prior knowledge when learning a new task, we make use of natural language. Our reasoning is that when goal descriptions are close in language space, they are potentially also close in policy space [8,14].

More specifically, we use a pre-trained word-embedding from [13]. This embedding was pre-trained on a set of tasks which are not tailored to our setting, utilizing the *OntoNotes 5* [26] dataset. Our embedding is trained [15] by taking as input a large corpus of texts, and outputs a vector space $\mathbb{R}^{300}$. Words that appear in similar contexts, are trained to also be close to each other in the resulting vector space.

When confronted with a new goal g_t, we calculate the cosine similarity of the resulting vector, after being processed through the word-embedding $\mathcal{M}(x)$ with all labels $\{z_0, ..., z_i\}$ attached to the available prior policies $\{\pi_{z_0}, ..., \pi_{z_i}\}$:

$$z_i = argmax_{z_i}(cos(\mathcal{M}(g_t), \mathcal{M}(z_i))) \tag{2}$$

We select the two policies whose labels are closest to the new goal in the word-embedding space as prior knowledge. Our method requires at least two policies in order to calculate a disagreement between their action distributions in the next phase. We use the minimum of two prior policies in the rest of this paper, as prior knowledge is often expensive to acquire. However, it's a straightforward extension to adapt our method to use more priors. The cosine similarity between goal objects used in our experiments is pictured in Fig. 1. For example, in an *ObjectGoal* task, when asked to navigate to a new goal object *shower*, policies attached to goals such as *bathtub* and *toilet* are most similar in the word-embedding space, and will be selected (if available) as most potent source tasks.

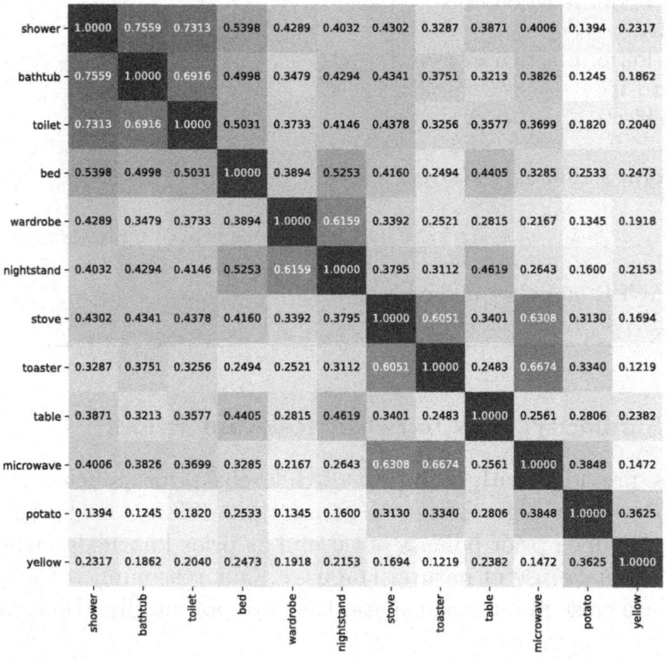

	shower	bathtub	toilet	bed	wardrobe	nightstand	stove	toaster	table	microwave	potato	yellow
shower	1.0000	0.7559	0.7313	0.5398	0.4289	0.4032	0.4302	0.3287	0.3871	0.4006	0.1394	0.2317
bathtub	0.7559	1.0000	0.6916	0.4998	0.3479	0.4294	0.4341	0.3751	0.3213	0.3826	0.1245	0.1862
toilet	0.7313	0.6916	1.0000	0.5031	0.3733	0.4146	0.4378	0.3256	0.3577	0.3699	0.1820	0.2040
bed	0.5398	0.4998	0.5031	1.0000	0.3894	0.5253	0.4160	0.2494	0.4405	0.3285	0.2533	0.2473
wardrobe	0.4289	0.3479	0.3733	0.3894	1.0000	0.6159	0.3392	0.2521	0.2815	0.2167	0.1345	0.1918
nightstand	0.4032	0.4294	0.4146	0.5253	0.6159	1.0000	0.3795	0.3112	0.4619	0.2643	0.1600	0.2153
stove	0.4302	0.4341	0.4378	0.4160	0.3392	0.3795	1.0000	0.6051	0.3401	0.6308	0.3130	0.1694
toaster	0.3287	0.3751	0.3256	0.2494	0.2521	0.3112	0.6051	1.0000	0.2483	0.6674	0.3340	0.1219
table	0.3871	0.3213	0.3577	0.4405	0.2815	0.4619	0.3401	0.2483	1.0000	0.2561	0.2806	0.2382
microwave	0.4006	0.3826	0.3699	0.3285	0.2167	0.2643	0.6308	0.6674	0.2561	1.0000	0.3848	0.1472
potato	0.1394	0.1245	0.1820	0.2533	0.1345	0.1600	0.3130	0.3340	0.2806	0.3848	1.0000	0.3625
yellow	0.2317	0.1862	0.2040	0.2473	0.1918	0.2153	0.1694	0.1219	0.2382	0.1472	0.3625	1.0000

Fig. 1. Similarity scores of different goals in the word-embedding space. These scores are used in order to decide what prior knowledge to use.

3.2   Task Adaptation: How Should We Use the Prior Knowledge?

Once we have selected the prior policies which we expect might be most useful, we can utilize these priors in order to solve the novel task. We treat the selected prior policies as options [24]. Thus, the agent now needs to decide when to use its primitive actions in order to explore, and when to follow the option policies in order to quickly reach new parts of the state-space.

This is a delicate balance, because when the agent would only follow the temporally extended actions greedily, it would not be capable of learning anything new. So, ideally, the agent should be capable of assessing when it should greedily follow the priors, and when it should explore. For example, when we are trying to locate a *toothbrush* object in a house, a temporally extended action that would take the agent to the bathroom is a useful prior. However, once we have entered the bathroom, the agent should explore it, in order to extend its capabilities.

Note that if the agent had access to a sensor that knows in which room the agent resides, this sensor could be used to steer the termination of the active option. Unfortunately, such a sensor is not trivially available, and we propose an alternative scheme based on disagreement between priors, to steer option termination.

In order to decide when to use prior knowledge, we utilize the action distributions of the selected prior policies. Given a state s_t these prior policies output different action distributions. We reason that when these distributions align, measured by the KL divergence between them, it is useful to greedily follow these policies as a temporally extended action. We call this score the *disagreement score*.

$$D_{KL}(\pi_{z_1}(s_t)||\pi_{z_2}(s_t)) = \sum_a \pi_{z_1}(a|s_t) \log \frac{\pi_{z_1}(a|s_t)}{\pi_{z_2}(a|s_t)} \tag{3}$$

Because the KL divergence is not symmetric, we calculate the disagreement score as follows:

$$d = min\left[D_{KL}(\pi_{z_1}(s_t)||\pi_{z_2}(s_t)), D_{KL}(\pi_{z_2}(s_t)||\pi_{z_1}(s_t))\right] \tag{4}$$

By using the minimum we slightly favor utilizing the prior knowledge, which experimentally yielded the best results.

When the two prior policies diverge on what the action of the agent should be, we terminate the temporally extended action and let the agent explore by itself. For example, two policies which pursue a *towel* and a *toothbrush* object, will have similar action distributions up until they reach the bathroom. Upon entering the bathroom the action distributions diverge, because their implicit high-level navigation target changed from reaching the bathroom to reaching the individual objects.

Because the action distributions of the prior policies can be noisy, we utilize a moving average of the disagreement scores B acquired over the last α steps. On each training step, we compare this moving average against a threshold β

in order to decide when to use our prior knowledge, and when to terminate the temporally extended action:

$$a_t = \begin{cases} \pi_{z_1}(s_t), & \text{if } \sum_i^{\alpha} B_i/\alpha > \beta \\ \pi(s_t), & \text{otherwise} \end{cases} \tag{5}$$

When the prior policies are in agreement, we randomly sample the recommended best action from one of the prior policies. As their divergence is small, they will output similar actions, so it does not matter which one to sample from. We take this action in the environment, and use it to update the new policy. In contrast, if there is disagreement, the agent uses the new policy to explore, by sampling an action from it.

While the disagreement window α and the disagreement threshold β are hyperparameters, which potentially are subject to an expensive search in order to get optimal values, we experimentally demonstrate that approximate optimal values can be found easily.

Because ACER has an experience replay buffer, and utilizes off-policy training, after a few iterations, prior knowledge will have found its way into the buffer, and thus also into the new policy. In order to gradually reduce the dependency on the priors, we only rely on the priors when the entropy of the action distribution of the new policy for the current observed state $\mathcal{H}(\pi(s_t, g_t))$ is still high. We assume this distribution entropy lowers as the new policy learns the new task. This is a realistic assumption in a deterministic environment in which an optimal policy will converge to assigning almost all probability to a single action given a state. The entropy measurement is used to gradually reduce the probability of invoking the temporally extended actions:

$$\mathcal{I}(s_t) = P(x \sim \mathcal{U}(0,1) < \mathcal{H}(\pi(s_t))) - 0.1 \tag{6}$$

We correct this probability with a small factor -0.1 in order to encourage exploration early on in training. Increasing this factor will reduce the usage of the priors.

3.3   Prior Policy Acquisition

We assume prior policies are provided a priori to the agent. A lot of different options are available to acquire such source policies. One could use any RL algorithm to train a policy. We especially envision RL methods that maximize entropy to be potent methods to acquire diverse prior policies. For example, VIC [9] tries to maximize the amount of different states the agent can reach by maximizing the mutual information between the set of skills and their termination states.

We also deem it possible to use an imitation learning approach [12,21] to bootstrap the agent, utilizing policies compiled from (human) expert demonstrations.

4   Experiments

We empirically show the effectiveness of our method in two different settings: a simple 3D *gridworld* and the photo-realistic Habitat simulator.

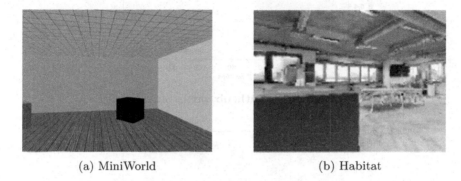

(a) MiniWorld   (b) Habitat

Fig. 2. Example ego-centric RGB states used in our experiments.

4.1   3D MiniWorld

The setting of our first set of experiments consists of a visually basic 3D world. In this environment we simulate a domestic apartment setting with three fixed different designated rooms: a bedroom, a kitchen and a bathroom. Each room has a visually distinct theme, and has multiple objects in it. The objects are represented using differently colored cubes in fixed positions. These three rooms are connected by a corridor. The agent always starts in a random position in this corridor. This setting is implemented as a custom level in the *MiniWorld* [6] environment.

In each episode the agent is tasked with finding an object in this environment. The state-space consists only of the ego-centric RGB render (e.g., Fig. 2a). Additionally, the agent observes a densely defined reward signal, which consists of the decrease of distance between the agent and the goal object. We also penalize the agent for slacking by subtracting a negative reward of -0.01 for each step taken. A positive reward of 10 is rewarded upon reaching a minimum distance to the goal object. The agent is allowed a maximum of 500 steps to reach the goal.

Room Sensor. In order to validate our hypothesis that prior knowledge can be useful to navigate the agent to the room with the goal object in it, we first equip the agent with a room sensor. This sensor informs the agent when it is positioned in the corridor, and thus should follow the prior policies greedily, in order to navigate to the room containing the goal object. We selected prior

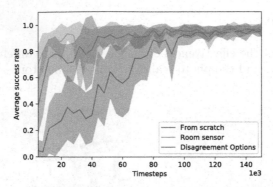

(a) New goal object: **bathtub**, priors: shower, toilet

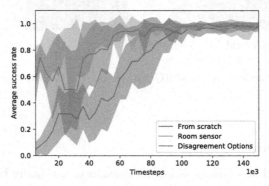

(b) New goal object: **nightstand**, priors: bed, wardrobe

Fig. 3. Average success rate of our disagreement agent (green) and our disagreement agent with a room sensor (orange) during training in the *MiniWorld* environments. We compare with learning the task from scratch (blue). Results are averaged over 10 runs and utilized window size $\alpha = 10$ and disagreement threshold $\beta = 0.1$. (Color figure online)

policies which were trained on goal objects that are in the same room as the new goal object. Once inside the correct room, the agent knows not to follow the prior anymore, but to explore by itself.

When utilizing this room sensor with prior policies capable of navigating to the *shower* and *toilet* goal objects, our results show that the agent almost instantly (50k training steps) is capable of adapting to reliably reach the new *bathtub* goal (Fig. 3a). Similarly, the agent is capable of quickly learning to navigate to the *nightstand* goal object using prior policies capable of reaching the *bed* and *wardrobe* (Fig. 3b). We plot an example trajectory followed during training in Fig. 4. In this trajectory the usage of prior knowledge that led the agent to the correct room is plotted in green, while the exploratory part of the trajectory is plotted in blue.

Fig. 4. Example trajectory of our disagreement agent followed during training. In the green part of the trajectory the agent follows the prior, in the blue part the agent explores the environment. In this case, the agent has access to a room sensor and only explores in the room of the goal object. (Color figure online)

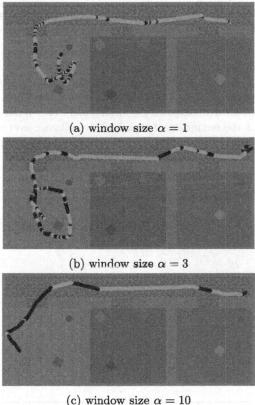

(a) window size $\alpha = 1$

(b) window size $\alpha = 3$

(c) window size $\alpha = 10$

Fig. 5. Example trajectories of our disagreement agent using different disagreement windows in the *MiniWorld* environment. Parts of the trajectory marked in green utilized the prior knowledge, in blue parts the agent explored. In this setting larger disagreement windows lead to more stable utilization of the prior knowledge. (Color figure online)

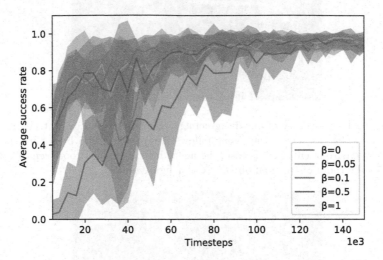

Fig. 6. Ablation study of the disagreement threshold in the *MiniWorld* environment (new goal: bathtub, priors: shower, toilet). A value of 0 never utilizes the prior knowledge, while a value of 1 does not explore the environment (when the action distribution entropy is still high at the beginning of training). Results are averaged over 10 runs.

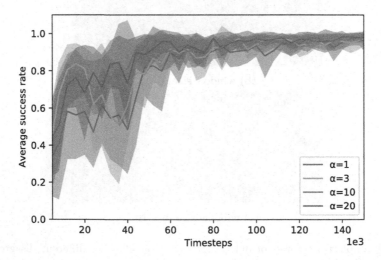

Fig. 7. Average success rate of our disagreement agent in the *MiniWorld* environment on the bathtub task. We compare different disagreement window sizes. Longer disagreement windows lead to more stable utilization of the temporally extended actions. Results are averaged over 10 runs.

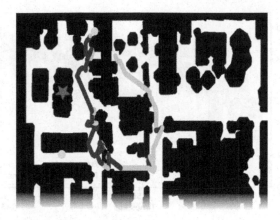

Fig. 8. An example trajectory of the agent in a scan of our office floor. The red star is the new goal, while the prior goals are marked with a yellow circle. (Color figure online)

Disagreement Options. However, a room sensor is not something an autonomous agent typically has access to. In the second set of experiments we wanted to validate whether the disagreement options provide a similar efficient usage of prior knowledge without such a sensor.

As plotted in Fig. 3, the agent is capable of efficiently utilizing the prior knowledge when using the disagreement scheme ($\alpha = 10$, $\beta = 0.1$), starting with a success rate averaging 60–80%, and quickly getting an average success rate of nearly 100%.

We also did an ablation study of our hyperparameters in this setting. In Fig. 6, we demonstrate the impact of the disagreement threshold β. When setting the value too high, the agent does not explore enough, while a too low β value will only limitedly benefit the task adaptation.

Figure 7 presents the impact of the disagreement window size α. In this setting, larger window sizes ($\alpha > 3$) are more efficient, as smaller window sizes lead to noisy trajectories, while a larger window size allows the agent to exploit the prior knowledge more systematically (Fig. 5).

4.2   Photorealistic Simulator

For our second set of experiments, we use the Habitat photo-realistic simulator [22] and a 3D scan of our office floor (Fig. 8). This environment is considerably more challenging than the *MiniWorld* environment, both structurally and visually. We use the same reward setting as in our *MiniWorld* experiments. Similar to the *MiniWorld* environment, the agent only has access to a visual RGB ego-centric observation of the current state. In this setting the agent starts in a completely random position, and is allowed to take 500 actions in order to reach a new goal in a fixed position (the main table in the office canteen). In order to master this novel task, the agent has access to two prior policies which are capable of navigating to two other goals within the canteen.

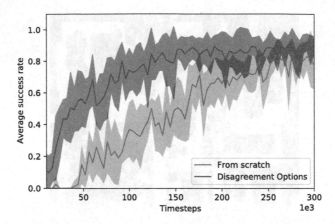

Fig. 9. Results of the disagreement agent (window size $\alpha = 10$, disagreement threshold $\beta = 0.1$) in the photorealistic Habitat simulator (green) compared to learning the task from scratch. Results are averaged over 10 runs. (Color figure online)

The results from using our disagreement options within this environment can be found in Fig. 9. While this task is considerably harder in terms of structure than the *MiniWorld* tasks, the agent is capable of utilizing the prior knowledge in order to reach a nearly perfect average success rate on the novel task considerably faster (150k training steps vs 250k training steps), than if the agent would have to start from scratch.

5   Towards Real-World Task Adaptation

Because our method only relies on goals formulated in natural language and egocentric visual observations, we can also potentially use our method in a real-world setting. In this setting we let the agent solve different tasks in simulation, and through *sim2real* techniques, utilize them in the real world. When confronted with a new task in the real world, the agent could use the prior knowledge gathered in simulation to solve the novel task considerably faster in the real world.

It is often not possible to define a dense reward signal in the real world. The use of prior knowledge allows our agent to efficiently reach states closer to the goal object, and thus increases the chance of the agent obtaining positive learning signals. This allows us to believe that it might be possible to learn only from sparse reward signals, which are more obtainable in real-world scenarios.

6   Related Work

Transfer learning has been utilized successfully in supervised learning tasks. In this setting, multiple levels of low-level learned features can often be re-used in order to speed up learning novel tasks.

A lot of research has been conducted on how prior knowledge can be utilized in RL as well. As learned lower level features in RL are often task specific, it is generally difficult to simply re-use them. Instead, prior knowledge has been utilized in RL as auxiliary reward signals, policy distillation, inter-task mapping and as temporally extended actions. An example algorithm using natural language as an auxiliary reward can be found in [3]. [20] proposes to use natural language as an intermediate channel to facilitate transfer between different domains. LamBERT [17] has a multi-modal visual and language representation which proved to be beneficial for transfer to novel tasks.

Word embeddings have been used in RL as an action-space reduction technique [8], and we examined the usefulness of word embeddings for task adaptation in prior work [14].

Some of the research which is closest to ours includes the Deep Q-learning from Demonstration (DQfD) architecture [11], which learns both from demonstrations and trial-and-error learning by introducing an additional replay buffer.

[7] proposed a probabilistic distribution over prior policies (the policy library) based on the expected performance gain of utilizing the prior policy.

7   Discussion

Our task-adaptation method is supported by the assumption that goals that are close in language-space should also be close in policy-space. However, this might not always be the case. If the agent selects prior goals which are physically located nowhere near the new goal, but in different locations, our method will not hinder progress as the priors will always disagree, and thus the agent will not use the priors. If however the wrong priors do agree on the next action, the agent will be steered in the wrong direction, and learning will be slower. In these settings the disagreement threshold could be lowered.

In our experiments we utilized a deterministic environment. If the environment is completely stochastic (e.g., all objects are randomly placed in random rooms) our method would not be able to utilize prior knowledge. However, if objects are placed in random positions, but always in the same rooms, our adaptation method would still be capable of adapting, and could even benefit from the learned ability of the priors to explore a certain room.

8   Conclusion

In this paper, we presented a novel method to transfer prior knowledge from prior tasks to a new task through temporally extended actions. We do this by selecting prior knowledge based on cosine similarity in a prior word-embedding space. In order to decide when to utilize our prior knowledge, and when to explore our environment, we rely on the disagreement between action distributions of the selected priors.

We demonstrate the effectiveness of our method in a visually simple 3D *MiniWorld* and a photorealistic simulator. We also hint at how our method might be used in the real world to expand the capabilities of a real-world embodied agent.

As future work, we would like to address the management and scaling of prior and novel policies. We also would like to examine how additional abstractions can be discovered using our disagreement method, potentially in a multi-level hierarchical system.

Acknowledgements. This research received funding from the Flemish Government under the "Onderzoeksprogramma Artificiële Intelligentie (AI) Vlaanderen" program.

References

1. Anderson, P., et al.: On evaluation of embodied navigation agents. arXiv:1807.06757 [cs] (2018)
2. Bacon, P.L., Harb, J., Precup, D.: The option-critic architecture. In: AAAI17 (2017)
3. Bahdanau, D., et al.: Learning to understand goal specifications by modelling reward. In: ICLR19 (2019)
4. Bengio, Y., Ducharme, R., Vincent, P., Jauvin, C.: A neural probabilistic language model. JMLR **3**(Feb), 1137–1155 (2003)
5. Chaplot, D.S., Gandhi, D., Gupta, A., Salakhutdinov, R.: Object goal navigation using goal-oriented semantic exploration. In: NeurIPS20 (2020)
6. Chevalier-Boisvert, M.: gym-miniworld environment for openai gym (2018). https://github.com/maximecb/gym-miniworld
7. Fernández, F., Veloso, M.: Probabilistic policy reuse in a reinforcement learning agent. In: AAMAS06 (2006)
8. Fulda, N., Ricks, D., Murdoch, B., Wingate, D.: What can you do with a rock? Affordance extraction via word embeddings. In: IJCAI17 (2017)
9. Gregor, K., Rezende, D.J., Wierstra, D.: Variational intrinsic control. arXiv:1611.07507 [cs] (2016)
10. Gupta, S., Tolani, V., Davidson, J., Levine, S., Sukthankar, R., Malik, J.: Cognitive mapping and planning for visual navigation. Int. J. Comput. Vision **128**(5), 1311–1330 (2020). https://doi.org/10.1007/s11263-019-01236-7
11. Hester, T., et al.: Deep Q-learning from demonstrations. In: AAAI18 (2017)
12. Ho, J., Ermon, S.: Generative adversarial imitation learning. In: Advances in Neural Information Processing Systems, vol. 29 (2016)
13. Honnibal, M., Montani, I., Van Landeghem, S., Boyd, A.: spaCy: industrial-strength natural language processing in Python (2020). https://doi.org/10.5281/zenodo.1212303
14. Hutsebaut-Buysse, M., Mets, K., Latré, S.: Pre-trained word embeddings for goal-conditional transfer learning in reinforcement learning. In: 1st Workshop on Language in Reinforcement Learning (2020)
15. Mikolov, T., Chen, K., Corrado, G., Dean, J.: Efficient estimation of word representations in vector space (2013)
16. Mirowski, P., et al.: Learning to navigate in complex environments. In: ICLR17 (2017)
17. Miyazawa, K., Aoki, T., Horii, T., Nagai, T.: lamBERT: language and action learning using multimodal BERT (2020)

18. Mousavian, A., Toshev, A., Fiser, M., Kosecka, J., Wahid, A., Davidson, J.: Visual representations for semantic target driven navigation. In: ICRA19, pp. 8846–8852. IEEE, Montreal (2019). https://doi.org/10.1109/ICRA.2019.8793493
19. Nachum, O., Gu, S., Lee, H., Levine, S.: Data-efficient hierarchical reinforcement learning. In: NIPS18 (2018)
20. Narasimhan, K., Barzilay, R., Jaakkola, T.: Grounding language for transfer in deep reinforcement learning. JAIR **63**, 849–874 (2018)
21. Ross, S., Gordon, G., Bagnell, D.: A reduction of imitation learning and structured prediction to no-regret online learning. In: Gordon, G., Dunson, D., Dudík, M. (eds.) Proceedings of the Fourteenth International Conference on Artificial Intelligence and Statistics. Proceedings of Machine Learning Research, vol. 15, pp. 627–635. PMLR, Fort Lauderdale (2011)
22. Savva, M., et al.: Habitat: a platform for embodied AI research. In: ICCV19 (2019)
23. Silver, D.L., Yang, Q., Li, L.: Lifelong machine learning systems: beyond learning algorithms. In: AAAI13 (2013)
24. Sutton, R.S., Precup, D., Singh, S.: Between MDPs and semi-MDPs: a framework for temporal abstraction in reinforcement learning. Artif. Intell. **112**(1–2), 181–211 (1999). https://doi.org/10.1016/S0004-3702(99)00052-1
25. Wang, Z., et al.: Sample efficient actor-critic with experience replay. In: ICLR17 (2017)
26. Weischedel, R., et al.: OntoNotes: a large training corpus for enhanced processing (2013)
27. Wijmans, E., et al.: DD-PPO: learning near-perfect pointgoal navigators from 2.5 billion frames. In: ICLR20 (2020)
28. Zhu, Y., et al.: Target-driven visual navigation in indoor scenes using deep reinforcement learning. In: ICRA17 (2017). https://doi.org/10.1109/ICRA.2017.7989381

Deep Adaptive Multi-intention Inverse Reinforcement Learning

Ariyan Bighashdel[✉], Panagiotis Meletis, Pavol Jancura, and Gijs Dubbelman

Eindhoven University of Technology, 5612 AZ Eindhoven, The Netherlands
{a.bighashdel,p.c.meletis,p.jancura,g.dubbelman}@tue.nl

Abstract. This paper presents a deep Inverse Reinforcement Learning (IRL) framework that can learn an *a priori* unknown number of nonlinear reward functions from unlabeled experts' demonstrations. For this purpose, we employ the tools from Dirichlet processes and propose an adaptive approach to simultaneously account for both complex and unknown number of reward functions. Using the conditional maximum entropy principle, we model the experts' multi-intention behaviors as a mixture of latent intention distributions and derive two algorithms to estimate the parameters of the deep reward network along with the number of experts' intentions from unlabeled demonstrations. The proposed algorithms are evaluated on three benchmarks, two of which have been specifically extended in this study for multi-intention IRL, and compared with well-known baselines. We demonstrate through several experiments the advantages of our algorithms over the existing approaches and the benefits of online inferring, rather than fixing beforehand, the number of expert's intentions.

Keywords: Inverse reinforcement learning · Multiple intentions · Deep learning

1 Introduction

The task of learning from demonstrations (LfD) lies in the heart of many artificial intelligence applications [29,38]. By observing the expert's behavior, an agent learns a mapping between world states and actions. This so-called *policy* enables the agent to select and perform an action, given the current world state. Despite the fact that this policy can be directly learned from expert's behaviors, inferring the *reward function* underlying the policy is generally considered the most succinct, robust, and transferable methodology for the LfD task [1]. Inferring the reward function, which is the objective of Inverse Reinforcement Learning (IRL), is often very challenging in real-world scenarios. The demonstrations come from multiple experts who can have different intentions, and their behaviors are consequently not well modeled with a single reward function. Therefore, in this study, we research and extend the concept of *mixture of conditional maximum entropy models* and propose a deep IRL framework to infer an *a priori* unknown number of reward functions from experts' demonstrations without intention labels.

© Springer Nature Switzerland AG 2021
N. Oliver et al. (Eds.): ECML PKDD 2021, LNAI 12975, pp. 206–221, 2021.
https://doi.org/10.1007/978-3-030-86486-6_13

Standard IRL can be described as the problem of extracting a reward function, which is consistent with the observed behaviors [34]. Obtaining the exact reward function is an ill-posed problem, since many different reward functions can explain the same observed behaviors [26,40]. Ziebart et al. [40] tackled this ambiguity by employing the principle of maximum entropy [16]. The principle states that the probability distribution, which best represents the current state of knowledge, is the one with the largest entropy [16]. Therefore, Ziebart et al. [40] chose the distribution with maximal information entropy to model the experts' behaviors. The maximum entropy IRL has been widely employed in various applications [17,35]. However, this method suffers from a strong assumption that the experts have one single intention in all demonstrations. In this study, we explore the principle of the mixture of maximum entropy models [31] that inherits the advantages of maximum entropy principle, while at the same time is capable of modeling multi-intention behaviors.

In many real-world applications, the demonstrations are often collected from multiple experts whose intentions are potentially different from each other [2,3,5, 10]. This leads to multiple reward functions, which is in direct contradiction with the single reward assumption in traditional IRL. To address this problem, Babes et al. [5] proposed a clustering-IRL scheme where the class of each demonstration is jointly learned via the respective reward function. Despite the recovery of multiple reward functions, the number of clusters in this method is assumed to be known *a priori*. To overcome this assumption, Choi et al. [10] presented a non-parametric Bayesian approach using the Dirichlet Process Mixture (DPM) to infer an unknown number of reward functions from unlabeled demonstrations. However, the proposed method is formulated based on the assumption that the reward functions are formed by a linear combination of a set of world state features. In our work, we do not make this assumption on linearity and model the reward functions using deep neural networks.

DPM is a stochastic process in the Bayesian non-parametric framework that deals with mixture models with a countably infinite number of mixture components [25]. In general, full Bayesian inference in DPM models is not feasible, and instead, approximate methods like Monte-Carlo Markov chain (MCMC) [4,20] and variational inference [8] are employed. When deep neural networks are involved in DPM (e.g. deep nonlinear reward functions in IRL), approximates methods may not be able to scale with high dimensional parameter spaces. MCMC sampling methods are shown to be slow in convergence [8,30] and variational inference algorithms suffer from restrictions in the distribution family of the observable data, as well as various truncation assumptions for the variational distribution to yield a finite dimensional representation [12,24]. These limitations apparently make approximate Bayesian inference methods inapplicable for DPM models with deep neural networks. Apart from that, the algorithms for maximum likelihood estimations like standard EM are no longer tractable when dealing with DPM models. The main reason is that the number of mixture components exponentially grows with non-zero probabilities, and after some iterations, the Expectation-step would be no longer available in a closed-form.

However, inspired by two variants of EM algorithms that cope with infeasible Expectation-step [9,37], we propose two solutions in which the Expectation-step is either estimated numerically with sampling (based on Monte Carlo EM [37]) or computed analytically and then replaced with a sample from it (based on stochastic EM [9]).

This study's main contribution is to develop an IRL framework where one can benefit from the strength of 1) maximum entropy principle, 2) deep nonlinear reward functions, and 3) account for an unknown number of experts' intentions. To the best of our knowledge, we are the first to present an approach that can combine all these three capabilities.

In our proposed framework, the experts' behavioral distribution is modeled as a mixture of conditional maximum entropy models. The reward functions are parameterized as a deep reward network, consisting of two parts: 1) a base reward model, and 2) an adaptively growing set of intention-specific reward models. The base reward model takes as input the state features and outputs a set of reward features shared in all intention-specific reward models. The intention-specific reward models take the reward features and output the rewards for the respective expert's intention. A novel adaptive approach, based on the concept of the Chinese Restaurant Process (CRP), is proposed to infer the number of experts' intentions from unlabeled demonstrations. To train the framework, we propose and compare two novel EM algorithms. One is based on stochastic EM and the other on Monte Carlo EM. In Sect. 3, this problem of multi-intention IRL is defined, following our two novel EM algorithms in Sect. 4. The results are evaluated on three available simulated benchmarks, two of which are extended in this paper for multi-intention IRL, and compared with two baselines [5,10]. These experimental results are reported in Sect. 5 and Sect. 6 is devoted to conclusions. The source code to reproduce the experiments is publicly available[1].

2   Related Works

In the past decades, a number of studies have addressed the problem of multi-intention IRL. A comparison of various methods for multi-intention IRL, together with our approach, is depicted in Table 1.

In an early work, Dimitrakakis and Rothkopf [11] formulated the problem of learning from unlabeled demonstrations as a multi-task learning problem. By generalizing the Bayesian IRL approach of Ramachandran and Amir [33], they assumed that each observed trajectory is responsible for one specific reward function, all of which shares a common prior. The same approach has also been employed by Noothigattu et al. [28], who assumed that each expert's reward function is a random permutation of one sharing reward function. Babes et al. [5] took a different approach and addressed the problem as a clustering task with IRL. They proposed an EM approach that clusters the observed trajectories by inferring the rewards function for each cluster. Using maximum likelihood, they estimated the reward parameters for each cluster.

---

[1] https://github.com/tue-mps/damiirl.

The main limitation in EM clustering approach is that the number of clusters has to be specified as an input parameter [5,27]. To overcome this assumption, Choi and Kim [10] employed a non-parametric Bayesian approach via the DPM model. Using MCMC sampler, they were able to infer an unknown number of reward functions, which are linear combinations of state features. Other authors have also employed the same methodology in the literature [2,23,32].

All above methods are developed on the basis of model-based reinforcement learning (RL), in which the model of the environment is assumed to be known. In the past few years, a couple of approximate, model-free methods have been developed for IRL with multiple reward functions [14,15,21,22]. Such methods aimed to solve large-scale problems by approximating the Bellman optimality equation with model-free RL.

In this study, we constrain ourselves to model-based RL and propose a multi-intention IRL approach to infer an unknown number of experts' intentions and corresponding nonlinear reward functions from unlabeled demonstrations.

Table 1. Comparison of proposed models for multi-intention IRL.

Models	Type		Features		
	Model based	Model free	Unlabeled demonstrations	Unknown # intentions	Non-linear reward fun.
Dimitrakakis and Rothkopf [11]	✓		✓		
Babes et al. [5]	✓		✓		
Nguyen et al. [27]	✓		✓		
Choi and Kim [10]	✓		✓	✓	
Rajasekaran et al. [32]	✓		✓	✓	
Li et al. [21]		✓	✓		✓
Hausman et al. [14]		✓	✓		✓
Lin and Zhang [22]		✓			✓
Hsiao et al. [15]		✓	✓		✓
Ours	✓		✓	✓	✓

3   Problem Definition

In this section, the problem of multi-intention IRL is defined. To facilitate the flow, we first formalize the multi-intention RL problem. For both problems, we follow the conventional modelling of the environment as a Markov Decision Process (MDP). A finite state MDP in a multi-intention RL problem is a tuple $(S, A, T, \gamma, b_0, R_1, R_2, ..., R_K)$ where S is the state space, A is the action space, $T : S \times A \times S \rightarrow [0,1]$ is the transition probability function, $\gamma \in [0,1)$ is the discount factor, $b_0(s)$ is the probability of staring in state s, and $R_k : S \rightarrow \mathbb{R}$ is the kth reward function with K to be the total number of intentions. A policy is a mapping function $\pi_k : S \rightarrow A \ \forall k \in \{1, 2, ..., K\}$. The value of policy π_k

with respect to the kth reward function is the expected discounted reward for following the policy and is defined as $V_{R_k}^{\pi} = \mathbb{E}[\sum_t \gamma^t R_k(s_t)|b_0]$. The optimal policy (π_k^*) for the kth reward function is the policy that maximizes the value function for all states and satisfies the respective Bellman optimality equation [36].

In multi-intention IRL, the context of this study, a finite-state MDP\R is a tuple $(S, A, T, \gamma, b_0, \tau^1, \tau^2, ..., \tau^M)$ where τ^m is the mth demonstration and M is the total number of demonstrations. In this work, it is assumed that there is a total of K intentions, each of which corresponds to one reward function, so that τ^m with length T_τ is generated from the optimal policy (π_k^*) of the kth reward function. It is further assumed that the demonstrations are without intention labels, i.e. they are *unlabeled*. Therefore, the goal is to infer the number of intentions K and the respective reward function of each intention. In the next section, we model the experts' behaviors as a mixture of conditional maximum entropy models, parameterize the reward functions via deep neural networks, and propose a novel approach to infer an unknown number of experts' intentions from unlabeled demonstrations.

4   Approach

In the proposed framework for multi-intention IRL, the experts' behavioral distribution is modeled as a mixture of conditional maximum entropy models. The Mixture of conditional maximum entropy models is a generalization of standard maximum entropy formulation for cases where the data distributions arise from a mixture of simpler underlying latent distributions [31]. According to this principal, a mixture of conditional maximum entropy models is a promising candidate to justify the multi-intention behaviors of the experts. The experts' behaviors with the kth intention is defined via a conditional maximum entropy distribution:

$$p(\tau|\eta_k = 1, \Psi) = exp(R_k(\tau, \Psi_k))/Z_k, \tag{1}$$

where $\eta = \{\eta_1, \eta_2, ..., \eta_K | \forall \eta_k \in \{0, 1\}, \sum_{k=1}^{K} \eta_k = 1\}$ is the latent intention vector, $R_k(\tau, \Psi_k) = \sum_{s \in \tau} R_k(s, \Psi_k)$ is the reward of the trajectory with respect to the kth reward function with $R_k(s, \Psi_k)$ as the state reward value, and Z_k is the kth partition function.

We define the kth reward function as: $R_k(s, \Psi_k) = R_{\Psi_k}(\boldsymbol{f}_s)$, where R_{Ψ_k} is a deep neural network with finite set of parameters $\Psi_k = \{\Theta_0, \Theta_k\}$ which consists of a base reward model R_{Θ_0} and an intention-specific reward model R_{Θ_k} (See Fig. 1). The base reward model with finite set of parameters Θ_0 takes the state feature vector $\boldsymbol{f}_s$ and outputs the state reward feature vector $\boldsymbol{r}_s$: $\boldsymbol{r}_s = R_{\Theta_0}(\boldsymbol{f}_s)$. The state reward feature vector $\boldsymbol{r}_s$ that is produced by the base reward model is input to all intention-specific reward models. The kth intention-specific reward model with finite set of parameters Θ_k, takes the state reward feature vector $\boldsymbol{r}_s$ and outputs the state reward value: $R_k(s, \Psi_k) = R_{\Theta_k}(\boldsymbol{r}_s)$. Therefore the total set of reward parameters is $\Psi = \{\Theta_0, \Theta_1, ..., \Theta_K\}$. The reward of the trajectory τ with respect to the kth reward function can be further obtained as:

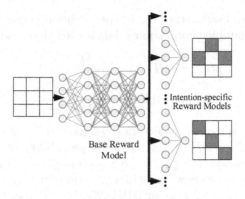

Fig. 1. Schematics of deep reward network.

$R_k(\boldsymbol{\tau}, \Psi_k) = \boldsymbol{\mu}(\boldsymbol{\tau})^\mathsf{T} \boldsymbol{R}_{\Psi_k}(\boldsymbol{\tau})$, where $\boldsymbol{\mu}(\boldsymbol{\tau})$ is the expected State Visitation Frequency (SVF) vector for trajectory $\boldsymbol{\tau}$ and $\boldsymbol{R}_{\Psi_k}(\boldsymbol{\tau}) = \{R_{\Psi_k}(\boldsymbol{f}_s) | \forall s \in S\}$ is the vector of reward values of all states with respect to the kth reward function.

In order to infer the number of intentions K, we propose an adaptive approach in which the number of intentions adaptively changes whenever a trajectory is visited/re-visited. For this purpose, at each iteration we first assume to have $M-1$ demonstrated trajectories $\{\boldsymbol{\tau}^1, \boldsymbol{\tau}^2, ..., \boldsymbol{\tau}^{m-1}, \boldsymbol{\tau}^{m+1}, ..., \boldsymbol{\tau}^M\}$ that are already assigned to K intentions with known latent intention vectors $\boldsymbol{H}^{-m} = \{\boldsymbol{\eta}^1, \boldsymbol{\eta}^2, ..., \boldsymbol{\eta}^{m-1}, \boldsymbol{\eta}^{m+1}, ..., \boldsymbol{\eta}^M\}$. Then, we visit/re-visit a demonstrated trajectory $\boldsymbol{\tau}^m$ and the task is to obtain the latent intention vector $\boldsymbol{\eta}^m$, which can be assigned to a new intention $K+1$, and update the reward parameters Ψ. As emphasized before, our work aims to develop a method in which K, the number of intentions, is a priori unknown and can, in theory, be arbitrarily large. Now we define the predictive distribution for the trajectory $\boldsymbol{\tau}^m$ as a mixture of conditional maximum entropy models:

$$p(\boldsymbol{\tau}^m | \boldsymbol{H}^{-m}, \Psi) = \sum_{k=1}^{K+1} p(\boldsymbol{\tau}^m | \eta_k^m = 1, \Psi) p(\eta_k^m = 1 | \boldsymbol{H}^{-m}) \qquad (2)$$

where $p(\eta_k^m = 1 | \boldsymbol{H}^{-m})$ is the prior intention assignment for trajectory $\boldsymbol{\tau}^m$, given all other latent intention vectors. In the case of K intentions, we define a multinomial prior distribution over all latent intention vectors $\boldsymbol{H} = \{\boldsymbol{H}^{-m}, \boldsymbol{\eta}^m\}$:

$$p(\boldsymbol{H} | \boldsymbol{\phi}) = \prod_{k=1}^{K} \phi_k^{M_k} \qquad (3)$$

where M_k is the number of trajectories with intention k and $\boldsymbol{\phi}$ is the vector of mixing coefficients $\boldsymbol{\phi} = \{\phi_1, \phi_2, ...\phi_K\}$ with Dirichlet prior distribution $p(\boldsymbol{\phi}) = \mathrm{Dir}(\alpha/K)$, where α is the concentration parameter. As $K \to \infty$ the main problematic parameters are the mixing coefficients. Marginalizing out the mixing coefficients and separating the latent intention vector for m^{th} trajectory yield (see Sect. 1 of supplementary materials for full derivation [6]):

$$p(\eta_k^m = 1|\mathbf{H}^{-m}) = \frac{M_k^{-m}}{M-1+\alpha}$$
$$p(\eta_{K+1}^m = 1|\mathbf{H}^{-m}) = \frac{\alpha}{M-1+\alpha}$$

(4)

where M_k^{-m} is the number of trajectories assigned to intention k excluding the mth trajectory, $p(\eta_k^m = 1|\mathbf{H}^{-m})$ is the prior probability of assigning the new trajectory m to intention $k \in \{1, 2, ..., K\}$, and $p(\eta_{K+1}^m = 1|\mathbf{H}^{-m})$ is the prior probability of assigning the new trajectory m to intention $K+1$. Equation (4) is known as the CRP representation for DPM [25]. Considering the exchangeability property [13], the following optimization problem is defined:

$$\max_{\Psi} L^m(\Psi) = \log \sum_{k=1}^{K+1} p(\boldsymbol{\tau}^m|\eta_k^m = 1, \Psi)p(\eta_k^m = 1|\mathbf{H}^{-m}) \quad \forall m \in \{1, 2, ..., M\}$$

(5)

The parameters Ψ can be estimated via Expectation Maximization (EM) [7]. Differentiating $L^m(\Psi)$ with respect to $\psi \in \Psi$ yields the following E-step and M-step (see Sect. 2 of supplementary materials for full derivation):

E-Step. Evaluation of the posterior distribution over the latent intention vector $\forall k \in \{1, 2, ..., K\}$:

$$\gamma_k^m = \frac{M_k^{-m} \prod_{t=0}^{T_\tau-1} \pi_k(a_t|s_t)}{\alpha \prod_{t=0}^{T_\tau-1} \pi_{K+1}(a_t|s_t) + \sum_{\hat{k}=1}^{K} M_{\hat{k}}^{-m} \prod_{t=0}^{T_\tau-1} \pi_{\hat{k}}(a_t|s_t)}$$

(6)

and for $k = K+1$:

$$\gamma_k^m = \frac{\alpha \prod_{t=0}^{T_\tau-1} \pi_k(a_t|s_t)}{\alpha \prod_{t=0}^{T_\tau-1} \pi_{K+1}(a_t|s_t) + \sum_{\hat{k}=1}^{K} M_{\hat{k}}^{-m} \prod_{t=0}^{T_\tau-1} \pi_{\hat{k}}(a_t|s_t)}$$

(7)

where we have defined $\gamma_k^m = p(\eta_k^m = 1|\boldsymbol{\tau}^m, \mathbf{H}^{-m}, \Psi)$.

M-Step update of the parameter value $\psi \in \Psi$ with gradient of:

$$\nabla_\psi L(\Psi) = \sum_{k=1}^{K+1} \gamma_k^m (\boldsymbol{\mu}(\boldsymbol{\tau}^m) - \mathbb{E}_{p(\boldsymbol{\tau}|\eta_k=1,\Psi)}[\boldsymbol{\mu}(\boldsymbol{\tau})])^\intercal \frac{d\boldsymbol{R}_{\Psi_k}(\boldsymbol{\tau})}{d\psi}$$

(8)

where $\mathbb{E}_{p(\boldsymbol{\tau}|\eta_k=1,\Psi)}[\boldsymbol{\mu}(\boldsymbol{\tau})]$ is the expected SVF vector under the parameterized reward function R_{Ψ_k} [40].

When K approaches infinity, the EM algorithm is no longer tractable since the number of mixture components exponentially grows with non-zero probabilities. As a result, after some iterations, the E-step would be no longer available in a closed-form. We propose two solutions for estimation of the reward parameters which are inspired by stochastic and Monte Carlo EM algorithms. Both proposed solutions are deeply evaluated and compared with in Sect. 5.

Algorithm 1: Adaptive multi-intention IRL based on stochastic EM

Initialize K, $\Theta_0, \Theta_1, \Theta_2, ..., \Theta_K$, $M_1, M_2, ..., M_K$;
while *iteration* $<$ *MaxIter* **do**
    Solve for $\pi_1, \pi_2, ..., \pi_K$;
    for $m = 1$ **to** M **do**
        Initialize Θ_{K+1} and solve for π_{K+1};
        E-step *Obtain* γ_k^m $\forall k \in \{1, 2, ..., K, K+1\}$;
        S-step *Sample* $\eta_k^m \sim \gamma_k^m$;
        if $\eta_{K+1}^m = 1$ **then**
            $K = K + 1$;
        end
        Remove K_u unoccupied intentions: $K = K - K_u$;
        Update $M_1, M_2, ..., M_K$;
        M-step *Update* $\psi \in \{\Theta_0, \Theta_1, \Theta_2, ..., \Theta_K\}$ *by (8)*;
    end
end

4.1 First Solution with Stochastic Expectation Maximization

Stochastic EM, introduces a stochastic step (S-step) after the E-step that represents the full expectation with a single sample [9]. Alg. 1 presents the summary of the first solution to multi-intention IRL via stochastic EM algorithm when the number of intentions is no longer known.

Given (6) and (7), first the posterior distribution over the latent intention vector $\boldsymbol{\eta}^m$ for trajectory $\boldsymbol{\tau}^m \in \{\boldsymbol{\tau}^1, \boldsymbol{\tau}^2, ..., \boldsymbol{\tau}^M\}$ is obtained. Then, the full expectation is estimated with a sample $\boldsymbol{\eta}^m$ from the posterior distribution. Finally, the reward parameters are updated via (8).

4.2 Second Solution with Monte Carlo Expectation Maximization

The Monte Carlo EM algorithm is a modification of the EM algorithm where the expectation in the E-step is computed numerically via Monte Carlo simulations [37]. As indicated, Algorithm 1 relies on the full posterior distribution which can be time-consuming. Therefore, another solution for multi-intention IRL is presented in which the E-step is performed through Metropolis-Hastings sampler (see Algorithm 2 for the summary).

First, a new intention assignment for m^{th} trajectory, $\boldsymbol{\eta}^{*m}$, is sampled from the prior distribution of (4), then $\boldsymbol{\eta}^m = \boldsymbol{\eta}^{*m}$ is set with the acceptance probability of $min\{1, \frac{p(\boldsymbol{\tau}^m|\boldsymbol{\eta}^{*m}, \Psi)}{p(\boldsymbol{\tau}^m|\boldsymbol{\eta}^m, \Psi)}\}$ where (see Sect. 3 of supplementary materials for full derivation):

$$\frac{p(\boldsymbol{\tau}^m|\eta_{k^*}^{*m} = 1, \Psi)}{p(\boldsymbol{\tau}^m|\eta_k^m = 1, \Psi)} = \frac{\prod_{t=1}^{T_\tau} \pi_{k^*}(a_t^m|s_t^m)}{\prod_{t=1}^{T_\tau} \pi_k(a_t^m|s_t^m)} \tag{9}$$

with $k \in \{1, 2, ..., K\}$ and $k^* \in \{1, 2, ..., K, K+1\}$.

Algorithm 2: Adaptive multi-intention IRL based on Monte Carlo EM

Initialize K, $\Theta_0, \Theta_1, \Theta_2, ..., \Theta_K$, $M_1, M_2, ..., M_K$;
while *iteration* $< MaxIter$ **do**

    Solve for $\pi_1, \pi_2, ..., \pi_K$;
    for $m = 1$ **to** M **do**

        Obtain $p(\boldsymbol{\eta}^m | \boldsymbol{H}^{-m}, \alpha)$;
        Sample $\boldsymbol{\eta}^{*m} \sim p(\boldsymbol{\eta}^m | \boldsymbol{H}^{-m}, \alpha)$;
        if $\eta_{K+1}^{*m} = 1$ **then**
          | Initialize Θ_{K+1} and solve for π_{K+1};
        end
        E-step *Assign* $\boldsymbol{\eta}^{*m} \to \boldsymbol{\eta}^m$ *by probability of* $min\{1, \frac{p(\boldsymbol{\tau}^m | \boldsymbol{\eta}^{*m}, \Psi)}{p(\boldsymbol{\tau}^m | \boldsymbol{\eta}^m, \Psi)}\}$;
        if $\eta_{K+1}^m = 1$ **then**
          | $K = K + 1$;
        end
        Remove K_u unoccupied intentions: $K = K - K_u$;
        Update $M_1, M_2, ..., M_K$;
        M-step *Update* $\psi \in \{\Theta_0, \Theta_1, \Theta_2, ..., \Theta_K\}$ *by (8)*;
    end
end

---

5 Experimental Results

In this section, we evaluate the performance of our proposed methods through several experiments with three goals: 1) to show the advantages of our methods in comparison with the baselines in environments with both linear and non-linear rewards, 2) to demonstrate the advantages of adaptively inferring the number of intentions rather than predefining a fixed number, and 3) to depict the strengths and weaknesses of our proposed algorithms with respect to each other.

5.1 Benchmarks

In order to deeply compare the performances of various models, the experiments are conducted on three different environments: GridWorld, Multi-intention ObjectWorld, and Multi-intention BinaryWorld. Variants of all three environments have been widely employed in IRL literature [19,39].

GridWorld [10] is a 8×8 environment with 64 states and four actions per state with 20% probability of moving randomly. The grids are partitioned into non-overlapping regions of size 2×2, and the feature function is defined by a binary indicator function for each region. Three reward functions are generated with linear combinations of state features and reward weights which are sampled to have a non-zero value with the probability of 0.2. The main idea behind using this environment is to compare all the models in aspects other than their capability of handling linear/non-linear reward functions.

Multi-intention ObjectWorld (M-ObjectWorld) is our extension of Object-World [19] for multi-intention IRL. ObjectWorld is a 32×32 grid of states with five actions per state with a 30% chance of moving in a different random direction. The objects with two different inner and outer colors are randomly placed, and the binary state features are obtained based on the Euclidean distance to the nearest object with a specific inner or outer color. Unlike ObjectWorld, M-ObjectWorld has six different reward functions, each of which corresponds to one intention. The intentions are defined for each cell based on three rules: 1) within 3 cells of outer color one and within 2 cells of outer color two, 2) Just within 3 cells of outer color one, and 3) everywhere else (see Table 2). Due to the large number of irrelevant features and the nonlinearity of the reward rules, the environment is challenging for methods that learn linear reward functions. Figure 2 (top three) shows a 8×8 zoom-in of M-ObjectWorld with three reward functions and respective optimal policies.

Multi-intention BinaryWorld (M-BinaryWorld) is our extension of BinaryWorld [39] for multi-intention IRL. Similarly, BinaryWorld has 32×32 states, five actions per state with a 30% chance of moving in a different random direction. But every state is randomly occupied with one of the two-color objects. The feature vector for each state consequently consists of a binary vector, encoding the color of each object in 3×3 neighborhood. Similar to M-ObjectWorld, six different intentions can be defined for each cell of M-BinaryWorld based on three rules: 1) four neighboring cells have color one, 2) five neighboring cells have color one, and 3) everything else (see Table 2). Since in M-BinaryWorld the reward depends on a higher representation for the basic features, the environment is arguably more challenging than the previous ones. Therefore, most of the experiments are carried in this environment. Figure 2 (bottom three) shows a 8×8 zoom-in of M-BinaryWorld with three different reward functions and policies.

In order to assess the generalizability of the models, the experiments are also conducted on *transferred* environments. In transferred environments, the learned reward functions are re-evaluated on new randomized environments.

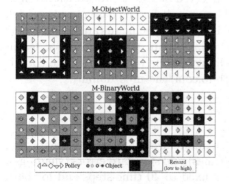

Fig. 2. 8×8 zoom-ins of M-ObjectWorld (top three) and M-BinaryWorld (bottom three) with three reward functions.

Table 2. Reward values in M-Object-World and M-BinaryWorld

Intention	Reward rule		
	1	2	3
A	+5	−10	0
B	−10	0	+5
C	0	+5	−10
D	−10	+5	0
E	+5	0	−10
F	0	−10	+5

5.2   Models

In this study, we compare our methods with existing approaches that can handle IRL with multiple intentions and constrain the experiments to model-based methods. The following models are evaluated on the benchmarks:

- EM-MLIRL(K), proposed by Babes et al. [5]. This method requires the number of experts' intentions K to be known. To research the influence on setting K for this method, we use $K \in \{2, 3, 4\}$.
- DPM-BIRL, a non-parametric multi-intention IRL method proposed by Choi and Kim [10].
- SEM-MIIRL, our proposed solution based on stochastic EM.
- MCEM-MIIRL, our proposed solution based on Monte Carlo EM.
- KEM-MIIRL, a simplified variant of our approach where the concentration parameter is zero and the number of intentions are fixed to $K \in \{2, 5\}$.

5.3   Metric

Following the same convention used in [10], the imitation performance is evaluated by the average of expected value difference (EVD). The EVD measures the performance difference between the expert's optimal policy and the optimal policy induced by the learned reward function. For $m \in \{1, 2, ..., M\}$, $\mathrm{EVD} = |V_{\tilde{R}^m}^{\tilde{\pi}^m} - V_{\tilde{R}^m}^{\pi^m}|$, where $\tilde{\pi}^m$ and $\tilde{R}^m$ are the true policy and reward function for m^{th} demonstration, respectively, and π^m is the predicted policy under the predicted reward function demonstration. In all experiments, a lower average-EVD corresponds to better imitation performance.

5.4   Implementations Details

In our experiments, we employed a fully connected neural network with five hidden layers of dimension 256 and a rectified linear unit for the base reward model, and a set of linear functions represents the intention-specific reward models. The reward network is trained for 200 epochs using Adam [18] with a fixed learning rate of 0.001. For easing the reproducibility of our work, the source code is shared with the community at https://github.com/tue-mps/damiirl.

5.5   Results

Each experiment is repeated for 6 times with different random environments, and the results are shown in the form of means (lines) and standard errors (shadings). The demonstration length for GridWorld is fixed to 40 time-steps and for both M-ObjectWorld and M-BinaryWorld is 8 time-steps.

Figure 3 and Fig. 4 show the imitation performances of our SEM-MIIRL and MCEM-MIIRL in comparison with two baselines, EM-MLIRL(K) and DPM-BIRL, for varying number of demonstrations per reward function in original and transferred environments, respectively. Each expert is assigned to one

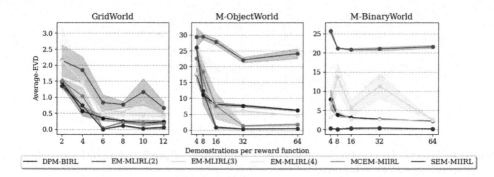

Fig. 3. Imitation performance in comparison with the baselines. Lower average-EVD is better.

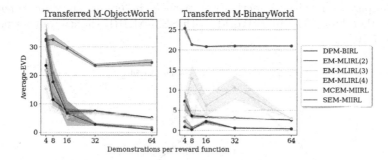

Fig. 4. Imitation performance in comparison with the baselines in transferred environments. Lower average-EVD is better.

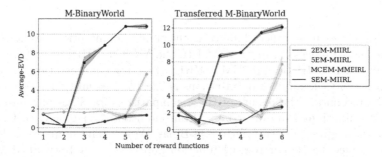

Fig. 5. Effects of overestimating/underestimating vs inferring the number of reward functions in original (left) and transferred (right) M-BinaryWorlds. Lower average-EVD is better

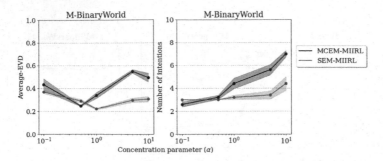

Fig. 6. Effects of α on Average-EVD (left) and number of predicted intentions (right). Lower average-EVD is better

out of three reward functions (intentions A, B, and C in M-ObjectWorld and M-BinaryWorld) and the concentration parameter is set to one. The results show clearly that our methods achieve significant lower average-EVD errors when compared to existing methods, especially in nonlinear environments of M-ObjectWorld and M-BinaryWorld, with SEM-MIIRL slightly outperforming MCEM-MIIRL.

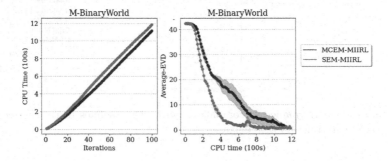

Fig. 7. Execution time (right) and Convergence (left). Lower average-EVD is better.

To address the importance of inferring the number of intentions, we have compared the performances of our SEM-MIIRL and MCEM-MIIRL with two simplified variants, 2EM-MIIRL and 5EM-MIIRL, where the concentration parameter is set to zero and the number of intentions is fixed and equal to 2 and 5, respectively. Figure 5 shows the results of these comparisons for a varying number of true reward functions from one to six (from intention: {A} to {A, B, C, D, E, F}) in both original and transferred M-BinaryWorld. The number of demonstrations is fixed to 16 per reward function and $\alpha = 1$ for both SEM-MIIRL and MCEM-MIIRL. As depicted, overestimation and underestimation of the number of reward functions, as happens frequently in both 2EM-MIIRL and 5EM-MIIRL, deteriorate the imitation performance. This while the adaptability in SEM-MIIRL and MCEM-MIIRL yields to less sensitivity with changes in the number of true reward functions.

Further experiments are conducted to deeply assess and compare MCEM-MIIRL and SEM-MIIRL. Figure 6 depicts the effects of the concentration parameter on both Average-EVD and number of predicted intentions. The number of demonstrations is fixed to 16 per reward function and intentions are {A, B, C}. As shown, the best value for the concentration parameter is between 0.5 to 1, with lower values leading to higher Average-EVD and lower number of predicted intentions, while higher values result in higher Average-EVD and higher number of predicted intentions for both MCEM-MIIRL and SEM-MIIRL. The final experiment is devoted to the convergence behavior of MCEM-MIIRL and SEM-MIIRL. The number of demonstrations is again fixed to 16 per reward function, intentions are {A, B, C} and the concentration parameter is set to 1. As shown in Fig. 7 (left image), the per-iteration execution time of MCEM-MIIRL is lower than SEM-MIIRL. The main reason is that SEM-MIIRL evaluates the posterior distribution over all latent intentions. However, this extra operation guarantees faster converges of SEM-MIIRL, making it overall the more efficient than MCEM-MIIRL as can be seen in Fig. 7 (right image).

6   Conclusions

We proposed an inverse reinforcement learning framework to recover complex reward functions by observing experts whose behaviors originate from an unknown number of intentions. We presented two algorithms that are able to consistently recover multiple, highly nonlinear reward functions and whose benefits were pointed out through a set of experiments. For this, we extended two complex benchmarks for multi-intention IRL in which our algorithms distinctly outperformed the baselines. We also demonstrated the importance of inferring rather than underestimating or overestimating the number of experts' intentions

Having shown the benefits of our approach in inferring the unknown number of experts' intention from a collection of demonstrations via model-based RL, we aim to extend the same approach in model-free environments by employing approximate RL methods.

Acknowledgments. This research has received funding from ECSEL JU in collaboration with the European Union's 2020 Framework Programme and National Authorities, under grant agreement no. 783190.

References

1. Abbeel, P., Coates, A., Quigley, M., Ng, A.: An application of reinforcement learning to aerobatic helicopter flight. In: Advances in Neural Information Processing Systems, pp. 1–8 (2007)
2. Almingol, J., Montesano, L., Lopes, M.: Learning multiple behaviors from unlabeled demonstrations in a latent controller space. In: International Conference on Machine Learning, pp. 136–144 (2013)
3. Almingol, J., Montesano, L.: Learning multiple behaviours using hierarchical clustering of rewards. In: 2015 IEEE/RSJ International Conference on Intelligent Robots And Systems (IROS), pp. 4608–4613 (2015)

4. Andrieu, C., De Freitas, N., Doucet, A., Jordan, M.: An introduction to MCMC for machine learning. Mach. Learn. **50**, 5–43 (2003)
5. Babes, M., Marivate, V., Subramanian, K., Littman, M.: Apprenticeship learning about multiple intentions. In: Proceedings of the 28th International Conference on Machine Learning (ICML-11), pp. 897–904 (2011)
6. Bighashdel, A., Meletis, P., Jancura, P., Dubbelman, G.: Supplementary materials (2020). https://github.com/tue-mps/damiirl/blob/main/Documents/DAMIIRL_SupMat.pdf
7. Bishop, C.: Pattern Recognition and Machine Learning. Springer, Heidelberg (2006)
8. Blei, D., Jordan, M.: Variational methods for the Dirichlet process. In: Proceedings of the Twenty-first International Conference on Machine Learning, p. 12 (2004)
9. Celeux, G.: The SEM algorithm: a probabilistic teacher algorithm derived from the EM algorithm for the mixture problem. Comput. Stat. Q. **2**, 73–82 (1985)
10. Choi, J., Kim, K.: Nonparametric Bayesian inverse reinforcement learning for multiple reward functions. In: Advances In Neural Information Processing Systems, pp. 305–313 (2012)
11. Dimitrakakis, C., Rothkopf, C.A.: Bayesian multitask inverse reinforcement learning. In: Sanner, S., Hutter, M. (eds.) EWRL 2011. LNCS (LNAI), vol. 7188, pp. 273–284. Springer, Heidelberg (2012). https://doi.org/10.1007/978-3-642-29946-9_27
12. Echraibi, A., Flocon-Cholet, J., Gosselin, S., Vaton, S.: On the variational posterior of Dirichlet process deep latent Gaussian mixture models. ArXiv Preprint ArXiv:2006.08993 (2020)
13. Gershman, S., Blei, D.: A tutorial on Bayesian nonparametric models. J. Math. Psychol. **56**, 1–12 (2012)
14. Hausman, K., Chebotar, Y., Schaal, S., Sukhatme, G., Lim, J.: Multi-modal imitation learning from unstructured demonstrations using generative adversarial nets. In: Advances in Neural Information Processing Systems, pp. 1235–1245 (2017)
15. Hsiao, F., Kuo, J., Sun, M.: Learning a multi-modal policy via imitating demonstrations with mixed behaviors. ArXiv Preprint ArXiv:1903.10304 (2019)
16. Jaynes, E.: Information theory and statistical mechanics. Phys. Rev. **106**, 620 (1957)
17. Jin, J., Petrich, L., Dehghan, M., Zhang, Z., Jagersand, M.: Robot eye-hand coordination learning by watching human demonstrations: a task function approximation approach. In: 2019 International Conference on Robotics and Automation (ICRA), pp. 6624–6630 (2019)
18. Kingma, D., Ba, J.: Adam: a method for stochastic optimization. ArXiv Preprint ArXiv:1412.6980 (2014)
19. Levine, S., Popovic, Z., Koltun, V.: Nonlinear inverse reinforcement learning with gaussian processes. In: Advances in Neural Information Processing Systems, pp. 19–27 (2011)
20. Li, Y., Schofield, E., Gönen, M.: A tutorial on Dirichlet process mixture modeling. J. Math. Psychol. **91**, 128–144 (2019)
21. Li, Y., Song, J., Ermon, S.: InfoGAIL: interpretable imitation learning from visual demonstrations. In: Advances in Neural Information Processing Systems, pp. 3812–3822 (2017)
22. Lin, J., Zhang, Z.: ACGAIL: imitation learning about multiple intentions with auxiliary classifier GANs. In: Geng, X., Kang, B.-H. (eds.) PRICAI 2018. LNCS (LNAI), vol. 11012, pp. 321–334. Springer, Cham (2018). https://doi.org/10.1007/978-3-319-97304-3_25

23. Michini, B., How, J.P.: Bayesian nonparametric inverse reinforcement learning. In: Flach, P.A., De Bie, T., Cristianini, N. (eds.) ECML PKDD 2012. LNCS (LNAI), vol. 7524, pp. 148–163. Springer, Heidelberg (2012). https://doi.org/10.1007/978-3-642-33486-3_10

24. Nalisnick, E., Smyth, P.: Stick-breaking variational autoencoders. In: 5th International Conference on Learning Representations, ICLR 2017, Toulon, France, 24–26 April 2017, Conference Track Proceedings (2017)

25. Neal, R.: Markov chain sampling methods for Dirichlet process mixture models. J. Comput. Graph. Stat. **9**, 249–265 (2000)

26. Ng, A., Russell, S., et al.: Algorithms for inverse reinforcement learning. In: Icml, vol. 1, p. 2 (2000)

27. Nguyen, Q., Low, B., Jaillet, P.: Inverse reinforcement learning with locally consistent reward functions. In: Advances in Neural Information Processing Systems, pp. 1747–1755 (2015)

28. Noothigattu, R., Yan, T., Procaccia, A.: Inverse reinforcement learning from like-minded teachers. Manuscript (2020)

29. Odom, P., Natarajan, S.: Actively interacting with experts: a probabilistic logic approach. In: Frasconi, P., Landwehr, N., Manco, G., Vreeken, J. (eds.) ECML PKDD 2016. LNCS (LNAI), vol. 9852, pp. 527–542. Springer, Cham (2016). https://doi.org/10.1007/978-3-319-46227-1_33

30. Papamarkou, T., Hinkle, J., Young, M., Womble, D.: Challenges in Bayesian inference via Markov chain Monte Carlo for neural networks. ArXiv Preprint ArXiv:1910.06539 (2019)

31. Pavlov, D., Popescul, A., Pennock, D., Ungar, L.: Mixtures of conditional maximum entropy models. In: Proceedings of the 20th International Conference on Machine Learning (ICML-03), pp. 584–591 (2003)

32. Rajasekaran, S., Zhang, J., Fu, J.: Inverse reinforce learning with nonparametric behavior clustering. ArXiv Preprint ArXiv:1712.05514 (2017)

33. Ramachandran, D., Amir, E.: Bayesian inverse reinforcement learning. IJCAI **7**, 2586–2591 (2007)

34. Russell, S.: Learning agents for uncertain environments. In: Proceedings of the Eleventh Annual Conference on Computational Learning Theory, pp. 101–103 (1998)

35. Shkurti, F., Kakodkar, N., Dudek, G.: Model-based probabilistic pursuit via inverse reinforcement learning. In: 2018 IEEE International Conference on Robotics And Automation (ICRA), pp. 7804–7811 (2018)

36. Sutton, R., Barto, A.: Reinforcement Learning: An Introduction. MIT Press, Cambridge (2018)

37. Wei, G., Tanner, M.: A Monte Carlo implementation of the EM algorithm and the poor man's data augmentation algorithms. J. Am. Stat. Assoc. **85**, 699–704 (1990)

38. Wei, H., Chen, C., Liu, C., Zheng, G., Li, Z.: Learning to simulate on sparse trajectory data. In: Dong, Y., Mladenić, D., Saunders, C. (eds.) ECML PKDD 2020. LNCS (LNAI), vol. 12460, pp. 530–545. Springer, Cham (2021). https://doi.org/10.1007/978-3-030-67667-4_32

39. Wulfmeier, M., Ondruska, P., Posner, I.: Maximum entropy deep inverse reinforcement learning. ArXiv Preprint ArXiv:1507.04888 (2015)

40. Ziebart, B., Maas, A., Bagnell, J., Dey, A.: Maximum entropy inverse reinforcement learning. In: Proceedings Of The 23rd National Conference on Artificial Intelligence, vol. 3, pp. 1433–1438 (2008)

Unsupervised Task Clustering
for Multi-task Reinforcement Learning

Johannes Ackermann[1](✉), Oliver Richter[2](✉), and Roger Wattenhofer[2]

[1] Technical University of Munich, Munich, Germany
johannes.ackermann@tum.de
[2] ETH Zurich, Zürich, Switzerland
{richtero,wattenhofer}@ethz.ch

Abstract. Meta-learning, transfer learning and multi-task learning have recently laid a path towards more generally applicable reinforcement learning agents that are not limited to a single task. However, most existing approaches implicitly assume a uniform similarity between tasks. We argue that this assumption is limiting in settings where the relationship between tasks is unknown a-priori. In this work, we propose a general approach to automatically cluster together similar tasks during training. Our method, inspired by the expectation-maximization algorithm, succeeds at finding clusters of related tasks and uses these to improve sample complexity. We achieve this by designing an agent with multiple policies. In the expectation step, we evaluate the performance of the policies on all tasks and assign each task to the best performing policy. In the maximization step, each policy trains by sampling tasks from its assigned set. This method is intuitive, simple to implement and orthogonal to other multi-task learning algorithms. We show the generality of our approach by evaluating on simple discrete and continuous control tasks, as well as complex bipedal walker tasks and Atari games. Results show improvements in sample complexity as well as a more general applicability when compared to other approaches.

1 Introduction

Imagine we are given an arbitrary set of tasks. We know that dissimilarities and/or contradicting objectives can exist. However, in most settings we can only guess these relationships and how they might affect joint training. Many recent works rely on such human guesses and (implicitly or explicitly) limit the generality of their approaches. This can lead to impressive results, either by explicitly modeling the relationships between tasks as in transfer learning [42], or by meta learning implicit relations [15]. However, in some cases an incorrect similarity assumption can slow training [19]. With this paper we provide an easy, straightforward approach to avoid human assumptions on task similarities.

J. Ackermann and O. Richter—Equal contribution. Johannes Ackermann did his part while visiting ETH Zurich.

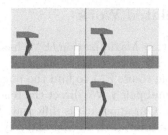

Fig. 1. Left: An agent (smiley) should reach one of 12 goals (stars) in a grid world. Learning to reach a goal in the top right corner helps it to learn about the other goals in that corner. However, learning to reach the green stars (bottom left corner) at the same time gives conflicting objectives, hindering training. **Right:** When all tasks are very similar, treating them as independent is disadvantageous. Task clustering allows us to perform well in both cases. (Color figure online)

An obvious solution is to train a separate policy for each task. However, this might require a large amount of experience to learn the desired behaviors. Therefore, it is desirable to have a single agent and share knowledge between tasks. This is generally known as multi-task learning, a field which has received a large amount of interest in both the supervised learning and reinforcement learning (RL) community [41]. If tasks are sufficiently similar, a policy that is trained on one task provides a good starting point for another task, and experience from each task will help training in the other tasks. This is known as *positive transfer* [19]. However, if the tasks are sufficiently dissimilar, *negative transfer* occurs and reusing a pre-trained policy is disadvantageous. Here using experience from the other tasks might slow training or even prevent convergence to a good policy. Most previous approaches to multi-task learning do not account for problems caused by negative transfer directly and either accept its occurrence or limit their experiments to sufficiently similar tasks. We present a hybrid approach that is helpful in a setting where the task set contains clusters of related tasks, amongst which transfer is helpful. To illustrate the intuition we provide a conceptualized example in Fig. 1 on the left. Note however that our approach goes beyond this conceptual ideal and can be beneficial even if the clustering is not perceivable by humans a-priori.

Our approach iteratively evaluates a set of policies on all tasks, assigns tasks to policies based on their respective performance and trains policies on their assigned tasks. This leads to policies naturally specializing to clusters of related tasks, yielding an interpretable decomposition of the full task set. Moreover, we show that our approach can improve the learning speed and final reward in multi-task RL settings. To summarize our contributions:

- We propose a general approach inspired by Expectation-Maximization (EM) that can find clusters of related tasks in an unsupervised manner.
- We provide an evaluation on a diverse set of multi-task RL problems that shows the improved sample complexity and reduction in negative transfer.
- We show the importance of meaningful clustering and the sensitivity to the assumed number of clusters in an ablation study.

2   Related Work

Expectation-Maximization (EM) has previously been used in RL to directly learn a policy. By reformulating RL as an inference problem with a latent variable, it is possible to use EM to find the maximum likelihood solution, corresponding to the optimal policy. We direct the reader to the survey on the topic by Deisenroth et al. [9]. Our approach is different: We use an EM-inspired approach to cluster tasks in a multi-task setting and rely on recent RL algorithms to learn the tasks.

In supervised learning, the idea of subdividing tasks into related clusters was proposed by Thrun and O'Sullivan [34]. They use a distance metric based on generalization accuracy to cluster tasks. Another popular idea related to our approach that emerged from supervised learning is the use of a mixture of experts [16]. Here, multiple sub-networks are trained together with an input dependent gating network. Jordan and Jacobs [18] also proposed an EM algorithm to learn the mixture of experts. While those approaches have been extended to the control setting [4,17,26,33], they rely on an explicit supervision signal. It is not clear how such an approach would work in an RL setting. A variety of other methods have been proposed in the supervised learning literature. For brevity we direct the reader to the survey by Zhang et al. [41], which provides a good overview of the topic. In contrast, we focus on RL, where no labeled data set exists.

In RL, task clustering has in the past received attention in works on transfer learning. Carroll and Seppi [5] proposed to cluster tasks based on a distance function. They propose distances based on Q-values, reward functions, optimal policies or transfer performance. They propose to use the clustering to guide transfer. Similarly, Mahmud et al. [25] propose a method for clustering Markov Decision Processes (MDPs) for source task selection. They design a cost function for their chosen transfer method and derive an algorithm to find a clustering that minimizes this cost function. Our approach differs from both in that we do not assume knowledge of the underlying MDPs and corresponding optimal policies. Furthermore, the general nature of our approach allows it to scale to complex tasks, where comparing properties of the full underlying MDPs is not feasible. Wilson et al. [38] developed a hierarchical Bayesian approach for multi-task RL. Their approach uses a Dirichlet process to cluster the distributions from which they sample full MDPs in the hope that the sampled MDP aligns with the task at hand. They then solve the sampled MDP and use the resulting policy to gather data from the environment and refine the posterior distributions for a next iteration. While their method is therefore limited to simple MDPs, our approach can be combined with function approximation and therefore has the potential to scale to MDPs with large or infinite state spaces which cannot be solved in closed form. Lazaric and Ghavamzadeh [20] use a hierarchical Bayesian approach to infer the parameters of a linear value function and utilize EM to infer a policy. However, as this approach requires the value function to be a linear function of some state representation, this approach is also difficult to scale to larger problems which we look at. Li et al. [22] note that believe states in partially observable MDPs can be grouped according to the decision they require. Their model infers the parameters of the corresponding decision state

MDP. Their approach scales quadratically with the number of decision states and at least linearly with the number of collected transitions, making it as well difficult to apply to complex tasks.

More recent related research on multi-task RL can be split into two categories: Works that focus on very similar tasks with small differences in dynamics and reward, and works that focus on very dissimilar tasks. In the first setting, approaches have been proposed that condition the policy on task characteristics identified during execution. Lee et al. [21] use model-based RL and a learned embedding over the local dynamics as additional input to their model. Yang et al. [39] train two policies, one that behaves in a way that allows the easy identification of the environment dynamics and another policy that uses an embedding over the transitions generated by the first as additional input. Zintgraf et al. [43] train an embedding over the dynamics that accounts for uncertainty over the current task during execution and condition their policy on it. Our approach is more general than these methods as our assumption on task similarity is weaker. In the second group of papers, the set of tasks is more diverse. Most approaches here are searching for a way to reuse representations from one task in the others. Riemer et al. [30] present an approach to learn hierarchical options, and use it to train an agent on 21 Atari tasks. They use the common NatureDQN network [27] with separate final layers for option selection policies, as well as separate output layers for each task to account for the different action spaces. Eramo et al. [11] show how a shared representation can speed up training. They then use a network strucuture with separate heads for each task, but shared hidden layers. Our multi-head baseline is based on these works. Bräm et al. [2] propose a method that addresses negative transfer between multiple tasks by learning an attention mechanism over multiple sub-networks, similar to a mixture of experts. However, as all tasks yield experience for one overarching network, their approach still suffers from interference between tasks. We limit this interference by completely separating policies. Wang et al. [36] address the problem of open-ended learning in RL by iteratively generating new environments. Similar to us, they use policy rankings as a measure of difference between tasks. However, they use this ranking as a measure of novelty to find new tasks, addressing a very different problem. Hessel et al. [14] present PopArt for multi-task deep RL. They address the issue that different tasks may have significantly different reward scales. Sharma et al. [31] look into active learning for multi-task RL on Atari tasks. They show that uniformly sampling new tasks is suboptimal and propose different sampling techniques. Yu et al. [40] propose Gradient Surgery, a way of projecting the gradients from different tasks to avoid interference. These last three approaches are orthogonal to our work and can be combined with EM-clustering. We see this as an interesting direction for future work.

Quality-Diversity (QD) algorithms [7,29] in genetic algorithms research aim to find a diverse set of good solutions for a given problem. One proposed benefit of QD is that it can overcome local optima by using the solutions as "stepping stones" towards a global optimum. Relatedly in RL, Eysenbach et al. [12] and Achiam et al. [1] also first identify diverse skills and then use the learned skills to solve a given task. While we do not explicitly encourage diversity in our approach, our approach is related in that our training leads to multiple good performing, distinct policies trained on distinct tasks. This can lead to a policy trained on one task becoming the best on a task that it was not trained on, similar to the "stepping stones" in QD. However, in our work this is more a side-effect than the proposed functionality.

3   Background and Notation

In RL [32] tasks are specified by a Markov Decision Process (MDP), defined as tuple (S, A, P, R, γ), with state space S, action space A, transition function $P(\cdot|s, a)$, reward function $R(s, a)$ and decay factor γ. As we are interested in reusing policies for different tasks, we require a shared state-space S and action-space A across tasks. Note however that this requirement can be omitted by allowing for task specific layers. Following prior work, we do allow for a task specific final layer in our Atari experiments to account for the different action spaces. In all other experiments however, tasks only differ in their transition function and reward function. We therefore describe a task as $\tau = (P_\tau, R_\tau)$ and refer to the set of given tasks as T. For each task $\tau \in T$ we aim to maximize the discounted return $G_\tau = \sum_{t=0}^{t=L} \gamma^t r_t^\tau$, where $r_t^\tau \sim R_\tau(s_t, a_t)$ is the reward at time step t and L is the episode length. Given a set of policies $\Pi = \{\pi_1, ..., \pi_n\}$, we denote the return obtained by policy π_i on task τ as $G_\tau(\pi_i)$.

4   Clustered Multi-task Learning

Before we introduce our proposed clustering approach, we first want to briefly discuss the straight forward, yet often disregarded limitation that exists when learning multiple task with a single policy.

Proposition 1. *The optimal policy of a jointly learned task set* $T = \{\tau_1, \tau_2\}$ *can be arbitrarily far from the optimal policy on task* τ_1.

To see this, consider task τ_2 given as $\tau_2 = (P_{\tau_1}, -2 \cdot R_{\tau_1})$. Optimizing a policy π to maximizing the joint objective $G_{\tau_1}(\pi) + G_{\tau_2}(\pi)$ is equivalent to optimizing π to minimize $G_{\tau_1}(\pi)$ as for any policy π we have $G_{\tau_1}(\pi) + G_{\tau_2}(\pi) = -G_{\tau_1}(\pi)$.

On the other hand, as the growing body of literature on meta-, transfer- and multi-task learning suggests, we can expect a gain through positive transfer if we train a single policy π_i on a subset of related tasks $T_k \subset T$.

We incorporate these insights into our algorithm by modeling the task set T as a union of K disjoint task clusters $T_1, ..., T_K$, i.e., $T = \bigcup_{k=1}^K T_k$ with $T_i \cap T_j = \emptyset$ for $i \neq j$. Tasks within a cluster allow for positive transfer while we do not assume any relationship between tasks of different clusters. Tasks in different clusters may therefore even have conflicting objectives. Note that the assignment of tasks to clusters is not given to us and therefore needs to be inferred by the algorithm. Note also that this formulation only relies on minimalistic assumptions. That is, we do not assume a shared transition function or a shared reward structure. Neither do we assume the underlying MDP to be finite and/or solvable in closed form. Our approach is therefore applicable to a much broader range of settings than many sophisticated models with stronger assumptions. As generality is one of our main objectives, we see the minimalistic nature of the model as a strength rather than a weakness.

Given this problem formulation, we note that it reflects a clustering problem, in which we have to assign each task $\tau \in \mathcal{T}$ to one of the clusters $\mathcal{T}_k$, $k \in \{1, \ldots, K\}$. At the same time, we want to train a set of policies $\Pi = \{\pi_1, \ldots, \pi_n\}$ to solve the given tasks. Put differently, we wish to infer a latent variable (cluster assignment of the tasks) while optimizing our model parameters (set of policies).

An EM [10] inspired algorithm allows us to do just that. On a high level, in the expectation step (E-step) we assign each of the tasks $\tau \in \mathcal{T}$ to a policy π_i, representing an estimated cluster $\tilde{\mathcal{T}}_i$. We then train the policies in the maximization step (M-step) on the tasks they got assigned, specializing the policies to their clusters. These steps are alternatingly repeated—one benefiting from the improvement of the other in the pre-

---

Algorithm 1: Task-Clustering

Initialize N policies $(\pi_1, \ldots, \pi_N)$
Initialize N buffers $(\boldsymbol{D}_1, \ldots, \boldsymbol{D}_N)$
while *not converged* **do**
  ▷ E-Step
  $\tilde{\mathcal{T}}_i \leftarrow \emptyset$ for $i \in \{1, \ldots, n\}$
  for $\tau \in \mathcal{T}$ **do**
    $k \leftarrow \arg\max_i G_\tau(\pi_i)$
    $\tilde{\mathcal{T}}_k \leftarrow \tilde{\mathcal{T}}_k \cup \tau$
  $\tilde{\mathcal{T}}_i \leftarrow \mathcal{T}$ where $\tilde{\mathcal{T}}_i = \emptyset$
  ▷ M-Step
  for $\pi_i \in \{\pi_1, \ldots, \pi_n\}$ **do**
    $t \leftarrow 0$
    while $t < T_M$ **do**
      $\tau \sim \tilde{\mathcal{T}}_i$
      Run π_i on τ for L steps, store transitions in $\boldsymbol{D}_i$
      Update π_i from $\boldsymbol{D}_i$
      $t \leftarrow t + L$

---

ceding step—until convergence. Given this general framework we are left with filling in the details. Specifically, how to assign tasks to which policies (E-step) and how to allocate training time from policies to assigned tasks (M-step).

For the assignment in the E-step we want the resulting clusters to represent clusters with positive transfer. Given that policy π_i is trained on a set of tasks $\tilde{\mathcal{T}}_i$ in a preceding M-step, we can base our assignment of tasks to π_i on the performance of π_i: Tasks on which π_i performs well likely benefited from the preceding training and therefore should be assigned to the cluster of π_i. Specifically, we can evaluate each policy $\pi_i \in \{\pi_1, \ldots, \pi_n\}$ on all tasks $\tau \in \mathcal{T}$ to get an estimate of $G_\tau(\pi_i)$ and base the assignment on this performance evaluation. To get to an implementable algorithm we state two additional desiderata for our assignment: (1) We do not want to constrain cluster sizes in any way as clusters can be of unknown, non-uniform sizes. (2) We do not want to constrain the diversity of the tasks. This implies that the assignment has to be independent of the reward scales of the tasks, which in turn limits us to assignments based on the relative performances of the policies $\pi_1, \ldots, \pi_n$. We found a greedy assignment—assigning each task to the policy that performs best—to work well. That is, a task τ_k is assigned to the policy $\pi = \arg\max_{\pi_i} G_{\tau_k}(\pi_i)$. A soft assignment based on the full ranking of policies might be worth exploring in future work. Given the greedy assignment, our method can also be seen as related to k-means [24], a special case of EM.

In the M-step, we take advantage of the fact that clusters reflect positive transfer, i.e., training on some of the assigned tasks should improve performance on the whole cluster. We can therefore randomly sample a task from the assigned tasks and train on it for one episode before sampling the next task. Overall we train each policy for a fixed number of updates T_M in each M-step with T_M independent of the cluster size. This independence allows us to save environment interactions as larger clusters benefit from positive transfer and do not need training time proportional to the number of assigned tasks.

Note that the greedy assignment (and more generally any assignment fulfilling desiderata 1 above) comes with a caveat: Some policies might not be assigned any tasks. In this case we sample the tasks to train these policies from all tasks $\tau \in \mathcal{T}$, which can be seen as a random exploration of possible task clusters. This also ensures that, early on in training, every policy gets a similar amount of initial experience. For reference, we provide a pseudo code of our approach in Algorithm 1. Note that we start by performing an E-Step, i.e., the first assignment is based on the performance of the randomly initialized policies.

4.1   Convergence Analysis

We now show that both, the E- and M-step yield a monotonic improvement. Thereby, our algorithm improves the objective monotonically in every iteration.

We denote our overall objective function that we aim to maximize as $o(\Pi, \tilde{\mathcal{T}}) = \sum_{\pi_i \in \Pi} \sum_{\tau_j \in \tilde{\mathcal{T}}_i} G_\tau(\pi_i)$, as a function of our policy set $\Pi = \{\pi_1, \ldots, \pi_n\}$ and their corresponding task assignments $\tilde{\mathcal{T}} = \{\tilde{\mathcal{T}}_1, \ldots, \tilde{\mathcal{T}}_n\}$. In the E-step, we evaluate all policies on all tasks to determine the returns $G_\tau(\pi_i)$. Using the greedy assignment strategy, we assign each task to the policy that achieves the respective highest return $\arg\max_i G_\tau(\pi_i)$ and obtain a new assignment set $\tilde{\mathcal{T}}'$. It is easy to see that this assignment step can only improve the objective, as

$$o(\Pi, \tilde{\mathcal{T}}') = \sum_{\tau \in \mathcal{T}} \max_{\pi_i \in \Pi} G_\tau(\pi_i) \geq \sum_{\tau \in \mathcal{T}} \sum_{\pi_i \in \Pi} \mathbf{1}_{[\tau \in \tilde{\mathcal{T}}_i]} G_\tau(\pi_i) = o(\Pi, \tilde{\mathcal{T}})$$

for any previous assignments $\tilde{\mathcal{T}}$, since the indicator function $\mathbf{1}_{[\tau \in \tilde{\mathcal{T}}_i]}$ will only indicate one cluster.[1] Note that this derivation relies on a deterministic evaluation of policies, i.e. deterministic task environments. For stochastic environments we can take the average over multiple evaluations, trading off the computational overhead with the accuracy of the evaluation. In our experiments we found that a relatively small number of evaluations is sufficient for the algorithm to converge.

---

[1] Note that assigning all tasks to a cluster that did not get any tasks assigned is only done for exploration. In the evaluation of our objective these clusters remain empty.

During the M-step the assignments are fixed, and every policy π_i is trained on its assigned tasks $\tau \in \tilde{T}_i$ by sampling from them uniformly. We derive the case for shared transition dynamics $P_\tau = P \; \forall \tau \in T$ here and extend it to the case of tasks with distinct transition dynamics in Appendix A.[2]

The value of policy π_i on task τ can be defined recursively as

$$V_\tau^{\pi_i}(s) = \mathbb{E}_{a \sim \pi_i}[R_\tau(s,a) + \gamma \mathbb{E}_{s' \sim P(\cdot|s,a)}[V_\tau^{\pi_i}(s')]]$$

such that $V_\tau^{\pi_i}(s_0) = G_\tau(\pi_i)$ for the starting state s_0. We further note that the expected value $V_{\mathcal{M}}^{\pi_i}(s) = \mathbb{E}_{\tau \sim \tilde{T}_i}[V_\tau^{\pi_i}(s)]$ is in itself a value function over an MDP $\mathcal{M}$ defined by the expected reward with

$$V_{\mathcal{M}}^{\pi_i}(s) = \mathbb{E}_{a \sim \pi_i}[\mathbb{E}_{\tau \sim \tilde{T}_i}[R_\tau(s,a)] + \gamma \mathbb{E}_{s' \sim P(\cdot|s,a)}[\mathbb{E}_{\tau \sim \tilde{T}_i}[V_\tau^{\pi_i}(s')]]]$$

$$= \mathbb{E}_{a \sim \pi_i}[\mathbb{E}_{\tau \sim \tilde{T}_i}[R_\tau(s,a)] + \gamma \mathbb{E}_{s' \sim P(\cdot|s,a)}[V_{\mathcal{M}}^{\pi_i}(s')]]$$

Policy iteration on $\mathcal{M}$ will yield an improved policy π_i' with $V_{\mathcal{M}}^{\pi_i'}(s) \geq V_{\mathcal{M}}^{\pi_i}(s) \; \forall s \in S$. More generally, any off-policy RL algorithm that samples uniformly over collected (s, r, a, s') transition tuples will implicitly optimize $\mathcal{M}$. Note that $V_{\mathcal{M}}^{\pi_i}(s_0) = \mathbb{E}_{\tau \sim \tilde{T}_i}[G_\tau(\pi_i)] = \frac{1}{|\tilde{T}_i|} \sum_{\tau \in \tilde{T}_i} G_\tau(\pi_i)$ for uniformly sampled tasks. Any improvement in $V_{\mathcal{M}}^{\pi_i}$ therefore directly translates into an improvement in our overall objective. While we focus on off-policy RL in this paper, we conjecture that a similar optimisation can be done on-policy.

5   Experiments

As a proof of concept we start the evaluation of our approach on two discrete tasks. The first environment consists of a chain of discrete states in which the agent can either move to the left or to the right. The goal of the agent is placed either on the left end or the right end of the chain. This gives rise to two task clusters, where tasks within a cluster differ in the frequency with which the agent is rewarded on its way to the goal. The second environment reflects the 2-dimensional grid-world presented in Fig. 1. Actions correspond to the cardinal directions in which the agent can move and the 12 tasks in the task set T are defined by their respective goal. We refer an interested reader to Appendix B.1 for a detailed description (See footnote 2).

We train policies with tabular Q-learning [37] and compare our approach to two baselines: In the first we train a single policy on all tasks. We refer to this as SP (Single Policy). In the other we train a separate policy per task and evaluate each policy on the task it was trained on. This is referred to as PPT (Policy per Task). Our approach is referred to as EM (Expectation-Maximization).

The results and task assignment over the course of training are shown in Fig. 2 and Fig. 3. Looking at the assignments, we see that in both environments our approach converges to the natural clustering, leading to a higher reward

---

[2] The Appendix and implementations of all our experiments can be found at https://github.com/JohannesAck/EMTaskClustering.

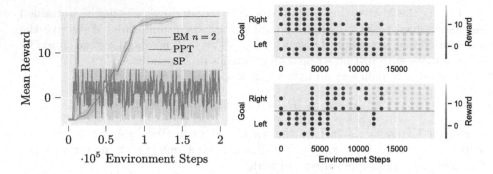

Fig. 2. Left: Mean reward and 95% confidence interval (shaded area) from 10 trials when training on the chain environment. **Right:** Task assignment (dots) and task specific reward (color) over the course of training the two policies in our approach. Each plot shows one of the policies/estimated clusters. The assignments converge to the natural clustering reflected by the goal location.

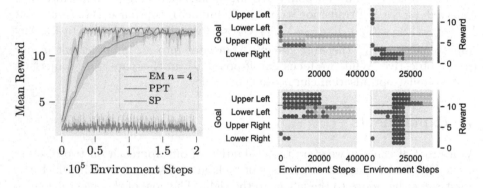

Fig. 3. Left: Mean reward and 95% confidence interval (shaded area) from 10 trials when training on the grid-world environment depicted in Fig. 1. **Right:** Task assignment (dots) and task specific reward (color) over the course of training for the $n = 4$ policies (estimated clusters) in our approach. The assignment naturally clusters the tasks of each corner together.

after finding these assignments. Both our EM-approach and PPT converge to an optimal reward in the chain environment, and a close to optimal reward in the corner-grid-world. However, PPT requires a significantly higher amount of environment steps to reach this performance, as it does not share information between tasks and therefore has to do exploration for each task separately. SP fails to achieve a high reward due to the different tasks providing contradicting objectives.

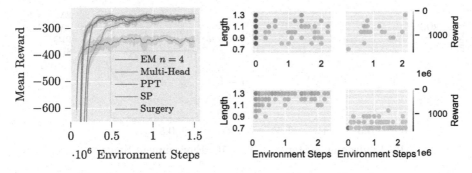

Fig. 4. Left: Mean reward and 95% confidence interval (shaded area) from 10 trials when training on the pendulum environment. The curves are smoothed by a rolling average to dampen the noise of the random starting positions. For [40] we used 12 trials out of which 3 failed to converge and were excluded. **Right:** Task assignment (dots) and task specific reward (color) from a sample run. Two policies focus on long and short, while the others focus on medium lengths.

5.1 Pendulum

Next we consider a simple continuous control environment where tasks differ in their dynamics. We use the pendulum gym task [3], in which a torque has to be applied to a pendulum to keep it upright. Here the environment is the same in all tasks, except for the length of the pendulum, which is varied in the range $\{0.7, 0.8, ..., 1.3\}$, giving a total of 7 tasks. Note that there is no obvious cluster structure here and the experiment therefore serves as an edge-case to test the applicability of our approach.

We use Twin Delayed Deep Deterministic Policy Gradient (TD3) [13] with hyperparameters optimized as discussed in Appendix B.2. By default, we use $n = 4$ policies and did not tune this hyperparameter. This was done to give a fair comparison to baseline approaches which do not have this extra degree of freedom. For application purposes the number of clusters can be treated as a hyperparameter and included in the hyperparameter optimization. We compare against SP, PPT, gradient surgery [40] and a multi-head network structure similar to the approach used by Eramo et al. [11]. Each policy in our approach uses a separate replay buffer. The multi-head network has a separate replay-buffer and a separate input and output layer per task. Surgery uses a separate replay-buffer and output layer per task. We adjust the network size of the multi-head baseline, surgery and SP to avoid an advantage of our method due to a higher parameter count, see Appendix B.2 for details. The results are shown in Fig. 4.

We observe that EM, PPT, multi-head and surgery all achieve a similar final performance, with EM and surgery achieving a high reward earlier than PPT or multi-head. The multi-head approach requires signficantly more experience to converge than even PPT in this setup. We believe this is due to the inherent interference of learning signals in the shared layers. Our approach manages to avoid this interference, as does surgery. SP is unable to achieve a high reward as

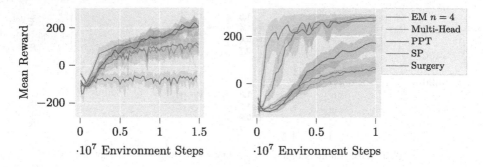

Fig. 5. Evaluation of the BipedalWalker experiments. The shaded areas show the 95% confidence interval on the mean task reward. **Left:** Track and field task set; 6 tasks with varying objectives. Results reflect 20 trials of each approach. **Right:** Task set with varying leg lengths and obstacles; 9 tasks with the same reward function. Results reflect 10 trials of each approach.

it cannot specialize to the tasks. In contrast to the surgery baseline, our approach can give further insights by producing intuitive cluster assignments, see Fig. 4.

5.2 Bipedal Walker

As a more complex continuous control environment we focus on *BipedalWalker* from the OpenAI Gym [3], which has previously been used in multi-task and generalization literature [28,35,36]. It consists of a bipedal robot in a two-dimensional world, where the default task is to move to the right with a high velocity. The action space consists of continuous torques for the hip and knee joints of the legs and the state space consists of joint angles and velocities, as well as hull angle and velocity and 10 lidar distance measurements. Examples are shown in Fig. 1 on the right.

To test our approach, we designed 6 tasks inspired by track and field sports: Jumping up at the starting position, jumping forward as far as possible, a short, medium and long run and a hurdle run. As a second experiment, we create a set of 9 tasks by varying the leg length of the robot as well as the number of obstacles in its way. This task set is inspired by task sets in previous work [28]. Note that we keep the objective—move forward as fast as possible—constant here. We again use TD3 and tune the hyperparameters of the multi-head baseline and our approach (with $n = 4$ fixed) with grid-search. Experiment details and hyperparameters are given in Appendix B.3.

The results in Fig. 5 (left) on the track and field tasks show a significant advantage in using our approach over multi-head TD3, surgery or SP and a better initial performance than PPT, with similar final performance. SP fails to learn a successful policy altogether due to the conflicting reward functions. In contrast, the results in Fig. 5 (right) from the second task set show that SP can learn a policy that is close to optimal on all tasks here. The multi-head, surgery and PPT approaches suffer in this setup as each head/policy only gets the experience

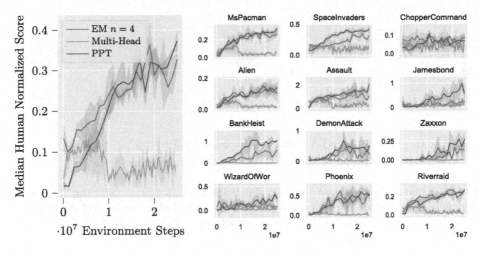

Fig. 6. The results of our experiments on a subset of the Atari Learning Environment games. The reward is averaged across 3 trials and the shaded region shows the standard deviation of the mean.

from its task and therefore needs more time to converge. Our approach can take advantage of the similarity of the tasks, converging significantly quicker. We note that the experiments presented here reflect two distinct cases: One in which it is advantageous to separate learning, reflected by PPT outperforming SP, and one where it is better to share experience between tasks, reflected by SP outperforming PPT. Our approach, unlike surgery or multi-head, demonstrates general applicability as it is the only one performing competitively in both. We provide an insight into the assignment of tasks to policies in Appendix C.1.

5.3   Atari

To test the performance of our approach on a more diverse set of tasks, we evaluate on a subset of the Arcade Learning Environment (ALE) tasks [23]. Our choice of tasks is similar to those used by [30], but we exclude tasks containing significant partial-observability. This is done to reduce the computational burden as those tasks usually require significantly more training data. We built our approach on top of the Implicit Quantile Network (IQN) implementation in the Dopamine framework [6,8]. We chose IQN due to its sample efficiency and the availability of an easily modifiable implementation. As the different ALE games have different discrete action spaces, we use a separate final layer and a separate replay buffer for each game in all approaches. We use the hyperparameters recommended by [6], except for a smaller replay buffer size to reduce memory requirements. As in the Bipedal Walker experiments we fix the number of policies in our approach without tuning to $n = 4$. We choose the size of the network such that each approach has the same number of total tunable parameters. We provide the details in Appendix B.4.

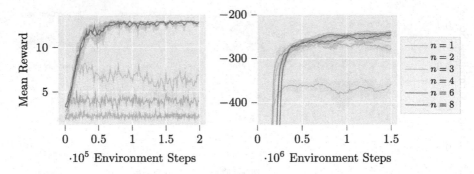

Fig. 7. Ablations for different number of policies n. Shaded areas show the 95% confidence interval of the mean reward from 10 trials each. **Left:** Corner-grid-world tasks. **Right:** Pendulum tasks, learning curves smoothed.

The results are given in Fig. 6. The multi-head approach is unable to learn any useful policy here due to negative transfer between tasks. This is in line with experiments in other research [14] and is due to the large variety of the tasks. On the other hand, both our EM-approach and PPT are able to achieve significantly higher reward. However, our approach does not perform better than PPT. Note that we can only expect a better performance than PPT if there are clusters of tasks that benefit from positive transfer. The diverse set of Atari games seems to violate this assumption. While we cannot benefit from positive transfer, our approach avoids the negative interference impacting the multi-head approach, even with just 4 clusters. Task assignments in our approach are given in Appendix C.2.

5.4 Ablations

To gain additional insight into our approach, we perform two ablation studies on the discrete corner-grid-world environment and the pendulum environment.

First, we investigate the performance of our approach for different numbers of policies n. The results in Fig. 7 show that using too few policies can lead to a worse performance, as the clusters cannot distinguish the contradicting objectives. On the other hand, using more policies than necessary increases the number of environment interactions required to achieve a good performance in the pendulum task, but does not significantly affect the final performance.

As a second ablation, we are interested in the effectiveness of the clustering. It might be possible that simply having fewer tasks per policy is giving our approach an advantage compared to SP or multi-head TD3. We therefore provide an ablation in which task-policy assignments are determined randomly at the start nd kept constant during the training. Results from this experiment can be seen in Fig. 8, with additional results in Appendix D. The results show that using random clusters performs significantly worse than using the learned clusters. This highlights the importance of clustering tasks meaningfully.

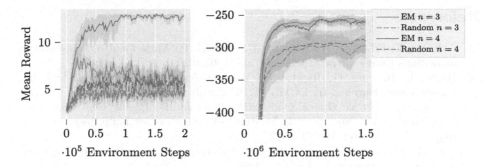

Fig. 8. Comparison of our approach against randomly assigning tasks to policies at the start of training. Shaded areas show the 95% confidence interval of the mean reward. **Left:** Corner-grid-world tasks, 10 trials each. **Right:** Pendulum tasks, 10 trials each, learning curves smoothed.

6 Conclusion

We present an approach for multi-task reinforcement learning (RL) inspired by Expectation-Maximization (EM) that automatically clusters tasks into related subsets. Our approach uses a set of policies and alternatingly evaluates the policies on all tasks, assigning each task to the best performing policy and then trains the policies on their assigned tasks. While the repeated evaluation of policies adds a small computational overhead, it provides an effective way to mitigate negative transfer. Our algorithm is straightforward and can easily be combined with a variety of state-of-the-art RL algorithms. We evaluate the effectiveness of our approach on a diverse set of environments. Specifically, we test its performance on sets of simple discrete tasks, simple continuous control tasks, two complex continuous control task sets and a set of Arcade Learning Environment tasks. We show that our approach is able to identify clusters of related tasks and use this structure to achieve a competitive or superior performance to evaluated baselines, while additionally providing insights through the learned clusters. We further provide an ablation over the number of policies in our approach and a second ablation that highlights the need to cluster tasks meaningfully.

Our approach offers many possibilities for future extensions. An adaption to on-policy learning and combination with orthogonal approaches could improve the applicability further. Another interesting direction would be hierarchical clustering. This could prove helpful for complicated tasks like the Atari games. It would also be interesting to see how our approach can be applied to multi-task learning in a supervised setting. Further, different assignment strategies with soft assignments could be investigated. Overall, we see our work as a good stepping stone for future work on structured multi-task learning.

References

1. Achiam, J., Edwards, H., Amodei, D., Abbeel, P.: Variational option discovery algorithms (2018). https://arxiv.org/abs/1807.10299
2. Bräm, T., Brunner, G., Richter, O., Wattenhofer, R.: Attentive multi-task deep reinforcement learning. In: ECML PKDD (2019)
3. Brockman, G., et al.: Openai gym (2016). http://arxiv.org/abs/1606.01540
4. Cacciatore, T.W., Nowlan, S.J.: Mixtures of controllers for jump linear and non-linear plants. In: NeurIPS (1993)
5. Carroll, J.L., Seppi, K.: Task similarity measures for transfer in reinforcement learning task libraries. In: IJCNN (2005)
6. Castro, P.S., Moitra, S., Gelada, C., Kumar, S., Bellemare, M.G.: Dopamine: a research framework for deep reinforcement learning (2018). http://arxiv.org/abs/1812.06110
7. Cully, A., Demiris, Y.: Quality and diversity optimization: a unifying modular framework. IEEE Trans. Evol. Comput. **22**, 245–259 (2018)
8. Dabney, W., Ostrovski, G., Silver, D., Munos, R.: Implicit quantile networks for distributional reinforcement learning. In: ICML (2018)
9. Deisenroth, M.P., Neumann, G., Peter, J.: A survey on policy search for robotics. Found. Trends Robot. **2**(1–2), 1–142 (2013)
10. Dempster, A.P., Laird, N.M., Rubin, D.B.: Maximum likelihood from incomplete data via the EM algorithm. J. Roy. Stat. Soc. Ser. B (Methodol.) **39**(1), 1–38 (1977)
11. Eramo, C.D., Tateo, D., Bonarini, A., Restelli, M., Milano, P., Peters, J.: Sharing knowledge in multi-task deep reinforcement learning. In: ICLR (2020)
12. Eysenbach, B., Gupta, A., Ibarz, J., Levine, S.: Diversity is all you need: learning skills without a reward function. In: ICLR (2018)
13. Fujimoto, S., van Hoof, H., Meger, D.: Addressing function approximation error in actor-critic methods. In: ICML (2018)
14. Hessel, M., Soyer, H., Espeholt, L., Czarnecki, W., Schmitt, S., Van Hasselt, H.: Multi-task deep reinforcement learning with PopArt. In: AAAI (2019)
15. Hospedales, T., Antoniou, A., Micaelli, P., Storkey, A.: Meta-learning in neural networks: a survey (2020). https://arxiv.org/abs/2004.05439
16. Jacobs, R.A., Jordan, M.I., Nowlan, S.E., Hinton, G.E.: Adaptive mixture of experts. Neural Comput. **3**, 79–87 (1991)
17. Jacobs, R., Jordan, M.: A competitive modular connectionist architecture. In: NeurIPS (1990)
18. Jordan, M.I., Jacobs, R.A.: Hierarchical mixtures of experts and the EM algorithm. In: IJCNN (1993)
19. Lazaric, A.: Transfer in reinforcement learning: a framework and a survey. In: Wiering, M., van Otterlo, M. (eds.) Reinforcement Learning - State of the Art, vol. 12, pp. 143–173. Springer, Heidelberg (2012). https://doi.org/10.1007/978-3-642-27645-3_5
20. Lazaric, A., Ghavamzadeh, M.: Bayesian multi-task reinforcement learning. In: ICML (2010)
21. Lee, K., Seo, Y., Lee, S., Lee, H., Shin, J.: Context-aware dynamics model for generalization in model-based reinforcement learning. In: ICML (2020)
22. Li, H., Liao, X., Carin, L.: Multi-task reinforcement learning in partially observable stochastic environments. J. Mach. Learn. Res. **10**, 1131–1186 (2009)

23. Machado, M.C., Bellemare, M.G., Talvitie, E., Veness, J., Hausknecht, M.J., Bowling, M.: Revisiting the arcade learning environment: evaluation protocols and open problems for general agents. JAIR **61**, 523–562 (2018)
24. MacQueen, J.: Some methods for classification and analysis of multivariate observations. In: Proceedings of the Fifth Berkeley Symposium on Mathematical Statistics and Probability, Volume 1: Statistics, pp. 281–297 (1967)
25. Mahmud, M.M.H., Hawasly, M., Rosman, B., Ramamoorthy, S.: Clustering Markov decision processes for continual transfer (2013). http://arxiv.org/abs/1311.3959
26. Meila, M., Jordan, M.I.: Learning fine motion by Markov mixtures of experts. In: NeurIPS (1995)
27. Mnih, V., et al.: Human-level control through deep reinforcement learning. Nature **518**, 529–533 (2015)
28. Portelas, R., Colas, C., Hofmann, K., Oudeyer, P.Y.: Teacher algorithms for curriculum learning of deep RL in continuously parameterized environments. In: CoRL (2019)
29. Pugh, J.K., Soros, L.B., Stanley, K.O.: Quality diversity: a new frontier for evolutionary computation. Front. Robot. AI **3**, 40 (2016)
30. Riemer, M., Liu, M., Tesauro, G.: Learning abstract options. In: NeurIPS (2018)
31. Sharma, S., Jha, A.K., Hegde, P.S., Ravindran, B.: Learning to multi-task by active sampling. ICLR 2018 - Conference Track (2018)
32. Sutton, R.S., Barto, A.G.: Reinforcement Learning: An Introduction. MIT Press (2017)
33. Tang, G., Hauser, K.: Discontinuity-sensitive optimal control learning by mixture of experts. In: ICRA (2019)
34. Thrun, S., O'Sullivan, J.: Discovering structure in multiple learning tasks : the TC algorithm. In: ICML (1996)
35. Wang, R., Lehman, J., Clune, J., Stanley, K.O.: Paired open-ended trailblazer (POET): endlessly generating increasingly complex and diverse learning environments and their solutions (2019). http://arxiv.org/abs/1901.01753
36. Wang, R., et al.: Enhanced POET: open-ended reinforcement learning through unbounded invention of learning challenges and their solutions. In: ICML (2020)
37. Watkins, C.J.C.H.: Learning from delayed rewards. Ph.D. thesis, King's College, Cambridge (1989)
38. Wilson, A., Fern, A., Ray, S., Tadepalli, P.: Multi-task reinforcement learning: a hierarchical Bayesian approach. In: ICML (2007)
39. Yang, J., Petersen, B., Zha, H., Faissol, D.: Single episode policy transfer in reinforcement learning. In: ICLR (2020)
40. Yu, T., Kumar, S., Gupta, A., Levine, S., Hausman, K., Finn, C.: Gradient surgery for multi-task learning (2020). http://arxiv.org/abs/2001.06782
41. Zhang, Y., Yang, Q.: A survey on multi-task learning (2017). https://arxiv.org/abs/1707.08114
42. Zhu, Z., Lin, K., Zhou, J.: Transfer learning in deep reinforcement learning: a survey (2020). http://arxiv.org/abs/2009.07888
43. Zintgraf, L., et al.: VariBAD: a very good method for bayes-adaptive deep RL via meta-learning. In: ICLR (2020)

Deep Model Compression via Two-Stage Deep Reinforcement Learning

Huixin Zhan[1], Wei-Ming Lin[2], and Yongcan Cao[2]($\boxtimes$)

[1] Texas Tech University, Lubbock, TX 79415, USA
huixin.zhan@ttu.edu
[2] The University of Texas at San Antonio, San Antonio, TX 78249, USA
{weiming.lin,yongcan.cao}@utsa.edu

Abstract. Besides accuracy, the model size of convolutional neural networks (CNN) models is another important factor considering limited hardware resources in practical applications. For example, employing deep neural networks on mobile systems requires the design of accurate yet fast CNN for low latency in classification and object detection. To fulfill the need, we aim at obtaining CNN models with both high testing accuracy and small size to address resource constraints in many embedded devices. In particular, this paper focuses on proposing a generic reinforcement learning-based model compression approach in a two-stage compression pipeline: pruning and quantization. The first stage of compression, i.e., pruning, is achieved via exploiting deep reinforcement learning (DRL) to co-learn the accuracy and the FLOPs updated after layer-wise channel pruning and element-wise variational pruning via information dropout. The second stage, i.e., quantization, is achieved via a similar DRL approach but focuses on obtaining the optimal bits representation for individual layers. We further conduct experimental results on CIFAR-10 and ImageNet datasets. For the CIFAR-10 dataset, the proposed method can reduce the size of VGGNet by 9× from 20.04 MB to 2.2 MB with a slight accuracy increase. For the ImageNet dataset, the proposed method can reduce the size of VGG-16 by 33× from 138 MB to 4.14 MB with no accuracy loss.

Keywords: Compression · Computer vision · Deep reinforcement learning

1 Introduction

CNN has shown advantages in producing highly accurate classification in various computer vision tasks evidenced by the development of numerous techniques, e.g., VGG [26], ResNet [9], DenseNet [15], and numerous automatic neural architecture search approaches [29,33]. Albeit promising, the complex structure and large number of weights in these neural networks often lead to explosive computation complexity. Real world tasks often aim at obtaining high accuracy under limited computational resources. This motivates a series of works towards

© Springer Nature Switzerland AG 2021
N. Oliver et al. (Eds.): ECML PKDD 2021, LNAI 12975, pp. 238–254, 2021.
https://doi.org/10.1007/978-3-030-86486-6_15

a light-weight architecture design and better speed-up ratio-accuracy trade-off, including Xception [5], MobileNet/MobileNet-V2 [13], ShuffleNet [34], and CondenseNet [14], where group and deep convolutions are crucial.

In addition to the development of the aforementioned efficient CNN models for fast inference, many results have been reported on the compression of large scale models, e.g., reducing the size of large-scale CNN models with little or no impact on their accuracies. Examples of the developed methods include low-rank approximation [7,20], network quantization [23,30], knowledge distillation [12], and weight pruning [8,11,16,21,36], which focus on identifying unimportant channels that can be pruned. However, one key limitation in these methods is the lack of automatic learning of the pruning policies or quantization strategies for reduced models.

Instead of identifying insignificant channels and then conducting compression during training, another potential approach is to use reinforcement learning (RL) based policies to determine the compression policy automatically. There are limited results on RL based model compression [10,31]. In particular, [10,31] proposed a deep deterministic policy gradient (DDPG) approach that uses reinforcement learning to efficiently sample the designed space for the improvement of model compression quality. While DDPG can provide good performance in some cases, it often suffers from performance volatility with respect to the hyperparameter setup and other tuning methods. Besides, these RL-based methods don't directly deal with leveraging the sparse features of CNN, i.e., pruning the small weight connections.

Recently, RL based search strategies have been developed to formulate neural architecture search. For example, [35,37] considered the generation of a neural architecture via considering agent's action space as the search space in order to model neural architecture search as a RL problem. Different RL approaches were developed to emphasize different representations of the agent's policies along with the optimization methods. In particular, [37] used a recurrent neural network based policy to sequentially sample a string that in turn encodes the neural architecture. Both REINFORCE policy gradient algorithm [28] and Proximal Policy Optimization (PPO) [25] were used to train the network. Differently, [3] used Q-learning to train a policy that sequentially chooses the type of each layer and its corresponding hyper-parameters. Note that [35,37] focuses on generating CNN models with efficient architectures, while not on the compression of large scale CNN models.

In this paper, we propose to develop a novel two-stage DRL framework for deep model compression. In particular, the proposed framework integrates layer-wise pruning rate learning based on testing accuracy and FLOPs, element-wise variational pruning, and per-layer bits representation learning. In the pruning stage, we first conduct channel pruning that will prune the input channel dimension (i.e., C dimension) with minimized accumulated error in feature maps with the obtained per-layer pruning rate. Then fine-tuning with element-wise pruning via information dropout is conducted to prune the weights in the kernel (i.e., from H and W dimensions).

Briefly, this paper has three main contributions:

1. We propose a novel DRL algorithm that can obtain stabilized policy and address Q-value overestimation in DDPG by introducing four improvements: (1) computational constrained PPO: Instead of collecting T timesteps of action advantages in each of M parallel actors and updating the gradient in each iteration based on MT action advantages in one iteration of the typical PPO, we propose to collect Q-values in each tilmestep of M parallel actors and update the gradient each timestep based on the M sampled Q-values; (2) PPO-Clip Objective: We propose to modify the expected return of the policy by clipping subject to policy change penalization. (3) smoothed policy update: Our algorithm first enables multiple agents to collect one minibatch of Q-values based on the prior policy and updates the policy while penalizing policy change. The target networks are then updated by slowly tracking the learned policy network and critic network; and (4) target policy regularization: We propose to smooth Q-functions along regularized actions via adding noise to the target action. The four improvements altogether can substantially improve performance of DRL to yield more stabilized layer-wise prune ratio and bit representations for deep compression, hence outperforming the traditional DDPG. We experimentally show the volatility of DDPG-based compression method in order to backup some common failure mode of policy exploitation in DDPG-based method as shown in Fig. 2.
2. Pruning: We propose a new **ppo** with variational pruning compression structure with element-wise variational pruning that can prune three dimensions of CNN. We further learn the *Pareto front* of a set of models with two-dimensional outputs, namely, model size and accuracy, such that at least one output is better than, or at least as good as, all other models by constraining the actions. More compressed models can be obtained with little or no accuracy loss.
3. Quantization: We propose a new quantization method that uses the same DRL-supported compression structure, where the optimal bit allocation strategy (layer-wise bits representation) is obtained in each iteration via learning the updated accuracy. Fine-tuning is further executed after each rollout.

2   A Deep Reinforcement Learning Compression Framework

In this section, we focus on presenting the proposed new generic reinforcement learning based model compression approach in a two-stage compression pipeline: pruning and quantization. Figure 1 shows the overall structure. The two-stage compression pipeline includes pruning and quantization. Adopting the pipeline can achieve a typical model compression rate between 4× and 33×. Investigating the Pareto front of candidate compression models shows little or no accuracy loss.

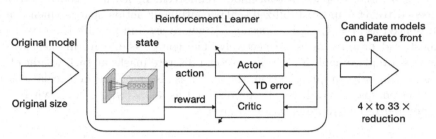

Fig. 1. The proposed deep reinforcement learning compression framework.

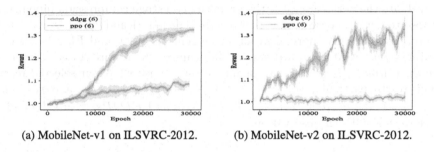

(a) MobileNet-v1 on ILSVRC-2012.   (b) MobileNet-v2 on ILSVRC-2012.

Fig. 2. Comparison of RL-based pruning methods, e.g., **ppo (modified)** and **ddpg**, for MobileNet-v1 and MobileNet-v2 on ILSVRC-2012 for 6 runs.

2.1 State

In both pruning and quantization, in order to discriminate each layer in the neural network, we use a 8-dimension vector space to model a continuous state space:

$$s_t = [N_{Lr}, N, C, H, W, Stride, A_H^t, FLOPs],   (1)$$

where N_{Lr} is the index of the layer, N and C are the dimension of, respectively, output channels and input channels, H is the kernel height, W is the kernel width, $Stride$ is the number of pixels shifts over the input matrix, A_H^t is the maximum pruning rate in pruning (respectively, the maximum and minimum bits representation in quantization) with respect to layer t, and $FLOPs$ is the number of floating point operations in each layer.

2.2 Action

In pruning, determining the compression policy is challenging because the pruning rate of each layer in CNN is related in an unknown way to the accuracy of the post-compression model. Since our goal is to simultaneously prune the C, H, and W dimensions. As the dimension of channels increases or the network goes deeper, the computation complexity increases exponentially. Instead of searching over a discrete space, a continuous reinforcement learning control strategy is needed to get a more stabilized scalar continuous action space, which can be represented as $a_t = \{pr_t | pr_t \in [pr_h, pr_l]\}$, where pr_l and pr_h are the lowest and highest and pruning rates, respectively. The compression rate in each layer is taken as a replacement of high-dimensional discrete masks at each weight of the kernels. Similarly, in quantization, the action is also modeled in a scalar continuous action space, which can be represented as $a_t = \{b_t | b_t \in \mathbb{N}^+\}$, where b_t is the number of bits representation in layer t.

2.3 Reward

To evaluate the performance of the proposed two-stage compression pipeline, we propose to construct two reward structures, labeled $r1$ and $r2$. $r1$ is a synthetic reward system as the normalization of current accuracy and FLOPs. $r2$ is an accuracy-concentrated reward system. In pruning, the reward for each layer can be chosen from $r_t \in \{r1, r2\}$. In quantization, we use $r2$ as our selected reward structure. In particular, $r1 = 1 - \frac{FLOPs_t - FLOPs_{low}}{FLOPs_{high} - FLOPs_{low}} + p_{ac}$ and $r2 = p_{ac}$, where p_{ac} is the current accuracy, $FLOPs_{high}$ and $FLOPs_{low}$ are the highest and lowest FLOPs in observation.

2.4 The Proposed DRL Compression Structure

In the proposed model compression method, we learn the Pareto front of a set of models with two-dimensional outputs (model size and accuracy) such that at least one output is better than (or at least as good as) all other outputs. We adopt a popular asynchronous actor critic [22] RL framework to compress a pre-trained network in each layer sequentially. At time step t, we denote the observed state by s_t, which corresponds to the per-layer features. The action set is denoted by $\mathcal{A}$ of size 1. An action, $a_t \in \mathcal{A}$, is drawn from a policy function distribution: $a_t \sim \mu(s_t | \theta^\mu) + \mathcal{N}_t \in \mathbb{R}^1$, referred to as an actor, where θ^μ is the current policy network parameter and the noise $\mathcal{N}_t \in \mathcal{N}(0, \epsilon)$. The actor receives the state s_t, and outputs an action a_t. After this layer is compressed with pruning rate or bits representation a_t, the environment then returns a reward r_t according to the reward function structure $r1$ or $r2$. The updated state s_{t+1} at next time step $t+1$ is observed by a known state transition function $s_{t+1} = f(s_t; a_t)$, governed by the next layer. In this way, we can observe a random minibatch of transitions consisting of a sequence of tuples $\mathbb{B} = \{(s_t; a_t; r_t; s_{t+1})\}$. In typical PPO, the surrogate objective is represented by $\hat{\mathbb{E}}_t[\frac{\pi^{\theta^\mu}(a|s_t)}{\pi^{\theta^{\mu^-}}(a|s_t)} \hat{A}_t]$, where the expectation $\hat{\mathbb{E}}_t[\cdot]$ is the empirical average over a finite batch of samples and θ^{μ^-} is the prior policy network parameter. If we compute the action advantage $\hat{A}_t$ in each layer, T-step time difference rewards are needed, which is computationally intensive. In resource constrained PPO, we propose to replace the action advantages by Q-functions given by $Q(s_t, a_t) = \mathbb{E}[\sum_{i=t}^{t+T} \gamma^{i-t} r_i | s_t, a_t]$, referred to as critic.

The policy network parameterized by θ^{μ} and the value function parameterized by θ^{Q} are then jointly modeled by two neural networks. Let $a = \mu(s_i|\theta^{\mu})$, we can learn θ^{Q} via Q-function regression, namely, Eq. (2), and learn θ^{μ} over the tuples $\mathbb{B}$ with PPO-Clip objective stochastic policy gradient, namely, Eq. (3) as

$$\theta^{Q} = \arg\min_{\theta} \frac{1}{|\mathbb{B}|} \sum_{(s_i,a_i,r_i,s_{i+1})\in\mathbb{B}} (y_i - Q(s_i,a_i|\theta))^2, \tag{2}$$

$$\theta^{\mu} = \arg\max_{\theta} \hat{\mathbb{E}}_{(s_i,a_i,\cdots)\in\mathbb{B}} \min\left\{ \frac{\pi^{\theta}(a|s_i)}{\pi^{\theta^{\mu^{-}}}(a|s_i)} Q(s_i,a|\theta^{Q}), \right.$$
$$\left. clip(\frac{\pi^{\theta}(a|s_i)}{\pi^{\theta^{\mu^{-}}}(a|s_i)}, 1-c, 1+c) \times Q(s_i,a|\theta^{Q}) \right\}, \tag{3}$$

where c is the probability ratio of the clipping. A pseudocode of DRL compression structure is shown in Algorithm 1.

Algorithm 1: The proposed DRL compression structure in pruning.

Data: Randomly initialize critic network $Q(s,a|\theta^{Q})$ and actor $\mu(s|\theta^{\mu})$ with weights θ^{Q} and θ^{μ}. Initialize target network Q' and μ' with weights $\theta^{Q'} \leftarrow \theta^{Q}$, $\theta^{\mu'} \leftarrow \theta^{\mu}$, the learning rate of the target network ρ, A_H^t, and empty replay buffer $\mathcal{D}$

Result: Weights θ^{Q} and θ^{μ}.

1 initialization;
2 **while** *Episode* $< M$ **do**
3   Initialize a random process $\mathcal{N}$ for action exploration;
4   Receive initial observation state s_1;
5      $M \leftarrow M + 1$;
6   **for** $t = 1, \cdots, T$ **do**
7   Select action $a_t = clip(\mu(s_t|\theta^{\mu}) + \mathcal{N}_t, A_H^t)$ according to the current policy and exploration noise;
8   Execute a_t;
9   Store (s_t, a_t, r_t, s_{t+1}) in replay buffer $\mathcal{D}$;
10   **for** $t = 1, \cdots$ **do**
11   Sample a random minibatch of $\mathbb{B}$ trajectories from $\mathcal{D}$;
12   Set $y_t = r_t + \gamma Q(s_{t+1}, \mu(s_{t+1})|\theta^{Q'})$;
13   Update the policy by maximizing the "surrogate" objective via stochastic gradient ascent with Adam in Eq. 3;
14   Pruning the C dimension of t-th layer with pruning rate a_t;
15   Executing the element-wise variational pruning in Algorithm 2;
16   Update the critic by minimizing the combinatorial loss via stochastic gradient descent in Eq. 2;
17   Update the target networks via $\theta^{Q'} \leftarrow \rho\theta^{Q'} + (1-\rho)\theta^{Q}$, $\theta^{\mu'} \leftarrow \rho\theta^{\mu'} + (1-\rho)\theta^{\mu}$;
18   **end**
19   **end**
20 **end**

---

3 Pruning

In this section, we present two schemes to compress CNN with little or no loss in accuracy by employing reinforcement learning to co-learn the layer-wise pruning rate and the element-wise variational pruning via information dropout. Similar to the aforementioned a3c framework, the layer-wise pruning rate is computed by the actor. After obtaining pruning rate a_t, layer s_t is pruned by a typical channel pruning method [11], whose detail will be given below, to select the most representative channels and reduce the accumulated error of feature maps. In other words, after we get the pruning rate, channel pruning can be used to determine which specific channels are less important or we can simply prune based on the weight magnitude. In each iteration, the CNN layer is further compressed by variational pruning. In particular, we start by learning the connectivity via normal network training. Then, we prune the small-weight connections: all connections with weights that create a representation of the data that is minimal sufficient for the task of reconstruction are remained. Finally, we retrain the network to learn the final weights for the remaining sparse connections.

In pruning, β is a vector whose dimension matches the 4D tensor with shape $N \times C \times W \times H$ in each layer. We also define β^i, the i-th entry of β, as a binary mask for each weight in the kernel. Figure 3 shows the pruning flow in our two schemes. In DRL compression framework, the scalar mask of the j-th weight w_j with mask β^j is set to zero if the weights are pruned based on LASSO regression, discussed in subsection. If pruned, these weights are moved to $\bar{s}^j$, defined as a set of pruned weights. Otherwise, if the weights are pruned based on information dropout, discussed in subsection, the scalar mask of the j-th weight w_j with mask β^j will be moved to $\bar{s}^j$ with probability $p_{a_\theta}(\xi^{(j)})$. The weights that play more important role in reducing the classification error are less likely to be pruned.

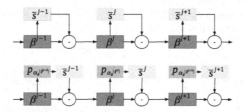

Fig. 3. Comparison of the pruning on two aforementioned schemes, i.e., channel pruning and variational pruning via information dropout.

3.1 Pruning from C Dimension: Channel Pruning

The C-dimension channel pruning can be formulated as:

$$\underset{\beta, W}{\arg\min} \frac{1}{2N} \left\| Y - \sum_{i=1}^{C} \beta X_i W_i^T \right\|_F^2 + \lambda \|\beta\|_1 \qquad (4)$$

$$subject\ to\ \|\beta\|_0 \leq pr \times C$$

$$\|W_i\|_F = 1, \forall i,$$

where pr is the pruning rate, X_i and Y are the input volume and the output volume in each layer, W_i is the weights, β is the coefficient vector of length C for channel selection, and λ is a positive weight to be selected by users. Then we assign $\beta_i \leftarrow \beta_i \|W_i\|_F$ and $W_i \leftarrow \frac{W_i}{\|W_i\|_F}$.

3.2 Pruning from H and W Dimensions: Variational Pruning

In information dropout, we propose a solution to: (1) efficiently approximate posterior inference of the latent variable $\mathbf{z}$ given an observed value $\mathbf{x}$ based on parameter θ, where $\mathbf{z}$ is a representation of $\mathbf{x}$ and defined as some (possibly nondeterministic) function of $\mathbf{x}$ that has some desirable properties in some coding tasks $\mathbf{y}$ and (2) efficiently approximate marginal inference of the variable $\mathbf{x}$ to allow for various inference tasks where a prior over $\mathbf{x}$ is required.

Without loss of generality, let us consider Bayesian analysis of some dataset $\mathcal{D} = \left\{(\mathbf{x}^{(i)}, \mathbf{y}^{(i)})\right\}_{i=1}^{N}$ consisting of N i.i.d samples of some discrete variable $\mathbf{x}$. We assume that the data are generated by some random process, involving an unobserved continuous random variable $\mathbf{z}$. Bayesian inference in such a scenario consists of (1) updating some initial belief over parameters $\mathbf{z}$ in the form of a prior distribution $p_{\theta^*}(\mathbf{z})$, and (2) a belief update over these parameters in the form of (an approximation to) the posterior distribution $p_\theta(\mathbf{z}|\mathbf{x})$ after observing data $\mathbf{x}$. In variational inference [17], inference is considered as an optimization problem where we optimize the parameters θ of some parameterized model $p_\theta(\mathbf{z})$ such that $p_\theta(\mathbf{z})$ is a close approximation of $p_\theta(\mathbf{z}|\mathbf{x})$ as measured by the KL divergence $D_{KL}(p_\theta(\mathbf{z}|\mathbf{x})|p_\theta(\mathbf{z}))$. The divergence between $p_\theta(\mathbf{z}|\mathbf{x})$ and the true posterior is minimized by minimizing the negative variational lower bound $\mathcal{L}(\theta)$ of the marginal likelihood of the data, namely,

$$\mathcal{L}(\theta, \theta^\star; \mathbf{x}^{(i)}) = - \sum_{(\mathbf{x}^{(i)}, \mathbf{y}^{(i)}) \in \mathcal{D}} \mathbb{E}_{p_\theta(\mathbf{z}|\mathbf{x}^{(i)})}[\log p(\mathbf{y}^{(i)}|\mathbf{z})]$$
$$+ \alpha D_{KL}(p_\theta(\mathbf{z}|\mathbf{x}^{(i)})|p_\theta^\star(\mathbf{z})). \tag{5}$$

As shown in [18], the neural network weight parameters θ are less likely to overfit the training data if adding input noise during training. We propose to represent $\mathbf{z}$ by computing a deterministic map of activations $f(\mathbf{x})$, and then multiply the result in an element-wise manner by a random noise ξ, drawn from a parametric distribution p_a with the variance that depends on the input data $\mathbf{x}$, as

$$\mathbf{z} = (\mathbf{x} \circ \xi)\theta,$$
$$\xi_{i,j} \sim p_{a_\theta}(\mathbf{x})(\xi_{i,j}), \tag{6}$$

where $\circ$ denotes the element product operation of two vectors. A choice for the distribution $p_{a_\theta(\mathbf{x})}(\xi_{i,j})$ is the log-normal distribution $\log(p_{a_\theta(\mathbf{x})}(\xi_{i,j})) = \mathcal{N}(0, a_\theta^2(\mathbf{x}))$ [2] that makes the normally fixed dropout rates p_a adaptive to the input data, namely,

$$\log(p_{a_\theta(\mathbf{x})}(\xi_{i,j})) \sim \mathcal{N}(\mathbf{z}; 0, a_\theta^2(\mathbf{x})I),$$
$$\log(p_{\theta^*}) \sim \mathcal{N}(\mathbf{z}; \mu, \sigma^2 I), \tag{7}$$

where $a_\theta(\mathbf{x})$ is an unspecified function of $\mathbf{x}$. The resulting estimator becomes

$$\mathcal{L}(\theta; \mathbf{x}^{(i)}) \sim \frac{1}{N} \sum_{j=1}^{N} [-\log p(\mathbf{y}^{(i)}|\mathbf{z}^{(i,j)})]$$
$$+ \alpha[\frac{1}{2\sigma^2}(a_\theta^2(\mathbf{x}^{(i)}) + \mu^2) - \log\frac{a_\theta^2(\mathbf{x}^{(i)})}{\sigma} - \frac{1}{2}], \tag{8}$$

where $\mathbf{z}^{(i,j)} \sim (\mathbf{x}^{(i)} \circ \xi^{(i,j)})\theta$ and $\xi^{(i,j)} \sim p_{a_\theta}(\xi) = \log \mathcal{N}(0, a_\theta^2(\mathbf{x}))$. This loss can be optimized using stochastic gradient descent. A pseudocode of this variational pruning is shown in Algorithm 2 and an illustrative experiment is given in Subsect. 5.4.

Algorithm 2: Variational pruning.

Data: Pruned model parameters at this iteration θ, the number of fine-tuning iterations $\mathcal{Z}$, learning rate γ and decay of learning rate τ.

Result: Further compressed and tuned model parameters θ.

1 **for** $\mathcal{Z}$ *iterations* **do**
2   Randomly choose a mini-batch of samples from the training set;
3   Compute gradient of $\mathcal{L}(\theta; \mathbf{x}^{(i)})$ by $\frac{\partial \mathcal{L}(\theta; \mathbf{x}^{(i)})}{\partial \theta}$, where $\mathcal{L}(\theta; \mathbf{x}^{(i)})$ is computed by Eq. (8);
4   Update θ using $\theta \leftarrow \theta - \gamma\frac{\partial \mathcal{L}(\theta; \mathbf{x}^{(i)})}{\partial \theta}$;
5     $\gamma \leftarrow \tau\gamma$
6 **end**

---

4 Quantization

In the proposed DRL-based quantization-aware training, the RL agent automatically searches for the optimal bit allocation representation strategy for each layer. The modeling of quantization state, action, and rewards are defined in Sect. 2. The DRL structure is the same as the one for pruning in Subect. 2.4. In the fine-tuning step of the quantized CNN, we apply Straight-Through Estimator (STE) [4]. The idea of this estimator of the expected gradient through stochastic neurons is simply to back-propagate through the hard threshold function, e.g., sigmoidal non-linearity function [32]. The gradient is 1 if the argument is positive and 0 otherwise.

5 Experiments

5.1 Settings

In all experiments, the MNIST and CIFAR-10 dataset are both divided by 50000 samples for training, 5000 samples for validation, and 5000 samples for evaluation. The ILSVRC-12 dataset is divided by 1281167 samples for training, 10000 samples for validation, and 50000 samples for evaluation. We adopt a neural network policy with one hidden layer of size 64 and one fully-connected layer using sigmoid as the activation function. We use the proximal policy optimization clipping algorithm with c = 0.2 as the optimizer. The critic also has one hidden layer of size 64. The discounting factor is selected as $\gamma = 0.99$. The learning rate of the actor and the critic is set as 1×10^{-3}. In CIFAR-10, the per GPU batch size for training is 128 and the batch size for evaluation is 100. In ILSVRC-12, the per GPU batch size for training is 64 and the batch size for evaluation is 100. The fine tuning steps for each layer are selected as 2000 in the quantization. The parameters are optimized using the SGD with momentum algorithm [27]. For MNIST and CIFAR-10, the initial learning rate is set as 0.1 for LeNet, ResNet, and VGGNet. For ILSVRC-12, the initial learning rate is set as 1 and divided by 10 at rollouts 30, 60, 80, and 90. The decay of learning rate is set to 0.99. All experiments were performed using TensorFlow, allowing for automatic differentiation through the gradient updates [1], on 8 NVIDIA Tesla K80 GPUs.

5.2 MNIST and CIFAR-10

The MNIST [6] and CIFAR-10 dataset [19] consists of images with a 32 × 32 resolution. Table 1 shows the performance of the proposed method. It can be observed that the proposed method can not only reduce model size but also improve the accuracy (i.e., red-

Table 1. Results on MNIST and CIFAR-10 dataset.

LeNet	Error (%)	Para	Pruned para. (%)	FLOPs (%)
LeNet-5 (DropPruning)	0.73	60K	87.0	–
LeNet-5	0.34	5.94K	90.1	16.4
CIFAR-10	Error (%)	Para	Pruned Para. (%)	FLOPs (%)
VGGNet(Baseline)	6.54	20.04M	0	100
VGGNet	6.33	2.20M	89.0	48.7
VGGNet	6.20	2.29M	88.6	49.1
ResNet-152 (Baseline)	5.37	1.70M	0	100
ResNet-152	5.19	1.30M	23.5	71.2
ResNet-152	5.33	1.02M	40.0	55.1

uce error rate). In MNIST, comparing with the most recent DropPruning method, our method for LeNet obtains 10× model compression with a slightly accuracy increase (0.68%). In CIFAR-10, comparing with the baseline model, our method for the VGGNet achieves 9× model compression with a slightly accuracy increase (0.34%). In addition, we compare our algorithm with the commonly adopted weight magnitude channel selection strategy and channel pruning strategy to demonstrate the importance of variational pruning. Please refer to the Subsect. 5.3 for more details.

5.3 ImageNet

Table 2. Results on ImageNet dataset for ResNet-18 and ResNet-50 with different speed-ups.

Model	Top-1/Top-5 error (%)	Pruned para. (%)	FLOPs (%) Pruning	Speed-up ×
				Pruning + Quantization
ResNet-18 (Baseline)	29.36/10.02	0	100	1
ResNet-18	30.29/10.43	30.2	71.4	11.4
ResNet-18	30.65/11.93	51.0	44.2	16.0
ResNet-18	33.40/13.37	76.7	29.5	28.2
ResNet-50 (Baseline)	24.87/6.95	0	100	1
ResNet-50	23.42/6.93	31.2	66.7	12.0
ResNet-50	24.21/7.65	52.1	47.6	16.0
ResNet-50	28.73/8.37	75.3	27.0	29.6

To evaluate the effect of different pruning rates A_H^t, we select 30%, 50%, and 70% for ResNet-18 and ResNet-50 and then evaluate the model pruning on ImageNet ILSVRC-2012 dataset [24]. Experimental results are shown in Table 2 while the per-layer weight bits policy for the quantization is shown in Fig. 4. From Table 2, it can be seen that the error increases as the pruning rate increases. However, our pruned ResNet-50 with 30% pruning rate outperforms the pre-trained baseline model in the top-1 accuracy and our pruned ResNet-50 with 30% and 50% pruning rate outperforms the pre-trained baseline model in the top-1 accuracy. In Fig. 4, the 8-bit uniform quantization strategy is shown in blue bar, and the **ppo** with variational pruning policy is shown in red bar. The DRL-supported policy generates a more compressed model with a faster inference speed. By observing the DRL-supported policy, the 3×3 layer is more important than the 1×1 layers because the 1×1 layers are represented by less bits naturally.

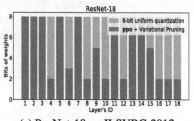
(a) ResNet-18 on ILSVRC-2012.

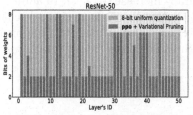
(b) ResNet-50 on ILSVRC-2012.

Fig. 4. ResNet-18 and ResNet-50 with different bit allocation strategies.

To show the importance of our DRL-supported compression structure with variational pruning, we compare RL with channel pruning and RL with variational pruning. Table 3 shows that **ppo** with channel pruning can find the optimal layer-wise pruning rates while **ppo** with variational pruning can further decrease the testing error of the compressed model. Another observation is with the same compression scope, e.g., 50% FLOPs, our model's accuracy outperforms **ddpg** based algorithms. A comparison of the reward $r1$ for AMC [10] (DDPG-based pruning) and our proposed method (PPO-based pruning) is also shown in Fig. 2 for $A_H^t = 50\%$ in 6 runs. A typical failure mode of **ddpg** training is the Q-value overestimation, which leads to a lower reward in $r1$.

Table 3. MobileNet-v1 and MobileNet-v2 on ILSVRC-12.

Model	FLOPs (%)	Δacc (%)
MobileNet-v1 (Baseline)	100	0
MobileNet-v1 (**ppo** + Channel Pruning)	50	−0.2
MobileNet-v1 (**ppo** + Channel Pruning)	40	−1.1
MobileNet-v1 (**ppo** + Variational Pruning)	47	+0.1
MobileNet-v1 (**ppo** + Variational Pruning)	40	−0.8
MobileNet-v1 (**ddpg**) [10]	50	−0.4
MobileNet-v1 (**ddpg**) [10]	40	−1.7
0.75 MobileNet-v1 (Uniform) [13]	56	−2.5
0.75 MobileNet-v1 (Uniform) [13]	41	−3.7
Model/Pruning	FLOPs (%)	Δacc(%)
MobileNet-v2 (Baseline)	100	0
MobileNet-v2 (**ppo** + Variational Pruning)	21	−2.4
MobileNet-v2 (**ppo** + Variational Pruning)	59	−1.0
MobileNet-v2 (**ppo** + Variational Pruning)	70	−0.8
MobileNet-v2 (**ddpg**) [10]	30	−3.1
MobileNet-v2 (**ddpg**) [10]	60	−2.1
MobileNet-v2 (**ddpg**) [10]	70	−1.0
0.75 MobileNet-v2 (Uniform) [13]	70	−2.0
Model/Quantization	Model size	Δacc(%)
MobileNet-v2 (**ppo** + Variational Pruning)	0.95M	−2.9
MobileNet-v2 (**ppo** + Variational Pruning)	0.89M	−3.2
MobileNet-v2 (**ppo** + Variational Pruning)	0.81M	−3.6
MobileNet-v2 (HAQ) [30]	0.95M	−3.3
MobileNet-v2 (Deep Compression) [8]	0.96M	−11.9

We also report the results for ILSVRC-12 on MobileNet-v2 on Table 3. Although for 70% FLOPs, our model's performance is competitive to the **ddpg** approach, we achieve lower accuracy decrease on smaller models such as 30% and 60% FLOPs.

We further examine the results when applying both pruning and quantization on ILSVRC-12. We use the VGG-16 model with 138 million parameters as the reference model. Table 4 shows that VGG-16 can be compressed to 3.0% of its original size (i.e., 33× speed-up) when weights in the convolution layers are represented with 8 bits, and fully-connected layers with 5 bits. Again, the compressed model outperforms the baseline model in both the top-1 and top-5 errors.

Table 4. Comparison with another non-RL two-stage compression method (VGG-16 on ILSVRC-12).

Model (MobileNet-v1)	Layer	Parameters	Pruned para. (%)	Weight bits Pruning + Quantization	Speed-up × Pruning + Quantization
ppo + Variational Pruning	conv1_1/conv1_2	2K/37K	42/89	8/8	2.5/10.2
	conv2_1/conv2_2	74K/148K	72/69	8/8	7.0/6.8
	conv3_1/conv3_2/conv3_3	295K/590K/590K	50/76/58	8/8/8	4.6/10.3/5.9
	conv4_1/conv4_2/conv4_3	1M/2M/2M	68/88/76	8/8/8	7.6/9.1/7.2
	conv5_1/conv5_2/conv5_3	1M/2M/2M	70/76/69	8/8/8	7.1/8.5/7.1
	fc_6/fc_7/fc_8	103M/17M/4M	96/96/77	5/5/5	62.5/66.7/14.1
	Total	138M	93.1	5	33×
Deep compression [8]	conv1_1/conv1_2	2K/37K	42/78	5/5	2.5/10.2
	conv2_1/conv2_2	74K/148K	66/64	5/5	6.9/6.8
	conv3_1/conv3_2/conv3_3	295K/590K/590K	47/76/58	5/5/5	4.5/10.2/5.9
	conv4_1/conv4_2/conv4_3	1M/2M/2M	68/73/66	5/5/5	7.6/9.2/7.1
	conv5_1/conv5_2/conv5_3	1M/2M/2M	65/71/64	5/5/5	7.0/8.6/6.8
	fc_6/fc_7/fc_8	103M/17M/4M	96/96/77	5/5/5	62.5/66.7/14.0
	Total	138M	92.5	5	31×

5.4   Variational Pruning via Information Dropout

The goal of this illustrative experiment is to validate the approach in subsection 3.2 and show that our regularized loss function $\mathcal{L}(\theta; \mathbf{x}^{(i)})$ shown in Equation (8) can automatically adapt to the data and can better exploit architectures for further compression. The random noise ξ is drawn from a distribution $p_{a_\theta(\mathbf{x})}(\xi)$ with a unit mean $u = 1$ and a variance $a_\theta(\mathbf{x})$ that depends on the input data $\mathbf{x}$. The variance $a_\theta(\mathbf{x})$ is parameterized by θ. To determine the best allocation of parameter θ to minimize the KL-divergence term $D_{KL}(p_\theta(\mathbf{z}|\mathbf{x})|p_{\theta^*}(\mathbf{z}))$, we still need to have a prior distribution $p_{\theta^*}(\mathbf{z})$. The prior distribution is identical to the expected distribution of the activation function $f(\mathbf{x})$, which represents how much data $\mathbf{x}$ lets flow to the next layer. For a network that is implemented using the softplus activation function, a log-normal distribution is a good fit for the prior distribution (Achille and Soatto 2018). After we fix this prior distribution as $\log(p_{\theta^*}(\mathbf{z})) \sim \mathcal{N}(0, 1)$, the loss can be computed using stochastic gradient descent to back-propagate the gradient through the sampling of $\mathbf{z}$ to obtain the optimized parameter θ. Even if $\log(p_{\theta^*}(\mathbf{z})) \sim \mathcal{N}(0, 1)$, the actual value of u is not necessarily equal to 1 during the runtime. Hence, the mean u and the variance $a_\theta(\mathbf{x})$ of the random noise ξ can be computed via solving the following two equations

$$E(\xi) = e^{u + \frac{a_\theta^2(\mathbf{x})}{2}}, \tag{9}$$

$$D(\xi) = (e^{a_\theta^2(\mathbf{x})} - 1)e^{a_\theta^2(\mathbf{x}) + 2u}, \tag{10}$$

where $E(\xi)$ is the mean of sampled ξ and $D(\xi)$ is the variance of sampled ξ. We add a constraint, $a_\theta(\mathbf{x}) \leq 0.8$, to avoid a large noise variance. Figure 5d shown the probability density function (PDF) of the noise parameter by experiment, which matches a log-normal distribution. The result shows that we optimize the parameters θ of the parameterized model $p_{\theta^*}(\mathbf{z})$ such that $p_{\theta^*}(\mathbf{z})$ is a close approximation of $p_\theta(\mathbf{z}|\mathbf{x})$ as measured by the KL divergence $D_{KL}(p_\theta(\mathbf{z}|\mathbf{x})|p_{\theta^*}(\mathbf{z}))$. After the noise distribution is known, the distribution of $p_\theta(\mathbf{z}|\mathbf{x})$ in Equation (5) can be obtained. In order to show how much information from images that information dropout is transmitting to the second layer, Fig. 5b shows the latent variable $\mathbf{z}$ while Fig. 5c shows the weights. As shown in Fig. 5b, the network lets through the input data (Fig. 5a).

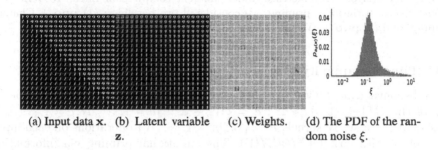

(a) Input data $\mathbf{x}$. (b) Latent variable $\mathbf{z}$. (c) Weights. (d) The PDF of the random noise ξ.

Fig. 5. An illustrative information dropout experiment. (a) shows the input data $\mathbf{x}$. (b) shows the plot of the latent variable $\mathbf{z}$ at a choice of parameter θ at each spatial location in the third information dropout layer of LeNet trained on MNIST with $\alpha = 1$. The resulting representation $\mathbf{z}$ is robust to nuisances, and provides good performance. (c) shows the weights. (d) shows the PDF of the noise parameter ξ.

5.5 Single Layer Acceleration Performance

In order to further show the importance of variational pruning after obtaining the optimized pruning rate based on the **ppo** algorithm, we test a simple 4-layer convolutional neural network, including 2 convolution (conv) layers and 2 fully connected (fc) layers, for image classification on the CIFAR-10 dataset. We evaluate single layer acceleration performance using the proposed **ppo** with variational pruning algorithm in Sect. 3 and compare it with the channel pruning strategy. A third typical weight magnitude pruning method is also tested for further comparison, i.e., pruning channels based on the weights' magnitude (**ppo** + Weight Magnitude Pruning).

Figure 6 shows the performance comparison measured by the error increase after a certain layer is pruned. By analyzing this figure, we can observe that our method (**ppo** + Variational Pruning) earns the best performance in all layers. Since, **ppo** + Channel Pruning applies a LASSO regression based channel selection to minimize the reconstruction error, it achieves a better performance than the weight magnitude method. Furthermore, the proposed policy considers the fully-connected layers more important than the convolutional layers because the error increase for fully-connected layers is typically larger under the same compression rate.

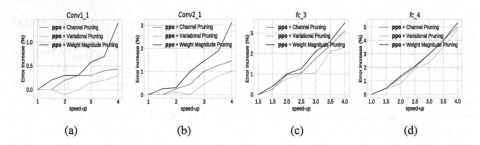

Fig. 6. Single layer error increase under different compression rates. To verify the importance of variational pruning, we considered two baselines: (1) **ppo** + Channel Pruning, and (2) **ppo** + Weight Magnitude Pruning.

5.6   Time Complexity

A single convolutional layer with N kernels requires evaluating a total number of NC of the $2D$ kernels $W_n^c * z^c : F = \{W_n^c \in \mathbb{R}^{d \times d} | n = 1, \cdots, N; c = 1, \cdots, C\}$. Note that there are N kernels $F = \{W_n^c | n = 1, \cdots, N\}$ operations on each input channel z^c with cost $O(NCd^2HW)$. The variational pruning via information dropout involves computing a total number of NC' of the $2D$ kernels $W_n^c * z^c$ with cost $O(NC'd^2HW)$, indicating that efficiency inference requires that $C' \ll C$. In subsection 3.2, we consider ameliorating the inference efficiency by information dropout. In the kernels $s^c = \{s_m^c | m = 1, \cdots, M\}$, the cost can be reduced to $O(NC'd^2HW)$.

6   Conclusion

Using hand-crafted features to get compressed models requires domain experts to explore a large design space and the trade-off among model size, speed-up, and accuracy, which is often suboptimal and time-consuming. This paper proposed a deep model compression method that uses reinforcement learning to automatically search the action space, improve the model compression quality, and use the FLOPs obtained from fine-tuning with information dropout pruning for the further adjustment of the policy to balance the trade-off among model size, speed-up, and accuracy. Experimental results were conducted on CIFAR-10 and ImageNet to achieve 4× - 33× model compression with limited or no accuracy loss, proving the effectiveness of the proposed method.

Acknowledgment. This work was supported in part by the Army Research Office under Grant W911NF-21-1-0103.

References

1. Abadi, M., et al.: TensorFlow: a system for large-scale machine learning. In: 12th USENIX Symposium on Operating Systems Design and Implementation, pp. 265–283 (2016)

2. Achille, A., Soatto, S.: Information dropout: learning optimal representations through noisy computation. IEEE Trans. Pattern Anal. Mach. Intell. **40**(12), 2897–2905 (2018)
3. Baker, B., Gupta, O., Naik, N., Raskar, R.: Designing neural network architectures using reinforcement learning. arXiv preprint arXiv:1611.02167 (2016)
4. Bengio, Y., Léonard, N., Courville, A.: Estimating or propagating gradients through stochastic neurons for conditional computation. arXiv preprint arXiv:1308.3432 (2013)
5. Chollet, F.: Xception: deep learning with depthwise separable convolutions. In: Proceedings of the IEEE Conference on Computer Vision and Pattern Recognition, pp. 1251–1258 (2017)
6. Cohen, G., Afshar, S., Tapson, J., Van Schaik, A.: EMNIST: extending MNIST to handwritten letters. In: 2017 International Joint Conference on Neural Networks, pp. 2921–2926 (2017)
7. Denton, E.L., Zaremba, W., Bruna, J., LeCun, Y., Fergus, R.: Exploiting linear structure within convolutional networks for efficient evaluation. In: Advances in Neural Information Processing Systems, pp. 1269–1277 (2014)
8. Han, S., Mao, H., Dally, W.J.: Deep compression: compressing deep neural networks with pruning, trained quantization and Huffman coding. arXiv preprint arXiv:1510.00149 (2015)
9. He, K., Zhang, X., Ren, S., Sun, J.: Deep residual learning for image recognition. In: Proceedings of the IEEE Conference on Computer Vision and Pattern Recognition, pp. 770–778 (2016)
10. He, Y., Lin, J., Liu, Z., Wang, H., Li, L.J., Han, S.: AMC: AutoML for model compression and acceleration on mobile devices. In: Proceedings of the European Conference on Computer Vision, pp. 784–800 (2018)
11. He, Y., Zhang, X., Sun, J.: Channel pruning for accelerating very deep neural networks. In: Proceedings of the IEEE International Conference on Computer Vision, pp. 1389–1397(2017)
12. Hinton, G., Vinyals, O., Dean, J.: Distilling the knowledge in a neural network. arXiv preprint arXiv:1503.02531 (2015)
13. Howard, A.G., et al.: MobileNets: efficient convolutional neural networks for mobile vision applications. arXiv preprint arXiv:1704.04861 (2017)
14. Huang, G., Liu, S., Van der Maaten, L., Weinberger, K.Q.: CondenseNet: an efficient densenet using learned group convolutions. In: Proceedings of the IEEE Conference on Computer Vision and Pattern Recognition, pp. 2752–2761 (2018)
15. Huang, G., Liu, Z., Van Der Maaten, L., Weinberger, K.Q.: Densely connected convolutional networks. In: Proceedings of the IEEE Conference on Computer Vision and Pattern Recognition, pp. 4700–4708 (2017)
16. Jia, H., et al.: Droppruning for model compression. arXiv preprint arXiv:1812.02035 (2018)
17. Kingma, D.P., Welling, M.: Auto-encoding variational bayes. arXiv preprint arXiv:1312.6114 (2013)
18. Kingma, D.P., Salimans, T., Welling, M.: Variational dropout and the local reparameterization trick. In: Advances in Neural Information Processing Systems, pp. 2575–2583 (2015)
19. Krizhevsky, A., Nair, V., Hinton, G.: The CIFAR-10 dataset, p. 55 (2014). http://www.cs.toronto.edu/kriz/cifar.html
20. Lebedev, V., Ganin, Y., Rakhuba, M., Oseledets, I., Lempitsky, V.: Speeding-up convolutional neural networks using fine-tuned CP-decomposition. arXiv preprint arXiv:1412.6553 (2014)

21. Liu, Z., Li, J., Shen, Z., Huang, G., Yan, S., Zhang, C.: Learning efficient convolutional networks through network slimming. In: Proceedings of the IEEE International Conference on Computer Vision, pp. 2736–2744 (2017)
22. Mnih, V., Badia, A., et al.: Asynchronous methods for deep reinforcement learning. In: Proceedings of the International Conference on Machine Learning, pp. 1928–1937 (2016)
23. Rastegari, M., Ordonez, V., Redmon, J., Farhadi, A.: XNOR-net: ImageNet classification using binary convolutional neural networks. In: Leibe, B., Matas, J., Sebe, N., Welling, M. (eds.) ECCV 2016. LNCS, vol. 9908, pp. 525–542. Springer, Cham (2016). https://doi.org/10.1007/978-3-319-46493-0_32
24. Russakovsky, O., et al.: ImageNet large scale visual recognition challenge. Int. J. Comput. Vis. **115**(3), 211–252 (2015)
25. Schulman, J., Wolski, F., Dhariwal, P., Radford, A., Klimov, O.: Proximal policy optimization algorithms. arXiv preprint arXiv:1707.06347 (2017)
26. Simonyan, K., Zisserman, A.: Very deep convolutional networks for large-scale image recognition. arXiv preprint arXiv:1409.1556 (2014)
27. Sutskever, I., Martens, J., Dahl, G., Hinton, G.: On the importance of initialization and momentum in deep learning. In: Proceedings of the International Conference on Machine Learning, pp. 1139–1147(2013)
28. Sutton, R.S., McAllester, D.A., Singh, S.P., Mansour, Y.: Policy gradient methods for reinforcement learning with function approximation. In: Advances in Neural Information Processing Systems, pp. 1057–1063 (2000)
29. Veit, A., Belongie, S.: Convolutional networks with adaptive inference graphs. In: Proceedings of the European Conference on Computer Vision, pp. 3–18 (2018)
30. Wang, K., Liu, Z., Lin, Y., Lin, J., Han, S.: HAQ: hardware-aware automated quantization with mixed precision. In: Proceedings of the IEEE Conference on Computer Vision and Pattern Recognition, pp. 8612–8620 (2019)
31. Wu, J., et al.: PocketFlow: an automated framework for compressing and accelerating deep neural networks (2018)
32. Yin, X., Goudriaan, J., Lantinga, E.A., Vos, J., Spiertz, H.J.: A flexible sigmoid function of determinate growth. Ann. Bot. **91**(3), 361–371 (2003)
33. Yu, X., Yu, Z., Ramalingam, S.: Learning strict identity mappings in deep residual networks. In: Proceedings of the IEEE Conference on Computer Vision and Pattern Recognition, pp. 4432–4440 (2018)
34. Zhang, X., Zhou, X., Lin, M., Sun, J.: ShuffleNet: an extremely efficient convolutional neural network for mobile devices. In: Proceedings of the IEEE Conference on Computer Vision and Pattern Recognition, pp. 6848–6856 (2018)
35. Zhong, Z., Yan, J., Wu, W., Shao, J., Liu, C.L.: Practical block-wise neural network architecture generation. In: Proceedings of the IEEE Conference on Computer Vision and Pattern Recognition, pp. 2423–2432 (2018)
36. Zhuang, Z., et al.: Discrimination-aware channel pruning for deep neural networks. In: Advances in Neural Information Processing Systems, pp. 875–886 (2018)
37. Zoph, B., Vasudevan, V., Shlens, J., Le, Q.V.: Learning transferable architectures for scalable image recognition. In: Proceedings of the IEEE Conference on Computer Vision and Pattern Recognition, pp. 8697–8710 (2018)

Dropout's Dream Land: Generalization from Learned Simulators to Reality

Zac Wellmer[✉] and James T. Kwok

Department of Computer Science and Engineering, Hong Kong University of Science and Technology, Kowloon, Hong Kong
{zwwellmer,jamesk}@cse.ust.hk

Abstract. A World Model is a generative model used to simulate an environment. World Models have proven capable of learning spatial and temporal representations of Reinforcement Learning environments. In some cases, a World Model offers an agent the opportunity to learn entirely inside of its own dream environment. In this work we explore improving the generalization capabilities from dream environments to real environments (Dream2Real). We present a general approach to improve a controller's ability to transfer from a neural network dream environment to reality at little additional cost. These improvements are gained by drawing on inspiration from Domain Randomization, where the basic idea is to randomize as much of a simulator as possible without fundamentally changing the task at hand. Generally, Domain Randomization assumes access to a pre-built simulator with configurable parameters but oftentimes this is not available. By training the World Model using dropout, the dream environment is capable of creating a nearly infinite number of *different* dream environments. Previous use cases of dropout either do not use dropout at inference time or averages the predictions generated by multiple sampled masks (Monte-Carlo Dropout). Dropout's Dream Land leverages each unique mask to create a diverse set of dream environments. Our experimental results show that Dropout's Dream Land is an effective technique to bridge the reality gap between dream environments and reality. Furthermore, we additionally perform an extensive set of ablation studies (The code is available at https://github.com/zacwellmer/DropoutsDreamLand).

1 Introduction

Reinforcement learning [30] (RL) has experienced a flurry of success in recent years, from learning to play Atari [20] to achieving grandmaster-level performance in StarCraft II [32]. However, in all these examples, the target environment is a simulator that can be directly trained in. Reinforcement learning is often not a practical solution without a simulator of the environment.

Sometimes the target environment is expensive, dangerous, or even impossible to interact with. In these cases, the agent is trained in a simulated source environment. Approaches that train an agent in a simulated environment with

© Springer Nature Switzerland AG 2021
N. Oliver et al. (Eds.): ECML PKDD 2021, LNAI 12975, pp. 255–270, 2021.
https://doi.org/10.1007/978-3-030-86486-6_16

the hopes of generalization to the target environment experience a common problem referred to as the *reality gap* [13]. One approach to bridge the reality gap is Domain Randomization [31]. The basic idea is that an agent which can perform well in an ensemble of simulations will also generalize to the real environment [2,21,24,31]. The ensemble of simulations is generally created by randomizing as much of the simulator as possible without fundamentally changing the task at hand. Unfortunately, this approach is only applicable when a simulator is provided and the simulator is configurable.

A recently growing field, World Models [9], focuses on the side of this problem when the simulation does not exist. World Models offer a general framework for optimizing controllers directly in *learned* simulated environments. The learned dynamics model can be viewed as the agent's dream environment. This is an interesting area because access to a learned dynamics model removes the need for an agent to train in the target environment. Some related approaches [10, 11,15,19,25,29] focus on an adjacent problem which allows the controller to continually interact with the target environment.

Despite the recent improvements [10,11,15,16,25] of World Models, little has been done to address the issue that World Models are susceptible to the reality gap. The learned dream environment can be viewed as the source domain and the true environment as the target domain. Whenever there are discrepancies between the source and target domains the reality gap can cause problems. Even though World Models suffer from the reality gap, none of the Domain Randomization approaches are directly applicable because the dream environment does not have easily configurable parameters.

In this work we present Dropout's Dream Land (DDL), a simple approach to bridge the reality gap from learned dream environments to reality (Dream2Real). Dropout's Dream Land was inspired by the first principles of domain randomization, namely, train a controller on a large set of *different* simulators which all adhere to the fundamental task of the target environment. We are able to generate a nearly infinite number of different simulators via the insight that dropout [27] can be understood as learning an ensemble of neural networks [3].

Our empirical results demonstrate that Dropout's Dream Land is an effective technique to cross the Dream2Real gap and offers improvements over baseline approaches [9,16]. Furthermore, we perform an extensive set of ablation studies which indicate the source of generalization improvements, requirements for the method to work, and when the method is most useful.

2 Related Works

2.1 Dropout

Dropout [27] was introduced as a regularization technique for feedforward and convolutional neural networks. In its most general form, each unit is dropped with a probability p during the training process. During training weights are scaled by $\frac{1}{1-p}$. Weight scaling ensures that for any hidden unit the *expected* output is the same as the actual output at test time [27]. Recurrent neural networks (RNNs) initially had issues benefiting from dropout. Zaremba *et al.* [35] suggests not to apply dropout to the hidden state units of the RNN cell. Gal *et al.* [7] shortly after show that the mask can also be applied to the hidden state units, but the mask must be fixed across the sequence during training.

In this work, we follow the dropout approach from [7] when training the RNN. More formally, for each sequence, the Boolean masks $\mathbf{m}_{xi}$, $\mathbf{m}_{xf}$, $\mathbf{m}_{xw}$, $\mathbf{m}_{xo}$, $\mathbf{m}_{hi}$, $\mathbf{m}_{hf}$, $\mathbf{m}_{hw}$, and $\mathbf{m}_{ho}$ are sampled, then used in the following LSTM update:

$$\mathbf{i}_t = \mathbf{W}_{xi}(\mathbf{x}_t \odot \mathbf{m}_{xi}) + \mathbf{W}_{hi}(\mathbf{h}_{t-1} \odot \mathbf{m}_{hi}) + \mathbf{b}_i, \tag{1}$$

$$\mathbf{f}_t = \mathbf{W}_{xf}(\mathbf{x}_t \odot \mathbf{m}_{xf}) + \mathbf{W}_{hf}(\mathbf{h}_{t-1} \odot \mathbf{m}_{hf}) + \mathbf{b}_f, \tag{2}$$

$$\mathbf{w}_t = \mathbf{W}_{xw}(\mathbf{x}_t \odot \mathbf{m}_{xw}) + \mathbf{W}_{hw}(\mathbf{h}_{t-1} \odot \mathbf{m}_{hw}) + \mathbf{b}_w, \tag{3}$$

$$\mathbf{o}_t = \mathbf{W}_{xo}(\mathbf{x}_t \odot \mathbf{m}_{xo}) + \mathbf{W}_{ho}(\mathbf{h}_{t-1} \odot \mathbf{m}_{ho}) + \mathbf{b}_o, \tag{4}$$

$$\mathbf{c}_t = \sigma(\mathbf{i}_t) \odot \tanh(\mathbf{w}_t) + \sigma(\mathbf{f}_t) \odot \mathbf{c}_{t-1}, \tag{5}$$

$$\mathbf{h}_t = \sigma(\mathbf{o}_t) \odot \tanh(\mathbf{c}_t), \tag{6}$$

where $\mathbf{x}_t$, $\mathbf{h}_t$, and $\mathbf{c}_t$ are the input, hidden state, and cell state, respectively, $\mathbf{W}_{xi}$, $\mathbf{W}_{xf}$, $\mathbf{W}_{xw}$, $\mathbf{W}_{xo} \in \mathbb{R}^{d \times r}$ $\mathbf{W}_{hi}$, $\mathbf{W}_{hf}$, $\mathbf{W}_{hw}$, $\mathbf{W}_{ho} \in \mathbb{R}^{d \times d}$ are the LSTM weight matrices, and $\mathbf{b}_i$, $\mathbf{b}_f$, $\mathbf{b}_w$, $\mathbf{b}_o \in \mathbb{R}^d$ are the LSTM biases. The masks are fixed for the entire sequence, but may differ between sequences in the mini-batch.

Monte-Carlo (MC) Dropout [6] runs multiple forward passes with independently sampled masks. In related works [14], Monte-Carlo (MC) Dropout [6] has been used to approximate the mean and variance of output predictions from an ensemble. We emphasize that Dropout's Dream Land does not use MC Dropout. Details are in Sect. 3.2.

2.2 Domain Randomization

The goal of Domain Randomization [24,31] is to create many different versions of the dynamics model with the hope that a policy generalizing to all versions of the dynamics model will do well on the true environment. Figure 1 illustrates many simulated environments ($\hat{e}^j$) overlapping with the actual environment (e^*). Simulated environments are often far cheaper to operate in than the actual environment. Hence, it is desirable to be able to perform the majority of interactions in the simulated environments.

Randomization has been applied on observations (e.g., lighting, textures) to perform robotic grasping [31] and collision avoidance of drones [24]. Randomization has also proven useful when applied to the underlying dynamics of simulators [23]. Often, both the observations and simulation dynamics are randomized [1].

Domain randomization generally uses some pre-existing simulator which then injects randomness into specific aspects of the simulator (e.g., color textures, friction coefficients). Each of the simulated environments in Fig. 1 can be thought of as a noisy sample of the pre-existing simulator. To the best of our knowledge, Domain Randomization has yet to be applied to entirely learned simulators.

Algorithm 1. World Models: Training in dreams.

1: Initialize parameters of V, M, and C
2: Collect N trajectories **o**, d, and **a** from e^*
3: Optimize V on observations **o**
4: Generate embeddings **z** for **o** with V
5: Optimize M on **z** and d
6: Generate dream environment $\hat{e}$ from M
7: **for** iteration=1, 2, ... **do**
8:     Optimize C via interactions with $\hat{e}$

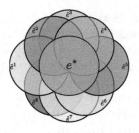

Fig. 1. e^* is the actual environment, and $\hat{e}^j$'s are randomized variants of the simulated environment.

2.3 World Models

The world model [9] has three modules trained separately: (i) vision module (V); (ii) dynamics module (M); and (iii) controller (C). A high-level view is shown in Algorithm 1. The vision module (V) is a variational autoencoder (VAE) [17], which maps an image observation (**o**) to a lower-dimensional representation $\mathbf{z} \in \mathbb{R}^n$.

The dynamics model (M) is a mixture density network recurrent neural network (MDN-RNN) [8,9]. The MDN-RNN models the dynamics of the environment, so modifying the parameters changes the dynamics of the learned simulated environment. It is implemented as an LSTM followed by a fully-connected layer outputting parameters for a Gaussian mixture model with k components. Each feature has k different π parameters for the logits of multinomial distribution, and (μ, σ) parameters for the k components in the Gaussian mixture. At each timestep, the MDN-RNN takes in the state **z** and action **a** as inputs and predicts $\boldsymbol{\pi}, \boldsymbol{\mu}, \boldsymbol{\sigma}$. To draw a sample from the MDN-RNN, we first sample the multinomial distribution parameterized by $\boldsymbol{\pi}$, which indexes which of the k normal distributions in the Gaussian mixture to sample from. This is then repeated for each of the n features. Depending on the experiments, Ha and Schmidhuber [9] also include an auxiliary head to the LSTM which predicts whether the episode terminates (d).

The controller (C) is responsible for deciding what actions to take. It takes features produced by the encoder V and dynamics model M as input (not the raw observations). The simple controller is a single-layer model which uses an evolutionary algorithm (CMA-ES [12]) to find its parameters. Depending on the problem setting, the controller (C) can either be optimized directly on the target environment (e^*) or on the dream environment ($\hat{e}$). This paper is focused on the case of optimizing exclusively in the dream environment.

3 Dropout's Dream Land

In this work we introduce Dropout's Dream Land (DDL). Dropout's Dream Land is the first work to offer a strategy to bridge the *reality gap* between learned neural network dynamics models and reality. Traditional Domain Randomization generates many *different* dynamics models by randomizing configurable parameters of a given simulation. This approach does not apply to neural network dynamics models because they generally do not have configurable parameters (such as textures and friction coefficients). In Dropout's Dream Land, the controller can interact with billions[1] of dream environments, whereas previous works [9,16] only use one dream environment. A naive way to go about this would be to train a population of neural network world models. However, this would be computationally expensive.

To keep the computational cost low, we go about this by applying dropout to the dynamics model in order to form different dynamics models. Crucially, dropout is applied at **both** training and inference of the dynamics model M. Each unique dropout mask applied to M can be viewed as a different environment. Similar to the spirit of Domain Randomization, an agent is expected to perform well in the real environment if it can perform well in a variety of simulated environments.

3.1 Learning the Dream Environment

The Dropout's Dream Land environments are built around the dynamics model M. The controller interactions during training are described by Fig. 2, in which $\hat{r}$, $\hat{d}$, and $\hat{\mathbf{z}}$ are generated entirely by M. In this work, M is an LSTM where $\mathbf{x} = [\mathbf{z}^\top, \mathbf{a}^\top]^\top$ from Eqs. (1)–(4). The LSTM is followed by multiple heads for predictions of the latent state ($\hat{\mathbf{z}}$), reward ($\hat{r}$) and termination ($\hat{d}$). The reward and termination heads are simple fully-connected layers. Latent state prediction is done with a MDN-RNN [8,9], but this could be replaced by any other neural network that supports dropout (e.g., GameGAN [16]).

Loss Function. The dynamics model M jointly optimizes all three heads. The loss of a single transition is defined as:

$$\mathcal{L}^M = \mathcal{L}^z + \alpha_r \mathcal{L}^r + \alpha_d \mathcal{L}^d. \tag{7}$$

Here, $\mathcal{L}^z = -\sum_{i=1}^n \log(\sum_{j=1}^k \hat{\pi}_{i,j} \mathcal{N}(z_i | \hat{\mu}_{i,j}, \hat{\sigma}_{i,j}^2))$ is a mixture density loss for the latent state predictions, where n is the size of the latent feature vector z, $\hat{\pi}_{i,j}$ is the jth component's probability for the ith feature, $\hat{\mu}_{i,j}, \hat{\sigma}_{i,j}$ are the corresponding mean and standard deviation. $\mathcal{L}^r = (r - \hat{r})^2$ is the square loss on rewards, where r and $\hat{r}$ are the true and estimated rewards, respectively. $\mathcal{L}^d = -d \log(\hat{d}) - (1 - d) \log(1 - \hat{d})$ is the cross-entropy loss for termination

---

[1] In practice we are bounded by the total number of steps instead of every possible environment.

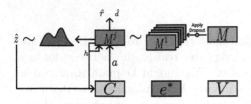
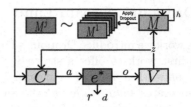

Fig. 2. Interactions with the dream environment. A dropout mask is sampled at every step yielding a new M^j.

Fig. 3. Interactions with the real environment. The controller being optimized only interacts with the real environment during the final testing phase.

prediction, where d and $\hat{d}$ are the true and estimated probabilities of the episode ending, respectively. Constants α_d and α_r in (7) are for trading off importance of the termination and reward objectives. The loss ($\mathcal{L}^M$) is aggregated over each sequence and averaged across the mini-batch.

Training Dynamics Model M with Dropout. At training time of M (Algorithm 1, Line 5), we apply dropout [7] to the LSTM to simulate different random environments. For each input and hidden unit, we first sample a Boolean indicator with probability p_{train}. If the indicator is 1, the corresponding input/hidden unit is masked. Masks $\mathbf{m}_{xi}$, $\mathbf{m}_{xf}$, $\mathbf{m}_{xw}$, $\mathbf{m}_{xo}$, $\mathbf{m}_{hi}$, $\mathbf{m}_{hf}$, $\mathbf{m}_{hw}$, and $\mathbf{m}_{ho}$ are sampled independently (Esq. (1)–(4)). When training the RNN, each mini-batch contains multiple sequences. Each sequence uses an independently sampled dropout mask. We fix the dropout mask for the entire sequence as this was previously found to be critically important [7].

Training the RNN with many different dropout masks is critical in order to generate multiple different dynamics models. At the core of Domain Randomization is the requirement that the randomizations do not fundamentally change the task. This constraint is violated if we do not train the RNN with dropout but apply dropout at inference (explored further empirically in Sect. 4.3). After optimizing the dynamics model M, we can use it to construct dream environments (Sect. 3.2) for controller training (Sect. 3.3).

In this work, we never sample masks to apply to the action (**a**). We do not zero out the action because in some environments this could imply the agent taking an action (e.g., moving to the left). This design choice could be changed depending on the environment, for example, when a zero'd action corresponds to a no-op or a sticky action.

3.2   Interacting with Dropout's Dream Land

Interactions with the dream environment (Algorithm 1, Line 8) can be characterized as training time for the controller (C) and inference time of the dynamics model (M). An episode begins by generating the initial latent state vector $\hat{z}$ by either sampling from a standard normal distribution or sampling from the starting points of the observed trajectories used to train M [9]. The hidden cell (**c**) and state (**h**) vectors are initialized with zeros.

The controller (C) decides the action to take based on $\hat{\mathbf{z}}$ and **h**. In Fig. 2, the controller also observes $\hat{r}$ and $\hat{d}$, but these are exclusively used for the optimization process of the controller. The controller then performs an action **a** on a dream environment.

A new dropout mask is sampled (with probability p_{infer}) and applied to M. We refer to the masked dynamics model as M^j and the corresponding Dropout's Dream Land environment as $\hat{e}^j$. The current latent state $\hat{\mathbf{z}}$ and action **a** are concatenated, and passed to M^j to perform a forward pass. The episode terminates based on a sample from a Bernoulli distribution parameterized by $\hat{d}$. The dream environment then outputs the latent state, LSTM's hidden state, reward, and whether the episode terminates.

It is crucial to apply dropout at inference time (of the dynamics model M) in order to create *different* versions of the dream environment for the controller C. Our experiments (Sects. 4.2 and 4.3) consider an extensive set of ablation studies as to how and when dropout should be applied.

Dropout's Dream Land Is Not Monte-Carlo Dropout. The only work we are aware of that applies dropout at inference time is Monte-Carlo Dropout [7]. In Sect. 4.1 we include a Monte-Carlo Dropout World Model baseline because DDL can easily be misinterpreted as an application of Monte-Carlo Dropout. This baseline passes the expected hidden ($\bar{h}_t$) and cell ($\bar{c}_t$) state to the next time-step, in which the expectation is over dropout masks from Eqs. (1)–(4). In practice we follow a similar approach to previous work [7] and approximate the expectation by performing multiple forward passes (each forward pass samples a new dropout mask), and averages the results. At each step, the expected Mixture Model parameters ($\bar{\pi}, \bar{\mu}, \bar{\sigma}$), reward ($\bar{r}$), and termination ($\bar{d}$) are used. Maximizing expected returns from the Monte-Carlo Dropout World Model is equivalent to maximizing expected returns on a *single* dream environment, the average dynamics model. On the other hand, the purpose of DDL's approach to dropout is to generate many *different* versions of the dynamics model. More explicitly, the controller is trained to maximize expected returns across many different dynamics models in the ensemble, as opposed to maximizing expected returns on the ensemble average.

Dropout has also traditionally been used as a model regularizer. Dropout as a model regularizer is only applied at training time but not at inference time. In this work, this approach would regularize the dynamics model M. The usual trade-off is lower test loss at the cost of higher training loss [7,27]. However, DDL's ultimate goal is not to lower test loss of the World Model (M). The ultimate goal is providing dream environments to a controller so that the optimal policy in Dropout's Dream Land also maximizes expected returns in the target environment (e^*).

3.3 Training the Controller

Training with CMA-ES. We follow the same controller optimization procedure as was done in World Models [9] and GameGAN [16] on their Doom-TakeCover experiments. We train the controller with CMA-ES [12]. At every generation CMA-ES [12] spawns a population (of size N_{pop}) of agents. Each agent in the population reports their mean returns on a set of N_{trials} episodes generated in the dream environments. As controllers in the population do not share a dream environment, the probability of controllers interacting with the same sequence of dropout masks is vanishingly small. Let $N_{max\_ep\_len}$ be the maximum number of steps in an episode. In a single CMA-ES iteration, the population as a whole can interact with $N_{pop} \times N_{trials} \times N_{max\_ep\_len}$ *different* environments. In our experiments, $N_{pop} = 64$, $N_{trials} = 16$, and $N_{max\_ep\_len}$ is 1000 for CarRacing and 2100 for DoomTakeCover. This potentially results in $> 1,000,000$ different environments at each generation.

Dream Leader Board. After every fixed number of generations (25 in our experiments), the best controller in the population (which received the highest average returns across its respective N_{trials} episodes) is selected for evaluation [9,16]. This controller is evaluated for another $N_{pop} \times N_{trials}$ episodes in the Dropout's Dream Land environments. The controller's mean across $N_{pop} \times N_{trials}$ trials is logged to the Dream Leader Board. After 2000 generations, the controller at the top of the Dream Leader Board is evaluated in the real environment.

Interacting with the Real Environment. In Fig. 3 we illustrate the controller's interaction with the real target environment (e^*). Interactions with e^* do not apply dropout to the input or hidden units of M. The controller only interacts with the target environment during testing. These interactions are never used to modify parameters of the controller. At test time r, d, and o are generated by e^*, and $\mathbf{z}$ is the embedding of o from the VAE (V). The only use of M when interacting with the target environment is producing $\mathbf{h}$ as a feature for the controller.

4 Experiments

Broadly speaking, our experiments are focused on either evaluating the dynamics model (M) or the controller (C). Architecture details of V, M, and C are in the Appendix. Experiments are performed on the DoomTakeCover-v0 [22] and CarRacing-v0 [18] environments from OpenAI Gym [4]. These have also been used in related works [9,16]. Even though both baseline target environments are simulators we still consider this "reality" because we do not leverage knowledge about the simulator mechanics to learn the source environment (M).

Quality of the dynamics model is evaluated against a training and testing set of trajectories (described below). Quality of the controller is measured by returns in the target environments. For all experiments the controller is trained exclusively in the dream environment (Sect. 3.2) for 2,000 generations. The controller only interacts with the target environments for testing (Sect. 3.3). The target environment is never used to update parameters of the controller. Means and standard deviations of returns achieved by the best controller (Sect. 3.3) in the target environment are reported based on 100 trials for CarRacing and 1000 trials for DoomTakeCover.[2]

DoomTakeCover Environment. DoomTakeCover is a control task in which the goal is to dodge fireballs for as long as possible. The controller receives a reward of $+1$ for every step it is alive. The maximum number of frames is limited to 2100.

For all tasks on this environment, we collect a training set of 10,000 trajectories and a test set of 100 trajectories. A trajectory is a sequence of state ($\mathbf{z}$), action ($\mathbf{a}$), reward (r), and termination (d) tuples. Both datasets are generated according to a random policy. Following the same convention as World Models [9], on the DoomTakeCover environment we concatenate $\mathbf{z}$, $\mathbf{h}$, and $\mathbf{c}$ as input to the controller. In (7), we set $\alpha_d = 1$ and $\alpha_r = 0$ because the Doom reward function is determined entirely based on whether the controller lives or dies.

CarRacing Environment. CarRacing is a continuous control task to learn from pixels. The race track is split up into "tiles". The goal is to make it all the way around the track (i.e., crossing every tile). We terminate an episode when all tiles are crossed or when the number of steps exceeds 1,000. Let N_{tiles} be the total number of tiles. The simulator [18] defines the reward r_t at each timestep as $\frac{100}{N_{\text{tiles}}} - 0.1$ if a new tile is crossed, and -0.1 otherwise. The number of tiles is not explicitly set by the simulator. We generated 10,000 tracks and observed that the number of tiles in the track appears to follow a normal distribution with mean 289. To simplify the reward function, we fix N_{tiles} to 289 in the randomly generated tracks, and call the modified environment CarRacingFixedN.

For all tasks on this environment, the training set contains 5,000 trajectories and the test set contains 100 trajectories. Both datasets are collected by following an expert policy with probability 0.9, and a random policy with probability 0.1. The expert policy was trained directly on the CarRacing environment and received an average return of 885 ± 63 across 100 trials. In comparison, the performance of the random policy is -53 ± 41. This is similar to the setup in GameGAN [16] on the Pacman environment which also used an expert policy. For this environment, we set $\alpha_d = \alpha_r = 1$ in (7).

4.1   Comparison with Baselines

Dropout's Dream Land (DDL) is compared against World Models (WM), Monte-Carlo Dropout World Models (MCD-WM), and a uniform random policy on the CarRacing and DoomTakeCover environments. The Monte-Carlo Dropout World Models baseline uses $p_{\text{train}} = 0.05$, $p_{\text{infer}} = 0.1$, and 10 samples. On the Doom

---

[2] 100 trials are used for the baselines GameGAN and Action-LSTM.

Table 1. Returns from baseline methods and DDL (p_{train} = 0.05 and p_{infer} = 0.1) on the DoomTakeCover environment.

	DoomTakeCover
random policy	210 ± 108
GameGAN	765 ± 482
Action-LSTM	280 ± 104
WM	849 ± 499
MCD-WM	798 ± 464
DDL	$\mathbf{933 \pm 552}$

Table 2. Returns from baseline methods and DDL (p_{train} = 0.05 and p_{infer} = 0.1) on the CarRacingFixedN and the original CarRacing environments.

	CarRacingFixedN	CarRacing
random policy	-50 ± 38	-53 ± 41
WM	399 ± 135	388 ± 157
MCD-WM	-56 ± 31	-53 ± 32
DDL	$\mathbf{625 \pm 289}$	$\mathbf{610 \pm 267}$

environment, we also compare with GameGAN [16] and Action-LSTM [5][3]. All controllers are trained entirely in dream environments.

Results on the target environments are in Tables 1 and 2. The CarRacing results appear different from those found in World Models [9] because we are not performing the same experiment. In this paper, we train the controller entirely in the dream environment and only interact with the target environment during testing. In World Models [9], the controller was trained directly in the CarRacing environment.

In Tables 1 and 2, we observe that DDL offers performance improvements over all the baseline approaches in the target environments. We suspect this is because the WM dream environments were easier for the controller to exploit errors between the simulator and reality. Forcing the controller to succeed in many different dropout environments makes it difficult to exploit discrepancies between the dream environment and reality. This leads us to the conclusion that forcing the controller to succeed in many different dropout environments is an effective technique to cross the Dream2Real gap.

The DoomTakeCover returns in the target environment as reported by the temperature-regulated variant[4] in [9] are higher than the returns we obtain from DDL, which does not use temperature. However, we emphasize that adjusting temperature is only useful for a limited set of dynamics models. For example, it would not be straightforward to apply temperature to any dynamics model which does not produce a probability density function (e.g., GameGAN); whereas the DDL approach of generating many *different* dynamics models is useful to any learned neural network dynamics model. Moreover, even though the temperature-regulated variant increases uncertainty of the dream environment, it is still only capable of creating *one* dream environment.

4.2   Inference Dropout and Dream2Real Generalization

In this experiment, we study the effects of dropout on the World Model. First, we evaluate the relationship between dropout and World Model accuracies. Second,

---

[3] Results on GameGAN and Action-LSTM returns are from [16].

[4] We were unable to reproduce the temperature results in [9].

Table 3. RNN's loss with and without dropout ($p_{\text{train}} = 0.05$ and $p_{\text{infer}} = 0$) during training.

	DoomTakeCover		CarRacingFixedN	
	Training loss	Test loss	Training loss	Test loss
Without dropout	0.89	**0.91**	2.36	**3.10**
With dropout	0.93	**0.91**	3.19	3.57

we evaluate the relationship between dropout and generalization from the World Model to the target environment. Model loss is measured by the loss in (7) on the test sets. Returns in the target environment are reported based on the best controller (Sect. 3.3) trained with varying levels of inference dropout. The same training and test sets described at the beginning of Sect. 4 are used.

Standard use cases of dropout generally observe a larger training loss but lower test loss relative to the same model trained without dropout [6,27]. In Table 3, we do not observe any immediate performance improvements of the World Model trained with dropout ($p_{\text{train}} = 0.05$ and $p_{\text{infer}} = 0$). In fact, we observe worse results on the test set. The poor performance of both DDL RNNs (Table 3) indicates a clear conclusion about the results from Tables 1 and 2. The improved performance of DDL relative to World Models comes from forcing the controller to operate in many different environments and not from a single more accurate dynamics model M.

Next we take a World Model trained with dropout and evaluate the model loss on a test set across varying levels of inference dropout (p_{infer}). As expected, in Fig. 4 we observe that as the inference dropout rate is increased the model loss increases. In Fig. 5 we observe that increasing the inference dropout rate improves generalization to the target environment. We believe that the boost in returns on the target environments comes from an increase in capacity to distort the dynamics model. Figures 4 and 5 suggest that we can sacrifice accuracy of the dream environments to better cross the Dream2Real gap between dream and target environments. However, this should only be useful up to the point where *the task at hand is fundamentally changed*. Figure 5 suggests this point is somewhere between 0.1 and 0.2 for p_{infer}, though we suspect in practice this will be highly dependent on network architecture and the environment.

In Fig. 5 we observe relatively weak returns on the real CarRacingFixedN environment when the inference dropout rate is zero. Recall from Table 3 that the dropout variant has a much higher test loss than the non-dropout variant on CarRacingFixedN. This means that when $p_{\text{infer}} = 0$, the *single* environment DDL is able to create is relatively inaccurate. It is easier for the controller to exploit any discrepancies between the dream environment and target environment because only a single dream environment exists. However, as we increase the inference dropout rate it becomes harder for the controller to exploit the dynamics model, suggesting that DDL is especially useful when it is difficult to learn an accurate World Model.

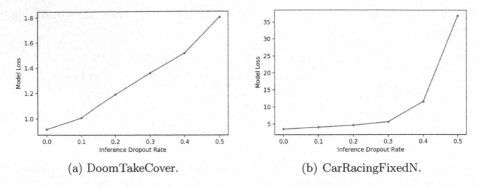

(a) DoomTakeCover.   (b) CarRacingFixedN.

Fig. 4. Loss of DDL dynamics model ($p_{\text{train}} = 0.05$) at different inference dropout rates.

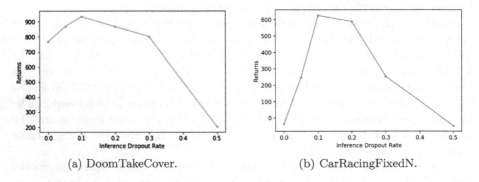

(a) DoomTakeCover.   (b) CarRacingFixedN.

Fig. 5. DDL ($p_{\text{train}} = 0.05$) returns at different inference dropout rates in the target environments.

4.3   When Should Dropout Masks Be Randomized During Controller Training?

In this ablation study we evaluate when the dropout mask should be randomized during training of C. We consider two possible approaches of when to randomize the masks. The first case only randomizes the mask at the beginning of an episode (*episode randomization*). The second case samples a new dropout mask at every step (*step randomization*). We also consider if it is effective to only apply dropout at inference time but not during M training (i.e., $p_{\text{infer}} > 0, p_{\text{train}} = 0$).

As can be seen in Table 4, randomizing the mask at each step offers better returns on both target environments. Better returns in the target environment when applying step randomization comes from the fact that the controller is exposed to a much larger number ($> 1000\times$) of dream environments. We also observe that applying step randomization without training the dynamics model with dropout yields a weak policy on the target environment. This is due to the randomization fundamentally changing the task. Training the dynamics model with dropout ensures that at inference time the masked model (M^j) is meaningful.

Table 4. Returns of the controller with different frequencies to randomize the dropout mask.

	DoomTakeCover	CarRacingFixedN
episode randomization ($p_{\text{train}} = 0.05$, $p_{\text{infer}} = 0.1$)	786 ± 469	601 ± 197
step randomization ($p_{\text{train}} = 0.05$, $p_{\text{infer}} = 0.1$)	$\mathbf{933 \pm 552}$	$\mathbf{625 \pm 289}$
step randomization ($p_{\text{train}} = 0$, $p_{\text{infer}} = 0.1$)	339 ± 90	-43 ± 52

4.4 Comparison to Standard Regularization Methods

In this experiment we compare Dropout's Dream Land with standard regularization methods. First, we consider applying the standard use case of dropout ($0 < p_{\text{train}} < 1$ and $p_{\text{infer}} = 0$). Second, we consider a noisy variant of M when training C. The Noisy World Model uses exactly the same parameters for M as the baseline World Model. When training the controller, a small amount of Gaussian noise is added to z at every step.

In Table 5, we observe that DDL is better at generalizing from the dream environment to the target environment than the standard regularization methods. Dropout World Models can be viewed as a regularizer on M. Noisy World Models can be viewed as a regularizer on the controller C. The strong returns on the target environment by DDL suggest that it is better at crossing the Dream2Real gap than standard regularization techniques.

Table 5. Returns from World Models, Dropout World Models ($p_{\text{train}} = 0.05$ and $p_{\text{infer}} = 0.0$), Noisy World Models, and DDL ($p_{\text{train}} = 0.05$ and $p_{\text{infer}} = 0.1$) on the CarRacingFixedN and the original CarRacing environments.

	CarRacingFixedN	CarRacing
World Models	399 ± 135	388 ± 157
Dropout World Models	-36 ± 19	-36 ± 20
Noisy ($\mathcal{N}(0,1)$) World Models	147 ± 121	180 ± 132
Noisy ($\mathcal{N}(0,10^{-2})$) World Models	455 ± 171	442 ± 171
Dropout's Dream Land	$\mathbf{625 \pm 289}$	$\mathbf{610 \pm 267}$

4.5 Comparison to Explicit Ensemble Methods

In this experiment we compare Dropout's Dream Land with two other approaches for randomizing the dynamics of the dream environment. We consider using an explicit ensemble of a population of dynamics models. Each environment in the population was trained on the same set of trajectories described at the beginning of Sect. 4 with a different initialization and different mini-batches. With the population of World Models we train a controller with Step Randomization and a controller with Episode Randomization. Note that the training cost of dynamics models and RAM requirements at inference time scale linearly with the population size. Due to the large computational cost we consider a population size of 2.

In Table 6, we observe that neither Population World Models (PWM) Step Randomization or Episode Randomization substantially close the Dream2Real gap. Episode Randomization does not dramatically improve results because the controller is forced to understand the hidden state ($\mathbf{h}$) representation of every M in the population. Step Randomization performs even worse than Episode Randomization because on top of the previously stated limitations, each dynamics model in the population is also forced to be compatible with the hidden state ($\mathbf{h}$) representation of all other dynamics models in the population. DDL does not suffer from any of the previously stated issues and is also computationally cheaper because only one M must be trained as opposed to an entire population.

Table 6. Returns from World Models, PWM Episode Randomization, PWM Step Randomization, and DDL ($p_{\text{train}} = 0.05$ and $p_{\text{infer}} = 0.1$) on the CarRacingFixedN and the original CarRacing environments.

	CarRacingFixedN	CarRacing
World Models	399 ± 135	388 ± 157
PWM Episode Randomization	398 ± 126	402 ± 142
PWM Step Randomization	-78 ± 14	-77 ± 13
Dropout's Dream Land	$\mathbf{625 \pm 289}$	$\mathbf{610 \pm 267}$

5   Conclusion

Dropout's Dream Land introduces a novel technique to improve controller generalization from dream environments to reality. This is accomplished by taking inspiration from Domain Randomization and training the controller on a large set of different simulators. A large set of different simulators are generated at little cost by the insight that dropout can be used to generate an ensemble of neural networks. To the best of our knowledge this is the first work to bridge the reality gap between learned simulators and reality. Previous work from Domain Randomization [31] is not applicable to learned simulators because they often do not have easily configurable parameters. Future direction for this work could be modifying the dynamics model parameters in a targeted manner [28,33,34]. This simple approach to generating different versions of a model could also be useful in committee-based methods [25,26].

References

1. Andrychowicz, M., et al.: Learning dexterous in-hand manipulation. Int. J. Robot. Res. **39**(1), 3–20 (2020)
2. Antonova, R., Cruciani, S., Smith, C., Kragic, D.: Reinforcement learning for pivoting task. Preprint arXiv:1703.00472 (2017)
3. Baldi, P., Sadowski, P.J.: Understanding dropout. In: Advances in Neural Information Processing Systems, pp. 2814–2822 (2013)
4. Brockman, G., et al.: Openai gym. Preprint arXiv:1606.01540 (2016)
5. Chiappa, S., Racaniere, S., Wierstra, D., Mohamed, S.: Recurrent environment simulators. Preprint arXiv:1704.02254 (2017)
6. Gal, Y., Ghahramani, Z.: Dropout as a Bayesian approximation: representing model uncertainty in deep learning. In: International Conference on Machine Learning, pp. 1050–1059 (2016)
7. Gal, Y., Ghahramani, Z.: A theoretically grounded application of dropout in recurrent neural networks. In: Advances in Neural Information Processing Systems, pp. 1019–1027 (2016)
8. Graves, A.: Generating sequences with recurrent neural networks. Preprint arXiv:1308.0850 (2013)
9. Ha, D., Schmidhuber, J.: Recurrent world models facilitate policy evolution. In: Advances in Neural Information Processing Systems, pp. 2450–2462 (2018)
10. Hafner, D., Lillicrap, T., Ba, J., Norouzi, M.: Dream to control: learning behaviors by latent imagination. In: International Conference on Learning Representations (2020)
11. Hafner, D., et al.: Learning latent dynamics for planning from pixels. In: International Conference on Machine Learning, pp. 2555–2565 (2019)
12. Hansen, N., Ostermeier, A.: Completely derandomized self-adaptation in evolution strategies. Evol. Comput. **9**(2), 159–195 (2001)
13. Jakobi, N., Husbands, P., Harvey, I.: Noise and the reality gap: the use of simulation in evolutionary robotics. In: Morán, F., Moreno, A., Merelo, J.J., Chacón, P. (eds.) ECAL 1995. LNCS, vol. 929, pp. 704–720. Springer, Heidelberg (1995). https://doi.org/10.1007/3-540-59496-5_337
14. Kahn, G., Villaflor, A., Pong, V., Abbeel, P., Levine, S.: Uncertainty-aware reinforcement learning for collision avoidance. Preprint arXiv:1702.01182 (2017)
15. Kaiser, Ł., et al.: Model based reinforcement learning for Atari. In: International Conference on Learning Representations (2020)
16. Kim, S.W., Zhou, Y., Philion, J., Torralba, A., Fidler, S.: Learning to simulate dynamic environments with GameGAN. In: IEEE Conference on Computer Vision and Pattern Recognition, pp. 1231–1240 (2020)
17. Kingma, D.P., Welling, M.: Auto-encoding variational Bayes. Preprint arXiv:1312.6114 (2013)
18. Klimov, O.: Carracing-v0 (2016). https://gym.openai.com/envs/CarRacing-v0/
19. Kurutach, T., Clavera, I., Duan, Y., Tamar, A., Abbeel, P.: Model-ensemble trust-region policy optimization. Preprint arXiv:1802.10592 (2018)
20. Mnih, V., et al.: Human-level control through deep reinforcement learning. Nature **518**(7540), 529–533 (2015)
21. Mordatch, I., Lowrey, K., Todorov, E.: Ensemble-CIO: full-body dynamic motion planning that transfers to physical humanoids. In: IEEE/RSJ International Conference on Intelligent Robots and Systems, pp. 5307–5314 (2015)

22. Paquette, P.: Doomtakecover-v0 (2017). https://gym.openai.com/envs/DoomTakeCover-v0/
23. Peng, X.B., Andrychowicz, M., Zaremba, W., Abbeel, P.: Sim-to-real transfer of robotic control with dynamics randomization. In: IEEE International Conference on Robotics and Automation, pp. 1–8 (2018)
24. Sadeghi, F., Levine, S.: CAD2RL: real single-image flight without a single real image. Preprint arXiv:1611.04201 (2016)
25. Sekar, R., Rybkin, O., Daniilidis, K., Abbeel, P., Hafner, D., Pathak, D.: Planning to explore via self-supervised world models. Preprint arXiv:2005.05960 (2020)
26. Settles, B.: Active learning literature survey. Technical report, University of Wisconsin-Madison Department of Computer Sciences (2009)
27. Srivastava, N., Hinton, G., Krizhevsky, A., Sutskever, I., Salakhutdinov, R.: Dropout: a simple way to prevent neural networks from overfitting. J. Mach. Learn. Res. **15**(1), 1929–1958 (2014)
28. Such, F.P., Rawal, A., Lehman, J., Stanley, K.O., Clune, J.: Generative teaching networks: accelerating neural architecture search by learning to generate synthetic training data. Preprint arXiv:1912.07768 (2019)
29. Sutton, R.S.: Integrated architectures for learning, planning, and reacting based on approximating dynamic programming. In: International Conference on Machine learning, pp. 216–224 (1990)
30. Sutton, R.S., Barto, A.G.: Reinforcement Learning: An Introduction. MIT Press, Cambridge (2018)
31. Tobin, J., Fong, R., Ray, A., Schneider, J., Zaremba, W., Abbeel, P.: Domain randomization for transferring deep neural networks from simulation to the real world. In: IEEE/RSJ International Conference on Intelligent Robots and Systems, pp. 23–30 (2017)
32. Vinyals, O., et al.: Grandmaster level in StarCraft II using multi-agent reinforcement learning. Nature **575**(7782), 350–354 (2019)
33. Wang, R., Lehman, J., Clune, J., Stanley, K.O.: Paired open-ended trailblazer (POET): endlessly generating increasingly complex and diverse learning environments and their solutions. Preprint arXiv:1901.01753 (2019)
34. Wang, R., et al.: Enhanced POET: open-ended reinforcement learning through unbounded invention of learning challenges and their solutions. Preprint arXiv:2003.08536 (2020)
35. Zaremba, W., Sutskever, I., Vinyals, O.: Recurrent neural network regularization. Preprint arXiv:1409.2329 (2014)

Goal Modelling for Deep Reinforcement Learning Agents

Jonathan Leung$^{(\boxtimes)}$, Zhiqi Shen, Zhiwei Zeng, and Chunyan Miao

School of Computer Science and Engineering, Nanyang Technological University,
50 Nanyang Avenue, Singapore 639798, Singapore
jonathan008@e.ntu.edu.sg, {zqshen,zhiwei.zeng,ascymiao}@ntu.edu.sg

Abstract. Goals provide a high-level abstraction of an agent's objectives and guide its behavior in complex environments. As agents become more intelligent, it is necessary to ensure that the agent's goals are aligned with the goals of the agent designers to avoid unexpected or unwanted agent behavior. In this work, we propose using Goal Net, a goal-oriented agent modelling methodology, as a way for agent designers to incorporate their prior knowledge regarding the subgoals an agent needs to achieve in order to accomplish an overall goal. This knowledge is used to guide the agent's learning process to train it to achieve goals in dynamic environments where its goal may change between episodes. We propose a model that integrates a Goal Net model and hierarchical reinforcement learning. A high-level goal selection policy selects goals according to a given Goal Net model and a low-level action selection policy selects actions based on the selected goal, both of which use deep neural networks to enable learning in complex, high-dimensional environments. The experiments demonstrate that our method is more sample efficient and can obtain higher average rewards than other related methods that incorporate prior human knowledge in similar ways.

Keywords: Deep reinforcement learning · Hierarchical reinforcement learning · Goal modelling

1 Introduction

Deep reinforcement learning (DRL) has enabled agents to achieve human-level, and in some cases superhuman-level, results in complex, high-dimensional environments. In many applications, agents are required to achieve multiple goals in complex environments. However, many DRL methods are limited in that they can only complete one task or goal. Kaelbling [13] proposed a method to train reinforcement learning agents to learn to achieve a wide variety of goals. This work forms the basis of recent goal-conditioned and multi-goal reinforcement learning methods that make use of deep neural networks [23].

As agents become more intelligent, it is necessary to ensure that the agent's goals are aligned with the goals of the agent designers, which has been referred

© Springer Nature Switzerland AG 2021
N. Oliver et al. (Eds.): ECML PKDD 2021, LNAI 12975, pp. 271–286, 2021.
https://doi.org/10.1007/978-3-030-86486-6_17

to as the agent alignment problem [16]. Although many recent deep learning methods reduce the amount of prior knowledge given to models to improve performance, the inclusion of such knowledge may improve agent alignment by providing more context to the agent. One way to leverage prior human knowledge in RL would be to model goals so that they can be understood and specified by agent developers and designers, regardless of their technical knowledge. Goal models, which originate from Goal-Oriented Requirements Engineering (GORE), can provide a way for agent designers to express the high-level behavior that they desire from their agents. GORE focuses on goals as a way to define system objectives and to communicate the rationale behind system requirements to stakeholders of varying technical knowledge [31]. In GORE, goal models have been used in agent design to support formal representation and reasoning with goals [4,33]. Goal models define goals and capture the relationships between them, such as AND/OR relationships between subgoals that conjunctively/disjunctively achieve a high-level goal.

In this work, we propose a hierarchical reinforcement learning (HRL) model that incorporates an agent designer's prior knowledge about an agent's overall goal within a Goal Net model. Goal Net is an agent modelling methodology that uses goal modelling to define agent behavior [27]. Unlike other goal models that only specify the decomposition of goals into subgoals, Goal Net allows agent designers to specify the sequential relationships between goals to allow agents to reason about goals at run time, which makes it a suitable choice for our work. Our model consists of a high-level goal selection policy that provides goals to a low-level action selection policy, as shown in Fig. 1. Given a high-level goal, an agent may select subgoals to achieve the goal which in turn affects the agent's actions. A Goal Net model is used to provide valid goal selection options to the high-level policy, and the low-level policy is a goal-conditioned policy that operates in a goal-augmented state space which incorporates a symbolic goal space. We propose an algorithm that trains a hierarchical Deep Q-Network (h-DQN) [15] combined with a Goal Net model. Then, we evaluate our model against other related methods in which Goal Net could be incorporated, namely deep abstract Q-networks (DAQN) [25] and reward machines (RM) [12]. The results suggest that our method is more sample efficient and can achieve higher average rewards in environments with randomized goal locations.

2   Background

Reinforcement Learning (RL) aims to train an agent to act optimally within an environment [28]. This problem is typically formulated as a Markov decision process (MDP), which is defined as a tuple $\mathcal{M} = \langle \mathcal{S}, \mathcal{A}, \mathcal{P}, r, \gamma \rangle$, where $\mathcal{S}$ is a set of states, $\mathcal{A}$ is a set of actions, $\mathcal{P}(s'|s, a)$ is a transition probability function, $r(s, a, s')$ is a reward function, and γ is a discount factor. At a time step t, an agent observes a state $s_t \in \mathcal{S}$ and takes an action $a_t \in \mathcal{A}$. After the action is executed, the agent observes a new state s_{t+1} and receives rewards r_{t+1} according to the reward function. The goal of the agent is to learn a policy $\pi : \mathcal{S} \rightarrow \mathcal{A}$ that maximizes the rewards the agent obtains while interacting with the environment.

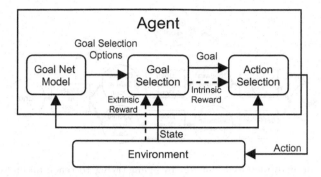

Fig. 1. An overview of the proposed model. The goal selection and action selection, in this work, are trained through reinforcement learning.

Many RL algorithms make use of value functions in order to learn the optimal policy. The Q-value function measures the expected future discounted rewards an agent can obtain by taking an action in a given state, and is defined as:

$$Q(s, a) = \mathbb{E}[\sum_{t=0}^{T} \gamma^t r_{t+1} | s_0 = s, a_0 = a]. \tag{1}$$

Deep Q-Networks (DQN) use deep neural networks to estimate the Q-value function [20]. DQN uses an experience replay buffer [18] that stores tuples containing information such as the states and actions the agent experiences while interacting with the environment. Experience tuples are sampled from the replay buffer to train the network, which enables data reuse and stabilizes the learning process.

Hierarchical Reinforcement Learning (HRL) involves training an agent to use multiple levels of policies where higher level policies may invoke or direct lower level policies to achieve subgoals. The options framework is a commonly used HRL formalism in which a high-level policy may use a temporally extended option, or macro-action, instead of a primitive action [29]. The framework makes use of the semi-Markov decision process (SMDP) that generalizes MDPs to the settings where actions may take a varying number of timesteps [24]. An option is a tuple $\langle \mathcal{I}_o, \pi_o, \beta_o \rangle$ where $\mathcal{I}_o \subseteq \mathcal{S}$ is an initiation set describing in which states the option can be invoked, $\pi_o : \mathcal{S} \rightarrow \mathcal{A}$ is an intra-option policy, and $\beta_o : \mathcal{S} \rightarrow [0, 1]$ is a termination function indicating when the option ends.

Goal-Conditioned Reinforcement Learning trains agents to learn a value function parametrized by the agent's goal g, which generalizes learning experience in achieving one goal to other goals [13]. Universal Value Function Approximators (UVFAs) make use of function approximators such as deep neural networks to enable generalization to new goals unseen at training time [26]. Hindsight Experience Replay (HER) [2] improves sample efficiency by adding relabelled experience tuples to the replay buffer. The relabelling process replaces the agent's original goal with the goal the agent actually reaches. Such works

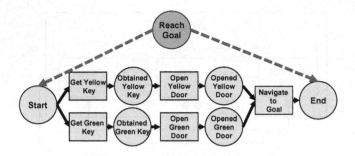

Fig. 2. An example Goal Net for an agent trying to reach an end goal.

have given rise to multi-goal RL [23], which trains agents to achieve a wide variety of goals.

Goal Net is a graphical model that defines agents' goals, subgoals, and the relationships between those goals [27]. Agent development using Goal Net involves the co-operation of agent designers who may be domain experts that can define high-level behavior and logic of an agent, and agent developers who have the development skills to implement the functions required by the agent. A Goal Net consists of goals and actions, which are represented graphically by circles and rectangles, respectively. Actions represent the transitions between goals and define any tasks that need to be completed in order to reach a goal. Goals can be composite, meaning that they can be decomposed into more goals, or atomic. Composite and atomic goals are represented as red and green circles, respectively. An example Goal Net is shown in Fig. 2, which shows a Goal Net for an agent attempting to reach an end goal by either obtaining a green key and opening a green door, or by using a yellow key to open a yellow door. We denote the set of goals within a Goal Net as G_{net}. Each Goal Net contains a root composite goal that indicates the overall goal to be achieved, a start goal, and an end goal. Arcs connect goals and actions together, indicating valid paths the agent may take to achieve the overall goal. At run time, the agent begins in the start goal, and uses goal selection algorithms to determine which goal to pursue and action selection algorithms to decide how goals should be achieved. The agent transitions to the next goal if it successfully achieves it.

Goal Net can also represent and define concurrent goal pursuit. A concurrency relation between goals represents a partially ordered goal achievement requirement where all goals in the concurrency relationship must be achieved. Concurrent goal paths will synchronize at a goal or action, which represents the point at which all paths must reach before transitioning to the next goal. Graphically, concurrent goals are represented using diamond-shaped arcs. Figure 3 shows an example Goal Net that contains a concurrent goal relation where the agent must reach both a yellow and blue subgoal before navigating to the final goal state, but the order in which the subgoals are reached does not matter.

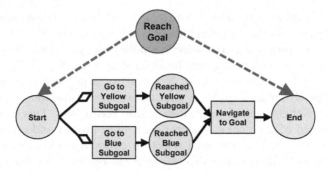

Fig. 3. Concurrent goal pursuit represented by diamond-shaped arcs, indicating that the agent must complete both subgoals before reaching the final goal.

3 Deep Reinforcement Learning with Goal Net

In this work, we utilize Goal Nets as models to define high-level agent behavior, which may be provided by agent designers. We consider the case where the goal and action selection within a Goal Net model are learned using reinforcement learning. This may be desirable when the environment is complex, or to reduce the workload of developers so that the goal and action selection algorithms do not need to be hand-engineered. We treat this setting as a hierarchical reinforcement learning problem with two policy levels: a high-level goal selection policy and a low-level action selection policy. Our hierarchical structure is based on the options framework, as well as h-DQN which trains two DQNs: a low-level controller and a high-level meta-controller [15].

In addition to the Goal Net model, we require agent designers and developers to create a goal space $\mathcal{G}$ that consists of symbolic attributes related to the goals and subgoals of the agent. For example, in a goal reaching task where an agent needs to reach a given position in a coordinate space, an agent designer may define the goal space as the agent's current coordinates. We will refer to goals within the Goal Net model as $g_{net} \in G_{net}$ to differentiate between points in the goal space $g \in \mathcal{G}$ and the Goal Net goals. Referring back to Fig. 1, a Goal Net goal g_{net} is passed to the goal selection policy and is used to select a target goal g'_{net}, and then this is converted to goal space $\mathcal{G}$. Similar to other related methods such as DAQN and QRM, this conversion is performed by a labelling function $\mathcal{F} : \mathcal{S} \rightarrow \mathcal{G}$, which we assume to be given by agent developers. The goal space allows us to take advantage of goal-conditioning and HER by augmenting the state space, inducing a state space $\mathcal{S}_{lo} = \mathcal{S} \times \mathcal{G}$. The low-level policy operates in this goal-augmented state space and is therefore defined as $\pi_{lo} : \mathcal{S}_{lo} \rightarrow \mathcal{A}$. The low-level policy selects actions using a goal-conditioned DQN that is trained to estimate the optimal Q-value function:

$$Q^*_{lo}(s_{lo}, a, g) = r_i + \gamma \sum_{s'_{lo} \in \mathcal{S}_{lo}} \mathcal{P}(s'_{lo}|s_{lo}, a, g) \max_{a'} Q^*_{lo}(s'_{lo}, a', g), \qquad (2)$$

where $g \in \mathcal{G}$ is a subgoal to achieve and $r_i \in \{0, 1\}$ is an intrinsic reward of 1 if the low-level policy reaches a given subgoal and 0 otherwise.

We denote the high-level goal selection policy as π_{hi}, which operates in the state space $\mathcal{S}_{hi} = \mathcal{S} \times G_{net}$ and can be defined as $\pi_{hi} : \mathcal{S}_{hi} \to G_{net}$. The goal selection policy selects the next Goal Net goal to target and does not need to select goals directly in goal space. A goal achievement function $\beta_g : \mathcal{S}_{hi} \to \mathcal{G}$ is used to generate the terminal conditions within goal space. This helps increase the training speed of the goal selection policy since invalid and unused goals in the goal space are pruned. The goal selection policy operates within a SMDP, and its associated DQN is trained to estimate the optimal Q-value function:

$$Q_{hi}^*(s_{hi}, g_{net}) = \sum_{s_{hi}', \tau} P(s_{hi}', \tau | s_{hi}, g_{net})[R_\tau + \gamma^\tau \max_{g_{net}'} Q_{hi}^*(s_{hi}', g_{net}')], \quad (3)$$

where $s_{hi} \in \mathcal{S}_{hi}$, τ is the number of timesteps taken by the low-level policy to complete the subgoal, and $R_\tau = \sum_{t=0}^{\tau} \gamma^t r_{t+1}$. The extrinsic rewards from the environment obtained while running the low-level policy are passed to the high-level policy.

Algorithm 1 shows the overall training procedure for our model. Line 6 is the start of the goal selection loop, and in lines 7–8 the next Goal Net goal for the agent to achieve is selected and converted to goal space. In lines 11–12, the action selection policy is used to select actions based on the target goal. Both the goal and action selection use ϵ-greedy style exploration strategies based on the exploration strategy used in h-DQN. Such strategies select a random action with probability ϵ, and select the action with the maximum Q-value otherwise. A common strategy to enable sufficient exploration is to initialize ϵ with a high value and to decay it, typically linearly, over the course of the training process. However, in complex, temporally extended problems, it is difficult to pick a decay rate that ensures that the agent adequately explores the environment. Therefore, we take advantage of the Goal Net model to determine the exploration rates for the action and goal selection policies. The exploration rate of the action selection policy scales according to the success rate of achieving the selected Goal Net goal:

$$\epsilon_{lo} = -(\epsilon_{max} - \epsilon_{min})(\text{success}_N(g_{net}, g_{net}')) + \epsilon_{max}, \quad (4)$$

where ϵ_{max} and ϵ_{min} are hyperparameters defining the maximum and minimum values of ϵ_{lo}, respectively, and $\text{success}_N(g_{net}, g_{net}')$ is the average success rate of achieving goal g_{net}' starting from g_{net} over the past N attempts at achieving the goal. The exploration rate for goal selection follows a standard ϵ-greedy strategy, but instead of randomly selecting a goal with probability ϵ_{hi}, we weigh the probability of selecting particular goals based on the success rate of achieving them. The formula for determining the probability of selecting a goal is:

$$p(g_{net}, g_{net}') = \frac{1 - \text{success}_N(g_{net}, g_{net}') + \rho}{\sum_{g_{net}^* \in G_{net}^*} 1 - \text{success}_N(g_{net}, g_{net}^*) + \rho}, \quad (5)$$

where $\rho < 1$ is a small number that ensures that all goals have a chance to be selected and to prevent any division by 0, and G_{net}^* is the set of goals that

---

Algorithm 1: h-DQN Training with Goal Net

    Input: Goal Net Model

1 Initialize DQNs Q_{lo}, Q_{hi}

2 Initialize experience replay buffers $\mathcal{R}_{lo}, \mathcal{R}_{hi}$

3 **for** $i = 0$ **to** num_episodes **do**

4      $g_{net} \leftarrow$ initial Goal Net goal

5      $s, s_{hi}, s_{lo}, g \leftarrow$ environment reset

6   **for** $j = 0$ **to** max_steps **do**

7         $g'_{net} \leftarrow$ SelectGoal(s_{hi})

8         $g_{target} \leftarrow \beta_g(s, g'_{net})$

9         $r_{total} \leftarrow 0, \text{steps} \leftarrow 0$

10   **for** $k = j$ **to** max_steps **do**

11           $a \leftarrow$ SelectAction(s_{lo}, g_{target})

12           $s', g', r, \text{done} \leftarrow$ ExecuteAction(a)

13           $r_{total} \leftarrow r_{total} + r$

14           $g^*_{net} \leftarrow$ GNetReached(s', g', g_{net})

15           $\text{done}_{lo} \leftarrow (g^*_{net} \; != g_{net})$ **or** done

16   **if** $g^*_{net} == g'_{net}$ **then**

17               $r_i \leftarrow 1$

18   **else**

19               $r_i \leftarrow 0$

20   Add $\langle s_{lo}, a, g_{target}, (s', g'), r_i, \text{done}_{lo} \rangle$ to $\mathcal{R}_{lo}$

21   Update Q_{lo} using $\mathcal{R}_{lo}$

22           $s_{lo} \leftarrow (s', g')$

23           $\text{steps} \leftarrow \text{steps} + 1$

24   **if** done_{lo} **then**

25   break

26   **end**

27   Add relabelled experience tuples to $\mathcal{R}_{lo}$, replacing g_{target} with g'

28   Add $\langle s_{hi}, g^*_{net}, (s', g^*_{net}), r_{total}, \text{done}, \text{steps} \rangle$ to $\mathcal{R}_{hi}$

29   Update Q_{hi} using $\mathcal{R}_{hi}$

30        $s_{hi} \leftarrow (s', g^*_{net})$

31   **if** done **then**

32   break

33   **end**

34 **end**

---

can be selected from g_{net} as defined by the Goal Net model. By using this goal exploration strategy, we attempt to ensure that the agent learns to achieve all goals by focusing on goals that the agent cannot reach consistently.

During training, it is likely that the action selection policy inadvertently achieves a different goal than the goal proposed by the goal selection policy. For example, an agent using the Goal Net in Fig. 2 may obtain the yellow key even though it tried to obtain the green key. Line 14 checks which Goal Net goal the agent has reached by comparing g' with $\beta_g(s, g'_{net})$ across all possible Goal Net goals reachable from g_{net}. An intrinsic reward is provided to the low-level policy if it reaches the proposed goal, and the loop breaks if the low-level policy transitions to a new Goal Net goal. In line 28, we add experience tuples to the high-level replay buffer using the Goal Net goal actually reached by the agent rather than the one proposed by the high-level policy.

To handle concurrent goals, Algorithm 1 uses a list of goal paths in the Goal Net model containing the goals the agent has currently reached. Then the available goals that can be selected consists of all possible goal selection options across all goal paths.

4   Experiments

In the experiments[1], we make use of the Minigrid environment [6], Miniworld environment [5], and AI2-THOR [14]. Some example images of the environments used are shown in Fig. 4. More details about each environment are provided in the following subsections. The extrinsic reward function used in our experiments is based on the default reward function provided by Minigrid, defined as:

$$R = 1 - 0.9\left(\frac{n_{steps}}{n_{max}}\right), \tag{6}$$

where n_{steps} is the number of steps taken by the agent to reach the goal, and n_{max} is the maximum episode length. We use this reward function as it integrates both the agent's success rate and steps taken to reach the goal.

The goal of the experiments is to compare our method with other related methods in which a Goal Net model could be incorporated and to highlight problems that they have. We compare 4 different models: the proposed model, a variant of our model where the low-level policy operates on the state space without goal-augmentation, a model based on DAQN, and a model based on Q-learning for Reward Machines (QRM). We will refer to these models as GNet, GNet without GA, DAQN, and QRM respectively. GNet without GA will be used to compare whether using a goal-augmented state space for the low-level policy provides any benefits. We use DAQN and QRM as comparisons as both methods provide ways for agent designers to provide knowledge to an agent so that they can achieve temporally extended goals in a similar manner as our proposed method. DAQN is a HRL method where the high-level policy operates in an abstract state space defined by an agent designer and may invoke a low-level policy that is trained to reach a specific abstract state. The high-level policy in the DAQN model uses a tabular Q-learning algorithm that selects goals according to the provided Goal Net model. To make comparisons fairer, we also provide the valid goal selection choices based on the Goal Net model to the high-level policy. Additionally, we use goal-conditioning on the low-level DAQN policy by providing the Cartesian coordinates of the target goal and use HER to relabel experience tuples using the coordinate reached by the agent. QRM provides a comparison to a flat, non-goal-conditioned model capable of training an agent

---

[1] Code available at: https://github.com/jleung1/goal_modelling_rl.

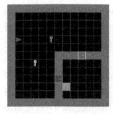  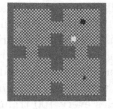 

(a) Two keys environ-
ment.
(b) 3D four rooms en-
vironment.
(c) Top view of the
3D four rooms envi-
ronment.
(d)   An   example
kitchen from AI2-
THOR.

Fig. 4. Sample images of the experimental environments. Positions of objects, subgoals, and the agent are randomized each episode.

to complete temporally extended tasks by having agent designers define a finite state machine that represents the reward function. The reward machines used in our experiments are based on the Goal Net models used in each experiment, where in most cases each goal acts as a reward machine state that provides the agent with a reward of 1. The transitions between reward machine states are determined by the goal spaces and β_g used in each experiment.

We attempt to use similar neural network architectures for all models. The low-level DQNs of the hierarchical models and the QRM model use convolutional layers that take the environment state as input. In GNet, the goal state and target goal are concatenated with the output of the convolutional layers and then passed through a set of linear layers. Only the target goal is used in GNet without GA. In DAQN and QRM, we pass one-hot encodings to the network to differentiate between multiple policies as opposed to the multi-headed or separate networks used in the respective works. We use Double DQN [11] for all models, as was done in both original DAQN and QRM works.

4.1 Two Keys

The first experimental environment, which we will refer to as the "two keys" environment, was created using Minigrid. In this environment, shown in Fig. 4a, the agent must reach the green goal coordinate. To do this, the agent needs to learn to acquire either the yellow or green key, then open the corresponding door to reach the goal room located in the bottom right quadrant. This environment tests the agent's ability to choose the key and door that lead it to the goal the fastest. Every episode, the agent and the keys are positioned randomly outside of the goal room, and the goal's position within the goal room is also randomized. In addition, the positions of the two doors may be randomly swapped. The actions available to the agent include moving forward, turning left or right, picking up a key, and opening a door. The Goal Net model used in this experiment is shown in Fig. 2. The goal space consists of the x and y coordinates of the agent's target goal, followed by a set of propositional symbolic features indicating whether the agent has the yellow or green key, and whether the yellow and green doors are opened or closed. We use a symbolic state space provided by the Minigrid environment, which is a $3 \times 13 \times 13$ grid that contains information about the object type, color, and status at each grid coordinate.

We train each agent for 50000 episodes, except for QRM which needed more training to converge, with a maximum episode length of 300 steps. We perform 100 evaluation episodes every 100 training episodes where the agent takes the greedy action at each time step. This process is repeated 5 times using the same set of random seeds across all models, and the means and standard deviations of the average rewards obtained are reported. The results are shown in Fig. 5a, and the rewards per frame are shown in Fig. 5b to illustrate the difference in sample efficiency between QRM and the other methods.

GNet is able to learn from the transitions within the goal-augmented state space and relate it to the target goals used in training a goal-conditioned DQN for the low-level policy. Since the state space used in GNet without GA does not directly include the goal space, the agent needs to learn the associations between the target goal and the state space itself, and thus is slightly less sample efficient. It should be noted that the goal space in this experiment contains information that is readily available within the state space given to the agent. This showcases the potential of providing an agent with a simplified representation of the state space alongside the full state space to improve learning efficiency.

DAQN converges to a lower average reward value because the abstract goal space does not provide enough information to the high-level policy about which key it should obtain. This problem was discussed by Gopalan et al. [10] and is demonstrated by this experiment. The agent cannot differentiate between obtaining the yellow of green key because both options lead the agent to the goal in the same number of steps with respect to the high-level policy. In order for DAQN to find a better policy, a goal space containing more information about the state space would be required. In contrast, the GNet models use the full state space at both policy levels and thus are more robust to the goal space design, imposing fewer restrictions on agent designers.

QRM can learn a policy that converges to higher rewards than DAQN, however it takes much longer to learn since it does not use HER. Whereas QRM only uses the goal space to determine reward machine state transitions, GNet allows the agent to actively select its next goal, which allows the use of HER. As QRM does not use a hierarchy, it is not clear how a goal selection mechanism would be incorporated into the method to allow for the use of goal-conditioning and HER. As described by Icarte et al. [12], QRM can share learning experience between RM state Q-value functions by using the RM to determine whether any RM state transitions have occurred when the agent interacts with the environment. However, this does not aid the agent in this environment because each RM state corresponds to separate sets of environment states. For example, if the agent opens the yellow door and then reaches the goal, this experience cannot be transferred to the case where the agent opens the green door and reaches the goal because the environment state is different in both cases.

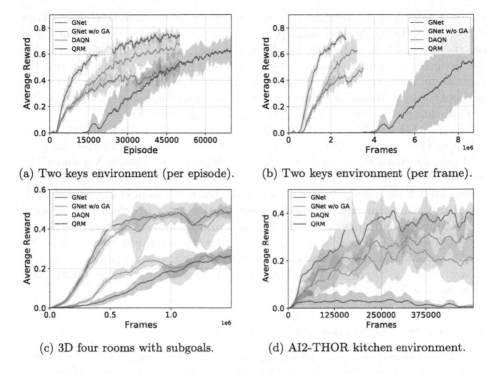

(a) Two keys environment (per episode).   (b) Two keys environment (per frame).

(c) 3D four rooms with subgoals.   (d) AI2-THOR kitchen environment.

Fig. 5. Average rewards obtained across all runs of the experiments.

4.2   3D Four Rooms with Subgoals

In this experiment, we modify an implementation of a 3D version of the four rooms environment provided by Miniworld [5]. The goal of the agent is to visit a blue and yellow subgoal before reaching the green goal, whose positions are all randomized at each episode. The order in which the agent visits the subgoals does not matter, and so this environment tests the agent's ability to handle partially ordered subgoals in a high-dimensional 3D environment with much randomness. If the agent reaches the final goal before reaching both subgoals, the agent receives a reward of 0. The agent views the environment in a first person perspective and receives RGB images as state observations, as shown in Fig. 4b. We provide the agent with the previous 4 frames to help the agent handle partial observability, making the size of the agent's observation $4 \times 3 \times 60 \times 80$. An overhead view of the environment is shown in Fig. 4c. The Goal Net model used for this experiment is shown in Fig. 3, however the equivalent RM contains an extra state that represents the agent having reached both subgoals. The agent can turn left or right by a random amount between $10°$ and $30°$ and move forward.

The goal space used in this experiment consists of the x and z positions of the goal, subgoals, and the agent, as well as whether the agent has reached the blue and yellow subgoals. To make comparisons fairer, we provide the goal, subgoal, and agent coordinates as additional state information to GNet without GA, DAQN, QRM. We use a deeper neural network for this experiment based on the one used by Espeholt et al., which uses residual connections [9]. We perform our evaluation similarly to the previous experiment where we run 100 evaluation episodes after every 100 training episodes and repeat the training process 5 times. We use a maximum episode length of 300 steps and use an ϵ of 0.05 during evaluation episodes. The results are shown in Fig. 5c.

Both GNet and GNet without GA perform similarly, with a bit more instability in GNet without GA. Since we provided the agent coordinates, subgoal, and goal locations to all methods, the only difference between GNet and GNet without GA is the inclusion of the subgoal completion statuses within the goal-augmented state space. However, as will be shown in the next experiment, this small difference can have a larger impact on the agent's performance in some environments. As in the two keys environment, DAQN performs worse because the high-level policy cannot determine whether visiting the yellow or blue subgoal first is better. Exploration in this experiment is easier than the two keys environment, as there are only three movement actions, making QRM learn quicker than in the previous experiment.

4.3   Kitchen Navigation and Interaction

In this experiment we use AI2-THOR, a 3D home environment created in the Unity game engine [14]. AI2-THOR provides various rooms where agents can interact with various objects. For this experiment, we use the 30 different kitchen environments provided by AI2-THOR and train the agent to first close the fridge door, and then turn off the light switch. An example of one of the kitchens is shown in Fig. 4d. The actions available to the agent are turning left and right, moving forward, closing an object, and toggling off an object. The episode ends when the agent turns off the light, with a reward of 0 being given if it turns off the light before closing the fridge door. This experiment tests the agent's use of the goal space to learn to navigate and complete tasks in high-dimensional environments that vary greatly between episodes. Since the sequence of subgoals is always the same in this experiment, the high-level policies of the hierarchical models do not need to be trained, which allows the low-level policy to be isolated and analyzed. The neural network architecture is similar to the previous experiment, using a deeper model with residual connections. The state observations given to the agent consists of the last four 100×100 RGB-depth images. The goal space consists of the agent's position, as well as the position of the fridge and light switch. Similar to the previous experiment, we give the x and z positions of the agent, fridge, and light switch to all models as extra state information to make comparisons fairer. Each episode, the agent is positioned randomly in one of the 30 kitchens and runs for a maximum of 200 steps. We perform evaluation every 50 episodes where we run the agent once through each kitchen using greedy actions. This process is repeated 5 times with different random seeds, and the results are shown in Fig. 5d.

As in the previous experiment, the only difference between GNet and GNet without GA is the subgoal completion statuses within the augmented state space. In the 3D four rooms environment, each action the agent took affected the x and z positions of the agent. However, this environment contains actions to close objects and toggle objects off, which do not have any effect if the agent is not near any object where these actions are applicable. This problem was demonstrated in the first experiment, where GNet without GA had to learn to associate the target goal with the state space. By using a goal-augmented state space, we provide a generalized way for agent designers to guide agents. DAQN performs similarly to the GNet models because when only considering the low-level policy, the methods are similar. Thus, a key benefit of our proposed method is the use of the full state space in the high-level policy.

5   Related Work

There have been proposed methods to incorporate prior knowledge in a RL agent. A closely related method is the hierarchy of abstract machines (HAM) [22], where partial policies can be defined using a hierarchy of finite state machines. Reward Machines also use finite state machines, but instead of directly defining an agent's behavior, they are used to define reward functions that may represent complex, temporally extended tasks [12]. Andreas et al. proposed a method to include agent designers' prior knowledge using policy sketches that are used to train an agent to complete tasks via subtask sequences [1]. However, a dataset of policy sketches is assumed to be available whereas our proposed method assumes a labelling function is defined. Additionally, policy sketches impose a specific ordering of subgoals whereas a Goal Net model also enables the definition of partially ordered subgoals. Roderick et al. proposed DAQN [25], which extends abstract MDPs [10] by combining tabular RL and DRL. Lyu et al. also propose a HRL method that combines tabular RL and DRL and uses symbolic planning to incorporate prior human knowledge [19]. Our method, however, uses HER to improve sample efficiency and can handle environments where goals may change between episodes. Icarte et al. use Linear Temporal Logic (LTL) formulae to describe tasks and decompose them into subtasks [30]. Our method uses Goal Net to provide a representation of an agent's objectives that is understandable to stakeholders who may have little technical knowledge, however a combination of LTL and Goal Net could be explored in the future. Zhang et al. propose a method where agents learn to plan in a human-defined attribute space, which is similar to the goal space of our method, and use count-based exploration to train agents in a task agnostic manner [34]. Unlike their method, our proposed method augments the state space using the goal space, making our method less reliant on how an agent designer defines the goal space.

Hierarchical reinforcement learning has roots in works such as the options framework [29], MAXQ value function decomposition [8], and Feudal Networks [7]. Many recent works in HRL incorporate deep neural networks. Bacon et al. extended the options framework by training agents to learn the intra-option and option termination functions in an end-to-end manner [3]. Vezhnevets et al. extend Feudal Networks to use deep neural networks and propose a model where a manager and worker are learned in parallel, and the manager learns to produce goals that represent a desired direction in a learned latent goal space [32]. Levy et al. proposed a method for training a hierarchical agent with potentially many levels of goal-conditioned policies [17]. Nachum et al. improved the sample efficiency of HRL methods using off-policy RL for the high-level policy by correcting transitions in the replay buffer with new goals according to the current low-level policy [21]. We note that our method is not necessarily orthogonal to other HRL methods and could potentially be integrated such that an agent designer proposes high-level subgoals via a Goal Net model and the agent learns to further decompose the subgoals through its own hierarchy.

6   Discussion and Conclusion

We proposed a goal-oriented model and algorithm which use agent designers' prior knowledge to train a hierarchical RL agent. We used Goal Net to accomplish this as goals provide an abstraction of agent behavior that is understandable by agent designers with diverse levels of technical knowledge. We compared our method to two related methods, DAQN and QRM, which make use of similar levels of prior knowledge. We demonstrated that the proposed method can make better use of the information provided to it by the agent designers and learn more quickly in various environments. We also showed that the agent is more robust to the goal space design because we augment the state space of the original MDP rather than reduce it. If the goal space is missing information that may help the agent achieve its goal more efficiently, the agent can still learn because it is not necessarily dependent on the goal space. If an agent designer provides redundant information to the agent, it can still leverage the goal space to learn more efficiently. However, information needed by the agent that is not contained in the state space should be included in the goal space.

A future direction may investigate methods of learning goal spaces to allow agents to have a better understanding of its goals. This may help apply our method to domains outside navigation and goal-reaching tasks, such as dialogue systems. Another direction could involve improving goal selection to handle partially observable environments where the subgoal locations may not be known to the agent. In such environments, the agent may need to change its goal based on new information. An investigation on the use of the proposed method to promote safe AI could be a future direction, as Goal Net can help create agents whose policies are controllable and interpretable. Incorporating other goal types, such as maintenance or avoidance goals, may help in this regard.

Acknowledgments. This research is supported, in part, by the National Research Foundation, Prime Minister's Office, Singapore under its NRF Investigatorship Programme (NRFI Award No. NRF-NRFI05-2019-0002). Any opinions, findings and conclusions or recommendations expressed in this material are those of the author(s) and do not reflect the views of National Research Foundation, Singapore.

References

1. Andreas, J., Klein, D., Levine, S.: Modular multitask reinforcement learning with policy sketches. In: International Conference on Machine Learning, pp. 166–175 (2017)
2. Andrychowicz, M., et al.: Hindsight experience replay. In: Advances in Neural Information Processing Systems, pp. 5048–5058 (2017)
3. Bacon, P.L., Harb, J., Precup, D.: The option-critic architecture. In: Thirty-First AAAI Conference on Artificial Intelligence (2017)
4. Bresciani, P., Perini, A., Giorgini, P., Giunchiglia, F., Mylopoulos, J.: Tropos: an agent-oriented software development methodology. Auton. Agent. Multi-Agent Syst. **8**(3), 203–236 (2004). https://doi.org/10.1023/B:AGNT.0000018806.20944. ef
5. Chevalier-Boisvert, M.: gym-miniworld environment for openai gym (2018). https://github.com/maximecb/gym-miniworld
6. Chevalier-Boisvert, M., Willems, L., Pal, S.: Minimalistic gridworld environment for openai gym (2018). https://github.com/maximecb/gym-minigrid
7. Dayan, P., Hinton, G.E.: Feudal reinforcement learning. In: Advances in Neural Information Processing Systems, pp. 271–278 (1993)
8. Dietterich, T.G.: Hierarchical reinforcement learning with the MAXQ value function decomposition. J. Artif. Intell. Res. **13**, 227–303 (2000)
9. Espeholt, L., et al.: IMPALA: scalable distributed deep-RL with importance weighted actor-learner architectures. In: International Conference on Machine Learning, pp. 1407–1416. PMLR (2018)
10. Gopalan, N., et al.: Planning with abstract Markov decision processes. In: Twenty-Seventh International Conference on Automated Planning and Scheduling (2017)
11. van Hasselt, H., Guez, A., Silver, D.: Deep reinforcement learning with double q-learning. In: Proceedings of the Thirtieth AAAI Conference on Artificial Intelligence, pp. 2094–2100 (2016)
12. Icarte, R.T., Klassen, T., Valenzano, R., McIlraith, S.: Using reward machines for high-level task specification and decomposition in reinforcement learning. In: International Conference on Machine Learning, pp. 2107–2116 (2018)
13. Kaelbling, L.P.: Learning to achieve goals. In: Proceedings of the Thirteenth International Joint Conference on Artificial Intelligence, pp. 1094–1099 (1993)
14. Kolve, E., et al.: AI2-THOR: An Interactive 3D Environment for Visual AI. arXiv (2017)
15. Kulkarni, T.D., Narasimhan, K., Saeedi, A., Tenenbaum, J.: Hierarchical deep reinforcement learning: integrating temporal abstraction and intrinsic motivation. In: Advances in Neural Information Processing Systems, pp. 3675–3683 (2016)
16. Leike, J., Krueger, D., Everitt, T., Martic, M., Maini, V., Legg, S.: Scalable agent alignment via reward modeling: a research direction. arXiv preprint arXiv:1811.07871 (2018)
17. Levy, A., Konidaris, G., Platt, R., Saenko, K.: Learning multi-level hierarchies with hindsight. arXiv preprint arXiv:1712.00948 (2017)
18. Lin, L.J.: Self-improving reactive agents based on reinforcement learning, planning and teaching. Mach. Learn. **8**(3–4), 293–321 (1992). https://doi.org/10.1007/BF00992699
19. Lyu, D., Yang, F., Liu, B., Gustafson, S.: SDRL: interpretable and data-efficient deep reinforcement learning leveraging symbolic planning. In: Proceedings of the AAAI Conference on Artificial Intelligence, vol. 33, pp. 2970–2977 (2019)

20. Mnih, V., et al.: Playing Atari with deep reinforcement learning. arXiv preprint arXiv:1312.5602 (2013)
21. Nachum, O., Gu, S.S., Lee, H., Levine, S.: Data-efficient hierarchical reinforcement learning. In: Advances in Neural Information Processing Systems, pp. 3303–3313 (2018)
22. Parr, R., Russell, S.J.: Reinforcement learning with hierarchies of machines. In: Advances in Neural Information Processing Systems, pp. 1043–1049 (1998)
23. Plappert, M., et al.: Multi-goal reinforcement learning: challenging robotics environments and request for research. arXiv preprint arXiv:1802.09464 (2018)
24. Puterman, M.L.: Markov Decision Processes: Discrete Stochastic Dynamic Programming. Wiley, Hoboken (2014)
25. Roderick, M., Grimm, C., Tellex, S.: Deep abstract q-networks. In: Proceedings of the 17th International Conference on Autonomous Agents and MultiAgent Systems, pp. 131–138 (2018)
26. Schaul, T., Horgan, D., Gregor, K., Silver, D.: Universal value function approximators. In: International Conference on Machine Learning, pp. 1312–1320 (2015)
27. Shen, Z., Miao, C., Gay, R., Li, D.: Goal-oriented methodology for agent system development. IEICE Trans. Inf. Syst. **89**(4), 1413–1420 (2006)
28. Sutton, R.S., Barto, A.G.: Reinforcement Learning: An Introduction. MIT Press, Cambridge (2018)
29. Sutton, R.S., Precup, D., Singh, S.: Between MDPs and semi-MDPs: a framework for temporal abstraction in reinforcement learning. Artif. Intell. **112**(1–2), 181–211 (1999)
30. Toro Icarte, R., Klassen, T.Q., Valenzano, R., McIlraith, S.A.: Teaching multiple tasks to an RL agent using LTL. In: Proceedings of the 17th International Conference on Autonomous Agents and MultiAgent Systems, pp. 452–461 (2018)
31. van Lamsweerde, A.: Goal-oriented requirements engineering: a guided tour. In: Proceedings Fifth IEEE International Symposium on Requirements Engineering, pp. 249–262 (2001)
32. Vezhnevets, A.S., et al.: Feudal networks for hierarchical reinforcement learning. arXiv preprint arXiv:1703.01161 (2017)
33. Yu, E.S.: Towards modelling and reasoning support for early-phase requirements engineering. In: Proceedings of ISRE 1997: 3rd IEEE International Symposium on Requirements Engineering, pp. 226–235. IEEE (1997)
34. Zhang, A., Sukhbaatar, S., Lerer, A., Szlam, A., Fergus, R.: Composable planning with attributes. In: International Conference on Machine Learning, pp. 5842–5851. PMLR (2018)

Time Series, Streams, and Sequence Models

Deviation-Based Marked Temporal Point Process for Marker Prediction

Anand Vir Singh Chauhan, Shivshankar Reddy, Maneet Singh$^{(\boxtimes)}$,
Karamjit Singh, and Tanmoy Bhowmik

AI Garage, Mastercard, New Delhi, India
{anandvirsingh.chauhan,shivshankar.reddy,maneet.singh,
karamjit.singh,tanmoy.bhowmik}@mastercard.com

Abstract. Temporal Point Processes (TPPs) are useful for modeling event sequences which do not occur at regular time intervals. For example, TPPs can be used to model the occurrence of earthquakes, social media activity, financial transactions, etc. Owing to their flexible nature and applicability in several real-world scenarios, TPPs have gained wide attention from the research community. In literature, TPPs have mostly been used to predict the occurrence of the next event (time) with limited focus on the *type/category* of the event, termed as the *marker*. Further, limited focus has been given to model the inter-dependency of the event time and marker information for more accurate predictions. To this effect, this research proposes a novel Deviation-based Marked Temporal Point Process (DMTPP) algorithm focused on predicting the marker corresponding to the next event. Specifically, the deviation between the estimated and actual occurrence of the event is modeled for predicting the event marker. The DMTPP model is explicitly useful in scenarios where the marker information is not known immediately with the event occurrence, but is instead obtained after some time. DMTPP utilizes a Recurrent Neural Network (RNN) as its backbone for encoding the historical sequence pattern, and models the dependence between the marker and event time prediction. Experiments have been performed on three publicly available datasets for different tasks, where the proposed DMTPP model demonstrates state-of-the-art performance. For example, an accuracy of 91.76% is obtained on the MIMIC-II dataset, demonstrating an improvement of over 6% from the state-of-the-art model.

Keywords: Temporal point processes · Marker prediction · Recurrent neural network

1 Introduction

The developments in technology and fast-paced lifestyle have resulted in the generation of large amount of temporal data containing *events* spanned across irregular time intervals. For example, activity on social media such as uploading images, post reacts, interactions with other users; utilizing public transportation

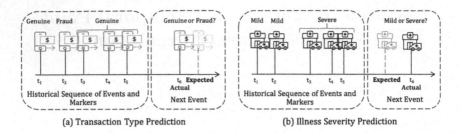

(a) Transaction Type Prediction   (b) Illness Severity Prediction

Fig. 1. Event marker prediction has wide-spread applicability in various real-world scenarios. The proposed Deviation-based Marked Temporal Point Process model focuses on predicting the marker of an event in real-time, while modeling the inter-dependence between the expected event time and the actual event time.

such as cabs, taxis, or buses; financial activity such as buying/selling stocks, online purchases; and dining out at restaurants or reviewing eating joints. Coupled with the advent of Machine Learning and the day-to-day usage of different deployed applications, developing algorithms for automated event prediction has garnered substantial research attention. Traditionally, event prediction referred to determining *when* the next event would happen. With several recent real-world applications, research has also focused on predicting the *type* of the event referred to as the *marker* corresponding to an event. Figure 1 presents sample real-world applications requiring event type prediction (often in real-time). Figure 1(a) presents a sample scenario where banks could utilize algorithms to identify whether the current transaction (event) was fraudulent or not (marker), and Fig. 1(b) presents another scenario where hospitals could identify the duration or severity (marker) of a patient's visit (event) based on their historical information. Event marker prediction thus has wide applicability in real-world scenarios across different domains, demanding dedicated attention.

Initial research on event prediction [21,27] utilized statistical techniques [2], followed by modeling the sequences as time series [12]. While earlier research focused primarily on events spaced evenly in time, as discussed previously, most of the above mentioned activities are uneven or irregular in terms of the inter-event time. The uneven characteristic of event sequences makes it appropriate to model them as Temporal Point Processes (TPP) [10,18], often defined by an intensity function modeling the inter-event duration. Generally, historical sequences are modeled to predict the occurrence of the next event, and a categorical value associated with it, referred to as the event marker. While event time prediction has been well studied in the past few years, limited attention has been given to the task of marker prediction. To this effect, this research proposes a novel *Deviation-based Marked Temporal Point Process (DMTPP)* model for predicting the event marker. The proposed model is specifically applicable in scenarios where the event marker is not available immediately after the event occurrence, but is instead computed/obtained after some time. For example, a fraudulent transaction (marker) might be reported after some time of the transaction (event) by the concerned person (Fig. 1(a)), the severity of an ill-ness (marker) is not known upon the immediate admission of a patient (event) into a hospital (Fig. 1(b)), and the impact of an online advertisement (event) on the subsequent sales (marker) is known after some time. By utilizing the real-time event occurrence, the DMTPP model presents high applicability in such scenarios, where the marker prediction is also performed in real-time.

The proposed DMTPP model focuses on explicitly modeling the inter-dependence between the marker prediction and the variation observed in the expected behavior. This enables the model to capture *anomalous* behavior with respect to the event occurrence in real-time, while utilizing the representation from the historical event sequence. Therefore, the contributions of this research are:

- A novel Deviation-based Marked Temporal Point Process (DMTPP) model has been proposed for marker prediction in real-time. The DMTPP algorithm models the dependence of the marker on the expected and actual time occur-rence of the next event. To the best of our knowledge, this is the first-of-a-kind model operating at real time, which explicitly models the dependence of the marker on the deviation in the expected and actual event time.
- The DMTPP model utilizes a Recurrent Neural Network (RNN) as its back-bone architecture. The RNN learns an embedding based on the past sequence of events and markers, while modeling the intensity function of the TPP as a non-linear function. The choice of RNN as the backbone architecture pro-vides more flexibility during sequence modeling, and also prevents learning of *user-specific* models/representations. Thus enabling the proposed DMTPP model to be useful in real-world scenarios of unseen test users as well.
- The proposed model has been evaluated on three marker prediction tasks: (i) retweet prediction on the Retweet dataset [28], (ii) illness type prediction on the MIMIC-II dataset [16], and (iii) badge prediction on the StackOver-flow dataset [4]. Comparison has been performed with recent state-of-the-art methods, where the proposed model demonstrates improved performance. For example, it achieves a classification accuracy of 91.76% on the challeng-ing MIMIC-II dataset. The improved performance promotes the utility of the proposed model for real-time marker prediction tasks.

2 Related Work

This section analyzes the related concepts and research in the area of marked temporal point process. Marked temporal point processes build upon the tradi-tional temporal point processes by associating a *marker* with the occurrence of each event. Here, *marker* can refer to the category of the event or some addi-tional information of the event that is mostly categorical in nature. Research in marked temporal point processes has focused on the next event and marker prediction based on the sequence of historical events which is measured by an intensity function. The intensity function measures the number of events that can be expected in a specific time interval.

The traditional methods in temporal point processes like Poisson [11], Self-exciting [6], and Self-correcting [9] processes estimate the conditional intensity function by making parametric assumptions. Such assumptions limit the flexibility of the model, thus making it challenging to apply in different real-world scenarios. The more recent models which utilize deep learning algorithms train the models by maximizing the log-likelihood of the desired loss function. One of the seminal algorithms at the intersection of marked temporal point processes and deep learning is the Recurrent Marked Temporal Point Processes (RMTPP) model [3], which uses a neural network to predict both next event time and event marker independently using a Recurrent Neural Network (RNN). Following this, Wang et al. [22] proposed an RNN network to build a marker-specific intensity function that considers the inter-dependency between the marker and time of the next event. Beyond RNNs, in 2018, Decoupled Learning for Factorial Marked Temporal Point Processes [23] was proposed, where a decoupling approach is presented for learning the factorial marked temporal point process, in which each event is represented by multiple markers. Recently, Türkmen et al. [20] leveraged both Hawkes processes and RNN to capture local and global temporal relationships. Shchur et al. [19] proposed a novel approach of using neural density estimation to estimate the conditional density instead of modeling the conditional intensity function. Further, in 2020, Transformer Hawkes Process (THP) [29] and Self-Attentive Hawkes Process (SAHP) [26] addressed the problem of long-term dependencies by using a self-attention mechanism to capture short-term and long-term dependencies in the past sequence of the event.

As demonstrated above, the field of marked temporal point processes has recently garnered substantial attention. TPPs have shown applicability in several real-time applications, and are successful in capturing the influence of past historical sequence information for the prediction of the next event time and marker. However, in most of the existing literature (Fig. 2(a)–(b)), the event time and marker are assumed to be independent, which might not be true in real-time applications where the event time and marker are inter-dependent. For example, as shown in Fig. 1(a), in scenarios of fraudulent transaction detection, a given transaction is required to be classified as fraudulent or not. In this scenario, unusual occurrence (time) of the actual event as compared to the predicted time by the learned model can help in identifying a fraudulent transaction. Similarly, in other domain applications such as predicting visits to the Intensive Care Unit (ICU) (Fig. 1(b)), illness severity prediction can be dependent on the deviation between the next predicted event time and the actual event time. Similar trend can be observed in the scenario of online advertisements, where variation between the actual and expected time of posting can often result in variation of the next marker type (impact of advertisement measured by subsequent sales).

Based on the above intuition, this research proposes utilizing the deviation in the predicted event time and actual event time for predicting the event marker. A novel Deviation-based Marked Temporal Point Process (DMTPP) model is proposed, which incorporates the inter-dependence between the event time and

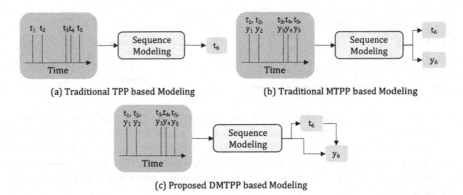

(a) Traditional TPP based Modeling   (b) Traditional MTPP based Modeling

(c) Proposed DMTPP based Modeling

Fig. 2. Existing literature in TPP based modeling has focused mostly on (a) predicting the next event time (t_6), or (b) predicting the next event and marker information (t_6, y_6) independently. (c) The proposed DMTPP based model learns the dependence of the marker information on the time prediction.

marker (Fig. 2(c)) by considering the real-time event occurrence information for predicting the next event marker.

3   Proposed Algorithm

Figure 3 presents a diagrammatic overview of the proposed Deviation-based Marked Temporal Point Process (DMTPP) model. The proposed model takes the historical sequence of time and marker, and predicts the marker for the next event. Further, the model also utilizes the actual time of the next event for predicting the corresponding marker. The proposed model thus demonstrates high applicability in scenarios where the marker is computed/derived after some time as opposed to being simultaneously available with the event occurrence. As shown in Fig. 3, a RNN based architecture is used for modeling the relationship between the lists of past event times and markers, which learns a non-linear hidden representation based on the past sequence. The next event time and marker are predicted by utilizing the hidden representation. The time deviation measures the variation between the predicted event time and actual event time, which is passed to a dense layer along with the RNN hidden representation to predict the next event marker. The following subsections elaborate upon the mathematical problem formulation, preliminaries for the proposed model, and the in-depth explanation of the DMTPP model, followed by the implementation details.

3.1   Problem Definition

The problem setting involves a sequence of events denoted by their time of occurrence and corresponding markers. Mathematically, each sequence is represented

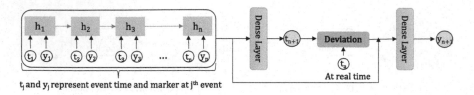

Fig. 3. Diagrammatic representation of the proposed DMTPP model. A sequence of past events $((t_j, y_j), j \in (1, n))$ is provided as input to the RNN model, which outputs the learned embedding for the sequence, which is then used for predicting the next event time (t_{n+1}). The deviation between the expected and the actual time event is calculated, followed by the combination of the embedding and the deviation for marker prediction (y_{n+1}).

by $S = \{(t_1, y_1), (t_2, y_2), \ldots, (t_n, y_n)\}$, where n refers to the total sequence length. Here, (t_j, y_j) refers to the jth event represented by the time of the event (t_j) and the corresponding marker (y_j). By default, the events are ordered in time, such that $t_{j+1} \geq t_j$. Given the sequence of last n events, often the task is to predict the next event time t_{n+1} and the corresponding marker y_{n+1}. In reference to Fig. 1(a), the event refers to a transaction, event time refers to the time of the transaction, and event marker refers to whether the transaction (event) was fraudulent or not.

3.2   Preliminaries

Temporal Point Process (TPP): A temporal point process is a stochastic process that models a sequence of discrete events occurring in a continuous-time interval [1]. Typically, a TPP is modeled by using a conditional intensity function, which measures the number of events that can be expected in a specific time interval, given the historical sequence of event information. Mathematically, the intensity function of a TPP is defined as the probability an event will occur in $[t, t + dt]$ time interval given the event history h_t till time t:

$$\lambda^*(t)dt = \lambda(t \mid h_t) = P(event\ in\ [t, t + dt] \mid h_t) \tag{1}$$

where, dt refers to a small window of time, and $P(.)$ refers to the probability function. As derived by Du et al. [3], the conditional density function $(f(t|h_t))$ of an event occurring at time t can thus be specified as:

$$f(t|h_t) = \lambda^*(t) \exp\left(-\int_{t_n}^{t} \lambda^*(\tau)d\tau\right) \tag{2}$$

where, t_n refers to the last event and τ corresponds to a very small value tending to zero. The conditional intensity function has been modelled using different parametric forms in the past. Some of the well known methods are:

- Poisson process [11]: Events are assumed to be independent of their history, such that $\lambda(t|h_t) = \lambda(t)$.

– Hawkes Process [6]: In the Hawkes process, the conditional intensity function constitutes a time decay kernel to take into consideration the events history. The intensity function is assumed to be a linear function ($\gamma(.)$) of the history along with the base intensity value (γ_0) and a weight parameter (α) as:

$$\lambda^*(t) = \gamma_0 + \alpha \sum_{t_j < t} \gamma(t, t_j) \tag{3}$$

As demonstrated above, the traditional temporal point process based techniques model the conditional intensity function by assuming that the data follows some parametric form, which can often be estimated using maximum likelihood estimation (MLE). The above assumption constraints the expressive power of the conditional intensity function, since the true form of the intensity function is unknown in real-time scenarios. This limitation often renders the above techniques unusable in several real-world applications having complex intensity functions.

Marked Temporal Point Process: A natural extension of the TPP based techniques is the inclusion of a marker information along with each event. In such scenarios, the model is expected to predict the next event time and corresponding marker, while having access to the history of past events. Therefore, the conditional intensity function for a marked temporal point process can be formulated as follows:

$$\lambda^*((t_j, y_j)) = \lambda((t_j, y_j) \mid h_t) \tag{4}$$

where, t_j and y_j refer to the event time and event marker, respectively. $\lambda^*((t_j, y_j))$ can take multiple forms, however, for mathematical simplicity it is mostly assumed that the event time and marker are conditionally independent given the event history i.e. $\lambda^*((t_j, y_j)) = \lambda^*(t_j)\lambda^*(y_j)$. The above assumption assumes independent marker and time occurrence, which often limits the model performance in scenarios where the event time and marker are interdependent.

3.3 Proposed Deviation-Based Marked Temporal Point Process

In the proposed Deviation-based Marked Temporal Point Process (DMTPP) model, the objective is to predict the next event marker given the history of the past events. The proposed DMTPP model addresses the discussed limitations by utilizing a universal approximator for learning the conditional intensity function, and by explicitly modeling the relationship between the event time and event marker. This is achieved by utilizing a Recurrent Neural Network (RNN) as the backbone model, and by incorporating the deviation between the actual event time and the predicted event time for marker prediction. Since RNNs are characterized by the property of being a universal approximator, they can thus be applied to model complex intensity functions, and the deviation component can be used to model the relationship between the event occurrence and the marker category. In literature, one of the seminal works involving the usage of RNNs for marked temporal point processes was presented by Du et al. [3]. The proposed model extends the current body of literature by modeling the inter-dependence of the marker on the event time as well.

Figure 3 presents the diagrammatic representation of the proposed technique. The model takes as input the past time sequence and marker sequence. It utilizes an RNN as the backbone architecture and learns an embedding based on the past events, followed by the prediction of the next event time. The deviation between the predicted time and the actual time of the event is then concatenated with the previously learned embedding for predicting the marker corresponding to the next event. The RNN is thus used to model the intensity function for the given sequences of events.

Mathematically, the last n events are passed as one sequence to the model $(\mathcal{S} = \left((t_j, y_j)_{j=1}^n\right))$. For processing, instead of the absolute time stamps, the sequence of the inter-event duration is provided to the algorithm for learning a model invariant to the absolute time. For the time sequence t_j, t_{j-1}, the inter-event duration is calculated as $d_j = t_j - t_{j-1}$. The inter-event duration sequence is calculated for all consecutive events in the sequence $\mathcal{S}$ and is provided to the model $(d_1, d_2, ..., d_n)$ along with the previous marker sequence $(y_1, y_2, ..., y_n)$. The marker information is converted into a sparse one-hot encoding for better representation, followed by learning a feature vector using an embedding layer:

$$y_j^{em} = \mathbf{W}_{em}^\top y_j + b_{em} \tag{5}$$

where, $\mathbf{W}_{em}$ is the weight matrix for the embedding layer and b_{em} is the bias vector. Thus, the input sequence consisting of the historical inter-event duration (d_j) and the past marker sequence (y_j^{em}) are provided as input to the RNN for learning an embedding capturing the relation between the event sequence and marker. The RNN utilizes the past historical representation (h_{j-1}) along with the other inputs and returns the updated hidden representation as follows:

$$h_j = ReLU(\mathbf{W}^y y_j^{em} + \mathbf{W}^d d_j + \mathbf{W}^h h_{j-1} + b_h) \tag{6}$$

where $\mathbf{W}^y$, $\mathbf{W}^d$, $\mathbf{W}^h$ and b_h denote the marker weight matrix, time duration weight matrix, RNN's representation (history) weight matrix, and the bias weight vector, respectively. Based on the hidden representation h_j, the next inter-event time duration and marker can be calculated as follows:

$$p(d_{j+1} \mid h_j) = f_t(d_{j+1} \mid h_j); \ p(y_{j+1} \mid h_j) = f_y(y_{j+1} \mid h_j, \delta_{j+1}) \tag{7}$$

where δ_{j+1} is the deviation for the $(j+1)^{th}$ event which corresponds to the difference between the predicted time and the actual time:

$$\delta_{j+1} = t_{j+1} - t_a \tag{8}$$

where t_a and t_{j+1} is the actual and predicted time of the $(j+1)^{th}$ event, respectively. Therefore, the proposed DMTPP model utilizes the learned embedding from the past sequence for predicting the event time, and also incorporates the deviation between the actual and expected time for marker prediction. Mathematically, for the time prediction, the conditional intensity function is calculated whereas for the marker prediction, the conditional distribution of the marker on hidden representation and deviation of time is calculated. Similar to Du et al. [3], the conditional intensity function for time prediction can be represented as:

$$\lambda^*(t) = \exp\left(w^{h^\top} \mathbf{h}_j + \beta (t - t_j) + b\right) \tag{9}$$

where w^h is a weight vector, while β and b are scalar values. The above equation ensures that the conditional intensity is dependent upon the inter-dependence of the past marker and time sequence obtained via the hidden representation from the RNN (first term), the influence of the current time (second term), and an offset base intensity value (third term). Given the above conditional intensity function, the conditional density function for TPPs (Eq. 2) can thus be updated. Therefore, the likelihood of the next event occurring at t_{j+1} given the history h_j can be given as:

$$f_t(t_{j+1} \mid h_j) = \lambda(t_{j+1} \mid h_j) \exp\left(-\int_{t_j}^{t_{j+1}} \lambda(t_j \mid h_j) dt\right) \tag{10}$$

The above equation is utilized for predicting the next event time, given the learned hidden representation from an RNN. Given the expected (predicted) and actual time occurrence, the corresponding marker can be predicted using the hidden representation h_j and the time deviation δ_{j+1}, by using the Softmax function on the conditional probability as:

$$f_y(y_{j+1} = k \mid h_j, \delta_{j+1}) = \frac{\exp\left(\mathbf{W}_k[h_j|\delta_{j+1}] + b_k^y\right)}{\sum_{i=1}^{K} \exp\left(\mathbf{W}_i[h_j|\delta_{j+1}] + b_i^y\right)} \tag{11}$$

where $[h_j|\delta_{j+1}]$ represents the concatenation of the embedding obtained via the RNN and the computed time deviation. Given a K class problem for marker prediction, $\mathbf{W}_i$ refers to the weight vector for the ith marker, and k refers to the correct marker for the next event. The inclusion of the deviation parameter enables the model to learn the inter-dependence between the user behavior (for event occurrence) and the marker. This is especially useful in scenarios where the marker is not known at real-time, but is instead computed after some time of the event occurrence. For example, earthquake intensity or influence of a retweet/advertisement. In such scenarios, the deviation from the expected time of the event can often impact the outcome of the event (marker). By modeling the time deviation, the proposed DMTPP model is thus able to capture the corresponding variations in the marker in real-time. The model is trained by maximizing the joint log-likelihood of the event prediction and marker prediction loss functions as follows:

$$\mathcal{L}(\{\mathcal{S}\}) = \sum_{i=1}^{n} \sum_{j=r-1}^{S_n^i - 1} \left(\underbrace{\lambda_1 \log f_y\left(y_{j+1}^i \mid [h_j^i, \delta_{j+1}]\right)}_{Marker\ Prediction} + \underbrace{\lambda_2 \log f_t\left(t_{j+1}^i \mid h_j^i\right)}_{Time\ Prediction}\right) \tag{12}$$

298 A. V. S. Chauhan et al.

Table 1. Details of the datasets used in this research demonstrating variability in terms of size, number of marks, number of events, and average sequence length.

Dataset	No. of Markers	No. of Events	Avg. Seq. Length
MIMIC-II [16]	75	2419	4
StackOverflow [4]	22	480414	72
Retweet [28]	3	2M	209

where, r refers to the first event that is being predicted for sequence i $(\mathcal{S}^i)$. n refers to the number of sequences in the training set $\mathcal{S}$, $\mathcal{S}_n^i$ refers to the number of events in sequence $\mathcal{S}^i$. λ_1 and λ_2 refer to the weight given to each component.

3.4 Implementation Details

The proposed DMTPP model has been implemented in the Pytorch environment [17] with a NVIDIA Quadro RTX6000 GPU. As demonstrated in Fig. 3, the DMTPP model consists of an initial embedding layer for the marker sequence, a RNN model, and modules for predicting the event time and marker. The embedding layer is of dimension 10, while the RNN model consists of a single layer Long Short-Term Memory module [8]. A 32 dimension representation is obtained via the RNN model, which is provided to a dense-layer for predicting the time, followed by another dense-layer for marker prediction. Dropout [7] has also been applied as a regularizer after the RNN layer. The weight parameters in Eq. 12 are initialized as follows: $\lambda_1 = 0.15$ and $\lambda_2 = 0.05$. The model is trained using the Adam optimizer [13] for 100 epochs with 1024 batch-size.

4 Experiments and Protocols

In order to evaluate the effectiveness of the proposed Deviation-based Marked Temporal Point Process model, experiments have been performed on three datasets corresponding to different tasks. Table 1 presents the dataset statistics demonstrating high variability across different parameters. Details regarding the datasets and protocols are as follows:

(i) Disease Type Prediction on the MIMIC II Dataset [16]: The MIMIC II dataset is a subset of the Electrical Medical Records Dataset which contains a collection of clinical visit records of Intensive Care Unit (ICU) patients over a period of seven years. Each event contains the time when a patient visits the ICU along with their type of disease (75 disease types). For this dataset, marker prediction corresponds to predicting the disease type. Similar to the existing protocol [29], 90% of the data has been used for training the model, while the remaining 10% corresponds to the test set.

(ii) Badge Prediction on the StackOverflow Dataset [4]: StackOverflow[1] is a popular question answering website where users are awarded with different

---
[1] https://archive.org/details/stackexchange.

Table 2. Marker prediction accuracy (%) on the MIMIC-II and StackOverflow datasets. Comparisons have been performed with the state-of-the-art algorithms. Owing to the same protocol, results have directly been taken from Zuo *et al.* [29].

Model	MIMIC-II	StackOverflow
Recurrent Marked Temporal Point Process [3]	81.2	45.9
Neural Hawkes Process [15]	83.2	46.3
Time Series Event Sequence [24]	83.0	46.2
Transformer Hawkes Process [29]	85.3	47.0
Deviation-based Marked TPP	**91.76**	**55.42**

badges (Guru, Great Answer, Stellar Question, etc.) for enhancing user engagement and popularity. The StackOverflow dataset contains sequence of badges (marker) received by a user along with the time when the badge is given. A similar protocol and pre-processing is followed as the existing manuscript [3]. The processed dataset contains 6,633 users and 480,414 events with total 22 badges. 90% of the data is used for training, and the remaining 10% is the test set.

(iii) Retweet Prediction on the Retweet Dataset [28]: The Retweet dataset is formed through the Seismic dataset[2]. A stream of retweets is available in which each event is a retweet with the time and number of followers of the user (who has retweeted). Markers are divided into three classes based on the number of followers (degree) of each user: (i) a normal user having degree lower than the median, (ii) an influencer having degree higher than or equal to the median and lower than 95 percentile, and (iii) a celebrity having degree higher than or equal to 95 percentile. Similar to the existing protocol [5], we randomly sample 10,000 streams of retweets and apply five-fold cross validation for experiments.

Consistent with the existing research, the following metrics have been used across different datasets:

- **Classification Accuracy/Micro-F1:** Micro-F1 measures the F1-score of the aggregated contributions of all classes. It is also defined as the overall accuracy which is the ratio of correctly classified samples out of all samples.
- **Macro-F1:** For a multi-class classification problem, Macro-Averaged F1-score or Macro-F1 score is defined as the average of F1-scores of each class.

5 Results and Analysis

Tables 2–3 present the performance of the proposed Deviation-based Marked Temporal Point Process model and other comparative techniques. Figures 4–5 present the analysis performed on the DMTPP based model. The following paragraphs elaborate the results and analysis of the proposed model:

---

[2] http://snap.stanford.edu/seismic/.

Table 3. Marker prediction performance on the Retweet Dataset for predicting the type of retweet. Owing to the same protocol, comparative results have directly been taken from the published manuscript [5].

Model	Micro-F1	Macro-F1
Noise-Contrastive Estimation Poisson (NCE-P)	0.52	0.28
Noise-Contrastive Estimation Gaussian (NCE-G)	0.49	0.30
MTPP with Discriminative Loss Function (DIS)	0.49	0.29
Maximum Likelihood Estimation (MLE)	0.50	0.29
Monte Carlo Maximum Likelihood Estimation (MCMLE)	0.49	0.28
INITIATOR [5]	0.57	0.35
Deviation-based Marked TPP	**0.58**	**0.45**

Comparison with State-of-the-Art Algorithms: Table 2 presents the performance on the MIMIC-II and StackOverflow datasets. It can be observed that the proposed technique achieves improved performance as compared to the state-of-the-art algorithm (Transformer Hawkes Process [29]). Due to the same protocol, results have directly been taken from the published manuscript. Specifically, on the MIMIC-II dataset, an improvement of at least 6% is observed from the existing results reported by the Transformer Hawkes Process. Further, comparison has also been made with the Recurrent Marked Temporal Point Process model (which forms the base for the proposed technique), where an improvement of over 10% is obtained. A similar improvement is observed on the StackOverflow dataset as well, wherein, the proposed technique achieves 55.42%, resulting in an improvement of over 8% from the current best results (Transformer Hawkes Process [29]). Further, Table 3 presents the performance obtained on the Retweet dataset, wherein results are reported using the standard Micro F-1 and Macro F-1 metrics. Due to the same protocol, results have directly been taken from the published manuscript [5]. From Table 3 it is observed that the proposed model's performance on both metrics is higher than the state-of-the-art model: the INITIATOR algorithm. Specifically, the proposed model achieves a Micro-F1 and Macro-F1 value of 0.58 and 0.45, respectively. Since the Retweet dataset is characterized by heavy class imbalance in the testing set, the above metrics provide a better understanding of the model performance as compared to traditional classification accuracy. We have also performed the Chi-Squared Statistic Test of Independence [14] on the Retweet dataset to evaluate the statistical association between the results obtained by the proposed model and the state-of-the-art INITIATOR model. A p-value of less than 0.01 is obtained between the Micro-F1 scores, which provides us with sufficient evidence to conclude that the models are disassociated. We believe that the explicit modeling of the dependence of the marker prediction on the time prediction enables the model to learn better features, thus resulting in improved marker prediction performance.

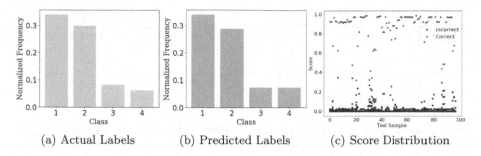

(a) Actual Labels   (b) Predicted Labels   (c) Score Distribution

Fig. 4. (a–b) Bar graphs demonstrating the distribution for the (a) ground-truth and (b) predicted marker labels. Distribution of the top four classes from the MIMIC-II dataset has been plotted. The predicted distribution follows a similar pattern as the ground-truth distribution, thus suggesting that the DMTPP model is able to capture the marker spread well. (c) Score distribution obtained from the DMTPP model on the test events of the MIMIC-II dataset. For each sample the actual (correct) class score and the other (incorrect) class scores obtained via the model have been plotted. For almost all samples, a clear distinction is seen between the correct and incorrect class scores.

Analysis of the proposed Deviation-based Marked TPP Model: Fig. 4(a–b) presents the marker distribution of the (a) ground-truth labels and the (b) labels predicted by the DMTPP model on the MIMIC-II dataset. A similar distribution is observed across the two graphs, thus suggesting that the proposed model is able to learn and simulate the varying occurrence of different marker types. Further, Fig. 4(c) presents the score distribution for different samples of the MIMIC-II dataset, where clear distinction can be observed between the scores of the correct class versus the scores of the incorrect class. A large number of incorrect class scores fall below the range of 0.1 which demonstrates the discriminative nature of the learned classifier. Experiments have also been performed to analyze the different components and hyper-parameters of the Deviation-based Marked TPP model. Discussions regarding different aspects are as follows:

(i) Effect of Sequence Length: Experiments have been performed on the MIMIC-II dataset for understanding the effect of the input sequence length on the model's performance. The sequence length determines the relevance of the length of the user's history for predicting the next marker. We observe that on higher sequence length the model's performance decreases. Specifically, the model achieves 72.3% and 86.1% with a sequence length of 5 and 4, respectively, while achieving 91.76% with sequence length 3. Reducing the length further to 2 results in an accuracy of 88.2%, thus demonstrating a slight drop, while still achieving improved results from the state-of-the-art technique.

(ii) Effect of Weight Hyper-parameters: Experiments have also been performed to understand the effect of the weight hyper-parameters (λ_1 and λ_2 in Eq. 12). Specifically, the MIMIC-II dataset has been used to see the impact of

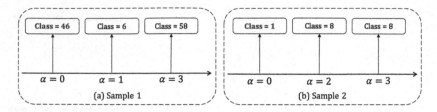

Fig. 5. Sample simulation demonstrating the effect of the deviation parameter for predicting the marker (class). Two sequences are presented where an alpha value (α) was added to the deviation, followed by marker prediction. Variation in the class prediction suggests dependency on the deviation parameter.

varying weights of time loss and marker loss on the model's performance. As mentioned earlier, best performance of 91.76% is achieved with the following combination: ($\lambda_1 = 0.05$ and $\lambda_2 = 0.15$). A drop in performance is observed upon varying the value of λ_1 or λ_2, respectively. For example, a classification accuracy of 90.73% is obtained with the weight pairs (0.05, 1), (0.05, 2), and (0.01, 1). The slight drop in performance suggests the model's robustness to variations in the hyper-parameter selection.

(iii) Performance on Time Prediction: While the aim of the DMTPP model is to perform accurate marker prediction, analysis has also been performed on the MIMIC-II dataset to understand its performance for time prediction. The proposed model obtains a Root Mean Squared Error (RMSE) value of 0.89 for time prediction, which is the second best in comparison to the state-of-the-art performance of 0.82 obtained by the Transformer Hawked Process [29]. Further, the proposed model performs better as compared to the other reported results, specifically, Recurrent Marked Temporal Point Process [3]: 6.12, Neural Hawkes Process [15]: 6.13, Time Series Event Sequence [24]: 4.70, Self-Attentive Hawkes Process [25]: 3.89. The accurate time prediction further supports the high marker prediction performance of the proposed model.

(iv) Effect of Deviation: Experiments have also been performed to understand the effect of the deviation parameter, and whether it contributes to the marker prediction or not. A simulation was performed on the MIMIC-II dataset, where, a small value (α) was added to the deviation obtained after the time prediction. The updated or perturbed deviation value was then provided with the learned embedding for predicting the corresponding marker. Figure 5 presents the output obtained on two sample sequences. In both the cases, the predicted marker was updated when the deviation was changed by an α value. Similar behavior was observed across different sequences as well. The simulation suggests that the DMTPP based model is able to learn the inter-dependence between the user behavior (event occurrence) and the corresponding marker, and updates its prediction based on the variability between the expected and actual event time.

6 Conclusion and Discussion

In various real-world scenarios, the marker information is not immediately known after the occurrence of an event. For example, the impact of an advertisement retweet on increased sales or online traffic generation, intensity of an earthquake, or identifying fraudulent transactions. In such scenarios, the marker information (advertisement impact, earthquake intensity, and fraud event) is known after some time of the event occurrence. It is our hypothesis that in such scenarios, the variation between the expected and actual event occurrence also impacts the corresponding marker. Therefore, in this research, a novel Deviation based Marked Temporal Point Process (DMTPP) model is proposed. The proposed model focuses on learning the dependency between the event time and marker information for predicting accurate markers. The DMTPP model builds upon the existing literature in the field of Marked Temporal Point Processes which has focused majorly on predicting the next event time and corresponding marker information without explicitly modeling the relationship between the two. The efficacy of the proposed model has been demonstrated on three different tasks and datasets (Table 2 and Table 3), where it achieves state-of-the-art performance. Further analysis on the model demonstrates its high performance for time prediction and impact of the deviation component as well. We believe that the research performed in this paper can act as a stepping stone to further explore the possibilities that Temporal Point Processes hold in terms of high accuracy of event type prediction and not just event time prediction. One of the key highlights of the DMTPP model is the requirement of real-time event occurrence (time) for marker prediction, which can further be improved in future algorithms. As part of future work, the proposed DMTPP model can also be extended to incorporate additional meta-information during training which can further boost the marker prediction performance.

References

1. Daley, D.J., Vere-Jones, D.: An Introduction to the Theory of Point Processes: Volume I: Elementary Theory and Methods. Springer, Heidelberg (2003)
2. Diggle, P.J.: Statistical Analysis of Spatial and Spatio-Temporal Point Patterns. CRC Press, Boca Raton (2013)
3. Du, N., Dai, H., Trivedi, R., Upadhyay, U., Gomez-Rodriguez, M., Song, L.: Recurrent marked temporal point processes: embedding event history to vector. In: Proceedings of ACM KDD, pp. 1555–1564 (2016)
4. Grant, S., Betts, B.: Encouraging user behaviour with achievements: an empirical study. In: Proceedings of MSR, pp. 65–68 (2013)
5. Guo, R., Li, J., Liu, H.: Initiator: noise-contrastive estimation for marked temporal point process. In: Proceedings of IJCAI, pp. 2191–2197 (2018)
6. Hawkes, A.G.: Spectra of some self-exciting and mutually exciting point processes. Biometrika **58**(1), 83–90 (1971)
7. Hinton, G.E., Srivastava, N., Krizhevsky, A., Sutskever, I., Salakhutdinov, R.R.: Improving neural networks by preventing co-adaptation of feature detectors. arXiv preprint arXiv:1207.0580 (2012)

8. Hochreiter, S., Schmidhuber, J.: Long short-term memory. Neural Comput. **9**(8), 1735–1780 (1997)
9. Isham, V., Westcott, M.: A self-correcting point process. Stoch. Process. Appl. **8**(3), 335–347 (1979)
10. Ji, Y., et al.: Temporal heterogeneous interaction graph embedding for next-item recommendation. In: Proceedings of ECML-PKDD (2020)
11. Kemp, A.: Poisson Processes. Wiley Online Library (1994)
12. Keogh, E., Chu, S., Hart, D., Pazzani, M.: Segmenting time series: a survey and novel approach. In: Data Mining in Time Series Databases, pp. 1–21 (2004)
13. Kingma, D.P., Ba, J.: Adam: A method for stochastic optimization. arXiv preprint arXiv:1412.6980 (2014)
14. McHugh, M.L.: The chi-square test of independence. Biochem. Med. **23**(2), 143–149 (2013)
15. Mei, H., Eisner, J.: The neural hawkes process: A neurally self-modulating multivariate point process. arXiv preprint arXiv:1612.09328 (2016)
16. Pan, Z., Du, H., Ngiam, K.Y., Wang, F., Shum, P., Feng, M.: A self-correcting deep learning approach to predict acute conditions in critical care. arXiv preprint arXiv:1901.04364 (2019)
17. Paszke, A., et al.: Pytorch: an imperative style, high-performance deep learning library. In: Proceedings of NeurIPS, pp. 8024–8035 (2019)
18. Rasmussen, J.G.: Temporal Point Processes: The Conditional Intensity Function. Lecture Notes, Jan (2011)
19. Shchur, O., Biloš, M., Günnemann, S.: Intensity-free learning of temporal point processes. arXiv preprint arXiv:1909.12127 (2019)
20. Türkmen, A.C., Wang, Y., Smola, A.J.: Fastpoint: scalable deep point processes. In: Proceedings of ECML-PKDD, pp. 465–480 (2019)
21. Verenich, I., Dumas, M., Rosa, M.L., Maggi, F.M., Teinemaa, I.: Survey and cross-benchmark comparison of remaining time prediction methods in business process monitoring. ACM TIST **10**(4), 1–34 (2019)
22. Wang, Y., Liu, S., Shen, H., Gao, J., Cheng, X.: Marked temporal dynamics modeling based on recurrent neural network. In: Proceedings of PAKDD, pp. 786–798 (2017)
23. Wu, W., Yan, J., Yang, X., Zha, H.: Decoupled learning for factorial marked temporal point processes. In: Proceedings of ACM KDD, pp. 2516–2525 (2018)
24. Xiao, S., Yan, J., Yang, X., Zha, H., Chu, S.: Modeling the intensity function of point process via recurrent neural networks. In: Proceedings of AAAI, vol. 31 (2017)
25. Zhang, Q., Lipani, A., Kirnap, O., Yilmaz, E.: Self-attentive hawkes processes. arXiv preprint arXiv:1907.07561 (2019)
26. Zhang, Q., Lipani, A., Kirnap, O., Yilmaz, E.: Self-attentive hawkes process. In: Proceedings of ICML, pp. 11183–11193 (2020)
27. Zhao, L.: Event Prediction in Big Data Era: A Systematic Survey. arXiv preprint arXiv:2007.09815 (2020)
28. Zhao, Q., Erdogdu, M.A., He, H.Y., Rajaraman, A., Leskovec, J.: Seismic: a self-exciting point process model for predicting tweet popularity. In: Proceedings of KDD, pp. 1513–1522 (2015)
29. Zuo, S., Jiang, H., Li, Z., Zhao, T., Zha, H.: Transformer hawkes process. In: Proceedings of ICML, pp. 11692–11702 (2020)

Deep Structural Point Process for Learning Temporal Interaction Networks

Jiangxia Cao[1,2], Xixun Lin[1,2(✉)], Xin Cong[1,2], Shu Guo[3], Hengzhu Tang[1,2], Tingwen Liu[1,2(✉)], and Bin Wang[4]

[1] Institute of Information Engineering, Chinese Academy of Sciences, Beijing, China
{caojiangxia,linxixun,congxin,tanghengzhu,liutingwen}@iie.ac.cn
[2] School of Cyber Security, University of Chinese Academy of Sciences, Beijing, China
[3] National Computer Network Emergency Response Technical Team/Coordination Center of China, Beijing, China
guoshu@cert.org.cn
[4] Xiaomi AI Lab, Xiaomi Inc., Beijing, China
wangbin11@xiaomi.com

Abstract. This work investigates the problem of learning temporal interaction networks. A temporal interaction network consists of a series of chronological interactions between users and items. Previous methods tackle this problem by using different variants of recurrent neural networks to model interaction sequences, which fail to consider the structural information of temporal interaction networks and inevitably lead to sub-optimal results. To this end, we propose a novel **Deep Structural Point Process** termed as **DSPP** for learning temporal interaction networks. DSPP simultaneously incorporates the *topological structure* and *long-range dependency structure* into the intensity function to enhance model expressiveness. To be specific, by using the topological structure as a strong prior, we first design a topological fusion encoder to obtain node embeddings. An attentive shift encoder is then developed to learn the long-range dependency structure between users and items in continuous time. The proposed two modules enable our model to capture the user-item correlation and dynamic influence in temporal interaction networks. DSPP is evaluated on three real-world datasets for both tasks of item prediction and time prediction. Extensive experiments demonstrate that our model achieves consistent and significant improvements over state-of-the-art baselines.

Keywords: Temporal interaction networks · Temporal point process · Graph neural networks

1 Introduction

Temporal interaction networks are useful resources to reflect the relationships between users and items over time, which have been successfully applied in many

N. Oliver et al. (Eds.): ECML PKDD 2021, LNAI 12975, pp. 305–320, 2021.
https://doi.org/10.1007/978-3-030-86486-6_19

real-world domains such as electronic commerce [3], online education [18] and social media [12]. A temporal interaction network naturally keeps a graph data structure with temporal characteristics, where each edge represents a user-item interaction marked with a concrete timestamp.

Representation learning on temporal interaction networks has gradually become a hot topic in the research of machine learning [24]. A key challenge of modeling temporal interaction networks is how to capture the evolution of user interests and item features effectively. Because users may interact with various items sequentially and their interests may shift in a period of time. Similarly, item features are also ever-changing and largely influenced by user behaviours. Recent works have been proposed to tackle this challenge by generating the dynamic embeddings of users and items [7,16,17,21]. Although these methods achieve promising results to some extent, they still suffer from the following two significant problems.

1) **Topological structure missing.** Most previous methods regard learning temporal interaction networks as a coarse-grained sequential prediction problem and ignore the topological structure information. In fact, instead of only treating a temporal interaction network as multiple interaction sequences, we can discover user similarity and item similarity from the view of the topological structure. Nevertheless, due to the bipartite nature of temporal interaction networks, each node is not the same type as its adjacent nodes, so that we have to develop a flexible method to capture such a meaningful topology. 2) **Long-range dependency structure missing.** Most current methods are built upon the variants of recurrent neural networks (RNNs) to learn interaction sequences. Hence, they typically pay more attention to short-term effects and miss the dependency structure in long-range historical information [8,25]. But learning the long-range dependency structure in temporal interaction networks is also critical, since it can better model the long-standing user preference and intrinsic item properties.

In this paper, we propose the **D**eep **S**tructural **P**oint **P**rocess termed as **DSPP** to solve above problems. Following the framework of Temporal Point Process (TPP) [29], we devise a novel intensity function which combines the topological structure and the long-range dependency structure to capture the dynamic influence between users and items. Specifically, we first design a topological fusion encoder (TFE) to learn the topological structure. TFE includes a two-steps layer to encourage each node to aggregate homogeneous node features. To overcome the long-range dependency issue, we then develop an attentive shift encoder (ASE) to recognize the complex dependency between each historical interaction and the new-coming interaction. Finally, we incorporate the learned embeddings from TFE and ASE into our intensity function to make time prediction and item prediction. The main contributions of our work are summarized as follows:

- We propose the novel DSPP to learn temporal interaction networks within the TPP paradigm, with the goal of solving two above structural missing problems simultaneously.

- DSPP includes the well-designed TFE and ASE modules. TFE utilizes the topological structure to generate steady embeddings, and ASE exploits the long-range dependency structure to learn dynamic embeddings. Furthermore, these two types of embeddings can be seamlessly incorporated into our intensity function to achieve future prediction.
- Extensive experiments are conducted on three public standard datasets. Empirical results show that the proposed method achieves consistent and significant improvements over state-of-the-art baselines[1].

2   Related Work

Previous studies for learning temporal interaction networks can be roughly divided into the following three branches: random walk based method, RNN based methods and TPP based method.

- Random walk based method. Nguyen *et al.* propose CTDNE [21] which models user and item dynamic embeddings by the random walk heuristics with temporal constraints. CTDNE first samples some time increasing interaction sequences, and then learns context node embeddings via the skip-gram algorithm [20]. However, it ignores the useful time information of the sampled sequences, e.g., a user clicks an item frequently may indicate that the user pays more attention to this item at the moment.
- RNN based methods. LSTM [11], RRN [27], LatentCross [2] and Time-LSTM [30] are pioneering works in this branch. For example, RRN provides a unified framework for combining the static matrix factorization features with the dynamic embeddings based on LSTM. Moreover, it provides a behavioral trajectory layer to project user and item embeddings over time. LatentCross is an extension of the architecture of GRU [5], which incorporates multiple types of context information. Time-LSTM develops the time gates for LSTM for modeling the interaction time information. Furthermore, JODIE [16] and DGCF [17] are the state-of-the-art methods for learning temporal interaction networks via the coupled variants of RNNs. JODIE defines two-steps embedding update operation and an embedding projection function to predict the target item embedding for each user directly. DGCF extends JODIE by considering the 1-hop neighboring information of temporal interaction networks.
- TPP based method. DeepCoevolve [7] is a promising work that applies TPPs to learn temporal interaction networks. It uses a multi-dimensional intensity function to capture the dynamic influence between users and items. However, DeepCoevolve maintains the same embeddings of user and item until it involves a new interaction, which is not consistent with real-world facts [16]. Furthermore, it uses a linear intensity function to describe the dynamic influence, leading to the limited model expressiveness.

---

[1] The source code is available from https://github.com/cjx96/DSPP.

3   Background

3.1   Temporal Interaction Network

A series chronological interactions can be represented as a temporal interaction network. Formally, a temporal interaction network on the time window $[0, T)$ can be described as $\mathcal{G}(T) = (\mathcal{U}, \mathcal{V}, \mathcal{E})$, where $\mathcal{U}$, $\mathcal{V}$ and $\mathcal{E}$ denote the user set, item set and interaction set, respectively. Each element $(u_i, v_j, t) \in \mathcal{E}$ is an interaction, describing that the user $u_i \in \mathcal{U}$ conducts an action with the item $v_j \in \mathcal{V}$ at the concrete timestamp t.

3.2   Temporal Point Process

TPPs are one of branches of stochastic processes for modeling the observed random discrete events, e.g. user-item interactions over time. Using conditional intensity function $\lambda(t)$ is a convenient way to describe TPPs. Given a interaction sequence that only has time information $\mathcal{T} := \{t_i\}_{i=1}^n$ and an infinitesimal time interval dt, where $t_i \in \mathbb{R}^+$ and $0 < t_1 < t_2 ... < t_n$, $\lambda(t)dt$ is the conditional probability of happening an interaction in the infinitesimal interval $[t, t + dt)$ based on $\mathcal{T}$. It can be also interpreted heuristically in the following way:

$$\lambda(t)dt := \mathbb{P}\{\text{an interaction occurs in } [t, t+dt)|\mathcal{T}\} = \mathbb{E}[N([t, t+dt))|\mathcal{T}],$$

where $N([t, t+dt))$ is used to count the number of interactions happened in the time interval $[t, t + dt)$. A general assumption made here is that there is either zero or one interaction happened in this infinitesimal interval, i.e., $N([t, t + dt)) \in \{0, 1\}$. Furthermore, given a future timestamp $t^+ > t_n$ and $\mathcal{T}$, we can formulate the conditional probability of no interactions happened during $[t_n, t^+)$ as $S(t^+) = \exp\left(-\int_{t_n}^{t^+} \lambda(t)dt\right)$. Therefore, the conditional probability density of the next interaction happened at the future timestamp t^+ can be defined as: $f(t^+) = S(t^+)\lambda(t^+)$, which means that no interaction happens in time interval $[t_n, t^+)$ and an interaction happens in infinitesimal interval $[t^+, t^+ + dt)$.

Multivariate Hawkes Process (MHP) is one of the most important TPPs for modeling interaction sequences [9]. We denote $\mathcal{S}_{u_i}(T) = (\mathcal{V}, \mathcal{H}_{u_i})$ as an interaction sequence of user u_i on the time window $[0, T]$, where $\mathcal{H}_{u_i}$ is the interaction sequence of user u_i. The h-th interaction of $\mathcal{H}_{u_i}$ is denoted as (v^h, t^h), which describes that the user $u_i \in \mathcal{U}$ has interacted with the item $v^h \in \mathcal{V}$ at $t^h \leq T$. The intensity function of an arbitrary item v_j in $\mathcal{S}_{u_i}(T)$ is defined as:

$$\lambda_{v_j}(t) = \mu_{v_j} + \sum_{t^h < t} \alpha_{(v_j, v^h)} \kappa(t - t^h),$$

where μ_{v_j} (a.k.a base intensity) is a positive parameter which is independent of the interaction sequence $\mathcal{S}_{u_i}(T)$, $\alpha_{(v_j, v^h)}$ is also a positive parameter that estimates the influence between item pair (v_j, v^h) and the $\kappa(t - t^h)$ is a triggering kernel function. The intensity function explicit models dynamic influence among interaction sequence. However, most existing approaches ignore to model the

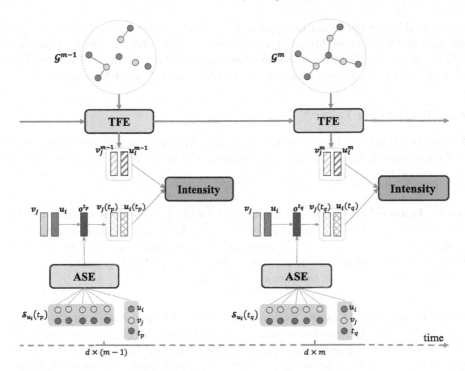

Fig. 1. A simple overview of DSPP. "TFE", "ASE", "Intensity" mean topological fusion encoder, attentive shift encoder and intensity function, respectively. $\mathcal{G}^{m-1}$ and $\mathcal{G}^m$ are two network snapshots. Noticeably, $\mathcal{G}^{m-1} \in \mathcal{G}^m$. (u_i, v_j, t_p), (u_i, v_j, t_q) denote two interactions, where $t_p \in [d \times (m-1), d \times m)$ and $t_q \in [d \times m, d \times (m+1))$. $\mathcal{S}_{u_i}(t_p)$ and $\mathcal{S}_{u_i}(t_q)$ are two interaction sequences.

topological structure between a series of interaction sequences. To fill this gap, DSPP includes a novel TFE module which provides a strong structure prior to enhance model expressiveness.

4   Proposed Model

4.1   Overview

For a temporal interaction network $\mathcal{G}(T) = (\mathcal{U}, \mathcal{V}, \mathcal{E})$, the adjacency matrix would change over time, because the emerge of a new interaction would introduce a new edge in the temporal interaction network, causing the huge occupation of memory. To sidestep this problem, we exploit an ordered snapshot sequence $\{\mathcal{G}^m\}_{m=0}^{M-1}$ with the same time interval $d = \frac{T}{M}$ to simulate the temporal interaction network $\mathcal{G}(T)$. Each snapshot $\mathcal{G}^m$ equals to $\mathcal{G}(d \times m)$, and M is a hype-parameter to control the number of snapshot.

Figure 1 shows the overview of our model. In DSPP, the TFE module aims to learn **steady embeddings** which represent the stable intentions of users and items in the corresponding time period $[d \times m, d \times (m+1))$ from $\mathcal{G}^m$. Here we denote the steady embeddings of an arbitrary user u_i and an item v_j as $\boldsymbol{u}_i^m$ and $\boldsymbol{v}_j^m$, respectively. The second module ASE aims to learn **dynamic embeddings** of users and items for describing their dynamic intentions at timestamp t.

4.2   Embedding Layer

The embedding layer is used to initialize node embeddings (all users and items) and time embedding. It consists of two parts: the node embedding layer and the time embedding layer.

Node Embedding Layer. The node embedding layer aims to embed users and items into a low-dimensional vector space. Formally, given a user u_i or an item v_j, we can obtain its D-dimensional representation ($\boldsymbol{u}_i \in \mathbb{R}^D$ or $\boldsymbol{v}_j \in \mathbb{R}^D$) from an initialization embedding matrix with a simple lookup operation, where D denotes the dimension number of embeddings.

Time Embedding Layer. Position embedding [13,22] is widely used to recognize the ordered information of sequences. However, the continuous time information of interaction sequence cannot be well reflected by the discrete position embedding. Therefore, we design a time embedding layer that encodes the discrete ordered information and continuous time information simultaneously. Concretely, given a future timestamp $t^+ \geq t$, user u_i and the h-th interaction (v^h, t^h) of its interaction sequence $\mathcal{S}_{u_i}(t)$ (detailed in Sect. 3.2), the h-th interaction time embedding $\boldsymbol{p}_{t^h}(t^+)$ of future timestamp t^+ can be formulated as follows:

$$[\boldsymbol{p}_{t^h}(t^+)]_j = \begin{cases} \cos(\omega_j(t^+ - t^h) + h/10000^{\frac{j-1}{D}}), & \text{if } j \text{ is odd,} \\ \sin(\omega_j(t^+ - t^h) + h/10000^{\frac{j}{D}}), & \text{if } j \text{ is even,} \end{cases}$$

where $[\boldsymbol{p}_{t^h}(t^+)]_j$ is the j-th element of the given vector $\boldsymbol{p}_{t^h}(t^+)$, and ω_j is a parameter to scale the time interval $t^+ - t^h$ in the j-th dimension.

4.3   Topological Fusion Encoder

In this section, we propose a novel topological fusion encoder (TFE) to learn the topological structure. The existing graph encoders [6,23] learn node embeddings by aggregating the features of neighboring nodes directly. However, in temporal interaction networks, it would easily lead to improper feature aggregations due to the fact that the user neighbors are items. In this paper, we introduce a topological aggregation layer (TAL) into TFE to alleviate this issue.

Topological Aggregation Layer. Different with homogeneous graphs, the distance between a user (item) and other users (items) is always an even number in temporal interaction network, e.g. 2, 4 and 6. This fact indicates that our encoder should aggregate homogeneous information through the even-number-hop. Based on this topological structure, we design a novel topological aggregation layer (TAL) as shown in Fig. 2. Concretely, given an interaction (u_i, v_j, t), we firstly compute its corresponding snapshot identifier $m = \lfloor \frac{t}{d} \rfloor$, where $\lfloor \cdot \rfloor$ denotes the floor function and d is the time interval of ordered snapshot sequence. Then, to generate the user representation $\boldsymbol{u}_i^k$ in the k-th TAL, we calculate its intermediate representation $\widehat{\boldsymbol{v}}_c^k$ as follows:

$$\widehat{\boldsymbol{v}}_c^k = \delta\Big(\widehat{W}_u^k \, \text{MEAN}(\{\boldsymbol{u}_q^{k-1} : u_q \in \mathcal{N}_m(v_c)\})\Big), \text{where } v_c \in \mathcal{N}_m(u_i), \tag{1}$$

where δ is the ReLU activity function, $\widehat{W}_u^k$ is a parameter matrix, and $\mathcal{N}_m(v_c)$ denotes the set of 1-hop neighbors (user-type) of item v_c in $\mathcal{G}^m$. Therefore, $\widehat{\boldsymbol{v}}_c^k$ can be consider as a user-type representation since it only aggregates the user-type features. After obtaining the intermediate representation $\widehat{\boldsymbol{v}}_c^k$, we leverage the attention mechanism [25] to learn different weights among the neighboring intermediate representations for user u_i. The final embedding $\boldsymbol{u}_i^k$ can be formulated as:

$$e_c = \delta\big((\overline{W}_u^k \boldsymbol{u}_i^{k-1})^\top \widehat{\boldsymbol{v}}_c^k\big), \text{where } v_c \in \mathcal{N}_m(u_i),$$

$$\alpha_c = \frac{\exp(e_c)}{\sum_{v_q \in \mathcal{N}_m(u_i)} \exp(e_q)},$$

$$\overline{\boldsymbol{u}}_i^k = \delta\big(\sum_{v_c \in \mathcal{N}_m(u_i)} \alpha_c \widehat{\boldsymbol{v}}_c^k\big),$$

$$\boldsymbol{u}_i^k = W_u^k \,[\overline{\boldsymbol{u}}_i^k | \boldsymbol{u}_i^{k-1}], \tag{2}$$

where the $\overline{W}_u^k$ and W_u^k are parameter matrices, $(\cdot)^\top$ is the transpose operation and $[\cdot|\cdot]$ is concatenation operation. Analogously, we can employ the same learning procedure to update $\boldsymbol{v}_j^{k-1}$. In TFE, we stack K TALs and denote the final outputs of $\boldsymbol{u}_i^K$ and $\boldsymbol{v}_j^K$ as our steady embeddings $\boldsymbol{u}_i^m$ and $\boldsymbol{v}_j^m$ for user u_i and item v_j in $\mathcal{G}^m$, respectively.

Temporal Fusion Layer. The proposed TAL can effectively deal with a single network snapshot, but it cannot capture the structural variations across the ordered snapshot sequence $\{\mathcal{G}^0, \mathcal{G}^1, ..., \mathcal{G}^{M-1}\}$. To mitigate this problem, after obtaining user and item embeddings (e.g. $\boldsymbol{u}_i^m$ and $\boldsymbol{v}_j^m$) for each discrete snapshot $\mathcal{G}^m$, we introduce a temporal fusion layer to encode these dynamical changes in the ordered snapshot sequence:

$$\boldsymbol{u}_i^m = f_u(\boldsymbol{u}_i^{m-1}, \boldsymbol{u}_i^m), \quad \boldsymbol{v}_j^m = f_v(\boldsymbol{v}_j^{m-1}, \boldsymbol{v}_j^m), \tag{3}$$

where f_u and f_v are temporal modeling functions. There are many alternative methods that can be used for concrete implementations. In our model, we choose two separate GRUs [5] to model f_u and f_v, respectively.

4.4   Attentive Shift Encoder

In this section, we develop an attentive shift encoder (ASE) for temporal interaction networks for capturing the long-range dependency structure. Previous works

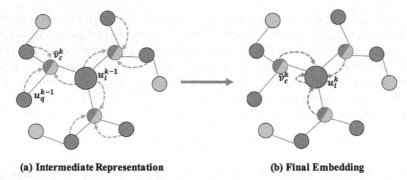

(a) Intermediate Representation   **(b) Final Embedding**

Fig. 2. Illustration of topological aggregation layer (TAL). Blue and yellow color nodes denote users and items, respectively. Nodes with two colors denote the intermediate representations, and gray lines denote that users have interacted with items. The subgraphs (a) and (b) show the learning procedures of u_i^{k-1} in k-th TAL. The green dotted lines (Eq. (1)) and orange dotted lines (Eq. (2)) describe how to derive the embedding u_i^k by considering the topological structure of temporal interaction network. (Color figure online)

employ different RNN variants which tend to forget the long history information, leading to the problem of long-range dependency structure missing. In contrast, our ASE module can explicitly learn the dependencies between each historical interaction and the new-coming interaction via the attention mechanism.

Attentive Interaction Layer. Considering a new-coming interaction (u_i, v_j, t) that user u_i has an interaction with item v_j at the timestamp t, we can use it to generate the dynamic embeddings of users and items and compute the correlation among historical interactions in $\mathcal{S}_{u_i}(t)$. The concrete implementation is given as:

$$e_h = \left(W_Q[\boldsymbol{u}_i|(\boldsymbol{v}_j + \boldsymbol{p}_{t|\mathcal{H}_{u_i}|}(t))]\right)^\top W_K[\boldsymbol{u}_i|(\boldsymbol{v}^h + \boldsymbol{p}_{t^h}(t))], \text{where } (v^h, t^h) \in \mathcal{S}_{u_i}(t)$$

$$\alpha_h = \frac{\exp(e_h)}{\sum_{(v^c, t^c) \in \mathcal{S}_{u_i}(t)} \exp(e_c)}, \tag{4}$$

$$\boldsymbol{o}^t = \delta\left(\sum_{(v^h, t^h) \in \mathcal{S}_{u_i}(t)} \alpha_h W_V \boldsymbol{v}^h\right),$$

where $|\mathcal{H}_{u_i}|$ is the number of interaction sequence $\mathcal{S}_{u_i}(t)$, and $\boldsymbol{o}^t$ is the new-coming interaction feature. W_Q, W_K and W_V are the query, key and value parameter matrices, respectively. Afterwards, we generate the embeddings of user u_i and item v_j at timestamp t via the following operations:

$$\boldsymbol{u}_i(t) = g_u(\boldsymbol{u}_i, \boldsymbol{o}^t), \quad \boldsymbol{v}_j(t) = g_v(\boldsymbol{v}_j, \boldsymbol{o}^t), \tag{5}$$

where g_u and g_v are embedding generation functions. In our model, we also use two separate GRUs for their implementations.

Temporal Shift Layer. Intuitively, the embeddings of user and item should be changed over time. For example, electronic products will gradually reduce their prices over time, and users may have different intentions when they returned to the E-commerce platform again. Hence, maintaining the same embeddings in a period cannot reflect the reality for the future prediction [2]. In our work, we devise a temporal shift layer to achieve dynamic embeddings over time. Specifically, after obtaining the embeddings of user u_i and item v_j at timestamp t, i.e., $u_i(t)$ and $v_j(t)$ in Eq. (5), their dynamic embeddings at future timestamp $t^+ \geq t$ can be calculated as follows:

$$u_i(t^+) = (1 + \Delta w_{u_i}) * u_i(t), \quad v_j(t^+) = (1 + \Delta w_{v_j}) * v_j(t), \tag{6}$$

where $\Delta = t^+ - t$ is the shift time interval, $*$ is the element-wise product, w_{u_i} and w_{v_j} are corresponding learnable shift vectors of user u_i and item v_j, respectively. We assume that the user or item embedding can shift in continuous space with its own trajectory, so each user or item has a specific shift vector.

4.5   Model Training

To explicitly capture dynamic influence between users and items, we devise a novel intensity function which is generated via the steady embeddings and dynamic embeddings.

Intensity Function. We model all possible interactions for all users with items via a multi-dimensional intensity function, where each user-item pair holds one dimension. Formally, based on the learned user and item embeddings, the intensity function of user-item pair (u_i, v_j) is defined as follows:

$$\lambda_{(u_i,v_j)}(t) = \sigma\Big(\underbrace{(u_i^{\lfloor \frac{t}{d} \rfloor})^\top v_j^{\lfloor \frac{t}{d} \rfloor}}_{\substack{\text{base intensity} \\ \text{(TFE)}}} + \underbrace{(u_i(t))^\top v_j(t)}_{\substack{\text{dynamic change} \\ \text{(ASE)}}} \Big), \tag{7}$$

where $\lfloor \frac{t}{d} \rfloor$ denotes the corresponding network snapshot identifier and σ is the softplus function for ensuring that the intensity function is positive and smooth. Our intensity function is similar with MHP (detailed in Sect. 3.2): 1) The former term $(u_i^{\lfloor \frac{t}{d} \rfloor})^\top v_j^{\lfloor \frac{t}{d} \rfloor}$ is provided by TFE, which uses the topological structure of temporal interaction network as a strong prior to generate the base intensity. 2) The latter term $(u_i(t))^\top v_j(t)$ is obtained by ASE, which describes the dynamic changes for this user-item pair.

Objective Function. Based on the proposed intensity function, we can train our model by maximizing the log-likelihood of these happened interactions during time window $[0, T)$:

$$\mathcal{L} = \sum_{(u_i,v_j,t) \in \mathcal{G}(T)} \log(\lambda_{(u_i,v_j)}(t)) - \underbrace{\int_0^T \lambda(t) \mathrm{d}t}_{\text{non-happened interactions}}, \tag{8}$$

Algorithm 1. The training procedure of DSPP.

Input: The training temporal interaction network $\mathcal{G}(T_{tr})$, the ordered snapshot sequence $\{\mathcal{G}^0, \mathcal{G}^1, ..., \mathcal{G}^{M-1}\}$, the time interval d, the user set $\mathcal{U}$, the item set $\mathcal{V}$, sampling number N.

1: Initialize model parameters.
2: **while** not convergence **do**
3:    Enumerate a batch of consecutive interactions from $\mathcal{G}(T_{tr})$ as B.
4:    $\nabla \leftarrow 0$ \\ Happened interactions.
5:    $\Lambda \leftarrow 0$ \\ Non-happened interactions.
6:    **for** each interaction $(u_i, v_j, t) \in B$ **do**
7:       $m \leftarrow \lfloor \frac{t}{d} \rfloor$ \\ Calculate the snapshot identifier.
8:       Calculate steady embedding $\boldsymbol{u}_i^m$ and $\boldsymbol{v}_j^m$ via TFE based on $\mathcal{G}^m$.
9:       Calculate dynamic embedding $\boldsymbol{u}_i(t)$ and $\boldsymbol{v}_j(t)$ by ASE.
10:       $\nabla \leftarrow \nabla + \log(\lambda_{(u_i, v_j)}(t))$
11:       **if** u_i has next interaction at future timestamp t^+ with item v_c **then**
12:          $\nabla \leftarrow \nabla + \log(\lambda_{(u_i, v_c)}(t^+))$
13:          Uniformly sample a timestamp set $t^s = \{t_k^s\}_{k=1}^N \leftarrow \text{Uniform}(t, t^+, N)$.
14:          Sample the negative item set Υ \\ Negative sampling.
15:          **for** $k \in \{2, ..., N\}$ **do**
16:             $\Lambda \leftarrow \Lambda + (t_k^s - t_{k-1}^s)\lambda(t_k^s)$ \\ Monte Carlo estimation.
17:          **end for**
18:       **end if**
19:    **end for**
20:    $\mathcal{L} \leftarrow \nabla - \Lambda$
21:    Update the model parameters by Adam optimizer.
22: **end while**

$$\lambda(t) = \sum_{u_i \in \mathcal{U}} \sum_{v_j \in \mathcal{V}} \lambda_{(u_i, v_j)}(t). \qquad (9)$$

Maximizing the likelihood function $\mathcal{L}$ can be interpreted intuitively in the following way: 1) The first term ensures that all happened interactions probabilities are maximized. 2) The second term penalizes the sum of the log-probabilities of infinite non-happened interactions, because the probability of no interaction happens during $[t, t + dt)$ is $1 - \lambda(t)dt$, and its log form is $-\lambda(t)dt$ [19].

Prediction Tasks. Beneficial from TPP framework, DSPP can naturally tackle the following two tasks:

- Item prediction: Given user u_i and a future time t^+, *what is the item that this user will interact at time t^+?* To answer this question, we rank all items and recommend the one that has the maximum intensity:

$$\text{argmax}_{v_j} \frac{\lambda_{(u_i, v_j)}(t^+)}{\sum_{v_c \in \mathcal{V}} \lambda_{(u_i, v_c)}(t^+)}, \qquad (10)$$

- Time prediction: Given the user u_i, item v_j and timestamp t_n, *how long will this user interact with this item again after timestamp t_n?* To answer this question, we estimate the following time expectation:

$$\Delta = \int_{t_n}^{\infty} (t - t_n) f_{(u_i, v_j)}(t) \mathrm{d}t, \tag{11}$$

where $f_{(u_i, v_j)}(t) = S_{(u_i, v_j)}(t) \lambda_{(u_i, v_j)}(t)$ is the conditional density (details in Sect. 3.2) and Δ is the expectation interaction time.

4.6 Model Analysis

Differences with Sequential Recommendation. Sequential recommendation methods [4,10,13,28] also focus on modeling sequential user preferences. Compared with them, DSPP has the following fundamental differences:

- In the task level, DSPP concentrates on modeling the dynamic evolution of users and items in continuous time. DSPP can not only predict the next item, but also explicitly estimates the time expectation of a given user-item interaction. In contrast, sequential recommendation aims to model interaction sequences in the discrete manner. Thus, most of them ignore the timestamp information and cannot model the time distribution.
- In the model level, DSPP simultaneously captures the topological structure and the long-range dependency structure via our TFE and ASE modules, but sequential recommendation methods usually ignore the topology information in temporal interaction networks.

Time Complexity. To accelerate the training process of DSPP, we adopt *t-batch* algorithm [16] to organize data for paralleling training. Moreover, we apply Monte Carlo Algorithm [1] with the negative sampling trick [7] to estimate our objective function Eq. (8). Hence, the main operations of DSPP fall into the proposed TFE and ASE modules. The computational complexity of TFE is $\mathcal{O}(K|\mathcal{E}|D)$, and the ASE is $\mathcal{O}(HBD)$, where K is the number of TAL, B is the batch size, and H is a hype-parameter to control the maximum length of historical interactions. In general, our model keeps an efficient training speed. Empirically, in the same running environment, JODIE [16], DGCF [17] and DSPP would cost about 5.1 min, 17.7 min, and 8.15 min per epoch on the Reddit dataset, respectively. The pseudo code of the training procedure is shown in Algorithm 1.

Table 1. Statistics of three datasets.

Datasets	$|\mathcal{U}|$	$|\mathcal{V}|$	Interactions	Action repetition
Reddit	10,000	1,000	672,447	79%
Wikipedia	8,227	1,000	157,474	61%
Last.FM	1,000	1,000	1,293,103	8.6%

5   Experiments

5.1   Datasets

To make a fair comparison, we evaluate DSPP on three pre-processed benchmark datasets [16], i.e., Reddit[2], Wikipedia[3] and Last.FM[4]. The concrete statistics of users, items, interactions and action repetition are listed in Table 1. Noticeably, these datasets are largely different in terms of action repetition rate, which can verify whether DSPP is able to capture the dynamic influence in various action repetition scenarios accurately.

5.2   Experiment Setting

Data Preprocessing: As used in JODIE [16] and DGCF [17], for each dataset, we first sort all interactions by chronological order. Then, we use the first 80% interactions to train, the next 10% interactions to valid, and the remaining 10% interactions for the test. In contrast with JODIE and DGCF, we generate a snapshot sequence $\{\mathcal{G}^m\}_{m=0}^{M-1}$. In our setting, the validation snapshots cover the training data, and the test snapshots also contain all training and validation data.

Evaluation Metrics: To evaluate our model performance, for each interaction, we first generate corresponding user and item steady and dynamic embeddings. Then, we rank all items by Eq. (10) and predict the future time by Eq. (11). Afterward, we evaluate the item prediction task with the following metrics: Mean Reciprocal Rank (MRR) and Recall@10. higher values for both metrics are better. For the time prediction task, we use Root Mean Square Error (RMSE) to measure model performance, and a lower value for RMSE is preferred.

Baselines: We compare DSPP with the following baselines.

- Random walk model: CTDNE [21].
- Recurrent network models: LSTM [11], RRN [27], LatentCross [2], Time-LSTM [30], JODIE [16] and DGCF [17].
- Temporal point process model: DeepCoevolve [7].

Implementation Details: In our experiments, we use the official implementations[5] of DeepCoevolve. Except from it, we directly report the experimental results in the original papers [16,17]. DSPP follows the same hyper-parameter setting with baselines: the embedding dimension D is fixed as 128, the batch size B is fixed as 128, the learning rate is fixed as 0.001, the model weight decay is fixed as 0.00001, the sampling number for Monte Carlo estimate is fixed as 64, the number of negative sampling is fixed as 10, the number of TAL K is

---

[2] http://snap.stanford.edu/jodie/reddit.csv.
[3] http://snap.stanford.edu/jodie/wikipedia.csv.
[4] http://snap.stanford.edu/jodie/lastfm.csv.
[5] https://hanjun-dai.github.io/supp/torch_coevolve.tar.gz.

Table 2. Performance (%) comparison of item prediction.

Model	Last.FM		Wikipedia		Reddit	
	Recall@10	MRR	Recall@10	MRR	Recall@10	MRR
CTDNE	1.0	1.0	5.6	3.5	25.7	16.5
LSTM	12.7	8.1	45.9	33.2	57.3	36.7
Time-LSTM	14.6	8.8	35.3	25.1	60.1	39.8
RRN	19.9	9.3	62.8	53.0	75.1	60.5
LatentCross	22.7	14.8	48.1	42.4	58.8	42.1
DeepCoevolve	33.6	21.3	60.6	48.5	78.7	65.4
JODIE	38.7	23.9	82.1	74.6	85.1	72.4
DGCF	45.6	32.1	85.2	78.6	85.6	72.6
DSPP	**47.1**	**34.3**	**90.5**	**82.1**	**86.7**	**74.5**

fixed as 2, the Attention is stacked 8 layers, the GRUs are 1 layer, the number of snapshots M is selected from $\{128, 256, 512, 1024\}$ and the maximum length of interaction sequence H is chosen from $\{20, 40, 60, 80\}$. The Adam [14] optimizer is used to update all model parameters.

5.3 Item Prediction

For item prediction, Table 2 shows the comparison results on the three datasets according to Recall@10 and MRR. From the experimental results, we have the following observations:

- DSPP consistently yields the best performances on all datasets for both metrics. Compared with state-of-the-art baselines, the most obvious effects are that DSPP achieves the 6.8% improvement in terms of MRR on Last.FM, the 6.2% improvement in terms of Recall@10 on Wikipedia and the 2.6% improvement in terms of MRR on Reddit. It reveals that incorporating the topological structure and long-range dependency structure can bring good robustness in different action repetition scenarios.
- DSPP outperforms DeepCoevolve significantly. This phenomenon demonstrates that DSPP has a more powerful intensity function that can better capture the dynamic influence between users and items.
- DSPP and DGCF are superior to other baselines on Last.FM, which indicates that it is critical to model the topological structure information for learning temporal interaction networks.

Table 3. Performance (hour^2) comparison of time prediction.

Model	Last.FM	Wikipedia	Reddit
DeepCoevolve	9.62	10.94	11.07
DSPP	7.78	8.71	9.06

Table 4. Performance (%) comparison of different model variants.

Model	Recall@10	MRR
DSPP	47.1	34.3
Remove TFE	40.2	25.3
Replace TAL with GCN	44.5	32.4
Replace TAL with GAT	45.2	32.7

5.4 Time Prediction

Table 3 shows the prediction performances of DeepCoevolve and DSPP. From it, we can observe that DSPP achieves more accurately time prediction. Specifically, our model achieves the 19.1% improvement on Last.FM, the 20.3% improvement on Wikipedia and the 18.1% improvement on Reddit. We suppose that the improvements owe to the following reasons: 1) DeepCoevolve uses a linear intensity function to model dynamic influence over time, which would reduce the model flexibility of intensity function. 2) DeepCoevolve remains the same user and item embeddings until it involves a new-coming interaction, so it limits model expressiveness. In contrast, our intensity function can learn the nonlinear dynamic influence, since the ASE module can provide time-aware dynamic embeddings.

5.5 Discussion of Model Variants

To investigate the effectiveness of our model components, we implement several variants of DSPP and conduct the experiment on Last.FM dataset for the task of item prediction. The experimental results are reported in Table 4. According to it, we can draw the following conclusions:

– Remove TFE. To verify whether our proposed TFE module is useful to enhance the expressiveness of the intensity function, we first remove it and only remain the second term of Eq. (7) as our intensity function. Then, we directly predict the item that is most likely to interact with each user via Eq. (10). As shown in the results, both metrics Recall@10 and MRR sharply drop 14.6% and 26.2%, respectively. It demonstrates that modeling the topological structure of temporal interaction networks can provide a powerful structural prior for enhancing the expressiveness of intensity function.

- Remove TFE. This variant can be also viewed as a non-graph based model, since it does not exploit the topological structure, and the remaining temporal attention shift encoder only provides the long-range dependency structure to model intensity function. Compared with DeepCoevolve, this variant yields 19.6% and 18.7% improvements on Recall@10 and MRR respectively. This observation shows that our proposed temporal attention shift encoder can further enhance the intensity function.
- Replace TAL with GCN/GAT. To verify whether our proposed TAL is superior to other graph encoders for capturing the topological structure of temporal interaction networks. We replace TAL by GCN [15] and GAT [26]. For the GCN variant, both Recall@10 and MRR drop 5.8%. For GAT variant, Recall@10 and MRR drop 4.0% and 4.6%, respectively. So, We suppose that our proposed TAL can better capture the information of the same type entity.

6   Conclusion

In this paper, we present the deep structural point process for learning temporal interaction networks. Our model includes two proposed modules, i.e., topological fusion encoder and attentive shift encoder to learn the topological structure and the long-range dependency structure in temporal interaction networks, respectively. On top of that, a novel intensity function, which combines the learned steady and dynamic embeddings, is introduced to enhance the model expressiveness. Empirically, we demonstrate the superior performance of our model on various datasets for both tasks of item prediction and time prediction.

Acknowledgement. This work is supported in part by the Strategic Priority Research Program of Chinese Academy of Sciences (Grant No. XDC02040400) and the Youth Innovation Promotion Association of CAS (Grant No. 2021153).

References

1. Aalen, O., Borgan, O., Gjessing, H.: Survival and Event History Analysis: A Process Point of View. Springer, Heidelberg (2008)
2. Beutel, A., et al.: Latent cross: making use of context in recurrent recommender systems. In: WSDM (2018)
3. Cao, J., Lin, X., Guo, S., Liu, L., Liu, T., Wang, B.: Bipartite graph embedding via mutual information maximization. In: WSDM (2021)
4. Chen, X., et al.: Sequential recommendation with user memory networks. In: WSDM (2018)
5. Cho, K., et al.: Learning phrase representations using RNN encoder-decoder for statistical machine translation. arXiv (2014)
6. Da, X., Chuanwei, R., Evren, K., Sushant, K., Kannan, A.: Inductive representation learning on temporal graphs. In: ICLR (2020)
7. Dai, H., Wang, Y., Trivedi, R., Song, L.: Deep coevolutionary network: embedding user and item features for recommendation. arXiv (2016)
8. Devlin, J., Chang, M.W., Lee, K., Toutanova, K.: BERT: pre-training of deep bidirectional transformers for language understanding. arXiv (2018)

9. Hawkes, A.G.: Spectra of some self-exciting and mutually exciting point processes. Biometrika (1971)
10. Hidasi, B., Karatzoglou, A., Baltrunas, L., Tikk, D.: Session-based recommendations with recurrent neural networks. In: ICLR (2016)
11. Hochreiter, S., Schmidhuber, J.: Long short-term memory. Neural Comput. (1997)
12. Iba, T., Nemoto, K., Peters, B., Gloor, P.A.: Analyzing the creative editing behavior of Wikipedia editors: through dynamic social network analysis. Procedia-Soc. Behav. Sci. (2010)
13. Kang, W.C., McAuley, J.: Self-attentive sequential recommendation. In: ICDM (2018)
14. Kingma, P.D., Ba, L.J.: Adam: a method for stochastic optimization. In: ICLR (2015)
15. Kipf, T.N., Welling, M.: Semi-supervised classification with graph convolutional networks. In: ICLR (2017)
16. Kumar, S., Zhang, X., Leskovec, J.: Predicting dynamic embedding trajectory in temporal interaction networks. In: KDD (2019)
17. Li, X., Zhang, M., Wu, S., Liu, Z., Wang, L., Yu, P.S.: Dynamic graph collaborative filtering. In: ICDM (2020)
18. Liyanagunawardena, T.R., Adams, A.A., Williams, S.A.: MOOCs: a systematic study of the published literature 2008–2012. In: International Review of Research in Open and Distributed Learning (2013)
19. Mei, H., Eisner, J.M.: The neural Hawkes process: a neurally self-modulating multivariate point process. In: NeurIPS (2017)
20. Mikolov, T., Chen, K., Corrado, G., Dean, J.: Efficient estimation of word representations in vector space. arXiv (2013)
21. Nguyen, G.H., Lee, J.B., Rossi, R.A., Ahmed, N.K., Koh, E., Kim, S.: Continuous-time dynamic network embeddings. In: WWW (2018)
22. Sankar, A., Wu, Y., Gou, L., Zhang, W., Yang, H.: DySAT: deep neural representation learning on dynamic graphs via self-attention networks. In: WSDM (2020)
23. Seo, Y., Defferrard, M., Vandergheynst, P., Bresson, X.: Structured sequence modeling with graph convolutional recurrent networks. In: ICONIP (2018)
24. Skarding, J., Gabrys, B., Musial, K.: Foundations and modelling of dynamic networks using dynamic graph neural networks: a survey. arXiv (2020)
25. Vaswani, A., et al.: Attention is all you need. In: NeurIPS (2017)
26. Veličković, P., Cucurull, G., Casanova, A., Romero, A., Liò, P., Bengio, Y.: Graph attention networks. In: ICLR (2018)
27. Wu, C.Y., Ahmed, A., Beutel, A., Smola, A.J., Jing, H.: Recurrent recommender networks. In: WSDM (2017)
28. Xu, C., et al.: Recurrent convolutional neural network for sequential recommendation. In: WWW (2019)
29. Yan, J., Xu, H., Li, L.: Modeling and applications for temporal point processes. In: KDD (2019)
30. Zhu, Y., et al.: What to do next: Modeling user behaviors by time-LSTM. In: IJCAI (2018)

Holistic Prediction for Public Transport Crowd Flows: A Spatio Dynamic Graph Network Approach

Bingjie He[1], Shukai Li[2], Chen Zhang[1(✉)], Baihua Zheng[3], and Fugee Tsung[4]

[1] Industrial Engineering, Tsinghua University, Beijing, China
hebj20@mails.tsinghua.edu.cn, zhangchen01@tsinghua.edu.cn
[2] State Key Laboratory of Rail Traffic Control and Safety, Beijing Jiaotong University, Beijing, China
shkli@bjtu.edu.cn
[3] School of Computing and Information Systems, Singapore Management University, Singapore, Singapore
bhzheng@smu.edu.sg
[4] Industrial Engineering and Decision Analytics, The Hong Kong University of Science and Technology, Kowloon, Hong Kong
season@ust.hk

Abstract. This paper targets at predicting public transport in-out crowd flows of different regions together with transit flows between them in a city. The main challenge is the complex dynamic spatial correlation of crowd flows of different regions and origin-destination (OD) paths. Different from road traffic flows whose spatial correlations mainly depend on geographical distance, public transport crowd flows significantly relate to the region's functionality and connectivity in the public transport network. Furthermore, influenced by commuters' time-varying travel patterns, the spatial correlations change over time. Though there exist many works focusing on either predicting in-out flows or OD transit flows of different regions separately, they ignore the intimate connection between the two tasks, and hence lose efficacy. To solve these limitations in the literature, we propose a Graph spAtio dynamIc Network (GAIN) to describe the dynamic non-geographical spatial correlation structures of crowd flows, and achieve holistic prediction for in-out flows of each region together with OD transit flow matrix between different regions. In particular, for spatial correlations, we construct a dynamic graph convolutional network for the in-out flow prediction. Its graph structures are dynamically learned from the prediction of OD transit flow matrix, whose spatial correlations are further captured via a multi-head graph attention network. For temporal correlations, we leverage three blocks of gated recurrent units, which capture minute-level, daily-level and weekly-level temporal correlations of crowd flows separately. Experiments on real-world datasets are used to demonstrate the efficacy and efficiency of GAIN.

Keywords: Crowd flows prediction · Origin-destination matrix · Dynamic spatial correlation · Public transport system · Graph attention network

© Springer Nature Switzerland AG 2021
N. Oliver et al. (Eds.): ECML PKDD 2021, LNAI 12975, pp. 321–336, 2021.
https://doi.org/10.1007/978-3-030-86486-6_20

1   Introduction

The public transport system is the backbone for human mobility in urban areas. Accurate prediction of public transport flow is greatly important to both system operators and passengers. It allows operators to conduct better train operation planning, detect potential abnormal traffic flows and render fast remedial strategies. It also provides real-time traffic information to passengers for better travel planning. Currently, the Automated Fare Collection System (AFC), widely used in urban public transport network, provides a convenient way for passenger travel data collection. The AFC data (aka smart card data) record passengers' each trip information, including the boarding and alighting time and corresponding stations, and offer big support for crowd flow analysis. Generally, two types of crowd flows are of interest. The first is the in-out flow of each region, which measures the number of passengers entering/leaving the region via public transport systems for each time step. The second is the finer-grain transit flow from each origin region to another destination region, i.e., Origin-Destination (OD) matrix prediction. How to utilize AFC data to predict these two levels of crowd flows has been raising researchers' interest in data mining for intelligent transportation system construction. There are several challenges involved.

The first and most critical is the complex spatial correlation of crowd flow data. Unlike road traffic flows, where the spatial correlations for different regions are based on the "first law of geography", i.e., near things are more related than distant things. However, for public transport systems, the spatial correlations are not fully based on geographical distance, but also the connectivity structure of the public transportation network and the region functionality. For example, though two non-adjacent regions are far away from each other on the map, they could be directly connected by a metro line or located in the same functional regions, and consequently share highly correlated crowd flow patterns. As such, traditional models trying to capture geographically neighboring spatial features are not suitable.

Furthermore, the spatial correlations are time-dynamic. For example, in the morning peak hour, the transport system carries hundreds of thousands of commuters from residential areas to the central business district (CBD), which leads to a high correlation between outflows of residential districts and inflows of CBD. This correlation dies down after the morning peak. In the evening, the outflows of CBD become more related to inflows of commercial regions, since people go for entertainment and leisure after work. At night, the outflows of commercial regions begin to impact inflows of residential regions. However, most current studies assume that spatial correlations are static, and hence lose the prediction accuracy.

Last but foremost, predictions on the two granularities of crowd flows, i.e., in-out flows and OD transit flows, have intimate connections with each other. The sum of passenger flows from one region to others is the outflow of that region. Likewise, destinations of OD transit flows determine the inflow of the corresponding regions. Consequently, accurate prediction of outflows in one region can help predict the transition flows from it to other regions more accurately, vice versa. Consequently, the OD transit flow matrix and in-out flows of regions mutually influence each other and a holistic prediction model with consideration of their connections is expected to have better prediction performance.

To address above issues, in this paper, we consider the connections between in-out flows and OD transit flows of different regions, and develop a genuine holistic framework for crowd flows prediction of the public transport system. In particular, we treat each region as a node in the graph, and propose a dynamic graph-based neural network framework to capture the dynamic spatial correlations of crowd flows of different nodes. First, we formulate a multi-head graph attention network (GAT) block for OD matrix prediction. The GAT model could dynamically leverage features from spatial correlated regions using the attention mechanism, and track the OD transit patterns of different regions accurately. Consider different kinds of spatial correlations may co-exist simultaneously, we apply the multi-head technique. In addition, the learned attention graphs of GAT can be really good representations of spatial correlation structures of different regions. Hence we further use them as the dynamic input graphs of the graph convolution network (GCN) block for in-out flow prediction of different regions. Consequently, the graph structure of GCN is not required to be predefined or dependent on any prior information, but is dynamically learned from the prediction process of the OD transit flow matrix. In this way, the two tasks are intimately interrelated with each other for joint training. Last, data from urban railway transit systems of Hong Kong and Shenzhen validate our proposed methodology. Extensive experiments and comparisons with state-of-the-art methods demonstrate the out-performance of our proposed method.

2 Literature Review

2.1 In-Out Flow Prediction

Generally, crowd flows prediction is referred as in-out flow prediction of a region. It has been extensively studied in many literature works, among which deep learning-based methods are the current mainstream tools since they could effectively model temporal and spatial correlations of crowd flows simultaneously. Various network structures have been proposed and applied to solve different problems.

For temporal correlation description with neural network models, recurrent neural network (RNN) and its variants, e.g. long short-term memory (LSTM) and gated recurrent unit (GRU) have been widely applied. For example, in [1], LSTM units are built to model peak-hour and post-accident traffic state. In [2], the authors extended the fully-connected LSTM to have convolutional structures such that it can handle spatio-temporal data. In [3], a periodically shifted attention mechanism is introduced to handle the long-term periodic temporal shifting. Different methods have demonstrated competitive performances in different data sets.

For spatial correlation description, lines of study adopt convolutional neural network (CNN)-based structures to model nonlinear spatial characteristics [4,5]. They learned traffic flows as heat map images and utilized convolutions to capture spatial correlations. The typical applications include taxi trajectory prediction, bike rent/return prediction, etc., where geographically nearby regions are important to help predict the target region. However, CNN is not suitable for public transport crowd flows data where correlation structures of flows between two regions are generally not only related to their geographical distance, but also a lot of other factors, such as spatial structures of the transportation network. Substantial research generalized the convolution operator to non-Euclidean data [6]. Among them, Graph Convolutional Neural Network (GCN) is a significant stride [7,8]. GCN-based methods assume each region as a node in the graph, and spatial correlations between different regions are denoted as edge weights between nodes. GCN has been an appealing choice for public transport flow forecasting, where the graph is defined based on station connectivity, geograph attributes, contextual features (point-of-interest) [7], flow profile similarity, etc. One limitation of these works is that the prediction performance is greatly influenced by the pre-defined graph. Yet how to choose these various kinds of graphs case-by-case is a practically difficult problem depending on the specific application purpose. It's also difficult to evaluate which kind of graph is better, and some specific types of graphs are not suitable in general cases. There are also some attempts using attention strategy [8]. However, they still rely on pre-defined graph structures.

Last but foremost, all the above methods only predict the in-out flows of different regions, yet ignore OD transit flows between different regions.

2.2 OD Transit Flow Matrix Prediction

OD matrix forecasting aims at predicting transit flows between different regions. In [9], the authors proposed a contextualized spatial-temporal network, which incorporated diverse contextual information to predict taxi OD demand. In [10], the authors formulated the OD matrix together with other geographical features as tensors and developed a multi-scale convolutional LSTM for predicting future OD traffic demand. In [11], Multi-Perspective Graph Convolutional Networks (MPGCN) with LSTM is proposed to extract temporal features for OD matrix prediction. In [12], a matrix factorization-embedded graph CNN is proposed for road OD matrix prediction.

Yet these methods only consider OD matrix prediction, without taking in-out flow prediction into account. In [13], the authors first considered multi-task learning of OD matrix and in-out flows together. It developed a grid-embedding based multi-task learning framework to predict OD passenger demands, together with in-out flows of each region. Yet the two tasks are merely added together as one objective function without any information sharing. In [14], the authors further proposed a better information fusion framework. It first designed two separate CNN modules to extract features of OD matrix and in-out flows. Then the two tasks' features were concatenated together in a fusion module. However, this simple concatenation does not consider task differences carefully and does not design which features are shared and which ones should be task-specific. In [15], an adversarial network is proposed for OD matrix and in-out flow prediction. It adopts a shared-private framework which contains both private and shared spatial-temporal encoders and decoders. A discriminative loss on task classification and an adversarial loss on shared feature extraction are incorporated to reduce information redundancy. However, these methods target at road traffic flow prediction and assume neighboring regions are more correlated. Consequently, they are not suitable for public transport flow prediction. And none of them consider dynamic spatial correlations and hence have limited prediction power. Last, these models define one region's inflow (outflow) as directed sum of OD transit flows over all the destination (origin) regions. This indicates only OD trips completed in the same time step are considered for counting in-out flows, and the learned model is forced to capture the concurrent spatial correlations between in-out flow data and OD transit data. However, for trips in public transport systems, they generally take longer time than one time step and simple summation of OD transit flows cannot substitute in-out flows, leading the above models to fail to give accurate prediction, as shown in our case studies later.

3   Problem Formulation

We first introduce some basic notations and define the public transport crowd flows prediction problem formally.

Definition 1 *(Region)*: We partition the city into N non-overlapping regions. Each region denoted as $g(n), n = 1, \ldots, N$, can have irregular figure and different size, depending on geography of the public transport system. The whole grid map is represented by $\mathcal{M}_N$.

This definition is a bit different from some previous studies [4], which assume each region should be a rectangular, with in total of $I \times J$ grids based on longitude and latitude dimensions. This is because our analysis focuses on public transport systems, whose spatial connectivity is not critically dependent on the geographical distance between different regions. The city map could be partitioned according to functionality of different regions, points of interest, volume of crowd flows, etc. Furthermore, we need to remove certain regions without public transport stations inside.

Definition 2 *(Node)*: Given the map $\mathcal{M}_N$ with partitioned regions, we define the city graph with $V = \{v_1, v_2, \ldots, v_N\}$ as the node set. Each node corresponds to one region in $g(n), n = 1, \ldots, N$.

Definition 3 *(Inflow/Outflow)*: Let (τ, l) be spatial-temporal coordinates, where τ denotes a timestamp and l denotes a location. Define $\mathcal{P}$ as a set of trip data. Each trip is denoted by its origination information $o = (\tau_o, l_o)$ and destination information $d = (\tau_d, l_d)$. Here τ_o and τ_d represent the trip starting time and ending time respectively; l_o and l_d represent the origin and destination location respectively. Given the corresponding city graph, the in-out flow of node $v_n \in V$, whose corresponding region in $\mathcal{M}_N$ is $g(n)$, is defined as $\boldsymbol{y}_t^n \in \mathbb{R}^2$:

$$(\boldsymbol{y}_t^n)_1 = |\{(o, d) \in \mathcal{P} : l_d \in v_n \wedge \tau_d \in t\}|, \tag{1}$$
$$(\boldsymbol{y}_t^n)_2 = |\{(o, d) \in \mathcal{P} : l_o \in v_n \wedge \tau_o \in t\}|, \tag{2}$$

where $(y_t^n)_1$ and $(y_t^n)_2$ represent the inflow and outflow of region $g(n)$ respectively. The symbol $|\cdot|$ denotes the cardinality of the set. With abuse of notation, we further define $y_t^{\text{in}} = [(y_t^1)_1, \ldots, (y_t^N)_1]^T \in \mathbb{R}^N$, $y_t^{\text{out}} = [(y_t^1)_2, \ldots, (y_t^N)_2]^T \in \mathbb{R}^N$, and $\mathbf{Y}_t = [y_t^{\text{in}}, y_t^{\text{out}}] \in \mathbb{R}^{N \times 2}$.

Definition 4 *(OD transit flow matrix):* Similarly, given data $\mathcal{P}$, and grid map $\mathcal{M}_N$ in time step t with corresponding city graph with nodes V, the OD transit flow matrix in time step t is defined as $\mathbf{S}_t \in \mathbb{R}^{N \times N}$:

$$(\mathbf{S}_t)_{mn} = |\{(o, d) \in \mathcal{P} : l_o \in v_n \wedge l_d \in v_m \wedge \tau_d \in t\}|, \tag{3}$$

where the corresponding regions of v_n and v_m in $\mathcal{M}_N$ are $g(n)$ and $g(m)$, respectively. $(\mathbf{S}_t)_{mn}$ represents OD transit flow from node v_n to node v_m. Abusing notation a bit, the mth row of OD transit flow matrix $\mathbf{S}_t$ is denoted as s_t^m, which describes the OD transit flows from all other nodes to node m in time step t.

Problem: Our goal is to provide a holistic prediction framework for the in-out flows of each region and the OD transit flow matrix. Specifically, given the city region nodes $V = \{v_1, v_2, \ldots, v_N\}$, current time step t, and historical data $\mathbf{Y}_{t-s}, \ldots, \mathbf{Y}_t, \mathbf{S}_{t-s}, \ldots, \mathbf{S}_t$ for $s = 0, \ldots, t-1$, we propose a model to collectively predict $\mathbf{Y}_{t+1}$ and $\mathbf{S}_{t+1}$.

4   Methodology

The framework of the proposed Graph spAtio dynamIc Network (GAIN) is shown in Fig. 1a. For spatial correlations, we use the multi-head GAT block for OD flow prediction. Its learned dynamic attention network can effectively capture non-adjacent spatial correlations of different regions, and hence can be fed into the GCN block for in-out flow prediction. The two blocks cooperate with each other and achieve joint prediction. For temporal correlations, we connect the above spatial blocks to three blocks of GRU, which capture the minute-level, daily-level and weekly-level correlations respectively. Last, the outputs of the three blocks are fused together in the output layer for final prediction.

4.1   Spatial Correlation

We utilize graph networks to capture non-adjacent spatial correlations. Figure 1b represents the structure of this block.

Spatial Correlation for OD Transit Flow Prediction. First, we propose a GAT module to capture the spatial correlations of s_t^i. To be specific, we first define a graph $G_t = \{V, E, \mathbf{A}_t\}$, where each node v_i is one region (as in Definition 2), E is the edge set, $\mathbf{A}_t$ is the adjacency matrix, and each weight represents the pairwise spatial relationship between two regions. We perform the information aggregation with the graph-based attention encoder to preserve high-order

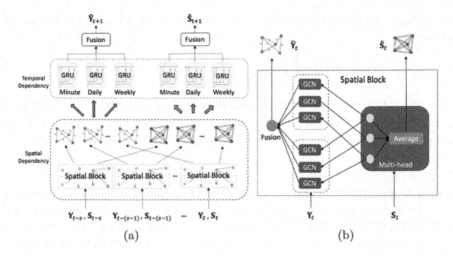

Fig. 1. (a) The structure of Graph spAtio dynamIc Network (GAIN); (b) Inner structure in Spatial Block.

region-wise crowd relations from a global perspective. Its general idea is to learn which regions are able to attend in terms of their crowd flow patterns in a dynamic way, i.e., how to aggregate both self-features and neighbor features for prediction for different time steps dynamically. Yet unlike GCN which requires to pre-define the adjacency matrix $\mathbf{A}_t$, GAT only requires the prior connectivity information E, i.e., whether there is an edge from v_i to v_j. As to the weight of the edge, it can be automatically learned via attention mechanisms.

In particular, to enhance the expressive power of feature representations during the graph-based aggregation process, we first perform linear transformation on the input feature $s_t^i \in \mathbb{R}^N$ of node v_i with a shared parameterized weight matrix $\mathbf{W} \in \mathbb{R}^{N \times N}$, i.e., $\mathbf{W}s_t^i$. Then we compute a pairwise attention coefficient between v_i and v_j by concatenating the projected embeddings $\mathbf{W}s_t^i$ and $\mathbf{W}s_t^j$, and taking a dot product of them with a weight vector c, i.e., $\alpha_{ij,t} = c^T \left[\mathbf{W}s_t^i \| \mathbf{W}s_t^j \right]$, where $\|$ is the concatenation operation. The activation function of LeakyReLU and the softmax are further applied to generate the attention coefficient:

$$a_{ij,t} = \text{softmax}_j \left(LeakyReLU(\alpha_{ij,t}) \right) = \frac{\exp \left(LeakyReLU(\alpha_{ij,t}) \right)}{\sum_{v_k \in neigh(i)} \exp \left(LeakyReLU(\alpha_{ik,t}) \right)}, \quad (4)$$

where $neigh(i)$ denotes the neighbour set of v_i defined by E. In this paper, we can suppose the graph is fully-connected without taking into account any prior information about the connectivity property of different regions. Alternatively, we can also use the connectivity structure of the public transport system, such as the urban railway transit map, as a reasonable prior of the connectivity property.

In addition, to capture different types of spatial correlations and improve the fitting ability of the self-attention, multi-head attention is employed in the mechanism, which uses the average of K parallel attention results as the updated features:

$$\tilde{s}_t^i = \sigma \left(\frac{1}{K} \sum_{k=1}^{K} \sum_{v_j \in neigh(i)} a_{ij,t}^k \mathbf{W}^k s_t^j \right), \tag{5}$$

where $a_{ij,t}^k$ are normalized attention coefficients computed by the kth attention mechanism. We denote the kth-head attention graph as $G_{k,t} = \{V, E, \mathbf{A}_t^k\}$, and the output from GAT that has captured the spatial correlations of $\mathbf{S}_t$ as $\tilde{\mathbf{S}}_t$, with $\tilde{s}_t^i$ as its ith row.

Dynamic Spatial Correlation for Inflow/Outflow Prediction. To capture dynamic and non-Euclidean spatial correlation structures, we design a dynamic GCN module. GCN conducts convolution over a graph with an adjacency matrix $\mathbf{A}$ where each element a_{ij} represents the spatial correlation between v_i and v_j. The general idea of GCN is to learn node representations by exchanging information among its correlated neighbours, and consequently extract the patterns hidden in the graphs.

It is noted that the learned edge weights $\alpha_{ij,t}^k$ from GAT can be regarded as good representations of dynamical spatial correlation structures of different regions. Thus we use $\mathbf{A}_t^k, k = 1, \ldots, K$ as graph inputs of GCN for in-out flow prediction. Specifically, we also employ the multi-head for the GCN block with the kth-head adjacency matrix as $(\mathbf{A}_t^k)_{ij} = \alpha_{ij,t}^k$. As the OD transit flows evolve over time, $\alpha_{ij,t}^k$ also changes over time. Consequently, the dynamic correlation structures of both $\mathbf{Y}_t$ and $\mathbf{S}_t$ have been successfully described in the collaborative model.

Furthermore, consider the correlation structure between inflows of regions is different from that between outflows of regions, two GCNs are conducted for inflow and outflow, respectively. Take inflow for example, the input node features are y_t^{in}, then the spectral convolutions on graph are defined as:

$$\tilde{y}_t^{in,k} = g_\theta *_{G_{k,t}} y_t^{in} = \mathbf{U}_t^k g_\theta(\mathbf{\Lambda}_t^k) \mathbf{U}_t^{k^T} y_t^{in}, \tag{6}$$

where $\mathbf{D}_t^k \in \mathbb{R}^{N \times N}$ is the diagonal degree matrix with the ith diagonal element as $(\mathbf{D}_t^k)_{ii} = \sum_j (\mathbf{A}_t^k)_{ij}$; $\mathbf{L}_t^k = (\mathbf{D}_t^k)^{-1}(\mathbf{D}_t^k - \mathbf{A}_t^k)$ is the Laplacian matrix; $\mathbf{\Lambda}_t^k \in \mathbb{R}^{N \times N}$ and $\mathbf{U}_t^k$ are results of the eigenvalue decomposition of $\mathbf{L}_t^k = \mathbf{U}_t^k \mathbf{\Lambda}_t^k \mathbf{U}_t^{k^T}$. $g_\theta(\mathbf{\Lambda}_t^k)$ is a function of the eigenvalues of $\mathbf{L}_t^k$, and can be localized in space and reduce learning complexity by a polynomial filter [6]. The GCN construction for outflow data y_t^{out} can be conducted in the same way, and get the output $\tilde{y}_t^{out,k}$. Then we fuse the results from these GCNs:

$$\mathbf{Y}_t = \text{ReLU} \left(\sum_{k=1}^{K} y_t^{in,k} \mathbf{W}^{in,k} + \sum_{k=1}^{K} \tilde{y}_t^{out,k} \mathbf{W}^{out,k} \right), \tag{7}$$

where $\tilde{\mathbf{Y}}_t \in \mathbb{R}^{N \times 2}$ are outputs from the GCNs, and $\mathbf{W}^{in,k}, \mathbf{W}^{out,k} \in \mathbb{R}^{1 \times 2}$ are parameters to be learned.

4.2   Temporal Correlation

Now we talk about temporal correlation modeling for $\mathbf{Y}_t$ and $\mathbf{S}_t$. Training long-term temporal information is a nontrivial task. To address this issue, we explicitly model relative historical time steps by capturing the minute-level, daily-level and weekly-level correlations separately [4]. For each time level, we construct the GRU cells using $\tilde{\mathbf{Y}}_t$ and $\tilde{\mathbf{S}}_t$ as input.

Take daily-level feature extraction of $\tilde{\mathbf{Y}}_t$ for example. After capturing spatial correlation features in the GCN block, we first use a flatten layer to transform $\tilde{\mathbf{Y}}_t \in \mathbb{R}^{N \times 2}$ to a feature vector $\tilde{\boldsymbol{y}}_t \in \mathbb{R}^{2N}$. The sequence to be inputted in GRU is $\{\tilde{\boldsymbol{y}}_{t+1-l_d \cdot d}, \tilde{\boldsymbol{y}}_{t+1-(l_d-1) \cdot d}, \cdots \tilde{\boldsymbol{y}}_{t+1-d}\}$, where d is the number of time steps in one day and l_d is the considered maximum lag for daily-level feature extraction. Then the GRU captures the temporal correlations of $\tilde{\boldsymbol{y}}_t$ as:

$$z_t^d = \sigma \left(\mathbf{W}_z^d \tilde{\boldsymbol{y}}_t + \mathbf{U}_z^d \boldsymbol{h}_{l-1}^d + \boldsymbol{b}_z^d \right), \tag{8}$$

$$r_t^d = \sigma \left(\mathbf{W}_r^d \tilde{\boldsymbol{y}}_t + \mathbf{U}_r^d \boldsymbol{h}_{t-1}^d + \boldsymbol{b}_r^d \right), \tag{9}$$

$$\tilde{\boldsymbol{h}}_t^d = \tanh \left(\mathbf{W}_h^d \tilde{\boldsymbol{y}}_t + \mathbf{U}_h^d \left(\boldsymbol{r}_t^d \circ \boldsymbol{h}_{t-1}^d \right) + \boldsymbol{b}_h^d \right), \tag{10}$$

$$\boldsymbol{h}_t^d = (1 - z_t) \circ \boldsymbol{h}_t^d + z_t \circ \tilde{\boldsymbol{h}}_{t-1}^d. \tag{11}$$

The output of the last layer GRU, denoted as $\boldsymbol{h}_{t+1}^d \equiv \hat{\boldsymbol{y}}_{t+1}^d$, represents the daily-level temporal feature. And then we reshape it into $\hat{\mathbf{Y}}_{t+1}^d \in \mathbb{R}^{N \times 2}$. Similarly, we can input the minute-level sequence $\{\tilde{\boldsymbol{y}}_{t+1-l_m}, \tilde{\boldsymbol{y}}_{t+1-(l_m-1)}, \cdots, \tilde{\boldsymbol{y}}_t\}$ where l_m is the considered maximum lag for minute-level feature extraction, and get $\hat{\mathbf{Y}}_{t+1}^m$. We input the weekly-level sequence $\{\tilde{\boldsymbol{y}}_{t+1-l_w \cdot w}, \tilde{\boldsymbol{y}}_{t+1-(l_w-1) \cdot w}, \cdots, \tilde{\boldsymbol{y}}_{t+1-w}\}$ where w equals the number of time steps in one week and l_w is the maximum lag for weekly-level feature extraction, and get $\hat{\mathbf{Y}}_{t+1}^w$. Likewise, we can construct another three GRU blocks to capture temporal relationship of $\mathbf{S}_{t+1}$ and get the minute-level, daily-level and weekly-level components $\hat{\mathbf{S}}_{t+1}^m$, $\hat{\mathbf{S}}_{t+1}^d$, $\hat{\mathbf{S}}_{t+1}^w$ respectively.

4.3   Fusion

Last, we combine results from the three GRUs together for final prediction by parametric-matrix-based fusion with tanh hyperbolic function:

$$\hat{\mathbf{Y}}_{t+1} = \tanh \left(\mathbf{W}_m^1 \circ \hat{\mathbf{Y}}_{t+1}^m + \mathbf{W}_d^1 \circ \hat{\mathbf{Y}}_{t+1}^d + \mathbf{W}_w^1 \circ \hat{\mathbf{Y}}_{t+1}^w \right), \tag{12}$$

$$\hat{\mathbf{S}}_{t+1} = \tanh \left(\mathbf{W}_m^2 \circ \hat{\mathbf{S}}_{t+1}^m + \mathbf{W}_d^2 \circ \hat{\mathbf{S}}_{t+1}^d + \mathbf{W}_w^2 \circ \hat{\mathbf{S}}_{t+1}^w \right), \tag{13}$$

where $\circ$ is element-wise multiplication, and $\mathbf{W}_m^1$, $\mathbf{W}_d^1$, $\mathbf{W}_w^1$, $\mathbf{W}_m^2$, $\mathbf{W}_d^2$, $\mathbf{W}_w^2$ are parameters to be learned to represent impacts of different components.

The final loss function adopts mean squared error between the true flows and the predicted ones:

$$\mathcal{L}(\theta) = \lambda_{\text{region}} \left\| \mathbf{Y}_{t+1} - \hat{\mathbf{Y}}_{t+1} \right\|^2 + \lambda_{OD} \left\| \mathbf{S}_{t+1} - \hat{\mathbf{S}}_{t+1} \right\|^2, \tag{14}$$

where λ_{region} and λ_{OD} are adjustable hyper-parameters, and θ indicates all parameters in GAIN.

Remark 1. Note that from Definitions 3 and 4, the outflow of a region can be computed by summing all the OD transit flows whose origin is that region. As such, we may add a regularization term to penalize the difference between the predicted $\widehat{\boldsymbol{y}}_{t+1}^{out}$ and $\widehat{\mathbf{S}}_{t+1}\mathbf{1}$ where $\mathbf{1} \in \mathbb{R}^{N \times 1}$ is a vector with all components equal to 1, i.e., $\lambda_{\text{lim}} \left\| \widehat{\boldsymbol{y}}_{t+1}^{out} - \widehat{\mathbf{S}}_{t+1}\mathbf{1} \right\|^2$ in (14) with adjustable hyper-parameter λ_{lim}.

5 Experiments and Results

5.1 Experimental Settings

Data. In our experiment, we consider two large-scale real-world datasets for performance evaluation, which contain smart card data from the corresponding AFC systems as follows.

- **Hong Kong Dataset** (HK): The dataset contains passengers' railway trip records in HK from Jan 1st 2017 to Feb 28th 2017. We use the first 52 days for training, and the remaining 7 days for testing. We split the whole city into 40×60 regions, $N = 92$ of which have at least one station. The length of each time step is set as 10 min.
- **Shenzhen Dataset** (SZ): The dataset contains passengers' railway trip records in SZ from Dec 1st 2015 to Dec 30st 2015. The previous 23 days are used for training, and the rest 7 days are for testing. We split the whole city into 10×10 regions, $N = 36$ of which have at least one station. The length of each time step is set as 10 min.

Evaluation Metric. Two metrics are used for performance evaluation: Rooted Mean Squared Error (RMSE) and Mean Absolute Error (MAE).

Baselines. We compared GAIN with the following state-of-the-art methods. The parameters for all the methods are well tuned with the best performance reported. It is noted that for GEML and MDL, they also aim at joint prediction of in-out flows and OD transit flows.

- **AR:** We build AR models for the inflow and outflow of each region, and transit flow of each OD path separately. Each model's lag order is tuned by Akaike information criterion (AIC).
- **ARIMA:** We build ARIMA models for the inflow and outflow of each region, and transit flow of each OD path separately. Each model's lag order is tuned by Akaike information criterion (AIC).
- **GRU:** All crowd flows, including in-out flows of each region and transit flows of all the OD paths, are stacked together as a matrix with rows as time step (the size of which equals look-back window $K = 6$) and columns as different crowd flow variables. The matrix is inputted into GRU for prediction.

- **CNN:** In-out flows of all the regions in each time step are viewed as two images inputted into CNN. The temporal information is modeled as features and we set look-back window K as 6.
- **ConvLSTM** [6]: In-out flows of all the regions are mapped into city grids. The LSTM structure is comprised with 2 ConvLSTM layers and 1 convolutional layer, and the look-back time window is set to 6.
- **DeepST** [4]: In-out flows of all the regions are mapped into city grids. 6 convolution layers are used, and the sequence lag length is set as 6, 3, 1 for modules of temporal closeness, period and trend dependencies, respectively.
- **ST-ResNet** [5]: Three residual units are stacked, and each is with two combinations of "ReLU+Convolution".
- **ASTGCN** [8]: Two ST blocks are stacked, and the look-back time window is set to 6. Inflow and outflow are fed in as features, and predicted respectively.
- **GEML** [13]: One layer GCN and Periodic-skip LSTM are conducted, with the length of the skipped time steps set as the number of time steps in one day.
- **MDL** [14]: All crowd flows are mapped into city grids. Three residual units and two convolution layers are stacked for OD transit flow network and in-out flow network, respectively.

Experiment Settings. For GAIN, tanh activation function is used. Min-Max normalization is used to standardize data into range $[-1, 1]$. In the evaluation, we apply inverse Min-Max transformation obtained on the training set to recover flow values. For temporal correlation, we set $l_m = 6$, $l_d = 3$ and $l_w = 1$. For the spatial block, we set $K = 3$ (the number of attention heads in GAT) and $L = 1$ (the number of GCN layers). The order of polynomials of the Laplacian is set as 1. The batch size is set to 10. 80% of the training samples are selected for training each model and the remaining are in validation set for parameter tuning. We use Adam as our optimizer and the epoch is set as 100. We also use early-stopping to avoid overfitting in all experiments, with patience set to be 20, and we reduce learning rate when a metric has stopped improving, with patience set to be 5 and factor be 0.1.

5.2 In-Out Flow Prediction

We first compare GAIN with baseline methods for in-out flow prediction. As shown in Table 1, GAIN achieves the best results among all approaches for both inflow and outflow prediction of the two datasets.

As for AR, ARIMA, and GRU, they perform poorly as they do not consider the spatial correlations into the model. CNN performs bad as it simply models the spatial information as features. Furthermore, the performance of ConvLSTM is quite unsatisfactory. One possible reason is that the temporal pattern in our data is not very complex. Yet LSTM is over complicated and tends to overfit the data a lot, leading to an even worse performance than AR, ARIMA and GRU. Two spatio-temporal deep-learning based models, i.e., DeepST and ST-ResNet, still perform worse than GAIN. This is because their spatial correlation

Table 1. In-out flow prediction of different methods.

Method	HK				SZ			
	Inflow		Outflow		Inflow		Outflow	
	RMSE	MAE	RMSE	MAE	RMSE	MAE	RMSE	MAE
AR	108.25	64.37	78.48	47.84	109.82	60.48	81.44	44.35
ARIMA	106.27	63.01	77.27	47.27	107.72	60.00	80.73	44.32
GRU	90.44	51.42	68.60	39.71	80.22	49.75	73.30	43.11
CNN	110.09	61.14	77.86	45.59	95.12	56.55	83.02	49.80
ConvLSTM	129.43	74.33	121.17	79.56	144.76	86.50	147.32	85.44
DeepST	94.31	54.87	74.54	45.17	94.11	56.22	90.67	55.28
ST-Resnet	82.27	48.98	64.73	40.24	83.31	52.60	76.20	48.33
ASTGCN	102.84	60.40	98.27	56.78	112.10	66.00	97.05	56.98
GEML	110.09	65.53	111.44	71.08	99.95	61.60	108.33	68.66
MDL	92.97	55.01	88.57	50.46	79.15	51.48	65.89	43.04
GAIN	**81.81**	**45.86**	**58.48**	**36.00**	**68.68**	**42.00**	**61.35**	**37.01**

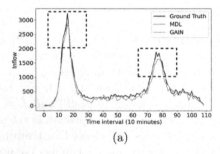 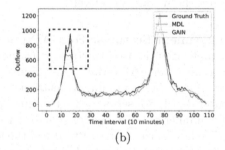

(a)   (b)

Fig. 2. Inflow and outflow results for 24 Dec. 2015 of SZ dataset: (a) Region 21 (Daxin, Taoyuan, Yuehaimen, Shenzhen University); (b) Region 14 (Honglang North, Xingdong, Liuxiandong).

is based on geographical distance, which works for road traffic prediction, such as for taxi or bicycles. However, they are not good at public transport crowd flow prediction. ASTGCN performs even worse than ST-ResNet. One reason is that its spatial correlation highly depends on the pre-defined network graph, which is not helpful for prediction.

As to the two joint prediction models, GEML and MDL, they perform worse than GAIN. For GEML, it is because its network structure mainly targets at OD flows. Yet the in-out flows are less conscientiously calculated by simply weighted sum of features of OD flows, and hence have lower prediction accuracy. For MDL, it is because it assumes geographically close regions are more correlated, and thus is not suitable for public transport flow prediction.

Table 2. OD flow prediction results of different methods.

Method	HK					SZ				
	ARIMA	GRU	GEML	MDL	**GAIN**	ARIMA	GRU	GEML	MDL	**GAIN**
RMSE	4.71	5.03	6.04	4.63	**4.83**	7.81	6.83	8.76	6.86	**6.78**
MAE	1.86	1.87	1.95	2.26	**1.80**	3.71	3.36	4.18	3.80	**3.33**

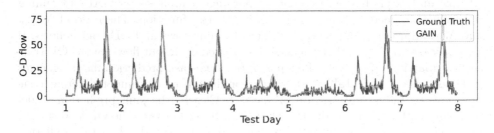

Fig. 3. The OD transit flow prediction from Region 73 (Quarry Bay) to Region 48 (Kowloon Bay) in HK dataset.

To better demonstrate the prediction performance, we randomly select one day and plot the prediction results of GAIN and MDL (the best baseline in SZ dataset) against the ground truth inflow and outflow. Figure 2 shows that GAIN is closer to the ground truth than MDL for most time steps. Especially, for peak hours with extreme high crowd flows, GAIN performs much better than MDL, as shown in the framed time windows in Fig. 2. In some time steps with sudden crowd flow changes, however, both methods could not predict well. One possible reason is that we ignore some external features like weather, due to lack of data.

5.3 OD Transit Flow Prediction

Now we evaluate the performance of OD transit flow matrix prediction of GAIN. Here we select some representative baselines: GEML and MDL which two are targeted at OD flow prediction, ARIMA and GRU which are basic models and can be easily applied into OD flow prediction. As to other works in the literature, their original papers aim at in-out flow prediction and cannot be easily extended for OD matrix prediction, so we do not compare with them. The results are shown in Table 2. Clearly, GAIN has overall the best performance. Though MDL outperforms GAIN a bit for HK dataset in terms of RMSE, their differences are insignificant, and GAIN even performs better in terms of MAE. Furthermore, GAIN also overwhelmingly outperforms MDL for in-out flow prediction. Combining results of Table 1, we can conclude the joint prediction framework of GAIN is more efficient than GEML and MDL. As to ARIMA, surprisingly it performs well for HK dataset, but generally ARIMA performs much worse than others for SZ dataset and for in-out flows. Figure 3 shows the ground truth and the prediction results of GAIN for one selected OD path in one week. We can see that the predicted curve can capture the various passenger flow patterns in different days accurately.

5.4 Sensitivity Analysis

To better evaluate the connection between in-out flow prediction and OD transit flow matrix prediction, we conduct parametric analysis for GAIN by tuning the hyper-parameters λ_{region} and λ_{OD} in the loss function. The ratio of λ_{region} and λ_{OD} adjusts the importance weight of in-out flows and OD transit flows. If $\lambda_{region} = 0$ or $\lambda_{OD} = 0$, then the model only predicts in-out flows, or the OD transit flow matrix, respectively. This means our dual-task prediction model changes to single-task model. Table 3 shows the prediction results under different combinations of λ_{region} and λ_{OD} on SZ dataset. Clearly, the joint prediction model performs better than the single-task model. It verifies that the OD matrix and in-out flows of regions mutually influence each other and a holistic prediction model with consideration of their intimate connections tends to increase prediction performance. Furthermore, when $\lambda_{OD}/\lambda_{region}$ increases, the prediction performance of both tasks improves. This means if we adjust more importance weights on OD transit flow prediction, its prediction accuracy becomes better and also results in better in-out flow prediction. This further demonstrates these two tasks mutually influence each other. In contrast, when $\lambda_{region}/\lambda_{OD}$ increases, the prediction performance of in-out flows improves insignificantly, while OD matrix prediction becomes much worse. As the ratio increases more, both tasks perform worse. This indicates the bottleneck of the multi-task prediction is OD matrix prediction, whose worse performance also deteriorates prediction of in-out flows.

We also evaluate the effect of attention head number on the performance. The number of heads represents how many different spatial correlations are captured in GAT. Figure 4a and b show when head number is 3 or 4, both RMSE and MAE achieve the lowest equivalently for in-out flow prediction. When head number is 3, both RMSE and MAE achieve the lowest for OD transit flow prediction. This indicates there may be three kinds of spatial correlations between different regions. We guess they represent the correlations between inflows of regions, correlations between outflows of regions, and interactive correlations between inflows and outflows of regions. This further demonstrates GAIN can extract dynamic and complex spatial features adaptively. As more than 3 heads are included, the model complexity increases and tends to be over-fitting. Last, Fig. 4c and d show the effect of batch size. Smallest batch size achieves the best generalization performance. This is because large batch size leads the model to make large gradient updates and consequently reach local minimum, while small batch size is noisy, offering more randomness and lower generation error.

Table 3. The impact of hyperparameter ratio for GAIN on SZ dataset.

Hyperparameter		Inflow		Outflow		Transition	
λ_{region}	λ_{OD}	RMSE	MAE	RMSE	MAE	RMSE	MAE
1	10	68.68	42.00	61.35	37.01	6.78	3.33
1	5	73.29	43.37	62.18	37.24	6.98	3.44
1	1	74.40	44.09	63. 65	37.91	7.41	3.62
0	1	\	\	\	\	8.13	3.80
1	0	85.94	52.98	78.60	48.36	\	\
5	1	73.18	44.66	63.58	38.31	7.96	3.66
10	1	74.05	45.09	65.34	39.32	8.16	3.71

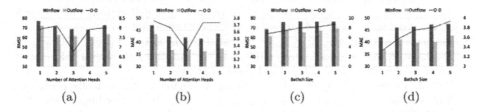

(a)   (b)   (c)   (d)

Fig. 4. Effect of parameter settings on SZ dataset: (a) RMSE and (b) MAE on different numbers of attention head; (c) RMSE and (d) MAE on different batch sizes.

6 Conclusions

This work proposes a holistic prediction framework for in-out flows and OD transit flow matrix for public transport network based on graph neural networks. It uses dynamic GCNs for in-out flow prediction. The graph structures are dynamically learned from the prediction process of OD transit flow matrix, where a multi-head GAT model is used to capture spatial correlations. The above spatial blocks are further inputted into three GRU blocks for minute-level, daily-level and weekly-level temporal correlation description separately. Experiments on two real-world datasets show that our model outperforms several state-of-the-arts.

Acknowledgments. The authors would like to show their great appreciation to the Metro Corporation for sharing this passenger flow data, and in the protection of privacy, all data have been desensitized. This work was supported in part by the NSFC Grant 71901131, 71932006 and 71931006, in part by the ASFC Grant 2020Z063058001, in part by the Tsinghua GuoQiang Research Center Grant 2020GQG1014 and Tsinghua University Intelligent Logistics & Supply Chain Research Center Grant THUCSL20182911756-001, in part by the Ministry of Education, Singapore, AcRF Tier 2 Funding Grant MOE2019-T2-2-116, and in part by the Hong Kong RGC GRF Grant 16216119 and 16201718.

References

1. Yu, R., Li, Y., Shahabi, C., Demiryurek, U., Liu, Y.: Deep learning: a generic approach for extreme condition traffic forecasting. In: Proceedings of the 2017 SIAM International Conference on Data Mining, pp. 777–785. SIAM (2017)
2. Shi, X., Chen, Z., Wang, H., Yeung, D.Y., Wong, W.K., Woo, W.C.: Convolutional LSTM network: a machine learning approach for precipitation nowcasting. In: Advances in Neural Information Processing Systems, pp. 802–810 (2015)
3. Yao, H., Tang, X., Wei, H., Zheng, G., Li, Z.: Revisiting spatial-temporal similarity: a deep learning framework for traffic prediction. In: Proceedings of the AAAI Conference on Artificial Intelligence, vol. 33, pp. 5668–5675 (2019)
4. Zhang, J., Zheng, Y., Qi, D., Li, R., Yi, X.: DNN-based prediction model for spatio-temporal data. In: Proceedings of the 24th ACM SIGSPATIAL International Conference on Advances in Geographic Information Systems, pp. 1–4 (2016)
5. Zhang, J., Zheng, Y., Qi, D.: Deep spatio-temporal residual networks for citywide crowd flows prediction. arXiv preprint arXiv:1610.00081 (2016)
6. Defferrard, M., Bresson, X., Vandergheynst, P.: Convolutional neural networks on graphs with fast localized spectral filtering. In: Advances in Neural Information Processing Systems, vol. 29, pp. 3844–3852 (2016)
7. Pan, Z., Liang, Y., Wang, W., Yu, Y., Zheng, Y., Zhang, J.: Urban traffic prediction from spatio-temporal data using deep meta learning. In: Proceedings of the 25th ACM SIGKDD International Conference on Knowledge Discovery & Data Mining, pp. 1720–1730 (2019)
8. Guo, S., Lin, Y., Feng, N., Song, C., Wan, H.: Attention based spatial-temporal graph convolutional networks for traffic flow forecasting. In: Proceedings of the AAAI Conference on Artificial Intelligence, vol. 33, pp. 922–929 (2019)
9. Liu, L., Qiu, Z., Li, G., Wang, Q., Ouyang, W., Lin, L.: Contextualized spatial-temporal network for taxi origin-destination demand prediction. IEEE Trans. Intell. Transp. Syst. 20(10), 3875–3887 (2019)
10. Chu, K.F., Lam, A.Y., Li, V.O.: Deep multi-scale convolutional LSTM network for travel demand and origin-destination predictions. IEEE Trans. Intell. Transp. Syst. 21(8), 3219–3232 (2019)
11. Shi, H., et al.: Predicting origin-destination flow via multi-perspective graph convolutional network. In: ICDE, pp. 1818–1821. IEEE (2020)
12. Hu, J., Yang, B., Guo, C., Jensen, C.S., Xiong, H.: Stochastic origin-destination matrix forecasting using dual-stage graph convolutional, recurrent neural networks. In: ICDE, pp. 1417–1428. IEEE (2020)
13. Wang, Y., Yin, H., Chen, H., Wo, T., Xu, J., Zheng, K.: Origin-destination matrix prediction via graph convolution: a new perspective of passenger demand modeling. In: Proceedings of the 25th ACM SIGKDD International Conference on Knowledge Discovery & Data Mining, pp. 1227–1235 (2019)
14. Zhang, J., Zheng, Y., Sun, J., Qi, D.: Flow prediction in spatio-temporal networks based on multitask deep learning. IEEE Trans. Knowl. Data Eng. 32(3), 468–478 (2019)
15. Wang, S., Miao, H., Chen, H., Huang, Z.: Multi-task adversarial spatial-temporal networks for crowd flow prediction. In: Proceedings of the 29th ACM International Conference on Information & Knowledge Management, pp. 1555–1564 (2020)

Reservoir Pattern Sampling in Data Streams

Arnaud Giacometti[ID] and Arnaud Soulet[(✉)][ID]

Université de Tours, LIFAT, Blois, France
{arnaud.giacometti,arnaud.soulet}@univ-tours.fr

Abstract. Many applications generate data streams where online analysis needs are essential. In this context, pattern mining is a complex task because it requires access to all data observations. To overcome this problem, the state-of-the-art methods maintain a data sample or a compact data structure retaining only recent information on the main patterns. This paper addresses online pattern discovery in data streams based on pattern sampling techniques. Benefiting from reservoir sampling, we propose a generic algorithm, named RESPAT, that uses a limited memory space and that integrates a wide spectrum of temporal biases simulating landmark window, sliding window or exponential damped window. For these three window models, we provide fast damping optimizations and we study their temporal complexity. Experiments show that the performance of RESPAT algorithms is particularly good. Finally, we illustrate the interest of our approach with online outlier detection in data streams.

1 Introduction

Many applications generate data streams, especially with the rise of network sensors [2] and the Internet of Things [8]. Beyond their operational utility, the analysis of these data streams raise strategic issue in mobile data stream mining [25], online transaction analysis [18] and so on. In most cases, it is not possible to consider storing these data to perform an off-line analysis because of their volume. In addition, the usefulness of certain analyzes like early outlier detection necessarily relies on online processing. Unfortunately, knowledge discovery in data streams remains a challenging task due to the continuous arrival of data observations that must be processed in a short time (*time constraint*) despite a limited memory space (*space constraint*) [19,24]. This problem is particularly exacerbated for pattern mining whose combinatorial complexity is costly by nature [11,16]. It aims to maintain a collection of interesting patterns extracted from the data stream while respecting these constraints. For this purpose, a first strategy consists in maintaining a compact data structure containing the information on the pattern occurrences appearing in each data observation [15,21,22,26]. In addition to being expensive to update, the size of this structure is not limited and sometimes requires significant memory space. A second strategy is to maintain a data sample representative of the data stream. It is then possible to mine the interesting patterns from this sample [1,4,23]. However, the expensive cost

© Springer Nature Switzerland AG 2021
N. Oliver et al. (Eds.): ECML PKDD 2021, LNAI 12975, pp. 337–352, 2021.
https://doi.org/10.1007/978-3-030-86486-6_21

of this mining step prevents it from being repeated at the arrival of each data observation and then, from having an up-to-date collection of patterns.

This paper revisits the pattern discovery in data streams at the light of *pattern sampling* [3,5]. Frequent pattern sampling consists in drawing patterns at random proportionally to their frequency. In our context, the principle is to maintain a sample of k patterns representative of the data stream. For example, a pattern twice as frequent will be twice as likely to be picked. To the best of our knowledge, no method exists to sample patterns in data streams. For this purpose, the key idea is to benefit from *reservoir sampling*. This family of randomized algorithms picks a random sample of k items from a population of unknown size in a single pass over the items [10,20,27]. Unfortunately, the existing reservoir sampling methods are not suitable to deal with two challenges: output space and temporal bias. First, existing reservoir sampling methods are designed to perform input space sampling (i.e., from data observations) and not output space sampling (i.e., from patterns covering the data observations). Of course, it is not possible to enumerate the exponential number of patterns contained in each data observation to build the population to be sampled. Second, there are reservoir sampling methods that take into account a static distribution on the population [10]. To the best of our knowledge, existing reservoir sampling methods do not incorporate a temporal bias to favor the most recent observations. However, the damping of the oldest data observations is important for pattern mining in data streams [17].

This paper provides the first pattern sampling method in data streams using reservoir sampling. The general principle is to generate a key for each occurrence of patterns and to keep in the reservoir the k occurrences with the largest keys. More specifically, our contributions are as follows:

- We present a generic algorithm RESPAT that performs exponential random jumps in the output space so as not to compute a key for each occurrence and that updates the value of the keys of the reservoir to integrate several window models. We also propose *fast damping* optimized algorithms (RESPAT$_{\text{no}}$, RESPAT$_{\text{win}}$ and RESPAT$_{\text{exp}}$) for three window models that avoid having to explicitly modify the keys of the reservoir.
- We demonstrate that the proposed RESPAT algorithm family based on reservoir sampling is exact and that it requires a memory space linear with the sample size k. Interestingly, our theoretical study proves its efficiency by computing the complexity of the number of insertions in the reservoir.
- We evaluate the effectiveness of our algorithm family by performing experiments on UCI benchmarks and synthetic data. This experimental study shows the important contribution of the exponential jump and fast damping optimizations. A use case also illustrates the interest of pattern sampling to detect outliers in data streams. In particular, our online and one-pass method rediscovers the outliers of an off-line method with a good accuracy.

The outline of this paper is as follows. Section 2 reviews some related work about pattern mining in data streams and pattern sampling methods. Section 3 introduces basic definitions and the formal problem statement. We present our reservoir pattern sampling algorithms for data streams in Sect. 4. We evaluate our approach in Sect. 5 and conclude in Sect. 6.

2 Related Work

Data mining over data streams is a daunting task [19,24], especially pattern mining [11,16]. Most of the existing methods aim to extract all the frequent patterns and more rarely, are limited to the top-k frequent patterns [28] or other measure like max-frequency [6]. Itemsets is the most popular pattern language and only few works are interested in particular forms like maximal patterns [18] or closed patterns [7,22]. Several static window models [17] are implemented to consider (i) the entire stream from a certain time (landmark window) [7,21,28], (ii) only the data observations inside a window (sliding window) [7,22,26] or (iii) the entire data stream by weighting the observations to favor the most recent ones (damped window) [15,23]. Clearly this latter is the more complex model and it is also the least dealt with in the literature. The majority of methods relies on a tree-like data structure in order to efficiently store and manipulate the current mined patterns [15,21,22,26]. This structure is updated according to the data stream to maintain a collection taking into account the considered window. Statistical techniques such as the Chernoff bound [28] are often used to estimate the frequency of the patterns in order to safely remove the less promising ones. Most of these techniques compute approximated collection of patterns contrary to our proposal, which guarantees an exact sampling: whatever the window model, the mined sample is equivalent to what would be mined if all the data observations were stored in memory.

Rather than incrementally maintaining the collection of interesting patterns, another approach consists in incrementally maintaining a data sample representative of the data stream benefiting from reservoir sampling [10,20,27]. The idea is then to extract the collection of patterns from this data sample by simulating different window models (e.g., sliding window [4], exponential bias [1] or tilted window [23]). Unfortunately, this approach makes it necessary to repeat the pattern discovery after each modification of the data sample, which is very costly (both for the frequent pattern mining and for the pattern sampling). For this reason, it would be more advantageous to directly sample the output space (i.e., pattern space) instead of the input space (i.e., data space). To the best of our knowledge, there are methods for sequential data [9] but not for sampling patterns in a data stream. First, stochastic methods [3] require evaluating the measure m on the entire data for selecting the next state of the random walk. This evaluation is impossible in a data stream because we do not have all the data observations. Second, multi-step random procedures [5,9] have the advantage of not directly evaluating the measure. They consist in drawing a data observation proportionally to the utility sum of patterns that it contains and then, in drawing a pattern from these patterns proportionally to its utility. Unfortunately, the essential normalization constant for drawing the transaction will only be known at the end of the stream. In short, all the pattern sampling methods require a full-access to data incompatible with the notion of data stream where not all the data observations can be stored. Thus, this work proposes to extend reservoir sampling methods dedicated to the input space so that they effectively deal with the output space with a time bias.

3 Preliminaries

Data Stream and Patterns. This paper exclusively addresses itemset language. Given a set of literals $\mathcal{I}$, a data stream is a sequence of transactions with timestamps: $\mathcal{D} = \langle(t_1, d_1), \ldots, (t_n, d_n)\rangle$ such that $d_j \subseteq \mathcal{I}$ for $j \in [1..n]$ and $t_j < t_{j+1}$ for $j \in [1..n-1]$. Without loss of generality, we consider that $t_1 = 0$. The itemset language $\mathcal{L}_\mathcal{I} = 2^\mathcal{I}$ is the set of all patterns (or itemsets). The cover of a pattern φ in a data stream $\mathcal{D}$, denoted by $\mathcal{D}[\varphi]$, is the set of data observations containing φ: $\mathcal{D}[\varphi] = \{(t, d) \in \mathcal{D} : \varphi \subseteq d\}$. For instance, Table 1 provides a data stream with 6 data observations described by 5 items $\mathcal{I} = \{A, B, C, D, E\}$. ABD is the transaction of the first data observation containing 8 itemsets: $2^{d_0} = \{\emptyset, A, B, D, AB, AD, BD, ABD\}$. Of course, a data stream evolves over the time with the addition of new data observations (e.g., a transaction will likely be added at timestamp 6).

Table 1. A running example with three damping functions

Time.	Items	ω_{no}	ω_{win}^2	$\omega_{exp}^{0.3}$
0	A B D	1	0	0.223
1	A B C D	1	0	0.301
2	A C E	1	0	0.406
3	A B C	1	1	0.549
4	C D E	1	1	0.741
5	C D E	1	1	1.000

(Data stream $\mathcal{D}$ — Damping function ω; left axis labeled "Time")

Interestingness Measure. A damping function $\omega : \Re^+ \rightarrow [0,1]$ is a decreasing function that assigns a lower weight to the oldest data observations such that $\omega(0) = 1$. This damping function enables us to consider the different existing window models [17]: **(i) Landmark window:** A landmark window ω_{no} considers all the data observations equally since a landmark point (as t_1 without loss of generality): $\omega_{no} : t \mapsto 1$ **(ii) Sliding window:** The (time-stamp based) sliding window ω_{win}^T considers only the most recent data observations [12]. Formally, $\omega_{win}^T : t \mapsto 1$ if $t \leq T$ and 0 otherwise, where $T > 0$ is the window time size. **(iii) Exponential damped window:** A popular damped window is the exponential bias [1] defined as $\omega_{exp}^\alpha : t \mapsto e^{-\alpha t}$ where α is the damping factor. For instance, Table 1 illustrates this three damping functions. It is easy to see that the damping functions ω_{win}^2 and $\omega_{exp}^{0.3}$ gives a larger weight to the most recent observations. This is suitable for applications where people are interested only in the most recent information of the data streams. For this purpose, we weight the support with the damping function:

Definition 1 (Damping support). *Given a damping function ω, the damped support of a pattern φ in $\mathcal{D} = \langle(t_1, d_1), \ldots, (t_n, d_n)\rangle$ is defined as below:*

$$supp_\omega(\varphi, \mathcal{D}) = \frac{\sum_{(t,d) \in \mathcal{D}[\varphi]} \omega(t_n - t)}{\sum_{(t,d) \in \mathcal{D}} \omega(t_n - t)}$$

Obviously, the damping function ω_{no} leads to the traditional support. For instance, $supp_{\omega_{no}}(AB, \mathcal{D}) = 3/6$ and $supp_{\omega_{no}}(CDE, \mathcal{D}) = 2/6$ meaning that AB occurs more frequently than CDE. On the latest data observations, the situation is reversed: $supp_{\omega_{win}^2}(AB, \mathcal{D}) = 1/3$ (or $supp_{\omega_{exp}^{0.3}}(AB, \mathcal{D}) = 1.073/3.220$) is lower than $supp_{\omega_{win}^2}(CDE, \mathcal{D}) = 2/3$ (or $supp_{\omega_{exp}^{0.3}}(CDE, \mathcal{D}) = 1.741/3.220$).

Problem Statement. Let Ω be a population and $f : \Omega \to [0, 1]$ be a measure, the notation $x \sim f(\Omega)$ means that the element x is drawn randomly from Ω with a probability distribution $\pi(x) = f(x)/Z$ where Z is a normalizing constant.

Given a data stream $\mathcal{D} = \langle(t_1, d_1), \ldots, (t_n, d_n)\rangle$, a language $\mathcal{L}_\mathcal{I}$ and a damping function ω, we aim at selecting k patterns $\varphi_1, \ldots, \varphi_k$ in $\mathcal{L}_\mathcal{I}$ where the probability of each pattern φ_i to be selected is determined by its relative weight $supp_\omega(\varphi_i, \mathcal{D})$: $\varphi_i \sim supp_\omega(\mathcal{L}_\mathcal{I}, \mathcal{D})$ for $i \in \{1, \ldots, k\}$.

As mentioned in the introduction, a method for processing data streams must comply with two constraints: **(i) Space constraint:** In most cases, it is not possible to store all the observations in the data stream $\mathcal{D}$. Therefore, the sampling method has to be done in a single pass to avoid disk storage and the space complexity has to be independent of the number of observations n. **(ii) Time constraint:** Each observation must be processed in a short time to avoid the accumulation of continuously arriving data, which would violate the above space constraint. Next sections address this problem using reservoir sampling.

4   Reservoir Algorithms for Pattern Sampling

4.1   Challenges and Key Ideas

First, we reformulate our *pattern* sampling problem as an *occurrence* sampling problem. Drawing a pattern according to the weighted support $supp_\omega$ is equivalent to drawing an occurrence according to the damping function ω: $\varphi \sim supp_\omega(\mathcal{L}_\mathcal{I}, \mathcal{D}) \Leftrightarrow \varphi \sim \omega(\mathcal{L}(\mathcal{D}))$ where the multi-set $\mathcal{L}(\mathcal{D}) = \bigcup_{(t,d) \in \mathcal{D}} 2^d$ gathers all the occurrences of patterns from $\mathcal{D}$:

$$\mathcal{L}(\mathcal{D}) = \{ \underbrace{\varphi_1^0, \varphi_1^1, \ldots}_{2^{d_1} \text{ with } \omega(t_n - t_1)}, \quad \underbrace{\varphi_2^0, \varphi_2^1, \ldots}_{2^{d_2} \text{ with } \omega(t_n - t_2)}, \ldots, \underbrace{\varphi_n^0, \varphi_n^1, \ldots}_{2^{d_n} \text{ with } \omega(t_n - t_n)} \}$$

and each occurrence $\varphi_j^i \subseteq d_j$ has a weight $\omega(\varphi_j^i) = \omega(t_n - t_j)$. Interestingly, this reformulation of the problem makes it possible to directly reuse reservoir sampling algorithms from the literature where the occurrences form the population to be sampled. More precisely, this family of algorithms selects a sample (often without replacement) of a population having an unknown size in a single pass. Some algorithms having the best complexity rely on a central result that we also use in the rest of this paper. Given two keys $key_1 = {u_1}^{1/\omega_1}$ and $key_2 = {u_2}^{1/\omega_2}$ where $\omega_i > 0$ and u_i is uniformly drawn from $[0, 1]$, we have:

$$P(key_1 \leq key_2) = \frac{\omega_2}{\omega_1 + \omega_2} \tag{1}$$

Based on this observation, [10] demonstrates that assigning each occurrence in $\mathcal{L}(\mathcal{D})$ a key ${u_i}^{1/\omega_i}$ where u_i is a random number and then, selecting the k patterns with the largest keys is equivalent to sampling k occurrences *without* replacement from $\mathcal{L}(\mathcal{D})$ proportionally to ω. Besides, as the number of occurrences is very large (i.e., $|\mathcal{L}(\mathcal{D})| \gg k$), a sampling method *without* replacement is equivalent to a sampling method *with* replacement. Consequently, this paper benefits from some principles of the sampling algorithm without replacement proposed by [10].

The context of pattern sampling raises two challenges with respect to the use of reservoir sampling. The first challenge is to address the output space $\mathcal{L}(\mathcal{D})$ rather than the input space $\mathcal{D}$. We could naively apply a reservoir sampling method by enumerating the output space (i.e., all the occurrences for each data observation). For the itemset language, this approach would lead to an exponential complexity $2^{|d|}$ for processing a data observation d, which the time constraint prevents. *Inserting step* in Sect. 4.2 shows how to avoid the enumeration of the output space by directly selecting the occurrence to insert into the sample. The second challenge is to take into account the damping function ω in the maintenance of the sample. In Eq. 1, the weights are static, while in our context, they are dynamic. At the insertion time t_{ins}, all occurrences have 1 as weight – by definition of the damping function ω, we have $\omega(t_{ins} - t_{ins}) = \omega(0) = 1$. When new observations arrive, the weight of the patterns in the sample decreases except for the landmark window. Whatever the damping function, *Damping step* in the next subsection shows how to integrate this modification relying on Eq. 1. Finally, Sect. 4.3 proposes fast damping optimizations for ω_{win}^{T} and $\omega_{\text{exp}}^{\alpha}$.

4.2   Generic Algorithm: RESPAT

Overview. This section presents our generic algorithm to address the two above challenges. Algorithm 1 takes a data stream $\mathcal{D}$, a damping function ω and a sample size k as inputs and returns a sample $\mathcal{S}$ containing k patterns randomly drawn in $\mathcal{L}_{\mathcal{I}}$ proportionally to the damped support in $\mathcal{D}$. Its general principle is to process each observation one by one. The inserting step (lines 7 to 11) inserts some occurrences of the jth observation without enumerating all the output space. The damping step (lines 5 and 6) modifies the keys of the occurrences contained in the sample to integrate the damping function ω.

Before detailing the two main steps, it is important to note that the reservoir $\mathcal{S}$ is a set of triples $\langle key, \varphi, t \rangle$ meaning that the occurrence φ was inserted at time t with the key key. The function MINKEY (lines 13–17) returns the smallest key of the sample if the sample contains k occurrences and 0 otherwise. The function UPDATESAMPLE (lines 18–23) inserts an occurrence into the sample and removes the occurrence with the smallest key (if necessary for maintaining $|\mathcal{S}| \leq k$). Note that the first k occurrences at the beginning of the stream are automatically added into the sample due to the smallest key that equals to zero (lines 20–21).

Algorithm 1. RESPAT: Pattern sampling in data streams with damping

Require: A data stream $\mathcal{D}$, a damping function ω and a number of patterns k
Ensure: A sample $\mathcal{S}$ containing k patterns randomly drawn in $\mathcal{L}_{\mathcal{I}}$ proportionally to the damped support in $\mathcal{D}$
1: $\mathcal{S} := \emptyset$
2: $jump := 0$
3: **for** $j \in \langle 1, \dots, n \rangle$ **do**
4:     $i := 0$
    // Damping step
5:     **if** $j \geq 2$ **then**
6:         Update the key of each element $\langle key, \varphi, t \rangle \in \mathcal{S}$ with $key^{\omega(t_{j-1}-t)/\omega(t_j-t)}$
    // Inserting step
7:     **while** $i + jump < |2^{d_j}|$ **do**
8:         $i := i + jump$
9:         UPDATESAMPLE$(\mathcal{S}, k, unif(\text{MINKEY}(\mathcal{S}, k), 1), \mathcal{L}(d_j)^i, t_j)$
10:         $jump := \lfloor \log(unif(0,1))/\log(\text{MINKEY}(\mathcal{S},k)) \rfloor + 1$
11:     $jump := (i + jump) - |2^{d_j}|$
12: **return** the sample $\mathcal{S}$
13: **function** MINKEY$(\mathcal{S}, k)$ // Return the minimum key in $\mathcal{S}$
14:     **if** $|\mathcal{S}| < k$ **then**
15:         **return** 0
16:     **else**
17:         **return** $\min_{\langle key, \varphi, t \rangle \in \mathcal{S}} key$
18: **procedure** UPDATESAMPLE$(\mathcal{S}, k, key, \varphi, t)$ // Add the pattern φ in $\mathcal{S}$
19:     $e := \langle key, \varphi, t \rangle$
20:     **if** $|\mathcal{S}| < k$ **then**
21:         Add the pattern e in $\mathcal{S}$
22:     **else**
23:         Replace the pattern with the minimum key in $\mathcal{S}$ by the pattern e

Inserting Step. A naive algorithm could draw a uniform number between 0 and 1, saying u, for each occurrence and it could insert this occurrence when $u^{1/\omega} = u$ exceeds the smallest key m (because the weight of each occurrence is 1). Instead of enumerating one by one all the occurrences, it is enough to calculate how many occurrences, saying $jump$, should be drawn before having the one that will be inserted in the reservoir [20]. Intuitively, to insert a pattern in the reservoir, it is sufficient to calculate the weight ω so that the key $u^{1/\omega}$ is larger than the smallest key in the reservoir, saying m. As each occurrence has a weight of 1, this weight ω simply corresponds to the number of occurrences skipped and the equation to solve is $m \leq u^{1/jump}$. This intuition is formalized by the below property:

Property 1 (Exponential random jump [20]). Given the smallest key m, the number of occurrences to jump before reaching the occurrence to be inserted in the sample is given by the random variable X_m:

$$X_m = \left\lfloor \frac{\log unif(0,1)}{\log m} \right\rfloor + 1$$

Property 1 is the first key ingredient of the inserting step at line 10 of Algorithm 1. Let us assume in our running example provided by Table 1 that after having processed the first two transactions, the smallest key of the reservoir is $m = 0.8$. If we draw $u = 0.1$, then we get $jump = \left\lfloor \frac{\log 0.1}{\log 0.8} \right\rfloor + 1 = 11$ and the 11th occurrence of $ACDE$ is inserted into the reservoir by replacing the occurrence having the smallest key m. With this technique, there is 1 single draw instead of 11 with a naive enumeration. We will measure this significant gain both theoretically (in Sect. 4.4) and practically (in Sect. 5).

For the random jump to be really useful, we must access the $jump$-th occurrence without listing all the previous ones. For this purpose, we introduce the notion of index operator. Given a data observation $(t, d) \in \mathcal{D}$, an *index operator* is a bijection mapping each number $i \in [0..|2^d| - 1]$ to a pattern $\varphi \in 2^d$:

Definition 2 (Itemset index operator). *Given a transaction $d = \{I_0, \ldots, I_{|d|-1}\}$ and an index number $i \in [0..2^{|d|} - 1]$, we consider its value $b_{|d|-1} \ldots b_1 b_0$ in binary system and then, the itemset X returned by $\mathcal{L}(d)^i$ contains all items I_j where $b_j = 1$: $I_j \in X \Leftrightarrow b_j = 1$.*

This index operator is the second key ingredient used at line 9 of Algorithm 1 for selecting directly the right occurrence to insert in the sample without enumerating all the patterns and applying a filter. For instance, $\mathcal{L}(ACDE)^{11}$ is ACE because the decimal value $(11)_{10}$ has $(1011)_2$ as binary value. This operator is efficient because its complexity is linear with the number of items in d.

Damping Step. All patterns are inserted into the reservoir with 1 as initial weight but after, their weight must be decreased to take into account the damping function ω. In other words, at every time t_j, a pattern inserted at t_{ins} must have a key $u^{1/\omega(t_j - t_{ins})}$. For this purpose, the key of each pattern is raised to the power $\frac{\omega(t_{j-1} - t_{ins})}{\omega(t_j - t_{ins})}$ at each iteration j. The below property formalizes this idea:

Property 2 (Key damping). Given a pattern φ inserted at time t_{ins} with a key key considering a damping function ω, we have for any $t_j \geq t_{ins}$:

$$key^{\frac{\omega(t_{ins} - t_{ins})}{\omega(t_{ins+1} - t_{ins}))}} \frac{\omega(t_{ins+1} - t_{ins})}{\omega(t_{ins+2} - t_{ins})} \cdots \frac{\omega(t_{j-1} - t_{ins})}{\omega(t_j - t_{ins})} = key^{1/\omega(t_j - t_{ins})}$$

Due to lack of space, the proofs are omitted. This property follows from the fact that the left hand-side of the equality can be rewritten as $key^{\prod_{i=ins+1}^{j} \frac{\omega(t_{i-1} - t_{ins})}{\omega(t_i - t_{ins})}}$ and the simplification of the exponent gives $key^{1/\omega(t_j - t_{ins})}$. Assume that ACE was inserted with the initial key $key = 0.9$ at time 2. Of course, this key will not be modified with the damping function ω_{no}. For the function ω_{win}^2, it decreases to 0 with the sixth observation $(5, CDE)$ because $key^{\omega_{win}^2(4-2)/\omega_{win}^2(5-2)} = key^{1/0}$ is equal to 0 (by convention). Finally, with the function $\omega_{exp}^{0.3}$, this weight is $0.9^{\omega_{exp}^{0.3}(2-2)/\omega_{exp}^{0.3}(3-2)} = 0.867$ at time 3, $0.867^{\omega_{exp}^{0.3}(3-2)/\omega_{exp}^{0.3}(4-2)} = 0.825$ at time 4 and $0.825^{\omega_{exp}^{0.3}(4-2)/\omega_{exp}^{0.3}(5-2)} = 0.772$ at time 5 that also corresponds to $0.9^{1/\omega_{exp}^{0.3}(5-2)} = 0.772$ as desired.

It is clear that the damping step decreases all the keys (except for ω_{no}) and therefore, the smallest key m in the sample. Therefore, the stronger the damping, the smaller the size of the jumps at the inserting step (see Property 1). Consequently, our approach is less efficient with the damping functions focusing on the latest data observations. Furthermore, the computational cost of decreasing the keys is an important defect of this damping step because this operation is done for each element in S at every iteration t_j. Fortunately, it is sometimes possible to skip this step for some damping functions as shown in the next section.

4.3   Fast Damping Algorithms: RESPAT_{no}, RESPAT_{win} and RESPAT_{exp}

This section improves the generic algorithm RESPAT by emulating the key damping without explicitly updating the key of each element contained in the reservoir. Of course, the damping step is useless for ω_{no} as the key value is not modified (indeed, $key^{\omega_{no}(t_{j-1}-t)/\omega_{no}(t_j-t)} = key^{1/1} = key$). Consequently, lines 5 and 6 can be safely removed leading to the algorithm denoted by RESPAT_{no}. But, we show below that it is also possible to remove these lines for ω_{win}^T and ω_{exp}^α by adapting the functions MINKEY and UPDATESAMPLE.

Sliding Window. Ideally, the patterns that are too old (i.e., $|t - t_{ins}| > T$) should be removed from the sample S at each damping step. Our key idea is to count them as being missing from S even if we do not take them out. Consequently, the number of patterns contained in the sample takes into account the too old patterns (see lines 2 and 8 of Algorithm 2). In practice, this number is maintained at each modification of S (lines 9 and 11) (with a circular array storing the number of insertions for the last T times). In particular, if the pattern in S with the minimum key (line 11) is too old, then this pattern is removed and the pattern with the next minimum key is considered. For instance, assume that ACE was inserted at time 2 with the initial key $key = 0.9$ and it remains in the reservoir S until the end of time 4. At the beginning of time 5, the function MINKEY will return 0 whatever the value of the smallest key in S because ACE is expired (due to $5 - 2 \geq 3$). The insertion of a new pattern will refill the reservoir with the correct number of unexpired patterns. Later, by taking care of timestamps, when the smallest key will be that of an expired itemset (here, ACE with 0.9), this element will be removed from S and the next smallest key will be considered.

Exponential Damping. Whatever the insertion time of the key, we can observe that the damping function ω_{exp}^α raises the key to the same power (because ω_{exp}^α is a memory-less bias function [1]). The below property formalizes this intuition:

Property 3 (Exponential key damping). Given a pattern φ inserted at time t_{ins} with a key key and an exponent α, we have for any $t_j \geq t_{ins}$:

$$key^{1/\exp(\alpha \times t_{ins})^{\exp(\alpha \times t_j)}} = key^{1/\omega_{exp}^\alpha(t_j - t_{ins})}$$

Algorithm 2. $\text{RESPAT}_{\text{win}}$: RESPAT with sliding window fast damping

1: **function** $\text{MINKEY-WIN}(S, k, t)$ // Return the minimum key in S
2:    **if** $|\{\langle key, \varphi, u \rangle \in S : |t - u| \leq T\}| < k$ **then**
3:        **return** 0
4:    **else**
5:        **return** $\min_{\langle key, \varphi, u \rangle \in S \text{ s.t. } |t-u| \leq T} key$
6: **procedure** $\text{UPDATESAMPLE-WIN}(S, k, key, \varphi, t)$ // Add the pattern φ in S
7:    $e := \langle key, \varphi, t \rangle$
8:    **if** $|\{\langle key, \varphi, u \rangle \in S : |t - u| \leq T\}| < k$ **then**
9:        Add the pattern e in S
10:   **else**
11:       Replace the pattern with the minimum key in S by the pattern e

---

This property follows from the fact that the two exponents can be rewritten as a single one $\frac{\exp(\alpha \times t_j)}{\exp(\alpha \times t_{ins})} = 1/\exp(-\alpha \times (t_j - t_{ins}))$. Property 3 means that the insertion time t_{ins} is useless for damping the key (if the initial insertion weight takes it into account). For instance, let us take again the example of the itemset ACE inserted at time 2 with the initial key $key = 0.9$. Benefiting from the exponential key damping for $\omega_{\exp}^{0.3}$, this itemset is inserted with the weight $key^{1/\exp(0.3 \times 2)} = 0.944$ and its damped key at time 5 is $0.944^{\exp(0.3 \times 5)} = 0.772$ (that equals to $0.9^{1/\omega_{\exp}^{0.3}(5-2)} = 0.9^{\exp(-0.3(5-2))}$ as desired). More generally, $\text{RESPAT}_{\exp}$ benefits from this damping strategy (see Algorithm 3). At line 7, we insert a new pattern at time t_{ins} with a weight greater than 1 so that the weight of the patterns already within the reservoir do not change. Then, when we consult the minimum key at line 5, we correct all the weights with the same power using Property 3.

Algorithm 3. $\text{RESPAT}_{\exp}$: RESPAT with exponential fast damping

1: **function** $\text{MINKEY-EXP}(S, k, t)$ // Return the minimum key in S
2:    **if** $|S| < k$ **then**
3:        **return** 0
4:    **else**
5:        **return** $\min_{\langle key, \varphi, t \rangle \in S} key^{\exp(\alpha \times t)}$
6: **procedure** $\text{UPDATESAMPLE-EXP}(S, k, key, \varphi, t)$ // Add the pattern φ in S
7:    $e := \langle key^{1/\exp(\alpha \times t)}, \varphi, t \rangle$
8:    **if** $|S| < k$ **then**
9:        Add the pattern e in S
10:   **else**
11:       Replace the pattern with the minimum key in S by the pattern e

---

4.4   Theoretical Analysis

This section studies the RESPAT algorithm family. First, the following property proves that these algorithms return a sample with the expected characteristics:

Property 4 (Correctness). Considering that $|\mathcal{L}(\mathcal{D})| \gg k$, the generic algorithm RESPAT and its fast damping variants (RESPAT$_{no}$, RESPAT$_{win}$ and RESPAT$_{exp}$) are correct: Given a data stream $\mathcal{D}$, a language $\mathcal{L}_\mathcal{I}$ and a damping function ω, each algorithm returns k patterns $\varphi_1, \ldots, \varphi_k$ such that $\varphi_i \sim supp_\omega(\mathcal{L}_\mathcal{I}, \mathcal{D})$.

This non-trivial property relies on [10] by proving that the temporal bias is correctly maintained thanks to the different key raising properties.

We now analyze the complexity of the algorithm family RESPAT. First, the space complexity of the different algorithms is linear with the sample size k. To the best of our knowledge, our proposal has the smallest space complexity for frequent pattern sampling and it is the first one-pass algorithm for frequent pattern sampling. Second, considering the time complexity, it is clear that the efficiency of fast damping algorithms depends essentially on the number of insertions (especially with the optimized versions without damping step). Of course, the lower this number, the more efficient the approach.

Property 5 (Number of insertions). Given a data stream $\mathcal{D}$, a language $\mathcal{L}_\mathcal{I}$ and a sample size k, the expected number of insertions into the reservoir is (after the filling phase): (i) $O(k \cdot \log\left(\frac{|\mathcal{L}(\mathcal{D})|}{k}\right))$ for the landmark window ω_{no} (see [10]), (ii) $O(k/T \cdot n)$ for the sliding window ω_{win}^T ($T \ll n$) and (iii) $O(k \cdot \alpha \cdot n)$ for the damped window ω_{exp}^α

Unlike other pattern methods in data streams [7,22,26], the weaker the damping is, the more efficient the approach is. In particular, for the sliding window, the larger the window T, the lower the number of insertions. For the exponential damping, the lower the exponent α, the lower the number of insertions.

5   Experimental Evaluation

This experimental study investigates the performance of our reservoir sampling approach in Sect. 5.1 and it shows its interest for outlier detection in Sect. 5.2. We use 12 benchmark datasets coming from the UCI Machine Learning repository and the FIMI repository, and 3 large synthetic datasets coming from [21]. Note that the condition $|\mathcal{L}(\mathcal{D})| \gg k$ is satisfied by these large datasets: $|\mathcal{L}(\mathcal{D})| \gg k$. The methods are implemented with the Java language[1]. All experiments are performed on a 2.5 GHz Xeon processor with the Linux operating system and 2 GB of RAM memory. Each of the reported measurements is the mean of 10 runs where the transactions were swapped.

5.1   Global and Longitudinal Performance Study

The first experiment assesses the overall efficiency by measuring its total execution time. For this purpose, Table 2 compares the RESPAT algorithm family with two state-of-the-art algorithms: the two step random procedure 2-STEP [5] and

---

[1] The source code is available: https://github.com/asoulet/ResPat.

Table 2. Running time in seconds for sampling 100k patterns

Dataset	No damping ω_{no}			Sliding window ω_{win}^{1000}			Exp. damping $\omega_{exp}^{0.003}$		
	2-STEP	A-RES	RESPAT$_{no}$	A-RES	RESPAT	RESPAT$_{win}$	A-RES	RESPAT	RESPAT$_{exp}$
abalone	0.079	0.8	0.8	75.9	81.1	2.1	79.5	84.4	2.9
chess	0.220	–	10.5	–	102.1	28.4	–	83.7	14.3
cmc	0.080	0.8	0.7	25.2	26.2	1.0	27.7	28.6	1.5
connect	0.725	–	14.3	–	2433.0	746.9	–	1569.5	20.0
crx	0.121	4.0	2.0	18.1	13.9	2.1	20.6	16.1	2.5
hypo	0.132	61.2	3.5	146.5	73.1	8.0	170.5	71.2	6.0
mushroom	0.201	–	5.6	–	218.3	30.8	–	184.9	14.3
retail *	6.669	–	115.9	–	2260.2	1025.6	–	1651.1	116.8
sick	0.153	851.8	4.4	938.4	71.7	10.4	964.2	65.1	7.3
T10I4D100K *	0.730	–	7.6	–	1948.3	334.0	–	2104.5	83.3
T10I4D1000K *	oom	–	5.2	–	17835.9	1615.2	–	19624.4	73.1
T15I6D1000K *	oom	–	9.1	–	19395.0	3367.6	–	19387.3	18.2
T30I20D1000K *	oom	–	6474.0	–	–	21229.8	–	29903.9	7008.3
vehicle	0.124	22.9	3.0	46.7	18.2	3.2	51.8	21.2	3.7
waveform	0.165	955.6	4.6	1104.3	131.0	17.6	1144.6	114.6	9.8

oom: out of memory/–: out of time ($\geq$ 10h)/*: variable size transactions

the baseline A-RES [10] that corresponds to RESPAT without the exponential random jump (see Property 1). We consider the execution times for a sample size $k =$ 100,000 and three damping functions: no damping ω_{no}, a sliding window ω_{win}^{1000} and an exponential damping $\omega_{exp}^{0.003}$. First, we observe that our RESPAT algorithm family manages to process large datasets with 1000K transactions, while the two step random procedure 2-STEP does not have enough memory (denoted by oom in Table 2). Of course, in return, our reservoir sampling approach is slower than 2-STEP. Second, it is clear that the fast damping algorithms RESPAT$_{no}$, RESPAT$_{win}$ and RESPAT$_{exp}$ outperform the baseline A-RES and the generic algorithm RESPAT. Indeed, as soon as the number of items per transaction is large, the exponential random jump becomes mandatory to maintain a reasonable processing time explaining timeouts for A-RES (denoted by -). Besides, the cost of the generic damping step is really prohibitive when the sample size is large because of the high number of key decreases for RESPAT. It is more efficient to simulate these key decreases as for the sliding window (see RESPAT$_{win}$ column) as well for the exponential damping (see RESPAT$_{exp}$ column). Finally, Fig. 1 plots the execution times of RESPAT$_{no}$, RESPAT$_{win}$ and RESPAT$_{exp}$ for the 6 largest datasets with respect to the sample size k. Unfortunately, our approach struggles for datasets containing very long transactions (here, retail and T30I20D1000K) regardless of the sample size k. Indeed, for very long transactions, the random jump is no longer enough to curb the combinatorial explosion of the occurrence space. In contrast, for the other four datasets, the asymptotically linear behavior is visible in accordance with Property 5.

This second experiment assesses the longitudinal efficiency of our algorithm by measuring the execution time that is required for processing each transaction. Figure 2 plots the execution time per transaction for a sample size

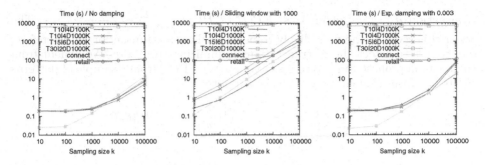

Fig. 1. Running time in seconds (y-axis) with respect to the sample size (x-axis)

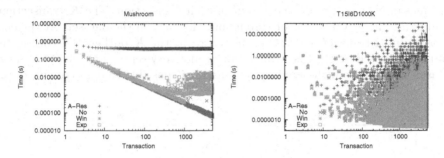

Fig. 2. Longitudinal performance of reservoir sampling algorithms

$k = 100,000$ and four algorithms: A-RES/RESPAT$_{no}$ for ω_{no}, RESPAT$_{win}$ for ω_{win}^{1000} and RESPAT$_{exp}$ for $\omega_{exp}^{0.003}$. We consider two datasets mushroom and T15I6D1000K that respectively represent the datasets with fixed size transactions and the datasets with variable size transactions (denoted by * in Table 2). First, we see again the strong impact of the exponential random jump that drastically reduces the execution time per transaction once the first transactions have been completed (magenta dots are above the others). Second, for mushroom, the execution time decreases steadily when there is no damping ω_{no}. For ω_{win}^{1000}, a disturbance is observed once the sliding window moves (after 1000) with values varying between a few milliseconds and several tens of milliseconds. For the exponential damping $\omega_{exp}^{0.003}$, the execution time per transaction stabilizes around a few milliseconds. Third, for T15I6D1000K, the situation is less visible because the execution time for each transaction depends strongly on its size. Because of the logarithmic scale and high dot density, one might imagine that the execution time increases, which is not true on average.

5.2 Use Case: One-Pass Frequent Pattern Outlier Detection

This section illustrates the interest of our sampling technique to detect outliers by performing a single pass on the data (as it is necessarily the case in a stream). We aim to rediscover the K outliers that we would have obtained with a multi-pass FPOF method. More precisely, the frequent pattern outlier factor of a transaction $d \in \mathcal{D}$ is defined as: $FPOF(d, \mathcal{D}) = \frac{\sum_{\varphi \subseteq d} supp(\varphi, \mathcal{D})}{\max_{d' \in \mathcal{D}} \sum_{\varphi \subseteq d'} supp(\varphi, \mathcal{D})}$. The lower this score is, the more likely to be an outlier the transaction is. Therefore, our goal is to detect the K transactions minimizing this score. We benefit from the formula proposed in [14] to approximate the FPOF from a sample of patterns $\mathcal{S}$ (drawn from $\mathcal{D}$ with respect to the support):

$$\lim_{|\mathcal{S}| = \infty} \underbrace{\frac{|\{\varphi \in \mathcal{S} \; : \; \varphi \subseteq d\}|}{\max_{d' \in \mathcal{D}} |\{\varphi \in \mathcal{S} \; : \; \varphi \subseteq d'\}|}}_{FPOF(d, \mathcal{S})} = FPOF(d, \mathcal{D})$$

Basically, the idea is to maintain a sample of frequent patterns $\mathcal{S}$ with our reservoir sampling approach. At the same time, we apply this formula using the current sample to estimate the FPOF of each transaction. The K transactions minimizing the FPOF are kept throughout the pass on the dataset. At the end, the remaining transactions are considered to be the K outliers. Of course, the sample computed on the first transactions is not very representative of the entire dataset (i.e., far from the final sample) and it is possible to miss true outliers.

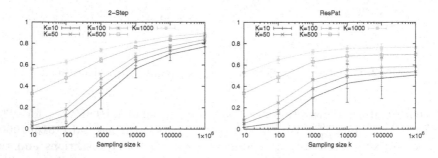

Fig. 3. Accuracy comparison between 2-STEP (left) and RESPAT (right)

Figure 3 plots the average accuracy of a multi-pass FPOF (2-STEP) and a one-pass FPOF (RESPAT$_{no}$) with the sample size k for retrieving the top-K outliers in all the datasets having a fixed size transaction except connect. Interestingly, we observe that our approach approximatively retrieves in a data stream the outliers that would be obtained by storing all the data observations (using 2-STEP). As expected, the accuracy of the two approaches increases rapidly with the sample size k. The gain of RESPAT is very strong between 10 and 1,000 but, much lower between 1000 and 1,000,000. The higher the number of outliers K, the more accurate the approach. On the one hand, the imprecision of the sampling only has an impact around the Kth outlier. On the other hand, a high K makes it possible to build a more representative sample from the first K transactions (which are all considered as outliers at the beginning of the pass). The latter explains why our approach is slightly less stable and less accurate than a non-streaming context with 2-STEP.

6   Conclusion

This paper presents the first frequent pattern sampling approach in data streams based on reservoir sampling. The strength of our generic algorithm is do deal with any damping function while having a space complexity only linear with the sample size. We have also shown how to optimize this algorithm for three damping functions usuallly considered in the state-of-the-art. Surprisingly, our theoretical analysis proves that these algorithms work best when the damping is low. Of course, they turn out to be slower than the two-step random procedure, but they require a limited memory space essential to process data streams or to process datasets that do not fit in memory. Finally, a use case illustrates the practical interest of an online pattern sample to detect outliers in one pass. Of course, this simple outlier detection method could be improved by keeping more transactions as candidate outliers and by using a bound to get statistical guarantees on rejected transactions as done in [14]. We would like to extend our approach to other languages and other interestingness measures. In both cases, the challenge lies in extending the index operator for mapping each value to a specific occurrence within a data observation. Finally, it would be interesting to consider a dynamic damping function for learning with drift detection [13].

References

1. Aggarwal, C.C.: On biased reservoir sampling in the presence of stream evolution. In: Proceedings of VLDB, pp. 607–618. VLDB Endowment (2006)
2. Aggarwal, C.C.: Managing and Mining Sensor Data. Springer, Heidelberg (2013). https://doi.org/10.1007/978-1-4614-6309-2
3. Al Hasan, M., Zaki, M.J.: Output space sampling for graph patterns. Proc. VLDB **2**(1), 730–741 (2009)
4. Babcock, B., Datar, M., Motwani, R.: Sampling from a moving window over streaming data. In: Proceedings of ACM-SIAM Symposium on Discrete Algorithms, pp. 633–634. Society for Industrial and Applied Mathematics (2002)
5. Boley, M., Lucchese, C., Paurat, D., Gärtner, T.: Direct local pattern sampling by efficient two-step random procedures. In: Proceedings of KDD, pp. 582–590. ACM (2011)
6. Calders, T., Dexters, N., Gillis, J.J., Goethals, B.: Mining frequent itemsets in a stream. Inf. Syst. **39**, 233–255 (2014)
7. Chi, Y., Wang, H., Yu, P.S., Muntz, R.R.: Moment: maintaining closed frequent itemsets over a stream sliding window. In: Proceedings of ICDM, pp. 59–66. IEEE (2004)
8. De Francisci Morales, G., Bifet, A., Khan, L., Gama, J., Fan, W.: IoT big data stream mining. In: Proceedings of KDD, pp. 2119–2120 (2016)
9. Diop, L., Diop, C.T., Giacometti, A., Li, D., Soulet, A.: Sequential pattern sampling with norm-based utility. Knowl. Inf. Syst. **62**(5), 2029–2065 (2019). https://doi.org/10.1007/s10115-019-01417-3
10. Efraimidis, P.S., Spirakis, P.G.: Weighted random sampling with a reservoir. Inf. Process. Lett. **97**(5), 181–185 (2006)
11. Gaber, M.M., Zaslavsky, A., Krishnaswamy, S.: Mining data streams: a review. ACM SIGMOD Rec. **34**(2), 18–26 (2005)

12. Gama, J.: A survey on learning from data streams: current and future trends. Progr. Artif. Intell. **1**(1), 45–55 (2012)
13. Gama, J., Medas, P., Castillo, G., Rodrigues, P.: Learning with drift detection. In: Bazzan, A.L.C., Labidi, S. (eds.) SBIA 2004. LNCS (LNAI), vol. 3171, pp. 286–295. Springer, Heidelberg (2004). https://doi.org/10.1007/978-3-540-28645-5_29
14. Giacometti, A., Soulet, A.: Frequent pattern outlier detection without exhaustive mining. In: Bailey, J., Khan, L., Washio, T., Dobbie, G., Huang, J.Z., Wang, R. (eds.) PAKDD 2016. LNCS (LNAI), vol. 9652, pp. 196–207. Springer, Cham (2016). https://doi.org/10.1007/978-3-319-31750-2_16
15. Giannella, C., Han, J., Pei, J., Yan, X., Yu, P.S.: Mining frequent patterns in data streams at multiple time granularities. Next Gener. Data Min. **212**, 191–212 (2003)
16. Jiang, N., Gruenwald, L.: Research issues in data stream association rule mining. ACM SIGMOD Rec. **35**(1), 14–19 (2006)
17. Jin, R., Agrawal, G.: Frequent pattern mining in data streams. In: Aggarwal, C.C. (ed.) Data Streams, vol. 31, pp. 61–84. Springer, Boston (2007). https://doi.org/10.1007/978-0-387-47534-9_4
18. Karim, M.R., Cochez, M., Beyan, O.D., Ahmed, C.F., Decker, S.: Mining maximal frequent patterns in transactional databases and dynamic data streams: a spark-based approach. Inf. Sci. **432**, 278–300 (2018)
19. Krempl, G., et al.: Open challenges for data stream mining research. ACM SIGKDD Explor. Newsl. **16**(1), 1–10 (2014)
20. Li, K.H.: Reservoir-sampling algorithms of time complexity O(n(1+log(N/n))). ACM Trans. Math. Softw. (TOMS) **20**(4), 481–493 (1994)
21. Manku, G.S., Motwani, R.: Approximate frequency counts over data streams. In: Proceedings of VLDB, pp. 346–357. Elsevier (2002)
22. Martin, T., Francoeur, G., Valtchev, P.: CICLAD: a fast and memory-efficient closed itemset miner for streams. In: Proceedings of KDD, pp. 1810–1818 (2020)
23. Raïssi, C., Poncelet, P.: Sampling for sequential pattern mining: from static databases to data streams. In: Proceedings of ICDM, pp. 631–636. IEEE (2007)
24. Ramírez-Gallego, S., Krawczyk, B., García, S., Woźniak, M., Herrera, F.: A survey on data preprocessing for data stream mining: current status and future directions. Neurocomputing **239**, 39–57 (2017)
25. ur Rehman, M.H., Liew, C.S., Wah, T.Y., Khan, M.K.: Towards next-generation heterogeneous mobile data stream mining applications: opportunities, challenges, and future research directions. J. Netw. Comput. Appl. **79**, 1–24 (2017)
26. Tanbeer, S.K., Ahmed, C.F., Jeong, B.S., Lee, Y.K.: Sliding window-based frequent pattern mining over data streams. Inf. Sci. **179**(22), 3843–3865 (2009)
27. Vitter, J.S.: Random sampling with a reservoir. ACM Trans. Math. Softw. (TOMS) **11**(1), 37–57 (1985)
28. Wong, R.C.W., Fu, A.W.C.: Mining top-k frequent itemsets from data streams. Data Min. Knowl. Disc. **13**(2), 193–217 (2006)

Discovering Proper Neighbors to Improve Session-Based Recommendation

Lin Liu[ID], Li Wang[✉][ID], and Tao Lian[ID]

Data Science College, Taiyuan University of Technology,
Jinzhong 030600, Shanxi, China
wangli@tyut.edu.cn

Abstract. Session-based recommendation shows increasing importance in E-commerce, news and multimedia applications. Its main challenge is to predict next item just using a short anonymous behavior sequence. Some works introduce other close similar sessions as complementary to help recommendation. But users' online behaviors are diverse and very similar sessions are always rare, so the information provided by such similar sessions is limited. In fact, if we observe the data at the high level of coarse granularity, we will find that they may present certain regularity of content and patterns. The selection of close neighborhood sessions at tag level can solve the problem of data sparsity and improve the quality of recommendation. Therefore, we propose a novel model CoKnow that is a collaborative knowledge-aware session-based recommendation model. In this model, we establish a tag-based neighbor selection mechanism. Specifically, CoKnow contains two modules: Current session modeling with item tag(Cu-tag) and Neighbor session modeling with item tag (Ne-tag). In Cu-tag, we construct an item graph and a tag graph based on current session, and use graph neural networks to learn the representations of items and tags. In Ne-tag, a memory matrix is used to store the representations of neighborhood sessions with tag information, and then we integrate these representations according to their similarity with current session to get the output. Finally, the outputs of these two modules are combined to obtain the final representation of session for recommendation. Extensive experiments on real-world datasets show that our proposed model outperforms other state-of-the-art methods consistently.

Keywords: Neighborhood sessions · Tag sequence graph · Memory network · Graph neural network · Session-based recommendation

1 Introduction

Session-based recommendation has become more and more popular in the field of recommendation systems, which aims to predict the next behavior based on anonymous user's behavior sequence in a short time. At the beginning of the research, item-to-item recommendations [1] are mainly used for this task. Since this method ignores the sequential information that plays a necessary role in the

© Springer Nature Switzerland AG 2021
N. Oliver et al. (Eds.): ECML PKDD 2021, LNAI 12975, pp. 353–369, 2021.
https://doi.org/10.1007/978-3-030-86486-6_22

session-based recommendation, Markov chain-based methods follow to predict the user's next behavior based on the previous one [2]. But this kind of method only models local sequential behavior, and when we try to include all sequences of the user, the state space will quickly become unmanageable.

With the development of deep learning, recurrent neural networks which are good at processing sequential data, gradually become active in session-based recommendation [3–5]. Hidasi et al. [3] apply Gated Recurrent Units (GRUs) to model behavior sequences. Li et al. [4] combines attention mechanism with RNN, while capturing the sequential behavior characteristics and user's main purpose. Although RNN-based methods can capture the sequential dependencies among behaviors in a session, it is difficult to model complex transition patterns in a session. To solve this problem, researchers begin to use graph neural networks for session-based recommendation. Wu et al. [6] use graph neural networks to model session sequences to capture complex transition mode between behaviors. On this basis, Liu et al. [7] introduce category-level information to enhance the expressive ability of the session. In addition, memory networks [8] have also been used in recommendation tasks to store the information of neighborhood sessions. Wang et al. [9] propose CSRM which applies memory network to store the representations of neighborhood sessions. However, this method only models from item level, the sparseness of neighbor information leads to biases in predictions.

Figure 1 shows an example of three sessions. On the one hand, if we analyze the sequence from the item level, we can find that the three sessions contain five, four and five completely different items respectively. The unique ID of each item makes the session sequence more confusing and complicated, which makes it difficult for us to dig out the user's real preferences. From the tag level, it can be found that these three sessions actually correspond to two, three and two types of items respectively, which greatly simplifies our judgment of user's preferences. On the other hand, looking at the three session sequences at the tag level, we can observe that each user has the most clicks on the phone. In other words, the behavior patterns of these three users are very similar. Therefore, we can refer to the first two sessions to find the real intent of the current session(session 3) when predicting the next action of current session. Combined with the information of neighborhood sessions with item tags, we can determine that the current user may want to buy a mobile phone. All brands and models may be selected by this user, rather than just limited to the products that appeared in neighborhood sessions. This approach alleviates the sparsity of neighbor information to a certain extent.

In this paper, we propose a model that improves session-based recommendation performance by discovering proper neighbors, namely CoKnow. We create Cu-tag module and Ne-tag module combined with tag information to model the current session and neighbor session respectively. The main contributions of our work can be summarized as follows:

(1) We establish a tag-aware neighborhood session selection mechanism to assist us in session-based recommendation. This method effectively reduces the sparsity of neighbor information.

(2) The expression of our current session is to integrate the information of the neighborhood session. Both the current session and neighborhood sessions are modeled in consideration of both the item level and the tag level, and the graph neural networks are used to capture complex transfer relationships.

(3) Extensive experiments conducted on real-world datasets indicate that CoKnow evidently outperforms the state-of-art methods.

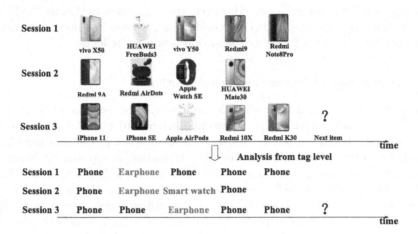

Fig. 1. An example of session sequences.

2 Related Work

2.1 Collaborative Filtering-Based Methods

Collaborative filtering-based recommendation is to discover user's preferences through the mining of user's historical behavior data and group users based on different preferences to recommend items with similar tastes. And it is mainly based on the following assumption: users with similar choices tend to have similar preferences. Recommendation based on collaborative filtering contains memory-based collaborative filtering and model-based collaborative filtering.

Memory-based collaborative filtering aims to find similar users or items by using a specific similarity function to generate recommendations. Such algorithms include user-based recommendations and item-based recommendations. User-based collaborative filtering calculates the similarity between users by analyzing their behaviors, and then recommendation items that similar users like. Jin et al. [10] present an optimization algorithm to automatically compute the weights for different items based on their ratings from training users. Item-based collaborative filtering method calculates the similarity between items, and recommends items that are more similar to the items the user likes. Sarwar et al. [1] analyze different item-based recommendation methods and different techniques of similarity calculation. Model-based collaborative filtering is to train a model with the help of technology such as machine learning, and then predict the target user's score for the item. Wu et al. [11] generalize collaborative filtering models by integrating a user-specific bias into an auto-encoder. He et al. [12] propose to leverage a multi-layer perceptron to learn the user-item interaction function. Wang et al. [9] propose CSRM to model collaborative filtering in session-based recommendation, using GRU to model the current session, and combining the collaborative information of neighborhood sessions to achieve gratifying results in session-based recommendation. However, the neighbor selection method based on item similarity adopted in this method obtains sparse neighbor information. Therefore, the sparsity of neighbor selection is still a challenge.

2.2 Graph Neural Networks-Based Methods

At present, graph neural networks [13] are widely used because of their strong ability to simulate relationships between different objects. Wu et al. [6] first construct each historical session sequence as a directed graph and use Grated Neural Networks to capture the complex transition mode among behaviors, which achieves better performance than RNN-based methods. Similarly, Xu et al. [14] use a multi-layer self-attention network to capture the global dependencies between items in a session. Qiu et al. [15] utilize graph neural networks to explore the inherent item transition patterns in a session, which is more than plain chronological order. Wang et al. [16] propose a multi-relational graph neural network model which models sessions using other types of user behavior beyond clicks. Liu et al. [7] introduce category-level representations and utilize graph neural networks to receive information at item level and category level. In a word, graph neural networks are good at modeling complex transfer relationships, but the above methods only simulate the behavior of the current session. In many cases, the information provided by the current session is far from enough.

2.3 Memory Network-Based Methods

Recently, memory networks for recommendation systems have received much attention because they achieve some good performance. Chen et al. [17] propose a memory-augmented neural network (MANN) integrated with the insights of collaborative filtering for recommendation. Huang et al. [18] propose a knowledge enhanced sequential recommendation method which integrates the RNN-based networks with Key-Value Memory Network (KV-MN). CSRM [9] utilizes memory networks to save the information of neighborhood sessions. These methods(such as CSRM) have improved recommendation performance through collaborative information. But it still has inherent limitations in dealing with the sparsity of neighbor information. Therefore, we try to use more coarse-grained tag information to make up for this defect.

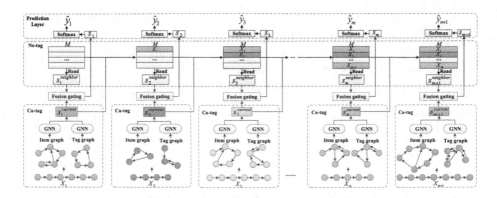

Fig. 2. The framework of CoKnow.

3   The Proposed Method: CoKnow

In this article, we propose CoKnow, which is a method to discover suitable neighbors with the help of tag information to improve session-based recommendation. The model framework is shown in Fig. 2. Specifically, the model contains two modules: Cu-tag and Ne-tag, modeling the current session and neighbor sessions respectively. In this section, we introduce these two modules in detail.

3.1   Problem Definition

Session-based recommendation aims to predict the next item that user will interact with, only based on user's anonymous interactive sequence. Here, we give some notations related to the task of session-based recommendations. Let $V = \{v_1, v_2, ..., v_M\}$ denote the set of all unique items involved in all sessions, $C = \{c_1, c_2, ..., c_N\}$ denote the set of item tag, where M and N represent the number of items and tags in the dataset respectively, and each item $v_i \in V$ corresponds to a tag $c_k \in C$. We define the session set in a dataset as $X = \{X_1, X_2, ..., X_Q\}$, where Q is the number of session sequences. And each session sequence X_i consists of two sequences: item sequence $S_i^m = [v_1, v_2, ..., v_n]$ and tag sequence $S_i^c = [c_1, c_2, ..., c_n]$. The goal of session-based recommendation is to calculate the recommendation probability $\widehat{y}_i$ of each candidate item, and get a list of probability $\widehat{y} = \{\widehat{y}_1, \widehat{y}_2, ..., \widehat{y}_M\}$.

3.2   Current Session Modeling with Item Tag (Cu-tag)

To model the current session, we jointly model from the item level and the tag level to get the representation of the current session, shown as Fig. 3. And the modeling of these two levels uses graph neural network for the same processing, so here is a unified description.

Learning the Embedding of Nodes in the Graph. Each item sequence $S_i^m = [v_1, v_2, ..., v_n]$ and its corresponding tag sequence $S_i^c = [c_1, c_2, ..., c_n]$ can

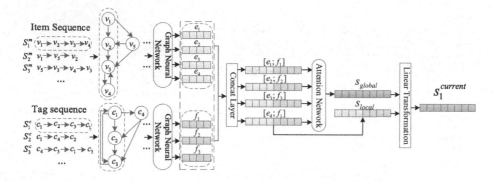

Fig. 3. The framework of the Cu-tag module.

be regarded as graph structure data, and they can be constructed as directed graphs. Here, we use $\mathcal{G}_m = (\nu_m, \varepsilon_m)$ to represent the directed graph constructed by the item sequence, each node refers to an item $v_i \in V$. And $\mathcal{G}_c = (\nu_c, \varepsilon_c)$ represents the directed graph constructed by the tag sequence, each node refers to a tag $c_i \in C$. For example, the item sequence is $S_i^m = [v_1, v_2, v_3, v_4, v_3, v_5]$ and the corresponding tag sequence is $S_i^c = [c_1, c_2, c_1, c_3, c_1, c_4]$. The constructed directed graph and connection matrix are Fig. 4(a), Fig. 4(b) and Fig. 4(c), Fig. 4(d). It should be noted that due to the repeated occurrence of a certain item or tag in an sequence, we assign a normalized weight to each edge, which is calculated as the occurrence of the edge divided by the outdegree of that edge's start node.

After constructing the direct graph and connection matrix, we obtain latent vectors of nodes via graph neural networks. At first, we embed every node into a united embedding space. The vector x_i denotes the latent vector of node learned via graph neural networks. The update functions are given as follows:

$$p^t = Concat(A_{In}^i([x_1^{t-1}, ..., x_n^{t-1}]^\top H_{In} + b_{In}), A_{Out}^i([x_1^{t-1}, ..., x_n^{t-1}]^\top H_{Out} + b_{Out})) \tag{1}$$

$$z^t = \sigma(W_z p^t + U_z x_i^{t-1}) \tag{2}$$

$$r^t = \sigma(W_r p^t + U_r x_i^{t-1}) \tag{3}$$

$$\widetilde{x_i^t} = tanh(W_o p^t + U_o(r^t \odot x_i^{t-1})) \tag{4}$$

$$x_i^t = (1 - z^t) \odot x_i^{t-1} + z^t \odot \widetilde{x_i^t} \tag{5}$$

where W_z, W_r, $W_o \in \mathbb{R}^{2d \times d}$, U_z, U_r, $U_o \in \mathbb{R}^{d \times d}$, H_{In}, $H_{Out} \in \mathbb{R}^{d \times d}$ are learnable parameters. b_{In}, $b_{Out} \in \mathbb{R}^d$ are the bias vectors. A_{In}^i, $A_{Out}^i \in \mathbb{R}^{1 \times n}$ are the corresponding rows in the matrices A_{In} and A_{Out}. $\sigma(\cdot)$ denotes the sigmoid function and $\odot$ represents element-wise multiplication. z^t, r^t are update gate and reset gate respectively, which decide what information to be preserved and discarded. When we update nodes in item graph and tag graph in this way, the obtained item and tag representations are expressed as e_i and f_i.

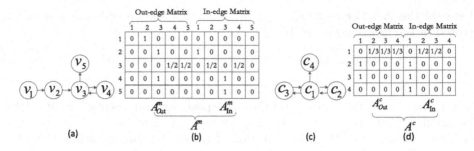

Fig. 4. Examples of item graph, tag graph and their connection matrices.

Current Session Embedding with Tag Information. The representation of the current session embedding consists of local representation and global representation. In our model, we suppose that the last interacting item plays a key role, so we use the representation of the last interacting item with tag information as local representation of the current session, i.e. $s_{local} = [e_n; f_n]$.

For global representation, considering the importance of each item in the session sequence is different, so we combine the attention mechanism to get a global representation of the session:

$$\alpha_i = r^\top \sigma(W_1[e_n; f_n] + W_2[e_i; f_i] + c) \tag{6}$$

$$s_{global} = \sum_{i=1}^{n} \alpha_i[e_i; f_i] \tag{7}$$

where $r \in \mathbb{R}^d$ and $W_1, W_2 \in \mathbb{R}^{d \times 2d}$ are learnable parameters.

Finally, we compute the current session embedding $s_i^{current}$ by taking linear transformation over the concatenation of the local embedding and global embedding:

$$s_i^{current} = W_3[s_{local}; s_{global}] \tag{8}$$

where $W_3 \in \mathbb{R}^{d \times 4d}$ is weight matrix.

3.3 Neighbor Session Modeling with Item Tag (Ne-tag)

The Cu-tag module only utilizes the information contained in the current session, which ignores the collaborative information in the neighborhood sessions that can supplement the expression of the current session. To address this omission, we design a neighbor session modeling method with tag participation. The Ne-tag module combines tag information to select neighborhood sessions that have the most similar behavior pattern to the current session and use them as auxiliary information of the current session, so as to better judge the user preference of the current session.

After using the Cu-tag module to obtain the representations of m last sessions, they are stored in the memory matrix M. Then we calculate the similarity between the global embedding s_{global} of the current session and each session m_i stored in the memory matrix M:

$$sim(s_{global}, m_i) = \frac{s_{global} \cdot m_i}{||s_{global}|| \times ||m_i||} \quad \forall m_i \in M \tag{9}$$

According to similarity values, we can get the k largest similarity values $[sim_1, sim_2, ..., sim_{k-1}, sim_k]$ and the corresponding k representations of neighborhood sessions $[m_1, m_2, ..., m_{k-1}, m_k]$. Since each neighborhood session contributes differently to the representation of current session, we utilize the softmax function to process the k similarity values, and more similar sessions will get greater weight. Finally, the ultimate neighborhood session embedding is obtained by integrating the information of k neighborhood sessions:

$$w_i = \frac{exp(\beta sim_i)}{\sum_{j=1} exp(\beta sim_j)} \tag{10}$$

$$s_i^{neighbor} = \sum_{i=1}^{k} w_i m_i \tag{11}$$

where β denotes the strength parameter.

3.4 Prediction Layer

Here, we apply a fusion gating mechanism to selectively combine the information of the current session and the neighborhood sessions to obtain the ultimate expression of the current session. The fusion gate g_i is given by:

$$g_i = \sigma(W_l s_{local} + W_g s_{global} + W_o s_i^{neighbor}) \tag{12}$$

Therefore, we can calculate the representation of the current session combined with the collaborative information of the neighborhood sessions:

$$s_i = g_i s_i^{current} + (1 - g_i) s_i^{neighbor} \tag{13}$$

Then, we compute the recommendation probability $\hat{y}$ by softmax function for each candidate item $v_i \in V$. The addition of tag information and neighbor information is beneficial to improve the score of candidate items that belong to same or similar tags as the items in the current session. The calculation of the recommendation probability is as follows:

$$\hat{y} = softmax(s_i^{\top}[e_i; f_i]) \tag{14}$$

We use cross-entropy loss as our loss function, which can be written as follows:

$$L(\hat{y}) = -\sum_{i=1}^{M} y_i log(\hat{y}_i) + (1 - y_i)log(1 - \hat{y}_i) + \lambda \parallel \theta \parallel^2 \tag{15}$$

where y is the one-hot encoding vector of the ground truth item, θ is the set of all learnable parameters.

Algorithm 1 elaborates the construction process of Cu-tag and Ne-tag in our proposed CoKnow.

Algorithm 1. CoKnow

Input: session sequence X_i, which contains item sequence S_i^m, tag sequence S_i^c; connection matrix A^m, A^c

Output: the probability list $\widehat{y} = \{\widehat{y}_1, \widehat{y}_2, ..., \widehat{y}_M\}$

1: Construct Cu-tag module:
2: **for** $S_i^m, S_i^c \in X_i$ **do**
3:     use (S_i^m, A^m) and (S_i^c, A^c) to construct item graph and tag graph
4:     obtain the representation of nodes e_i^t and f_i^t in the item graph and tag graph respectively using GNN based on Eq.(1)-(5)
5:     concatenate e_n and f_n to obtain the local embedding s_{local}
6:     obtain global embedding s_{global} based on attention mechanism(Eq. (6)-(8))
7:     combine s_{local} with s_{global} to obtain the current session embedding $s_i^{current}$
8:     store $s_i^{current}$ in the outer memory matrix M
9: **return**
10: Construct Ne-tag module:
11: **for** $m_i \in M$ **do**
12:     calculate the similarity between s_{global} and m_i
13:     obtain the neighborhood session embedding $s_i^{neighbor}$ based on Eq.(9)-(11)
14: **return**
15: combine $s_i^{current}$ with $s_i^{neighbor}$ to obtain the ultimate current session embedding s_i (Eq.(12)-(13))
16: compute the probability $\widehat{y}$ based on Eq.(14)

---

4   Experiments

4.1   Research Questions

In this section, we will answer the following research questions:

(RQ1) How is the performance of CoKnow compared with other state-of-the-art methods?

(RQ2) How does CoKnow perform on sessions with different lengths?

(RQ3) Does the existence of each module in CoKnow has some significance?

(RQ4) Will the prediction performance of CoKnow be influenced when the number of neighborhood sessions is different?

4.2   Datasets

We evaluate the proposed model on the following two real e-commerce datasets:

- **Cosmetics[1]:** This is a user behavior record of a medium cosmetics online store in October and November 2019. It is a public dataset published on kaggle competition platform. Each record contains attribute information, such as session_id, item_id, item_category and timestamp. It is worth mentioning that we are first to try to perform a session-based recommendation task on this dataset.

---

[1] https://www.kaggle.com/mkechinov/ecommerce-events-history-in-cosmetics-shop.

- **UserBehavior[2]:** A Taobao user behavior dataset provided by Alimama. It contains a series of purchase behavior records of users during November 25 to December 03, 2017. Each record contains attribute information, such as userID, itemID, item_category and timestamp. The userID is directly used as session_id.

Table 1. Statistics of datasets used in the experiments.

Statistics	Cosmetics		UserBehavior
	2019-Oct	2019-Nov	
# of training sessions	33087	20127	2816
# of test sessions	6322	4765	706
# of items	22063	16175	14641
# of categories	103	98	1403
Average length	6.9938	9.1128	18.6063

In this paper, we filter out all session sequences with a length shorter than 2 or longer than 50 items [5] and items appearing less than 5 times. In addition, similar to [19], we use the data augmentation method to generate sequences and corresponding labels by splitting the input session. Then, a session sequence of length n is divided into $n - 1$ sub-session sequences. For an input item sequence $S_i^m = [v_1, v_2, ..., v_n]$, we generate the sequences and corresponding labels $([v_1], v_2), ([v_1, v_2], v_3), ..., ([v_1, v_2, ..., v_{n-1}], v_n)$ for training and testing on all datasets, where $[v_1, v_2, ..., v_{n-1}]$ is the generated sequence and v_n is the label of the sequence, i.e. the next interactive item. Similarly, the same processing is performed for the tag sequence. The data statistics are shown in Table 1.

4.3   Baselines

To evaluate the prediction performance, we compare CoKnow with the following representative baselines:

- **Pop:** Pop always recommends the most popular items in the training set. It is a simple method that still performs well in some scenarios.
- **Item-KNN** [1]: A baseline method based on the cosine similarity to recommend items that are most similar to the candidate item within the session.
- **FPMC** [20]: This is a method that combines Markov chain model and matrix factorization for the next-basket recommendation. Here, we ignore user latent representations to make it suitable for session-based recommendation.
- **GRU4Rec-topK** [21]: Based on deep learning, this method uses RNN to model the user's interaction sequence. And it improved GRU4Rec with a top-K based ranking loss.

---

[2] https://tianchi.aliyun.com/dataset/dataDetail?dataId=649&userId=1.

- **NARM** [4]: This model combines the attention mechanism with RNN, while capturing sequential behavior characteristics and main purpose of users.
- **STAMP** [22]: STAMP not only considers the general interest from long-term historical behavior, but also considers the user's last click to mine the immediate interest
- **NEXTITNET** [23]: It is a CNN-based generative model. In this model, a stack of dilated convolutional layers is applied to increase the receptive field when modeling long-range sequences.
- **SR-GNN** [6]: A recently proposed algorithm for session-based recommendation applying graph neural network. It constructs the item sequence as a directed graph, and combines attention mechanism to obtain the embedding of each session sequence.
- **CSRM** [9]: This is the model most relevant to our model. It utilizes GRU to model the current session at the item level, and applies an outer memory module to model neighborhood sessions.
- **CaSe4SR** [7]: It is a deep learning model that utilizes graph neural network to model item sequence and category sequence respectively. The difference from the model in this paper is that it only models the current session and does not consider collaborative information.

4.4   Evaluation Metrics and Experimental Setup

Evaluation Metrics. We use the most commonly used metrics Recall@20 and MRR@20 for session-based recommendation to evaluate model performance. Recall@20 is the proportion of ground-truth items appearing in the top-20 positions of the recommendation list. It does not consider the rank of the item that user actually clicked. MRR@20 is the average of the inverse of the ground-truth item ranking. If the rank of an item is greater than 20, the value is set to 0. This indicator takes the position of the item in the recommendation list into account, and is usually important in some order-sensitive tasks.

Experimental Setup. During training, we randomly initialize all parameters with a Gaussian distribution with a mean of 0 and a standard deviation of 0.1. The dimension of the embedding vector is set to $d = 100$. The mini-batch Adam optimizer is exerted to optimize these parameters, where the initial learning rate is set to 0.001. In addition, the training batch size is set to 512 and the L2 penalty is 10^5. We vary the number of neighborhood sessions from [128, 256, 512] to study the effects of Ne-tag.

4.5   Results and Analysis

Comparison with Baseline Methods (RQ1). To demonstrate the performance of the proposed model, we compare CoKnow with other state-of-art session-based recommendation methods and the results can be seen in Table 2. It shows that CoKnow consistently achieves the best performance in terms of both Recall@20 and MRR@20 on two datasets. Overall, the recommendation performance of all methods on UserBehavior is low. This may because the time interval

Table 2. The performance comparison of CoKnow with other baseline methods over two datasets.

Method	Cosmetics				UserBehavior	
	2019-Oct		2019-Nov			
	Recall@20	MRR@20	Recall@20	MRR@20	Recall@20	MRR@20
Pop	0.0420	0.0104	0.0470	0.0109	0.0148	0.0038
Item-KNN	0.1519	0.0601	0.1657	0.0659	0.0428	0.0143
FPMC	0.2422	0.1877	0.2092	0.1623	0.0711	0.0478
GRU4Rec-topK	0.4007	0.2447	0.4048	0.2210	0.0855	0.0318
NARM	0.4103	0.2271	0.4149	0.2141	0.0806	0.0328
STAMP	0.3738	0.1967	0.4148	0.2294	0.0621	0.0240
NEXTITNET	0.3303	0.1741	0.3488	0.1639	0.0353	0.0164
SR-GNN	0.3325	0.1655	0.3377	0.1610	0.0481	0.0181
CSRM	<u>0.4515</u>	0.2530	0.4367	0.2283	0.0906	0.0431
CaSe4SR	0.4396	<u>0.2556</u>	<u>0.4551</u>	<u>0.2585</u>	<u>0.1865</u>	<u>0.0880</u>
CoKnow(ours)	**0.4558**	**0.2669**	**0.4669**	**0.2694**	**0.1989**	**0.1094**

in the sessions on this dataset is longer than other datasets, and as can be seen from Table 1, there are more item categories in this dataset, which means that the user's behavior is more complicated, making it harder to judge user's real intentions. Our model has been greatly improved compared to other models, which also confirms the ability of CoKnow to handle complex behavior sequences. In conventional methods, Item-KNN achieves some improvement over Pop which only considers the popularity of the items, which shows that considering the similarity of items in the session can improve the accuracy rate. The biggest difference among FPMC and the above two models is that it models user's historical behavior records. Experimental results show the contribution of sequential information to recommendation performance. Furthermore, deep learning-based methods generally outperform the conventional algorithms. GRU4Rec-topK and NARM use recurrent units to capture the general interests of users and have achieved more prominent prediction results, which indicates the effectiveness of RNN in sequence modeling. STAMP also obtains better results by combining general interests with current interests(the last-clicked item), which indicates the importance of the last click. In comparison, the prediction performance of NEXTITNET based on CNN is still slightly worse, indicating that CNN is still not as capable of capturing sequential information as RNN on our dataset. CSRM receives higher recommendation performance than SR-GNN, reflecting the role of neighborhood sessions. Looking at these baseline methods, CaSe4SR is the best recommendation method, which fully shows the effectiveness of category information for session-based recommendation.

Analysis on Sessions with Different Lengths (RQ2). In addition to verifying the prediction ability of our model on all sessions, we also group sessions in the datasets according to the length to verify the validity of our model. Specif-

Table 3. The performance of different methods on sessions with different lengths.

Method		Cosmetics				UserBehavior	
		2019-Oct		2019-Nov			
		Recall@20	MRR@20	Recall@20	MRR@20	Recall@20	MRR@20
Short	GRU4Rec-topK	0.4898	0.3518	0.5066	0.2879	–	–
	NARM	0.6207	0.4826	0.5788	0.3832	–	–
	STAMP	0.5462	0.4149	0.3859	0.2400	–	–
	NEXTITNET	0.4653	0.3272	0.4673	0.2819	–	- -
	SR-GNN	0.5396	0.3919	0.5110	0.3333	–	–
	CSRM	0.5962	0.4294	0.5685	0.3789	–	–
	CaSe4SR	**0.6356**	0.5049	0.5916	**0.4274**	–	–
	CoKnow(ours)	0.6223	**0.5161**	**0.6022**	0.4239	–	–
Medium	GRU4Rec-topK	0.3481	0.1491	0.3835	0.1753	0.1576	0.0733
	NARM	0.4012	0.2092	0.4342	0.2305	0.2342	0.1179
	STAMP	0.2889	0.1412	0.3422	0.1797	0.1171	0.0501
	NEXTITNET	0.2462	0.1175	0.2658	0.1229	0.0563	0.0363
	SR-GNN	0.3599	0.1813	0.3917	0.2017	0.1136	0.0455
	CSRM	0.3978	0.2091	0.4215	0.2246	0.2036	0.0815
	CaSe4SR	0.4130	0.2360	0.4438	0.2475	0.2827	0.1309
	CoKnow(ours)	**0.4192**	**0.2677**	**0.4441**	**0.2874**	**0.3092**	**0.1476**
Long	GRU4Rec-topK	0.3155	0.1412	0.3268	0.1584	0.0826	0.0293
	NARM	0.2425	0.0948	0.2802	0.1182	0.0467	0.0190
	STAMP	0.2694	0.1134	0.2939	0.1331	0.0479	0.0173
	NEXTITNET	0.2233	0.0864	0.2420	0.0966	0.0251	0.0125
	SR-GNN	0.2341	0.0946	0.2525	0.1033	0.0682	0.0255
	CSRM	0.3085	0.1358	0.3297	0.1545	0.0846	0.0364
	CaSe4SR	0.3503	0.1897	0.3707	0.2045	0.1742	0.0769
	CoKnow(ours)	**0.3666**	**0.2133**	**0.3788**	**0.2268**	**0.1815**	**0.0932**

ically, we divide each dataset into three groups based on the average length of sessions in datasets in Table 1, that is, the length is less than or equal to 5, the length is less than or equal to 15 and the length is greater than 15. We named these three groups "Short", "Medium" and "Long" respectively. It is worth noting that because the number of sessions in the "Short" group of UserBehavior is too small, we have not conducted experiments on this group of UserBehavior here.

It is obvious that as the length of session increases, the performance on all two datasets in terms of Recall@20 and MRR@20 decreases to varying degrees. This may be because as the length of the session increases, the number of items that users click on out of curiosity or accident will increase, resulting in an increase in irrelevant items in the session sequence, which is not conducive to the extraction of user preferences. Compared with other groups, the prediction performance of all models on short sessions is much better than other groups. In "Short" group,

NARM achieves the best performance among RNN-based baseline methods on all datasets, but as the length of the session increasing, its performance drops quickly. This may explain that RNN has difficulty coping with long sessions. SR-GNN and CSRM have also achieved good recommendations on the two datasets. This shows the role of graph neural networks and neighborhood information in capturing user preferences. The performance of CaSe4SR is more prominent.

According to the results, we can observe that our model performs best on different groups of all datasets, which fully proves that the information of tag and neighbors is an effective complement to the capture of user's intention in long and short sessions. Although the result on CaSe4SR in short sessions slightly exceeds CoKnow, this may be because the model cannot accurately capture user's preferences due to too little behavior, and the joining of neighborhood sessions has misled the current session.

Impact of Different Modules (RQ3). In order to illustrate more clearly the validity of each party of our model, we compare CoKnow with its three variants: (1) $CoKnow_{cu}$: CoKnow without the Ne-tag module and only graph neural networks are used to model the current session. It is actually CaSe4SR. (2) $CoKnow_{ne}$: CoKnow without the Cu-tag module which only uses the Ne-tag module to model neighborhood sessions. (3) $CoKnow_{item}$: CoKnow only constructs a model at item level, that is, it removes the tag information.

It can be seen from Table 4 that the prediction result of CoKnow combing two modules is the best. $CoKnow_{cu}$ outperforms $CoKnow_{ne}$, which indicates that a session's own information is more important than collaborative information of the neighborhood sessions. Besides, the prediction effect of $CoKnow_{item}$ is slightly worse than CoKnow, which shows that the tag information can supplement the information of the session to a certain degree, so as to capture user's real intention more accurately. Especially on the more complicated User-Behavior, the prediction performance of our proposed model CoKnow has been greatly improved compared to $CoKnow_{item}$. This fully indicates that tag information can simplify the sequence with various behaviors, which is beneficial to the capture of user's intentions.

Influence of the Number of Neighbors k (RQ4). The number of neighborhood sessions also has a certain effect on the prediction performance of the model. Therefore, we compare the performance of the models when the num-

Table 4. Performance comparison of different variants of CoKnow.

Method	Cosmetics				UserBehavior	
	2019-Oct		2019-Nov			
	Recall@20	MRR@20	Recall@20	MRR@20	Recall@20	MRR@20
$CoKnow_{cu}$	0.4396	0.2556	0.4551	0.2585	0.1865	0.0880
$CoKnow_{ne}$	0.3385	0.1812	0.3252	0.1624	0.0656	0.0217
$CoKnow_{item}$	0.4543	0.2666	0.4655	0.2688	0.1209	0.0754
CoKnow(ours)	**0.4558**	**0.2669**	**0.4669**	**0.2694**	**0.1989**	**0.1094**

Table 5. Performance comparison of CoKnow with different number of neighborhood sessions k.

Method	Cosmetics				UserBehavior	
	2019-Oct		2019-Nov			
	Recall@20	MRR@20	Recall@20	MRR@20	Recall@20	MRR@20
$k = 128$	**0.4558**	0.2669	**0.4669**	0.2694	**0.1989**	**0.1094**
$k = 256$	0.4499	**0.2706**	0.4658	0.2714	0.1949	0.1085
$k = 512$	0.4506	0.2655	0.4665	**0.2722**	0.1940	0.1059

ber of neighbors takes different values. It can be seen from Table 5 that not all datasets are consistent with the concept that the prediction effect increases with the number of neighborhood sessions. On 2019-Nov, it is true that as the number of neighbors k increases, the values of the MRR@20 metrics are increasing. Other datasets don't exactly follow this pattern. For example, the value of Recall@20 on 2019-Oct is the best when $k = 128$. This may be because an increase in the number of neighbor sessions may bring more irrelevant information to interfere with the expression of the current session for some datasets. Considering the results of all datasets comprehensively, our model CoKnow can achieve the optimal results when $k = 128$. Therefore, in this paper, the number of neighbors is set to 128.

5   Conclusion

In this paper, we propose a method that utilizes tag information to discover proper neighbors to improve session-based recommendation. We establish Cu-tag module and Ne-tag module. The former uses graph neural networks to model the item sequence and tag sequence of the current session, and the latter utilizes memory network to store information of neighborhood sessions to supplement the current session. These two modules are combined via a fusion gating mechanism to achieve better recommendations. Extensive experiments verify that the information of tags and neighbors play a good role in the expression of current session.

The limitation of this work is that we can only consider a limited number of neighborhood sessions, which may miss some more similar neighborhood sessions and affect prediction performance. In future work, we will explore solutions to this problem.

Acknowledgments. This work is supported by the National Natural Science Foundation of China (Grant No. 61872260), particularly supported by Science and Technology Innovation Project of Higher Education Institutions in Shanxi Province (No. 2020L0102).

References

1. Sarwar, B., Karypis, G., Konstan, J., Riedl, J.: Item-based collaborative filtering recommendation algorithms. In: Proceedings of the 10th international conference on World Wide Web (2001)
2. Shani, G., Heckerman, D., Brafman, R.I.: An MDP-based recommender system. J. Mach. Learn. Res. **6**(12/1/2005), 1265–1295 (2005)
3. Hidasi, B., Karatzoglou, A., Baltrunas, L., Tikk, D.: Session-based recommendations with recurrent neural networks. In: International Conference on Learning Representations 2016 (ICLR) (2016)
4. JLi, J., Ren, P., Chen, Z., Ren, Z., Lian, T., Ma, J.: Neural attentive session-based recommendation. In: Proceedings of the 2017 ACM on Conference on Information and Knowledge Management, pp. 1419–1428 (2017)
5. Ren, P., Chen, Z., Li, J., Ren, Z., Ma, J., De Rijke, M. : RepeatNet: a repeat aware neural recommendation machine for session-based recommendation. In: Proceedings of the AAAI Conference on Artificial Intelligence, pp. 4806–4813 (2019)
6. Wu, S., Tang, Y., Zhu, Y., Wang, L., Xie, X., Tan, T. : Session-based recommendation with graph neural networks. In: Proceedings of the AAAI Conference on Artificial Intelligence, pp. 346–353 (2019)
7. Liu, L., Wang, L., Lian, T.: CaSe4SR: using category sequence graph to augment session-based recommendation. Knowl. Based Syst. **212**, 106558 (2021)
8. Weston, J., Chopra, S., Bordes, A.: Memory network. In: International Conference on Learning Representations 2015 (ICLR) (2015)
9. Wang, M., Ren, P., Mei, L., Chen, Z., Ma, J., de Rijke, M. : A collaborative session-based recommendation approach with parallel memory modules. In: Proceedings of the 42nd International ACM SIGIR Conference on Research and Development in Information Retrieval, pp. 345–354 (2019)
10. Jin, R., Chai, J.Y., Si, L.: An automatic weighting scheme for collaborative filtering. In: SIGIR 2004 (2004)
11. Wu, Y., DuBois, C., Zheng, A.X., Ester, M.: collaborative denoising auto-encoders for top-n recommender systems. In: WSDM 2016 (2016)
12. He, X., Liao, L., Zhang, H., Nie, L., Hu, X., Chua, T.S.: Neural collaborative filtering. In: Proceedings of the 26th International Conference on World Wide Web (2017)
13. Qiao, J., Wang, L., Duan, L.: Sequence and graph structure co-awareness via gating mechanism and self-attention for session-based recommendation. Int. J. Mach. Learn. Cybern. **12**(9), 2591–2605 (2021). https://doi.org/10.1007/s13042-021-01343-3
14. Xu, C., et al.: Graph contextualized self-attention network for session-based recommendation. In: IJCAI, pp. 3940–3946 (2019)
15. Qiu, R., Li, J., Huang, Z., Yin, H.: Rethinking the item order in session-based recommendation with graph neural networks. In: Proceedings of the 28th ACM International Conference on Information and Knowledge Management, pp. 579–588 (2019)
16. Wang, W., et al.: Beyond clicks: modeling multi-relational item graph for session-based target behavior prediction. In: Proceedings of The Web Conference 2020 (2020)
17. Chen, X., et al.: Sequential recommendation with user memory networks. In: Proceedings of the Eleventh ACM International Conference on Web Search and Data Mining, pp. 108–116 (2018)

18. Huang, J., Zhao, W.X., Dou, H., Wen, J.R., Chang, E.Y.: Improving sequential recommendation with knowledge-enhanced memory networks. In: The 41st International ACM SIGIR Conference on Research & Development in Information Retrieval, pp. 505–514 (2018)

19. Tan, Y.K., Xu, X., Liu, Y.: Improved recurrent neural networks for session-based recommendations. In: Proceedings of the 1st Workshop on Deep Learning for Recommender Systems, pp. 17–22 (2016)

20. Rendle, S., Freudenthaler, C., Schmidt-Thieme, L. : Factorizing personalized markov chains for next-basket recommendation. In: Proceedings of the 19th international conference on World wide web, pp. 811–820 (2010)

21. Hidasi, B., Karatzoglou, A.: Recurrent neural networks with top-k gains for session-based recommendations. In: Proceedings of the 27th ACM international conference on information and knowledge management, pp. 843–852 (2018)

22. Liu, Q., Zeng, Y., Mokhosi, R., Zhang, H.: STAMP: short-term attention/memory priority model for session-based recommendation. In: Proceedings of the 24th ACM SIGKDD International Conference on Knowledge Discovery & Data Mining, pp. 1831–1839 (2018)

23. Yuan, F., Karatzoglou, A., Arapakis, I., Jose, J.M., He, X.: A simple convolutional generative network for next item recommendation. In: Proceedings of the Twelfth ACM International Conference on Web Search and Data Mining, pp. 582–590 (2019)

Continuous-Time Markov-Switching GARCH Process with Robust State Path Identification and Volatility Estimation

Yinan Li and Fang Liu[✉][iD]

University of Notre Dame, Notre Dame, IN 46530, USA
Yinan.Li.267@alumni.nd.edu, fang.liu.131@nd.edu

Abstract. We propose a continuous-time Markov-switching generalized autoregressive conditional heteroskedasticity (COMS-GARCH) process for handling irregularly spaced time series with multiple volatility states. We employ a Gibbs sampler in the Bayesian framework to estimate the COMS-GARCH model parameters, the latent state path and volatilities. To improve the computational efficiency and robustness of the identified state path and estimated volatilities, we propose a multi-path sampling scheme and incorporate the Bernoulli noise injection in the computational procedure. We provide theoretical justifications for the improved stability and robustness with the Bernoulli noise injection through the concept of ensemble learning and the low sensitivity of the objective function to external perturbation in the time series. The experiment results demonstrate that our proposed COMS-GARCH process and computational procedure are able to predict volatility regimes and volatilities in a time series with satisfactory accuracy.

Keywords: Bernoulli noise injection · Ensemble learning · Gibbs sampler · Irregularly (unevenly) spaced time series · Maximum a posterior (MAP) · Stability and robustness

1 Introduction

1.1 Motivation and Problem

Heteroskedasticity is a common issue in time series (TS) data. The generalized autoregressive conditional heteroskedasticity (GARCH) model is a popular discrete-time TS model that accommodates heteroskeasticity and estimates the underlying stochastic volatility. When there is variation in the presence of regime changes in the volatility dynamics, the Markov-switching GARCH (MS-GARCH) model can be employed and when collected TS data are irregularly spaced in time, continuous-time GARCH (CO-GARCH) can be applied.

In practice, TS data may exhibit heteroskedasticity and multiple regimes and are collected in irregularly spaced timepoints or on a near-continuous time scale. For example, heart rate variability TS are typically recorded on a millisecond

© Springer Nature Switzerland AG 2021
N. Oliver et al. (Eds.): ECML PKDD 2021, LNAI 12975, pp. 370–387, 2021.
https://doi.org/10.1007/978-3-030-86486-6_23

scale and can have multiple volatility regimes corresponding to different activities or stress levels. Seismic waves TS, used for studying the earth's interior structure and predicting earthquakes, consist of wave types of different magnitudes and are recorded on a millisecond scale. Financial data are often irregularly spaced in time due to weekend and holiday effects and known to exhibit different volatility states, such as changing behavior drastically from steadily trending to extremely volatile after a major event or news.

To the best of our knowledge, there does not exist a MS-GARCH model for irregularly spaced TS nor a CO-GARCH model to handle multiple states. To fill the methodological gap and respond to the practical needs to handle irregularly-spaced and multi-state TS data, we propose a new COMS-GARCH process to analyze such data and develop a robust and efficient computational procedure to identify multiple states and estimate volatilities.

1.2   Related Work

The GARCH process has been extensively studied [13,25,26,28,32]. [27] derives the conditions under which the discretized-time GARCH model converges in distribution to a bivariate non-degenerate diffusion process as the length of the discrete-time intervals goes to zero. The fact that the limiting process consisting of two independent Brownian motions (that drives the underlying volatility process and the accumulated TS, respectively) contradicts the GARCH model's intuition that large volatilities are the feedback of large innovations. [10] applies different parameterizations as a function of the discrete-time interval to GARCH(1, 1) and obtained both degenerate and non-degenerate diffusion limits. [31] further shows the asymptotic non-equivalence between the GARCH model and the continuous-time bivariate diffusion limit except for the degenerate case. [18] proposes a COntinuous-time GARCH (CO-GARCH) model that replaces the Brownian motions by a single Lévy process and incorporates the feedback mechanism by modeling the squared innovation as the quadratic variation of the Lévy process. For the inference of the CO-GARCH process, there exist quasi-likelihood [7], method of moments (MoM) [18], pseudo-likelihood [20,21] approaches, and Markov chain Monte Carlo (MCMC) procedures [24].

Volatility predictions by GARCH-type models may fail to capture the true variation in the presence of regime changes in the volatility dynamics [3,19,23]. The MS-GARCH model [14] solves this issue by employing a hidden discrete Markov chain to assign a state to each timepoint. For the inference in the MS-GARCH model, regular likelihood-based approaches require summing over exponentially many possible paths and can be computationally unfeasible. Several alternatives exist to deal with the problem. The collapsing procedures [11,14,16,17] use simplified versions of the MS-GARCH model and incorporate recombination mechanisms of the state space. [15] proposes a new MS-GARCH model that is analytically tractable and allows the derivation of stationarity condition and the process properties. [2] employs a Markov Chain Expectation Maximization approach. [5] proposes a Bayesian MCMC method but it can be slow in convergence. Recent methods focus on the efficient sampling of state paths.

[12] introduces a Viterbi-based technique to sample state paths; [4] proposes a particle MCMC algorithm; [6] uses a multi-point sampler in combination with the forward filtering backward sampling technique. Both the likelihood-based and MCMC estimations have been implemented in software (e.g., R package MSGARCH [1]).

For multi-state irregularly-spaced TS, simple and model-free approaches such as the realized volatility can be used to estimate the historical volatility, but they cannot systematically identify different volatility states. In addition, it is not always meaningful to aggregate measures across timepoints to calculate the realized volatilities. Though the existing MS-GARCH model can also be used to identify different volatility regimens, it cannot analyze irregularly spaced TS. This methodological gap motives our work.

1.3   Our Contribution

We propose a COMS-GARCH process, employing the Lévy process to model volatility in each state and the continuous-time hidden Markov chain to model state switching. The estimate volatilities via the COMS-GARCH process are expected to be more robust than those obtained via the realized volatility, largely due to the computational procedure we design specifically for the state path and volatility estimation. Furthermore, the COMS-GARCH process can be used to forecast volatilities and states. Other contributions are listed below.

- We propose a Bayesian Gibbs sampler to obtain inference of the COMS-GARCH parameters and maximum a posterior (MAP) estimation for state path and volatilities.
- We develop a computational procedure with a multi-path sampling scheme and the Bernoulli noise injection (NI) to accelerate the optimization and improve the robustness of the predicted state path and volatilities.
- We provide theoretical justifications for the Bernoulli NI from the perspectives of ensemble learning of the state path and lowered sensitivity of the objective function to small random external perturbation in the TS.
- We run experiments in both simulated and real TS data to demonstrate the application of the COMS-GARCH procedures and the computational procedure and to show satisfactory accuracy in predicting states and volatilities.

2   COMS-GARCH Process

We develop the COMS-GARCH process to handle multiple states, extending the CO-GARCH(1,1) process [18,20]. To the best of our knowledge, CO-GARCH(1,1) is the only Lévy-process driven CO-GARCH model that has analytical solutions for the model parameters from the stochastic differential equations and is inference-capable in the context of pseudo-likelihood. [9] theoretically analyzes the CO-GARCH(p, q) model driven by the Lévy process for general p and q values, but unable to obtain inferences for the model parameters.

Let G_t for $t \in (0, T)$ denote the observed TS, L refer to the innovation that is modeled by a Lévy process, s_t be the state $\{s_t\}$ and σ_t^2 is the underlying volatility process at time t. Our proposed COMS-GARCH process on $(G, \sigma^2, S) = (\{G_t\}, \{\sigma_t^2\}, \{s_t\})$ is the solution to the following set of stochastic differential equations

$$
\begin{cases}
dG_t = Y_t = \sigma_t dL_t(s_t) & (1) \\
d\sigma_t^2 = \alpha(s_t)dt - \beta(s_t)\sigma_{t-}^2 dt + \lambda(s_t)\sigma_{t-}^2 d[L, L]_t & (2) \\
\Pr(s_t = j | s_{t-} = k) = \eta_{jk} dt + o(dt) \text{ for } j \neq k & (3) \\
\Pr(s_t = k | s_{t-} = k) = 1 - \sum_{j \neq k} \eta_{jk} dt + o(dt). & (4)
\end{cases}
$$

$t-$ stands for $t - dt$ and (σ_{t-}^2, s_{t-}) refers to the volatility and associated state at time $t-$. α, β, and λ in Eq. (2) quantify how much the change in time (dt), the volatility at time $t-$, and the innovation at time $t-$ affects the volatility, respectively. The increment of the Lévy process $dL_t(s_t)$ in Eq. (1) is assumed standardized with mean 0 and variance 1, and $[L, L]_{t-}$ in Eq. (2) is its quadratic variation process. Equations (3) and (4) represent the hidden continuous-time Markov chain with ν discrete states and transition parameters $\eta = \{\eta_{jk}\}$ that model the regime switching in the TS for $j, k \in \{1, \cdots, \nu\}$ (note that η_{jk} in Eqs. (7) and (8) is not a probability, and the parameter space for η_{jk} is $(0, \infty)$ instead of $\in (0, 1)$).

Next, we define a family of discrete-time processes that approximates the above continuous-time process (G, σ^2, S), following the methodological framework in [20]. There are a couple of reasons for obtaining a discretized process. First, real-life observed TS data are often recorded in discrete-time, whether irregularly spaced or regardless of how fine the time scale is. Second, the discretization allows us to take advantage of the well-developed inferential approaches for discrete-time GARCH processes. We will show the discretized process converges to the COMS-GARCH process.

The discretization is defined over a finite time interval $[0, T]$ for $T > 0$. Let $0 = t_0 < t_1 < \cdots < t_i < \cdots < t_n = T$ be a deterministic sequence that divides $[0, T]$ into n sub-intervals of lengths $\Delta t_i = t_i - t_{i-1}$ for integers $i = 1, \ldots, n$. Let $G_0 = 0$ and ϵ_i be a first-jump approximation of the Lévy process [20]. A discretized COMS-GARCH process $(G_n, \sigma_n^2, s_n) = (\{G_i\}, \{\sigma_i^2\}, \{s_i\})$ satisfies

$$
\begin{cases}
G_i - G_{i-1} = Y_i = \sigma_{i-1}(\Delta t_i)^{1/2} \epsilon_i, & (5) \\
\sigma_i^2 = \alpha(s_i)\Delta t_i + (\sigma_{i-1}^2 + \lambda(s_i)Y_i^2) \exp(-\beta(s_i)\Delta t_i), & (6) \\
\Pr(s_i = j | s_{i-1} = k) = 1 - \exp(-\eta_{jk}\Delta t_i) \text{ for } j \neq k, & (7) \\
\Pr(s_i = k | s_{i-1} = k) = \sum_{j \neq k} \exp(-\eta_{jk}\Delta t_i). & (8)
\end{cases}
$$

374 Y. Li and F. Liu

Equations (5) and (6) model the "CO" component in a similar manner to [20] and Eqs. (7) and (8) model the "MS" part of the COMS-GARCH process. Since Y_i is obtained by differencing the observed G_i, it is also observed. To ensure the positivity of Eq. (6), we require $\alpha(k)$ and $\lambda(k)$ to be non-negative for all states $k = 1, \ldots, \nu$. To reflect the general belief that dependence between two quantities at two timepoints diminishes as the time gap increases, we also impose positivity on $\beta(k)$ for all states. As $n \to \infty$, $\Delta t_i \to 0$ and the discretized COMS-GARCH process in Eqs. (5) to (8) converges in probability to the COMS-GARCH process defined in Eqs. (1) to (4), as stated in Lemma 1.

Lemma 1 (Convergence of discretized COMS-GARCH process). Let (G, σ^2, s) be the COMS-GARCH process on time interval $[0, T]$, and (G_n, σ_n^2, s_n) be its discretized process. As $n \to \infty$, $\Delta t_i \to 0$ for $i = 1, \ldots, n$ and (G_n, σ_n^2, s_n) converges in probability to (G, σ^2, s) in that the Skorokhod distance $D_S((G_n, \sigma_n^2, S_n), (G, \sigma^2, s)) \xrightarrow{p} 0$ as $n \to \infty$.

The converges in probability in Lemma 1 also implies the convergence of (G_n, σ_n^2, s_n) in distribution to (G, σ^2, s). Lemma 1 is an extension of the theorem on the convergence of a discretized CO-GARCH process [20]. The added complexity in COMS-GARCH, that is, multiple states and state-dependent GARCH parameters, has no material impact on the discretization of the process and the underlying conditions that leads to the convergence. Therefore, the theoretical result of the convergence of the discretized CO-GARCH process can be directly extended to the discretized COMS-GARCH process. In fact, the CO-GARCH model can be regarded as a special case of the COMS-GARCH process when the number of states ν is 1. Therefore, the inferential approaches and theoretical results for COMS-GARCH in the next section also apply to CO-GARCH.

3 Inference for COMS-GARCH Process

The parameters in the COMS-GARCH process include $\Theta = \{\alpha(k), \beta(k), \lambda(k)\}$ $\forall k = 1, \ldots, \nu$ and transition parameters η. In addition, we are also interested in learning the latent state s_i and volatility σ_i^2 for $i = 1, \ldots, n$ so to better understand an observed TS and to aid prediction of future states and volatilities. We propose a Bayesian Gibbs sampler coupled with the pseudo-likelihood that is defined as follows.

$$f(Y_i | Y_1, \ldots, Y_{i-1}, s_1, \ldots, s_i) = N(0, \rho_i^2), \text{ where} \tag{9}$$

$$\rho_i^2 = \left(\sigma_{i-1}^2 - \frac{\alpha(s_i)}{\beta(s_i) - \lambda(s_i)} \right) \left(\frac{\exp((\beta(s_i) - \lambda(s_i))\Delta t_i) - 1}{\beta(s_i) - \lambda(s_i)} \right) + \frac{\alpha(s_i)\Delta t_i}{\beta(s_i) - \lambda(s_i)} \tag{10}$$

$$\approx \sigma_{i-1}^2 \Delta t_i$$

$$= \alpha(s_{i-1})\Delta t_{i-1} \Delta t_i + \Delta t_i \left(\sigma_{i-2}^2 + \lambda(s_{i-1})Y_{i-1}^2 \right) e^{-\beta(s_{i-1})\Delta t_{i-1}}. \tag{11}$$

Equation (11) is obtained by taking the first-order Taylor expansion of Eq. (10) around $\Delta t_i = 0$ and substituting σ_{i-1}^2 in Eq. (6). Equations (9) to (11) suggest that

$$\mathrm{E}(Y_i^2|Y_1,\ldots,Y_{i-1},s_1,\ldots,s_i)=\mathrm{V}(Y_i|Y_1,\ldots,Y_{i-1},s_1,\ldots,s_i)=\rho_i^2\approx\sigma_{i-1}^2\Delta t_i. \quad (12)$$

3.1   Gibbs Sampler for Bayesian Inference on Model Parameters

Since we are interested in obtaining inferences for Θ and $\boldsymbol{\eta}$ given the pseudo-likelihood and predicting states and volatilities, methods that rely on integrating out the latent states, such as the EM and MC-EM algorithms (we provide their steps in the expanded paper for interested readers), do not work well. In contrast, the Bayesian framework provides a more convenient and straightforward approach to reach the inferential goal. Below we propose a Gibbs sampler to obtain Bayesian inferences for the COMS-GARCH process.

Define $\boldsymbol{\Delta t} = (\Delta t_1,\ldots,\Delta t_n), \mathbf{Y} = (Y_1,\ldots,Y_n), S = (s_1,\ldots,S_n)$, and $\mathcal{S}$ is the set of all possible state paths. Denote the priors for Θ and $\boldsymbol{\eta}$ by $\pi(\Theta,\boldsymbol{\eta})$ and assume $\pi(\Theta,\boldsymbol{\eta})=\pi(\Theta)\pi(\boldsymbol{\eta})$. The conditional posterior distributions of $\Theta,\boldsymbol{\eta}$, and the states are respectively

$$f(\Theta|\boldsymbol{\eta},\mathbf{Y},\boldsymbol{\Delta t},S)\propto\pi(\Theta)L(\Theta,\boldsymbol{\eta}|\mathbf{Y},S)=\pi(\Theta)\prod_{i=1}^{n}\rho_i^{-1}\exp\left(-Y_i^2/(2\rho_i^2)\right), \quad (13)$$

$$f(\eta_{1k},\ldots,\eta_{\nu k}|\Theta,\mathbf{Y},\boldsymbol{\Delta t},S) = f(\eta_{1k},\ldots,\eta_{\nu k}|S,\boldsymbol{\Delta t}) \text{ for } k = 1,\ldots,\nu,$$

$$\propto\pi(\eta_{1k},\ldots,\eta_{\nu k})\prod_{\substack{s_{i+1}=k\\s_i=k}}^{n-1}\left(2-\nu+\sum_{j\neq\nu}e^{-\eta_{jk}\Delta t_{i+1}}\right)\prod_{j\neq k}\prod_{\substack{s_{i+1}=j\\s_i=k}}^{n-1}(1-e^{-\eta_{jk}\Delta t_{i+1}}), \quad (14)$$

$$f(s_i|S_{-i},\Theta,\boldsymbol{\eta},\mathbf{Y},\boldsymbol{\Delta t}) \propto \xi_{s_i,s_{i-1}}\xi_{s_{i+1},s_i}\prod_{t=i}^{n}\rho_t^{-1}\exp(-Y_t^2/(2\rho_t^2)), \quad (15)$$

where $\sum_{j=1}^{\nu}\eta_{jk} = 1$ in Eq. (14), and

$$\xi_{s_i,s_{i-1}}=\begin{cases}2-\nu+\sum_{k\neq s_{i-1}}\exp(-\eta_{k,s_{i-1}}\Delta t_i) \text{ when } s_i = s_{i-1}\\1-\exp(-\eta_{s_i,s_{i-1}}\Delta t_i) \text{ when } s_i \neq s_{i-1}\end{cases};$$

similarly for ξ_{s_{i+1},s_i}. When there are two states ($\nu = 2$), Eqs. (14) and (15) can be simplified to

$$f(\eta_{21}|S,\mathbf{Y},\boldsymbol{\Delta t})\propto\pi(\eta_{21})\prod_{\substack{s_{i+1}=1\\s_i=1}}^{n-1}e^{-\eta_{21}\Delta t_i}\prod_{\substack{s_{i+1}=2\\s_i=1}}^{n-1}(1-e^{-\eta_{21}\Delta t_{i+1}}) \quad (16)$$

$$f(\eta_{12}|S,\mathbf{Y},\boldsymbol{\Delta t})\propto\pi(\eta_{12})\prod_{\substack{s_{i+1}=2\\s_i=2}}^{n-1}e^{-\eta_{12}\Delta t_i}\prod_{\substack{s_{i+1}=1\\s_i=2}}^{n-1}(1-e^{-\eta_{12}\Delta t_{i+1}}) \quad (17)$$

$$f(s_i|S_{-i},\Theta,\boldsymbol{\eta},,\boldsymbol{\Delta t})\propto\xi_{1,s_{i-1}}^{2-s_i}\xi_{2,s_{i-1}}^{s_i-1}\xi_{1,s_i}^{2-s_{i+1}}\xi_{2,s_i}^{s_{i+1}-1}\prod_{t=i}^{n}\rho_t^{-1}\exp\left(-\frac{Y_t^2}{2\rho_t^2}\right), \quad (18)$$

where $\xi_{1,s_{i-1}}=e^{-\eta_{21}\Delta t_i}$ if $s_{i-1}=1$, and $1-e^{-\eta_{21}\Delta t_i}$ if $s_{i-1}=2$; $\xi_{1,s_i}=e^{-\eta_{21}\Delta t_{i+1}}$ if $s_i = 1$, and $1 - e^{-\eta_{21}\Delta t_{i+1}}$ if $s_i = 2$.

The Gibbs sampler draws samples on Θ, η and s_i for $i = 1, \ldots, n$ alternatively from Eqs. (13), (14), and (15). Upon convergence, after burning and thinning, we will have multiple, say M, sets of posterior samples of Θ, η, based on which their posterior inferences can be obtained. We will also have M sets of samples on state s_i and can calculate the posterior volatility σ_i^2 at each timepoint via Eq. (6). Connecting the states across the n times points from each set of the state posterior samples leads to a state path. Due to the large sample space (totally ν^n possible paths), it is difficult to identify the MAP estimate for the state path out of the M paths with acceptable accuracy unless $M \gg \nu^n$ and a significant portion of paths have close-to-0 posterior probabilities with a few paths having significantly higher probabilities compared to the rest. To solve this issue, we design a new computational algorithm (reSAVE) as detailed next.

3.2   Estimation for State Path and Volatility

To deal with the computational challenge in obtaining the MAP estimates for state path via the Gibbs sampler in Sect. 3.1, we propose an inferentially **R**obust and computationally **E**fficient iterative procedure for **S**tate path **A**nd **V**olatility **E**stimation (reSAVE). The steps of the procedure are listed in Algorithm 1.

The inferential robustness for the MAP estimates from the reSAVE procedure is brought by the Bernoulli NI implemented in each iteration of the procedure, leading to both ensemble learning and improved stability of the object function from which the MAP estimates are obtained (Sect. 3.4). The computational efficiency of the reSAVE can be attributed to a couple of factors: the Bernoulli NI that generates a sub-TS (smaller data size) in each iteration, of sampling of a small set of state m to calculate MAP estimates for the state path, and the employment of a maximization-maximization scheme to obtain the MAP estimates of the parameters Θ and η and for the state path in each iteration.

The number of iterations N in the procedure can be prespecified or determined using a convergence criterion, such as the l_1 distances on the MAP estimates (e.g., $\{|\hat{\eta}^{(l+1)} - \hat{\eta}^{(l)}|, |\hat{\Theta}^{(l+1)} - \hat{\Theta}^{(l)}|, |S^{(l+1)} - S^{(l)}|\}$) or the objective functions between two consecutive iterations. If the criterion reaches a prespecified threshold, the procedure stops. Regarding the number of sampled state paths $m > 1$, m too small will not lead to stable MAP estimates; m too large would increase the computational costs. In the experiments presented in Sect. 4, we used $m = 6$, which seems good enough. The ensemble size b refers to the number of observations following a given timepoint i that are used to update the conditional posterior distribution of s_i (Proposition 1). The specification of b is mainly for computational efficiency consideration and is optional.

Algorithm 1: The reSAVE Optimization Procedure

---

    input : data $(\mathbf{Y}, \Delta \mathbf{t})$, initial values $\Theta^{(0)}, \boldsymbol{\eta}^{(0)}, S^{(0)} = \left(s_1^{(0)}, \ldots, s_n^{(0)}\right)$, # of
               iterations N, # of sampled state paths m, ensemble size b

    output: MAP estimates $\hat{S}_{\text{MAP}}, \hat{\boldsymbol{\sigma}}^2_{\text{MAP}} = (\hat{\sigma}_1^2, \ldots, \hat{\sigma}_n^2), \hat{\Theta}_{\text{MAP}}, \hat{\boldsymbol{\eta}}_{\text{MAP}}$.

1   **for** $l = 1$ *to* N **do**

2   Apply Bernoulli NI in Alg. 2 to obtain a sub-TS $(\tilde{\mathbf{Y}}^{(l)}, \Delta \tilde{\mathbf{t}}^{(l)})$ of length $\tilde{n}^{(l)}$.
          Denote by $\mathcal{T}^{(l)}$ the set of the original timepoints retained in the sub-TS;

3   Calculate $\hat{\Theta}_{\text{MAP}}^{(l)} = \arg\max_{\Theta} f(\Theta | S^{(l-1)}, \tilde{\mathbf{Y}}^{(l)}, \Delta \tilde{\mathbf{t}}^{(l)})$ and

$$\hat{\boldsymbol{\eta}}_{\text{MAP}}^{(l)} = \arg\max_{\eta} f(\boldsymbol{\eta} | S^{(l-1)}, \tilde{\mathbf{Y}}^{(l)}, \Delta \tilde{\mathbf{t}}^{(l)});$$

4   **for** $j = 1$ *to* m **do**

5   **for** $i = 2$ *to* $n - 1$ **do**

6   sample $s_i^{(j)}$ for $i \in \mathcal{T}^{(l)}$ given $\Theta_{\text{MAP}}^{(l)}, \boldsymbol{\eta}_{\text{MAP}}^{(l)} \tilde{\mathbf{Y}}^{(l)}, \Delta \tilde{\mathbf{t}}^{(l)}, S_{-i}^{*(l-1)}$ per Eq.
             (15);

7   **end**

8   Let $\tilde{S}^{(j)} = \left(s_1^{(j)}, s_2^{(j)}, \ldots, s_{\tilde{n}^{(l)}}^{(j)}\right)$.

9   **end**

10   Let $\tilde{\mathcal{S}} = \left(\tilde{S}^{(1)}, \ldots, \tilde{S}^{(m)}\right)$, $\tilde{S}^{(l)} = \arg\max_{S \in \tilde{\mathcal{S}}} f(S | \Theta_{\text{MAP}}^{(l)}, \boldsymbol{\eta}_{\text{MAP}}^{(l)}, \tilde{\mathbf{Y}}^{(l)}, \Delta \tilde{\mathbf{t}}^{(l)})$, and

$$S^{(l)} = \{\tilde{s}_i^{(l)}\}_{i \in \mathcal{T}^{(l)}} \bigcup \{\tilde{s}_i^{(l-1)}\}_{i \notin \mathcal{T}^{(l)}};$$

11 **end**

12 Calculate MAP estimate for volatility $\{\sigma_i^2\}_{i=1,\ldots,n}$ given
     $S_{\text{MAP}} = S^{(N)}, \hat{\Theta}_{\text{MAP}} = \hat{\Theta}_{\text{MAP}}^{(N)}, \hat{\boldsymbol{\eta}}_{\text{MAP}} = \hat{\boldsymbol{\eta}}_{\text{MAP}}^{(N)}$ via Eq. (6).

The MAP estimates of Θ and $\boldsymbol{\eta}$ can be obtained either through direct optimization of their respective conditional posterior distributions or via MC approaches using samples from the conditional posterior distributions. The estimate of the state path in each iteration is defined as the path, out of the sampled m paths, that maximizes the conditional posterior distribution of S, which is proportional to $\prod_{i=1}^n \rho_i^{-1} \exp(-y_i^2/(2\rho_i^2))\eta_{i,i-1}$ [5], given the latest MAP estimates of Θ and $\boldsymbol{\eta}$. Though the Bernoulli NI is designed more for achieving ensemble learning and improving the stability of the objective functions for the state path optimization (see Sect. 3.4), we expect its usage also makes the inference for Θ and $\boldsymbol{\eta}$ more robust, especially if ν is relatively large.

3.3 Bernoulli Noise Injection

A key step in Algorithm 1 is generating sub-TS via Bernoulli NI. The rationale for sub-TS when estimating the state path and volatilities is that the estimation can be sensitive to the TS data for the COMS-GARCH process. The Bernoulli NI can create an ensemble of sub-TS' of considerable diversity among the ensemble members across iterations to reduce the sensitivity, to reduce the inference sensitivity. Algorithm 2 lists the steps of the Bernoulli NI. It outputs a sub-TS $\{\tilde{\mathbf{G}}, \Delta \tilde{\mathbf{t}}\}$, which is then fed to Algorithm 1; only the states $\tilde{S}$ at the retained timepoints are updated in each iteration, and the states of the dropped timepoints are kept at the values from the previous iteration, saving costs computationally. For the choice of Bernoulli rate p, a k-fold cross-validation (CV) procedure can be used, the detail of which are provided in the expanded paper.

Algorithm 2: Bernoulli Noise Injection

---

    input : Original TS $\mathbf{G}$; Bernoulli NI rate p.
    output: sub-TS $(\tilde{\mathbf{Y}}, \varDelta\tilde{\mathbf{t}}, \tilde{n})$.

1 Draw e_i from $\mathrm{Bern}(1-p)$ for $i=2,\ldots,n-1$. Set $e_0=e_1=e_n=1$;

2 Let $\tilde{\mathbf{G}} = \{\mathbf{G} : \mathbf{G}\cdot\mathbf{e} \neq 0\}$, where $\mathbf{e} = \{e_i\}_{i=0}^n$, $\tilde{n} = \sum_{i=1}^n e_i$;

3 Obtain $\tilde{\mathbf{Y}} = \{\tilde{Y}_1, \ldots, \tilde{Y}_{\tilde{n}}\} = \mathrm{diff}(\tilde{\mathbf{G}})$;

4 Set $\varDelta\tilde{\mathbf{t}} = \varDelta\mathbf{t}$. For $0\leq i \leq n-1$, let $\begin{cases} \varDelta\tilde{t}_{i+1} = \varDelta i + \varDelta\tilde{t}_{i+1} \text{ and } \varDelta i = 0 \text{ if } e_i = 0 \\ \varDelta\tilde{t}_{i+1} \leftarrow \varDelta\tilde{t}_{i+1} \text{ if } e_i = 1 \end{cases}$;

    $\varDelta\tilde{\mathbf{t}} = \{\varDelta\tilde{t} : \varDelta\tilde{t} \neq 0\}$.

Claim 1. The differenced $\tilde{\mathbf{Y}}$ in a sub-TS after Bernoulli NI is a summation of a sequence of differenced $\mathbf{Y}$ with the dropped observations in the original TS.

Claim 1 is a simple but interesting fact. For example, if G_{i+1} gets dropped from the sequence of $\ldots, G_i, G_{i+1}, G_{i+2}, \ldots$, then $\tilde{Y}_{i'} = G_{i+2} - G_i = (G_{i+2} - G_{i+1}) + (G_{i+1} - G_i) = Y_{i+2} + Y_{i+1}$; say r observations are dropped between G_i and G_{i+r+1}, then $\tilde{Y}_{i'} = G_{i+r+1} - G_i = (G_{i+r+1} - G_{i+r}) + (G_{i+r} - G_{i+r-1}) + \cdots + (G_{i+1} - G_i) = Y_{i+r+1} + \cdots + Y_{i+1}$. This fact is used in the proof of Proposition 2 in Sect. 3.4. Since the NI rate p is usually small and the timepoints are dropped from the original TS randomly, with the fine time scale on which the TS is collected, the COMS-GARCH process can "digests" these "missing" timepoints effortlessly, without needing an ad-hoc approach to handle these dropped timepoints. The full conditional distributions of $\Theta, \boldsymbol{\eta}$, and states $\{s_i\}$ given the sub-TS in each iteration are given in Eqs. (13) and (15) by replacing the original TS $(\mathbf{Y}, \varDelta\mathbf{t})$ with the sub-TS $(\tilde{\mathbf{Y}}, \varDelta\tilde{\mathbf{t}})$.

The Bernoulli NI for COMS-GARCH is inspired by the dropout technique for regularizing neural networks (NNs) [30], which injects Bernoulli noises to input and hidden nodes during training, leading to model regularization. The Bernoulli NI we propose here is different procedurally in that it is applied to the observed data and drops random timepoints in the original TS in each iteration, rather than generating sub-models; in other words, the COMS-GARCH model remains as is during training. The benefits of the Bernoulli NI here include reduced computational cost, its connection with ensemble learning and inferential stability and robustness. The Bernoulli NI also bears some similarity to bagging [8], a well-known ensemble learning algorithm, but differs from bagging in two aspects. First, the Bernoulli NI leads to a random sub-TS (without replacement) of the original TS in each iteration of Algorithm 1 whereas bagging often generates a bootstrapped sample set with replacement that is of the same size as the training data. Second, bagging often generates multiple sets of samples, trains a model on each set in parallel, and then ensembles them into a meta-model, whereas the ensemble learning brought by the Bernoulli NI to the MAP estimation for COMS-GARCH is implicit, iterative, and realized sequentially.

3.4   Theoretical Analysis on Inferential Benefits of Bernoulli NI

In this section, we investigate theoretically the reasons behind the inferential benefits (improved efficiency and robustness in the MAP estimates of state path and volatilities) of the Bernoulli NI in two aspects.

First, we show, through the iterative Algorithm 1, that the Bernoulli NI leads to sequential and implicit ensemble learning of the parameters and states for the COMS-GARCH model. The formal results are given in Proposition 1.

Proposition 1 (ensemble learning of state path). Let $\mathbf{Y}_{j-1} = (Y_1, \ldots, Y_{j-1})$, $\mathbf{S}_1 = (s_1, \ldots, s_i = k_1, \ldots, s_j)$ and $\mathbf{S}_2 = (s_1, \ldots, s_i = k_2, \ldots, s_j)$ $\forall k_1 \neq k_2 \in \{1, \ldots, \nu\}, i \leq n-b$. Assume that for $\forall\, \epsilon > 0$, $\exists b \in N+$ such that

$$\left| \frac{\prod_{j=i}^{n} f(Y_j|\mathbf{Y}_{j-1}, \mathbf{S}_1)}{\prod_{j=i}^{n} f(Y_j|\mathbf{Y}_{j-1}, \mathbf{S}_2)} - \frac{\prod_{j=i}^{i+b} f(Y_j|\mathbf{Y}_{j-1}, \mathbf{S}_1)}{\prod_{j=i}^{i+b} f(Y_j|\mathbf{Y}_{j-1}, \mathbf{S}_2)} \right| < \epsilon. \tag{19}$$

There exist C_{k-1}^{b-1} ways to yield a set of b observations from a sequence of $k \in [b, n-i]$ consecutive observations. Denote the ensemble of the resultant C_{k-1}^{b-1} sub-TS' by $\tilde{\mathcal{Y}}$. Given a Bernoulli NI rate p, the conditional posterior distribution of s_i given the ensemble $\tilde{\mathcal{Y}}$ is

$$\sum_{k=b}^{n-i} \left(p^{k-b}(1-p)^{b-1} \sum_{\tilde{\mathbf{Y}} \in \tilde{\mathcal{Y}}} f(\tilde{s}_i|\tilde{S}_{-i}, \Theta, \boldsymbol{\eta}, \tilde{\mathbf{Y}}) \right). \tag{20}$$

The proof of Proposition 1 is straightforward. Equation (6) implies that the conditional distribution of Y_j depends only on its variance since its mean is fixed at 0. Equation (9) suggests that the impact of state s_i on σ_{j-1}^2 (and thus ρ_j^2) decreases as i departs from j given the recursive formula on σ^2. Taken together, it implies that the state at time t_i has minimal effect on the distribution of Y_j at a future timepoint t_j once the distance $t_j - t_i$ surpasses a certain threshold, which we use b to denote. Mathematically, it means the ratio between $\prod_{j=i+b+1}^{n} f(Y_j|\mathbf{Y}_{j-1}, \mathbf{S}_1)$ and $\prod_{j=i+b+1}^{n} f(Y_j|\mathbf{Y}_{j-1}, \mathbf{S}_2)$ is arbitrarily close to 1, or

$$\frac{\prod_{j=i}^{i+b} f(Y_j|\mathbf{Y}_{j-1}, \mathbf{S}_1)}{\prod_{j=i}^{i+b} f(Y_j|\mathbf{Y}_{j-1}, \mathbf{S}_2)} \left| \frac{\prod_{j=i+b+1}^{n} f(Y_j|\mathbf{Y}_{j-1}, \mathbf{S}_1)}{\prod_{j=i+b+1}^{n} f(Y_j|\mathbf{Y}_{j-1}, \mathbf{S}_2)} - 1 \right| < \epsilon$$

for any $\epsilon > 0$, leading to Eq. (19). k given b and p follows a negative binomial distribution, leading directly to Eq. (20).

Taken together with Eq. (15), Eq. (19) implies the posterior distribution of s_i can be almost surely determined by the b observations in TS $\mathbf{Y}$ that immediately follow t_i; that is, $f(s_i|S_{-i}, \Theta, \boldsymbol{\eta}, \mathbf{Y}) = f(s_i|S_{-i}, \Theta, \boldsymbol{\eta}, Y_i, Y_{i+1}, \ldots, Y_{i+b})$. This narrow focus on just b observations is undesirable especially when b is small because the inference about the state path can become unstable and highly sensitive to even insignificant fluctuation in the TS. The Bernoulli NI helps mitigate this concern by diversifying the set of the b observations. After the Bernoulli NI, the posterior probability of s_i is a weighted average of the posterior distributions over multiple sets of b observations with different compositions across iterations, as suggested by Eq. (20), leading to more robust state estimation.

Figure 1 provides a visual illustration of the ensemble effect achieved through the Bernoulli NI. The dashed line in each plot corresponds to the right y-axis that presents the size of an ensemble. The solid lines correspond to the left y-axis, representing the weights assigned to ensembles of different sizes. When there is no Bernoulli NI ($p = 0$ and $k = b$), the ensemble is of size 1 (the first point on the dashed line in each plot). For $p > 0$, we have more than one way of generating the set of b observations; and the actual ensemble size depends on p and k. In brief, for a fixed b, as k increases, the size of the ensemble set $\tilde{\mathcal{Y}}$, C_{k-1}^{b-1}, increases dramatically (the dashed line within each plot), implying more sub-TS' are involved to obtain the posterior distribution of s_i. In addition, the ensemble set also increases dramatically with b for a fixed $k - b$ value (the trend of the dashed lines across the 3 plots). The separated lines for different p suggest that ensembles of different sizes are not weighted equally toward the conditional posterior distribution of s_i: the larger an ensemble, the smaller the weight it carries, especially for small p. Figure 1 implies that p as small as $O(0.01)$ can create an ensemble of sub-TS' of enough diversity among the ensemble members to bring more robustness to the inference. A large p leads to a larger ensemble, but also a higher computational cost which could overshadow the improved diversity. In addition, a large p may drop too many timepoints and lead to too much fluctuation in the sub-TS' from iteration to iteration, resulting in possibly large bias or large variance in the estimation.

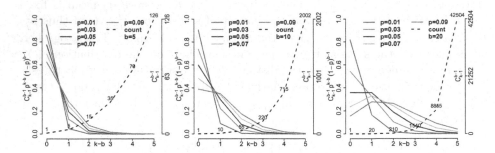

Fig. 1. Size of ensemble $\tilde{\mathcal{Y}}$ (right y-axis) and weights $w(k; p, b)$ assigned to ensembles of different sizes (left y-axis) for different p and b

Second, we establish that the Bernoulli NI also improves the stability of the objective function from which the MAP estimation is obtained, in the presence of random external perturbation in the TS. With a more stable objective function, the MAP estimates of $(\Theta, \eta, S, \sigma^2)$ are also expected to be more stable. The formal result is presented in Proposition 2 (the proof is in the expanded paper).

Proposition 2 (improved stability of objective function). Let $Y_i' = Y_i + z_i$, with $z_i \sim N(0, \varepsilon^2)$ independently for $i = 1, \ldots, n$, be an externally perturbed observation to the original observation Y_i from TS $\mathbf{Y}$; and Y_i' comprises the perturbed TS $\mathbf{Y}'$. Let $\tilde{\mathbf{Y}}$ and $\tilde{\mathbf{Y}}'$ denote a sub-TS of $\mathbf{Y}$ and $\mathbf{Y}'$, respectively, after implementing the Bernoulli NI in an iteration of Algorithm 1. The difference in the objection function (negative log-likelihood function or negative log-posterior distributions of $(\Theta, \boldsymbol{\eta})$, S and $\boldsymbol{\sigma}^2$) given $\tilde{\mathbf{Y}}'$ vs. that given $\tilde{\mathbf{Y}}$ after the Bernoulli NI is on average smaller than the difference obtained without the Bernoulli NI.

4 Experiments

We run 4 experiments to demonstrate the applications of the COMS-GARCH process and the reSAVE algorithm. Experiment 1 shows the improved robustness of the MAP estimates of volatilities brought by the Bernoulli NI in the reSAVE procedure in a one-state COMS-GARCH process. Experiment 2 demonstrates the inferential robustness and computational efficiency of the reSAVE procedure in a two-state COMS-GARCH process and to compare with the MSGARCH(1,1) process. Experiments 3 and 4 apply the COMS-GARCH process and the reSAVE procedure to a real exchange rate TS (https://www.histdata.com) and a real blood volume amplitude (BVA) TS (https://archive.ics.uci.edu/ml/datasets/PPG-DaLiA) – to identify multiple states and estimate volatility.

4.1 Experiment Setting

In experiment 1, the TS data $\mathbf{Y}$ is simulated from a single-state CO-GARCH model and has 500 time-points. The time gap Δt between two consecutive observations is characterized by a Poisson process with rate ζ, i.e., $E(\Delta t) = \zeta^{-1}$. We examined 4 ζ values (2.5, 5, 10, 20). In experiment 2, the simulated TS $\mathbf{Y}$ from the COMS-GARCH process has 2 states and 1,000 timepoints. The transitions among the states between two adjacent time-points are modeled by a hidden Markov process transition parameters. We examined 2 set of β: $\eta_{12} = \eta_{21} = 0.1$ and 0.25 ($\eta_{11} = \eta_{22} = 0.9$ and 0.75). A similar Poisson process as in experiment 1 was used to simulate the TS data in the two states at $\zeta = 10, 40$, respectively. In experiment 3, to keep the data at a manageable size, we took every 90-th observation and performed a log transformation on the exchange rate TS between US dollar and Canadian dollar. The final TS $\mathbf{G}$ contains 1501

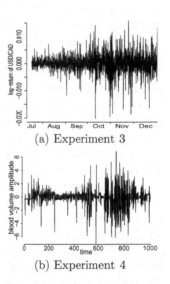

(a) Experiment 3

(b) Experiment 4

Fig. 2. Observed TS $\mathbf{Y}$

times points over half a year. In experiment 4, we extracted the measurements from the photoplethysmograph of the blood volume pulse (64 Hz; i.e., 64 times per second) by taking the valley and peak pulse values in each cycle and scaling them by 0.01. We then took the first 1,000 measurements from a random patient as the input $\mathbf{Y}$. The TS' in experiments 3 and 4 have multiple regimens and the timepoints are irregularly spaced.

In all 4 experiments, we imposed non-informative priors on $\Theta = (\alpha, \beta, \lambda)$, set $b = 20$ and $m = 6$ for multiple path sampling. We examined a range of the Bernoulli NI rates in in experiments 1 and 2 and set p at 0.02 in experiments 3 and 4. The convergence of the reSAVE procedure was examined by visual inspection of the trace plots of the MAP estimates for $(\Theta, \boldsymbol{\eta})$ and $\log(\Theta, \boldsymbol{\eta}, \sigma^2, S | \mathbf{Y}, \boldsymbol{\Delta t})$. MS-GARCH(1, 1) in experiment 2 was implemented using R package MSGARCH. Since the MS-GARCH model assumes evenly-space TS data, we first applied linear interpolation to each simulated TS to obtain the equally spaced TS data before fitting the model.

4.2   Results

Due to space limitation, we present selected main results; more results can be found in the expanded paper. Figure 3 presents examples on the estimated volatilities and states (a randomly chosen repeat out of 50 is shown in experiments 1 and 2). The main observations are as follows. In Figs. 3(a), (c) to (f)), the estimates of volatilities and states almost completely overlap with their true values at all timepoints, suggesting high accuracy in the estimation. The results in experiment 2 further suggest that when there is no external perturbation in the TS, the Bernoulli NI does not negatively impact the volatility and state estimation but improves the accuracy in the estimation when there is, implying that the Bernoulli NI is an "intelligent" technique, only doing its tricks when needed and is silent otherwise. By contract, MS-GARCH tend to over-estimate volatilities and there are more mis-predicted states (Fig. 3(b)). In experiments 3 and 4 (Fig. 3(g) and (h)), the estimated states and volatilities via COMS-GARCH and the reSAVE procedure reflect the two expected states in each case: a change in the state somewhere between September and October in experiment 3, and different BVA states corresponding to different physical conditions or emotional episodes of subjects in experiment 4.

Figure 4 summarizes the relative $\%|\text{bias}|$ of the estimated volatilities (l_1-distance scaled by the true volatility) and the state mis-prediction rate, both averaged across the time points in each TS and across the 50 repeats in experiments 1 and 2. Figure 4(a) suggests the volatility estimation can be sensitive to even mild fluctuation in the TS data as the bias with externally perturbed TS (crosses at $p = 0$) is larger than that at no external perturbation (circles at $p = 0$). Bernoulli NI helps bring the bias down at all the examined ζ values but there is not much of a difference across p. In Fig. 4(b) to 4(e), the accuracy of the state identification and volatility estimation is significantly improved with a proper Bernoulli NI rate p than without NI. The smallest mis-prediction rate (8%–17%) and volatility estimation bias (12%–25%) are achieved around p at 0.01–0.02 in most scenarios; and further increasing p does not seem to improve the prediction accuracy. In addition, how much the Bernoulli NI helps in reducing the prediction bias relates to ζ, $\boldsymbol{\eta}$, and Θ.

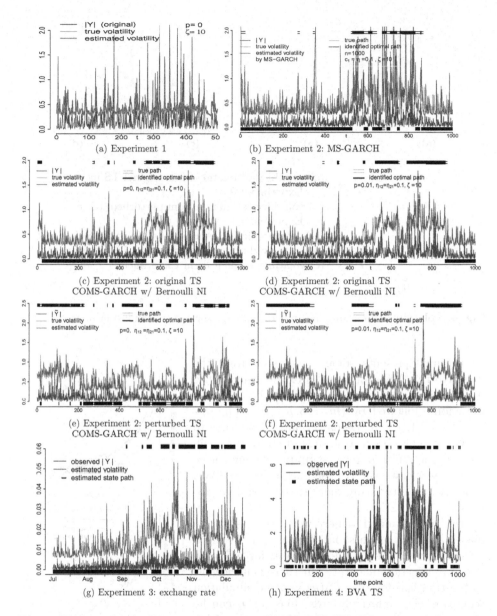

Fig. 3. Estimated volatilities and states via COMS-GARCH. Experiment 1 has only one state so there is no state path estimation; (b) shows the results from MS-GARCH, a comparison method to COMS-GARCH

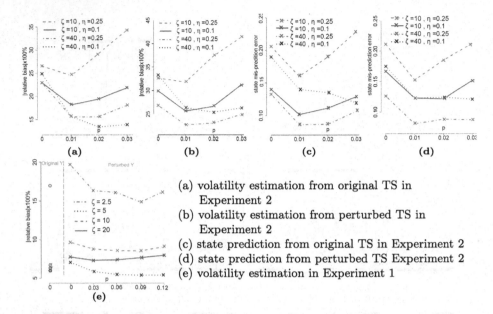

(a) volatility estimation from original TS in Experiment 2
(b) volatility estimation from perturbed TS in Experiment 2
(c) state prediction from original TS in Experiment 2
(d) state prediction from perturbed TS Experiment 2
(e) volatility estimation in Experiment 1

Fig. 4. Volatility estimation bias and state mis-prediction rate in Experiments 1 and 2

Figure 5 presents some trace plots from the iterative reSAVE procedure in experiment 2. The setting $(m = 6, p = 0.02)$ converges with the least iterations, followed by $(m = 6, p = 0)$. The setting $(m = 1, p = 0)$ needs the most iterations to converge. This observation suggests both Bernoulli NI and the multiple path sampling scheme (m = 6) can accelerate the convergence of the reSAVE procedure in estimating the state path and volatilities.

In the expanded paper, we also present the biases and root mean squared errors for the MAP estimates of Θ and η. The estimates for $\alpha(k)$ and $\beta(k)$ are generally accurate but there is noticeable estimation bias for η and $\lambda(k)$ in some simulation scenarios. The relatively large bias for η can be at least partially attributed to the low transition probabilities between different states, leading to data sparsity in estimating η. Estimation bias of Θ is rather a common problem and exists in the GARCH, CO-GARCH, and MS-GARCH estimation, rather than something unique to the Gibbs sampler or the reSAVE procedure we propose for the COMS-GARCH process. [7] suggests that the

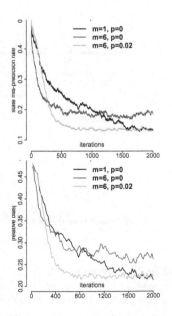

Fig. 5. Trace plots of state mis-prediction rate (top) and |relative bias| of estimated volatility (bottom) in a single TS repetition in Experiment 2

MLEs for the parameters from the GARCH model are biased; the estimation bias for the GARCH parameters for both the quasi-MLE and the constrained M-estimators (more robust) can be as large as 20% [22] and as large as 30% for the parameters of the CO-GARCH process [20].

5 Discussion

We propose the COMS-GARCH process for handling irregularly spaced TS data with multiple volatility states. We also introduce the reSAVE procedure with the Bernoulli NI for obtaining the MAP estimates for model parameters, state path, and volatilities. The computational efficiency and inferential robustness of the reSAVE procedure are established and illustrated theoretically or empirically. Foresting is often of major interest in TS analysis as they provide insights into future trends and are useful for decision making (e.g., developing option trading strategies in financial markets, predicting earthquakes). We present in the expanded paper an algorithm for the hth-step-ahead prediction of future volatilities and states through a trained COMS-GARCH process.

We conjecture that the reSAVE procedure is applicable not only to the COMS-GARCH and CO-GARCH processes but also to other solvable CO-*-GARCH processes. For example, it will make an interesting future topic to develop the COMS-Exponential-GARCH and COMS-Integrated-GARCH processes and examine the performance of the reSAVE procedure in these settings. We also expect that the reSAVE procedure can be used in the COMS-ARMA process for trend estimation, yielding some types of weighted l_2 regularization on the ARMA parameters. More work is needed to prove conjectures formally.

Regarding when to use the COMS-GARCH process, TS' that are of high-frequency or irregularly spaced are potential candidates (the "CO" component); we may plot the TS and visually examine whether there is any sign for multiple states (the "MS" component) or there may exist domain or prior knowledge suggesting state multiplicity (e.g., experiments 3 and 4 in Sect. 4). For a more quantitative approach, one may fit the candidate models (e.g., COGARCH vs. COMS-GARCH, MS-GARCH and COMS-GARCH, ν_1 vs ν_2 states in COMS-GARCH) to the TS in the Bayesian framework, and compare the deviance information criterion (DIC) [29], a Bayesian measure of model fit and choose the model with the smaller (or) smallest DIC.

Finally, as discussed briefly in the experiments, there lacks in-depth theoretical investigation on the asymptotic properties of the MLE and MAP estimation for the MS-GARCH and CO-GARCH processes, and the estimation bias for *-GARCH parameters is well documented. We will look into the asymptotic properties of the COMS-GARCH process as $n \to \infty$ and $T \to \infty$ and develop more accurate inferential procedures for COMS-GARCH in the future.

References

1. Ardia, D., Bluteau, K., Boudt, K., Catania, L., Trottier, D.A.: Markov-switching GARCH models in R: the MSGARCH package. J. Stat. Softw. **91**, 4 (2019)

2. Augustyniak, M.: Maximum likelihood estimation of the Markov-switching GARCH model. Comput. Stat. Data Anal. **76**, 61–75 (2014)
3. Bauwens, L., De Backer, B., Dufays, A.: A Bayesian method of change-point estimation with recurrent regimes: application to GARCH models. J. Empir. Financ. **29**, 207–229 (2014)
4. Bauwens, L., Dufaysa, A., Romboutsb, J.V.K.: Marginal likelihood for Markov-switching and change-point GARCH models. Lect. Notes Econ. Math. Syst. **178**(3), 508–522 (2014)
5. Bauwens, L., Preminger, A., Rombouts, J.V.K.: Theory and inference for a Markov switching GARCH model. Econometr. J. **13**, 218–244 (2010)
6. Billioa, M., Casarina, R., Osuntuyi, A.: Efficient Gibbs sampling for Markov switching GARCH models. Comput. Stat. Data Anal. **100**, 37–57 (2016)
7. Bollerslev, T., Engle, R.F., Nelson, D.B.: Arch models. Handb. Econ. **4**, 2959–3038 (1994)
8. Breiman, L.: Bagging predictors. Mach. Learn. **24**(2), 123–140 (1996)
9. Brockwell, P., Chadraa, E., Lindner, A.: Continuous-time GARCH processes. Ann. Appl. Probab. **16**, 790–826 (2006)
10. Corradi, V.: Reconsidering the continuous time limit of the GARCH (1, 1) process. J. Econometr. **96**(1), 145–153 (2000)
11. Dueker, M.J.: Markov switching in GARCH processes and mean-reverting stock-market volatility. J. Bus. Econ. Stat. **15**(1), 26–34 (1997)
12. Elliott, R., Lau, J., Miao, H., Siu, T.: Viterbi-based estimation for Markov switching GARCH models. Appl. Math. Financ. **19**(3), 1–13 (2012)
13. Glosten, L.R., Jagannathan, R., Runkle, D.E.: On the relation between the expected value and the volatility of the nominal excess return on stocks. J. Financ. **48**(5), 1779–1801 (1993)
14. Gray, S.F.: Modeling the conditional distribution of interest rates as a regime-switching process. J. Financ. Econ. **42**(1), 27–62 (1996)
15. Haas, M., Mittnik, S., Paolella, M.S.: A new approach to Markov-switching GARCH models. J. Financ. Economet. **2**(4), 493–530 (2004)
16. Kim, C.J.: Dynamic linear models with Markov-switching. J. Econometr. **60**(1–2), 1–22 (1994)
17. Klaassen, F.: Improving GARCH volatility forecasts with regime-switching GARCH. In: Hamilton, J.D., Raj, B. (eds.) Advances in Markov-Switching Models, pp. 223–254. Springer, Heidelberg (2002). https://doi.org/10.1007/978-3-642-51182-0_10
18. Kluppelberg, C., Lindner, A., Maller, R.: A continuous-time GARCH process driven by a levy process: stationarity and second-order behavior. J. Appl. Probab. **41**, 601–622 (2004)
19. Lamoureux, C.G., Lastrapes, W.D.: Persistence in variance, structural change, and the GARCH model. J. Bus. Econ. Stat. **8**(2), 225–234 (1990)
20. Maller, R.A., Müller, G., Szimayer, A.: GARCH modelling in continuous time for irregularly spaced time series data. Bernoulli **14**(2), 519–542 (2008)
21. Marín, J.M., Rodríguez-Bernal, M.T., Romero, E.: Data cloning estimation of GARCH and COGARCH models. J. Stat. Comput. Simul. **85**(9), 1818–1831 (2015)
22. Mendes, M., De, B.V.: Assessing the bias of maximum likelihood estimates of contaminated GARCH models. J. Stat. Comput. Simul. **67**(4), 359–376 (2000)
23. Mikosch, T., Starica, C.: Nonstationarities in financial time series, the long-range dependence, and the IGARCH effects. Rev. Econ. Stat. **86**(1), 378–390 (2004)
24. Müller, G.: MCMC estimation of the COGARCH(1, 1) model. J. Financ. Economet. **8**(4), 481–510 (2010)

25. Nelson, D.B.: Stationarity and persistence in the GARCH(1, 1) model. Economet. Theor. **6**, 318–334 (1990)
26. Nelson, D.B.: Conditional heteroskedasticity in asset returns: a new approach. Econometrica **59**, 347–370 (1991)
27. Nelson, D.: Arch models as diffusion approximations. J. Econ. **45**, 7–38 (1990)
28. Sentana, E.: Quadratic ARCH models. Rev. Econ. Stud. **62**(4), 639–661 (1995)
29. Spiegelhalter, D.J., Best, N.G., Carlin, B.P., Van Der Linde, A.: Bayesian measures of model complexity and fit. J. Roy. Stat. Soc.: Ser. B (Stat. Methodol.) **64**(4), 583–639 (2002)
30. Srivastava, N., Hinton, G., Krizhevsky, A., Sutskever, I., Salakhutdinov, R.: Dropout: a simple way to prevent neural networks from overfitting. J. Mach. Learn. Res. **15**, 1929–1958 (2014)
31. Wang, Y.: Asymptotic nonequivalence of GARCH models and diffusions. Ann. Stat. **30**(3), 754–783 (2002)
32. Zakoian, J.M.: Threshold heteroskedastic models. J. Econ. Dyn. Control **18**(5), 931–955 (1994)

Dynamic Heterogeneous Graph Embedding via Heterogeneous Hawkes Process

Yugang Ji[1], Tianrui Jia[1], Yuan Fang[2], and Chuan Shi[1(✉)]

[1] Beijing University of Posts and Telecommunications, Beijing, China
{jiyugang,jiatianrui,shichuan}@bupt.edu.cn
[2] Singapore Management University, Singapore, Singapore
yfang@smu.edu.sg

Abstract. Graph embedding, aiming to learn low-dimensional representations of nodes while preserving valuable structure information, has played a key role in graph analysis and inference. However, most existing methods deal with static homogeneous topologies, while graphs in real-world scenarios are gradually generated with different-typed temporal events, containing abundant semantics and dynamics. Limited work has been done for embedding dynamic heterogeneous graphs since it is very challenging to model the complete formation process of heterogeneous events. In this paper, we propose a novel **H**eterogeneous **H**awkes **P**rocess based dynamic **G**raph **E**mbedding (**HPGE**) to handle this problem. HPGE effectively integrates the Hawkes process into graph embedding to capture the excitation of various historical events on the current type-wise events. Specifically, HPGE first designs a heterogeneous conditional intensity to model the base rate and temporal influence caused by heterogeneous historical events. Then the heterogeneous evolved attention mechanism is designed to determine the fine-grained excitation to different-typed current events. Besides, we deploy the temporal importance sampling strategy to sample representative events for efficient excitation propagation. Experimental results demonstrate that HPGE consistently outperforms the state-of-the-art alternatives.

Keywords: Dynamic heterogeneous graph · Graph embedding · Heterogeneous Hawkes process · Heterogeneous evolved attention mechanism

1 Introduction

Graphs, such as social networks, e-commerce platforms and academic graphs, occur naturally in various real-world applications. Recently, graph embedding, whose goal is to encode high-dimensional non-Euclidean structures into low-dimensional vector space [2,10], has shown great popularity in tackling graph analytic problems such as node classification and link predictions.

© Springer Nature Switzerland AG 2021
N. Oliver et al. (Eds.): ECML PKDD 2021, LNAI 12975, pp. 388–403, 2021.
https://doi.org/10.1007/978-3-030-86486-6_24

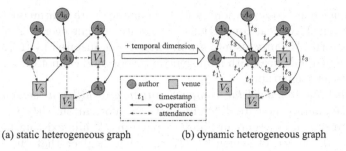

(a) static heterogeneous graph   (b) dynamic heterogeneous graph

Fig. 1. Toy examples of static and dynamic heterogeneous graphs.

Most existing graph embedding methods focus on modeling static homogeneous graphs, where both edges and nodes are of the same type and never change over time. However, in the real world, complex systems are commonly associated with multiple temporal interactions between different-typed nodes, forming the so-called dynamic heterogeneous graphs. Taking Fig. 1(b) as an example, there are two types of interactions ("co-operation" and "attendance") between two types of nodes (authors and venues) and each interaction is marked with a continuous timestamp to describe when it happened, compared to the static one in Fig. 1(a). Dynamic heterogeneous graphs indeed describe richer semantics and dynamics besides structural information, indicating the multiple evolutions of node representations, compared to static homogeneous graphs.

Paying attention to the abundant semantics, there have been several heterogeneous graph embedding methods [5,12,27,34], taking into account both types of nodes and edges when learning representations. While earlier approaches [5,6] employ shallow skip-gram models on heterogeneous sequences generated by meta-paths [24], recent studies [7,12,27,34] apply deeper graph neural networks (GNNs) which usually gather information from heterogeneous neighborhoods to enhance node representations. On the other line, to capture the temporal evolution of dynamic graphs, it is general to split the whole graph into several snapshots and generate representations by inputting all snapshot-based embeddings into sequential models like Long-Short Term Memory (LSTM) and Gated Recurrent Units (GRU) [8,19,22]. Recently, aware of the fact that historical events (i.e., temporal edges) consistently influence and excite the generation of current interactions, recent researchers [17,29,36] attempt to introduce temporal point process, especially Hawkes process, into graph embedding to model the formation process of dynamic graphs.

However, limited work has been done for embedding dynamic heterogeneous graphs. The semantics and dynamics introduce two essential challenges:

First, how to model the continuous dynamics of heterogeneous interactions? Although several works attempt to describe the formation process as sequential heterogeneous snapshots [1,18,32], the heterogeneous dynamics can only be reflected via the number of snapshots, while different-typed edges are indeed continuously generated over time. For instance, as shown in Fig. 1(a), heterogeneous events like "co-operation" and "attendance" are continuously generated over time and historical connections can excite current events. A naïve idea is to integrate Hawkes process into graph embedding, inspired by [17,29,36]. However, these methods deal with homogeneous events and cannot directly introduce into heterogeneous graphs.

Second, how to model the complex influence of different semantics? While different semantics indicate different views of information, they usually impact current various interactions in different patterns. While existing methods only model the difference of semantics [27,32], they neglect that the influence to different-typed current or future events could be very different. For example, in Fig. 1(b), the co-operation between A_1 and A_5 at T_3 could be excited more from historical co-operation events of A_4 and A_5, rather than the attendance between A_4 and V_3. Meanwhile, the attendance between A_1 and V_3 at time t_4 would be affected more from the historical attendance events of A_4. In a word, different-typed historical events would excite different-typed current events in different patterns.

Motivated by these challenges, we propose the **H**eterogeneous Hawkes **P**rocess for Dynamic Heterogeneous **G**raph **E**mbedding (**HPGE**). To handle the continuous dynamics, we treat heterogeneous interactions as multiple temporal events, which gradually occur over time, and introduce Hawkes process into heterogeneous graph embedding by designing a *heterogeneous conditional intensity* to model the excitation of historical heterogeneous events to current events. To handle the complex influence of semantics, we further design the *heterogeneous evolved attention mechanism* which considers both the intra-typed temporal importance of historical events but also the inter-typed temporal impacts from multiple historical events to current type-wise events. Moreover, as current events are influenced more by past important interactions, we adopt the temporal importance sampling strategy to select representative events from historical candidates, balancing their importance and recency. The contributions of this work are summarized as follows.

- We introduce Hawkes process into dynamic heterogeneous graph embedding, which can preserve both semantics and dynamics by learning the formation process of all heterogeneous temporal events. Although few works [17,36] attempt to model the formation process of graphs, they pay no attention to types of either historical or current events.
- Our proposed approach HPGE not only integrates complex evolved excitation of events but also enables efficient extraction of representative past events. To these ends, we respectively design the heterogeneous evolved attention mechanism and the temporal importance sampling strategy.
- We study the effectiveness and efficiency of HPGE empirically on three public datasets and the experimental results of node classification and temporal link prediction demonstrate that HPGE consistently outperforms the state-of-the-art alternatives.

2 Related Work

We discuss the related work on two lines, namely, static graph embedding and dynamic graph embedding, taking both homogeneous and heterogeneous methods into consideration.

Static Graph Embedding. This line of methods are to embed non-Euclidean structures into low-dimensional vector space. Earlier methods [9,23] input random walk-based contextual sequences into skip-gram framework to preserve relevance of connected nodes. Recently, graph neural networks (GNNs) [11,16,25] have attached much attention for their ability to integrate neighborhood influence via message passing. However, they neglect the types of either edges or nodes, and thus fail to model the abundant semantics in real-world graphs. Focus on dealing with heterogeneity, previous Metapath2Vec [5] and HIN2Vec [6] associate nodes by their local proximity through heterogeneous sequences, while current works focus on heterogeneous GNNs [7,35] to better exploit structures and semantics over the whole graph. In these methods, various heterogeneous attention mechanisms are designed to enhance traditional information aggregation [3,7,12,27,34]. More detailed discussions are summarized in [2,26]. However, all the above methods cannot deal with dynamic heterogeneous graphs because of overlooking evolution within interactions.

Dynamic Graph Embedding. On another line, there is significant research interest in dynamic graph embedding (also called temporal network embedding) during the past decade. CTDNE [21] considers dynamics as temporal bias and deploy temporal random walks to learn nodes. TGAT [30] designs a temporal encoder to project continuous timestamps as temporal vectors. Aware of the dynamic evolution of graphs, recent works prefer to split a graph into several snapshots and integrate deep auto-encoders [8] or recurrent neural networks [20,22] to learn the evolving embeddings. Focusing on handle both dynamics and semantics, dynamic heterogeneous graph embedding has also been explored to some extent [12,13,31,32]. Nevertheless, the performance of these methods is often limited as the timestamps of interactions in a snapshot are removed, whereas the formation process of graphs remains unknown. Recently, temporal point processes, most notably the Hawkes process, have become popular for their ability to simulate the formation history [17,36]. However, they are designed for homogeneous graphs while the heterogeneity introduces essential challenges to learn and inference.

3 Preliminaries

In this section, we introduce the definition of dynamic heterogeneous graphs, the problem of dynamic heterogeneous graph embedding as well as the general Hawkes process framework.

Definition 1 _Dynamic Heterogeneous Graph._ _A dynamic heterogeneous graph is_ $\mathcal{G} = (\mathcal{V}, \mathcal{E}, \mathcal{T}, \mathcal{O}, \mathcal{R})$ _where_ $\mathcal{V}$ _denotes the set of nodes,_ $\mathcal{E}$ _denotes the_ *temporal edges (i.e., events), $\mathcal{T}$ denotes the set of timestamps, $\mathcal{O}$ and $\mathcal{R}$ respectively denote node and edge types. In addition, there are two corresponding type mapping functions including $\phi : \mathcal{V} \rightarrow \mathcal{O}$ and $\psi : \mathcal{E} \rightarrow \mathcal{R}$. Notice that, each event is a quad $e = (v_i, v_j, t, r)$ where v_i and v_j are source and target nodes, $t \in \mathcal{T}$ is the continuous timestamp and $r \in \mathcal{R}$ is the event type.*

For instance, the academic graph in Fig. 1(b) consists of two types of nodes (i.e., authors and venues), two types of events (i.e., "co-operation" and "attendance") as well as the continuous timestamps t_1, t_2, t_3, t_4 and t_5 of these heterogeneous events, naturally forming a dynamic heterogeneous graphs. Obviously, heterogeneous events gradually happen and excite future interactions over time, expressing abundant semantics and dynamics, compared to static graphs.

Definition 2 *Dynamic Heterogeneous Graph Embedding.* *Given a dynamic heterogeneous graph $\mathcal{G}$, the goal of dynamic heterogeneous graph embedding is to learn a representation function $\mathcal{H}$ to project such a high-dimensional non-Euclidean structures into low-dimensional vector space, namely, $\mathcal{H}(\mathcal{G}) \rightarrow H, H \in \mathcal{R}^{|\mathcal{V}| \times d}$ where $|\mathcal{V}|$ and d are the size and dimension of nodes, $d \ll |\mathcal{V}|$. Meanwhile, both the dynamics and semantics besides structural information should be preserved as well.*

Definition 3 *Hawkes process.* *Hawkes process is a typical temporal point process with the assumption that historical events can influence the occurrence of the current event. Given historical events $\{e_h | t_h < t\}$ before current time t, a conditional intensity function is defined to characterizes the arrival rate of current event e, namely,*

$$\lambda(e) = \mu(e) + \sum_{e_h : t_h < t} \kappa(t - t_h), \tag{1}$$

where $\mu(e)$ is the base intensity (i.e., spontaneous arrival rate) of current event e, $\kappa(\cdot)$ is a time decay effect of historical events on the current e.

Obviously, the temporal excitation is well modeled and there are several works [17,36] attempt to embed dynamic graphs with Hawkes process. Nevertheless, these methods cannot handle the heterogeneity. In this paper, we focus on introducing Hawkes process into dynamic heterogeneous graph embedding, to learn the complete temporal formation process of heterogeneous events, keeping both semantics and dynamics.

4   The Proposed HPGE Model

In this section, we propose our model called HPGE. We begin with an overview, before zooming into the details.

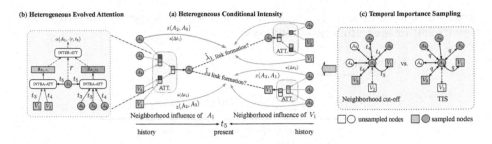

Fig. 2. The overall architecture of HPGE. (a) Heterogeneous conditional intensity function to model the heterogeneous temporal influence of A_1, A_3 or V_1, (b) Heterogeneous evolved attention to measure the relevance and evolution from historical neighbors to current type-wise event, consisting of intra- and inter-typed temporal attention, (c) Temporal importance sampling of heterogeneous events where q denotes the sampling probability and the nodes in white are unsampled, in comparison to a naïve cut-off strategy.

4.1 Overview

There are three main components of HPGE, namely, the heterogeneous conditional intensity function to learn the semantics and dynamics within the formation process of heterogeneous temporal events, the heterogeneous evolved attention mechanism to measure the importance and evolution from historic neighborhoods to current type-wise event, and the temporal importance sampling to handle the efficient extraction of representative events. First, as shown in Fig. 2(a), given the respective temporal heterogeneous neighbors of A_1, A_3, and V_1, HPGE evaluates the affinity between each node and its neighbors with a type-wise influence measure. Subsequently, hinged on a heterogeneous conditional intensity function, it accumulates the influence from historical heterogeneous neighbors, which characterizes the arrival rate at present. Second, an attentive manner is designed in 2(b) to capture both the temporal importance of same-typed neighborhoods (intra-att) and the evolution from historical types to the current type (inter-att). Third, as the graph evolves, in Fig. 2(c), the number of events gradually grows. For effective and efficient HPGE, we adopt a Temporal Importance Sampling (TIS) strategy to extract representative neighbors in both temporal and structural dimensions, instead of using the full neighborhood which is inefficient, or the traditional cut-off strategy based on recency only.

4.2 Heterogeneous Conditional Intensity Modeling

On a dynamic heterogeneous graph, various kinds of interactions are constantly being established over time, which can be regarded as a series of observed heterogeneous events. Intuitively, the current events are influenced by past events, and the heterogeneity of events implies different strengths of influence. For instance, attendance in a conference at present is influenced by different historical views, including the past attendance view and author collaboration view. Therefore, given current event $e = (v_i, v_j, t, r)$, we introduce the general heterogeneous conditional intensity function as follows:

$$\tilde{\lambda}(e) = \underbrace{\mu_r(v_i, v_j)}_{\text{base rate}}$$

$$+ \underbrace{\gamma_1 \sum_{r' \in \mathcal{R}} \sum_{p \in \mathcal{N}_{i,r',<t}} \alpha(p, e) z(v_p, v_j) \kappa_i(t - t_p)}_{\text{neighborhood influence on source } v_i},$$

$$+ \underbrace{\gamma_2 \sum_{r'' \in \mathcal{R}} \sum_{q \in \mathcal{N}_{j,r'',<t}} \alpha(q, e) z(v_q, v_i) \kappa_j(t - t_q)}_{\text{neighborhood influence on target } v_j}$$

(2)

where γ_1 and γ_2 are the balance parameters. This conditional intensity function consists of three major parts, including the type-wise base rate, the heterogeneous neighborhood on source node v_i and on target node v_j. At first, given v_i and v_j as well as event type r, the base rate $\mu_r(v_i, v_j)$ is defined as:

$$\mu_r(v_i, v_j) = -\sigma(f(\boldsymbol{h}_i \boldsymbol{W}_{\phi(v_i)} - \boldsymbol{h}_j \boldsymbol{W}_{\phi(v_j)}) \boldsymbol{W}_r + b_r),$$

(3)

where $\boldsymbol{h}_i \in \mathbb{R}^d$ and $\boldsymbol{h}_j \in \mathbb{R}^d$ are the embedding of v_i and v_j, d is the dimension of node embedding, $\boldsymbol{W}_{\phi(\cdot)} \in \mathbb{R}^{d \times d}$ denotes the type-$\phi(\cdot)$ projection matrix, $f(\cdot)$ denotes the element-level non-negative operation to measure the symmetrical similarity of v_i and v_j, and we adopt self Hadamard product in this paper, namely $f(\boldsymbol{X}) = \boldsymbol{X} \odot \boldsymbol{X}$, $\boldsymbol{W}_r$ and b_r are the projection and bias of type-r events, $\sigma(\cdot)$ is the non-linear activate function. In the base rate evaluation, both the types of nodes and edges are taken into consideration.

Besides, historical neighbors can continuously excite the occurrence of the current event. Taking the neighborhood influence on source node as an example, given its historical neighborhoods $\{\mathcal{N}_{i,r',<t} | r' \in \mathcal{R}\}$ the excitation is indeed associated with three aspects, (1) the time span to the current time, (2) the relevant historical neighbors to target node v_j and (3) the importance of historical neighbors to source node v_i. As the time decay to different nodes are different, we design $\kappa_i(\Delta t)$ as $\exp(-\delta_i(\Delta t))$, where $delta_i$ is the learnable personalized parameter and the influence become exponentially weak over time. The relevance between historical neighbors and target nodes are related to their types as well, namely,

$$z(v_p, v_j) = -\|\boldsymbol{h}_p \boldsymbol{W}_{\phi(p)} - \boldsymbol{h}_j \boldsymbol{W}_{\phi(j)}\|_2^2,$$

(4)

where $\| \cdot \|_2^2$ denotes the Euclidean distance measure, and the negative symbol indicates that closer nodes could affect greater. To measure the importance to source node, attention mechanisms [7,12,27] have shown powerful performance on static heterogeneous graphs. However, when dealing with the heterogeneous formation process, the complex temporal influence between different semantics remains an essential challenge. To handle the second challenge, we design the heterogeneous evolved attention mechanism in Sect. 4.3.

4.3   Heterogeneous Evolved Attention Mechanism

As mentioned in Sect. 1, the excitation of historical interactions not only associate with types of historical events but also depend on types of current events. Thus, the importance to current event $\alpha(p, e)$ is defined as

$$\alpha(p, e) = \xi(v_p, t_p | r', v_i, t)\beta(r | r', v_i, t), \tag{5}$$

where r' and r respectively denote the type of historical and current event, t_p and t are the corresponding timestamps, $\xi(v_p, t_p | r', v_i, t)$ is the intra-type heterogeneous temporal attention, calculated by

$$\xi(v_p, t_p | r', v_i, t) = \text{softmax}(\sigma(\kappa_i(t - t_p)[\boldsymbol{h}_i \boldsymbol{W}_{\phi(v_i)} \oplus \boldsymbol{h}_j \boldsymbol{W}_{\phi(v_j)}]\boldsymbol{W}_\xi)), \tag{6}$$

where $\boldsymbol{W}_\xi \in \mathbb{R}^{2d \times 1}$ denotes the attention projection matrix need to learn, $\oplus$ denotes the concatenation operation, $\text{softmax}(x)$ is in the form of $\exp(x)/\sum_{x'} \exp(x')$. Both the heterogeneity and time decay are taken into consideration. Furthermore, we design the inter-typed $\beta(r | r', v_i, t)$ to model the relevance from historical types to current types, namely

$$\beta(r | r', v_i, t) = \text{softmax}(\tanh(\tilde{\boldsymbol{g}}_i \boldsymbol{W}_r)\boldsymbol{w}_r)^{\mathrm{T}}, \tag{7}$$

where $\boldsymbol{W}_r \in \mathbb{R}^{d|\mathcal{R}| \times d_m}$ and $\boldsymbol{w}_r \in \mathbb{R}^{d_m \times 1}$ are the projection matrices need to learn, d_m is the length of latent dimension and we set $d_m = 0.5d$ here. $\tilde{\boldsymbol{g}}_i$ is the concatenation of historical excitation, namely $\tilde{\boldsymbol{g}}_i = [\tilde{\boldsymbol{g}}_{i,1} \oplus \tilde{\boldsymbol{g}}_{i,2} \oplus \cdots \oplus \tilde{\boldsymbol{g}}_{i,|\mathcal{R}|}]$, and the sub-excitation from type-r' neighbors is calculated by

$$\tilde{\boldsymbol{g}}_{i,r'} = \sigma\left(\left[\sum_p \xi(v_p, t_p | r', v_i, t)\boldsymbol{h}_p \boldsymbol{W}_{\phi(v_p)}\kappa_i(t - t_p)\right] \boldsymbol{W}_{\beta,r'} + b_{\beta,r'}\right), \tag{8}$$

where $\boldsymbol{W}_{\beta,r'} \in \mathbb{R}^{d \times d}$ and $b_{\beta,r'}$ are the projection matrix and bias need to learn. It is naturally a intra-typed attention based temporal excitation aggregation.

4.4   Temporal Importance Sampling

As more events are accumulated over time, it becomes expensive to materialize the heterogeneous conditional intensity function. For efficiency, existing Hawkes process on homogeneous graphs cut off events happened far away in the past, and only focus on the most recent events. However, the cut-off point is often arbitrary and difficult to set. Furthermore, the recency-only strategy risks in omitting structurally important neighbors that have frequent interactions over time. As illustrated in Fig. 2(b), A_5 would be cut off based on recency only, but it is desirable to retain A_5 for modeling due to its frequent interaction with A_1.

To efficiently extract representative candidates with both recency and structural importance, inspired by importance sampling [4,14], we propose the strategy of Temporal Importance Sampling (TIS). TIS considers both temporal and structural information to extract representation neighbors. Weighed by the excitation rate and the time decay function, we design the sampler of TIS as follows,

$$q(v_p | v_i, r', t) = \frac{\kappa_i(t - t_p)N_i(v_p)}{\sum_{v_{p'} \in \mathcal{N}_{i,r',<t}} \kappa_i(t - t'_p)N_i(v'_p)}, \tag{9}$$

where $q(v_p|v_i, r', t)$ denotes the sampling probability, depending on the importance of node v_h relating to event type r', times of historical occurrence $N_i(v_p)$ as well as time t. Thus, the estimator of the sampled neighbor influence is given by

$$z(\hat{v}_p, v_j) = \frac{1}{n} \cdot \frac{z(\hat{v}_p, v_j)}{q(\hat{v}_p|v_i, r', t)}, \quad \hat{v}_p \sim q(v_p|v_i, r', t) \tag{10}$$

where n is the sample size, $\hat{v}_p$ denotes a sampled historical neighbor. Thus, both temporal and structural importance of the neighbors can be retained for influence modeling. In particular, the estimator ensures the expectation of weighted sampled excitation is equal to propagate all historical influences.

4.5   Optimization Objective

By modeling the temporal heterogeneous event formation with heterogeneous Hawkes process, the current neighbor formation events can be inferred from the heterogeneous conditional intensity. Given all the historical neighborhoods $\mathcal{N}_{i,<t}$ of v_i and $\mathcal{N}_{j,<t}$ of v_j before time t, the probability of forming type-r connection between v_i and v_j at time t can be inferred as

$$p(e_{i,j,r}|\mathcal{N}_{i,t}, \mathcal{N}_{j,t}) = \frac{\lambda(e_{i,j,r})}{\sum_{r' \in \mathcal{R}} \left(\sum_{j' \in \mathcal{N}_{i,t}^{r'}} \lambda(e_{i,j',r'}) + \sum_{i' \in \mathcal{N}_{j,t}^{r'}} \lambda(e_{i',j,r'}) \right)}, \tag{11}$$

where $\lambda(e_{i,j,r}) = \exp(\lambda(\tilde{e}_{i,j,r}))$ denotes the positive intensity. As directed likelihood optimization would suffer from the heavily computational complexity of $p(e_{i,j,r}|\mathcal{N}_{i,t}, \mathcal{N}_{j,t})$, we consider Eq. (11) as the softmax normalization of $\tilde{\lambda}(e_{i,j,r})$, and adopt negative sampling to accelerate learning, thus, the loss of the current event e is defined as follows,

$$\mathcal{L}_{hp}(e) = -\sum_{e \in \mathcal{E}} \log \sigma(\tilde{\lambda}(e)) - \sum_{k} \mathbb{E}_{j'} \log \sigma(-\tilde{\lambda}(e_{j'})) - \sum_{k} \mathbb{E}_{i'} \log \sigma(-\tilde{\lambda}(e_{i'})), \tag{12}$$

where $e_{i'}$ and $e_{j'}$ are the abbreviations of $e_{i',j,r,t}$ and $e_{i,j',r,t}$, k is the size of negative samples, and $\mathcal{L}_{hp} = \frac{1}{|\mathcal{E}|} \sum_{e \in \mathcal{E}} \mathcal{L}_{hp}(e)$.

Besides, focusing on the downstream tasks like node classification and temporal link prediction, we design the unified loss function as follows:

$$\mathcal{L} = \mathcal{L}_{hp} + \omega_1 \mathcal{L}_{task} + \omega_2 \Omega(\boldsymbol{\Theta}), \tag{13}$$

where $\Omega(\boldsymbol{\Theta})$ is the l2-norm regularization of learnt parameters, $\mathcal{L}_{task}$ is the loss of specific tasks. For node classification and temporal link prediction, we input node embedding or the concatenation of embedding pair into a Multi-Layer Perception to extract the distribution of classifications or the probability of connections, and then evaluate the cross-entropy loss values, i.e., $\mathcal{L}_{task}$. ω_1 and ω_2 are the weights. We adopt Adam optimizer [15] to minimize the loss function for each mini-batch.

Table 1. Statistics of the three public datasets.

Datasets	Node types	#Nodes	Event types	#Events	Time span
Aminer	Author (A)	23,037	A-A	71,121	16 years
	Conference (C)	22	A-C	52,399	
DBLP	Author (A)	34,766	A-A	133,684	10 years
	Venue (V)	20	A-V	98,262	
Yelp	User (U)	494,524	BrU	1,145,070	60 quarters
	Business (B)	13,507	BtU	226,728	

5 Experiments

In this section, we conduct extensive experiments on three public real-world dynamic heterogeneous graphs to demonstrate the effectiveness of HPGE.

5.1 Experimental Settings

Datasets. The three real-world datasets are the academic Aminer and DBLP graphs and the Yelp business graph. The details are introduced as follows and the statistics are listed in Table 1. (1) **Aminer**[1]. This is a benchmark bibliographic graph, which consists of two types of nodes, namely, authors (A) and conferences (C), as well as two types of temporal events, namely "co-operation" (A-A) and "attendance" (A-C). Notice that each author is labeled by one of the five research domains including data mining, database, medical informatics, theory, and visualization. (2) **DBLP**[2]. This is another bibliographic graph, which also consists of two types of temporal events between authors (A) and venues (V), namely, A-A and A-V. We follow previous work [27] to extract 20 venues in four areas, namely, database, data mining, machine learning, information retrieval. The authors are labeled by the research area they focus on. (3) **Yelp**[3]. This is a business review dataset, containing timestamped user reviews and tips on businesses. There are two types of nodes, users (U) and businesses (B), and four types of temporal events including "reviewed" (UrB), "tipped" (UtB), "reviewed by" (BrU) and "tipped by" (BtU). We extract interactions of three categories of businesses, including "Fast Food", "Sushi" and "American (New) Food", to construct the dynamic graph. Each business is labeled with its most related category.

Baselines. We compare the proposed HPGE with three groups of graph embedding models, namely, heterogeneous graph embedding (Metapath2vec [5], HEP [35], HAN [27] and HGT [12]), dynamic graph embedding (CTDNE [21],

---

[1] Available at Aminer website.
[2] Available at DBLP website.
[3] Available at Yelp website.

EvolveGCN [22], and M^{2}DNE [17]), and dynamic heterogeneous graph embedding (DHNE [33], DyHNE [28], and DyHATR [31]).

- **Metapath2vec** [5] and **HEP** [35]: They are two heterogeneous graph embedding models, where the former learns node embedding with sequences generated by a meta-path, and the latter propagates embedding information among different-typed interactions.
- **HAN** [27] and **HGT**: They are two attentive heterogeneous GNNs, where the former designs a hierarchical attention considering both node- and semantic-levels while the latter takes into account both the types of nodes and edges to design a heterogeneous mutual attention.
- **CTDNE** [21], **EvolveGCN** [22] and **M^{2}DNE** [17]: They are three typical dynamic homogeneous graph embedding approaches. CTDNE is a skip-gram model based on temporal random walks; EvolveGCN learns the evolution among snapshots by integrating with RNNs to sequentially update convolutional parameters; and M^{2}DNE introduces Hawkes process into modeling the formation process of dynamic graphs where neighbor influence of both source and target nodes are simultaneously extracted.
- **DHNE** [33], **DyHNE** [28] and **DyHATR** [31]: These are three representative temporal heterogeneous graph embedding models. DHNE performs metapath-based random walk between historical snapshots and the current snapshot and design a dynamic heterogeneous skip-gram model to capture representations of nodes; DyHNE splits graphs into several snapshots and employs eigenvalue perturbation to derive the updated embeddings between different snapshots; DyHATR uses hierarchical attention to learn heterogeneous information and incorporates RNNs with temporal attention to capture evolutionary patterns between different snapshots.

Parameter Settings. For all methods, we set the embedding dimension $d = 128$, batch size as 1024, learning rate as 0.001, regularization weight $\omega_2 = 0.01$ (if any), and negative sampling size as $k = 5$ (if any). These values give robust performance and are consistent with guidelines from the literature. For HAN, HGT, M^{2}DNE, DyHATR and our HPGE, we respectively limit the size of neighboring candidates to 5, 5 and 10 on the three datasets, using TIS for our method, recency cut-off for M^{2}DNE and random sampling for others. For dynamic homogeneous baselines, we treat events as homogeneous. For Metapath2Vec and DHNE, we sample sequences via A-A, A-A and B-U-B on the three datasets, respectively. The other parameters of all baselines follow their original papers. For our HPGE, we set $\gamma_1 = 0.5$ and $\gamma_2 = 0.5$, $\omega_1 = 1$. In addition, the max iteration is set as 500, 500 and 50 on the three datasets.

5.2   Effectiveness Analysis

Node Classification. This task is to predict the research area of authors on Aminer and DBLP and the category of businesses on Yelp. The train/test ratio

Table 2. Performance evaluation (with standard deviation) on node classification. The best performance is bolded and the second best is underlined.

Dataset	Aminer		DBLP		Yelp	
Metric	Micro-F1	Macro-F1	Micro-F1	Macro-F1	Micro-F1	Macro-F1
M2V	0.824(0.029)	0.853(0.032)	0.874(0.024)	0.885(0.029)	0.537(0.023)	0.642(0.017)
HEP	0.949(0.016)	0.952(0.013)	0.903(0.022)	0.913(0.018)	0.622(0.012)	0.694(0.009)
HAN	0.967(0.008)	0.970(0.009)	0.912(0.014)	0.914(0.007)	0.621(0.019)	0.691(0.025)
HGT	0.963(0.007)	0.971(0.011)	0.920(0.002)	0.927(0.001)	0.633(0.026)	0.705(0.022)
CTDNE	0.897(0.038)	0.895(0.025)	0.872(0.001)	0.892(0.005)	0.512(0.011)	0.639(0.011)
E.GCN	0.952(0.020)	0.955(0.018)	0.887(0.009)	0.881(0.010)	0.611(0.009)	0.687(0.008)
M2DNE	0.969(0.015)	0.972(0.018)	0.891(0.022)	0.909(0.027)	0.619(0.003)	0.693(0.005)
DHNE	0.901(0.010)	0.913(0.009)	0.888(0.007)	0.909(0.008)	0.578(0.001)	0.665(0.001)
DyHNE	0.970(0.008)	0.978(0.007)	0.922(0.003)	0.922(0.004)	0.622(0.011)	0.721(0.015)
DyHATR	0.973(0.002)	0.969(0.003)	0.933(0.011)	0.935(0.010)	0.627(0.008)	0.717(0.007)
HPGE	**0.988(0.002)**	**0.984(0.003)**	**0.951(0.005)**	**0.952(0.004)**	**0.649(0.010)**	**0.731(0.012)**

is set to 80%/20%. We run all methods five times and evaluate the average Micro-F1 and Macro-F1 scores.

As shown in Table 2, our proposed HPGE consistently outperforms all baselines on the three datasets. We make the following observations. (1) Compared with heterogeneous graph embedding approaches (Metapath2vec, HEP, HAN and HGT), HPGE is able to model the temporal dynamics of heterogeneous events. Similarly, compared to dynamic graph embedding approaches (CTDNE, EvolveGCN and M^2DNE), HPGE benefits from integrating the abundant semantic information within heterogeneous events. Not surprisingly, the performance gains of HPGE are larger relative to these baselines. (2) Compared with the best competitor DyHATR, which considers both the temporal and heterogeneous information, our HPGE can still achieve substantial improvements. The stable improvements demonstrate that modeling the formation process of DIIGs can embed evolving nodes better than just paying attention to the evolution between snapshots. (3) Compared with Aminer and DBLP, our model improves more on Yelp. The potential reason is that Yelp is a larger dataset, such that our temporal importance sampling strategy can benefit more.

Temporal Link Prediction. This task is to predict the type-r interaction at time t. Given all temporal heterogeneous events before time t and two nodes v_i and v_j. We treat all events at time t as the positive link, and randomly sample 2 negative instances for both v_i and v_j as the negative links. Subsequently, we test all baselines and our HPGE five times and report the average performance of Accuracy, F1 score, and ROC-AUC in Table 3. Obviously, HPGE still achieves the best performance on all datasets. Besides the observations on node classification, HPGE evaluates node proximity based on event types and continuously propagates the influence of types via the temporal point process, while traditional type-wise projections can only model the heterogeneity rather than the interactivity. In addition, HAN, HEP, HGT, DyHNE, DyHATR and our HPGE always performs better than CTDNE, EvolveGCN and M^2DNE. This

Table 3. Performance evaluation on temporal link prediction. The best performance is bolded and the second best is underlined.

Dataset	Aminer			Yelp			DBLP		
Metric	ACC	F1	AUC	ACC	F1	AUC	ACC	F1	AUC
M2V	0.806	0.359	0.759	0.790	0.419	0.702	0.798	0.375	0.656
HEP	0.921	0.814	0.944	0.853	0.566	0.829	0.910	0.753	0.934
HAN	0.923	0.811	0.955	0.855	0.591	0.833	0.903	0.751	0.940
HGT	0.938	0.822	0.963	0.859	0.588	0.833	0.899	0.761	0.941
CTDNE	0.824	0.382	0.763	0.806	0.342	0.635	0.713	0.345	0.653
E.GCN	0.904	0.767	0.922	0.822	0.526	0.785	0.853	0.714	0.905
M2DNE	0.929	0.790	0.951	0.854	0.547	0.818	0.896	0.734	0.939
DHNE	0.875	0.634	0.827	0.831	0.504	0.717	0.821	0.668	0.808
DyHNE	0.928	**0.838**	0.959	0.861	0.592	0.831	0.909	0.767	0.940
DyHATR	<u>0.941</u>	0.832	<u>0.966</u>	<u>0.870</u>	<u>0.598</u>	<u>0.843</u>	<u>0.914</u>	<u>0.773</u>	<u>0.936</u>
HPGE	**0.953**	<u>0.835</u>	**0.976**	**0.873**	**0.603**	**0.850**	**0.938**	**0.793**	**0.957**

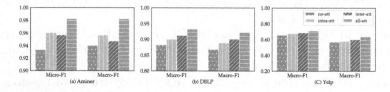

Fig. 3. Effect of hierarchical attention mechanism on node classification.

phenomenon indicates that integrating semantics into link formation can benefit temporal link prediction more, compared with simply preserving evolving structures.

5.3   Model Analysis

Effect of Heterogeneous Evolved Attention Mechanism. We further discuss the effect of heterogeneous evolved attention mechanism by comparing with three model variants including no attention (no-att), intra-type temporal attention (intra-att) and inter-type temporal attention (inter-att), as well as HPGE (all-att). The results for the node classification task are shown in Fig. 3. We observe the following. (1) Simultaneously modeling both intra- and inter-type temporal attention achieves the most improvements, while the no-attention variant performs the worst on all datasets. (2) Compared with the intra-attention variant, HPGE has the ability to evaluate the importance of influence of different types of historical events to current type of interactions. Meanwhile, HPGE can filter the neighborhoods via intra-typed attention, compared with the inter-typed variant. These observations demonstrate the effectiveness of our heterogeneous evolved attention mechanism.

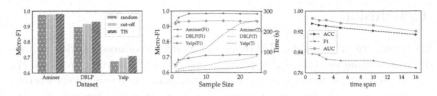

(a) sampling strategies   (b) effective sample size (c) varying the dynamics

Fig. 4. Efficacy of TIS and the ability of evolution modeling.

Efficacy of Temporal Importance Sampling. The other key design is our temporal importance sampling (TIS), which considers both structural importance and time decay. We analyze the effectiveness of TIS by comparing with the often used random sampling and recency-based cut-off, as well as the efficiency of TIS under the effective sample size. (1) Comparison of sampling strategies. Figure 4(a) reports the Micro-F1 scores of different sampling strategies for the node classification task. Notice that the sample size is set as 5, 5 and 10 for all strategies on the three datasets, respectively. Among the three sampling strategies, it is clear that our TIS strategy performs the best, especially on the larger datasets DBLP and Yelp. The results are intuitive since the cut-off strategy ignores structurally important neighbors, while the random sampling, which performs the worst, pays no attention to either structure or dynamics. (2) Effective sample size. Effective sample size plays an important role in sampling to achieve the balance between effectiveness and efficiency. As shown in Fig. 4(b), we increase the sample size from 5 to 25 and showcase both the Micro-F1 score (solid lines) and time cost (dotted lines). A larger sample size gradually increases Micro-F1, which converges quickly around 5 or 10. Here 5 or 10 is the effective samples size, which is much smaller than the full neighborhoods. In particular, when using a larger sample size (e.g., 25 or even the full size), the time cost becomes unbearable.

Ability of Modeling Evolution. As the dynamics of graphs are in the form of timestamps, we "coarsen" the timestamps by considering time spans of varying size. In Fig. 4(c), on the Aminer dataset, we vary the size of time span from every 1 year (i.e., finest time units) to 16 year (i.e., the entire graph consists of a single time span of 16 years, which effectively become a static graph), and showcase the performance on temporal link prediction. The performance of HPGE consistently degrades with the increasing size of time span, indicating that modeling evolving dynamics with finer granularity (i.e., smaller time span) lead to better performance. Notice that when the time span is 16, the graph becomes a static graph and our HPGE also degrades to a static model. Overall, the results further illustrate the effectiveness of HPGE in handling evolution.

6 Conclusion

In this paper, we propose the HPGE model which introduces Hawkes process to handle the challenging dynamic heterogeneous graph embedding problem. Focusing on modeling the formation process of temporal heterogeneous events, we respectively design the heterogeneous conditional intensity function to capture the excitation from historical multiple events, the heterogeneous evolved attention mechanism to learn fine-grained representations considering both intra- and inter-typed temporal influences. HPGE hinges on a novel temporal importance sampling strategy, to enable efficient extraction of representative events. Experimental results on three public datasets demonstrate that HPGE outperforms the alternatives on fundamental graph tasks.

Acknowledgements. This work is supported in part by the National Natural Science Foundation of China (No. U20B2045, 61772082, 62002029). This research is also supported by the Agency for Science, Technology and Research (A*STAR) under its AME Programmatic Funds (Grant No. A20H6b0151).

References

1. Bian, R., Koh, Y.S., Dobbie, G., Divoli, A.: Network embedding and change modeling in dynamic heterogeneous networks. In: SIGIR, pp. 861–864 (2019)
2. Cai, H., Zheng, V.W., Chang, K.C.: A comprehensive survey of graph embedding: problems, techniques, and applications. IEEE TKDE **30**(9), 1616–1637 (2018)
3. Cen, Y., Zou, X., Zhang, J., Yang, H., Zhou, J., Tang, J.: Representation learning for attributed multiplex heterogeneous network. In: ACM SIGKDD, pp. 1358–1368 (2019)
4. Chen, J., Ma, T., Xiao, C.: FastGCN: fast learning with graph convolutional networks via importance sampling. In: ICLR (2018)
5. Dong, Y., Chawla, N.V., Swami, A.: metapath2vec: scalable representation learning for heterogeneous networks. In: ACM SIGKDD, pp. 135–144 (2017)
6. Fu, T., Lee, W., Lei, Z.: HIN2Vec: explore meta-paths in heterogeneous information networks for representation learning. In: CIKM, pp. 1797–1806 (2017)
7. Fu, X., Zhang, J., Meng, Z., King, I.: MAGNN: metapath aggregated graph neural network for heterogeneous graph embedding. In: WWW, pp. 2331–2341 (2020)
8. Goyal, P., Kamra, N., He, X., Liu, Y.: DynGEM: deep embedding method for dynamic graphs. CoRR abs/1805.11273 (2018)
9. Grover, A., Leskovec, J.: node2vec: scalable feature learning for networks. In: SIGKDD, pp. 855–864 (2016)
10. Hamilton, W.L., Ying, R., Leskovec, J.: Representation learning on graphs: methods and applications. IEEE Data Eng. Bull. **40**(3), 52–74 (2017)
11. Hamilton, W.L., Ying, Z., Leskovec, J.: Inductive representation learning on large graphs. In: NeuIPS, pp. 1024–1034 (2017)
12. Hu, Z., Dong, Y., Wang, K., Sun, Y.: Heterogeneous graph transformer. In: WWW, pp. 2704–2710 (2020)
13. Ji, Y., et al.: Temporal heterogeneous interaction graph embedding for next-item recommendation. In: ECML-PKDD (2020)

14. Ji, Y., et al.: Accelerating large-scale heterogeneous interaction graph embedding learning via importance sampling. ACM TKDD **15**(1), 1–23 (2020)
15. Kingma, D.P., Ba, J.: Adam: a method for stochastic optimization. In: ICLR (2015)
16. Kipf, T.N., Welling, M.: Semi-supervised classification with graph convolutional networks. In: ICLR (2017)
17. Lu, Y., Wang, X., Shi, C., Yu, P.S., Ye, Y.: Temporal network embedding with micro- and macro-dynamics. In: CIKM, pp. 469–478 (2019)
18. Luo, W., et al.: Dynamic heterogeneous graph neural network for real-time event prediction. In: ACM SIGKDD, pp. 3213–3223 (2020)
19. Ma, Y., Guo, Z., Ren, Z., Tang, J., Yin, D.: Streaming graph neural networks. In: SIGIR, pp. 719–728 (2020)
20. Manessi, F., Rozza, A., Manzo, M.: Dynamic graph convolutional networks. Pattern Recogn. **97**, 107000 (2020)
21. Nguyen, G.H., Lee, J.B., Rossi, R.A., Ahmed, N.K., Koh, E., Kim, S.: Continuous-time dynamic network embeddings. In: WWW, pp. 969–976 (2018)
22. Pareja, A., et al.: EvolveGCN: evolving graph convolutional networks for dynamic graphs. In: AAAI, vol. 34, pp. 5363–5370 (2020)
23. Perozzi, B., Al-Rfou, R., Skiena, S.: DeepWalk: online learning of social representations. In: ACM SIGKDD, pp. 701–710 (2014)
24. Sun, Y., Han, J., Yan, X., Yu, P.S., Wu, T.: PathSim: meta path-based top-k similarity search in heterogeneous information networks. VLDB **4**(11), 992–1003 (2011)
25. Veličković, P., Cucurull, G., Casanova, A., Romero, A., Liò, P., Bengio, Y.: Graph attention networks. In: ICLR (2018)
26. Wang, X., Bo, D., Shi, C., Fan, S., Ye, Y., Yu, P.S.: A survey on heterogeneous graph embedding: methods, techniques, applications and sources. arXiv preprint arXiv:2011.14867 (2020)
27. Wang, X., et al.: Heterogeneous graph attention network. In: WWW, pp. 2022–2032 (2019)
28. Wang, X., Lu, Y., Shi, C., Wang, R., Cui, P., Mou, S.: Dynamic heterogeneous information network embedding with meta-path based proximity. TKDE (2020)
29. Wu, W., Liu, H., Zhang, X., Liu, Y., Zha, H.: Modeling event propagation via graph biased temporal point process. IEEE TNNLS (2020)
30. Xu, D., Ruan, C., Körpeoglu, E., Kumar, S., Achan, K.: Inductive representation learning on temporal graphs. In: ICLR (2020)
31. Xue, H., Yang, L., Jiang, W., Wei, Y., Hu, Y., Lin, Y.: Modeling dynamic heterogeneous network for link prediction using hierarchical attention with temporal RNN. arXiv preprint arXiv:2004.01024 (2020)
32. Yang, L., Xiao, Z., Jiang, W., Wei, Y., Hu, Y., Wang, H.: Dynamic heterogeneous graph embedding using hierarchical attentions. In: Jose, J.M., et al. (eds.) ECIR 2020. LNCS, vol. 12036, pp. 425–432. Springer, Cham (2020). https://doi.org/10.1007/978-3-030-45442-5_53
33. Yin, Y., Ji, L.X., Zhang, J.P., Pei, Y.L.: DHNE: network representation learning method for dynamic heterogeneous networks. IEEE Access **7**, 134782–134792 (2019)
34. Zhao, J., Wang, X., Shi, C., Hu, B., Song, G., Ye, Y.: Heterogeneous graph structure learning for graph neural networks. In: AAAI (2021)
35. Zheng, V.W., et al.: Heterogeneous embedding propagation for large-scale e-commerce user alignment. In: ICDM, pp. 1434–1439 (2018)
36. Zuo, Y., Liu, G., Lin, H., Guo, J., Hu, X., Wu, J.: Embedding temporal network via neighborhood formation. In: ACM SIGKDD, pp. 2857–2866 (2018)

Explainable Online Deep Neural Network Selection Using Adaptive Saliency Maps for Time Series Forecasting

Amal Saadallah[(✉)], Matthias Jakobs, and Katharina Morik

Artificial Intelligence Group, Department of Computer Science, TU Dortmund,
Dortmund, Germany
{amal.saadallah,matthias.jakobs, katharina.morik}@tu-dortmund.de

Abstract. Deep neural networks such as Convolutional Neural Networks (CNNs) have been successfully applied to a wide variety of tasks, including time series forecasting. In this paper, we propose a novel approach for online deep CNN selection using saliency maps in the task of time series forecasting. We start with an arbitrarily set of different CNN forecasters with various architectures. Then, we outline a gradient-based technique for generating saliency maps with a coherent design to make it able to specialize the CNN forecasters across different regions in the input time series using a performance-based ranking. In this framework, the selection of the adequate model is performed in an online fashion and the computation of saliency maps responsible for the model selection is achieved adaptively following drift detection in the time series. In addition, the saliency maps can be exploited to provide suitable explanations for the reason behind selecting a specific model at a certain time interval or instant. An extensive empirical study on various real-world datasets demonstrates that our method achieves excellent or on par results in comparison to the state-of-the-art approaches as well as several baselines.

Keywords: Deep neural networks · Time series forecasting · Model selection · Grad-CAM · Explainability

1 Introduction

Both the complex and time-evolving nature of time series make forecasting one of the most challenging tasks in time series analysis [26].

This work is supported by the Deutsche Forschungsgemeinschaft (DFG) within the Collaborative Research Center SFB 876 and the Federal Ministry of Education and Research of Germany as part of the competence center for machine learning ML2R (01–S18038A).

Electronic supplementary material The online version of this chapter (https://doi.org/10.1007/978-3-030-86486-6_25) contains supplementary material, which is available to authorized users.

ⓒ Springer Nature Switzerland AG 2021
N. Oliver et al. (Eds.): ECML PKDD 2021, LNAI 12975, pp. 404–420, 2021.
https://doi.org/10.1007/978-3-030-86486-6_25

Several machine learning methods have been proposed to solve this task either by dealing with the data as ordered sequences of observations in an online or a streaming manner, or by using time series embeddings which map a set of target observations to a k-dimensional feature space corresponding to the k past lagged values of the observation [8, 26]. In particular, Artificial Neural Networks (ANNs) have been widely applied to solve the forecasting task [24, 28]. Nowadays, deep ANNs (DNNs) have shown some improvements over previous shallow ANN architectures [24]. In fact, DNNs have shown the ability to automatically learn new, complex and enriched feature representation from input data [29], thus achieving good performance in solving a wide variety of task. Recurrent-based NNs such as Long Short-Term Memory Networks (LSTMs), as well as Convolutional Neural Networks (CNNs), have been widely used as state-of-the-art NN methods in the context of forecasting [13, 24]. Many improvements over these network architectures have been proposed in literature, ranging from optimizing the architecture structure to combining these networks together in one single forecasting task [13, 19, 20]. However, it is generally accepted that none of the proposed machine learning forecasting methods is universally valid for every application, and even within the same application, models have varying relative performance over time [9, 22, 25, 26]. Hence, different forecasting models have different areas of expertise and a varying relative performance [8, 9, 22]. Therefore, adequate and adaptive model selection in real-time is required to cope with the time evolving nature of time series and the fact that models have certain expected level of expertise in predicting a given sequence in the time series. While some works focused on online single model selection, others have been based on the assumption that no single model is expert the whole time and suggested to combine several single models in an ensemble framework by adaptively combining single models into one [8, 9, 25, 26]. Given a set of candidate models for performing a well-defined forecasting task, different tactics ranging from statistical estimations to applying meta-learning to learning the adequate selection strategy have been suggested. The approaches for single model selection can be divided into three main families. The first family of methods is based on approximating a posterior over the expected error of the different candidates using parametric [6] or non-parametric estimation methods [2]. These methods are not practical in the context of forecasting since continuous composite densities for the error function of the target and estimated time series values have to be approximated. The results depends largely on the quality of approximation. The second family consists of using empirical estimation of the unseen error of a given model using a independent validation/calibration dataset. Models with lowest estimated error are selected subsequently [23]. These methods are quite ineffective in practice since the estimated empirical error is usually lower than the true error. The third family is based on the meta-learning paradigm, where the selection of the adequate method is decided by another machine learning model which learns from previous selection realizations characterized by a set of devised meta features [8, 26].

Meta-learning can also be used for model selection by specializing the set of candidate models over different parts of the input so that each part gets assigned to one expert model based on comparison of predicted candidate model performances [8,9]. These parts are called **Region of Competence** (RoC) of a model [22]. The RoCs of one model are either stored individually or clustered and cluster centres are stored. At test time, the distance of the current input (i.e. in our case time series input sequence) to the RoCs or the RoCs cluster centres are computed, selecting the model with the lowest distance to perform the prediction. Generally, the computation of the RoCs is done by considering a static division of the time series into equally sized intervals. One way is to consider each time series observation t and its corresponding k lagged values as one interval and sliding the time window by one time step, whereas another way is to split the training or validation set into equally sized intervals [22].

Generally, the selection is performed in a static manner [25], i.e. the decision is made once at a time in favour of one model and this model is used subsequently to forecast all the required values at test time. The selection can also be updated continuously (i.e. blindly at each time instant or periodically) [8,9]. However, this is usually expensive in terms of time and resources [26], especially when the candidate models include DNNs. Few works in the literature performed the selection of single or ensemble models in an informed manner following drift detection of the relative performance of candidate models [25,26].

Candidate models in model selection can be the result of considering different parameter settings of the same model or by training different models belonging to different families of models. In the former case, the first family of model selection methods is widely used [2,6], while in the latter, a wide variety of selection approaches have been proposed [8,23,26]. However, the search for optimal network architectures for a given application is still an open research question [15]. This is even more challenging in the case of forecasting, where the decision for the adequate architecture have to be made in real-time. Due to their high training run-time and general resource consumption it is usually impractical to search for the adequate architecture at test time at each time instant or even in a periodic manner. Therefore, we focus in this work on approaching this problem by considering different candidate models of DNNs from different architectures (i.e. based on CNNs combined with other NNs models) and we perform the selection of the adequate architecture in real-time in an adaptive informed manner using concept drift detection in the time series.

We start by computing the RoCs of candidate CNNs using saliency maps. Saliency maps are usually used to establish a relationship between the output and the input of a neural net given fixed weights. They are widely used in the context of computer vision with CNNs to create a class-specific heatmap based on a particular input image and a chosen class of interest [32]. These maps are used for visualizing the regions of input that are "important" for prediction by the model and for understanding a model's prediction [27]. We suggest not only to transfer the class-activation maps from the context of classification to forecasting, but also to establish a mapping between the input time series and the performance so that dynamic RoCs are computed for each single CNN. Opposingly to the aforementioned approaches, the RoCs are considered as dynamic since their size is automatically decided and changed over time by the saliency map depending on the input time series sequence and the CNN performance. The RoCs are computed using a time-sliding window over a validation set. At test time, we produce forecasts step by step. At each time step, the distance between the recent observed window of time series observations (i.e. lagged values used to compute the forecast) and the pre-computed RoCs is determined. The model corresponding to the RoC with the lowest distance is selected to perform the forecasting. Additionally, the pre-computed RoCs are adaptively updated in case a concept drift is detected in the time series by sliding the validation set to take into account the probable presence of new concepts in the data when computing RoCs. The saliency maps can also be exploited to provide explanations for the reason behind selection one particular model for a given sequence of input data.

We further conduct comprehensive empirical analysis to validate our framework using 102 real-world time series datasets from various domains. The obtained results demonstrate that our method achieves excellent results in comparison to the SoA approaches for DNN selection as well as several baselines for time series forecasting. We note that all the experiments are fully reproducible, and both code and datasets are publicly available[1].

The main contributions of this paper are thus summarized as follows.

- We present a novel method for online CNNs selection for time series forecasting by computing RoCs for a set of candidate CNN-based models using an adaption of saliency maps.
- We update the RoCs in an informed manner following concept drift detection in the time series data.
- We exploit the saliency maps to provide suitable explanations for the reason behind selecting a specific model at a certain time instant or interval.
- We provide a comparative empirical study with state-of-the-art methods, and discuss their implications in terms of predictive performance and scalability.

2   Literature Review

Over the recent years, deep learning methods have been successfully applied in a wide variety of real-world learning tasks, including time series forecasting [10,19,20]. Currently, Recurrent Neural Networks (RNNs), and particularly Long-Short Term Memory (LSTM) nets, are considered to be the state-of-the-art in time series forecasting [7,19,20]. Thanks to their design based on recurrent connections, these networks have the ability to learn from the entire history of previous time series values. Another alternative for the use of DNNs in the forecasting task is to employ a Convolutional Neural Network (CNN) with multiple layers of dilated convolutions [31]. The layered structure of CNNs enables them to work well on noisy series, by removing the noise within each subsequent layer

---

[1] https://github.com/MatthiasJakobs/os-pgsm/tree/ecml2021.

and extracting only the meaningful patterns, performing thus similarly to neural networks which use wavelet transform on input time series [7]. This also allows for the receptive field of the network to expand exponentially, hereby making the network, similarly to RNNs, access a wide range of historical data. Some works have focused on improving CNN based architectures by combining CNN and LSTM in one single model to take advantage of the ability of LSTMs to cope with long temporal correlations [19,20]. In [21], the authors propose an undecimated convolutional network for time series forecasting using the undecimated wavelet transform. An autoregressive weighting schema for forecasting financial time series is presented in [5] where the weights are learnt through a CNN. However, convolutional architectures in literature are much more commonly applied to time series classification problems compared to forecasting [7,11].

The aforementioned works have focused on searching the most suitable network architecture for a well-defined application. At test time, the architecture and the learned weights are kept fixed and used to produce forecasts. However, to cope with the time-evolving nature of time series data, the forecasting schema has to be designed in a dynamic adaptive manner [25,26]. Since the same model can't be guaranteed to hold the same performance over time [8,9,22,25,26], online adequate model selection is required. This is usually hard to achieve with DNNs in general since the architecture tuning and the re-training of such models are intensively time consuming operations [7,20]. We suggest to mitigate this problem by training different CNN-based models with various architectures offline and decide for the online selection of the adequate network at each time instant at test time. The selection is achieved using a computation of the so-called RoCs using saliency maps (i.e. known also as attribution heat-maps) [32].

Saliency maps (in the form of *class activation maps (CAMs)*) were originally designed for computer vision classification tasks to visualize which part of an image is of high relevance to the network to make its decision [27,32]. They are considered tools for better understanding a models behaviour, e.g. by providing insight into model failure modes [32]. Many saliency map generation methods are post-hoc methods, in the sense that they are applied to an already-trained model. CAM uses the feature maps produced by the last convolutional layer of a CNN. This is motivated by the fact that the last convolution layer is expected to contain both high-level semantic and detailed spatial information [27]. More recently, CAMs have been applied in the context of time series classification to explain which features and which joint contribution of all the features during which time interval are responsible for a given time series class [3]. However, to the best of our knowledge, ours is the first work to apply saliency maps for online CNN-based model selection for time series forecasting. The generation of these maps is performed in an informed adaptive fashion. In addition, they are exploited to provide explanation for particular timely model selection.

3   Methodology

This section introduces our method and its main stages. In a first stage, we train different candidate CNN-based models with various architectures offline. The second stage consists of determining the RoCs for these models using sliding windows over a validation set. The RoCs are computed using a modified version of saliency maps, in the sense that instead of using these maps as class activation maps (CAM), we employ them to establish a relation between a relative "good" performance of a given candidate CNN and a particular pattern within the input time-sliding window sequences. We base our method on a gradient-based technique for generating saliency maps called Grad-CAM [27]. We call our modified version in the following "Performance Gradient-based Saliency Map (PGSM)". In the third stage, in order to produce a forecast at a given time instant t_f, the distance of the current input time sequence (i.e. time series observations from t_{f-k} to t_{f-1}, $f > k$) to the computed RoCs of each model is measured. The model corresponding to RoC with the lowest distance is selected to forecast the time series value at t_f. A concept drift detection mechanism in the time series is employed at test time (i.e. during forecasting). Once a drift is detected, an alarm is triggered to update the validation set by taking into account the new observed time series values and to subsequently update the RoCs. The PGSMs can be used to provide suitable explanations for the reason behind selecting a specific model at a certain time interval or instant. Practical examples of explanations are shown with details in Sect. 4. Our framework is denoted in the rest of the paper, OS-PGSM: Online CNN-based models Selection using Performance Gradient-based Saliency Maps.

3.1 Preliminaries

A time series X is a temporal sequence of values, where $X_t = \{x_1, x_2, \cdots, x_t\}$ is a sequence of X until time t and x_i is the value of X at time i. Denote with $P_{CNN} = \{C_0, C_1, \cdots, C_{N-1}\}$ the pool of trained CNN-based models. Let $\hat{x} = (\hat{x}^{C_0}, \hat{x}^{C_2}, \cdots, \hat{x}^{C_{N-1}})$ be the vector of forecast values of X at time instant $t + f, f \geq 1$ (i.e. x_{t+f}) by each of the models in P_{CNN}. The goal of the dynamic online selection is to identify which $\hat{x}^{C_j}$ should be used to produce this forecast.

We divide the time series X_t into $X_\omega^{train} = \{x_1, x_2, \cdots, x_{t-\omega}\}$ and $X_\omega^{val} = \{x_{t-\omega+1}, x_{t-\omega+2}, \cdots, x_t\}$, with ω a provided window size. X_ω^{train} is used for training the models in P_{CNN} and X_ω^{val} is used to compute the RoCs using the PGSMs, since to measure models performance both true and predicted values of the time series are required. The RoCs for each model $C_j, j \in \{0, \cdots, N-1\}$ are obtained by performing time-sliding window operations of size $n_\omega, n_\omega < \omega$ over X_ω^{val} either by one step or by z steps.

3.2 Candidate CNN Architectures

The candidate models are CNN-based models that share more less the same basic types of layers. The common basic structure consists of sequence of 1D-convolutional layers with different filter and kernel sizes, followed by a batch normalization layer, in some cases a LSTM layer and an output layer of one neuron. The different architectures are obtained by varying the number of the convolutional layers and their corresponding parameters (i.e. the size of filters

and kernels) and in some cases adding or removing another neural network type
to the last convolutional layer, like a LSTM layer. To obtain further architectures
variations, the number of units in the LSTM are also varied.

3.3   Online Model Selection

Performance Gradient-Based Saliency Maps. The PGSMs are inspired
from the class activation saliency maps, more specifically, Grad-CAM [27]. This
method has been proven to successfully pass commonly used sanity checks, which
are devised to check whether the saliency map is truly providing insights into
what the model is doing or not [1]. However, instead of using these maps to
derive the importance of certain features for a given class, we use them to map
the performance of a given forecasting model to a specific time interval. The performance of each model $C_j, j \in \{0, \cdots, N-1\}$ is evaluated using an error-related
measure, namely the Mean Squared Error, ϵ_j^i on $X_{n_\omega}^{val,i}$: the ith time interval window of X_ω^{val} of size n_ω. Our goal is to estimate the importance of each value in
$X_{n_\omega}^{val,i}$ to the measured error ϵ_j^i of C_j. This can be interpreted similarly to Grad-
CAM exploiting the spatial information that is preserved through convolutional
layers, in order to understand which parts of an input image are important for a
classification decision. However, we are focused here on the temporal information
explaining certain behaviour/performance of C_j. To do so, the last layer which
has produced the last feature maps f_{maps} is considered. For each activation unit
u at each generic feature map A, an importance weight w^ϵ associated with ϵ_j^i,
is obtained. This is done by computing the gradient of the ϵ_j^i with respect to A.
Subsequently, a global average over all the units in A is computed:

$$w^\epsilon = \frac{1}{U} \sum_u \frac{\partial \epsilon_j^i}{\partial A_u} \tag{1}$$

where U is the total number of units in A. We use w^ϵ to compute a weighted
combination between all the feature maps for a given measured value of the error
ϵ_j^i. Since we are mainly interested in highlighting temporal features contributing
most to ϵ_j^i a ReLU is used to remove all the negative contributions by:

$$L_j^i = ReLU\left(\sum_{f_{maps}} w^\epsilon A \right) \tag{2}$$

$L_j^i \in \mathbb{R}^U$ is used to find the regions in $X_{n_\omega}^{val,i}$ that have mainly contributed to ϵ_j^i
of the network C_j. Note that the candidates are designed such that $U < n_\omega$.

RoCs Computation. Our goal is to determine the region of competences of
each model on X_ω^{val}. However, one single evaluation of the models on X_ω^{val} obviously lead to just one best model. Therefore, we need to split X_ω^{val} into equally
sized time intervals of size n_ω, so that different evaluations of the candidate
models are performed and different rankings are derived. To increase this number of evaluations, the intervals of size n_ω can be obtained using a time-sliding

window approach over X_ω^{val} where the sliding operations are performed each z-steps. The lower z, the higher the number of evaluations is. If z is set to 1, the sliding window approach is performed in a step-wise manner. After evaluating the models on each of the $X_{n_\omega}^{val,i}$, the RoC of the model with the lowest error is computed using L_j^i of the PGSMs, where j the index of the candidate model C_j satisfying: $C_j = \mathrm{argmin}_{z \in \{1, \cdots, N\}} \epsilon_z^i$.

To obtain one continuous region RoC R^j within the time series sequence $X_{n_\omega}^{val,i}$, a smoothing operation is applied to L_j^i. This is achieved by normalizing L_j^i values between 0 and 1 and applying a threshold $\tau = 0.5$ to filter out smaller values (i.e. these values are set to 0). Further smoothing using a moving-average of size 3, is applied where each point is compared to the previous and the subsequent value. Whenever R^j of model C_j is computed, it is added to a corresponding RoC^j buffer which includes all collected RoCs for the model C_j (i.e. since C_j can be the best performing model on different $X_{n_\omega}^{val,i}$).

Online Forecasting. For forecasting the value of X at $t + f$ (assume $f = 1$ for simplicity), the candidate CNNs are devised such that they use the same k-lagged values of the time series as input, $p_t^k = \{x_{t-k+1}, \cdots, x_t\}$, $(t \geq k)$. To perform the selection, the distance of the input pattern p_t^k to the RoCs for each model in P_{CNN} is measured. The RoCs of a given model $C_j, j \in \{0, \cdots, N-1\}$ are already collected in $RoC^j = \{R_1^j, R_2^j, \cdots, R_{M^j}^j\}$, where M^j is the total number of regions of competence that have been determined by the PGSMs. Since the length of each RoC can be different from k (i.e. length of p_t^k), Dynamic Time Wrapping (DTW) [4] is used to measure the similarity between p_t^k and each $R_z^j, z \in \{1, \cdots, M^j\}$ within each $RoC^j, j \in \{0, \cdots, N-1\}$. The model C_b satisfying:

$$C_b = \underset{\substack{j \in \{0, \cdots, N-1\}; \\ z \in \{1, \cdots, M^j\}}}{\mathrm{argmin}} DTW(R_z^j, p_t^k) \qquad (3)$$

is selected to forecast $t + 1$.

RoCs Update. As explained above, the ROCs are computed offline using the validation set X_ω^{val}. However, due to the dynamic behaviour of time series, streaming upcoming values can be subject to significant changes, more specifically to concept drifts [12]. As a result, the ROCs have to be updated to take into account the possible presence of new patterns after the occurrence of such drifts and also to gain knowledge of which models are more adequate to handle these patterns if they ever reoccur again (i.e. note that the already computed RoCs are preserved and enriched with the new ones). Once a drift is detected, an alarm is triggered to update of the ROCs by sliding X_ω^{val} to include the new recent observations. The detection of concept drifts is performed by monitoring the deviation Δm_{t_f} in the mean of the time series [26]: $\Delta m_{t_f} = \mathbb{E}(X_{t_f}) - \mu$, with $\mu = \mathbb{E}(X_t), t \leq t_f$, the initial computed mean of X up to time t, a drift is assumed to take place at t_f if the true mean of Δm_{t_f} diverges in a significant way from 0. We propose to detect the validity of this using the well-known

Parameters: size of the validation set: ω; size of time windows within the validation set: n_ω; CNNs Pool: P_{CNN}.

1: **Models Training and RoCs Computation**:
2: Train each $C_j \in P_{CNN}, j \in \{0, \cdots, N-1\}$ on X_ω^{train}.
3: Initialize RoC buffers RoC^j for each $C_j, j \in \{0, \cdots, N-1\}$
4: **for** each $X_{n_\omega}^{val,i} \in X_\omega^{val}$ **do**
5:      Determine the best performing C_j.
6:      Compute the corresponding R_i^j using PGSMs (i.e. L_j^i Eq. 2)
7:      Add R_i^j to the corresponding buffer RoC^j
8: **end for**
9: **Online Forecasting**: Forecasting next N_f values :
10: predict x_{t+1} by the model C_b selected using Eq. 3
11: **for** $j \in \{2, \cdots, N_f\}$ **do**
12:     **if** an alarm is triggered (concept drift detected) **then**
13:         Update $X_\omega^{val} = \{x_{t-\omega+j}, x_{t-\omega+2}, \cdots, x_{t+j-1}\}$
14:         Recompute and add new RoCs (steps: 4–7)
15:     **end if**
16:     predict x_{t+j} by the model C_b selected using Eq. 3
17: **end for**

Algorithm 1: OS-PGSM

Hoeffding-Bound [16], which states that after W independent observations of a real-value random variable with range r, its true mean has not diverged if the sample mean is contained within $\pm\xi_m$:

$$\xi_m = \sqrt{\frac{r^2 \ln(1/\delta)}{2W}} \tag{4}$$

with a probability of $1-\delta$ (a user-defined hyperparameter). Once $|\Delta m_{t_f}|$ exceeds ξ_m, an alarm is triggered and the reference mean μ is reset by setting $t = t_f$. This checking procedure is continuously applied online at forecasting time. All the steps of OS-PGSM are summarized in Algorithm 1.

4   Experiments

We present the experiments carried out to validate OS-PGSM and to answer these research questions: **Q1**: How does OS-PGSM perform compared to the state-of-the-art (SoA) and existing online model selection methods for time series forecasting?; **Q2**: What is the advantage of reducing the step size z for sliding the window of size n_ω over X_ω^{val} on the performance of OS-PGSM? **Q3**: What is the advantage of updating the RoCs in an informed fashion (i.e. following drift detection)?; **Q4**: How scalable is OS-PGSM in terms of computational resources compared to the most competitive online model selection methods? and what is the computational advantage of the **drift-aware** adaption of the models' RoCs?; **Q5**: How can OS-PGSM be exploited to provide suitable explanations for the reason behind selecting a specific model at a certain time interval or instant?

4.1 Experimental Setup

The methods used in the experiments were evaluated using the root mean squared error (RMSE). The used time series was split using 50% for training (X_ω^{train}), and 25% for validation (X_ω^{val}) and 25% for testing. We use 102 real-world time series. A full list of the used datasets, together with a description, is given in the code repository[2].

4.2 OS-PGSM Setup and Baselines

We construct a pool P_{CNN} of CNN-based candidate models using different parameter settings (e.g. number of filters varies in $\{32, 64, 128\}$, kernel size varies in $\{1, 3\}$), like explained in Sect. 3.2. By combining these different parameters and adding/removing LSTM layer, 12 different CNNs with various architectures are created. OS-PGSM has also a number hyper-parameters that are summarized in Table 1. In our experiments, k is set equal to n_ω and the RoCs (after computation and smoothing) result in even smaller size than k. However, this is not problematic since we are interested in extracting distinctive patterns that are responsible for good performance of a given candidate. The difference in lengths of the RoC and the input sequence (k) are handled by the DTW measure. We compare OS-PGSM against the following approaches which include SoA methods for forecasting and model selection methods devised in the context of forecasting. Some of them operate in an online fashion. First, we compare against **ARIMA** and Exponential Smoothing (**ETS**) [18]. Next, we add the two best performing candidate models, denoted as **CNN** and **CNN-LSTM** [24]. Additionally, a simple **LSTM** is added for comparison [14]. **KNN-RoC** [22] which computes static RoCs using complete intervals of the validation set as input and the rank of the individual candidates on each interval as labels for a *KNN* classifier, using DTW distance and $K = 3$, is also used for comparison. At test time, the *KNN* predicts which candidate should be selected. Next, we

Table 1. Hyperparameters of OS-PGSM and their values for the experiments.

Parameter	Description	Value
k	Number of lagged values (size of the input to CNNs p_t^k)	5
ω	Size of validation set	25% of the dataset length
n_ω	Size of time windows within the validation set	5
z	Number of time steps with which time windows Within the validation set are slided	1
δ	Hoeffding-Bound parameter	0.05

---

[2] https://github.com/MatthiasJakobs/os-pgsm/tree/ecml2021.

also compare ourselves against a simple validation procedure where the CNNs are evaluated offline and the best model is selected to forecast all the upcoming data points [23]. **ADE** [8,9] was recently developed for online dynamic ensemble of forecasters construction. However, instead of selecting many models, we select the best performing model using the same principle. A Random Forest is used for estimating each candidate error and select the best model based on the lowest predicted error at test time. As **Stacking**, we denote a method where a meta-learner (Random Forest) is trained to predict which model to select using a set of meta-features consisting of input time sequence statistical characteristics and performance-based features [30]. Finally, we compare ourselves against **Adaptive Mixture** [17], which consists of some experts (Shallow CNNs) and a gating network. The gating network acts as a selector by performing a single output stochastic switch to select a given expert with the estimated switch probability.

We also compare OS-PGSM with some variants of itself. Note that OS-PGSM uses the Hoeffding-based drift detection mechanism to update the RoCs.

OS-PGSM-Int: Same as our method, but the time windows of size n_ω are slided with step size $z = n_\omega$.

OS-PGSM-Euc: Instead of using DTW as similarity measure, we use Euclidean distance. However, values in the RoC corresponding to 0 are not taken into consideration in the k-lagged values sequence (p_t^k).

OS-PGSM-Int-Euc: It is a combination of **OS-PGSM-Int** and **OS-PGSM-Euc**.

OS-PGSM-St: Same as our method, but the RoCs are not updated using the drift detection mechanism. The RoCs are computed and stored offline, only the selection takes place online.

OS-PGSM-Per: Same as our method, but the RoCs are update periodically in a blind manner (i.e. without taking into account the occurrence of concept drifts) with periodicity each upcoming 10% data points.

4.3   Results

Table 2 presents the average ranks and their deviation for all methods. For the paired comparison, we compare our method OS-PGSM against each of the other methods. We counted wins and losses for each dataset using the RMSE scores. We use the non-parametric Wilcoxon Signed Rank test to compute significant wins and losses, which are presented in parenthesis (significance level 0.05).

In the results in Table 2, OS-PGSM outperforms the baseline methods in terms of wins/loses in pairwise comparison. The online model selection approaches, e.g., KNN-RoC, ADE-Single and Adaptive Mixture, show inferior performance compared to OS-PGSM. ARIMA, ETS, LSTM, and CNN, SoA methods for forecasting, are considerably worse in average rank compared to OS-PGSM. CNN-LSTM shows slightly better performance, but is still worse than OS-PGSM. The most competitive SoA approach to OS-PGSM is ADE-Single. Nevertheless, it has a higher average rank and a lower performance than all the variants of our method. These results address the research question **Q1**.

Table 2. Comparison of OS-PGSM to different SoA for 102 time series. The rank column presents the average rank and its standard deviation across different time series. A rank of 1 means the model was the best performing on all time series

	ARIMA	ETS	LSTM	CNN	CNN-LSTM	Valid	KNN-RoC	ADE-single
Losses	7(6)	5(4)	17(8)	18(6)	23(6)	20(12)	22(16)	30(19)
Wins	95(93)	97(96)	17(8)85(83	84(80)	79(71)	82(72)	80(73)	72(61)
Avg. Rank	12.90	13.04	9.97	9.00	7.30	7.84	7.41	4.28
±	4.35	4.26	3.60	3.60	3.59	4.18	3.95	2.90

	Stacking	Adapt. Mixture	OS-PGSMInt	Euc	Int-Euc	St	Per	OS-PGSM
Losses	11(10)	17(8)	40(7)	35(10)	33(9)	49(10)	45(7)	–
Wins	91(80)	85(84)	62(54)	66(56)	69(66)	53(44)	57(46)	
Avg. Rank	12.11	10.93	3.62	3.86	3.95	2.93	3.09	**2.78**
±	4.12	5.78	2.80	3.10	3.13	2.62	3.05	**2.63**

Comparing OS-PGSM to different variants of our method, we see a clear advantage in using all the choices in our method. First, the DTW distance is better in sketching the similarities between the input sequences and the RoCs, especially when both have different lengths as explained above. This explains why the variants using Euclidean distance have worse performance. In addition, by setting $z = 1$, higher number of windows of size n_ω are created and as a result, a higher number of RoCs are computed (See Sect. 3.3). This contributes to creating richer information about RoCs of different candidates, compared to setting $z = n_\omega$ in OS-PGSM-Int and OS-PGSM-Int-Euc. Finally, OS-PGSM-St is even better then OS-PGSM-Per, which shows that unnecessary updates are not always beneficial. Opposingly, OS-PGSM which relies on informed adaption of the RoCs using concept drift detection is better than OS-PGSM-Per and OS-PGSM-St. This can be explained by the fact that the update of the RoCs is only beneficial for datasets where concept drifts can be detected and more probably taking into account these new appearing concepts is helpful for the selection of models since a knowledge of which models are more adequate to handle these patterns if they ever reoccur again is gained and the old sets of RoCs are enriched. Figure 1 show an illustrative example of the RoCs of C_7 before and after drift detection. New patterns are added as new RoCs to the old RoCs of C_7. This answers research questions **Q2–Q3**.

In the next experiment, we compare the runtime of OS-PGSM and its variants against the most competitive SoA method, ADE-Single, in Table 3. The reported runtime for OS-PGSM and OS-PGSM-Per takes into account the computation of the new RoCs. ADE-Single [9] relies on periodic update of the meta-learning strategy behind the selection (same periodicity as OS-PGSM-Per). All the reported runtimes concern only the online predictions and any operation computed offline is not taken into account. The results demonstrate that OS-PGSM and its variants have lower average runtime than ADE-Single. OS-PGSM-Int is faster than OS-PGSM-St since fewer evaluation windows of size n_ω are created and as a result, a lower number of RoCs is generated for distance comparisons. OS-PGSM has lower runtime than OS-PGSM-Per. This is due to using drift

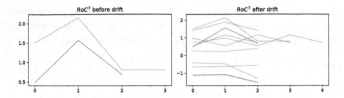

Fig. 1. RoCs for the candidate model C_7 before and after drift detection.

detection to update the RoCs only when necessary. This results in faster predictions and less computational requirements. The high deviation of the runtime of OS-PGSM is due to the different numbers of drifts per time series. This answers question **Q4**.

Table 3. Empirical runtime comparison between different methods in Seconds.

Method	ADE-single	OS-PGSM	OS-PGSMSt	OS-PGSMInt	OS-PGSMPer
Avg. runtime	167.87	8.42	2.33	0.90	154.11
±	56.40	18.30	4.80	1.65	204.22

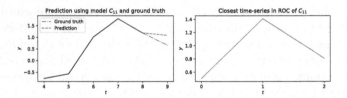

Fig. 2. Comparison of the current input pattern to the closest RoC (C_{11}). (Color figure online)

Last but not least, we provide some insights how OS-PGSM can be used to provide suitable explanations for the reason behind model selection. Figure 2 shows a comparison between the current input time series pattern p_t^k (left part in black) with the RoC of the selected model to perform the forecast. A clear similarity between both patterns can be observed which justifies the choice of this model since it has been proven to show some degree of competence in forecasting using similar patterns as input. This is further validated when also comparing between the true time series value (ground truth, green) and the predicted value (red). While these two values differ slightly, an evaluation of all the candidates in this point showed that our selected model C_{11} has the smallest error. A more general overview over the RoCs for AbnormalHeartbeat dataset is shown in Fig. 3.

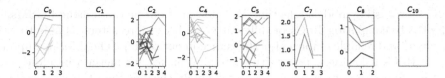

Fig. 3. Visualization of RoCs for AbnormalHeartbeat data using OS-PGSM.

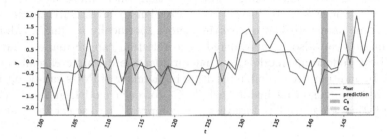

Fig. 4. A visualization of model selection one the AbnormalHeartbeat dataset.

Some models have quite similar RoCs (patterns). For example, C_0 appears to be expert in increasing linear trend patterns, or in peaks followed by a slight plateau, while C_7 is the best in dealing with sharp peaks. As it can be seen with the varied amount of transparency of lines of the RoCs, many identical RoCs are collected for each models, confirming thus the assumption that certain models are experts on specific input regions of the time series. Some models do not have any RoC. This can be explained by the fact that they never get selected in the validation process or their PGSMs are too small to form a pattern that it is why they get filtered out in the smoothing procedure. This also leads to better explainability in the sense of sparseness, since not every model is forced to contribute to the forecasting. Hence, models that are poorly designed or not well-trained, get ignored during the selection. Practitioners of our method can then use this insight to focus their attention on improving certain, poorly performing models, or remove the unused models entirely to save runtime. Another visualization aspect of the forecasting is shown in Fig. 4. We focus for visualization clarity on the two models (C_8 and C_0). We highlight regions where they are selected by OS-PGSM. Notice that preceding every decision where C_0 is chosen, the time series exhibits a peak, which corresponds to the model's RoCs in Fig. 3. The same conclusion can be drawn for C_8, which is picked after valley-shaped parts. While the two models are not picked for all peaks and valleys, our method clearly aligns certain time series regions with specific models. All the shown aspects address clearly research question **Q5**.

4.4 Discussion and Future Work

The empirical results indicate that OS-PGSM has performance advantages compared to popular forecasting methods and the most recent SoA approaches for online forecasting models selection. We show that our method, using PGSMs for adaptively computing RoCs of different CNN-based forecasters, is able to gain excellent and reliable empirical performance in our setting. The informed update of the RoCs following concept drift detection makes our method in addition to better predictive performance, computationally cheaper than the most competitive SoA, namely ADE-single. OS-PGSM can also successfully be used for providing useful explanations behind model selection which can be used to optimize further our framework. As future work, we plan to investigate the impact of varying some parameters in our setting, more specifically n_ω and k. More candidate models can also be considered. The selection can be made in favour of top-M best performing models, so that an adaptive dynamic ensemble can be created. In addition, the possible resulting big number of RoCs can be optimized further using a clustering inside each RoC^j for each model, and only clusters representatives are considered for distance computation to select the best model.

5   Concluding Remarks

This paper introduces OS-PGSM: a novel, practically useful online CNN-based models selection using saliency maps framework for time series forecasting. OS-PGSM uses gradient-based saliency maps to derive Region of Competences RoCs of a set of candidate models. These RoCs are updated in an informed-manner using concept drift detection in the time series. An exhaustive empirical evaluation, including 102 real-world datasets and multiple comparison algorithms showed the advantages of OS-PGSM in terms of performance and scalability.

References

1. Adebayo, J., Gilmer, J., Muelly, M., Goodfellow, I., Hardt, M., Kim, B.: Sanity checks for saliency maps. arXiv preprint arXiv:1810.03292 (2018)
2. Argiento, R., Guglielmi, A., Pievatolo, A.: Bayesian density estimation and model selection using nonparametric hierarchical mixtures. Comput. Stat. Data Anal. **54**(4), 816–832 (2010)
3. Assaf, R., Schumann, A.: Explainable deep neural networks for multivariate time series predictions. In: IJCAI, pp. 6488–6490 (2019)
4. Berndt, D.J., Clifford, J.: Using dynamic time warping to find patterns in time series. In: KDD Workshop, vol. 10, pp. 359–370 (1994)
5. Binkowski, M., Marti, G., Donnat, P.: Autoregressive convolutional neural networks for asynchronous time series. In: International Conference on Machine Learning, pp. 580–589. PMLR (2018)
6. Birgé, L., Massart, P.: Gaussian model selection. J. Eur. Math. Soc. **3**(3), 203–268 (2001)
7. Borovykh, A., Bohte, S., Oosterlee, C.W.: Conditional time series forecasting with convolutional neural networks. arXiv preprint arXiv:1703.04691 (2017)
8. Cerqueira, V., Torgo, L., Pinto, F., Soares, C.: Arbitrated ensemble for time series forecasting. In: Ceci, M., Hollmén, J., Todorovski, L., Vens, C., Džeroski, S. (eds.) ECML PKDD 2017. LNCS (LNAI), vol. 10535, pp. 478–494. Springer, Cham (2017). https://doi.org/10.1007/978-3-319-71246-8_29

9. Cerqueira, V., Torgo, L., Pinto, F., Soares, C.: Arbitrage of forecasting experts. Mach. Learn. **108**(6), 913–944 (2018). https://doi.org/10.1007/s10994-018-05774-y
10. Demertzis, K., Iliadis, L., Anezakis, V.D.: A deep spiking machine-hearing system for the case of invasive fish species. In: INISTA, pp. 23–28 (2017)
11. Ismail Fawaz, H., Forestier, G., Weber, J., Idoumghar, L., Muller, P.-A.: Deep learning for time series classification: a review. Data Min. Knowl. Disc. **33**(4), 917–963 (2019). https://doi.org/10.1007/s10618-019-00619-1
12. Gama, J., Žliobaitė, I., Bifet, A., Pechenizkiy, M., Bouchachia, A.: A survey on concept drift adaptation. ACM comput. Surv. (CSUR) **46**(4), 1–37 (2014)
13. Gamboa, J.C.B.: Deep learning for time-series analysis. arXiv preprint arXiv:1701.01887 (2017)
14. Gers, F.A., Eck, D., Schmidhuber, J.: Applying LSTM to time series predictable through time-window approaches. In: Tagliaferri, R., Marinaro, M. (eds.) Neural Nets. Perspectives in Neural Computing, pp. 193–200. Springer, London (2002). https://doi.org/10.1007/978-1-4471-0219-9_20
15. Goodfellow, I., Bengio, Y., Courville, A., Bengio, Y.: Deep Learning, vol. 1. MIT Press, Cambridge (2016)
16. Hoeffding, W.: Probability inequalities for sums of bounded random variables. In: Fisher, N.I., Sen, P.K. (eds.) The Collected Works of Wassily Hoeffding, pp. 409–426. Springer, New York (1994). https://doi.org/10.1007/978-1-4612-0865-5_26
17. Jacobs, R.A., Jordan, M.I., Nowlan, S.J., Hinton, G.E.: Adaptive mixtures of local experts. Neural Comput. **3**(1), 79–87 (1991)
18. Jain, G., Mallick, B.: A study of time series models ARIMA and ETS. Available at SSRN 2898968 (2017)
19. Kim, T.Y., Cho, S.B.: Predicting residential energy consumption using CNN-LSTM neural networks. Energy **182**, 72–81 (2019)
20. Livieris, I.E., Pintelas, E., Pintelas, P.: A CNN-LSTM model for gold price time-series forecasting. Neural Comput. Appl. **32**(23), 17351–17360 (2020)
21. Mittelman, R.: Time-series modeling with undecimated fully convolutional neural networks. arXiv preprint arXiv:1508.00317 (2015)
22. Priebe, F.: Dynamic model selection for automated machine learning in time series (2019)
23. Rivals, I., Personnaz, L.: On cross validation for model selection. Neural Comput. **11**(4), 863–870 (1999)
24. Romeu, P., Zamora-Martínez, F., Botella-Rocamora, P., Pardo, J.: Time-series forecasting of indoor temperature using pre-trained deep neural networks. In: Mladenov, V., Koprinkova-Hristova, P., Palm, G., Villa, A.E.P., Appollini, B., Kasabov, N. (eds.) ICANN 2013. LNCS, vol. 8131, pp. 451–458. Springer, Heidelberg (2013). https://doi.org/10.1007/978-3-642-40728-4_57
25. Saadallah, A., Moreira-Matias, L., Sousa, R., Khiari, J., Jenelius, E., Gama, J.: Bright-drift-aware demand predictions for taxi networks. IEEE Trans. Knowl. Data Eng. **32**(2), 234–245 (2020)
26. Saadallah, A., Priebe, F., Morik, K.: A drift-based dynamic ensemble members selection using clustering for time series forecasting (2019)
27. Selvaraju, R.R., Cogswell, M., Das, A., Vedantam, R., Parikh, D., Batra, D.: Grad-CAM: visual explanations from deep networks via gradient-based localization. In: ICCV, pp. 618–626 (2017)
28. Taieb, S.B., Bontempi, G., Atiya, A.F., Sorjamaa, A.: A review and comparison of strategies for multi-step ahead time series forecasting based on the NN5 forecasting competition. Expert Syst. Appl. **39**(8), 7067–7083 (2012)

29. Utgoff, P.E., Stracuzzi, D.J.: Many-layered learning. Neural Comput. **14**(10), 2497–2529 (2002)
30. Wolpert, D.H.: Stacked generalization. Neural Netw. **5**(2), 241–259 (1992)
31. Yu, F., Koltun, V.: Multi-scale context aggregation by dilated convolutions. arXiv preprint arXiv:1511.07122 (2015)
32. Zhou, B., Khosla, A., Lapedriza, A., Oliva, A., Torralba, A.: Learning deep features for discriminative localization. In: CVPR, pp. 2921–2929 (2016)

Change Detection in Multivariate Datastreams Controlling False Alarms

Luca Frittoli[1](✉), Diego Carrera[2], and Giacomo Boracchi[1]

[1] DEIB, Politecnico di Milano, via Ponzio 34/5, Milan, Italy
{luca.frittoli,giacomo.boracchi}@polimi.it
[2] STMicroelectronics, via Camillo Olivetti 2, Agrate Brianza, Italy
diego.carrera@st.com

Abstract. We introduce QuantTree Exponentially Weighted Moving Average (QT-EWMA), a novel change-detection algorithm for multivariate datastreams that can operate in a nonparametric and online manner. QT-EWMA can be configured to yield a target Average Run Length (ARL_0), thus controlling the expected time before a false alarm. Control over false alarms has many practical implications and is rarely guaranteed by online change-detection algorithms that can monitor multivariate datastreams whose distribution is unknown. Our experiments, performed on synthetic and real-world datasets, demonstrate that QT-EWMA controls the ARL_0 and the false alarm rate better than state-of-the-art methods operating in similar conditions, achieving comparable detection delays.

Keywords: Online change detection · Nonparametric monitoring · Multivariate datastreams · Histograms · False alarms

1 Introduction

Detecting changes in datastreams is a frequently encountered problem in industrial [13] and traffic monitoring [12], security [31], cryptographic attacks [9], and finance [29], to name a few examples. Change detection is also relevant in machine learning, where changes are also known as *concept drifts*, and classifiers have to be adapted as the data-generating process changes [10]. In many of these applications, datastreams have to be processed *online*, i.e., while acquiring observations, and this condition poses crucial, sometimes conflicting, challenges from both algorithmic and technological standpoints. On the one hand, online change-detection algorithms have to analyze virtually unlimited datastreams, while storing a limited amount of data and performing a fixed number of operations per incoming sample. On the other hand, to detect even subtle changes,

Electronic supplementary material The online version of this chapter (https://doi.org/10.1007/978-3-030-86486-6_26) contains supplementary material, which is available to authorized users.

© Springer Nature Switzerland AG 2021
N. Oliver et al. (Eds.): ECML PKDD 2021, LNAI 12975, pp. 421–436, 2021.
https://doi.org/10.1007/978-3-030-86486-6_26

online change-detection algorithms are requested to perform a sequential analysis, leveraging at time t all the data samples observed until t, in contrast with *one-shot* detectors, which perform independent tests over batches of data.

This paper addresses the crucial problem of controlling false alarms in online change detection, which is particularly important in industrial monitoring and unsupervised drift detection, where each detected change possibly triggers a costly intervention or supervision. Operating at controlled false alarm rates enables allocating the right amount of resources for the supervision, as well as identifying issues in the monitored process when there are more detections than expected. Unfortunately, most online change-detection algorithms that can monitor multivariate datastreams, especially nonparametric ones, fail to control false alarms effectively.

Most change-detection algorithms feature two main ingredients: i) a *statistic* with a known response to data following the initial distribution ϕ_0, and ii) a *decision rule* to analyze the values of the statistic and report changes. In online monitoring, the statistic T_t considers all the data acquired until time t, and the decision rule typically compares the statistic T_t against a threshold h, which is defined to control false alarms. In offline tests and one-shot detectors analyzing a fixed amount of data, the threshold h is set as a specific quantile of T and computed analytically [23], or by bootstrap [3]. In online monitoring, setting thresholds becomes more complicated, as one typically wants to control the *Average Run Length* (ARL$_0$), i.e., the average time before raising a false alarm [2]. Controlling the ARL$_0$ requires defining a sequence of thresholds $\{h_t\}_t$ where each threshold h_t is set as a specific quantile of the statistic T_t conditioned on the fact that the algorithm has not detected any change before t. In both offline and online monitoring, tests based on *nonparametric* statistics are preferable because their thresholds $\{h_t\}_t$ can be set without any information about ϕ_0.

In this paper, we propose *QT-EWMA*, a novel online change-detection algorithm for multivariate datastreams combining a QuantTree histogram (QT) [3] as a general and flexible model for ϕ_0, with an online statistic based on the Exponentially Weighted Moving Average (EWMA) [28]. In particular, we define a novel EWMA statistic to monitor the proportion of incoming samples falling in each bin of the histogram, and use this to define an efficient and practical online change-detection algorithm. The theoretical background of QuantTree guarantees that this test statistic does not depend on ϕ_0 nor the data dimension [3]. Thus we derive an efficient Monte Carlo scheme to compute thresholds controlling ARL$_0$. By design, these thresholds guarantee a constant false alarm probability over time and, consequently, a fixed false alarm rate at each time instant. Thus, QT-EWMA controls both the ARL$_0$ and the false alarm rate. Our main contributions are:

- We develop QT-EWMA, an online change-detection algorithm for multivariate datastreams based on a novel statistic that monitors the bin probabilities of a QuantTree histogram (Sect. 4.1).
- We design an efficient Monte Carlo scheme to compute thresholds controlling the ARL$_0$ in QT-EWMA, as in [29]. Thanks to the theoretical properties of

QuantTree, our thresholds control both the ARL_0 and the probability of false alarms at a given time, for any distribution ϕ_0 (Sect. 4.2).
- We propose two simple yet theoretically sound procedures to extend a generic one-shot detector to monitor datastreams while controlling the ARL_0 (Sect. 5), which we employ as baselines in our experiments.

Our experiments performed on both simulated and real-world datastreams show that QT-EWMA controls the ARL_0 better than the baselines, and Scan-B [22], a competing algorithm based on a *Maximum Mean Discrepancy* (MMD) statistic. Our results also show that QT-EWMA maintains the expected false alarm rate, which Scan-B does not guarantee. Moreover, in the considered real-world datasets, QT-EWMA achieves detection delays on par with online and nonparametric tests implementing powerful statistics based on kernels, which cannot control false alarms. QT-EWMA implementation and thresholds are available for download at https://github.com/diegocarrera89/quantTree.

The rest of the paper is organized as follows: in Sect. 2 we illustrate the literature of online change detection algorithms for multivariate datastreams, while in Sect. 3 we provide a formal definition of the online change-detection problem. In Sect. 4 we introduce our proposed solution, together with our procedure to compute the thresholds controlling ARL_0. In Sect. 5 we illustrate how to extend one-shot change detectors to monitor datastream controlling the ARL_0, and discuss the theoretical guarantees and limitations of these approaches. Section 6 analyzes the computational complexity of QT-EWMA compared to alternative methods and our experiments in Sect. 7 demonstrate the effectiveness of QT-EWMA on both simulated and real-world datastreams.

2 Related Work

Most change-detection algorithms in the literature are designed to monitor univariate datastreams [13,28,29]. The vast majority of these methods cannot be extended to monitor multivariate datastreams, especially those leveraging nonparametric statistics based on ranks [29]. Change detection in multivariate datastreams has often been addressed in multi-stream (multi-channel) settings, i.e., by separately analyzing each input component [8,25,32]. However, the hypotheses underpinning the multi-stream monitoring setting are fundamentally different from those of our multivariate datastream change-detection problem. In particular, [8,25,32] assume the components of the data points are generated by separate random variables rather than a single multivariate random vector. Moreover, in multi-stream settings, changes typically affect the distribution of a subset of these random variables [32], while, in multivariate settings, more general kinds of distribution changes are admissible [5]. In particular, changes affecting the correlation between components or subtle changes involving the whole vector can be hard to detect by separately analyzing the components.

The first change-detection algorithms for multivariate datastreams assume parametric hypotheses: a remarkable example [33] leverages the Hotelling test statistic [15] to perform online monitoring while controlling the ARL_0. A popular semi-parametric approach consists of reducing the data dimensionality by

monitoring the log-likelihood of the observations with respect to a density model
fitted on a training set. The most common models for the initial distribution ϕ_0
are Gaussian [12], and Gaussian mixture models [1,17]. The main limitation of
these solutions is the implicit assumption that ϕ_0 can be approximated well by
a probability distribution from a known family, which is not guaranteed in general. Other approaches reduce the dimensionality of the data by PCA [18,27],
or by a *strangeness measure* [14,26]. After reducing the dimensionality, online
change detection boils down to monitoring a univariate datastream, for instance
by Martingale-based permutation tests [14]. However, none of these methods can
be configured before deployment to operate at a target ARL_0.

Kernel methods based on Maximum Mean Discrepancy (MMD) have been
introduced for nonparametric one-shot change detection [11]. Recently, the MMD
statistic has been employed for online change detection [16,22], following a
sliding-window approach to compare the new observations to a training set.
Among these methods, Scan-B [22] is the only where ARL_0 can be set for an
unknown distribution ϕ_0. However, the thresholds for this method are defined
by analyzing the asymptotic behavior of the ARL_0 when the threshold tends to
infinity [22]. As we show in our experiments, this approach does not guarantee
an accurate control of the ARL_0 and the false alarm rate. NEWMA [16] also
employs a sliding-window monitoring scheme and analyzes the relation between
two EWMA statistics with different forgetting factors based on MMD to detect
changes online. Unfortunately, thresholds controlling the ARL_0 can only be set
when ϕ_0 is known [16], which limits the applicability of this solution.

A very general nonparametric approach consists of modeling ϕ_0 by a histogram [4], as in QuantTree [3], an algorithm to construct histograms over the
input domain that are adaptively defined over the distribution ϕ_0 of a training
set. A one-shot change-detection method is presented in [3], leveraging a nonparametric statistic to test whether a single batch of data follows ϕ_0. This test cannot
be directly used for online change detection. As we show in Sect. 5, QuantTree
and other one-shot detectors can be extended to online change detection controlling false alarms, but these solutions have substantial limitations in maintaining
the ARL_0 and in terms of detection power. Another change-detection algorithm
based on histograms is presented in [19], and can operate online controlling the
ARL_0, but has been tested only on univariate datastreams. An extension to multivariate data is infeasible since the number of bins scales exponentially with the
data dimension. Moreover, this algorithm requires to know the analytical expression of ϕ_0, or an accurate approximation, which would require a large training
set to be estimated when the data dimension is high [19].

The proposed QT-EWMA overcomes all these limitations, being an online
change-detection algorithm controlling the ARL_0 in any practical condition,
without requiring to know ϕ_0.

3   Problem Formulation

We address the online change-detection problem in a virtually unlimited multivariate datastream $x_1, x_2 \ldots \in \mathbb{R}^d$. We assume that, as long as there are no

changes, all the data samples are i.i.d. realizations of a random variable having unknown distribution ϕ_0. We define the *change point* τ as the unknown time instant when a change $\phi_0 \rightarrow \phi_1$ takes place:

$$x_t \sim \begin{cases} \phi_0 & \text{if } t < \tau \\ \phi_1 & \text{if } t \geq \tau \end{cases}. \tag{1}$$

We assume that both ϕ_0 and $\phi_1 \neq \phi_0$ are unknown, and that a training set TR containing N realizations of ϕ_0 is provided.

Online change-detection algorithms assess, for each new incoming sample x_t, whether the sequence $\{x_1, \ldots x_t\}$ contains a change point. Typically, a statistic T_t is computed at each incoming x_t, then a decision rule is applied. Usually, the rule consists in controlling whether $T_t > h_t$ for an appropriate threshold h_t, and the detection time t^* is defined as the first time index when this happens:

$$t^* = \min\{t : T_t > h_t\}. \tag{2}$$

The detection time t^* is the first time instant when there is enough statistical evidence to claim that the datastream $\{x_1, \ldots x_t\}$ contains a change point.

A fundamental issue in change detection is to define a sequence of thresholds $\{h_t\}_t$ to control false alarms. We measure false alarms by the Average Run Length [2], defined as $ARL_0 = \mathbb{E}[t^*]$, where the expectation is taken assuming that the whole datastream is drawn from ϕ_0. Thus, the ARL_0 is the average time before a false alarm. Ideally, the ARL_0 of an online change-detection method should be set *a priori*, similarly to Type I error probability in hypothesis testing.

4 Proposed Solution

Here we introduce our novel algorithm QT-EWMA (Sect. 4.1) and describe the procedure to define its thresholds controlling the ARL_0 (Sect. 4.2).

4.1 QuantTree Exponentially Weighted Moving Average

The QT-EWMA algorithm (Algorithm 1) leverages a novel online statistic T_t defined over a QuantTree histogram, which is constructed from the training set TR and K target probabilities $\{\pi_j\}_{j=1}^K$ (line 2). The QuantTree algorithm returns a histogram defined by K bins $\{S_j\}_{j=1}^K$, where each $S_j \subset \mathbb{R}^d$ is set to contain $\pi_j N$ training samples. Further details on QuantTree – including how to define $\{(S_j, \pi_j)\}_{j=1}^K$ when TR cannot be exactly split to match target probabilities – can be found in [3].

The QT-EWMA statistic T_t monitors the proportion of samples in the datastream that fall in each bin S_j. In particular, for each x_t we define K binary statistics $\{y_{j,t}\}_j$ as the indicator functions of each bin S_j, namely

$$y_{j,t} = \mathbb{1}(x_t \in S_j), \quad j \in \{1, \ldots, K\} \tag{3}$$

Algorithm 1: QT-EWMA

> **input** : datastream $x_1, x_2, \ldots$, target $\{\pi_j\}_{j=1}^K$, thresholds $\{h_t\}_t$, TR
> **output**: detection flag ChangeDetected, detection time t^*
>
> 1 ChangeDetected $\leftarrow$ False, $t^* \leftarrow \infty$;
> 2 estimate QT histogram $\{(S_j, \pi_j)\}_{j=1}^K$ from TR and define $\{\hat{\pi}_j\}_{j=1}^K$ as in (4);
> 3 $Z_{j,0} \leftarrow \hat{\pi}_j \; \forall j = 1, \ldots, K$;
> 4 **for** $t = 1, \ldots$ **do**
> 5 $\quad$ $y_{j,t} \leftarrow \mathbb{1}(x_t \in S_j)$;
> 6 $\quad$ $Z_{j,t} \leftarrow (1 - \lambda)Z_{j,t-1} + \lambda y_{j,t}, \quad j = 1 \ldots, K$;
> 7 $\quad$ $T_t \leftarrow \sum_{j=1}^K (Z_{j,t} - \hat{\pi}_j)^2 / \hat{\pi}_j$;
> 8 $\quad$ **if** $T_t > h_t$ **then**
> 9 $\quad\quad$ ChangeDetected $\leftarrow$ True, $t^* \leftarrow t$;
> 10 $\quad\quad$ break;
> 11 $\quad$ **end**
> 12 **end**
> 13 **return** ChangeDetected, t^*

---

to track in which bin the input sample x_t falls. It is possible to show that, when $x_t \sim \phi_0$ and $TR \sim \phi_0$ then:

$$\mathbb{E}[y_{j,t}] \approx \hat{\pi}_j := \frac{N\pi_j}{N+1}, \quad j < K \quad \text{and} \quad \mathbb{E}[y_{K,t}] \approx \hat{\pi}_K := \frac{N\pi_K + 1}{N+1}. \quad (4)$$

We evaluate these statistics $y_{j,t}$ for each incoming sample x_t (line 5), and then we compute the EWMA statistic [28] $Z_{j,t}, \; j \in \{1, \ldots, K\}$ (line 6), to monitor the proportion of data that falls in each bin S_j:

$$Z_{j,t} = (1 - \lambda)Z_{j,t-1} + \lambda y_{j,t} \quad \text{where} \quad Z_{j,0} = \hat{\pi}_j. \quad (5)$$

Since, under ϕ_0, the expected value $\mathbb{E}[Z_{j,t}] \approx \hat{\pi}_j$ for $j = 1, \ldots, K$, we define the QT-EWMA change-detection statistic as follows:

$$T_t = \sum_{j=1}^K \frac{(Z_{j,t} - \hat{\pi}_j)^2}{\hat{\pi}_j}. \quad (6)$$

Similarly to the Pearson statistic [20], T_t measures the overall difference between the proportion of points in each bin S_j, represented by $Z_{j,t}$, and their approximated expected values $\hat{\pi}_j$ under ϕ_0. This difference naturally increases as a consequence of a change $\phi_0 \to \phi_1$ that modifies the probability of some bin S_j. The QT-EWMA statistic is computed at each incoming sample (line 7) and then compared against the corresponding threshold h_t to detect changes (line 9).

The QT-EWMA algorithm inherits from QuantTree the fundamental property [3] that the distribution of the statistics (5) and (6) – like any other statistic entirely defined over QuantTree bins – does not depend on ϕ_0, so the thresholds $\{h_t\}_t$ can be defined *a priori* to guarantee the ARL$_0$ on any datastream.

4.2   Computing Thresholds Controlling the ARL$_0$

The sequence of thresholds $\{h_t\}_t$ has to be properly defined to guarantee a given ARL$_0$. According to the definition in Sect. 3, the ARL$_0 = \mathbb{E}[t^*]$, where the expected value is computed assuming that the whole datastream is drawn from ϕ_0. Following Margavio et al. [24], we adopt the relevant design choice of setting the thresholds $\{h_t\}_t$ to guarantee a constant false alarm probability α at each time instant t. When this property is satisfied, the detection time t^* is a Geometric random variable [24] with parameter α and expected value

$$\text{ARL}_0 = \mathbb{E}[t^*] = \frac{1}{\alpha}. \tag{7}$$

However, defining a set of thresholds $\{h_t\}_t$ that guarantee a constant false alarm probability α in online monitoring is not straightforward. As noted in [24], the thresholds must satisfy the following equation:

$$\mathbb{P}(T_t > h_t \mid T_k \le h_k \; \forall k < t) = \alpha \quad \forall t \ge 1. \tag{8}$$

Since it is infeasible to exactly compute the conditional probabilities in (8), we resort to Monte Carlo simulations as in [29]. An important consequence of our design choice is that QuantTree properties [3] ensure that the thresholds for T_t do not depend on ϕ_0 nor on the data dimension d. Thus, we can conveniently generate 1-dimensional Gaussian streams and perform Monte Carlo simulations to compute the thresholds $\{h_t\}_t$ very efficiently.

In particular, we generate 1,000,000 training sets of $N = 4096$ normal realizations $x_t \sim \mathcal{N}(0, 1)$. For each TR we construct a QuantTree histogram and then generate 5000 samples from $\mathcal{N}(0, 1)$ that we use to compute the statistics $\{T_t\}_{t=1}^{5000}$ as in (6). Then, we define the threshold h_1 yielding the target ARL$_0$ as the empirical $(1 - \alpha)$-quantile of T_1 values, bearing in mind that $\alpha = 1/\text{ARL}_0$. Similarly, all the thresholds h_t are computed as the $(1 - \alpha)$-quantiles of the values T_t, but when $t > 1$ we compute the empirical quantiles only considering those sequences whose statistic has never exceeded any of the previous thresholds $h_1, \ldots, h_{t-1}$, namely having $T_k \le h_k$, $\forall k < t$. Thresholds $\{h_t\}_t$ computed in this way guarantee (8) to hold, so the target ARL$_0$ is preserved.

We compute all the thresholds $\{h_t\}_{t=1}^{5000}$ and then fit a polynomial to these values, as suggested in [29]. This allows both to estimate h_t for $t > 5000$ and to improve the estimates $\{h_t\}_{t=1}^{5000}$ by leveraging correlation among thresholds. In particular, we estimate a polynomial in powers of $1/t$ that returns h_t for a given t. In our experiments we employ the following target ARL$_0$ values: 500, 1000, 2000, 5000, but we also compute and provide in the supplementary material the polynomial expressions of the thresholds for higher ARL$_0$ values (10000, 20000), which can be very useful to control false alarms in high-throughput applications.

An important consequence of setting a constant false alarm probability in (8) is that, being t^* a Geometric random variable with parameter α, the probability of having a false alarm before t by the geometric sum:

$$\mathbb{P}(t^* \le t) = \sum_{k=1}^{t} \alpha(1 - \alpha)^{k-1} = \alpha \cdot \frac{1 - (1 - \alpha)^t}{\alpha} = 1 - (1 - \alpha)^t, \tag{9}$$

where the probability $\mathbb{P}$ is computed under ϕ_0. This fact has relevant practical implications: since false alarms are controlled by (9), any substantial increase in their occurrence might indicate a drift in the data-generating process, which requires to update the change-detection algorithm. Other strategies based on asymptotic approximations cannot guarantee (9), as shown in our experiments.

5   Datastream Monitoring by One-Shot Detectors

In this section we present how to adapt one-shot change-detection algorithms to monitor datstreams by controlling ARL$_0$. We focus on both algorithms that operate batch-wise (Sect. 5.1), and element-wise (Sect. 5.2).

5.1   Monitoring Datastreams by Batch-Wise Detectors

Several change-detection algorithms operates dividing the datastream $x_1, x_2, \ldots$ in non-overlapping batches of ν samples:

$$W_t = [x_{(t-1)\nu+1}, \ldots, x_{t\nu}], \tag{10}$$

and computing a batch-wise test statistic $T^\nu(\cdot)$ over each W_t. This test statistic is typically defined upon a model fit over TR. For example, QuantTree [3] builds a histogram, while SPLL [17] fits a Gaussian Mixture $\widehat{\phi_0}$ to approximate ϕ_0. One-shot algorithms detect a change as soon as $T^\nu(W_t) > h^\nu$, and the threshold h^ν is set to control the probability of false alarm over each individual batch.

   The very same monitoring scheme can be adopted to monitor datastreams at a controlled ARL$_0$, as long as the threshold h^ν is accordingly modified. This result is based on the following proposition

Proposition 1. *Let the threshold h^ν be such that*

$$\mathbb{P}(T^\nu(W) > h^\nu) = \alpha, \tag{11}$$

where W is a batch of ν samples drawn from ϕ_0. Then, the monitoring scheme $T^\nu(W) > h^\nu$ yields ARL$_0 \geq \nu/\alpha$.

Proof. Reported in the supplementary material due to space limitations.

   Proposition 1 implies that, when we set $\alpha = \nu/\text{ARL}_0$, our online change-detection algorithm is conservative, since its ARL$_0$ can be greater than the target. In some cases, however, we can provide guarantees that $\alpha = \nu/\text{ARL}_0$, to exactly control the ARL$_0$, as shown in the following proposition.

Proposition 2. *Let the threshold h^ν be such that*

$$\mathbb{P}(T^\nu(W) > h^\nu \mid TR) = \alpha, \tag{12}$$

where W is a batch of ν samples drawn from ϕ_0. Then, the monitoring scheme $T^\nu(W) > h^\nu$ yields ARL$_0 = \nu/\alpha$.

Proof. Reported in the supplementary material due to space limitations.

We use the above propositions to adapt two relevant examples of batch-wise change-detection algorithm to monitor datastreams controlling ARL_0: Quant-Tree (QT) [3] and Semi-Parametric Log-Likelihood (SPLL) [17]. The theoretical property of QuantTree allows setting h^ν to guarantee (11) for any α regardless the data generating distribution ϕ_0. Thanks to Proposition 1, this can be extended to provide a lower bound on ARL_0 by instead setting the threshold h^ν as to guarantee that $\mathbb{P}(T^\nu(W) > h^\nu) = \nu/ARL_0$.

In the batch-wise SPLL detector, we compute the thresholds to guarantee the false positive rate using bootstrap over the training set TR since the distribution of the statistic depends on ϕ_0. In this case, we are under the hypothesis of Proposition 2, since the false positive probability is conditioned on the training set realization. Therefore, by adopting the same monitoring scheme and setting h^ν to guarantee $\mathbb{P}(T^\nu(W) > h^\nu|TR) = \nu/ARL_0$ we can obtain the target ARL_0. These two different guarantees will become very apparent in the experiments.

5.2   Monitoring Datastreams by Element-Wise Detectors

As we remarked in Sect. 2, a popular approach to perform online monitoring in multivariate datastreams consists in reducing the dimensionality of the data, so that it is possible to monitor a 1-dimensional datastream using online change-detection algorithms designed for univariate data. Here we consider a dimensionality reduction method based on SPLL. In particular, we fit a Gaussian Mixture Model (GMM) $\widehat{\phi}_0$ on TR. Then, we reduce the dimensionality of each incoming sample x_t by computing $-\log(\widehat{\phi}_0(x_t))$, building a new, univariate sequence. Finally, we monitor this sequence using a nonparametric online CPM [29] leveraging the Lepage test statistic [21]. Remarkably, the resulting algorithm, which we call SPLL-CPM, operates at the target ARL_0 thanks to the CPM, whose thresholds have been computed for this purpose.

6   Computational Complexity

Here we analyze the computational complexity and memory requirements of QT-EWMA, Scan-B [22] (presented in Sect. 2), and of the modified one-shot detectors QuantTree [3], SPLL [17] and SPLL-CPM discussed in Sect. 5. The results of this analysis are summarized in Table 1.

QT-EWMA and QuantTree: Both algorithms are extremely efficient from both computational and memory points of view. Both place the incoming sample x_t in the corresponding bin of the QuantTree histogram, resulting in $\mathcal{O}(K)$ operations [3], where K is the number of bins. Then, QT-EWMA computes the test statistics (3), (5), (6) and these operations have a constant cost that falls within $\mathcal{O}(K)$. The QuantTree algorithm computes the Pearson statistic at the end of each batch, and this does not increase the order of computational complexity either, resulting in $\mathcal{O}(K)$ operations as in QT-EWMA. In terms of memory requirements, QT-EWMA updates the statistics $Z_{j,t}$ (5) at each new sample

Table 1. Computational complexity for each update of the statistic and memory requirement of QT-EWMA compared to the other considered methods. In our experiments we set $K = 32$ for QT-EWMA and QT [3], $w = 1000$ for SPLL-CPM as in [29], and $B = 100, n = 5$ for Scan-B [22].

Algorithm	QT-EWMA	QT [3]	SPLL [17]	SPLL-CPM	Scan-B[22]
Complexity	$\mathcal{O}(K)$	$\mathcal{O}(K)$	$\mathcal{O}(md)$	$\mathcal{O}(md + w \log w)$	$\mathcal{O}(nBd)$
Memory	K	K	1	w	$(n+1)Bd$

x_t, which requires storing only the K values $Z_{j,t-1}, j = 1, \ldots, K$. Similarly, the Pearson statistic in QuantTree requires storing the proportions of points from the batch falling in each of the K bins, thus K values. Hence, both algorithms have the same, constant, memory requirement of K values.

SPLL and SPLL-CPM. Both these tests need to compute the log-likelihood of each sample with respect to each of the m components of a GMM $\widehat{\phi}_0$ fit on TR. This results in $\mathcal{O}(md)$ operations per sample [17]. In case of batch-wise monitoring (SPLL), averaging the log-likelihood over the batch, falls within the $\mathcal{O}(md)$ operations per sample, since it can be computed incrementally. Hence, only 1 value has to be stored. The SPLL-CPM algorithm instead leverages the CPM framework [29] to monitor the log-likelihood of each new observation x_t with respect to $\widehat{\phi}_0$. The Lepage statistic [21] used in the CPM requires sorting the entire sequence of log-likelihood values, which results in additional $\mathcal{O}(t \log t)$ operations. In this case, all the t values of the sequence have to be analyzed and stored, thus computational complexity and memory requirement steadily increase over time. Since this is not appropriate for online monitoring, [29] quantizes and stores old observations in a histogram, operating over a window of most recent w samples. Thus, both operations and memory requirements relate to w.

Scan-B. The Scan-B algorithm [22] operates in a sliding-window fashion, with a fixed window size B and a fixed number n of reference windows from the training set. At each time step t, this algorithm requires to update n Gram matrices by computing $\mathcal{O}(B)$ times the MMD statistic, resulting in $\mathcal{O}(nBd)$ operations [16]. The Scan-B algorithm need to store n reference windows of B samples each from the training set, on top of the current window, yielding $(n+1)Bd$ values in memory [16] for $d-$dimensional samples.

7   Experiments

Here we present the experimental evaluation of the proposed solution. Our aim is to show that QT-EWMA controls the false alarms better than competing methods while achieving lower or comparable detection delays. In QT-EWMA and QT we set $K = 32$ and uniform target probabilities $\pi_j = 1/K$, as in [3], since uniform histograms have been shown to be very effective for change detection purposes [4]. In both QT and SPLL we set $\nu = 32$, as in [3], and we employ the original configuration of the Scan-B algorithm [22].

7.1   Considered Datasets

We generate synthetic datastreams in different dimension $d \in \{2, 4, 8, 16, 32\}$ by choosing an initial Gaussian distribution ϕ_0 with random covariance matrix, and as alternative distribution a random roto-translation of ϕ_0 computed as $\phi_1 = \phi_0(Q \cdot + v)$. The roto-translation parameters Q and v are computed using the CCM framework [5] to guarantee a fixed symmetric Kullback-Leibler divergence $sKL(\phi_0, \phi_1) \in \{0.5, 1, 1.5, 2, 2.5, 3\}$, which is useful to compare the detection performance obtained in different dimensions [1].

We also test our change-detection method on seven traditional multivariate classification datasets: Credit Card Fraud Detection ("credit", $d = 28$) from [6], Sensorless Drive Diagnosis ("sensorless", $d = 48$), MiniBooNE particle identification ("particle", $d = 50$), Physicochemical Properties of Protein Ternary Structure ("protein", $d = 9$), El Niño Southern Oscillation ("niño", $d = 5$), and two of the Forest Covertype datasets ("spruce" and "lodgepole", $d = 10$) from the UCI Machine Learning Repository [7]. As in [3], we standardize the datasets and sum to each component of "sensorless", "particle", "spruce" and "lodgepole" imperceptible Gaussian noise to avoid repeated values, which harm the construction of QuantTree histograms. The distribution of these datasets is typically considered stationary [18], so we randomly sample the datastreams from the datasets and introduce changes by shifting the post-change samples by a random vector drawn from a d-dimensional Gaussian scaled by the total variance of the dataset, as in [3,18]. Here we report only the results on Gaussian datastreams with $d \in \{4, 32\}$ and the average results over these seven datasets (which we indicate by "UCI+credit"), leaving results on Gaussian data with $d \in \{2, 8, 16\}$ and over each individual dataset in the supplementary material.

We also test our method on the recently published INSECTS dataset [30] ($d = 33$), which describes the wing-beat frequency of different species of flying insects. This dataset is meant as a classification benchmark in datastreams affected by concept drift, i.e., the data is acquired under different environmental conditions affecting the insects' behavior. The dataset contains six concepts referring to different distributions. We assemble data from different concepts to form datastreams that include 30 realistic changes: we start sampling observations from one concept (ϕ_0) and switch to another (ϕ_1) introducing a change.

7.2   Figures of Merit

Empirical ARL_0. To assess whether QT-EWMA and the other considered methods maintain the target ARL_0, we compute the empirical ARL_0 as the average time before raising a false alarm. To this purpose, we run the considered methods on 5000 datastreams drawn from ϕ_0, setting the target $ARL_0 \in \{500, 1000, 2000, 5000\}$. In this experiment, we consider datastreams of length $L = 6 \cdot ARL_0$ to have a detection in each datastream. Since, by construction, the detection time t^* of our method under ϕ_0 is a Geometric random variable with parameter $\alpha = 1/ARL_0$, (9) indicates that the probability of having a false alarm before L is $\mathbb{P}(t^* \leq L) \approx 0.9975$.

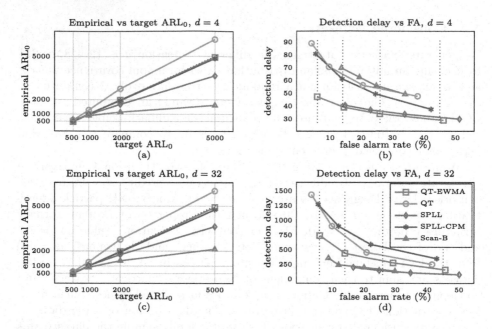

Fig. 1. Experimental results obtained on d-dimensional Gaussian datastreams ($d \in \{4, 32\}$). (a, c) show the empirical ARL_0 of the considered methods, and the target $ARL_0 \in \{500, 1000, 2000, 5000\}$ is maintained when the line is close to the dotted diagonal. (b, d) show the average detection delay against the percentage of false alarms, which should approach the dotted false alarm rates computed by (9).

Detection Delay. We evaluate the detection performance of QT-EWMA and the other considered methods by their detection delay, i.e. $ARL_1 = \mathbb{E}[t^* - \tau]$, where the expectation is taken assuming that a change point τ is present [2]. We run the methods configured with target $ARL_0 \in \{500, 1000, 2000, 5000\}$ on 1000 datastreams of length 10000, each containing a change point at $\tau = 300$. We estimate the ARL_1 as the average difference $t^* - \tau$, excluding false alarms.

False Alarm Rate. To assess whether the considered methods maintain the target false alarm probability, we compute the percentage of false alarms obtained on the datastreams used to evaluate the detection delay, i.e., those in which a detection occurs at some $t^* < \tau$. Also in this case, we set the the target $ARL_0 \in \{500, 1000, 2000, 5000\}$, which according to (9) yield a false alarm in 45%, 26%, 14% and 6% of the datastreams, respectively.

7.3   Results and Discussion

Empirical ARL_0. The comparison of empirical and target ARL_0 on simulated Gaussian datastreams is reported in Fig. 1(a, c), which show that QT-EWMA and SPLL-CPM control the ARL_0 very accurately, regardless of the data dimension, while the empirical ARL_0 of QT is higher than the target, as we expected from Proposition 1. Figures 2(a, c) show that we obtain the same result on

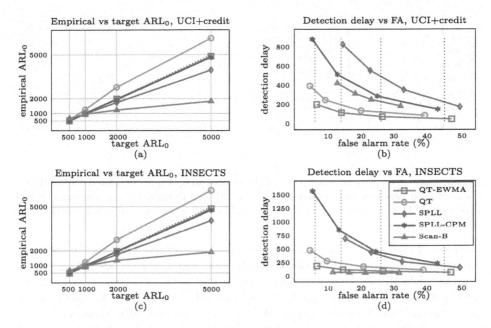

Fig. 2. Experimental results obtained on datastreams sampled from real-world datasets. (a, c) show the empirical ARL_0 of the considered methods respectively on the UCI+credit and INSECTS datasets. The target $ARL_0 \in \{500, 1000, 2000, 5000\}$ is maintained when the line is close to the dotted diagonal. (b, d) show the detection delay against the percentage of false alarms achieved on the same datasets, which should approach the dotted false alarm rates by (9).

datastreams sampled from the considered real-world dataset, confirming the non-parametric nature of QuantTree. In contrast, SPLL and Scan-B cannot control high target values of ARL_0 (on both simulated and real-world datastreams), due to detection thresholds inaccurately estimated. In particular, the SPLL statistic strongly depends on ϕ_0, so the thresholds have to be computed by bootstrap on a relatively small training set, which might yield inaccurate estimates. Thresholds for Scan-B are defined by an asymptotic approximation that strongly depends on the sliding window size B [22]. In particular, this approximation is more accurate for higher target ARL_0 when B is large, which increases the computational complexity and memory usage (Table 1). Thus, Scan-B with a fixed window size B can only maintain low ARL_0 values.

Detection Delay vs False Alarms. We plot the percentage of false alarms against the average detection delay, setting different ARL_0 values, to assess the trade-off between these two quantities. Figures 1(b, d) show the performance of the considered methods on simulated Gaussian datastreams ($d \in \{4, 32\}$, respectively) containing a change point at $\tau = 300$ such that $sKL(\phi_0, \phi_1) = 2$.

In terms of detection delay, QT-EWMA is the best method when $d = 4$, while SPLL outperforms all the others when $d = 32$, which is expected since its parametric assumptions are met (ϕ_0 is a Gaussian). All methods decrease their

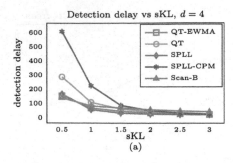 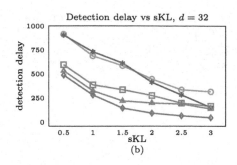

Fig. 3. Detection delay as a function of the change magnitude computed on simulated Gaussian datastreams with dimension $d \in \{4, 32\}$ containing a change point at $\tau = 300$ with target $ARL_0 = 1000$ (which is maintained by all methods, see Fig. 1(a, c)).

power as d increases, which is also expected due to *detectability loss* [1]. Scan-B seems to be more robust to detectability loss, yielding lower detection delays than QT-EWMA on Gaussian data when d increases, as shown in Figs. 1(b, d) and in supplementary material. Statistics defined on histograms are known to be less powerful than those based on MMD, as they perceive only changes affecting bin probabilities (and are for instance totally blind to distribution changes inside each bin). This effect can be mitigated by increasing the number of bins, thus the computational complexity, which is reasonable when d increases (e.g., when $K = d = 32$ there is on average a single split per dimension). However, in the considered real-world datasets, QT-EWMA turns out to be more effective than Scan-B in even larger dimensions, as shown in Figs. 2(b, d) and supplementary material. On average, QT-EWMA is clearly the best method in terms of detection delay on the UCI+credit datasets, and achieves delays similar to Scan-B on the INSECTS dataset. The fact that QT-EWMA consistently outperforms QT on simulated and real-world data indicates that our sequential statistic is more powerful than the original QuantTree statistic (designed for batch-wise monitoring) when detecting changes online.

In terms of false alarm rate, QT-EWMA and SPLL-CPM approach the target values computed by (9), while QT and SPLL have fewer and more false alarms, respectively, as a consequence of their empirical ARL_0, and this happens in all the considered monitoring scenarios. The false alarms of Scan-B, instead, exhibit a completely different behavior which also depends on the data distribution, due to the fact that its thresholds do not yield a constant false alarm probability.

Detection Delay. Figure 3 shows the detection delays of the considered methods as a function of the change magnitude on Gaussian datastreams. To enable a fair comparison, we configured all methods by setting the target $ARL_0 = 1000$, which all methods can maintain (see Figs. 1(a, c)). The best method on Gaussian datastreams is SPLL, as can be expected since its parametric assumptions are met. The best nonparametric method is QT-EWMA when $d = 4$, and Scan-B when $d = 32$. As observed also in Fig. 1(b, d), Scan-B seems to be more robust to detectability loss than QT-EWMA. As expected, all methods decrease their detection delays when the change magnitude increases.

8   Conclusions

We introduce QT-EWMA, a novel nonparametric online change-detection algorithm for multivariate datastreams. Our monitoring scheme is computationally light and can effectively maintain the target ARL_0 and the expected false alarm rates. Such accurate control over false alarms is very useful in practical applications. Our experiments on simulated and real-world data show that alternative solutions do not provide such guarantees in nonparametric settings, and that QT-EWMA turns out to be very effective in real-world datasets. Future work concerns incremental monitoring schemes based on QuantTree, to start monitoring with small training sets, and further investigation on how to set the optimal number of bins in QuantTree histograms for online change detection.

References

1. Alippi, C., Boracchi, G., Carrera, D., Roveri, M.: Change detection in multivariate datastreams: likelihood and detectability loss. In: Proceedings of the International Joint Conference on Artificial Intelligence (IJCAI), vol. 2, pp. 1368–1374 (2016)
2. Basseville, M., Nikiforov, I.V., et al.: Detection of Abrupt Changes: Theory and Application, vol. 104. Prentice Hall, Englewood Cliffs (1993)
3. Boracchi, G., Carrera, D., Cervellera, C., Macciò, D.: QuantTree: histograms for change detection in multivariate data streams. In: International Conference on Machine Learning, pp. 639–648 (2018)
4. Boracchi, G., Cervellera, C., Macciò, D.: Uniform histograms for change detection in multivariate data. In: 2017 IEEE International Joint Conference on Neural Networks (IJCNN), pp. 1732–1739. IEEE (2017)
5. Carrera, D., Boracchi, G.: Generating high-dimensional datastreams for change detection. Big Data Res. **11**, 11–21 (2018)
6. Dal Pozzolo, A., Boracchi, G., Caelen, O., Alippi, C., Bontempi, G.: Credit card fraud detection: a realistic modeling and a novel learning strategy. IEEE Trans. Neural Netw. Learn. Syst. **29**(8), 3784–3797 (2017)
7. Dua, D., Graff, C.: UCI machine learning repository (2017). http://archive.ics.uci.edu/ml
8. Fellouris, G., Tartakovsky, A.G.: Multichannel sequential detection-part I: non-I.I.D. data. IEEE Trans. Inform. Theory **63**(7), 4551–4571 (2017)
9. Frittoli, L., et al.: Strengthening sequential side-channel attacks through change detection. IACR Trans. Cryptogr. Hardw. Embed. Syst. **3**, 1–21 (2020)
10. Gama, J., Žliobaitė, I., Bifet, A., Pechenizkiy, M., Bouchachia, A.: A survey on concept drift adaptation. ACM Comput. Surv. (CSUR) **46**(4), 44 (2014)
11. Gretton, A., Borgwardt, K., Rasch, M., Schölkopf, B., Smola, A.J.: A kernel method for the two-sample-problem. In: Advances in Neural Information Processing Systems, pp. 513–520 (2007)
12. Guralnik, V., Srivastava, J.: Event detection from time series data. In: 5th ACM SIGKDD International Conference on Knowledge Discovery and Data Mining, pp. 33–42 (1999)
13. Hawkins, D.M., Qiu, P., Kang, C.W.: The changepoint model for statistical process control. J. Qual. Technol. **35**(4), 355–366 (2003)

14. Ho, S.S.: A martingale framework for concept change detection in time-varying data streams. In: Proceedings of the 22nd International Conference on Machine Learning, pp. 321–327 (2005)

15. Hotelling, H.: A generalized t test and measure of multivariate dispersion. In: Proceedings of the Second Berkeley Symposium on Mathematical Statistics and Probability. University of California (1951)

16. Keriven, N., Garreau, D., Poli, I.: NEWMA: a new method for scalable model-free online change-point detection. IEEE Trans. Signal Process. **68**, 3515–3528 (2020)

17. Kuncheva, L.I.: Change detection in streaming multivariate data using likelihood detectors. IEEE Trans. Knowl. Data Eng. **25**(5), 1175–1180 (2011)

18. Kuncheva, L.I., Faithfull, W.J.: PCA feature extraction for change detection in multidimensional unlabeled data. IEEE Trans. Neural Netw. Learn. Syst. **25**(1), 69–80 (2013)

19. Lau, T.S., Tay, W.P., Veeravalli, V.V.: A binning approach to quickest change detection with unknown post-change distribution. IEEE Trans. Signal Process. **67**(3), 609–621 (2018)

20. Lehmann, E.L., Romano, J.P.: Testing Statistical Hypotheses. Springer, Heidelberg (2006). https://doi.org/10.1007/0-387-27605-X

21. Lepage, Y.: A combination of Wilcoxon's and Ansari-Bradley's statistics. Biometrika **58**(1), 213–217 (1971)

22. Li, S., Xie, Y., Dai, H., Song, L.: M-statistic for kernel change-point detection. In: Advances in Neural Information Processing Systems, vol. 28, pp. 3366–3374 (2015)

23. Lung-Yut-Fong, A., Lévy-Leduc, C., Cappé, O.: Robust changepoint detection based on multivariate rank statistics. In: 2011 IEEE International Conference on Acoustics, Speech and Signal Processing (ICASSP), pp. 3608–3611. IEEE (2011)

24. Margavio, T.M., Conerly, M.D., Woodall, W.H., Drake, L.G.: Alarm rates for quality control charts. Stat. Probab. Lett. **24**(3), 219–224 (1995)

25. Mei, Y.: Efficient scalable schemes for monitoring a large number of data streams. Biometrika **97**(2), 419–433 (2010)

26. Mozafari, N., Hashemi, S., Hamzeh, A.: A precise statistical approach for concept change detection in unlabeled data streams. Comput. Math. Appl. **62**(4), 1655–1669 (2011)

27. Qahtan, A.A., Alharbi, B., Wang, S., Zhang, X.: A PCA-based change detection framework for multidimensional data streams. In: 21st ACM SIGKDD International Conference on Knowledge Discovery and Data Mining, pp. 935–944 (2015)

28. Roberts, S.: Control chart tests based on geometric moving averages. Technometrics **1**(3), 239–250 (1959)

29. Ross, G.J., Tasoulis, D.K., Adams, N.M.: Nonparametric monitoring of data streams for changes in location and scale. Technometrics **53**(4), 379–389 (2011)

30. Souza, V.M.A., dos Reis, D.M., Maletzke, A.G., Batista, G.E.A.P.A.: Challenges in benchmarking stream learning algorithms with real-world data. Data Min. Knowl. Disc. **34**(6), 1805–1858 (2020). https://doi.org/10.1007/s10618-020-00698-5

31. Tartakovsky, A.G., Rozovskii, B.L., Blazek, R.B., Kim, H.: A novel approach to detection of intrusions in computer networks via adaptive sequential and batch-sequential change-point detection methods. IEEE Trans. Signal Process. **54**(9), 3372–3382 (2006)

32. Xie, Y., Siegmund, D.: Sequential multi-sensor change-point detection. In: 2013 Information Theory and Applications Workshop, pp. 1–20. IEEE (2013)

33. Zamba, K., Hawkins, D.M.: A multivariate change-point model for statistical process control. Technometrics **48**(4), 539–549 (2006)

Approximation Algorithms for Confidence Bands for Time Series

Nikolaj Tatti[(✉)] [iD]

University of Helsinki, Helsinki, Finland
nikolaj.tatti@helsinki.fi

Abstract. Confidence intervals are a standard technique for analyzing data. When applied to time series, confidence intervals are computed for each time point separately. Alternatively, we can compute confidence bands, where we are required to find the smallest area enveloping k time series, where k is a user parameter. Confidence bands can be then used to detect abnormal time series, not just individual observations within the time series. We will show that despite being an **NP**-hard problem it is possible to find optimal confidence band for some k. We do this by considering a different problem: discovering regularized bands, where we minimize the envelope area minus the number of included time series weighted by a parameter α. Unlike normal confidence bands we can solve the problem exactly by using a minimum cut. By varying α we can obtain solutions for various k. If we have a constraint k for which we cannot find appropriate α, we demonstrate a simple algorithm that yields $\mathcal{O}(\sqrt{n})$ approximation guarantee by connecting the problem to a minimum k-union problem. This connection also implies that we cannot approximate the problem better than $\mathcal{O}\left(n^{1/4}\right)$ under some (mild) assumptions. Finally, we consider a variant where instead of minimizing the area we minimize the maximum width. Here, we demonstrate a simple 2-approximation algorithm and show that we cannot achieve better approximation guarantee.

1 Introduction

Confidence intervals are a common tool to summarize the underlying distribution, and to indicate outlier behaviour. In this paper we will study the problem of computing confidence intervals for time series.

Korpela et al. [11] proposed a notion for computing confidence intervals: instead of computing point-wise confidence intervals, the authors propose computing confidence bands. More formally, given n time series T, we are asked to find k time series $U \subseteq T$ that minimize the envelope area, that is, the sum $\sum_i \left(\max_{t \in U} t(i)\right) - \left(\min_{t \in U} t(i)\right)$. The benefit, as argued by Korpela et al. [11], of using confidence bands instead of point-wise confidence intervals is better family-wise error control: if we were to use point-wise intervals we can only say that a time series *at some fixed point* is an outlier and require a correction for

© Springer Nature Switzerland AG 2021
N. Oliver et al. (Eds.): ECML PKDD 2021, LNAI 12975, pp. 437–452, 2021.
https://doi.org/10.1007/978-3-030-86486-6_27

multiple testing (such as Bonferroni correction) if we want to state with a certain probability that the *whole* time series is normal.

In this paper we investigate the approximation algorithms for finding confidence bands. While Korpela et al. [11] proved that finding the optimal confidence band is an **NP**-hard problem, they did not provide any approximation algorithms nor any inapproximability results.

We will first show that despite being an **NP**-hard problem, we can solve the problem for *some k*. We do this by considering a different problem, where instead of having a hard constraint we have an objective function that prefers selecting time series as long as they do not increase the envelope area too much. The objective depends on the parameter α, larger values of α allow more increase in the envelope area. We will show that this problem can be solved exactly in polynomial time and that each α correspond to a certain value of k. We will show that there are at most $n + 1$ of such bands, and that we can discover all of them in polynomial time by varying α.

Next, we provide a simple algorithm for approximating confidence bands by connecting the problem to the weighted k-MinUnion problem. We will provide a variant of an algorithm by Chlamtáč et al. [2] that yields $\sqrt{n}+1$ guarantee. We also argue that—under certain conjecture—we cannot approximate the problem better than $\mathcal{O}(n^{1/4})$.

Finally, we consider a variant of the problem where instead of minimizing the envelope area, we minimize the width of the envelope, that is, we minimize the maximum difference between the envelope boundaries. We show that a simple algorithm can achieve 2-approximation. This approximation provides interesting contrast to the inapproximability results when minimizing the envelope area. Surprisingly this guarantee is tight: we will also show that the there is no polynomial-time algorithm with smaller guarantee unless $\mathbf{P} = \mathbf{NP}$.

The remainder of the paper is organized as follows. We define the optimization problems formally in Sect. 2. We solve the regularized band problem in Sect. 3, approximate minimization of envelope area in Sect. 4, and approximate minimization of envelope width in Sect. 5. Section 6 is devoted to the related work. We present our experiments in Sect. 7 and conclude with discussion in Sect. 8.

2   Preliminaries and Problem Definitions

Assume that we are given time series T with each time series $f : D \to \mathbb{R}$ mapping from domain D to a real number. We will often write $n = |T|$ to be the number of given time series, and $m = |D|$ to mean the size of the domain.

Given a set of time series T, we define the upper and lower *envelopes* as

$$ub(T, i) = \max_{t \in T} t(i) \quad \text{and} \quad \ell b(T, i) = \min_{t \in T} t(i).$$

Our main goal is to find k time series that minimize the envelope area.

*Problem 1 (*SUMBAND*).* Given a set of n time series $T = (t_1, \ldots, t_n)$, an integer $k \le n$, and a time series $x \in T$ find k time series $U \subseteq T$ containing x minimizing

$$s_1(U) = \sum_i ub(U, i) - \ell b(U, i).$$

We will refer to U as *confidence bands.*

Note that the we also require that we must specify at least one sequence $x \in T$ that must be included in the input whereas the original definition of the problem given by Korpela et al. [11] did not require specifying x. As we will see later, this requirement simplifies the computational problem. On the other hand, if we do not have x at hand, then we can either test every $t \in T$ as x, or we can use the mean or the median of T. We will use the latter option as it does not increase the computational complexity and at the same time is a reasonable assumption. Note that in this case most likely $x \notin T$, so we define $T' = T \cup \{x\}$, increase $k' = k + 1$, and solve SUMBAND for T' and k' instead.

We can easily show that the area function $s_1(\cdot)$ is a submodular function for all non-empty subsets, that is,

$$s_1(U \cup \{t\}) - s_1(U) \le s_1(W \cup \{t\}) - s_1(W),$$

where $U \supseteq W \ne \emptyset$. In other words, adding t to a larger set U increases the cost less than adding t to W.

We also consider a variant of SUMBAND where instead of minimizing the area of the envelope, we will minimize the maximum width.

*Problem 2 (*INFBAND*).* Given a set of n time series $T = (t_1, \ldots, t_n)$, an integer $k \le n$, and a time series $x \in T$, find k time series $U \subseteq T$ containing x minimizing

$$s_\infty(U) = \max_i ub(U, i) - \ell b(U, i).$$

We will show that we can 2-approximate INFBAND and that the ratio is tight.

Finally, we consider a regularized version of SUMBAND, where instead of requiring that the set has a minimum size k, we add a term $-\alpha|U|$ into the objective function. In other words, we will favor larger sets as long as the area $s_1(U)$ does not increase too much.

*Problem 3 (*REGBAND*).* Given a set of n time series $T = (t_1, \ldots, t_n)$, a number $\alpha > 0$, and a time series $x \in T$, find a subset $U \subseteq T$ containing x minimizing

$$s_{reg}(U; \alpha) = s_1(U) - \alpha|U|.$$

In case of ties, use $|U|$ as a tie-breaker, preferring larger values.

We refer to the solutions of REGBAND as *regularized bands.* It turns out that REGBAND can be solved in polynomial time. Moreover, the solutions we obtain from REGBAND will be useful for approximating SUMBAND.

3   Regularized Bands

In this section we will list useful properties of REGBAND, show how can we solve REGBAND in polynomial time for a single α, and finally demonstrate how we can discover *all* regularized bands by varying α.

3.1   Properties of Regularized Bands

Our first observation is that the output of REGBAND also solves SUMBAND for certain size constraints.

Proposition 1. *Assume time series T and $\alpha > 0$. Let U be a solution to* REGBAND(α). *Then U is also a solution for* SUMBAND *with $k = |U|$.*

The proof of this proposition is trivial and is omitted.

Our next observation is that the solutions to REGBAND form a chain.

Proposition 2. *Assume time series T and $0 < \alpha < \beta$. Let V be a solution to* REGBAND(T, α) *and let U be a solution to* REGBAND(T, β). *Then $V \subseteq U$.*

Proof. Assume otherwise. Let $W = V \setminus U$. Due to the optimality of V,

$$0 \geq s_{reg}(V; \alpha) - s_{reg}(V \cap U; \alpha) = s_1(V) - s_1(V \cap U) - \alpha|W|.$$

Since s_1 is a submodular function, we have

$$s_1(V) - s_1(V \cap U) = s_1(W \cup (V \cap U)) - s_1(V \cap U) \geq s_1(W \cup U) - s_1(U).$$

Combining these inequalities leads to

$$
\begin{aligned}
0 &\geq s_1(W \cup U) - s_1(U) - \alpha|W| \\
&\geq s_1(W \cup U) - s_1(U) - \beta|W| \\
&= s_{reg}(W \cup U; \beta) - s_{reg}(U; \beta),
\end{aligned}
$$

which contradicts the optimality of U.   □

This property is particularly useful as it allows clean visualization: the envelopes resulting from different values of α will not intersect. Moreover, it allows us to stored all regularized bands by simply storing, per each time series, the index of the largest confidence band containing the time series.

Interestingly, this result does not hold for SUMBAND.

Example 1. Consider 4 constant time series $t_1 = 0$, $t_2 = -1$ and $t_3 = t_4 = 2$. Set the seed time series $x = t_1$. Then the solution for SUMBAND with $k = 2$ is $\{t_1, t_2\}$ and the solution SUMBAND with $k = 3$ is $\{t_1, t_3, t_4\}$.

3.2   Computing Regularized Band for a Single α

Our next step is to solve REGBAND in polynomial time. Note that since $s_1(\cdot)$ is submodular, then so is $s_{reg}(\cdot)$. Minimizing submodular function is solvable in polynomial-time [15]. Solving REGBAND using a generic solver for minimizing submodular functions is slow, so instead we will solve the problem by reducing it to a minimum cut problem. In such a problem, we are given a weighted directed graph $G = (V, E, W)$, two nodes, say $\theta, \eta \in V$, and ask to partition V into $X \cup Y$ such that $\theta \in X$ and $\eta \in Y$ minimizing the total weight of edges from X to Y.

In order to define G we need several definitions. Assume we are given n time series T, a real number α and a seed time series $x \in T$. Let m be the size of the domain. For $i \in [m]$, we define $p_i = \{t_j(i) \mid j \in [n]\}$ to be the set (with no duplicates) sorted, smallest values first. In other words, p_{ij} is the jth smallest distinct observed value in T at i. Let P be the collection of all p_i.

We also define c_{ij} to be the number of time series at i smaller than or equal to p_{ij}, that is, $c_{ij} = |\{\ell \in [n] \mid t_\ell(i) \leq p_{ij}\}|$. We also write $c_{i0} = 0$.

We are now ready to define our graph. We define a weighted directed graph $G = (V, E, W)$ as follows. The nodes V have three sets A, B, and C. The set A has $|P|$ nodes, a node $a_{ij} \in A$ corresponding to each entry $p_{ij} \in P$. The set $B = \{b_j\}$ has n nodes, and the set C has two nodes, θ and η. Here, θ acts as a source node and η acts as a terminal node.

The edges and the weights are as follows: For each $a_{ij} \in A$ such that $p_{ij} > x(i)$, we add an edge $(a_{i(j-1)}, a_{ij})$ with the weight

$$w(a_{i(j-1)}, a_{ij}) = n - c_{i(j-1)} + \frac{m}{\alpha}(p_{i(j-1)} - x(i)).$$

For each $a_{ij} \in A$ such that $p_{ij} < x(i)$, we add an edge $(a_{i(j+1)}, a_{ij})$ with the weight

$$w(a_{i(j+1)}, a_{ij}) = c_{ij} + \frac{m}{\alpha}(x(i) - p_{i(j+1)}).$$

For each $a_{ij} \in A$ such that $p_{ij} = x(i)$, we add an edge (θ, a_{ij}) with the weight ∞. For each $i \in [m]$ and $\ell = |p_i|$, we add two edges $(a_{i\ell}, \eta)$ and (a_{i1}, η) with the weights

$$w(a_{i\ell}, \eta) = \frac{m}{\alpha}(p_{i\ell} - x(i)) \quad \text{and} \quad w(a_{i1}, \eta) = \frac{m}{\alpha}(x(i) - p_{i1}).$$

In addition, for each $i \in [m]$, $\ell \in [n]$, let j be such that $p_{ij} = t_\ell(i)$ and define two edges (a_{ij}, b_ℓ) and (b_ℓ, a_{ij}) with the weights,

$$w(a_{ij}, b_\ell) = 1 \qquad\qquad w(b_\ell, a_{ij}) = \infty.$$

Our next proposition states the minimum cut of G also minimizes REGBAND.

Proposition 3. *Let X, Y be a (θ, η)-cut of G with the optimal cost. Define $f(i) = \min_j \{p_{ij} \mid a_{ij} \in X\}$ and $g(i) = \max_j \{p_{ij} \mid a_{ij} \in X\}$.*
Then the cost of the cut is equal to

$$nm - m|\{j \mid b_j \in X\}| + \frac{m}{\alpha}\sum_i g(i) - f(i).$$

Moreover, if $b_\ell \in X$, then $g(i) \leq b_\ell(i) \leq f(i)$, for all i.

Proof. The last claim follows immediately as otherwise there is a cross-edge with infinite cost making the cut suboptimal.

Define $u(i) = \arg\min_j \{p_{ij} \mid a_{ij} \in X\}$ and $v(i) = \arg\max_j \{p_{ij} \mid a_{ij} \in X\}$ to be the indices yielding f and g. Define also

$$d_i = |\{j \mid u(i) \leq t_j(i) \leq v(i)\}| = c_{iv(i)} - c_{i(u(i)-1)}$$

to be the number of time series between $u(i)$ and $v(i)$ at i.

Note that $a_{ij} \in X$ whenever $u(i) \leq j \leq v(j)$ as otherwise we can move a_{ij} to X and decrease the cost.

The cut consists of the cross-edges originating from $a_{iv(i)}$ and $a_{iu(i)}$, and cross-edges between A and B. The cost of the former is equal to

$$\sum_i n - c_{iv(i)} + m\frac{p_{iv(i)} - x(i)}{\alpha} + c_{i(u(i)-1)} + m\frac{x(i) - p_{iu(i)}}{\alpha}$$
$$= nm + \sum_i \frac{m}{\alpha}(g(i) - f(i)) - \sum_i d_i$$

while the cost of the latter is

$$\sum_i |\{j \mid a_{ij} \in X, b_j \notin X\}| = \sum_i |\{j \mid u(i) \leq t_j(i) \leq v(i) \in X, b_j \notin X\}|$$
$$= \sum_i d_i - |\{j \mid u(i) \leq t_j(i) \leq v(i) \in X, b_j \in X\}|$$
$$= \sum_i d_i - m|\{j \mid b_j \in X\}|.$$

Combining the two equations proves the claim.   □

Corollary 1. *Let U' be the solution to REGBAND(α). Let (X, Y) be a minimum (θ, η)-cut of G. Set $U = \{t_\ell \mid b_\ell \in X\}$. Then $s_{reg}(U; \alpha) = s_{reg}(U'; \alpha)$.*

Proof. Proposition 3 states that the cost of the minimum cut is $nm + \frac{m}{\alpha}s_{reg}(U; \alpha)$.

Construct a cut (X', Y') from U' by setting X' to be the nodes from A and B that correspond to the time series U'. The proof of Proposition 3 now states that the cut is equal to $nm + \frac{m}{\alpha}s_{reg}(U'; \alpha)$.

The optimality of (X, Y) proves the claim.   □

We may encounter a pathological case, where we have multiple cuts with the same optimal cost. REGBAND requires that in such case we use largest solution. This can be enforced by modifying the weights: first scale the weights so that they are all multiples of $nm + 1$, then add 1 to the weight of each (θ, α_{ij}). The cut with the modified graph yields the largest band with the optimal cost.

The constructed graph G has $\mathcal{O}(nm)$ nodes and $\mathcal{O}(nm)$ edges. Consequently, we can compute the minimum cut in $\mathcal{O}((nm)^2)$ time [13]. In practice, solving minimum cut is much faster.

3.3 Computing All Regularized Bands

Now that we have a method for solving REGBAND(α) for a fixed α, we would like to find solutions for all α. Note that Proposition 2 states that we can have at most $n + 1$ different bands.

We can enumerate the bands with the divide-and-conquer approach given in Algorithm 1. Here, we are given two, already discovered, regularized bands $U \subsetneq V$, and we try to find a middle band W with $U \subsetneq W \subsetneq V$. If W exists, we recurse on both sides. To enumerate all bands, we start with ENUMREG($\{x\}, V$).

Algorithm 1: ENUMREG(U, V) finds all regularized bands between U and V

---

1   $\gamma \leftarrow \frac{s_1(V) - s_1(U)}{|V| - |U|} - \frac{\Delta}{n^2}$;

2   $W \leftarrow$ solution to REGBAND(γ);

3   **if** $U \neq W$ **then**

4   report W;

5   ENUMREG(U, W); ENUMREG(W, V);

The following proposition proves the correctness of the algorithm: during each split we will always find a new band if such exist.

Proposition 4. *Assume time series T with n time series. Let $\{U_i\}$ be all the possible regularized confidence bands ordered using inclusion. Define*

$$\Delta = \min \{|t(i) - u(i)| \mid t, u \in T, i, t(i) \neq u(i)\}.$$

Let $i < j$ be two integers and define

$$\gamma = \frac{s_1(U_j) - s_1(U_i)}{|U_j| - |U_i|} - \frac{\Delta}{n^2}.$$

Let U_ℓ be the solution for REGBAND(γ). Then $i \leq \ell < j$. If $j > i+1$, then $i < \ell$, otherwise $\ell = i$.

For simplicity, let us define $f(x, y) = \frac{s_1(U_y) - s_1(U_x)}{|U_y| - |U_x|}$.

In order to prove the result we need the following technical lemma.

Lemma 1. *Assume time series T with n time series. Let $\{U_i\}$ be all the possible regularized confidence bands ordered using inclusion. Let $\alpha > 0$. Let U_i be the solution for REGBAND(α). Then $f(i - 1, i) \leq \alpha < f(i, i+1)$.*

Proof. Due to the optimality of U_i,

$$s_1(U_i) - \alpha|U_i| = s_{reg}(U_i; \alpha) < s_1(U_{i+1}) - \alpha|U_{i+1}|.$$

Solving for α gives us the right-hand side of the claim. Similarly,

$$s_1(U_i) - \alpha|U_i| = s_{reg}(U_i; \alpha) \leq s_1(U_{i-1}) - \alpha|U_{i-1}|.$$

Solving for α gives us the left-hand side of the claim. $\square$

Proof (of Proposition 4). It is straightforward to see that Lemma 1 implies that $f(a,b) \leq f(x,y)$ for $a \leq x$ and $b \leq y$. Moreover, the equality holds only if $x = a$ and $y = b$, in other cases $f(a,b) + \frac{A}{n^2} \leq f(x,y)$.

If $\ell \geq j$, then Lemma 1 states that $f(i,j) \leq f(\ell-1,\ell) \leq \gamma$, which contradicts the definition of γ. Thus $\ell < j$.

Since $f(i,j) - f(i-1,i) \geq \frac{A}{n^2}$, we have $f(i-1,i) \leq \gamma$. If $\ell < i$, then Lemma 1 states that $\gamma < f(i-1,i)$, which is a contradiction. Thus, $\ell \geq i$.

If $j = i + 1$, then immediately $\ell = i$.

Assume that $j > i+1$. Since $f(i,j) - f(i,i+1) \geq \frac{A}{n^2}$, we have $f(i,i+1) \leq \gamma$. If $\ell = i$, then according to Lemma 1 $\gamma < f(i,i+1)$, which is a contradiction. Thus, $\ell > i$. □

Lemma 1 reveals an illuminating property of regularized bands, namely each band minimizes the ratio of additional envelope area and the number of new time series.

Proposition 5. *Let U be a regularized band. Define $g(X) = \frac{s_1(X) - s_1(U)}{|X| - |U|}$. Let $V \supsetneq U$ be the adjacent regularized band. Then $g(V) = \min_{X \supsetneq U} g(X)$.*

Proof. Let $O = \arg\min_{X \supsetneq U} g(X)$, and set $\beta = g(O)$. We will prove that $g(V) \leq \beta$. Let $W = \text{REGBAND}(\beta)$. Let α be the parameter for which $U = \text{REGBAND}(\alpha)$. Assume that $\alpha \geq \beta$. We can rewrite the equality $\beta = g(O)$ as

$$0 = s_{reg}(O;\beta) - s_{reg}(U;\beta) \geq s_{reg}(O;\alpha) - s_{reg}(U;\alpha),$$

which violates the optimality of U. Thus $\alpha < \beta$. Proposition 2 states that $U \subseteq W$. Moreover, due to submodularity,

$$s_{reg}(O \cup W;\beta) - s_{reg}(W;\beta) \leq s_{reg}(O \cup U;\beta) - s_{reg}(U;\beta) = 0,$$

which due to the optimality of W implies that $O \subseteq W$. Thus $W \neq U$ and $V \subseteq W$. Lemma 1, possibly applied multiple times, shows that $g(V) \leq g(W) \leq \beta$. □

Proposition 2 states that there are at most $n + 1$ bands. Queries done by ENUMREG yield the same band at most twice. Thus, ENUMREG performs at most $\mathcal{O}(n)$ queries, yielding computational complexity of $\mathcal{O}(n^3 m^2)$. In practice, ENUMREG is faster: the number of bands is significantly smaller than n and the minimum cut solver scales significantly better than $\mathcal{O}(n^2 m^2)$. Moreover, we can further improve the performance with the following observation: Proposition 2 states that when processing ENUMREG(U,V), the bands will be between U and V. Hence, we can ignore the time series that are outside V, and we can safely replace U with its envelope $\ell b(U)$ and $ub(U)$.[1]

4   Discovering Confidence Bands Minimizing s_1

In this section, we will study SUMBAND. Korpela et al. [11] showed that the problem is **NP**-hard. We will argue that we can approximate the problem and establish a (likely) lower bound for the approximation guarantee.

---

[1] We need to make sure that the envelope is always selected. This can be done by connecting θ to the envelope with edges of infinite weight.

Algorithm 2: FINDSUM(T, k, x), approximates SUMBAND

---

1 $\{B_i\} \leftarrow$ ENUMREG$(\{x\}, T)$;
2 $j \leftarrow$ largest index for which $|B_j| \leq k$;
3 **if** $|B_j| \leq k - \sqrt{n}$ **then** $W \leftarrow B_{j+1} \setminus B_j$ **else** $W \leftarrow T \setminus B_j$;
4 $U \leftarrow B_j$;
5 greedily add $k - |U|$ entries from W to U, minimizing s_1 at each step;
6 **return** U;

As a starting point, note that SUMBAND is an instance of k-MINUNION, weighted minimum k-union problem. In k-MINUNION we are given n sets over a universe with weighted points, and ask to select k sets minimizing the weighted union. In our case, the universe is the set P described in Sect. 3, the weights are the distances between adjacent points, and a set consists of all the points between a time series and x.

The *unweighted* k-MINUNION problem has several approximation algorithms: a simple algorithm achieving $\mathcal{O}(\sqrt{n})$ guarantee by Chlamtáč et al. [2] and an algorithm achieving lower approximation guarantee of $\mathcal{O}(n^{1/4})$ by Chlamtáč et al. [3]. We will use the former algorithm due to its simplicity and the fact that it can be easily adopted to handle weights.

The pseudo-code for the algorithm is given in Algorithm 2. The algorithm first looks for the largest possible regularized band, say B_j, whose size at most k. The remaining time series are then selected greedily from a set of candidates W. The set W depends on how many additional time series is needed: if we need at most $\sqrt{n}$ additional time series, we set W to be the remaining time series $T \setminus B_j$, otherwise we select the time series from the next regularized band, that is, we set $W = B_{j+1} \setminus B_j$.

Proposition 6. FINDSUM *yields* $\sqrt{n} + 1$ *approximation guarantee.*

Proof. Let O be the optimal solution for SUMBAND(k), and let $r = s_1(O)$. Let U be the output of FINDSUM. Assume that $B_j \neq O$, as otherwise we are done. We split the proof in two cases.

First, assume that $|B_j| \leq k - \sqrt{n}$. Since s_1 is submodular we have $s_1(O \cup B_j) - s_1(B_j) \leq s_1(O) - s_1(\{x\}) = r$, leading to

$$\frac{s_1(B_{j+1}) - s_1(B_j)}{|B_{j+1}| - |B_j|} \leq \frac{s_1(O \cup B_j) - s_1(B_j)}{|O \cup B_j| - |B_j|} \leq \frac{r}{k - |B_j|} \leq \frac{r}{\sqrt{n}},$$

where the first inequality is due to Proposition 5. Rearranging the terms gives us

$$s_1(B_{j+1}) - s_1(B_j) \leq \frac{r(|B_{j+1}| - |B_j|)}{\sqrt{n}} \leq r\frac{n}{\sqrt{n}} = r\sqrt{n}.$$

Finally,

$$s_1(U) = s_1(B_j) + (s_1(U) - s_1(B_j))$$
$$\leq s_1(B_j) + (s_1(B_{j+1}) - s_1(B_j)) \leq s_1(B_j) + r\sqrt{n} \leq r(1 + \sqrt{n}),$$

where the last inequality is implied by Proposition 1 and the fact that $|B_j| \leq k$.

Assume that $|B_j| > k - \sqrt{n}$, and let $q = k - |B_j|$. Note that $q < \sqrt{n}$. Let $c_1, \ldots, c_q$ be the additional time series added to U. Write $U_i = B_j \cup \{c_1, \ldots, c_i\}$.

Let c_i' be the closest time series to x outside U_{i-1}. Note that $s_1(\{x, c_i'\}) - s_1(\{x\}) = s_1(\{x, c_i'\}) \leq r$ for $i = 1, \ldots, q$ as otherwise r has to be larger. In addition, Proposition 1 and $|B_j| \leq k$ imply that $s_1(B_j) \leq r$. Consequently,

$$
s_1(U_q) = s_1(B_j) + \sum_{i=1}^{q} s_1(U_i) - s_1(U_{i-1})
$$

$$
\leq s_1(B_j) + \sum_{i=1}^{q} s_1(\{x, c_i'\}) - s_1(\{x\}) \leq (1 + \sqrt{n})r,
$$

where the first inequality is due to the submodularity of s_1.   □

FINDSUM resembles greatly the algorithm given by Chlamtáč et al. [2] but has few technical differences: we select B_j as our starting point whereas the algorithm by Chlamtáč et al. [2] constructs the starting set by iteratively finding and adding sets with the smallest average area, $s_1(X)/|X|$, that is, solving the problem given in Proposition 5.[2] Such sets can be found with a linear program. Proposition 5 implies that both approaches result in the same set B_j but our approach is faster.[3] Moreover, this modification allows us to prove a tighter approximation guarantee: the authors prove that their algorithm yields $2\sqrt{n}$ guarantee whereas we show that we can achieve $\sqrt{n} + 1$ guarantee. Additionally, we select additional time series iteratively by selecting those time series that result in the smallest increase of the current area, whereas the original algorithm would simply select time series that are closest to $\{x\}$.

Chlamtáč et al. [3] argued that under some mild but technical conjecture there is no polynomial-time algorithm that can approximate k-MINUNION better than $\mathcal{O}(n^{1/4})$. Next we will show that we can reduce k-MINUNION to SUMBAND while preserving approximation.

Proposition 7. *If there is an $f(n)$-approximation polynomial-time algorithm for* SUMBAND, *then there is an $f(n+1)$-approximation polynomial-time algorithm for k-MINUNION.*

Proof. Assume that we are given an instance of k-MINUNION with n sets $\mathcal{S} = (S_1, \ldots, S_n)$. Let $D = \bigcup_i S_i$ be the union of all S_i.

Define T containing $n+1$ time series over the domain D. The first n time series correspond to the sets S_i, that is, given $i \in D$, we set $t_j(i) = 1$ if $i \in S_j$, and 0 otherwise. The remaining single time series, named x, is set to be 0.

---

[2] The original algorithm is described using set/graph terminology but we use our terminology to describe the differences.

[3] The computational complexity of the state-of-the-art linear program solver is $\mathcal{O}((nm)^{2.37} \log(nm/\delta))$, where δ is the relative accuracy [4]. We may need to solve $\mathcal{O}(n)$ such problems, leading to a total time of $\mathcal{O}(n(nm)^{2.37} \log(nm/\delta))$.

Assume that we have an algorithm estimating $\textsc{SumBand}(T, x, k+1)$, and let U be the output of this algorithm. Note that since $x \in U$, we have $\ell b(U, i) = 0$.

Let $\mathcal{V}$ be the subset of $\mathcal{S}$ corresponding to the non-zero time series in U. Let $C = \bigcup_{S \in \mathcal{V}} S$ be the union of sets in $\mathcal{V}$. Since $\ell b(U, i) = 0$, and $ub(U, i) = 1$ if and only if $i \in C$, we have $s_1(U) = |C|$.   □

The above result implies that unless the conjecture suggested by Chlamtáč et al. [3] is false, we cannot approximate $\textsc{SumBand}$ better than $\mathcal{O}(n^{1/4})$. This proposition holds even if we replace $s_1(\cdot)$ with an ℓ_p^p norm, $\sum_i |t(i) - u(i)|^p$, where $1 \le p < \infty$, or any norm that reduces to hamming distance if t is a binary sequence and u is 0. Interestingly, we will show in the next section that we can achieve a tighter approximation if we use s_∞.

5 Discovering Confidence Bands Minimizing s_∞

In this section we consider the problem $\textsc{InfBand}$. Namely, we will show that a straightforward algorithm 2-approximates the problem, and more interestingly we show that the guarantee is tight.

The algorithm for $\textsc{InfBand}(T, x, k)$ is simple: we select k time series that are closest to x according to the norm $\|t(i) - x(i)\|_\infty = \max_i |t(i) - x(i)|$. We will refer to this algorithm as $\textsc{FindInf}$.

It turns out that this simple algorithm yields 2-approximation guarantee.

Proposition 8. $\textsc{FindInf}$ *yields 2-approximation for* $\textsc{InfBand}$.

Proof. Let U be the optimal solution for $\textsc{InfBand}$. Let V be the result produced by $\textsc{FindInf}$. Define $c = \max_{t \in V} \|t - x\|_\infty$. Then

$$c = \max_{t \in V} \|t - x\|_\infty \le \max_{t \in U} \|u - x\|_\infty \le s_\infty(U),$$

where the first inequality holds since V contains the closest time series and the second inequality holds since $x \in U$.

Let i be the index such that $s_\infty(V) = ub(V, i) - \ell b(V, i)$. Then

$$s_\infty(V) = ub(V, i) - \ell b(V, i) = (ub(V, i) - t(i)) + (t(i) - \ell b(V, i)) \le 2c.$$

Thus, $s_\infty(V) \le 2c \le 2s_\infty(U)$, proving the claim.   □

While $\textsc{FindInf}$ is trivial, surprisingly it achieves the best possible approximation guarantee for a polynomial-time algorithm.

Proposition 9. *There is no polynomial-time algorithm for* $\textsc{InfBand}$ *that yields* $\alpha < 2$ *approximation guarantee unless* $\boldsymbol{P = NP}$.

Proof. To prove the claim we will show that we can solve k-$\textsc{Clique}$ in polynomial time if we can α-approximate $\textsc{InfBand}$ with $\alpha < 2$. Since k-$\textsc{Clique}$ is an $\textbf{NP}$-complete problem, this is a contradiction unless $\textbf{P} = \textbf{NP}$.

The goal of k-CLIQUE is given a graph $G = (V, E)$ with n nodes and m edges to detect whether there is a k-clique, a fully connected subgraph with k nodes, in G. We can safely assume that G has no nodes that are fully-connected.

Fix an order for nodes $V = (v_1, \ldots, v_n)$ and let F be all the edges that are not in E, that is, $F = \{(v_x, v_y) \mid (v_x, v_y) \notin E, x < y\}$.

Next, we will define an instance of INFBAND. The set of time series $T = (t_1, \ldots, t_n) \cup \{x\}$ consists of n time series t_i corresponding to the node v_i, and a single time series x which we will use a seed. We set the domain to be F. Each time series t_i maps an element of $e = (v_x, v_y) \in F$ to an integer,

$$t_i(e) = 1, \text{ if } i = x, \quad t_i(e) = -1, \text{ if } i = y, \quad t_i(e) = 0, \text{ otherwise.}$$

We also set $x = 0$. First note that since $t_i(e)$ is an integer between -1 and 1, the score $s_\infty(U)$ is either 0, 1, or 2 for any $U \subseteq T$.

Since we do not have any fully-connected nodes in G, there is no non-zero t_i in T. Since $x \in U$ for any solution of INFBAND, then $s_\infty(U) = 0$ implies $|U| = 1$.

Let $W \subseteq V$ be a subset of nodes, and let U be the corresponding time series. We claim that $s_\infty(U) = 1$ if and only if W is a clique. To prove the claim, first observe that if $v_i, v_j \in W$ such that $e = (v_i, v_j) \in F$, then $t_i(e) = 1$ and $t_j(e) = -1$, thus $s_\infty(U) = 2$. On the other hand, if W is a clique, then for every $t_i, t_j \in U$ and $e \in F$ such that $t_i(e) \neq 0$, we have $t_j(e) = 0$ since otherwise $(v_i, v_j) \notin E$. Thus, $s_\infty(U) = 1$ if and only if W is a clique.

Let O be the solution for INFBAND$(T, k+1, x)$. Note that $s_\infty(O) = 1$ if and only if G has a k-clique, and $s_\infty(O) = 2$ otherwise.

Let S be the output of α-approximation algorithm. Since $k > 1$, we know that $s_\infty(O)$ is either 1 or 2. If $s_\infty(O) = 2$, then $s_\infty(S) = 2$. If $s_\infty(O) = 1$, then $s_\infty(S) \leq \alpha s_\infty(O) < 2 \times 1$. Thus, $s_\infty(S) = 1$. In summary, $s_\infty(O) = s_\infty(S)$.

We have shown that $s_\infty(S) = 1$ if and only if G has a k-clique. This allows us to detect k-clique in G in polynomial time proving our claim.   □

6   Related Work

Confidence bands are envelopes for which confidence intervals of individual points hold simultaneously. Davison and Hinkley [5,12] proposed a non-parametric approach for finding simultaneous confidence intervals. Here, time series are ordered based on its *maximum* value, and α-confidence intervals are obtained by removing $\alpha/2$ portions from each tail. Note that unlike SUMBAND and INFBAND this definition is not symmetric: if we flip the sign of time series we may get a different interval.

There is a strong parallel between finding regularized bands and finding dense subgraphs. Proposition 5 states that the inner-most regularized band has the smallest average envelope area, or alternatively it has the highest ratio of time series per envelope area. A related graph-theoretical concept is a dense subgraph, a subgraph H of a given subgraph G with the largest ratio $|E(H)|/|V(H)|$. The method proposed by Goldberg [7] for finding dense subgraphs in polynomial time is based on maximizing $|E(H)| - \alpha|V(H)|$ and selecting α to be as small as possible without having an empty solution. Moreover, Tatti [16] extended the notion of dense subgraphs to density-friendly core decomposition, which essentially consists of the subgraphs minimizing $|E(H)| - \alpha|V(H)|$ for various values of α, the algorithm for finding the decomposition is similar to the algorithm for enumerating all regularized bands. In addition, Tsourakakis [17] extended the notion of dense subgraphs to triangle-density and hypergraphs, and also used minimum cut to find the solutions. As pointed out in Sect. 4 is that we can view time series as sets of points in P. In fact, the minimum cut used in Sect. 3 share some similarities with the minimum cut proposed by Tsourakakis [17]. Finally, the algorithm proposed by Korpela et al. [11] to find confidence bands resembles the algorithm by Charikar [1] for approximating the densest subgraph: in the former we delete the time series that reduce the envelope area the most while in the latter we delete vertices that have the smallest degree.

We assume that we are given a seed time series x. If such series is not given then we need to test every $t \in T$ as a seed. If we consider a special case of $k = 2$, then the problem of finding regularized band reduces to the closest pair problem: find two time series with the smallest distance: a well-studied problem in computational geometry. A classic approach by Dietzfelbinger et al. [14], Khuller and Matias [6], Rabin [10] allows to solve the closest pair problem in $\mathcal{O}(n)$ time but the analysis treats the size of the domain, m, as a constant; otherwise, the computational complexity has an exponential factor in m and can be only used for very small values of m. For large values of m, Indyk et al. [9] proposed an algorithm for solving the closest pair problem minimizing $s_1(\cdot)$ in $\mathcal{O}(n^{2.687})$ time and minimizing $s_\infty(\cdot)$ in $\mathcal{O}(n^{2.687} \log \Delta)$ time, where Δ is the width of the envelope of the whole data.

7   Experimental Evaluation

In this section we describe our experimental evaluation.

We implemented ENUMREG and FINDSUM using C++ and used a laptop with Intel Core i5 (2.3 GHz) to conduct our experiments.[4] As a baseline we used the algorithm by Korpela et al. [11], which we will call PEEL. We implemented PEEL also with C++, and modified it to make sure that the seed time series x is always included. Finally, we implemented FINDINF with Python. In all algorithms we used the median as the seed time series.

Datasets: We used 4 real-world datasets as benchmark datasets. The first dataset, *Milan*, consists of monthly averages of maximum daily temperatures in Milan between the years 1763–2007.[5] The second dataset, *Power*, consists of hourly power consumption (variable `global_active_power`) of a single household over almost 4 years, a single time series representing a day.[6] Our last 2

---

[4] The code is available at https://version.helsinki.fi/DACS.
[5] https://www.ncdc.noaa.gov/.
[6] http://archive.ics.uci.edu/ml/datasets/Individual+household+electric+power+consumption.

Table 1. Basic characteristics of the datasets and performance measures of the algorithms. Here, n stands for the number of time series, m stands for the domain size, $|B_1|$ is the size of the smallest non-trivial regularized band, $|\mathcal{B}|$ is the number of regularized bands, and time is the required time to execute ENUMREG in seconds. The scores s_1 for the algorithms FINDSUM, FINDINF, and PEEL are normalized with the envelope area of the whole data and multiplied by 100.

Dataset	n	m	$|B|$	$|\mathcal{B}|$	Time	s_1 for $k = \lfloor 0.9n \rfloor$			s_1 for $k = \lfloor 0.95n \rfloor$		
						SUM	PEEL	INF	SUM	PEEL	INF
Milan	245	12	209	17	0.03	70.34	72.49	74.1	75.31	76.99	78.45
Power	1 417	24	1 102	56	3.68	70.94	72.83	77.06	78.89	81.17	82.31
ECG-normal	1 507	253	1 289	72	39.44	51.72	52	72.97	57.22	57.51	73.15
ECG-pvc	520	253	484	19	6.67	80.28	80.02	91.97	83.84	83.92	95.69

Table 2. Scores s_∞ of discovered confidence bands. The scores are normalized with the envelope width of the whole data and multiplied by 100.

Dataset	s_∞ for $k = \lfloor 0.9n \rfloor$			s_∞ for $k = \lfloor 0.95n \rfloor$		
	SUM	PEEL	INF	SUM	PEEL	INF
Milan	72.54	79.78	67.35	78.04	79.78	73.95
Power	79.08	82.16	73.13	82.16	98.71	79.08
ECG-normal	64.78	64.78	54.81	65.64	64.78	57.39
ECG-pvc	93.24	93.24	66.41	93.24	93.24	81.9

datasets *ECG-normal* and *ECG-pvc* are heart beat data [8]. We used MLII data of a single patient (id 106) from the MIT-BIH arrhythmia database,[7] and split the measurements into normal beats (*ECG-normal*) and abnormal beats with premature ventricular contraction (*ECG-pvc*). Each time series represent measurements between -300 ms and 400 ms around each beat. The sizes of the datasets are given in Table 1.

Results: First let us consider ENUMREG. From the results given Table 1 we see that the number of distinct regularized bands $|\mathcal{B}|$ is low: about 4%–7% of n, the

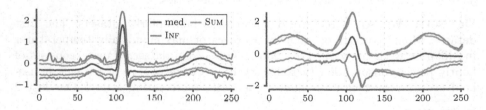

Fig. 1. Envelopes for *ECG-normal* (left) and *ECG-pvc* (right) and $k = \lfloor 0.9n \rfloor$.

---

[7] https://physionet.org/content/mitdb/1.0.0/.

number of time series. Having so few bands in practice reduces the computational cost of ENUMREG since the algorithm tests at most $2|\mathcal{B}|$ values of α Interestingly, the smallest non-trivial band B_1 is typically large, containing about 70%–90% of the time series. Note that Proposition 5 states that B_1 has the smallest ratio of $s_1(B_1)/|B_1|$. For our benchmark datasets, B_1 is large suggesting that most time series are equally far away from the median while the remaining the time series exhibit outlier behaviour.

The algorithms are fast for our datasets: Table 1 show that ENUMREG requires at most 40 s. Additional steps required by FINDSUM are negligible, completing in less than a second. The baseline algorithm is also fast, requiring less than a second to complete.

Let us now compare FINDSUM against PEEL. We compared the obtained areas by both algorithms with $k = \lfloor 0.9n \rfloor$ and $k = \lfloor 0.95n \rfloor$. We see from the results in Table 1, that FINDSUM performs slightly better than PEEL. The improvement in score is modest, 1%–2%. We conjecture that in practice PEEL is close to the optimal, so any improvements are subtle. Interestingly, enough PEEL performs better than FINDSUM for ECG-pvc and $\gamma = 0.1$. The reason for this is that the inner band B_1 contains more than 90% of the time series. In such a case FINDSUM will reduce to a simple greedy method, starting from $\{x\}$. Additional testing revealed that PEEL outperforms FINDSUM when $k \leq |B_1|$ about 50%–90%, depending on the dataset, suggesting that whenever $k \leq |B_1|$ it is probably better to run both algorithms and select the better envelope.

Next let us compare FINDINF against the other methods. The results in Tables 1–2 show that FINDINF yields inferior s_1 scores but superior s_∞ scores. This is expected as FINDINF optimizes s_∞ while FINDSUM and PEEL optimize s_1. The differences are further highlighted in the envelopes for ECG datasets shown in Fig. 1: FINDINF yields larger envelopes but provides a tighter bound under the peak (R wave).

8   Concluding Remarks

In this paper we consider the approximation algorithms for discovering confidence bands. Namely, we proposed a practical algorithm that approximates SUMBAND with a guarantee of $\mathcal{O}(n^{1/2})$. We also argued that the lower bound for the guarantee is most likely $\mathcal{O}(n^{1/4})$. In addition, we showed that we can 2-approximate INFBAND, a variant of SUMBAND problem, with a simple algorithm and that the guarantee is tight.

Our experiments showed that FINDSUM outperforms the original baseline method for large values of k, that is, as long as k is larger than the smallest regularized band. Our experiments suggest that this condition usually holds, if we are interested, say in, 90%–95% confidence.

Interesting future line of work is to study the case for time series with multiple modes, that is, a case where instead of a single seed time series, we are given a set of time series, and we are asked to find confidence bands around each seed.

References

1. Charikar, M.: Greedy approximation algorithms for finding dense components in a graph. APPROX (2000)
2. Chlamtáč, E., Dinitz, M., Konrad, C., Kortsarz, G., Rabanca, G.: The densest k-subhypergraph problem. SIAM J. Discret. Math. **32**(2), 1458–1477 (2018)
3. Chlamtáč, E., Dinitz, M., Makarychev, Y.: Minimizing the union: tight approximations for small set bipartite vertex expansion. In: Proceedings of the Twenty-Eighth Annual ACM-SIAM Symposium on Discrete Algorithms, pp. 881–899. SIAM (2017)
4. Cohen, M.B., Lee, Y.T., Song, Z.: Solving linear programs in the current matrix multiplication time. J. ACM (JACM) **68**(1), 1–39 (2021)
5. Davison, A.C., Hinkley, D.V.: Bootstrap Methods and Their Application. Cambridge University Press, Cambridge (1997)
6. Dietzfelbinger, M., Hagerup, T., Katajainen, J., Penttonen, M.: A reliable randomized algorithm for the closest-pair problem. J. Algorithms **25**(1), 19–51 (1997)
7. Goldberg, A.V.: Finding a maximum density subgraph. University of California Berkeley Technical report (1984)
8. Goldberger, A.L., et al.: Physiobank, physiotoolkit, and physionet: components of a new research resource for complex physiologic signals. Circulation **101**(23), e215–e220 (2000)
9. Indyk, P., Lewenstein, M., Lipsky, O., Porat, E.: Closest pair problems in very high dimensions. In: Díaz, J., Karhumäki, J., Lepistö, A., Sannella, D. (eds.) ICALP 2004. LNCS, vol. 3142, pp. 782–792. Springer, Heidelberg (2004). https://doi.org/10.1007/978-3-540-27836-8_66
10. Khuller, S., Matias, Y.: A simple randomized sieve algorithm for the closest-pair problem. Inf. Comput. **118**(1), 34–37 (1995)
11. Korpela, J., Puolamäki, K., Gionis, A.: Confidence bands for time series data. Data Min. Knowl. Discov. 1530–1553 (2014). https://doi.org/10.1007/s10618-014-0371-0
12. Mandel, M., Betensky, R.A.: Simultaneous confidence intervals based on the percentile bootstrap approach. Comput. Stat. Data Anal. **52**(4), 2158–2165 (2008)
13. Orlin, J.B.: Max flows in $O(nm)$ time, or better. In: Proceedings of the Forty-Fifth Annual ACM Symposium on Theory of Computing, pp. 765–774 (2013)
14. Rabin, M.O.: Probabilistic algorithms. In: Traub, J.F. (ed.) Algorithms and Complexity: New Directions and Recent Results. Academic Press, New York (1976)
15. Schrijver, A.: A combinatorial algorithm minimizing submodular functions in strongly polynomial time. J. Comb. Theory Ser. B **80**(2), 346–355 (2000)
16. Tatti, N.: Density-friendly graph decomposition. ACM Trans. Knowl. Discov. Data (TKDD) **13**(5), 1–29 (2019)
17. Tsourakakis, C.: The k-clique densest subgraph problem. In: Proceedings of the 24th International Conference on World Wide Web, pp. 1122–1132 (2015)

A Mixed Noise and Constraint-Based Approach to Causal Inference in Time Series

Charles K. Assaad[1,2]($\boxtimes$), Emilie Devijver[1]($\boxtimes$), Eric Gaussier[1]($\boxtimes$), and Ali Ait-Bachir[2]($\boxtimes$)

[1] Univ. Grenoble Alpes, CNRS, Grenoble INP, LIG, Grenoble, France
`emilie.devijver@univ-grenoble-alpes.fr`, `eric.gaussier@imag.fr`
[2] Coservit, Grenoble, France
`ali.ait-bachir@coservit.com`

Abstract. We address, in the context of time series, the problem of learning a summary causal graph from observations through a model with independent and additive noise. The main algorithm we propose is a hybrid method that combines the well-known constraint-based framework for causal graph discovery and the noise-based framework that gained much attention in recent years. Our method is divided into two steps. First, it uses a noise-based procedure to find the potential causes of each time series. Then, it uses a constraint-based approach to prune all unnecessary causes. A major contribution of this study is to extend the standard causation entropy measure to time series to handle lags bigger than one time step, and to rely on a lighter version of the faithfulness hypothesis, namely the *adjacency faithfulness*. Experiments conducted on both simulated and real-world time series show that our approach is fast and robust wrt to different causal structures and yields good results over all datasets, whereas previously proposed approaches tend to yield good results on only few datasets.

Keywords: Causal graph discovery · Noise-based approach · Constraint-based approach · Time series

1 Introduction

Identifying causal structure from observational data is an important but also challenging task in many applications. Most causal graph discovery algorithms assume that causal relations can be described within causal graphs, where arrows encode causal information. For time series, the true complete causal graph $\mathcal{G} = (V, E)$ with V the set of vertices and E the set of edges, is called a *full time causal graph* and represents a complete graph of the dynamic system, through infinite vertices. In practice, inferring an infinite graph is unfeasible, so most algorithms assume that causal relations are consistent throughout time, *i.e.* for two time

© Springer Nature Switzerland AG 2021
N. Oliver et al. (Eds.): ECML PKDD 2021, LNAI 12975, pp. 453–468, 2021.
https://doi.org/10.1007/978-3-030-86486-6_28

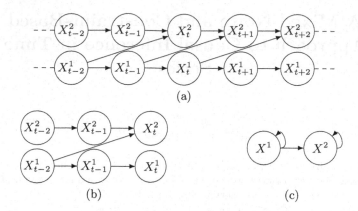

(a)

(b)   (c)

Fig. 1. Different causal graphs that one can infer from two time series X^1 and X^2: full time causal graph (a), window causal graph of size $\tau = 2$ (b) and summary causal graph (c).

series X^p and X^q, if X^p_{t-i} causes X^q_t, denoted $X^p_{t-i} \to X^q_t$, then $X^p_{t-i-j} \to X^q_{t-j}$ for all j. Under this assumption, and given the maximal lag τ between cause and effect that can be present in the system, the full time causal graph can be contracted to give a finite graph which we call *window causal graph*, with $\tau + 1$ nodes for each time series [15]. However, sometimes, in practice, knowing only the causes of a given time series without necessarily knowing the time delay between the cause and the effect is all we need, so the true complete graph of the system is compressed even more via the so-called *summary graph* which represents causal relations between time series without referencing to lags [12]. Those notions are illustrated in Fig. 1. Algorithms that detect causal relations can be classified according to the type of graph they look for.

Whatever the type of graph, the algorithms in question rely also on additional assumptions. The *Causal Markov Condition* states that in a causal graph, each node is independent of all other nodes given its parents, except its children [17]. The *causal sufficiency* states that there are no hidden common causes. One of the best known approach for causal discovery methods is the constraint-based approach that relies on conditional independencies and assume *faithfulness*, which states that the joint distribution P over V is faithful to the true causal Directed Acyclic Graph (DAG) $\mathcal{G}$ over V in the sense that every conditional independence statement satisfied by P is entailed by $\mathcal{G}$ [17].

Our contribution is two-fold. We introduce a new measure of dependence between two time series called the temporal causation entropy, which is an extension of the standard causation entropy measure [18] to time series to handle instantaneous relations and lags bigger than one. Based on it, we develop an algorithm to infer a summary causal graph from observational time series that is not limited to the Markov equivalent class even for instantaneous relations (which are common in practice due to the discretization of the time), that assumes causal Markov condition and a weaker version of faithfulness described in Sect. 3.1, and which is proved to be complete. The algorithm we propose is hybrid in the sense that it combines two different families of causal graph discovery methods: the noise-based family to find the potential causes of each time series and the constraint-based family to prune all unnecessary causes by looking at possible confounders and therefore end-up with only genuine cause. Remarkably, this is to our knowledge the first algorithm hybrid between constraint-based and noise-based methods for time series. Our evaluation, conducted on several datasets, illustrates the efficacy and efficiency of our approach.

The remainder of the paper is organized as follows: Sect. 2 describes related work. Section 3 then introduces the main causal discovery algorithm, called NBCB for Noise-Based/Constraint-Based approach. It relies on weak assumptions that are reminded first, and on the temporal causation entropy that is also introduced. The causal graph discovery algorithm we propose is illustrated and evaluated on several datasets in Sect. 4. Finally, Sect. 5 concludes the paper.

2   State of the Art

Granger Causality is one of the oldest methods proposed to detect causal relations between time series. However, this approach is known to handle a restricted version of causality that focuses on causal priorities as it assumes that the past of a cause is necessary and sufficient for optimally forecasting its effect [5]. The simplicity constitutes its advantage but also its limitations: for instance, it cannot deal with instantaneous effects. More recently, [11] exploited deep learning to learn causal relations between time series using an attention mechanism within convolutional networks. It infers a potential set of causes by analysing the estimated coefficients and then applies a validation step that is to some extent comparable to conditional Granger causality. However, in our experiments, we observe particularly bad results for TCDF.

Constraint-based approaches for time series are usually extended from causal graph discovery algorithm for nontemporal data. The main idea is to eliminate potential causes by finding conditional independencies in the data. The PC algorithm [17] is known to be the representative of this family of methods in case of i.i.d. data, which optimize the search of the smallest conditioning set needed to achieve separation between each pair of nodes. PCMCI [15] is an extension of PC for time series where a window causal graph is constructed, using temporal priority constraint to reduce the search space of the causal structure. oCSE [18] takes a different procedure compared to PC: instead of limiting as much as possible the size of its conditioning set, it conditions since the start on all potential causes which constitute the past of all available nodes. However, it limits its search for causal relations with a lag of one to find a summary graph. These methods usually assume causal Markov condition and faithfulness. Moreover, in general, graphs can only be recovered up to Markov equivalence[1] classes. In the context of time series, the notion of time can make such algorithms go beyond the Markov equivalence class but only for lagged relations [15], *i.e.* instantaneous relations are always limited to the Markov equivalence class.

---

[1] Two DAGs are Markov equivalent if and only if they have the same skeleton and the same v-structures [19].

Lastly, noise-based methods assume that the causal system can be defined by a set of equations that explain each variable by its direct causes and an additional noise. Causal relations are in this case discovered using footprints produced by the causal asymmetry in the data. For time series, the most well known algorithms in this family are tsLiNGAM [7], which is an extension of LiNGAM through autoregressive models, and TiMINo [12], which discovers a causal relationship by looking at independence between the noise and the potential causes. However, self-causation is not discovered within TiMINo, and the summary graph is assumed to be acyclic. These methods usually assume causal Markov condition and causal minimality, which is a weaker assumption than faithfulness. The main drawbacks are that such methods usually do not scale well [4] and might need a large sample size to achieve good performance [8].

The method we propose, as an hybrid method, takes benefit of the two approaches: it is not limited to a Markov equivalence class and provides a specific graph, scales better and needs a smaller sample size.

3   Causal Graph Discovery Between Two Time Series

Let us consider d univariate time series $X^1, \cdots, X^d$. Our goal is to find a summary causal graph between them, as represented in Fig. 1(c).

3.1   Assumptions

The faithfulness assumption is difficult to check in practice, and it has been debated for a long time. It assumes that there are no accidental conditional independence relations in the true distribution, that is, no conditional independence relations unless entailed by the true causal structure. The faithfulness assumption is mainly used in constraint-based methods, where it is used at two different stages, skeleton construction and edge orientation. As such, it can be decomposed into two assumptions, as proposed in [14], namely adjacency faithfulness and orientation faithfulness. As the orientation process we rely on differs from the one used in PC-like algorithms, we dispense here with the second assumption and solely rely on adjacency faithfulness, which is defined as follows:

Definition 1 (Adjacency faithfulness [14]**).** *For every $X^p, X^q \in V$, if X^p and X^q are adjacent in $\mathcal{G}$, then they are not conditionally independent given any subset of $V \backslash \{X^p, X^q\}$.*

As shown in [14], the relaxation of the faithfulness assumption still leads to provably correct skeletons.

Finally, the approach we propose discovers causal relations from time series under the causal Markov condition, the causal minimality condition (needed in the causal ordering, when using the noise-based method) and adjacency faithfulness (needed in the pruning step, when using the constraint-based method). We also assume that time series satisfy a causal ordering, meaning that we assume that the summary graph is acyclic. The inferred summary graph can however be cyclic, as opposed to [12], with loops between at least 3 time series.

3.2   Method

Our approach is a hybrid method which is decomposed into two parts. The first part, a noise-based approach, is described in Algorithm 1. It is based on a Gaussian process to map the past of the time series to the present, and a dependency measure between its input and its residuals to infer which time series potentially causes the other. The second part, a constraint-based approach, is described in Algorithm 2. It prunes the graph being constructed to remove spurious causes by considering the set of potential parents. The two parts are detailed below.

Causal Ordering. The first step relies on noise-based approaches, which were initially introduced for i.i.d. data. However, they gained much attention in recent years [1,6,9,10], and have also been extended for time series [12].

In this paper, we focus on Additive Noise Models (ANMs), which are defined as follows:

$$X_t^q = f\left([Par(X_t^q)_{t'}]_{t-\tau \leq t' \leq t}\right) + \xi_t^q \tag{1}$$

where f is a potentially nonlinear function, $Par(X_t^q)_{t'}$ is the set of parents of X_t^q at time-point t', $(\xi_t^q)_{q,t}$ are jointly independent; futhermore, for each q, ξ_t^q are identically distributed in t and the finite dimensional distributions for the time series $(X^q)_{1 \leq q \leq d}$ are absolutely continuous wrt a product measure. Note that this model allows instantaneous relations. ANMs belong to the Identifiable Functional Model Class (IFMOC) [13], even in case of non-faithful causal models, for which conditional independence-based methods, as constraint-based, usually fail [13].

Similarly to the bivariate case [6,10], the independence between the signal and the residuals allows one to detect the most probable cause from a set of variables through the following principle.

Principle 1 (Multivariate additive noise principle). *Suppose we are given a joint distribution $P(X^1, \cdots, X^d)$. If it satisfies an identifiable Additive Noise Model such that $\{(X_{t-j}^p)_{1 \leq p \neq q \leq d, 0 \leq j \leq \tau}, (X_{t-j}^q)_{1 \leq j \leq \tau}\} \to X^q$, then it is likely that $\{(X_{t-j}^p)_{1 \leq p \neq q \leq d, 0 \leq j \leq \tau}, (X_{t-j}^q)_{1 \leq j \leq \tau}\}$ precedes X^q in the causal order.*

Similarly to [10], when considering a suitable regression estimator and a suitable dependency estimator, the true causal order will be inferred. If we consider the fully connected graph given by this causal ordering (an edge between each node and its parents), it leads to a graph that contains the real graph as all true causal relations are in the inferred graph.

In practice, we first estimate for all $q \in \{1, \ldots, d\}$,

$$f_q : \{(X_{t-j}^p)_{1 \leq p \neq q \leq d, 0 \leq j \leq \tau}, (X_{t-j}^q)_{1 \leq j \leq \tau}\} \mapsto X_t^q$$

by a Gaussian Process and deduce the residuals

$$\hat{\xi}_t^q = X_t^q - \hat{f}\{(X_{t-j}^p)_{1 \leq p \neq q \leq d, 0 \leq j \leq \tau}, (X_{t-j}^q)_{1 \leq j \leq \tau}\}.$$

Algorithm 1. NBCB Part I: noise-based approach to order causes

---

Result: $\mathcal{G}$
X a d-dimensional time series, τ a window size;
$\mathcal{G}$ an empty graph with nodes $\{X^1, \ldots, X^d\}$; $S = \{1, \ldots, d\}$;
while $length(S) > 1$ **do**
  for $j \in S$ **do**
    Learn $\hat{f}^j : \{(X_{t-j}^p)_{p \in S, p \neq q, 0 \leq j \leq \tau}, (X_{t-j}^q)_{1 \leq j \leq \tau}\} \mapsto X_t^j$;
    Deduce $\hat{\xi}_t^j$ and compute;
    c_j from Eq. (2)
  Choose $j^* = \operatorname{argmin} c_j$;
  $S = S \backslash j^*$;
  for $s \in S$ **do**
    Add $X^s \to X^{j^*}$ in $\mathcal{G}$;

The last place in the causal ordering (which belongs to the most probable effect of all other time series) is given to the time series which yields the residuals that are more independent to the other time series. The dependency between the residuals and the input is estimated with

$$c_q = C\left(\{(X_{t-j}^p)_{1 \leq p \neq q \leq d, 0 \leq j \leq \tau}, (X_{t-j}^q)_{1 \leq j \leq \tau}\}, \hat{\xi}_t^q\right), \tag{2}$$

where C is a dependence measure[2].

However, this method is not capable to detect independence between two time series, and thus it is susceptible to treat indirect causes as direct causes. To remove indirect causes or detect independencies, we complement this procedure with a second step that prunes spurious relations from the graph. It necessitates an exact estimation of the lag between two time series (through a maximum window of size τ).

Since this procedure uses a regression function estimator, it is subject to the curse of dimensionality when d is large compared to n. So we also consider a pairwise version of the procedure which consists on estimating for each pair of time series X^q, X^p two regression functions

$$f^q : \{(X_{t-j}^p)_{0 \leq j \leq \tau}, (X_{t-j}^q)_{1 \leq j \leq \tau}\} \mapsto X_t^q,$$
$$f^p : \{(X_{t-j}^q)_{0 \leq j \leq \tau}, (X_{t-j}^p)_{1 \leq j \leq \tau}\} \mapsto X_t^p.$$

We then compare the dependency of the residuals of those two functions with their inputs, and as before the potential cause is the one that is mapped by the function that yields the higher dependency, i.e. we choose the causal direction that yields the best bivariate ANM. While one cannot prove that the inferred graph contains the real one, numerical experiments show good performances for this method.

---

[2] As motivated in [12], we use the partial correlation to measure the dependence, but one can use our procedure with any measure.

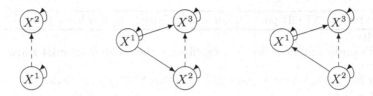

Fig. 2. Wrong causal relations potentially inferred in the first step of our algorithm. Dashed lines represents wrong causal relations. On the left, we show a spurious cause, whereas on the middle and on the right, we provide two indirect causes.

Pruning Using Temporal Causation Entropy. Knowing the list of potential parents of each time series, as detected in the previous step, one way to prune the causes that are not genuine is to conduct conditional independence tests between time series. Indeed, suppose X^p is a potential cause of X^q but X^p and X^q are conditionally independent, as illustrated in Fig. 2. Then, we can conclude that X^p is not a cause of X^q.

In order to capture the dependencies (and conditional dependencies) between two time series, one needs to take into account the lag between them, as the true causal relations might not be instantaneous. Several studies have acknowledged the importance of taking into account lags to measure (conditional) dependencies between time series [5,18]. Causation entropy, introduced in [18], is an asymmetric measure that detects the uncertainty reduction of the future states of X^q as a result of knowing the past states of X^p given that the past of X^R is already known, where $\mathbf{R}$ is a subset of $\{1, \cdots, d\}$. However, it only considers causation with a lag of size one, whereas it can take any values in practice.

In addition to lags, a window-based representation may be necessary to fully capture the dependencies between the two time series. So it may be convenient to consider them together when assessing whether the time series are dependent or not. We thus introduce the temporal causation entropy, that extends the causation entropy to general lags and window representation of time series.

Definition 2 (Temporal causation entropy). *We first define the optimal lag γ_{pq} between time series X^p and X^q and $(\lambda_{pq}, \lambda_{qp})$ the optimal windows of time series X^p regarding X^q and of time series X^q regarding X^p respectively as:*

$$\gamma_{pq}, \lambda_{pq}, \lambda_{qp} = \underset{\gamma \geq 0, \lambda_1, \lambda_2}{\mathrm{argmax}} \ h(X^q_{t:t+\lambda_2} \mid X^q_{t-1}, X^p_{t-\gamma-1}))$$
$$- h(X^q_{t:t+\lambda_2} \mid X^p_{t-\gamma-1:t-\gamma+\lambda_1}, X^q_{t-1}),$$

where h denotes the entropy. The temporal causation entropy *from time series X^p to time series X^q conditioned on a set $X^R = \{X^{r_1}, \cdots, X^{r_K}\}$ is given by:*

$$TCE(X^p \to X^q \mid X^R) = \underset{\Gamma_{r_i} \geq 0, \ 1 \leq i \leq K}{\min} \ h(X^q_{t:t+\lambda_{qp}} \mid (X^{r_i}_{t-\Gamma_{pq|r_i}})_{1 \leq i \leq K}, X^q_{t-1}, X^p_{t-\gamma_{pq}-1}))$$
$$- h(X^q_{t:t+\lambda_{qp}} \mid (X^{r_i}_{t-\Gamma_{pq|r_i}})_{1 \leq i \leq K}, X^p_{t-\gamma_{pq}-1:t-\gamma_{pq}+\lambda_{pq}}, X^q_{t-1}),$$

where $\Gamma_{pq|r_1}, \cdots, \Gamma_{pq|r_K}$ are the lags between X^R and X^q.

Algorithm 2. NBCB part II: constraint-based approach for pruning

---

Result: $\mathcal{G}$
X d-dimensional time series, α a significance threshold, $\mathcal{G}$ a causal graph;
n = 0;
while *there exists* $X^q \in V$ *such that* $card(Par(X^q, \mathcal{G})) \geq n+1$ **do**
    $\mathbf{D} = list()$;
    for $X^q \in V$ *such that* $card(Par(X^q, \mathcal{G})) \geq n+1$ **do**
        for $X^p \in Par(X^q, \mathcal{G})$, $X^{\mathbf{R}} \subset Par(X^q, \mathcal{G}) \setminus \{X^p\}$ *with* $card(X^{\mathbf{R}}) = n$ **do**
            $y_{q,p,\mathbf{R}} = \text{TCE}(X^p; X^q \mid X^{\mathbf{R}})$;
            append($\mathbf{D}, \{X^q, X^p, X^{\mathbf{R}}\}$));
    Sort $\mathbf{D}$ by increasing order of y;
    while D *is not empty* **do**
        $\{X^q, X^p, X^{\mathbf{R}}\} = $ pop($\mathbf{D}$);
        if $X^p \in Par(X^q, \mathcal{G})$ *and* $X^{\mathbf{R}} \subset Par(X^q, \mathcal{G})$ **then**
            Compute z the p-value given by Eq. (3);
            if $z > \alpha$ **then**
                Remove edge $X^p \rightarrow X^q$ from $\mathcal{G}$;
    n=n+1;

First, the lag between X^p and X^q is detected by maximizing the dependency between X^p and X^q. As we measure the amount of information brought by the observations of one variable on the observations of another variable, taking the maximum ensures that one does not miss any possible information contributing to relating the two time series. In a second step, we find the lags between (X^p, X^q) and $X^{\mathbf{R}}$ that minimize the conditional dependency between X^p and X^q conditioned on $X^{\mathbf{R}}$. Taking the minimum ensures that we search for the lags that break the maximal dependence. Following the temporal priority principle, which states that causes precede their effects in time, we also ensure while finding only nonnegative lags that X^p as well as the conditional variables should precede in time X^q. If $\gamma = 1$ and $\lambda_{pq} = \lambda_{qp} = 1$, then the temporal causation entropy is equivalent to causation entropy when the latter is conditioned on the past.

In practice, the success of temporal causation entropy (and in fact, any entropy-based approaches) depends crucially on reliable estimation of the relevant entropies from data. This leads to two practical challenges. The first one is based on the fact that entropies must be estimated from finite time series data. To do so, we rely here on the k-NN estimator introduced in [3]. We denote by $\epsilon_{ik}/2$ the distance from

$$(X^p_{t-\gamma_{pq}:t-\gamma_{pq}+\lambda_{pq}}, X^q_{t:t+\lambda_{pq}}, ((X^{r_i}_{t-\Gamma_{pq|r_i}})_{1 \leq i \leq K}, X^q_{t-1}, X^r_{t-\gamma_{pq}}))$$

to its k-th neighbor, $n_i^{1,3}$, $n_i^{2,3}$ and n_i^3 the numbers of points with distance strictly smaller than $\epsilon_{ik}/2$ in the subspace

$$(X^p_{t-\gamma_{pq}:t-\gamma_{pq}+\lambda_{pq}}, ((X^{r_i}_{t-\Gamma_{pq|r_i}})_{1\leq i\leq K}, X^q_{t-1}, X^p_{t-\gamma_{pq}})),$$

$$(X^q_{t:t+\lambda_{pq}}, ((X^{r_i}_{t-\Gamma_{pq|r_i}})_{1\leq i\leq K}, , X^q_{t-1}, X^p_{t-\gamma_{pq}}))$$

and

$$((X^{r_i}_{t-\Gamma_{pq|r_i}})_{1\leq i\leq K}, X^q_{t-1}, X^p_{t-\gamma_{pq}})$$

respectively, and $n_{\gamma_{r,p},\gamma_{r,q}}$ the number of observations. The estimate of the temporal causation entropy is then given by:

$$\widehat{TCE}(X^p \to X^q \mid X^{\mathbf{R}}) = \psi(k) + \frac{1}{n_{\gamma_{r,p},\gamma_{r,q}}} \sum_{i=1}^{n_{\gamma_{r,p},\gamma_{r,q}}} \psi(n_i^3) - \psi(n_i^{1,3}) - \psi(n_i^{2,3})$$

where ψ denotes the digamma function. The second problem is the following: to detect independence, we need a statistical test to check if the temporal causation entropy is equal to zero. We rely here on a permutation test:

Definition 3 (Permutation test of TCE). *Given X^p, X^q and $X^{\mathbf{R}}$, the p-value associated to the permutation test of TCE is given by:*

$$p = \frac{1}{B} \sum_{b=1}^{B} \mathbb{1}_{\widehat{TCE}(b(X^p)\to X^q|X^R)\geq \widehat{TCE}(X^p\to X^q|X^R)}, \qquad (3)$$

where $b(X^p)$ is a permuted version of X^p, $\mathbb{1}$ denotes the indicator function and B the maximum number of bootstrap sampling.

The method, detailed in Algorithm 2, can be summarized as follows. Starting with a fully directed graph (with one sided edges coming from a causal ordering), the first step consists in removing edges between nodes that are unconditionally independent: for each pair of nodes, a test of TCE is computed an edge is removed if the dependency, measured by TCE, is not significant given a threshold α. Once this is done, the algorithm checks, for the remaining oriented edges, whether two time series are conditionally independent or not given a set of parents of the arrow side node: in the first iteration the set of parents is of size one and then it gradually increases until either the edge between X^p and X^q is removed or all subsets of parents of X^q have been considered. Note that we make use of the same strategy as the one used in PC-stable [2], which consists in sorting time series according to their TCE scores and, when an independence is detected, removing all other occurrences of the time series. This leads to an order-independent procedure.

The following theorem states that the graph obtained by the above procedure is the true one.

Theorem 4. *Given the true ordering of the causal process, Algorithm 2 is complete.*

Proof. Similarly to PC, Algorithm 2 prunes all unnecessary edges by removing edges that are conditionally independent given a subset S. Thanks to the causal order, the possible subsets space is reduced. By removing all links that are conditionally independent, by causal Markov condition, adjacency faithfulness and causal sufficiency, we are left with links that are directly causal and which are oriented wrt causal ordering.

Self Causes. Finally, given the graph $\mathcal{G}$ inferred with the above procedure, one can verify for each node X^q in $\mathcal{G}$ if it is self causal by checking if there exists a $\gamma > 0$ such that for all t, $X_t^q \not\perp\!\!\!\perp X_{t-\gamma}^q \mid Par(X^q)$ in $\mathcal{G}$.

3.3   Complexity Analysis

Our proposed methods benefit from a smaller number of tests compared to constraint-based methods that infer the full temporal graph. In the worst case, the complexity of PC in a temporal graph is bounded by:

$$\frac{(d \cdot \tau)^2 (d \cdot \tau - 1)^{k-1}}{(k-1)!}$$

where k represents the maximal degree of any vertex and each operation consists in conducting significance test to a conditional independence measure. Algorithms more adapted to time series, such as PCMCI [15], use the notion of time to reduce the number of tests. In those cases, the complexity would be divided by 2 (if instantaneous relations are not taken into account). NBCB is inferring a summary graph, which limits the number of decisions that need to be taken. NBCB's complexity in the worst case (when all relations are instantaneous) is bounded by:

$$d^2.f(n,d) + \frac{d^2(d-1)^{k-1}}{(k-1)!}$$

where $f(n,d)$ is the complexity of the user-specific regression method.

4   Experiments

To illustrate the behavior of our method, we test it on several artificial and real datasets.

NBCB[3] and its pairwise version denoted pwNBCB are fitting a Gaussian Process with zero mean and squared exponential covariance function. The hyperparameters are automatically chosen by marginal likelihood optimization.

We compare NBCB with seven state-of-the-art methods: the constraint-based methods PCMCI[4] [15], where two variations of PCMCI are considered, varying

---

[3] Python code available at https://github.com/kassaad/causal_discovery_for_time _series.

[4] Python code available at https://github.com/jakobrunge/tigramite.

Table 1. Structures of simulated data.

Pair	Pair-sc	Fork	V-structure	Mediator	Diamond
X^1 → X^2	X^1 → X^2	X^2, X^1 → X^3	X^1, X^3 → X^2	X^1 → X^2 → X^3	X^1 → X^2, X^3 → X^4

the measure of independence between the mutual information for PCMCI-MI and the linear partial correlation for PCMCI-PC, and oCSE[5] [18]; the noise-based methods TiMINo[6] [12] with the linear time series model and VarLiNGAM[7] [7] where the regularization parameter in the adaptive Lasso is selected using BIC; the multivariate version of Granger Causality denoted GC[8] [5] and the Neural Network based method TCDF[9] [11] with default hyperparameters as introduced in the original paper. For all the methods, the best time lag is determined with the Akaike Information Criterion, the window size is set to $\tau = 5$ and the significant threshold for hypothesis testing to $\alpha = 0.05$.

In the different experimental settings, we compare the results wrt the *F1-score* denoted F1 of the orientations in the graph obtained without considering self causes, as it is treated differently depending on the methods.

4.1 Simulated Data

We first test our method on simulated data generated from five different causal structures (pair, fork, V-structure, mediator, diamond) presented in Table 1. We distinguish pairs when the time series are self caused (Pair-sc) or not (Pair). For each benchmark, we generate randomly 10 data sets with 1000 observations. The data generating process is the following: for all q, $X_0^q = 0$ and for all $t > 0$,

$$X_t^q = a_{t-1}^{qq} X_{t-1}^q + \sum_{\substack{(p,\gamma) \\ X_{t-\gamma}^p \in Par(X_t^q)}} a_{t-\gamma}^{pq} f(X_{t-\gamma}^p) + 0.1\xi_t^q,$$

where $\gamma \geq 0$, a_t^{jq} are random coefficients chosen uniformly in $\mathcal{U}([-1; -0.1] \cup [0.1; 1])$ for all $1 \leq j \leq d$, $\xi_t^q \sim \mathcal{N}(0, \sqrt{15})$ and f is a non linear function chosen at random uniformly between absolute value, tanh, sine, cosine. Two scenarios

---

[5] Python code available at https://github.com/kassaad/causal_discovery_for_time _series.

[6] R code available at http://web.math.ku.dk/~peters/code.html.

[7] Python code available at https://github.com/cdt15/lingam.

[8] Matlab code available at https://github.com/SacklerCentre/MVGC1.

[9] Python code available at https://github.com/M-Nauta/TCDF.

Table 2. Results obtained on the simulated data for the different structures with 1000 observations. We report the mean and the standard deviation of the F1 score. The best results are in bold.

	Pair	Pair-sc	V-struct	Fork	Mediator	Diamond
NBCB	0.7 ± 0.46	0.7 ± 0.46	0.67 ± 0.28	0.67 ± 0.38	0.66 ± 0.32	0.71 ± 0.16
pwNBCB	0.7 ± 0.46	0.7 ± 0.46	0.75 ± 0.18	0.67 ± 0.38	0.7 ± 0.30	0.83 ± 0.12
PCMCI-PC	0.57 ± 0.47	0.6 ± 0.49	0.61 ± 0.33	0.53 ± 0.39	0.75 ± 0.24	0.63 ± 0.26
PCMCI-MI	0.9 ± 0.16	0.8 ± 0.4	0.67 ± 0.37	0.78 ± 0.17	0.84 ± 0.09	0.82 ± 0.16
oCSE	$\mathbf{1.0 \pm 0.0}$	$\mathbf{1.0 \pm 0.0}$	$\mathbf{0.90 \pm 0.16}$	$\mathbf{0.8 \pm 0.12}$	$\mathbf{0.95 \pm 0.08}$	$\mathbf{0.88 \pm 0.09}$
TiMINo	0.57 ± 0.49	0.5 ± 0.5	0.65 ± 0.37	0.52 ± 0.44	0.80 ± 0.19	0.60 ± 0.25
VarLiNGAM	0.54 ± 0.49	0.5 ± 0.5	0.0 ± 0.0	0.0 ± 0.0	0.0 ± 0.0	0.03 ± 0.08
GC	0.67 ± 0.44	0.4 ± 0.49	0.37 ± 0.25	0.44 ± 0.38	0.83 ± 0.22	0.66 ± 0.26
TCDF	0.0 ± 0.0	0.1 ± 0.3	0.13 ± 0.26	0.26 ± 0.32	0.05 ± 0.15	0.16 ± 0.19

are considered: all the coefficients are random, or some coefficients are fixed to not be faithful to the true causal graph. Results are summarized in Table 2 for faithful data and in Table 3 for unfaithful data.

From Table 2, one can note that methods from the constraint-based family consistently outperform all other methods. However, our proposed algorithm is able to compete against pure constraint-based approaches. Specifically for the fork structure, the pairwise version outperforms all other methods, and for the others structures it performs better than most of the methods. When pwNBCB outperforms NBCB, one can expect that a larger sample size would improve the performance for NBCB, as the bivariate analysis is seen as a lower dimensional proxy of the full regression model. Furthermore, the similarity of results of our methods obtained by the F1 scores regarding to the structures illustrates the stability of our method. VarLINGAM performs particularly bad, but all the assumptions are violated in this design (Gaussian noise, non linearity). TCDF has also bad performances, whereas Granger Causality is surprisingly good, particularly for Mediator.

In Table 3 we consider two unfaithful datasets. The first one is a mediator, where $a^{13} = -a^{12}a^{23}$, without self cause, and all relations are instantaneous. Following [20], the second dataset is a linear unfaithful diamond without self causes, where we set the coefficient $a^{34} = -a^{12}a^{23}/a^{13}$ and all relations are instantaneous. From Table 3, we can see that PCMCI and oCSE perform poorly for unfaithful data, as expected. VarLINGAM has still bad results, again due to the simulation process. Our proposed algorithm comes out as one of the best algorithms in terms of performance, and there is an improvement with the full NBCB instead of its pairwise version.

Figure 3 provides an empirical illustration of the algorithmic complexity. We compare NBCB to oCSE, PCMCI-MI and TiMINo on four structures (v-structure, fork, mediator, diamond), sorted according to their number of nodes, their maximal out-degree and their maximal in-degree. The time is given in seconds. As one

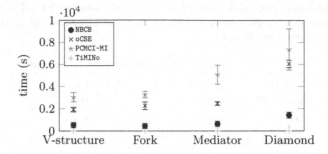

Fig. 3. Time computation (in second) for `NBCB`, `oCSE` and `PCMCI-MI` for four basic causal structures (V-structure, Fork, Mediator, Diamond). We report the mean and the standard deviation.

can note, `NBCB` is always faster than `oCSE` and `PCMCI-MI`, the difference being more important when the structure to be inferred is more complex. TiMINo is faster than NBCB, because the vector autoregressive model has been used instead of Gaussian process (there were not available in the online package) and the pruning in TiMINo relies on the noise-based approach, so less steps have to be considered. This illustrates again the trade-off between constraint-based and noise-based approaches.

4.2 Real Data

Three different real datasets are considered in this study. We detail the performance of each method in the following paragraphs, but the results are summarized in Table 4.

Temperature. This bivariate time series[10] of length 168 is about indoor X^{in} and outdoor X^{out} measurements. We expect that there is the following causal link: $X^{\mathrm{out}} \rightarrow X^{\mathrm{in}}$. VarLiNGAM wrongly infers no causal relation, Granger infers a bidirected arrow and TiMINo remains undecided. PCMCI-PC, PCMCI-MI, oCSE, NBCB and NBCBk correctly infer $X^{\mathrm{out}} \rightarrow X^{\mathrm{in}}$.

Diary. This dataset[11] provides10 years (from 09/2008 to 12/2018) of monthly prices for milk X^m, butter X^b and cheddar cheese X^c, so the three time series are of length 124. We expect that the price of milk is a common cause of the price of butter and the price of cheddar cheese: $X^b \leftarrow X^m \rightarrow X^c$. VarLiNGAM wrongly infers X^b as common cause of X^m and X^c, Granger wrongly infers $X^m \leftrightarrow X^b \rightarrow X^c \rightarrow X^m$ and TiMINo only infers one wrong causal relation $X^c \rightarrow X^m$. TCDF infers no causal relation. PCMCI-PC and PCMCI-MI wrongly infer the causal chain $X^c \rightarrow X^m \rightarrow X^b$. oCSE, NBCB and pwNBCB correctly infer the causal relations but also add a wrong causal $X^c \rightarrow X^b$.

---

[10] Data is available at https://webdav.tuebingen.mpg.de/cause-effect/.
[11] Data is available at http://future.aae.wisc.edu.

Table 3. Results obtained on the unfaithful simulated data for the different structures with 1000 observations. We report the mean and the standard deviation of the F1 score. The best results are in bold.

	unfaith. Mediator	unfaith. Diamond
NBCB	0.56 ± 0.26	$\mathbf{0.5 \pm 0.31}$
pwNBCB	0.46 ± 0.23	0.39 ± 0.22
PCMCI-PC	0.21 ± 0.21	0.19 ± 0.16
PCMCI-MI	0.05 ± 0.15	0.20 ± 0.22
oCSE	0.05 ± 0.15	0.08 ± 0.16
TiMINo	$\mathbf{0.64 \pm 0.08}$	0.49 ± 0.03
VarLiNGAM	0.0 ± 0.0	0.02 ± 0.06
GC	0.12 ± 0.27	0.14 ± 0.23
TCDF	0.4 ± 0.22	0.33 ± 0.17

Table 4. Results for real datasets. We report the mean and the standard deviation of the F1 score.

	Temperature	Diary	FMRI
NBCB	1	0.8	0.40 ± 0.21
pwNBCB	1	0.8	0.39 ± 0.21
PCMCI-PC	1	0.5	0.29 ± 0.19
PCMCI-MI	1	0.5	0.22 ± 0.18
oCSE	1	0.8	0.16 ± 0.20
TiMINo	0	0.0	0.32 ± 0.11
VarLiNGAM	0	0.0	0.49 ± 0.28
GC	0.66	0.33	0.24 ± 0.18
TCDF	0	0.0	0.07 ± 0.13

FMRI. The last real-world dataset benchmark is about FMRI[12] (Functional Magnetic Resonance Imaging) that contains BOLD (Blood-oxygen-level dependent) datasets[16] for 28 different underlying brain networks. It measures the neural activity of different regions of interest in the brain based on the change of blood flow. There are 50 regions in total, each with its own associated time series. Since not all existing methods can handle 50 time series, datasets with more than 10 time series are excluded. In total we are left with 26 datasets containing between 5 and 10 brain regions. NBCB and VarLINGAM clearly outperforms other methods. All other methods are comparable, except TCDF which performs very poorly. Interestingly, PCMCI-PC performs better than PCMCI-MI,

[12] Original data is available at https://www.fmrib.ox.ac.uk/datasets/netsim/index.html, a preprocessed version is available at https://github.com/M-Nauta/TCDF/tree/master/data/fMRI.

and VarLINGAM outperforms TiMINo which suggests the existence of linear causal relations.

5 Conclusion

We have addressed in this study the problem of learning a summary causal graph on time series without being restricted to the Markov equivalent class even in the case of instantaneous relations. To do so, we followed a hybrid strategy. First we used a noise-based method to find the causal ordering between the time series under the assumption of additive noise models. Second, we used a constraint-based method to prune unnecessary parents and therefore ending up with an oriented causal graph. The second step heavily relies on a new temporal causation entropy measure that generalizes the causation entropy by removing the restriction of one time lag. Experiments conducted on different benchmark datasets and involving previous state-of-the-art proposals showed that the algorithm we have introduced outperforms previous proposals. In particular, we have illustrated and compared the behavior of our algorithm robustness wrt to different causal structures which yielded good results over all datasets, particularly on real ones.

In the future, we would like to test the method on large datasets, increasing both the number of time series d and the number of timepoints n. In particular, it would be interesting to study the quality of estimation in the regime $d \gg n$.

Acknowledgements. This work has been partially supported by MIAI@Grenoble Alpes (ANR-19-P3IA-0003).

References

1. Assaad, C.K., Devijver, E., Gaussier, E., Ait-Bachir, A.: Scaling causal inference in additive noise models. In: Le, T.D., Li, J., Zhang, K., Cui, E.K.P., Hyvärinen, A. (eds.) Proceedings of Machine Learning Research. Proceedings of Machine Learning Research, Anchorage, Alaska, USA, vol. 104, pp. 22–33. PMLR, 05 August 2019
2. Colombo, D., Maathuis, M.H.: Order-independent constraint-based causal structure learning. J. Mach. Learn. Res. **15**(116), 3921–3962 (2014)
3. Frenzel, S., Pompe, B.: Partial mutual information for coupling analysis of multivariate time series. Phys. Rev. Lett. **99**, 204101 (2007)
4. Glymour, C., Zhang, K., Spirtes, P.: Review of causal discovery methods based on graphical models. Front. Genet. **10**, 524 (2019)
5. Granger, C.W.J.: Time series analysis, cointegration, and applications. Am. Econ. Rev. **94**(3), 421–425 (2004)
6. Hoyer, P.O., Janzing, D., Mooij, J.M., Peters, J., Schölkopf, B.: Nonlinear causal discovery with additive noise models. In: Advances in Neural Information Processing Systems 21. ACM Press (2009)
7. Hyvärinen, A., Shimizu, S., Hoyer, P.O.: Causal modelling combining instantaneous and lagged effects: an identifiable model based on non-gaussianity. In: Proceedings of the 25th International Conference on Machine Learning, ICML 2008, pp. 424–431. ACM, New York (2008)

8. Malinsky, D., Danks, D.: Causal discovery algorithms: a practical guide. Philos. Compass **13**(1) (2018)

9. Mooij, J., Janzing, D., Peters, J., Schölkopf, B.: Regression by dependence minimization and its application to causal inference in additive noise models. In: Proceedings of the 26th International Conference on Machine Learning, pp. 745–752. Max-Planck-Gesellschaft, ACM Press, New York (2009)

10. Mooij, J.M., Peters, J., Janzing, D., Zscheischler, J., Schölkopf, B.: Distinguishing cause from effect using observational data: methods and benchmarks. J. Mach. Learn. Res. **17**(1), 1103–1204 (2016)

11. Nauta, M., Bucur, D., Seifert, C.: Causal discovery with attention-based convolutional neural networks. Mach. Learn. Knowl. Extr. **1**(1), 312–340 (2019)

12. Peters, J., Janzing, D., Schölkopf, B.: Causal inference on time series using restricted structural equation models. In: Advances in Neural Information Processing 26, pp. 154–162 (2013)

13. Peters, J., Mooij, J.M., Janzing, D., Schölkopf, B.: Identifiability of causal graphs using functional models. In: Proceedings of the Twenty-Seventh Conference on Uncertainty in Artificial Intelligence, UAI 2011, Arlington, Virginia, USA, pp. 589–598. AUAI Press (2011)

14. Ramsey, J., Spirtes, P., Zhang, J.: Adjacency-faithfulness and conservative causal inference. In: Proceedings of the Twenty-Second Conference on Uncertainty in Artificial Intelligence, UAI 2006, Arlington, Virginia, USA, pp. 401–408. AUAI Press (2006)

15. Runge, J., Nowack, P., Kretschmer, M., Flaxman, S., Sejdinovic, D.: Detecting and quantifying causal associations in large nonlinear time series datasets. Sci. Adv. **5**(11) (2019)

16. Smith, S.M., et al.: Network modelling methods for FMRI. Neuroimage **54**, 875–891 (2011)

17. Spirtes, P., Glymour, C., Scheines, R.: Causation, Prediction, and Search, 2nd edn. MIT Press, Cambridge (2001)

18. Sun, J., Taylor, D., Bollt, E.: Causal network inference by optimal causation entropy. SIAM J. Appl. Dyn. Syst. **14**(1), 73–106 (2015)

19. Verma, T., Pearl, J.: Equivalence and synthesis of causal models. In: Proceedings of the Sixth Annual Conference on Uncertainty in Artificial Intelligence, UAI 1990, pp. 255–270. Elsevier Science Inc., New York (1991)

20. Zhalama, Zhang, J., Mayer, W.: Weakening faithfulness: some heuristic causal discovery algorithms. Int. J. Data Sci. Anal. **3**, 93–104 (2016)

Estimating the Electrical Power Output of Industrial Devices with End-to-End Time-Series Classification in the Presence of Label Noise

Andrea Castellani[1]([✉])(iD), Sebastian Schmitt[2](iD), and Barbara Hammer[1](iD)

[1] Bielefeld University, Bielefeld, Germany
{acastellani,bhammer}@techfak.uni-bielefeld.de
[2] Honda Research Institute Europe, Offenbach, Germany
sebastian.schmitt@honda-ri.de

Abstract. In complex industrial settings, it is common practice to monitor the operation of machines in order to detect undesired states, adjust maintenance schedules, optimize system performance or collect usage statistics of individual machines. In this work, we focus on estimating the power output of a Combined Heat and Power (CHP) machine of a medium-sized company facility by analyzing the total facility power consumption. We formulate the problem as a time-series classification problem, where the class label represents the CHP power output. As the facility is fully instrumented and sensor measurements from the CHP are available, we generate the training labels in an automated fashion from the CHP sensor readings. However, sensor failures result in mislabeled training data samples which are hard to detect and remove from the dataset. Therefore, we propose a novel multi-task deep learning approach that jointly trains a classifier and an autoencoder with a shared embedding representation. The proposed approach targets to gradually correct the mislabelled data samples during training in a self-supervised fashion, without any prior assumption on the amount of label noise. We benchmark our approach on several time-series classification datasets and find it to be comparable and sometimes better than state-of-the-art methods. On the real-world use-case of predicting the CHP power output, we thoroughly evaluate the architectural design choices and show that the final architecture considerably increases the robustness of the learning process and consistently beats other recent state-of-the-art algorithms in the presence of unstructured as well as structured label noise.

Keywords: Time-series · Deep learning · Label noise · Self-supervision · Non-intrusive load monitoring · Time-series classification

Electronic supplementary material The online version of this chapter (https://doi.org/10.1007/978-3-030-86486-6_29) contains supplementary material, which is available to authorized users.

1   Introduction

It is common to monitor multiple machines in complex industrial settings for many diverse reasons, such as to detect undesired operational states, adjust maintenance schedules or optimize system performance. In situations where the installation of many sensors for individual devices is not feasible due to cost or technical reasons, Non-Intrusive Load Monitoring (NILM) [18] is able to identify the utilization of individual machines based on the analysis of cumulative electrical load profiles. The problem of the generation of labelled training data sets is a cornerstone of data-driven approaches to NILM. In this context, industry relies mostly on manually annotated data [11] and less often the training labels can be automatically generated from the sensors [13], which is often unreliable because of sensor failure and human misinterpretation. Data cleaning techniques are often hard to implement [42] which unavoidably leads to the presence of wrongly annotated instances in automatically generated datasets i.e. *label noise* [12]. Many machine learning methods, and in particular deep neural networks, are able to overfit training data with noisy labels [48], thus it is challenging to apply data-driven approaches successfully in complex industrial settings.

We consider a medium-sized company facility and target the problem of estimating the electrical power output of a Combined Heat and Power (CHP) machine by only analyzing the facility electrical power consumption. The electrical power output of the CHP is sufficient to supply a substantial share of the total electricity demand of the facility. Therefore, knowing the CHP's electrical power output is very helpful for distributing the electrical energy in the facility, for example when scheduling the charging of electrical vehicles (EVs) or reducing total peak-load [28]. We propose a data-driven deep learning-based approach to this problem, which is modelled as time-series classification challenge in the presence of label noise, where the class label of each time series represents the estimated CHP power output level. As the facility is fully instrumented, and sensor measurements from the CHP are available, we generate the training labels in an automated fashion from the CHP sensor readings. However, these sensors fail from time to time which resulting wrong labels. To tackle this problem, we propose a novel multi-task deep learning approach named Self-Re-Labeling with Embedding Analysis (SREA), which targets the detection and re-labeling of wrongly labeled instances in a self-supervised training fashion.

In the following, after a formal introduction to SREA, we empirically validate it with several benchmarks data sets for time-series classification and compare against state-of-the-art (SotA) algorithms. In order to evaluate the performance, we create a training data set from clean sensor readings and we corrupt it by introducing three types of artificial noise in a controlled fashion. After that, we apply the proposed method to a real-world use-case including real sensor failures and show that those are properly detected and corrected by our algorithm. Finally, we perform an extensive ablation study to investigate the sensitivity of the SREA to its hyper-parameters.

2   Related Work

Deep learning based techniques constitute a promising approach to solve the NILM problem [30, 34]. Since the requirement of large annotated data sets is very challenging, the problem is often addressed as a semi-supervised learning task [4, 20, 46], where only a part of the data is correctly labeled, and the other is left without any label. Here we take a different stance to the problem of energy disaggregation: we assume that labels for all data are given, but not all labels are correct, as is often the case in complex sensor networks [12].

Learning noisy labels is an important topic in machine learning research [16, 38]. Several approaches are based on the fact that deep neural networks tend to first learn clean data statistics during early stages of training [2, 36, 48]. Methods can be based on a loss function which is robust to label noise [40, 50], or they can introduce an explicit [5] or implicit [35] regularization term. Another adaptation of the loss function is achieved by using different bootstrapping methods based on the predicted class labels [16, 33, 41]. Some other common approaches are based on labeling samples with smaller loss as clean ones [1, 15, 23, 27], and fitting a two-component mixture model on the loss distribution, or using cross-validation techniques [7] in order to separate clean and mislabeled samples. Several existing methods [15, 29, 39] require knowledge about the percentage of noisy samples, or estimate this quantity based on a given. The suitability of these approaches is unclear for real-world applications. Since our proposed method, SREA, targets the correction of mislabeled training samples based on their embedding representation, we do not rely on specific assumptions on the amount of label noise in advance. Our modeling approach is based on the observation that self-supervision has proven to provide an effective representation for downstream tasks without requiring labels [17, 19], this leading to an improvement of performance in the main supervised task [22].

Most applications of models which deal with label noise come from the domain of image data, where noise can be induced by e.g. crowd-working annotations [16, 25]. Some approaches consider label noise in other domains such as human activity detection [3], sound event classification [10], or malware detection [14]. An attempt to analyze the effect of noisy data on applications for real-world data sets is made in [44], but the authors do not compare to the SotA. Up to the authors knowledge, we are the first to report a detailed evaluation of time-series classification in the presence of label noise, which is a crucial data characteristics in the domain of NILM.

3   CHP Electrical Power Output Estimation

The CHP is a complex industrial machinery which burns natural gas in order to produce heat and electrical power. It is controlled by an algorithm where only some aspects are known, so its behavior is mostly unclear and not well predictable. It is known that important control signals are the ambient outside temperature T_{amb}, the internal water temperature T_{water}, the generated electrical power output P_{CHP}, and the total electricity demand of the facility P_{tot}.

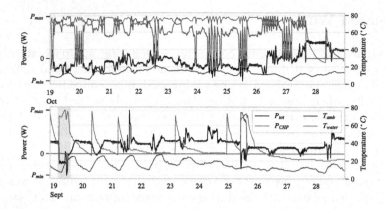

Fig. 1. Sensory data from the CHP and total electrical power. Upper: normal operation. Lower: an example of P_{CHP} sensor malfunction is highlighted in yellow.

Figure 1 shows examples of recorded data from the CHP as well as P_{tot} of the facility. T_{amb} has a known strong influence as the CHP is off in the summer period and more or less continuously on in winter and cold periods. In the transition seasons (spring and fall), the CHP sometimes turns on (night of 25th Sept.), sometimes just heats up its internal water (nights of 20st, 21st, and 23rd Sept.), or exhibits a fast switching behavior (e.g. 20st, and 27th Oct.). Even though the CHP usually turns on for a couple of hours (e.g. 25th Sept.) at rare instances it just turns on for a very short time (e.g. 28th Oct.).

Due to the complicated operational pattern, it is already hard to make a detector for the CHP operational state even with full access to the measurement data. Additionally, the sensors measuring the CHP output power are prone to failure which can be observed in the yellow highlighted area in the bottom side of Fig. 1. During that 10-hour period, the CHP did produce electrical power, even though the sensor reading does not indicate this. The total electrical power drawn from the grid, P_{tot}, provides a much more stable measurement signal. The signature of the CHP is clearly visible in the total power signal and we propose to estimate P_{CHP} from this signal, but many variables also affects the total load, e.g. PV system, changing workloads, etc.

We focus on estimating the power output but where the estimate power value should only be accurate within a certain range. Thus, the problem is formulated as time-series classification instead of regression, where each class represents a certain range of output values. The class labels are calculated directly from the P_{CHP} sensor measurement as the mean power output of a fixed-length sliding window. Due to frequent sensor malfunctioning, as displayed in Fig. 1, the resulting classification problem is subject to label noise.

4   Self-Re-Labeling with Embedding Analysis (SREA)

Architecture and Loss Function. In this work, column vectors are denoted in bold (e.g. $\boldsymbol{x}$). As described in Sect. 3, we treat the challenge to model time

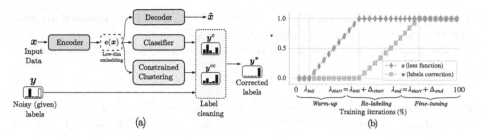

Fig. 2. SREA processing architecture (a) and the dynamics of the parameters α and w during the training epochs (b).

series data as a classification problem to predict averaged characteristics of the process using windowing techniques. Hence, we deal with a supervised k-class classification problem setting with a dataset of n training examples $\mathcal{D} = \{(\boldsymbol{x}_i, \boldsymbol{y}_i), i = 1, ..., n\}$ with $\boldsymbol{y}_i \in \{0, 1\}^k$ being the one-hot encoding label for sample $\boldsymbol{x}_i$. Thereby, **label noise** is present, i.e. we expect that $\boldsymbol{y}_i$ is wrong for a substantial (and possibly skewed) amount of instances in the training set.

The overall processing architecture of the proposed approach is shown in Fig. 2(a). The *autoencoder* (f_{ae}), represented by the *encoder* (e) and the *decoder*, provides a strong surrogate supervisory signal for feature learning [22] that is not affected by the label noise. Two additional components, a *classifier* network (f_c) and a *constrained clustering* module (f_{cc}) are introduced, which independently propose class labels as output. Each of the three processing pipelines share the same embedding representation, created by the encoder, and each output module is associated with one separate contribution to the total loss function.

For the autoencoder, a typical reconstruction loss is utilized:

$$\mathcal{L}_{ae} = \frac{1}{n} \sum_{i=1}^{n} (\hat{\boldsymbol{x}}_i - \boldsymbol{x}_i)^2, \tag{1}$$

where $\hat{\boldsymbol{x}}_i$ is the output of the autoencoder given the input $\boldsymbol{x}_i$.

Cross entropy is used as loss function for the classification network output,

$$\mathcal{L}_c = -\frac{1}{n} \sum_{i=1}^{n} \boldsymbol{y}_i^T \cdot \log(\boldsymbol{p}_i^c), \tag{2}$$

where $\boldsymbol{p}_i^c$ are the k-class softmax probabilities produced by the model for the training sample i, i.e. $\boldsymbol{p}_i^c = \text{softmax}(f_c(\boldsymbol{x}_i))$.

For the constraint clustering loss, we first initialize the cluster center $\boldsymbol{C} \in \mathbb{R}^{d \times k}$ in the d-dimensional embedding space, with the k-means clustering of the training samples. Then, inspired by the good results achieved in [47], we constrain the embedding space to have small intra-class and large inter-class distances, by iteratively adapting $\boldsymbol{C}$. The resulting clustering loss is given by:

$$\mathcal{L}_{cc} = \frac{1}{n} \sum_{i=1}^{n} \left[\underbrace{\|e(\boldsymbol{x}_i) - \boldsymbol{C}_{\boldsymbol{y}_i}\|_2^2}_{\text{intra-class}} + \underbrace{\log \sum_{j=1}^{k} \exp\left(-\|e(\boldsymbol{x}_i) - \boldsymbol{C}_j\|_2\right)}_{\text{inter-class}} \right] + \ell_{reg}. \tag{3}$$

The entropy regularization $\ell_{reg} = -\sum_i^k \min_{i \neq j} \log \|C_i - C_j\|_2$ is aimed to create well separated embedding for different classes [37].

The final total loss function is given by the sum of those contributions,

$$\mathcal{L} = \mathcal{L}_{ae} + \alpha \left(\mathcal{L}_c + \mathcal{L}_{cc} + \mathcal{L}_\rho\right), \tag{4}$$

where we introduced the dynamic parameter $0 \leq \alpha \leq 1$ which changes during the training (explained below). We also add the regularization loss $\mathcal{L}_\rho$ to prevent the assignment of all labels to a single class, $\mathcal{L}_\rho = \sum_{j=1}^k h_j \cdot \log \frac{h_j}{p_j^\rho}$, where h_j denotes the prior probability distribution for class j, which assume to be uniformly distributed to $1/k$. The term p_j^ρ is the mean softmax probability of the model for class j across all samples in dataset which we approximate using mini-batches as done in previous works [1].

Re-Labeling Strategy. We do not compute the total loss function in Eq. (4) with the given (noisy) label y_i. But, we estimate the true label for each data sample y_i^* by taking the weighted average of the given training label y_i, and the pseudo-labels proposed by the classifier y_i^c and the constraint clustering y_i^{cc}.

In order to increase robustness of the labels proposed by the classifier (y_i^c), we take for each data sample the exponentially averaged probabilities during the last five training epochs, with the weighting factor $\tau_t \sim e^{\frac{t-5}{2}}$:

$$y_i^c = \sum_{\text{last 5 epochs } t} \tau_t \, [p_i^c]_t. \tag{5}$$

The label from constraint clustering (y_i^{cc}) is determined by the distances of the samples to the cluster centers in the embedding space:

$$y_i^{cc} = \text{softmin}_j \left(\|e(x_i) - C_j\|_2\right) \tag{6}$$

Then, the corrected label y_i^* (in one hot-encoding) is produced by selecting the class corresponding to the maximum entry:

$$y_i^* = \text{argmax}\left[(1-w)\, y_i + w\, (y_i^c + y_i^{cc})\right] \tag{7}$$

where the dynamic weighting factor $0 \leq w \leq 1$ is function of the training epoch t and to be discussed below when explaining the training dynamics.

Training Dynamics. A key aspect of our proposed approach is to dynamically change the loss function, as well as the label correction mechanism, during the training. This is achieved by changing the parameters α (loss function) and w (label correction mechanism) as depicted in Fig. 2(b). The training dynamics is completely defined by the three hyper-parameters λ_{init}, Δ_{start} and Δ_{end}.

Initially, we start with $\alpha = 0$ and $w = 0$, and only train the autoencoder from epoch $t = 0$ to epoch $t = \lambda_{init}$. At the training epoch $t = \lambda_{init}$, α is ramped up linearly until it reaches $\alpha = 1$ in training epoch $t = \lambda_{start} = \lambda_{init} + \Delta_{start}$.

Algorithm 1: SREA: Self-Re-Labeling with Embedding Analysis.

Require: Data $\{(\boldsymbol{x}_i, \boldsymbol{y}_i)\}_n$, autoencoder f_{ae}, classifier f_c, constraint clustering f_{cc}, hyper-parameters: $\lambda_{init}, \Delta_{start}, \Delta_{end}$.

1 Init hyper-parameter ramp-up functions w_t and α_t *// see Fig. 2(b)*
2 **for** *training epoch* $t = 0$ **to** t_{end} **do**
3   Fetch mini-batch data $\{(\boldsymbol{x}_i, \boldsymbol{y}_i)\}_b$ at current epoch t
4   **for** $i = 1$ **to** b **do**
5   **if** $t == \lambda_{init}$ **then**
6            $f_{cc} \leftarrow k\text{-means}(e(\boldsymbol{x}_i))$ *// Initialize constraint clustering*
7   **end**
8         $\hat{\boldsymbol{x}}_i = f_{ae}(\boldsymbol{x}_i)$ *// Auto-encoder forward pass*
9         $\boldsymbol{y}_i^c \leftarrow$ Eq.(5)   *// Classifier forward pass*
10        $\boldsymbol{y}_i^{cc} \leftarrow$ Eq.(6)   *// Constraint clustering output*
11   Adjust $w \leftarrow w_t, \alpha \leftarrow \alpha_t$ *// Label correction and loss parameters*
12        $\boldsymbol{y}_i^* \leftarrow$ Eq.(7)   *// Re-labeling*
13        $\mathcal{L} \leftarrow$ Eq.(4)   *// Evaluate loss function*
14   Update f_{ae}, f_c, f_{cc} by SGD on $\mathcal{L}$
15   **end**
16 **end**

The purpose of this first *warm-up* period is an unsupervised initialization of the embedding space with slowly turning on the supervision of the given labels. The dominant structure of the clean labels is learned, as neural networks tend to learn the true labels, rather than overfit to the noisy ones, at early training stages [2,48]. Then, we also increase w linearly from zero to one, between epochs $t = \lambda_{start}$ to $t = \lambda_{start} + \Delta_{end} = \lambda_{end}$, thereby turning on the label correction mechanism (*re-labeling*). After training epoch $t = \lambda_{end}$ until the rest of the training, we keep $\alpha = 1$ and $w = 1$ which means we are fully self-supervised where the given training labels do not enter directly anymore (*fine-tuning*). We summarize the SREA and display the pseudo-code in Algorithm 1.

5 Experimental Setup

Label Noise. True labels are corrupted by a *label transition matrix* T [38], where T_{ij} is the probability of the label i being flipped into label j. For all the experiments, we corrupt the labels with *symmetric* (unstructured) and *asymmetric* (structured) noise with noise ratio $\epsilon \in [0, 1]$. For symmetric noise, a true label is randomly assigned to other labels with equal probability, i.e. $T_{ii} = 1 - \epsilon$ and $T_{ij} = \frac{\epsilon}{k-1}$ $(i \neq j)$, with k the number of classes. For asymmetric noise, a true label is mislabelled by shifting it by one, i.e. $T_{ii} = 1 - \epsilon$ and $T_{(j+1 \bmod k)j} = \epsilon$ $(i \neq j)$. For the estimation of the CHP power, we also analyze another kind of structured noise which we call *flip* noise, where a true label is only flipped to zero, i.e. $T_{ii} = 1 - \epsilon$ and $T_{i0} = \epsilon$. This mimics sensor failures, where a broken sensor produces a constant output regardless of the real value. Note that learning with structured noise is much harder than with unstructured noise [12].

Network Architecture. Since SREA is model agnostic, we use CNNs in the experiments, as these are currently the SotA deep learning network topology for time-series classification [9, 43]. The encoder and decoder have a symmetric structure with 4 convolutional blocks. Each block is composed by a 1D-conv layer followed by batch normalization [21], a ReLU activation and a dropout layer with probability 0.2. The dimension of the shared embedding space is 32. For the classifier we use a fully connected network with 128 hidden units and #classes outputs. We use the Adam optimizer [26] with an initial learning rate of 0.01 for 100 epochs. Such high value for the initial learning rate helps to avoid overfitting of noisy data in the early stages of training [2, 48]. We halve the learning rate every 20% of training (20 epochs). In the experiments, we assume to not have access to any clean data, thus it is not possible to use a validation set, and the models are trained without early stopping. The SREA hyper-parameters $\lambda_{init} = 0$, $\Delta_{start} = 25$ and $\Delta_{end} = 30$ are used if not specified otherwise. Further implementation details are reported in the supplementary material [6], including references to the availability of code and data sets.

Comparative Methods. In order to make a fair comparison, we use the same neural network topology throughout all the experiments. A baseline method, which does not take in accout any label noise correction criteria, is a CNN classifier [43] trained with cross-entropy loss function of Eq. (2), which we refer to as *CE*. We compare to *MixUp* [49], which is a data augmentation technique that exhibits strong robustness to label noise. In *MixUp-BMM* [1] a two-component beta-mixture model is fitted to the loss distribution and training with bootstrapping loss is implemented. *SIGUA* [15] implements stochastic gradient ascent on likely mislabeled data, thereby trying to reduce the effect of noisy labels. Finally, in *Co-teaching* [39] two networks are simultaneously trained which inform each other about which training examples to keep. The algorithms *SIGUA* and *Co-teaching* assume that the noise level ϵ is known. In our experiments, we use the true value of ϵ for those approaches, in order to create an upper-bound of their performance. All the hyper-parameters of the investigated algorithm are set to their default and recommended values.

Implementation Details. For the problem of estimating CHP power output, the raw data consists of 78 days of measurement with a sampling rate of 1 sample/minute. As preprocessing, we do a re-sampling to 6 samples/hour. The CHP should have a minimal on-time of one hour, in order to avoid too rapid switching which would damage the machine. However, during normal operation, the CHP is controlled in a way that on- and off-time periods are around 4 to 8 h. Due to these time-scales, we are interested in the power output on a scale of 6 h, which means we use a sliding window with a size of 6 h (36 samples) and a stride of 10 min (1 sample). Therefore, the preprocessing of the three input variables P_{tot}, T_{water}, and T_{amb} lead to $\mathbb{R}^{(36 \times 3)}$-dimensional data samples. For generating the labels, we use 5 power output levels, linearly spaced from 0 to $P_{CHP,max}$, and correspondingly to a five-dimensional one-hot encoded label vector $\boldsymbol{y}_i \in \mathbb{R}^5$. For every dataset investigated, we normalize the dataset to have zero mean and unit standard deviation, and we randomly divide the total available data in train-set and test-set with a ratio 80:20.

Evaluation Measures. To evaluate the performance, we report the *averaged $\mathcal{F}_1$-score* on the test-set, where the well-known F_1-scores are calculated for each class separately and then averaged via arithmetic mean, $\mathcal{F}_1 = \frac{1}{k}\sum_{j=1}^{k} F_{1,j}$. This formulation of the F_1-score results in a larger penalization when the model do not perform well in the minority class, in cases with class imbalance. Other metrics, such as the accuracy, show qualitatively similar results and are therefore not reported in this manuscript. In order to get performance statistics of the methods, all the experiments have been repeated 10 times with different random initialization. The non-parametric statistical *Mann-Whitney U Test* [32] is used to compare the SREA against the SotA algorithms.

6   Results and Discussion

Benchmarks Datasets. We evaluate the proposed SREA on publicly available time-series classification datasets from UCR repository [8]. We randomly choose 10 datasets with different size, length, number of classes and dimensions in order to try to avoid bias in the data. A summary of the datasets is given in Table 1.

We show a representative selection of all comparisons of the results in Table 2. Without any label noise, CE is expected to provide very good results. We observe that our SREA gives similar or better $\mathcal{F}_1$ scores than the CE method on 9 out of 10 datasets without label noise. Considering all algorithms and datasets, with symmetric noise we achieve statistically significantly better scores in 62, similar scores in 105, and worse scores in 33 experiments out of a total of 200 experiments[1]. For the more challenging case of asymmetric noise, the SREA-results are 86 times significantly better, 97 times equal, and 17 times worse than SotA algorithms.

Table 1. UCR Single-variate and Multi-variate dataset description.

Dataset	Size	Length	#classes	#dim
ArrowHead[†]	211	251	3	1
CBF	930	128	3	1
Epilepsy	275	206	4	3
FaceFour[†]	112	350	4	1
MelbourneP	3633	24	10	1
NATOPS	360	51	6	24
OSULeaf[†]	442	427	6	1
Plane	210	144	7	1
Symbols[†]	1020	398	6	1
Trace[†]	200	275	4	1

† results reported in the supplementary material [6].

CHP Power Estimation. Table 3 shows the results of the estimation of the CHP output power level. Without labels noise, the proposed approach has comparable performance to CE, with an average $\mathcal{F}_1$-score of 0.979. This implies that

---

[1] Number of experiments: 10 datasets × 4 noise levels × 5 algorithms = 200. Each experiment consists of 10 independent runs.

Table 2. F_1 test scores on UCR datasets. The best results per noise level are underlined. In parenthesis the results of a Mann–Whitney U test with $\alpha = 0.05$ of SREA against the other approaches: SREA F_1 is significantly higher (+), lower (−) or not significant (≈).

Dataset	Noise	%	CE	MixUp	M-BMM	SIGUA	Co-teach	SREA
CBF	–	0	1.000 (+)	0.970 (+)	0.886 (+)	1.000 (+)	0.997 (+)	<u>1.000</u>
	Symm	15	0.943 (+)	0.923 (+)	0.941 (+)	0.976 (+)	0.923 (+)	<u>1.000</u>
		30	0.780 (+)	0.799 (+)	0.932 (+)	0.923 (+)	0.833 (+)	<u>0.998</u>
	Asymm	10	0.973 (+)	0.956 (+)	0.920 (+)	0.989 (+)	0.963 (+)	<u>1.000</u>
		20	0.905 (+)	0.897 (+)	0.949 (+)	0.980 (+)	0.900 (+)	<u>1.000</u>
Epilepsy	–	0	<u>0.974</u> (≈)	0.955 (+)	0.926 (+)	<u>0.978</u> (≈)	0.971 (+)	0.973
	Symm	15	<u>0.890</u> (≈)	<u>0.913</u> (≈)	<u>0.899</u> (≈)	<u>0.884</u> (≈)	<u>0.861</u> (≈)	0.861
		30	0.784 (−)	0.823 (−)	0.805 (−)	<u>0.741</u> (≈)	<u>0.744</u> (≈)	0.708
	Asymm	10	<u>0.919</u> (≈)	<u>0.930</u> (≈)	<u>0.847</u> (≈)	<u>0.905</u> (≈)	<u>0.919</u> (≈)	0.888
		20	<u>0.861</u> (≈)	0.894 (−)	0.891 (−)	<u>0.826</u> (≈)	<u>0.863</u> (≈)	0.825
Melbourne	–	0	<u>0.923</u> (≈)	0.879 (+)	0.773 (+)	<u>0.918</u> (≈)	<u>0.913</u> (≈)	0.911
	Symm	15	0.869 (+)	0.870 (+)	0.856 (+)	<u>0.883</u> (≈)	<u>0.886</u> (≈)	0.883
		30	0.826 (+)	<u>0.858</u> (≈)	<u>0.870</u> (≈)	<u>0.855</u> (≈)	0.876 (−)	0.862
	Asymm	10	0.898 (+)	0.877 (+)	0.860 (+)	0.899 (+)	0.897 (+)	<u>0.911</u>
		20	0.865 (+)	0.861 (+)	0.851 (+)	0.858 (+)	<u>0.893</u> (≈)	0.903
NATOPS	–	0	<u>0.858</u> (≈)	<u>0.801</u> (≈)	0.711 (+)	<u>0.848</u> (≈)	<u>0.835</u> (≈)	0.866
	Symm	15	<u>0.779</u> (≈)	0.718 (+)	0.702 (+)	<u>0.754</u> (≈)	<u>0.761</u> (≈)	0.796
		30	<u>0.587</u> (≈)	0.580 (+)	0.602 (+)	0.593 (+)	0.673 (+)	<u>0.670</u>
	Asymm	10	0.798 (+)	<u>0.822</u> (≈)	0.756 (+)	0.764 (+)	0.790 (+)	<u>0.829</u>
		20	<u>0.703</u> (≈)	<u>0.763</u> (≈)	<u>0.762</u> (≈)	<u>0.698</u> (≈)	<u>0.733</u> (≈)	0.762
Plane	–	0	<u>0.995</u> (≈)	0.962 (+)	0.577 (+)	0.981 (+)	<u>0.990</u> (≈)	0.998
	Symm	15	0.930 (+)	0.953 (+)	0.873 (+)	<u>0.971</u> (≈)	<u>0.981</u> (≈)	0.983
		30	0.887 (+)	0.902 (+)	<u>0.943</u> (≈)	0.862 (+)	<u>0.941</u> (≈)	0.944
	Asymm	10	<u>0.981</u> (≈)	<u>0.986</u> (≈)	0.648 (+)	<u>0.986</u> (≈)	<u>0.990</u> (≈)	0.976
		20	<u>0.952</u> (≈)	<u>0.923</u> (≈)	0.751 (+)	<u>0.976</u> (≈)	<u>0.990</u> (≈)	0.966

we successfully solve the CHP power estimating problem by analyzing the total load with an error rate less than 2%. When the training labels are corrupted with low level of symmetric label noise, $\epsilon \leq 0.4$, SREA consistently outperforms the other algorithms. With higher level of symmetric noise we achieve a comparable performance to the other algorithms. Under the presence of asymmetric label noise, SREA shows a performance comparable to the other SotA algorithms. Only for a highly unrealistic asymmetric noise level of 40%, the performance is significantly worse than the SotA. This indicates that, during the warm-up and relabelling phase, the network is not able to learn the true labels but also learns the wrong labels induced by the structured noise. During the fine-tuning phase, the feedback of the wrongly labeled instances is amplified and the wrong labels a reinforced. For flip noise, which reflects sensors failures, SREA retains a high performance and outperforms all other SotA algorithms up to noise levels of 40%. For low noise levels up to 20%, SREA has similar performance to Co-teaching, but without the need to know the amount of noise.

Table 3. $\mathcal{F}_1$ test scores of the CHP power estimation. Same notation as Table 2.

Noise	%	CE	MixUp	M-BMM	SIGUA	Co-teach	SREA
–	0	0.980 ($\approx$)	0.957 (+)	0.882 (+)	0.979 ($\approx$)	0.974 ($\approx$)	0.979
Symmetric	15	0.931 (+)	0.934 (+)	0.903 (+)	0.954 ($\approx$)	0.950 ($\approx$)	0.960
	30	0.856 (+)	0.910 (+)	0.897 (+)	0.912 (+)	0.920 (+)	0.938
	45	0.763 (+)	0.883 (+)	0.895 (+)	0.867 (+)	0.886 (+)	0.918
	60	0.661 (+)	0.761 ($\approx$)	0.692 (+)	0.817 ($\approx$)	0.839 ($\approx$)	0.800
Asymmetric	10	0.954 ($\approx$)	0.945 (+)	0.893 (+)	0.959 ($\approx$)	0.964 ($\approx$)	0.961
	20	0.924 (+)	0.925 (+)	0.899 (+)	0.935 ($\approx$)	0.938 ($\approx$)	0.946
	30	0.895 ($\approx$)	0.909 ($\approx$)	0.873 (+)	0.916 ($\approx$)	0.923 ($\approx$)	0.919
	40	0.807 (−)	0.848 (−)	0.784 (−)	0.836 (−)	0.876 (−)	0.287
Flip	10	0.970 ($\approx$)	0.950 (+)	0.860 (+)	0.965 (+)	0.973 ($\approx$)	0.971
	20	0.942 (+)	0.942 (+)	0.860 (+)	0.945 (+)	0.962 ($\approx$)	0.963
	30	0.908 (+)	0.919 (+)	0.880 (+)	0.923 (+)	0.792 (+)	0.956
	40	0.868 (+)	0.791 (+)	0.696 (+)	0.779 (+)	0.623 (+)	0.945

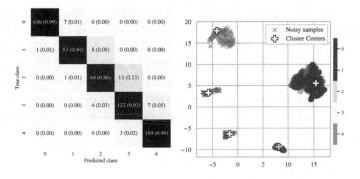

Fig. 3. Confusion matrix of the corrected labels (left) and embedding space of the train-set (right) of the CHP power estimation, corrupted with 30% flip noise.

In Fig. 3 (left) we show the label confusion matrix (with in-class percentage in parentheses) of SREA for the resulting corrected labels for the case of 30% of flip label noise. The corrected labels have 99% and 98% accuracy for the fully off- (0) and on-state (4), respectively. The intermediate power values' accuracies are 90% for state 1, 86% for state 2 and 92% for state 3. As an example, a visualization of the 32 dimensional embedding using the UMAP [31] dimension reduction technique is shown in Fig. 3 for the cases of 30% flip noise. The clusters representing the classes are very well separated, and we can see that the majority of the noisy label samples have been corrected and assigned to their correct class. Similar plots for other noise types as well as critical distance plots can be found in the supplementary material [6].

Finally, we run SREA on a different real-world test-set which includes a sensor failure, and the corresponding noisy label, as shown in Fig. 1. The method was able to correctly re-label the period of the sensor failure.

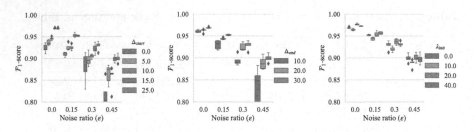

Fig. 4. SREA sensitivity to hyper-parameters Δ_{start} (left), Δ_{end} (middle), λ_{init} (right) for the CHP data and symmetric noise.

Table 4. Ablation studies on loss function components of SREA.

Noise	%	$\mathcal{L}_c$	$\mathcal{L}_c + \mathcal{L}_{ae}$	$\mathcal{L}_c + \mathcal{L}_{cc}$	$\mathcal{L}_c + \mathcal{L}_{ae} + \mathcal{L}_{cc}$
–	0	0.472 ± 0.060	0.504 ± 0.012	0.974 ± 0.003	$\underline{0.980 \pm 0.003}$
Symmetric	15	0.388 ± 0.027	0.429 ± 0.023	0.943 ± 0.004	$\underline{0.957 \pm 0.007}$
	30	0.355 ± 0.038	0.366 ± 0.041	0.919 ± 0.006	$\underline{0.930 \pm 0.008}$
	45	0.290 ± 0.014	0.318 ± 0.010	0.892 ± 0.009	$\underline{0.902 \pm 0.008}$
Asymmetric	10	0.400 ± 0.026	0.407 ± 0.012	0.949 ± 0.003	$\underline{0.957 \pm 0.003}$
	20	0.348 ± 0.003	0.358 ± 0.003	0.930 ± 0.009	$\underline{0.941 \pm 0.009}$
	30	0.342 ± 0.002	0.349 ± 0.004	0.901 ± 0.007	$\underline{0.922 \pm 0.010}$
Flip	10	0.460 ± 0.030	0.468 ± 0.030	0.961 ± 0.009	$\underline{0.973 \pm 0.006}$
	20	0.415 ± 0.028	0.424 ± 0.037	0.951 ± 0.008	$\underline{0.962 \pm 0.005}$
	30	0.405 ± 0.030	0.411 ± 0.040	0.943 ± 0.007	$\underline{0.957 \pm 0.003}$

6.1 Ablation Studies

Hyper-parameter Sensitivity. We investigate the effect of the hyper-parameters of SREA on both, benchmarks and CHP datasets. The observed trends were similar in all datasets, and therefore we only report the result for the CHP dataset. The effect of the three hyper-parameters related to the training dynamics (λ_{init}, Δ_{start}, and Δ_{end}) are reported in Fig. 4 for the cases of unstructured symmetric label noise (results for the other noise types can be found in the supplementary material [6]). For the variation of Δ_{start} and Δ_{end} there is a clear pattern for every noise level as the performance increases with the values of the hyper-parameters, and best performance is achieved by $\Delta_{start} = 25$ and $\Delta_{end} = 30$. This shows that both the *warm-up* and *re-labeling* periods should be rather long and last about 55% of the training time, before fully self-supervised training. The effect of λ_{init} is not as clear, but it seems that either a random initialization of the cluster centers (λ_{init} close to zero) or an extended period of unsupervised training of the autoencoder (λ_{init} between 20 and 40) is beneficial.

Loss Function Components. We study what effect each of the major loss function components of Eq. 4 has on the performance of SREA and report the

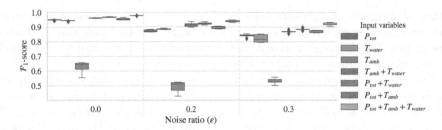

Fig. 5. SREA sensitivity to the input variables for CHP data and asymmetric noise.

results in Table 4. It can be observed, that not including the constrained clustering, i.e. using only $\mathcal{L}_c$ or $\mathcal{L}_c + \mathcal{L}_{ae}$, gives rather poor performance for all noise types and levels. This is explicitly observed as the performance decreases again during training in the self-supervision phase without $\mathcal{L}_{cc}$ (not shown). This seems understandable as the label correction method repeatedly bootstraps itself by using only the labels provided by the classifier, without any anchor to preserve the information from the training labels. This emphasizes the necessity of constraining the data in the embedding space during self-supervision.

Input Variables. We investigate the influence of the selection of the input variables on the estimation of the CHP power level. The results for all possible combination of the input signals are reported in Fig. 5. Unsurprisingly, using only the ambient temperature gives by far the worst results, while utilizing all available inputs results in the highest scores. Without the inclusion of the P_{tot}, we still achieve a $\mathcal{F}_1$-score above 0.96 with only using the T_{water} as input signal.

7   Conclusion and Future Work

In this work, we presented the problem of estimating the electrical power output of a Combined Heat and Power (CHP) machine by analyzing the total electrical power consumption of a medium size company facility. We presented an approach to estimate the CHP power output by analyzing the total load, the ambient temperature and the water temperature of the CHP, all of which are known to be control variables of the CHP. The training dataset for the deep-learning based approach was automatically derived from sensor measurements of the CHP power output, and sensor failures create noisy samples in the generated class labels. The proposed Self-Re-Labeling with Embedding Analysis (SREA) incorporates an autoencoder, a classifier and a constraint clustering which all share and operate on a common low-dimensional embedding representation. During the network training, the loss function and the label correction mechanism are adjusted in a way that a robust relabeling of noisy training labels is possible. We compare SREA to five SotA label noise correction approaches on ten time-series classification benchmarks and observe mostly comparable or better performance for various noise levels and types. We also observe superior performance on the CHP use-case for a wide range of noise levels and all studied noise types. We thoroughly analyzed the dependence of the proposed methods on the (hyper-)parameters and architecture choices.

The proposed approach is straight-forward to realize without any (hyper-)parameter tuning, as there are clear insights on how to set the parameters and the method is not sensitive to details. It also has the strong benefit that the amount of label noise need not be known or guessed. We used CNNs as building blocks of the proposed algorithm, but since SREA is model agnostic, it is possible to utilize other structures, such as recurrent neural networks [24] or transformers [45], which would also utilize the time-structure of the problem. The application of such dynamic models is left for future work.

Estimating the CHP output as shown in this work will be used in the future in energy optimization scenarios to arrive at more reliable and robust EV charging schedules. But, due to the robustness of the proposed method and the ability to exchange the neural networks with arbitrary other machine learning modules, we see a high potential for this architecture to be used for label noise correction in other application domains. We also see a high potential for an application in anomaly detection scenarios where sensor failures need to be detected. A thorough evaluation in these application areas is left for future work.

References

1. Arazo, E., Ortego, D., Albert, P., O'Connor, N., McGuinness, K.: Unsupervised label noise modeling and loss correction. In: International Conference on Machine Learning, pp. 312–321 (2019)
2. Arpit, D., et al.: A closer look at memorization in deep networks. In: International Conference on Machine Learning, pp. 233–242 (2017)
3. Atkinson, G., Metsis, V.: Identifying label noise in time-series datasets. In: Adjunct Proceedings of the 2020 ACM International Joint Conference on Pervasive and Ubiquitous Computing and Proceedings of the 2020 ACM International Symposium on Wearable Computers, pp. 238–243 (2020)
4. Barsim, K.S., Yang, B.: Toward a semi-supervised non-intrusive load monitoring system for event-based energy disaggregation. In: 2015 IEEE Global Conference on Signal and Information Processing (GlobalSIP), pp. 58–62 (2015)
5. Berthelot, D., Carlini, N., Goodfellow, I., Papernot, N., Oliver, A., Raffel, C.: Mixmatch: a holistic approach to semi-supervised learning. arXiv:1905.02249 (2019)
6. Castellani, A., Schmitt, S., Hammer, B.: Supplementary material for: estimating the electrical power output of industrial devices with end-to-end time-series classification in the presence of label noise. arXiv:2105.00349 (2021)
7. Chen, P., Liao, B.B., Chen, G., Zhang, S.: Understanding and utilizing deep neural networks trained with noisy labels. In: International Conference on Machine Learning, pp. 1062–1070 (2019)
8. Dau, H.A., et al.: The UCR time series classification archive (2018)
9. Ismail Fawaz, H., Forestier, G., Weber, J., Idoumghar, L., Muller, P.-A.: Deep learning for time series classification: a review. Data Min. Knowl. Disc. **33**(4), 917–963 (2019). https://doi.org/10.1007/s10618-019-00619-1

10. Fonseca, E., Plakal, M., Ellis, D.P., Font, F., Favory, X., Serra, X.: Learning sound event classifiers from web audio with noisy labels. In: ICASSP 2019-2019 IEEE International Conference on Acoustics, Speech and Signal Processing (ICASSP), pp. 21–25 (2019)
11. Fredriksson, T., Mattos, D.I., Bosch, J., Olsson, H.H.: Data labeling: an empirical investigation into industrial challenges and mitigation strategies. In: Morisio, M., Torchiano, M., Jedlitschka, A. (eds.) PROFES 2020. LNCS, vol. 12562, pp. 202–216. Springer, Cham (2020). https://doi.org/10.1007/978-3-030-64148-1_13
12. Frénay, B., Verleysen, M.: Classification in the presence of label noise: a survey. IEEE Trans. Neural Netw. Learn. Syst. **25**(5), 845–869 (2013)
13. Gan, O.P.: Automatic labeling for personalized IoT wearable monitoring. In: IECON 2018-44th Annual Conference of the IEEE Industrial Electronics Society, pp. 2861–2866 (2018)
14. Gavrilut, D., Ciortuz, L.: Dealing with class noise in large training datasets for malware detection. In: 2011 13th International Symposium on Symbolic and Numeric Algorithms for Scientific Computing, pp. 401–407 (2011)
15. Han, B., et al.: SIGUA: forgetting may make learning with noisy labels more robust. In: International Conference on Machine Learning, pp. 4006–4016 (2020)
16. Han, B., et al.: A survey of label-noise representation learning: past, present and future. arXiv:2011.04406 (2020)
17. Hendrycks, D., Mazeika, M., Kadavath, S., Song, D.: Using self-supervised learning can improve model robustness and uncertainty. arXiv:1906.12340 (2019)
18. Holmegaard, E., Kjærgaard, M.B.: NILM in an industrial setting: a load characterization and algorithm evaluation. In: 2016 IEEE SMARTCOMP, pp. 1–8 (2016)
19. Huang, L., Zhang, C., Zhang, H.: Self-adaptive training: bridging the supervised and self-supervised learning. arXiv:2101.08732 (2021)
20. Humala, B., Nambi, A.S.U., Prasad, V.R.: UniversalNILM: a semi-supervised energy disaggregation framework using general appliance models. In: Proceedings of the Ninth International Conference on Future Energy Systems, pp. 223–229 (2018)
21. Ioffe, S., Szegedy, C.: Batch normalization: accelerating deep network training by reducing internal covariate shift. In: International Conference on Machine Learning, pp. 448–456 (2015)
22. Jawed, S., Grabocka, J., Schmidt-Thieme, L.: Self-supervised learning for semi-supervised time series classification. In: Pacific-Asia Conference on Knowledge Discovery and Data Mining, pp. 499–511 (2020)
23. Jiang, L., Zhou, Z., Leung, T., Li, L.J., Fei-Fei, L.: MentorNet: learning data-driven curriculum for very deep neural networks on corrupted labels. In: International Conference on Machine Learning, pp. 2304–2313 (2018)
24. Karim, F., Majumdar, S., Darabi, H., Chen, S.: LSTM fully convolutional networks for time series classification. IEEE Access **6**, 1662–1669 (2018)
25. Karimi, D., Dou, H., Warfield, S., Gholipour, A.: Deep learning with noisy labels: exploring techniques and remedies in medical image analysis. Med. Image Anal. **65**, 101759 (2020)
26. Kingma, D.P., Ba, J.: Adam: a method for stochastic optimization. arXiv:1412.6980 (2014)
27. Li, J., Socher, R., Hoi, S.C.: DivideMix: learning with noisy labels as semi-supervised learning. arXiv:2002.07394 (2020)
28. Limmer, S.: Evaluation of optimization-based EV charging scheduling with load limit in a realistic scenario. Energies **12**(24), 4730 (2019)

29. Mandal, D., Bharadwaj, S., Biswas, S.: A novel self-supervised re-labeling approach for training with noisy labels. In: Proceedings of the IEEE/CVF Winter Conference on Applications of Computer Vision, pp. 1381–1390 (2020)

30. Massidda, L., Marrocu, M., Manca, S.: Non-intrusive load disaggregation by convolutional neural network and multilabel classification. Appl. Sci. **10**(4), 1454 (2020)

31. McInnes, L., Healy, J.: UMAP: uniform manifold approximation and projection for dimension reduction. arXiv:1802.03426 (2018)

32. McKnight, P.E., Najab, J.: Mann-Whitney u test. In: The Corsini Encyclopedia of Psychology, p. 1 (2010)

33. Nguyen, D.T., Mummadi, C.K., Ngo, T.P.N., Nguyen, T.H.P., Beggel, L., Brox, T.: Self: Learning to filter noisy labels with self-ensembling. In: International Conference on Learning Representations (2019)

34. Paresh, S., Thokala, N., Majumdar, A., Chandra, M.: Multi-label auto-encoder based electrical load disaggregation. In: 2020 International Joint Conference on Neural Networks (IJCNN), pp. 1–6 (2020)

35. Reed, S.E., Lee, H., Anguelov, D., Szegedy, C., Erhan, D., Rabinovich, A.: Training deep neural networks on noisy labels with bootstrapping. In: ICLR (2015)

36. Rolnick, D., Veit, A., Belongie, S., Shavit, N.: Deep learning is robust to massive label noise. arXiv:1705.10694 (2017)

37. Sablayrolles, A., Douze, M., Schmid, C., Jégou, H.: Spreading vectors for similarity search. In: ICLR 2019–7th International Conference on Learning Representations, pp. 1–13 (2019)

38. Song, H., Kim, M., Park, D., Lee, J.G.: Learning from noisy labels with deep neural networks: a survey. arXiv:2007.08199 (2020)

39. Sugiyama, M.: Co-teaching: robust training of deep neural networks with extremely noisy labels. In: NeurIPS (2018)

40. Van Rooyen, B., Menon, A.K., Williamson, R.C.: Learning with symmetric label noise: the importance of being unhinged. arXiv:1505.07634 (2015)

41. Wang, J., Ma, Y., Gao, S.: Self-semi-supervised learning to learn from noisylabeled data. arXiv:2011.01429 (2020)

42. Wang, X., Wang, C.: Time series data cleaning: a survey. IEEE Access **8**, 1866–1881 (2020)

43. Wang, Z., Yan, W., Oates, T.: Time series classification from scratch with deep neural networks: a strong baseline. In: 2017 International Joint Conference on Neural Networks (IJCNN), pp. 1578–1585 (2017)

44. Wang, Z., Yi Luo, X., Liang, J.: A label noise robust stacked auto-encoder algorithm for inaccurate supervised classification problems. Math. Probl. Eng. **2019**, 1–19 (2019)

45. Wu, N., Green, B., Ben, X., O'Banion, S.: Deep transformer models for time series forecasting: the influenza prevalence case. arXiv:2001.08317 (2020)

46. Yang, Y., Zhong, J., Li, W., Gulliver, T.A., Li, S.: Semisupervised multilabel deep learning based nonintrusive load monitoring in smart grids. IEEE Trans. Ind. Inform. **16**(11), 6892–6902 (2019)

47. Zeghidour, N., Grangier, D.: Wavesplit: end-to-end speech separation by speaker clustering. arXiv:2002.08933 (2020)

48. Zhang, C., Bengio, S., Hardt, M., Recht, B., Vinyals, O.: Understanding deep learning requires rethinking generalization. arXiv:1611.03530 (2016)

49. Zhang, H., Cisse, M., Dauphin, Y.N., Lopez-Paz, D.: mixup: beyond empirical risk minimization. arXiv:1710.09412 (2017)

50. Zhang, Z., Sabuncu, M.R.: Generalized cross entropy loss for training deep neural networks with noisy labels. arXiv:1805.07836 (2018)

Multi-task Learning Curve Forecasting Across Hyperparameter Configurations and Datasets

Shayan Jawed[1]([✉]), Hadi Jomaa[1], Lars Schmidt-Thieme[1], and Josif Grabocka[2]

[1] University of Hildesheim, Hildesheim, Germany
{shayan,jomaah,schmidt-thieme}@ismll.uni-hildesheim.de
[2] University of Freiburg, Freiburg, Germany
grabocka@informatik.uni-freiburg.de

Abstract. The computational challenges arising from increasingly large search spaces in hyperparameter optimization necessitate the use of performance prediction methods. Previous works have shown that approximated performances at various levels of fidelities can efficiently early terminate sub-optimal model configurations. In this paper, we design a Sequence-to-sequence learning curve forecasting method paired with a novel objective formulation that takes into account earliness, multi-horizon and multi-target aspects. This formulation explicitly optimizes for forecasting shorter learning curves to distant horizons and regularizes the predictions with auxiliary forecasting of multiple targets like gradient statistics that are additionally collected over time. Furthermore, via embedding meta-knowledge, the model exploits latent correlations among source dataset representations and configuration trajectories which generalizes to accurately forecasting partially observed learning curves from unseen target datasets and configurations. We experimentally validate the superiority of the method to learning curve forecasting baselines and several ablations to the objective function formulation. Additional experiments showcase accelerated hyperparameter optimization culminating in near-optimal model performance.

Keywords: Neural forecasting · Learning curves · Hyperparameter optimization · Sequence-to-sequence neural networks · Multi-task learning

1 Introduction

Hyperparameter optimization is a vital process in machine learning workflows. Practitioners commonly either rely on brute-force search over long grids, or via treating the loss surface in a black-box optimization framework [15]. Even so, given the configuration evaluation times, both of these methods fail to scale for large search spaces [14]. This motivates the research problem of designing novel methods to tackle this characteristic complexity and speeding up optimization.

© Springer Nature Switzerland AG 2021
N. Oliver et al. (Eds.): ECML PKDD 2021, LNAI 12975, pp. 485–501, 2021.
https://doi.org/10.1007/978-3-030-86486-6_30

The prominent theme has been to exploit cheap-to-evaluate fidelities (or prox-
ies) to the actual validation metrics. The simplest example of it is of a Learning curve in iterative learning algorithms which can be considered an iterative
fidelity to the final performance of the hyperparameter configuration. Learning
curve forecasting methods speed up optimization by extrapolating the performance metric from short runs to arrive at keep-or-kill decisions faster. Specific
works [2,6,10,13] that model this extrapolation as a fidelity have exploited the
partially observed time-series from validation metrics and hyperparameter configuration features (batch size, learning rate etc.) to predict the asymptote (final
performance) or forecast multiple steps ahead till the asymptote. We provide an
overview of the related work in the accompanying Appendix 1.[1]

Considering neural network training, there can be several statistical properties for e.g. μ, σ for $layer_i$ associated with the weights that dynamically change
throughout training and can also be modeled as fidelities to the final performance [16]. In this paper, we propose a Sequence-to-sequence learning model
that can model this inherent multivariate aspect present in a multi-task learning
problem setting. We propose to forecast these additional channels together with
the target validation accuracy for all timesteps till the asymptote. This leads
to an interesting multi-task problem formulation with a rich output space to be
modeled. Specifically, we formulate the main tasks to be the future points needed
to be forecasted for a target channel and in contrast, all other channel's future
value predictions as auxiliary tasks. An additional aspect to the learning curve
forecasting problem relates to earliness in the prediction of the learning curve.
The intuition behind catering for earliness is that ideally we wish to extrapolate
performance of the underlying architecture from noting only a few timesteps of
its performance. On the other hand, predicting the asymptote value with input
of a longer length curve is comparatively trivial and not useful, since training
might have converged already. We model for this aspect in the training of the
forecasting network with task-specific weighting that incentivizes early forecasting of the learning curve.

A related stream of works considers meta-learning for the purpose of sample-
efficiency when proceeding with hyperparameter optimization for new datasets
[7,8,12,17,18]. The intuition is to exploit past optimization runs to expedite
search for new datasets. Generally, the optimization runs are first gathered in
a meta-dataset; a dataset describing datasets. Besides containing multivariate
time series information from the iterative optimization of various architectures on
various datasets, the meta-dataset also contains descriptive statistics about the
underlying individual datasets, termed as meta-features. Common meta-features
include, the number of attributes, classes and the instances. Hence, a meta-
dataset can provide sufficient data enabling learning of a deep neural network
model like we propose above, and allow transfer learning possibilities considering forecasting of a new dataset's partial learning curve. What is normally
referred to as the cold-start problem in the literature [9], can be hence tackled
for a new dataset by exploiting meta-features that capture dataset relations
and the underlying patterns relating different hyperparameter configuration

---

[1] Appendix available via arXiv.

performances across datasets [7,12]. In summary, our core contributions can be listed as follows:

- We propose a novel Sequence-to-sequence learning curve forecasting method that incorporates various additional dynamic gradient statistics in a multi-task problem formulation.
- We design and optimize a novel corresponding multi-task loss function inspired by the problem setting that enforces early forecasting and incorporates biased weighted regularization for target performance metric tasks in contrast to counterpart weaker fidelity auxiliary tasks from various gradient metrics.
- We demonstrate that the method can be meta-learned and exploit configuration and meta features that generalize forecasting across hyperparameter configurations and datasets.
- We also show that our method is capable of accelerating Hyperparameter optimization when carrying out early stopping of sub-optimal configurations when integrated with model-free and meta-learned baselines.
- A thorough ablation study grounded on rigorously validating the effect of separate building components of the proposed method. Ultimately, proving the method on whole is well-founded.

2 Problem Setting

We consider a set of datasets $\mathcal{D} \in \mathbb{R}^{P \times S \times F}$, where each of the P datasets is an independent and identical set of S samples and F features upon which a supervised classification task is defined. The datasets can be differentiated on underlying data generating processes and different data modalities (tabular data, images etc.), however, parallels can be drawn based on a set of meta-features denoted as $\phi \in \mathbb{R}^{P \times M}$. Further, we consider a set of hyperparameter configurations $\Lambda \subset \mathbb{R}^{L \times K}$, where each $\Lambda_{1:L} \in \Lambda$ is a hyperparameter configuration of K hyperparameters[2]. Formally, we can define the meta-dataset $X \in \mathbb{R}^{N \times C \times T}$ as a Cartesian product $\Lambda \times \mathcal{D}$, that is the result of training Neural networks [3] with L hyperparameter configurations $\Lambda_{1:L}$ on each of the P datasets. We can describe X as the set of multivariate time-series of C metrics/channels (training loss, gradient norms, validation loss, etc.) across T epochs with $N = P \times L$. For notation ease, we assume the last channel C represents a particular metric of interest (target metric), which typically is the validation accuracy. To fix ideas, the problem definition with respect to the main-task:

Given the observed metrics from the conditioning range $[1 : \tau]$ of the n-th experiment, denoted as $X_{n,:,1:\tau} \in \mathbb{R}^{C \times \tau}$ using the slicing notation;
Given the hyperparameter configuration $\Lambda_l \in \mathbb{R}^K$ and the dataset meta-features $\phi_p \in \mathbb{R}^M$ of the n-th experiment;
Predict the value of the C-th metric (validation accuracy) at the final epoch of the n-th experiment, i.e. estimate $X_{n,C,T}$.

---

[2] Each of the $[1 : K]$ hyperparameter is sampled from a domain of valid values.
[3] In this work, we only consider Neural networks as the algorithm class.

3   Multi-LCNet: Multivariate Multi-step Forecasting with Meta-features

Our proposed model is dubbed Multi-LCNet. It is based on the encoder-decoder framework with several auxiliary tasks of predicting multiple channels for multi-step ahead. We let both the encoder and decoder networks be multi-layered Gated Recurrent Unit Networks (GRUs) [5]. A basic premise of our modeling objective is to exploit the configuration and meta-feature embeddings jointly with the multivariate time-series channels. However, incorporating these embeddings is not straight-forward given the fact that the rest of the data has a natural ordering with respect to time. In order to still exploit the embeddings denoted as $\xi \in \mathbb{R}^{Q \times \tau}$ jointly with the rest of the sequence modeling, we resort to repeating the embeddings on the time-axis to form additional Q channels that are concatenated with the rest of the multivariate time-series. The GRU encoder updates the hidden state recursively in the conditioning range $[1 : \tau]$. We let all channels share the same hidden-state parameters given existing correlations.

The last hidden-state from the encoding is generally referred to as the context vector [1]. Most prior approaches linearly extrapolate for one-step ahead from the context vector maximizing one-step likelihood. However, we can exploit the context vector to initialize a decoder network for multi-step forecasting. By having another decoder network, we can forecast for an arbitrarily long horizon ahead $H \in \mathbb{N}$. In addition to granting the model, the capacity to model across a wide range of fidelities, this also has a regularization effect given the pattern leading up to the asymptote can be covered in the modeling phase.

We simply initialize the decoder network's initial hidden state by copying the context from the encoder network, and feeding in the last timestep from the conditioning range i.e. τ as its first input. The decoder network has the same hidden dimensionality and number of layers as the encoder network, making this trivially possible. However, we note the discrepancy of feeding in the ground truth element at each timestep to the encoder whereas the decoder is trained in an auto-regressive manner that is consuming its own generated output at each successive timestep to compute the next hidden state and output. The output at each timestep is a $\mathbb{R}^{C+Q}$ dimensional extrapolation from the hidden state during decoding. During decoding, the model outputs $Q < M + L$ static features $[\hat{\Lambda}_l \circ \hat{\phi}_p]$ on which a reconstruction loss is defined. We noted experimentally that reconstructing static features during decoding lead to a regularization effect, improving modeling accuracy than otherwise. Additionally, we incorporate the attention mechanism [1] which allows the decoder to focus on the entirety of encoder outputs instead of solely relying on the last encoder hidden state.

We refer to Appendix 2 for a more detailed description of modeling above.

3.1   Optimizing Multi-LCNet

We have formulated the problem with respect to the main-task and explained how we can generate multivariate multi-step forecasts by modeling auxiliary tasks as well. Below, we formulate objective functions with respect to both. For simplicity, let Multi-LCNet be $f(X_{n,:,1:\tau}, \Lambda_l, \phi_p, H; \theta)$ where arguments denote availability and respective ranges and θ all learnable parameters.

Standard Objective. The standard approach is to train the model that predicts the target (C-th) metric at the final epoch T after observing τ observations of the metrics. The corresponding objective:

$$\arg\min_{\theta} \sum_{n=1}^{N} \| X_{n,C,T} - f(X_{n,:,1:\tau}, \Lambda_l, \phi_p, T; \theta)_C \|_\rho \tag{1}$$

However, we could improve this objective further, as it does not use the remaining targets $c \in \{1, ..., C-1\}$, the observations after the index τ till $T-1$ and lastly does not give more importance to the first observations. Given the practical importance associated with early decision-making regarding a hyperparameter configuration, it is important to accurately estimate the target metric after only a few epochs, otherwise convergence is already reached and curve plateaued.

An *Early, Multivariate and Multi-step* Forecasting Objective. To address the aforementioned drawbacks, we can optimize Multi-LCNet's parameters using the objective listed below:

$$\arg\min_{\theta} \sum_{n=1}^{N} \sum_{c=1}^{C} \sum_{t=1}^{\tau} \sum_{z=\tau+1}^{T} w_{ctz} \| X_{n,c,\tau+z} - \tag{2}$$
$$f(X_{n,:,1:\tau}, \Lambda_l, \phi_p, T; \theta)_c \|_\rho$$

In contrast to Eq. (1), this incorporates several additional auxiliary tasks in a weighted multi-task loss. The intuition is that these auxiliary tasks induce a strong regularization effect on the main-task learning. Differentiation between auxiliary and main-tasks is defined through task-specific weighting $w_{ctz} \subset (0,1) \subset \mathbb{R}^{+4, 5}$. These task weights are hyperparameters in the objective function formulation. Manual tuning of these weights is computationally infeasible given the number of tasks could explode in the case of predicting for a decent sized horizon in standard multivariate setting. Therefore, we propose a novel factorization of the weights customized with respect to the sub-objectives relating to inducing earliness, balancing multiple channels and their point forecasts ahead:

$$\arg\min_{\theta} \sum_{n=1}^{N} \sum_{c=1}^{C} \sum_{t=1}^{\tau} \sum_{z=\tau+1}^{T} \alpha_c \beta_t \gamma_z \| X_{n,c,\tau+z} - \tag{3}$$
$$f(X_{n,:,1:\tau}, \Lambda_l, \phi_p, T; \theta)_c \|_\rho$$

Where $\alpha_c, \beta_t, \gamma_z \in (0,1) \subset \mathbb{R}^+$ can be chosen with regard to the following insights:

---

[4] We also normalize all meta-data in $(0,1)$ unit interval.
[5] We overload the notation, in this subsection w defines task-weight.

i) Predicting the target metric (validation accuracy) is more important than predicting other metrics i.e. $\alpha_C > \alpha_c$. On the other hand, $\alpha_c > 0, \forall c \in 1, ..., C$ meaning we do not want to avoid predicting the other metrics since correlated channels have a beneficial regularization effect. We emphasize that such an objective formulation is a multi-task setting, where we have a target task/metric (the validation accuracy) and a set of auxiliary tasks/metrics (training loss, gradient norms, etc.).

ii) Correct forecasts with few observations $\tau \ll T$ are more important than estimations close to the converged epoch $\tau \approx T$. Practically speaking, we should be able to predict the performance of a poorly-performing hyperparameter configuration Λ_l after as few epochs as possible. Therefore, the weights $\beta_{1:\tau}$ can be set as exponentially decaying, which incorporates stronger penalization towards the errors made with small t values in the objective. Concretely, $\beta_{t=1} \approx 1$ and $\beta_\tau \approx 0$.

iii) Predicting the metric values at the last epoch is more important than the next immediate epoch after τ, in the desired case when $\tau \ll T$. Therefore, the forecasts indexed higher in the prediction range $[\tau + 1 : T]$ and their corresponding loss terms need to be penalized stronger. In that regard, the horizon task weights $\gamma_{\tau+1:T}$ can be formulated with the decay rate inverted and generated similarly from the exponential function. Concretely, $\gamma_{\tau+1} \approx 0$ and $\gamma_T \approx 1$.

We make the effort to elaborate more on the earliness aspect of the objective formulation given its distinct and novel formulation. In order to model for earliness, we generate what are normally called roll-outs after each input timestep observed from the learning curves. All roll-outs are multivariate multi-step forecasts till the asymptote of the curves. During training, we set $\tau = T - 1$, to utilize the full-extent of the curves and the model subsequently generates forecasts of different H length adjusted accordingly. Once all roll-outs are made, that is when $\tau = T - 1$ we can weight the errors based on the combined weighting scheme motivated above.

Exponential Weighting. The exponential weighting is defined as follows:

$$\beta_{1:\tau} = \exp\left(\frac{-|j - center|}{g}\right) \tag{4}$$

$$g = -\left(\frac{\tau - 1}{\log(u)}\right)$$

Where, j defines the index of roll-out and *center* is the parameter defining center location of the weighting function. g defines the decay. We fix $center = 0$, u is then the fraction of window remaining at the very end, that is the weight for last indexed roll-out. Setting the value for $u \in (0, 1) \subset \mathbb{R}^+$ defines the entire set of weights $\beta_{1:\tau}$. We can generate the weights $\gamma_{\tau+1:T}$ by defining another u value, replacing τ with H and inverting the weights generated through Eq.(4).

4 Experiments[6]

4.1 Datasets, Meta-Datasets and Evaluation Protocol

Meta-dataset. We use the dataset created in [21]. Each sample contains multivariate training logs of a configuration trained on a particular underlying classification dataset. All datasets used to evaluate the configurations came from the AutoML benchmark [11], in total numbering to 35. The overall meta-level distribution can be considered diverse in terms of underlying dataset characteristics such as number of samples, features and classes. Exhaustive sets of meta-features for each dataset are also available that besides these characteristics note additional many such. We also shed light on the configuration space that is used to sample valid hyperparameter configurations through in Appendix 3. We note that all architectures are funnel-shaped feed-forward networks, defined with respect to number of layers and initial units. A total of 2000 configurations are sampled from this configuration space and trained/validated/tested on the corresponding splits of each underlying dataset for a total of 52 epochs. The resulting multivariate channels also include global and layer-wise gradient statistics (max, mean, median, norm, standard deviation, and quartiles Q10, Q25, Q75, Q90), learning rate, runtime and balanced accuracies, up-to a total 54 channels. This results into X, ϕ_p, Λ_l with shapes ($N = 70000 \times C = 54 \times T = 52$), ($P = 35 \times M = 107$) and ($L = 2000 \times K = 7$) respectively. We label encoded, normalized all channels besides validation accuracy between 0 and 1 and zero-padded in case of missing values due to conditionally undefined layer-wise statistics, hyperparameters or meta-features across these tensor and matrices.

Evaluation Protocol. The evaluation protocol is aligned to a realistic meta-learning setting where prior meta-data across datasets is considered available and the goal would be to warm start hyperparameter optimization for new datasets as tackled in [8,17,18]. In light of this, we divide the meta-dataset into meta-train, meta-validation and meta-test splits covering 25, 5 and 5 datasets each. We highlight important characteristics of the validation and test split datasets in Appendix 3. We refer the remaining 25 train split datasets and a more thorough summary of data-set characteristics to [11,21]. We split the above noted tensor and data matrices accordingly. We now proceed to define evaluation metrics that shall quantify success from different lenses. Firstly, we rely on measuring the mean-squared-error on the prediction of the last timestep (final performance) for the target metric i.e. our main-task as formulated earlier in Sect. 2. In alignment with previous works [2,13], we judge the predictions based on $\approx 20\%$ of the curve as input. Nevertheless, as motivated earlier, for the purpose of realistic hyperparameter optimization it is necessary to quantify how early the predictions match the ground-truth asymptotic performance of the curves as well. Therefore, we also evaluate our results as the average error of all final-performance errors till observing $\approx 20\%$ of the curve as input. And for completion's sake, we report the average of all final-performance errors made till $T - 1$.

---

[6] github.com/super-shayan/multi-lcnet; Baseline Implementation Details in Appendix.

We also report simple Regret, the difference between optimal accuracy (pre-computed from meta-test data) and one achieved over a set of trials [8,17,18].

4.2   Baselines

Learning Curve Baselines
Last Value [14] propagates the last observed value for H timesteps.

LCNet [13] is a Bayesian Neural network that estimates parameters and creates weighted ensembles of increasing and saturating functions from power law or sigmoidal family to model learning curves. As input, however the model only takes into account configuration features by repeating these along the time-axis and learns joint embeddings via hidden layers. Hence, straight-forward application would prevent meta-learning where we wish to forecast accuracy across datasets. In light of this, we propose an extension of this model with meta-features which we refer to as **LCNet(MF)** where we simply concatenate the configuration and meta-features before joint embeddings are learned as in the standard setting.

ν-**SRM** is the model from [2]. The modeling for learning curves is based on training $T-1$ many feature engineered models, where each successive ν-Support Vector Machine Regression (SVR) model takes an additional timestep of the learning curve. All $T-1$ many SVR models only predict for last timestep, the validation accuracy at T. We train and validate the baseline **SRM** and its extension with multivariate channels **SRM(M)** and with multivariate channels plus meta-features **SRM(MM)** on meta-train and meta-validation datasets.

LCRankNet [18] learns latent features for learning curves via stack of non-linear Convolutional layers and architectural embeddings via Sequence-to-sequence networks. It embeds Dataset IDs for modeling learning curves across Datasets which are however generated randomly for modeling across datasets and therefore we propose to embed meta-features instead. We focus on only the ablation reported on learning $L2$-loss based pairwise rankings, which proved to be better for early predictions across all datasets and makes learning comparable to models in this paper. We remove the Sequence-to-sequence based embeddings that might be more applicable to deeper network topologies as tackled originally in that paper. Extension of **LCRankNet** with multivariate channels is termed **LCRankNet(M)**, and like before we also craft **LCRankNet(MM)**.

Multivariate Multi-step Forecasting Baselines
TT-RNNs Tensor-train RNN is a sequence-to-sequence model [20]. The working principle is to replace the first-order markovian dynamics abiding hidden states in RNNs with a polynomial expansion computed over the last many hidden states. Tensor decomposition is used for dimensionality reduction for this new state. We train the baseline on all meta-train datasets but unlike above baselines, repeated the static features including configuration and meta-features on the time-axis to form additional channels for **TT-RNN (MM)**.

MCNN is a multivariate time-series forecasting baseline crafted through heuristically searched parameter sharing between stacks of convolutional layers and exponentially decaying weighted schemes that tackle scale changes in long-term forecasting. We also designed another meta-feature based extension named

MCNN(MM) where we stack non-linear embedded meta and configuration features directly with the latent convolutional features before feeding to the stack of fully-connected layers predicting for multiple channels and time-steps ahead.

Multi-Task LASSO induces shared sparsity among parameter vectors of multiple regularized linear regression models. In our setting, we can consider all timesteps to be forecasted as separate tasks and instead of solving multiple lasso models independently, feature selection is stabilized by shared sparsity induced via block-regularization schemes.

Model-Free Hyperparameter Optimization Baselines
Random Search [3] samples hyperparameter configurations randomly from the space of configurations defined in the meta-split.

Hyperband [14] is a bandit-based method that samples configurations randomly and terminates sub-optimal configurations according to predefined downsampling rates at each round, only advancing better performing ones to be run for more iterations. We report results for different downsampling rates in brackets Table 2 with fixed max-iterations i.c. 52 from meta-data.

Meta Learning Hyperparameter Optimization Baselines For the purpose of hyperparameter optimization, most work has focused on meta learned Bayesian Optimization (BO). Hence, we benchmark and propose orthogonal extensions to:

TAF from [19], given the same intuition of ours, transfers knowledge between tasks in a meta-setting. Transferable Acquisition Function (TAF) incorporates source and target relationships in the acquisition function during BO. The acquisition function scores the next configuration based on expected improvement on the target dataset and predicted improvement over the source datasets. We also orthogonally integrate Multi-LCNet as a meta-learned forecasting model within BO. The Gaussian Process (GP) surrogate updates its parameters sequentially on early terminated estimations of Multi-LCNet instead of on final-performances of configuration trained fully. We refer to the extension as **TAF-MLCNet**. Also, early terminating only applies to meta-testing, source GPs remain unaltered.

MBO from [17] is the recent state-of-the-art baseline for Meta-learning in Bayesian Optimization (MBO) that acquires next configurations efficiently in a meta reinforcement learning setting and modeling the acquisition function with a neural network. We also orthogonally integrate Multi-LCNet similar to above crafting **MBO-MLCNet**.

Table 1. Comparison against learning curve and multivariate forecasting baselines in terms of MSE $\cdot 10^{-2}$ on validation accuracy scaled to [0–1]. Columnar least is boldfaced, second-least is underlined.

Methods	Segment			Shuttle			Sylvine			Vehicle			Volkert		
	$\tau=9$	A(9)	A(51)	$\tau=9$	A(9)	A(51)	$\tau=9$	A(9)	A(51)	$\tau=9$	A(9)	A(51)	$\tau=9$	A(9)	A(51)
LCNet(MF)	1.83	3.57	2.98	32.24	27.1	35.44	3.28	2.38	3.25	2.29	5.39	3.21	2.69	3.03	5.34
SRM	**0.95**	2.04	0.46	2.02	4.45	1.09	<u>0.42</u>	1.08	0.22	**0.4**	<u>0.83</u>	**0.19**	0.24	<u>0.38</u>	<u>0.1</u>
SRM(M)	1.37	3.21	0.69	2.55	5.35	1.38	0.87	2.58	0.5	0.67	1.24	0.29	<u>0.2</u>	0.39	0.1
SRM(MM)	1.29	2.75	0.6	2.9	6.07	1.43	1.05	2.51	0.52	0.63	1.22	0.28	**0.19**	0.3	**0.08**
LCRankNet	1	1.7	<u>0.4</u>	**1.24**	<u>2.67</u>	**0.63**	**0.27**	<u>0.79</u>	<u>0.16</u>	0.58	1.39	0.3	0.6	1.83	0.36
LCRankNet(M)	1.34	2.81	0.63	1.86	3.94	0.96	0.76	2.32	0.48	<u>0.55</u>	1.28	0.29	0.21	0.56	0.13
LCRankNet(MM)	<u>0.99</u>	<u>1.62</u>	0.49	<u>1.43</u>	2.79	<u>0.94</u>	0.62	1.25	0.35	0.75	1.35	0.44	0.49	0.84	0.2
Last Value	1.51	3.57	0.76	2.76	6.15	1.61	0.69	1.72	0.35	0.73	1.6	0.35	0.31	0.78	0.17
TT-RNN	1.56	2.29	–	4.95	5.72	–	0.97	1.5	–	1.15	1.79	–	1.2	2.64	–
TT-RNN(MM)	5.63	4.62	–	10.08	11.47	–	5.12	4.99	–	7.84	7.06	–	5.15	2.86	–
MCNN(M)	7.25	7.04	7.1	5.05	5.28	5.18	2.42	2.5	2.46	7.51	7.26	7.31	9.54	9.14	9.25
MCNN(MM)	7.14	7.1	7.04	5.13	5.29	5.2	2.47	2.5	2.48	7.38	7.33	7.23	9.32	9.21	9.14
MTL-LASSO(M)	1.35	1.99	–	6.32	7.85	–	1.49	1.95	–	0.96	1.31	–	0.77	0.94	–
Multi-LCNet(M)	1.02	**1.47**	**0.38**	1.57	**2.27**	1.42	**0.27**	**0.68**	**0.15**	0.92	1.53	0.38	1.46	3.27	0.7
Multi-LCNet(MM)	1.07	2.09	0.5	2.68	3.9	1.43	0.61	1.38	0.32	**0.4**	**0.75**	<u>0.21</u>	0.32	0.94	0.2

Multi-LCNet Ablations
Multi-LCNet(M) does not embed meta-features. However, it uses configuration features repeated and forecasted as channels.

Multi-LCNet(MM) embeds meta-features jointly with configuration features and forms channels with these joint embeddings as noted earlier.

4.3   Forecasting Results

We report the comparison of our proposed Multi-LCNet with the learning curve and multivariate forecasting baselines in Table 1. We ran all baselines described earlier with 3 different seeds and report mean performances across standard objective with $\tau=9$ and the aggregated metrics $A(9), A(51)$ that quantify earliness and dynamic performance throughout the curve length. $A(9), A(51)$ denote the Average of final-performance errors (main-task) made observing curves till $\tau=9$ and $\tau=51$ respectively. A number of interesting observations can be drawn from these results. Firstly, we can see that Multi-LCNet is able to outperform the baselines across multiple metrics on all datasets besides one. Most interesting are the lifts on $A(9)$ compared to other metrics, since it quantifies performance with regard to earliness. We can credit the auxiliary supervision provided to the model through multivariate multi-step forecasting as the basis for these leads. This stands to reason, provided learning curve baselines already rely on deep learning primitives such as feed-forward layers and convolutions in LCNet and LCRankNet. Crucially, the baselines are all provided input of the same dimensionality with their respective extensions.

Another set of observations can be derived from benchmarking the performance of machine learning baselines to the naive last value forecasting baseline. We can validate the finding from prior works about this baseline's exceptionally strong performance on learning curves, which have a natural tendency to plateau rather early. Nevertheless, we can see that majority of learned baselines outperform it especially on the first two metrics that quantify earliness.

With regard to auxiliary supervision through multivariate multi-step forecasting baselines, we observe that MTL-LASSO and TT-RNN perform equally well across datasets and metrics, but indeed are less generalizable than counterpart learning curve baselines. We hypothesize that the reason for this sub-par performance is due to the rather fixed dimensionality forecasts that prohibit all multivariate forecasting baselines considered to exploit training on dynamic length input curves. These baselines were initially proposed for long range input as compared to extremely short learning curves where data cannot be generated through rolling windows and rather every curve needs to be partitioned into a fixed conditioning and prediction range beforehand. This explains why despite MCNN and LCRankNet being both convolutional neural networks, the LCRankNet baseline and extensions can perform much better by modeling for only 1 fixed window but exploiting the same curves multiple times with dynamic conditioning history. This is also the reason why one joint model across all input length for LCNet, SRM, MTL-LASSO, TT-RNN, MCNN is not possible and we did hyperparameter tuning for validating their performance for only $\tau = 9$ and re-trained the baselines for all other $\tau = [2...51]$. We also dropped the comparison given scalability challenges with TT-RNN, MTL-LASSO for metric $A(51)$ as noted by '–' in Table 1. Except for these baselines, we tuned the hyperparameters for all other models on the metric $A(51)$. On the other hand, this highlights yet another advantage of Multi-LCNet which can be trained on dynamic conditioning history as well as exploit auxiliary regularization through multivariate forecasting as it can generate dynamic length forecasts from any window of the curve.

Additional observations can be made with regard to multivariate channels, hyperparameter configuration features and lastly meta-features. SRM baseline is unable to cater for both multivariate gradient statistics and meta-features, as evident by higher errors made throughout the metrics and datasets by respective extensions. This could be because the model is considered rather shallow and unable to learn non-linear feature interactions as the deep learning counterpart methods are able to. In fact, we see that LCRankNet and Multi-LCNet both benefit from additional multivariate information plus meta-features comparatively more so. On the other hand, we can observe that directly feeding all static features as channels through repetition on time axis lead to downgrade in modeling accuracy for both the TT-RNN and MCNN baselines. This is the reason why for Multi-LCNet we explored another way to embed meta-features to a more fine-grained representation explained earlier in the method section.

4.4 Accelerating Hyperparameter Optimization

In this section, our aim is to firstly formulate a predictive termination criterion based on the predictions from Multi-LCNet. We follow the lead of [2, 4, 6], and model a similar criteria that is essentially based on the heuristic that if at any given time the forecasted accuracy from a partially observed configuration's curve falls below a certain best observed accuracy in a given set of configurations, then this configuration can be early terminated to save valuable compute and time resources. Specifically, we adapt the criteria from [2] to a meta-setting. To ground the termination decision in probabilistic terms, one can model the forecast as a Gaussian perturbation around the original estimate to safeguard against poor out-of-sample generalization. To fix ideas, we randomly sample $M \ll N$ configurations where $m \in 1, ..., M$ and at each successive epoch τ, we generate forecasts, $\hat{X}_{1:M,C,T}$ for all M configurations. To model the uncertainty associated with the forecasts we estimate the standard deviation σ by leave-p-out cross-validation[7]. Specifically, we account for the uncertainty by keeping dropout active and noting the σ among $\{(\tau - p)...\tau\}$ forecasts of $\hat{X}_{1:M,C,T}$. With the uncertainty, we can estimate the forecasts as a gaussian perturbation $\hat{y}_{1:M,C,T} = \mathcal{N}(\hat{X}_{1:M,C,T}, \sigma)$. Finally, the probabilistic termination criterion: $p(\hat{y}_{1:M,C,T} \leq \max(X_{1:M,C,1:\tau})) = \Phi(\max(X_{1:M,C,1:\tau}); \hat{y}_{1:M,C,T}, \sigma)$, where $\Phi(.; \mu, \sigma)$ is the Cumulative distribution function (CDF) of the Normal distribution. For configuration m if probability $p(\hat{y}_{m,C,T} \leq \max(X_{1:M,C,1:\tau})) \geq \Delta$ does not hold, we can early terminate it. Where, Δ balances the tradeoff between early terminating configurations for more significant acceleration or on the other hand the risk of observing higher regret. Additionally, for ensembling's sake, one can let the top-η confs to complete training. For our experiments we set Δ, p and η via cross-validation on the validation datasets based on observing the regret. We note that this criteria despite sharing characteristic similarities differs from the one in [2], given the cross-dataset setting. This setting does not require any burn-in period to observe learning curves completely for new datasets. As a downside however, we track the maximum observed accuracy from $[1..\tau]$ for all new configurations instead of the accuracy at T from the burn-in period. Other notable differences include a more robust estimation of uncertainty given dropout and multi-horizon recursive forecasts and tuning of Δ, p and η on meta-validation. We also refer to an example termination in Appendix 4.

4.5   Acceleration Results

This section reports the results on accelerating hyperparameter optimization through early termination. For our first set of experiments we accelerate Random Search (RS-MLCNet) given its simplicity and vast utility. We randomly sample two sets of confs. i.e. M = 50 and M = 100 and report corresponding time(m) and regret in Table 2. We report the average of 10 runs. We set $\delta = 0.99$, $\eta = 5$ and $p = 5$ for both Multi-LCNet(RS-MLCNet) and SRM based early stopping (RS-SRM). The regret is stated in terms of percentage classification accuracy. We first note the comparison between RS and its counterpart acceleration with Multi-LCNet. The results indicate huge gains with regard to time saved with very little to no harm in regret. We also note that the standard deviation is on a similar scale. Since initial random selection of 50 or 100 configurations from 2000

---

[7] We overload notation σ to denote standard deviation, p in cross-validation.

is bound to affect the final regret, we keep these same for RS, and its acceleration through Multi-LCNet & SRM across all runs.

We also compare these gains with SRM based early termination. In terms of retrieving the optimal model among the initial trials, both RS accelerations lead to similar regret given the same early stopping criteria. Our initial assumption was that the difference in MSE would result in better acceleration performance, but however in terms of regret computed for optimal configuration the gains in forecasting accuracy did not transfer gracefully.

We also compare accelerated RS (RS-MLCNet) to Hyperband. Hyperband also randomly samples configurations and uses the last value based extrapolation to early terminate. However, Hyperband dynamically selects the configurations to evaluate, which prohibits reporting results for 50 or 100 trials. To report a fair comparison, we report results for Hyperband initialized with three different downsampling rates. Among the two trial settings for RS-MLCNet and Hyperband variants, we can see that on Segment, Shuttle and Volkert we are able to outperform Hyperband in terms of balance between regret and time taken.

Lastly, we benchmark the meta-learned approaches TAF and MBO. Firstly, we note that both these baselines stand out due to consistent minimal possible regret across both trial settings[8]. This is consistent with known superiority of Bayesian optimization to RS. The efficacy of RL framework from MBO saves even more time compared to BO based TAF. We observed that MBO can ask to run the same configuration repeatedly, in contrast to TAF if it discovers the optimal configuration early on. Hence, we count the regret and time taken for only unique configurations among the 50 or 100 specified initially[9] and can see that the difference in time for the two initial trial sets remains similar for MBO. We turn to report the results for Multi-LCNet integrated counterparts MBO-MLCNet and TAF-MLCNet that enable early termination for both these methods. Given the drawback that TAF and MBO are both sequential approaches, we modify the early stopping criteria to consider the values until current timestep for only the single incumbent configuration. This puts the early stopping criteria at a disadvantage, but nevertheless we observe a clear lift across all datasets without loss in regret. Equally worth noting is the fact that early terminated objectives do not generally interfere with acquisitions within the context of either BO nor RL. This is important because one might worry that the sequential chain of successive configuration acquisitions might be affected if the underlying GP parameters are updated on the early stopped performance (objectives) for target dataset configurations instead of on their final performance. Nevertheless, when provided with early stopped objectives, the number unique confs. did arise for MBO-MLCNet@100 for Shuttle and Sylvine datasets notably. We hypothesize this is due to comparable fewer differences between configurations on these datasets compared to other datasets as evident in lower standard deviation for RS regret too. Still, with higher number of configurations, the times were less.

---

[8] The results for MBO and TAF are not averaged across runs given the stationarity of GP modeling and meta-data; based on personal correspondence with the authors.

[9] Optimization is not terminated when regret is 0 to simulate real-world testing where regret is unknown apriori.

Table 2. Accelerated Hyperparameter Optimization results.

Methods	Segment		Shuttle		Sylvine		Vehicle		Volkert	
	Time	Regret	Time	Regret	Time	Regret	Time	Regret	Time	Regret
MBO@50	35.29	0.0	164.61	0.0	40.76	0.0	25.84	0.0	75.09	0.0
MBO@100	46.43	0.0	170.5	0.0	47.86	0.0	27.38	0.0	82.18	0.0
TAF@50	78.02	0.0	186.59	0.0	66.3	0.0	72.52	0.0	268.16	0.0
TAF@100	160.77	0.0	364.68	0.0	146.86	0.0	125.33	0.0	447.14	0.0
MBO-MLCNet@50	24.39	0.0	129.29	0.0	31.9	0.0	18	0.0	52.14	0.0
MBO-MLCNet@100	30.81	0.0	145.47	0.0	36.08	0.0	18.72	0.0	68.89	0.0
TAF-MLCNet@50	43.22	0.0	117.68	0.0	55.01	0.0	47.07	0.5319	169.24	0.0
TAF-MLCNet@100	93.97	0.0	200.75	0.0	110.45	0.0	84.24	0.0	322.75	0.0
Hyperband(2)	24.9 ± 3.2	4.3 ± 0.7	$57. \pm 9.1$	1.0 ± 0.5	38.1 ± 5.1	0.5 ± 0.4	26.6 ± 3.0	3.0 ± 3.1	62.6 ± 6.3	7.2 ± 3.2
Hyperband(2.5)	15.4 ± 2.0	4.4 ± 1.4	32.9 ± 3.4	1.2 ± 0.5	20.5 ± 1.8	0.7 ± 0.5	$15. \pm 1.3$	3.4 ± 2.2	37.6 ± 5.0	7.8 ± 4.3
Hyperband(3)	9.1 ± 0.9	5.6 ± 0.9	22.5 ± 5.5	1.5 ± 0.3	16.3 ± 3.4	0.6 ± 0.5	$10. \pm 1.3$	4.3 ± 1.5	21.8 ± 2.7	8.7 ± 3.3
RS@50	57.4 ± 4.4	5.6 ± 2.4	126.1 ± 11.6	0.5 ± 0.6	60.2 ± 7.1	1.1 ± 0.7	53.1 ± 5.8	6.3 ± 1.4	188.1 ± 24	8.2 ± 3.8
RS@ 100	135.5 ± 12.5	3.3 ± 0.7	298.6 ± 24.1	0.4 ± 0.2	146.4 ± 11.7	0.4 ± 0.4	$131. \pm 7.7$	5.0 ± 0.7	384.3 ± 14.9	5.3 ± 3.3
RS-SRM@50	8.0 ± 0.7	5.6 ± 2.4	25.7 ± 3.9	1.1 ± 0.7	$9. \pm 1.3$	1.5 ± 1.1	7.8 ± 1.0	6.7 ± 1.3	24.1 ± 1.1	8.9 ± 4.5
RS-SRM@100	13.2 ± 0.8	3.4 ± 0.9	38.6 ± 5.3	1.0 ± 0.6	16.2 ± 1.5	1.1 ± 0.9	13.2 ± 1.1	5.0 ± 0.7	43.3 ± 3.5	5.3 ± 3.3
RS-MLCNet@50	8.0 ± 0.7	5.7 ± 2.5	25.5 ± 4.0	1.1 ± 0.7	9.1 ± 1.1	1.5 ± 1.1	8.0 ± 1.3	6.7 ± 1.3	35.1 ± 3.1	8.5 ± 4.0
RS-MLCNet@100	$13. \pm 0.7$	3.4 ± 0.9	36.9 ± 3.9	1.1 ± 0.6	15.9 ± 2.0	1.2 ± 0.8	13.4 ± 1.2	5.0 ± 0.7	64.3 ± 3.7	5.3 ± 3.3

4.6 Ablation Study on the Meta-validation Set

The above objective function comprehensively captures the entirety of the multi-task output space, with sub-objectives exploiting inherent characteristics of the learning curve forecasting problem setting. However, there exist possibilities of designing the objective functions in between the two extremities, as given by Eq. (1) and Eq. (3). Specifically, if we consider either of the earliness, multi-target or the multi-step outer loop as either present or discarded leads to $2^3 = 8$ possibilities with respect to objective formulation. We can study if either of the unstated 6 combinations, for example the standard function in Eq. (1) equipped with earliness and associated hyperparameters $\beta_{1:\tau}$ dictating the exponentially decaying scheme for task weights models the main-task more accurately than the objective in Eq. (3). Moreover, we can also study whether any of the associated components in the input space, channels $c \in \{1, ..., C - 1\}$, the configuration features Λ_l and the meta-features ϕ_p lead to improvement or on the contrary decline in modeling accuracy with respect to the main-task. To fix ideas, we term the changes to the objective function and removal of input configuration or meta-features as ablations and refer to number of recurrent layers, number of hidden units, number of fully connected layers and respective units, activation functions, dropout, batch sizes, learning rate as standard hyperparameters to the network that need to be tuned regardless the ablation. Additionally, we consider the auxiliary task weights $\alpha_c, \beta_t, \gamma_z$ as conditional hyperparameters that are only defined when the corresponding ablation is chosen. Attention is considered as additional conditional hyperparameter that is defined only for ablations considering encoder-decoder modeling of the entire horizon. We tune the hyperparameters of each ablation together with well-defined associated hyperparameter configurations, since one hyperparameter config. might not generalize to another ablation.

For all ablations besides the proposed objective formulation and input space, we introduce another model termed as Standard-Net. This model is characterized as an encoder-only network with an output fully-connected-layer whose dimen-

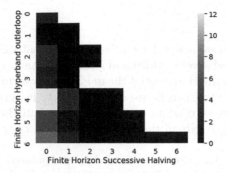

Fig. 1. In each cell we plot the ratio of StandardNet configurations selected to Multi-LCNet configurations across different successive halving iterations in Hyperband. We can observe that Hyperband increasingly selects Multi-LCNet configurations as successive halving continues and hence the ratio decreases

sionality corresponds to the dimensionality of the target space. The target space can vary from a single point to forecast for the main channel (as in standard objective) or all channels at the last horizon and lastly forecasting all channels for the entire horizon. Interestingly, the Standard-Net cannot be trained with an Earliness sub-objective when impeded by fixed dimensionality of output. Also, conditional upon the ablation Standard-Net can incorporate all input channels, meta-features and configuration features, but however cannot incorporate attention.

Including root-level binary valued hyperparameters that define presence or absence of ablative sub-objectives and input features, hyperparameters conditioned upon these ablations (α, β, γ) and standard network hyperparameters leads to 14-dimensional hyperparameter configurations. Given this relatively large search space, we rely on Hyperband [14], to conduct a thorough analysis in order to better judge whether the proposed formulation of Multi-LCNet leads to a gain in predictive accuracy over Standard-Net. The working principle behind Hyperband also qualitatively expresses whether a particular configuration is iteratively selected consecutively in various levels of Successive Halving. We define a large search space to randomly sample configurations from and observe that configurations trained with the proposed objective formulation are given increasingly higher budget, which testifies modeling accuracy of the proposed method to be higher than counterpart ablations. We note for Hyperband that all initial search spaces and following number of successive halving rounds are defined with respect to maximum number of iterations. By setting this to 1000 and default downsampling rate (=3), we allow for the possibility of multiple rounds and larger initial search spaces before these rounds. In these spaces ablations outnumber the Multi-LCNet configurations by 8x, however, across all successive halving iterations (y-axis of Fig. 1), the ratio converges to 0 (x-axis) showcasing that Hyperband spends more budget on selected Multi-LCNet configurations.

5   Conclusion

In this work, we propose a novel meta-learned forecasting model that models validation accuracy and several additional gradient statistics in a weighted multi-task loss. Empirical evaluation showed the model outperformed multiple forecasting baselines and forecasts can be used to accelerate hyperparameter optimization in the simple case of random search and also meta Bayesian optimization. As future work, we shall extend the modeling in novel meta-learning directions.

Acknowledgements. This work is co-funded by the industry project "Data-driven Mobility Services" of ISMLL and Volkswagen Financial Services; also through "IIP-Ecosphere: Next Level Ecosphere for Intelligent Industrial Production".

References

1. Bahdanau, D., Cho, K., Bengio, Y.: Neural machine translation by jointly learning to align and translate. arXiv preprint arXiv:1409.0473 (2014)
2. Baker, B., Gupta, O., Raskar, R., Naik, N.: Accelerating neural architecture search using performance prediction. arXiv preprint arXiv:1705.10823 (2017)
3. Bergstra, J., Bengio, Y.: Random search for hyper-parameter optimization. J. Mach. Learn. Res. **13**(2) (2012)
4. Chandrashekaran, A., Lane, I.R.: Speeding up hyper-parameter optimization by extrapolation of learning curves using previous builds. In: Ceci, M., Hollmén, J., Todorovski, L., Vens, C., Džeroski, S. (eds.) ECML PKDD 2017. LNCS (LNAI), vol. 10534, pp. 477–492. Springer, Cham (2017). https://doi.org/10.1007/978-3-319-71249-9_29
5. Chung, J., Gulcehre, C., Cho, K., Bengio, Y.: Empirical evaluation of gated recurrent neural networks on sequence modeling. arXiv preprint arXiv:1412.3555 (2014)
6. Domhan, T., Springenberg, J.T., Hutter, F.: Speeding up automatic hyperparameter optimization of deep neural networks by extrapolation of learning curves. In: Twenty-Fourth International Joint Conference on Artificial Intelligence (2015)
7. Feurer, M., Klein, A., Eggensperger, K., Springenberg, J.T., Blum, M., Hutter, F.: Auto-sklearn: efficient and robust automated machine learning. In: Hutter, F., Kotthoff, L., Vanschoren, J. (eds.) Automated Machine Learning. TSSCML, pp. 113–134. Springer, Cham (2019). https://doi.org/10.1007/978-3-030-05318-5_6
8. Feurer, M., Letham, B., Bakshy, E.: Scalable meta-learning for Bayesian optimization. arXiv preprint arXiv:1802.02219 (2018)
9. Gantner, Z., Drumond, L., Freudenthaler, C., Rendle, S., Schmidt-Thieme, L.: Learning attribute-to-feature mappings for cold-start recommendations. In: 2010 IEEE International Conference on Data Mining, pp. 176–185. IEEE (2010)
10. Gargiani, M., Klein, A., Falkner, S., Hutter, F.: Probabilistic rollouts for learning curve extrapolation across hyperparameter settings. arXiv preprint arXiv:1910.04522 (2019)
11. Gijsbers, P., LeDell, E., Thomas, J., Poirier, S., Bischl, B., Vanschoren, J.: An open source AutoML benchmark. arXiv preprint arXiv:1907.00909 (2019)
12. Jomaa, H.S., Schmidt-Thieme, L., Grabocka, J.: Dataset2Vec: learning dataset meta-features. arXiv preprint arXiv:1905.11063 (2019)
13. Klein, A., Falkner, S., Springenberg, J.T., Hutter, F.: Learning curve prediction with Bayesian neural networks (2016)

14. Li, L., Jamieson, K., DeSalvo, G., Rostamizadeh, A., Talwalkar, A.: Hyperband: a novel bandit-based approach to hyperparameter optimization. J. Mach. Learn. Res. **18**(1), 6765–6816 (2017)
15. Shahriari, B., Swersky, K., Wang, Z., Adams, R.P., De Freitas, N.: Taking the human out of the loop: a review of Bayesian optimization. Proc. IEEE **104**(1), 148–175 (2015)
16. Unterthiner, T., Keysers, D., Gelly, S., Bousquet, O., Tolstikhin, I.: Predicting neural network accuracy from weights. arXiv preprint arXiv:2002.11448 (2020)
17. Volpp, M., et al.: Meta-learning acquisition functions for transfer learning in Bayesian optimization. arXiv preprint arXiv:1904.02642 (2019)
18. Wistuba, M., Pedapati, T.: Learning to rank learning curves. arXiv preprint arXiv:2006.03361 (2020)
19. Wistuba, M., Schilling, N., Schmidt-Thieme, L.: Scalable gaussian process-based transfer surrogates for hyperparameter optimization. Mach. Learn. **107**(1), 43–78 (2018). https://doi.org/10.1007/s10994-017-5684-y
20. Yu, R., Zheng, S., Anandkumar, A., Yue, Y.: Long-term forecasting using higher order tensor RNNs. arXiv preprint arXiv:1711.00073 (2017)
21. Zimmer, L., Lindauer, M., Hutter, F.: Auto-PyTorch tabular: multi-fidelity MetaLearning for efficient and robust AutoDL. arXiv preprint arXiv:2006.13799 (2020)

Streaming Decision Trees for Lifelong Learning

Łukasz Korycki[(✉)] and Bartosz Krawczyk

Department of Computer Science, Virginia Commonwealth University,
Richmond, VA, USA
koryckil@vcu.edu, bkrawczyk@vcu.edu

Abstract. Lifelong learning models should be able to efficiently aggregate knowledge over a long-term time horizon. Comprehensive studies focused on incremental neural networks have shown that these models tend to struggle with remembering previously learned patterns. This issue known as catastrophic forgetting has been widely studied and addressed by several different approaches. At the same time, almost no research has been conducted on online decision trees in the same setting. In this work, we identify the problem by showing that streaming decision trees (i.e., Hoeffding Trees) fail at providing reliable long-term learning in class-incremental scenarios, which can be further generalized to learning under temporal imbalance. By proposing a streaming class-conditional attribute estimation, we attempt to solve this vital problem at its root, which, ironically, lies in leaves. Through a detailed experimental study we show that, in the given scenario, even a rough estimate based on previous conditional statistics and current class priors can significantly improve the performance of streaming decision trees, preventing them from catastrophically forgetting earlier concepts, which do not appear for a long time or even ever again.

Keywords: Lifelong learning · Continual learning · Catastrophic forgetting · Data streams · Decision trees

1 Introduction

Modern machine learning calls for algorithms that are able not only to generalize patterns from a provided data set but also to continually improve their performance while accumulating knowledge from constantly arriving data [12]. Lifelong learning aims at developing models that will be capable of working on constantly expanding problems over a long-time horizon [18]. Such learning models should keep utilizing new instances (i.e., like online learning), new classes (i.e., like class-incremental learning), or even new tasks (i.e., like multi-task learning). Whenever new information becomes available it must be incorporated into the lifelong learning model to expand its knowledge base and make it suitable for predictive analytics over a new, more complex view on the analyzed problem [15].

© Springer Nature Switzerland AG 2021
N. Oliver et al. (Eds.): ECML PKDD 2021, LNAI 12975, pp. 502–518, 2021.
https://doi.org/10.1007/978-3-030-86486-6_31

This requires a flexible model structure capable of continual storage of incrementally arriving data. At the same time, adding a new class or task to the model may cause an inherent bias towards this newly arrived distribution, leading to a decline of performance over previously seen classes/tasks [22]. This phenomenon is known as catastrophic forgetting and must be avoided at all costs, as robust lifelong learning models should be capable of both accumulating new knowledge and retaining the previous one [10]. Most of the research done in this domain focuses on deep neural network architectures. However, lifelong learning has many parallels with data stream mining domain, where other models (especially decision trees and their ensemble versions) are highly effective and popular [12]. Therefore, adapting streaming decision trees is an attractive potential solution to the considered issues, due to their advantages, such as lightweight structure and interpretability.

Research Goal. To propose a lifelong learning version of streaming decision trees that will be enhanced with a modified splitting mechanism offering robustness to catastrophic forgetting, while maintaining all the advantages of this popular streaming classifier.

Motivation. Streaming decision trees are highly popular and effective algorithms for learning from continuously arriving data. They offer a combination of a lightweight model, adaptiveness, and interpretability while being able to handle ever-growing streams of instances. Streaming decision trees have not been investigated from the perspective of lifelong learning problems that impose the need for not only integrating new knowledge into the model, but also retaining the previously learned one. This calls for modifications of the streaming decision tree induction algorithms that will make them robust to catastrophic forgetting when creating new splits over newly appearing classes or tasks.

Overview. We offer a detailed analysis of Hoeffding Trees in the lifelong learning set-up. We show that neither these trees, nor any streaming ensemble technique using them, can retain useful knowledge over time. Their success in data stream mining can be attributed to their ability to adapt to the newest information, but no research so far has addressed the fact that they cannot memorize learned concepts well over a long-term time horizon. We identify this a fundamental problem can be found at leaves of the streaming decision trees, as they are not able to maintain information about distributions of previously seen classes, and propose a potential solution to the problem.

Main Contributions. This paper offers the following contributions to the lifelong learning domain.

– **Streaming decision trees for lifelong learning.** We propose the first approach for using streaming decision trees for lifelong learning tasks, introducing a modification of Hoeffding Tree that is capable of both incremental addition of new knowledge, as well as retaining the previously learned concepts over a long-term time horizon.

- **New splitting mechanism robust to catastrophic forgetting.** We show that splitting procedure for creating new leaves in Hoeffding Tree directly contribute to the occurrence of catastrophic forgetting. To alleviate this problem, we enhance the streaming tree induction with the propagation of class-conditional attribute estimators and utilization of the class priors during entropy calculation and Bayesian classification.
- **Decision tree ensembles for lifelong learning.** We show that the proposed modification of Hoeffding Tree can be used to create highly effective ensembles robust to catastrophic forgetting, allowing us to introduce Incremental Random Forest for lifelong learning.
- **Detailed experimental study.** We evaluate the robustness of the proposed streaming decision trees through a detailed experimental study in the lifelong learning setting. We evaluate not only the global and per-class accuracy over time, but additionally the propagation of errors and model retention after being exposed to multiple new classes.

2   Related Works

Data Streams. Learning from data stream focuses on developing algorithms capable of batch-incremental or online processing of incoming instances [4]. Due to the high velocity of data, time and memory constraints are important, as algorithms should be lightweight and capable of fast decision-making [12]. The focus is put on adaptation to the current state of the stream, as concept drift may dynamically impact the properties of data [13]. Thus, streaming algorithms offer high-speed and adaptive learners that provide powerful capabilities for learning from new information [3]. At the same time, knowledge aggregation and retaining mechanisms are not commonly investigated, making streaming algorithms unsuitable for lifelong learning.

Catastrophic Forgetting. Lifelong learning focuses on preserving knowledge learned over a long-term time horizon, mainly with the usage of deep neural networks [18]. It has been observed that these models are biased toward the newest class, while gradually dropping their performance on older classes, which is known as catastrophic forgetting [21]. Several interesting solutions have been proposed to make neural networks robust to this phenomenon, such as experience replay [6], masking [14] or hypernetworks [16]. Despite the fact that not only neural networks suffer from catastrophic forgetting, the research on avoiding its occurrence in other learning models is still very limited.

3   Decision Trees and Lifelong Learning

Typical scenarios of lifelong learning and catastrophic forgetting involve cases in which classes arrive subsequently one after another. This means that once a given class was presented it may never appear again. Extensive works on using neural networks in such scenarios showed that such settings lead to severe learning problems for them, as mentioned in Sect. 2. While very little attention has been given to decision trees in similar scenarios, our preliminary studies of hybridizing convolutional networks with tree-based classifiers for lifelong learning indicated that streaming decision trees may struggle with exactly the same problems as neural networks. In this section, we want to emphasize this issue and propose a possible solution.

3.1 Forgetting in Streaming Decision Trees

Online decision trees have been proven to be excellent algorithms for learning from stationary and non-stationary data streams [2]. However, a more in-depth analysis of the conducted experimental research may reveal that algorithms like Hoeffding Tree [5] and Adaptive Random Forest [7] have been evaluated mainly in scenarios where incoming data per class is generally uniformly distributed over time, which means that instances of different classes are reasonably mixed with each other, without long delays between them [12]. Although researchers usually take into consideration the dynamic imbalance of analyzed streams [11], they still assume that instances of all classes appear rather frequently, even if ratios between them are skewed. The class-incremental scenarios are edge cases of extreme temporal imbalance, where the older classes do not appear ever again and the newer ones completely dominate the learning process. Let us introduce the main components of the state-of-the-art streaming decision trees and analyze what consequences the given scenario has for them.

Entropy and Splits. The Hoeffding Tree model is built upon two fundamental components used at leaves: (i) Hoeffding bound that determines when we should split a node, and (ii) node statistics that are used for finding the best splits. The former is defined as:

$$\epsilon = \sqrt{\frac{R^2 \ln(1/\delta)}{2n}}, \tag{1}$$

where R is a value range, equal to $R = \log C$ for information gain calculations (C is the total number of classes), n is a number of examples seen at a node and δ is a confidence parameter. If a difference between the best potential split and the current state of the node is greater than ϵ, then there is a $1 - \delta$ confidence that the attribute introduces superior information gain and it should be used to create a split. We can express it using the following condition:

$$\Delta G(x_i, s_j) = E(x_i, s_j) - E_0 > \epsilon, \tag{2}$$

where the best potential information gain $\Delta G(x_i, s_j)$ is equal to the difference between the entropy after the best possible split $E(x_i, s_j)$ on an attribute x_i using a split value s_j, and before the split E_0. Although the condition alone is not directly related to the forgetting problem, the entropy values are, as we will show in the next steps.

The entropy for a given binary split s_j on an attribute x_i can be calculated as:

$$E(x_i, s_j) = \sum_{k=1}^{C} -p(c_k|x_i \leq s_j) \log(p(c_k|x_i \leq s_j)) - p(c_k|x_i > s_j) \log(c_k|x_i > s_j) \tag{3}$$

which simply boils down to the entropy on the left $(x_i \leq s_j)$ from the split s_j and on the right $(x_i > s_j)$. For the current entropy E_0 at the node we simply have:

$$E_0 = \sum_{k=1}^{C} -p(c_k) \log(p(c_k)). \tag{4}$$

Based on the given formulas, in order to find the best potential splits over all attributes and classes, we need to maintain two groups of estimators at leaves: (i) class priors $p(c_k)$, and (ii) conditional class probabilities $p(c_k|x_i)$. The former estimations can be easily obtained by counting occurrences of each class:

$$p(c_k) = \frac{n_k}{n}, \tag{5}$$

where n_k is the number of instances of class k counted for a node and n is the total number of examples received. For the latter values we use the fact that we have discrete classes and apply the conditional probability formula:

$$p(c_k|x_i) = \frac{p(x_i|c_k)p(c_k)}{p(x)}, \tag{6}$$

where $p(x)$ is the normalizing constant for all classes. The prior probability $p(c_k)$ can be omitted here, as a part of the prior scaling, to alleviate the class imbalance problems [11]. The required class-conditional attribute probabilities $p(x_i|c_k)$ are modeled using Gaussian estimators, which provide a quick and memory efficient way of obtaining the required values [19]. We use triplets consisting of a count $n_{k,i}$, mean $\mu_{k,i}$ and variance $\sigma_{k,i}$ for all pairs of classes c_k and attributes x_i. By having those models we can easily apply Eq. 6 to obtain $p(c_k|x_i \leq s_j)$ and $p(c_k|x_i > s_j) = 1.0 - p(c_k|x_i \leq s_j)$. We end up with $p(x_i \leq s_j|c_k)$, which can be calculated using the cumulative distribution function for the standard normal distribution $\Phi_k(s_j)$. It can be expressed using the error function:

$$p(x_i \leq s_j|c_k) = \Phi_k(s_j) = 0.5(1 + erf_k(s_j/\sqrt{2}), \tag{7}$$

where the value of the error function can be calculated using the stored triplets.

Finally, after finding the best possible split s_j for an attribute x_i that minimizes the entropy after a split (Eq. 3) and passing the Hoeffding bound test (Eq. 2) we can split the node and estimate the total number of instances that will go to the left and right child:

$$p_l(c_k) = p(c_k)p(c_k|x_i \leq s_j) = 1 - p_r(c_k), \tag{8}$$

where $p_l(c_k)$ and $p_r(c_k)$ are priors for the left and right child for the given class c_k, and x_i is the selected split attribute.

By default, we omit the estimation of all $p(c_k|x_i)$ after the split as it is a non-trivial task, which most likely cannot be quickly solved in the current form of the algorithm. This fact has a crucial impact on the streaming decision trees in the class-incremental scenario as we will show in the subsequent paragraphs.

Classification at Leaves. After forwarding an incoming instance to a leaf in the decision tree, it is classified using majority voting based on the class priors. To improve the classification process the simple procedure is often combined with a naive Bayes classifier [1], which can be easily applied using the already stored estimators:

$$p(c_k|\mathbf{x}) = \frac{p(\mathbf{x}|c_k)p(c_k)}{p(\mathbf{x})}, \tag{9}$$

where $\mathbf{x}$ is the vector of input attributes and $p(\mathbf{x}|c_k)$ is equal to:

$$p(\mathbf{x}|c_k) = \prod_{i=1}^{m} p(x_i|c_k), \tag{10}$$

where m is the number of features. Each $p(x_i|c_k)$ can be obtained using the Gaussian density function.

Forgetting Scenario. After the introduction of the leaf components and required calculations, let us now consider what will happen in the class-incremental scenario after subsequent splits. In Fig. 1 we can see an example of a sequence of 3 class batches. At the beginning, there are only instances of the first class (C0) for which the algorithm accumulates values for the prior count (Eq. 5) and conditional estimators (Eq. 6) only at the root, since there is no need for a split.

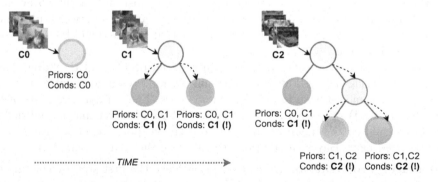

Fig. 1. Catastrophic forgetting in streaming decision trees learning from a class-incremental sequence.

Next, the second class (C1) starts arriving and at some point the Hoeffding Tree algorithm finds a good split, which creates two additional nodes and distributes priors accordingly to Eq. 8. After this step, the child nodes have some smaller priors for C0 and C1, however, the conditional estimators have been reset by default. Although we can assume that after the split some instances of class C1 can still appear and rebuild the conditional estimators, there is no chance that the same will happen for C0, which means that while its priors will be good for now, its conditional estimators (Eq. 6) will remain equal to zero, resulting in an inability of the naive Bayes classifier (Eq. 9) to recognize this class.

When the next class starts coming (C2) we can already observe a problem – since there are no instances of C0, its $p(c_0|x_i)$ is still equal to zero, which leads to the situation in which the older class is completely ignored during the entropy calculations when looking for a split (Eq. 3). Finally, once a new split is created, there will be no prior for the class at the newest leaves, since based on Eq. 8 it has to be zeroed. This concludes the learning process for class C0 which has been completely erased at the third level of the tree, and which may very likely disappear from the model completely. Even worse is the fact that the same will most likely happen to C1 and C2 as soon as new classes arrive.

Based on the analysis, we can conclude that in the class-incremental scenario, catastrophic forgetting in streaming decision trees manifests itself in three ways: **(i)** by excluding older classes from a meaningful contribution to the best split criterion, **(ii)** by disabling the conditional classification, and finally **(iii)** by erasing priors which leads to complete class forgetting at a given node.

3.2   Overcoming Catastrophic Forgetting

The observations from the previous section clearly indicate that the source of the problem with forgetting can be found at leaves and their conditional estimators. It is worth emphasizing that this issue practically does not exist in most of the commonly used data stream benchmarks, which provide instances of different classes for most of the time during the learning process. In such a case, the estimators can always rebuild themselves after new instances arrive, preventing them from forgetting most of the classes. The longer are gaps between subsequent instances of one class, the higher is the chance that the class will be temporarily or forever forgotten.

To make a step towards solving the introduced problem in Hoeffding Trees, we propose using a rough class-conditional attribute estimation after the split to prevent the model from forgetting older classes. The approach consists of two modifications: **(i)** propagating class-conditional attribute estimators (needed for Eq. 7) to children of a node being split, and **(ii)** keeping the class priors in the entropy and naive Bayes calculations to calibrate the rough estimation.

Estimator Propagation. We can simply achieve the first step by copying the Gaussian parameters of each class-conditional distribution $p_{t-1}(x_i|c_k)$ before split at time step t to the left node with $p_{t,l}(x_i|c_k)$ and to the right one with $p_{t,r}(x_i|c_k)$, which results in:

$$p_{t,l}(x_i|c_k) = p_{t,r}(x_i|c_k) = p_{t-1}(x_i|c_k), \tag{11}$$

for each class c_k and attribute x_i. This is obviously a very rough estimate, however, since we assume simple Gaussian distributions, the error does not have to be critical and may provide more benefits than obstructions. Most likely, providing any platform for an older class is more important than the risk of making the estimation error. In addition, the estimate may still be fine-tuned by instances that come to this node before the class batch ends.

Prior Scaling. By sticking to the prior probabilities $p(c_k)$ in the entropy calculations (Eq. 3) and Bayesian classification (Eq. 9), we attempt to somehow adjust the rough estimate from the previous step. Since the split class priors are relatively well-estimated, we can utilize them to softly scale the class-conditional distributions to become more adequate to the state after the split. Although this step does not change the shape of the distribution horizontally, it may increase or decrease the influence of the distribution by scaling it vertically based on the formula:

$$p_t(x_i|c_k) = p_{t-1}(x_i|c_k)p_t(c_k). \tag{12}$$

Ensembles. Finally, the modified Hoeffding Tree can be simply used as a base learner of the Incremental Random Forest, which is an Adaptive Random Forest without change detectors and node replacement mechanisms. The only difference between the standard forest and the ensemble using our modified tree is that we have to keep statistics for all attributes at leaves, not only for those within a random subspace, since we do not know which attributes will be needed at a lower level. By combining the robustness of ensemble techniques with improvements of the base learner we may potentially alleviate the catastrophic forgetting problem even more.

4   Experimental Study

In the following experiments, we aim at proving that our proposed modifications of the Hoeffding Tree algorithm are capable of alleviating the catastrophic forgetting in decision trees learning from class-incremental streams, allowing for the application of these models in such scenarios. Our goal was to answer the following research questions.

- **RQ1**: Does the proposed algorithm effectively address the problem of catastrophic forgetting in streaming decision trees?
- **RQ2**: Can the presented decision tree be utilized as a base learner of a random forest? Does it further improve the classification performance?
- **RQ3**: Is it possible to solve the presented problem by using a different already available ensemble technique?

In order to improve reproducibility of this work, all of the presented algorithms and details of the evaluation have been made available in a public repository: github.com/lkorycki/lldt.

4.1   Data

To evaluate the baseline and proposed models in the scenario of lifelong learning and catastrophic forgetting, we used popular visual data sets commonly used for the given task. The first three were used as simpler sequences consisting of 10 classes: **MNIST**, **FASHION**, **SVHN**. Next, we utilized 20 superclasses of the CIFAR100 data set (**CIFAR20**), as well as we extracted two 20-class subsets of the IMAGENET: **IMAGENET20A** and **IMAGENET20B**. All of the sets were transformed into class-incremental sequences in which each batch contained only one class and each class was presented to a classifier only once. All of the evaluated models were processing the incoming batches in a streaming manner, one instance after another.

The MNIST and FASHION data sets were transformed into a series of flattened arrays (from raw images), which provided us with feature vectors of size 784. The rest of the used benchmarks were pre-processed using pre-trained feature extractors. For SVHN and CIFAR20 we used ResNeXt-29 with its cardinality equal to 8 and using widen factor equal to 4. We extracted the output of the last 2D average pooling and processed it with an additional 1D average pooling, which resulted in a feature vector consisting of 512 values. For the IMAGENET-based sets we directly utilized the output of the last average pooling layer of the ResNet18 model, which once again gave us 512-element vectors.

4.2   Algorithms

In our experiments, we compared the proposed single tree (**HT+AE**) with the original streaming algorithm (**HT**) [5], as well as the incremental random forest using our base learner (**IRF+AE**) with its baseline (**IRF**) to answer the first two research questions. Next, we evaluated other ensemble techniques to check whether it is possible that a solution to the introduced problem lies solely in a different committee design (the last research question). We investigated drift-sensitive Adaptive Random Forest (**ARF**) [7], online bagging without random subspaces per node (**BAG**) [17], online random subspaces per tree (**RSP**) [8] and the ensemble of 1-vs-all classifiers (**OVA**) [9].

All of the algorithms used Hoeffding Trees as base learners with confidence set to $\delta = 0.01$, bagging lambda equal to $\lambda = 5$, split step $s = 0.1$ (10% of a difference between the maximum and minimum attribute value) and split wait equal to $w = 100$ for all sets except for the slightly smaller IMAGENET-based ones for which we set $w = 10$. All of the ensembles used $n = 40$ base learners.

4.3   Evaluation

Firstly, for all of the considered sequences, we measured **hold-out accuracy** [20] per each class after each class batch and used it to calculate the average accuracy per batch and the overall average for a whole sequence. Secondly, we collected data for **confusion matrices** after each batch to generate the average matrices which could help us illustrate the bias related to catastrophic forgetting.

Finally, we measured the **retention** of the baseline and improved algorithms to show how well the given models remember previously seen concepts.

4.4 Results

Analysis of the Average Predictive Accuracy. Table 1 presents the average accuracy over all classes for all six used class–incremental benchmarks. This is the bird's eye view on the problem and the performance of the analyzed methods, allowing us to assess the general differences among the algorithms. We can see that the standard HT and IRF were significantly outperformed by the proposed HT+AE and IRF+AE approaches. For HT the proposed propagation of class-conditional attribute estimators and storing the class priors led to very significant improvements on all data sets, which is especially visible on CIFAR20 (almost 0.3) and IMAGENET20A (0.2). Similar improvements can be observed for IRF, especially for CIFAR20 where the modifications led to 0.28 improvement. The SVHN benchmark shows the smallest improvements out of all six data sets, which can be explained by the extractor potentially being very strongly fine-tuned for this problem. Thus extracting well-separated class embeddings may slightly alleviate the catastrophic forgetting on its own (although the proposed modifications still help).

The Impact of Different Ensemble Architectures. To truly understand the impact of catastrophic forgetting on HT and IRF, we decided to see if other ensemble architectures may behave better in class-incremental lifelong learning scenarios. Table 1 presents results for four other popular streaming ensemble architectures. We can see that all of them performed poorly on every data set, offering inferior predictive accuracy to the baseline IRF. This shows that the choice of an ensemble architecture on its own does not offer improved robustness to catastrophic forgetting. As a result, we have a good indication that our modifications of the HT splitting procedure are the sole source of the achieved impressive gains in accuracy. However, a more in-depth analysis of these models will allow us to gain better insights into the nature of catastrophic forgetting in streaming decision trees.

Table 1. The average accuracy on all class-incremental sequences.

Model	MNIST	FASHION	SVHN	CIFAR20	IMGN20A	IMGN20B
HT	0.6283	0.5720	0.8845	0.3511	0.4589	0.5301
HT+AE	**0.8398**	**0.7037**	**0.9510**	**0.6497**	**0.6530**	**0.6730**
IRF	0.8662	0.7355	0.9334	0.4467	0.6890	0.7500
IRF+AE	**0.9645**	**0.8698**	**0.9733**	**0.7298**	**0.7777**	**0.8121**
ARF	0.2929	0.2929	0.2929	0.1799	0.2411	0.2849
OVA	0.3416	0.2929	0.5033	0.1805	0.3842	0.3847
BAG	0.7096	0.6446	0.9029	0.3737	0.5709	0.6635
RSP	0.6202	0.5898	0.9087	0.3734	0.6337	0.6995

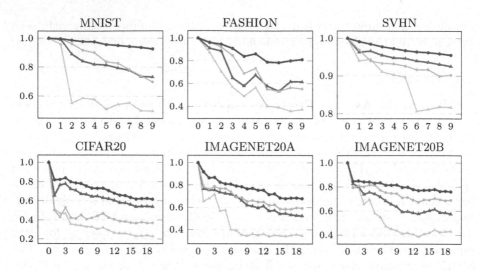

Fig. 2. The average class accuracy for the baseline tree-based models (▲ HT, ● IRF) and the proposed ones (▲ HT+AE, ● IRF+AE) after each class batch.

Analysis of the Class-Batch Performance. Figure 2 depicts the average accuracy after each class appearing incrementally. This allows us to visually analyze the stability of the examined methods and their response to the increasing model size (when more and more classes need to be stored and remembered). We can see that both proposed HT+AE and IRF+AE offered significantly improved stability over the baseline approaches, maintaining their superior predictive accuracy regardless of the number of classes. Additionally, we can see that the baseline models tended to deteriorate faster when the number of classes became higher (e.g., HT and IRF on MNIST and FASHION). At the same time, the proposed modifications could accommodate all the classes from the used benchmarks without destabilization of their performance. It is worth noting that HT+AE was often capable of outperforming IRF. This is a very surprising observation, as the modification of class-conditional estimators allows a single decision tree to outperform a powerful ensemble classifier. This shows that the proposed introduction of robustness to catastrophic forgetting into streaming decision trees is a crucial improvement of their induction mechanisms.

Analysis of the Class-Based Performance. Figure 3 depicts the accuracy per batch on selected classes. This allows us to understand how the appearance of new classes affects the performance on previously seen ones. We can clearly see that both HT and IRF were subject to catastrophic forgetting, very quickly forgetting the old classes. While they were very good at learning the newest concept, their performance degraded with every newly arriving class, showing their capabilities of aggressively adapting to new knowledge, but not retaining it over time. This was especially vivid for the first class for each data set (C0), where it was completely forgotten (i.e., accuracy on it drops to zero) as

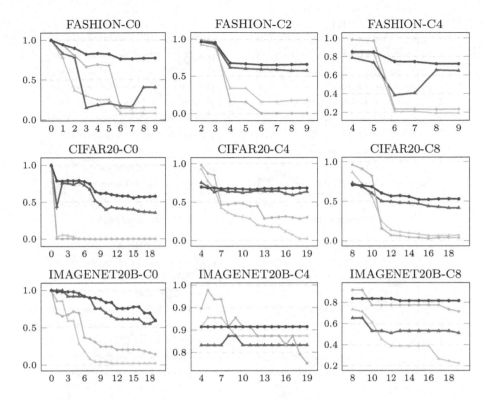

Fig. 3. The average accuracy for selected classes of FASHION, CIFAR20 and IMA-GENET20B for the baseline tree-based models (▲ HT, ● IRF) and the proposed ones (▲ HT+AE, ● IRF+AE) after subsequent class batches.

soon as 1–2 new classes appeared. The proposed propagation of class-conditional attribute estimators and storing the class priors in HT+AE and IRF+AE led to a much better retaining of knowledge extracted from old classes. In some cases (e.g., CIFAR20 or IMAGENET20) we can see that the accuracy for old classes remained almost identical through the entire duration of the lifelong learning process. This is a highly sought-after property and attests to the effectiveness of our proposed modifications.

Analysis of the Confusion Matrices. Figure 4 depicts the confusion matrices averaged over all examined data sets (10 classes from each for the visualization sake). Based on that we can directly compare how errors are distributed among classes for HT vs. HT+AE and IRF vs. IRF+AE. We can see that the proposed modifications in HT+AE and its ensemble version led to a much more balanced lifelong learning procedure that both avoided the bias towards the newest class (i.e., is robust to catastrophic forgetting) and the bias towards older classes (i.e., offers capabilities for incorporating new information into the model in an effective manner). These confusion matrices further confirm our observations

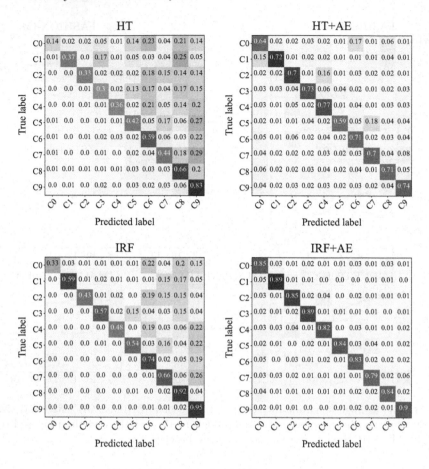

Fig. 4. The average confusion matrices.

made in previous points of this discussion that our proposed modifications lead
to robust streaming decision tree induction for lifelong learning.

Analysis of the Retention of Information. Figure 5 depicts the average
retention of information about a class after +k new classes appeared. This helps
us analyze how each of examined models manages its knowledge base and how
flexible it is to add new information to it. An ideal model would perfectly retain
the performance on every previously seen class, regardless of how many new
classes it has seen since then. We can see that the baseline HT and IRF offered
very good performance on the newest class, but drastically dropped it after see-
ing as little as 2 new classes. This further enforces our hypothesis that standard
decision trees and their ensembles cannot avoid catastrophic forgetting and thus
cannot be directly used for lifelong learning. However, when we enhance HT with
the proposed propagation of class-conditional attribute estimators and storing
the class priors, we obtain a streaming decision tree that can learn new infor-

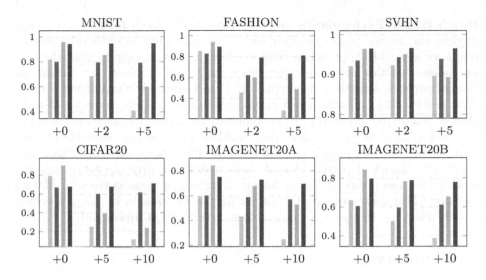

Fig. 5. The average retention after +k class batches since the moment a class appeared for: HT, HT+AE, IRF, IRF+AE

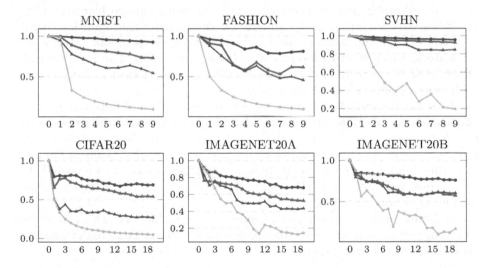

Fig. 6. The average class accuracy for other baseline models (▲ OVA, ▲ BAG) and the proposed ones (● HT+AE, ● IRF+AE) after each class batch.

mation almost as effectively as its standard counterpart, while offering excellent robustness to catastrophic forgetting (**RQ1 answered**). Furthermore, we can see that HT+AE can be utilized as a base learner for ensemble approaches, leading to even further improvements in its accuracy and information retention (**RQ2 answered**).

L. Korycki and B. Krawczyk

Batch-Based Performance of the Ensemble Architectures. Figure 6 depicts the average accuracy after each class appearing incrementally for the reference ensemble approaches. This confirms our observations from the earlier point that the ensemble architecture itself does not have any impact on the catastrophic forgetting occurrence. Reference methods use different ways of data partitioning (subsets of instances, features, or classes), but none of them allowed for better retention of old information. What is highly interesting is that HT+AE (a single decision tree) could outperform any ensemble of trees that do not use our proposed modifications. This shows the importance and significant impact of propagation of class-conditional attribute estimators and storing the class priors on the usefulness of streaming decision trees for lifelong learning. Therefore, catastrophic forgetting can be avoided by using a robust base learner, not changing the ensemble structure (**RQ3 answered**).

5 Summary

In this work, we identified and emphasized the issue of catastrophic forgetting that occurs when traditional streaming decision trees attempt to learn in class-incremental lifelong learning scenarios. Through an in-depth analysis of the Hoeffding Tree algorithm, we found out that the source of the algorithm's weakness comes from the lack of additional support for class-conditional attribute estimators, which tend to forget older classes after splits. The issue critically affects different aspects of tree-based learning, ranging from the procedure for finding new splits to classification on leaves.

To solve the introduced problem, we proposed a rough estimation of the conditional distributions after a split, based on distributions and priors aggregated at a node before it is divided. Our extensive experimental study has shown that this simple yet effective approach is capable of providing excellent improvements for both single trees and incremental forests. As a result, we proved that the proposed method turns the standard streaming trees into learners suitable for lifelong learning scenarios.

In future works, we plan to find more precise estimators, which may need to be supported by some local experience replay utilizing small buffers of either input instances or prototypes.

References

1. Bifet, A., Gavaldà, R.: Adaptive learning from evolving data streams. In: Adams, N.M., Robardet, C., Siebes, A., Boulicaut, J.-F. (eds.) IDA 2009. LNCS, vol. 5772, pp. 249–260. Springer, Heidelberg (2009). https://doi.org/10.1007/978-3-642-03915-7_22
2. Bifet, A., et al.: Extremely fast decision tree mining for evolving data streams. In: Proceedings of the 23rd ACM SIGKDD International Conference on Knowledge Discovery and Data Mining, Halifax, NS, Canada, 13–17 August 2017, pp. 1733–1742. ACM (2017)

3. Cano, A., Krawczyk, B.: Kappa Updated Ensemble for drifting data stream mining. Mach. Learn. **109**(1), 175–218 (2019). https://doi.org/10.1007/s10994-019-05840-z

4. Ditzler, G., Roveri, M., Alippi, C., Polikar, R.: Learning in nonstationary environments: a survey. IEEE Comput. Intell. Mag. **10**(4), 12–25 (2015)

5. Domingos, P.M., Hulten, G.: Mining high-speed data streams. In: Ramakrishnan, R., Stolfo, S.J., Bayardo, R.J., Parsa, I. (eds.) Proceedings of the Sixth ACM SIGKDD International Conference on Knowledge Discovery and Data Mining, Boston, MA, USA, 20–23 August 2000, pp. 71–80. ACM (2000)

6. Fujimoto, S., Meger, D., Precup, D.: An equivalence between loss functions and non-uniform sampling in experience replay. In: Advances in Neural Information Processing Systems 33: Annual Conference on Neural Information Processing Systems 2020, NeurIPS 2020, 6–12 December 2020, virtual (2020)

7. Gomes, H.M., et al.: Adaptive random forests for evolving data stream classification. Mach. Learn. **106**(9), 1469–1495 (2017). https://doi.org/10.1007/s10994-017-5642-8

8. Gomes, H.M., Read, J., Bifet, A.: Streaming random patches for evolving data stream classification. In: 2019 IEEE International Conference on Data Mining, ICDM 2019, Beijing, China, 8–11 November 2019, pp. 240–249. IEEE (2019)

9. Hashemi, S., Yang, Y., Mirzamomen, Z., Kangavari, M.R.: Adapted one-versus-all decision trees for data stream classification. IEEE Trans. Knowl. Data Eng. **21**(5), 624–637 (2009)

10. Korycki, Ł., Krawczyk, B.: Class-incremental experience replay for continual learning under concept drift. CoRR abs/2104.11861 (2021). arXiv:2104.11861

11. Korycki, Ł., Krawczyk, B.: Online oversampling for sparsely labeled imbalanced and non-stationary data streams. In: 2020 International Joint Conference on Neural Networks, IJCNN 2020, Glasgow, United Kingdom, 19–24 July 2020, pp. 1–8. IEEE (2020)

12. Krawczyk, B., Minku, L.L., Gama, J., Stefanowski, J., Wozniak, M.: Ensemble learning for data stream analysis: a survey. Inf. Fusion **37**, 132–156 (2017)

13. Lu, J., Liu, A., Dong, F., Gu, F., Gama, J., Zhang, G.: Learning under concept drift: a review. IEEE Trans. Knowl. Data Eng. **31**(12), 2346–2363 (2019)

14. Mallya, A., Davis, D., Lazebnik, S.: Piggyback: adapting a single network to multiple tasks by learning to mask weights. In: Ferrari, V., Hebert, M., Sminchisescu, C., Weiss, Y. (eds.) ECCV 2018. LNCS, vol. 11208, pp. 72–88. Springer, Cham (2018). https://doi.org/10.1007/978-3-030-01225-0_5

15. Mishra, M., Huan, J.: Learning task grouping using supervised task space partitioning in lifelong multitask learning. In: Proceedings of the 24th ACM International Conference on Information and Knowledge Management, CIKM 2015, Melbourne, VIC, Australia, 19–23 October 2015, pp. 1091–1100. ACM (2015)

16. von Oswald, J., Henning, C., Sacramento, J., Grewe, B.F.: Continual learning with hypernetworks. In: 8th International Conference on Learning Representations, ICLR 2020, Addis Ababa, Ethiopia, 26–30 April 2020. OpenReview.net (2020)

17. Oza, N.C.: Online bagging and boosting. In: Proceedings of the IEEE International Conference on Systems, Man and Cybernetics, Waikoloa, Hawaii, USA, 10–12 October 2005, pp. 2340–2345. IEEE (2005)

18. Parisi, G.I., Kemker, R., Part, J.L., Kanan, C., Wermter, S.: Continual lifelong learning with neural networks: a review. Neural Netw. **113**, 54–71 (2019)

19. Pfahringer, B., Holmes, G., Kirkby, R.: Handling numeric attributes in Hoeffding trees. In: Washio, T., Suzuki, E., Ting, K.M., Inokuchi, A. (eds.) PAKDD 2008. LNCS (LNAI), vol. 5012, pp. 296–307. Springer, Heidelberg (2008). https://doi.org/10.1007/978-3-540-68125-0_27
20. Raschka, S.: Model evaluation, model selection, and algorithm selection in machine learning. CoRR arXiv:1811.12808 (2018)
21. Yao, X., Huang, T., Wu, C., Zhang, R., Sun, L.: Adversarial feature alignment: avoid catastrophic forgetting in incremental task lifelong learning. Neural Comput. **31**(11), 2266–2291 (2019)
22. Zaidi, N.A., Webb, G.I., Petitjean, F., Forestier, G.: On the inter-relationships among drift rate, forgetting rate, bias/variance profile and error. CoRR abs/1801.09354 (2018). arXiv:1801.09354

Transfer and Multi-task Learning

Unifying Domain Adaptation and Domain Generalization for Robust Prediction Across Minority Racial Groups

Farzaneh Khoshnevisan[1]([✉]) and Min Chi[2]

[1] Intuit Inc., San Diego, USA
farzaneh_khoshnevisan@intuit.com
[2] Department of Computer Science, North Carolina State University, Raleigh, USA
mchi@ncsu.edu

Abstract. In clinical deployment, the performance of a model trained from one or more medical systems often deteriorates on another system and such deterioration is especially evident among minority patients who often have limited data. In this work, we present a multi-source adversarial domain separation (MS-ADS) framework which unifies domain adaptation and domain generalization. MS-ADS is designed to address two types of discrepancies: *covariate shift* stemming from differences in patient populations, and *systematic bias* on account of differences in data collection procedures across medical systems. We evaluate MS-ADS for early prediction of *septic shock* on three tasks. On *a task of domain adaptation* across three medical systems, we show that by leveraging data from multiple systems while accounting for both types of discrepancies, MS-ADS improves the prediction performance across all three systems; on *a task of domain generalization to an unseen medical system*, we show that MS-ADS can perform better than or close to the gold standard supervised models built for the system; last but not least, on a task that involves both domain adaptation and domain generalization. *generalization to unseen racial groups across medical systems*, MS-ADS shows robust out-performance by addressing covariate shift across different racial groups and systematic bias across medical systems simultaneously.

Keywords: Domain adaptation · Domain generalization · Cross-racial transfer · Septic shock

1 Introduction

Machine learning is used increasingly in clinical care to improve diagnosis, treatment policy, and healthcare efficiency. Because machine learning models learn from historically collected data, electronic health records (EHRs), populations that are under-represented in the training data are often vulnerable to harm by incorrect predictions. For example, between the two medical systems involved

© Springer Nature Switzerland AG 2021
N. Oliver et al. (Eds.): ECML PKDD 2021, LNAI 12975, pp. 521–537, 2021.
https://doi.org/10.1007/978-3-030-86486-6_32

in this work, the percentages of White vs. African American are 71% vs. 22.5% in Christiana Care whereas 91% vs. 3% in Mayo clinic. For certain diseases like sepsis, different racial groups often exhibit distinct progression patterns [35]. Therefore, a model that can leverage EHRs across multiple medical systems to improve prediction among minority racial groups is needed. However, EHRs across medical systems can vary dramatically because different systems serve different demographic populations and often employ different infrastructure, workflows and administrative policies [1]. For this work, we refer to the discrepancies caused by the heterogeneous patient populations as *covariate shift* and those caused by incompatible data collection procedures as *systematic bias*.

We propose a multi-source adversarial domain separation (MS-ADS) framework which unifies domain adaptation and domain generalization. More specifically, MS-ADS separates the local representation of each domain from the global latent representation across all domains to address *systematic bias* and leverages multi-domain discriminator in conjunction with gradient reversal layer to address the *covariate shift* across each pair of domains. More specifically, our MS-ADS is built atop variational recurrent neural networks (VRNN) [5] due to VRNN's ability to handle variabilities in EHRs, such as missing data, and its ability to capture complex conditional and temporal dependencies [26,39]; it is shown that VRNN significantly outperforms commonly-used variations of RNN such as long short-term memory (LSTM) on EHRs [16,39]. The effectiveness of MS-ADS is compared against another strong VRNN-based domain adaptation framework called VRADA [28] for early prediction of a challenging condition in hospitals, septic shock. Sepsis is a life-threatening condition caused by a dysregulated body response to infection [32]. Septic shock is the most severe complication of sepsis, associated with high mortality rate and prolonged length of hospitalization [32]. Timing is critical for this condition as every hour delay in antibiotic treatment leads to 8% increase in the chance of mortality. Early prediction of septic shock is challenging due to vague symptoms and subtle body responses [19]. Also, sepsis, like cancer, involves various disease etiologies that span a wide range of syndromes, and different patient groups may show vastly different symptoms [35].

To investigate the early prediction of septic shock, we leverage EHRs collected from three large medical systems located in different parts of the US. The effectiveness of MS-ADS is evaluated on three tasks involving domain adaptation (DA), domain generalization (DG), or both. First, on a task of *DA across three medical systems*, we compare MS-ADS against VRADA and a VRNN model trained on all three domains and show that MS-ADS improves the prediction performance across the three domains and outperforms all baselines. Further, through visualization we show that MS-ADS indeed capture both covariate shift and systematic bias. Second, on a task of *DG to an unseen system*, we evaluate the performance of MS-ADSs trained with two medical systems on a third target system. The results suggest that MS-ADS can perform as well as or better than the gold standard: supervised model trained on the target domain. Finally, probably the most important, we evaluate MS-ADS on the task of *generalization to an unseen racial group across medical systems.* We demonstrate that by treating each medical system and each racial group as a separate domain, our MS-ADS is capable of addressing both covariate shifts across different racial groups and systematic bias across medical systems. Our results suggest that MS-ADS significantly improves generalization performance to African American population in Mayo as compared to the other baselines. Our contributions are:

- By tackling two different types of discrepancies, MS-ADS can effectively leverage EHRs from multiple *medical systems* to improve prediction performance on each system individually and also combined.
- Domain-invariant representations generated by MS-ADS are generalizable to new domains such that they perform close to or better than the gold standard supervised models trained on those systems.
- By unifying DA and DG, as far as we know, MS-ADS is the first framework that shows great potential on generalization to unseen racial groups across medical systems.

2   Methodology

Problem Description. We have K domains: $\{\mathcal{D}_1, \mathcal{D}_2, ..., \mathcal{D}_K\}$ and a domain contains n patient visits represented as $X = \{\mathbf{x}^1, ..., \mathbf{x}^n\}$. Each visit $\mathbf{x}^i$ is a multivariate time-series that is composed of T^i medical events and can be denoted as $\mathbf{x}^i = (\boldsymbol{x}_t^i)_{t=1}^{T^i}$ where $\boldsymbol{x}_t^i \in \mathbb{R}^D$. Additionally, each visit has a visit-level outcome label represented as $Y = \{\mathrm{y}^1, ..., \mathrm{y}^n\}$ where $\mathrm{y}^i \in \{1, 0\}$ indicates the outcome of visit i: septic shock or non-septic shock. By combining X and Y for each domain, we have: $\mathcal{D}_k = \{\mathbf{x}_{\mathcal{D}_k}^i, \mathrm{y}_{\mathcal{D}_k}^i\}_{i=1}^{n_k}$, where n_k is the number of visits in $\mathcal{D}_k$; Here we assume each $\{\mathbf{x}_{\mathcal{D}_k}^i, \mathrm{y}_{\mathcal{D}_k}^i\}_{i=1}^{n_k}$ is drawn from distribution $p_k(\mathbf{x}, \mathrm{y})$ that is different from $\{p_j(\mathbf{x}, \mathrm{y}) : j \neq k\}$. Our objective is to minimize the discrepancies between these K domains in a common latent space by aligning their latent representations: $\mathbf{z}_{\mathcal{D}_1}, ..., \mathbf{z}_{\mathcal{D}_K}$, so that to create a unified, generalizable classifier $C : \mathbf{z} \mapsto \mathrm{y}$ that predicts the outcome optimally in *all* K domains. To do so, we adversarially learn $\binom{K}{2}$ discriminators to minimize the distance between global latent representations of each pair of domain $\mathbf{z}_{\mathcal{D}_i}$ and $\mathbf{z}_{\mathcal{D}_j}$. We describe this framework in detail in the following.

Multi-Source Adversarial Domain Separation (MS-ADS). Figure 1 illustrates MS-ADS architecture: it separates one globally-shared latent representation for all domains from domain-specific (local) information. This architecture would allow global information to be purified so that the discrepancies caused by systematic bias are addressed. MS-ADS employs VRNN as the base model to process sequential input EHRs. VRNN has an encode-decoder structure where its four internal operations interact with each other to capture dependencies between latent random variables across time steps (please see [5] for more details). MS-ADS ensures that the global latent representations are different from the local ones by maximizing a dissimilarity measure. Additionally, multiple domain discriminators and a label predictor are employed to ensure domain-invariant and class-discriminative projection. In the following, we briefly describe the two steps for training the MS-ADS framework.

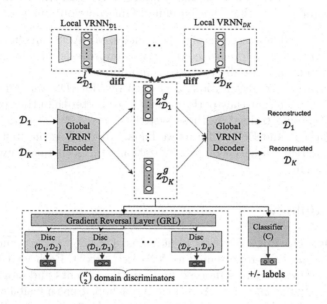

Fig. 1. Multi-Source Adversarial Domain Separation (MS-ADS) Framework

Step 1: Pre-train Source and Target VRNNs. Optimal local latent representations $\mathbf{z}^l_{\mathcal{D}_1}, ..., \mathbf{z}^l_{\mathcal{D}_K}$ are obtained by pre-training a local VRNN per each domain separately. The VRNN's loss objective ($\mathcal{L}^l_{\text{vrnn}}$) optimizes the inference (encoder) and the generative (decoder) processes to minimize the reconstruction loss [5].

Step 2: Discriminative Adversarial Separation. As shown in Fig. 1, MS-ADS is composed of the K pre-trained local VRNNs from step 1, and one global VRNN that takes the concatenation of all K domains as input. The global Encoder will generate global latent representations, and the global decoder reconstructs the input for each domain. The set of discriminators align the global latent representations between every two domains from $\mathcal{D}_i$ and $\mathcal{D}_j$. Finally, the unified classifier learns to predict the outcome labels using all latent representations regardless of their source domain. Following formalizes each component's loss objective.

1. **Global and Local VRNNs**: The parameters of local $\text{VRNN}_{\mathcal{D}_1}, ..., \text{VRNN}_{\mathcal{D}_K}$ are initialized based on the K pre-trained VRNNs to generate local representations: $\mathbf{z}^l_{\mathcal{D}_1}, ..., \mathbf{z}^l_{\mathcal{D}_K}$. The global VRNN also takes concatenation of all

domain's data as input and the global encoder generates $\mathbf{z}^g_{\mathcal{D}_1}, ..., \mathbf{z}^g_{\mathcal{D}_K}$. Further, for optimizing reconstruction loss in each of the local and global VRNNs we follow the original VRNN loss as follows:

$$\mathcal{L}^l_{\text{vrnn}}(\mathbf{x}_{\mathcal{D}_1}, ..., \mathbf{x}_{\mathcal{D}_K}; \Theta^l) = \sum_{i=1}^{K} \mathcal{L}_{\text{vrnn}}(\mathbf{x}_{\mathcal{D}_i}; \theta_{e_i}, \theta_{d_i}) \tag{1}$$

$$\mathcal{L}^g_{\text{vrnn}}([\mathbf{x}_{\mathcal{D}_1}, ..., \mathbf{x}_{\mathcal{D}_K}]; \Theta^g) = \mathcal{L}_{\text{vrnn}}([\mathbf{x}_{\mathcal{D}_1}, ..., \mathbf{x}_{\mathcal{D}_K}]; \Theta^g) \tag{2}$$

where $\Theta^l = \bigcup_{i=1}^{K} (\theta_{e_i}, \theta_{d_i})$ and $\Theta^g = (\theta^g_e, \theta^g_d)$ indicate the local and global VRNN parameters, respectively.

The main novelty of MS-ADS is to *separate* local and global features by maximizing the distance between them so that they extract system specific features such as systematic bias. To do so, we add a dissimilarity measurement between $(\mathbf{z}^g_{\mathcal{D}_i}, \mathbf{z}^l_{\mathcal{D}_i})$ for all $\mathcal{D}_i, i \in \{1, ..., K\}$ for each sample, defined by a *Frobenius norm* which measures the orthogonality between global and local representation from each domain. Let us denote matrix $\mathbf{Z}^g_{\mathcal{D}_i}$ as global matrix of $\mathcal{D}_i$ where each row j of it is composed of $\mathbf{z}^g_{\mathcal{D}_i}$ for sample j in this domain. Similarly, $\mathbf{Z}^l_{\mathcal{D}_i}$ indicates local matrix of $\mathcal{D}_i$. Therefore, the difference loss is defined as:

$$\mathcal{L}_{\text{diff}}(\mathbf{x}_{\mathcal{D}_1}, ..., \mathbf{x}_{\mathcal{D}_K}; \Theta^l, \Theta^g) = \sum_{i=1}^{K} \left\| {\mathbf{Z}^g_{\mathcal{D}_i}}^\top \mathbf{Z}^l_{\mathcal{D}_i} \right\|^2_F \tag{3}$$

where $\|\cdot\|^2_F$ refers to the squared Frobenius norm where zero indicates orthogonal vectors. Finally, the overall separation loss is:

$$\begin{aligned} \mathcal{L}_{\text{sep}}(\mathbf{x}_{\mathcal{D}_1}, ..., \mathbf{x}_{\mathcal{D}_K}; \Theta) = &\mathcal{L}^l_{\text{vrnn}}(\mathbf{x}_{\mathcal{D}_1}, ..., \mathbf{x}_{\mathcal{D}_K}; \Theta^l) + \mathcal{L}^g_{\text{vrnn}}([\mathbf{x}_{\mathcal{D}_1}, ..., \mathbf{x}_{\mathcal{D}_K}]; \Theta^g) \\ &+ \alpha \mathcal{L}_{\text{diff}}(\mathbf{x}_{\mathcal{D}_1}, ..., \mathbf{x}_{\mathcal{D}_K}; \Theta^l, \Theta^g). \end{aligned} \tag{4}$$

2. **Classifier:** A simple fully connected neural network is used as a classifier that consumes the global latent representations from the last time step T. This network is optimized based on the binary cross-entropy loss ($\mathcal{L}_B$) for all domains as:

$$\mathcal{L}_{\text{clf}}(\mathbf{x}_{\mathcal{D}_1}, ..., \mathbf{x}_{\mathcal{D}_K}; \theta_c, \theta^g_e) = \sum_{i=1}^{K} \mathcal{L}_B(C_{\theta_c}(E_g(\mathbf{x}_{\mathcal{D}_i}; \theta^g_e)_T), \mathbf{y}_{\mathcal{D}_i}) \tag{5}$$

where θ_c indicates the classifier parameters.

3. **Discriminator:** To minimize the difference between source domains, we propose to build a domain discriminator for each pair of domains. Therefore, each discriminator $D_{i,j}$ is a fully connected neural network that takes the last time step from global representations $\mathbf{z}^g_{\mathcal{D}_i}$ and $\mathbf{z}^g_{\mathcal{D}_j}$ as input to infer a domain label.

This will result in $\binom{K}{2}$ binary classifiers and the total discriminator loss would become:

$$\mathcal{L}_{\text{disc}}(\mathbf{z}^g_{\mathcal{D}_1},...,\mathbf{z}^g_{\mathcal{D}_K};\theta_{\text{disc}}) = \binom{K}{2}^{-1} \sum_{i=1}^{K-1} \sum_{j=i+1}^{K} \mathcal{L}_B(D_{i,j}(\mathbf{z}^g_{\mathcal{D}_i},\mathbf{z}^g_{\mathcal{D}_j});\theta^{i,j}_{\text{disc}}) \quad (6)$$

where $\theta^{i,j}_{\text{disc}}$ indicates the parameters of discriminator $D_{i,j}$. The discriminator's objective is to minimize this loss while the global VRNN tries to maximize this loss. Therefore, the adversarial learning process captures the notion of invariant latent representations between different domains. We have explored multiple other discriminative adversarial learning designs for multi-source problems such as a single discriminator with one vs. rest discrimination or with accumulated gradients [33,38], but the results show that the pairwise architecture performs the best.

Inspired by Ganin et al. [10] we use the gradient reversal layer (GRL) to effectively combine and optimize all three loss components using backpropagation. GRL can be represented as $\mathcal{R}(x)$ with different forward and backward propagation behavior, where I is the identity matrix and λ is a constant (a specified schedule during training can be used):

$$\mathcal{R}(x) = x; \frac{\partial \mathcal{R}}{\partial x} = -\lambda I \quad (7)$$

The GRL would handle the gradients from the discriminators that should be optimized in the reverse order and the overall optimization becomes:

$$\arg\min_{\Theta,\theta_c,\theta_{\text{disc}}} \mathcal{L}_{\text{sep}}(\mathbf{x}_{\mathcal{D}_1},...,\mathbf{x}_{\mathcal{D}_K};\Theta) + \mathcal{L}_{\text{clf}}(\mathbf{x}_{\mathcal{D}_1},...,\mathbf{x}_{\mathcal{D}_K};\theta_c,\theta^g_e)+$$
$$\mathcal{L}_{\text{disc}}(\mathcal{R}(\mathbf{z}^g_{\mathcal{D}_1}),...,\mathcal{R}(\mathbf{z}^g_{\mathcal{D}_K});\theta_{\text{disc}}) \quad (8)$$

Equation 8 yields a multi-source domain adaptation framework that can separate domain-specific features from the globally-shared latent representations and adversarially learn an invariant representation between each pair of the source domains. We hypothesize that MS-ADS will address both systematic bias and covariate shift effectively in a multi-source learning environment and builds a unified classifier that is robust across all source domains. We assess this hypothesis through experimentation in the following sections.

3   Experimental Setup

Three EHR Datasets: 210,289 visits of adult patients (i.e. age > 18) admitted to *Christiana Care Health System (CCHS)* in Newark, Delaware (07/2013-12/2015); 106,844 adult patient visits from *Mayo* Clinic in Rochester, Minnesota (07/2013-12/2015); and 53,423 ICU visits of patients admitted to Beth Israel Deaconess Medical Center in Boston, Massachusetts (2001–2012), *MIMIC-III* [15]. *Note that the nature of MIMIC-III data is different from CCHS and Mayo.*

To be consistent among all datasets, we define our target population as *suspected of infection*, identified by administration of any anti-infectives, or a positive PCR test result. This definition and the following data pre-processing steps are determined by three leading clinicians with extensive experience on this subject.

Labeling: We adopt the agreement between International Classification of Diseases, Ninth Revision (ICD-9) codes recorded in EHRs, and our expert-defined rules based on the Third International Consensus Definitions for Sepsis and Septic Shock [32] to achieve the most reliable population across all datasets. Our clinicians identify septic shock at event-level as having received vasopressor(s) or persistent hypotension for more than 1 h (systolic blood pressure (SBP)<90; or mean arterial pressure<65; or drop in SBP>40 in an 8-h window).

Sampling: Using the agreement criteria results in 2,963 positive cases in CCHS, 3,499 in Mayo, and 2,459 cases in MIMIC-III. To balance the number of positive and negative cases, we perform a stratified random sampling by 1) maintaining the same underlying age, gender, ethnicity, and length of stay distribution, and 2) having the same level of severity as positive samples. The severity of septic shock visits is identified as the presence of different stages of sepsis in their visits: infection, inflammation, and organ failure as defined by experts.

Aggregation: To align the sampling frequency across all datasets, we use a 30-min aggregation window to summarized all records into a single event and missing if none. Our feature set includes 7 vital signs (e.g.: SBP, Temperature), 2 oxygen information (FIO2 and OxygenFlow), and 10 lab results (e.g.: WBC, BUN). To handle the remaining missing values, we first use expert rules to carry forward vital signs (for 8 h) and lab results (for 24 h), then we apply the mean imputation along with the missing indicator. Our experiments show that this strategy will help VRNN address such variabilities in data more efficiently.

Prediction Task: Figure 2 shows our prediction task setup: using EHRs in an observation window to predict whether a patient is going to develop septic shock n hours later; n varies from 24 to 72 h denoted as *prediction window* and *observation window* is set to be capped at 48 h as suggested by the leading physicians. All the sequences are aligned by their end time, which is the shock onset for pos-

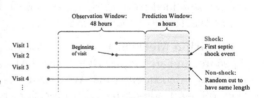

Fig. 2. Septic shock early prediction task

itive visits and a truncated time point for non-shock visits. To prevent the potential bias in models, negative visits are truncated such that they have the same distribution of length as positives. As the prediction window expands, the number of visits remaining in the observation window will drop. For a fair comparison, we sample the same number of positive/negative visits in all domains. This results in 1,315 total visits from each domain for 24 h early prediction and 620

Table 1. Multi-source DA performance (± std) evaluated on integration of ALL domains and each domain separately for 24 h early prediction task.

Test domain	Model	Accuracy	Precision	Recall	F_1 Score	AUC
ALL	1. VRNN(CCHS)	0.735(±0.012)	**0.823**(±0.019)	0.6(±0.048)	0.692(±0.026)	**0.815**(±0.014)
	2. VRNN(Mayo)	0.741(±0.017)	0.753(±0.021)	0.718(±0.053)	0.734(±0.026)	0.81(±0.015)
	3. VRNN(MIMIC)	0.732(±0.016)	0.677(±0.023)	**0.894****(±0.037)	0.**769**(±0.009)	0.814(±0.015)
	4. VRNN(Separate)	**0.803**‡(±0.012)	**0.817**‡(±0.017)	0.781(±0.037)	**0.797**‡(±0.017)	0.864(±0.01)
	5. VRNN(All)	0.795(±0.004)	0.791(±0.014)	**0.801**(±0.022)	0.796(±0.006)	**0.882**‡(±0.003)
	6. Multi VRADA	0.78(±0.029)	0.778(±0.046)	0.766(±0.034)	0.771(±0.021)	0.855(±0.031)
	7. MS-ADS	**0.81****(±0.011)	**0.828****(±0.018)	0.782‡(±0.027)	**0.804****(±0.014)	**0.893****(±0.009)
CCHS	1. VRNN(CCHS)	**0.778**(±0.008)	**0.833**(±0.022)	0.698(±0.034)	0.759(±0.014)	0.837(±0.012)
	7. MS-ADS	0.777(±0.012)	0.791(±0.016)	**0.75**(±0.028)	**0.77**(±0.014)	**0.862**(±0.013)
Mayo	2. VRNN(Mayo)	**0.731**(±0.011)	0.732(±0.016)	**0.729**(±0.04)	**0.73**(±0.017)	0.795(±0.004)
	7. MS-ADS	0.73(±0.022)	**0.752**(±0.034)	0.688(±0.046)	0.718(±0.028)	**0.796**(±0.02)
MIMIC	3. VRNN(MIMIC)	0.9(±0.018)	0.888(±0.014)	0.917(±0.038)	0.902(±0.02)	0.961(±0.014)
	7. MS-ADS	**0.921**(±0.015)	**0.935**(±0.018)	0.907(±0.018)	**0.921**(±0.016)	**0.974**(±0.005)

· The *best* and the *second best* models are labeled with ** and ‡, respectively.

visits for 72 h. Therefore, as the number of samples decreases, it is more crucial to integrate different domains to build more robust classifiers.

Parameters and Training: As illustrated in Eq. 8, there are three sets of parameters: discriminator (θ_{disc}), classifier (θ_c), and VRNN (Θ) parameters optimized through a GRL for adversarial training using NAdam optimizer [34], with learning rate $\alpha_{\text{total}} = 8e^{-4}$. Then the classifier and VRNN models are optimized in an additional step to compete against the gradients from the discriminator, with learning rates: $\alpha_c = 10e^{-4}$, $\alpha_{\text{VRNN}} = 10e^{-4}$. In every epoch, the order of optimization between the three optimizers is altered from the previous epoch to prevent over-training of a specific network. All the models are implemented in Tensorflow using mini-batch with batch size 32. The same experimental setup is used for all the models with 160 epochs and early stopping. The VRNN's hidden size is set to 30 and the latent size is defined as 50. All the sequences are zero-padded to have the same length and the zero-paddings are masked for reconstruction loss calculation.

Evaluation Metrics: Our evaluation metrics include accuracy, recall, precision, F_1 score, and area under ROC curve (AUC) obtained from 2-fold cross-validation in three independent runs. We mainly use F_1 and AUC as the main metrics as they offer a trade-off between precision, recall, and specificity.

4   Multi-Source DA Across the Three Medical Systems

By leveraging data from multiple medical systems while accounting for both *covariate shift* and *systematic bias* across them, we expect MS-ADS would improve the prediction performance across all three systems. Therefore, in this task, the test set is composed of an equal number of visits from CCHS, Mayo, and MIMIC. MS-ADS is compared against six baselines:

Table 2. Multi-source DA performance evaluated for 24–72 h early prediction.

Model	Accuracy	Precision	Recall	F_1 Score	AUC
1. VRNN(CCHS)	0.674(±0.04)	0.734(±0.055)	0.559(±0.052)	0.63(±0.041)	0.728(±0.046)
2. VRNN(Mayo)	0.66(±0.031)	0.678(±0.037)	0.624(±0.075)	0.645(±0.044)	0.712(±0.032)
3. VRNN(MIMIC)	0.689(±0.014)	0.633(±0.015)	**0.909****(±0.029)	0.745(±0.009)	0.759(±0.016)
4. VRNN(Separate)	**0.755‡**(±0.025)	**0.775‡**(±0.031)	0.719(±0.053)	0.742(±0.031)	0.804(±0.023)
5. VRNN(All)	0.749(±0.012)	0.757(±0.021)	0.743(±0.051)	**0.747‡**(±0.019)	**0.835‡**(±0.01)
6. 3-d VRADA	0.746(±0.021)	0.751(±0.034)	0.739(±0.043)	0.743(±0.022)	0.829(±0.024)
7. MS-ADS	**0.771****(±0.016)	**0.782****(±0.016)	0.75‡(±0.033)	**0.765****(±0.02)	**0.85****(±0.012)

· The *best* and the *second best* models are labeled with ** and ‡, respectively.

1. *VRNN(CCHS)*: a VRNN trained on CCHS only.
2. *VRNN(Mayo)*: a VRNN trained on Mayo only.
3. *VRNN(MIMIC)*: a VRNN trained on MIMIC-III only.
4. *VRNN(Separate)*: will use the individual VRNN trained above to predict the corresponding test data.
5. *VRNN(All)*: a VRNN trained on a combined data from CCHS, Mayo, MIMIC.
6. *Multi VRADA* [28]: a modified version of VRADA to address multi-source DA by changing the domain classifier loss to categorical cross-entropy loss.

24 H Early Prediction: Table 1 presents the DA results for 24 h early prediction on the combined test data (ALL) first (top) and then on test data in each system separately. The top row shows that 1) among the five non-adaptive baselines (1–5), VRNN(All) outperforms all single-domain VRNNs and VRNN(Separate). This result suggests that a more effective classifier can be achieved by leveraging more training samples; 2) By comparing the two multi-source DA models against VRNN(All), we show that VRADA is not able to outperform VRNN(All) while MS-ADS performs robustly and achieves the best performance on all measures except on recall. The highest recall is achieved by VRNN(MIMIC) at a cost of low precision.

The bottom three rows in Table 1 show whether the performance of these models differ across different medical systems (domains). Due to the space limitation, for each domain, we only listed the performance of the corresponding VRNN trained on the same domain compared with the best of the remaining six models. Table 1 shows MS-ADS consistently to be the best model on CCHS and MIMIC but for Mayo, VRNN (Mayo) has a higher F1 score and very close AUC score to MS-ADS. Additionally, MIMIC data has extremely good results while the performance on Mayo is the worst. Such results suggest that early sepsis shock prediction is relatively trivial for MIMIC dataset probably because MIMIC only includes ICU visits. Therefore, in the following, we mainly focus on generalization to CCHS and Mayo only.

Varying 24–72 H Early Prediction: Table 2 shows the average performance by varying the prediction window from 24 to 72 h, with every 12 h interval. For each prediction window, our test set has an equal number of visits from each domain. Table 2 shows MS-ADS significantly outperforms all other baselines

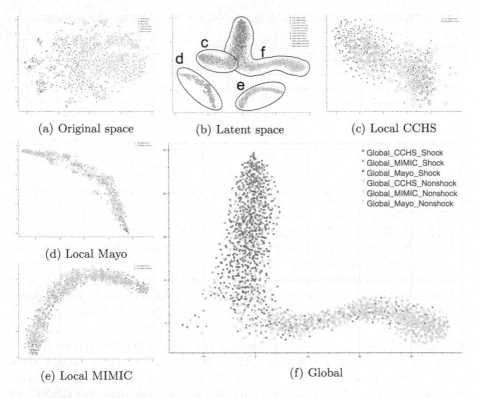

(a) Original space   (b) Latent space   (c) Local CCHS

* Global_CCHS_Shock
* Global_MIMIC_Shock
* Global_Mayo_Shock
  Global_CCHS_Nonshock
  Global_MIMIC_Nonshock
  Global_Mayo_Nonshock

(d) Local Mayo

(e) Local MIMIC   (f) Global

Fig. 3. Visit-level t-SNE visualization of (a) Original vs. (b) Latent space of MS-ADS. (c)–(e) show domain-specific representations while (f) illustrates the globally-shared representation. Solid dots represent septic shock visits.

including VRADA and VRNN(All) for *all* metrics except recall. VRNN(MIMIC) performs with the highest recall across all domains, but at the cost of very low precision. Comparing VRNN(Separate) with MS-ADS shows a ∼3% improvement for recall and ∼4.5% improvement for AUC across all three domains. This result demonstrates that by integrating EHRs across medical systems, MS-ADS can address insufficient labeled data problems and by adopting an effective domain adaptation architecture, MS-ADS can address both systematic bias and covariate shift across medical systems.

Visit-Level Visual Investigation. Figure 3 illustrates t-SNE visualization of the original and latent representation of all visits for 24 h early prediction. In all graphs, different colors represent different medical systems and solid and hollow points represent shock and non-shock visits, respectively. Figure 3a illustrates the original space and 3b shows that the latent space generated by MS-ADS can separate the three local representations (c), (d), (e) (enlarged in Fig. 3c–3e) from the global ones (f) (enlarged in Fig. 3f). Figure 3c–3e suggests that MS-ADS can address systematic bias effectively while Fig. 3f shows that in the global space,

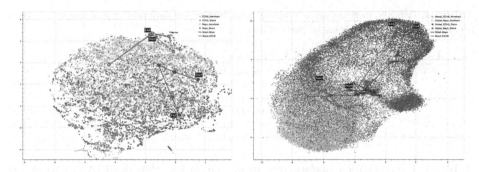

Fig. 4. Event-level t-SNE visualization of Original (left) vs. Global Latent (right) representation of MS-ADS on CCHS and Mayo. Red (CCHS) and Blue (Mayo) traces show sepsis progression of two similar patients. (Color figure online)

Table 3. DG performance to unseen target domains for 24 h early prediction.

Source	Unseen target	Model	Accuracy	Precision	Recall	F_1 Score	AUC
MIMIC + Mayo	CCHS	VRNN	0.747(±0.02)	0.739‡(±0.041)	↑0.77(±0.028)	0.753(±0.011)	0.831(±0.015)
MIMIC + Mayo		VRADA	0.763‡(±0.021)	0.739(±0.043)	↑**0.826**(±0.048)	↑**0.778**(±0.013)	↑0.846‡(±0.017)
MIMIC + Mayo		MS-ADS	**0.764**(±0.017)	**0.74**(±0.027)	↑0.813‡(±0.023)	↑0.774‡(±0.013)	↑**0.851**(±0.008)
CCHS		VRNN	0.778(±0.008)	0.833(±0.022)	0.698(±0.034)	0.759(±0.014)	0.837(±0.012)
CCHS + MIMIC	Mayo	VRNN	0.698(±0.014)	0.725‡(±0.036)	**0.644**(±0.066)	0.678‡(±0.028)	0.763‡(±0.018)
CCHS + MIMIC		VRADA	0.678‡(±0.039)	0.702(±0.035)	0.608(±0.072)	0.65(±0.055)	0.739(±0.039)
CCHS + MIMIC		MS-ADS	↑**0.73**(±0.012)	↑**0.796**(±0.025)	0.62‡(±0.011)	**0.697**(±0.01)	↑**0.8**(±0.014)
Mayo		VRNN	0.731(±0.011)	0.732(±0.016)	0.729(±0.04)	0.73(±0.017)	0.795(±0.004)

·In each block, the best performance is in **bold**; Models that outperform the gold standard (bottom) are labeled with ↑.

samples from different domains are close together and mostly aligned and mixed. This shows that MS-ADS can address covariate shift effectively as well.

Event-Level Visual Investigation. Further, we look at the original and global latent space at the event level to validate if covariate shift is addressed by MS-ADS along the temporal axis. We select two similar septic shock visits across CCHS and Mayo such that both develop inflammation and multiple organ failure symptoms within the observation window. Figure 4 shows these two traces (CCHS (red) and Mayo (blue)) in the original and global latent spaces. Despite their similarity, their progression deviates in the original space while in the latent representation their temporal progression is aligned. This further demonstrates the effectiveness of MS-ADS in addressing covariate shift at a temporal level.

5   Domain Generalization to Unseen Medical System

In the second task, MS-ADS is trained on EHRs from two medical systems and evaluated on an unseen target system: CCHS or Mayo. MS-ADS is compared against two baselines: a VRNN trained on the combination of two source domains and the original VRADA applied for DA across the two domains. Table 3

532   F. Khoshnevisan and M. Chi

Table 4. VRNN performance trained and tested on different racial groups across medical systems for 24 h early prediction.

Train domain	Test Domain	Accuracy	Precision	Recall	F_1 Score	AUC
CCHS(WA)	CCHS(WA)	0.888(±0.014)	0.869(±0.025)	0.916(±0.017)	0.891(±0.013)	0.956(±0.008)
	CCHS(AA)	0.885(±0.011)	0.874(±0.018)	0.9(±0.004)	0.887(±0.01)	0.946(±0.007)
Mayo(WA)	Mayo(WA)	0.841(±0.038)	0.83(±0.043)	0.86(±0.039)	0.844(±0.036)	0.909(±0.03)
	Mayo(AA)	0.809(±0.025)	0.821(±0.042)	0.813(±0.038)	0.816(±0.024)	0.847(±0.038)
CCHS(AA)	Mayo(AA)	0.715(±0.031)	0.751(±0.031)	0.68(±0.061)	0.712(±0.038)	0.811(±0.037)
CCHS(WA+AA)	Mayo(AA)	0.792(±0.032)	0.834(±0.048)	0.753(±0.054)	0.79(±0.034)	0.872(±0.021)

Table 5. Generalization performance to unseen African American (AA) patients in Mayo using 2-domains and 3-domains.

Train domains	Model	Accuracy	Precision	Recall	F_1 Score	AUC
CCHS(WA + AA), Mayo(WA)	VRNN(All)	0.844(±0.031)	0.895(±0.012)	0.793(±0.075)	0.839(±0.04)	0.913(±0.01)
	2-d VRADA	0.87(±0.025)	0.87(±0.034)	0.873(±0.063)	0.87(±0.029)	0.922(±0.021)
	2-d MS-ADS	0.854(±0.017)	**0.901**(±0.033)	0.813(±0.068)	0.852(±0.026)	0.914(±0.012)
CCHS(WA), CCHS(AA), Mayo(WA)	3-d VRADA	0.847(±0.016)	0.861(±0.033)	0.847(±0.049)	0.852(±0.017)	0.917(±0.034)
	3-d MS-ADS	**0.87**(±0.012)	0.876(±0.007)	**0.876**(±0.035)	**0.875**(±0.014)	**0.947**(±0.005)

· The best overall performance is in **bold**.

presents the generalization performance of all three models for 24 h early prediction. Table 3 shows MS-ADS outperforms the two baselines for most metrics in both target domains. Finally, we also compared them against the gold standard supervised VRNN model trained on the target domain (last row in each section), Table 3 shows that MS-ADS outperforms the supervised VRNN for AUC metric in both target domains.

6   Unseen Racial Group Across Medical Systems

In this task, we focus on two racial groups: White American (WA) and African American (AA). Table 4 compares the performance of models that are trained and tested on different racial groups across medical systems. Table 4 shows that while the model trained on CCHS(WA) performs equally well on CCHS(WA) and CCHS(AA); the model trained on Mayo(WA) performs much better on Mayo(WA) than Mayo(AA). This is probably because the percentages of WA and AA are more balanced than those of Mayo: 71% vs. 22.5% in CCHS while 91% vs. 3% in Mayo. The last block in Table 4 shows transfer across medical systems. The model trained on CCHS(AA) does not perform well on Mayo(AA) probably due to systematic bias across medical systems while adding CCHS(WA) to the training population can help predictions on Mayo(AA) probably because of more training data. As a result, our training domain settings will involve WA in Mayo and WA and AA in CCHS.

Table 5 compares the generalization performance on Mayo(AA) by using two training domains: CCHS (WA+AA) and Mayo (WA) (upper) vs. three training domains: CCHS (WA), CCHS(AA), and Mayo (WA) (bottom). For the two-domain generalization, MS-ADS is compared against the best non-DA baseline: VRNN(All) and VRADA. Table 5 shows that both VRADA and MS-ADS outperform the VRNN(All) and VRADA achieves the best F1 and AUC. When we conduct the same task by using three domains, the bottom block in Table 5 shows that the performance of VRADA suffered while the performance of MS-ADS improved. Indeed, Table 5 shows that the F1 and AUC of 3-domain MS-ADS on Mayo(AA) are 0.875 and 0.947, catching up with the other three racial groups across the two systems. We argue the effectiveness of 3-domain MS-ADS over 2-domain MS-ADS is probably because the former can leverage 1) Mayo(WA) (same system different race), 2) CCHS(AA) (same race, different system), 3) the DA mechanism learned from unifying AA and WA in CCHS (addressing covariate shift within the same system), and 4) the DA learned from unifying WA between CCHS and Mayo (addressing systematic bias in the same racial group).

7 Related Work

Septic Shock Early Prediction: A variety of machine learning models have been developed to predict septic shock several hours before the onset. Among traditional approaches, multivariate logistic regression and survival analysis models have been proposed for early detection [14,31]. Moreover, sequential pattern mining approaches have shown to be effective for early prediction of septic shock while producing explainable patterns [12,17]. Recently, various deep learning-based approaches have been proposed, especially variations of recurrent neural networks such as LSTM, and they have shown promising power in predicting septic shock several hours before the onset [22,40,41]. Despite the great power of LSTM models, they are not designed to address the high missing rate in EHR [18]. Variational recurrent neural network (VRNN) [5] is recently proposed to model complex temporal and conditional dependencies in sequential data and account for variabilities, like missing data, and has shown great promise [4,26,39].

Multi-source Domain Adaptation: The majority of existing DA work either addresses this problem by generating an invariant feature space for all pairs of source-target distributions [27,43] or constructs the target distribution as a weighted combination of source distributions [23,37]. For example, VRADA is a VRNN-based DA that has been applied to EHRs from different groups of patients and it has shown significant improvement in creating domain-invariant representations using adversarial learning [28]. In this work, we further expanded VRADA's architecture to address multi-source DA problems and use it as a baseline. These studies treat multiple medical systems as "source" domains to improve the prediction performance in a specific "target" medical system. By treating all domains equally, a group of DA studies aim at learning a unified minimal risk model from multiple domains [6,8,30]. While the majority of such

DA research is conducted in computer vision and text classification, a few studies have proposed DA approaches to integrate EHRs across multiple medical systems and improve prediction in a target domain by addressing covariate shift and feature mismatch [36,38]. All existing approaches have shown great power in accounting for the covariate shift but not domain-specific characteristics or systematic bias that *should not* be unified across domains. Our MS-ADS model is capable of integrating multiple medical systems to build a robust unified model that improves prediction across *all* systems while accounting for the covariate shift and systematic bias simultaneously but differently.

Domain Generalization aims to learn a model from an arbitrary number of source domains such that it can generalize to previously unseen target domains [9,13,29]. One class of approaches proposes to learn domain-invariant representations by minimizing domain mismatch across source domains, similar to DA approaches [11,21,25]. For example, Motiian et al. propose a unified DA and DG model exploiting Siamese architecture using a contrasting loss to minimize the distance between samples from the same class but in different domains [24]. Another type of DG method utilizes meta-learning techniques to synthesize domain shift and directly learn and optimize for generalization task [3,7,20]. Despite the critical application of DG in clinical deployment, especially in presence of limited data, this problem is still under-explored.

Further, to close the gap between performances among different groups of patients, previous studies have explored DA approaches to account for the covariate shift between groups within a medical system [2,28]. For example, Zhang et al. proposed a time-aware adversarial LSTM network to transfer knowledge across different racial, age, and gender groups and improve prediction for minority groups [42]. As far as we know, this study is the first that investigates DG to simultaneously address the covariate shift across different racial groups and systematic bias across medical systems to generalize robustly and improve prediction among minority groups.

8   Conclusion

In this work, we propose a multi-source adversarial domain separation (MS-ADS) framework that unifies domain adaptation (DA) and domain generalization (DG) by accounting for systematic bias across medical systems and covariate shift among different patient groups to achieve a robust generalization. In specific, MS-ADS separates the global representation of each domain from the local ones to address systematic bias and leverage a multi-domain discriminator with Gradient Reversal Layer (GRL) to account for the covariate shift. We evaluate MS-ADS in three tasks for septic shock early prediction using EHR from three medical systems. First, on a task of DA across three medical systems, we show that the effective adaptation under MS-ADS leads to performance improvement in all three domains. Second, on a task of DG to an unseen medical system, we demonstrate the generalization power brought by MS-ADS architecture by comparing and showing its robustness against a gold standard supervised model on the target domain. Lastly, on a task of generalization to unseen racial groups across the medical system, we show that unifying DA and DG MS-ADS can significantly improve prediction among minority racial groups.

Acknowledgments. This research was supported by the NSF Grants #2013502, #1726550, #1651909, and #1522107.

References

1. Agniel, D., et al.: Biases in electronic health record data due to processes within the healthcare system: retrospective observational study. BMJ **361** (2018)
2. Alves, T., Laender, A., Veloso, A., Ziviani, N.: Dynamic prediction of ICU mortality risk using domain adaptation. In: 2018 IEEE International Conference on Big Data (Big Data), pp. 1328–1336. IEEE (2018)
3. Balaji, Y., Sankaranarayanan, S., Chellappa, R.: Metareg: towards domain generalization using meta-regularization. In: NeurIPS, pp. 998–1008 (2018)
4. Chien, J.T., Kuo, K.T., et al.: Variational recurrent neural networks for speech separation. In: Interspeech, VOLS 1–6: Situated Interaction, pp. 1193–1197 (2017)
5. Chung, J., Kastner, K., Dinh, L., Goel, K., Courville, A.C., Bengio, Y.: A recurrent latent variable model for sequential data. In: NeurIPS, pp. 2980–2988 (2015)
6. Ding, X., Shi, Q., Cai, B., Liu, T., Zhao, Y., Ye, Q.: Learning multi-domain adversarial neural networks for text classification. IEEE Access **7**, 40323–40332 (2019)
7. Dou, Q., Castro, D.C., Kamnitsas, K., Glocker, B.: Domain generalization via model-agnostic learning of semantic features. arXiv:1910.13580 (2019)
8. Dredze, M., Kulesza, A., Crammer, K.: Multi-domain learning by confidence-weighted parameter combination. Mach. Learn. **79**(1–2), 123–149 (2010). https://doi.org/10.1007/s10994-009-5148-0
9. Du, Y., et al.: Learning to learn with variational information bottleneck for domain generalization. In: Vedaldi, A., Bischof, H., Brox, T., Frahm, J.-M. (eds.) ECCV 2020. LNCS, vol. 12355, pp. 200–216. Springer, Cham (2020). https://doi.org/10.1007/978-3-030-58607-2_12
10. Ganin, Y., Lempitsky, V.: Unsupervised domain adaptation by backpropagation. arXiv:1409.7495 (2014)
11. Ghifary, M., Bastiaan Kleijn, W., Zhang, M., Balduzzi, D.: Domain generalization for object recognition with multi-task autoencoders. In: CVPR, pp. 2551–2559 (2015)
12. Ghosh, S., Li, J., Cao, L., Ramamohanarao, K.: Septic shock prediction for ICU patients via coupled hmm walking on sequential contrast patterns. JBI **66**, 19–31 (2017)
13. Gong, R., Li, W., Chen, Y., Gool, L.V.: DLOW: domain flow for adaptation and generalization. In: CVPR, pp. 2477–2486 (2019)
14. Henry, K.E., et al.: A targeted real-time early warning score (TREWScore) for septic shock. Sci. Transl. Med. **7**(299), 299ra122 (2015)
15. Johnson, A.E., et al.: MIMIC-III, a freely accessible critical care database. Sci. Data **3**, 160035 (2016)
16. Khoshnevisan, F., Chi, M.: An adversarial domain separation framework for septic shock early prediction across EHR systems. arXiv:2010.13952 (2020)
17. Khoshnevisan, F., Ivy, J., Capan, M., Arnold, R., Huddleston, J., Chi, M.: Recent temporal pattern mining for septic shock early prediction. In: IEEE ICHI, pp. 229–240. IEEE (2018)

18. Kim, Y.J., Chi, M.: Temporal belief memory: imputing missing data during RNN training. In: IJCAI (2018)
19. Kumar, A., et al.: Duration of hypotension before initiation of effective antimicrobial therapy is the critical determinant of survival in human septic shock. Crit. Care Med. **34**(6), 1589–1596 (2006)
20. Li, D., Yang, Y., Song, Y.Z., Hospedales, T.M.: Learning to generalize: meta-learning for domain generalization. arXiv:1710.03463 (2017)
21. Li, Y., et al.: Deep domain generalization via conditional invariant adversarial networks. In: Ferrari, V., Hebert, M., Sminchisescu, C., Weiss, Y. (eds.) ECCV 2018. LNCS, vol. 11219, pp. 647–663. Springer, Cham (2018). https://doi.org/10.1007/978-3-030-01267-0_38
22. Lin, C., et al.: Early diagnosis and prediction of sepsis shock by combining static and dynamic information using convolutional-LSTM. In: IEEE ICHI, pp. 219–228. IEEE (2018)
23. Mansour, Y., Mohri, M., Rostamizadeh, A.: Domain adaptation with multiple sources. In: NeurIPS, pp. 1041–1048 (2009)
24. Motiian, S., Piccirilli, M., Adjeroh, D.A., Doretto, G.: Unified deep supervised domain adaptation and generalization. In: CVPR, pp. 5715–5725 (2017)
25. Muandet, K., Balduzzi, D., Schölkopf, B.: Domain generalization via invariant feature representation. In: ICML, pp. 10–18 (2013)
26. Mulyadi, A.W., Jun, E., Suk, H.I.: Uncertainty-aware variational-recurrent imputation network for clinical time series. arXiv:2003.00662 (2020)
27. Peng, X., Bai, Q., Xia, X., Huang, Z., Saenko, K., Wang, B.: Moment matching for multi-source domain adaptation. In: CVPR, pp. 1406–1415 (2019)
28. Purushotham, S., Carvalho, W., Nilanon, T., Liu, Y.: Variational recurrent adversarial deep domain adaptation (2016)
29. Qiao, F., Zhao, L., Peng, X.: Learning to learn single domain generalization. In: Proceedings of the CVPR, pp. 12556–12565 (2020)
30. Schoenauer-Sebag, A., Heinrich, L., Schoenauer, M., Sebag, M., Wu, L.F., Altschuler, S.J.: Multi-domain adversarial learning. arXiv:1903.09239 (2019)
31. Shavdia, D.: Septic shock: providing early warnings through multivariate logistic regression models. Ph.D. thesis, Massachusetts Institute of Technology (2007)
32. Singer, M., et al.: The third international consensus definitions for sepsis and septic shock (sepsis-3). JAMA **315**(8), 801–810 (2016)
33. Tasar, O., Tarabalka, Y., Giros, A., Alliez, P., Clerc, S.: StandardGAN: multi-source domain adaptation for semantic segmentation of very high resolution satellite images by data standardization. In: CVPR Workshops, pp. 192–193 (2020)
34. Tato, A., Nkambou, R.: Improving adam optimizer (2018)
35. Tintinalli, J., Stapczynski, J., Ma, O.J., Cline, D., Cydulka, R., Meckler, G.: Septic Shock (chap. 146). In: Tintinallis Emergency Medicine A Comprehensive Study Guide, 7 edn., pp. 1003–1014. McGraw-Hill Education (2011)
36. Wiens, J., Guttag, J., Horvitz, E.: A study in transfer learning: leveraging data from multiple hospitals to enhance hospital-specific predictions. J. Am. Med. Inform. Assoc. **21**(4), 699–706 (2014)
37. Xu, R., Chen, Z., Zuo, W., Yan, J., Lin, L.: Deep cocktail network: multi-source unsupervised domain adaptation with category shift. In: CVPR, pp. 3964–3973 (2018)
38. Yoon, J., Jordon, J., van der Schaar, M.: RadialGAN: leveraging multiple datasets to improve target-specific predictive models using generative adversarial networks. arXiv:1802.06403 (2018)

39. Zhang, S., Xie, P., Wang, D., Xing, E.P.: Medical diagnosis from laboratory tests by combining generative and discriminative learning. arXiv:1711.04329 (2017)
40. Zhang, Y., Lin, C., Chi, M., Ivy, J., Capan, M., Huddleston, J.M.: LSTM for septic shock: adding unreliable labels to reliable predictions. In: IEEE Big Data, pp. 1233–1242. IEEE (2017)
41. Zhang, Y., Yang, X., Ivy, J., Chi, M.: Attain: attention-based time-aware LSTM networks for disease progression modeling. In: IJCAI, pp. 10–16 (2019)
42. Zhang, Y., Yang, X., Ivy, J., Chi, M.: Time-aware adversarial networks for adapting disease progression modeling. In: IEEE ICHI, pp. 1–11. IEEE (2019)
43. Zhao, H., Zhang, S., Wu, G., Moura, J.M., Costeira, J.P., Gordon, G.J.: Adversarial multiple source domain adaptation. In: NeurIPS, pp. 8559–8570 (2018)

Deep Multi-task Augmented Feature Learning via Hierarchical Graph Neural Network

Pengxin Guo[1], Chang Deng[3], Linjie Xu[4], Xiaonan Huang[1], and Yu Zhang[1,2(✉)]

[1] Department of Computer Science and Engineering, Southern University of Science and Technology, Shenzhen, China
12032913@mail.sustech.edu.cn
[2] Peng Cheng Laboratory, Shenzhen, China
[3] Committee on Computational and Applied Mathematics, University of Chicago, Chicago, USA
changdeng@uchicago.edu
[4] Game AI Group, Queen Mary University of London, London, UK
linjie.xu@qmul.ac.uk

Abstract. Deep multi-task learning attracts much attention in recent years as it achieves good performance in many applications. Feature learning is important to deep multi-task learning for sharing common information among tasks. In this paper, we propose a Hierarchical Graph Neural Network (HGNN) to learn augmented features for deep multi-task learning. The HGNN consists of two-level graph neural networks. In the low level, an intra-task graph neural network is responsible of learning a powerful representation for each data point in a task by aggregating its neighbors. Based on the learned representation, a task embedding can be generated for each task in a similar way to max pooling. In the second level, an inter-task graph neural network updates task embeddings of all the tasks based on the attention mechanism to model task relations. Then the task embedding of one task is used to augment the feature representation of data points in this task. Moreover, for classification tasks, an inter-class graph neural network is introduced to conduct similar operations on a finer granularity, i.e., the class level, to generate class embeddings for each class in all the tasks using class embeddings to augment the feature representation. The proposed feature augmentation strategy can be used in many deep multi-task learning models. Experiments on real-world datasets show the significant performance improvement when using this strategy.

Keywords: Multi-task learning · Feature learning · Graph neural network

1 Introduction

Multi-task learning [8,38] aims to leverage useful information contained in multiple learning tasks to improve their performance simultaneously. During

N. Oliver et al. (Eds.): ECML PKDD 2021, LNAI 12975, pp. 538–553, 2021.
https://doi.org/10.1007/978-3-030-86486-6_33

past decades, many multi-task learning models have been proposed to identify the shared information which can take a form of the instance, feature, and model, leading to three categories including instance-based multi-task learning [5], feature-based multi-task learning [3, 19, 21, 23, 26, 28, 31, 41], and model-based multi-task learning [2, 6, 11–14, 17, 39, 40].

At present, using the output of shared hidden layer as the representation of hidden features shared by tasks is the mainstream approach in deep multi-task learning, which has achieved good results in many problems. However, these methods can not learn the task-specific feature representation of each task, which limits the further improvement of the performance. At another extreme, some works [9, 26] use a neural network as a base model for each task in multi-task learning. One advantage of this approach is that different tasks can learn their own feature representation, however, a major problem is that the model parameters of the entire multi-task learning model are linear with respect to the number of tasks, making this approach not scalable to a large number of tasks in multi-task learning.

In this paper, we study deep multi-task learning between the two extremes and hope to learn task-specific feature representation to improve the learning performance but without increasing the number of parameters as well as the model complexity too much. To achieve that, we propose to use the task representation, which is also called the task embedding, as a type of an augmented feature representation to improve the expressiveness of the feature representation as well as the performance, since the task embedding contains the unique characteristics of a task. To derive the task embedding of each task, the training dataset of that task is used in terms of a graph where nodes represent data points and edges denote the similarities between data points. Besides, the relationship between tasks can also be represented in a graph. Inspired by this idea, in this paper, we propose a Hierarchical Graph Neural Network (HGNN) to further improve the performance of multi-task learning models by learning augmented features. The HGNN consists of two-level graph neural networks. In the first level, an intra-task graph neural network is to learn a powerful representation for each data point in a task by aggregating its neighbored data points in this task. Based on the representation learned in the first level, we can generate the task embedding, which is a representation for this task, in a way similar to max pooling. For classification tasks, we can generate the class embedding for each class in this task based on max pooling. Based on task embeddings of all the tasks generated in the first level, an inter-task graph neural network in the second level updates all the task embeddings based on the attention mechanism. For classification tasks, an inter-class graph neural network is introduced in the second level to update all the class embeddings based on neighbored class embeddings. Finally, each of the learned task embeddings as well as the class embeddings for classification tasks is used to augment the feature representation of all the data points in the corresponding task. The proposed HGNN can be used in many multi-task learning models. We analyze the use of HGNN in terms of both the training loss and generalization loss. Extensive experiments show the effectiveness of the proposed HGNN. Our contributions are summarized as follows:

- We propose a Hierarchical Graph Neural Network (HGNN) to learn augmented features for deep multi-task learning. The proposed feature augmentation strategy can be used in many deep multi-task learning models for regression and classification tasks.
- We provide some analyses for the proposed feature augmentation strategy, which can help understand why the incorporation of the task embedding can improve the performance.
- We conduct extensive experiments on four benckmark datasets to demonstrate that HGNN can improve the performance of many deep multi-task learning models.

2 Related Works

Liu [20] explores the problem of learning the relationship between multiple tasks dynamically and formulate this problem as a message passing process over a graph neural network. Meng [25] solves relative attribute learning via a message passing scheme on a graph and the main idea is that relative attribute learning naturally benefits from exploiting the dependency graph among different relative attributes of images. The multi-task attention network proposed in [21] consists of a single shared network containing a global feature pool and a soft-attention module for each task that allows to learn task-specific feature-level attentions. Lu [24] presents a graph star net which utilizes the message-passing and attention mechanisms for multiple prediction tasks, including node classification, graph classification, and link prediction. Though the aforementioned works propose GNN or the attention mechanism for multi-task learning, none of them use a hierarchical version of GNN as well as the attention mechanism to learn augmented features for multi-task learning, which is the focus of this paper.

Kim [15] proposes a hierarchical attention network for stock prediction which can selectively aggregate information on different relation types and add the information to each representation of the company for the stock market prediction. However, after obtaining the additional information to each representation, this work only uses the neighbored nodes to aggregate the information, which is different from our work that aggregates all the nodes in the graph. Moreover, this method adds the additional information to the original feature representation, which is different from the concatenation method used in this paper. Ryu [30] proposes a Hierarchical graph Attention-based Multi-Agent actor-critic (HAMA) method, which employs a hierarchical graph neural network to effectively model the inter-agent relationships in each group of agents and inter-group relationships among groups, and additionally employ inter-agent and inter-group attentions to adaptively extract state-dependent relationships among agents. However, similar to [15], the HAMA method, a network stacking multiple Graph Attention Networks (GAT) [33] hierarchically, only processes local observations of each agent but not all the information in the graph. Different from these two methods, we first use an intra-task graph neural network to generate a task embedding for each task by using all the data in this task and then use the inter-task graph neural network to update task embeddings of all the tasks based on the inter-task structure. To the best of our knowledge, we are the first to use the feature augmentation strategy in multi-task learning.

3   Hierarchical Graph Neural Network

In this section, we introduce the proposed architecture, the Hierarchical Graph Neural Network (HGNN), for deep multi-task learning. Whilst the architecture can be incorporated into any multi-task learning network, in the following sections we show how to build the HGNN upon a multi-task network.

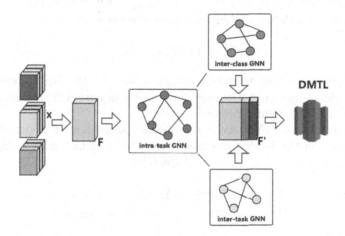

Fig. 1. An illustration of the hierarchical graph neural network for multi-task learning, where **F** is the hidden feature representation, 'intra-task GNN' is the first level GNN to aggregate all the information contained in the data of a task to generate the task embedding, 'inter-task GNN' and 'inter-class GNN' (for classification task) are the second level GNN to update all the task embeddings and class embeddings by sharing information among all the tasks and classes, and **F'** is the augmented feature representation used to do prediction. **F'** is the concatenated of the hidden feature representation **F** and its corresponding task embedding and class embedding (for classification task).

3.1   Overview of the Architecture

The HGNN consists of two-level GNNs. The first-level GNN is an intra-task GNN to aggregate all the information contained in the data of a task to generate a task representation, which is called the task embedding. In the second level, based on the generated task embeddings in the first level, an inter-task GNN is used

542   P. Guo et al.

to update all the task embeddings by sharing information among all the tasks. Finally the task embeddings are used to augment the feature representation of the data to improve the learning performance. For classification tasks, we can learn augmented features in a fine granularity - the class level. That is, the intra-task GNN is also used to aggregate all the information in a class of a task to generate a class embedding. Then based on class embeddings in all the tasks, an inter-class GNN is used to update them. Finally, both task embeddings and class embeddings are used to augment the feature representation. The whole architecture of HGNN is shown in Fig. 1.

3.2 The Model

Suppose that there are m multi-class classification tasks where each task has k classes. The training dataset of the ith task consists of n_i pairs of data samples and corresponding labels, i.e., $\mathcal{D}^i = \{(\mathbf{x}_p^i, y_p^i)\}_{p=1}^{n_i}$ where $y_p^i \in \{1, \ldots, k\}$.

For $\mathbf{x}_p^i$, we first define its hidden representation as

$$\hat{\mathbf{h}}_p^i = \sigma_s(\hat{\mathbf{W}}_s \mathbf{x}_p^i + \hat{\mathbf{b}}_s), \tag{1}$$

where $\sigma_s(\cdot)$ can be any activation function such as the ReLU function, and $\hat{\mathbf{W}}_s, \hat{\mathbf{b}}_s$ are shared parameter among all the tasks. Equation (1) defines the shared layer in a multi-task neural network.

For the intra-task GNN, we first construct an adjacency matrix $\mathbf{G}^i$ for the ith task based on the hidden representation and label information. Specifically, the (p, q)th entry in $\mathbf{G}^i$, g_{pq}^i, can be defined as

$$g_{pq}^i = \begin{cases} \exp\{-\|\hat{\mathbf{h}}_p^i - \hat{\mathbf{h}}_q^i\|_2^2\} & \text{if } y_p^i = y_q^i \\ -\exp\{-\|\hat{\mathbf{h}}_p^i - \hat{\mathbf{h}}_q^i\|_2^2\} & \text{otherwise} \end{cases},$$

where $\|\cdot\|_2$ denotes the ℓ_2 norm of a vector. Then the intra-task GNN can be defined as

$$\mathbf{H}^i = \sigma_h(\mathbf{W}_h^i \mathbf{X}^i + \hat{\mathbf{H}}^i \mathbf{G}^i + \mathbf{b}_h^i \mathbf{1}), \tag{2}$$

where $\sigma_h(\cdot)$ can be any activation function such as the ReLU function, $\mathbf{X}^i = (\mathbf{x}_1^i, \ldots, \mathbf{x}_{n_i}^i)$, $\hat{\mathbf{H}}^i = (\hat{\mathbf{h}}_1^i, \ldots, \hat{\mathbf{h}}_{n_i}^i)$, $\mathbf{1}$ denotes a vector of all ones with an appropriate size, $\mathbf{W}_h$ and $\mathbf{b}_h$ are the parameters in the GNN. $\mathbf{G}^i$ in Eq. (2) can make similar data points in the same class have similar representations in $\mathbf{H}^i$ and dissimilar data points from different classes have dissimilar representations. The intra-task GNN can have two or more layers each of which is defined as in Eq. (2).

Based on the intra-task GNN, the task embedding of the ith task is defined as

$$\mathbf{e}_t^i = \max_p\{\mathbf{h}_p^i\},$$

where the max operation is conducted elementwisely, $\mathbf{h}_p^i$ is the pth column in $\mathbf{H}^i$. So the task embedding is obtained via the max pooling on all the data points in the ith task based on the hidden representation learned by the intra-task GNN. Similarly, the class embedding of the rth class in ith task is defined as

$$\mathbf{e}_c^{i,r} = \max_{p:y_p^i=r} \{\mathbf{h}_p^i\},$$

which means that the class embedding of the rth class in the ith task is obtained via the max pooling on all the data points in the rth class of the ith task based on the intra-task GNN. We have tried other pooling methods such as the mean pooling but the performance is inferior to the max pooling. One reason is that the max pooling can bring some nonlinearity but the mean pooling is a linear operation.

Then m task embeddings $\{\mathbf{e}_t^i\}_{i=1}^m$ for the m tasks can form a graph. The inter-task GNN is responsible of learning for the graph constructed by task embeddings $\{\mathbf{e}_t^i\}_{i=1}^m$ to generate new task embeddings $\{\hat{\mathbf{e}}_t^i\}_{i=1}^m$ by exchanging information among tasks. Here we use GAT as an implementation of the inter-task GNN. In order to learn powerful task embeddings based on the inter-task relation, each task embedding $\mathbf{e}_t^i$ is first transformed by a weight matrix $\mathbf{W}$. Then we perform *self-attention* on the task embeddings. That is, an attentional mechanism computes *attention coefficients* as

$$d_{ij} = a(\mathbf{W}\mathbf{e}_t^i, \mathbf{W}\mathbf{e}_t^j),$$

where the attentional mechanism $a(\cdot, \cdot)$ we use is the cosine function, which is different from the original GAT. To normalize coefficients, we transform d_{ij} via the softmax function as

$$\alpha_{ij} = \text{softmax}_j(d_{ij}) = \frac{\exp(d_{ij})}{\sum_l \exp(d_{il})}.$$

Attention values can be viewed as a measure of task relations between each pair of tasks. Once obtained, the normalized attention coefficients are used as combination coefficients to compute the updated task embeddings via a nonlinear activation function σ as

$$\hat{\mathbf{e}}_t^i = \sigma\left(\sum_{j=1}^m \alpha_{ij} \mathbf{W}\mathbf{e}_t^j\right).$$

According to this equation, we can see that $\hat{\mathbf{e}}_t^i$ contains useful information from embeddings of other tasks. In experiments, the inter-task GNN adopts two such layers to generate the new task embeddings.

Similarly, the mk class embeddings $\{\mathbf{e}_c^{i,r}\}$ also can form a graph. We use another inter-class GNN to generate new class embeddings $\{\hat{\mathbf{e}}_c^{i,r}\}$ in a similar way to the inter-task GNN.

The learned task embeddings and class embeddings can be used to augment the data feature representation to form a more expressive one as $\tilde{\mathbf{h}}_p^i = $ concat$(\hat{\mathbf{h}}_p^i, \hat{\mathbf{e}}_t^i, \hat{\mathbf{e}}_c^{i,r})$, where concat$(\cdot, \cdot, \cdot)$ denotes the concatenation operation. Then data in such augmented representation can be fed into a deep multi-task learning model to predict class labels. To see that, the objective to be minimized in our proposed learning framework for multi-task learning can be formulated as

$$\mathcal{L} = \sum_{i=1}^{m} \frac{1}{n_i} \sum_{p=1}^{n_i} \mathcal{L}_C(f(\text{concat}(\gamma(\mathbf{x}_p^i), \hat{\mathbf{e}}_t^i, \hat{\mathbf{e}}_c^{i,y_p^i})), y_p^i) + \lambda r(\boldsymbol{\Theta}), \qquad (3)$$

where $\mathcal{L}_C$ denotes the classification loss on the available labeled data, $f(\cdot)$ denotes the learning function of a multi-task neural network starting from the second layer, the function $\gamma(\cdot)$ denotes the function defined in Eq. (1) by omitting the parameters, $r(\boldsymbol{\Theta})$ denotes a regularization function on $\boldsymbol{\Theta}$ that includes all the parameters of the model, and λ is a regularization parameter.

3.3   Testing Process

At the testing process, we do not know the true label, hence we cannot directly concatenate the class embedding to the hidden representation. We use the following method to solve this problem. For each testing sample, we concatenate the class embedding of each class c to its hidden representation as its new hidden representation and then compute the prediction probability that the testing sample belongs to class c via the softmax function used in the multi-task neural network. Finally we choose class c with the largest prediction probability as the predicted label. In mathematics, we predict the class label of a testing sample as

$$c^* = \arg \max_{r \in [k]} \mathbb{P}(y = r | f(\text{concat}(\gamma(\mathbf{x}_*^i), \hat{\mathbf{e}}_t^i, \hat{\mathbf{e}}_c^{i,r}))), \qquad (4)$$

where $[k]$ denotes a set of positive integers no larger than k and $\mathbf{x}_*^i$ denotes the test data point from the ith task. Note that in the prediction rule (4), the concatenated class embedding $\hat{\mathbf{e}}_c^{i,r}$ changes with r.

3.4   Extension to Regression Tasks

For the regression problems, there are only continuous labels and we cannot define class embeddings. So we only use task embeddings as the augmented feature representation. Furthermore, the adjacency matrix $\mathbf{G}^i$ for the ith task is constructed differently from classification tasks. Specifically, the (p, q)th entry in $\mathbf{G}^i$, g_{pq}^i, for a regression task is defined as

$$g_{pq}^i = \exp\{-\|\hat{\mathbf{h}}_p^i - \hat{\mathbf{h}}_q^i\|_2^2\}.$$

Since there is no class embedding, we do not need the prediction rule as in Eq. (4). The rest is identical to classification tasks.

3.5   Analysis

The proposed approach to augment the feature representation based on HGNN is interesting and here we provide some analyses to give insights into this model.

To understand why the incorporation of the task embedding can improve the performance, we use single-task learning as an example to make an illustration, which also works for multi-task learning. We denote the learning function without the use of the task embedding as $g(\mathbf{x})$ and that with the task embedding as $\hat{g}(\text{concat}(\mathbf{x}, \mathbf{e}))$, where $\mathbf{e}$ denotes the task embedding. If those two learning functions are within the same family such as the linear model or the neural network, we can see that $\hat{g}(\text{concat}(\mathbf{x}, \mathbf{e}))$ can reduce to $g(\mathbf{x})$ when all the parameters related to $\mathbf{e}$ in $\hat{g}(\cdot)$ are set to 0, making the expressiveness of $g(\mathbf{x})$ lower than that of $\hat{g}(\text{concat}(\mathbf{x}, \mathbf{e}))$. Hence, given the same training dataset, the training loss of $\hat{g}(\text{concat}(\mathbf{x}, \mathbf{e}))$ is usually lower than that of $g(\mathbf{x})$. Of course, the model complexity of $\hat{g}(\text{concat}(\mathbf{x}, \mathbf{e}))$ is larger than that of $g(\mathbf{x})$. Based on generalization bounds derived in single-task learning based on for example the Rademacher complexity [4], the generalization loss is upper-bounded by the sum of the training loss and the model complexity. So if the decrease of the training loss of $\hat{g}(\text{concat}(\mathbf{x}, \mathbf{e}))$ compared with $g(\mathbf{x})$ is larger than the increasing of the model complexity in $\hat{g}(\text{concat}(\mathbf{x}, \mathbf{e}))$, then $\hat{g}(\text{concat}(\mathbf{x}, \mathbf{e}))$ is likely to have a lower generalization loss than $g(\mathbf{x})$. Such analysis also holds for multi-task learning. In experiments, we find that when the dimension of the task embedding is small, leading to a small number of additional parameters incurred as well as a low model complexity, the generalization performance is better than that with a larger dimension for the task embedding, which verifies the above analysis.

The input space, which is a subset of a vector space, is denoted by $\mathcal{X}$ and the output space is denoted by $\mathcal{Y}$. Training samples $\{(\mathbf{x}_i, \mathbf{y}_i)_{i=1}^n\} \in \mathcal{X} \times \mathcal{Y}$ are distributed according to some unknown distribution P. Let $\ell : \mathbb{R}^k \times \mathcal{Y} \to \mathbb{R}^+$ be the loss function, where k denotes the dimension of the label space. The learning function is defined as $f(\mathbf{x}) = \mathbf{W}^\mathsf{T}\mathbf{x}$ where the superscript $^\mathsf{T}$ denotes the transpose and $\mathbf{W}$ is abused to denote the parameter in this linear learner. The expected loss is defined as $\mathcal{L}(\mathbf{W}) = \mathbb{E}[\ell(\mathbf{W}^\mathsf{T}\mathbf{x}, \mathbf{y})]$. The empirical loss is defined as $\hat{\mathcal{L}}(\mathbf{W}) = \frac{1}{n}\sum_{i=1}^n \ell(\mathbf{W}^\mathsf{T}\mathbf{x}_i, \mathbf{y}_i)$. The data matrix $\mathbf{X}$ is defined as $\mathbf{X} = (\mathbf{x}_1, \ldots, \mathbf{x}_n) \in \mathbb{R}^{p \times n}$ and the label matrix $\mathbf{Y}$ is defined as $\mathbf{Y} = (\mathbf{y}_1, \ldots, \mathbf{y}_n) \in \mathbb{R}^{k \times n}$. $\mathbf{e} \in \mathbb{R}^{q \times 1}$ denotes the task embedding and $\mathbf{E} = \mathbf{e}\mathbf{1}^\mathsf{T} \in \mathbb{R}^{q \times n}$ is the task embedding matrix for all the training data, where $\mathbf{1}$ denotes a column vector of all ones with an appropriate size.

Let us consider two models. The objective function of model 1 is formulated as

$$\hat{\mathbf{W}}_1 = \operatorname*{argmin}_{\mathbf{W}_1} \|\mathbf{Y} - \mathbf{W}_1^\mathsf{T}\mathbf{X}\|_2^2 + \lambda\|\mathbf{W}_1\|_2^2, \tag{5}$$

and that of model 2 is

$$\hat{\mathbf{W}}_2 = \operatorname*{argmin}_{\mathbf{W}_2} \|\mathbf{Y} - \mathbf{W}_2^\mathsf{T}\hat{\mathbf{X}}\|_2^2 + \lambda\|\mathbf{W}_2\|_2^2, \tag{6}$$

where $\mathbf{W}_1 \in \mathbb{R}^{p \times k}$, $\mathbf{W}_2 \in \mathbb{R}^{(q+p) \times k}$, $\hat{\mathbf{x}}_i = (\mathbf{x}_i^\mathsf{T}, \mathbf{e}^\mathsf{T})^\mathsf{T}$, $\hat{\mathbf{X}} = (\hat{\mathbf{x}}_1, \ldots, \hat{\mathbf{x}}_n) = (\mathbf{X}^\mathsf{T}, \mathbf{E}^\mathsf{T})^\mathsf{T} \in \mathbb{R}^{(q+p) \times n}$. So model 1 is a ridge regression model which can be applied to both classification and regression tasks and model 2 is a variant of model 1 with the task embedding incorporated. For training losses of those two models, we have the following result, whose proof is in the appendix.

Theorem 1. *If* $\mathbf{X}$ *and* $\mathbf{E}$ *satisfy* $\mathbf{X}^{\mathsf{T}}\mathbf{X}\mathbf{E}^{\mathsf{T}}\mathbf{E}+\mathbf{E}^{\mathsf{T}}\mathbf{E}\mathbf{X}^{\mathsf{T}}\mathbf{X}+2\lambda\mathbf{E}^{\mathsf{T}}\mathbf{E}+\mathbf{E}^{\mathsf{T}}\mathbf{E}\mathbf{E}^{\mathsf{T}}\mathbf{E} \succeq 0$ *where* $\mathbf{M}_1 \succeq \mathbf{M}_2$ *means that* $\mathbf{M}_1 - \mathbf{M}_2$ *is positive semidefinite, then the training loss of model 2 with the task embedding is always lower than that of model 1 without the task embedding. That is, we have*

$$\|\mathbf{Y} - \hat{\mathbf{W}}_1^{\mathsf{T}}\mathbf{X}\|_2^2 \geq \|\mathbf{Y} - \hat{\mathbf{W}}_2^{\mathsf{T}}\hat{\mathbf{X}}\|_2^2. \tag{7}$$

Remark 1. Theorem 1 implies that for a model, incorporating the task embedding to augment the feature representation will incur a lower training loss than that without the task embedding. From the perspective of the model capacity, model 1 is a reduced version of model 2 by setting the task embedding to be zero and hence mode 2 has a larger capacity than model 1, making model 2 possess a large chance to have a lower training loss. The condition proposed in Theorem 1 is very easy to check and we can adjust λ to ensure the positive semidefiniteness of the condition.

We also analyze the generalization bound of the two models. We first rewrite problems (5) and (6) into equivalent formulations as

$$\hat{\mathbf{W}}_1 = \underset{\|\mathbf{W_1}\|_2 \leq W_*}{\operatorname{argmin}} \ \|\mathbf{Y} - \mathbf{W}_1^{\mathsf{T}}\mathbf{X}\|_2^2$$

$$\hat{\mathbf{W}}_2 = \underset{\|\mathbf{W_2}\|_2 \leq W_*}{\operatorname{argmin}} \ \|\mathbf{Y} - \mathbf{W}_2^{\mathsf{T}}\hat{\mathbf{X}}\|_2^2.$$

For the above two problems, we have the following result, whose proof is in the appendix.

Theorem 2. *Suppose* $\|\mathbf{x}_i\|, \|\hat{\mathbf{x}}_i\| \leq X_*$, *the task embedding satisfies the condition in Theorem 1. Then for any* $\delta > 0$, *with probability at least* $1 - \delta$, *we have*

$$\mathbb{E}_{\mathbf{x},\mathbf{y}}(\|\mathbf{y} - \hat{\mathbf{W}}_1^{\mathsf{T}}\mathbf{x}\|_2^2) \leq \frac{1}{n}\sum_{i=1}^{n}\|\mathbf{y}_i - \hat{\mathbf{W}}_1^{\mathsf{T}}\mathbf{x}_i\|_2^2 + 4X_*\beta_* \sqrt{\frac{1}{n}} + 2X_*\beta_* \sqrt{\frac{\log(1/\delta)}{2n}}$$

$$\mathbb{E}_{\mathbf{x},\mathbf{y}}(\|\mathbf{y} - \hat{\mathbf{W}}_2^{\mathsf{T}}\hat{\mathbf{x}}\|_2^2) \leq \frac{1}{n}\sum_{i=1}^{n}\|\mathbf{y}_i - \hat{\mathbf{W}}_2^{\mathsf{T}}\hat{\mathbf{x}}_i\|_2^2 + 4X_*\beta_* \sqrt{\frac{1}{n}} + 2X_*\beta_* \sqrt{\frac{\log(1/\delta)}{2n}}.$$

Remark 2. According to Theorem 2, the generalization upper-bound of model 2 with the use of the task embedding is lower than that without the task embedding because of the lower training loss of model 2 which has been proved in Theorem 1. This may imply that there is a large chance that the expected loss of model 2 is lower than that of model 1, which can be verified in empirical studies.

4   Experiments

In this section, we conduct empirical studies to test the performance of the proposed HGNN.

4.1   Experimental Settings

We conduct experiments on several benchmark datasets, including **ImageCLEF** [7], **Office-Caltech-10** [10], **Office-Home** [34], and **SARCOS** [39].

The **ImageCLEF** dataset is the benchmark for Image-CLEF domain adaptation challenge which contains about 2,400 images from 12 common categories shared by four tasks including *Caltech-256* (**C**), *ImageNet ILSVRC* (**I**), *Pascal VOC 2012* (**P**), and *Bing* (**B**). There are 50 images in each category and 600 images in each task.

The **Office-Caltech-10** dataset includes 10 common categories shared by the Office-31 and Caltech-256 datasets. It contains four domains: *Caltech* (**C**) that is sampled from Caltech-256 dataset, *Amazon* (**A**) that contains images collected from the amazon website, *Webcam* (**W**) and *DSLR* (**D**) that are taken by the web camera and DSLR camera under the office environment. In our experiment, we regard each domain as a task.

The **Office-Home** dataset has 15,500 images across 65 classes in the office and home settings from four domains with a large domain discrepancy: *Artistic images* (**Ar**), *Clip art* (**Cl**), *Product images* (**Pr**), and *Real-world images* (**Rw**). In our experiment, we regard each domain as a task.

The **SARCOS** dataset studies a multi-output problem of learning the inverse dynamics of 7 SARCOS anthropomorphic robot arms, each of which corresponds to a task, based on 21 features, including seven joint positions, seven joint velocities, and seven joint accelerations. By following [39], we treat each output as a task and randomly sample 2000 data points from each output to construct the dataset.

Since the proposed HGNN can be combined with many deep multi-task learning models as discussed before, we incorporate the HGNN into the Deep Multi-Task Learning (DMTL) which shares the first several layers as the common hidden feature representation for all the tasks as did in [8,18,22,27,37], Deep Multi-Task Representation Learning (DMTRL) [35], and Trace Norm Regularised Deep Multi-Task Learning (TNRMTL) [36], respectively, to show the benefit of the learned augmented features.

In experiments, we use the Tensorflow package [1] to implement all the models and leverage the VGG-19 network [32] pretrained on the ImageNet dataset [29] as the backbone of the feature extractor. After that, all the multi-task learning models adopt a two-layer fully-connected architecture (#data_dim × 600 × #classes) and the ReLU activation function is used. The first layer is shared by all tasks to learn a common representation corresponding to Eq. (1) and the second layer is for task-specific outputs.

For optimization, we use the Adam method [16] with the learning rate varying as $\eta = \frac{0.02}{1+p}$, where p is the number of the iteration. By following GAT, the dimension of task embeddings is set to 8, i.e., $F'_t = 8$. Similarly, we also set the dimension of class embeddings to 8, i.e., $F'_c = 8$. The size of mini-batch is set to 32. Each experiment repeats for 5 times and we report the average performance as well as the standard deviation.

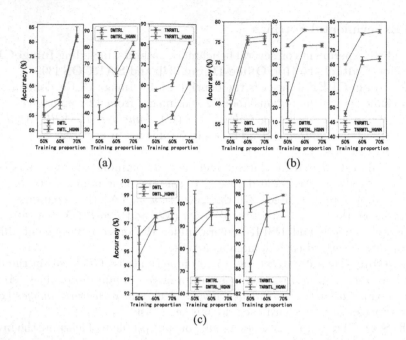

Fig. 2. Performance of different models on different datasets when varying with the training proportion: (a) ImageCLEF; (b) Office-Home; (c) Office-Caltech-10.

4.2 Experimental Results

Results on Classification Tasks. For classification tasks, the performance measure is the classification accuracy. To investigate the effect of the size of the training dataset on the performance, we vary the proportion of training data from 50% to 70% at an interval of 10% and plot the average test accuracy of different methods in Figs. 2(a)–2(c). According to results reported in these figures, we can see that the incorporation of the HGNN into baseline models improves the classification accuracy of all baseline models especially when the training proportion is small. As reported in Figs. 2(b) and 2(c), the incorporation of the HGNN boosts the performance of all the baseline on the Office-Caltech-10 and Office-Home datasets. For the DMTRL and TNRMTL models, the improvement is significant with the use of the HGNN. Moreover, when using augmented features learned by the HGNN, the standard deviation becomes smaller than the corresponding baseline model without using the HGNN under every experimental setting, which implies that the HGNN can improve the stability of baseline models to some extent.

Results on Regression Tasks. For regression tasks, we use the mean squared error to measure the performance. The test errors on the SARCOS dataset are shown in Fig. 3 where the training proportion varies from 50% to 70% at an interval of 10%. As shown in Fig. 3, after using the HGNN, the test error of each baseline model has a significant decrease at each training proportion, which demonstrate the effectiveness of augmented features learned in the HGNN method.

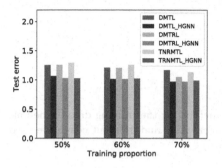 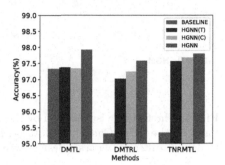

Fig. 3. Performance of different models on the SARCOS dataset.

Fig. 4. The ablation study on the Office-Caltech-10 dataset.

4.3   Ablation Study

To study the effectiveness of task embeddings and class embeddings in the HGNN model, we study two variants of HGNN, including HGNN(T) that only augments with the task embedding and HGNN(C) that only augments with the class embedding. The comparison among baseline models, HGNN, variants of HGNN on the Office-Caltech-10 dataset is shown in Fig. 4. According to the results, we can see that the use of only the class embedding in HGNN(C) or the task embedding in HGNN(T) can improve the performance over baseline models, which shows that augmented features learned in two ways are effective. HGNN(C) seems better than HGNN(T) in this experiment. One reason is that class embeddings may contain more discriminative features for the classification task. Figure 4 also indicates that using both task embeddings and class embeddings achieves the best performance, which again verifies the usefulness of the HGNN.

4.4   Visualization

To dive deeper into the learned features, we plot in Figs. 5 and 6 the t-SNE embeddings of the feature representations learned for the four tasks on the Office-Caltech-10 dataset by TNRMTL and TNRMTL_HGNN, respectively, at the training and testing processes. We observe that the data based on the representation derived by the HGNN model are more separable among classes in each task during either the training process or the testing process. This phenomenon verifies the effectiveness of the augmented features learned in the HGNN to help discriminate data points in different classes of all the tasks.

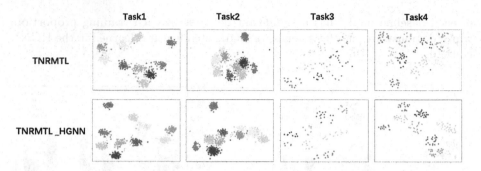

Fig. 5. The feature visualization by the t-SNE method for the training data in the four tasks on the Office-Caltech-10 dataset. Different markers and different colors are used to denote different classes. (Best viewed in color.)

4.5   Sensitivity Analysis

We conduct the sensitivity analysis of the performance with respect to the dimension of task embedding (denoted by F'_t) and class embedding (denoted by F'_c), respectively, on the ImageCLEF dataset. The results are shown in Tables 1 and 2. According to the results, we can see that $F'_t = 8$ and $F'_c = 8$ are a good choice in most cases, though in some case, a lower value (i.e., 4) performs better. When the dimension is not so large (e.g., not large than 32), the performance changes a little, making the choice of the dimension insensitive. However, when using a larger dimension (e.g., 64), the classification accuracy drops significantly, implying that the HGNN prefers a small dimension.

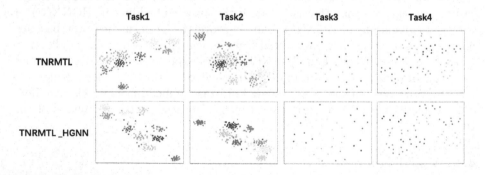

Fig. 6. The feature visualization by the t-SNE method for the testing data in the four tasks on the Office-Caltech-10 dataset. Different markers and different colors are used to denote different classes. (Best viewed in color.)

Table 1. The classification accuracy (%) on ImageCLEF when varying F_t' and fixing F_c' as 8.

F_t'	4	8	16	32	64
DMTL_HGNN	81.22±1.27	**82.03±1.98**	81.18±1.43	80.41±0.88	79.89±0.65
DMTRL_HGNN	81.21±0.80	**82.07±1.47**	81.64±0.77	81.15±2.35	81.48±1.78
TNRMTL_HGNN	**82.35±1.66**	81.11±0.94	82.27±1.19	81.74±0.91	81.20±0.40

Table 2. The classification accuracy (%) on ImageCLEF when varying F_c' and fixing F_t' as 8.

F_c'	4	8	16	32	64
DMTL_HGNN	**82.37±1.13**	82.03±1.98	81.75±0.56	80.71±1.30	81.65±2.84
DMTRL_HGNN	81.73±1.39	**82.07±1.47**	80.40±1.57	81.94±2.03	80.74±2.16
TNRMTL_HGNN	80.07±3.36	**81.11±0.94**	80.64±1.11	80.17±0.98	80.47±0.80

5 Conclusion

In this paper, we propose a hierarchical graph neural network to learn augmented features for deep multi-task learning. The proposed HGNN has two levels. In the first level, the intra-task graph neural network is used to learn a powerful representation for each data point in a task by aggregating information from its neighbors in this task. Based on the learned representation, we can learn the task embedding for each task as well as the class embedding if any. The inter-task graph neural network as well inter-class graph neural network is used to update each task embedding and each class embedding. Finally the learned task embedding and class embedding can be used to augment the data representation. Extensive experiments show the effectiveness of the proposed HGNN. In our future work, we are interested in applying the HGNN to other multi-task learning models.

Acknowledgements. This work is supported by NSFC grant 62076118.

References

1. Abadi, M., et al.: Tensorflow: a system for large-scale machine learning. In: Proceedings of the 12th USENIX Symposium on Operating Systems Design and Implementation, pp. 265–283 (2016)
2. Ando, R.K., Zhang, T.: A framework for learning predictive structures from multiple tasks and unlabeled data. J. Mach. Learn. Res. **6**, 1817–1853 (2005)
3. Argyriou, A., Evgeniou, T., Pontil, M.: Multi-task feature learning. Adv. Neural. Inf. Process. Syst. **19**, 41–48 (2006)
4. Bartlett, P.L., Mendelson, S.: Rademacher and Gaussian complexities: Risk bounds and structural results. J. Mach. Learn. Res. **3**, 463–482 (2002)
5. Bickel, S., Bogojeska, J., Lengauer, T., Scheffer, T.: Multi-task learning for HIV therapy screening. In: Proceedings of the Twenty-Fifth International Conference on Machine Learning, pp. 56–63 (2008)

6. Bonilla, E., Chai, K.M.A., Williams, C.: Multi-task Gaussian process prediction. In: Advances in Neural Information Processing Systems 20, pp. 153–160. Vancouver, British Columbia, Canada (2007)
7. Caputo, B., et al.: ImageCLEF 2014: overview and analysis of the results. In: Kanoulas, E., et al. (eds.) CLEF 2014. LNCS, vol. 8685, pp. 192–211. Springer, Cham (2014). https://doi.org/10.1007/978-3-319-11382-1_18
8. Caruana, R.: Multitask learning. Mach. Learn. **28**(1), 41–75 (1997)
9. Gao, Y., Bai, H., Jie, Z., Ma, J., Jia, K., Liu, W.: MTL-NAS: task-agnostic neural architecture search towards general-purpose multi-task learning. In: Proceedings of IEEE Conference on Computer Vision and Pattern Recognition (2020)
10. Gong, B., Shi, Y., Sha, F., Grauman, K.: Geodesic flow kernel for unsupervised domain adaptation. In: 2012 IEEE Conference on Computer Vision and Pattern Recognition, pp. 2066–2073. IEEE (2012)
11. Han, L., Zhang, Y.: Learning multi-level task groups in multi-task learning. In: Proceedings of the 29th AAAI Conference on Artificial Intelligence (2015)
12. Han, L., Zhang, Y.: Multi-stage multi-task learning with reduced rank. In: Proceedings of the 30th AAAI Conference on Artificial Intelligence (2016)
13. Jacob, L., Bach, F., Vert, J.P.: Clustered multi-task learning: a convex formulation. Adv. Neural. Inf. Process. Syst. **21**, 745–752 (2008)
14. Jalali, A., Ravikumar, P.D., Sanghavi, S., Ruan, C.: A dirty model for multi-task learning. In: Advances in Neural Information Processing Systems 23, pp. 964–972. Vancouver, British Columbia, Canada (2010)
15. Kim, R., So, C.H., Jeong, M., Lee, S., Kim, J., Kang, J.: HATS: A hierarchical graph attention network for stock movement prediction. CoRR abs/1908.07999 (2019)
16. Kingma, D.P., Ba, J.: Adam: A method for stochastic optimization. arXiv preprint arXiv:1412.6980 (2014)
17. Kumar, A., III, H.D.: Learning task grouping and overlap in multi-task learning. In: Proceedings of the 29th International Conference on Machine Learning. Edinburgh, Scotland, UK (2012)
18. Li, S., Liu, Z., Chan, A.B.: Heterogeneous multi-task learning for human pose estimation with deep convolutional neural network. IJCV **113**(1), 19–36 (2015)
19. Liu, H., Palatucci, M., Zhang, J.: Blockwise coordinate descent procedures for the multi-task lasso, with applications to neural semantic basis discovery. In: Proceedings of the 26th Annual International Conference on Machine Learning (2009)
20. Liu, P., Fu, J., Dong, Y., Qiu, X., Cheung, J.C.K.: Multi-task learning over graph structures. CoRR abs/1811.10211 (2018)
21. Liu, S., Johns, E., Davison, A.J.: End-to-end multi-task learning with attention. In: Proceedings of IEEE Conference on Computer Vision and Pattern Recognition, pp. 1871–1880 (2019)
22. Liu, W., Mei, T., Zhang, Y., Che, C., Luo, J.: Multi-task deep visual-semantic embedding for video thumbnail selection. In: Proceedings of IEEE Conference on Computer Vision and Pattern Recognition, pp. 3707–3715 (2015)
23. Lozano, A.C., Swirszcz, G.: Multi-level lasso for sparse multi-task regression. In: Proceedings of the 29th International Conference on Machine Learning. Edinburgh, Scotland, UK (2012)
24. Lu, H., Huang, S.H., Ye, T., Guo, X.: Graph star net for generalized multi-task learning. CoRR abs/1906.12330 (2019)
25. Meng, Z., Adluru, N., Kim, H.J., Fung, G., Singh, V.: Efficient relative attribute learning using graph neural networks. In: Proceedings of the 15th European Conference on Computer Vision (2018)

26. Misra, I., Shrivastava, A., Gupta, A., Hebert, M.: Cross-stitch networks for multi-task learning. In: Proceedings of IEEE Conference on Computer Vision and Pattern Recognition, pp. 3994–4003 (2016)
27. Mrksic, N., et al.: Multi-domain dialog state tracking using recurrent neural networks. In: Proceedings of the 53rd Annual Meeting of the Association for Computational Linguistics, pp. 794–799 (2015)
28. Obozinski, G., Taskar, B., Jordan, M.: Multi-task feature selection. Technical report, Department of Statistics, University of California, Berkeley (June 2006)
29. Russakovsky, O., et al.: Imagenet large scale visual recognition challenge. Int. J. Comput. Vis. **115**(3), 211–252 (2015)
30. Ryu, H., Shin, H., Park, J.: Multi-agent actor-critic with hierarchical graph attention network. CoRR abs/1909.12557 (2019)
31. Shinohara, Y.: Adversarial multi-task learning of deep neural networks for robust speech recognition. In: Proceedings of the 17th Annual Conference of the International Speech Communication Association, pp. 2369–2372 (2016)
32. Simonyan, K., Zisserman, A.: Very deep convolutional networks for large-scale image recognition. arXiv preprint arXiv:1409.1556 (2014)
33. Veličković, P., Cucurull, G., Casanova, A., Romero, A., Lio, P., Bengio, Y.: Graph attention networks. arXiv preprint arXiv:1710.10903 (2017)
34. Venkateswara, H., Eusebio, J., Chakraborty, S., Panchanathan, S.: Deep hashing network for unsupervised domain adaptation. In: Proceedings of the IEEE Conference on Computer Vision and Pattern Recognition, pp. 5018–5027 (2017)
35. Yang, Y., Hospedales, T.M.: Deep multi-task representation learning: a tensor factorisation approach. In: Proceedings of the 6th International Conference on Learning Representations (2017)
36. Yang, Y., Hospedales, T.M.: Trace norm regularised deep multi-task learning. In: Proceedings of the 6th International Conference on Learning Representations, Workshop Track (2017)
37. Zhang, W., et al.: Deep model based transfer and multi-task learning for biological image analysis. In: Proceedings of the 21th ACM SIGKDD International Conference on Knowledge Discovery and Data Mining, pp. 1475–1484 (2015)
38. Zhang, Y., Yang, Q.: A survey on multi-task learning. CoRR abs/1707.08114 (2017)
39. Zhang, Y., Yeung, D.Y.: A convex formulation for learning task relationships in multi-task learning. In: Proceedings of the 26th Conference on Uncertainty in Artificial Intelligence, pp. 733–742 (2010)
40. Zhang, Y., Wei, Y., Yang, Q.: Learning to multitask. In: Advances in Neural Information Processing Systems 31, pp. 5776–5787 (2018)
41. Zhang, Y., Yeung, D., Xu, Q.: Probabilistic multi-task feature selection. Probabilistic multi-task feature selection. In: Advances in Neural Information Processing Systems 23, pp. 2559–2567 (2010)

Bridging Few-Shot Learning and Adaptation: New Challenges of Support-Query Shift

Etienne Bennequin[1,2]([✉]), Victor Bouvier[1,3]([✉]), Myriam Tami[1],
Antoine Toubhans[2], and Céline Hudelot[1]

[1] CentraleSupélec, Mathématiques et Informatique pour la Complexité et les
Systèmes, Université Paris-Saclay, 91190 Gif-sur-Yvette, France
{etienne.bennequin,victor.bouvier,myriam.tami,
celine.hudelot}@centralesupelec.fr
[2] Sicara, 48 boulevard des Batignolles, 75017 Paris, France
etienneb@sicara.com
[3] Sidetrade, 114 Rue Gallieni, 92100 Boulogne-Billancourt, France
vbouvier@sidetrade.com

Abstract. *Few-Shot Learning* (FSL) algorithms have made substantial progress in learning novel concepts with just a handful of labelled data. To classify *query* instances from novel classes encountered at test-time, they only require a *support set* composed of a few labelled samples. FSL benchmarks commonly assume that those queries come from the same distribution as instances in the support set. However, in a realistic setting, data distribution is plausibly subject to change, a situation referred to as *Distribution Shift* (DS). The present work addresses the new and challenging problem of ***F****ew-Shot Learning under **S**upport/**Q**uery **S**hift* (**FSQS**) *i.e.*, when support and query instances are sampled from related but different distributions. Our contributions are the following. First, we release a testbed for FSQS, including datasets, relevant baselines and a protocol for a rigorous and reproducible evaluation. Second, we observe that well-established FSL algorithms unsurprisingly suffer from a considerable drop in accuracy when facing FSQS, stressing the significance of our study. Finally, we show that transductive algorithms can limit the inopportune effect of DS. In particular, we study both the role of Batch-Normalization and Optimal Transport (OT) in aligning distributions, bridging *Unsupervised Domain Adaptation* with FSL. This results in a new method that efficiently combines OT with the celebrated *Prototypical Networks*. We bring compelling experiments demonstrating the advantage of our method. Our work opens an exciting line of research by providing a testbed and strong baselines. Our code is available at https://github.com/ebennequin/meta-domain-shift.

E. Bennequin and V. Bouvier—Equal contribution.

Electronic supplementary material The online version of this chapter (https://doi.org/10.1007/978-3-030-86486-6_34) contains supplementary material, which is available to authorized users.

© Springer Nature Switzerland AG 2021
N. Oliver et al. (Eds.): ECML PKDD 2021, LNAI 12975, pp. 554–569, 2021.
https://doi.org/10.1007/978-3-030-86486-6_34

Keywords: Few-shot learning · Distribution shift · Adaptation · Optimal transport

1   Introduction

In the last few years, we have witnessed outstanding progress in supervised deep learning [15]. As the abundance of labelled data during training is rarely encountered in practice, ground-breaking works in *Few-Shot Learning* (FSL) have emerged [12,28,31], particularly for image classification. This paradigm relies on a straightforward setting. At test-time, given a set of not seen during training and *few* (typically 1 to 5) labelled examples for each of those classes, the task is to classify query samples among them. We usually call the set of labelled samples the *support set*, and the set of query samples the *query set*. Well-adopted FSL benchmarks [25,30,31] commonly sample the support and query sets from the same distribution. We stress that this assumption does not hold in most use cases. When deployed in the real-world, we expect an algorithm to infer on data that may shift, resulting in an acquisition system that deteriorates, lighting conditions that vary, or real world objects evolving [1].

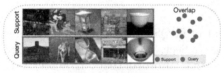 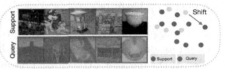

   (a) Standard FSL   (b) FSL under Support / Query Shift

Fig. 1. Illustration of the FSQS problem with a 5-way 1-shot classification task sampled from the miniImageNet dataset [31]. In (a), a standard FSL setting where support and query sets are sampled from the same distribution. In (b), the same task but with shot-noise and contrast perturbations from [16] applied on support and query sets (respectively) that results in a support-query shift. In the latter case, a similarity measure based on the Euclidean metric [28] may become inadequate.

The situation of *Distribution Shift* (DS) *i.e.*, when training and testing distributions differ, is ubiquitous and has dramatic effects on deep models [16], motivating works in *Unsupervised Domain Adaptation* [22], *Domain Generalization* [14] or *Test-Time Adaptation* [32]. However, the state of the art brings insufficient knowledge on few-shot learners' behaviours when facing distribution shift. Some pioneering works demonstrate that advanced FSL algorithms do not handle cross-domain generalization better than more naive approaches [5]. Despite its great practical interest, FSL under distribution shift between the support and query sets is an under-investigated problem and attracts a very recent attention [11]. We refer to it as *Few-Shot Learning under Support/Query Shift* (**FSQS**) and provide an illustration in Fig. 1. It reflects a more realistic situation where the algorithm is fed with a support set at the time of deployment and infers continuously on data subject to shift. The first solution is to re-acquire a support set that follows the data's evolution. Nevertheless, it implies human intervention to select and annotate data to update an already deployed model, reacting to a potential drop in performances. The second solution consists in designing an algorithm that is robust to the distribution shift encountered during inference. This is the subject of the present work. Our contributions are:

1. `FewShiftBed`: a testbed for FSQS available at https://github.com/ ebennequin/meta-domain-shift. The testbed includes 3 challenging benchmarks along with a protocol for fair and rigorous comparison across methods as well as an implementation of relevant baselines, and an interface to facilitate the implementation of new methods.
2. We conduct extensive experimentation of a representative set of few-shot algorithms. We empirically show that *Transductive* Batch-Normalization [3] mitigates an important part of the inopportune effect of FSQS.
3. We bridge *Unsupervised Domain Adaptation* (UDA) with FSL to address FSQS. We introduce *Transported Prototypes*, an efficient transductive algorithm that couples *Optimal Transport* (OT) [23] with the celebrated *Prototypical Networks* [28]. The use of OT follows a long-standing history in UDA for aligning representations between distributions [2,13]. Our experiments demonstrate that OT shows a remarkable ability to perform this alignment even with only a few samples to compare distributions and provide a simple but strong baseline.

In Sect. 2 we provide a formal statement of FSQS, and we position this new problem among existing learning paradigms. In Sect. 3, we present `FewShiftBed`. We detail the datasets, the chosen baselines, and a protocol that guarantees a rigorous and reproducible evaluation. In Sect. 4, we present a method that couples Optimal Transport with Prototypical Networks [28]. Finally, in Sect. 5, we conduct an extensive evaluation of baselines and our proposed method using the testbed.

2　The Support-Query Shift Problem

2.1　Statement

Notations. We consider an input space $\mathcal{X}$, a representation space $\mathcal{Z} \subset \mathbb{R}^d$ $(d > 0)$ and a set of classes C. A representation is a learnable function from $\mathcal{X}$ to $\mathcal{Z}$ and is noted $\varphi(\cdot; \theta)$ with $\theta \in \Theta$ for Θ a set of parameters. A dataset is a set $\Delta(\mathsf{C}, \mathsf{D})$ defined by a set of classes C and a set of domains D *i.e.*, a domain $\mathcal{D} \in \mathsf{D}$ is a set of IID realizations from a distribution noted $p_{\mathcal{D}}$. For two domains $\mathcal{D}, \mathcal{D}' \in \mathsf{D}$, the distribution shift is characterized by $p_{\mathcal{D}} \neq p_{\mathcal{D}'}$. For instance, if the data consists of images of letters handwritten by several users, $\mathcal{D}$ can consist of samples from a specific user. Referring to the well known UDA terminology of source/target [22], we define a couple of source-target domains as a couple $(\mathcal{D}_s, \mathcal{D}_t)$ with $p_{\mathcal{D}_s} \neq p_{\mathcal{D}_t}$, thus presenting a distribution shift. Additionally, given $\mathcal{C} \subset \mathsf{C}$ and $\mathcal{D} \in \mathsf{D}$, the restriction of a domain $\mathcal{D}$ to images with a label that belongs to $\mathcal{C}$ is noted $\mathcal{D}^{\mathcal{C}}$.

Dataset Splits. We build a split of $\Delta(\mathsf{C}, \mathsf{D})$, by splitting D (respectively C) into $\mathsf{D}_{\text{train}}$ and D_{test} (respectively $\mathsf{C}_{\text{train}}$ and C_{test}) such that $\mathsf{D}_{\text{train}} \cap \mathsf{D}_{\text{test}} = \emptyset$

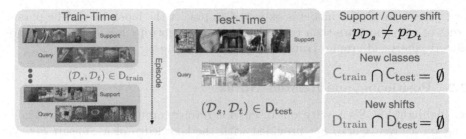

Fig. 2. During meta-learning (Train-Time), each episode contains a support and a query set sampled from different distributions (for instance, illustrated by noise and contrasts as in Fig. 1(b)) from a set of *training domains* (D_{train}), reflecting a situation that may potentially occurs at test-time. When deployed, the FSL algorithm using a trained backbone is fed with a support set sampled from new classes. As the algorithm is subject to infer continuously on data subject to shift (Test-Time), we evaluate the algorithm on data with an unknown shift (D_{test}). Importantly, both classes ($C_{train} \cap C_{test} = \emptyset$) and shifts ($D_{train} \cap D_{test} = \emptyset$) are not seen during training, making the FSQS a challenging problem of generalization.

and $D_{train} \cup D_{test} = D$ (respectively $C_{train} \cap C_{test} = \emptyset$ and $C_{train} \cup C_{test} = C$). This gives us a train/test split with the datasets $\Delta_{train} = \Delta(C_{train}, D_{train})$ and $\Delta_{test} = \Delta(C_{test}, D_{test})$. By extension, we build a validation set following the same protocol.

Few-Shot Learning under Support-Query Shift (FSQS). Given:

- $D' \in \{D_{train}, D_{test}\}$ and $C' \in \{C_{train}, C_{test}\}$,
- a couple of source-target domains $(\mathcal{D}_s, \mathcal{D}_t)$ from D',
- a set of classes $\mathcal{C} \subset C'$;
- a small labelled support set $\mathcal{S} = (x_i, y_i)_{i=1,...,|\mathcal{S}|}$ (named *source support set*) such that for all i, $y_i \in \mathcal{C}$ and $x_i \in \mathcal{D}_s$ i.e., $\mathcal{S} \subset \mathcal{D}_s^{\mathcal{C}}$;
- an unlabelled query set $\mathcal{Q} = (x_i)_{i=1,...,|\mathcal{Q}|}$ (named *target query set*) such that for all i, $y_i \in \mathcal{C}$ and $x_i \in \mathcal{D}_t$ i.e., $\mathcal{Q} \subset \mathcal{D}_t^{\mathcal{C}}$.

The task is to predict the labels of query set instances in $\mathcal{C}$. When $|\mathcal{C}| = n$ and the support set contains k labelled instances for each class, this is called an n-way k-shot FSQS classification task. Note that this paradigm provides an additional challenge compared to classical Few-shot classification tasks, since at test time, the model is expected to generalize to both new classes and new domains while support set and query set are sampled from different distributions. This paradigm is illustrated in Fig. 2.

Episodic Training. We build an episode by sampling some classes $\mathcal{C} \subset C_{train}$, and a source and target domain $\mathcal{D}_s, \mathcal{D}_t$ from D_{train}. We build a support set $\mathcal{S} = (x_i, y_i)_{i=1...|\mathcal{S}|}$ of instances from source domain $\mathcal{D}_s^{\mathcal{C}}$, and a query set $\mathcal{Q} = (x_i, y_i)_{i=|\mathcal{S}|+1,...,|\mathcal{S}|+|\mathcal{Q}|}$ of instances from target domain $\mathcal{D}_t^{\mathcal{C}}$, such that $\forall i \in [1, |\mathcal{S}| + |\mathcal{Q}|]$, $y_i \in \mathcal{C}$. Using the labelled examples from $\mathcal{S}$ and unlabelled instances from $\mathcal{Q}$, the model is expected to predict the labels of $\mathcal{Q}$. The parameters of the model are then trained using a cross-entropy loss between the predicted labels and ground truth labels of the query set.

2.2 Positioning and Related Works

To highlight FSQS's novelty, our discussion revolves around the problem of inferring on a given *Query Set* provided with the knowledge of a *Support Set*. We refer to this class of problems as *SQ problems*. Intrinsically, FSL falls into the category of SQ problems. Interestingly, *Unsupervised Domain Adaptation* [22] (UDA), defined as labelling a dataset sampled from a target domain based on labelled data sampled from a source domain, is also a SQ problem. Indeed, in this case, the source domain plays the role of support, while the target domain plays the query's role. Notably, an essential line of study in UDA leverages the target data distribution for aligning source and target domains, reflecting the importance of transduction in a context of adaptation [2,13] *i.e.*, performing prediction by considering all target samples together. Transductive algorithms also have a special place in FSL [10,21,25] and show that leveraging a query set as a whole brings a significant boost in performances. Nevertheless, UDA and FSL exhibit fundamental differences. UDA addresses the problem of distribution shift using important source data and target data (typically thousands of instances) to align distributions. In contrast, FSL focuses on the difficulty of learning from few samples. To this purpose, we frame UDA as both SQ problem with *large* transductivity and Support/Query Shift, while Few-Shot Learning is a SQ problem, eventually with *small* transductivity for transductive FSL. Thus, FSQS combines both challenges: distribution shift and small transductivity. This new perspective allows us to establish fruitful connections with related learning paradigms, presented in Table 1, that we review in the following. A thorough review is available in Appendix A[1].

Adaptation. Unsupervised Domain Adaptation (UDA) requires a whole target dataset for inference, limiting its applications. Recent pioneering works, referred to as Test-Time Adaptation (TTA), adapt at test-time a model provided with a batch of samples from the target distribution. The proposed methodologies are test-time training by self-supervision [29], updating batch-normalization statistics [27] or parameters [32], or meta-learning to condition predictions on the whole batch of test samples for an *Adaptative Risk Minimization* (ARM) [33]. Inspired from the principle of invariant representations [2,13], the seminal work [7] brings *Optimal Transport* (OT) [23] as an efficient framework for aligning data distributions. OT has been recently applied in a context of transductive FSL [17] and our proposal (TP) is to provide a simple and strong baseline following the principle of OT as it is applied in UDA. In this work, following [3], we also study the role of Batch-Normalization for SQS, that points out the role of transductivity. Our conviction was that the batch-normalization is the first lever for aligning distributions [27,32].

---

[1] https://arxiv.org/abs/2105.11804.

Table 1. An overview of SQ problems. We divide SQ problems into two categories, presence or not of **Support-Query** shift; **No SQS** *vs* **SQS**. We consider three classes of transductivity: point-wise transductivity that is equivalent to inductive inference, small transductivity when inference is performed at batch level (typically in [32,33]), and large transductivity when inference is performed at dataset level (typically in UDA). New classes (resp. new domains) describe if the model is evaluated at test-time on novel classes (resp. novel domains). Note that we frame UDA as a fully test-time algorithm. Notably, Cross-Domain FSL (CDFSL) [5] assumes that the support set and query set are drawn from the same distribution, thus No SQS.

SQ problems		Train-time				Test-time			New	New
		Support		Query		Support		Query	classes	domains
		Size	Labels	Size	Labels	Size	Labels	Transductivity		
No SQS	FSL [12,28]	Few	✓	Few	✓	Few	✓	Point-wise	✓	✗
	TransFSL [21,25]	Few	✓	Few	✓	Few	✓	Small	✓	✗
	CDFSL [5]	Few	✓	Few	✓	Few	✓	Point-wise	✓	✓
SQS	UDA [22,24]					Large	✓	Large		
	TTA [27,29,32]	Large	✓					Small		✓
	ARM [33]	Large	✓	Few	✓			Small		✓
	Ind FSQS	Few	✓	Few	✓	Few	✓	Point-wise	✓	✓
	Trans FSQS	Few	✓	Few	✓	Few	✓	Small	✓	✓

Few-Shot Classification. We usually frame Few-Shot Classification methods [5] as either metric-based methods [28,31], or optimization-based methods that learn to fine-tune by adapting with few gradient steps [12]. A promising line of study leverages *transductivity* (using the query set as unlabelled data while inductive methods predict individually on each query sample). Transductive Propagation Network [21] meta-learns label propagation from the support to query set concurrently with the feature extractor. Transductive Fine-Tuning [10] minimizes the prediction entropy of all query instances during fine-tuning. Evaluating cross-domain generalization of FSL (FSCD), *i.e.*, a distributional shift between meta-training and meta testing, attracts the attention of a few recent works [5]. Zhao *et al.* propose a Domain-Adversarial Prototypical Network [34] in order to both align source and target domains in the feature space while maintaining discriminativeness between classes. Sahoo *et al.* combine Prototypical Networks with adversarial domain adaptation at the task level [26]. Notably, Cross-Domain Few-Shot Learning [5] (CDFSL) addresses the distributional shift between meta-training and meta-testing assuming that the support set and query set are drawn from the same distribution, not making it a SQ problem with support-query shift. Concerning the novelty of FSQS, we acknowledge the very recent contribution of Du *et al.* [11] which studies the role of learnable normalization for domain generalization, in particular when support and query sets are sampled from different domains. Note that our statement is more ambitious: we evaluate algorithms on both source and target domains that were unseen during training, while in their setting the source domain has already been seen during training.

Benchmarks in Machine Learning. Releasing benchmark has always been an important factor for progress in the *Machine Learning* field, the most

outstanding example being ImageNet [9] for the Computer Vision community. Recently, `DomainBed` [14] aims to settle Domain Generalization research
into a more rigorous process, where `FewShiftBed` takes inspiration from it.
`Meta-Dataset` [30] is an other example, this time specific to FSL.

3   FewShiftBed: A Pytorch Testbed for FSQS

3.1   Datasets

We designed three new image classification datasets adapted to the FSQS problem. These datasets have two specificities.

1. They are dividable into groups of images, assuming that each group corresponds to a distinct domain. A key challenge is that each group must contain
   enough images with a sufficient variety of class labels, so that it is possible
   to sample FSQS episodes.
2. They are delivered with a train/val/test split ($\Delta_{\text{train}}, \Delta_{\text{val}}, \Delta_{\text{test}}$), along both
   the class and the domain axis. This split is performed following the principles
   detailed in Sect. 2. Therefore, these datasets provide true few-shot tasks at
   test time, in the sense that the model will not have seen any instances of test
   classes and domains during training. Note that since we split along two axes,
   some data may be discarded (for instance images from a domain in D_{train}
   with a label in C_{test}). Therefore it is crucial to find a split that minimizes this
   loss of data.

Meta-CIFAR100-Corrupted (MC100-C). CIFAR-100 [19] is a dataset of 60k
three-channel square images of size 32×32, evenly distributed in 100 classes.
Classes are evenly distributed in 20 superclasses. We use the same method used
to build CIFAR-10-C [16], which makes use of 19 image perturbations, each one
being applied with 5 different levels of intensity, to evaluate the robustness of a
model to domain shift. We modify their protocol to adapt it to the FSQS problem: (i) we split the classes with respect to the superclass structure, and assign
13 superclasses (65 classes) to the training set, 2 superclasses (10 classes) to the
validation set, and 5 superclasses (25 classes) to the testing set; (ii) we also split
image perturbations (acting as domains), following the split of [33]. We obtain
2,184k transformed images for training, 114k for validation and 330k for testing.
The detailed split is available in the documentation of our code repository.

miniImageNet-Corrupted (mIN-C). *mini*ImageNet [31] is a popular benchmark
for few-shot image classification. It contains 60k images from 100 classes from
the ImageNet dataset. 64 classes are assigned to the training set, 16 to the
validation set and 20 to the test set. Like MC100-C, we build mIN-C using the
image perturbations proposed by [16] to simulate different domains. We use the
original split from [31] for classes, and use the same domain split as for MC100-
C. Although the original *mini*ImageNet uses 84×84 images, we use 224×224
images. This allows us to re-use the perturbation parameters calibrated in [16]

for ImageNet. Finally, we discard the 5 most time-consuming perturbations. We obtain a total of 1.2M transformed images for training, 182k for validation and 228k for testing. The detailed split in the documentation of our code repository.

FEMNIST-FewShot (FEMNIST-FS). EMNIST [6] is a dataset of images of handwritten digits and uppercase and lowercase characters. Federated-EMNIST [4] is a version of EMNIST where images are sorted by writer (or user). FEMNIST-FS consists in a split of the FEMNIST dataset adapted to few-shot classification. We separate both users and classes between training, validation and test sets. We build each group as the set of images written by one user. The detailed split is available in the code. Note that in FEMNIST, many users provide several instances for each digits, but less than two instance for most letters. Therefore it is hard to find enough samples from a user to build a support set or a query set. As a result, our experiments are limited to classification tasks with only one sample per class in both the support and query sets.

3.2   Algorithms

We implement in `FewShiftBed` two representative methods of the vast literature of FSL, that are commonly considered as strong baselines: Prototypical Networks (**ProtoNet**) [28] and Matching Networks (**MatchingNet**) [31]. Besides, for transductive FSL, we also implement with Transductive Propagation Network (**TransPropNet**) [21] and Transductive Fine-Tuning (**FTNet**) [10]. We also implement our novel algorithm *Transported Prototypes* (**TP**) which is detailed in Sect. 4. `FewShiftBed` is designed for favoring a straightforward implementation of a new algorithm for FSQS. To add a new algorithm, we only need to implement the `set_forward` method of the class `AbstractMetaLearner`. We provide an example with our implementation of the Prototypical Network [28] that only requires few line of codes:

```
class ProtoNet(AbstractMetaLearner):
    def set_forward(self, support_images, support_labels, query_images):
        z_support, z_query = self.extract_features(support_images, query_images)
        z_proto = self.get_prototypes(z_support, support_labels)
        return - euclidean_dist(z_query, z_proto)
```

3.3   Protocol

To prevent the pitfall of misinterpreting a performance boost, we draw three recommendations to isolate the causes of improvement rigorously.

– **How important is episodic training?** Despite its wide adoption in meta-learning for FSL, in some situation episodic training does not perform better than more naive approaches [5]. Therefore we recommend to report both the result obtained using episodic training and standard ERM (see the documentation of our code repository).

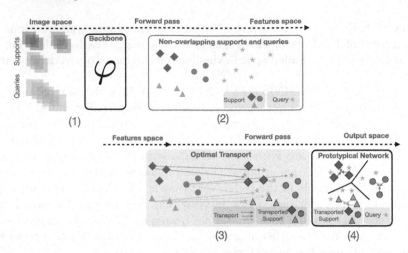

Fig. 3. Overview of *Transported Prototypes*. (1) A support set and a query set are fed to a trained backbone that embeds images into a feature space. (2) Due to the shift between distributions, support and query instances are embedded in non-overlapping areas. (3) We compute the Optimal Transport from support instances to query instances to build the transported support set. Note that we represent the transport plan only for one instance per class to preserve clarity in the schema. (4) Provided with the transported support, we apply the Prototypical Network [28] *i.e.*, L^2 similarity between transported support and query instances.

- **How does the algorithm behave in the absence of Support-Query Shift?** In order to assess that an algorithm designed for distribution shift does not provide degraded performance in an ordinary concept, and to provide a top-performing baseline, we recommend reporting the model's performance when we do not observe, at test-time, a support-query shift. Note that it is equivalent to evaluate the performance in cross-domain generalization, as firstly described in [5].
- **Is the algorithm transductive?** The assumption of transductivity has been responsible of several improvements in FSL [3,25] while it has been demonstrated in [3] that MAML [12] benefits strongly from the Transductive Batch-Normalization (TBN). Thus, we recommend specifying if the method is transductive and adapting the choice of the batch-normalization accordingly (Conventional Batch Normalization [18] and Transductive Batch Normalization for inductive and transductive methods, respectively) since transductive batch normalization brings a significant boost in performance [3].

4   Transported Prototypes: A Baseline for FSQS

4.1   Overall Idea

We present a novel method that brings UDA to FSQS. As aforementioned, FSQS presents new challenges since we no longer assume that we sample the support set and the query set from the same distribution. As a result, it is unlikely that the support set and query sets share the same representation space region (non-overlap). In particular, the L^2 distance, adopted in the celebrated Proto-typical Network [28], may not be relevant for measuring similarity between query and support instances, as presented in Fig. 1. To overcome this issue, we develop a two-phase approach that combines Optimal Transport (Transportation Phase) and the celebrated Prototypical Network (Prototype Phase). We give some back-ground about Optimal Transport (OT) in Sect. 4.2 and the whole procedure is presented in Algorithm 1.

4.2 Background

Definition. We provide some basics about Optimal Transport (OT). A thorough presentation of OT is available at [23]. Let p_s and p_t be two distributions on $\mathcal{X}$, we note $\Pi(p_s, p_t)$ the set of joint probability with marginal p_s and p_t i.e., $\forall \pi \in \Pi(p_s, p_t), \forall x \in \mathcal{X}, \pi(\cdot, x) = p_s, \pi(x, \cdot) = p_t$. The *Optimal Transport*, associated to cost c, between p_s and p_t is defined as:

$$W_c(p_s, p_t) := \min_{\pi \in \Pi(p_s, p_t)} \mathbb{E}_{(x_s, x_t) \sim \pi} [c(x_s, x_t)] \tag{1}$$

with $c(\cdot, \cdot)$ any metric. We note $\pi^*(p_s, p_t)$ the joint distribution that achieves the minimum in Eq. 1. It is named the *transportation plan* from p_s to p_t. When there is no confusion, we simply note π^*. For our applications, we will use as metric the euclidean distance in the representation space obtained from a representation $\varphi(\cdot; \theta)$ i.e., $c_\theta(x_s, x_t) := ||\varphi(x_s; \theta) - \varphi(x_t; \theta)||_2$.

Discrete OT. When p_s and p_t are only accessible through a finite set of samples, respectively $(x_{s,1}, ..., x_{s,n_s})$ and $(x_{t,1}, ..., x_{t,n_t})$ we introduce the empirical distri-butions $\hat{p}_s := \sum_{i=1}^{n_s} w_{s,i} \delta_{x_{s,i}}$, $\hat{p}_t := \sum_{j=1}^{n_t} w_{t,j} \delta_{x_{t,j}}$, where $w_{s,i}$ ($w_{t,j}$) is the mass probability put in sample $x_{s,i}$ ($x_{t,j}$) i.e., $\sum_{i=1}^{n_s} w_{s,i} = 1$ ($\sum_{j=1}^{n_t} w_{t,j} = 1$) and δ_x is the Dirac distribution in x. The discrete version of the OT is derived by intro-ducing the set of couplings $\mathbf{\Pi}(p_s, p_t) := \{\pi \in \mathbb{R}^{n_s \times n_t}, \pi \mathbf{1}_{n_s} = \mathbf{p}_s, \pi^\top \mathbf{1}_{n_t} = \mathbf{p}_t\}$ where $\mathbf{p}_s := (w_{s,1}, \cdots, w_{s,n_s})$, $\mathbf{p}_t := (w_{t,1}, \cdots, w_{1,n_t})$, and $\mathbf{1}_{n_s}$ (respectively $\mathbf{1}_{n_t}$) is the unit vector with dim n_s (respectively n_t). The discrete transportation plan π^*_θ is then defined as:

$$\pi^*_\theta := \underset{\pi \in \mathbf{\Pi}(p_s, p_t)}{\mathrm{argmin}} \ \langle \pi, \mathbf{C}_\theta \rangle_F \tag{2}$$

where $\mathbf{C}_\theta(i, j) := c_\theta(x_{s,i}, x_{t,j})$ and $\langle \cdot, \cdot \rangle_F$ is the Frobenius dot product. Note that π^*_θ depends on both p_s and p_t, and θ since $\mathbf{C}_\theta$ depends on θ. In practice, we use Entropic regularization [8] that makes OT easier to solve by promoting smoother transportation plan with a computationally efficient algorithm, based on Sinkhorn-Knopp's scaling matrix approach (see the Appendix C).

4.3 Method

Transportation Phase. At each episode, we are provided with a source support set $\mathcal{S}$ and a target query set $\mathcal{Q}$. We note respectively $\mathbf{S}$ and $\mathbf{Q}$ their represen-tations from a deep network $\varphi(\cdot; \theta)$ i.e., $z_s \in \mathbf{S}$ is defined as $z_s := \varphi(x_s; \theta)$ for $x_s \in \mathcal{S}$, respectively $z_q \in \mathbf{Q}$ is defined as $z_q := \varphi(x_q; \theta)$ for $x_q \in \mathcal{Q}$. As these two sets are sampled from different distributions, $\mathbf{S}$ and $\mathbf{Q}$ are likely to lie in different

Algorithm 1. Transported Prototypes. Blue lines highlight the OT's contribution in the computational graph of an episode compared to the standard Prototypical Network [28].

Input: Support set $\mathcal{S} := (x_{s,i}, y_{s,i})_{1 \leq i \leq n_s}$, query set $\mathcal{Q} := (x_{q,j}, y_{q,j})_{1 \leq j \leq n_q}$, classes $\mathcal{C}$, backbone φ_θ.

Output: Loss $\mathcal{L}(\theta)$ for a randomly sampled episode.

1: $z_{s,i}, z_{q,j} \leftarrow \varphi(x_{s,i}; \theta), \varphi(x_{q,j}; \theta)$, for i, j ▷ Get representations.
2: $\mathbf{C}_\theta(i, j) \leftarrow \|z_{s,i} - z_{q,j}\|^2$, for i, j ▷ Cost-matrix.
3: $\pi_\theta^\star \leftarrow$ Solve Equation 2   ▷ Transportation plan.
4: $\hat{\pi}_\theta^\star(i, j) \leftarrow \pi_\theta^\star(i, j) / \sum_j \pi_\theta^\star(i, j)$, for i, j ▷ Normalization.
5: $\hat{\mathbf{S}} = (\hat{z}_{s,i})_i \leftarrow$ Given by Equation 3   ▷ Get transported support set.
6: $\hat{\mathbf{c}}_k \leftarrow \frac{1}{|\hat{\mathbf{S}}_k|} \sum_{\hat{z}_s \in \hat{\mathbf{S}}_k} \hat{z}_s$, for $k \in \mathcal{C}$.   ▷ Get transported prototypes.
7: $p_\theta(y|x_{q,j}) \leftarrow$ From Equation 4, for j
8: **Return:** $\mathcal{L}(\theta) := \frac{1}{n_q} \sum_{j=1}^{n_q} - \log p_\theta(y_{q,j}|x_{q,j})$.

---

regions of the representation space. In order to adapt the source support set $\mathcal{S}$ to the target domain, which is only represented by the target query set $\mathcal{Q}$, we follow [7] to compute $\hat{\mathbf{S}}$ the *barycenter mapping* of $\mathcal{S}$, that we refer to as the *transported support set*, defined as follows:

$$\hat{\mathbf{S}} := \hat{\pi}_\theta^\star \mathbf{Q} \tag{3}$$

where $\pi_\theta^\star$ is the transportation plan from $\mathbf{S}$ to $\mathbf{Q}$ and $\hat{\pi}_\theta^\star :=$ $\pi_\theta^\star(i, j) / \sum_{j=1}^{n_t} \pi_\theta^\star(i, j)$. The *transported* support set $\hat{\mathbf{S}}$ is an estimation of labelled examples in the target domain using labelled examples in the source domain. The success relies on the fact that transportation conserves labels, *i.e.*, a query instance close to $\hat{z}_s \in \hat{\mathbf{S}}$ should share the same label with x_s, where $\hat{z}_s$ is the barycenter mapping of $z_s \in \mathbf{S}$. See step (3) of Fig. 3 for a visualization of the transportation phase.

Prototype Phase. For each class $k \in \mathcal{C}$, we compute the *transported prototypes* $\hat{\mathbf{c}}_k := \frac{1}{|\hat{\mathbf{S}}_k|} \sum_{\hat{z}_s \in \hat{\mathbf{S}}_k} \hat{z}_s$ (where $\hat{\mathbf{S}}_k$ is the transported support set with class k and $\mathcal{C}$ are classes of current episode). We classify each query x_q with representation $z_q = \varphi(x_q; \theta)$ using its euclidean distance to each transported prototypes;

$$p_\theta(y = k|x_q) := \frac{\exp\left(-\|z_q - \hat{\mathbf{c}}_k\|^2\right)}{\sum_{k' \in \mathcal{C}} \exp\left(-\|z_q - \hat{\mathbf{c}}_{k'}\|^2\right)} \tag{4}$$

Crucially, the standard Prototypical Networks [28] computes euclidean distance to each prototypes while we compute the euclidean to each *transported* prototypes, as presented in step (4) of Fig. 3. Note that our formulation involves the query set in the computation of $(\hat{\mathbf{c}}_k)_{k \in \mathcal{C}}$.

Genericity of OT. FewShiftBed implements OT as a stand-alone module that can be easily plugged into any FSL algorithm. We report additional baselines in Appendix B where other FSL algorithms are equipped with OT. This technical choice reflects our insight that OT may be ubiquitous for addressing FSQS and makes its usage in the testbed straightforward.

Table 2. Top-1 accuracy of few-shot learning models in various datasets and numbers of shots with 8 instances per class in the query set (except for FEMNIST-FS: 1 instance per class in the query set), with 95% confidence intervals. The top half of the table is a comparison between existing few-shot learning methods and Transported Prototypes (TP). The bottom half is an ablation study of TP. OT denotes Optimal Transport, TBN is Transductive Batch-Normalization, OT-TT refers to the setting where Optimal Transport is applied at test time but not during episodic training, and ET means episodic training *i.e.*, w/o ET refers to the setting where training is performed through standard Empirical Risk Minimization. TP w/o SQS reports model's performance in the absence of support-query shift. † flags if the method is transductive. For each setting, the best accuracy among existing methods is shown in bold, as well as the accuracy of an ablation if it improves TP.

	Meta-CIFAR100-C		miniImageNet-C		FEMNIST-FS
	1-shot	5-shot	1-shot	5-shot	1-shot
ProtoNet [28]	30.02 ± 0.40	42.77 ± 0.47	36.37 ± 0.50	47.58 ± 0.57	84.31 ± 0.73
MatchingNet [31]	30.71 ± 0.38	41.15 ± 0.45	35.26 ± 0.50	44.75 ± 0.55	84.25 ± 0.71
TransPropNet† [21]	**34.15 ± 0.39**	47.39 ± 0.42	24.10 ± 0.27	27.24 ± 0.33	86.42 ± 0.76
FTNet† [10]	28.91 ± 0.37	37.28 ± 0.40	39.02 ± 0.46	51.27 ± 0.45	86.13 ± 0.71
TP† (ours)	**34.00 ± 0.46**	**49.71 ± 0.47**	**40.49 ± 0.54**	**59.85 ± 0.49**	**93.63 ± 0.63**
TP w/o OT †	32.47 ± 0.41	48.00 ± 0.44	40.43 ± 0.49	53.71 ± 0.50	90.36 ± 0.58
TP w/o TBN †	33.74 ± 0.46	49.18 ± 0.49	37.32 ± 0.55	55.16 ± 0.54	92.31 ± 0.73
TP w. OT-TT †	32.81 ± 0.46	48.62 ± 0.48	**44.77 ± 0.57**	**60.46 ± 0.49**	**94.92 ± 0.55**
TP w/o ET †	**35.94 ± 0.45**	48.66 ± 0.46	42.46 ± 0.53	54.67 ± 0.48	94.22± 0.70
TP w/o SQS †	*85.67 ± 0.26*	*88.52 ± 0.17*	*64.27 ± 0.39*	*75.22 ± 0.30*	*92.65 ± 0.69*

5   Experiments

We compare the performance of baseline algorithms with *Transported Prototypes* on various datasets and settings. We also offer an ablation study in order to isolate the source to the success of *Transported Prototypes*. Extensive results are detailed in Appendix B. Instructions to reproduce these results can be found in the code's documentation.

Setting and Details. We conduct experiments on all methods and datasets implemented in **FewShiftBed**. We use a standard 4-layer convolutional network for our experiments on Meta-CIFAR100-C and FEMNIST-FewShot, and a ResNet18 for our experiments on miniImageNet. Transductive methods are equipped with a Transductive Batch-Normalization. All episodic training runs contain 40k episodes, after which we retrieve model state with best validation accuracy. We run each individual experiment on three different random seeds. All results presented in this paper are the average accuracies obtained with these random seeds.

Analysis. The top half of Table 2 reveals that Transported Prototypes (TP) outperform all baselines by a strong margin on all datasets and settings. Importantly, baselines perform poorly on FSQS, demonstrating they are not equipped to address this challenging problem, stressing our study's significance. It is also interesting to note that the performance of transductive approaches, which is significantly better in a standard FSL setting [10,21], is here similar to inductive methods (notably, TransPropNet [21] fails loudly without Transductive Batch-Normalization showing that propagating label with non-overlapping support/query can have a dramatic impact, see Appendix B). Thus, FSQS deserves a fresher look to be solved. Transported Prototypes mitigate a significant part of the performance drop caused by support-query shift while benefiting from the simplicity of combining a popular FSL method with a time-tested UDA method. This gives us strong hopes for future works in this direction.

Ablation Study. Transported Prototypes (TP) combines three components: Optimal Transport (OT), Transductive Batch-Normalization (TBN) and episode training (ET). Which of these components are responsible for the observed gain? Following recommendations from Sect. 3.3, we ablate those components in the bottom half of Table 2. We observe that both OT and TBN individually improve the performance of ProtoNet for FSQS, and that the best results are obtained when the two of them are combined. Importantly, OT without TBN performs better than TBN without OT (except for 1-shot mIN-C), demonstrating the superiority of OT compared to TBN for aligning distributions in the few samples regime. Note that the use of TaskNorm [3] is beyond the scope of the paper[2]; we encourage future work to dig into that direction and we refer the reader to the very recent work [11]. We observe that there is no clear evidence that using OT at train-time is better than simply applying it at test-time on a ProtoNet trained without OT. Additionally, the value of Episodic Training (ET) compared to standard Empirical Risk Minimization (ERM) is not obvious. For instance, simply training with ERM and applying TP at test-time is better than adding ET on 1-shot MC100-C, 1-shot mIN-C and FEMNIST-FS, making it an another element to add to the study [20] who put into question the value of ET. Understanding why and when we should use ET or only OT at test-time is interesting for future works. Additionally, we compare TP with MAP [17] which implements an OT-based approach for transductive FSL. Their approach includes a power transform to reduce the skew in the distribution, so for fair comparison we also implemented it into Transported Prototypes for these experiments[3]. We also used the OT module only at test-time and compared with two backbones, respectively trained with ET and ERM. Interestingly, our experiments in Table 3 show that MAP is able to handle SQS. Finally, in order to evaluate the performance drop related to Support-Query Shift compared to a setting with support and query instances sampled from the same distribution, we test Transported Prototypes on few-shot classification tasks without SQS (TP w/o SQS in Table 2), making a setup equivalent to CDFSL. Note that in both cases, the model is trained in an episodic fashion on tasks presenting a Support-Query Shift. These results show that SQS presents a significantly harder challenge than CDFSL, while there is considerable room for improvements.

---

[2] These normalizations are implemented in `FewShiftBed` for future works.
[3] Therefore results in Table 3 differ from results in Table 2.

Table 3. Top-1 accuracy with 8 instances per class in the query set when applying Transported Prototypes and MAP on two different backbones: ⋆ is standard ERM (*i.e.*, without Episodic Training) and † is ProtoNet [28]. Transported Prototypes performs equally or better than MAP [17]. Here TP includes power transform in the feature space.

	Meta-CIFAR100-C		miniImageNet-C		FEMNIST-FS
	1-shot	5-shot	1-shot	5-shot	1-shot
TP⋆	36.17 ± 0.47	50.45 ± 0.47	45.41 ± 0.54	57.82 ± 0.48	93.60 ± 0.68
MAP⋆	35.96 ± 0.44	49.55 ± 0.45	43.51 ± 0.47	56.10 ± 0.43	92.86 ± 0.67
TP†	32.13 ± 0.45	46.19 ± 0.47	45.77 ± 0.58	59.91 ± 0.48	94.92 ± 0.56
MAP†	32.38 ± 0.41	45.96 ± 0.43	43.81 ± 0.47	57.70 ± 0.43	87.15 ± 0.66

6 Conclusion

We release `FewShiftBed`, a testbed for the under-investigated and crucial problem of Few-Shot Learning when the support and query sets are sampled from related but different distributions, named FSQS. `FewShiftBed` includes three datasets, relevant baselines and a protocol for reproducible research. Inspired from recent progress of Optimal Transport (OT) to address Unsupervised Domain Adaptation, we propose a method that efficiently combines OT with the celebrated Prototypical Network [28]. Following the protocol of `FewShiftBed`, we bring compelling experiments demonstrating the advantage of our proposal compared to transductive counterparts. We also isolate factors responsible for improvements. Our findings suggest that Batch-Normalization is ubiquitous, as described in related works [3,11], while episodic training, even if promising on paper, is questionable. As a lead for future works, `FewShiftBed` could be improved by using different datasets to model different domains, instead of using artificial transformations. Since we are talking about domain adaptation, we also encourage the study of accuracy as a function of the size of the target domain, *i.e.*, the size of the query set. Moving beyond the transductive algorithm, as well as understanding when meta-learning brings a clear advantage to address FSQS remains an open and exciting problem. `FewShiftBed` brings the first step towards its progress.

Acknowledgements. Etienne Bennequin is funded by Sicara and ANRT (France), and Victor Bouvier is funded by Sidetrade and ANRT (France), both through a CIFRE collaboration with CentraleSupélec. This work was performed using HPC resources from the "Mésocentre" computing center of CentraleSupélec and École Normale Supérieure Paris-Saclay supported by CNRS and Région Île-de-France (http://mesocentre.centralesupelec.fr/).

References

1. Amodei, D., et al.: Concrete problems in AI safety. arXiv preprint arXiv:1606.06565 (2016)
2. Ben-David, S., et al.: Analysis of representations for domain adaptation. In: Advances in Neural Information Processing Systems, pp. 137–144 (2007)
3. Bronskill, J., et al.: Tasknorm: rethinking batch normalization for meta-learning. In ICML, pp. 1153–1164. PMLR (2020)
4. Caldas, S., et al.: Leaf: a benchmark for federated settings. arXiv preprint arXiv:1812.01097 (2018)
5. Chen, W.-Y., et al.: A closer look at few-shot classification. In: International Conference on Learning Representations (2019)
6. Cohen, G., et al.: EMNIST: extending MNIST to handwritten letters. In: IJCNN. IEEE (2017)
7. Courty, N., et al.: Optimal transport for domain adaptation. IEEE Trans. Pattern Anal. Mach. Intell. **39**(9), 1853–1865 (2016)
8. Cuturi, M.: Sinkhorn distances: lightspeed computation of optimal transport. In: Advances in Neural Information Processing Systems, vol. 26, pp. 2292–2300 (2013)
9. Deng, J., et al.: ImageNet: a large-scale hierarchical image database. In: 2009 IEEE Conference on Computer Vision and Pattern Recognition, pp. 248–255. IEEE (2009)
10. Dhillon, G.S., et al.: A baseline for few-shot image classification. In: ICLR (2020)
11. Du, Y., et al.: MetaNorm: learning to normalize few-shot batches across domains. In: International Conference on Learning Representations (2021). https://openreview.net/forum?id=9z_dNsC4B5t
12. Finn, C., et al.: Model-agnostic meta-learning for fast adaptation of deep networks. In: ICML. JMLR (2017)
13. Ganin, Y., Lempitsky, V.: Unsupervised domain adaptation by backpropagation. In: International Conference on Machine Learning, pp. 1180–1189 (2015)
14. Gulrajani, I., Lopez-Paz, D.: In search of lost domain generalization. In: International Conference on Learning Representations (2021)
15. He, K., et al.: Deep residual learning for image recognition. In: Proceedings of the IEEE Conference on Computer Vision and Pattern Recognition, pp. 770–778 (2016)
16. Hendrycks, D., Dietterich, T.: Benchmarking neural network robustness to common corruptions and perturbations. In: ICLR (2019)
17. Hu, Y., et al.: Leveraging the feature distribution in transfer-based few-shot learning. arXiv preprint arXiv:2006.03806 (2020)
18. Ioffe, S., Szegedy, C.: Batch normalization: accelerating deep network training by reducing internal covariate shift. In: ICML. PMLR (2015)
19. Krizhevsky, A., et al.: Learning multiple layers of features from tiny images. Citeseer (2009)
20. Laenen, S., Bertinetto, L.: On episodes, prototypical networks, and few-shot learning. arXiv preprint arXiv:2012.09831 (2020)
21. Liu, Y., et al.: Learning to propagate labels: transductive propagation network for few-shot learning. In: ICLR (2019)
22. Pan, S.J., Yang, Q.: A survey on transfer learning. IEEE Trans. Knowl. Data Eng. **22**(10), 1345–1359 (2009)
23. Peyré, G., et al.: Computational optimal transport: with applications to data science. Found. Trends® Mach. Learn. **11**(5–6), 355–607 (2019)

24. Quionero-Candela, J., et al.: Dataset Shift in Machine Learning. The MIT Press, Cambridge (2009)
25. Ren, M., et al.: Meta-learning for semi-supervised few-shot classification. In: ICLR (2019)
26. Sahoo, D., et al.: Meta-learning with domain adaptation for few-shot learning under domain shift (2019)
27. Schneider, S., et al.: Improving robustness against common corruptions by covariate shift adaptation. In: Advances in Neural Information Processing Systems, vol. 33 (2020)
28. Snell, J., et al.: Prototypical networks for few-shot learning. In: Advances in Neural Information Processing Systems, pp. 4077–4087 (2017)
29. Sun, Y., et al.: Test-time training with self-supervision for generalization under distribution shifts. In: ICML (2020)
30. Triantafillou, E., et al.: Meta-dataset: a dataset of datasets for learning to learn from few examples. In: ICLR (2020)
31. Vinyals, O., et al.: Matching networks for one shot learning. In: NIPS (2016)
32. Wang, D., et al.: Fully test-time adaptation by entropy minimization. In: ICLR (2021)
33. Zhang, M., et al.: Adaptive risk minimization: a meta-learning approach for tackling group shift. In: ICLR (2021)
34. Zhao, A., et al.: Domain-adaptive few-shot learning. arXiv preprint arXiv:2003.08626 (2020)

Source Hypothesis Transfer for Zero-Shot Domain Adaptation

Tomoya Sakai$^{(\boxtimes)}$ (iD)

NEC Corporation, Tokyo, Japan
tomoya_sakai@nec.com

Abstract. Making predictions in target unseen domains without training samples is frequent in real-world applications, such as new products'
sales predictions. Zero-shot domain adaptation (ZSDA) has been studied to achieve this important but difficult task. An approach to ZSDA
is to use multiple source domain data and domain attributes. Several
recent domain adaptation studies have mentioned that source domain
data are not often available due to privacy, technical, and contractual
issues in practice. To address these issues, hypothesis transfer learning
(HTL) has been gaining attention since it does not require access to
source domain data. It has shown its effectiveness in supervised/unsu-
pervised domain adaptation; however current HTL methods cannot be
readily applied to ZSDA because we have no training data (even unlabeled data) for target domains. To solve this problem, we propose an
HTL-based ZSDA method that connects multiple source hypotheses by
domain attributes. Through theoretical analysis, we derive the convergence rate of the estimation error of our proposed method. Finally, we
numerically demonstrate the effectiveness of our proposed HTL-based
ZSDA method.

Keywords: Hypothesis transfer learning · Zero-shot domain
adaptation · Unseen domains · Domain adaptation

1 Introduction

In real-world applications, training data of our task of interest are not often
available. To name a few, sales prediction of new products, preference prediction
of new users, and energy consumption prediction of new sites are applications
in which labeled training data in a target domain does not exist. The task of
making predictions in *unseen* domains (e.g., new products) without any target
training data is known as *zero-shot domain adaptation* (ZSDA) [25,26]. To enable
ZSDA, these studies proposed an approach that uses multiple source data (i.e.,
training datasets obtained from multiple domains) and *domain attributes* (i.e.,
descriptions of domains). For sales prediction of food, domain attributes can be
colors, size, and nutritional components. Intuitively, the relation between sales of
existing products and domain attributes can be regarded as a clue of estimating

© Springer Nature Switzerland AG 2021
N. Oliver et al. (Eds.): ECML PKDD 2021, LNAI 12975, pp. 570–586, 2021.
https://doi.org/10.1007/978-3-030-86486-6_35

Table 1. Comparison of our method and related work. SHOT [16] does not require source data but cannot be applied to unseen domains. In contrast, MDMT [25] can handle unseen domains but requires source data. Our method addresses both issues.

	Proposed method	SHOT [16]	MDMT [25]
HTL	✓	✓	–
ZSDA	✓	–	✓

sales of new products, since a combination of domain attributes of a new product is new but each element already appears in existing products.

Another crucial issue that has emerged is that source domain data are not always available due to legal, technical, and contractual constraints between data owners and data customers [4]. It is common for decision-making rules to be only available, e.g., learned prediction functions are accessible but not source domain data. To handle this situation, *hypothesis transfer learning* (HTL) [14,16] is promising because it does not require source data for training a new model. Since HTL does not require access to source domain data, it secures private information in the source domain data and saves memory and computation time for training a new target model [16].

In the existing ZSDA methods, multiple source domain data are used for training. This is not suitable for applications that are privacy sensitive and require expensive computational resources to store source domain data. The method proposed by Mansour et al. [17] allows us to train a target model from multiple source hypotheses. However, the method requires training data obtained from target domains. Thus, HTL-based ZSDA methods should enable to solve sales prediction of new products while maintaining privacy and reducing storage costs.

In this paper, we propose a ZSDA method that is based on HTL. The main challenge is that we cannot use current HTL methods since target training data are not available in ZSDA. To tackle this challenge, we introduce a new learning objective that connects hypotheses for existing domains through domain attributes to train a prediction model for unseen domains. An advantage of our method is that it can be easily implemented with *Scikit-learn* [20] in Python. Through theoretical analysis, we derive the convergence rate of an estimation error of our method. To the best of our knowledge, this is the first study to present HTL for ZSDA (see also Table 1) and the convergence rate of an estimation error in ZSDA. We then conducted numerical experiments to show that our proposed method achieved comparable or sometimes superior performance to a non-HTL method for ZSDA.

2 Related Work

Domain adaptation without target samples has been studied from several aspects. For example, *domain generalization* (DG) [2,15] obtains predictions without any sample obtained from target domains. In DG, samples from multiple source domains are assumed available. However, in some applications, it is difficult to assume multiple source domains. To address this issue, *zero-shot deep domain adaptation* [21] (ZDDA) has been proposed.

Compared with DG and ZDDA, our method requires domain attributes, similarly to the methods in [25,26]. While the requirement might restrict applications of our method, there is a trade-off between having and not having domain attributes. The use of domain attributes incurs annotation cost but enables us to handle a response that depends on both input features and domain attributes. However, the approach without domain attributes cannot handle such a case. ZSDA is beneficial when discriminating domains from input features is difficult or almost impossible.

While several studies have considered making predictions in unseen target domains without any target training data, they relied on the availability of source domain data. From the viewpoint of HTL, the method proposed by Mansour et al. [17] can be regarded as a method from multiple source hypotheses but requires target training data.

3   Problem Setting and Background

In this section, we explain our problem setting and background knowledge.

3.1   Problem Setting

Let a covariate $x^{(t)} \in \mathbb{R}^d$ and its corresponding response $y^{(t)} \in \mathbb{R}$, where t denotes the task index and d is a positive integer. Let us denote a set of seen domain indices by $\mathcal{T}_S = \{1, \ldots, T_S\}$, where T_S is the number of seen domains. Similarly, let $\mathcal{T}_U := \{T_S + 1, \ldots, T_S + T_U\}$ be a set of unseen target domain indices, where T_U denotes the number of unseen domains. As a signature of a domain, we assume an m-dimensional vector $a^{(t)} \in \mathbb{R}^m$ is available for each domain and call a domain attributes (domain-attribute vector). Let us define a set of attribute vectors for seen and unseen domain as

$$\mathcal{A}_S := \{a^{(1)}, \ldots, a^{(T_S)}\},$$
$$\mathcal{A}_U := \{a^{(T_S+1)}, \ldots, a^{(T_S+T_U)}\},$$

respectively.

Let $h^{(t)} \colon \mathbb{R}^d \to \mathbb{R}$ be a source hypothesis for a domain t and $h_S := (h^{(1)}, \ldots, h^{(T_S)})^\top$. In this paper, we assume that the (learned) source hypotheses

$$\widehat{h}_S := (\widehat{h}^{(1)}, \ldots, \widehat{h}^{(T_S)})^\top$$

are available. The source hypotheses $\widehat{h}_S$ can be obtained by supervised learning independently or by *multi-task learning* (MTL) [5] jointly from multiple source domain data. Note that as long as the input-output constraint is satisfied, any class of model (e.g., linear model, tree model, and neural networks) can be used with our method, while neural networks are assumed as a class of hypotheses with SHOT [16].

Our goal is to obtain a prediction of a test sample x' in an unseen target domain $t' \in \mathcal{T}_\mathrm{U}$ by using $a^{(t')}$, $\mathcal{A}_\mathrm{S}$, and $\widehat{h}_\mathrm{S}$ without source domain data.

3.2 Ordinary Supervised Learning

Ordinary supervised learning does not handle unseen domains, but we review the method of standard supervised learning since it can be used for obtaining source hypotheses h_S.

Suppose that we have a set of labeled samples for a seen domain t, i.e., source domain data:

$$\mathcal{D}^{(t)} := \big\{(x_i^{(t)}, y_i^{(t)})\big\}_{i=1}^{n^{(t)}},$$

where $n^{(t)}$ is the number of labeled samples on t. A simple approach to obtain predictions in a seen domain is to train a predictor with the corresponding labeled samples $\mathcal{D}^{(t)}$. Specifically, for each $t \in \mathcal{T}_\mathrm{S}$, a predictor $h^{(t)}$ is trained to minimize the training error plus a regularization functional:

$$\underset{h^{(t)}}{\text{minimize}} \; \frac{1}{n^{(t)}} \sum_{i=1}^{n^{(t)}} \ell\big(h^{(t)}(x_i^{(t)}), y_i^{(t)}\big) + \lambda W\big(h^{(t)}\big), \tag{1}$$

where W is the regularization functional, $\lambda \geq 0$ is the regularization parameter, and $\ell: \mathbb{R} \times \mathbb{R} \to \mathbb{R}_{\geq 0}$ is the loss function such as the squared loss: $\ell_\mathrm{sq}(y, y') := (y - y')^2$. With the learned source hypotheses $\widehat{h}_\mathrm{S}$, we obtain a prediction for seen domains. However, it is not possible to make a prediction in an unseen domain because we only have $\widehat{h}_\mathrm{S}$ for seen domains.

3.3 ZSDA with Source Domain Data

An approach to make a prediction in unseen domains is to include attribute vectors into the prediction model. We review one of the state-of-the-art ZSDA methods on the basis of domain attributes [25], the usefulness of which was also investigated, e.g., [10,23,26]. Let $F: \mathbb{R}^d \times \mathcal{T} \to \mathbb{R}$ be an *attribute-aware* predictor. An example of F is a bilinear function defined as

$$F(x, t) = x^\top W a^{(t)},$$

where $W \in \mathbb{R}^{d \times m}$ is the parameter matrix to be learned.

Suppose we have training data $\mathcal{D} := \{\mathcal{D}^{(t)}\}_{t \in \mathcal{T}_\mathrm{S}}$ and a set of seen attributes $\mathcal{A}_\mathrm{S}$. We then train F by labeled samples and attribute vectors for all seen domains. That is, we solve the following optimization problem:

$$\underset{F}{\text{minimize}} \; \frac{1}{\mathcal{T}_\mathrm{S}} \sum_{t \in \mathcal{T}_\mathrm{S}} \frac{1}{n^{(t)}} \sum_{i=1}^{n^{(t)}} \ell\big(F(x_i^{(t)}, t), y_i^{(t)}\big) + \widetilde{\lambda} \widetilde{W}(F), \tag{2}$$

where $\widetilde{W}$ is a regularization functional and $\widetilde{\lambda} \geq 0$ is the regularization parameter. After obtaining a learned predictor, denoted as $\widehat{F}$, a prediction of a test sample x' in an unseen domain t' can be obtained by $\widehat{F}(x', t')$.

Although this approach can make predictions for a sample on an unseen domain, it requires access to a source training dataset $\mathcal{D}$, which is not always possible in practice [4].

4 Proposed Method

We explain how to make predictions in unseen domains by using multiple source hypotheses $\widehat{h}_{\mathrm{S}}$.

4.1 Model Collaboration

To obtain predictions in unseen domains, we make the learned predictors take domain attributes into account. While it is difficult to design the predictors to handle domain attributes after their parameters are fixed, our approach can connect the learned predictors with domain attributes.

Our key idea is to make an implicit connection between the learned predictors and another prediction model that can take domain attributes into account. Instead of training the new prediction model, we compute the prediction of input in an unseen domain at *test* time. That is, the computation time of our method is zero until a test sample comes in. Fortunately, this does not cause any problems with ZSDA because we do not know the information of unseen domains in advance.

More specifically, let x' be a test input and $g \colon \mathcal{A} \to \mathbb{R}$ be a prediction function for x'. We refer to g as a *fixed-input* model because g is in charge of the prediction of x' only. We then connect g with $\widehat{h}_{\mathrm{S}}$ by minimizing the model collaboration (MC) error defined as

$$\widehat{R}_{\mathrm{MC}}(g) := \frac{1}{T_{\mathrm{S}}} \sum_{t=1}^{T_{\mathrm{S}}} \ell\big(g(a^{(t)}), \widehat{h}^{(t)}(x')\big). \tag{3}$$

In practice, we add a regularization functional $\widetilde{W}$ and solve the optimization problem expressed as

$$\underset{g \in \mathcal{G}}{\text{minimize}} \ \widehat{R}_{\mathrm{MC}}(g) + \widetilde{\lambda}\widetilde{W}(g),$$

where $\widetilde{\lambda}$ is the regularization parameter and $\mathcal{G}$ is a function class, such as linear models, tree models, and neural networks. Minimization of the MC error connects g with $\widehat{h}_{\mathrm{S}}$. From the perspective of generalization, $\widehat{h}^{(t)}$ handles features while $\widehat{g}$ handles domain attributes.

At a glance, this approach might seem heuristic, e.g., one may think the MC error just connects hypotheses $\widehat{h}_{\mathrm{S}}$ and a new model g through a loss function.

However, this is a theoretically justified method. In Sect. 5, we show that the proposed method is theoretically valid.

After solving the above optimization problem, we obtained the learned fixed-input model $\widehat{g}$. The prediction of $\boldsymbol{x}'$ in an unseen domain $\boldsymbol{a}^{(t')}$ is given by $\widehat{g}(\boldsymbol{a}^{(t')})$. Since g is only in charge of $\boldsymbol{x}'$, for another test input $\boldsymbol{x}''$, we retrain g to obtain a prediction. However, as we explain in Sect. 4.3 and show through experimentation, this computation in an inference phase can be efficiently done.

Our method can be regarded as *transductive* inference [3,24], where the task is to estimate labels of test instances included in the training samples and abandon the ability of prediction for new test instances in the future. Transductive inference is known as an easier task than *inductive* inference [3,24]. In this sense, although our method is affected by the accuracy of learned predictors for seen domains, the advantage of transductive inference might neutralize the effect of using learned predictors.

Table 2 summarizes the required input to the proposed method and a (non-HTL) ZSDA method [25], a method of multi-domain and multi-task learning, called as MDMT. A notable difference is that our proposed method does not require source training data $\mathcal{D}$.

Table 2. Required input to each method. MDMT requires training data $\mathcal{D}$ from multiple source domains. Our method does not require using $\mathcal{D}$ and enables us to make predictions by leveraging source hypotheses $\widehat{\boldsymbol{h}}_{\mathrm{S}}$.

	Proposed method	MDMT [25]
Training	$\boldsymbol{x}' \in \mathbb{R}^d$, $\mathcal{A}_{\mathrm{S}}$	$\mathcal{D}$, $\mathcal{A}_{\mathrm{S}}$
Inference	$\boldsymbol{a}' \in \mathcal{A}_{\mathrm{U}}$	$\boldsymbol{x}' \in \mathbb{R}^d$, $\boldsymbol{a}' \in \mathcal{A}_{\mathrm{U}}$
Prediction model	$g \colon \mathcal{A} \to \mathbb{R}$	$F \colon \mathbb{R}^d \times \mathcal{A} \to \mathbb{R}$
Source hypotheses	$\widehat{\boldsymbol{h}}_{\mathrm{S}}$	–

4.2 Hyperparameter Tuning

To tune a hyperparameter such as the regularization parameter, we use domain-wise dataset split. For example, if we have 100 domains, by 8:2 domain-wise split, we use 80 domains for training data and 20 for test. Similarly to class-wise cross-validation [22], we can use domain-wise cross-validation. In our implementation with linear ridge regression, we can use a computationally efficient implementation of leave-one-domain-out cross-validation (LOOCV) to tune the regularization parameter.

4.3 Implementation

General Implementation: An example of g is the linear model defined as $g(\boldsymbol{a}^{(t)}) = \boldsymbol{\beta}^{\top} \boldsymbol{a}^{(t)}$, where $\boldsymbol{\beta} \in \mathbb{R}^m$ is a parameter vector. For the linear model,

```
def predict(x_test, h_S, a_S, a_U):
    """
    @x_test: a test data point (1 times d)
    @h_S: a list of seen predictors (T_S size)
    @a_S: a matrix of domain attributes (T_S times m)
    @a_U: a vector of domain-attributes for target (1 times m)
    """
    yh_S = [h.predict(x_test) for h in h_S]
    reg = Any_ScikitLearn_Regressor()
    reg.fit(a_S, yh_S)
    return reg.predict(a_U)
```

Code 1. Example of Python implementation

if we use the squared loss and ℓ_2-regularizer, the optimization problem becomes the linear ridge regression [8] and the solution can be obtained analytically.

A prediction in an unseen domain associated with the attribute vector $\boldsymbol{a}^{(t')}$ can generally be obtained as follows. We first feed $\{\boldsymbol{a}^{(t)}, \widehat{h}^{(t)}(\boldsymbol{x}')\}_{t \in \mathcal{T}_S}$ as training data into a function of a regression method then obtain prediction by feeding $\boldsymbol{a}^{(t')}$ into the trained model. Note that, in the above procedure, we can use any regression method.

Our method can be implemented by a few lines of Python code with the *Scikit-learn* [20] package. We show an example of Python implementation in Code 1. As shown in this code, our method is model-independent. We can thus use any method for both $\widehat{h}^{(t)}$ and g.

Computationally-Efficient Implementation: An apparent drawback of our method is the necessity of calculation for each test sample. Although we might not need to handle millions of test samples in one second in practice, it is better that the computation time of our method be short.

In this section, we explain a computationally-efficient implementation based on the linear ridge regression with LOOCV [8]. For example, we can use the *RidgeCV* in Scikit-learn as an implementation of the linear ridge regression with LOOCV. In LOOCV, the eigendecomposition of $\boldsymbol{A}_S = (\boldsymbol{a}^{(1)}, \ldots, \boldsymbol{a}^{(T_S)})$ is necessary but only once unless we add new seen domains or change the representation of domain attributes. That is, after eigendecomposition, the computation time of prediction consists of several multiplications of vectors and matrices.

More specifically, the matrix multiplication of $\boldsymbol{A}_S^\top \boldsymbol{A}_S$ takes $\mathcal{O}(m^2 T_S)$ time. The eigendecomposition of $\boldsymbol{A}_S^\top \boldsymbol{A}_S$ takes $\mathcal{O}(m^3)$ time. Once we obtain the eignedecomposition, we can reuse the result for any test point as long as the representation of domain attributes is fixed. If we can compute the eigendecomposition of $\boldsymbol{A}_S^\top \boldsymbol{A}_S$ in advance, $\mathcal{O}(m^3 + m^2 T_S)$ does not matter in prediction.

Let $\mathcal{O}(H)$ be the computational complexity of computing prediction for an in-service predictor, i.e., inference time. For example, H becomes d if we use the linear models as $\widehat{h}^{(t)}$. Then, obtaining the outputs of all learned predictors for a single test data point requires $\mathcal{O}(HT_{\mathrm{S}})$ time. For each hyperparameter, we can compute the score of LOOCV in $\mathcal{O}(T_{\mathrm{S}}^2)$ time. Let L be the number of candidates of the regularization parameters. The total computation time of the hyperparameter tuning and parameter estimation takes $\mathcal{O}(LT_{\mathrm{S}}^2)$. After we determine the hyperparameter, we then compute the prediction in $\mathcal{O}(mT_{\mathrm{S}})$ time.

5   Theoretical Analysis

Our ultimate goal is to obtain a prediction function that minimizes error to the ground truth function. In contrast, the MC error measures the average loss between a prediction function and learned hypotheses. In this sense, one may think that MC error minimization is just a heuristic. However, it is not. In this section, we investigate the estimation error for evaluating the difference between a minimizer of the MC error and an optimal one that is as close as to the ground truth function, and we elucidate the convergence rate of the estimation error bound.

5.1   Notations and Assumptions

In this analysis, we assume a set of attribute vectors of size T is drawn independently from the distribution with density η:

$$\mathcal{A} = \{\boldsymbol{a}^{(t)}\}_{t=1}^{T} \sim \eta^{T}(\boldsymbol{a}).$$

This assumption would be natural as long as observations of new domains is independent to past observations.

We then define the (expected) risk, i.e., the error over target domains as

$$R_{\mathrm{U}}(g) := \mathrm{E}_{\eta}\big[\ell(g(\boldsymbol{a}), f(\boldsymbol{a}))\big].$$

Let us define two minimizers as

$$g^{*} = \underset{g \in \mathcal{G}}{\mathrm{argmin}}\, R_{\mathrm{U}}(g),$$

$$\widehat{g} = \underset{g \in \mathcal{G}}{\mathrm{argmin}}\, \widehat{R}_{\mathrm{MC}}(g),$$

where $\mathcal{G}$ is a function class. Note that unlike the standard setting, $\widehat{R}_{\mathrm{MC}}(g)$ is not a sample approximation of $R_{\mathrm{U}}(g)$.

In this section, we investigate the estimation error defined as

$$R_{\mathrm{U}}(\widehat{g}) - R_{\mathrm{U}}(g^{*}).$$

More precisely, let us assume that $|g(\boldsymbol{a}) - f(\boldsymbol{a})| \leq M$ for all $g \in \mathcal{G}$ and $\boldsymbol{a} \in \mathbb{R}^m$, where $f\colon \mathcal{A} \to \mathbb{R}$ is the labeling function. For ℓ, we consider the ℓ_p loss defined as $\ell_p(y, y') = |y - y'|^p$ for $p \geq 1$, which includes the squared loss if $p = 2$.

Let us also assume that there exists a constant $C_a > 0$ such that $\|a\| \leq C_a$ for all $a \in \mathbb{R}^m$, i.e., the attribute vector is bounded.

Let $\mathcal{G}$ be the function class for prediction models. For example, the function class of the linear model can be expressed as $\mathcal{G} = \{w^\top a \mid w \in \mathbb{R}^m; \|w\| \leq C_w\}$. Let $\mathfrak{R}_T(\mathcal{G}) := \mathrm{E}_{\mathcal{A} \sim \eta^T}[\mathrm{E}_{\sigma}[\sup_{g \in \mathcal{G}} \frac{1}{T} \sum_{t=1}^{T} \sigma_t g(a^{(t)})]]$ be the Rademacher complexity, where $\sigma := (\sigma_1, \ldots, \sigma_T)^\top$, with σ_is independent uniform random variables taking values in $\{-1, +1\}$.

We then define the empirical risk as

$$\widehat{R}_S(g) := \frac{1}{T_S} \sum_{t=1}^{T_S} \ell(g(a^{(t)}), f(a^{(t)})).$$

Additionally, let us define the empirical risk for h_S as

$$\xi_{\mathrm{MTL}}(h_S) := \frac{1}{T_S} \sum_{t=1}^{T_S} \ell(h^{(t)}(x'), f(a^{(t)})).$$

5.2   Results

First, we have the following lemma:

Lemma 1. *For any $\delta > 0$, we have with probability at least $1 - \delta$, the following inequality:*

$$\sup_{g \in \mathcal{G}} |R_U(g) - \widehat{R}_{\mathrm{MC}}(g)| \leq \xi_{\mathrm{MTL}}(h_S) + 2pM^{p-1}\mathfrak{R}_{T_S}(\mathcal{G}) + M^p \sqrt{\frac{\ln(2/\delta)}{2T_S}}. \tag{4}$$

Proof. From the triangle inequality, we have

$$\ell(g(a^{(t)}), f(a^{(t)})) \leq \ell(g(a^{(t)}), \widehat{h}^{(t)}(x')) + \ell(\widehat{h}^{(t)}(x'), f(a^{(t)}))$$

for any $\widehat{h}^{(t)}$ at x'. Thus,

$$\widehat{R}_S(g) \leq \widehat{R}_{\mathrm{MC}}(g) + \xi_{\mathrm{MTL}}(h_S). \tag{5}$$

Similarly,

$$\widehat{R}_{\mathrm{MC}}(g) \leq \widehat{R}_S(g) + \xi_{\mathrm{MTL}}(h_S). \tag{6}$$

Next, on the basis of the standard Rademacher complexity analysis (see, e.g., [19, Theorem 10.3]), for any $\delta > 0$, we have with probability at least $1 - \delta/2$, the following inequality for all $g \in \mathcal{G}$:

$$R_U(g) - \widehat{R}_S(g) \leq 2pM^{p-1}\mathfrak{R}_{T_S}(\mathcal{G}) + M^p \sqrt{\frac{\ln(2/\delta)}{2T_S}},$$

$$\widehat{R}_S(g) - R_U(g) \leq 2pM^{p-1}\mathfrak{R}_{T_S}(\mathcal{G}) + M^p \sqrt{\frac{\ln(2/\delta)}{2T_S}}.$$

From Eqs. (5) and (6), we thus have with probability at least $1 - \delta/2$, the following inequality for all $g \in \mathcal{G}$:

$$R_{\mathrm{U}}(g) - \widehat{R}_{\mathrm{MC}}(g) \leq \xi_{\mathrm{MTL}}(\widehat{\boldsymbol{h}}_{\mathrm{S}}) + 2pM^{p-1}\mathfrak{R}_{T_{\mathrm{S}}}(\mathcal{G}) + M^p \sqrt{\frac{\ln(2/\delta)}{2T_{\mathrm{S}}}},$$

$$\widehat{R}_{\mathrm{MC}}(g) - R_{\mathrm{U}}(g) \leq \xi_{\mathrm{MTL}}(\widehat{\boldsymbol{h}}_{\mathrm{S}}) + 2pM^{p-1}\mathfrak{R}_{T_{\mathrm{S}}}(\mathcal{G}) + M^p \sqrt{\frac{\ln(2/\delta)}{2T_{\mathrm{S}}}},$$

which concludes the proof. $\qquad\square$

On the basis of Lemma 1, we have the following estimation error bound:

Theorem 2. *For any $\delta > 0$, we have with probability at least $1-\delta$, the following inequality:*

$$R_{\mathrm{U}}(\widehat{g}) - R_{\mathrm{U}}(g^*) \leq 2\xi_{\mathrm{MTL}}(\widehat{\boldsymbol{h}}_{\mathrm{S}}) + 4pM^{p-1}\mathfrak{R}_{T_{\mathrm{S}}}(\mathcal{G}) + 2M^p \sqrt{\frac{\ln(2/\delta)}{2T_{\mathrm{S}}}} \qquad (7)$$

Proof. By definition of $\widehat{g}$, we have $\widehat{R}_{\mathrm{MC}}(\widehat{g}) \leq \widehat{R}_{\mathrm{MC}}(g^*)$. We then derive the upper bound of the estimation error:

$$R_{\mathrm{U}}(\widehat{g}) - R_{\mathrm{U}}(g^*) \leq R_{\mathrm{U}}(\widehat{g}) - \widehat{R}_{\mathrm{MC}}(\widehat{g}) + \widehat{R}_{\mathrm{MC}}(\widehat{g}) - R_{\mathrm{U}}(g^*)$$

$$\leq \left(\sup_{g \in \mathcal{G}} R_{\mathrm{U}}(g) - \widehat{R}_{\mathrm{MC}}(g) \right) + \widehat{R}_{\mathrm{MC}}(g^*) - R_{\mathrm{U}}(g^*)$$

$$\leq 2 \sup_{g \in \mathcal{G}} \left| R_{\mathrm{U}}(\widehat{g}) - \widehat{R}_{\mathrm{MC}}(\widehat{g}) \right|.$$

Combining the above with Lemma 1, we obtain the theorem. $\qquad\square$

For $\mathfrak{R}_{T_{\mathrm{S}}}(\mathcal{G})$, the Rademacher complexity for various models are known to be bounded [1,19]. To observe the effect of the Rademacher complexity, let us assume the linear model as the function class:[1]

$$\mathcal{G} = \{\boldsymbol{w}^{\top}\boldsymbol{a} \mid \boldsymbol{w} \in \mathbb{R}^m; \|\boldsymbol{w}\| \leq C_w\}.$$

From Theorem 2, we then have the following corollary:

Corollary 3. *Assume that the linear model is used as $\mathcal{G}$. For any $\delta > 0$, we have with probability at least $1 - \delta$, the following inequality:*

$$R_{\mathrm{U}}(\widehat{g}) - R_{\mathrm{U}}(g^*) \leq 2\xi_{\mathrm{MTL}}(\widehat{\boldsymbol{h}}_{\mathrm{S}}) + \frac{C_{a,w,\delta}}{\sqrt{T_{\mathrm{S}}}}, \qquad (8)$$

where $C_{a,w,\delta} := 4pM^{p-1}C_a C_w + M^p \sqrt{2\ln(2/\delta)}$.

---

[1] The linear-in-parameter model and kernel model can be handled in a similar manner.

Proof. Since the linear models are used for $\mathcal{G}$, we can prove (see [19, Theorem 4.3] for details)

$$\mathfrak{R}_{T_\mathrm{S}}(\mathcal{G}) \leq \frac{C_a C_w}{\sqrt{T_\mathrm{S}}}.$$

Plugging the above equation into Eq. (7) and defining $C_{a,w,\delta} := 4pM^{p-1}C_a C_w + M^p\sqrt{2\ln(2/\delta)}$, we conclude the proof. $\square$

Corollary 3 shows that the second term on the right-hand side in Eq. (8) decreases in $\mathcal{O}_p(1/\sqrt{T_\mathrm{S}})$, meaning that the term becomes small if the number of seen domains T_S is large.[2] If we have accurate source hypotheses $\widehat{h}_\mathrm{S}$, $\xi_\mathrm{MTL}(\widehat{h}_\mathrm{S})$ will be also small. Thus, minimizing $\widehat{R}_\mathrm{MC}$ leads to a smaller estimation error in Eq. (8).

It should be noted that in a standard supervised learning setting, the estimation error converges with $\mathcal{O}_p(1/\sqrt{N})$, where N is the number of training samples. The convergence rate is known as optimal under certain mild conditions in empirical risk minimization [18]. Since our analysis is also based on a tool for empirical risk minimization, the connection indicates that $\mathcal{O}_p(1/\sqrt{T_\mathrm{S}})$ convergence derived from our analysis is optimal under mild conditions.

Table 3. Statistics of datasets. T denotes number of domains.

Dataset	n	d	m	T
Synth (T)	$100T$	10	20	T
Coffee	1,161	63	36	23
School	4,593	21	6	23
Book	7,282	64	77	169
Wine	32,906	751	8	75
Sushi	50,000	31	15	100

6 Experiments

We evaluated our proposed method on various synthetic and benchmark datasets. We used a PC equipped with Intel Xeon Gold 6142 and NVIDIA Quadro RTX 5000.

6.1 Datasets

Synthetic Datasets: We generated a synthetic dataset, called Synth (T). By varying the number of seen domains T, we confirmed the performance change of ZSDA methods in terms of T. We generated Synth (T) on the basis of the following procedure:

---

[2] $\mathcal{O}_p$ denotes the order in probability.

1. Prepare the Gaussian basis functions $\{\phi_\ell(\boldsymbol{x}) = \exp(-\|\boldsymbol{x} - \boldsymbol{x}_\ell\|^2)\}_{\ell=1}^b$;
2. Create b-dimensional parameter vector, $\boldsymbol{w}^{(t)} \in \mathbb{R}^b$, each element of which is drawn from the standard normal distribution $\mathcal{N}(0, 1^2)$;
3. Make an m-dimensional attribute vector by $\boldsymbol{a}^{(t)} = \boldsymbol{Q}\boldsymbol{w}^{(t)}$, where $\boldsymbol{Q} = (\boldsymbol{q}_1, \ldots, \boldsymbol{q}_b) \in \mathbb{R}^{m \times b}$ and $\{\boldsymbol{q}_i \in \mathbb{R}^m\}_{i=1}^b$ is the set of m-dimensional orthonormal vectors;
4. Generate feature vectors, each element of which is drawn from the uniform distribution $\mathcal{U}(0, 1)$;
5. Observe the paired samples of size $n^{(t)}$ from $y^{(t)} = \boldsymbol{w}^{(t)\top}\phi(\boldsymbol{x}) + 0.1\varepsilon$, where ε is drawn from the standard normal distribution.

We set $d = 10$, $m = 20$, $b = 20$, and $n^{(t)} = 100$ (thus, $n = 100T$).

Benchmark Datasets: We used the goodbooks-10k (Book), coffee quality (Coffee), School [7], SUSHI preference [11] (Sushi), and wine reviews (Wine) datasets as benchmark datasets.[3] Table 3 summarizes the statistics of the datasets used in our experiments.

The Coffee dataset consists of reviews of the coffee beans for several farms. We used "Country of Origin", "Certification Body", and "Altitude" as features of farms and "Species", "Processing Method", and "Variety" as domain attributes of the coffee beans. The "Total Cup Points" was used as the score. We removed coffee beans that received less than ten reviews.

The Book dataset is a collection of book ratings from readers. We used "Age" and "Country" as features of readers and the tags of books annotated by users in the book-rating platform as domain attributes. We manually extracted book tags that are likely to be relevant to rating.

The School dataset contains examination scores of 15,362 students from 139 schools. Similarly to Yang and Hospedales [25], we chose the school in which each year group had more than 50 students. After preprocessing, we have 4,593 samples and 23 domains. We used the school gender (Mixed, Male, and Female) and school denomination (Maintained, Church of England, and Roman Catholic) as domain attributes.

In the Wine dataset, we had 32,906 wine reviews after preprocessing. We used "Variety", "Country", and "Price" as features, and color and taste information extracted from the description as domain attributes.

The SUSHI dataset consists of ratings of 100 types of sushi from 5,000 people. We used the information of users as features and the information of sushi as domain attributes.

6.2   Setting

We used the 8:2 domain-wise dataset split to create data for seen and unseen domains. We left the data for unseen domains for evaluation of our method's

---

[3] Book: https://github.com/zygmuntz/goodbooks-10k. Coffee: https://github.com/jldbc/coffee-quality-database. Sushi: http://www.kamishima.net/sushi/. Wine: https://www.kaggle.com/zynicide/wine-reviews.

Table 4. Average and standard error of relative mRMSE (α/β) over 20 trials, where α and β are mRMSE$_U$ of proposed and MDMT, respectively. When α/β was less than 1, our proposed method was more accurate than MDMT. Even though our method does not use source domain data, its performance was often comparable or sometimes superior to that of MDMT.

Dataset	Ridge+MC vs		LGBM+MC vs	
	MDMT1	MDMT2	MDMT1	MDMT2
Synth (50)	1.01 ± 0.03	1.00 ± 0.04	1.00 ± 0.03	0.99 ± 0.04
Synth (100)	0.91 ± 0.02	0.96 ± 0.04	1.01 ± 0.02	1.07 ± 0.04
Coffee	0.06 ± 0.01	0.10 ± 0.00	0.06 ± 0.01	0.10 ± 0.01
School	0.51 ± 0.07	0.73 ± 0.03	0.49 ± 0.06	0.72 ± 0.03
Book	0.98 ± 0.00	0.92 ± 0.01	0.99 ± 0.00	0.92 ± 0.00
Wine	1.00 ± 0.01	0.96 ± 0.02	0.93 ± 0.01	0.89 ± 0.02
Sushi	0.98 ± 0.00	0.83 ± 0.01	0.98 ± 0.00	0.83 ± 0.01

performance. We further split the data for seen domains into 80% training and 20% test data.

For our proposed method, we used the linear ridge regression (Ridge) [8] as the fixed-input predictor g. As the trained predictor for a seen domain $\widehat{h}^{(t)}$, we used two methods: Ridge and LightGBM [12]. The former is denoted as Ridge+MC and the latter as LGBM+MC.

As a baseline, we used MDMT [25]. For the attribute-aware predictor F, we used $F(\boldsymbol{x}, t) = \boldsymbol{\phi}(\boldsymbol{x})^\top \boldsymbol{W} \boldsymbol{a}^{(t)} + c$, where $\boldsymbol{W} \in \mathbb{R}^{r \times m}$, $c \in \mathbb{R}$ is the intercept and $\boldsymbol{\phi} \colon \mathbb{R}^d \to \mathbb{R}^r$ is a feature transformation.

We prepared two architectures, i.e., MDMT1 and MDMT2. For MDMT1, $\boldsymbol{\phi}$ is the identity transformation and $r = d$, meaning that F is a bilinear function, and $\boldsymbol{W}$ and c are the parameters to be learned. For MDMT2, $\boldsymbol{\phi}$ is the two-layer neural network consisting of a $d \times r$ linear layer, batch normalization [9], and rectified linear unit (ReLU) activation [6]. We thus learn $\boldsymbol{W}$, c, and $\boldsymbol{\phi}$ in MDMT2. For both MDMT1 and MDMT2, we used Adam optimizer [13] to solve the optimization problem in Eq. (2). The number of epochs was set to 500.

6.3   Evaluation Measure

To evaluate the performance of our proposed method, we first defined the mean of the *root-mean-square error* (mRMSE) over *unseen* domains:

$$\mathrm{mRMSE_U} := \frac{1}{T_U} \sum_{t \in \mathcal{T}_U} \sqrt{\mathrm{MSE}(t)},$$

where $\mathrm{MSE}(t) := (1/n^{(t)}) \sum_{i=1}^{n^{(t)}} (y_i^{(t)} - \widehat{y}_i^{(t)})^2$ and $\widehat{y}_i^{(t)}$ is the prediction of a test sample $\boldsymbol{x}_i^{(t)}$ in t.

To measure the performance of our proposed method, we used *relative* mRMSE. Specifically, let α and β be the mRMSE$_U$ of our method and MDMT, respectively. We reported α/β (smaller is better), the relative mRMSE. When α/β was close to one, our method performed comparably with the baseline that can access source domain data.

6.4   Results

Theory and Practice: We first show the mRMSE$_U$ of the proposed method as a function of the number of seen domains T_S on Synth (T_S). In the theoretical analysis discussed in Sect. 5, we proved that the generalization error in terms of domains converges at the rate of $\mathcal{O}(1/\sqrt{T_S})$. This theoretical result indicates that the expected error over unseen domains decreases with the number of seen domains.

Figure 2 shows the mRMSE$_U$ of our method and the curve of $1/\sqrt{T_S}$ plus a constant.[4] The results indicate that the mRMSE$_U$ of our method decreased as the number of seen domains increased. Compared with the curve $\mathcal{O}(1/\sqrt{T_S})$, the mRMSE$_U$ of our method behaved similarly, indicating that our theoretical analysis can be a guideline on how many seen domains are necessary to obtain a certain performance improvement.

Prediction Performance: Table 4 lists the average and standard errors of relative mRMSE (α/β) over 20 trials, where α and β are the mRMSE$_U$ of the proposed method and MDMT, respectively. For example, α is the mRMSE$_U$ of Ridge+MC and β is that of MDMT1. When α/β was less than 1, our proposed method was more accurate than MDMT. Table 4 shows that even though our method does not use source domain data, its performance was often comparable or sometimes superior to that of MDMT. We thus conclude that HTL for ZSDA is possible with our method.

Computation Time: Since the training time of our method is zero (see Sect. 4), we are thus interested in the inference time. Figure 1 shows the average inference times (in microseconds). The inference time of our method was less than 1 *ms* (i.e., 1000 μs) except for the Book dataset. Since the total number of domains of the Book dataset is 169, which is slightly larger than the other datasets, the computation time of our method took slightly longer. This is because the computational complexity of our method depends on the number of seen domains, as discussed in Sect. 4.3.

In summary, even if our method requires a certain amount of computation time in an inference phase, it is computationally efficient.

---

[4] The constant is calculated such that the value of the curve at $T_S = 200$ is equivalent to that of Ridge+MC.

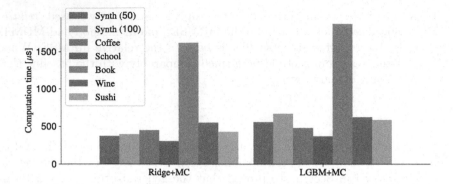

Fig. 1. Average computation time of Ridge+MC and LGBM+MC over 10 trials.

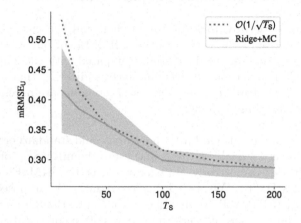

Fig. 2. Average and standard deviation of mRMSE$_U$ of Ridge+MC over 100 trials. Curve $\mathcal{O}(1/\sqrt{T_S})$, derived from our theoretical analysis, is $1/\sqrt{T_S}$ plus constant. mRMSE$_U$ of our method decreased as T_S increased and shape was similar to $\mathcal{O}(1/\sqrt{T_S})$.

7 Conclusions

We proposed a hypotheses transfer method for zero-shot domain adaptation that can work with source hypotheses without accessing source domain data. We argued that our method can be easily implemented with Scikit-learn in Python. When linear models are used, we can make predictions very efficiently, as confirmed from both computational complexity analysis and experiments. We investigated the estimation error bound of our proposed method and revealed that our method is theoretically valid. Finally, through numerical experiments, we demonstrated the effectiveness of our proposed method.

Acknowledgments. We thank the anonymous reviewers for their helpful comments.

References

1. Bartlett, P.L., Mendelson, S.: Rademacher and Gaussian complexities: risk bounds and structural results. J. Mach. Learn. Res. **3**, 463–482 (2002)
2. Blanchard, G., Lee, G., Scott, C.: Generalizing from several related classification tasks to a new unlabeled sample. In: Advances in Neural Information Processing Systems, pp. 2178–2186 (2011)
3. Chapelle, O., Schölkopf, B., Zien, A.: Semi-Supervised Learning. The MIT Press, Cambridge (2006)
4. Chidlovskii, B., Clinchant, S., Csurka, G.: Domain adaptation in the absence of source domain data. In: Proceedings of the 22nd ACM SIGKDD International Conference on Knowledge Discovery and Data Mining, pp. 451–460 (2016)
5. Evgeniou, T., Pontil, M.: Regularized multi-task learning. In: Proceedings of the Tenth ACM SIGKDD International Conference on Knowledge Discovery and Data Mining (2004)
6. Glorot, X., Bordes, A., Bengio, Y.: Deep sparse rectifier neural networks. In: Proceedings of the Fourteenth International Conference on Artificial Intelligence and Statistics, pp. 315–323 (2011)
7. Rabe-Hesketh, S., Skrondal, A.: Multilevel modelling of complex survey data. J. R. Stat. Soc. Ser. A (Stat. Soc.) **169**(4), 805–827 (2006). https://www.jstor.org/stable/3877401
8. Hastie, T., Tibshirani, R., Friedman, J.: The Elements of Statistical Learning. Springer, New York (2009). https://doi.org/10.1007/978-0-387-84858-7
9. Ioffe, S., Szegedy, C.: Batch normalization: accelerating deep network training by reducing internal covariate shift. In: Proceedings of the 32nd International Conference on Machine Learning, vol. 37, pp. 448–456 (2015)
10. Ishii, M., Takenouchi, T., Sugiyama, M.: Zero-shot domain adaptation based on attribute information. In: Proceedings of The Eleventh Asian Conference on Machine Learning, pp. 473–488 (2019)
11. Kamishima, T., Akaho, S.: Efficient clustering for orders. In: Zighed, D.A., Tsumoto, S., Ras, Z.W., Hacid, H. (eds.) Mining Complex Data, vol. 165, pp. 261–279. Springer, Heidelberg (2009). https://doi.org/10.1007/978-3-540-88067-7_15
12. Ke, G., et al.: LightGBM: a highly efficient gradient boosting decision tree. Adv. Neural Inf. Process. Syst. **30**, 3146–3154 (2017)
13. Kingma, D.P., Ba, J.: Adam: a method for stochastic optimization. In: Proceedings of the 32nd International Conference on Machine Learning (2015)
14. Kuzborskij, I., Orabona, F.: Stability and hypothesis transfer learning. In: Proceedings of the 30th International Conference on Machine Learning, pp. 942–950 (2013)
15. Li, D., Yang, Y., Song, Y.Z., Hospedales, T.M.: Deeper, broader and artier domain generalization. In: Proceedings of the IEEE International Conference on Computer Vision, pp. 5542–5550 (2017)
16. Liang, J., Hu, D., Feng, J.: Do we really need to access the source data? Source hypothesis transfer for unsupervised domain adaptation. In: Proceedings of the 37th International Conference on Machine Learning, pp. 6028–6039 (2020)
17. Mansour, Y., Mohri, M., Rostamizadeh, A.: Domain adaptation with multiple sources. In: Advances in Neural Information Processing Systems, pp. 1041–1048 (2009)

18. Mendelson, S.: Lower bounds for the empirical minimization algorithm. IEEE Trans. Inf. Theory **54**(8), 3797–3803 (2008)
19. Mohri, M., Rostamizadeh, A., Talwalkar, A.: Foundations of Machine Learning. MIT Press, Cambridge (2012)
20. Pedregosa, F., et al.: Scikit-learn: machine learning in Python. J. Mach. Learn. Res. **12**, 2825–2830 (2011)
21. Peng, K.C., Wu, Z., Ernst, J.: Zero-shot deep domain adaptation. In: Proceedings of the European Conference on Computer Vision, pp. 764–781 (2018)
22. Romera-Paredes, B., Torr, P.: An embarrassingly simple approach to zero-shot learning. In: Proceedings of the 32nd International Conference on Machine Learning (2015)
23. Sakai, T., Ohasaka, N.: Predictive optimization with zero-shot domain adaptation. In: Proceedings of the 2021 SIAM International Conference on Data Mining, pp. 369–377 (2021)
24. Vapnik, V.: The Nature of Statistical Learning Theory. Springer, New York (1995). https://doi.org/10.1007/978-1-4757-2440-0
25. Yang, Y., Hospedales, T.M.: A unified perspective on multi-domain and multi-task learning. In: Proceedings of 3rd International Conference on Learning Representations (2015)
26. Yang, Y., Hospedales, T.M.: Zero-shot domain adaptation via kernel regression on the Grassmannian. In: International Workshop on Differential Geometry in Computer Vision for Analysis of Shapes, Images and Trajectories (2015)

FedPHP: Federated Personalization with Inherited Private Models

Xin-Chun Li[1] , De-Chuan Zhan[1](✉) , Yunfeng Shao[2], Bingshuai Li[2],
and Shaoming Song[2]

[1] State Key Laboratory for Novel Software Technology, Nanjing University,
Nanjing, China
lixc@lamda.nju.edu.cn, zhandc@nju.edu.cn
[2] Huawei Noah's Ark Lab, Beijing, China
{shaoyunfeng,libingshuai,shaoming.song}@huawei.com

Abstract. Federated Learning (FL) generates a single global model via collaborating distributed clients without leaking data privacy. However, the statistical heterogeneity of non-iid data across clients poses a fundamental challenge to the model personalization process of each client. Our significant observation is that the *newly downloaded global model* from the server may perform poorly on local clients, while it could become better after adequate personalization steps. Inspired by this, we advocate that the *hard-won personalized model* in each communication round should be rationally exploited, while standard FL methods directly overwrite the previous personalized models. Specifically, we propose a novel concept named *"inHerited Private Model" (HPM)* for each local client as a temporal ensembling of its historical personalized models and exploit it to supervise the personalization process in the next global round. We explore various types of knowledge transfer to facilitate the personalization process. We provide both theoretical analysis and abundant experimental studies to verify the superiorities of our algorithm.

1 Introduction

Federated Learning (FL) [6,17,23] has been proposed as an efficient decentralized training method under data privacy constraints. In FL, clients' data are not permitted to send out, and only models or parameters could be transmitted. Usually, FL contains two fundamental stages: personalization and aggregation. During personalization, a small subset of clients download the global model, which is referred to as the *"newly downloaded model"*, and then personalize it on their private data to obtain the *"personalized model"*; during aggregation, the server receives the personalized models from these clients and aggregates them. A global communication round contains these two stages, and amounts of rounds will be taken until convergence. FL faces many challenges [12,17]. The statistical heterogeneity caused by the non-iid data among clients could dispel the clients' incentive to participate in FL, because the globally aggregated model may be worse than the locally trained model [9,26]. Hence, it is necessary to design effective personalization strategies in non-iid scenes.

© Springer Nature Switzerland AG 2021
N. Oliver et al. (Eds.): ECML PKDD 2021, LNAI 12975, pp. 587–602, 2021.
https://doi.org/10.1007/978-3-030-86486-6_36

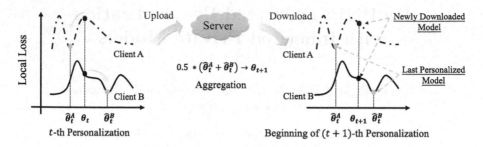

Fig. 1. Motivation: the local performance degradation in the beginning of FedAvg. The y-axis shows the local loss. We draw the curves of client A and B without overlapping for better visualization. The local losses of θ_{t+1} are higher than $\hat{\theta}_t^A$ and $\hat{\theta}_t^B$.

As the most standard FL algorithm, FedAvg [17] aims to generate a single global model, which is hard to capture heterogeneous local distributions simultaneously. An empirical observation in FedAvg is that the newly downloaded global model could perform poorly on local data. As shown in Fig. 1, during the personalization stage of the tth global round, two clients first finetune the global model θ_t according to their own data distributions and obtain personalized models $\hat{\theta}_t^A$, $\hat{\theta}_t^B$ respectively. During aggregation, the server collects the personalized models and takes a direct parameter averaging as $\theta_{t+1} \leftarrow (\hat{\theta}_t^A + \hat{\theta}_t^B)/2$. At the beginning of the next global round, the aggretated model θ_{t+1} may perform worse than the last personalized models correspondingly.

We apply FedAvg to several FL benchmarks, i.e., Cifar10-100-5, Cifar100-100-20, and FeMnist. For each scene, we take several groups of hyper-parameters, varying the local epoch E, the client selection ratio Q, and the learning rate η. The details of benchmarks and hyper-parameters can be found in Sect. 4.1. The performance measures could be found in Sect. 3.1. The observations are shown in Fig. 2. In each personalization stage, we first record the local test accuracy of the newly downloaded model (marked by "x"). Then, we personalize this model and record the personalized model's performance (marked by subsequent "+"). We can observe that the local performances will improve during the personalization (the solid segments), while the newly downloaded models' performances could be much worse than the last personalized model (the dotted segments). This conforms to the motivation in Fig. 1.

A fundamental problem here is that the hard-won personalized models are directly overwritten by the newly downloaded global model, and the clients have to personalize the global model from scratch in a new round. On one hand, in the tth round, only a fraction of clients S_t could be selected for personalization due to stragglers in FL. The tth aggregation will take the average of personalized models from S_t. In the $(t+1)$th round, another subset of clients will be selected. If $c \in S_{t+1}$ but $c \notin S_t$, it is manifest that the aggregated model from S_t may perform worse on c due to the distribution shift. On the other hand, even we could select all clients, such as in cross-silo FL scenes [6], the data heterogeneity

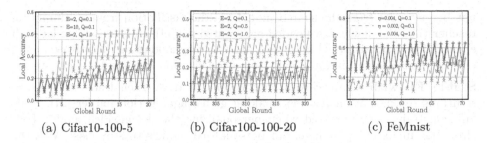

(a) Cifar10-100-5   (b) Cifar100-100-20   (c) FeMnist

Fig. 2. Observation: the local performance degradation in FedAvg on three FL benchmarks. In each curve, marker "x" and "+" show the performances of the newly downloaded model and the personalized model, respectively. The performances will degrade a lot once receiving the newly downloaded model and overwriting the last personalized model (the dotted segments).

still induces the performance degradation as shown in Fig. 2 when $Q = 1.0$. To solve this, we propose a novel concept named *"inHerited Private Model"* *(HPM)* to keep the moving average of historical personalized models in each client and utilize it to supervise the newly downloaded model in next round. We denote our algorithm as "Federated Personalization with inHerited Private models" *(FedPHP)* and briefly introduce it with three progressive explanations: (1) HPM is a novel kind of private-shared models in FL; (2) it keeps and transfers the historical valuable personalized knowledge to the newly downloaded global model; (3) it takes advantage of temporal ensembling, leading to better and stable personalization results. The illustration can be found in Fig. 3.

2   Related Works

Private-Shared Models for Federated Personalization. The major incentive of clients to participate in FL is to obtain better models with limited data or computation budget [9,16]. Simultaneously keeping private components on local clients is a natural solution for effective personalization. FedPer [1] splits models into base and personalization layers and only aggregates the transferable base layers; FedL2G [14] keeps representation learning private to learn useful and compact features for heterogeneous tasks; FedFu [25] fuses the features extracted by the fixed global model into the local models as a feature-level rectification; FLDA [19] directly combines a complete local private model and the global model as a mixture-of-experts for per-user domain adaptation. Although the specific problems solved by these methods are slightly different, they can all be regarded as variants of private-shared models for FL, where only the shared components participate in the aggregation procedure. These methods do not explicitly exploit the knowledge transfer between shared and private models, which may be less effective in some special FL cases. FedDML [20] also takes a similar architecture as FLDA, while it additionally uses knowledge distillation [5] to share the knowledge between private and shared models. However, the private model will not benefit a lot from the shared model via the distillation from an immature teacher (i.e., the newly downloaded aggregated model). In contrast, we let each client inherit the historical hard-won personalized models as the private model (HPM), which are utilized to facilitate personalization in the subsequent stages.

Stable FL with Constraints. Heterogeneous data often leads to diverged solutions during personalization, making global aggregation harder [27]. Various constraints are proposed when personalizing the downloaded model. Fed-Prox [12] limits the parameters of personalized models to stay close with the global model via a proximal term; FedCurv [21] avoids catastrophic forgetting via elastic weight consolidation; FedMMD [24] aims to mitigate the feature discrepancy between local and global models. In FL with amounts of clients, only selecting a subset of clients in each round is a solution to stragglers due to communication delay or computation limitation. However, this introduces additional randomness and slows the convergence. Similar to SGD, whose randomness is mainly resulted from batch data sampling and can be mitigated with momentum, updating global model with momentum can accelerate the convergence in FL, e.g., Scaffold [7]. Although these methods can partly lead to stable updates in FL, they do not take advantage of the hard-won personalized models and are almost proposed for better aggregation. Different from them, we propose Fed-PHP to enhance the personalization ability of local clients and aim to obtain better personalization performances compared with these related FL methods.

3   Our Methods

Suppose we have K distributed clients, and each client has a local data distribution $\mathcal{D}^k = \mathcal{P}^k(\mathbf{x}, y)$. The kth client has the optimization target:

$$\min_{\theta^k} F^k(\theta^k) \triangleq E_{(\mathbf{x}^k, y^k) \sim \mathcal{D}^k} \left[\ell \left(f(\mathbf{x}^k; \theta^k), y^k \right) \right], \tag{1}$$

where $f(\cdot; \theta^k)$ is the prediction function based on parameters θ^k, $f(\mathbf{x}^k; \cdot)$ returns the "logits" before softmax operation, and $\ell(\cdot, \cdot)$ is the loss function, e.g., the softmax cross-entropy loss. Owing to lack of enough labeled data, individually training could not generate well-performed models. Standard FL algorithms, e.g., FedAvg [17], collaborate local clients via: $\min_\theta \sum_{k=1}^{K} p_k F^k(\theta)$, where p_k is often set as $n_k / \sum_k n_k$ and n_k is the number of samples on the kth client. This can be solved by rounds of personalization and aggregation stages. We denote θ_t^k as the model parameters of the kth client in tth global round. Without additional declaration, we omit the superscript "k" and use θ_t to represent the global parameters on the server.

During the personalization stage in tth round, a subset of clients S_t is selected, and the selected clients download the global model, i.e., $\theta_t^k \leftarrow \theta_t$ for $k \in S_t$, and train this model with private data. Formally, for a data batch $\left\{ (\mathbf{x}_i^k, y_i^k) \right\}_{i=1}^{B}$, the empirical loss according to Eq. 1 is calculated as:

$$\mathcal{L}\left(\theta_t^k\right) = \frac{1}{B} \sum_{i=1}^{B} \ell\left(f(\mathbf{x}_i^k; \theta_t^k), y_i^k \right), \tag{2}$$

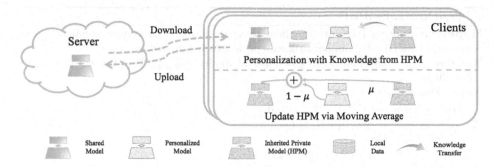

Fig. 3. Illustration of the proposed FedPHP. Each client stores the hard-won personalized models via moving average, i.e., the "inHerited Private Model" (HPM). During subsequent personalization stages, HPM could transfer the historical personalization knowledge to the newly downloaded model.

where B is the batch size. Then θ_t^k can be personalized with deep learning optimization methods, e.g., SGD with momentum. We denote as $\hat{\theta}_t^k$ the *personalized model*. During aggregation, the server collects these personalized models and takes a parameter averaging as: $\theta_{t+1} \leftarrow \sum_{k \in S_t} \frac{1}{|S_t|} \hat{\theta}_t^k$. Iteratively, the global model θ_{t+1} will be sent to another subset of clients for next round of personalization and aggregation. For the local clients, the received $\theta_t^k \leftarrow \theta_t$, $\theta_{t+1}^k \leftarrow \theta_{t+1}$ are called as *newly downloaded models*. In FedAvg, we can observe that the personalized model $\hat{\theta}_t^k$ is directly overwritten by the θ_{t+1}, and the local client has to personalize θ_{t+1} from scratch.

3.1 Empirical Observation and Goal

In FedAvg, the newly downloaded global model may perform poorly on local test data. We have a local test set $\{(\mathbf{x}_i^k, y_i^k)\}_{i=1}^{m_k}$ with m_k samples on the kth client. Formally, we denote as

$$\text{Acc}^k(\theta) \triangleq \frac{1}{m_k} \sum_{i=1}^{m_k} \mathcal{I}\left\{f(\mathbf{x}_i^k; \theta), y_i^k\right\} \tag{3}$$

the local test accuracy of kth client, where $\mathcal{I}\{\cdot, \cdot\}$ returns 1 if the prediction is right and 0 otherwise. Empirically, a performance degradation appears at the beginning of $(t+1)$th personalization stage:

$$\delta_t^k \triangleq \text{Acc}^k(\theta_{t+1}) - \text{Acc}^k\left(\hat{\theta}_t^k\right) < 0. \tag{4}$$

That is, the previous personalized model could be better than the newly downloaded global model as illustrated and shown in Fig. 1 and Fig. 2.

Due to the performance degradation, each client has to personalize the newly downloaded models from scratch, leading to slower convergence. Our goal is to

---

Algorithm 1. FedPHP

HyperParameters (partial): Q: client selection ratio; T: maximum number of communication rounds; E: number of local epochs; B: batch size

Return: $\theta_{s,T+1}$: the final aggregated global model; $\{\theta_{p,T+1}^k\}_{k=1}^K$: the HPM for each local client

ServerProcedure:

1: **for** global round $t = 0, 1, 2, \ldots, T$ **do**
2:    $S_t \leftarrow$ sample $\max(Q \cdot K, 1)$ clients
3:    **for** $k \in S_t$ **do**
4:        $\hat{\theta}_{s,t}^k \leftarrow$ ClientProcedure$(k, \theta_{s,t})$
5:    **end for**
6:    $\theta_{s,t+1} \leftarrow \sum_{k \in S_t} \frac{1}{|S_t|} \hat{\theta}_{s,t}^k$
7: **end for**

ClientProcedure$(k, \theta_{s,t})$:

1: $\theta_{s,t}^k \leftarrow \theta_{s,t}$
2: **for** local epoch $e = 1, 2, \ldots, E$ **do**
3:    **for** each batch $\{(\mathbf{x}_i^k, y_i^k)\}_{i=1}^B$ sampled from $\mathcal{D}^k$ **do**
4:        Calculate loss as in Eq. 10
5:        Update $\theta_{s,t}^k$ using, e.g., SGD with momentum
6:    **end for**
7: **end for**
8: Denote the personalized model as $\hat{\theta}_{s,t}^k$
9: $\theta_{p,t+1}^k \leftarrow (1 - \mu_t^k)\hat{\theta}_{s,t}^k + \mu_t^k \theta_{p,t}^k$
10: Adjust μ_t^k as in Sect. 3.2
11: **Return**: $\hat{\theta}_{s,t}^k$

---

accelerate the personalization in FL. Specifically, we record the *local test accuracy (Eq. 3) of the personalized model in each personalization stage and report the mean accuracy averaged among selected clients, i.e.,* $\frac{1}{|S_t|} \sum_{k=1}^{|S_t|} \text{Acc}^k \left(\hat{\theta}_t^k \right)$, as the personalization performance.

3.2   Inherited Private Models

Inspired by the empirical observation, we aim to preserve the previously hard-won personalized model on each client as private models. Specifically, we divide the whole model parameters into the global shared parameters θ_s and private parameters for local clients $\{\theta_p^k\}_{k=1}^K$. Different from existing private-shared models for FL, we mainly exploit the private model for preservation of historical hard-won personalized models, and name it as *"inHerited Private Model" (HPM)*. In contrast, the selected clients download the global model and directly overwrite the hard-won personalized models in FedAvg.

Formally, at the beginning of the tth personalization stage, the clients download the global shared model: $\theta_{s,t}^k \leftarrow \theta_{s,t}$ and obtain the personalized model $\hat{\theta}_{s,t}^k$. The specific personalization process will be introduced later. At the end, we update the HPM via:

$$\theta_{\mathrm{p},t+1}^k \leftarrow (1 - \mu_t^k)\hat{\theta}_{\mathrm{s},t}^k + \mu_t^k \theta_{\mathrm{p},t}^k, \tag{5}$$

which keeps a moving average of historical personalized models. $\mu_t^k \in [0,1]$ is the momentum term for the kth client. When $\mu_t^k = 0$, the HPM only keeps the current personalized model; when $\mu_t^k = 1$, the HPM degenerates into an independent private model. Because only a fraction of clients are selected in each round, the update frequency and learning speed of local clients are slightly distinct. Hence, the momentum should be client-specific and dynamically adjusted. We assign a counter z_t^k as the number of times that the kth client has been selected. If $k \in S_t$, then $z_t^k = z_{t-1}^k + 1$. Then we linearly set $\mu_t^k = \mu * z_t^k/(Q * T)$, where Q is the client selection ratio, T is the maximum number of global rounds, and $Q * T$ denotes the expected times of being selected. μ is the macro momentum that controls the change of μ_t^k, and we take $\mu = 0.9$ in this paper by default. Finally, we limit μ_t^k in a range of $[0,1]$. We will verify several possible ways to adjust μ_t^k in experimental studies, i.e., Sect. 4.3.

3.3   FedPHP

With the observation in Fig. 2 and Eq. 4, the HPM commonly perform better than the newly downloaded model. Hence, we exploit the HPM to supervise the personalization process of the newly downloaded model. As categorized in transfer learning [18], what to transfer refers to the specific content of the knowledge, which could be the outputs, features, or parameters. Hence, we explore several specific forms of knowledge transfer. For a specific sample $\mathbf{x}_i^k$, we denote the outputs of $\theta_{\mathrm{s},t}^k$ and $\theta_{\mathrm{p},t}^k$ as $\mathbf{g}_{\mathrm{s},i}$ and $\mathbf{g}_{\mathrm{p},i}$, respectively; and their intermediate features as $\mathbf{h}_{\mathrm{s},i}$ and $\mathbf{h}_{\mathrm{p},i}$, respectively. The outputs are "logits" without softmax, while the intermediate features are d-dimension vectors extracted by an extractor. We omit the index of k and t for simplification.

First, we could transfer the knowledge contained in the outputs. The knowledge distillation [5] could be utilized to enhance the information transfer via:

$$\mathcal{L}_{\mathrm{kd}}\left(\theta_{\mathrm{s},t}^k\right) = \tau^2 \frac{1}{B} \sum_{i=1}^{B} D_{KL}\left(\sigma(\mathbf{g}_{\mathrm{p},i}/\tau) \| \sigma(\mathbf{g}_{\mathrm{s},i}/\tau)\right), \tag{6}$$

where D_{KL} refers to the KL-divergence, $\sigma(\cdot)$ is the softmax operation, and τ is the temperature. Different from FedDML [20], we take an asymmetric distillation way and view the HPM as the teacher. We can also transfer the knowledge contained in the intermediate features. We can take a simple L2 regularization or a maximum mean discrepancy (MMD) [3] to align the feature distributions:

$$\mathcal{L}_{\mathrm{l2}}\left(\theta_{\mathrm{s},t}^k\right) = \frac{1}{2B} \sum_{i=1}^{B} \|\mathbf{h}_{\mathrm{s},i} - \mathbf{h}_{\mathrm{p},i}\|_2^2, \tag{7}$$

$$\mathcal{L}_{\mathrm{mmd}}\left(\theta_{\mathrm{s},t}^k\right) = \left\| \frac{1}{B} \sum_{i=1}^{B} \Phi(\mathbf{h}_{\mathrm{s},i}) - \frac{1}{B} \sum_{i=1}^{B} \Phi(\mathbf{h}_{\mathrm{p},i}) \right\|_{\mathcal{H}}^2, \tag{8}$$

where $\Phi(\cdot)$ is a feature map induced by a specific kernel function, i.e., $k(\mathbf{h}_i, \mathbf{h}_j) = \langle \Phi(\mathbf{h}_i), \Phi(\mathbf{h}_j) \rangle$. We use multiple Gaussian kernels with different bandwidths as in [15]. Finally, we can also transfer the knowledge directly from the parameters as in [12,13]:

$$\mathcal{L}_{\text{prox}} \left(\theta_{s,t}^k \right) = \left\| \theta_{s,t}^k - \theta_{p,t}^k \right\|_2^2. \tag{9}$$

With these types of knowledge transfer, and combined with Eq. 2, the total personalization loss is denoted as:

$$\mathcal{L}_{\text{total}} \left(\theta_{s,t}^k \right) = (1 - \lambda)\mathcal{L} \left(\theta_{s,t}^k \right) + \lambda \mathcal{L}_{\text{kt}} \left(\theta_{s,t}^k \right), \tag{10}$$

where $\mathcal{L}_{\text{kt}}$ could be $\mathcal{L}_{\text{kd}}$, $\mathcal{L}_{\text{mmd}}$, $\mathcal{L}_{\text{l2}}$, or $\mathcal{L}_{\text{prox}}$.

We will investigate these types of knowledge transfer and the coefficient λ in our experimental studies, i.e., Sect. 4.3. During the whole personalization stage, we only utilize HPM to facilitate the learning process of $\theta_{s,t}^k$ and do not update the HPM, which is efficient to implement. Once the personalization ends, we update the HPM as in Eq. 5. The illustration of FedPHP is shown in Fig. 3, and the complete pseudo-code of FedPHP is in Algorithm 1.

3.4   Discussion

A Novel Private-Shared Model in FL. Our proposed HPM can be viewed as a novel way to keep private models on local clients, which is majorly motivated by the empirical observation in Fig. 2. We take advantage of historical hard-won personalized models in each client and fully exploit them in subsequent personalization stages.

A Different Type of Regularization. As in previous FL studies [12,21,24], restricting the personalized model not go far away from the newly downloaded model is a natural solution to obtain stable FL. However, they are designed to make the aggregation more stable without considering personalization. Our proposed FedPHP takes a novel regularization way, utilizing the HPM to improve the personalization.

Temporal Ensembling and Mean Teacher. The moving average in Eq. 5 is inherently one type of temporal ensembling [10]. Utilizing the model $(1 - \mu_t^k)\hat{\theta}_{s,t}^k + \mu_t^k \theta_{p,t}^k$ to supervise the learning process of $\theta_{s,t+1}^k$ in next round works similarly as the self-ensembling and mean teacher in [22], which could lead to stable and better results.

3.5   Theoretical Analysis

We provide a macroscopic analysis of the failure in FedAvg and the advantages of FedPHP. Similar to FedBoost [4], we utilize the Bregman Divergence B_F as the loss function and assume that F is strictly convex and B_F is jointly convex. The loss of the kth client is $B_F \left(\mathcal{D}^k \| h^k \right) = F(\mathcal{D}^k) - F(h^k) - \langle \nabla F(h^k), \mathcal{D}^k - h^k \rangle$, where h^k is the learned estimator with a little abuse of notations. As shown in

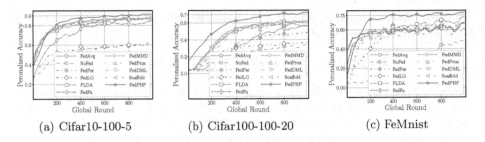

(a) Cifar10-100-5   (b) Cifar100-100-20   (c) FeMnist

Fig. 4. Performance comparisons of the proposed FedPHP with previous FL methods. FedPHP could obtain faster convergence and better personalization results.

FedBoost [4], FedAvg aims to minimize a uniform combination of local losses: $\arg\min_h \sum_{k=1}^{K} p_k B_F \left(\mathcal{D}^k || h\right)$, and the optimal solution is $h^* = \mathcal{D}_g$, where $\mathcal{D}_g \triangleq \sum_{k=1}^{K} p_k \mathcal{D}^k$ is the global distribution. Directly applying this solution to local clients leads to a loss $B_F \left(\mathcal{D}^k || \mathcal{D}_g\right)$, which could be very large due to non-iid data. This theoretically explains the major observation that the newly downloaded model performs poorly on local clients.

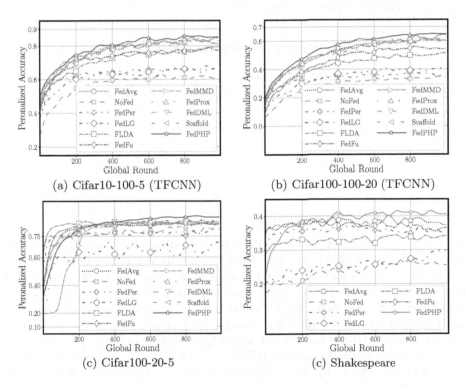

(a) Cifar10-100-5 (TFCNN)   (b) Cifar100-100-20 (TFCNN)

(c) Cifar100-20-5   (c) Shakespeare

Fig. 5. Performance comparisons on various settings. The (a) and (b) take another base model (i.e., TFCNN) for Cifar10-100-5 and Cifar100-100-20, respectively. The (c) takes a cross-silo FL scene, i.e., Cifar100-20-5. The (d) takes the Shakespeare benchmark.

With a combination of private and shared models: $h^k = (1-\alpha)h_s + \alpha h_p^k$, the local loss function on the kth client is:

$$B_F\left(\mathcal{D}^k||h^k\right) \leq (1-\alpha)B_F\left(\mathcal{D}^k||h_s\right) + \alpha B_F\left(\mathcal{D}^k||h_p^k\right),$$

implying that the personalization error of local clients is bounded by two components: the error of shared model with a shrinkage factor $1-\alpha$, $\alpha \in [0,1]$; the error of private models. The updated HPM in the tth round is actually an interpolation as: $h_{p,t+1}^k = (1-\alpha)\hat{h}_{s,t}^k + \alpha h_{p,t}^k$, with $\alpha = \mu_t^k$. We also have:

$$B_F\left(\mathcal{D}^k||h_{p,t+1}^k\right) \leq (1-\alpha)B_F\left(\mathcal{D}^k||\hat{h}_{s,t}^k\right) + \alpha B_F\left(\mathcal{D}^k||h_{p,t}^k\right),$$

which bounds the personalization error via two components: $B_F\left(\mathcal{D}^k||\hat{h}_{s,t}^k\right)$ denotes the error of after personalizing $h_{s,t}^k$, which can be smaller than $B_F\left(\mathcal{D}^k||h_{s,t}^k\right)$ with appropriate fine-tuning; $B_F\left(\mathcal{D}^k||h_{p,t}^k\right)$ denotes the error of the tth HPM, which can be deduced similarly to the $(t-1)$th round. To be brief, the macroscopic theoretical analysis shows that HPM is a special kind of private-shared model that can inherit the historical personalized models' ability, leading to smaller personalization errors.

Table 1. Statistical information of the investigated benchmarks.

	K	C	Loc.C	No.Tr	No.Te		K	C	Loc.C	No.Tr	No.Te
C10-100-5	100	10	5	400	100	C100-100-20	100	100	20	400	100
FeMnist	3550	62	62	181	45						
C100-20-5	20	100	5	2k	500	Shakespeare	1129	81	81	2994	749

4 Experiments

4.1 Scenes and Basic Settings

We verify the superiorities of FedPHP on several non-iid scenes: Cifar10-100-5, Cifar100-100-20, FeMnist. The Cifar10-100-5 is constructed via distributing the Cifar10 dataset [8] onto 100 clients according to labels, where each client only could observe 5 classes, and each client owns 400/100 samples for training/testing. Similarly, for Cifar100-100-20, it is constructed by distributing Cifar100 dataset [8] onto 100 clients, where each client only observes 20 classes. Such partitions can also be found in previous works [11,27]. FeMnist is the benchmark recommended by LEAF [2], which is to classify the mixture of digits and characters with data from 3550 writers. We resize the images in FeMnist into 28×28 ones. We list the detailed statistics in Table 1, which shows the number of clients K, the number of total classes C, the number of seen classes of each local client on average Loc.C, the number of local training samples for training/testing on average No.Tr/No.Te. For Cifar10-100-5 and Cifar100-100-20, we take $T = 1000$ as the number of global rounds, $E = 2$ as the number of local epochs, $B = 64$ as the batch size, $Q = 0.1$ as the client selection ratio. We use VGG11 without BN in PyTorch[1] as the base model. For FeMnist, we take $T = 1000$, $E = 10$, $B = 50$, $Q = 0.01$, and utilize the model with two convolutional layers and two fully-connected layers as in LEAF [2]. We use SGD with momentum 0.9 as optimizer and a constant learning rate $\eta = 0.03$ for Cifar10-100-5 and Cifar100-100-20, $\eta = 4e - 3$ for FeMnist. We take the features fed into the fully-connected layer as the intermediate features. For FedPHP, we use the MMD regularization as in Eq. 8 with $\lambda = 0.01$ for Cifar10-100-5 and Cifar100-100-20, and $\lambda = 0.1$ for FeMnist, as in Eq. 10. We only take the predictions from the HPM, i.e., $\theta_{p,t+1}^k$, to calculate the personalization accuracy. Our comparison methods can be divided into four categories: individual training (denoted as NoFed) and FedAvg [17]; existing FL methods with private-shared models: FedPer [1], FedL2G [14], FedFu [25], FLDA [19], and FedDML [20]; FL methods with constraints: FedProx [12], FedMMD [24]; FL aggregation with momentum: SCAFFOLD [7].

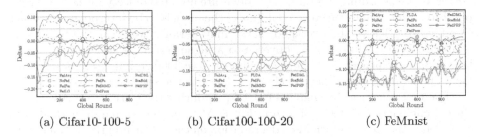

(a) Cifar10-100-5   (b) Cifar100-100-20   (c) FeMnist

Fig. 6. Comparisons of δ_t^k (Eq. 4) with existing FL methods. FedPHP could omit the local performance degradation with δ_t^k being nearly zero.

4.2   Experimental Results

We record the personalized models' local test accuracy as the personalization performance as introduced in Sect. 3.1. The personalized accuracy curves are plotted in Fig. 4. We can find that FedPHP can obtain the best performances. NoFed obtains the worst performance due to few local data. FedPer [1] and FedLG [14] also perform worse due to the possible feature mismatch between the downloaded shared model and the private model. Compared with FedAvg [17], FedMMD [24] leads to slightly faster convergence on Cifar100-100-20, while FLDA [19] could obtain higher personalization performance when converged. Other methods perform similarly. We can find that our methods could surpass the compared methods with a large margin, which is exciting and inspiring.

---

[1] https://pytorch.org/docs/stable/torchvision/models.html.

To further verify the superiorities of our methods, we vary several settings. First, we want to explore whether the improvement is related to the used network. We apply the TFCNN (Tensorflow CNN)[2] used in FedAvg [17] to Cifar10-100-5 and Cifar100-100-20. Then, we vary the number of clients and construct a cross-silo FL scene [6]. Specifically, we distribute the Cifar100 dataset [8] onto 20 clients with disjoint classes, i.e., with each client owning 5 classes. We denote this scene as Cifar100-20-5. For this scene, we use VGG11 again as the base model and select all clients in each round, i.e., $Q = 1.0$. We also investigate another benchmark recommended by LEAF [2], i.e., Shakespeare. It is a next-character prediction task and contains 1129 clients. We take the CharLSTM model used in FedAvg [17], and set $Q = 0.01$, $\eta = 1.47$, $E = 2$, and $B = 10$. For this scene, we only compare parts of previous methods due to computation limitation.

The results are shown in Fig. 5. From Fig. 5 (a) and (b), we can deduce that our proposed FedPHP is robust to the base models. However, FedPHP could converge slower on Cifar100-20-5 than most of the compared methods. This results from two reasons. First, in this scene, each client owns 2000 training samples with 400 samples for each class. It is enough for the compared FL methods to train a well-performed model on local clients, e.g., even NoFed could perform better at the beginning. Second, the models are updated quickly initially, and the possibly induced high variance could make the models to be interpolated distinct significantly. Hence, the averaged model may perform worse at the beginning. However, FedPHP could still converge to a higher result as in Fig. 5 (c). Similarly, on Shakespeare, each client can almost have 3000 training samples on each local client. Also, these may be closely related to the specific FL task.

Then we compare the change of δ_t^k (Eq. 4) during the learning process on the three benchmarks, which is shown in Fig. 6. Most of the compared methods will experience the local performance degradation on these benchmarks, i.e., negative δ_t^k. However, FedPHP could make δ_t^k nearly zero, which omits the degradation when personalization.

4.3   Ablation Studies

As introduced in Sect. 3.3, the knowledge transfer from the HPM to the newly downloaded model could have various types and could be applied with different levels of forces with various λ. Hence, we first explore the comparisons of these types of knowledge and the settings of $\lambda \in \{0.0, 0.001, 0.01, 0.1\}$. We compare the performances on Cifar100-100-20 and FeMnist. The results are shown in Fig. 7. To show the convergence speed, we report both the personalization performance around the 200th round and the 1000th round, denoted as "Begin" and "End", respectively. The types of knowledge transfer, i.e., the "KD", "MMD", "L2", and "Prox", are shown along with the x-axis. The y-axis shows the λ. For better comparison, we also report the corresponding best accuracy of the compared methods in Fig. 4, which is listed in the "[]". On Cifar100-100-20, "MMD" could effectively accelerate the convergence speed at the beginning, and these

---

[2] https://www.tensorflow.org/tutorials/images/cnn.

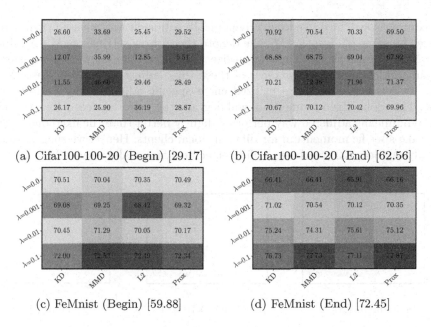

Fig. 7. Ablation studies on the types of knowledge transfer and the coefficient λ (Eq. 10). "Begin" means the average performance around the round $t = 200$; "End" means the final converged performance, averaged across the last 10 rounds' results. The number in "[]" denotes the best performance of the compared methods in Fig. 4.

types of knowledge transfer perform nearly equally when converged. On FeMnist, the knowledge transfer does not impact the personalization significantly, while the coefficient λ influences the results a lot. Anyway, these settings could almost nearly obtain better results than the compared methods, which verifies the superiorities of FedPHP again. Additionally, utilizing "MMD" to align feature distributions between the HPM and the newly downloaded model should be prioritized for a novel FL non-iid scene.

Table 2. Comparisons on various ways of adjusting the momentum, i.e., μ_t^k in Eq. 5.

	Begin			End		
	Mu-A	Mu-B	Mu-C	Mu-A	Mu-B	Mu-C
Cifar10-100-5	**80.35**	70.24	79.24	**91.04**	84.37	85.56
Cifar100-100-20	**46.60**	11.82	30.45	**72.46**	47.67	59.81
FeMnist	**72.53**	58.92	62.88	**77.73**	69.87	71.29

For the momentum in Eq. 5, we can take several settings. The first is the client-specific dynamically adjusted way as introduced in Sect. 3.2. Second, we could also take a global μ, i.e., setting $\mu_t^k = \mu, \forall t \in [1, T], k \in S_t$. We set $\mu = 0.9$

for comparison. Third, we take a dynamically adjusted way with all clients in S_t share the same $\mu_t^k = \mu_t = t/T$. We denote these three settings as Mu-A, Mu-B, and Mu-C, respectively. Then we utilize "MMD" as the knowledge transfer by default and list the performances of these three adjusting ways in Table 2. We can obviously find that the utilized client-specific adjusted way performs the best. The global fixed momentum will degrade the performances a lot. Although the dynamically adjusted way performs slightly better than the fixed one, it does not take specific momentum for different local clients. Hence, we suggest taking the Mu-A (Sect. 3.2) to obtain a client-specific dynamically adjusted momentum.

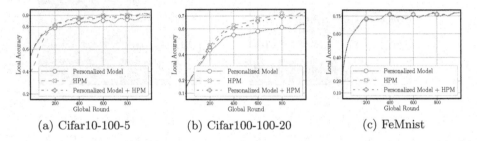

(a) Cifar10-100-5   (b) Cifar100-100-20   (c) FeMnist

Fig. 8. Comparisons of personalization accuracies with different inference methods on local clients.

The final question that we want to explore is the inference method with the HPM. For each client, we can only use the HPM to predict, i.e., the $\theta_{p,t}^k$, which is used in the above experiments by default. However, we could also use the personalized model to make predictions, i.e., $\hat{\theta}_{s,t}^k$. Also, we could ensemble their predictions via an averaging of their returned probability vectors. We show these performances as in Fig. 8. We can find that the performances of the personalized models could not reach the performance of the HPM on Cifar10-100-5 and Cifar100-100-20, while they could perform similarly on FeMnist. These result from two reasons. First, Cifar10-100-5 and Cifar100-100-20 take a larger client selection ratio $Q = 0.1$, where the HPM could be updated 100 times on average, while FeMnist takes $Q = 0.01$ and the HPM is updated not so frequently. Hence, the performance gap will be enlarged on Cifar10-100-5 and Cifar100-100-20. This may also be related to the natural property of these benchmarks. Anyway, we only take the predictions from the HPM in FedPHP.

5   Conclusion

Based on an empirical observation that the newly downloaded model in FedAvg could perform poorly and the hard-won personalized models in previous rounds are overwritten, we propose a novel concept named "inHerited Private Model" (HPM) to keep the historical valuable personalization knowledge in FL. Specifically, we take a moving average of personalized models on each client and exploit them to supervise the newly-downloaded model in the next global round. Our proposed FedPHP possesses the advantage of temporal ensembling, leading to better and stable performances. We advocate our work's main contribution is the exploitation of historical hard-won personalized models on local clients. We also explore various types of knowledge transfer and find that aligning feature distributions via MMD performs better. Searching for other advanced techniques to better exploit the hard-won personalized models are future works.

Acknowledgments. This research was partially supported by National Natural Science Foundation of China (Grant Nos. 61773198, 61632004 and 61921006), and NSFC-NRF Joint Research Project under Grant 61861146001, and Collaborative Innovation Center of Novel Software Technology and Industrialization. Thanks to Huawei Noah's Ark Lab NetMIND Research Team for funding this research. Professor De-Chuan Zhan is the corresponding author.

References

1. Arivazhagan, M.G., Aggarwal, V., Singh, A.K., Choudhary, S.: Federated learning with personalization layers. CoRR abs/1912.00818 (2019)
2. Caldas, S., et al.: LEAF: A benchmark for federated settings. CoRR abs/1812.01097 (2018)
3. Gretton, A., Borgwardt, K.M., Rasch, M.J., Schölkopf, B., Smola, A.J.: A kernel method for the two-sample-problem. In: Advances in Neural Information Processing Systems 19, pp. 513–520 (2006)
4. Hamer, J., Mohri, M., Suresh, A.T.: FedBoost: a communication-efficient algorithm for federated learning. In: Proceedings of the 37th International Conference on Machine Learning, pp. 3973–3983 (2020)
5. Hinton, G.E., Vinyals, O., Dean, J.: Distilling the knowledge in a neural network. CoRR abs/1503.02531 (2015)
6. Kairouz, P., et al.: Advances and open problems in federated learning. CoRR abs/1912.04977 (2019)
7. Karimireddy, S.P., Kale, S., Mohri, M., Reddi, S.J., Stich, S.U., Suresh, A.T.: SCAFFOLD: stochastic controlled averaging for federated learning. In: Proceedings of the 37th International Conference on Machine Learning, pp. 5132–5143 (2020)
8. Krizhevsky, A.: Learning multiple layers of features from tiny images (2012)
9. Kulkarni, V., Kulkarni, M., Pant, A.: Survey of personalization techniques for federated learning. CoRR abs/2003.08673 (2020)
10. Laine, S., Aila, T.: Temporal ensembling for semi-supervised learning. In: 5th International Conference on Learning Representations (2017)
11. Li, D., Wang, J.: FedMD: Heterogenous federated learning via model distillation. CoRR abs/1910.03581 (2019)
12. Li, T., Sahu, A.K., Zaheer, M., Sanjabi, M., Talwalkar, A., Smith, V.: Federated optimization in heterogeneous networks. In: Proceedings of Machine Learning and Systems (2020)
13. Li, X., Grandvalet, Y., Davoine, F.: Explicit inductive bias for transfer learning with convolutional networks. In: Proceedings of the 35th International Conference on Machine Learning, vol. 80, pp. 2830–2839 (2018)

14. Liang, P.P., Liu, T., Liu, Z., Salakhutdinov, R., Morency, L.: Think locally, act globally: Federated learning with local and global representations. CoRR abs/2001.01523 (2020)
15. Long, M., Cao, Y., Wang, J., Jordan, M.I.: Learning transferable features with deep adaptation networks. In: Proceedings of the 32nd International Conference on Machine Learning, vol. 37, pp. 97–105 (2015)
16. Mansour, Y., Mohri, M., Ro, J., Suresh, A.T.: Three approaches for personalization with applications to federated learning. CoRR abs/2002.10619 (2020)
17. McMahan, B., Moore, E., Ramage, D., Hampson, S., y Arcas, B.A.: Communication-efficient learning of deep networks from decentralized data. In: Proceedings of the 20th International Conference on Artificial Intelligence and Statistics, pp. 1273–1282 (2017)
18. Pan, S.J., Yang, Q.: A survey on transfer learning. IEEE Trans. Knowl. Data Eng. **22**(10), 1345–1359 (2010)
19. Peterson, D., Kanani, P., Marathe, V.J.: Private federated learning with domain adaptation. CoRR abs/1912.06733 (2019)
20. Shen, T., et al.: Federated mutual learning. CoRR abs/2006.16765 (2020)
21. Shoham, N., et al.: Overcoming forgetting in federated learning on non-iid data. CoRR abs/1910.07796 (2019)
22. Tarvainen, A., Valpola, H.: Mean teachers are better role models: weight-averaged consistency targets improve semi-supervised deep learning results. In: 5th International Conference on Learning Representations (2017)
23. Yang, Q., Liu, Y., Chen, T., Tong, Y.: Federated machine learning: concept and applications. ACM TIST **10**(2), 12:1–12:19 (2019)
24. Yao, X., Huang, C., Sun, L.: Two-stream federated learning: reduce the communication costs. In: IEEE Visual Communications and Image Processing, pp. 1–4 (2018)
25. Yao, X., Huang, T., Wu, C., Zhang, R., Sun, L.: Towards faster and better federated learning: a feature fusion approach. In: IEEE International Conference on Image Processing, pp. 175–179 (2019)
26. Yu, T., Bagdasaryan, E., Shmatikov, V.: Salvaging federated learning by local adaptation. CoRR abs/2002.04758 (2020)
27. Zhao, Y., Li, M., Lai, L., Suda, N., Civin, D., Chandra, V.: Federated learning with non-iid data. CoRR abs/1806.00582 (2018)

Rumour Detection via Zero-Shot Cross-Lingual Transfer Learning

Lin Tian[1], Xiuzhen Zhang[1(✉)], and Jey Han Lau[2]

[1] RMIT University, Melbourne, Australia
s3795533@student.rmit.edu.au, xiuzhen.zhang@rmit.edu.au
[2] The University of Melbourne, Melbourne, Australia

Abstract. Most rumour detection models for social media are designed
for one specific language (mostly English). There are over 40 languages
on Twitter and most languages lack annotated resources to build rumour
detection models. In this paper we propose a zero-shot cross-lingual
transfer learning framework that can adapt a rumour detection model
trained for a *source language* to another *target language*. Our framework utilises pretrained multilingual language models (e.g. multilingual
BERT) and a self-training loop to iteratively bootstrap the creation
of "silver labels" in the target language to adapt the model from the
source language to the target language. We evaluate our methodology on
English and Chinese rumour datasets and demonstrate that our model
substantially outperforms competitive benchmarks in *both* source and
target language rumour detection.

Keywords: Rumour detection · Cross-lingual transfer · Zero-shot

1 Introduction

Online social media platforms provide an alternative means for the general public
to access information. The ease of creating a social media account has the impli-
cation that rumours—*stories or statements with unverified truth value* [1]—can
be fabricated by users and spread quickly on the platform.

To combat misinformation on social media, one may rely on fact checking
websites such as `snopes.com` and `emergent.info` to dispel popular rumours.
Although manual evaluation is the most reliable way of identifying rumours, it
is time-consuming.

Automatic rumour detection is therefore desirable [14,36]. Content-based
methods focus on rumour detection using the textual content of messages
and user comments. Feature-based models exploit features other than text
content, such as author information and network propagation features, for
rumour detection. [19,20,22]. Most rumour detection models, however, are
built for English [29,30], and most annotated rumour datasets are also in
English [13,20,28].

Rumours can spread in different languages and across languages. Table 1
shows an example (untruthful) rumour about Bill Gates circulated on Twitter

© Springer Nature Switzerland AG 2021
N. Oliver et al. (Eds.): ECML PKDD 2021, LNAI 12975, pp. 603–618, 2021.
https://doi.org/10.1007/978-3-030-86486-6_37

Table 1. An illustration of a COVID-19 rumour being circulated in English, French and Italian on Twitter.

Date	Language	Tweet
04-02-2020	English	Bill Gates admits the vaccine will no doubt kill 700000 people. The virus so far has killed circa 300000 globally. Can anyone explain to me why you would take a vaccine that kills more people than the virus it's desgined to cure?
17-04-2020	French	et si bill gates etait le seul manipulateur de ce virus.. il veut moinsvde gens sur terre. veut vous vacciner et parle de pandemie depuis des années c est quand meme fou cette citation, non? #covid #BillGates
06-05-2020	Italian	Bill Gates:"la cosa più urgente nel mondo ora è il vaccino contro il Covid-19." I bambini africani che hanno ricevuto i vaccini di Bill Gates o sono morti o sono diventati epilettici. I vaccini di Bill Gates sono più pericolosi di qualsiasi coronavirus. #BillGates #Coronavirus

during the COVID-19 pandemic.[1] The rumour is found not only in English but also in Spanish and French.[2]

There are over 40 languages on Twitter[3] and most languages lack annotated resources for building rumour detection models. Although we have seen recent successes with deep learning based approaches for rumour detection [4,15,18,19, 29,39] most systems are monolingual and require annotated data to train a new model for a different language.

In this paper, we propose a zero-shot cross-lingual transfer learning framework for building a rumour detection system without requiring annotated data for a new language. Our system is cross-lingual in the sense that it can detect rumours in two languages based on one model. Our framework first fine-tunes a multilingual pretrained language model (e.g. multilingual BERT) for rumour detection using annotated data for a source language (e.g. English), and then uses it to classify rumours on another target language (zero-shot prediction) to create "silver" rumour labels for the target language. We then use these silver labels to fine-tune the multilingual model further to adapt it to the target language.

At its core, our framework is based on MultiFiT [8] which uses a multilingual model (LASER; [2]) to perform zero-shot cross-lingual transfer from one language to another. An important difference is that we additionally introduce a self-training loop—which iteratively refines the quality of the silver labels—that

---

[1] https://www.bbc.com/news/52847648.

[2] The following article clarifies several rumours surrounding Bill Gates: https://www.reuters.com/article/uk-factcheck-gates-idUSKBN2613CK.

[3] https://semiocast.com/downloads/Semiocast_Half_of_messages_on_Twitter_are_not_in_English_20100224.pdf.

can substantially improve rumour detection in the *target language*. Most interestingly, we also found that if we include the original gold labels in the source language in the self-training loop, detection performance in the *source language* can also be improved, creating a rumour detection system that excels in *both* source and target language detection.

To summarise, our contributions are: (1) we extend MultiFiT, a zero-shot cross-lingual transfer learning framework by introducing a self-training loop to build a cross-lingual model; and (2) we apply the proposed framework to the task of rumour detection, and found that our model substantially outperforms benchmark systems in both source and target language rumour detection.

2   Related Work

Rumour detection approaches can be divided into two major categories according to the types of data used: text-based and non-text based. Text-based methods focus on rumour detection using the textual content, which may include the original source document/message and user comments/replies. [18] proposed a recursive neural network model to detect rumours. Their model first clusters tweets by topics and then performs rumour detection at the topic level. [29] introduced linguistic features to represent writing styles and other features based on sensational headlines from Twitter and to detect misinformation. To detect rumours as early as possible, [39] incorporated reinforcement learning to dynamically decide how many responses are needed to classify a rumour. [30] explored the relationship between a source tweet and its comments by transferring stance prediction model to classify the veracity of a rumour. Non-text-based methods utilise features such as user profiles or propagation patterns for rumour detection [15, 19]. In this paper, we adopt the text-based approach to rumour detection.

Most studies on rumour detection focus on a specific social media platform or language (typically English). Still there are a few exceptions that explore cross-domain/multilingual rumour detection or related tasks. [31] proposed a set of 10 hand-crafted cross-lingual and cross-platform features for rumour detection by capturing the similarity and agreement between online posts from different social media platforms. [24] introduced a contrastive learning-based model for cross-lingual stance detection using memory networks. Different to these studies, we specifically focus on how to transfer learned knowledge from a source language to a target language for automatic rumour detection.

Transfer learning has been successfully applied to many natural language processing (NLP) tasks, where modern pretrained language models (e.g. BERT) are fine-tuned with annotated data for down-stream tasks [7, 16, 38]. Multilingual pretrained language models have also been explored. For example, BERT has a multilingual version trained using 104 languages of Wikipedia,[4] and [6] incorporate RoBERTa's training procedure for pretraining a multilingual language model that produces sentence embeddings for 100 languages. [25] found that multilingual BERT is surprisingly good at zero-shot cross-lingual transfer, i.e. it

---

[4] https://github.com/google-research/bert/blob/master/multilingual.md

can be fine-tuned for a particular task in one language and used to make predictions in another language without any further training. [8] proposed MultiFiT, a zero-shot cross-lingual transfer framework that uses predicted labels from a fine-tuned multilingual model to train a monolingual model on the same task in a target language; their transfer learning objective is only to optimise the model for the target language. Different from MultiFiT, our objective is to optimise models for both the target and source languages.

Self-training [27] is an early semi-supervised learning approach that has been explored for a variety of NLP tasks, such as neural machine translation [11], semantic segmentation [40]. Self-training involves teacher and student models, where the teacher model is trained with labelled data and then used to make predictions on unlabelled data to create more training data for training a student model. The process is repeated for several iterations with the student model replacing the original teacher model at the end of each iteration, and through iterative refinement of the predicted labels the student model improves over time. We apply self-training in a novel way to fine-tune pre-trained multilingual language models for cross-lingual rumour detection, and show that the student model improves over time during the transfer.

3   Methodology

We are interested in the task of rumour detection, and particularly how to do zero-shot cross-lingual transfer to build a multilingual rumour detection model. That is, we assume we have labelled rumours in one language (*source*) where we can build a supervised rumour detection model, and the goal is to transfer the model to detect rumours in a second language (*target*) without any labelled data in that second language. After transfer, it should have the ability to detect rumours in both languages (hence a multilingual model). We first describe the rumour classifier in Sect. 3.1, and return to detail the cross-lingual transfer learning framework in Sect. 3.2.

3.1   Rumour Classifier

We focus on binary rumour detection, and follow previous studies to use crowd comments to classify whether a microblog post constitutes a rumour or not [18, 30,39].

Given an initial post s_i and its reactions r_i,[5] we feed them to a pre-trained multilingual language model (we use multilingual BERT [7] and XML-RoBERTa [6] in our experiments) as:[6]

$$[CLS] + s_i + [SEP] + r_i + [SEP]$$

---

[5] Reactions are replies and quotes. r_i represents all reactions that can fit the maximum sequence length (384) for the pretrained model, concatenated together as a long string.

[6] For XLM-RoBERTa, we have 2 $[SEP]$ symbols between s_i and r_i, following https://huggingface.co/transformers/model_doc/xlmroberta.html#transformers. XLMRobertaTokenizer.build_inputs_with_special_tokens.

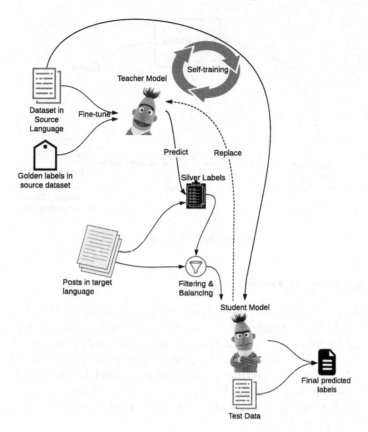

Fig. 1. Proposed cross-lingual transfer framework.

where $[CLS]$ and $[SEP]$ are special symbols used for classification and separating sequences [7].

We then take the contextual embedding of $[CLS]$ ($h_{[CLS]}$) and feed it to a fully-connected layer to perform binary classification of the rumour.

$$y_i = softmax\left(W_i h_{[CLS]} + b_s\right)$$

Given ground truth rumour labels, the model is fine-tuned with standard binary cross-entropy loss. All parameters are updated except for the word embeddings (rationale detailed in the following section).

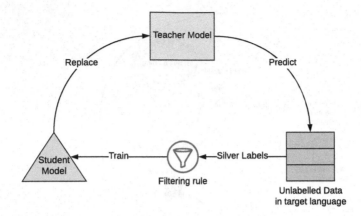

Fig. 2. Self-training loop.

3.2 Cross-Lingual Transfer

Our zero-shot cross-lingual transfer learning framework is based on MultiFiT [8]. MultiFiT works by first fine-tuning a multilingual model (e.g. LASER [2] is used in the original paper) for a task in a source language, and then applying it (zero-shot) to the same task in a target language to create silver labels. These silver labels are then used to fine-tune a monolingual model in the target language. MultiFiT is shown to substantially improve document classification compared to zero-shot predictions by a series of multilingual models trained using only gold labels in the source language.[7]

We present our zero-shot cross-lingual transfer learning framework in Fig. 1. One key addition that we make is a self-training loop that *iteratively refines the quality of the adapted model*. In the original MultiFiT framework, the teacher model is a multilingual model, and the student model is a monolingual model in the target language. As we are interested in multilingual rumour detection, the student model is a multilingual model in our case, although in our experiments (Sect. 4.3) we also present variations where the student model is a monolingual model.

Figure 2 illustrates the self-training loop. The student model is initialised using the teacher model (so both are multilingual models). Once the student is trained, the teacher model in the next iteration will be replaced by the student model.

To reduce noise in the silver labels, we introduce a filtering and balancing procedure in the self-training loop. The procedure was originally introduced to image classification and shown to improve performance [34]. With the filtering procedure, instances with prediction confidence/probability lower than a threshold p are filtered. The balancing procedure effectively drops some high

---

[7] Silver labels refer to the predicted labels in the target language, while gold labels refer to the real labels in the source language.

confidence instances that pass the threshold to ensure that an equal number of positive (rumour) and negative (non-rumour) instances.

Following [10], we perform adaptive pretraining on the teacher model before fine-tuning it for the rumour detection task. That is, we take the off-the-shelf pretrained multilingual model and further pretrain it using the masked language model objective on data in our rumour detection/social media domain. In terms of pretraining data we use both the unlabelled rumour detection data ("task adaptive") and externally crawled microblog posts ("domain adaptive") in the target language.

The degree of overlap in terms of vocabulary between the source and target language varies depending on the language pair. If the overlap is low, after fine-tuning the source language subword embeddings would have shifted while the target language subword embeddings remained the same (due to no updates), creating a synchronisation problem between the subword embeddings and intermediate layers. We solve this issue by freezing the subword embeddings when we first fine-tune the teacher model; subsequent fine-tuning in the self-training loop, however, updates all parameters. We present ablation tests to demonstrate the importance of doing this in Sect. 4.4.

Aiming for a model that performs well for both the target and source languages, we also introduce gold labels in the source language during self-training, i.e. we train the student using both the silver labels in the target language and the gold labels in the source language. This approach produces a well-balanced rumour detection model that performs well in both source and target languages, as we will see in Sect. 4.3.

4   Experiments and Results

We evaluated our cross-lingual transfer learning framework for rumour detection using three English and Chinese datasets. We formulate the problem as a binary classification task to distinguish rumours from non-rumours.

4.1   Datasets

Two English datasets Twitter15/16 [20] and PHEME [13], and one Chinese dataset WEIBO [18] were used in our experiments. PHEME and WEIBO have two class labels, rumour and non-rumour. For the Chinese WEIBO dataset, rumours are defined as "a set of known rumours from the Sina community management center (http://service.account.weibo.com), which reports various misinformation" [18]. The original Twitter15/16 dataset [20] has four classes, true rumour, false rumour, unverified rumour and non-rumour. We therefore extract tweets with labels "false rumour" and "non-rumour" from Twitter15/16 to match the definition of rumours and non-rumours of WEIBO and use the extracted data for experiments. Table 2 shows statistics of the experiment datasets.

Table 2. Rumour datasets.

	T15/16	PHEME	WEIBO
#initial posts	1,154	2,246	4,664
#all posts	182,535	29,387	3,805,656
#users	122,437	20,529	2,746,818
#rumours	575	1,123	2,313
#non-rumours	579	1,123	2,351
Avg. # of reactions	279	26	247
Max. # of reactions	3,145	289	2,313
Min. # of reactions	74	12	10

To ensure fair comparison of the performance across all models, for each dataset we reserved 20% data as test and we split the rest in a ratio of 4:1 for training and validation partitions. The validation set was used for hyperparameter tuning and early-stopping. For the PHEME dataset, to be consistent with the experiment set up in the literature, we followed the 5-fold split from [21]. For adaptive pretraining (Sect. 3.2), we used an external set of microblogs data for English (1.6M posts; [28]) and Chinese (39K posts).[8]

4.2 Experiment Setup

We used multilingual BERT [7] and XLM-RoBERTa [6] for the multilingual models, and implemented in PyTorch using the HuggingFace Libraries.[9]

For adaptive pretraining, we set batch size = 8. For the fine-tuning, we set batch size = 16, maximum token length = 384, and dropout rate = 0.1. Training epochs vary between 3–5 and learning rate in the range of $\{1e\text{-}5, 2e\text{-}5, 5e\text{-}5\}$; the best configuration is chosen based on the development data. We also tuned the number of self-training iterations and p, the threshold for filtering silver labels (Sect. 3.2), based on development.[10] All experiments were conducted using $1 \times$ V100 GPU.

4.3 Results

For our results, we show cross-lingual transfer performance from English to Chinese and vice versa. As we have two English datasets (T15/16 and PHEME) and one Chinese dataset (WEIBO), we have four sets of results in total: T15/16→ WEIBO, PHEME→WEIBO, WEIBO→T15/16 and WEIBO→PHEME. We evaluate rumour detection performance using accuracy, and present the

---

[8] https://archive.ics.uci.edu/ml/datasets/microblogPCU.
[9] https://github.com/huggingface.
[10] For p we search in the range of 0.94–0.96.

Table 3. Rumour detection results (Accuracy (%)) for English to Chinese transfer. Each result is an average over 3 runs, and subscript denotes standard deviation. monoBERT is a Chinese BERT model in this case. Boldfont indicates optimal zero-shot performance.

Model	T15/16 →	WEIBO	PHEME →	WEIBO
	Source	Target	Source	Target
Supervised				
multiBERT+source	$95.8_{0.1}$	—	$83.7_{0.5}$	—
multiBERT+target	—	$93.9_{0.2}$	—	$93.9_{0.2}$
multiBERT+both	$94.8_{0.1}$	$93.0_{0.3}$	$82.1_{0.8}$	$95.2_{0.2}$
XLMR+source	$96.3_{0.4}$	—	$82.8_{0.5}$	—
XLMR+target	—	$94.8_{0.1}$	—	$94.8_{0.1}$
XLMR+both	$95.5_{0.1}$	$92.2_{0.2}$	$85.8_{1.9}$	$95.4_{0.1}$
[18]+source	$83.5_{0.7}$	—	$80.8_{0.4}$	—
[15]+source	$85.4_{0.4}$	—	$64.5_{1.0}$	—
[30]+source	$87.2_{0.9}$	—	$86.7_{1.5}$	—
[4]+source	$96.3_{0.7}$	—	—	—
Zero-shot				
multiBERT	—	$64.3_{2.1}$	—	$65.9_{1.1}$
XLMR	—	$64.7_{1.1}$	—	$68.1_{1.0}$
MF [8]	—	$70.6_{0.4}$	—	$61.1_{1.0}$
MF-monoBERT	—	$67.5_{0.5}$	—	$68.2_{0.4}$
MF-monoBERT+ST	—	$\mathbf{81.3_{0.1}}$	—	$\mathbf{79.0_{0.2}}$
MF-multiBERT+ST	$61.3_{0.4}$	$78.6_{3.9}$	$66.3_{1.5}$	$72.6_{1.9}$
MF-multiBERT+ST+GL	$\mathbf{96.6_{0.2}}$	$78.3_{0.8}$	$83.0_{0.5}$	$74.3_{3.3}$
MF-XLMR+ST	$57.6_{1.0}$	$81.2_{0.1}$	$62.1_{1.3}$	$77.4_{0.3}$
MF-XLMR+ST+GL	$96.2_{0.1}$	$80.2_{0.2}$	$\mathbf{85.3_{0.7}}$	$77.2_{0.8}$

English→Chinese results in Table 3 and Chinese→English in Table 4 respectively. All performance is an average over 3 runs with different random seeds.

We include both supervised and zero-shot baselines in our experiments. For the supervised benchmarks, we trained multilingual BERT and XLM-RoBERTa using: (1) source labels; (2) target labels; and (3) both source and target labels. The next set of supervised models are state-of-the-art monolingual rumour detection models: (1) [18] is a neural model that processes the initial post and crowd comments with a 2-layer gated recurrent units; (2) [15] uses recurrent and convolutional networks to model user metadata (e.g. followers count) in the crowd responses;[11] (3) [30] uses BERT to encode comments (like our model) but it is pre-trained with stance annotations; and (4) [4] uses bidirectional graph convolu-

---

[11] Following the original paper, only a maximum of 100 users are included.

Table 4. Rumour detection results (Accuracy (%)) for Chinese to English transfer. Each result is an average over 3 runs, and subscript denotes standard deviation. monoBERT is an English BERT model here.

Model	WEIBO → T15/16		WEIBO → PHEME	
	Source	Target	Source	Target
Supervised				
multiBERT+source	$93.9_{0.1}$	—	$93.9_{0.2}$	—
multiBERT+target	—	$95.8_{0.1}$	—	$83.7_{0.5}$
multiBERT+both	$93.0_{0.3}$	$94.8_{0.1}$	$95.2_{0.2}$	$82.1_{0.8}$
XLMR+source	$94.8_{0.1}$	—	$94.8_{0.1}$	—
XLMR+target	—	$96.3_{0.4}$	—	$82.8_{0.5}$
XLMR+both	$92.2_{0.2}$	$95.5_{0.1}$	$95.4_{0.1}$	$85.8_{1.9}$
[18]+source	$91.0_{0.1}$	—	$91.0_{0.1}$	—
[15]+source	$92.1_{0.2}$	—	$92.1_{0.2}$	—
[4]+source	$96.1_{0.4}$	—	$96.1_{0.4}$	—
Zero-shot				
multiBERT	—	$60.8_{1.3}$	—	$67.2_{0.4}$
XLMR	—	$73.9_{0.8}$	—	$69.0_{1.5}$
MF [8]	—	$73.4_{1.4}$	—	$64.1_{2.0}$
MF-monoBERT	—	$64.7_{0.1}$	—	$70.7_{0.7}$
MF-monoBERT+ST	—	$\mathbf{85.7_{0.4}}$	—	$\mathbf{78.9_{0.6}}$
MF-multiBERT+ST	$55.1_{1.8}$	$82.2_{1.2}$	$66.0_{1.0}$	$72.6_{1.5}$
MF-multiBERT+ST+GL	$97.0_{0.1}$	$80.9_{1.5}$	$95.8_{0.5}$	$73.4_{0.4}$
MF-XLMR+ST	$52.4_{0.3}$	$83.0_{1.0}$	$62.7_{0.5}$	$75.4_{0.4}$
MF-XLMR+ST+GL	$\mathbf{97.6_{0.1}}$	$81.3_{0.1}$	$\mathbf{95.9_{0.5}}$	$77.9_{1.1}$

tional networks to model crowd responses in the propagation path. Note that we only have English results (T15/16 and PHEME) for [30] as it uses stance annotations from SemEval-2016 [23], and only T15/16 and WEIBO results for [4] as PHEME does not have the propagation network structure. To get user metadata for [15], we crawled user profiles via the Twitter API.[12]

For the zero-shot baselines, multilingual BERT and XLM-RoBERTa were trained using the source labels and applied to the target language (zero-shot predictions); subword embeddings are frozen during fine-tuning for these zero-shot models. We also include the original MultiFiT model [8], which uses LASER [2] as the multilingual model (teacher) and a pretrained quasi-recurrent neural network language model [5] as the monolingual model (student).[13]

---

[12] https://developer.twitter.com/en/docs/twitter-api/v1.
[13] The monolingual student model is pretrained using Wikipedia in the target language.

Table 5. Rumour detection results (F_1 score (%)) for both the source and target languages. "R" and "NR" denote the rumour and non-rumour classes respectively.

	T15/16 →	WEIBO	PHEME →	WEIBO	WEIBO →	T15/16	WEIBO →	PHEME
	Source	Target	Source	Target	Source	Target	Source	Target
MF-multiBERT+ST+GL								
R	96.6	79.1	84.2	75.5	96.9	83.3	94.3	75.2
NR	96.1	77.2	79.5	73.1	97.0	76.8	94.4	70.3
MF-XLMR+ST+GL								
R	96.6	83.1	85.2	81.6	97.4	82.9	96.9	74.4
NR	95.6	76.1	85.7	74.5	96.8	81.1	95.0	81.4

We first look at the supervised results. XLM-RoBERTa ("XLMR") is generally better (marginally) than multilingual BERT ("multiBERT"). In comparison, for the monolingual rumour detection models, [4] has the best performance overall (which uses network structure in addition to crowd comments), although XLM-RoBERTa and multilingual BERT are not far behind.

Next we look at the zero-shot results. Here we first focus on target performance and baseline models. The zero-shot models ("multiBERT" and XLMR") outperform the MultiFiT baseline ("MF") in 2–3 out of 4 cases, challenging the original findings in [8]. When we replace the teacher model with multilingual BERT and the student model with monolingual BERT ("MF-monoBERT"), we found mixed results compared to MultiFiT ("MF"): 2 cases improve but the other 2 worsen. When we incorporate the self-training loop ("MF-monoBERT+ST"), however, we see marked improvement in all cases—the largest improvement is seen in WEIBO→T15/16 (Chinese to English, Table 4), from 64.7% to 85.7%—demonstrating the benefits of iteratively refining the transferred model. These results set a new state-of-the-art for zero-shot cross-lingual transfer learning for our English and Chinese rumour detection datasets. That said, there is still a significant gap (10+ accuracy points) compared to supervised models, but as we see in Sect. 4.6 the gap diminishes quickly as we introduce some ground truth labels in the target domain.

We now discuss the results when we use a multilingual model for the student model, i.e. replacing it with either multilingual BERT ("MF-multiBERT+ST") or XLM-RoBERTa ("MF-XLMR+ST"), which turns it into a *multilingual* rumour detection system (i.e. after fine-tuned it can detect rumours in both source and target language). Similar to the supervised results, we see that the latter ("MF-XLMR+ST") is a generally better multilingual model. Comparing our best multilingual student model ("MF-XLMR+ST") to the monolingual student model ("MF-monoBERT+ST") we see only a small drop in the target performance (about 1–4 accuracy points depending on domain), demonstrating that the multilingual rumour detection system is competitive to the monolingual detection system in the target language.

For the source performance, we see a substantial drop (20–40 accuracy points) after cross-lingual transfer (e.g. "XLMR+source" vs. "MF-XLMR+ST"), implying there is catastrophic forgetting [9,12,32,35]—the phenomenon where adapted neural models "forget" and perform poorly in the original domain/task. When we incorporate gold labels in the source domain in the self-training loop ("MF-multiBERT+ST+GL" or "MF-XLMR+ST+GL"), we found a surprising observation: not only was catastrophic forgetting overcame, but the source performance actually surpasses some supervised monolingual models, e.g. "MF-multiBERT+ST+GL" and "MF-XLMR+ST+GL" outperform [4] in T15/16 (96.6% vs. 96.3%) and WEIBO (97.6% vs. 96.1%) respectively, creating a new state-of-the-art for rumour detection in these two domains. One explanation is that the transfer learning framework maybe functioning like a unique data augmentation technique that creates additional data in a different language (unique in the sense it works only for improving multilingual models). Note that incorporating the gold labels generally does not hurt the *target performance* – e.g. comparing "MF-XLMR+ST" with "MF-XLMR+ST+GL" we see a marginal dip in 2 cases, but in 2 other cases we see similar or improved performance—which shows that this is an effective approach for building multilingual models.

We further examine class-specific performance of our best models. The F_1 scores of rumour and non-rumour classes are presented in Table 5. For this binary classification task with relatively balanced class distributions, not surprisingly we observe that our models have reasonably good performance in both the rumour and non-rumour classes; lowest F_1 score is 70.3% of MF-multiBERT+ST+GL (Chinese to English transfer) for non-rumours in PHEME. That said, performance of the rumour class is generally better than that of the non-rumour class in both the source and target languages (the only exception is Chinese to English transfer on PHEME).

4.4 Adaptive Pretraining and Layer Freezing

To understand the impact of adaptive pretraining and layer freezing, we display zero-shot multilingual BERT results (test set) in Table 6. We can see that there are clear benefits for adaptive pretraining (top-3 vs. bottom-3 rows). For layer freezing, we have 3 options: no freezing ("∅"), only freezing the subword embeddings ("*") and freezing the first 3 layers ("**"). The second option (subword embedding frozen) consistently produces the best results (irrespective of whether adaptive pretraining is used), showing that this approach is effective in tackling the synchronisation issue (Sect. 3.2) that arises when we fine-tune a multilingual model on one language.

Table 6. Influence of adaptive pretraining ("Ad. Pt.") and layer freezing ("Frz.") for results (Accuracy (%)). "∅" denotes no freezing of any layers; "*" freezing the subword embedding layer; and "**" freezing the first 3 layers.

Ad. Pt.	Frz.	T15/16→ WEIBO	PHEME→ WEIBO	WEIBO→ T15/T16	WEIBO→ PHEME
N	∅	53.2	58.4	50.1	54.5
	*	61.3	60.0	52.6	63.8
	**	57.5	60.9	52.9	58.5
Y	∅	56.6	61.8	50.9	61.6
	*	**64.3**	**65.9**	**60.8**	**67.2**
	**	60.3	63.8	55.0	61.8

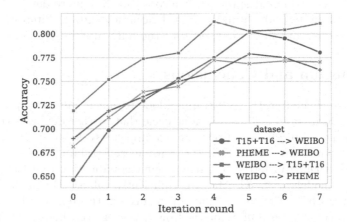

Fig. 3. Accuracy over iteration during self-training.

4.5 Self-training

To measure the influence of the self-training loop, we present target performance (test set) of our multilingual model ("MF-XLMR+ST+GL") over different iterations in the self-training loop in Fig. 3. We can see the performance improves rapidly in the first few iterations, and gradually converges after 4–7 iterations. These results reveal the importance of refining the model over multiple iterations during cross-lingual transfer.

4.6 Semi-supervised Learning

Here we explore feeding a proportion of ground truth labels in the target domain to our zero-shot model ("MF-XLMR+ST+GL") and compare it to supervised multilingual model ("XLMR+both"). We present T15/16→WEIBO results (test set) in Table 7. We can see that the gap shrinks by more than half (12.0 to 5.9 accuracy difference) with just 20% ground truth target label. In general our

Table 7. T15/16→WEIBO results (Accuracy(%)) as we incorporate more ground truth target labels ("GT Label").

% GT Label	Supervised	Zero-shot
0%	—	80.2
20%	79.8	86.3
40%	83.3	89.2
60%	89.3	92.9
80%	91.0	93.5
100%	92.2	—

unsupervised cross-lingual approach is also about 20% more data efficient (e.g. supervised accuracy@40% ≈ unsupervised accuracy@20%). Interestingly, with 60% ground truth our model outperforms the fully supervised model.

5   Discussion and Conclusions

One criticism of the iterative self-training loop is that it suffers from poor initial prediction which could lead to a vicious cycle that further degrades the student model. The poor initial predictions concern appears to less of a problem in our task, as the pure zero-shot models (i.e. without self-training) appear to do reasonably well when transferred to a new language, indicating that the pretrained multilingual models (e.g. XLMR) are sufficiently robust. By further injecting the gold labels from the source domain during self-training, we hypothesise it could also serve as a form of regularisation to prevent continuous degradation if the initial predictions were poor. Also, although our proposed transfer learning framework has only been applied to multilingual rumour detection, the architecture of the framework is general and applicable to other tasks.

To conclude, we propose a zero-shot cross-lingual transfer learning framework to build a multilingual rumour detection model using only labels from one language. Our framework introduces: (1) a novel self-training loop that iteratively refines the multilingual model; and (2) ground truth labels in the source language during cross-lingual transfer. Our zero-shot multilingual model produces strong rumour detection performance in both source and target language.

Acknowledgments. This research is supported in part by the Australian Research Council Discovery Project DP200101441.

References

1. Allport, G.W., Postman, L.: The psychology of rumor (1947)
2. Artetxe, M., Schwenk, H.: Massively multilingual sentence embeddings for zero-shot cross-lingual transfer and beyond. Trans. Assoc. Comput. Linguist. **7**, 597–610 (2019)

3. Bengio, Y.: Deep learning of representations for unsupervised and transfer learning. In: Proceedings of ICML Workshop on Unsupervised and Transfer Learning (2012)
4. Bian, T., et al.: Rumor detection on social media with bi-directional graph convolutional networks. In: AAAI (2020)
5. Bradbury, J., Merity, S., Xiong, C., Socher, R.: Quasi-recurrent neural networks (2016)
6. Conneau, A., et al.: Unsupervised cross-lingual representation learning at scale (2019)
7. Devlin, J., Chang, M.W., Lee, K., Toutanova, K.: BERT: pre-training of deep bidirectional transformers for language understanding. In: NAACL (2019)
8. Eisenschlos, J., Ruder, S., Czapla, P., Kardas, M., Gugger, S., Howard, J.: MultiFiT: efficient multi-lingual language model fine-tuning (2019)
9. French, R.M.: Catastrophic forgetting in connectionist networks. Trends Cogn. Sci. **3**(4), 128–135 (1999)
10. Gururangan, S., et al.: Don't stop pretraining: adapt language models to domains and tasks. In: ACL (2020)
11. He, J., Gu, J., Shen, J., Ranzato, M.: Revisiting self-training for neural sequence generation (2019)
12. Kirkpatrick, J., et al.: Overcoming catastrophic forgetting in neural networks. Proc. Natl. Acad. Sci. **114**(13), 3521–3526 (2017)
13. Kochkina, E., Liakata, M., Zubiaga, A.: All-in-one: multi-task learning for rumour verification. In: COLING (2018)
14. Liu, X., Nourbakhsh, A., Li, Q., Fang, R., Shah, S.: Real-time rumor debunking on Twitter. In: CIKM (2015)
15. Liu, Y., Wu, Y.F.B.: Early detection of fake news on social media through propagation path classification with recurrent and convolutional networks. In: AAAI (2018)
16. Liu, Y., et al.: Roberta: a robustly optimized bert pretraining approach (2019)
17. Long, Y., Lu, Q., Xiang, R., Li, M., Huang, C.R.: Fake news detection through multi-perspective speaker profiles. In: IJCNLP (2017)
18. Ma, J., et al.: Detecting rumors from microblogs with recurrent neural networks. In: IJCAI (2016)
19. Ma, J., Gao, W., Wei, Z., Lu, Y., Wong, K.F.: Detect rumors using time series of social context information on microblogging websites. In: CIKM (2015)
20. Ma, J., Gao, W., Wong, K.F.: Detect rumors in microblog posts using propagation structure via kernel learning. In: ACL (2017)
21. Ma, J., Gao, W., Wong, K.F.: Detect rumors on Twitter by promoting information campaigns with generative adversarial learning. In: WWW (2019)
22. Mendoza, M., Poblete, B., Castillo, C.: Twitter under crisis: can we trust what we rt? In: Proceedings of the first workshop on social media analytics (2010)
23. Mohammad, S., Kiritchenko, S., Sobhani, P., Zhu, X., Cherry, C.: Semeval-2016 task 6: detecting stance in tweets. In: Proceedings of the 10th International Workshop on Semantic Evaluation (SemEval-2016) (2016)
24. Mohtarami, M., Glass, J., Nakov, P.: Contrastive language adaptation for cross-lingual stance detection. In: EMNLP-IJCNLP (2019)
25. Pires, T., Schlinger, E., Garrette, D.: How multilingual is multilingual bert? (2019)
26. Ren, Y., Zhang, J.: HGAT: Hierarchical Graph Attention Network for Fake News Detection (2020)
27. Scudder, H.: Probability of error of some adaptive pattern-recognition machines. IEEE Trans. Inf. Theory **11**(3), 363–371 (1965)

28. Shu, K., Mahudeswaran, D., Wang, S., Lee, D., Liu, H.: Fakenewsnet: a data repository with news content, social context and dynamic information for studying fake news on social media. Big Data **8**(3), 171–188 (2020)

29. Shu, K., Sliva, A., Wang, S., Tang, J., Liu, H.: Fake news detection on social media: a data mining perspective. ACM SIGKDD Explor. Newsl. **19**(1), 22–36 (2017)

30. Tian, L., Zhang, X., Wang, Y., Liu, H.: Early detection of Rumours on Twitter via stance transfer learning. In: ECIR (2020)

31. Wen, W., Su, S., Yu, Z.: Cross-lingual cross-platform rumor verification pivoting on multimedia content. In: EMNLP (2018)

32. Wiese, G., Weissenborn, D., Neves, M.: Neural domain adaptation for biomedical question answering (2017)

33. Wu, L., Liu, H.: Tracing fake-news footprints: characterizing social media messages by how they propagate. In: WSDM (2018)

34. Xie, Q., Luong, M.T., Hovy, E., Le, Q.V.: Self-training with noisy student improves imagenet classification. In: CVPR (2020)

35. Xu, Y., Zhong, X., Yepes, A.J.J., Lau, J.H.: Forget me not: reducing catastrophic forgetting for domain adaptation in reading comprehension. In: IJCNN (2020)

36. Yang, F., Liu, Y., Yu, X., Yang, M.: Automatic detection of rumor on sina weibo. In: Proceedings of the ACM SIGKDD Workshop on Mining Data Semantics (2012)

37. Yang, S., Shu, K., Wang, S., Gu, R., Wu, F., Liu, H.: Unsupervised fake news detection on social media: a generative approach. In: AAAI (2019)

38. Yin, D., Meng, T., Chang, K.W.: Sentibert: a transferable transformer-based architecture for compositional sentiment semantics. In: ACL (2020)

39. Zhou, K., Shu, C., Li, B., Lau, J.H.: Early rumour detection. In: NAACL (2019)

40. Zou, Y., Yu, Z., Vijaya Kumar, B., Wang, J.: Unsupervised domain adaptation for semantic segmentation via class-balanced self-training. In: ECCV (2018)

Continual Learning with Dual Regularizations

Xuejun Han[1]($\boxtimes$) and Yuhong Guo[1,2]($\boxtimes$)

[1] Carleton University, Ottawa, Canada
xuejunhan@cmail.carleton.ca, yuhong.guo@carleton.ca
[2] Canada CIFAR AI Chair, Amii, Edmonton, Canada

Abstract. Continual learning (CL) has received a great amount of attention in recent years and a multitude of continual learning approaches arose. In this paper, we propose a continual learning approach with dual regularizations to alleviate the well-known issue of catastrophic forgetting in a challenging continual learning scenario – domain incremental learning. We reserve a buffer of past examples, dubbed memory set, to retain some information about previous tasks. The key idea is to regularize the learned representation space as well as the model outputs by utilizing the memory set based on interleaving the memory examples into the current training process. We verify our approach on four CL dataset benchmarks. Our experimental results demonstrate that the proposed approach is consistently superior to the compared methods on all benchmarks, especially in the case of small buffer size.

Keywords: Continual learning · Representation regularization · Functional regularization

1 Introduction

Ideally, the intelligent system should be capable of coping with and adapting to continually changing environments like humans. Unfortunately, most of existing machine learning algorithms are not provided with such ability. To address this shortcoming, a problem setting known as *continual learning* [18] or *lifelong learning* [22] came into being in recent years, prompting machine learners to behave more like humans by fast acquiring new knowledge without forgetting what has been learned in the past. In this setting, the learner is presented with a sequence of similar or dissimilar tasks and the goal is to train a learner to work well on all seen tasks.

Generally, continual learning can be split into three scenarios: *task incremental learning*, *domain incremental learning* and *class incremental learning* [7,24]. In *task incremental learning*, the task identities are always available and hence it admits model architectures with task specific components such as a multi-head output layer. In contrast, the task identities are unknown at test time in *incremental domain learning*. Single-head networks are typically exploited for

© Springer Nature Switzerland AG 2021
N. Oliver et al. (Eds.): ECML PKDD 2021, LNAI 12975, pp. 619–634, 2021.
https://doi.org/10.1007/978-3-030-86486-6_38

this scenario and the output layer stays the same with exposure to new tasks. Different from the above two scenarios, *class incremental learning* deals with the circumstance where each task in the sequence only contains a subset of classes and new classes progressively emerge with new tasks arriving. In the case that the output spaces of sequential tasks can be treated as an identical setting, e.g., all binary tasks, class incremental learning has been viewed as a special case of domain incremental learning, where the domain shifts between different tasks are relatively substantial and the task identities are no long available.

The most critical issue in continual learning nevertheless is the *catastrophic forgetting* [5,11] for previous learned tasks when the model is retrained for new ones. To address this problem, many continual learning methods have been developed. Some methods overcome the forgetting problem by regularizing the model parameters or outputs [1,2,8,9,26], which is known as *regularization-based methods*. Some methods resort to a memory set consisting of previous seen examples to preserve certain knowledge about past tasks [3,4,10,13–15,17,19], which is called *memory-based methods*. These methods typically keep a constant network architecture during the overall learning phase. By contrast, *model-based methods* propose to revise the network architecture by dynamically adding new neurons for new tasks or revising some specific neurons for early tasks. However, such kind of methods may result in a network with excessive size as the number of tasks increases.

Among the exiting regularization-based and memory-based approaches, it is worth noting that some regularization-based models such as [8] can completely fail even on the simple Split MNIST dataset in the domain incremental learning scenario, while many memory-based models can work extremely well in the task incremental learning scenario but perform poorly in the domain incremental learning setting. Some memory-based models such as [10] behave well in both settings but are computational inefficient.

In this paper, we propose a novel continual learning approach for the domain incremental learning setting, which can be viewed as a hybrid of regularization and memory based methods and does not require much computational and memory cost. The key idea is to regularize the model outputs as well as the learned representation space by use of a set of past examples, known as memory set. First, we interleave memory examples with current data throughout model training to learn shared feature representations. On this basis, we enforce the model outputs on both current and memory data to be close to previous ones via knowledge distillation so as to preserve information about previous model and thereby overcome forgetting. Meanwhile, feature selection is implemented in the learned representation space to further minimize the domain discrepancy of current and memory datasets. Finally, mixup of current and memory data on the representation level is exploited to yield a beneficial effect on the generalization performance. We verify our approach on four dataset benchmarks – Split MNIST, Permuted MNIST, Split CIFAR-10 and Split CIFAR-100 and the proposed approach consistently outperforms the compared methods on all the benchmarks. Notably, our approach is empirically much more effective when the size of memory set is small.

2   Related Work

Generally speaking, continual learning methods can be grouped into three categories: regularization-based methods, memory-based methods and model-based methods. Our approach is a combination of regularization-based and memory-based methods. We hence will briefly review prior work for these two categories.

Regularization-based Methods. To alleviate the issue of catastrophic forgetting, the regularization-based methods equip the loss function with a regularization, either on weights or functions outputs, to protect the model from unwanted changes. *Elastic Weight Consolidation (EWC)* [8] penalizes the parameter changes in terms of the importance of parameters for old tasks measured by Fisher information matrix. In a similar way, *Synaptic Intelligence (SI)* [26] proposed a regularization penalty but in an online manner and along the entire training trajectory. *Online EWC* [20] is a modified variant of *EWC* by treating all old tasks equivalently which are able to gracefully forget old tasks when the model is out of capacity. Based on *Online EWC, Riemmanian Walk (RWalk)* [2] further replaces the Euclidean distance between parameters by KL divergence, i.e. distance in the Riemannian manifold. *Memory Aware Synapses (MAS)* [1] measures the importance of parameters based on the sensitivity to model predictions in an unsupervised and online setting. In addition to weight-regularization approaches, there also exists a group of functional-regularization methods, such as *Learning without Forgetting (LwF)* [9] which enforces the model outputs to be close to the previous ones by use of *knowledge distillation*.

Memory-based Methods. The memory-based continual learning methods can be divided into two types based on whether interleaving memory examples with current data in the model training. *Gradient Episodic Memory (GEM)* [10] does not absorb the memory set into the training data. Specifically, It incorporates the memory set into the objective as $t-1$ inequality constraints where $t-1$ is the number of seen tasks, such that the loss on previous tasks cannot increase. *Averaged GEM (A-GEM)* [3] is a simplified version of *GEM* by combining $t-1$ task-specific constraints into one constraint for all old tasks together. In contrast, the manner that the memory data participates directly in the model training is named as *Experience Replay (ER)* [15,19], which we believe makes the best use of all data on hand. [4] gave a comprehensive study for *experience replay* and evaluated a group of different memory selection strategies. Based on ER, many CL techniques were put forward. *Variational Continual Learning (VCL)* [12] extends online variational inference to deal with continual learning tasks. *Meta-Experience Replay (MER)* [17] combines experience replay with the optimization-based meta-learning. *Dark Experience Replay (DER)* [14] encourages the model outputs to mimic the previous ones on memory examples by minimizing their KL divergence. *Functional Regularization of Memorable Past (FROMP)* [13] regularizes the model outputs at a few memory examples with a Gaussian process formulation of deep neural networks. Moreover, instead of reserving a subset from previous datasets directly, *Deep Generative Replay*

(DGR) [21] and *Replay through Feedback (RtF)* [23] train generative models on old tasks and rehearsal pseudo-examples during new task training. However, it is difficult to make them work well on complicated datasets.

In a nutshell, the proposed approach falls into the category of a hybrid of regularization-based and memory-based CL methods, and is expected to integrate strengths of both.

3   Methodology

3.1   Problem Formulation

Continual learning copes with a sequence of similar or dissimilar tasks and the model is optimized on one task at a time without accessing to previous data in general. The goal is to train a model that works well on all tasks, which naturally involves a severe problem of catastrophic forgetting. In practice, a small buffer is realizable to preserve some information about old tasks, which is crucial to alleviate the forgetting problem. Formally, let the task index be $t \in \{1, \ldots, T\}$, with corresponding dataset $\mathcal{D}_t$. In addition, we keep a memory set $\mathcal{M}$ of size m consisting of examples from previous tasks, and allocate $m/(t-1)$ memories to each of the $t-1$ previous tasks. By default we assume $m \geq T$. Furthermore, to make the notation of memory set more explicitly, we denote the memory set used during task t by $\mathcal{M}_{t-1}$, which contains $t-1$ subsets, specifically, $\mathcal{M}_{t-1} = \bigcup_{i=1}^{t-1} \mathcal{M}_{t-1}^i$.

When the task t arrives, the model f_θ attempts to minimize the loss over current data $\mathcal{D}_t$. Out of sample efficiency, we also incorporate memory data $\mathcal{M}_{t-1}$ into the training dataset. Thus, the loss function is defined as follows,

$$\mathcal{L}_{CE} = \mathbb{E}_{(\mathbf{x},y) \sim \mathcal{D}_t \bigcup \mathcal{M}_{t-1}} l(f_\theta(\mathbf{x}), y) \tag{1}$$

where l is the classification loss function, in particular the cross-entropy loss.

It is worth noticing that the memory set is equally allocated to each seen tasks, so as to ensure there is as few as one example for each previous task. Notably, the joint training on current data and stored previous data falls into the scope of *experience repay* [15,19], which was comprehensively explored by [4] and demonstrated to be a very simple and strong CL baseline.

3.2   Proposed Approach

The key idea of the proposed approach is to regularize the model outputs as well as the learned representation space by use of the memory set through knowledge distillation and feature selection, respectively. Moreover, the representation regularization and the memory set usage are further strengthened by deploying a manifold mixup component. In this section, we provide a progressive and comprehensible exposition of the proposed approach.

Knowledge Distillation. The phenomenon of catastrophic forgetting usually occurs when the model changes drastically after retraining for new tasks. For this reason, preventing the model from dramatic drifts is one of coping strategies, which was investigated in [8,26] by regularizing the model parameters. Besides, a more straightforward and effective alternative is to directly regularize the model outputs [9,13,14,16], since what ultimately matters is model predictions rather than learned representations. To achieve this goal, it is effective to make the outputs of new model to be close to the ones of previous model by means of knowledge distillation.

We use $f_{\theta_{t-1}}$ to denote the prediction model the model after learning tasks $\{1 \ldots, t-1\}$. To preserve the knowledge about these previous tasks, a knowledge distillation loss can be incorporated into the objective of the classification loss:

$$\mathcal{L}_{KD} = \mathbb{E}_{(\mathbf{x},y) \sim \mathcal{D}_t \bigcup \mathcal{M}_{t-1}} l(f'_\theta(\mathbf{x}), f'_{\theta_{t-1}}(\mathbf{x})) \tag{2}$$

where l denotes the cross-entropy loss, $f'_\theta(\mathbf{x})$ and $f'_{\theta_{t-1}}(\mathbf{x})$ are modified versions of current and previous model predictions, respectively. Explicitly, let $h_\theta(\mathbf{x})$ and $h_{\theta_{t-1}}(\mathbf{x})$ be pre-softmax outputs (i.e. logits) of of $f_\theta(\mathbf{x})$ and $f_{\theta_{t-1}}(\mathbf{x})$ on example $\mathbf{x}$, then

$$f'_\theta(\mathbf{x}) = \text{softmax}(h_\theta(\mathbf{x})/\mathcal{T}) \tag{3}$$

where $\mathcal{T}$ is a temperature parameter to adjust soft target distributions. Same definition applies for $f'_{\theta_{t-1}}(\mathbf{x})$.

We empirically found that letting $\mathcal{T} = 2$ is slightly better than just leaving $\mathcal{T} = 1$, therefore, we will use $\mathcal{T} = 2$ during experiments in this paper, which aligns with the choices of [6,9]. It is worth noting that, different from [9], we adopted a memory set and encouraged the similar predictions of current and old models not only on current data but also on memory examples. Moreover, [9] pointed out the similar performance between the cross-entropy loss and other reasonable loss functions for knowledge distillation, as long as trying to force the outputs of previous model to stay steady.

Feature Selection via Sparse Regularization. The performance deterioration of old tasks arising from model retaining for new tasks is mainly caused by the domain drifts between the new and old tasks. By interleaving the memory data with current data in the training stage, the learned representation space are to some extent shared across the current and previous tasks. Ideally, the distribution gap between current and previous tasks is expected to be eliminated in the shared representation space. However, such aspiration is strenuous to realize especially when the memory set is small. Therefore, an elaborately filtration among features in the representation space is worthy of consideration to alleviate such issue. On this account, we incorporate a regularization on the classifier to implicitly select features that behave similarly in both current and previous tasks, so that in the selected feature space, the discrepancy between the current and previous task domains is able to be further diminished.

To begin with, let us rephrase the model parameters $\theta = [\theta^G, \theta^F]$, where θ^G and θ^F are parameters for the feature extractor G and the classifier F, respectively. The dimension of the feature space is d.

To realize feature selection, we first resort to a statistic – Pearson correlation coefficient (PCC) – that measures the correlation between two variables such as X and Y. The definition is as follows,

$$\rho_{X,Y} = \frac{\mathbb{E}[(X - \mu_X)(Y - \mu_Y)]}{\sigma_X \sigma_Y}$$

It is worth noticing that $\rho_{X,Y} \in [-1, 1]$ with a value of 1 if X and Y are perfectly correlated and -1 otherwise.

Here we aim to capture the correlation between features and the class labels via PCC, which can reflect the prediction power of the corresponding features. Specifically, we denote the PCC between the prediction of the classifier based on the feature ψ of datapoint $\mathbf{x}$ and the label y for all pairs $(\mathbf{x}, y)$ from $\mathcal{D}_t$ as $\rho_{\mathcal{D}_t}(F_\psi(G(\mathbf{x})), y)$, where F_ψ uses only the parameters related to feature ψ. The larger the value of $\rho_{\mathcal{D}_t}(F_\psi(G(\mathbf{x})), y)$, the greater the compatibility degree of the prediction based on the feature ψ and the label. Similarly, let us denote the PCC for all pairs in $\mathcal{M}_{t-1}$ as $\rho_{\mathcal{M}_{t-1}}(F_\psi(G(\mathbf{x})), y)$. Ideally, we desire to select features that are similarly predictive in both the current and memory datasets. For this purpose, let us define the *compatibility* of the prediction based on the feature ψ and the label in two datasets as the product $\rho_{\mathcal{D}_t}(F_\psi(G(\mathbf{x})), y) \cdot \rho_{\mathcal{M}_{t-1}}(F_\psi(G(\mathbf{x})), y)$. while the *incompatibility* is

$$\Delta(\psi) = 1 - \rho_{\mathcal{D}_t}(F_\psi(G(\mathbf{x})), y)\rho_{\mathcal{M}_{t-1}}(F_\psi(G(\mathbf{x})), y) \qquad (4)$$

which represents the undesirableness of a specific feature ψ for bridging the domain discrepancy in terms of the classification problem. It is straightforward to see that $\Delta(\psi) \in [0, 2]$. By using the incompatibility values of features as weights for l_1 norm regularization, we obtain the following sparse regularizer of interest:

$$\mathcal{R}(\theta) = \sum_{\psi=1}^{d} \Delta(\psi)|\theta_\psi^F| \qquad (5)$$

which enforces stronger regularization on more incompatible features so as to achieve the goal of selecting features that are more compatible across the current and previous tasks.

Manifold Mixup Enhancement. We reserve a memory set to preserve some knowledge about previous tasks, however, such memory set merely provides partial information especially when the buffer size is small. In the selected representation space via feature selection, we expect the domain discrepancy between current and old tasks to be diminished. However, oscillations are inevitable when predicting the examples from old tasks outside the memory set. Out of concern for this issue, we recall a data augmentation technique – *mixup* [27], which is

Algorithm 1. Training Protocol

Input: dataset $\mathcal{D}_t$, memory set $\mathcal{M}_{t-1}$, previous model θ_{t-1}, batch size b, number of iterations num_iter, learning rate τ

Output: model θ

1: Initialize $\theta \leftarrow \theta_{t-1}$
2: **for** $iter = 1 : num\_iter$ **do**
3:     $B_D \leftarrow$ random sample$(\mathcal{D}, b)$
4:     $B_M = \bigcup_{i=1}^{t-1} B_{M_i}$ and $B_{M_i} \leftarrow$ random sample$(\mathcal{M}_{t-1}^i, \frac{b}{t-1})$
5:     $B_{mix} = \text{Mixup}_G(B_D, B_M)$ acc. to Eq. (6).
6:     compute loss l acc. to Eq. (8)
7:     update $\theta \leftarrow \theta - \tau \nabla_\theta \mathcal{L}_\theta$
8: **end for**
9: **return** θ

to extend the training dataset with convex combinations of pairs of datapoints and their corresponding labels and was shown to have a favourable effect on generalization performance.

Different from vanilla mixup, we construct the mixup samples on a representation level by linearly interpolating between current and memory data representations, dubbed $\text{Mixup}_G(\mathcal{D}_t, \mathcal{M}_{t-1})$. Specifically,

$$
\begin{aligned}
(\mathbf{g}_{mix}, y_{mix}) &:= \text{Mixup}_G((\mathbf{x}_D, y_D), (\mathbf{x}_M, y_M)) \\
\mathbf{g}_{mix} &= \lambda G(\mathbf{x}_D) + (1-\lambda)G(\mathbf{x}_M) \\
y_{mix} &= \lambda y_D + (1-\lambda)y_M
\end{aligned}
\tag{6}
$$

where $(\mathbf{x}_D, y_D) \sim \mathcal{D}_t$, $(\mathbf{x}_M, y_M) \sim \mathcal{M}_{t-1}$, $\lambda \sim Beta(\alpha_0, \alpha_0)$ and G is the feature extractor. We take $\alpha_0 = 3$ in the experiments. Notably, Eq. (6) can be viewed as a simplified variant of *manifold mixup* [25], where the mixup is performed in the hidden states at each layer of the network whereas ours is only in the representation space right before the classifier.

The classification loss computed on such mixup samples is defined as follows,

$$
\mathcal{L}_{MM} = \mathbb{E}_{(\mathbf{x}_D, y_D) \sim \mathcal{D}_t, (\mathbf{x}_M, y_M) \sim \mathcal{M}_{t-1}} l(F(\mathbf{g}_{mix}), y_{mix})
\tag{7}
$$

where $(\mathbf{g}_{mix}, y_{mix})$ is derived from Eq. (6) and F is the classifier of the proposed model. By embracing mixup samples between the memory set and current task, we expect smoother decision boundaries and benefits to classification performance when the undesirable test oscillation occurs. By performing mixup in the extracted feature space, we expect this mixup based classification loss can work together with the feature selection regularization above to bridge domain discrepancy and improve generalization performance.

Overall Learning Problem. By integrating the knowledge distillation based classification loss in Eq. (2), the mixup based classification loss in Eq. (7), and the

Table 1. The characteristics of datasets.

Datasets	#Tasks	#Classes/task	#Training/task	#Test/task
Split MNIST	5	2	12000	2000
Permuted MNIST	10	10	60000	10000
Split CIFAR-10	5	2	10000	2000
Split CIFAR-100	10	10	5000	1000

sparse regularization in Eq. (5) together, we obtain the following optimization problem for the current task t:

$$\min_{\theta} \mathcal{L}_{\theta}(\mathcal{D}_t, \mathcal{M}_{t-1}) = \mathcal{L}_{CE} + \alpha\mathcal{L}_{KD} + \beta\mathcal{L}_{MM} + \gamma\mathcal{R}(\theta) \qquad (8)$$

where α, β and γ are trade-off parameters. We solve it using a batch-wise gradient descent algorithm, which is summarized in Algorithm 1. After training for task t, the memory set $\mathcal{M}_{t-1}$ can be updated to $\mathcal{M}_t$ through task-wise random sampling before going to the next task.

4 Experiments

We conducted experiments on four benchmark continual learning datasets in the domain incremental learning scenario. In this section, we report the experimental setting and results.

4.1 Experimental Setting

Datasets. We used four benchmark continual learning datasets: Split MNIST, Permuted MNIST, Split CIFAR-10, and Split CIFAR-100. Split MNIST [7] is generated by splitting the source MNIST dataset into five binary-class subsets in sequence (0/1, 2/3, 4/5, 6/7, 8/9). Permuted MNIST [7] is a variant of MNIST dataset, by applying a certain random pixel-level permutations to the entire MNIST dataset. We consider 10 tasks for this dataset. Split CIFAR-10 [16] is a sequential split of CIFAR-10 with five binary classification tasks. Split CIFAR-100 [16] has 10 ten-class classification tasks. The characteristics of all datasets are summarized in Table 1.

Comparison Methods. We compared the proposed approach with nine comparison methods, two baselines and seven continual learning competitors: (1) *Joint* is a baseline that jointly trains on the data of all tasks. It serves as the upper bound for continual learning techniques. (2) *Finetune* is a baseline that simply fine-tunes the model from previous tasks on the current task dataset. (3) *EWC* is the regularization based continual learning method, Elastic Weight Consolidation [8]. (4) *SI* is another regularization based method, Synaptic Intelligence [26]. (5) *LwF* is a Learning without Forgetting method [9]. (6) *GEM* is

Table 2. The average accuracy ± standard deviation (%) on test data of all tasks across 5 runs with different random seeds. The result of joint training, i.e. the upper bound, and the best accuracies of different buffer size are marked in bold. Note that '5t/10t' means after training 5/10 tasks.

Buffer	Model	MNIST		CIFAR-10	CIFAR-100	
		S-MNIST	P-MNIST	S-CIFAR10	S-CIFAR100-5t	S-CIFAR100-10t
–	Joint	$98.59_{\pm0.15}$	$97.90_{\pm0.09}$	$90.67_{\pm0.22}$	$51.30_{\pm0.42}$	$43.80_{\pm0.83}$
	Finetune	$56.72_{\pm2.01}$	$74.38_{\pm1.63}$	$72.21_{\pm0.53}$	$25.60_{\pm0.30}$	$16.76_{\pm0.43}$
	EWC	$57.04_{\pm1.46}$	$86.60_{\pm1.62}$	$72.12_{\pm1.23}$	$25.41_{\pm0.48}$	$16.79_{\pm0.25}$
	SI	$68.06_{\pm2.18}$	$93.59_{\pm0.77}$	$75.16_{\pm0.87}$	$25.84_{\pm1.22}$	$17.60_{\pm1.30}$
	LwF	$77.93_{\pm0.53}$	$53.25_{\pm1.84}$	$78.90_{\pm0.81}$	$31.66_{\pm0.34}$	$19.41_{\pm0.23}$
200	GEM	$91.08_{\pm0.70}$	$86.67_{\pm0.48}$	$79.26_{\pm0.31}$	$28.29_{\pm1.60}$	$18.72_{\pm0.37}$
	A-GEM	$86.56_{\pm1.01}$	$69.84_{\pm2.01}$	$76.40_{\pm1.02}$	$27.37_{\pm0.56}$	$16.86_{\pm0.19}$
	ER-Res	$86.71_{\pm1.50}$	$86.73_{\pm0.44}$	$78.58_{\pm0.84}$	$28.69_{\pm0.62}$	$18.78_{\pm0.36}$
	FROMP	$71.75_{\pm1.31}$	$76.71_{\pm0.75}$	$75.20_{\pm1.12}$	$24.42_{\pm1.09}$	$12.35_{\pm2.37}$
	Ours	$94.37_{\pm0.35}$	$93.11_{\pm0.43}$	$84.62_{\pm0.28}$	$34.34_{\pm0.59}$	$21.15_{\pm0.41}$
500	GEM	$94.67_{\pm0.32}$	$92.89_{\pm0.86}$	$81.83_{\pm0.61}$	$32.04_{\pm0.77}$	$20.88_{\pm0.31}$
	A-GEM	$89.53_{\pm1.12}$	$83.06_{\pm1.78}$	$77.38_{\pm1.70}$	$28.33_{\pm0.38}$	$17.20_{\pm0.27}$
	ER-Res	$90.91_{\pm0.98}$	$91.30_{\pm0.41}$	$81.79_{\pm0.48}$	$31.9_{\pm0.58}$	$20.24_{\pm0.34}$
	FROMP	$78.71_{\pm0.45}$	$90.81_{\pm0.47}$	$74.84_{\pm2.29}$	$23.17_{\pm2.86}$	$15.13_{\pm2.58}$
	Ours	$96.41_{\pm0.19}$	$94.72_{\pm0.19}$	$85.69_{\pm0.25}$	$37.70_{\pm0.40}$	$23.08_{\pm0.53}$
1000	GEM	$95.72_{\pm0.81}$	$94.36_{\pm0.25}$	$84.66_{\pm0.44}$	$35.46_{\pm0.82}$	$23.05_{\pm0.16}$
	A-GEM	$95.31_{\pm1.48}$	$88.34_{\pm0.41}$	$79.12_{\pm0.47}$	$28.21_{\pm0.39}$	$17.37_{\pm0.27}$
	ER-Res	$95.09_{\pm0.22}$	$93.00_{\pm0.11}$	$83.77_{\pm0.37}$	$34.06_{\pm0.54}$	$22.23_{\pm0.23}$
	FROMP	$88.34_{\pm1.11}$	$93.09_{\pm0.09}$	$74.75_{\pm2.85}$	$24.04_{\pm3.36}$	$16.62_{\pm2.88}$
	Ours	$97.35_{\pm0.25}$	$95.60_{\pm0.18}$	$86.72_{\pm0.30}$	$40.20_{\pm0.52}$	$25.62_{\pm0.22}$

the Gradient Episodic Memory method [10]. (7) *A-GEM* is a lightweight variant of GEM, Averaged GEM (A-GEM) [3]. (8) *ER-Reservoir* is an Experience Replay method based on reservoir sampling [4]. (9) *FROMP* is the Functional Regularization of Memorable Past method from [13] .

Architecture and Hyperparameter Selection. As for model architectures, we employ a single-head multi-layer perceptron with two hidden layers following [7,23,24] for the MNIST datasets, and a CNN with 4 convolutional layers and 2 dense layers with dropouts following [13,26] for CIFAR-10/100 datasets. For fair comparison, all approaches share the same model architectures. The hyperparameters are selected by a coarse grid search for all approaches and the best results are reported. For the proposed approach, we recommend $\alpha = 2$ and $\beta = 0.1$ for three Split datasets. For Permuted MNIST, we have $\alpha = 1$ and $\beta = 0.001$. For all datasets, $\gamma \in [10^{-4}, 10^{-1}]$. Generally, the greater buffer size results in smaller γ.

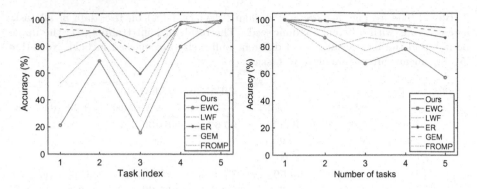

Fig. 1. Split MNIST dataset with buffer size of 200. **Left:** The test accuracy of each task after learning all tasks. **Right:** The evolution of average test accuracy of all seen tasks as more tasks are learned. All results are obtained via 5 runs with different random seeds.

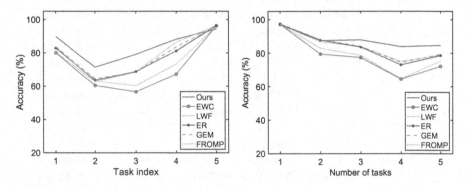

Fig. 2. Split CIFAR-10 dataset with buffer size of 200. **Left:** The test accuracy of each task after learning all tasks. **Right:** The evolution of average test accuracy of all seen tasks as more tasks are learned. All results are obtained via 5 runs with different random seeds.

4.2   Experimental Results

We experimented with three different buffer sizes, {200, 500, 1000}, for all benchmark datasets. Each experiment for each approach is repeated 5 runs with different random seeds. The random seeds work for the network initialization, training data shuffle and memories selection. The order of sequential tasks in Split MNIST and Split CIFAR-10/100 datasets remain unchanged throughout the experiments. For Permuted MNIST, once the random permutations for each task are generated, the task order is fixed across all runs. The results in terms of the average accuracies evaluated on test data of all tasks for the proposed approach and comparison methods are reported in Table 2. It is worth noting that our proposed approach achieves the best results compared to memory-based competitors across all dataset benchmarks over different buffer sizes and

Table 3. The average BWT ± standard deviation (%) across 5 runs with different random seeds. The best BWT of different buffer size are marked in bold. Note that the larger BWT means the better performance.

Buffer	Model	S-MNIST	S-CIFAR-10
–	EWC	-53.24 ± 1.81	-27.67 ± 1.53
	LwF	$\mathbf{-5.26 \pm 1.73}$	$\mathbf{-16.73 \pm 1.20}$
200	GEM	-9.49 ± 1.01	-18.71 ± 0.34
	ER-Res	-16.15 ± 1.86	-19.42 ± 1.12
	FROMP	-34.72 ± 1.06	-22.63 ± 1.01
	Ours	$\mathbf{-2.29 \pm 0.61}$	$\mathbf{-8.08 \pm 0.29}$
500	GEM	-4.55 ± 0.43	-16.34 ± 0.71
	ER-Res	-10.86 ± 1.25	-15.25 ± 0.37
	FROMP	-24.68 ± 1.33	-22.65 ± 2.02
	Ours	$\mathbf{-1.41 \pm 0.39}$	$\mathbf{-5.76 \pm 0.37}$

significantly outperforms these competitors when the buffer size is small, like 200. The proposed approach also outperforms the regularization-based methods in most cases except on the Permuted MNIST with buffer size 200, where the accuracy of the proposed approach is approximately 0.48% lower than *SI*. In the split datasets of either MNIST or CIFAR-10/100, *EWC* and *SI* perform poorly. This usually happens to most of parameter-regularization models in the domain incremental learning scenario. A memory set contrarily is favourable in this scenario. However, some memory-based models still produce poor performance when the buffer size is small, such as *FROMP* in most datasets and *A-GEM* in the permuted MNIST dataset with buffer size of 200. By contrast, our proposed approach demonstrate good performance across all datasets and scenarios.

Figure 1 and Fig. 2 show the test accuracy results of the multiple tasks after and during continue learning on Split MNIST and Split CIFAR-10, respectively, when the buffer size is 200. Specifically, the figures on the left side of Fig. 1 and Fig. 2 report the test accuracy of each learned task after finishing the training on all tasks, while the figures on the right side report the evolution of average test accuracy of all seen tasks as new task arrives. It is clear to observe that after learning all tasks, the proposed approach still keep a competent memory of previously learned tasks compared to other competitors, where *EWC* and *FROMP* almost thoroughly forget task 1 and 3 of Split MNIST and also perform pretty poorly on task 3 of Split CIFAR-10. In the figures of evolution of average accuracy, the proposed approach maintain the best performance from task 3 of both datasets, which demonstrates the outstanding ability of our approach to overcome the catastrophic forgetting even with a small memory set.

Table 4. The effectiveness of knowledge distillation (KD), feature selection (FS) and manifold mixup (MM) on Split MNIST and Split CIFAR-10 datasets. CE represents the cross-entropy loss (Eq. 1) over current and memory datasets. The average accuracy ± standard deviation (%) across 5 runs are reported.

S-MNIST				Buffer Size	
CE	KD	FS	MM	200	500
✓				88.71 ± 1.73	92.73 ± 0.51
✓	✓			93.03 ± 0.41	95.70 ± 0.33
✓	✓	✓		94.10 ± 0.35	96.27 ± 0.23
✓	✓	✓	✓	94.37 ± 0.35	96.41 ± 0.19
S-CIFAR-10				Buffer Size	
CE	KD	FS	MM	200	500
✓				76.05 ± 0.55	79.08 ± 0.53
✓	✓			83.06 ± 0.59	84.91 ± 0.29
✓	✓	✓		84.43 ± 0.28	85.44 ± 0.32
✓	✓	✓	✓	84.62 ± 0.28	85.69 ± 0.25

Backward Transfer (BWT). In addition to test accuracy, another metric, called *Backward Transfer (BWT)* [10], has been used to measure the influence on previous tasks after learning the new task, which is defined as follows,

$$\text{BWT} = \frac{1}{T-1} \sum_{i=1}^{T-1} R_{T,i} - R_{i,i}$$

where $R_{i,j}$ is the classification accuracy of the model on test data of task j after learning the training data from task i. The positive backward transfer means the improvement of performance on some previous tasks after learning new ones, whereas the negative backward transfer is known as forgetting. The larger BWT value indicates the better ability of the model to overcome forgetting. When two models have similar test accuracies, the one with larger BWT value would be preferable. Here we report our BWT results on the Split MNIST and Split CIFAR-10 datasets in Table 3. We can see that the proposed approach achieves much larger BWT values compared to other models on both datasets, which again validated its efficacy.

Discussion. Following the literature work that uses memory set, we assume $m \geq T$ in this paper. What if the memory set is out of capacity with the increase of the number of tasks? This $m < T$ issue can be addressed by elaborately selecting memory examples or exploiting mixup to generate examples that are able to best stand for all past tasks.

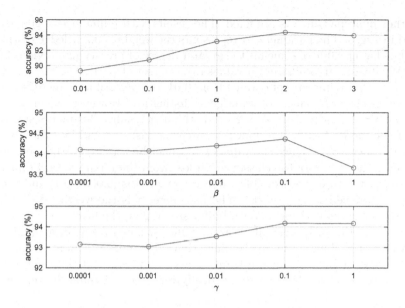

Fig. 3. Sensitivity to α, β and γ on Split MNIST dataset with the buffer size of 200.

4.3 Ablation Study

We further conducted an ablation study to verify the impact of each component in our proposed approach on the Split MNIST and Split CIFAR-10 datasets with the buffer size of 200 and 500. The results are presented in Table 4.

As demonstrated in [4], experience replay is a very simple and strong baseline, which here is treated as base model represented by 'CE' - the cross-entropy loss on current and memory datasets (Eq. (1)). We empirically observed that the experience replay with knowledge distillation (KD) is a much stronger method, which is also simple and effective and can beat the majority of elaborate CL techniques. Similar phenomenon was as well explored in [14]. As shown in Table 4, by adding the term of knowledge distillation, the performance obtains remarkable improvement. On the basis of knowledge distillation, we further implement the feature selection (FS) in the representation space by equipping model with a weighted l_1 regularization, so that only shared features across current and old tasks are utilized. Table 4 shows that by adding the feature regularization, the results get further improved. Finally, the simplified manifold mixup (MM) of current and memory data in the selected feature space is employed to further improve the generalization performance which is also verified by the slight but consistent improvement demonstrated in Table 4.

Hyperparameter Sensitivity Analysis. We also conducted experiments on the Split MNIST dataset to evaluate the sensitivity of the proposed model to hyperparameters α, β and γ. When evaluating one hyperparameter, the values of the other hyperparameters are fixed. The results with buffer size of 200 are presented in Fig. 3. The parameter α represents the weight of knowledge distillation loss. The larger value of α means the greater degree of preserving previous knowledge. We investigated different α values from the set of $\{0.01, 0.1, 1, 2, 3\}$. We can see that with the increase of α value from 0.01, the performance increases due to the increasing contribution of knowledge distillation. However, when $\alpha = 3$ the performance degrades while $\alpha = 2$ yields the best result. This makes sense since that if the model focuses too much on precious tasks, it will lose the ability to adapt to a new task. Hence the value of α cannot be too large. The parameter β represents the weight of the classification loss from the mixup data. From Fig. 3, we can observe that a reasonable value range for β is $[0.0001, 0.1]$ and the best result is produced when $\beta = 0.1$. The parameter γ is the weight of the feature selection regularization term. With buffer size as 200, the best result is achieved at $\gamma = 0.1$ in Fig. 3. In general, a larger buffer size can result in a smaller suitable value of γ. This is due to the fact that with more previous samples from the memory set involved in the training loss, they will contribute more to the feature representation learning to naturally reduce the domain discrepancy.

5   Conclusions

In this paper, we proposed a simple but effective continual learning approach with representational and functional regularizations for the domain incremental learning scenario. To avoid catastrophic forgetting, we enforced the model predictions on both current and memory data to approach the previous ones by means of knowledge distillation. Additionally, to alleviate the domain discrepancy in the feature representation space between the old task domain and the current task domain, we further resorted to feature selection so as to minimize the domain gap in the selected representation space. Finally, mixup between current and memory data representations are incorporated to improve the generalization performance. We demonstrated the effectiveness of the proposed approach through extensive experiments on four benchmark continual learning datasets.

Acknowledgments. This research was supported in part by the NSERC Discovery Grant, the Canada Research Chairs Program, and the Canada CIFAR AI Chairs Program.

References

1. Aljundi, R., Babiloni, F., Elhoseiny, M., Rohrbach, M., Tuytelaars, T.: Memory aware synapses: learning what (not) to forget. In: Proceedings of the European Conference on Computer Vision (ECCV) (2018)
2. Chaudhry, A., Dokania, P.K., Ajanthan, T., Torr, P.H.: Riemannian walk for incremental learning: understanding forgetting and intransigence. In: European Conference on Computer Vision (ECCV) (2018)

3. Chaudhry, A., Ranzato, M., Rohrbach, M., Elhoseiny, M.: Efficient lifelong learning with a-gem. In: International Conference on Learning Representations (ICLR) (2019)

4. Chaudhry, A., et al.: Continual learning with tiny episodic memories. arXiv preprint arXiv:1902.10486 (2019)

5. Goodfellow, I.J., Mirza, M., Xiao, D., Courville, A., Bengio, Y.: An empirical investigation of catastrophic forgetting in gradient-based neural networks. In: International Conference on Learning Representations (ICLR) (2014)

6. Hinton, G., Vinyals, O., Dean, J.: Distilling the knowledge in a neural network. In: NIPS Deep Learning and Representation Learning Workshop (2015)

7. Hsu, Y.C., Liu, Y.C., Ramasamy, A., Kira, Z.: Re-evaluating continual learning scenarios: a categorization and case for strong baselines. In: NeurIPS Continual Learning Workshop (2018)

8. Kirkpatrick, J., et al.: Overcoming catastrophic forgetting in neural networks. Natl. Acad. Sci. (PNAS) **114**(13), 3521–3526 (2017)

9. Li, Z., Hoiem, D.: Learning without forgetting. In: European Conference on Computer Vision (ECCV) (2016)

10. Lopez-Paz, D., Ranzato, M.: Gradient episodic memory for continual learning. In: Advances in Neural Information Processing Systems (NIPS) (2017)

11. McCloskey, M., Cohen, N.J.: Catastrophic interference in connectionist networks: the sequential learning problem. Psychol. Learn. Motiv. Adv. Res. Theory **24**(C), 109–165 (1989)

12. Nguyen, C.V., Li, Y., Bui, T.D., Turner, R.E.: Variational continual learning. In: International Conference on Learning Representations (ICLR) (2018)

13. Pan, P., Swaroop, S., Immer, A., Eschenhagen, R., Turner, R.E., Khan, M.E.: Continual deep learning by functional regularisation of memorable past. In: Advances in Neural Information Processing Systems (NeurIPS) (2020)

14. Pietro, B., Matteo, B., Angelo, P., Davide, A., Simone, C.: Dark experience for general continual learning: a strong, simple baseline. In: Advances in Neural Information Processing Systems (NeurIPS) (2020)

15. Ratcliff, R.: Connectionist models of recognition memory: constraints imposed by learning and forgetting functions. Psychol. Rev. **97**(2), 285–308 (1990)

16. Rebuffi, S., Kolesnikov, A., Sperl, G., Lampert, C.H.: iCaRL: incremental classifier and representation learning. In: Conference on Computer Vision and Pattern Recognition (CVPR) (2017)

17. Riemer, M., et al.: Learning to learn without forgetting by maximizing transfer and minimizing interference. In: International Conference on Learning Representations (ICLR) (2019)

18. Ring, M.B.: Continual Learning in Reinforcement Environments. Ph.D. thesis, The University of Texas at Austin (1994)

19. Robins, A.: Catastrophic forgetting, rehearsal and pseudorehearsal. Connect. Sci. **7**(2), 123–146 (1995)

20. Schwarz, J., et al.: Progress & compress: a scalable framework for continual learning. In: International Conference on Machine Learning (ICML) (2018)

21. Shin, H., Lee, J.K., Kim, J., Kim, J.: Continual learning with deep generative replay. In: Advances in Neural Information Processing Systems (NIPS) (2017)

22. Thrun, S.: A lifelong learning perspective for mobile robot control. In: IEEE/RSJ International Conference on Intelligent Robots and Systems (IROS) (1994)

23. van de Ven, G.M., Tolias, A.S.: Generative replay with feedback connections as a general strategy for continual learning. arXiv preprint arXiv:1809.10635 (2018)

24. van de Ven, G.M., Tolias, A.S.: Three scenarios for continual learning. arXiv preprint arXiv:1904.07734 (2019)
25. Verma, V., et al.: Manifold mixup: better representations by interpolating hidden states. In: International Conference on Machine Learning (ICML) (2019)
26. Zenke, F., Poole, B., Ganguli, S.: Continual learning through synaptic intelligence. In: International Conference on Machine Learning (ICML) (2017)
27. Zhang, H., Cisse, M., Dauphin, Y.N., Lopez-Paz, D.: Mixup: beyond empirical risk minimization. In: International Conference on Learning Representations (ICLR) (2018)

EARLIN: Early Out-of-Distribution Detection for Resource-Efficient Collaborative Inference

Sumaiya Tabassum Nimi[✉], Md Adnan Arefeen, Md Yusuf Sarwar Uddin, and Yugyung Lee

University of Missouri-Kansas City, Kansas City, MO, USA
{snvb8,aa4cy}@mail.umkc.edu, {muddin,LeeYu}@umkc.edu

Abstract. Collaborative inference enables resource-constrained edge devices to make inferences by uploading inputs (e.g., images) to a server (i.e., cloud) where the heavy deep learning models run. While this setup works cost-effectively for successful inferences, it severely underperforms when the model faces input samples on which the model was not trained (known as Out-of-Distribution (OOD) samples). If the edge devices could, at least, detect that an input sample is an OOD, that could potentially save communication and computation resources by not uploading those inputs to the server for inference workload. In this paper, we propose a novel lightweight OOD detection approach that mines important features from the *shallow* layers of a pretrained CNN model and detects an input sample as ID (In-Distribution) or OOD based on a distance function defined on the reduced feature space. Our technique (a) works on pretrained models without any retraining of those models, and (b) does not expose itself to any OOD dataset (all detection parameters are obtained from the ID training dataset). To this end, we develop EARLIN (**EARL**y OOD detection for Collaborative **IN**ference) that takes a pretrained model and partitions the model at the OOD detection layer and deploys the considerably small OOD part on an edge device and the rest on the cloud. By experimenting using real datasets and a prototype implementation, we show that our technique achieves better results than other approaches in terms of overall accuracy and cost when tested against popular OOD datasets on top of popular deep learning models pretrained on benchmark datasets.

Keywords: Out-of-distribution detection · Collaborative inference · Novelty detection · Neural network

1 Introduction

With the emergence of Artificial Intelligence (AI), applications and services using deep learning models, especially Convolutional Neural Networks (CNN), for per-

Electronic supplementary material The online version of this chapter (https://doi.org/10.1007/978-3-030-86486-6_39) contains supplementary material, which is available to authorized users.

© Springer Nature Switzerland AG 2021
N. Oliver et al. (Eds.): ECML PKDD 2021, LNAI 12975, pp. 635–651, 2021.
https://doi.org/10.1007/978-3-030-86486-6_39

forming intelligent tasks, such as image classification, have become prevalent. However, several issues have been observed in deployment of the deep learning models for real-life application. First, since the models tend to be very large in size (100's of MB in many cases), they require higher computation, memory, and storage to run, which makes it difficult to deploy them on end-user/edge devices. Second, these models usually predict with high confidence, even for those input samples that are supposed to be unknown to the models (called out-of-distribution (OOD) samples) [18,20]. Since both in-distribution (ID) and OOD input samples are likely to appear in real-life settings, OOD detection has emerged as a challenging research problem.

The first issue, the deployment of deep learning models in end/edge devices, has been studied well in the literature [9]. One solution is to run collabora-tive inference, in which the end devices do not run the heavy model on-board, instead offload the inference task by uploading the input to a nearby server (or to the cloud in appropriate cases) and obtain the inference/prediction results from there. Other recent works propose doing edge-cloud collaboration [5], model compression [19] or model splitting [8] for faster inference. The second issue, the OOD detection, has received much attention in the deep learning research community [3,4,6,14,22]. We note several gaps in these research works, partic-ularly their suitability of deployment in collaborative inference setup. Firstly, in most of these works, the input data were detected as an OOD sample using the outputs from the *last* [7,10,14,17] or *penultimate* [12] layer of the deep learn-ing classifiers. We argue that detecting an input sample as OOD after these many computations are done by the model is inefficient. Secondly, most of the OOD detection approaches rely on full retraining the original classifier model to enable the OOD detection [10,11,17], which is computationally very expensive. Thirdly, in most of these works [12,14,17], several model hyperparameters for the detection task need to be tuned based on a validation dataset of OOD samples. The fitted model is then tested, thereby inducing bias towards those datasets. Finally, some OOD detector requires computationally expensive pre-processing of the input samples [12,14].

In this paper, we tackle the above two discussed issues jointly. We propose a novel OOD detection approach, particularly for Convolutional Neural Networks (CNN) models, that detects an input sample as OOD early into the network pipeline, using the portion of the feature space obtained from the shallow layers of the pretrained model. It is documented that early layers in CNN models usu-ally pick up some salient features representing the overall input space whereas the deeper layers progressively capture more discriminant features toward classifying the input samples to the respective classes [16]. This, therefore, suggests that these salient feature maps extracted from a designated *early* layer will be different for ID and OOD samples. This is the principle observation based on which we attempt to build our OOD detector model. However, the space spanned by the obtained feature maps is in most of the cases too big to make any significant partitioning between ID and OOD samples. Hence we compress the high dimensional feature space by mining the *most significant* information out of the space. We apply a series of "feature selection" operations on the extracted feature maps, namely *indexed-*

pooling and max-pooling, to reduce the large feature space to a manageable size. After the reduction, we construct a distance function defined on the reduced feature space so that the distance measure can differentiate ID and OOD samples. For deployment in edge-cloud collaboration setup, we partition the model around the selected layer to obtain a super-small OOD detection model and readily deploy the lightweight model on an edge device. With that, the edge device can detect an incoming input sample as OOD and if detected, does not upload the sample to the server/cloud (thus saves communication and computation resources).

To this end, we develop EARLIN (EARLy OOD Detection for Collaborative INference) based on the our proposed OOD detection technique. We evaluate EARLIN on a set of popular CNN models for image classification, namely Densenet, ResNet34, ResNet44, and VGG16 models pretrained on benchmark image datasets CIFAR-10 and CIFAR-100 [13]. We also compare our OOD detection algorithm with state-of-the-art other OOD detection techniques discussed in the literature. Furthermore, we design and develop an OOD-aware collaborative inference system and show that this setup results in faster and more precise inference in edge devices. To the best of our knowledge, ours is the first work to propose such OOD-aware collaborative inference framework. Furthermore, we define a novel performance metric, the *joint accuracy* of a model combined with its detector, to quantify the performance of the model and detector combination, and formally characterize EARLIN's performance and cost using that metric.

We summarize our contributions as follows:

- We propose a novel OOD detection approach called EARLIN that enables detection of OOD samples early in the computation pipeline, with minimal computation.
- Our technique does not require retraining the neural network classifier and thus can be implemented as an external module on top of available pretrained classifiers.
- We do not exploit samples from unknown set of OOD data for tuning hyperparamters, thereby reducing bias towards any subset of the unknown set of OOD samples.
- We propose a novel OOD-aware edge-cloud collaborative setup based on our proposed detector for precise and resource-efficient inference at edge devices, along with characterizations of its performance and cost.

2   Related Work

Deep Learning based methods have been designed to achieved huge success in recent years in recognition tasks but they have their limitations. The problem of reporting high confidence for all input samples, even those outside the domain of training data is inherent in the general construct of the popular deep learning models. In order to deploy the deep learning models in real-life applications, this issue should be mitigated. Hence in the recent years, a large number of research works have been conducted towards this direction. In [6], confidence of the deep learning classifiers in the form of output softmax probability for the predicted class was used to differentiate between ID and OOD samples. Later, in [12,14], OOD detection approaches were proposed that worked without making any change

Table 1. Comparison of approaches.

	Baseline [6]	ODIN [14]	Mahalanobis [12]	MALCOM [23]	DeConf [7]	EARLIN
Without Retraining?	✓	✓	✓	✓	×	✓
Before Last layer?	×	×	×	✓	×	✓
Use One Layer Output?	✓	✓	×	×	✓	✓
Without OOD Exposed?	✓	×	×	✓	✓	✓
Without Input Preprocessing?	✓	×	×	✓	×	✓

to the original trained deep learning models. We note several limitations in the works. Firstly, samples are detected as OOD at the very last layer of the classifier, thereby wasting computational resources on unnecessary computations done on input samples, that is eventually going to be identified unsuitable for classification. Secondly, the hyperparameters were tuned while being exposed to subset of OOD samples that the approach was tested on, inducing bias towards those samples. Also, due to this exposure, it can not be guaranteed that the approach will be as successful on any completely different set of OOD data. Thirdly, this approach required computationally heavy preprocessing of the input samples for the approach to work. The preprocessing involved two forward and one backward passes over the classifier model, rendering the approach completely unsuitable for real-time deployment. In the recent literature [7,10,17] also, OODs were not detected on the top of readily available deep learning models that were pretrained with traditional cross entropy loss, instead retraining was required. A better approach was proposed in [23], that did not require retraining the classifier. Although this approach achieved good performance on OOD detection, we note that best reported results were obtained when feature maps from the deeper layers were used.

3   Proposed OOD Detection Approach: EARLIN

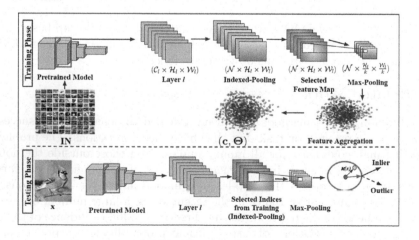

Fig. 1. Framework of the training and inference using EARLIN.

We propose an OOD detection approach, called EARLIN (**EARL**y OOD detection for Collaborative **IN**ference), that enables OOD detection from the shallow layers of a neural network classifier, without requiring to retrain the classifier and without exposure to any OOD sample during training. EARLIN infers a test input sample to be ID or OOD as follows. It first feeds the input to the classification CNN model and computes up to a designated shallow layer of the model and extracts feature maps from that layer. The output of the intermediate layer is a stack of 2D feature maps, out of which a small subset of them are selected. The selected maps are ones that supposedly contain the most information entailed by all of those maps. This process is called *indexed-pooling* the parameters of which (i.e., the positions of 2D maps to be selected) are determined from the training ID dataset during the training phase of the process. We then do *max-pooling* for downsampling the feature space even further. With that, we obtain a vector representation of the original input in some high dimensional feature space. During training, we do this for a large number of samples drawn from the training ID dataset, aggregate them in a single cluster, and find the centroid of the ID space. Consequently, we define a distance function from the ID samples to the centroid such that at a certain level of confidence, it can be asserted that the sample is ID if the distance is less than a threshold. Since the threshold is a measure of distance, its value is expected to be low for ID samples. During inference, we use this obtained value of threshold to differentiate between ID and OOD samples. The framework of our proposed ID detection approach is shown in Fig. 1.

Feature Selection and Downsampling: Indexed-Pooling and Max-Pooling: We at first select a subset of 2D feature maps from a designated shallow intermediate layer of pretrained neural network classifier based on a quantification of the amount of information each 2D feature map contains. We denote the chosen layer by ℓ. From this layer, we choose the most informative $\mathcal{N}$ feature maps. We know from studies done previously that feature maps at shallow layers of the classifier capture useful properties out of input images [16], but the space spanned by the feature maps is too big to capture properties inherent to the ID images out of this space. Hence we reduce the feature space by selecting a subset of the features. Visual observation reveals that some of the maps, the ones for which we obtain almost monochromatic plots, do not carry significant observation about the input image. Whereas, there are some maps that capture useful salient features from the image. We consider *variance* of a 2D feature map as the quantification of the amount of information contained in the map.

Suppose at layer ℓ, feature map, $f \in \mathbb{R}^{\mathcal{C}_\ell \times \mathcal{H}_\ell \times \mathcal{W}_\ell}$ is obtained. So we have a total of $\mathcal{C}_\ell$ 2D maps (also known as channels), each of which with a dimension of $\mathcal{H}_\ell \times \mathcal{W}_\ell$. Our goal is to select $\mathcal{N}$ most informative 2D feature maps out of these $\mathcal{C}_\ell$ maps. For finding the most informative feature maps using this assumption, we choose a subset of ID training data, $\mathbf{D}_{in}$. Using each data sample from $\mathbf{D}_{in}$, we calculate feature map f, with shape $\mathcal{C}_\ell \times \mathcal{H}_\ell \times \mathcal{W}_\ell$ from layer ℓ and finally obtain collection of feature maps, $\mathcal{F}$, with shape $|\mathbf{D}_{in}| \times \mathcal{C}_\ell \times \mathcal{H}_\ell \times \mathcal{W}_\ell$ for $\mathbf{D}_{in}$.

We define information contained in each feature map j, denoted as $\psi(j)$, as the summation of variance (aggregate variance) of 2D maps obtained from all input sample in the training ID dataset $(\mathbf{D}_{in})$. This collectively measures how important map j in layer ℓ is with respect to the entire ID population. Formally, we compute:

$$\psi_j = \sum_{\mathbf{x} \in \mathbf{D}_{in}} Var(m_\ell^{\mathcal{M}}(\mathbf{x})[j]) \tag{1}$$

where $m_\ell^{\mathcal{M}}$ denotes layer ℓ of model $\mathcal{M}$ with a tensor of size $\mathcal{C}_\ell \times \mathcal{H}_\ell \times \mathcal{W}_\ell$ and $m_\ell^{\mathcal{M}}[j]$ denote j-th 2D map in that layer having the size of size $\mathcal{H}_\ell \times \mathcal{W}_\ell$. Once the ψ values are obtained from all $\mathcal{C}_\ell$ maps, we find the order statistic of ψ values (sort the values in the descending order) as such:

$$\psi_{(1)} \geq \psi_{(2)} \geq \cdots \geq \psi_{(\mathcal{N})} \geq \cdots \geq \psi_{(\mathcal{C}_\ell)}$$

We then find the *indices* of top $\mathcal{N}$ channels that have the largest aggregate variance across the ID training dataset and populate a binary index vector γ to denote whether a certain map from that layer ℓ is selected or not. More precisely,

$$\gamma_j = \begin{cases} 1, & \text{if } \psi_j \leq \psi_{(\mathcal{N})} \\ 0, & \text{otherwise.} \end{cases}$$

Obviously, $\sum_{j=1}^{\mathcal{C}_\ell} \gamma_j = \mathcal{N}$. Given this binary index-vector, γ, and the layer output of $m_\ell^{\mathcal{M}}(\mathbf{x})$ for an input sample $\mathbf{x}$, the *indexed-pooling* operation takes out only those feature maps (channels) as specified by the index-vector thus effectively reduces the feature space dimension from $\mathcal{C}_\ell \times \mathcal{H}_\ell \times \mathcal{W}_\ell$ to $\mathcal{N} \times \mathcal{H}_\ell \times \mathcal{W}_\ell$. Consequently, we define the indexed-pooling operator as $\varGamma_1 : \mathbb{R}^{\mathcal{C}_\ell \times \mathcal{H}_\ell \times \mathcal{W}_\ell} \rightarrow \mathbb{R}^{\mathcal{N} \times \mathcal{H}_\ell \times \mathcal{W}_\ell}$ as follows:

$$\varGamma_1(\mathbf{x}) = \|_{j=1}^{\mathcal{C}_\ell} m_\ell^{\mathcal{M}}(\mathbf{x})[j] \text{ if } \gamma_j = 1 \tag{2}$$

where $\|$ indicates concatenation.

We note that the feature space spanned by the chosen $\mathcal{N}$ feature maps from layer ℓ is still too large to capture useful information. Hence, we downsample the space by (k, k) max-pooling. Max-pooling is an operation that is traditionally done within the deep leaning model architectures for downsampling the feature space, so that only the most relevant information out of a bunch of neighboring values is retained. We follow the same practice for downsampling our feature space. The max-pooling operator is denoted as $\varGamma_2 : \mathbb{R}^{\mathcal{N} \times \mathcal{H}_\ell \times \mathcal{W}_\ell} \rightarrow \mathbb{R}^{\mathcal{N} \times \frac{\mathcal{H}_\ell}{k} \times \frac{\mathcal{W}_\ell}{k}}$. We usually use $k = 4$.

Let $\phi(\mathbf{x})$ be the vector representation for an input, $\mathbf{x}$ obtained from layer ℓ, constructed by applying two pooling operators on the extracted features maps, *indexed-pooling* and *max-pooling*. Using the above two pooling operators $\varGamma_1$ and $\varGamma_2$, therefore, the construction of $\phi(\mathbf{x})$, for an input $\mathbf{x}$, can be written as:

$$\phi(\mathbf{x}; \ell, \gamma) = (\varGamma_2 \circ \varGamma_1)(\mathbf{x}) \tag{3}$$

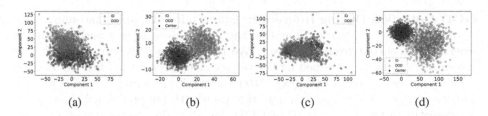

(a)   (b)   (c)   (d)

Fig. 2. Effect of the proposed Feature Pooling strategy on differentiating between ID and OOD (LSUN) in feature space of ResNet model, visualized in 2D using PCA. (a) Original feature space and (b) Pooled feature space obtained for ID dataset CIFAR-10, (c) Original feature space and (d) Pooled feature space obtained for ID dataset CIFAR-100. The black dots represent the center of the feature space.

where $\circ$ means $(f \circ g)(x) = f(g(x))$ [composite function]. We show in Fig. 2, the segregation between ID and OOD samples obtained in feature space after executing the above feature selection operations.

Feature Aggregation: We compute $\phi(\mathbf{x})$ for input all inputs $\mathbf{x} \in \mathbf{D}_{in}$ to represent the entire ID space as $\phi(\mathbf{D}_{in})$. We then find a set of aggregated information for the entire ID space. For that, we at first find the aggregated cluster *centroid*, denoted by $\mathbf{c}$, of the feature space. The centroid is defined as follows:

$$\mathbf{c} = \text{MEAN}(\phi(\mathbf{x})) \text{ for } \mathbf{x} \in \mathbf{D}_{in} \tag{4}$$

where MEAN computes the element-wise mean of the the collection of vectors obtained from $\mathbf{D}_{in}$. The centroid, $\mathbf{c}$, ultimately designates a center position of the ID space around which all ID samples position themselves in a close proximity. In that, the distance between the center and $\phi(\mathbf{x})$ for any ID sample $\mathbf{x}$ should follow a low-variance distribution. This distance is denoted as $\kappa(\mathbf{x})$, which is defined as the Euclidean distance between $\phi(\mathbf{x})$ and $\mathbf{c}$:

$$\kappa(\mathbf{x}) = \|\phi(\mathbf{x}) - \mathbf{c}\| \tag{5}$$

We hypothesize that since the centroid is a pre-determined value calculated using features of ID samples, distance, $\kappa(\mathbf{x})$, will take smaller value for any an ID sample than the distance obtained for an OOD sample. We observe from Fig. 2 that feature space $\phi(\mathbf{x})$, when visualized in 2D, validates our hypothesis. That requires us to find a suitable a threshold value on this distance value based on which ID and OOD sample can be separated out. This is what we do next.

We empirically find a threshold value, Θ, that detects ID samples with some confidence p, such that p fraction of the ID training samples have $\kappa(\mathbf{x}) \leq \Theta$. That is:

$$p = \frac{\sum_{\mathbf{x} \in \mathbf{D}_{in}} [\kappa(\mathbf{x}) \leq \Theta]}{|\mathbf{D}_{in}|} \tag{6}$$

We usually set $p = 0.95$. This is actually the expected TPR (True Positive Rate) of the OOD detector that we expect (the detector's capability to detect a true ID sample as ID). We note that, tuning the value of this hyperparameter Θ does not require exposure to any a priori known OOD samples.

OOD Detection During Inference: During inference, for an incoming input sample $\mathbf{x}$, we first pass the input into the model up to layer ℓ and extract the intermediate output from that layer. We then choose $\mathcal{N}$ best maps from that intermediate output (specified by the binary index-vector γ). Then we do max-pooling on that space to find $\phi(\mathbf{x})$. After that we find the distance of $\phi(\mathbf{x})$ from the centroid, $\mathbf{c}$, as $\kappa(\mathbf{x})$ and compare this value with the predefined threshold Θ for detecting if the sample is ID or OOD. Let $\mathcal{D}(\mathbf{x})$ denote the detector output for input $\mathbf{x}$, which can be obtained as:

$$\mathcal{D}(\mathbf{x}) = \begin{cases} 1, & \text{if } \kappa(\mathbf{x}) \leq \Theta, \\ 0, & \text{otherwise.} \end{cases} \tag{7}$$

So the inference is very fast and since our chosen layer ℓ is very shallow (unlike [6,12,14] that detect OOD samples at the last layer), we can reject the extraneous OOD samples, way before lots of unnecessary computations are done on the sample, which would lead nowhere. Hence our ID detection approach gives higher throughput during batch inference. Besides we do not retrain the classifier model unlike [7]. Also, we detect using features collected from a single layer only unlike [12,23], without preprocessing input samples unlike [7,12,14] and without using OOD samples for validation unlike [12,14]. The comparison of EARLIN with other approaches is summarized in Table 1.

4   Collaborative Inference Based on EARLIN

Based on our proposed OOD detection technique, we develop a setup for collaborative inference as a collaboration between an edge device and a server (this server can be in the cloud or can be a nearby edge resources, such as Cloudlet [21], Mobile Edge Cloud (MEC) [15], we generically refer to it as "server"). Deep learning models usually have large memory and storage requirements, and hence are difficult to deploy in the constrained environment of the edge devices. Thus, edge devices make remote call to the server devices for inference. If the incoming image is Out-of-Distribution, making such call is useless since the model would not be able to classify the image. Hence,

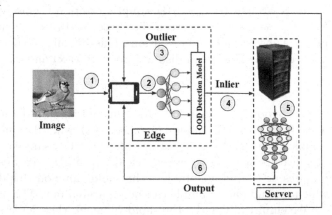

Fig. 3. Collaborative Inference Scheme.

we both save resources and make more precise recognition by not allowing to call when input image is OOD. Since our detection model, consisting of the first few layers of the network architecture, is very lightweight, we deploy the detection pipeline in the edge device. Then, if the image is detected as ID, we send the image to the server for classification. Otherwise, we report that the image is OOD and hence not classifiable by the model. We thus save resource by not sending the OOD images to the server. The schematic diagram of the framework appears in Fig. 3. We note that we send the original image, instead of the intermediate layer output to the server, when the sample is detected as ID. This is because intermediate layer outputs from deep learning models at the shallower layers are often significantly higher in dimension than the original images. With that, we save considerable upload bandwidth. As the servers are usually high-end machines, repeating the same computation up to layer ℓ adds very nominal overhead compared to the volume of data to be uploaded. Moreover, ℓ is a very shallow layer (below 10% from the input layer) as reported in Table 2. We also note that all model parameters are estimated/trained in the cloud using base model and the training datasets, and the resultant detector model is deployed on the edge device.

Overall Accuracy of OOD Detector and Deep Learning Classifier: Traditionally the performance of deep learning classification models are reported in terms of accuracy to establish how well they perform. *Accuracy* is defined as the ratio or percentage of samples classified correctly by the model, *given that* every sample comes from In-Distribution (ID). Let us denote $\mathbf{x} \in ID$ to indicate if an input $\mathbf{x}$ *truly* belongs to ID and $\mathbf{x} \in OOD$ to denote if $\mathbf{x}$ truly belongs to OOD (ideally, $\mathbf{x} \in OOD$ is logically equivalent to $\mathbf{x} \notin ID$). In terms of probability expression, the *Accuracy* (written as *acc* in short) of a model, $\mathcal{M}$, can be written as:

$$acc_{\mathcal{M}} \triangleq \mathcal{P}(\mathcal{M}(\mathbf{x}) = y(\mathbf{x}) \mid \mathbf{x} \in ID)$$

where $\mathcal{M}(\mathbf{x})$ represents the classification output of model $\mathcal{M}$ and $y(\mathbf{x})$ represents the true class label of input $\mathbf{x}$, $\mathcal{P}(E)$ denotes probability of event E. On the other hand, the performance of an OOD detector can be expressed in term of two metrics: True Positive Rate (TPR) and True Negative rate (TNR). TPR is the ratio of ID samples correctly classified as ID by the detector where TNR is the ratio of the true OOD samples detected as OODs. Let $\mathcal{D}$ denote a detector and $\mathcal{D}(\mathbf{x})$ denote a binary output of the detector to indicate that whether input $\mathbf{x}$ is detected as ID or OOD ($\mathcal{D}(\mathbf{x}) = 1$ if detected as ID else 0). Consequently, the TPR and TNR values of a detector, $\mathcal{D}$, can be expressed as:

$$TPR_{\mathcal{D}} \triangleq \mathcal{P}(\mathcal{D}(\mathbf{x}) = 1 \mid \mathbf{x} \in ID) \tag{8}$$

$$TNR_{\mathcal{D}} \triangleq \mathcal{P}(\mathcal{D}(\mathbf{x}) = 0 \mid \mathbf{x} \in OOD) \tag{9}$$

While the capability of the classification model ($\mathcal{M}$) and the OOD detector model ($\mathcal{D}$) can be expressed individually in terms of their respective performance metrics (that is, model classification accuracy, TPR and TNR), it is interesting to

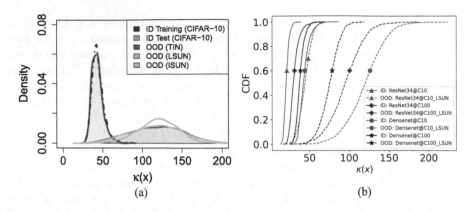

Fig. 4. (a) Histogram of $\kappa(\mathbf{x})$ that differentiates between ID and OOD samples (drawn from benchmark datasets) for Densenet pretrained on CIFAR-10. (b) CDF of $\kappa(\mathbf{x})$ that differentiates between ID and OOD samples with CIFAR-10 (C10) and CIFAR-100 (C100) as ID and the TinyImageNet (TIN), LSUN and iSUN dataset as OOD.

note how these three terms play a role in measuring the accuracy of the model and the detector combined. We refer to this as the joint accuracy or *overall accuracy*. We define the **Overall Accuracy** as the success rate of assigning *correct* class labels to test inputs. That is, for an ID sample, this corresponds to assigning correct the class label to the input whereas, for an OOD sample, this corresponds to detecting it as an OOD (OOD samples do not have any correct class label other than being flagged as OOD). Let us use $\mathcal{M} \oplus \mathcal{D}$ to denote the classification model and detector combined and we are interested to determine the accuracy of $\mathcal{M} \oplus \mathcal{D}$ as a function of its constituents. We observe that in addition to the above three metrics, the overall accuracy of the model and OOD detector combined is dependent on what fraction of inputs are actually OOD as opposed to ID as inputs are passed to the model. Let this ratio be denoted as ρ. Formally,

$$\rho = \mathcal{P}(\mathbf{x} \in OOD) \text{ and } 1 - \rho = \mathcal{P}(\mathbf{x} \in ID)$$

More specifically, given the accuracy of model $\mathcal{M}$ and the TPR and TNR values of the associated OOD detector, $\mathcal{D}$, the overall accuracy of $\mathcal{M} \oplus \mathcal{D}$ is given by:

$$acc_{\mathcal{M} \oplus \mathcal{D}} = acc_{\mathcal{M}} \times TPR_{\mathcal{D}} \times (1 - \rho) + TNR_{\mathcal{D}} \times \rho \tag{10}$$

The proof of the above equation is based on the fact that a correct output occurs when either of the two mutually exclusive events happen with respect to an input sample: (a) the input sample truly belongs to ID, and the detector also detects it as ID and the model correctly classifies it, (b) the input sample belongs to OOD and the detector detects this as OOD (detail appears in the supplementary document).

Performance and Cost Characteristics of the Collaborative Setup:
As per Eq. (10), the overall accuracy depends on four quantities: accuracy of
the original model, TPR and TNR of the detector, and ρ (fraction of samples
being OOD in the inference workload). Without any detector in place (when
TPR becomes 1 and TNR is 0), the overall accuracy of the model $acc_{\mathcal{M} \oplus \mathcal{D}} =$
$acc_{\mathcal{M}} \times (1 - \rho)$, sharply declines with ρ. With the detector combined, the overall
accuracy of the model, in fact, improves at a rate of $TNR - acc_{\mathcal{M}} \times TPR$ with
respect to ρ (actually, the accuracy grows only when the slope is positive, that
is, $TNR > acc_{\mathcal{M}} \times TPR$). In Sect. 6, we demonstrate this.

In EARLIN, as shown in Fig. 3, we send inputs to the server only when
they are detected as ID by the lightweight OOD detector deployed at the edge.
Let T_E be the time required for OOD detection at the edge, T_C be the round-
trip communication delay between the edge and the server, and T_S be the time
required for classifying the image at the server when sent. In that, when we
encounter a sample that is detected as OOD (when $\mathcal{D}(\mathbf{x}) = 0$), the time required
is only T_E (no communication to the server nor processing at the server). On
the other hand, when an incoming sample is detected as ID (when $\mathcal{D}(\mathbf{x}) = 1$),
the inference latency becomes $T_E + T_C + T_S$. So, the time required for inference
is closely associated with the ratio of OOD samples, ρ and the precision with
which the detector detects input samples as ID vs OOD. We can characterize the
cost, in terms of latency, involved with each inference using $\mathcal{M} \oplus D$ as follows:

$$T_{\mathcal{M} \oplus D} = T_E + (T_C + T_S)(TPR_{\mathcal{D}}(1 - \rho) + (1 - TNR_{\mathcal{D}})\rho) \qquad (11)$$

Similar to our performance indicator, *Overall Accuracy*, the cost characteris-
tics of the setup, $T_{\mathcal{M} \oplus D}$, can also be approximated as a linear function of ρ (OOD
ratio). In general, the end-to-end inference latency declines as ρ grows as OOD
samples are intercepted by the OOD detector at edge thus reducing inference
latency and saving communication resources. In particular, the inference latency
declines at a rate of $(T_C + T_S) \times (FPR_{\mathcal{D}} - TPR_{\mathcal{D}})$, where $FPR = 1 - TNR$
with respect to ρ. More detailed performance characterization can be found in
the supplementary section.

5   Experimental Evaluation of EARLIN

In this section, we show how our proposed OOD detector, EARLIN, performs on
standard pretrained models and benchmark datasets compared to the previously
proposed approaches for OOD detection.

Evaluation Metrics of OOD Detection: *TNR and FPR at 95% TPR:* This
is the rate of detecting an OOD sample as OOD. Hence, $TNR = TN/(FP+TN)$,
where FP is the number of OOD samples detected as ID and TN is the number of
OOD samples detected as OOD. We report TNR values obtained when $TPR = TP/(FN + TP)$, TP being the number of ID samples detected as ID, is as high
as 95%. And FPR is defined as (1-TNR).

Detection Accuracy and Detection Error: This depicts the overall accuracy of detection and is calculated using formula $0.5 \times (TPR + TNR)$, assuming that both ID and OOD samples are equally likely to be encountered by the classifier during inference. And Detection Error is (1-Detection Accuracy).

Table 2. Chosen Layer and Size of corresponding OOD detection models in pretrained Models

Model	# of Layers	Chosen Layer	Size of Detector Model n KB)	Training Dataset
ResNet	34	BN (5th)	112	CIFAR-10
ResNet	34	BN (2nd)	55	CIFAR-100
DenseNet	100	BN (10th)	256	CIFAR-10
DenseNet	100	BN (10th)	256	CIFAR-100

AUROC: This evaluates area under the ROC curve.

Results: We conduct experiments on Densenet with 100 layers (growth rate = 12) and ResNet with 34 layers pretrained on CIFAR-10 and CIFAR-100 datasets. Each of the ID datasets contains 50,000 training images and 10,000 test images. Summary of the pretrained models used in terms of their total number of layers, chosen layer ℓ for OOD detection, size of detector model $\mathcal{D}$, ID dataset on which the model was trained and the classification accuracy of the corresponding model are shown in Table 2.

In Table 3, we show the TNR (at 95% TPR) and Detection Accuracy of our approach. We compare our results with those obtained using previously proposed approaches, Baseline [6], ODIN [14] Mahalanobis Detector [12] and MAL-COM [23] on benchmark datasets [1] TinyImagenet, LSUN and iSUN, popularly used for testing OOD detection techniques. It is to be noted that, we did not implement the earlier approaches (except Baseline), rather compare with the results reported in [23] by using the same experimental setup. We see from the results in Table 3 that EARLIN performs better than the previous approaches in most of the cases. We report another set of results in Table 4, where we compare performance of EARLIN against DeConf [7] on DenseNet model pretrained on CIFAR-10 and CIFAR-100 datasets, in terms of metrics TNR at 95% TPR and AUROC. We note that we did not obtain results in the experimental setting on ResNet34 pretrained models in [7]. We see from the results in Table 4 also that EARLIN performs better than the previous approaches in most of the cases. We report yet another set of experimental results on VGG16 and ResNet44 models pretrained on CIFAR-10 and CIFAR-100 in the supplementary file.

Table 3. OOD detection performance on different datasets and pretrained models. Here MLCM stands for MALCOM [23], BASE for Baseline [6], ODIN for ODIN [14] and MAHA for Mahalanobis [12]. **bold** indicates best result.

ID Dataset	Model	OOD	TNR at 95% TPR					Detection Accuracy					AUROC				
			MLCM	BASE	ODIN	MAHA	EARLIN	MLCM	BASE	ODIN	MAHA	EARLIN	BASE	ODIN	MAHA	MLCM	EARLIN
CIFAR-10	Densenet	TinyImagenet	95.50	81.20	87.59	93.61	**97.50**	95.33	88.10	92.34	94.38	**96.25**	94.10	97.69	98.29	99.06	**99.14**
		LSUN	96.78	85.40	94.53	96.21	**99.30**	96.07	90.20	94.91	95.78	97.15	95.50	98.85	98.91	99.23	**99.85**
		iSUN	95.59	83.30	91.81	93.21	**97.60**	95.41	89.15	93.82	94.17	96.30	94.80	98.40	97.98	99.04	**99.37**
CIFAR-10	Resnet34	TinyImagenet	**98.10**	71.60	70.39	97.53	93.92	**96.92**	83.30	85.80	96.55	94.46	91.00	91.88	99.43	**99.56**	97.54
		LSUN	**99.04**	71.70	81.94	98.83	98.00	**97.65**	83.35	90.01	97.58	96.5	91.10	95.55	99.64	**99.70**	99.55
		iSUN	**98.25**	71.90	77.89	97.64	95.19	**96.94**	83.45	88.4	96.66	95.09	91.00	94.26	99.47	**99.59**	98.74
CIFAR-100	Densenet	TinyImagenet	87.12	47.90	53.88	80.37	**92.60**	91.65	61.45	81.32	88.40	**93.80**	71.60	89.16	93.64	97.21	**98.04**
		LSUN	90.46	49.70	60.77	85.74	**98.10**	92.87	62.35	84.51	90.85	**96.55**	70.80	92.06	95.82	97.61	**99.96**
		iSUN	88.29	47.30	54.85	81.78	**94.00**	92.04	61.15	82.51	89.30	**94.50**	69.60	90.29	94.81	97.34	**98.61**
CIFAR-100	Resnet34	TinyImagenet	92.88	31.00	64.48	91.76	**95.40**	94.10	58.00	85.77	93.56	**95.20**	67.10	93.06	98.28	98.54	**98.55**
		LSUN	94.76	35.30	64.95	95.31	**99.20**	94.92	55.15	86.09	95.22	**97.10**	65.60	93.39	98.81	98.71	**99.74**
		iSUN	92.36	36.70	63.03	91.98	**97.10**	93.81	55.85	85.33	93.76	**96.05**	65.60	92.76	98.27	98.24	**99.22**

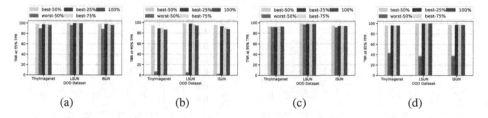

(a) (b) (c) (d)

Fig. 5. TNR at 95% TPR for different combinations of feature selection. (a) DenseNet and (b) ResNet34 pretrained on CIFAR-10, (c) DenseNet and (d) ResNet34 pretrained on CIFAR-100

To demonstrate the clear separation of ID and OOD samples based on the estimated distance, $\kappa(\mathbf{x})$, in Fig. 4, we show the density and the corresponding CDF of $\kappa(\mathbf{x})$ obtained from various test ID and test OOD datasets. We observe that ID and OOD samples have separable distribution based on $\kappa(\mathbf{x})$.

Table 4. OOD detection performance of EARLIN compared to DeConf [7]. **bold** indicates best result.

ID Dataset	Model	OOD	TNR at 95% TPR		AUROC	
			DeConf [7]	EARLIN	DeConf [7]	EARLIN
CIFAR-10	DenseNet	TinyImagenet	95.80	**97.50**	99.10	**99.14**
		LSUN	97.60	**99.30**	99.40	**99.85**
		iSUN	97.50	**97.60**	**99.40**	99.37
CIFAR-100	DenseNet	TinyImagenet	93.30	**93.80**	**98.60**	98.60
		LSUN	93.80	**98.10**	98.70	**99.96**
		iSUN	92.50	**94.00**	98.40	**98.61**

Ablation Studies: In order to detect samples as OOD as early as possible, we explore top (shallowest) 10% layers of the pretrained models to find the separation between ID and OOD samples and report the layer ℓ that performed the best. It is to be noted that we consider only the Batch Normalization (BN) layers of the pretrained models as in these layers parameters sensitive to the ID dataset are learned during training. We show in Fig. 6 how our end result, TNR at 95% TPR varies for different choices of the shallow BN layers in ResNet34 and DenseNet models pretrained on CIFAR-10. In Table 2 we show our choice of layers for the models we considered. We observe that in all cases, the chosen layer is quite early in the network pipeline.

For finding the other hyperparameters, such as the number of maps $\mathcal{N}$, centroid $\mathbf{c}$, and threshold Θ, for each pretrained model, 20% of the training ID samples were used as $\mathbf{D}_{in}$ without using their corresponding classification labels. For each pretrained model, we set $\mathcal{N}$ to be half of the number of 2D feature maps at layer ℓ. The threshold, Θ, is set to 95% ID detection confidence. In Fig. 5, we show the effect of selecting different number of feature maps ($\mathcal{N}$), other than the default 50% (half). Figure 5 shows the TNR values at 95% TPR for different datasets on different pretrained models, for different combinations of selecting 2D features: best 50%, best 25%, best 75%, worst 50% based on ψ (Eq. (1)) and also all 100%. We see that in almost all cases, selecting worst 50% leads to the worst TNR for all datasets (more noticeable for ResNet34 models). Selecting top 50% of the maps leads to either better or equivalent TNR, compared to selecting top 25%, top 75% and all 100%. The choice of top 50% of the 2D feature maps apparently produces the best results.

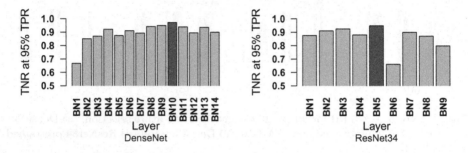

Fig. 6. TNR at 95% TPR for different choice of shallow BN layers in DenseNet and ResNet34 models pretrained on CIFAR-10. Random 1000 iSUN samples have been used as validation OOD data. Our chosen layer is shown in red in each case.

6   Prototype Implementation and Results

Experimental Setup: We build a collaborative inference testbed where a client program with our EARLIN OOD detector runs on an edge device and the deep learning models are deployed on a server machine. Our client program runs on a desktop computer with a moderate CPU-only configuration (Intel®Core™ i7-9750H@2.60 GHz CPU) and 32 GB RAM, a configuration similar to the edge setup described in [2]). The server program, developed using Flask and TensorFlow framework in Python, is deployed at the Google Cloud and is powered by Nvidia K80 GPU devices. For demonstrating the effectiveness of EARLIN, we deploy two CNN models in the cloud: (a) DenseNet with 100 layers and (b) ResNet with 34 layers (both are pretrained on CIFAR-100 with 70% classification accuracy). We deploy their corresponding OOD detection part on the edge device. In all experiments, TinyImageNet dataset is used as OOD. We set a threshold to have 95% prediction confidence on ID samples, the condition we considered while reporting the results on EARLIN in Sect. 5. Hence all the TPR, TNR, and *Accuracy* (detector accuracy) values match those reported in Table 3. We note that the mean latency for computations done at the edge (T_E)

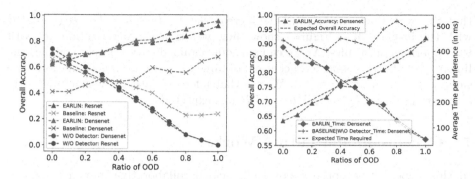

Fig. 7. Change in (a) Performance and (b) Cost vs Performance of Collaborative Setup with ratio of OOD samples using models pretrained on CIFAR-100 dataset and Tiny-Imagenet as OOD dataset.

is 32.8 ± 15 ms and at the server (T_S), it is 47.8 ± 25 ms. The mean communication delay (T_C) is 186.5 ± 52.12 ms. We observe that latency at both edge and server is quite small. At the edge, we deploy a small portion of the model hence the latency is low. On the other hand, the server runs models on GPU resources so the inference time is small there. The communication delay to the server, which accounts round-trip delay between the edge and the server and all other request processing delays before hitting the inference model, seems to the the heavy part of the latency. In our EARLIN-based setup, we improve this latency by not sending to the server when not required and thus getting rid of the communication delay.

Experimental Results: We show a set of aggregated results in Fig. 7. We show the accuracy results for varying degree of OOD samples for EARLIN, Baseline, and "no detector". We observe that as the OOD ratio ρ rises, the accuracy drops sharply if no OOD detector is applied. The overall accuracy of Baseline also declines whereas the accuracy of EARLIN grows as the OOD ratio grows. This is because EARLIN has considerably higher TNR value and higher detection accuracy than the Baseline detector. It is to note that when ρ is close to 0 (very few samples are OOD compared to ID), the accuracy of EARLIN is slightly worse than that of when no detector is used. This is because EARLIN detects, in the worst case, 5% ID samples as OODs (since TPR is 95%), which contributes to reducing the overall accuracy.

Figure 7a shows the performance of EARLIN as we increase ρ. We observe that, as we increase ρ, the overall accuracy increases and the inference latency decreases. The decline in inference latency is due to the fact that as more OOD inputs are fed, they are detected at the edge as OOD. The samples being detected as ID are uploaded to the cloud accounting all three components of delay and the number of those samples decline as ρ grows. We note that time required per inference when model is not associated with any detector is equivalent to the case when OOD samples are detected at the last layer of the model, as in both cases input will be sent to the server for classification and OOD detection. In Fig. 7b, we show the average time required per inference in this case. In Sect. 4, we showed that the overall accuracy of a model increases at a rate of $TNR - acc_{\mathcal{M}} \times TPR$ with the increase of ρ. Figure 7 shows how well that characterization fits with the experimental results. As we can see, our obtained curve closely matches the linear curve for the expected accuracy obtained based on our formulation. The same is true for cost (latency). We see that inference latency decreases linearly at a rate of $(T_C + T_S) \times (FPR - TPR)$, as expected.

7   Conclusion and Future Works

In this paper, we propose a novel edge-cloud collaborative inference system, EARLIN, based on a proposed Out-of-Distribution (OOD) detection technique. EARLIN enables the detection of OOD samples using feature maps obtained from the shallow layers of the pretrained deep learning classifiers. We exploit the advantage of early detection to design at OOD-aware edge-cloud collaborative inference framework as we deploy the small foot-print detector part on an edge device and the full model in the cloud. During inference, the edge detects if an input sample is ID. If it is, the sample is sent to the cloud for classification. Otherwise, the sample is reported as OOD and the edge starts processing the next sample in the pipeline. In this way, we make the inference at the edge faster and more precise. We characterize the performance and cost of the setup. Experimental results on benchmark datasets show that EARLIN performs well on OOD detection. Moreover, when deployed on a prototype implementation, results obtained show that expected improvement in cost and performance is achieved using proposed EARLIN-based setup. In future, we plan to investigate more on building a context-aware adaptive OOD detection setup that takes advantage of choosing from multiple candidate OOD detectors based on desired cost-accuracy trade offs.

References

1. OOD datasets. https://github.com/facebookresearch/odin
2. Canel, C., et al.: Scaling video analytics on constrained edge nodes. In: Proceedings of Machine Learning and Systems 2 (MLSys 2020) (2019)
3. Cardoso, D.O., Gama, J., França, F.M.: Weightless neural networks for open set recognition. Mach. Learn. **106**(9), 1547–1567 (2017)
4. Deecke, L., Vandermeulen, R., Ruff, L., Mandt, S., Kloft, M.: Image anomaly detection with generative adversarial networks. In: Berlingerio, M., Bonchi, F., Gärtner, T., Hurley, N., Ifrim, G. (eds.) ECML PKDD 2018. LNCS (LNAI), vol. 11051, pp. 3–17. Springer, Cham (2019). https://doi.org/10.1007/978-3-030-10925-7_1
5. Gazzaz, S., Nawab, F.: Collaborative edge-cloud and edge-edge video analytics. In: Proceedings of the ACM Symposium on Cloud Computing, p. 484 (2019)
6. Hendrycks, D., Gimpel, K.: A baseline for detecting misclassified and out-of-distribution examples in neural networks. In: ICLR (2017)
7. Hsu, Y.C., Shen, Y., Jin, H., Kira, Z.: Generalized odin: detecting out-of-distribution image without learning from out-of-distribution data. In: Proceedings of the IEEE/CVF Conference on Computer Vision and Pattern Recognition, pp. 10951–10960 (2020)

8. Kang, Y., et al.: Neurosurgeon: collaborative intelligence between the cloud and mobile edge. ACM SIGARCH Comput. Archit. News **45**(1), 615–629 (2017)

9. Laskaridis, S., Venieris, S.I., Almeida, M., Leontiadis, I., Lane, N.D.: Spinn: synergistic progressive inference of neural networks over device and cloud. In: Proceedings of the 26th Annual International Conference on Mobile Computing and Networking, pp. 1–15 (2020)

10. Lee, D., Yu, S., Yu, H.: Multi-class data description for out-of-distribution detection. In: Proceedings of the 26th ACM SIGKDD International Conference on Knowledge Discovery & Data Mining, pp. 1362–1370 (2020)

11. Lee, K., Lee, H., Lee, K., Shin, J.: Training confidence-calibrated classifiers for detecting out-of-distribution samples. In: ICLR (2018)

12. Lee, K., Lee, K., Lee, H., Shin, J.: A simple unified framework for detecting out-of-distribution samples and adversarial attacks. In: Advances in Neural Information Processing Systems, pp. 7167–7177 (2018)

13. Krizhevsky, A., Hinton, G., et al.: Learning multiple layers of features from tiny images (2009)

14. Liang, S., Li, Y., Srikant, R.: Enhancing the reliability of out-of-distribution image detection in neural networks. In: International Conference on Learning Representations (2018)

15. Liu, H., Eldarrat, F., Alqahtani, H., Reznik, A., De Foy, X., Zhang, Y.: Mobile edge cloud system: architectures, challenges, and approaches. IEEE Syst. J. **12**(3), 2495–2508 (2017)

16. Matthew, D., Fergus, R.: Visualizing and understanding convolutional neural networks. In: Proceedings of the 13th European Conference Computer Vision and Pattern Recognition, Zurich, Switzerland, pp. 6–12 (2014)

17. Mohseni, S., Pitale, M., Yadawa, J., Wang, Z.: Self-supervised learning for generalizable out-of-distribution detection. In: AAAI, pp. 5216–5223 (2020)

18. Nguyen, A., Yosinski, J., Clune, J.: Deep neural networks are easily fooled: high confidence predictions for unrecognizable images. In: Proceedings of the IEEE Conference on Computer Vision and Pattern Recognition, pp. 427–436 (2015)

19. Schindler, G., Zöhrer, M., Pernkopf, F., Fröning, H.: Towards efficient forward propagation on resource-constrained systems. In: Berlingerio, M., Bonchi, F., Gärtner, T., Hurley, N., Ifrim, G. (eds.) ECML PKDD 2018. LNCS (LNAI), vol. 11051, pp. 426–442. Springer, Cham (2019). https://doi.org/10.1007/978-3-030-10925-7_26

20. Szegedy, C., et al.: Intriguing properties of neural networks. arXiv preprint arXiv:1312.6199 (2013)

21. Verbelen, T., Simoens, P., De Turck, F., Dhoedt, B.: Cloudlets: bringing the cloud to the mobile user. In: Proceedings of the third ACM Workshop on Mobile Cloud Computing and Services, pp. 29–36 (2012)

22. Xie, C., Koyejo, O., Gupta, I.: SLSGD: secure and efficient distributed on-device machine learning. In: Brefeld, U., Fromont, E., Hotho, A., Knobbe, A., Maathuis, M., Robardet, C. (eds.) ECML PKDD 2019. LNCS (LNAI), vol. 11907, pp. 213–228. Springer, Cham (2020). https://doi.org/10.1007/978-3-030-46147-8_13

23. Yu, H., Yu, S., Lee, D.: Convolutional neural networks with compression complexity pooling for out-of-distribution image detection. In: Proceedings of the Twenty-Ninth International Joint Conference on Artificial Intelligence (IJCAI-20). IJCAI (2020)

Semi-supervised and Few-Shot Learning

LSMI-Sinkhorn: Semi-supervised Mutual Information Estimation with Optimal Transport

Yanbin Liu[1](✉), Makoto Yamada[2,3], Yao-Hung Hubert Tsai[4], Tam Le[3], Ruslan Salakhutdinov[4], and Yi Yang[1]

[1] AAII, University of Technology Sydney, Ultimo, Australia
[2] Kyoto University, Kyoto, Japan
[3] RIKEN AIP, Tokyo, Japan
[4] Carnegie Mellon University, Pittsburgh, USA

Abstract. Estimating mutual information is an important statistics and machine learning problem. To estimate the mutual information from data, a common practice is preparing a set of paired samples $\{(x_i, y_i)\}_{i=1}^{n} \overset{\text{i.i.d.}}{\sim} p(x, y)$. However, in many situations, it is difficult to obtain a large number of data pairs. To address this problem, we propose the semi-supervised Squared-loss Mutual Information (SMI) estimation method using a small number of paired samples and the available unpaired ones. We first represent SMI through the density ratio function, where the expectation is approximated by the samples from marginals and its assignment parameters. The objective is formulated using the optimal transport problem and quadratic programming. Then, we introduce the **Least-Squares Mutual Information with Sinkhorn (LSMI-Sinkhorn)** algorithm for efficient optimization. Through experiments, we first demonstrate that the proposed method can estimate the SMI without a large number of paired samples. Then, we show the effectiveness of the proposed LSMI-Sinkhorn algorithm on various types of machine learning problems such as image matching and photo album summarization. Code can be found at https://github.com/csyanbin/LSMI-Sinkhorn.

Keywords: Mutual information estimation · Density ratio · Sinkhorn algorithm · Optimal transport

1 Introduction

Mutual information (MI) represents the statistical independence between two random variables [4], and it is widely used in various types of machine learning applications including feature selection [20, 21], dimensionality reduction [19], and causal inference [23]. More recently, deep neural network (DNN) models have started using MI as a regularizer for obtaining better representations from data

---

Y. Liu and M. Yamada—Equal contribution.

N. Oliver et al. (Eds.): ECML PKDD 2021, LNAI 12975, pp. 655–670, 2021.
https://doi.org/10.1007/978-3-030-86486-6_40

such as infoVAE [26] and deep infoMax [9]. Another application is improving the generative adversarial networks (GANs) [8]. For instance, Mutual Information Neural Estimation (MINE) [1] was proposed to maximize or minimize the MI in deep networks and alleviate the mode-dropping issues in GANS. In all these examples, MI estimation is the core of all these applications.

In various MI estimation approaches, the probability density ratio function is considered to be one of the most important components:

$$r(\boldsymbol{x}, \boldsymbol{y}) = \frac{p(\boldsymbol{x}, \boldsymbol{y})}{p(\boldsymbol{x})p(\boldsymbol{y})}.$$

A straightforward method to estimate this ratio is the estimation of the probability densities (i.e., $p(\boldsymbol{x}, \boldsymbol{y})$, $p(\boldsymbol{x})$, and $p(\boldsymbol{y})$), followed by calculating their ratio. However, directly estimating the probability density is difficult, thereby making this two-step approach inefficient. To address the issue, Suzuki *et al.* [21] proposed to directly estimate the density ratio by avoiding the density estimation [20, 21]. Nonetheless, the abovementioned methods requires a large number of paired data when estimating the MI.

Under practical setting, we can only obtain a small number of paired samples. For example, it requires a massive amount of human labor to obtain one-to-one correspondences from one language to another. Thus, it prevents us to easily measure the MI across languages. Hence, a research question arises:

Can we perform mutual information estimation using unpaired samples and a small number of data pairs?

To answer the above question, in this paper, we propose a semi-supervised MI estimation approach, particularly designed for the Squared-loss Mutual Information (SMI) (a.k.a., χ^2-divergence between $p(\boldsymbol{x}, \boldsymbol{y})$ and $p(\boldsymbol{x})p(\boldsymbol{y})$) [20]. We first formulate the SMI estimation as the optimal transport problem with density-ratio estimation. Then, we propose the **Least-Squares Mutual Information with Sinkhorn (LSMI-Sinkhorn)** algorithm to optimize the problem. The algorithm has the computational complexity of $O(n_x n_y)$; hence, it is computationally efficient. Through experiments, we first demonstrate that the proposed method can estimate the SMI without a large number of paired samples. Then, we visualize the optimal transport matrix, which is an approximation of the joint density $p(\boldsymbol{x}, \boldsymbol{y})$, for a better understanding of the proposed algorithm. Finally, for image matching and photo album summarization, we show the effectiveness of the proposed method.

The contributions of this paper can be summarized as follows:

- We proposed the semi-supervised Squared-loss Mutual Information (SMI) estimation approach that does not require a large number of paired samples.
- We formulate mutual information estimation as a joint density-ratio fitting and optimal transport problem, and propose an efficient **LSMI-Sinkhorn** algorithm to optimize it with a monotonical decreasing guarantee.
- We experimentally demonstrate the effectiveness of the proposed LSMI-Sinkhorn for MI estimation, and further show its broader applications to the image matching and photo album summarization problems.

2 Problem Formulation

In this section, we formulate the problem of Squared-loss Mutual Information (SMI) estimation using a small number of paired samples and a large number of unpaired samples.

Formally, let $\mathcal{X} \subset \mathbb{R}^{d_x}$ be the domain of random variable $\boldsymbol{x}$ and $\mathcal{Y} \subset \mathbb{R}^{d_y}$ be the domain of another random variable $\boldsymbol{y}$. Suppose we are given n independent and identically distributed (i.i.d.) *paired* samples:

$$\{(\boldsymbol{x}_i, \boldsymbol{y}_i)\}_{i=1}^n,$$

where the number of paired samples n is small. Apart from the paired samples, we also have access to n_x and n_y i.i.d. samples from the marginal distributions:

$$\{\boldsymbol{x}_i\}_{i=n+1}^{n+n_x} \overset{\text{i.i.d.}}{\sim} p(\boldsymbol{x}) \text{ and } \{\boldsymbol{y}_j\}_{j=n+1}^{n+n_y} \overset{\text{i.i.d.}}{\sim} p(\boldsymbol{y}),$$

where the number of unpaired samples n_x and n_y is much larger than that of paired samples n (e.g., $n = 10$ and $n_x = n_y = 1000$). We also denote $\boldsymbol{x}'_i = \boldsymbol{x}_{i-n}, i \in \{n+1, n+2, \ldots, n+n_x\}$ and $\boldsymbol{y}'_j = \boldsymbol{y}_{j-n}, j \in \{n+1, n+2, \ldots, n+n_y\}$, respectively. Note that the input dimensions d_x, d_y and the number of samples n_x, n_y may be different.

This paper aims to estimate the SMI [20] (a.k.a., χ^2-divergence between $p(\boldsymbol{x}, \boldsymbol{y})$ and $p(\boldsymbol{x})p(\boldsymbol{y})$) from $\{(\boldsymbol{x}_i, \boldsymbol{y}_i)\}_{i=1}^n$ with the help of the extra unpaired samples $\{\boldsymbol{x}_i\}_{i=n+1}^{n+n_x}$ and $\{\boldsymbol{y}_j\}_{j=n+1}^{n+n_y}$. Specifically, the SMI between random variables X and Y is defined as

$$\text{SMI}(X, Y) = \frac{1}{2} \iint (r(\boldsymbol{x}, \boldsymbol{y}) - 1)^2 p(\boldsymbol{x})p(\boldsymbol{y}) \mathrm{d}\boldsymbol{x}\mathrm{d}\boldsymbol{y}, \tag{1}$$

where $r(\boldsymbol{x}, \boldsymbol{y}) = \frac{p(\boldsymbol{x}, \boldsymbol{y})}{p(\boldsymbol{x})p(\boldsymbol{y})}$ is the density-ratio function. *SMI takes 0 if and only if X and Y are independent (i.e., $p(\boldsymbol{x}, \boldsymbol{y}) = p(\boldsymbol{x})p(\boldsymbol{y})$), and takes a positive value if they are not independent.*

Naturally, if we know the estimation of the density-ratio function, then we can approximate the SMI in Eq. 1 as

$$\widehat{\text{SMI}}(X, Y) = \frac{1}{2(n+n_x)(n+n_y)} \sum_{i=1}^{n+n_x} \sum_{j=1}^{n+n_y} (r_\alpha(\boldsymbol{x}_i, \boldsymbol{y}_j) - 1)^2,$$

where $r_\alpha(\boldsymbol{x}, \boldsymbol{y})$ is an estimation of the true density ratio function $r(\boldsymbol{x}, \boldsymbol{y})$ parameterized by $\boldsymbol{\alpha}$. More details are discussed in Sect. 3.1.

However, in many real applications, it is difficult or laborious to obtain sufficient paired samples for density ratio estimation, which may result in high variance and bias when computing the SMI. In this paper, the key idea is to align the unpaired samples under this limited number of paired samples setting, and propose an objective to incorporate both the paired samples and aligned samples for a better SMI estimation.

3   Methodology

In this section, we propose the SMI estimation algorithm with limited number of paired samples and large number of unpaired samples.

3.1   Least-Squares Mutual Information with Sinkhorn Algorithm

We employ the following density-ratio model. It first samples two sets of basis vectors $\{\widetilde{\boldsymbol{x}}_i\}_{i=1}^b$ and $\{\widetilde{\boldsymbol{y}}_i\}_{i=1}^b$ from $\{\boldsymbol{x}_i\}_{i=1}^{n+n_x}$ and $\{\boldsymbol{y}_j\}_{j=1}^{n+n_y}$, then computes

$$r_\alpha(\boldsymbol{x},\boldsymbol{y}) = \sum_{\ell=1}^b \alpha_\ell K(\widetilde{\boldsymbol{x}}_\ell,\boldsymbol{x})L(\widetilde{\boldsymbol{y}}_\ell,\boldsymbol{y}) = \boldsymbol{\alpha}^\top\boldsymbol{\varphi}(\boldsymbol{x},\boldsymbol{y}), \tag{2}$$

where $\boldsymbol{\alpha}\in\mathbb{R}^b$, $K(\cdot,\cdot)$ and $L(\cdot,\cdot)$ are kernel functions, $\boldsymbol{\varphi}(\boldsymbol{x},\boldsymbol{y})=\boldsymbol{k}(\boldsymbol{x})\circ\boldsymbol{l}(\boldsymbol{y})$ with $\boldsymbol{k}(\boldsymbol{x})=[K(\widetilde{\boldsymbol{x}}_1,\boldsymbol{x}),\dots,K(\widetilde{\boldsymbol{x}}_b,\boldsymbol{x})]^\top\in\mathbb{R}^b$, $\boldsymbol{l}(\boldsymbol{y})=[L(\widetilde{\boldsymbol{y}}_1,\boldsymbol{y}),\dots,L(\widetilde{\boldsymbol{y}}_b,\boldsymbol{y})]^\top\in\mathbb{R}^b$.

In this paper, we optimize $\boldsymbol{\alpha}$ by minimizing the squared error loss between the true density-ratio function $r(\boldsymbol{x},\boldsymbol{y})$ and its parameterized model $r_\alpha(\boldsymbol{x},\boldsymbol{y})$:

$$\text{Loss} = \frac{1}{2}\iint\left(r_\alpha(\boldsymbol{x},\boldsymbol{y})-\frac{p(\boldsymbol{x},\boldsymbol{y})}{p(\boldsymbol{x})p(\boldsymbol{y})}\right)^2 p(\boldsymbol{x})p(\boldsymbol{y})\mathrm{d}\boldsymbol{x}\mathrm{d}\boldsymbol{y}$$

$$= \frac{1}{2}\iint r_\alpha(\boldsymbol{x},\boldsymbol{y})^2 p(\boldsymbol{x})p(\boldsymbol{y})\mathrm{d}\boldsymbol{x}\mathrm{d}\boldsymbol{y} - \iint r_\alpha(\boldsymbol{x},\boldsymbol{y})p(\boldsymbol{x},\boldsymbol{y})\mathrm{d}\boldsymbol{x}\mathrm{d}\boldsymbol{y} + \text{const.} \tag{3}$$

For the first term of Eq. (3), we can approximate it by using a large number of unpaired samples as it only involves $p(\boldsymbol{x}),p(\boldsymbol{y})$. However, to approximate the second term, paired samples from the joint distribution (i.e., $p(\boldsymbol{x},\boldsymbol{y})$) are required. Since we only have a limited number of paired samples in our setting, the approximation of the second term may have high bias and variance.

To deal with this issue, we leverage the abundant unpaired samples to help the approximation of the second term. Since we have no access to the true pair information for these unpaired samples, we propose a practical way to estimate their pair information. Specifically, we introduce a matrix $\boldsymbol{\Pi}$ ($\pi_{ij}\geq 0$, $\sum_{i=1}^{n_x}\sum_{j=1}^{n_y}\pi_{i,j}=1$) that can be regarded as a parameterized estimation of the joint density function $p(\boldsymbol{x},\boldsymbol{y})$. Then, we approximate the second term of Eq. (3)

$$\iint r_\alpha(\boldsymbol{x},\boldsymbol{y})p(\boldsymbol{x},\boldsymbol{y})\mathrm{d}\boldsymbol{x}\mathrm{d}\boldsymbol{y} \approx \frac{\beta}{n}\sum_{i=1}^n r_\alpha(\boldsymbol{x}_i,\boldsymbol{y}_i)+(1-\beta)\sum_{i=1}^{n_x}\sum_{j=1}^{n_y}\pi_{ij}r_\alpha(\boldsymbol{x}_i',\boldsymbol{y}_j'), \tag{4}$$

where $0\leq\beta\leq 1$ is a parameter to balance the terms of paired and unpaired samples. Ideally, if we can set $\pi_{ij}=\delta(\boldsymbol{x}_i',\boldsymbol{y}_j')/n'$ where $\delta(\boldsymbol{x}_i',\boldsymbol{y}_j')$ is 1 for all paired $(\boldsymbol{x}_i',\boldsymbol{y}_j')$ and 0 otherwise, and n' is the total number of pairs, then we can recover the original empirical estimation (i.e., $\pi_{ij}=p(\boldsymbol{x}_i',\boldsymbol{y}_j')$ ideally).

Now, we can substitute Eq. (2) and Eq. (4) back into the squared error loss function Eq. (3) to obtain the final loss function as

$$J(\boldsymbol{\Pi},\boldsymbol{\alpha}) = \frac{1}{2}\boldsymbol{\alpha}^\top H\boldsymbol{\alpha} - \boldsymbol{\alpha}^\top h_{\boldsymbol{\Pi},\beta},$$

Algorithm 1: LSMI-Sinkhorn Algorithm.

Initialize $\mathbf{\Pi}^{(0)}$ and $\mathbf{\Pi}^{(1)}$ such that $\|\mathbf{\Pi}^{(1)} - \mathbf{\Pi}^{(0)}\|_F > \eta$ (η is the stopping parameter), and $\boldsymbol{\alpha}^{(0)}$, set the regularization parameters ϵ and λ, the number of maximum iterations T, and the iteration index $t = 1$.

while $t \leq T$ *and* $\|\mathbf{\Pi}^{(t)} - \mathbf{\Pi}^{(t-1)}\|_F > \eta$ **do**

   | $\boldsymbol{\alpha}^{(t+1)} = \operatorname{argmin}_{\alpha} J(\mathbf{\Pi}^{(t)}, \boldsymbol{\alpha})$.
   | $\mathbf{\Pi}^{(t+1)} = \operatorname{argmin}_{\Pi} J(\mathbf{\Pi}, \boldsymbol{\alpha}^{(t+1)})$.
   | $t = t + 1$.

return $\mathbf{\Pi}^{(t-1)}$ and $\boldsymbol{\alpha}^{(t-1)}$.

where

$$H = \frac{1}{(n + n_x)(n + n_y)} \sum_{i=1}^{n+n_x} \sum_{j=1}^{n+n_y} \boldsymbol{\varphi}(\boldsymbol{x}_i, \boldsymbol{y}_j) \boldsymbol{\varphi}(\boldsymbol{x}_i, \boldsymbol{y}_j)^{\top},$$

$$\boldsymbol{h}_{\Pi,\beta} = \frac{\beta}{n} \sum_{i=1}^{n} \boldsymbol{\varphi}(\boldsymbol{x}_i, \boldsymbol{y}_i) + (1 - \beta) \sum_{i=1}^{n_x} \sum_{j=1}^{n_y} \pi_{ij} \boldsymbol{\varphi}(\boldsymbol{x}'_i, \boldsymbol{y}'_j).$$

Since we want to estimate the density-ratio function by minimizing Eq. (3), the optimization problem is then given as

$$\min_{\Pi,\alpha} \quad J(\mathbf{\Pi}, \boldsymbol{\alpha}) = \frac{1}{2} \boldsymbol{\alpha}^{\top} H \boldsymbol{\alpha} - \boldsymbol{\alpha}^{\top} \boldsymbol{h}_{\Pi,\beta} + \epsilon H(\mathbf{\Pi}) + \frac{\lambda}{2} \|\boldsymbol{\alpha}\|_2^2$$

$$\text{s.t.} \quad \mathbf{\Pi} \mathbf{1}_{n_y} = n_x^{-1} \mathbf{1}_{n_x} \text{ and } \mathbf{\Pi}^{\top} \mathbf{1}_{n_x} = n_y^{-1} \mathbf{1}_{n_y}. \tag{5}$$

Here, we add several regularization terms. $H(\mathbf{\Pi}) = \sum_{i=1}^{n_x} \sum_{j=1}^{n_y} \pi_{ij}(\log \pi_{ij} - 1)$ is the negative entropic regularization to ensure $\mathbf{\Pi}$ non-negative, and $\epsilon > 0$ is the corresponding regularization parameter. $\|\boldsymbol{\alpha}\|_2^2$ is the regularization on $\boldsymbol{\alpha}$, and $\lambda \geq 0$ is the corresponding regularization parameter.

3.2 Optimization

The objective function $J(\mathbf{\Pi}, \boldsymbol{\alpha})$ is not jointly convex. However, if we fix one variable, it becomes a convex function for the other. Thus, we employ the alternating optimization approach (see Algorithm 1) on $\mathbf{\Pi}$ and $\boldsymbol{\alpha}$, respectively.

1) Optimizing $\mathbf{\Pi}$ Using the Sinkhorn Algorithm. When fixing $\boldsymbol{\alpha}$, the term in our objective relating to $\mathbf{\Pi}$ is

$$\sum_{i=1}^{n_x} \sum_{j=1}^{n_y} \pi_{ij} \boldsymbol{\alpha}^{\top} \boldsymbol{\varphi}(\boldsymbol{x}'_i, \boldsymbol{y}'_j) = \sum_{i=1}^{n_x} \sum_{j=1}^{n_y} \pi_{ij} [C_{\alpha}]_{ij},$$

where $C_{\alpha} = K^{\top} \operatorname{diag}(\boldsymbol{\alpha}) L \in \mathbb{R}^{n_x \times n_y}$, $K = (\boldsymbol{k}(\boldsymbol{x}'_1), \boldsymbol{k}(\boldsymbol{x}'_2), \dots, \boldsymbol{k}(\boldsymbol{x}'_{n_x})) \in \mathbb{R}^{b \times n_x}$, and $L = (\boldsymbol{l}(\boldsymbol{y}'_1), \boldsymbol{l}(\boldsymbol{y}'_2), \dots, \boldsymbol{l}(\boldsymbol{y}'_{n_y})) \in \mathbb{R}^{b \times n_y}$. This formulation can be considered as an optimal transport problem if we maximize it with respect to $\mathbf{\Pi}$ [5]. It is worth noting that the rank of C_{α} is at most $b \ll \min(n_x, n_y)$ with b being a constant (e.g., $b = 100$), and the computational complexity of the cost matrix C_α is $O(n_x n_y)$. The optimization problem with fixed α becomes

$$\min_{\Pi} \quad -\sum_{i=1}^{n_x}\sum_{j=1}^{n_y} \pi_{ij}(1-\beta)[C_\alpha]_{ij} + \epsilon H(\Pi)$$

$$\text{s.t.} \quad \Pi 1_{n_y} = n_x^{-1} 1_{n_x} \text{ and } \Pi^\top 1_{n_x} = n_y^{-1} 1_{n_y}, \tag{6}$$

which can be efficiently solved using the Sinkhorn algorithm [5,17][1]. When α is fixed, problem (6) is convex with respect to Π.

2) Optimizing α. Next, when we fix Π, the optimization problem becomes

$$\min_{\alpha} \quad \frac{1}{2}\alpha^\top H \alpha - \alpha^\top h_{\Pi,\beta} + \frac{\lambda}{2}\|\alpha\|_2^2. \tag{7}$$

Problem (7) is a quadratic programming and convex. It has an analytical solution

$$\hat{\alpha} = (H + \lambda I_b)^{-1} h_{\Pi,\beta}, \tag{8}$$

where $I_b \in \mathbb{R}^{b \times b}$ is an identity matrix. Note that the H matrix does not depend on either Π or α, and it is a positive definite matrix.

Convergence Analysis. To optimize $J(\Pi, \alpha)$, we alternatively solve two convex optimization problems. Thus, the following property holds true.

Proposition 1. *Algorithm 1 will monotonically decrease the objective function $J(\Pi, \alpha)$ in each iteration.*

Proof. We show that $J(\Pi^{(t+1)}, \alpha^{(t+1)}) \leq J(\Pi^{(t)}, \alpha^{(t)})$. First, because $\alpha^{(t+1)} = \text{argmin}_\alpha J(\Pi^{(t)}, \alpha)$ and $\alpha^{(t+1)}$ is the globally optimum solution, we have

$$J(\Pi^{(t)}, \alpha^{(t+1)}) \leq J(\Pi^{(t)}, \alpha^{(t)}).$$

Moreover, because $\Pi^{(t+1)} = \text{argmin}_\Pi J(\Pi, \alpha^{(t+1)})$ and $\Pi^{(t+1)}$ is the globally optimum solution, we have

$$J(\Pi^{(t+1)}, \alpha^{(t+1)}) \leq J(\Pi^{(t)}, \alpha^{(t+1)}).$$

Therefore,

$$J(\Pi^{(t+1)}, \alpha^{(t+1)}) \leq J(\Pi^{(t)}, \alpha^{(t)}).$$

$\square$

Model Selection. Algorithm 1 is dubbed as LSMI-Sinkhorn algorithm since it utilizes Sinkhorn algorithm for LSMI estimation. It includes several tuning parameters (i.e., λ and β) and determining the model parameters is critical to obtain a good estimation of SMI. Accordingly, we use the cross-validation with the hold-out set to select the model parameters.

---

[1] In this paper, we use the log-stabilized Sinkhorn algorithm [16].

First, the paired samples $\{(\boldsymbol{x}_i, \boldsymbol{y}_i)\}_{i=1}^n$ are divided into two subsets $\mathcal{D}_{\mathrm{tr}}$ and $\mathcal{D}_{\mathrm{te}}$. Then, we train the density-ratio $r_\alpha(\boldsymbol{x}, \boldsymbol{y})$ using $\mathcal{D}_{\mathrm{tr}}$ and the unpaired samples: $\{\boldsymbol{x}_i\}_{i=n+1}^{n+n_x}$ and $\{\boldsymbol{y}_j\}_{j=n+1}^{n+n_y}$. The hold-out error can be calculated by approximating Eq. (3) using the hold-out samples $\mathcal{D}_{\mathrm{te}}$ as

$$\widehat{J}_{\mathrm{te}} = \frac{1}{2|\mathcal{D}_{\mathrm{te}}|^2} \sum_{\boldsymbol{x}, \boldsymbol{y} \in \mathcal{D}_{\mathrm{te}}} r_{\widehat{a}}(\boldsymbol{x}, \boldsymbol{y})^2 - \frac{1}{|\mathcal{D}_{\mathrm{te}}|} \sum_{(\boldsymbol{x}, \boldsymbol{y}) \in \mathcal{D}_{\mathrm{te}}} r_{\widehat{a}}(\boldsymbol{x}, \boldsymbol{y}),$$

where $|\mathcal{D}|$ denotes the number of samples in the set $\mathcal{D}$, $\sum_{\boldsymbol{x}, \boldsymbol{y} \in \mathcal{D}_{\mathrm{te}}}$ denotes the summation over all possible combinations of $\boldsymbol{x}$ and $\boldsymbol{y}$ in $\mathcal{D}_{\mathrm{te}}$, and $\sum_{(\boldsymbol{x}, \boldsymbol{y}) \in \mathcal{D}_{\mathrm{te}}}$ denotes the summation over all pairs of $(\boldsymbol{x}, \boldsymbol{y})$ in $\mathcal{D}_{\mathrm{te}}$. We select the parameters that lead to the smallest $\widehat{J}_{\mathrm{te}}$.

3.3 Discussion

Relation to Least-Squares Object Matching (LSOM). In this section, we show that the LSOM algorithm [22,24] can be considered as a special case of the proposed framework. If $\boldsymbol{\Pi}$ is a permutation matrix and $n' = n_x = n_y$,

$$\boldsymbol{\Pi} = \{0, 1\}^{n' \times n'}, \ \boldsymbol{\Pi}\mathbf{1}_{n'} = \mathbf{1}_{n'}, \text{ and } \boldsymbol{\Pi}^\top \mathbf{1}_{n'} = \mathbf{1}_{n'},$$

where $\boldsymbol{\Pi}^\top \boldsymbol{\Pi} = \boldsymbol{\Pi}\boldsymbol{\Pi}^\top = \mathbf{I}_{n'}$. Then, the estimation of SMI using the permutation matrix can be written as

$$\widehat{\mathrm{SMI}}(X, Y) = \frac{\beta}{2n} \sum_{i=1}^n r_\alpha(\boldsymbol{x}_i, \boldsymbol{y}_i) + \frac{1}{2n'} \sum_{i=1}^{n'} (1 - \beta) r_\alpha(\boldsymbol{x}'_i, \boldsymbol{y}'_{\pi(i)}) - \frac{1}{2},$$

where $\pi(i)$ is the permutation function. In order to calculate $\widehat{\mathrm{SMI}}(X, Y)$, the optimization problem is written as

$$\min_{\boldsymbol{\Pi}, \boldsymbol{\alpha}} \ \frac{1}{2} \boldsymbol{\alpha}^\top \boldsymbol{H} \boldsymbol{\alpha} - \boldsymbol{\alpha}^\top \boldsymbol{h}_{\boldsymbol{\Pi}, \beta} + \frac{\lambda}{2} \|\boldsymbol{\alpha}\|_2^2$$

$$\text{s.t. } \boldsymbol{\Pi}\mathbf{1}_{n'} = \mathbf{1}_{n'}, \ \boldsymbol{\Pi}^\top \mathbf{1}_{n'} = \mathbf{1}_{n'}, \ \boldsymbol{\Pi} \in \{0, 1\}^{n' \times n'}.$$

To solve this problem, LSOM uses the Hungarian algorithm [10] instead of the Sinkhorn algorithm [5] for optimizing $\boldsymbol{\Pi}$. It is noteworthy that in the original LSOM algorithm, the permutation matrix is introduced to permute the Gram matrix (i.e., $\boldsymbol{\Pi}\boldsymbol{L}\boldsymbol{\Pi}^\top$) and $\boldsymbol{\Pi}$ is also included within the $\boldsymbol{H}$ computation. However, in our formulation, the permutation matrix depends only on $\boldsymbol{h}_{\boldsymbol{\Pi}, \beta}$. This difference enables us to show a monotonic decrease for the loss function of the proposed algorithm.

Since LSOM aims to seek the alignment, it is more suitable to find the exact matching among samples when the exact matching exists. In contrast, the proposed LSMI-Sinkhorn is reliable even when there is no exact matching. Moreover, LSOM assumes the same number of samples (i.e., $n_x = n_y$), while our

LSMI-Sinkhorn does not have this constraint. For computational complexity, the Hungarian algorithm requires $O(n'^3)$ while the Sinkhorn requires $O(n'^2)$.

Computational Complexity. First, the computational complexity of estimating Π is based on the computation of the cost matrix C_α and the Sinkhorn iterations. The computational complexity of C_α is $O(n_x n_y)$ and that of Sinkhorn algorithm is $O(n_x n_y)$. Therefore, the computational complexity of the Sinkhorn iteration is $O(n_x n_y)$. Second, for the α computation, the complexity to compute H is $O((n + n_x)^2 + (n + n_y)^2)$ and that for $h_{\Pi,\beta}$ is $O(n_x n_y)$. In addition, estimating α has the complexity $O(b^3)$, which is negligible with a small constant b. To conclude, the total computational complexity of the initialization needs $O((n + n_x)^2 + (n + n_y)^2)$ and the iterations requires $O(n_x n_y)$. In particular, for small n and large $n_x = n_y$, the computational complexity is $O(n_x^2)$.

As a comparison, for another related algorithm, Gromove-Wasserstein [11, 14], the time complexity of computing the objective function is $O(n_x^4)$ for general cases and $O(n_x^3)$ for some specific losses (e.g. L_2 loss, Kullback-Leibler loss) [14].

4   Related Work

In this paper, we focus on the mutual information estimation problem. Moreover, the proposed LSMI-Sinkhorn algorithm is related to Gromov-Wasserstein [11,14] and kernelized sorting [6,15].

Mutual Information Estimation. To estimate the MI, a straightforward approach is to estimate the probability density $p(\boldsymbol{x}, \boldsymbol{y})$ from the paired samples $\{(\boldsymbol{x}_i, \boldsymbol{y}_i)\}_{i=1}^n$, $p(\boldsymbol{x})$ from $\{\boldsymbol{x}_i\}_{i=1}^n$, and $p(\boldsymbol{y})$ from $\{\boldsymbol{y}_i\}_{i=1}^n$, respectively.

Because the estimation of the probability density is itself a difficult problem, this straightforward approach does not work well. To handle this, a density-ratio based approach can be promising [20,21]. More recently, deep learning based mutual information estimation algorithms have been proposed [1,12]. However, these approaches still require a large number of paired samples to estimate the MI. Thus, in real world situations when we only have a limited number of paired samples, existing approaches are not effective to obtain a reliable estimation.

Gromov-Wasserstein and Kernelized Sorting. Given two set of vectors in different spaces, the Gromov-Wasserstein distance [11] can be used to find the optimal alignment between them. This method considers the pairwise distance between samples in the same set to build the distance matrix, then it finds a matching by minimizing the difference between the pairwise distance matrices:

$$\min_{\Pi} \sum_{i=1}^{n_x}\sum_{j=1}^{n_y}\sum_{i'=1}^{n_x}\sum_{i'=1}^{n_y} \pi_{ij}\pi_{i'j'}(D(\boldsymbol{x}_i, \boldsymbol{x}_{i'}) - D(\boldsymbol{y}_j, \boldsymbol{y}_{j'}))^2,$$

$$\text{s.t.} \quad \Pi \mathbf{1}_{n_y} = \boldsymbol{a}, \Pi^\top \mathbf{1}_{n_x} = \boldsymbol{b}, \pi_{ij} \geq 0,$$

where $\boldsymbol{a} \in \Sigma_{n_x}$, $\boldsymbol{b} \in \Sigma_{n_y}$, and $\Sigma_n = \{p \in \mathbb{R}_n^+; \sum_i p_i = 1\}$ is the probability simplex.

Computing Gromov-Wasserstein distance requires solving the quadratic assignment problem (QAP), and it is generally NP-hard for arbitrary inputs [13,14]. In this work, we estimate the SMI by simultaneously solving the alignment and fitting the distribution ratio by efficiently leveraging the Sinkhorn algorithm and properties of the squared-loss. Recently, semi-supervised Gromov-Wasserstein-based Optimal transport has been proposed and applied to the heterogeneous domain adaptation problems [25]. However, their method cannot be directly used to measure the independence between two sets of random variables. In contrast, we can achieve this by the estimation of the density-ratio function.

Kernelized sorting methods [6,15] are highly related to Gromov-Wasserstein. Specifically, the kernelized sorting determines a set of paired samples by maximizing the Hilbert-Schmidt independence criterion (HSIC) between samples. Similar to LSOM [24], the kernelized sorting also has the assumption of the same number of samples (i.e., $\{x_i'\}_{i=1}^{n'}$ and $\{y_i'\}_{j=1}^{n'}$). This assumption prohibits both LSOM and kernelized sorting from being applied to a broader range of applications, such as photo album summarization in Sect. 5.5. To the contrary, since the proposed LSMI-Sinkhorn does not rely on this assumption, it can be applied to more general scenarios when $n_x \neq n_y$.

5 Experiments

In this section, we first estimate the SMI on both the synthetic data and benchmark datasets. Then, we apply our algorithm to real world applications, i.e., deep image matching and photo album summarization.

5.1 Setup

For the density-ratio model, we utilize the Gaussian kernels:

$$K(x, x') = \exp\left(-\frac{\|x - x'\|_2^2}{2\sigma_x^2}\right), L(y, y') = \exp\left(-\frac{\|y - y'\|_2^2}{2\sigma_y^2}\right),$$

where σ_x and σ_y denote the widths of the kernel that are set using the median heuristic [18] as $\sigma_x = 2^{-1/2}\text{median}(\{\|x_i - x_j\|_2\}_{i,j=1}^{n_x})$, $\sigma_y = 2^{-1/2}\text{median}(\{\|y_i - y_j\|_2\}_{i,j=1}^{n_y})$. We set the number of basis $b = 200$, $\epsilon = 0.3$, the maximum number of iterations $T = 20$, and the stopping parameter $\eta = 10^{-9}$. β and λ are chosen by cross-validation.

5.2 Convergence and Runtime

We first demonstrate the convergence of the loss function and the estimated SMI value. Here, we generate synthetic data from $y = 0.5x + \mathcal{N}(0, 0.01)$ and randomly choose $n = 50$ paired samples and $n_x = n_y = 500$ unpaired samples. The convergence curve is shown in Fig. 1. The loss value and SMI value converge quickly (<5 iterations), which is consistent with Proposition 1.

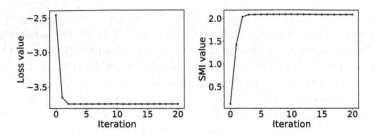

Fig. 1. Convergence curves of the loss and SMI values.

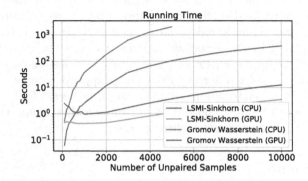

Fig. 2. Runtime comparison of LSMI-Sinkhorn and Gromov-Wasserstein. A base-10 log scale is used for the Y axis.

Then, we perform a comparison between the runtimes of the proposed LSMI-Sinkhorn and Gromov-Wasserstein for CPU and GPU implementations. The data are sampled from two 2D random measures, where $n_x = n_y \in \{100, 200, \ldots, 9000, 10000\}$ is the number of unpaired data and $n = 100$ is the number of paired data (only for LSMI-Sinkhorn). For Gromov-Wasserstein, we use the CPU implementation from Python Optimal Transport toolbox [7] and the Pytorch GPU implementation from [2]. We use the squared loss function and set the entropic regularization ϵ to 0.005 according to the original code. For LSMI-Sinkhorn, we implement the CPU and GPU versions using numpy and Pytorch, respectively. For fair comparison, we use the log-stabilized Sinkhorn algorithm and the same early stopping criteria and the same maximum iterations as in Gromov-Wasserstein. As shown in Fig. 2, in comparison to the Gromov-Wasserstein, LSMI-Sinkhorn is more than one order of magnitude faster for the CPU version and several times faster for the GPU version. This is consistent with our computational complexity analysis. Moreover, the GPU version of our algorithm costs only 3.47 s to compute 10,000 unpaired samples, indicating that it is suitable for large-scale applications.

5.3 SMI Estimation

For SMI estimation, we set up four baselines:

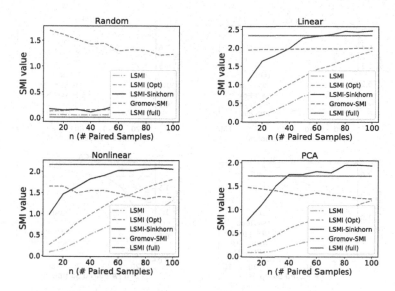

Fig. 3. SMI estimation on synthetic data ($n_x = n_y = 500$).

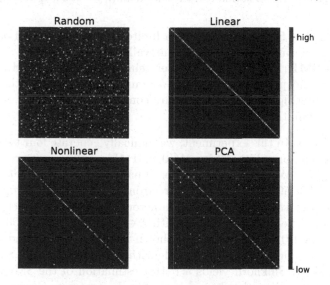

Fig. 4. Visualization of the matrix Π.

- **LSMI (full):** 10,000 paired samples are used for cross-validation and SMI estimation. It is considered as the ground truth value.
- **LSMI:** Only n (usually small) paired samples are used for cross-validation and SMI estimation.
- **LSMI (opt):** n paired samples are used for SMI estimation. However, we use the optimal parameters from LSMI (full) here. This can be seen as the

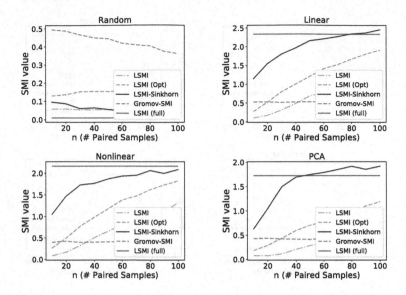

Fig. 5. SMI estimation on synthetic data ($n_x = 1000, n_y = 500$).

upper bound of SMI estimation with limited number of paired data because the optimal parameters are usually unavailable.

- **Gromov-SMI':** The Gromov-Wasserstein distance is applied on unpaired samples to find potential matching ($\hat{n} = \min(n_x, n_y)$). Then, the $\hat{n}$ matched pairs and existing n paired samples are combined to perform cross-validation and SMI estimation.

Synthetic Data. In this experiment, we manually generate four types of paired samples: random normal, $y = 0.5x + \mathcal{N}(0, 0.01)$ (Linear), $y = \sin(x)$ (Nonlinear), and $y = \text{PCA}(x)$. We change the number of paired samples $n \in \{10, 20, \dots, 100\}$ while fixing $n_x = 500$ and $n_y = 500$ for Gromov-SMI and the proposed LSMI-Sinkhorn, respectively. The model parameters λ and β are selected by cross-validation using the paired examples with $\lambda \in \{0.1, 0.01, 0.001, 0.0001\}$ and $\beta \in \{0.2, 0.4, 0.6, 0.8, 1.0\}$. The results are shown in Fig. 3. In the random case, the data are nearly independent and our algorithm achieves a small SMI value. In other cases, LSMI-Sinkhorn yields a better estimation of the SMI value and it lies near the ground truth when n increases. In contrast, Gromov-SMI has a small estimation value, which may be due to the incorrect potential matching. We further show the heatmaps of the matrix Π in Fig. 4. For the random case, Π distributes uniformly as expected. For all other cases, Π concentrate on the diagonal, indicating good estimation for the unpaired samples.

To show the flexibility of the proposed LSMI-Sinkhorn algorithm, we set $n_x = 1000, n_y = 500$ and fix all other settings. The results are shown in Fig. 5. Similarly, LSMI-Sinkhorn achieves the best performance among all methods. We also notice that Gromov-SMI achieves even worse estimation than $n_x = n_y$ case, which means it is not as stable as our algorithm to handle sophisticated situations ($n_x \neq n_y$).

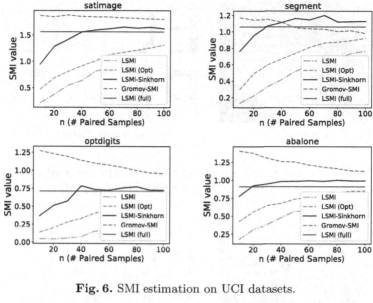

Fig. 6. SMI estimation on UCI datasets.

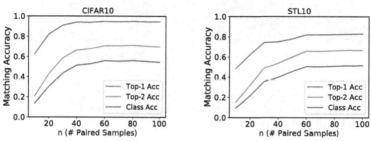

Fig. 7. Deep image matching.

UCI Datasets. We selected four benchmark datasets from the UCI machine learning repository. For each dataset, we split the features into two sets as paired samples. To ensure high dependence between these two subsets of features, we utilized the same splitting strategy as [15] according to the correlation matrix. The experimental setting is the same as the synthetic data experiment. We show the SMI estimation results in Fig. 6. Similarly, LSMI-Sinkhorn obtains better estimation values in all four datasets. Gromov-SMI tends to overestimate the value by a large margin, while other baselines underestimate the value.

5.4   Deep Image Matching

Next, we consider an image matching task with deep convolution features. We use two commonly-used image classification benchmarks: CIFAR10 and STL10 [3]. We extracted 64-dim features from the last layer (after pooling) of ResNet20

(a) 16 × 20   (b) "ECML PKDD"

Fig. 8. Photo album summarization on Flickr dataset. In (a), we fixed the corners with *blue, orange, green,* and *black* images. In (b), we fixed the center of each character with a different image. (Color figure online)

(a) 16 × 20   (b) "ECML PKDD"

Fig. 9. Photo album summarization on CIFAR10 dataset. In (a), we fixed the corners with *automobile, airplane, dog,* and *horse* images. In (b), we fixed the center of each character with a different image.

pretrained on the training set of CIFAR10. The features are divided into two 32-dim parts denoted by $\{x_i\}_{i=1}^N$ and $\{y_i\}_{i=1}^N$. We shuffle the samples of y and attempt to match x and y with limited pair samples ($n \in \{10, 20, \ldots, 100\}$) and unpaired samples ($n_x = n_y = 500$). Other settings are the same as the above experiments.

To evaluate the matching performance, we used top-1 accuracy, top-2 accuracy (correct matching is achieved in the top-2 highest scores), and class accuracy (matched samples are in the same class). As shown in Fig. 7, LSMI-Sinkhorn obtains high accuracy with only a few tens of supervised pairs. Additionally, the high class matching performance implies that our algorithm can be applied to further applications such as semi-supervised image classification.

5.5 Photo Album Summarization

Finally, we apply the proposed LSMI-Sinkhorn to the photo album summarization problem, where images are matched to a predefined structure according to the Cartesian coordinate system.

Color Feature. We first used 320 images from Flickr [15] and extracted the RGB pixels as color feature. Figure 8 depicts the semi-supervised summarization to the 16×20 grid with the corners of the grid fixed to *blue, orange, green,* and *black* images. Similarly, we show the summarization results on an "ECML PKDD" grid with the center of each character fixed. It can be seen that these layouts show good color topology according to the fixed color images.

Semantic Feature. We then used CIFAR10 with the ResNet20 feature to illustrate the semantic album summarization. Figure 9 shows the layout of 1000 images into the same 16×20, and "ECML PKDD" grids. In Fig. 9a, we fixed corners of the grid to *automobile, airplane, dog,* and *horse* images. In Fig. 9b, we fixed the eight character centers. It can be seen that objects are aligned together by their semantics rather than colors according to the fixed images.

Compared with previous summarization algorithms, LSMI-Sinkhorn has two advantages. (1) The semi-supervised property enables interactive album summarization, while kernelized sorting [6,15] and object matching [22] can not. (2) We obtained a solution for general rectangular matching (both $n_x = n_y$ and $n_x \neq n_y$), e.g., 320 images to a 16×20 grid, 1000 images to a 16×20 grid, while most previous methods [15,22] relied on the Hungarian algorithm [10] to obtain square matching ($n_x = n_y$) only.

6 Conclusion

In this paper, we proposed the Least-Square Mutual Information with Sinkhorn (LSMI-Sinkhorn) algorithm to estimate the SMI from a limited number of paired samples. To the best of our knowledge, this is the first semi-supervised SMI estimation algorithm. Experiments on synthetic and real data show that the proposed algorithm can successfully estimate SMI with a small number of paired samples. Moreover, we demonstrated that the proposed algorithm can be used for image matching and photo album summarization.

Acknowledgements. Yanbin Liu and Yi Yang are supported by ARC DP200100938. Makoto Yamada was supported by MEXT KAKENHI 20H04243 and partly supported by MEXT KAKENHI 21H04874. Tam Le acknowledges the support of JSPS KAK-ENHI Grant number 20K19873. Yao-Hung Hubert Tsai and Ruslan Salakhutdinov were supported in part by the NSF IIS1763562, IARPA D17PC00340, ONR Grant N000141812861, and Facebook PhD Fellowship.

References

1. Belghazi, M.I., et al.: Mutual information neural estimation. In: ICML (2018)
2. Bunne, C., Alvarez-Melis, D., Krause, A., Jegelka, S.: Learning generative models across incomparable spaces. In: ICML (2019)
3. Coates, A., Ng, A., Lee, H.: An analysis of single-layer networks in unsupervised feature learning. In: AISTATS (2011)
4. Cover, T.M., Thomas, J.A.: Elements of Information Theory, 2nd edn. Wiley, Hoboken (2006)

5. Cuturi, M.: Sinkhorn distances: lightspeed computation of optimal transport. In: NIPS (2013)
6. Djuric, N., Grbovic, M., Vucetic, S.: Convex kernelized sorting. In: AAAI (2012)
7. Flamary, R., Courty, N.: Pot python optimal transport library (2017). https://github.com/rflamary/POT
8. Goodfellow, I., et al.: Generative adversarial nets. In: NIPS (2014)
9. Hjelm, R.D., et al.: Learning deep representations by mutual information estimation and maximization. In: ICLR (2019)
10. Kuhn, H.: The Hungarian method for the assignment problem. Naval Res. Logist. Q. **2**(1–2), 83–97 (1955)
11. Mémoli, F.: Gromov-Wasserstein distances and the metric approach to object matching. Found. Comput. Math. **11**(4), 417–487 (2011)
12. Ozair, S., Lynch, C., Bengio, Y., Oord, A.V.D., Levine, S., Sermanet, P.: Wasserstein dependency measure for representation learning. In: NeurIPS (2019)
13. Peyré, G., Cuturi, M.: Computational optimal transport. Found. Trends® Mach. Learn. **11**(5–6), 355–607 (2019)
14. Peyré, G., Cuturi, M., Solomon, J.: Gromov-Wasserstein averaging of kernel and distance matrices. In: ICML (2016)
15. Quadrianto, N., Smola, A., Song, L., Tuytelaars, T.: Kernelized sorting. IEEE Trans. Pattern Anal. Mach. Intell. **32**, 1809–1821 (2010)
16. Schmitzer, B.: Stabilized sparse scaling algorithms for entropy regularized transport problems. SIAM J. Sci. Comput. **41**(3), A1443–A1481 (2019)
17. Sinkhorn, R.: Diagonal equivalence to matrices with prescribed row and column sums. In: Proceedings of the American Mathematical Society, vol. 45, no. 2, pp. 195–198 (1974)
18. Sriperumbudur, B.K., Fukumizu, K., Gretton, A., Lanckriet, G.R., Schölkopf, B.: Kernel choice and classifiability for RKHS embeddings of probability distributions. In: NIPS (2009)
19. Suzuki, T., Sugiyama, M.: Sufficient dimension reduction via squared-loss mutual information estimation. In: AISTATS (2010)
20. Suzuki, T., Sugiyama, M., Kanamori, T., Sese, J.: Mutual information estimation reveals global associations between stimuli and biological processes. BMC Bioinform. **10**(S52), 1–12 (2009)
21. Suzuki, T., Sugiyama, M., Tanaka, T.: Mutual information approximation via maximum likelihood estimation of density ratio. In: ISIT (2009)
22. Yamada, M., Sigal, L., Raptis, M., Toyoda, M., Chang, Y., Sugiyama, M.: Cross-domain matching with squared-loss mutual information. IEEE Trans. Pattern Anal. Mach. Intell. **37**(9), 1764–1776 (2015)
23. Yamada, M., Sugiyama, M.: Dependence minimizing regression with model selection for non-linear causal inference under non-gaussian noise. In: AAAI (2010)
24. Yamada, M., Sugiyama, M.: Cross-domain object matching with model selection. In: AISTATS (2011)
25. Yan, Y., Li, W., Wu, H., Min, H., Tan, M., Wu, Q.: Semi-supervised optimal transport for heterogeneous domain adaptation. In: IJCAI (2018)
26. Zhao, S., Song, J., Ermon, S.: InfoVAE: balancing learning and inference in variational autoencoders. In: AAAI (2019)

Spatial Contrastive Learning
for Few-Shot Classification

Yassine Ouali[(✉)], Céline Hudelot, and Myriam Tami

Université Paris-Saclay, CentraleSupélec, MICS, 91190 Gif-sur-Yvette, France
{yassine.ouali,celine.hudelot,myriam.tami}@centralesupelec.fr

Abstract. In this paper we explore contrastive learning for few-shot classification, in which we propose to use it as an additional auxiliary training objective acting as a data-dependent regularizer to promote more general and transferable features. In particular, we present a novel attention-based spatial contrastive objective to learn locally discriminative and class-agnostic features. As a result, our approach overcomes some of the limitations of the cross-entropy loss, such as its excessive discrimination towards seen classes, which reduces the transferability of features to unseen classes. With extensive experiments, we show that the proposed method outperforms state-of-the-art approaches, confirming the importance of learning good and transferable embeddings for few-shot learning. Code: https://github.com/yassouali/SCL.

Keywords: Few-shot learning · Contrastive learning · Deep learning

1   Introduction

Few-shot learning [21] has emerged as an alternative to supervised learning to simulate more realistic settings that mimic human capabilities, and in particular, it consists of reproducing the learner's ability to rapidly and efficiently adapt to novel tasks. In this paper, we tackle the problem of few-shot image classification, which aims to equip a learner with the ability to learn novel visual concepts and recognize unseen classes with limited supervision.

A popular paradigm to solve this problem is meta-learning [25,41] consisting of two disjoint stages, meta-training and meta-testing. During meta-training, the goal is to acquire transferable knowledge from a set of tasks sampled from the meta-training tasks so that the learner is equipped with the ability to adapt to novel tasks quickly. This fast adaptability to unseen classes is evaluated at test time by the average test accuracy over several meta-testing tasks.

Recently, a growing line of works [4,5,43] show that learning good representations results in fast adaptability at test time, suggesting that feature reuse [30] plays a more important role in few-shot classification than the meta-learning

Electronic supplementary material The online version of this chapter (https:// doi.org/10.1007/978-3-030-86486-6_41) contains supplementary material, which is available to authorized users.

© Springer Nature Switzerland AG 2021
N. Oliver et al. (Eds.): ECML PKDD 2021, LNAI 12975, pp. 671–686, 2021.
https://doi.org/10.1007/978-3-030-86486-6_41

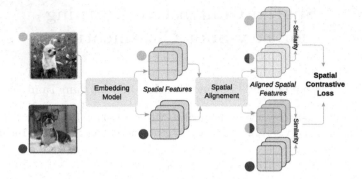

Fig. 1. Spatial Contrastive Learning (SCL). To learn more locally class-independent discriminative features, we propose to measure the similarity between a given pair of samples using their spatial features as opposed to their global features. We first apply an attention-based alignment, aligning each input with respect to the other. Then, we measure the one-to-one spatial similarities and compute the Spatial Contrastive (SC) loss.

aspect of existing algorithms. Such methods consider an extremely simple transfer learning baseline, in which the model is first pre-trained using the standard cross-entropy (CE) loss on the meta-training set. Then, at test time, a linear classifier is trained on the meta-testing set on top of the pre-trained model. The pre-trained model can either be fine-tuned [1,5] together with the classifier, or fixed and used as a feature extractor [4,43]. While promising, we argue that using the CE loss during the pre-training stage hinders the quality of the learned representations since the model only acquires the necessary knowledge to solve the classification task over seen classes at train time. As a result, the learned visual features are excessively discriminative against the training classes, rendering them sub-optimal for test time classification tasks constructed from an arbitrary set of unseen and novel classes.

To alleviate these limitations, we propose to leverage contrastive representation learning [3,14,51] as an auxiliary objective, where instead of only mapping the inputs to fixed targets, we also optimize the features, pulling together semantically similar (*i.e.*, positive) samples in the embedding space while pushing apart dissimilar (*i.e.*, negative) samples. By integrating the contrastive loss into the learning objective, we give rise to discriminative representations between dissimilar instances while maintaining an invariance towards visual similarities. Subsequently, the learned representations are more transferable and capture more prevalent patterns outside of the seen classes. Additionally, by combining both losses, we leverage the stability of the CE loss and its effectiveness on small datasets and small batch sizes, while taking benefit of the contrastive loss as a data-dependent regularizer promoting more general-purpose embeddings.

Specifically, we propose a novel attention-based spatial contrastive loss (see Fig. 1) as the auxiliary objective to further promote class-agnostic visual features and avoid suppressing local discriminative patterns. It consists of measuring the local similarity between the spatial features of a given pair of samples after an attention-based spatial alignment mechanism, instead of the global features (*i.e.*, avg. pooled spatial features) used in the standard contrastive loss. We also adopt the supervised formulation [19] of the contrastive loss to leverage the provided label information when constructing the positive and negative samples.

However, directly optimizing the features and promoting the formation of clusters of similar instances in the embedding space might result in extremely disentangled representations. Such an outcome can be undesirable for few-shot learning, where the testing tasks can be notably different from the tasks encountered during training, *e.g.*, training on generic categories, and testing on fine-grained sub-categories. To solve this, we propose contrastive distillation to reduce the compactness of the features in the embedding space and provide additional refinement of the representations.

Contributions. To summarize, our contributions are: (1) We explore contrastive learning as an auxiliary pre-training objective to learn more transferable features. (2) We propose a novel Spatial Contrastive (SC) loss with an attention-based alignment mechanism to spatially compare a pair of features, further promoting class-independent discriminative patterns. (3) We employ contrastive distillation to avoid excessive disentanglement of the learned embeddings and improve the performances. (4) We demonstrate the effectiveness of the proposed method with extensive experiments on standard and cross-domain few-shot classification benchmarks, achieving state-of-the-art performances. (5) We show the universality of the proposed method by applying it to a standard metric learning approach, resulting in a notable performance boost.

2   Related Work

Few-Shot Classification aims at learning to recognize unseen novel classes with a few labeled example in each class. Meta-learning remains the most popular paradigm to tackle this problem with two principal approaches: (1) optimization-based, or *learning to learn* methods [8,22,31,39], that integrate the fine-tuning process in the meta-training algorithm. (2) metric-based, or *learning to compare* methods [6,37,40,47], that learn a common embedding space in which the similarities between the data can help distinguish between different novel categories with a given distance metric. Most relevant to our work are the methods that follow the standard transfer learning strategy [1,4,5,43], which despite their apparent simplicity, yield state-of-the-art results on standard benchmarks.

Contrastive Learning acts directly on the low-dimensional representations with contrastive losses [13], which measure the similarities of different samples in the embedding space. Recently, contrastive learning based methods have emerged as the state-of-the-art approaches for self-supervised representation learning. The main difference between them is the way they construct and choose the positive samples. In this work, we differentiate between self-supervised contrastive methods [3,14,15,27,42,51] that leverage data augmentations to construct the positive pairs, and supervised contrastive methods [18,19,35,50] that leverage the provided labels to sample the positive examples.

Most relevant to our work are methods that try to build on the insights and advances in contrastive learning, or more broadly self-supervised learning, to improve the few-shot classification task. Such methods [6,9,11,23,38] integrates various types of self-supervised training objective into different few-shot learning frameworks in order to learn transferable features and improve the few-shot classification performance.

3   Preliminaries

3.1   Problem Definition

Few-shot classification usually involves a meta-training set $\mathcal{T}$ and a meta-testing set $\mathcal{S}$ with disjoint label spaces. The meta-training set discerns *seen* classes, while the meta-testing set discerns novel and *unseen* classes. Each one of the meta sets consists of a number of classification tasks where each task describes a pair of training (*i.e.*, support) and testing (*i.e.*, query) sets with few examples, *i.e.*, $\mathcal{T} = \{(\mathcal{D}_t^{\text{train}}, \mathcal{D}_t^{\text{test}})\}_{t=1}^T$ and $\mathcal{S} = \{(\mathcal{D}_q^{\text{train}}, \mathcal{D}_q^{\text{test}})\}_{q=1}^Q$, with each dataset containing pairs of images $\mathbf{x}$ and their ground-truth labels y.

The goal of few-shot classification is to learn a classifier f_θ parametrized by θ capable of exploiting the few training examples provided by the dataset $\mathcal{D}^{\text{train}}$ to correctly predict the labels of the test examples from $\mathcal{D}^{\text{test}}$ for a given task. However, given the high dimensionality of the inputs and the limited number of training examples, the classifier f_θ suffers from high variance. As such, the training and testing inputs are replaced with their corresponding features, which are produced by an embedding model f_ϕ parametrized by ϕ and then used as inputs to the classifier f_θ.

To this end, the objective of meta-training algorithms is to learn a good embedding model f_ϕ so that the average test error of the classifier f_θ is minimized. This usually involves two stages: first, a meta-training stage inferring the parameters ϕ of the embedding model using the meta-training set $\mathcal{T}$, followed by a meta-testing stage evaluating the embedding model's performance on meta-testing set $\mathcal{S}$.

3.2   Transfer Learning Baseline

In this work, we consider the simple transfer learning baseline of [43], in which the embedding model f_ϕ is first pre-trained on the merged tasks from the meta-training set using the CE loss. Then, the model is carried over to the meta-testing stage and fixed during evaluation.

Concretely, we start by merging all the meta-training tasks $\mathcal{D}_t^{\text{train}}$ from $\mathcal{T}$ into a single training set $\mathcal{D}^{\text{new}}$ of seen classes:

$$\mathcal{D}^{\text{new}} = \cup\{\mathcal{D}_1^{\text{train}}, \dots, \mathcal{D}_t^{\text{train}}, \dots, \mathcal{D}_T^{\text{train}}\}. \tag{1}$$

Then, during the meta-training stage, the embedding model f_ϕ can be pre-trained on the resulting set of seen classes using the standard CE loss L_{CE}:

$$\phi = \arg\min_\phi L_{\text{CE}}(\mathcal{D}^{\text{new}}; \phi). \tag{2}$$

Fig. 2. Analysis of the learned representations. (a) k-Nearest Neighbors Analysis. (b) GradCAM results.

The pre-trained model f_ϕ is then fixed (*i.e.*, no fine-tuning is performed) and leveraged as a feature extractor during the meta-testing stage. For a given task $(\mathcal{D}_q^{train}, \mathcal{D}_q^{test})$ sampled from $\mathcal{S}$, a linear classifier f_θ is first trained on top of the extracted features to recognize the unseen classes using the training dataset $\mathcal{D}_q^{train}$:

$$\theta = \arg\min_\theta L_{CE}(\mathcal{D}_q^{train}; \theta, \phi) + \mathcal{R}(\theta), \tag{3}$$

where $\mathcal{R}$ is a regularization term, and the parameters $\theta = \{\mathbf{W}, \mathbf{b}\}$ consist of weight and bias terms, respectively. The predictor f_θ can then be used on the features of the test dataset $\mathcal{D}_q^{test}$ to obtain the class predictions and evaluate f_ϕ.

3.3   Analysis of the Learned Representations

Although the baseline of Sect. 3.2 delivers impressive results, we hypothesis that the usage of the CE loss during the meta-training stage can hinder the performances. Our intuition is that the learned representations lack general discriminative visual features since the CE loss induces embeddings tailored for solving the classification task over the seen classes. As a results, their transferability to novel domains with unseen classes is reduced, and especially if the domain gap between the training and testing stages is significant.

To empirically validate such a hypothesis, we conduct a k-nearest neighbor search [17] on the learned embedding space. First, we train a model with the CE loss on the meta-training set of *mini*-ImageNet [47] as in Eq. (2). Then, for a given test image, we search for its neighbors from the meta-testing set. The results are shown in Fig. 2. For a fast test-time adaptation of the predictor f_θ, the desired outcome is to have visually and semantically similar images adjacent in the embedding space. However, we observe that the neighboring images are semantically dissimilar. Using Grad-CAM [36], we notice that dominant discriminative features acquired during training might not be useful for discriminating between unseen classes at test time. In the case of *mini*-ImageNet, this observation is reinforced by the fact that the meta-training and meta-testing sets are closely related, in which better transferability of the learned features in expected when compared to other benchmarks. We note that similar behavior was also observed by [6] for metric-learning based approaches.

To further investigate this behavior, we conduct a spectral analysis of the learned features. As shown in Fig. 3, we inspect the variance explained by a varying number of principal components and notice that almost all of the variance can be captured with a limited number of components, indicating that the CE loss only preserves the minimal amount of information required to solve the classification task.

4   Methodology

4.1   Contrastive Learning

We explore contrastive learning as an auxiliary pre-training objective to learn general-purpose visual embeddings capturing discriminative features usable outside of the meta-training set. It thus facilitate the test time recognition of unseen classes. Specifically, given that in a few-shot classification setting we are provided with the class labels, we examine the usage of the supervised formulation [19] of the contrastive loss which leverages the label information to construct positive and negative samples.

Formally, let f_ϕ be an embedding model mapping the inputs $\mathbf{x}$ to *spatial* features $\mathbf{z}^s \in \mathbb{R}^{HW \times d}$, followed by an average pooling operation to obtain the *global* features $\mathbf{z}^g \in \mathbb{R}^d$, which are then mapped into a lower dimensional space using a projection head p, *i.e.*, $\mathbf{f} = p(\mathbf{z}^g)$ with $\mathbf{f} \in \mathbb{R}^{d'}$, and let a global similarity function sim_g be denoted as the cosine similarity between a pair of projected global features $\mathbf{f}_i$ and $\mathbf{f}_j$ (*i.e.*, dot product between the ℓ_2 normalized features). First, we sample a batch of N pairs of images and labels from the merged meta-training set $\mathcal{D}^{\mathrm{new}}$ and augment each example in the batch, resulting in $2N$ data points. Then, the supervised contrastive loss [19], referred to as the Global Contrastive (GC) loss, can be computed as follows:

$$L_{\mathrm{GC}} = \sum_{i=1}^{2N} \frac{1}{2N_{y_i} - 1} \sum_{j=1}^{2N} \mathbb{1}_{i \neq j} \cdot \mathbb{1}_{y_i = y_j} \cdot \ell_{ij} , \tag{4}$$

$$\text{where } \ell_{ij} = -\log \frac{\exp(\mathrm{sim}_g(\mathbf{f}_i, \mathbf{f}_j)/\tau)}{\sum_{k=1}^{2N} \mathbb{1}_{i \neq k} \cdot \exp(\mathrm{sim}_g(\mathbf{f}_i, \mathbf{f}_k)/\tau)} ,$$

with $\mathbb{1}_{\mathrm{cond}} \in \{0, 1\}$ as an indicator function evaluating to 1 iff cond is satisfied, N_{y_i} as the total number of images with the same label y_i, and τ as a scalar temperature parameter. By using the GC loss of Eq. (4) as an additional pre-training objective with the CE loss, we push the embedding model f_ϕ to learn the visual similarities between instances of the same class, instead of only maintaining the useful features for the classification task over the seen classes, which results in more useful and transferable embeddings.

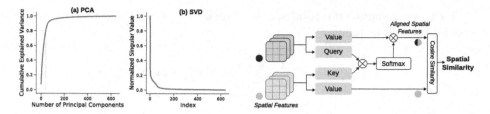

Fig. 3. Spectral analysis of the embedding matrix. (a) Principal component analysis. (b) Singular value decomposition.

Fig. 4. Attention-based spatial alignment mechanism.

4.2 Spatial Contrastive Learning

Although the GC loss is capable of producing good embeddings, using the global features $\mathbf{z}^g$ might suppress some local discriminative features present in the spatial features $\mathbf{z}^s$ that can be informative for down-stream tasks (*e.g.*, suppressing object specific features while overemphasizing the irrelevant background features). As an alternative, we propose a novel Spatial Contrastive (SC) loss that leverages the spatial features $\mathbf{z}^s$ to compute the similarity between a given pair of examples. However, to locally compare a pair of spatial features $\mathbf{z}_i^s$ and $\mathbf{z}_j^s$ and compute the SC loss, we first need to define a mechanism to align them spatially. To this end, we employ the attention mechanism [46] to compute the spatial attention weights to align the features $\mathbf{z}_i^s$ with respect to $\mathbf{z}_j^s$ and vice-versa. Then, we measure the one-to-one spatial similarity as illustrated in Fig. 4, and finally, compute the SC loss.

Attention-based Spatial Alignment. Let h_v, h_q and h_k denote the value, query and key projection heads, taking as input the spatial features $\mathbf{z}^s$ and outputting the value $\mathbf{v}$, query $\mathbf{q}$ and key $\mathbf{k}$ of d'-dimensional features, *i.e.*, $\mathbf{v}, \mathbf{q}, \mathbf{k} \in \mathbb{R}^{HW \times d'}$. Given a pair of spatial features $\mathbf{z}_i^s$ and $\mathbf{z}_j^s$ of two instances i and j, we want to compute the aligned values of i with respect to j, denoted as $\mathbf{v}_{i|j}$. Such an alignment can be obtained using the key $\mathbf{k}_i$ and the query $\mathbf{q}_j$ to compute the attention weights $\mathbf{a}_{ij} \in \mathbb{R}^{HW \times HW}$, which can then be applied to $\mathbf{v}_i$ to obtain $\mathbf{v}_{i|j}$. Concretely, this can be computed as follows:

$$\mathbf{v}_{i|j} = \mathbf{a}_{ij}\mathbf{v}_i \quad \text{where} \quad \mathbf{a}_{ij} = \text{softmax}\left(\frac{\mathbf{q}_j \mathbf{k}_i^\top}{\sqrt{d'}}\right). \tag{5}$$

Similarly, we compute $\mathbf{v}_{j|i}$ aligning the value of j with respect to i using the key $\mathbf{k}_j$ and the query $\mathbf{q}_i$.

Spatial Similarity. Given a pair of values $\mathbf{v}_i$ and $\mathbf{v}_j$, together with their two aligned versions $\mathbf{v}_{i|j}$ and $\mathbf{v}_{j|i}$ computed using the attention mechanism detailed above, and with $\mathbf{v}_*^r$ denoting a feature vector at a spatial location $r \in [1, HW]$, we first perform an ℓ_2 normalization step of the values $\mathbf{v}_*^r$ at each spatial location

r. Then, we compute the total spatial similarity $\text{sim}_s(\mathbf{z}_i^s, \mathbf{z}_j^s)$ between a pair of spatial features as follows:

$$\text{sim}_s(\mathbf{z}_i^s, \mathbf{z}_j^s) = \frac{1}{HW} \sum_{r=1}^{HW} \left[(\mathbf{v}_i^r)^\top \mathbf{v}_{j|i}^r + (\mathbf{v}_j^r)^\top \mathbf{v}_{i|j}^r \right]. \tag{6}$$

Spatial Contrastive Learning. With the spatial similarity function sim_s defined in Eq. (6), and similar to the GC loss in Eq. (4), the SC loss can be computed as follows:

$$L_{SC} = \sum_{i=1}^{2N} \frac{1}{2N_{y_i} - 1} \sum_{j=1}^{2N} \mathbb{1}_{i \neq j} \cdot \mathbb{1}_{y_i = y_j} \cdot \ell_{ij}, \tag{7}$$

$$\text{where } \ell_{ij} = -\log \frac{\exp(\text{sim}_s(\mathbf{z}_i^s, \mathbf{z}_j^s)/\tau')}{\sum_{k=1}^{2N} \mathbb{1}_{i \neq k} \cdot \exp(\text{sim}_s(\mathbf{z}_i^s, \mathbf{z}_k^s)/\tau')},$$

with τ' as a scalar temperature parameter.

4.3 Pre-training Objective

Based on the contrastive objectives in Eq. (4) and Eq. (7), the pre-training objective can take different forms. We mainly consider the case where the pre-training objective L_T is the summation of the CE and SC losses, with λ_{CE} and λ_{SC} as scaling weights to control the contribution of each term:

$$L_T = \lambda_{CE} L_{CE} + \lambda_{SC} L_{SC}. \tag{8}$$

However, we also explore other alternatives such as replacing L_{SC} with L_{GC} or training with both L_{GC} and L_{SC} as auxiliary losses with their corresponding weighting terms. Additionally, we also consider the self-supervised formulations of the GC and SC losses, where the label information is discarded and the only positives considered are the augmented versions of each example (*i.e.*, $y_i = i$ mod N). We refer to them as SS-GC and SS-SC (Self-Supervised Global and Spatial Contrastive) losses respectively.

Using the total loss L_T, the embedding model f_ϕ can be trained together with the projection head and the attention modules during the meta-training stage. Specifically, let ψ represent the parameters of the projection head p and the attention modules h_v, h_q and h_k. The parameters are obtained as follows:

$$\{\phi, \psi\} = \underset{\{\phi, \psi\}}{\arg \min} \, L_T(\mathcal{D}^{\text{new}}; \{\phi, \psi\}). \tag{9}$$

After the pre-training stage, the parameters ψ are discarded, and the embedding model f_ϕ is then fixed and carried over from meta-training to meta-testing.

4.4   Avoiding Excessive Disentanglement

Since the contrastive objectives encourage closely aligned embeddings of instances of the same class while distributing all of the normalized features uniformly on the hypersphere [48], we have to consider a possible over-clustering of the features of the same class. Such an outcome can be desired for closed-set recognition, but in a few-shot setting, in which the discrepancy between the meta-training and meta-testing domain might differ greatly from one case to the other (*e.g.*, training on coarse seen categories, and testing on fine-grained unseen sub-categories), this might lead to sub-optimal performances. As such, to avoid an excessive disentanglement of the learned features and to further improve the generalization of the embedding model, we propose Contrastive Distillation (CD) to reduce the compactness of the features in embeddings space.

Contrastive Distillation. Given a teacher model f_{ϕ_t} pre-trained with the objective in Eq. (8), we transfer its knowledge to a student model f_{ϕ_s} using the standard knowledge distillation [16] objective L_{KL} (*i.e.*, the Kullback-Leibler (KL) divergence between the student's predictions and the soft targets predicted by the teacher), but with an additional contrastive distillation loss L_{CD}. This loss consists of maximizing the inner dot product between the ℓ_2 normalized global features of the teacher $\mathbf{z}^{gt}$ and that of the student $\mathbf{z}^{gs}$, which corresponds to minimizing the squared Euclidean distance, formally:

$$L_{CD} = \frac{1}{N} \sum_{i=1}^{N} \|\mathbf{z}_i^{gt} - \mathbf{z}_i^{gs}\|_2^2. \tag{10}$$

To summarize, the student's parameters are learned as follows:

$$\phi_s = \arg\min_{\phi_s} \lambda_{CD} L_{CD}(\mathcal{D}^{new}; \phi_s, \phi_t) + \lambda_{KL} L_{KL}(\mathcal{D}^{new}; \phi_s, \phi_t). \tag{11}$$

This way, by only maximizing the similarity between the pairs of features without using any negative samples, we relax the uniformity constraint of the contrastive loss and reduce the disentanglement of the learned embeddings.

5   Experiments

For the experimental section, we base our implementation on the publicly available code of [43] and conduct experiments on ImageNet derivatives: *mini*-ImageNet [47] and *tiered*-ImageNet [33], and CIFAR-100 derivatives: CIFAR-CS [2] and FC100 [28]. Additionally, we present experiments on cross-domain few-shot benchmarks introduced by [44]. We note that additional experimental details and results are presented in the supplementary material.

5.1   Experimental Details

Architecture. For the embedding model f_ϕ, we follow [43] and use a ResNet-12 consisting of 4 residual blocks with Dropblock as a regularizer and

Table 1. Comparison of the mean acc. with different training objectives. "Aug." indicates the usage of SimCLR type augmentations.

Loss function	Aug.	mini-ImageNet, 5-way		CIFAR-CS, 5-way	
		1-shot	5-shot	1-shot	5-shot
CE		61.8 ± 0.7	79.7 ± 0.6	71.3 ± 0.9	86.1 ± 0.6
CE	✓	61.8 ± 0.8	78.6 ± 0.5	71.9 ± 0.9	86.3 ± 0.5
CE + SS-GC	✓	62.7 ± 0.7	81.0 ± 0.6	70.9 ± 0.9	84.5 ± 0.6
CE + SS-SC	✓	64.0 ± 0.8	81.5 ± 0.5	72.1 ± 0.8	86.2 ± 0.6
CE + SS-GC + SS-SC	✓	62.8 ± 0.8	81.1 ± 0.6	69.0 ± 0.9	85.0 ± 0.6
CE + GC	✓	65.0 ± 0.8	81.6 ± 0.5	74.0 ± 0.8	87.3 ± 0.6
CE + SC	✓	65.7 ± 0.8	82.5 ± 0.5	75.0 ± 0.9	87.4 ± 0.6
CE + GC + SC	✓	65.0 ± 0.8	81.3 ± 0.5	76.0 ± 0.7	87.5 ± 0.5

Table 2. Comparison of the mean acc. with different evaluation setting.

Features used	mini-ImageNet, 5-way		CIFAR-CS, 5-way	
	1-shot	5-shot	1-shot	5-shot
Spatial	64.5 ± 0.8	82.1 ± 0.5	75.0 ± 0.9	87.1 ± 0.6
Global	65.7 ± 0.8	82.5 ± 0.5	75.0 ± 0.9	87.4 ± 0.6
Glo. & Spa. (Max)	65.6 ± 0.8	82.1 ± 0.5	74.2 ± 0.8	87.3 ± 0.5
Glo. & Spa. (Sum)	65.7 ± 0.8	83.1 ± 0.5	75.6 ± 0.9	87.6 ± 0.6

Table 3. Comparison of the mean acc. distillation objectives.

Loss function	mini-ImageNet, 5-way		CIFAR-CS, 5-way	
	1-shot	5-shot	1-shot	5-shot
Teacher	65.7 ± 0.8	82.5 ± 0.5	75.0 ± 0.9	87.4 ± 0.6
KL	66.0 ± 0.8	82.5 ± 0.5	75.9 ± 0.9	87.4 ± 0.6
KL+CD	67.4 ± 0.8	82.7 ± 0.5	76.5 ± 0.9	87.6 ± 0.6

640-dimensional output features (*i.e.*, $d = 640$). For the projection head and the attention modules, we use an MLP with one hidden layer and a ReLU non-linearity similar to SimCLR, outputting 80-dimensional features (*i.e.*, $d' = 80$).

Training Setup. For optimization, we use SGD with a momentum of 0.9, a weight decay of 5×10^{-4}, a learning rate of 5×10^{-2} and a batch size of 64. For the loss functions, we set the temperature parameters τ and τ' to 0.1 and the scaling weights λ_{CE}, λ_{SC}, and λ_{GC} to 1.0, except for CIFAR-FS where we set them to 0.5. For distillation, we set λ_{CD} to 10.0 and λ_{KL} to 1.0 and use a temperature of 4.0 for the KL loss. For data augmentations, we use standard augmentations for the first N instances, while the remaining N instances are generated using SimCLR type augmentations, resulting in $2N$ augmented examples.

Evaluation Setup. During meta-testing, and given a pre-trained embedding model f_ϕ, we follow [43] and consider a linear classifier as the predictor f_θ, implemented in scikit-learn and trained on the ℓ_2 normalized features produced by f_ϕ. Specifically, we sample a number of C-way K-shot testing classification tasks constructed from the unseen classes of the meta-testing set, with C as the number of classes and K as the number of training examples per class. After training f_θ on the train set, the predictor is then applied to the features of the test set to obtain the prediction and compute the accuracy. In our case, we evaluate the model over 600 randomly sampled tasks and report the median accuracy over 3 runs with 95% confidence intervals, where in each run, the accuracy is the mean accuracy of the 600 sampled tasks.

5.2 Ablation Studies

Loss Functions. We evaluate the performances obtained with various loss functions as detailed in Sect. 4.3. The results are shown in Table 1. We observe a notable gain in performance when adopting auxiliary contrastive losses, be it supervised or self-supervised, with better gains when using the supervised formulation, highlighting the benefits of using the label information when constructing the positives and negatives samples. More importantly, the SC loss outperforms the standard GC loss, confirming the effectiveness of using the spatial features

Table 4. Comparison with prior few-shot classification works. [†]results obtained by training on both train and validation sets.

Method	Backbone	mini-ImageNet, 5-way		tiered-ImageNet, 5-way		CIFAR-FS, 5-way		FC100, 5-way	
		1-shot	5-shot	1-shot	5-shot	1-shot	5-shot	1-shot	5-shot
MAML [8]	32-32-32-32	48.70 ± 1.84	63.11 ± 0.92	51.67 ± 1.81	70.30 ± 1.75	58.9 ± 1.9	71.5 ± 1.0	–	–
Matching Networks [47]	64-64-64-64	43.56 ± 0.84	55.31 ± 0.73	–	–	–	–	–	–
Prototypical Networks[†] [37]	64-64-64-64	49.42 ± 0.78	68.20 ± 0.66	53.31 ± 0.89	72.69 ± 0.74	55.5 ± 0.7	72.0 ± 0.6	35.3± 0.6	48.6 ± 0.6
Relation Networks [40]	64-96-128-256	50.44 ± 0.82	65.32 ± 0.70	54.48 ± 0.93	71.32 ± 0.78	55.0 ± 1.0	69.3 ± 0.8	–	–
R2D2 [2]	96-192-384-512	51.20 ± 0.60	68.80 ± 0.10	-	-	65.3 ± 0.2	79.4 ± 0.1	–	–
SNAIL [24]	ResNet-12	55.71 ± 0.99	68.88 ± 0.92	–	–	–	–	–	–
TADAM [28]	ResNet-12	58.50 ± 0.30	76.70 ± 0.30	–	–	–	–	40.1 ± 0.4	56.1 ± 0.4
Shot-Free [32]	ResNet-12	59.04 ± n/a	77.64 ± n/a	63.52 ± n/a	82.59 ± n/a	69.2 ± n/a	84.7 ± n/a	–	–
TEWAM [29]	ResNet-12	60.07 ± n/a	75.90 ± n/a	–	–	70.4 ± n/a	81.3 ± n/a	–	–
Diversity w/ Coop. [7]	ResNet-18	59.48 ± 0.65	75.62 ± 0.48	–	–	–	–	–	–
Boosting [11]	WRN-28-10	63.77 ± 0.45	80.70 ± 0.33	70.53 ± 0.51	84.98 ± 0.36	73.6 ± 0.3	86.0 ± 0.2	–	–
Fine-tuning [5]	WRN-28-10	57.73 ± 0.62	78.17 ± 0.49	66.58 ± 0.70	85.55 ± 0.48	–	–	–	–
LEO-trainval[†] [34]	WRN-28-10	61.76 ± 0.08	77.59 ± 0.12	66.33 ± 0.05	81.44 ± 0.09	–	–	–	–
Prototypical Networks[†] [37]	ResNet-12	–	–	–	–	72.2 ± 0.7	83.5 ± 0.5	37.5 ± 0.6	52.5 ± 0.6
MetaOptNet [22]	ResNet-12	62.64 ± 0.61	78.63 ± 0.46	65.99 ± 0.72	81.56 ± 0.53	72.6 ± 0.7	84.3 ± 0.5	41.1 ± 0.6	55.5 ± 0.6
RFS [43]	ResNet-12	62.02 ± 0.63	79.64 ± 0.44	69.74 ± 0.72	84.41 ± 0.55	71.5 ± 0.8	86.0 ± 0.5	42.6 ± 0.7	59.1 ± 0.6
RFS-Distill [43]	ResNet-12	64.82 ± 0.60	82.14 ± 0.43	71.52 ± 0.69	86.03 ± 0.49	73.9 ± 0.8	86.9 ± 0.5	44.6 ± 0.7	60.9 ± 0.6
Ours	ResNet-12	65.69 ± 0.81	83.10 ± 0.52	71.48 ± 0.89	**86.88 ± 0.53**	75.6 ± 0.9	87.6 ± 0.6	44.4 + 0.8	60.8 ± 0.8
Ours-Distill	ResNet-12	**67.40 ± 0.76**	**83.19 ± 0.54**	**71.98 ± 0.01**	86.10 ± 0.59	**76.5 ± 0.9**	**88.0 + 0.6**	**44.8 ± 0.7**	**61.4 ± 0.7**

Table 5. Comparison with prior works on cross-domain few-shot classification benchmarks.

Method	CUB, 5-way		Cars, 5-way		Places, 5-way		Plantae, 5-way	
	1-shot	5-shot	1-shot	5-shot	1-shot	5-shot	1-shot	5-shot
MatchingNet [47]	35.89 ± 0.5	51.37 ± 0.7	30.77 ± 0.5	38.99 ± 0.6	49.86 ± 0.8	63.16 ± 0.8	32.70 ± 0.6	46.53 ± 0.6
MatchingNet w/ FT [44]	36.61 ± 0.6	55.23 ± 0.8	29.82 ± 0.4	41.24 ± 0.6	51.07 ± 0.7	64.55 ± 0.7	34.48 ± 0.5	41.69 ± 0.6
RelationNet [40]	42.44 ± 0.7	57.77 ± 0.7	29.11 ± 0.6	37.33 ± 0.7	48.64 ± 0.8	63.32 ± 0.8	33.17 ± 0.6	44.00 ± 0.6
RelationNet w/ FT [44]	44.07 ± 0.7	59.46 ± 0.7	28.63 ± 0.6	39.91 ± 0.7	50.68 ± 0.9	66.28 ± 0.7	33.14 ± 0.6	45.08 ± 0.6
GNN [10]	45.69 ± 0.7	62.25 ± 0.6	31.79 ± 0.5	44.28 ± 0.6	53.10 ± 0.8	70.84 ± 0.6	35.60 ± 0.5	52.53 ± 0.6
GNN w/ FT [44]	47.47 ± 0.6	66.98 ± 0.7	31.61 + 0.5	44.90 ± 0.6	55.77 ± 0.8	73.94 ± 0.7	35.95 ± 0.5	53.85 ± 0.6
Ours	49.58 ± 0.7	67.64 ± 0.7	34.46 ± 0.6	**52.22 ± 0.7**	59.37 ± 0.7	76.46 ± 0.6	**40.23 ± 0.6**	59.38 ± 0.6
Ours-Distill	**50.09 ± 0.7**	**68.81 ± 0.6**	**34.93 ± 0.6**	51.72 ± 0.7	**60.32 ± 0.8**	**76.51 ± 0.6**	39.75 ± 0.8	**59.91 ± 0.8**

rather than the global features. Additionally, using both the SC and GC losses does not result in distinct gains over the SC loss. Thus, for the rest of this section, we adopt the SC as a sole auxiliary loss.

Distillation. To improve the generalization of the embedding model, we investigate the effect of knowledge distillation by training a new (*i.e.*, student) model using a pre-trained (*i.e.*, teacher) network with various training objectives. Table 3 shows a clear performance gain with the proposed CD objective as an additional loss term, confirming the benefits of optimizing the learned features and relaxing the compactness of the embedding space.

Evaluation. Instead of only training the linear classifier on top of the global features during the meta-testing stage, we compare the performance when training over the global features, the spatial features, or both, where we train two classifiers and aggregate their predictions. Table 2 shows the evaluation results. Overall, using the global features to train the linear classifier offers slightly better results than the spatial features. We suspect this might result from slight overfitting of the classifier given that the spatial features increase the number of parameters to be learned, which negatively impacts the performances. However, when leveraging both the spatial and global features, we obtain better results confirming the usefulness of the spatial feature even during the meta-testing stage.

5.3   Few-Shot Classification

Based on the ablation studies, we fix the training objective as SC+CE during the meta-training stage and use both the spatial and global features during the meta-testing stage with a sum aggregate, and compare our approach with other popular few-shot classification methods. The results of 5-way classification are summarized in Table 4 for ImageNet and CIFAR derivatives. Our method outperforms previous works and achieves state-of-the-art performances across different datasets and evaluation settings. This suggests that our attention-based SCL approach coupled with the CE loss improves the transferability of the learned embeddings without any meta-learning techniques, with additional improvements using a contrastive distillation step. These results also show the potential of integrating contrastive losses as auxiliary objectives for various few-shot learning scenarios.

5.4   Cross-Domain Few-Shot Classification

To further affirm the improved transferability of the learned embedding with our approach, we explore the effects of an increased domain difference between the seen and unseen classes, $i.e.$, the discrepancy between the meta-training and meta-testing stages. Precisely, we follow the same procedure as [44] where we first train on the whole $mini$-ImageNet dataset using the same setting as detailed above. Then, we evaluate the embeddings model on four different domains: CUB [49], Cars [20], Places [52], and Plantae [45]. We show the obtained results in Table 5, and see a notable gain in performance using the proposed method, from 2% gain on CUB dataset, up to 7% gain on Cars dataset, indicating a clear enhancement in terms of the generalization of the embedding model.

6   ProtoNet Experiments

To demonstrate the generality of the proposed approach and its applicability in different settings, we provide additional metric-learning based experiments in which we integrate the contrastive losses into the ProtoNet [37] framework. ProtoNet is a distance-based learner trained in an episodic manner so that both the meta-training and meta-testing stages have matching conditions. During meta-training, for a C-way K-shot setting, we construct a meta-training set $\mathcal{T} = \{(\mathcal{D}_t^{\text{train}}, \mathcal{D}_t^{\text{test}})\}_{t=1}^{T}$ where each given task t depicts C randomly chosen classes from the seen classes, with K images per class for the training ($i.e.$, support) set $\mathcal{D}_t^{\text{train}}$, and M images per classes for the test ($i.e.$, query) set $\mathcal{D}_t^{\text{test}}$. At each training iteration, after sampling a given task from $\mathcal{T}$, we first compute the class prototypes for classification using the support set. Then, the embeddings model is trained to minimize the CE loss where each query example is classified based on the distances to the class prototypes. In order to add the contrastive

Table 6. The obtained improvement when adding the contrastive objectives as auxiliary losses. We show the mean acc. and 95% confidence interval for 5-way 5-shot classification across ImagetNet derivatives.

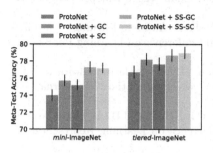

Table 7. Comparison with prior works on *mini*-ImageNet for 5-shot 5-way classification.

Method	Image size	Backbone	Aux. Loss	Acc. (%)
MAML	84 × 84	Conv4-64	–	63.1
ProtoNet		Conv4-64	–	68.2
RelationNet		Conv4-64	–	65.3
ProtoNet [4]	84 × 84	Conv4-64	–	64.2
	224 × 224	ResNet-18	–	73.7
ProtoNet [11]	84 × 84	Conv4-64	–	70.0
		Conv4-64	Rotation	71.7
		Conv4-512	–	71.6
		Conv4-512	Rotation	74.0
		WRN-28-10	–	68.7
		WRN-28-10	Rotation	72.1
ProtoNet [38]	224 × 224	ResNet-18	–	75.2
			Rotation	76.0
			Jigsaw	76.2
			Rot.+Jig	76.6
Ours	224 × 224	ResNet-18	–	74.0
			GC	75.2
			SC	75.2
			SS-GC	77.3
			SS-SC	77.2
			SS-GC+SS-SC	**77.6**

objectives as auxiliary losses to the ProtoNet training objective, we simply merge the query and support set, augment each exampled within it, and compute the contrastive losses detailed in Sect. 4 over this merged and augmented set. The experimental details of this section are presented in the supplementary material.

Results. To investigate the impact of the contrastive losses on the performances of ProtoNet, we report the mean acc. for 5-way 5-shot classification on ImageNet derivatives with different training objectives. The results in Table 6 show a notable performance gain over the ProtoNet baseline. Additionally, we compare the performances of our approach with other self-supervised auxiliary losses, *i.e.*, rotation prediction [12] and jigsaw puzzle [26], for which [38] provided their integration into the ProtoNet framework. As shown in Table 7, we observe that a larger performance gain can be obtained with the contrastive objectives as auxiliary losses compared to other self-supervised objectives, especially when using both the SS-SC and SS-GC losses with a 3.6% gain over the baseline, which further confirms the effectiveness of the proposed SC loss.

7 Conclusion

In this paper, we investigated contrastive losses as auxiliary training objectives along the CE loss to compensate for its drawbacks and learn richer and more transferable features. With extensive experiments, we showed that integrating contrastive learning into existing few-shot learning frameworks results in a notable boost in performances, especially with our spatial contrastive learning objective. Future work could investigate the spatial contrastive method extension for other few-shot learning scenarios and adapt it for other visual tasks such as unsupervised representation learning.

References

1. Afrasiyabi, A., Lalonde, J.-F., Gagné, C.: Associative alignment for few-shot image classification. In: Vedaldi, A., Bischof, H., Brox, T., Frahm, J.-M. (eds.) ECCV 2020. LNCS, vol. 12350, pp. 18–35. Springer, Cham (2020). https://doi.org/10.1007/978-3-030-58558-7_2
2. Bertinetto, L., Henriques, J.F., Torr, P.H., Vedaldi, A.: Meta-learning with differentiable closed-form solvers. In: International Conference on Learning Representations (2019)
3. Chen, T., Kornblith, S., Norouzi, M., Hinton, G.: A simple framework for contrastive learning of visual representations. In: Proceedings of the 37th International Conference on Machine Learning (2020)
4. Chen, W.Y., Liu, Y.C., Kira, Z., Wang, Y.C.F., Huang, J.B.: A closer look at few-shot classification. In: International Conference on Learning Representations (2019)
5. Dhillon, G.S., Chaudhari, P., Ravichandran, A., Soatto, S.: A baseline for few-shot image classification. In: International Conference on Learning Representations (2020)
6. Doersch, C., Gupta, A., Zisserman, A.: CrossTransformers: spatially-aware few-shot transfer. In: Advances in Neural Information Processing Systems (2020)
7. Dvornik, N., Schmid, C., Mairal, J.: Diversity with cooperation: ensemble methods for few-shot classification. In: IEEE International Conference on Computer Vision (2019)
8. Finn, C., Abbeel, P., Levine, S.: Model-agnostic meta-learning for fast adaptation of deep networks. In: Precup, D., Teh, Y.W. (eds.) Proceedings of the 34th International Conference on Machine Learning. Proceedings of Machine Learning Research, vol. 70, pp. 1126–1135. PMLR (2017)
9. Gao, Y., Fei, N., Liu, G., Lu, Z., Xiang, T., Huang, S.: Contrastive prototype learning with augmented embeddings for few-shot learning. arXiv preprint arXiv:2101.09499 (2021)
10. Garcia, V., Bruna, J.: Few-shot learning with graph neural networks. In: International Conference on Learning Representations (2018)
11. Gidaris, S., Bursuc, A., Komodakis, N., Pérez, P., Cord, M.: Boosting few-shot visual learning with self-supervision. In: Proceedings of the IEEE International Conference on Computer Vision, pp. 8059–8068 (2019)
12. Gidaris, S., Singh, P., Komodakis, N.: Unsupervised representation learning by predicting image rotations (2018)
13. Hadsell, R., Chopra, S., LeCun, Y.: Dimensionality reduction by learning an invariant mapping. In: 2006 IEEE Computer Society Conference on Computer Vision and Pattern Recognition (CVPR'06), vol. 2, pp. 1735–1742. IEEE (2006)
14. He, K., Fan, H., Wu, Y., Xie, S., Girshick, R.: Momentum contrast for unsupervised visual representation learning. In: Proceedings of the IEEE/CVF Conference on Computer Vision and Pattern Recognition (CVPR) (2020)
15. Henaff, O.: Data-efficient image recognition with contrastive predictive coding. In: Proceedings of the 37th International Conference on Machine Learning. Proceedings of Machine Learning Research, vol. 119, pp. 4182–4192. PMLR (2020)
16. Hinton, G., Vinyals, O., Dean, J.: Distilling the knowledge in a neural network. arXiv preprint arXiv:1503.02531 (2015)
17. Johnson, J., Douze, M., Jégou, H.: Billion-scale similarity search with GPUs. IEEE Trans. Big Data (2019)

18. Kamnitsas, K., et al.: Semi-supervised learning via compact latent space clustering. In: International Conference on Machine Learning, pp. 2459–2468. PMLR (2018)
19. Khosla, P., et al.: Supervised contrastive learning. In: Advances in Neural Information Processing Systems (2020)
20. Krause, J., Stark, M., Deng, J., Fei-Fei, L.: 3D object representations for fine-grained categorization. In: Proceedings of the IEEE International Conference on Computer Vision Workshops, pp. 554–561 (2013)
21. Lake, B., Salakhutdinov, R., Gross, J., Tenenbaum, J.: One shot learning of simple visual concepts. In: Proceedings of the Annual Meeting of the Cognitive Science Society, vol. 33 (2011)
22. Lee, K., Maji, S., Ravichandran, A., Soatto, S.: Meta-learning with differentiable convex optimization. In: Proceedings of the IEEE Conference on Computer Vision and Pattern Recognition, pp. 10657–10665 (2019)
23. Medina, C., Devos, A., Grossglauser, M.: Self-supervised prototypical transfer learning for few-shot classification. arXiv preprint arXiv:2006.11325 (2020)
24. Mishra, N., Rohaninejad, M., Chen, X., Abbeel, P.: A simple neural attentive meta-learner. In: International Conference on Learning Representations (2018)
25. Naik, D.K., Mammone, R.J.: Meta-neural networks that learn by learning. In: Proceedings 1992 IJCNN International Joint Conference on Neural Networks, vol. 1, pp. 437–442. IEEE (1992)
26. Noroozi, M., Favaro, P.: Unsupervised learning of visual representations by solving jigsaw puzzles. In: Leibe, B., Matas, J., Sebe, N., Welling, M. (eds.) ECCV 2016. LNCS, vol. 9910, pp. 69–84. Springer, Cham (2016). https://doi.org/10.1007/978-3-319-46466-4_5
27. Oord, A.v.d., Li, Y., Vinyals, O.: Representation learning with contrastive predictive coding. arXiv preprint arXiv:1807.03748 (2018)
28. Oreshkin, B., López, P.R., Lacoste, A.: Tadam: task dependent adaptive metric for improved few-shot learning. In: Advances in Neural Information Processing Systems, pp. 721–731 (2018)
29. Qiao, L., Shi, Y., Li, J., Wang, Y., Huang, T., Tian, Y.: Transductive episodic-wise adaptive metric for few-shot learning. In: IEEE International Conference on Computer Vision (2019)
30. Raghu, A., Raghu, M., Bengio, S., Vinyals, O.: Rapid learning or feature reuse? Towards understanding the effectiveness of MAML. In: International Conference on Learning Representations (2019)
31. Ravi, S., Larochelle, H.: Optimization as a model for few-shot learning. In: International Conference on Learning Representations (2017)
32. Ravichandran, A., Bhotika, R., Soatto, S.: Few-shot learning with embedded class models and shot-free meta training. In: Proceedings of the IEEE International Conference on Computer Vision, pp. 331–339 (2019)
33. Ren, M., et al.: Meta-learning for semi-supervised few-shot classification. In: International Conference on Learning Representations (2018)
34. Rusu, A.A., et al.: Meta-learning with latent embedding optimization. In: International Conference on Learning Representations (2019)
35. Salakhutdinov, R., Hinton, G.: Learning a nonlinear embedding by preserving class neighbourhood structure. In: Artificial Intelligence and Statistics, pp. 412–419 (2007)
36. Selvaraju, R.R., Cogswell, M., Das, A., Vedantam, R., Parikh, D., Batra, D.: Grad-CAM: visual explanations from deep networks via gradient-based localization. In: Proceedings of the IEEE International Conference on Computer Vision, pp. 618–626 (2017)

37. Snell, J., Swersky, K., Zemel, R.: Prototypical networks for few-shot learning. In: Advances in Neural Information Processing Systems, pp. 4077–4087 (2017)
38. Su, J.-C., Maji, S., Hariharan, B.: When does self-supervision improve few-shot learning? In: Vedaldi, A., Bischof, H., Brox, T., Frahm, J.-M. (eds.) ECCV 2020. LNCS, vol. 12352, pp. 645–666. Springer, Cham (2020). https://doi.org/10.1007/978-3-030-58571-6_38
39. Sun, Q., Liu, Y., Chua, T.S., Schiele, B.: Meta-transfer learning for few-shot learning. In: Proceedings of the IEEE Conference on Computer Vision and Pattern Recognition, pp. 403–412 (2019)
40. Sung, F., Yang, Y., Zhang, L., Xiang, T., Torr, P.H., Hospedales, T.M.: Learning to compare: relation network for few-shot learning. In: Proceedings of the IEEE Conference on Computer Vision and Pattern Recognition, pp. 1199–1208 (2018)
41. Thrun, S.: Lifelong learning algorithms. In: Thrun, S., Pratt, L. (eds.) Learning to Learn, pp. 181–209. Springer, Boston (1998). https://doi.org/10.1007/978-1-4615-5529-2_8
42. Tian, Y., Krishnan, D., Isola, P.: Contrastive multiview coding. In: Vedaldi, A., Bischof, H., Brox, T., Frahm, J.-M. (eds.) ECCV 2020. LNCS, vol. 12356, pp. 776–794. Springer, Cham (2020). https://doi.org/10.1007/978-3-030-58621-8_45
43. Tian, Y., Wang, Y., Krishnan, D., Tenenbaum, J.B., Isola, P.: Rethinking few-shot image classification: a good embedding is all you need? In: Vedaldi, A., Bischof, H., Brox, T., Frahm, J.-M. (eds.) ECCV 2020. LNCS, vol. 12359, pp. 266–282. Springer, Cham (2020). https://doi.org/10.1007/978-3-030-58568-6_16
44. Tseng, H.Y., Lee, H.Y., Huang, J.B., Yang, M.H.: Cross-domain few-shot classification via learned feature-wise transformation. In: International Conference on Learning Representations (2020)
45. Van Horn, G., et al.: The iNaturalist species classification and detection dataset. In: Proceedings of the IEEE Conference on Computer Vision and Pattern Recognition, pp. 8769–8778 (2018)
46. Vaswani, A., et al.: Attention is all you need. In: Advances in Neural Information Processing Systems, pp. 5998–6008 (2017)
47. Vinyals, O., Blundell, C., Lillicrap, T., Wierstra, D., et al.: Matching networks for one shot learning. In: Advances in Neural Information Processing Systems, pp. 3630–3638 (2016)
48. Wang, T., Isola, P.: Understanding contrastive representation learning through alignment and uniformity on the hypersphere. In: Proceedings of the 37th International Conference on Machine Learning (2020)
49. Welinder, P., et al.: Caltech-UCSD birds 200 (2010)
50. Wu, Z., Efros, A.A., Yu, S.X.: Improving generalization via scalable neighborhood component analysis. In: Proceedings of the European Conference on Computer Vision (ECCV), pp. 685–701 (2018)
51. Wu, Z., Xiong, Y., Yu, S.X., Lin, D.: Unsupervised feature learning via non-parametric instance discrimination. In: Proceedings of the IEEE Conference on Computer Vision and Pattern Recognition, pp. 3733–3742 (2018)
52. Zhou, B., Lapedriza, A., Khosla, A., Oliva, A., Torralba, A.: Places: a 10 million image database for scene recognition. IEEE Trans. Pattern Anal. Mach. Intell. 40(6), 1452–1464 (2017)

Ensemble of Local Decision Trees for Anomaly Detection in Mixed Data

Sunil Aryal$^{(\boxtimes)}$ and Jonathan R. Wells

School of Information Technology, Deakin University, Geelong, VIC, Australia
{sunil.aryal,j.wells}@deakin.edu.au

Abstract. Anomaly Detection (AD) is used in many real-world applications such as cybersecurity, banking, and national intelligence. Though many AD algorithms have been proposed in the literature, their effectiveness in practical real-world problems are rather limited. It is mainly because most of them: (i) examine anomalies globally w.r.t. the entire data, but some anomalies exhibit suspicious characteristics w.r.t. their local neighbourhood (local context) only and they appear to be normal in the global context; and (ii) assume that data features are all numeric, but real-world data have numeric/quantitative and categorical/qualitative features. In this paper, we propose a simple robust solution to address the above-mentioned issues. The main idea is to partition the data space and build local models in different regions rather than building a global model for the entire data space. To cover sufficient local context around a test data instance, multiple local models from different partitions (an ensemble of local models) are used. We used classical decision trees that can handle numeric and categorical features well as local models. Our results show that an Ensemble of Local Decision Trees (ELDT) produces better and more consistent detection accuracies compared to popular state-of-the-art AD methods, particularly in datasets with mixed types of features.

Keywords: Anomaly detection · Mixed data · LOF · IForest · Ensemble anomaly detection · Decision trees

1 Introduction

Anomaly Detection (AD) is a machine learning task of identifying anomalous data instances automatically using algorithms. Anomalies (also refer to as outliers) are data instances that are significantly different from most of the other data causing suspicions that they are generated from a different mechanism from the one that is normal or expected. AD has many applications such as intrusion detection in computer networks, fraud detection in banking, detecting illegal activities (e.g., drug trafficking, money laundering) in national intelligence/security. In the literature, AD problems have been solved using three learning approaches [8,12]: (i) **Supervised learning:** A classification model

© Springer Nature Switzerland AG 2021
N. Oliver et al. (Eds.): ECML PKDD 2021, LNAI 12975, pp. 687–702, 2021.
https://doi.org/10.1007/978-3-030-86486-6_42

is learned using training instances from both normal and anomalous classes to make predictions for test data; (ii) **Unsupervised learning:** Given data instances (which may have anomalies) are ranked directly based on some outlier scores, i.e., no training involved; and (iii) **Semi-supervised learning:** A profile of normal/expected behaviour is learned from labelled training samples of normal data only, and test data are ranked based on how well they comply with the learned profile of normal data.

Regardless of the learning approaches used, existing AD methods have some limitations/issues that restrict their wide applicability in practice. Supervised methods have the following major issues [12]: (i) it might be very expensive or even impossible to obtain labelled training anomalous samples in many real-world applications; (ii) even if possible, they are infinitesimally rare resulting in the class imbalanced problem; and (iii) a few known anomalies are not enough to generalise characteristics of all possible anomalous patterns because anomalies can be anywhere in the feature space. Though techniques like minority class (anomalies) oversampling, majority class (normal) under-sampling, and algorithmic adjustments [18] are used to alleviate the above-mentioned issues, their effectiveness in practice are limited. It is because they assume that unseen/future anomalies are generated from the same distribution as previously seen/observed anomalies. Often, it is not the case in practice. New anomalies can be very different from previously seen anomalies.

Un/semi-supervised approaches do not require labelled training anomalous samples. Unsupervised approaches do not require training samples at all. Assuming anomalies are few and different, they use distance/density based scores to rank given data (which may have anomalies) directly. They may perform poorly when the assumption does not hold, i.e., when there are far too many anomalies [8,12]. Semi-supervised approaches do not make such assumption. Because a vast majority of observed data are normal, normal training data can be obtained easily. Thus, we focus on semi-supervised AD approach in this paper.

Most existing un/semi-supervised AD methods assume that data have numeric features. However, in many real-world applications, data have both numeric (e.g., age, height) and categorical (e.g., gender, nationality) features. The common practice is to convert categorical features into numeric features using technique like one-hot encoding [15]. Each categorical label (e.g., Australian for nationality) is converted into a binary feature with the value of 1 (if the nationality is Australian) or 0 (otherwise) and treated as a numeric feature. A categorical feature with n possible values is converted into n binary numeric features, out of which only one has the value of 1 for each instance. Because of this, each original categorical feature and original numeric feature contribute differently to AD models, which can degrade the performances of AD methods. There are methods proposed for categorical data only [26]. Numeric features can be converted into categorical features through discretisation [15]. Most AD methods developed for categorical data have high computational complexities limiting their use in large real-world datasets.

Most existing AD methods examine data instances globally, i.e., w.r.t. the entire dataset. They can detect global anomalies that exhibit anomalous characteristics in the entire dataset. However, they cannot detect local anomalies that appear to be normal when examined globally but exhibit anomalous characteristics w.r.t. their local neighbourhood (i.e., in the local context). For example, in Fig. 1(a), a_4 and a_5 have significantly lower density than normal cluster C_1 in their neighbourhood but have the same density as many instances in normal cluster C_2. Most existing methods fail to detect them as anomalies. Real-world data have complex structures and instances may exhibit characteristics that look normal in the global perspective but suspicious in their local contexts. There are some methods that examine anomalies w.r.t. their localities (e.g., [5,11]), but they are limited to numeric data only.

To summarise, most existing AD methods do not work well in practical applications due to the following three main issues:

- **Lack of sufficient examples of known anomalies:** It is not possible to have a good representative sample of known anomalies to generalise characteristics of all possible anomalies.
- **Global view of anomalies:** Data often exhibit suspicious characteristics w.r.t. their neighbourhood (in local context) that can appear to be normal in the global context.
- **Limitations to handle mixed types of data features:** Most real-world applications have numeric and categorical features, but most existing methods can not handle mixed types of features well.

In this paper, we present a simple idea to address the above-mentioned issues and introduce a new semi-supervised AD method. Instead of using one global model, we propose to partition the data space into many regions and build an AD model in each region using data falling in the region only, i.e., many local AD models are built. To make prediction for a test instance, the AD model learned on the region where it falls is used. Instead of just relying on a local region from one partitioning of the space, we propose to create multiple partitions of the data space and use ensemble of multiple local models learned on local regions from each partition. It exploits the benefits of ensemble learning to consider sufficient locality around the test instance. Though there are not many AD models that can work well with mixed data, there are classifiers such as traditional Decision Tree (DT) [22] that can handle numeric and categorical features directly. We used DTs in local regions for AD without using labelled anomalies by adding synthetic data. Our results show that an Ensemble of Local Decision Trees (ELDT) produces better and more consistent detection results compared to popular state-of-the-art AD methods, particularly in datasets with mixed types of features.

2 Related Work

In the semi-supervised approach, a model is learned from a training set D of N instances belonging to the normal class only and evaluated on a test set Q, which is a mixture of normal and anomalous data. Let $\mathbf{x}$ be a data instance represented as an M-dimensional vector $\langle x_1, x_2, \cdots, x_M \rangle$, where each component represents its value of a feature that can be either numeric $x_i \in \mathbb{R}$ ($\mathbb{R}$ is a real domain) or categorical $x_i \in \{v_{i_1}, v_{i_2}, \cdots, v_{i_w}\}$ (where v_{i_j} is a label out of w possible labels for feature i). Let $F = \{A_1, A_2, \cdots, A_M\}$ be a set of data features, also called as attributes of data.

In this section, we review prior work related to this paper that includes AD methods for numeric and categorical data, and ensemble approaches for AD.

2.1 Methods for Numeric Data

Because anomalies are few and different, they are expected to have feature values that are significantly different from most data and lie in low density regions. Most of them use distance/density-based anomaly scores to rank data according to their degrees of outlying behaviour, e.g., Nearest Neighbours (NNs) or Support Vectors (SVs) based methods.

In the NN-based methods, the anomaly score of $\mathbf{x} \in Q$ is estimated based on the distances to its kNNs in D, where k is a user defined neighbourhood parameter. Local Outlier Factor (LOF) [11] and k^{th} NN distance [6] are the most widely used NNs-based methods. Being different from normal instances, anomalies are expected to have larger distances to their kNNs than normal instances. They require to compute distances of $\mathbf{x}$ with all instances in D, which can be computationally expensive when D is large. Though the nearest neighbour search can be speed up by using indexing schemes such as k-d tree [7], their effectiveness reduces as the number of dimension increases and become useless in high dimensional problems [19]. Sugiyama and Borgwardt (2013) [25] showed that the nearest neighbour search in a small subset $\mathcal{D} \subset D$ ($|\mathcal{D}| = \psi \ll N$) is enough. They proposed a simple, but very fast, anomaly detector called Sp where the anomaly score of $\mathbf{x}$ is its distance to the nearest neighbor (1NN) in $\mathcal{D}$. It has been shown that Sp with ψ as small as 25 produces competitive results to LOF but runs several orders of magnitude faster [25].

The SV-based methods define the boundary around normal (expected) data and identify a set of data instances lying in the boundary called Support Vectors (SVs). They compute the pairwise similarities of data using a kernel function. Gaussian kernel that uses Euclidean distance is a popular choice. In the testing phase, the anomaly score of $\mathbf{x} \in Q$ is estimated based on its kernel similarities with the SVs. One-Class Support Vector Machine (OCSVM) [23] and Support Vector Data Description (SVDD) [27] are widely used methods in this class. The training process is computationally expensive in the case of large D because of the pairwise similarity calculations.

2.2 Methods for Categorical Data

Despite the widespread prevalence of categorical/qualitative data in real-world applications, AD in categorical data has not received much attention in the research community [26]. The common practice is to convert categorical features into numeric features and use methods designed for numeric data. There are some methods proposed in the literature specifically for categorical data based on frequencies of categorical labels, information theory and data compression/encoding [26]. They are computationally expensive to run in large datasets and do not perform better than using methods for numeric data by converting categorical data into numeric data [4].

He et al. (2005) [17] proposed a method for categorical data based on frequent patterns. The intuition is that an instance is more likely to be an anomaly if it has a few or none of the frequent patterns. Akoglu et al. (2012) [2] proposed a pattern-based compression technique called COMPREX. The intuition is that the higher the cost of encoding $\mathbf{x}$, the more likely it is to be an anomaly. Aryal et al. (2016) [4] revisited the Simple Probabilistic AD (SPAD) where multi-dimensional probability is estimated as the product of one-dimensional probability and show that it works quite well compared to more complex state-of-the-art methods, such as LOF, One-Class SVM, in datasets with categorical only and mixed types of features. It uses the frequencies of categorical label in each feature individually assuming features are independent to each other. Most of these methods for categorical datasets except SPAD have high time and/or space complexities limiting their use in small and low-dimensional datasets only. SPAD is simple and arguably the fastest AD method.

2.3   Ensemble Approaches

To solve a given task, the ensemble methods build multiple models by using an algorithm on different subsets of given data (data sampling or feature sampling) or using different parameter settings of the algorithm [13]. The final decision of the ensemble is an aggregation of decisions by its individual models. The main idea is that models are different and they make different errors so that they compensate each other's weaknesses and results in better overall performance than any individual model. Ensemble learning is widely studied for classification problems and various frameworks have been proposed that can be used with different base classifiers [9,10,28]. However, the use of ensemble learning to solve the AD problem is rather limited [1]. Ensemble based AD methods build multiple models using subsamples of data and/or subsets of features, e.g., Lazarevic and Kumar (2005) [20] and Zimek et al. (2013) [29] used LOF using random subsets of features (i.e., subspaces) and data (i.e., subsamples), respectively. AD techniques such as iForest [21] and usfAD [3] used a collection of random trees to partition the data space using small subsamples of data until the instances are isolated. Each tree is using a small subset of features. The main idea of these methods is that anomalies are expected to isolate early in the trees and lie in leaves with low heights. They run very fast as they do not require pairwise distance calculations. All these ensemble-based methods assume that data have numeric features only. Most of them are not applicable to data with categorical only or mixed features. Also, many of them build multiple global models, they can not detect local anomalies.

3   Our Proposal: An Ensemble of Local Decision Trees

To address the three limitations of existing AD methods in practical real-world applications discussed in Sect. 1, we develop a new ensemble learning framework for anomaly detection based on the of idea of Feating [28]. First, we explain Feating for classification as used by Ting et al. (2011) [28] (Subsect. 3.1) and then discuss how we can adapt it for anomaly detection (Subsect. 3.2).

3.1   Feating for Classification

Feature-Subspace Aggregating (Feating) [28] is an ensemble framework developed for classification that uses an ensemble of local models. It is a feature bagging approach, ensemble learning using subsets of features of fixed size $m < M$ (i.e., using m-dimensional subspaces). In each subspace $S \subset F$ with $|S| = m$, rather than building a global model trained on the entire training set, it first partitions the subspace using a tree structure called "Level Tree" (LT). At each node of the tree, the space is partitioned using one of the m features in S. Each feature in S is used only once in the tree, resulting in the maximum tree height of m. LTs can handle both numeric and categorical features. For numeric feature, the space is divided into two regions by the cut-point selected in the same manner as in ordinary decision tree [22] based on information gain. For categorical feature with w possible values, the space is partition into w regions, one for each categorical label. Further partitioning of a node stops when the node is either pure (i.e., has instances belonging to the same class), there are less than $minPts$ data instances or reaches the maximum height of m. In each impure leaf node with more than $minPts$ samples, a classifier is learned from the training samples falling in the node only, i.e., a Local Model (LM) is built. For rest of the other leaves (with less than $minPts$ instances or pure), class probabilities are recorded based on the training samples they have. In the testing phase, a test instance is traversed from the root to a leaf in each LT. If a LM was built in the leaf node, the class probabilities are the predicted probabilities of the LM. Otherwise, the recorded class probabilities are used. The final prediction is based on the aggregated class probabilities from multiple LTs. The enumerated version of Feating builds $\binom{M}{m}$ LTs, which has a large space complexity making it infeasible in problems with large M (high-dimensional applications). To overcome this issue, Ting et al. (2011) introduced a randomised version, where only $t \ll \binom{M}{m}$ random subspaces of size m are used. It significantly improves the time and space requirements without any significant compromise in accuracy.

3.2   Feating for Anomaly Detection

We propose the following adjustments to use Feating in semi-supervised AD, where there are no labelled anomalies. LT building process uses class information but in this case we do not have labelled anomalies. We are given D which is a set of normal data only. To build each LT, we propose to consider the given data D as "+ve" class and the same number of synthetic points are added as "-ve" class

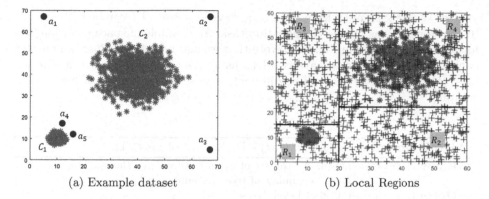

(a) Example dataset   (b) Local Regions

Fig. 1. An example of dataset and definition of local regions. (a) C_1 and C_2 are clusters of normal data, whereas $a_1, \cdots, a_5$ are anomalies. (b) Note only half of the normal data (red points) are used in the training process as "**+ve**" class samples. Blue points, which are uniformly generated synthetic points, are "**-ve**" class samples. R_1, R_2, R_3, and R_4 are local regions created by a Level Tree. Note that local classifiers are built in regions R_1 and R_4 only. (Color figure online)

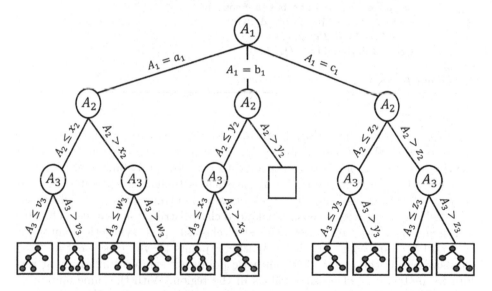

Fig. 2. An example of a Level Tree in 3-dimensional subspace $S = \{A_1, A_2, A_3\}$. Note that A_1 is a categorical feature with three possible values (a_1, b_1, c_1), and A_2 and A_3 are numeric features.

as done by Shi and Horvath (2006) [24] to use Random Forest in unsupervised problems. The values of synthetic points in each feature are selected uniformly at random from the range of possible values, i.e., for a numeric feature, values are selected uniformly randomly between the possible range defined by instances in D, and for a categorical feature, values are selected randomly from the possible values. With the samples from "+ve" and "−ve" classes, a LT can be built exactly in the same way as for the classification task. An example of space partition into local region is shown in Fig. 1(b). Note that we add new "−ve" class samples for each LT. It is possible to have two LTs using the same subset of m features if $t > \binom{M}{m}$, but the two LTs will be different because of the new set of "−ve" class samples.

Algorithm 1: Feating(D, A, m, t) - Build a set of Level Trees

---

    Input   : D - Given data, A - the set of given features, m - the maximum level
                of a level tree. t - number of trees to build

    Output: E - a collection of Level Trees

1 $N = |D|$ (#training samples);
2 $M \leftarrow |A|$ (#features);
3 $minPts = \lfloor \log_2(N) \rfloor + 1$;
4 $m = \lfloor \log_2(M) \rfloor + 1$;
5 $E \leftarrow \emptyset$;
6 **for** $i = 1$ *to* t **do**
7   // get a set of m attributes from A
8      $L \leftarrow randomSetOfAttributes(A, m)$;
9      $D_s \leftarrow addSyntheticPoints(D)$;
10    $E \leftarrow E \bigcup$ BuildLevelTree(D_s, L, 0);
11 **end**
12 **return** E;

Once the space is partitioned into regions, the idea is to build a classifier to separate "+ve" class (given training data, which are normal) from "−ve" class (synthetic data). Anomalies are expected to have low probabilities of belonging to "+ve" class (normal data). Local classifier is built in each leaf with more than $minPts$ samples from the "+ve" class and current class probabilities are recorded in other leaves. To ensure balanced class distribution to build a classifier in a local region, we first remove all "−ve" class samples (synthetic points) and then add the same amount of new synthetic points ("−ve" class) as the "+ve" class samples in the region. The synthetic points are sampled uniformly at random from the range of possible values in the region. With the same amount of "+ve" class and "−ve" class (newly added) samples, local classifier is built. We use Decision Tree (DT) [22] that can handle numeric and categorical features directly as local classifier. An example of a Level Tree in 3-dimensional subspace is provided in Fig. 2 and the procedures to build LTs and local DTs are provided in Algorithms 1, 2 and 3. In the testing phase the anomaly score of a test instance $\mathbf{x}$ is estimated as the aggregated $P(+ve|\mathbf{x})$ over t local models. Anomalies are expected to have low aggregated score than normal data. We call the proposed method as "**Ensemble of Local Decision Trees (ELDT)**".

Algorithm 2: BuildLevelTree(D_s, L, j) - Build a single Level Tree recursively

---

Input : D_s - Data with synthetic points to build a tree, L - Attribute list, j - Current tree level

Output: *node* - Level Tree node

```
1  // Check if we have enough positive samples
```
2 **if** $|D_s^+| < minPts$ **then**
3 $\quad$ **return** As a leaf with $P^+ = \frac{|D_s^+|}{N}$ and $P^- = 1.0 - P^+$;
4 **end**
```
5  // Check if we have a pure node with all -ve class samples
```
6 **if** $|D_s^-| = |D_s|$ **then**
7 $\quad$ **return** As a leaf with $P^+ = 0.0$ and $P^- = 1.0$;
8 **end**
```
9  // Check if we have a pure node with all +ve class samples
```
10 **if** $|D_s^+| = |D_s|$ **then**
11 $\quad$ `// Build a local DT in the node.`
12 $\quad$ **return** BuildLocalDecisionTree(D_s);
13 **end**
14 **if** $j = m$ **then** `// m is the maximum level of the Level Tree`
15 $\quad$ `// Build a local DT in the node.`
16 $\quad$ **return** BuildLocalDecisionTree(D_s);
17 **end**
```
18  // retrieve the next attribute from L based on current level, j
```
19 $a \leftarrow nextAttribute(L, j)$;
```
20  // Construct a node with attribute a
```
21 **if** a *is a numeric attribute* **then**
22 $\quad$ `// cut-point selection based on information gain`
23 $\quad$ $node.splitpoint \leftarrow$ findSplitPoint(a, D_t);
24 $\quad$ $D_1 \leftarrow$ filter(D_t, $a > node.splitpoint$);
25 $\quad$ $D_2 \leftarrow$ filter(D_t, $a \leq node.splitpoint$);
26 $\quad$ $node.branch(1) \leftarrow$ BuildLevelTree(D_1, L, $j + 1$);
27 $\quad$ $node.branch(2) \leftarrow$ BuildLevelTree(D_2, L, $j + 1$);
28 **else**
29 $\quad$ `// split according to categorical values`
30 $\quad$ let $\{v_1, \ldots, v_w\}$ be possible values of a;
31 $\quad$ **for** $i = 1$ *to* w **do**
32 $\quad\quad$ $D_i \leftarrow$ filter(D_t, $a == v_i$);
33 $\quad\quad$ $node.branch(i) \leftarrow$ BuildLevelTree(D_i, L, $j + 1$);
34 $\quad$ **end**
35 **end**
36 **return** *node*;

Algorithm 3: BuildLocalDecisionTree(D_s) - Build a local Decision Tree

Input : D_s - Training set
Output: *node* - Level Tree node

1 RemoveOldSyntheticPoints(D_s);
2 $D_S \leftarrow addSyntheticPoints(D_s)$;
3 // **Learn a Decision Tree**
4 *node.localModel* $\leftarrow$ BuildDecisionTree(D_S);
5 **return** *node*;

The ensemble of local DT based on the idea of Feating addresses the three limitations of existing AD approaches in practical problems discussed in Sect. 1. It does not require labelled anomalies. It examines anomalies with respect to their local context or locality defined by multiple local regions. This is useful to detect local anomalies. Using DT that can handle categorical and/or mixed features directly at the local regions, it works well with categorical and mixed data.

4 Experimental Results

In this section, we present the results of our experiments conducted to evaluate the performance of ELDT. The three parameters of ELDT were set as default to: $minPts = \lfloor \log_2(|D|) \rfloor + 1$, (note that D is the training normal data); $m = \lfloor \log_2(|F|) \rfloor + 1$ (note that F is the set of features of D), and $t = 100$. We compared the performance of ELDT with Bagging using DT (Bag.DT) [9] and Random Forest (RF) [10], where each model in the ensemble is a global DT for the entire data space. They also used D as "+ve" class and the same amount of synthetic points as "-ve" class as did in building level trees in ELDT. Each tree in the ensemble has different "-ve" class samples to ensure diversity between trees. We considered the following three state-of-the-art AD methods as main baselines:

1. iForest [21]: It is an ensemble-based AD method. It uses a collection of t random trees, where each tree T_i is constructed from a small random subsample of data $\mathcal{D}_i \subset D$, $|\mathcal{D}_i| = \psi$ (=256 by default). The idea is to isolate each instance in $\mathcal{D}_i$. Anomalies are expected to have shorter average path lengths over the collection of random trees. It produces good results and runs significantly fast. It works only with numeric features, so categorical features are converted into numeric features using one-hot encoding. It is unable to detect local anomalies [5].
2. LOF [11]: It is the most widely used AD method based on kNN ($k = \lfloor \sqrt{N} \rfloor$ by default) search. It compares the density of a test instance with the average densities of its kNNs. It examines anomalies w.r.t. to their locality defined by the kNNs. It is a local model-based existing AD method. It is also mainly for numeric data, categorical features have to be first converted into numeric features. It is computationally very expensive when D is large.

3. SPAD [4]: It is a simple probabilistic AD method, where multi-dimensional probability is estimated as the product of one-dimensional probabilities assuming features are independent. It works with discrete or categorical data. Numerical features are converted into categorical features through equal-width discretisation [15] with the number of bins b ($=\lfloor \log_2(N) \rfloor + 1$ by default). Despite its simplicity, it has been shown to perform better than more complex methods such as LOF, One-Class SVM and iForest [4].

We used 10 benchmark datasets with categorical only, mixed (categorical and numeric) and numeric only features. The characteristics of datasets used in terms of data size, dimensionality (numeric and categorical) and proportion of anomalies are provided in Table 1. Most of these datasets are from the UCI Machine Learning Repository [14][1]. All methods are implemented in JAVA using the WEKA platform [16]. We used **A**rea **U**nder the Receiver Operating Characteristic (ROC) **C**urve (AUC) as the performance evaluation metric. We conducted 10 trials of different train (D) and test (Q) sets and presented the average AUC over 10 runs.

Table 1. Characteristics of data sets. #Inst: data size, #Feat: num. of features, #NFeat: num. of numeric features, #CFeat: num. of categorical feaures, anomaly%: percentage of anomalies

Name	#Inst	#Feat	#NFeat	#CFeat	anomaly%
Census	299285	40	7	33	6.0
Covertype	287128	12	10	2	1.0
Kddcup99	64759	41	34	7	6.5
U2r	60821	41	34	7	0.5
Mnist	20444	96	96	0	3.5
Annthyroid	7200	21	6	15	7.5
Chess	4580	6	0	6	0.5
Mushroom	4429	22	0	22	5.0
Hypothyroid	3772	29	7	22	7.5
Spambase	2964	57	57	0	6.0

The average AUC results of contending methods are provided in Table 2. The results show that ELDT produced best results overall with the average AUC of 0.918 and average rank of 2.0 over 10 datasets used. It had the best AUC in four out of 10 datasets followed by Bag.DT in three datasets, RF and SPAD in two datasets each, and iFoest and LOF in only one dataset each. The closest contender in terms of consistent performance across datasets is SPAD the average AUC of 0.864 and the average rank of 3.3. Though Bag.DT produced the best results in three datasets, it performed worst in the other four datasets, whereas

---

[1] http://archive.ics.uci.edu/ml.

ELDT was ranked second in three datasets, third in two datasets and forth in the remaining one dataset. This results show that ELDT produced more consistent results across different datasets with numeric only, categorical only and mixed attributes and those with local and global anomalies. Among the three baselines, SPAD has the best overall performance. These results are consistent with those claimed by the authors in [4].

The runtime results in the five largest datasets with more than 10,000 instances are presented in Table 3. These results show that ELDT ran slower than all contender except LOF, but it had the runtimes in the same order of magnitudes with them. It was at last one order of magnitude faster than LOF, two orders of magnitude faster in the largest dataset.

Table 2. Average AUC over 10 runs. The best performance in each dataset is highlighted on bold.

Dataset	ELDT	Bag.DT	RF	iForest	LOF	SPAD
Census	**0.713**	0.561	0.57	0.589	0.491	0.684
Covertype	**0.995**	0.974	0.984	0.945	0.992	0.966
Kddcup99	0.993	0.504	0.636	**0.998**	0.896	**0.998**
U2r	**0.988**	0.515	0.576	0.978	0.931	0.988
Mnist	0.815	0.691	0.771	0.841	**0.880**	0.824
Annthyroid	0.921	**0.975**	0.802	0.771	0.612	0.705
Chess	0.998	0.997	**1.000**	0.889	0.968	0.995
Mushroom	0.999	**1.000**	**1.000**	0.791	0.996	0.977
Hypothyroid	0.953	**0.977**	0.894	0.694	0.607	0.723
Spambase	**0.808**	0.549	0.718	0.805	0.659	0.781
Avg. AUC	**0.918**	0.774	0.795	0.830	0.803	0.864
Avg. Rank	**2.0**	3.9	3.4	3.8	4.3	3.3

Table 3. Average runtime (in seconds) over 10 runs in the five largest datasets with more than 10,000 instances.

Dataset	Ft.DT	Bag.DT	RF	iForest	LOF	SPAD
Census	420	346	119	180	58,140	180
Covertype	1,621	1,572	187	39	10,429	150
Kddcup99	39	88	20	6	1,424	7
U2r	32	79	20	4	1,431	8
Mnist	84	282	19	2	160	6

4.1   Sensitivity of Parameters

In this section, we present the results of experiments conducted to assess the sensitivity of the three parameters, m (the size of subspaces that determines the maximum height of Level Trees), $minPts$ (minimum points required at leaf nodes to build local AD models) and t (ensemble size), in the performance of ELDT. We varied one parameter at a time setting the other two parameters to default values. For this experiments, we used two datasets - Annthyroid and Mnist. The results are presented in Figs. 3 and 4. The results show that the performance of ELDT can be improved by setting m and $minPts$ properly. In terms of t, higher the better. In both cases, performance improved when t was increased and started to flatten. Increasing t also increases time and space complexities linearly. Therefore, there has to be a trade-off between performance and complexities.

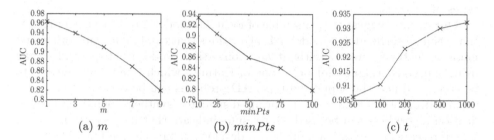

(a) m (b) $minPts$ (c) t

Fig. 3. Annthyroid: Effect of parameter in ELDT

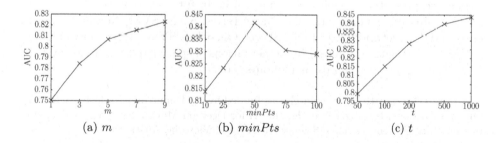

(a) m (b) $minPts$ (c) t

Fig. 4. Mnist: Effect of parameter in ELDT

5   Conclusions and Future Work

In this paper, we presented a simple idea to address the three main limitations of existing Anomaly Detection (AD) methods in practical applications: (i) lack of sufficient examples of known anomalies; (ii) unable to detect local anomalies; and (iii) inability to handle mixed attributes well. Instead of using one global model, we propose to partition the data space into many regions and build an AD model in each region using data falling in the region only, i.e., many local AD models are built. To make prediction for a test instance, the AD model learned on the region where it falls is used. Instead of just relying on a local region from one partitioning of the space, we proposed to create multiple partitions of the data space and use ensemble of multiple local models learned on local regions from each partition. It exploits the benefits of ensemble learning to consider sufficient locality around the test instance. We used the idea of Feating to partition the data space. Though there are not many AD models that can work well with mixed data, there are classifiers such as traditional Decision Tree (DT) that can handle numeric and categorical features directly. We used DTs in local regions for AD without using labelled anomalies by adding synthetic data. We presented a new AD method called Ensemble of Local Decision Trees (ELDT). Our results show that ELDT produces better and more consistent detection results compared to popular state-of-the-art AD methods, particularly in datasets with mixed types of features.

Our results suggest that ensemble of local AD models produces better results than using a single global model. AD algorithms that can handle categorical and numeric features directly without any conversion produce better results than using methods designed for only type of features, which require all features to be converted into the supported type. AD problems can be converted into classification problems by adding uniformly distributed synthetic points and classification algorithms can be used. Our results indicate that it is a very promising line of research to investigate further to develop a flexible and robust AD framework for practical use. It can lead to a general ensemble learning framework for AD, where different space partitioning techniques can be used to define local regions and any classifier or AD algorithm can be used in local regions. In this paper, we presented one simple variant of it. In future, we would like to focus on: (i) using other classifiers (e.g., Naive Bayes, KNN, SVM, Neural Networks, etc.) and AD methods (e.g., LOF, SPAD, One-Class SVM, etc.) as local models; and (ii) investigating different implementations of space partitioning: using trees (e.g., Feating), grids, nearest neighbours, etc.

Acknowledgement. This research was funded by the Department of Defence and the Office of National Intelligence under the AI for Decision Making Program, delivered in partnership with the Defence Science Institute in Victoria, Australia.

References

1. Aggarwal, C.C., Sathe, S.: Outlier Ensembles: An Introduction. Springer, Cham (2017). https://doi.org/10.1007/978-3-319-54765-7
2. Akoglu, L., Tong, H., Vreeken, J., Faloutsos, C.: Fast and reliable anomaly detection in categorical data. In: Proceedings of the 21st ACM Conference on Information and Knowledge Management (CIKM), pp. 415–424 (2012)
3. Aryal, S.: Anomaly detection technique robust to units and scales of measurement. In: Phung, D., Tseng, V.S., Webb, G.I., Ho, B., Ganji, M., Rashidi, L. (eds.)

PAKDD 2018. LNCS (LNAI), vol. 10937, pp. 589–601. Springer, Cham (2018). https://doi.org/10.1007/978-3-319-93034-3_47

4. Aryal, S., Ting, K.M., Haffari, G.: Revisiting attribute independence assumption in probabilistic unsupervised anomaly detection. In: Chau, M., Wang, G.A., Chen, H. (eds.) PAISI 2016. LNCS, vol. 9650, pp. 73–86. Springer, Cham (2016). https://doi.org/10.1007/978-3-319-31863-9_6

5. Aryal, S., Ting, K.M., Wells, J.R., Washio, T.: Improving iForest with relative mass. In: Tseng, V.S., Ho, T.B., Zhou, Z.-H., Chen, A.L.P., Kao, H.-Y. (eds.) PAKDD 2014. LNCS (LNAI), vol. 8444, pp. 510–521. Springer, Cham (2014). https://doi.org/10.1007/978-3-319-06605-9_42

6. Bay, S.D., Schwabacher, M.: Mining distance-based outliers in near linear time with randomization and a simple pruning rule. In: Proceedings of the Ninth ACM International Conference on Knowledge Discovery and Data Mining, pp. 29–38 (2003)

7. Bentley, J.L., Friedman, J.H.: Data structures for range searching. ACM Comput. Surv. **11**(4), 397–409 (1979)

8. Boriah, S., Chandola, V., Kumar, V.: Similarity measures for categorical data: a comparative evaluation. In: Proceedings of the Eighth SIAM International Conference on Data Mining, pp. 243–254 (2008)

9. Breiman, L.: Bagging predictors. Mach. Learn. **24**(2), 123–140 (1996). https://doi.org/10.1007/BF00058655

10. Breiman, L.: Random forests. Mach. Learn. **45**(1), 5–32 (2001). https://doi.org/10.1023/A:1010933404324

11. Breunig, M.M., Kriegel, H.P., Ng, R.T., Sander, J.: LOF: identifying density-based local outliers. In: In Proceedings of ACM SIGMOD International Conference on Management of Data, pp. 93–104 (2000)

12. Chandola, V., Banerjee, A., Kumar, V.: Anomaly detection: a survey. ACM Comput. Surv. **41**(3), 15:1–15:58 (2009)

13. Dietterich, T.G.: Ensemble methods in machine learning. In: Kittler, J., Roli, F. (eds.) MCS 2000. LNCS, vol. 1857, pp. 1–15. Springer, Heidelberg (2000). https://doi.org/10.1007/3-540-45014-9_1

14. Dua, D., Graff, C.: UCI machine learning repository (2019). http://archive.ics.uci.edu/ml

15. Duda, R.O., Hart, P.E., Stork, D.G.: Pattern Classification, 2nd edn. Wiley, Hoboken (2000)

16. Hall, M., Frank, E., Holmes, G., Pfahringer, B., Reutemann, P., Witten, I.H.: The WEKA data mining software: an update. SIGKDD Explor. Newsl. **11**(1), 10–18 (2009)

17. He, Z., Xu, X., Huang, J.Z., Deng, S.: FP-outlier: frequent pattern based outlier detection. Comput. Sci. Inf. Syst. **2**(1), 103–118 (2005)

18. Hilario, A.F., López, S.C., Galar, M., Prati, R., Krawczyk, B., Herrera, F.: Learning from Imbalanced Data Sets. Springer, Heidelberg (2018). https://doi.org/10.1007/978-3-319-98074-4

19. Hinneburg, A., Aggarwal, C.C., Keim, D.A.: What is the nearest neighbor in high dimensional spaces? In: Proceedings of the 26th International Conference on Very Large Data Bases, VLDB '00, pp. 506–515 (2000)

20. Lazarevic, A., Kumar, V.: Feature bagging for outlier detection. In: Proceedings of the Eleventh ACM SIGKDD International Conference on Knowledge Discovery in Data Mining (KDD), pp. 157–166 (2005)

21. Liu, F., Ting, K.M., Zhou, Z.H.: Isolation forest. In: In Proceedings of the Eighth IEEE International Conference on Data Mining, pp. 413–422 (2008)

22. Quinlan, J.R.: Induction of decision trees. Mach. Learn. **1**(1), 81–106 (1986). https://doi.org/10.1007/BF00116251
23. Schölkopf, B., Platt, J.C., Shawe-Taylor, J.C., Smola, A.J., Williamson, R.C.: Estimating the support of a high-dimensional distribution. Neural Comput. **13**(7), 1443–1471 (2001)
24. Shi, T., Horvath, S.: Unsupervised learning with random forest predictors. J. Comput. Graph. Stat. **15**(1), 118–138 (2006)
25. Sugiyama, M., Borgwardt, K.M.: Rapid distance-based outlier detection via sampling. In: Proceedings of the 27th Annual Conference on Neural Information Processing Systems, pp. 467–475 (2013)
26. Taha, A., Hadi, A.S.: Anomaly detection methods for categorical data: a review. ACM Comput. Surv. **52**(2), 38:1–38:35 (2019)
27. Tax, D.M., Duin, R.P.: Support vector data description. Mach. Learn. **54**, 45–66 (2004). https://doi.org/10.1023/B:MACH.0000008084.60811.49
28. Ting, K.M., Wells, J.R., Tan, S.C., Teng, S.W., Webb, G.I.: Feature-subspace aggregating: ensembles for stable and unstable learners. Mach. Learn. **82**(3), 375–397 (2011). https://doi.org/10.1007/s10994-010-5224-5
29. Zimek, A., Gaudet, M., Campello, R.J., Sander, J.: Subsampling for efficient and effective unsupervised outlier detection ensembles. In: Proceedings of KDD, pp. 428–436 (2013)

Learning Algorithms and Applications

Optimal Teaching Curricula with Compositional Simplicity Priors

Manuel Garcia-Piqueras[1,2]([envelope]) [iD] and José Hernández-Orallo[3] [iD]

[1] Department of Mathematics, Faculty of Education of Albacete, University of
Castilla-La Mancha, 02071 Albacete, Spain
manuel.gpiqueras@uclm.es
[2] Laboratory for the Integration of ICT in the Classroom Research Group,
University of Castilla-La Mancha, 02071 Albacete, Spain
[3] VRAIN, Universitat Politècnica de València, València, Spain
jorallo@upv.es

Abstract. Machine teaching under strong simplicity priors can teach
any concept in universal languages. Remarkably, recent experiments suggest that the teaching sets are shorter than the concept description itself.
This raises many important questions about the complexity of concepts
and their teaching size, especially when concepts are taught incrementally. In this paper we put a bound to these surprising experimental
findings and reconnect *teaching size* and concept complexity: complex
concepts do require large teaching sets. Also, we analyse teaching curricula, and find a new *interposition* phenomenon: the teaching size of a
concept can increase because examples are *captured* by simpler concepts
built on previously acquired knowledge. We provide a procedure that not
only avoids interposition but builds an *optimal curriculum*. These results
indicate novel curriculum design strategies for humans and machines.

Keywords: Machine teaching · Interposition · Kolmogorov complexity

1 Introduction

A *teacher* instructing a series of concepts to a *learner* using examples would ideally design a curriculum such that the whole teaching session is shortest. For one
concept, the field of *machine teaching* has analysed the efficiency of the teacher,
the learner or both, for different representation languages and teaching settings
[5,16,28,37,42]. For more than one concept, however, we need to consider different sequences of concepts, or *curricula*, to make learning more effective. While
there has been extensive experimental work in curriculum learning [36], the theoretical analysis is not abundant and limited to continuous models [12,26,40]. It
is not well understood how curriculum learning can be optimised when concepts
are *compositional*, with the underlying representation mechanisms being rich
languages, even Turing-complete. Also, in a curriculum learning situation where
a *teacher* chooses the examples sequentially, it is surprising that the connection with machine teaching has not been made explicit at a general conceptual

© Springer Nature Switzerland AG 2021
N. Oliver et al. (Eds.): ECML PKDD 2021, LNAI 12975, pp. 705–721, 2021.
https://doi.org/10.1007/978-3-030-86486-6_43

level, with only a specific minimax approach for gradient-based representations [10,11,41]. In other words, to our knowledge, a theoretical framework has not yet been articulated for curriculum learning in machine teaching, or *curriculum teaching*, when dealing with universal languages, as a counterpart to incremental inductive inference based on simplicity [34,35].

While the teaching dimension has been the traditional metric for determining how easy it is to teach a concept [42], the *teaching size* [38] is a new metric that is more reasonably related to how easy it is to teach an infinite compositional concept class. It is also more appropriate to understand 'prompting' of language models as a kind of teaching, where users need to think of the shortest prompts that make a language model such as BERT, GPT-2 or GPT-3 achieve a task by few-shot learning [4,6,27]. However, as far as we know, the following issues are not clear yet: (1) *What is the relationship between the Kolmogorov complexity of a concept and how difficult it is to be taught under the teaching size paradigm?* and (2) *Is there a way to extend machine teaching, and teaching size in particular, to consider the notion of* optimal teaching curricula?

Theorem 1 addresses the first question and shows that concepts with *high* complexity are *difficult to teach*, putting a limit to the surprising experimental finding recently reported in [38], where teaching a concept by examples was usually more economical (in total number of bits) than showing the shortest program for the concept. This connection suggests that the second question may rely on a strong relation between incremental learning using simplicity priors and curriculum teaching. For instance, consider the concepts c_+ for addition, $c_\times$ for multiplication, $c_\wedge$ for exponentiation and c_θ for the removal of zeros (Fig. 1). If the concept of c_+ is useful to allow for a shorter description of $c_\times$, is it also reasonable to expect that c_+ would also be useful to *teach* $c_\times$ from examples? Or even $c_\wedge$? In general, is the conditional algorithmic complexity $K(c_2|c_1)$ related to the minimal size of the examples needed to teach c_2 after having acquired c_1?

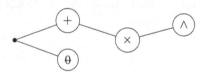

Fig. 1. Curriculum teaching for a set of concepts.

Our perspective studies the sequence of learning a set of concepts, instead of learning a sequence of instances under the same concept. In the general case, we define a teaching curriculum as a set of partial alternative sequences, such as the top and bottom branches in Fig. 1. The order between branches is irrelevant, but the order of concepts inside each branch is crucial. This tree structure is proposed as future work in [26]. Given a set of concepts, is there a curriculum that minimises the overall teaching size?

Our second group of contributions turns around this new concept of teaching curriculum. We provide a definition of *conditional teaching size*, given some other concepts already taught, $TS(c|c_1, \ldots, c_n)$. We show that, in general, $K(c_1|c_2) < K(c_2|c_1)$, for conditional Kolmogorov complexities, does not imply $TS(c_1|c_2) < TS(c_2|c_1)$, and vice versa. Furthermore, given a concept c, it is not true that $TS(c|B) \leq TS(c)$, $\forall B$. We find a new *interposition* phenomenon: acquired concepts may increase the teaching size of new concepts. We give conditions to avoid or provoke interposition. Theorems 3 and 4 are key results in this direction, providing an explicit range where interposition might happen. Finally, we present an effective procedure, $\mathbb{I}$-*search*, to design *optimal* curricula, minimising overall teaching size, for a given set of concepts.

2   Notation and Background

Let us consider a machine M and a universal (i.e., Turing complete) language L. We assume that L is formed by a finite set of instructions in an alphabet Υ, each of them been coded with the same number of bits. Hence, each program p in language L can simply be represented as a string in $\Sigma = \{0,1\}^*$, whose length is denoted by $\dot{\ell}(p)$ (in number of instructions) and denoted by $\ell(p)$ (in bits). There is a total order, $\prec$, over programs in language L defined by two criteria: (i) length and (ii) lexicographic order over Υ only applied when two programs have equal size. Programs map binary strings in Σ to $\Sigma \cup \bot$, denoted by $p(\mathbf{i}) = \mathbf{o}$, with $p(\mathbf{i}) = \bot$ representing that p does not halt for $\mathbf{i}$. Two programs are equivalent if they compute the same function.

We say that c is an L-concept if it is a total or partial function $c : \Sigma \to \Sigma \cup \bot$ computed by at least a program in language L. The class of concepts defined by all programs in L is denoted by C_L; $[p]_L$ denotes the equivalence class of program p. Given $c \in C_L$, we denote $[c]_L$ as the equivalence class of programs in L that compute the function defined by c. Examples are just pairs of strings, and their space is the infinite set $X = \{\langle \mathbf{i}, \mathbf{o} \rangle : \langle \mathbf{i}, \mathbf{o} \rangle \in \Sigma \times (\Sigma \cup \bot)\}$. A *witness* can be any finite example subset of X, of the form $S = \{\langle \mathbf{i}_1, \mathbf{o}_1 \rangle, \ldots, \langle \mathbf{i}_k, \mathbf{o}_k \rangle\}$. In order to calculate the *size* of these sets, we consider self-delimiting codes. Let δ be the number of bits needed to encode S, using certain prefix code. For instance, if we consider Elias coding [7], the string 01010010001001000101 (size = 20) expresses the example set $\{\langle 1, 010 \rangle, \langle 0, 1 \rangle\}$ unambiguously. The size of an example set is the size of its encoding (e.g., $\delta(\{\langle 1, 010 \rangle, \langle 0, 1 \rangle\} = 20$ in Elias coding). For output strings, the natural number to be encoded is increased by 1, to accommodate for $\bot$. We also define a total order $<$ on X, i.e., $\forall S, S'$ such that $S < S'$ then $\delta(S) \leq \delta(S')$ with any preference (e.g., lexicographic) for equal size.

A concept c defines a unique subset of the example space X and we call any element in that subset a *positive* example. A concept c satisfies example set S, denoted by $c \vDash S$, if S is a subset of the positive examples of c. For instance, a witness set for the concept c_θ (*remove zeros*) is $\{\langle 10011, 111 \rangle, \langle 001, 1 \rangle\}$. Example sets cannot have different outputs for equal inputs: $\{\langle 1, 00 \rangle, \langle 1, 01 \rangle\}$ is not valid.

A program p is compatible with $S = \{\langle \mathbf{i}_j, \mathbf{o}_j \rangle\}_{j=1}^k \subset X$, denoted by $p \vDash S$, if $p_S(\mathbf{i}_j) = \mathbf{o}_j$ for every $j \in \{1, \ldots, k\}$. For a finite example set S, there is always a program, denoted by $\dot{p}_S$, that implements a conditional hard-coded structure of if-then-elses (trie) specifically designed for S. If we know the number of bits of input $\mathtt{i}$ and the set of examples in S, the number of comparisons using a trie-data structure is linearly time-bounded. Namely, for any $\ddot{p}_S$, there exists a constant, ρ, such that $\rho \cdot \min\{\ell(\mathtt{i}), \ell(\mathtt{i}_{max})\} + \ell(\mathtt{o}_{max})$ is an upper bound of time steps for each input $\mathtt{i}$, where $\ell(\mathtt{i}_{max})$, $\ell(\mathtt{o}_{max})$ are the lengths of the longest input string and output string in S, respectively. In general, for any program that employs a trie-data structure for S, there exists a time-bound linear function, denoted by $\lambda_L(\mathtt{i}, S)$, that represents an upper bound in time steps on every input $\mathtt{i}$.

Complexity functions $\mathsf{f} : \mathbb{N} \to \mathbb{N}$ act as time bounds. We say that a program p is f-compatible with the example set $S = \{\langle \mathtt{i}_j, \mathtt{o}_j \rangle\}_{j=1}^k \subset X$, denoted by $p \vDash_\mathsf{f} S$, if $p(\mathtt{i}_j) = \mathtt{o}_j$ within $\max\{\mathsf{f}(\ell(\mathtt{i}_j)), \lambda_L(\mathtt{i}_j, S)\}$ time steps (time-bound) for each $j \in \{1, \ldots, k\}$. In other words, within time bound, for each pair $\langle \mathtt{i}, \mathtt{o} \rangle \in S$ the program p on input $\mathtt{i}$: (1) outputs $\mathtt{o}$ when $\mathtt{o} \neq \perp$ or (2) does not halt when $\mathtt{o} = \perp$. Note that: (i) For any complexity function f and any example set S, there is always[1], a program f-compatible with S, (ii) there may be programs p such that $p \nvDash_\mathsf{f} S \wedge p \vDash S$, if f and S do not guarantee enough time bound and (iii) larger complexity functions distinguish more programs.

3   Absolute Teaching Size and Complexity

Now we can study how a non-incremental teacher-learner setting works and the relationship between teaching size and Kolmogorov complexity.

Following the K-dimension [2,3], seen as preference-based teaching using simplicity priors [8,15], we assume that the learner is determined to find the shortest program (according to the prior $\prec$). Namely, the learner Φ returns the first program, in order $\prec$, for an example set S and a complexity function f as follows:

$$\Phi_\ell^\mathsf{f}(S) = \arg\min_p{}^\prec \{\ell(p) : p \vDash_\mathsf{f} S\}$$

Note that the f-bounded Kolmogorov complexity of an example set S, $K^\mathsf{f}(S)$, is the length of the program returned by the learner $K^\mathsf{f}(S) = \ell(\Phi_\ell^\mathsf{f}(S))$. We say that S is a *witness set* of concept c for learner Φ if S is a finite example set such that $p = \Phi_\ell^\mathsf{f}(S)$ and $p \in [c]_L$.

The teacher selects the *simplest* witness set that allows the learner to identify the concept, according to set size (δ) and associated total order $<$, as follows:

$$\Omega_\ell^\mathsf{f}(c) = \arg\min_S{}^\lessdot \{\delta(S) : \Phi_\ell^\mathsf{f}(S) \in [c]_L\}$$

The K^f-teaching size of a concept c is $TS_\ell^\mathsf{f}(c) = \delta(\Omega_\ell^\mathsf{f}(c))$.

Every program the teacher picks defines a concept c. The teacher-learner protocol is computable for any complexity function f and able to create pairs (p_c, w_c), where p_c defines a concept c and w_c is a witness set of c. We can think of these pairs as if they were inserted sequentially in the so-called f-*Teaching Book*

---

[1] Note that this $\ddot{p}_S$ is ensured by the max with time costs.

ordered by w_c, with no repeated programs or witness sets. For example, if we consider the concept $a \in C_L$ for swapping ones and zeros in a binary string, there will be a pair (p_a, w_a) in the f-Teaching Book, e.g., containing a witness set like $w_a = \{\langle 10, 01 \rangle, \langle 110, 001 \rangle\}$ that the teacher would provide with which the learner would output p_a, a program that swaps 1 and 0. Theorem 1 in [38] shows that *for any concept $c \in C_L$, there exists a complexity function f such that there is a pair (p_c, w_c) in the f-Teaching Book*. The teaching size makes more sense than the traditional teaching dimension (the smallest cardinality of a witness set for the concept) because some concepts could be taught by very few examples, but some of them could be extremely large. Also, the use of size instead of cardinality allows us to connect teaching size and Kolmogorov complexity, as we do next.

Our first result[2] shows an equipoise between teaching size and data compression, an extra support for machine teaching; the compressing performance of the learner and the minimisation of the teaching size go in parallel.

Proposition 1. Let f be a complexity function and Φ_ℓ^f the learner. There exist two constants $k_1, k_2 \in \mathbb{N}$, such that for any given pair $(w, p) \in$ f-Teaching Book we have that:[3]

$$K(p) \leq \delta(w) + k_1 \text{ and } K(w) \leq \ell(p) + k_2 \tag{1}$$

Proposition 1 is a key result ensuring that the size difference between programs and witness sets is bounded: a short witness set would not correspond with an arbitrarily complex concept and vice versa. This puts a limit to the surprising empirical observation in [38], where the size of the witness sets in bits was usually smaller than the size of the shortest program for that set, i.e., in terms of information it was usually cheaper to teach by example than sending the shortest description for a concept.

There is another close relationship between the Kolmogorov complexity of a concept and its teaching size. First we need to define the complexity of a concept through the *first program* of a concept in language L.

$$p_c^* = \arg\min_p{}^{\prec} \{\ell(p) : p \in [c]_L\}$$

For every concept $c \in C_L$, we will simply refer to the Kolmogorov complexity of a concept c with respect to the universal language L as $K_L(c) = \ell(p_c^*)$. Now,

Theorem 1. Let L be a universal language, M be a universal machine and k_M be a constant that denotes the length of a program for Φ in M.[4] For any concept $c \in C_L$, there exists a complexity function f, such that $K_L(c) \leq TS_\ell^f(c) + k_M$.

---

[2] The proofs can be found in [9].
[3] We use the standard definition of K using a monotone universal machine U [19] (we will drop U when the result is valid for any U), applied to binary strings (where programs and example sets are encoded as explained in the previous section). With K^f we refer to a non-universal version where the descriptional machine is the learner.
[4] For any universal Turing machine M, a finite program can be built coding an interpreter for Φ in M and taking w_c as input. The length of this 'glued' program does not depend on the concept c but on the machine M to glue things together and how many bits of the program instructions are required to code Φ, i.e., $K_M(\Phi)$.

This gives an upper bound (the teaching size) for the Kolmogorov complexity of a concept. On the other hand, this theorem implies that concepts with *high* complexity are *difficult to teach* in this setting. The surprising observation found in [38] of some concepts having shorter TS than K has a limit.

4   Conditional Teaching Size

In this section we introduce the notion of conditional teaching size and the *curriculum teaching problem*. We now assume that the learner can reuse any already learnt concept to *compose* other concepts. The curriculum teaching problem is to determine the optimal *sequential* way of teaching a set of concepts $Q = \{c_1, c_2, \ldots, c_n\}$, in terms of minimum total teaching size. Let $TS(c_i|c_j, c_k \ldots)$ be the conditional teaching size of concept c_i, given the set of concepts $\{c_j, c_k \ldots\}$ previously distinguished by the learner. The challenge is to minimise $TS(c_1) + TS(c_2|c_1) + TS(c_3|c_1, c_2) + \ldots$.

In this new setting we need a definition of $TS(c_i|c_j)$ that considers that (1) a concept c has infinitely many programs that generate it, so which one the learner has identified may be important, and (2) the learner must have some *memory*, where that program is stored. Interestingly, if we assume that memory is implemented by storing the identified programs in a library, where the learner can only make calls to—but not reuse its parts—, then it is irrelevant which program has been used to capture concept c, since the learner only reuses the functional *behaviour* of the program[5].

4.1   Conditional Teaching Size and Minimal Curriculum

We define a library $B = \{p_1, \ldots, p_k\}$, as a set of programs in the universal language used by the learner. Let $|B| = k$ the number of *primitives*. We assume that Υ always includes an instruction @ for making static[6] library calls. We use @i to denote the instruction that calls the primitive that is indexed as i in the library. If $|B| = 1$, then @ needs no index. Accordingly, the length of a call to the library is $\ell(@i) = \ell(@) + \log_2(|B|) = \log_2(|\Upsilon|) + \log_2(|B|)$ bits.

Let p, p' be programs in the universal language L and B a library. We say that a program p *contains a call to* p' when @i is a substring of p and i is the index of $p' \in B$. L_B denotes a language L that implements static calls to a library B. Even with static calls, the flow of the program may never reach @ for an input. Interestingly, we can avoid this undecidable question when dealing with programs in the teaching book by considering @ as the last instruction regarding lexicographical order.

---

[5] If the learner uses a complexity function f, then we may have that a particular program p_1 identifies c_1 and c_1 is very useful for c_2, but p_1 is too slow to be used in any reasonably efficient program for c_2, so becoming useless incrementally. Computational time has also been considered in other machine teaching frameworks [21,43].

[6] There is no loss of generality here, since every program that uses dynamic calls can be rewritten only using static calls [1].

Lemma 1. Let f be a complexity function and B a library. For any $(w, p) \in$ f-Teaching Book, if p has a call to B then p effectively reaches @ and executes a primitive on at least one input of w.

Let us use $\dot{p}$ to denote program @i, where i is the index of p in the library.

Lemma 2. Let B be a library. The language L_B satisfies: $\dot{p} \prec p'$, $\forall p'$ such that $p' \notin [p]_L \wedge p'$ has a call to p.

Now, we are able to redefine the learner, Φ_ℓ^f, and the time-bounded Kolmogorov complexity for a given library.

Definition 1. Let f be a complexity function, B a library and S an example set. The learner Φ calculates the *first program* for S in language L_B:

$$\Phi_\ell^f(S|B) = \arg \min_{p \in L_B}^{\prec} \{\ell(p) : p \vDash_f S\}$$

The f-bounded Kolmogorov complexity of S, denoted by $K^f(S|B)$, is the length of the program returned by the learner: $K^f(S|B) = \ell\left(\Phi_\ell^f(S|B)\right)$. The extension of the teacher, denoted by $\Omega_\ell^f(c|B)$, also selects the shortest witness set that makes the learner distinguish the concept:

$$\Omega_\ell^f(c|B) = \arg \min_{S}^{\prec} \{\delta(S) : \Phi_\ell^f(S|B) \in [c]_{L_B})\}$$

And the definition of the K^f-teaching size of a concept c is $TS_\ell^f(c|B) = \delta(\Omega_\ell^f(c|B))$.

We can also extend Theorem 1 in [38].

Corollary 1. Let L be a universal language and B a library. For any concept c in C_{L_B}, there is a complexity function f so that the f-Teaching Book will contain some (p_c, w_c) with $p_c \in [c]_{L_B}$ and $TS_\ell^f(c|B) = \delta(w_c)$.

Sometimes we will refer to the original L OR the augmented L_B depending on whether we see it conditional to B or not. We are now in position to give a formal definition of the conditional teaching size given a set of concepts.

Definition 2. Let $a \in C_L$, $\{c_i\}_{i=1}^n \subset C_L$ and let $p_i = \Phi(\Omega_\ell^f(c_i))$, for each $i = 1, \dots, n$. Let $B = \{p_i\}_{i=1}^n$. We define the conditional teaching size of concept a given the concepts $\{c_i\}_{i=1}^n$, denoted by $TS_\ell^f(a|c_1, \dots, c_n)$, as

$$TS_\ell^f(a|c_1, \dots, c_n) = TS_\ell^f(a|B)$$

The programs that identify the concepts are in the same f-Teaching Book.

We now give a definition of *curriculum*. Given a set of concepts, a curriculum is a set of disjoint sequences covering all the concepts. Our notion of curriculum is more general than just a simple sequence. If some branches are unrelated, a curriculum should not specify which branch comes first, and are considered

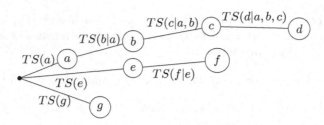

Fig. 2. Curriculum $\{a \to b \to c \to d, e \to f, g\}$ for a set of concepts $\{a, b, c, d, e, f, g\}$.

independent 'lessons'. We will see how this flexibility is handled by the algorithm that finds the optimal curriculum in Sect. 5. For instance, Fig. 2 shows how a set of concepts $\{a, b, c, d, e, f, g\}$ is partitioned into three branches: $\{a \to b \to c \to d, e \to f, g\}$, where $a \to b$ means that b must come after a in the curriculum. For each *branch*, there is no background knowledge or library at the beginning. The library grows as the teacher-learner protocol progresses in each branch.

Definition 3. Let $Q = \{c_i\}_{i=1}^n$ a set of n labelled concepts. A curriculum $\pi = \{\sigma_1, \sigma_2, \cdots, \sigma_m\}$ is a full partition of Q where each of the m subsets $\sigma_j \subset Q$ has a total order, becoming a sequence.

We denote $\overline{Q}$ as the set of all the curricula in Q. The order in which the subsets are chosen does not matter, but the order each subset is traversed does. For example, the curriculum $\pi = \{a \to b \to c \to d, e \to f, g\}$ can have many paths, such as $abcdefg$ or $gabcdef$. But note that π is different from $\pi' = \{b \to a \to c \to d, f \to e, g\}$. It is easy to check that, for any Q with n concepts, the number of different curricula is $|\overline{Q}| = n! \cdot \left(\sum_{k=0}^{n-1} \binom{n-1}{k} \cdot \frac{1}{(k+1)!} \right)$.

In what follows we will consider that all concepts are in the original f-Teaching Book, so they can be taught independently. This is not an important constraint, given Theorem 1 in [38] and Corollary 1. With this we ensure the same f for all of them. Now we can define the teaching size of a curriculum:

Definition 4. *Let* f *be a complexity function and let* Q *be a set of concepts that appear in the original* f-*Teaching Book. Let* $\pi = \{\sigma_1, \sigma_2, \cdots, \sigma_m\}$ *a curriculum in* Q. *We define the teaching size of each sequence* $\sigma = \{c_1, c_2, ..., c_k\}$ *as* $TS_\ell^f(\sigma) = TS_\ell^f(c_1) + \sum_{j=2}^k TS_\ell^f(c_j | c_1, \ldots, c_{j-1})$. *The overall teaching size of* π *is just* $TS_\ell^f(\pi) = \sum_{i=1}^m TS_\ell^f(\sigma_i)$.

We say that a curriculum in Q is minimal, denoted by π^*, if no other has less overall teaching size.

4.2   Interposition and Non-monotonicity

We now show a teaching phenomenon called *interposition*: new acquired concepts may lead to an increase in teaching size. The phenomenon might not even preserve the relationship established between two concepts, in terms of conditional Kolmogorov complexity, when considering conditional teaching size.

Definition 5. We say that B is an *interposed* library for concept c if $TS(c|B) > TS(c)$; if $B = \{p'\}$ we say that p' is an interposed program for c.

Proposition 2. For any $(w_c, p_c) \in$ f-Teaching Book, such that $@ \prec p_c$, there is an interposed library for concept c.

The above proposition means that virtually every concept (represented in the teaching book by a program of more than one instruction) may be interposed by a primitive that makes the witness set lead to another concept. Not only may some concepts be useless for the concepts yet to come in the curriculum, but that they may even be harmful. This will have important implications when we look for minimal curricula in the following section.

This contrasts with conditional Kolmogorov complexity, where for every a and b we have that $K(a|b) \leq K(a)$. Given this, we can study the monotonicity between concept complexity and teaching size. Namely, *is there any relationship between $K(a|b) \leq K(b|a)$ and $TS(a|b) \leq TS(b|a)$?* We now show that, for any universal language, the inequalities aforementioned have, in general, different directions. First, we give the following definition.

Definition 6. Let $c \in C_L$ and let B be a library. We define the Kolmogorov conditional complexity of a concept c given a library B as $K_{L_B}(c) = \ell(p_c^*)$ where p_c^* is calculated using L_B. We use the notation $K(c|B) = K_{L_B}(c)$

We now extend the conditional Kolmogorov complexity to a set of concepts through programs that identify the concepts given in the same f-Teaching Book:

Definition 7. Let $a \in C_L$, the set $\{c_i\}_{i=1}^n \subset C_L$ and $p_i = \Phi(\Omega_\ell^f(c_i))$, for each $i = 1, \ldots, n$. Let $B = \{p_i\}_{i=1}^n$. We define the Kolmogorov complexity of concept a given the concepts $\{c_i\}_{i=1}^n$, denoted by $K(a|c_1, \ldots, c_n)$, as

$$K(a|c_1, \ldots, c_n) = K(a|B)$$

In words, the conditional complexity of a concept given a set of concepts is equal to the conditional complexity of the concept given the canonical programs for those concepts as extracted from the original teaching book.

We now show the non-monotonicity between K and TS:

Theorem 2. There exist two concepts $a, b \in C_L$ and a complexity function, f, such that $K(a|b) < K(b|a)$ and $TS_\ell^f(a|b) > TS_\ell^f(b|a)$.

When considering conditional teaching size for curriculum learning, we need general conditions to avoid interposition. For instance, an important reduction of program size in language L_B usually minimises the risk of interposition.

Corollary 2. Let $(w_c, p_c) \in$ f-Teaching Book, with $p_c \in [c]_L$. If there exists a library B and a witness set w, verifying the following conditions (1) $\delta(w) < \delta(w_c)$ and (2) the first program $p_c' \in [c]_{L_B}$, using order $\prec$, such that $p_c' \vDash_f w$, precedes any other program p in language L_B, satisfying $p \vDash_f w$, then $TS_\ell^f(c|B) < TS_\ell^f(c)$.

These conditions to avoid interposition are strong, since we shall elucidate, e.g., whether a program is the shortest one, using a time complexity bound f.

5   Minimal Curriculum: Interposition Range and $\mathbb{I}$-search

One key reason why interposition is hard to avoid is the existence of programs (and concepts) with *parallel behaviour*, i.e., programs with equal inputs-outputs up to large sizes of the inputs, e.g., one implementing the even function, and the other doing the same except for the input 2^{300}. However, in practice, the concepts we use in the break-out for a curriculum do not have this problem. For instance, we can use addition to teach multiplication. They coincide in a few cases, $2+2 = 4$ and $2 \times 2 = 4$, but they clearly differ in many other short inputs.

Thus, let a, b be distinct concepts such that $\exists (w_a, p_a), (w_b, p_b) \in$ f−Teaching Book, with p_a, p_b in L verifying $w_a \nvDash_f p_b$ and $w_b \vDash_f p_a$. Assume that we use w_a first and the learner outputs p_a, and adds it to $B = \{p_a\}$. With this increased L_B, if we give w_b to the learner, it does not output p_a since $p_a \nvDash_f w_b$. However, there might still be interposition. For instance, suppose that L has four instructions: x, y, z and t. Let $B = \{xx\}$ and suppose that $p_b = $ zytxz is f-compatible with w_b. Suppose that there exists $p = $ xxytxx, expressed as $p = $ @yt@ in L_B, such that $p \vDash_f w_b$. Program p would interpose to p_b. It would be important to know about such programs p, i.e., the ones that precede p_b in L_B and are posterior in L.

5.1   Interposition Range: $\mathbb{I}$-sets

Firstly, we define the set of *interposed programs*.

Definition 8. Let w be a witness set and B be a library. Let p be a program in language L_B such that $p \vDash_f w$. We define the $\mathbb{I}$-set of *interposed programs* in language L_B for p and w as $\mathbb{I}^f_w(p|B) = \{q \text{ in } L_B : q \vDash_f w \text{ and } q \prec p\}$.

We now show how large the $\mathbb{I}$-sets can be. To do that, we use the size of a program when its library calls are *unfolded*, i.e., given a program p and a library B, we use $\circ(p)$ to denote the program that is equivalent to p (as it worked in L_B), where each primitive call @ has been replaced by the instructions of the called primitive in B.

Given an $\mathbb{I}$-set, we call *size-range*, denoted as $[i_{min}, i_{max}]$, to the range of $i = \dot{\ell}(\circ(q))$, $\forall q \in \mathbb{I}$-set. The *call-range*, denoted as $[j_{min}, j_{max}]$, is the range of the number of library calls, j, $\forall q \in \mathbb{I}$-set. We call *s/c-ranges* to both ranges; interposition occurs within them. The following theorem gives the *s/c-ranges* explicitly and provides a bound for the cardinality of the $\mathbb{I}$-set.

Theorem 3. Let (w_a, p_a), $(w_b, p_b) \in$ f-Teaching Book, with p_a, p_b in L and $p_a \nvDash_f w_b$. Consider the library $B = \{p_a\}$. Let p'_b an equivalent program to p_b for L_B. Then, the cardinal of $\mathbb{I}^f_{w_b}(p'_b|B)$ is bounded by $\sum_i \left(\sum_j \binom{i - \dot{\ell}(p_b) \cdot j + j}{j} \cdot (|\Upsilon| - 1)^{(i - j \cdot \dot{\ell}(p_b))} \right)$ with $i, j \in \mathbb{N}$ ranging in the intervals: (1) $i_{min} = \dot{\ell}(p_b)$, $i_{max} = 1 + (\dot{\ell}(p'_b) - 1) \cdot \dot{\ell}(p_a)$, $j_{min} = \lceil \frac{i - \dot{\ell}(p'_b)}{\dot{\ell}(p_a) - 1} \rceil$ and $j_{max} = \lfloor \frac{i}{\dot{\ell}(p_a)} \rfloor$, when $1 < \dot{\ell}(p_a) < \dot{\ell}(p_b)$; (2) $i_{min} = \dot{\ell}(p_a) + 1$ and the rest is as (1), when $\dot{\ell}(p_a) \geq \dot{\ell}(p_b)$.

Could we identify an empty $\mathbb{I}$*-set, based just on the sizes of the programs involved?* It happens when the *s/c-ranges* define an *empty region.* In Theorem 3 (1), it occurs whenever $i_{max} < i_{min}$. Namely, we have $\mathbb{I}^f_{w_b}(p_b'|B) = \emptyset$, when:

$$\dot{\ell}(p_b) > 1 + (\dot{\ell}(p_b') - 1) \cdot \dot{\ell}(p_a) \tag{2}$$

For instance, if $\dot{\ell}(p_a) = 4$, $\dot{\ell}(p_b) = 8$ and we know that $\dot{\ell}(p_b') = 2$, then $i_{min} = 8$ and $i_{max} = 1 + (2-1) \cdot 4 = 5$. We see that this becomes more likely as p_b is much greater than p_a and the program for b using B, i.e., p_b', is significantly reduced by the use of $B = \{p_a\}$.

Let p' be the first program in $[b]_{L_B}$ such that $p' \vDash_f w_b$. With the conditions of Theorem 3 (1), p' must be equivalent to p_b and operating with Eq. 2 we get $\dot{\ell}(p') < \frac{\dot{\ell}(p_b)-1}{\dot{\ell}(p_a)} + 1$, which means there is no interposition for any program for b by including $B = \{a\}$ and $TS^f_\ell(b|a) \leq TS^f_\ell(b)$. But, since $\ell(p_b') \geq K(b|a)$ we also have that Eq. 2 is impossible when $K(b|a) \geq (\frac{\dot{\ell}(p_b)-1}{\dot{\ell}(p_a)} + 1) \cdot \log_2 |\Upsilon|$.

We now consider a library with more than one primitive. We cannot extend Theorem 3 as a Corollary, since the relationships involved change completely, but we can connect both cases through the *s/c-ranges.*

Theorem 4. Let $\{(w_m, p_m)\}^n_{m=1}$, $(w_c, p_c) \in$ f-Teaching Book, with p_c, p_m in L, $\forall m$. Consider $B = \{p_m\}^n_{m=1}$ with $p_m \nvDash_f w_c, \forall m$, and $1 < |B|$. Let p_c' be an equivalent program to p_c for L_B. Let $D, r \in \mathbb{N}$ such that $\ell(p_c') = D \cdot \ell(@i) + r$, i.e., they are the *divisor* and the *remainder* of the division $\ell(p_c')/\ell(@i)$. Note that $\ell(@i) = \log_2 |\Upsilon| + \log_2 |B|$. Let $p_{max} = \max^\prec\{p_m\}^n_1$ and $p_{min} = \min^\prec\{p_m\}^n_1$. Then, the cardinal of $\mathbb{I}^f_{w_c}(p_c'|B)$ is bounded by $|B| \cdot \sum^{\dot{\ell}(p_c')}_{s=2} (\sum^s_{t-1}(|\Upsilon|-1)^{s-t} \cdot |B|^{t-1})$ and the *s/c-intervals* are: (1) if $1 < \dot{\ell}(p_{min}) \leq \dot{\ell}(p_c)$, then $i_{min} = \dot{\ell}(p_c)$, $i_{max} = D \cdot \dot{\ell}(p_{max}) + \lfloor r/\log_2 |\Upsilon| \rfloor$, $j_{min} = \lfloor \frac{\dot{\ell}(p_c')-\dot{\ell}(\circ(q))}{\dot{\ell}(@i)-\dot{\ell}(p_{max})} \rfloor$ and $j_{max} = \min\{D, \lfloor \frac{\dot{\ell}(\circ(q))}{\dot{\ell}(p_{min})} \rfloor\}$; (2) if $\dot{\ell}(p_c) < \dot{\ell}(p_{min})$, then $i_{min} = \dot{\ell}(p_{min}) + 1$ and the rest is as in (1).

We need $D \cdot \dot{\ell}(p_{max}) + \lfloor r/\log_2 |\Upsilon| \rfloor < \dot{\ell}(p_{min}) + 1$, to avoid interposition directly, in the same conditions as in Theorem 4 (1). It entails $\ell(p_c') < \ell(@i)$ when $\lfloor r/\log_2 |\Upsilon| \rfloor = 0$ in the extreme case. For Theorem 4 (2), an unfeasible *s-range* implies $D < \frac{\dot{\ell}(p_c)-\lfloor r/\log_2 |\Upsilon| \rfloor}{\dot{\ell}(p_{max})}$, which is restrictive.

5.2 Teaching Size Upper Bounds: $\mathbb{I}$-safe

In practice, we deal with a program p that has the desired behaviour for a given witness set, but there may be interposition. If we know which the interposed programs are, then it is possible to get an upper bound of the teaching size of the concept that defines p, by *deflecting* interposition, refining the witness sets.

We employ $\mathbb{I}$-*safe* witnesses: example sets attached to input/output pairs. For instance, if we want to teach exponentiation, a set of examples might be $\{(3,1) \to 3, (2,2) \to 4\}$. This witness set is compatible with exponentiation, but also compatible with multiplication. To avoid multiplication being interposed, we can add another example to distinguish both concepts: $\{(3,1) \to 3, (2,2) \to$

$4, (2,3) \rightarrow 8\}$. We can always replace the original witness set by an $\mathbb{I}$-safe witness set, where, in general, we need to add examples to avoid interposition.

Proposition 3. Let f be a complexity function and (w,p), $\{(w_m, p_m)\}_{m=1}^{n} \in$ f-Teaching Book, with p, p_m in L, $\forall m$. Let $B = \{p_m\}_{m=1}^{n}$ be a library such that $p_m \nVdash_f w$, $\forall m$. Let $c \in C_L$ such that $c \vDash w$. Let $p_c' \in [c]_{L_B}$ be the first program, using order $\prec$, such that $p_c' \nVdash_f w$. If $n = |\mathbb{I}_w^{f(p_c'|B)}|$, there exist $\{\langle i_k, o_k \rangle\}_{k=1}^{n}$ such that $TS_\ell^f(c|B) \leq \delta(w \bigcup_{k=1}^{n} \{\langle i_k, o_k \rangle\})$.

For a library B, if we find an example set w that can be converted into an $\mathbb{I}$-safe witness set $\overline{w} = w \bigcup_{k=1}^{n} \{\langle i_k, o_k \rangle\}$ with $\delta(\overline{w}) < TS_\ell^f(c)$ using B, then we reduce the teaching size. This is a sufficient and necessary condition to avoid interposition and get $TS_\ell^f(c|B) \leq TS_\ell^f(c)$.

Finally, given these general bounds: *how can we find minimal curricula?* Let us consider, for example, the set of concepts $Q = \{a, b\}$, where (w_a, p_a) and (w_b, p_b) are in the f-Teaching Book. We also know that their behaviours are not parallel, i.e., $p_a \nVdash_f w_b$ and $p_b \nVdash_f w_a$. There are three different curricula $\{a, b\}$, $\{a \rightarrow b\}$ or $\{b \rightarrow a\}$. There is an $\mathbb{I}$-safe witness set $\overline{w}$, such that $\delta(\overline{w}) \leq TS_\ell^f(b|a)$ (or $\delta(\overline{w}) \leq TS_\ell^f(a|b)$). Thus, we can choose a curriculum, with less overall teaching size than the non-incremental version.

5.3 Minimal Curriculum Algorithm: $\mathbb{I}$-search

We now *search* minimal curricula. For example, let $Q = \{c_+, c_\times\}$ be a set of two concepts from Fig. 1, which appear in the non-incremental f-Teaching Book as (w_+, p_+) and $(w_\times, p_\times)$. The set of possible curricula, $\overline{Q}$, is $\pi_0 = \{c_+, c_\times\}$, $\pi_1 = \{c_+ \rightarrow c_\times\}$ and $\pi_2 = \{c_\times \rightarrow c_+\}$.

The starting point for our algorithm will be π_0, the non-incremental curriculum, and its overall teaching size TS_ℓ^f. Then, we generate another curriculum: π_1. We know $TS_\ell^f(c_+) = \delta(w_+)$ and we need to add $TS_\ell^f(c_\times | c_+)$. We compare this total size to the best TS so far. We explore all the curricula in $\overline{Q}$ but, in order to save computational steps, we generate successive witness sets w_k, using order $\prec$, such that $c_\times \vDash w_k$ (Fig. 3).

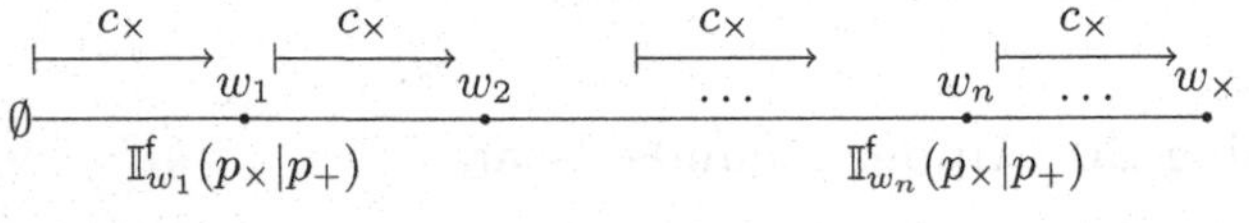

Fig. 3. Non-decreasing sequence of witness sets w_k, through $c_\times$ with $\delta(w_k) \leq \delta(w_\times)$.

For each w_k, we get the first program p_k of $\mathbb{I}_{w_k}^f(p_\times | p_+)$. We then investigate whether $p_k \in [p_\times]_{L_B}$ or not. If p_k acts like $p_\times$ to certain witness size limit, H, then we can identify p_k and $p_\times$. The $\mathbb{I}$-*search algorithm* (5.3) shown below extends this strategy.

Note that the *s/c-ranges* reduce, *drastically*, the computational effort of executing the teacher-learner protocol (calculating teaching book and TS).

Algorithm: I-search
Input: $Q = \{a, b, \dots\}$; f-Teaching Book (w_a, p_a), $(w_b, p_b)\dots$; Witness size limit H

1. **For each** distinct pair of concepts $\langle x, y \rangle \in Q \times Q$:
   (a) **If** $[TS^f_\ell(y|x) \leq TS^f_\ell(y) \wedge TS(x|y)^f_\ell \geq TS^f_\ell(x)]$
      then $\overline{Q} = \overline{Q} \setminus \{\pi : \exists$ a branch starting as $y \to x\}$
2. $\pi^* = \{a, b, \dots\}$, $TS^f_\ell(\pi^*) = \sum_{x \in Q} TS^f_\ell(x)$ and $\overline{Q} = \overline{Q} \setminus \{\pi^*\}$
3. **For each** $\pi \in \overline{Q}$:
   (a) $TS^f_\ell(\pi) = 0$
   (b) **For each** branch $\sigma \in \pi$:
      i. **For each** concept $x \in \sigma$ (ordered by σ):
         - $B = \{p_y : (y \in \sigma) \wedge (y \text{ precedes } x)\}$
         - Let p'_x be the first program equivalent to p_x in L_B, using order $\prec$
         - **For each** $w_k \in \{w \subset X : p'_x \vDash_f w_k\}$, using order $\lessdot$:
            - **If** $[TS^f_\ell(\pi^*) \leq TS^f_\ell(\pi) + \delta(w_k)]$ **then break** to 3
            - $p = \min^{\prec}\{\mathbb{I}^f_{w_k}(p'_x|B)\}$; use *s/c ranges* to refine the calculation
            - **If** $[p \vDash_f w \longleftrightarrow p_x \vDash_f w, \forall w \text{ such that } \delta(w) < H]$
               then $[\ TS^f_\ell(\pi) = TS^f_\ell(\pi) + \delta(w_k)\ $ **and break** to 3(b)i $]$
   (c) $\pi^* = \pi$ and $TS^f_\ell(\pi^*) = TS^f_\ell(\pi)$
Output: π^* and $TS^f_\ell(\pi^*)$

In the previous example, e.g., if there is a w_n such that $TS^f_\ell(c_\times|c_+) = \delta(w_n) < TS^f_\ell(c_\times)$, then we set $\pi^* = \pi_1$ (and $TS^f_\ell(\pi^*) = \delta(w_+) + \delta(w_n)$). Finally, we test π_2 and follow the same steps as with π_1. If, at some stage, there is a witness set w_m such that $TS^f_\ell(c_\times) + \delta(w_m) \geq TS^f_\ell(\pi^*)$, then π_1 is minimal and we stop.

The algorithm is complete but the search is not *exhaustive*, since we can *discard* curricula that contain a *branch* starting in a way that does not decrease the overall teaching size for sure. For example, if $TS^f_\ell(c_\times|c_+) \leq TS^f_\ell(c_\times)$ and $TS^f_\ell(c_+|c_\times) \geq TS^f_\ell(c_+)$, the branch $\sigma = \{c_+ \to c_\times \to c_\wedge\}$ has less or equal overall teaching size than $\sigma' = \{c_\times \to c_+ \to c_\wedge\}$. Consequently, we can remove all branches starting with $c_\times \to c_+$. We can test this for every pair of distinct concepts at the beginning of the branches.

The I-search algorithm (5.3) satisfies the following theorem.

Theorem 5. Let H be certain witness size limit, f be a complexity function and Q be a set of concepts registered in the f-Teaching Book. We also assume, for each $c \in Q$, that $c \vDash w \to p_c \vDash_f w, \forall w$ verifying $\delta(w) \leq \sum_{x \in Q} TS^f_\ell(x)$. Then, the I-search algorithm expressed in algorithm 5.3 returns a minimal curriculum.

The I-search algorithm shows that: (1) We should create curricula containing concepts that significantly reduce the complexity of another ones. For instance, if concepts $c_\times$ and c_+ (Fig. 1) satisfy $K(c_\times|c_+) < K(c_\times)$, then the chances to minimise the teaching size increase significantly. (2) Given a set of concepts, it may be useful to implement some kind of *isolation* (or even forgetting by separating concepts in different branches[7]). For instance, c_θ might be f-compatible with a considerable number of witness sets w_k and it may cause *interposition* with c_+, $c_\times$ or $c_\wedge$. This is why we should allocate c_θ in a different branch. (3) The branches (or lessons) could simply suggest ways in which we can arrange, *classify* and organise large sets of concepts. The tree-structure for curricula proposed here is a solution for the problem posed in [26].

6   Conclusions and Future Work

The teaching *size*—rather than teaching dimension—opened a new avenue for a more realistic and powerful analysis of machine teaching [38], its connections with information theory (both programs and examples can be measured in bits) and a proper handling of concept classes where examples and programs are compositional and possibly universal, such as natural language.

The intuitive concept of how much of the description of a concept is reused for the definition of another dates back to Leibniz's *règle pour passer de pensée en pensée* [18], and has been vindicated in cognitive science since Vigotsky's zone of proximal development [29,39], to more modern accounts of compositionality based on what has been learnt previously [22,24,30].

In mathematical terms, a gradient-based or continuous account of this view of incremental teaching, and the reuse of concepts, is not well accommodated. Incremental teaching is usually characterised as a compositional process, which is a more appropriate view for the acquisition of high-level concepts. The learning counterpart is still very elegantly captured by conditional Kolmogorov complexity, and some incremental learning schemata have followed this inspiration [13,17,20,23,31]. However, even if the concept of teaching *size* suggests that a mapping was possible, we have had to face a series of phenomena in order to translate some of these intuitions to the machine teaching scenario, and a new setting for curriculum teaching.

The absence of monotonicity because of interposition presents some difficulties for implementing curriculum teaching for compositional languages. Theorems 3 and 4 and its consequences make possible such an implementation: either through sufficient conditions to avoid interposition, by implementing 𝕀-safe witness sets or through the 𝕀-search.

Given the theoretical bounds and the algorithms for the optimal curricula, we can now start exploring novel algorithms and strategies for curriculum teaching that are suboptimal, but more efficient, such as (1) greedy algorithms introducing the next concept as the one with maximum local TS reduction, (2) approximations based on Vigotsky's zone of proximal development principles [29,39] where each step is bounded by some teaching length Z, i.e., such that $TS(c_{i+1}|c_1,\ldots,c_i) \leq Z, \forall i$; or (3) variations of the *incremental combinatorial optimal path* algorithm [32]. All these new research possibilities in curriculum teaching, and even others, are now wide open to exploration.

---

[7] Forgetting may simply refer to a lesson not using primitives that are considered out of the context of a "lesson".

Because of the fundamental (re-)connection we have done between K and TS in this paper, another novel possibility for curriculum teaching would be the combination of teaching by examples *and* descriptions of the concepts themselves. This is actually the way humans teach other humans, combining examples and descriptions, but it is nevertheless unprecedented in the application of machine teaching in natural language processing [25,33]. However, it is beginning to become common with language models, with prompts that combine examples and some indications of the task to perform [4,14].

Acknowledgements. This work was funded by the EU (FEDER) and Spanish MINECO under RTI2018-094403-B-C32, G. Valenciana under PROMETEO/2019/098 and EU's Horizon 2020 research and innovation programme under grant 952215 (TAILOR).

References

1. Antoniol, G., Di Penta, M.: Library miniaturization using static and dynamic information. In: International Conference on Software Maintenance, pp. 235–244 (2003)
2. Balbach, F.J.: Models for algorithmic teaching. Ph.D. thesis, U. of Lübeck (2007)
3. Balbach, F.J.: Measuring teachability using variants of the teaching dimension. Theoret. Comput. Sci. **397**(1–3), 94–113 (2008)
4. Brown, T.B., Mann, B., Ryder, N., et al.: Language models are few-shot learners. arXiv:2005.14165 (2020)
5. Cicalese, F., Laber, E., Molinaro, M., et al.: Teaching with limited information on the learner's behaviour. In: ICML, pp. 2016–2026. PMLR (2020)
6. Devlin, J., Chang, M.W., Lee, K., Toutanova, K.: Bert: pre-training of deep bidirectional transformers for language understanding. arXiv: 1810.04805 (2018)
7. Elias, P.: Universal codeword sets and representations of the integers. IEEE Trans. Inf. Theory **21**(2), 194–203 (1975)
8. Gao, Z., Ries, C., Simon, H.U., Zilles, S.: Preference-based teaching. J. Mach. Learn. Res. **18**(1), 1012–1043 (2017)
9. Garcia-Piqueras, M., Hernández-Orallo, J.: Conditional teaching size. arXiv: 2107.07038 (2021)
10. Gong, C.: Exploring commonality and individuality for multi-modal curriculum learning. In: AAAI, vol. 31 (2017)
11. Gong, C., Yang, J., Tao, D.: Multi-modal curriculum learning over graphs. ACM Trans. Intell. Syst. Technol. (TIST) **10**(4), 1–25 (2019)
12. Gong, T., Zhao, Q., Meng, D., Xu, Z.: Why curriculum learning & self-paced learning work in big/noisy data: a theoretical perspective. BDIA **1**(1), 111 (2016)
13. Gulwani, S., Hernández-Orallo, J., Kitzelmann, E., Muggleton, S.H., Schmid, U., Zorn, B.: Inductive programming meets the real world. Commun. ACM **58**(11), 90–99 (2015)
14. Hendrycks, D., Burns, C., Basart, S., Zou, A., Mazeika, M., Song, D., Steinhardt, J.: Measuring massive multitask language understanding. In: ICLR (2021)
15. Hernández-Orallo, J., Telle, J.A.: Finite and confident teaching in expectation: Sampling from infinite concept classes. In: ECAI (2020)
16. Kumar, A., Ithapu, V.: A sequential self teaching approach for improving generalization in sound event recognition. In: ICML, pp. 5447–5457 (2020)

17. Lake, B.M., Salakhutdinov, R., Tenenbaum, J.B.: Human-level concept learning through probabilistic program induction. Science **350**(6266), 1332–1338 (2015)
18. Leibniz, G.W., Rabouin, D.: Mathesis universalis: écrits sur la mathématique universelle. Mathesis (Paris, France) Librairie philosophique J. Vrin (2018)
19. Li, M., Vitányi, P.M.: An Introduction to Kolmogorov Complexity and Its Applications, 3rd edn. Springer, Heidelberg (2008). https://doi.org/10.1007/978-3-030-11298-1
20. Li, Y., Mao, J., Zhang, X., Freeman, W.T., Tenenbaum, J.B., Wu, J.: Perspective plane program induction from a single image. In: CVPR, pp. 4434–4443 (2020)
21. Liu, W., et al.: Iterative machine teaching. In: ICML, pp. 2149–2158 (2017)
22. Manohar, S., Zokaei, N., Fallon, S., Vogels, T., Husain, M.: Neural mechanisms of attending to items in working memory. Neurosci. Biobehav. Rev. **101**, 1–12 (2019)
23. Nye, M.I., Solar-Lezama, A., Tenenbaum, J.B., Lake, B.M.: Learning compositional rules via neural program synthesis. arXiv: 2003.05562 (2020)
24. Oberauer, K., Lin, H.Y.: An interference model of visual working memory. Psychol. Rev. **124**(1), 21 (2017)
25. Peng, B., Li, C., Li, J., Shayandeh, S., Liden, L., Gao, J.: Soloist: building task bots at scale with transfer learning and machine teaching. arXiv: 2005.05298 (2020)
26. Pentina, A., Sharmanska, V., Lampert, C.H.: Curriculum learning of multiple tasks. In: Proceedings of Computer Vision and Pattern Recognition (2015)
27. Radford, A., Wu, J., Child, R., Luan, D., Amodei, D., Sutskever, I.: Language models are unsupervised multitask learners. OpenAI Blog **1**(8), 9 (2019)
28. Rakhsha, A., Radanovic, G., Devidze, R., Zhu, X., Singla, A.: Policy teaching via environment poisoning: training-time adversarial attacks against reinforcement learning. In: ICML, pp. 7974–7984 (2020)
29. Salkind, N.: An Introduction to Theories of Human Development. Sage P. (2004)
30. Schneider, W.X., Albert, J., Ritter, H.: Enabling cognitive behavior of humans, animals, and machines: a situation model framework. ZiF **1**, 21–34 (2020)
31. Shi, Y., Mi, Y., Li, J., Liu, W.: Concept-cognitive learning model for incremental concept learning. IEEE Trans. Syst. Man Cybern. Syst. (2018)
32. Shindyalov, I., Bourne, P.: Protein structure alignment by incremental combinatorial extension of the optimal path. Prot. Eng. Des. Sel. **11**(9), 739–747 (1998)
33. Shukla, S., et al.: Conversation learner-a machine teaching tool for building dialog managers for task-oriented dialog systems. arXiv: 2004.04305 (2020)
34. Solomonoff, R.J.: A formal theory of inductive inference I. IC **7**(1), 1–22 (1964)
35. Solomonoff, R.J.: A system for incremental learning based on algorithmic probability. In: Proceedings of the Sixth Israeli Conference on AICVPR, pp. 515–527 (1989)
36. Soviany, P., Ionescu, R.T., Rota, P., Sebe, N.: Curriculum learning: a survey. arXiv:2101.10382 (2021)
37. Such, F.P., Rawal, A., Lehman, J., Stanley, K., Clune, J.: Generative teaching networks: accelerating neural architecture search by learning to generate synthetic training data. In: ICML, pp. 9206–9216 (2020)
38. Telle, J.A., Hernández-Orallo, J., Ferri, C.: The teaching size: computable teachers and learners for universal languages. Mach. Learn. **108**(8), 1653–1675 (2019). https://doi.org/10.1007/s10994-019-05821-2
39. Vygotsky, L.S.: Mind in Society: Development of Higher Psychological Processes. Harvard University Press, Cambridge (1978)
40. Weinshall, D., Cohen, G., Amir, D.: Curriculum learning by transfer learning: theory and experiments with deep networks. In: ICML, pp. 5235–5243 (2018)

41. Zhou, T., Bilmes, J.A.: Minimax curriculum learning: machine teaching with desirable difficulties and scheduled diversity. In: ICLR (Poster) (2018)
42. Zhu, X.: Machine teaching: an inverse problem to machine learning and an approach toward optimal education. In: AAAI, pp. 4083–4087 (2015)
43. Zhu, X., Singla, A., Zilles, S., Rafferty, A.: An overview of machine teaching. arXiv: 1801.05927 (2018)

FedDNA: Federated Learning with Decoupled Normalization-Layer Aggregation for Non-IID Data

Jian-Hui Duan, Wenzhong Li$^{(\boxtimes)}$ (ID), and Sanglu Lu

Nanjing University, Nanjing, China
{lwz,sanglu}@nju.edu.cn

Abstract. In the federated learning paradigm, multiple mobile clients train their local models independently based on the datasets generated by edge devices, and the server aggregates the model parameters received from multiple clients to form a global model. Conventional methods aggregate gradient parameters and statistical parameters without distinction, which leads to large aggregation bias due to cross-model distribution covariate shift (CDCS), and results in severe performance drop for federated learning under non-IID data. In this paper, we propose a novel decoupled parameter aggregation method called FedDNA to deal with the performance issues caused by CDCS. With the proposed method, the gradient parameters are aggregated using the conventional federated averaging method, and the statistical parameters are aggregated with an importance weighting method to reduce the divergence between the local models and the central model to optimize collaboratively by an adversarial learning algorithm based on variational autoencoder (VAE). Extensive experiments based on various federated learning scenarios with four open datasets show that FedDNA achieves significant performance improvement compared to the state-of-the-art methods.

Keywords: Federated learning · Deep learning · Machine learning

1 Introduction

Federated learning (FL) has emerged as a novel distributed machine learning paradigm that allows a global machine learning model to be trained by multiple mobile clients collaboratively while protecting their private data in local devices. In such a paradigm, mobile clients train local models based on datasets generated by edge devices such as sensors and smartphones, and the server is responsible to aggregate parameters from local models to form a global model without transferring data to a central server. Federated learning has been drawn much attention in mobile-edge computing with its advantages in preserving data privacy [9,35] and enhancing communication efficiency [22,28].

Parameter aggregation is the key technology of federated learning, which typically involves the following three steps repeated periodically during the training process: (1) the involved clients train the same type of models with their

© Springer Nature Switzerland AG 2021
N. Oliver et al. (Eds.): ECML PKDD 2021, LNAI 12975, pp. 722–737, 2021.
https://doi.org/10.1007/978-3-030-86486-6_44

local data independently; (2) when the server sends an aggregation signal to the clients, the clients transmit their model parameters to the server; (3) after receiving the local models' parameters, the server applies an aggregation method to the received parameters to form a global model, and broadcast the global model's parameters to the involved clients for the next round of federated training. The standard aggregation method FedAvg [22] and its variants such as q-FedSGD [19] applied a synchronous parameter averaging method to form the global model. Several efforts had been made to deal with non-IID data in federated learning. Zhao et al. proposed to use a globally shared dataset for training to address data heterogeneity [34]. FedProx [18] modified FedAvg by adding a heterogeneity bound on local datasets to tackle the non-IID condition. FedMA [28] demonstrated that permutations of layers could affect the parameter aggregation results and proposed a layer-wise parameter-permutation aggregation method to improve the accuracy of the global model.

Despite the efforts that have been made, applying the existing parameter aggregation methods for a large number of heterogeneous clients in federated learning suffers from performance degrade. It was reported in [34] that the accuracy of a convolutional neural network (CNN) model trained by FedAvg reduced by up to 55% for a highly skewed heterogeneous dataset. The work of [28] showed that the accuracy of FedProx [18] dropped over 11% when the client number increases from 5 to 20 under non-IID data partition. A key issue to cause the performance drop in federated learning could be the covariate shift of data distribution among clients due to non-IID data, which is known as *cross-model distribution covariate shift (CDCS)*. Such issue has not been addressed appropriately by the previous parameter aggregation methods. We use the following example to illustrate the impact of CDCS in federated learning.

Aggregation Bias Due to CDCS: A CNN model typically consists of the *convolutional (Conv) layers*, the *full connected (FC) layers*, and the *normalization layers*. Those layers are formulated by two different types of parameters to be trained: (1) the *gradient parameters* that represent the weights of the CNN model, which are commonly contained in all layers; (2) the *statistical parameters* that represent the statistical information such as mean and variance of the feature maps, which are solely contained in the normalization layers (e.g., Batch-Normalization and Layer-Normalization). Conventional federated learning approaches such as FedAvg simply average the local model parameters indiscriminately to form a global model, which will lead to bias on the statistical information for non-IID data. Figure 1(a) illustrates the KL-divergence [10] between the parameters of a FedAvg-aggregated model and that of a centrally-trained model with varying number of clients on non-IID data. It is shown that with the increasing of communication rounds, the divergence of gradient parameters approaches to 0, but that of statistical parameters increases to a large value. The more heterogeneous clients involved, the higher divergence is observed. The divergence is mainly caused by the model aggregation methods such as FedAvg that fail to address the CDCS of non-IID local datasets. As a result, the aggregated models' test accuracy decreases dramatically, as shown in Fig. 1(b).

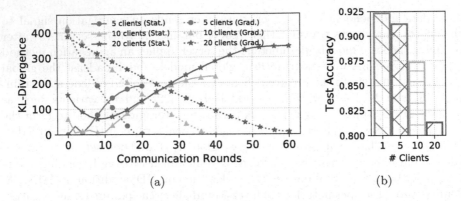

(a)   (b)

Fig. 1. Illustration of divergence and performance drop caused by CDCS on non-IID data. (a) The KL-divergence of gradient parameters and statistical parameters between a federated learning model and a central model of ResNet18@CIFAR-10. (b) The test accuracy of ResNet18 on CIFAR-10 with different #clients.

In this paper, we propose a novel decoupled model aggregation method called FedDNA (shorten for <u>fe</u>derated learning with <u>d</u>ecoupled <u>n</u>ormalization-layer parameter <u>a</u>ggregation) to address the performance issues caused by CDCS on non-IID data for federated learning. FedDNA aggregates gradient parameters and statistical parameters in a decoupled way. The gradient parameters are aggregated using the conventional distributed stochastic gradient descent (SGD) method, which is theoretically converged during distributed training. The statistical parameters are aggregated with an importance weighting method to reduce the divergence between the local models and the central model, and they are optimized collaboratively by an adversarial learning algorithm based on variational autoencoder (VAE). Extensive experiments based on a variety of federated learning scenarios with four open datasets show that FedDNA significantly outperforms the state-of-the-art methods.

The contributions of our work are summarized as follows.

- We illustrate that cross-model distribution covariate shift (CDCS) can cause large divergence in the aggregated statistical parameters of multiple heterogeneous local models, which is a key problem of performance drop for federated learning under non-IID data.
- We propose a novel decoupled model aggregation method to deal with the performance issues caused by CDCS, where the gradient parameters and statistical parameters are aggregated separately, aiming to reduce the divergence between the local models and the central model.
- We propose an adversarial learning algorithm to derive the optimal probabilistic weights for statistical parameters aggregation. It enables a data-free solution for the federated server with a variational autoencoder (VAE) to minimize the divergence of unknown distributions based on limited information received from the clients.

– We conduct extensive experiments using five mainstream CNN models based on three federated datasets under non-IID conditions. Compared to the de facto standard FedAvg, and the state-of-the-art for non-IID data (FedProx, FedMA), the proposed FedDNA has the lowest divergence of the aggregated parameters, and the test accuracy improves up to 9%.

The rest of the paper is organized as follows: Sect. 2 presents the related works on federated learning and optimizing federated learning under non-IID data. Section 3 proposes the detailed decoupled mechanism of FedDNA. Section 4 describes the adversarial learning algorithm on optimizing FedDNA, inference and optimization on VAE and the detailed algorithm of FedDNA. Section 5 shows the performance evaluation for FedDNA with 5 state-of-the-art baselines. And the paper is concluded in Sect. 6.

2 Related Work

Federated learning [14,21,24,26,31,32] is an emerging distributed machine learning paradigm that aims to build a global model based on datasets distributing across multiple clients. One of the standard parameter aggregation methods is FedAvg [22], which combined local stochastic gradient descent (SGD) on each client with a server that performs parameter averaging. Later, the lazily aggregated gradient (Lag) method [2] allowed clients running multiple epochs before model aggregation to reduce communication cost. The q-FedSGD [19] method improved FedAvg with a dynamic SGD update step using a scale factor to achieve fair resources allocation among heterogeneous clients. The FedDyn [1] method proposed a dynamic regularizer for each round of aggregation, so that different models are aligned to alleviate the inconsistency between local and global loss.

Several works focused on optimizing federated learning under non-IID data. Zhao et al. used the earth mover's distance (EMD) to quantify data heterogeneity and proposed to use globally shared data for training to deal with non-IID [34]. FedProx [18] modified FedAvg by adding a heterogeneity bound on local datasets to tackle heterogeneity. The RNN-based method in [8] adopted a meta-learning method to learn a new gradient from the received gradients and then applied it to update the global model. The FedMA [28] method, derived from AFL [23] and PFNM [33], demonstrated that permutations of layers can affect the parameter aggregation results, and proposed a layer-wise parameter-permutation aggregation method to solve the problem. FedBN [20] suggested keeping the local Batch Normalization parameters not synchronized with the global model to mitigate feature shifts in Non-IID data. FedGN [5] replaced Batch Normalization with Group Normalization to avoids the accuracy loss induced by the skewed distribution of data labels. SCAFFOLD [11] used variance reduction to correct the "client-drift" in each client's local updates to prevent unstable and slow convergence for heterogeneous data. FedRobust [25] adopted a distributed optimization method via gradient descent ascent to address affine distribution shifts across users in federated settings. Instead of averaging the cumulative local gradient, FedNova [29] aggregated normalized local gradients to eliminate objective inconsistency while preserving fast error convergence.

To the best of our knowledge, the problem of performance drop of federated learning caused by CDCS has not been explored in the literature. In this paper, we make the first attempt to deal with CDCS by decoupling the aggregation of gradient parameters and statistical parameters, and adopt an adversarial learning algorithm to optimize model aggregation to achieve high accuracy in non-IID conditions.

3   Decoupled Federated Learning with CDCS

3.1   Optimization Objectives

As discussed in Sect. 1, conventional federated learning methods suffers from noteworthy bias when aggregating statistical parameters and gradient parameters without distinction. To address this issue, we propose a decoupled method to optimize model parameters aggregation.

Consider a federated learning scenario with K clients that train their local deep neural network (DNN) models independently based on local datasets $\mathbf{x}_1, \mathbf{x}_2, \ldots, \mathbf{x}_K$, and report their model parameters to a central server. The objective of the server is to form an aggregate global DNN model to minimize the following loss function.

$$\min_{\mathbf{w}} \mathcal{L}(\mathbf{w}, \mathbf{x}) := \sum_{k=1}^{K} \frac{|\mathbf{x}_k|}{|\mathbf{x}|} \mathcal{L}_k(\hat{\mathbf{w}}_k, \tilde{\mathbf{w}}_k, \mathbf{x}_k), \tag{1}$$

where $\mathbf{x} = \{\mathbf{x}_1, \mathbf{x}_2, \ldots, \mathbf{x}_K\}$ is the total dataset; $\hat{\mathbf{w}}_k, \tilde{\mathbf{w}}_k$ are the gradient parameters and statistical parameters of local model received from the k-th client; $\mathcal{L}_k(\cdot)$ indicates the loss functions of the k-th local model; $\mathbf{w} = F(\hat{\mathbf{w}}_1, \tilde{\mathbf{w}}_1, \hat{\mathbf{w}}_2, \tilde{\mathbf{w}}_2, \cdots, \hat{\mathbf{w}}_K, \tilde{\mathbf{w}}_K)$ are the global model's parameters and F is the aggregation method to be derived that maps K local models to a global model.

To deal with the CDCS problem, we aggregate the gradient parameters and statistical parameters separatively. Different from the gradient parameters, which can be optimized with the conventional distributed stochastic gradient descent (SGD), the statistical parameters should be optimized to eliminate their covariate shifts. Since the real distribution of global data is unknown to the server, we propose a *collaborative optimization* approach that enables multiple clients to update their statistical parameters to gradually reduce their distribution divergence.

For the k-th client, we use $\tilde{\mathbf{w}}_k$ to denote its own statistical parameters, and $\tilde{\mathbf{w}}_{\neg k}$ to denote the averaged statistical parameters of the other clients except client-k. We refer to $\tilde{\mathbf{w}}_k$ and $\tilde{\mathbf{w}}_{\neg k}$ as the *twin statistics*. We assume that $\tilde{\mathbf{w}}_k$ is drawn from client-k's local distribution $\tilde{\mathbf{w}}_k \sim p_k(\tilde{\mathbf{w}})$, and $\tilde{\mathbf{w}}_{\neg k}$ is drawn from the distribution without client-k, i.e., $\tilde{\mathbf{w}}_{\neg k} \sim q_{\neg k}(\tilde{\mathbf{w}})$, where $\tilde{\mathbf{w}}$ represent the statistical parameters of the global model. Aggregation of statistical parameters should eliminate the discrepancy between $\tilde{\mathbf{w}}_k$ and $\tilde{\mathbf{w}}_{\neg k}$ so that they converge to the same objective distribution, which can be represented by the following optimization problem:

$$\min_{\tilde{\mathbf{w}}} \tilde{\mathcal{L}}(\tilde{\mathbf{w}}) := \frac{1}{K} \sum_{k=1}^{K} \mathbb{D}[p_k(\tilde{\mathbf{w}}), q_{\neg k}(\tilde{\mathbf{w}})], \tag{2}$$

where $\mathbb{D}[\cdot]$ represents the divergence of two distributions.

3.2 FedDNA Mechanism

Based on the optimization objectives, we proposed a decoupled method called FedDNA to optimize parameters aggregation for federated learning. The federated server received model parameters from the clients periodically. In the t-th communication round, the received parameters are:

- $\hat{\mathbf{w}}_k^t$: the gradient parameters of client-k in round t;
- $\tilde{\mathbf{w}}_k^t = [\tilde{\mathbf{mean}}_k^t, \tilde{\mathbf{var}}_k^t]$: the statical parameters of client-k in round t, where $\tilde{\mathbf{mean}}_k^t$ and $\tilde{\mathbf{var}}_k^t$ represent the parameters of statical mean and variance accordingly.

The parameter update process is as follows.

(1) Aggregation of Gradient Parameters: The gradient parameters can be aggregated using the principle of distributed stochastic gradient descent (SGD), the same as the method of FedAvg [22]:

$$\hat{\mathbf{w}}^{t+1} = \sum_{k=1}^{K} \frac{|\mathbf{x}_k|}{|\mathbf{x}|} \hat{\mathbf{w}}_k^t. \tag{3}$$

(2) Aggregation of Statistical Parameters: The statistical parameters are aggregated collaboratively to reduce the divergence between the individual's distribution and the overall distribution, which are updated with the following reweighting manner:

$$\tilde{\mathbf{mean}}_k^{t+1} = \gamma_k^t \tilde{\mathbf{mean}}_k^t + (1 - \gamma_k^t) \sum_{i=1, i \neq k}^{K} \frac{1}{K-1} \tilde{\mathbf{mean}}_i^t,$$

$$\tilde{\mathbf{var}}_k^{t+1} = \gamma_k^t \tilde{\mathbf{var}}_k^t + (1 - \gamma_k^t) \sum_{i=1, i \neq k}^{K} \frac{|\mathbf{x}_i| - 1}{|\mathbf{x}| - K - 1} \tilde{\mathbf{var}}_i^t. \tag{4}$$

In the above equations, the term $\sum_{i=1, i \neq k}^{K} \frac{1}{K-1} \tilde{\mathbf{mean}}_i^t$ is the average of other statistical means, and the term $\sum_{i=1, i \neq k}^{K} \frac{|\mathbf{x}_i| - 1}{|\mathbf{x}| - K - 1} \tilde{\mathbf{var}}_i^t$ is the weighted pooled variance [12] which gives an unbias estimation of the variance of the overall dataset without client-k. The adjustable parameter γ_k^t is called the *importance weight*, which tends to reduce the discrepancy between the twin statistics $\tilde{\mathbf{w}}_k$ and $\tilde{\mathbf{w}}_{\neg k}$. The weight γ_k^t can be formulated by

$$\gamma_k^t = f(\mathbb{D}[p_k(\tilde{\mathbf{w}}^t), q_{\neg k}(\tilde{\mathbf{w}}^t)]), \tag{5}$$

where $f\colon \mathbb{R}^+ \to [0,1)$ is a function that projects the divergence into the interval $[0,1)$. Generally speaking, the mapping f could be non-linear, and it can be implemented by a neural network in practice.

The proposed method minimize the loss function in Eq. (2) by gradually updating the statistical parameters to approach the overall distribution with an importance weight. By minimizing Eq. (2), the importance weight γ_k^t approaches to 0, and the statistical parameters of the ith client converges to that of the other clients. Thereafter, the final statistical parameters of the global model can be represented by $\tilde{\mathbf{w}}_k = [\frac{1}{K}\sum_{k=1}^K \tilde{\mathbf{mean}}_k^T, \sum_{k=1}^K \frac{(|\mathbf{x}_k|-1)}{|\mathbf{x}|-K}\tilde{\mathbf{var}}_k^T]$ where T is the total communication rounds.

To determine the value of importance weight γ_k^t, it needs to derive the divergence $\mathbb{D}$ and the mapping function f, which can be solved by the adversarial learning algorithm discussed in the following section.

4   Adversarial Learning Algorithm

FedDNA aggregates the statistical parameters with an importance weight γ_k^t intending to eliminate the discrepancy between the twin statistics $\tilde{\mathbf{w}}_k$ and $\tilde{\mathbf{w}}_{\neg k}$. The updating process can be viewed as sampling from $p_k(\tilde{\mathbf{w}})$ to match $q_{\neg k}(\tilde{\mathbf{w}})$ with acceptance rate $\alpha(\tilde{\mathbf{w}}) \in [0,1]$. Here $\alpha(\tilde{\mathbf{w}})$ can be interpreted as a binary classifier between "from $p_k(\tilde{\mathbf{w}})$" and "from $q_{\neg k}(\tilde{\mathbf{w}})$".

More formally, we let $y = 1$ to represent "from $p_k(\tilde{\mathbf{w}})$" and $y = 0$ to represent "from $q_{\neg k}(\tilde{\mathbf{w}})$". The overall dataset can be represent by a joint distribution $p(\tilde{\mathbf{w}}, y)$, which can be written as $p(\tilde{\mathbf{w}}, y) = p(\tilde{\mathbf{w}}|y = 1)p(y = 1) + p(\tilde{\mathbf{w}}|y = 0)p(y = 1)$. So the classifier can be represented by $\alpha(\tilde{\mathbf{w}}) = p(y = 1|\tilde{\mathbf{w}}_k)$. By Bayes' theorem, we have:

$$
\begin{aligned}
\alpha(\tilde{\mathbf{w}}) &= \frac{p(\tilde{\mathbf{w}}|y=1)p(y=1)}{p(\tilde{\mathbf{w}}, y)} = \frac{p_k(\tilde{\mathbf{w}})p(y=1)}{p(\tilde{\mathbf{w}}|y=1)p(y=1) + p(\tilde{\mathbf{w}}|y=0)p(y=0)} \\
&= \frac{p_k(\tilde{\mathbf{w}})p(y=1)}{p_k(\tilde{\mathbf{w}})p(y=1) + q_{\neg k}(\tilde{\mathbf{w}})p(y=0)} \propto \frac{p_k(\tilde{\mathbf{w}})}{p_k(\tilde{\mathbf{w}}) + q_{\neg k}(\tilde{\mathbf{w}})}.
\end{aligned}
\tag{6}
$$

Therefore, the $\alpha(\tilde{\mathbf{w}})$ is only related with $p_k(\tilde{\mathbf{w}})$ and $q_{\neg k}(\tilde{\mathbf{w}})$.

By introducing the classifier α, the derivation of the importance weight γ_k^t can be described as an adversarial learning framework, which is illustrated in Fig. 2(a). The classifier α works as an "adversarial discriminator" to distinguish $p_k(\tilde{\mathbf{w}})$ and $q_{\neg k}(\tilde{\mathbf{w}})$ as possible, which forms a *divergence measure* of the twin statistics. Based on the divergence from α, the project function f generates an importance weight γ_k^t, which is used to adjust the statistical parameters with

$\tilde{w}_k$ and $\tilde{w}_{\neg k}$ to reduce their discrepancy. The intuition is that the twin statistics are more likely to converge to the objective distribution if they are harder to be differentiated by a classifier.

4.1   Adversarial Training

Based on the adversarial learning framework with f and α, the training object to optimize Eq. (2) can be written as $\min_f \frac{1}{K} \sum_{k=1}^{K} \mathbb{D}(p_k(\tilde{w}, f, \alpha), q_{\neg k}(\tilde{w}, f, \alpha))$, where the divergence measure can be expressed by:

$$\mathbb{D}(p_k(\tilde{w}, f, \alpha), q_{\neg k}(\tilde{w}, f, \alpha)) = \max_{\alpha} \mathbb{E}_{\tilde{w} \sim p_k}[\log \alpha(\tilde{w})] + \mathbb{E}_{\tilde{w} \sim q_{\neg k}}[\log(1 - \alpha(\tilde{w}))].$$
(7)

Therefore f and α can be derived by solving the following min-max optimization:

$$\min_{f} \max_{\alpha} \frac{1}{K} \sum_{k=1}^{K} \mathbb{E}_{\tilde{w} \sim p_k}[\log \alpha(\tilde{w})] + \mathbb{E}_{\tilde{w} \sim q_{\neg k}}[\log(1 - \alpha(\tilde{w}))].$$
(8)

To form a learning model to solve the min-max problem, $f(\cdot)$ can be parameterized by a fully-connected neural network, and $\alpha(\cdot)$ can be parameterized by a variational auto-encoder (VAE) as illustrated in Fig. 2(a), where the latent variable $\mathbf{z}$ represents the divergence between $p_k(\tilde{w})$ and $q_{\neg k}(\tilde{w})$ in priori distribution. Next we inference $\alpha(\cdot)$ with VAE.

4.2   Inference with VAE

We construct a variational encoder-decoder model that takes the observed statistical parameters as input, encodes them to a latent variable, and decodes the statistical information from the latent variable to minimize the reconstruction error. The plate notions of the generative model are shown in Fig. 2(b), which are explained as follows.

- $\tilde{w}_k^t$ is the observed statistical parameters from the local model of client-k. In communication round t, the server receives a set statistical parameters $\tilde{w}^t = \{\tilde{w}_1^t, \ldots, \tilde{w}_K^t\}$, and they are used to train the VAE to generate the latent representation of the twin statistics.
- $\mathbf{z}_k^t$ is a latent variable whose prior is the joint distribution of $p(\tilde{w}, \mathbf{y}) \sim Gaussian(\boldsymbol{\mu}, \boldsymbol{\sigma})$, where $\boldsymbol{\mu}$ and $\boldsymbol{\sigma}$ are inferred parameters that are used to generate $\mathbf{z}_k^t$.
- $\boldsymbol{\theta}$ are the generative model parameters (decoder), and $\boldsymbol{\phi}$ are the variational parameters (encoder).

The solid lines in Fig. 2(b) denote the generative process $p_\theta(\mathbf{z}_k^t) p_\theta(\tilde{w}_k^t | \mathbf{z}_k^t)$, and the dashed lines denote the variational approximation $q_\phi(\mathbf{z}_k^t | \tilde{w}_k^t)$ to the intractable posterior $p_\theta(\mathbf{z}_k^t | \tilde{w}_k^t)$. We approximate $p_\theta(\mathbf{z}_k^t | \tilde{w}_k^t)$ with $q_\phi(\mathbf{z}_k^t | \tilde{w}_k^t)$ by minimizing their divergence:

$$\phi^*, \theta^* = arg \min_{\theta, \phi} \mathbb{D}(q_{\phi_k}(\mathbf{z}_k^t | \tilde{w}_k^t) \, || \, p_{\theta_k}(\mathbf{z}_k^t | \tilde{w}_k^t)).$$
(9)

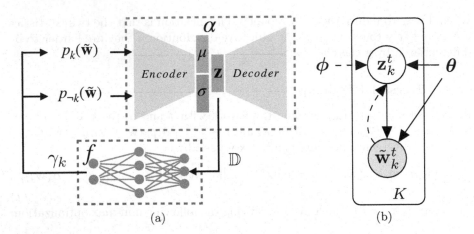

Fig. 2. Illustration of adversarial learning. (a) The framework. (b) The plate notations of VAE.

To derive the optimal value of the parameters ϕ and θ, we compute the marginal likelihood of $\tilde{\mathbf{w}}_k^t$:

$$\log p(\tilde{\mathbf{w}}_k^t) = \mathbb{D}_{KL}(q_\phi(\mathbf{z}_k^t|\tilde{\mathbf{w}}_k^t) \,\|\, p_\theta(\mathbf{z}_k^t|\tilde{\mathbf{w}}_k^t)) + \mathbb{E}_{q_\phi(\mathbf{z}_k^t|\tilde{\mathbf{w}}_k^t)}\left[\log \frac{p_\theta(\mathbf{z}_k^t, \tilde{\mathbf{w}}_k^t)}{q_\phi(\mathbf{z}_k^t|\tilde{\mathbf{w}}_k^t)}\right]. \quad (10)$$

In Eq. (10), the first term is the KL-divergence [10] of the approximate distribution and the posterior distribution; the second term is called the ELBO (Evidence Lower BOund) on the marginal likelihood of dataset in the k-th client.

Since $\log p(\tilde{\mathbf{w}}_k^t)$ is non-negative, the minimization problem of Eq. (9) can be converted to maximize the ELBO. To solve the problem, we change the form of ELBO as:

$$\mathbb{E}_{q_\phi(\mathbf{z}_k^t|\tilde{\mathbf{w}}_k^t)}\left[\log \frac{p_\theta(\mathbf{z}_k^t, \tilde{\mathbf{w}}_k^t)}{q_\phi(\mathbf{z}_k^t|\tilde{\mathbf{w}}_k^t)}\right] =$$
$$\underbrace{\mathbb{E}_{q_\phi(\mathbf{z}_k^t|\tilde{\mathbf{w}}_k^t)}\left[log \frac{p(\mathbf{z}_k^t)}{q_\phi(\mathbf{z}_k^t|\tilde{\mathbf{w}}_k^t)}\right]}_{\text{Encoder}} + \underbrace{\mathbb{E}_{q_\phi(\mathbf{z}_k^t|\tilde{\mathbf{w}}_k^t)}[\log p_\theta(\tilde{\mathbf{w}}_k^t|\mathbf{z}_k^t)]}_{\text{Decoder}}. \quad (11)$$

The above form is a variational encoder-decoder structure: the model $q_\phi(\mathbf{z}_k^t|\tilde{\mathbf{w}}_k^t)$ can be viewed as a probabilistic encoder that given an observed statistics $\tilde{\mathbf{w}}_k^t$ it produces a distribution over the possible values of the latent variables $\mathbf{z}_k^t$; The model $p_\theta(\tilde{\mathbf{w}}_k^t|\mathbf{z}_k^t)$ can be refered to as a probabilistic decoder that reconstructs the value of $\tilde{\mathbf{w}}_k^t$ based on the code $\mathbf{z}_k^t$. According to the theory of variational inference [13], the problem in Eq. (11) can be solved with the SGD method using a fully-connected neural network to optimize the mean squared error loss function.

After training the model, we feed the twin statistics $\{\tilde{\mathbf{w}}_k^t, \tilde{\mathbf{w}}_{\neg k}^t\}$ into the VAE and obtain the corresponding latent variables $\{\mathbf{z}_k^t, \mathbf{z}_{\neg k}^t\}$. The latent variables are

---

Algorithm 1. FedDNA

Initialize $\mathbf{w}^0$ and γ_k.
for each round $t = 0, 1, \ldots, T - 1$ do
    $S^t :=$ (random set of m clients)
    Send $\mathbf{w}^t$ to client in S^t as $\mathbf{w}_k^t$
    for each client $k \in S^t$ **in parallel** do
        for each local epoch $e = 0, 1, \ldots, E - 1$ do
            $\mathbf{w}_k^t := \mathbf{w}_k^t - \eta \nabla l(\mathbf{w}_k^t; \mathbf{x}_k)$
        end for
    end for
    Receive $\mathbf{w}_k^t$ from client in S^t
    for each client $k \in S^t$ do
        $\tilde{\mathbf{w}}_k^t := [\tilde{\mathbf{mean}}_k^t, \tilde{\mathbf{var}}_k^t]$.
    end for
    Update $\hat{\mathbf{w}}_k^t$ by Eq. (3)
    Compute $\tilde{\mathbf{w}}_{\neg k}^t$ based on Eq. (4)
    repeat
        $\boldsymbol{\mu}, \boldsymbol{\sigma} := \phi(\tilde{\mathbf{w}}_k^t), \phi(\tilde{\mathbf{w}}_{\neg k}^t)$
        Sample $\mathbf{z}_k^t$ and $\mathbf{z}_{\neg k}^t$ from $\mathcal{N}(\boldsymbol{\mu}, \boldsymbol{\sigma}^2)$
        $\gamma_k^t := f(\mathbb{D}[\mathbf{z}_k, \mathbf{z}_{\neg k}])$
        $(\boldsymbol{\phi}, \boldsymbol{\theta}) := (\boldsymbol{\phi}, \boldsymbol{\theta}) - \eta \nabla l(\boldsymbol{\phi}, \boldsymbol{\theta}; \tilde{\mathbf{w}}_k^t)$
        $f = f - \eta \nabla l(f; \mathbb{D}[\mathbf{z}_k, \mathbf{z}_{\neg k}])$
        Update every client's statistical parameters based on Eq. (4)
    until α and f Converge.
    Form $\mathbf{w}_k^t$ by $\hat{\mathbf{w}}_k^t$ and $\tilde{\mathbf{w}}_k^t$.
end for

---

further used by the nueral network of f to derive their divergence and output the important weight γ_k^t, which is used to update the statistical parameters in Eq. (4). The pseudo-code of FedDNA is shown in Algorithm 1.

5 Performance Evaluation

5.1 Experimental Setup

Implementation. We implement FedDNA[1] and the considered baselines in PyTorch. We train the models in a simulated federated learning environment consisting of one server and a set of clients with wireless network connections. Unless explicitly specified, the default number of clients is 20 as FedMA [28], and the learning rate $\beta = 0.01$. We conduct experiments on a GPU-equipped personal computer (CPU: Intel Core i7-8700 3.2 GHz, GPU: Nvidia GeForce RTX 2070, Memory: 32 GB DDR4 2666MHz, and OS: 64-bit Ubuntu 16.04).

Models and Datasets. We conduct experiments based on 6 mainstream neural network models: ResNet18 [4], LeNet [16], DenseNet121 [6], MobileNetV2 [27]

---

[1] The source code will be publicly available after acceptance.

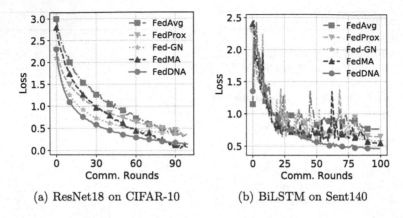

(a) ResNet18 on CIFAR-10   (b) BiLSTM on Sent140

Fig. 3. Convergence of different algorithms.

and BiLSTM [7]. The first 5 models used Batch-Normalization (BN), which are commonly applied in computer vision (CV), and the last model used Layer-Normalization (LN), which is applied in natural language processing (NLP).

We use 4 real world datasets: MNIST [17], Fashion-MNIST [30], CIFAR-10 [15] and Sentiment140 [3]. MNIST is a dataset for hand written digits classification with 60000 samples of 28×28 greyscale image. Fashion-MNIST is an extended version of MNIST for benchmarking machine learning algorithms. CIFAR-10 is a large image dataset with 10 categories, each of which has 6000 samples of size 32×32. Sentiment140 is a natural language process dataset containing 1,600,000 extracted tweets annotated in scale 0 to 4 for sentiment detection.

We generate non-IID data partition according to the work [22]. For each dataset, we use 80% as training dada to form non-IID local datasets as follows. We sort the data by their labels and divide each class into 200 shards. Each client draw samples from the shards to form a local dataset with probability $pr(x) = \begin{cases} \eta \in [0, 1], & \text{if } x \in class_j, \\ \mathcal{N}(0.5, 1), & \text{otherwise.} \end{cases}$ It means that the client draws samples from a particular class j with a fixed probability η, and from other classes based on a Gaussian distribution. The larger η is, the more likely the samples concentrate on a particular class, and the more heterogeneous the datasets are.

5.2 Performance Analysis

We compare the performance of FedDNA with 5 state-of-the-art methods: FedAvg [22], FedProx [18], Fed-GN [5], SCAFFOLD [11], and FedMA [28] with $\eta = 0.5$. The results are analyzed as follows.

Convergence: In this experiment, we study the convergence of FedDNA and all baselines by showing the total communication rounds versus train loss with local training epoch $E = 10$. Figure 3(a) and Fig. 3(b) show the result of ResNet18 on

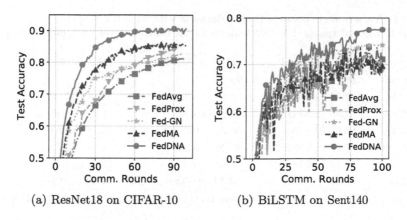

(a) ResNet18 on CIFAR-10   (b) BiLSTM on Sent140

Fig. 4. Training efficiency of different algorithms.

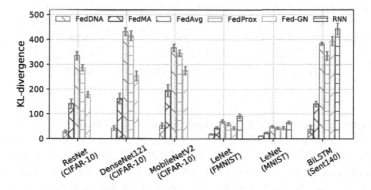

Fig. 5. Comparison of KL-divergence of statistical parameters of different models with different algorithms.

CIFAR-10 and BiLSTM on Sent140. It is shown that the loss of all algorithms tends to be stable after a number of epochs. Clearly, FedDNA has the lowest loss among all algorithms, and it converges faster than all baselines.

Training Efficiency: In this experiment, we study the test accuracy versus communication rounds during training with local training epoch $E = 10$. Figure 4(a) and Fig. 4(b) show the results of training ResNet18 on CIFAR-10 and BiLSTM on Sent140. For ResNet18, it is shown that FedDNA reaches 0.8 accuracy after 16 communication rounds, while the others take 30 to 80 communication rounds to reach the same accuracy. FedDNA exceeds 0.9 accuracy after 63 communication rounds, while the accuracy of other algorithms is below 0.86. Similar results are found for BiLSTM, where FedDNA achieves the highest accuracy at the same communication round. It suggests that FedDNA trains much faster than the baseline algorithms, and it can reach higher accuracy with less communication cost.

Table 1. Average test accuracy (%) on datasets with local training epoch $E = 10$. The "Central" method trains the CNN models in the central server with global dataset.

Algorithm	LeNet@MNIST	LeNet@F-MNIST	BiLSTM@Sent140
Central	98.95	90.42	81.47
FedAvg	97.32 (±0.09)	87.41 (±0.29)	72.14 (±0.93)
FedProx	97.55 (±0.14)	88.33 (±0.35)	71.08 (±1.28)
Fed-GN	95.88 (±0.25)	88.21 (±0.27)	74.44 (±1.04)
SCAFFOLD	97.47 (±0.19)	89.36 (±0.21)	73.83 (±0.79)
FedMA	97.86 (±0.18)	89.02 (±0.46)	72.81 (±1.38)
FedDNA	**98.49** (±0.12)	**90.11** (±0.18)	**77.51** (±0.87)
Algorithm	ResNet18@CIFAR-10	DenseNet121@CIFAR-10	MobileNetV2@CIFAR-10
Central	92.33	93.24	92.51
FedAvg	81.29 (±0.83)	81.86 (±0.46)	80.11 (±0.87)
FedProx	83.47 (±0.68)	85.03 (±0.65)	80.68 (±0.79)
Fed-GN	83.08 (±0.63)	83.43 (±0.53)	82.82 (±0.61)
SCAFFOLD	85.72 (±0.45)	86.94 (±0.39)	85.27 (±0.85)
FedMA	86.44 (±0.59)	87.12 (±0.71)	85.59 (±1.07)
FedDNA	**90.31** (±0.45)	**90.29** (±0.42)	**89.17** (±0.72)

Divergence: We compare the KL-divergence of statistical parameters between the global model aggregated by different algorithms and the central model, which are shown in Fig. 5. It is shown that FedMA, FedProx, Fed-GN, and FedAvg have exceptional high divergence varying from 140.4 to 447.1 on the CIFAR-10 dataset, while **FedDNA** has significantly lower divergences than all baselines for all models and datasets. It suggests that the statistical parameters aggregated by **FedDNA** are much more close to the central model in non-IID settings.

Global Model Accuracy: In this experiment, we compare the global model accuracy of different federated parameter aggregation algorithms after training to converge. We repeat the experiment for 20 rounds and show the average results in Table 1. As shown in the table, the central method yields the highest accuracy. In the comparison of different federated learning methods, **FedDNA** significantly outperforms the other algorithms in global model accuracy. It performs better than the state-of-the-art method FedMA with 3.87%, 3.17%, 3.58%, and 4.09% accuracy improvement on ResNet18, DenseNet121, MobileNetV2, and 4-L CNN respectively for CIFAR-10; 1.09% and 0.63% improvement on LeNet for F-MNIST and MNIST; 4.70% improvement on BiLSTM for Sent140. Compared to Fed-GN, **FedDNA** achieves accuracy improvement with 7.32%, 6.86%, 6.35%, and 3.08% on ResNet18, DenseNet121, MobileNetV2, and 4-L CNN respectively for CIFAR-10; 1.90% and 2.61% on LeNet for F-MNIST and MNIST; 3.07% on BiLSTM for Sent140 accordingly. Compared to FedAvg, **FedDNA** improves the test accuracy up to 9.02% for ResNet18 on CIFAR-10. In summary, **FedDNA**

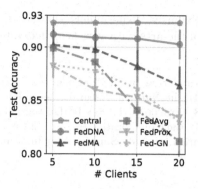

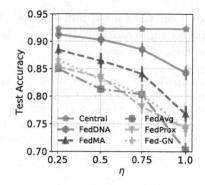

Fig. 6. Test accuracy with different number of clients (ResNet18 on CIFAR-10).

Fig. 7. Test accuracy on different level of heterogeneity (ResNet18 on CIFAR-10).

achieves the highest accuracy among all baselines, and it performs very close to the centralized method, whose accuracy drop is within 4% in all cases.

Hyperparameter Analysis: We further analyze the influence of two hyperparameters in federated learning: the number of clients and the heterogeneity of local datasets.

Figure 6 compares the test accuracy of the global model for a different number of involved clients. According to the figure, the performance of FedDNA remains stable. When the number of clients increases from 5 to 20, the test accuracy slightly decreases from 0.912 to 0.903. The other algorithms yield significant performance drop, and the accuracy of most baselines is below 0.85 for 20 clients. FedDNA achieves the highest test accuracy among all federated learning algorithms in all cases, and it performs very close to the central model.

In the experiment, the heterogeneity of local datasets is represented by η, the probability that a client tends to sample from a particular class. The more η approaches to 1, the more heterogeneous the local datasets are. Figure 7 shows the test accuracy under different levels of heterogeneity. As η increases, the test accuracy of all models decreases. FedDNA yields the highest test accuracy among all algorithms, and its performance much slower than that of the baselines. It verifies the effectiveness of the proposed decoupled aggregation approach under non-IID conditions.

6 Conclusion

Parameter aggregation played an important role in federated learning to form a global model. To address the problem of aggregation bias in federated learning for non-IID data, we proposed a novel parameter aggregation method called FedDNA that decoupled gradient parameters and statistical parameters to aggregate them separately with stochastic gradient descent and importance weighting method to reduce the divergence between the local models and the central model. FedDNA optimized parameter aggregation by an adversarial learning algorithm based on variational autoencoder (VAE). Extensive experiments showed that FedDNA significantly outperforms the state-of-the-arts on a variety of federated learning scenarios.

Acknowledgment. This work was partially supported by the National Key R&D Program of China (Grant No. 2018YFB1004704), the National Natural Science Foundation of China (Grant Nos. 61972196, 61832008, 61832005), the Key R&D Program of Jiangsu Province, China (Grant No. BE2018116), the Collaborative Innovation Center of Novel Software Technology and Industrialization, and the Sino-German Institutes of Social Computing.

References

1. Acar, D.A.E., Zhao, Y., Matas, R., Mattina, M., Whatmough, P., Saligrama, V.: Federated learning based on dynamic regularization. In: Proceedings of ICLR (2021)
2. Chen, T., Giannakis, G., Sun, T., Yin, W.: Lag: lazily aggregated gradient for communication-efficient distributed learning. In: Proceedings of NIPS, pp. 5050–5060 (2018)
3. Go, A., Bhayani, R., Huang, L.: Twitter sentiment classification using distant supervision (2009). http://help.sentiment140.com/home
4. He, K., Zhang, X., Ren, S., Sun, J.: Deep residual learning for image recognition. In: Proceedings of CVPR, pp. 770–778 (2016)
5. Hsieh, K., Phanishayee, A., Mutlu, O., Gibbons, P.B.: The Non-IID data quagmire of decentralized machine learning. In: Proceedings of ICML (2020)
6. Huang, G., Liu, Z., van der Maaten, L., Weinberger, K.Q.: Densely connected convolutional networks. In: Proceedings of CVPR, pp. 2261–2269 (2017)
7. Huang, Z., Xu, W., Yu, K.: Bidirectional LSTM-CRF models for sequence tagging (2015). http://arxiv.org/abs/1508.01991
8. Ji, J., Chen, X., Wang, Q., Yu, L., Li, P.: Learning to learn gradient aggregation by gradient descent. In: Proceedings of IJCAI, pp. 2614–2620 (2019)
9. Jiang, L., Tan, R., Lou, X., Lin, G.: On lightweight privacy-preserving collaborative learning for internet-of-things objects. In: Proceedings of IoTDI, pp. 70–81 (2019)
10. Joyce, J.M.: Kullback-Leibler Divergence, pp. 720–722 (2011)
11. Karimireddy, S.P., Kale, S., Mohri, M., Reddi, S., Stich, S., Suresh, A.T.: SCAFFOLD: stochastic controlled averaging for federated learning. In: Proceedings of ICML (2020)
12. Killeen, P.R.: An alternative to null-hypothesis significance tests. Psychol. Sci. **16**(5), 345–53 (2005)
13. Kingma, D., Welling, M.: Auto-encoding variational bayes. In: Proceedings of ICLR (2014)
14. Konecny, J., McMahan, H.B., Ramage, D.: Federated optimization: Distributed optimization beyond the datacenter. In: NIPS Optimization for Machine Learning Workshop 2015, p. 5 (2015)
15. Krizhevsky, A.: Learning multiple layers of features from tiny images. Technical report (2009)
16. Lecun, Y., Bottou, L., Bengio, Y., Haffner, P.: Gradient-based learning applied to document recognition. In: Proceedings of the IEEE, pp. 2278–2324 (1998)

17. LeCun, Y., Cortes, C., Burges, C.: MNIST handwritten digit database. ATT Labs (2010). http://yann.lecun.com/exdb/mnist
18. Li, T., Sahu, A.K., Zaheer, M., Sanjabi, M., Talwalkar, A., Smith, V.: Federated optimization in heterogeneous networks. In: Proceedings of MLSys, pp. 429–450 (2020)
19. Li, T., Sanjabi, M., Smith, V.: Fair resource allocation in federated learning. In: Proceedings of ICLR (2020)
20. Li, X., Jiang, M., Zhang, X., Kamp, M., Dou, Q.: FedBN: federated learning on Non-IID features via local batch normalization. In: Proceedings of ICLR (2021)
21. Malinovsky, G., Kovalev, D., Gasanov, E., Condat, L., Richtarik, P.: From local SGD to local fixed-point methods for federated learning. In: Proceedings of ICML (2020)
22. McMahan, B., Moore, E., Ramage, D., Hampson, S., Arcas, B.A.: Communication-efficient learning of deep networks from decentralized data. In: Proceedings of AISTATS, vol. 54, pp. 1273–1282 (2017)
23. Mohri, M., Sivek, G., Suresh, A.T.: Agnostic federated learning. In: ICML (2019)
24. Pathak, R., Wainwright, M.J.: FedSplit: an algorithmic framework for fast federated optimization. In: Proceedings of NeurIPS, vol. 33 (2020)
25. Reisizadeh, A., Farnia, F., Pedarsani, R., Jadbabaie, A.: Robust federated learning: the case of affine distribution shifts. In: Proceedings of NeurIPS (2020)
26. Rothchild, D., et al.: FetchSGD: communication-efficient federated learning with sketching. In: Proceedings of ICML (2020)
27. Sandler, M., Howard, A., Zhu, M., Zhmoginov, A., Chen, L.: MobilenetV2: inverted residuals and linear bottlenecks. In: Proceedings of CVPR, pp. 4510–4520 (2018)
28. Wang, H., Yurochkin, M., Sun, Y., Papailiopoulos, D., Khazaeni, Y.: Federated learning with matched averaging. In: Proceedings of ICLR (2020)
29. Wang, J., Liu, Q., Liang, H., Joshi, G., Poor, H.V.: Tackling the objective inconsistency problem in heterogeneous federated optimization. In: Proceedings of NeurIPS (2020)
30. Xiao, H., Rasul, K., Vollgraf, R.: Fashion-MNIST: a novel image dataset for benchmarking machine learning algorithms (2017)
31. Yu, F.X., Rawat, A.S., Menon, A., Kumar, S.: FedAwS: federated learning with only positive labels. In: Proceedings of ICML (2020)
32. Yuan, H., Ma, T.: Federated accelerated stochastic gradient descent. In: Proceedings of NeurIPS, vol. 33 (2020)
33. Yurochkin, M., Agarwal, M., Ghosh, S., Greenewald, K., Hoang, N., Khazaeni, Y.: Bayesian nonparametric federated learning of neural networks. In: Proceedings of ICML (2019)
34. Zhao, Y., Li, M., Lai, L., Suda, N., Civin, D., Chandra, V.: Federated learning with Non-IID data. arXiv abs/1806.00582 (2018)
35. Zhu, H., Jin, Y.: Multi-objective evolutionary federated learning. IEEE Trans. Neural Netw. Learn. Syst. **31**(4), 1310–1322 (2020)

The Curious Case of Convex Neural Networks

Sarath Sivaprasad[1,2]([✉]), Ankur Singh[1,3], Naresh Manwani[1],
and Vineet Gandhi[1]

[1] KCIS, IIIT Hyderabad, Hyderabad, India
{naresh.manwani,vgandhi}@iiit.ac.in
[2] TCS Research, Pune, India
sarath.s@research.iiit.ac.in
[3] IIT Kanpur, Kanpur, India
ankuriit@iitk.ac.in

Abstract. This paper investigates a constrained formulation of neural networks where the output is a convex function of the input. We show that the convexity constraints can be enforced on both fully connected and convolutional layers, making them applicable to most architectures. The convexity constraints include restricting the weights (for all but the first layer) to be non-negative and using a non-decreasing convex activation function. Albeit simple, these constraints have profound implications on the generalization abilities of the network. We draw three valuable insights: (a) Input Output Convex Neural Networks (IOC-NNs) self regularize and significantly reduce the problem of overfitting; (b) Although heavily constrained, they outperform the base multi layer perceptrons and achieve similar performance as compared to base convolutional architectures and (c) IOC-NNs show robustness to noise in train labels. We demonstrate the efficacy of the proposed idea using thorough experiments and ablation studies on six commonly used image classification datasets with three different neural network architectures. The appendix and codes for this paper are available at: https://github.com/sarathsp1729/Convex-Networks.

1 Introduction

Deep Neural Networks use multiple layers to extract higher-level features from the raw input progressively. The ability to automatically learn features at multiple levels of abstractions makes them a powerful machine learning system that can learn complex relationships between input and output. Seminal work by Zhang *et al.* [30] investigates the expressive power of neural networks on finite sample sizes. They show that even when trained on completely random labeling of the true data, neural networks achieve zero training error, increasing training time and effort by only a constant factor. Such potential of brute force memorization makes it challenging to explain the generalization ability of deep neural

---

Work done while at IIIT Hyderabad.

N. Oliver et al. (Eds.): ECML PKDD 2021, LNAI 12975, pp. 738–754, 2021.
https://doi.org/10.1007/978-3-030-86486-6_45

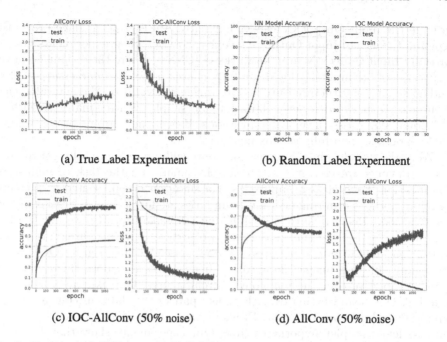

(a) True Label Experiment   (b) Random Label Experiment

(c) IOC-AllConv (50% noise)   (d) AllConv (50% noise)

Fig. 1. Training of AllConv and IOC-AllConv on CIFAR-10 dataset. (a) Loss curve while training with true labels. AllConv starts overfitting after few epochs. IOC-AllConv does not exhibit overfitting, and the test loss nicely follows the training loss. (b) Accuracy plots while training with randomized labels (labels were randomized for all the training images). If sufficiently trained, even a simple network like MLP achieves 100% training accuracy and gives around 10% test accuracy. IOC-MLP resists any learning on the randomized data and gives 0% generalization gap. (c) and (d) Loss and accuracy plots on CIFAR-10 data when trained with 50% labels randomized in the training set.

networks. They further illustrate that the phenomena of neural network fitting on random labeling of training data is largely unaffected by explicit regularization (such as weight decay, dropout, and data augmentation). They suggest that explicit regularization may improve generalization performance but is neither necessary nor by itself sufficient for controlling generalization error. Moreover, recent works show that generalization (and test) error in neural networks reduces as we increase the number of parameters [22,23], which contradicts the traditional wisdom that overparameterization leads to overfitting. These observations have given rise to a branch of research that focuses on explaining the neural network's generalization error rather than just looking at their test performance [24].

We propose a principled and reliable alternative that tries to affirmatively resolve the concerns raised in [30]. More specifically, we investigate a novel constrained family of neural networks called Input Output Convex Neural Networks (IOC-NNs), which learn a convex function between input and output. Convexity in machine learning typically refers to convexity in terms of the parameters w.r.t to the loss [3], which is not the case in our work. We use an IOC prefix to indicate the Input Output Convexity explicitly. Amos *et al.* [1] have previously explored the idea of Input Output convexity; however, their experiments limit to Partially Input Convex Neural Networks (PICNNs), where the output is convex w.r.t some of the inputs. They deem fully convex networks *unnecessary* in their studied setting of structured prediction, *highly restricted* on the allowable class of models, *highly limited*, even failing to do simple identity mapping without additional skip (pass-through) connections. Hence, they do not present even a single experiment on fully convex networks.

We wake this sleeping giant up and thoroughly investigate fully convex networks (outputs are convex w.r.t to all the inputs) on the task of multi-class classification. Each class in multi-class classification is represented as a convex function, and the resulting decision boundaries are formed as an `argmax` of convex functions. Being able to train IOC with NN-like capacity, we, for the first time, discover the beautiful underlying properties, especially in terms of generalization abilities and robustness to label noise. We investigate IOC-NNs on six commonly used image classification benchmarks and pose them as a preferred alternative over the non-convex architectures. Our experiments suggest that IOC-NNs avoid fitting over the noisy part of the data, in contrast to the typical neural network behavior. Previous work shows that [2] neural networks tend to learn simpler hypotheses first. Our experiments show that IOC-NNs tend to hold on to the simpler hypothesis even in the presence of noise, without overfitting in most settings.

A motivating example is illustrated in Fig. 1, where we train an All Convolutional network (AllConv) [28] and its convex counterpart IOC-AllConv on the CIFAR-10 dataset. AllConv starts overfitting the train data after a few epochs (Fig. 1(a)). In contrast, IOC-AllConv shows no signs of overfitting and flattens at the end (the test loss values pleasantly follow the training curve). Such an observation is consistent across all our experiments on IOC-NNs across different datasets and architectures, suggesting that IOC-NNs have lesser reliance on explicit regularization like early stopping. Fig. 1(b) presents the accuracy plots for the randomized test where we train Multi-Layer Perceptron (MLP) and IOC-MLP on a copy of the data where the true labels were replaced by random labels. MLP achieves 100% accuracy on the train set and gives a random chance performance on the test set (observations are coherent with [30]). IOC-MLP resists any learning and gives random chance performance (10% accuracy) on both train and test sets. As MLP achieves zero training error, the test error is the same as generalization error, i.e., 90% (the performance of random guessing on CIFAR10). In contrast, the IOC-MLP has a near 0% generalization error. We further present experiment with 50% noisy labels Fig. 1(c). The neural network training profile concurs with the observation of Krueger *et al.* [17], where the network learns a simpler hypothesis first and then starts memorizing. On the other hand, IOC-NN converges to the simpler hypothesis, showing strong resistance to fit the noise labels.

Input Output Convexity shows a promising paradigm, as any feed-forward network can be re-worked into its convex counterpart by choosing a non-decreasing (and convex) activation function and restricting its weights to be non-negative (for all but the first layer). Our experiments suggest that activation functions that allow negative outputs (like leaky ReLU or ELU) are more suited for the task as they help retain negative values flowing to subsequent layers in the network. We show that IOC-MLPs outperforms traditional MLPs in terms of test accuracy on five of the six studied datasets and IOC-NNs almost recover the performance of the base network in case of convolutional networks. In almost all studied scenarios, IOC networks achieve multi-fold improvements in terms of generalization error over unconstrained Neural Networks. Overall, our work makes the following contributions:

- We bring to light the little known idea of Input Output Convexity in neural networks. We propose a revised formulation to efficiently train IOC-NNs, retaining adequate capacity (with changes like using ELU, increasing nodes in the first layer, whitening transform at the input, etc.). To the best of our knowledge, we for the first time explore a usable form of IOC-NNs, and shows that they can be trained with NN like capacity.
- Through a set of intuitive experiments, we detail its internal functioning, especially in terms of its self regularization properties and decision boundaries. We show that how sufficiently complex decision boundaries can be learned using an `argmax` over a set of convex functions (where each class is represented by a single convex function). We further propose a framework to learn the ensemble of IOC-NNs.
- With a comprehensive set of quantitative and qualitative experiments, we demonstrate IOC-NN's outstanding generalization abilities. IOC-MLPs achieve near zero generalization error in all the studied datasets and a negative generalization error (test accuracy is higher than train accuracy) in a couple of them, even at convergence. Such never seen behaviour opens up a promising avenue for more future explorations.
- We explore the robustness of IOC-NNs to label noise and find that it strongly resists fitting the random labels. Even while training, IOC-NNs show no signs of fitting on noisy data and efficiently learns patterns from non noisy data. Our findings ignites explorations towards tighter generalization bounds for neural networks.

2 Related Work

Simple Convex Models: Our work relates to parameter estimation on models that are guaranteed to be convex by its construction. For regression problems, Magnani and Boyd [19] study the problem of fitting a convex piecewise linear function to a given set of data points. For classification problems, this traditionally translates to polyhedral classifiers. A polyhedral classifier can be described as an intersection of a finite number of hyperplanes. There have been several attempts to address the problem of learning polyhedral classifiers [15,20]. However, these algorithms require the number of hyperplanes as an input, which is a major constraint. Furthermore, these classifiers do not give completely smooth boundaries (at the intersection of hyperplanes). As another major limitation, these classifiers cannot model the boundaries in which each class is distributed over the union of non-intersecting convex regions (e.g., XOR problem). The proposed IOC-NN (even with a single hidden layer) supersedes this direction of work.

Convex Neural Networks: Amos *et al.* [1] mentions the possibility of fully convex networks, however, does not present any experiments with it. The focus of their work is to achieve structured predictions using partially convex network (using convexity w.r.t to some of the inputs). They propose a specific architecture called FICNN which is fully convex and has fully connected layers with skip connections. The skip connections are a must because their architecture cannot even achieve identity mapping without them. In contrast, our work can take any given architecture and derive its convex counterpart (we use the IOC suffix to suggest model agnostic nature of our work). The work by Kent *et al.* [16] analyze the links between polynomial functions and input convex neural networks to understand the trade-offs between model expressiveness and ease of optimization. Chen *et al.* [7,8] explore the use of input convex neural network in a variety of control applications like voltage regulation. The literature on input convex neural networks has been limited to niche tailored scenarios. Two key highlights of our work are: (a) to use activations that allow the flow of negative values (like ELU, leaky ReLU, etc.), which enables a richer representation (retaining fundamental properties like identity mapping which are not achievable using ReLU) and (b) to bring a more in-depth perspective on the functioning of convex networks and the resulting decision boundaries. Consequently, we present IOC-NNs as a preferred option over the base architectures, especially in terms of generalization abilities, using experiments on mainstream image classification benchmarks.

Generalization in Deep Neural Nets: Conventional machine learning wisdom says that overparameterization leads to poor generalization performance owing to overfitting. Counter-intuitively, empirical evidence shows that neural networks give better generalization with an increased number of parameters even without any explicit regularization [25]. Explaining how neural networks generalize despite being overparameterized is an important question in deep learning [22,25].

Neyshabur *et al.* [23] study different complexity measures and capacity bounds based on the number of parameters, VC dimension, Rademacher complexity etc., and conclude that these bounds fail to explain the generalization behavior of neural networks on overparameterization. Neyshabur *et al.* [24] suggest that restricting the hypothesis class gives a generalization bound that decreases with an increase in the number of parameters. Their experiments show that restricting the spectral norm of the hidden layer leads to tighter generalization bounds.

The above discussion implies that a hypothetical neural network that can fit any hypothesis will have a worse generalization than the practical neural networks which span a restricted hypothesis class. Inspired by this idea, we propose a principled way of restricting the hypothesis class of neural networks (by convexity constraints) that improves their generalization ability in practice. In the previous efforts to train fully input output convex networks, they were shown to have a limited capacity compared to its neural network counterpart [1,3], making their generalization capabilities ineffective in practice. To our knowledge, we for the first time present a method to formulate and efficiently train IOC-NNs opening an avenue to explore their generalization ability.

3   Input Output Convex Networks

We first consider the case of an MLP with k hidden layers. The output of ith neuron in the lth hidden layer will be denoted as $h_i^{(l)}$. For an input $\mathbf{x} = (x_1, \ldots, x_d)$, $h_i^{(l)}$ is defined as:

$$h_i^{(l)} = \phi(\sum_j w_{ij}^{(l)} h_j^{l-1} + b_i^{(l)}), \tag{1}$$

where, $h_j^{(0)} = x_j$ ($j = 1 \ldots d$) and $h_j^{(k+1)} = y_j$ (j^{th} output). The first hidden layer represents an affine mapping of input and preserves the convexity (i.e. each neuron in $h^{(1)}$ is convex function of input). The subsequent layers are a weighted sum of neurons from the previous layer followed by an activation function. The final output $\mathbf{y}$ is convex with respect to the input $\mathbf{x}$ by ensuring two conditions: (a) $w_{ij}^{(2:k+1)} \geq 0$ and (b) ϕ is convex and a non-decreasing function. The proof follows from the operator properties [5] that the non-negative sum of convex functions is convex and the composition $f(g(x))$ is convex if g is convex and f is convex and non-decreasing.

A similar intuition follows for convolutional architectures as well, where each neuron in the next layer is a weighted sum of the previous layer. Convexity can be assured by restricting filter weights to be non-negative and using a convex and non decreasing activation function. Filter weights in the first convolutional layer can take negative values, as they only represent an affine mapping of the input. The maxpool operation also preserves convexity since point-wise maximum of convex functions is convex [5]. Also, the skip connection does not violate Input Output Convexity, since the input to each layer is still a non-negative weighted sum of convex functions.

We use an ELU activation to allow negative values; this is a minor but a key change from the previous efforts that rely on ReLU activation. For instance, with non-negativity constraints on weights ($w_{ij}^{(2:k+1)} \geq 0$), ReLU activations restrict the allowable use of hidden units that mirror the identity mapping. Previous works rely on passthrough/skip connections to address [1] this concern. The use of ELU enables identity mapping and allows us to use the convex counterparts of existing networks without any architectural changes.

3.1   Convexity as Self Regularizer

We define self regularization as the property in which the network itself has some functional constraints. Inducing convexity can be viewed as a self regu-

744   S. Sivaprasad et al.

Fig. 2. Decision boundaries of different networks trained for two class classification. (a) Original data: one class shown by blue and the other orange. (b) Decision boundary learnt using MLP. (c) Decision boundary learnt using IOC-MLP with single node in the output layer. (d) Decision boundary learnt using IOC-MLP with two nodes in the output layer (ground truth as one hot vectors) (Color figure online)

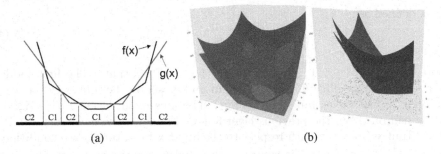

Fig. 3. (a) Using two simple 1-D functions we illustrate that `argmax` of two convex functions can result into non-convex decision boundaries. (b) Two convex functions whose `argmax` results into the decision boundaries shown in Fig. 2(d). The same plot is shown from two different viewpoints.

larization technique. For example, consider a quadratic classifier in $\mathbb{R}^2$ of the form $f(x_1, x_2) = w_1 x_1^2 + w_2 x_2^2 + w_3 x_1 x_2 + w_4 x_1 + w_5 x_2 + w_0$. If we want the function f to be convex, then it is required that the network imposes following constraints on the parameters, $w_1 \geq 0$, $w_2 \geq 0$, $-2\sqrt{w_1 w_2} \leq w_3 \leq 2\sqrt{w_1 w_2}$, which essentially means that we are restricting the hypothesis space.

Similar inferences can be drawn by taking the example of polyhedral classifiers. Polyhedral classifiers are a special class of Mixture of Experts (MoE) network [13,26]. VC-dimension of a polyhedral classifier in d-dimension formed by the intersection of m hyperplanes is upper bounded by $2(d+1)m \log(3m)$ [29]. On the other hand, VC-dimension of a standard mixture of m binary experts in d-dimension is $O(m^4 d^2)$ [14]. Thus, by imposing convexity, the VC-dimension becomes linear with the data dimension d and $m \log(m)$ with the number of experts. This is a huge reduction in the overall representation capacity compared to the standard mixture of binary experts.

Furthermore, adding non-negativity constraints alone can lead to regularization. For example, the VC dimension of a sign constrained linear classifier in $\mathbb{R}^d$ reduces from $d+1$ to d [6,18]. The proposed IOC-NN uses a combination of sign constraints and restrictions on the family of activation functions for induc-

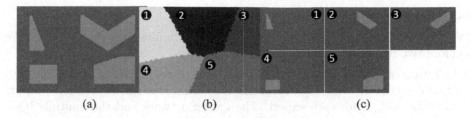

Fig. 4. (a) Original Data. (b) Output of the gating network, each color represents picking a particular expert. (c) Decision boundaries of the individual IOC-MLPs. We mark the correspondences between each expert and the segment for which it was selected. Notice how the V-shape is partitioned and classified using two different IOC-MLPs. (Color figure online)

ing convexity. The representation capacity of the resulting network reduces, and therefore, regularization comes into effect. This effectively helps in improving generalization and controlling overfitting, as clearly observed in our empirical studies (Sect. 4.1).

3.2 IOC-NN Decision Boundaries

Consider a scenario of binary classification in 2D space as presented in Fig. 2(a). We train a three-layer MLP with a single output and a sigmoid activation for the last layer. The network comfortably learns to separate the two classes. The learned boundaries by the MLP are shown in Fig. 2(b). We then train an IOC-MLP with the same architecture. The learned boundary is shown in Fig. 2(c). IOC-MLP learns a single convex function as output w.r.t the input and its contour at the value of 0.5 define the decision boundary. The use of non-convex activation like sigmoid in the last layer does not distort convexity of decision boundary (Appendix A).

We further explore IOC-MLP with a variant architecture where the ground truth is presented as a one-hot vector (allowing two outputs). The network learns two convex functions f and g representing each class, and their `argmax` defines the decision boundary. Thus, if $g(\mathbf{x}) - f(\mathbf{x}) > 0$, then $\mathbf{x}$ is assigned to class $C1$ and $C2$ otherwise. Therefore, it can learn non-convex decision boundaries as shown in Fig. 3. Please note that $g - f$ is no more convex unless $g'' - f'' \geq 0$. In the considered problem of binary classification in Fig. 2, using one-hot output allows the network to learn non-convex boundaries (Fig. 2 (d)). The corresponding two output functions (one for each class) are illustrated in Fig. 3 (b). We can observe that both the individual functions are convex; however, their arrangement is such that the `argmax` leads to a reasonably complex decision boundary. This happens due to the fact that the sets $S_1 = \{\mathbf{x} \mid g(\mathbf{x}) - f(\mathbf{x}) > 0\}$ and $S_2 = \{\mathbf{x} \mid g(\mathbf{x}) - f(\mathbf{x}) \leq 0\}$ can both be non-convex (even though functions $f(.)$ and $g(.)$ are convex).

3.3   Ensemble of IOC-NN

We further explore the ensemble of IOC-NN for multi-class classification. We explore two different ways to learn the ensembles:

1. Mixture of IOC-NN Experts: Training a mixture of IOC-NNs and an additional gating network [13]. The gating network can be non-convex and outputs a scalar weight for each expert. The gating network and the multiple IOC-NNs (experts) are trained in an Expectation-Maximization (EM) framework, i.e., training the gating network and the experts iteratively.
2. Boosting + Gating: In this setup, each IOC-NN is trained individually. The first model is trained on the whole data, and the consecutive models are trained with exaggerated data on the samples on which the previous model performs poorly. For bootstrapping, we use a simple re-weighting mechanism as in [10]. A gating network is then trained over the ensemble of IOC-NNs. The weights of the individual networks are frozen while training the gating network.

We detail the idea of ensembles using a representative experiment for binary classification on the data presented in Fig. 4(a). We train a mixture of **p** IOC-MLPs with a gating network using the EM algorithm. The gating network is an MLP with a single hidden layer, the output of which is a **p** dimensional vector. Each of the IOC-MLP is a three-layer MLP with a single output. We keep a single output to ensure that each IOC-MLP learns a convex decision boundary. The output of the gating network is illustrated in Fig. 4(b). A particular IOC-MLP was selected for each partition and led to five partitions. The decision boundaries of individual IOC-MLPs are shown in Fig. 4(c). It is interesting to note that the MoE of binary IOC-MLPs fractures the input space into sub-spaces where a convex boundary is sufficient for classification.

4   Experiments

Dataset and Architectures: To show the significance of enhanced performance of IOC-MLP over traditional NN, we train them on six different datasets: MNIST, FMNIST, STL-10, SVHN, CIFAR-10, and CIFAR-100. We use an MLP with three hidden layers and 800 nodes in each layer. We use batch normalization between every layer, and it's activation in all hidden layers. ReLU and ELU are used as activations for NN and IOC respectively, and softmax is used in the last layer. We use Adam optimizer with an initial learning rate of 0.0001 and use validation accuracy for early stopping.

We perform experiments that involve two additional architectures to extend the comparative study between IOC and NN on CIFAR-10 and CIFAR-100 datasets. We use a fully convolutional [28], and a densely connected architecture [12]. We choose DenseNet with growth rate $k-12$, for our experiments. We term the convex counterparts as IOC-AllConv, IOC-DenseNet, respectively, and compare against their base neural network counterparts [12,28]. In all comparative studies, we follow the same training and augmentation strategy to train IOC-NNs, as used by the aforementioned neural networks.

Training on Duplicate Free Data: The test sets of CIFAR-10 and CIFAR-100 datasets have 3.25% and 10% duplicate images, respectively [4]. Neural networks show higher performance on these datasets due to the bias created by this duplicate data (neural networks have been shown to memorize the data). CIFAIR-10 and CIFAIR-100 datasets are variants of CIFAR-10 and CIFAR-100 respectively, where all the duplicate images in the test data are replaced with new images. Barz *et al.* [4] observed that the performance of most neural architectures drops when trained and tested on bias-free CIFAIR data. We train IOC-NN and their neural network counterparts on CIFAIR-10 data with three different architectures: a fully connected network (MLP), a fully convolutional network (AllConv) [28] and a densely connected network (DenseNet) [12].

Training IOC Architectures: We tried four variations for weight constraints to enforce convexity constraints: clipping negative weights to zero, taking absolute of weights, exponentiation of negative weights and shifting the weights after each iteration. We use exponentiation strategy in all experiments, as it gave the best results. We exponentiate the negative weights after every update. The IOC constrained optimization algorithm differs only by a single step from the traditional algorithms (Appendix B).

To conserve convexity in the batch-normalization layer, we also constrain the gamma scaler with exponentiation. However, in practice we found that the IOC networks retains all desirable properties without constraining the gamma scalar. We make few additional modifications to facilitate the training of IOC-NNs. Such changes do not affect the performance of the base neural networks. We use ELU as an activation function instead of ReLU in IOC-NNs. We apply the whitening transformation to the input so that it is zero-centered, decorrelated, and spans over positive and negative values equally. We also increase the number of nodes in the first layer (the only layer where parameters can take negative values). We use a slower schedule for learning rate decay than the base counterparts. The IOC-NNs have a softmax layer at the last layer and are trained with cross-entropy loss (same as neural networks).

Training Ensembles of Binary Experts: We divide CIFAR-10 dataset into 2 classes, namely: 'Animal' (CIFAR-10 labels: 'Bird', 'Cat', 'Deer', 'Dog', 'Frog' and 'Horse') and 'Not Animal'. We train an ensemble of IOC-MLP, where each expert is a three-layer MLP with one output (with sigmoid activation at the output node). The gating network in the EM approach is a one layer MLP which takes an image as input and predicts the weights by which the individual expert predictions get averaged. We report test results of ensembles with each additional expert. This experiment resembles the study shown in Fig. 4.

Training Boosted Ensembles: The lower training accuracy of IOC-NNs makes them suitable for boosting (while the training accuracy saturates in non-convex counterparts). For bootstrapping, we use a simple re-weighting mechanism as in [10]. We train three experts for each experiment. The gating network is a regular neural network, which is a shallow version of the actual experts. We train

Table 1. Table shows train accuracy, test accuracy and generalization gap for MLP and IOC-MLP on six different datasets.

	NN			IOC-NN		
	Train	Test	Gen. gap	Train	Test	Gen. gap
MNIST	99.34	99.16	0.19	98.77	**99.25**	**−0.48**
FMNIST	94.8	**90.61**	3.81	90.41	90.58	**−0.02**
STL-10	81	52.32	28.68	62.3	**54.55**	**7.75**
SVHN	91.76	86.19	5.57	81.18	**86.37**	**−5.19**
CIFAR-10	97.99	63.83	34.16	73.27	**69.89**	**3.38**
CIFAR-100	84.6	32.68	51.92	46.9	**41.08**	**5.82**

Table 2. Train accuracy, test accuracy and generalization gap of three neural architectures and their IOC counterparts

	CIFAR-10						CIFAR-100					
	NN			IOC-NN			NN			IOC-NN		
	Train	Test	Gen. gap	Train	Test	Gen. gap	Train	Test	Gen. gap	Train	Test	Gen. gap
MLP	99.17	63.83	35.34	73.27	69.89	**3.3**	84.6	32.68	51.9	46.9	41.08	**5.8**
AllConv	99.31	92.8	6.5	93.2	90.6	**2.6**	97.87	69.5	28.4	67.07	65.08	**1.9**
DenseNet	99.46	94.06	5.4	94.22	91.12	**3.1**	98.42	75.36	23.06	74.9	68.53	**6.3**

an MLP with only one hidden layer, a four-layer fully convolutional network, and a DenseNet with two dense-blocks as the gate for the three respective architectures. We report the accuracy of the ensemble trained in this fashion as well as the accuracy if we would have used an oracle instead of the gating network.

Partially Randomized Labeling: Here, we investigate IOC-NN's behavior in the presence of partial label noise. We do a comparative study between IOC and neural networks using All-Conv architecture, similar to the experiment performed by [30]. We use CIFAR-10 dataset and make them noisy by systematically randomizing the labels of a selected percentage of training data. We report the performance of All-Conv, and it's IOC counterpart on 20, 40, 60, 80 and 100% noise in the train data. We report train and test scores at peak performance (performance if we had used early stopping) and at convergence (if loss goes below 0.001 or at 2000 epochs).

4.1 Results

IOC as a Preferred Alternative for Multi-Layer-Perceptrons: MLP is most basic and earliest explored form of neural networks. We compare the train and test scores of MLP and IOC-MLP in Table 1. With a sufficient number of parameters, MLP (a basic NN architecture) perfectly fits the training data. However, it fails to generalize well on the test data owing to brute force memorization. The results in Table 1 indicate that IOC-MLP gives a smaller generalization gap (the difference between train and test accuracies) compared to MLP. The generalization gap even goes to negative values on three of the datasets. MLP (being poorly optimized for parameter utilization) is one of the architectures prone to overfitting the most, and IOC constraints help retain test performance resisting the tendency to overfit. Obtaining negative or almost zero generalization error even at convergence is a never seen behaviour in deep networks and the results clearly suggest the profound generalization abilities of Input Output Convexity, especially when applied to fully connected networks.

Furthermore, having the IOC constraints significantly boosts the test accuracy on datasets where neural network gives a high generalization gap (Table 1). This trend is clearly visible in Fig. 5(b). For the CIFAR-10 dataset, unconstrained MLP gives 34.16% generalization gap, while IOC-NN brings down the generalization gap by more than ten folds and boosts the test performance by about 6%. Even in scenarios where neural networks give a smaller generalization gap (like MNIST and SVHN), IOC-NN marginally outperforms regular NN and gives an advantage in generalization. Overall, the results in Table 1 highlight that IOC constraints are extremely beneficial when training Multi Layer Perceptrons for image classification, giving comprehensive advantages in terms of generalization and test performance.

Better Generalization: We investigate the generalization capability of IOC-NN on other architectures. The results of the base architectures and their convex counterparts on CIFAR-10 and CIFAR-100 datasets are presented in Table 2. IOC-NN outperforms base NN on MLP architecture and gives comparable test accuracies for convolutional architectures. The train accuracies are saturated in the base networks (reaching above 99% in most experiments). The lower train accuracy in IOC-NNs suggests that there might still be room for improvement, possibly through better design choices tailored for IOC-NNs. In Table 2, the difference in train and test accuracy across all the architectures (generalization gap) demonstrates the better generalization ability of IOC-NNs. The generalization gap of base architectures is at least twofold more than IOC-NNs on the CIFAR-100 dataset. For instance, the generalization error of IOC-AllConv on CIFAR-100 is only 1.99%, in contrast to 28.4% in AllConv. The generalization ability of IOC-NNs is further qualitatively reflected using the training and validation loss profiles (e.g., Fig. 1(a)). We present a table showing the confidence intervals of prediction across all three architectures with repeated runs in Appendix C.

Table 5 shows the train and test performance of the three architectures on CIFAR-10 dataset and the drop incurred when trained on CIFAIR-10. The drop in test performance of IOC-NNs is smaller than the typical neural network. This further strengthens the claim that IOC-NNs are not memorizing the training data but learning a generic hypothesis.

Table 3. Results for systematically randomized labels at peak and at convergence for both IOC-NN and NN. The IOC constraints bring huge improvements in generalization error and test accuracy at convergence.

	NN					IOC-NN				
	Peak		Convergence			Peak		Convergence		
	Train	Test	Train	Test	Gen. gap	Train	Test	Train	Test	Gen. gap
100	98.63	10.53	97.80	10.1	**87.7**	9.98	10.62	10.21	9.94	**0.27**
80	22.40	60.24	97.83	27.75	**70.08**	21.93	61.48	23.80	56.20	**−32.4**
60	38.52	75.80	97.80	46.71	**51.09**	37.90	75.91	39.31	71.75	**−32.44**
40	56.48	80.47	97.96	61.83	**36.13**	55.01	81.58	54.63	81.01	**−26.38**
20	72.8	85.72	98.73	76.31	**22.42**	69.92	85.85	70.22	83.61	**−13.39**

Table 4. Results comparing FICNN [1] with IOC-NN on CIFAR-10 using MLP architecture. First column shows base MLP results. Second column presents results with a convex MLP using ReLU activation. Third and final columns show the accuracies of FICNN and IOC-NN, respectively.

	Base MLP	Constrained MLP	FICNN	IOC-NN
Train	99.17	46.81	62.8	73.27
Test	63.83	27.36	53.07	69.89
Gen-gap	35.34	19.45	9.73	3.38

Comparison with FICNN: Table 4 shows the results of IOC-NN and FICNN [1] on CIFAR-10 data. For comparison, we use a three layer MLP with 800 nodes in each layer, for both IOC-NN and FICNN. FICNN uses a skip connection from input layer to each of the intermittent layers. This enables each layer to learn identity mapping inspite of non-negative constraint. The number of parameters in FICNN model is almost twice compared to the base MLP and IOC models but still the test performance drops by more than 10%. The results clearly shows that IOC-NN gives better test accuracy and lower generalization gap compared to FICNN, while using the same number of parameters as the base MLP architecture.

Robustness to Random Label Noise: Robustness of IOC-NNs on partial and fully randomized labels (Fig. 1 (b, c, and d)) is one of its key properties. We further investigate this property by systematically randomizing increasing portion of labels. We report the results of neural networks and their convex counterparts with percentage of label noise varying from 20% to 100% in Table 3. The train performance of neural networks at convergence is near 100% across all noise levels. It is interesting to note that IOC-NN gives a large negative generalization gap, where the train accuracy is almost equal to the percentage of true labels in the data. This observation shows that IOC-NNs significantly resist learning noise in labels as compared to neural networks. Both neural network and it's convex

The Curious Case of Convex Neural Networks 751

Table 5. Results on CIFAIR-10 dataset

	NN			IOC-NN		
	C-10	CIFAIR	Gap	C-10	CIFAIR	Gap
MLP	63.6	63.08	0.52	69.89	69.51	**0.38**
AllConv	92.8	91.14	0.66	90.6	90.47	**0.13**
DenseNet	94.06	93.28	0.78	91.12	90.73	**0.39**

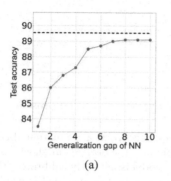
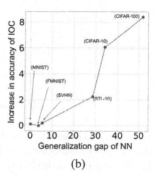

(a)   (b)

Fig. 5. (a) shows the test accuracy of IOC-MLP with increasing number of experts in the binary classification setting. Average performance of normal MLP is shown in red since it does not change with increase in number of experts. (b) The generalization gap of MLP plotted against the improvement gained by the IOC-MLP for the six different datasets (represented by every point on the plot). The performance gain with IOC constraints increase with the increase in generalization gap of MLP.

counterpart learns the simple hypothesis first. While IOC-NN holds on to this, in later epochs, the neural network starts brute force memorization of noisy labels. The observations are coherent with findings in [17,27], demonstrating neural network's heavy reliance on early stopping. IOC-AllConv outperforms test accuracy of AllConv + early stopping with a much-coveted generalization behavior. It is clear from this experiment that IOC-NN performs better in the presence of random label noise in the data in terms of test accuracy both at peak and convergence.

Leverage IOC Properties to Train Ensembles: We train binary MoE on the modified two-class setting of CIFAR-10 as described in Sect. 4. The result is shown in Fig. 5 (a). Traditional neural network gives a test accuracy of 89.63% with a generalization gap of 10%. Gated MoE of NNs does not improve the test performance as we increase the number of experts. In contrast, the performance of ensemble of IOC-NNs goes up with the addition of each expert and moves closer to the performance of neural networks. It is interesting to note that even in the higher dimensional space (like CIFAR-10 images), the intuitions derived from Fig. 4 holds. We also note that gate fractures the space into **p** partitions (where p is the number of experts). Moreover, in the binary case for a single expert, the generalization gap is almost zero. This can be attributed to the convex hull like

Table 6. Result for single expert, gated MoE and with oracle on CIFAR-10 for three architectures

	Single expert	Gate	Oracle
MLP	69.89	71.8	85.47
All-Conv	90.6	92.83	96.3
DenseNet	91.12	93.25	97.19

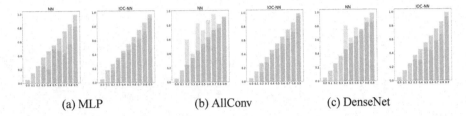

(a) MLP   (b) AllConv   (c) DenseNet

Fig. 6. These diagrams show expected sample accuracy as a function of confidence [9]. The blue bar shows the confidence of the bin and the orange bar shows the percentage correctness of prediction in that bin. If the model is perfectly calibrated, the bars align to form identity function. Any deviation from a perfect diagonal is a miscalibration. (Color figure online)

smooth decision boundary that the network predicts in the binary setting with a single output.

The results with the boosted ensembles of IOC-NNs are presented in Table 6. The boosted ensemble improves the test accuracies of IOC-NNs, matching or outperforming the base architectures. However, this performance gain comes at the cost of increased generalization error (still lower than the base architectures). In the boosted ensemble, the performance significantly improves if the gating network is replaced by an oracle. This observation suggests that there is a scope of improvement in model selection ability, possibly by using a better gating architecture.

Confidence Calibration of IOC-NNs: In a classification setting, given an input, the neural network predicts probability-like scores towards each class. The class with the maximum score is considered the predicted output, and the corresponding score to be the confidence. The confidence and accuracy being correlated is a desirable property, especially in high-risk applications like self-driving cars, medical diagnoses, etc. However, many modern multi-class classification networks are poorly calibrated, i.e., the probability values that they associate with the class labels they predict overestimate the likelihoods of those class labels being correct in the real world [11]. Recent works have explored methods to improve the calibration of neural networks [11,21].

We observe that adding IOC constraints improve calibration error on the base NN architecture. We present the reliability diagrams [9] (presenting accuracy as a function of confidence) of three neural architectures and their convex counterparts in Fig. 6. The sum of the difference between the blue bars and the orange bars represents the Expected Calibration Error. IOC constraints show improved calibration in all three architectures (with notable improvements in the case of MLP and AllConv). Better calibration further strengthens the case for IOC-NNs from the application perspective.

5   Conclusions

We present a subclass of neural networks, where the output is a convex function of the input. We show that with minimal constraints, existing neural networks can be adopted to this subclass called Input Output Convex Neural Networks. With a set of carefully chosen experiments, we unveil that IOC-NNs show outstanding generalization ability and robustness to label noise while retaining adequate capacity. We show that in scenarios where the neural network gives a large generalization gap, IOC-NN can give better test performance. An alternate interpretation of our work can be self regularization (regularization through functional constraints). IOC-NN puts to rest the concerns around brute force memorization of deep neural networks and opens a promising horizon for the community to explore. We show that in the case of Multi-Layer-Perceptrons, IOC constraints improve accuracy, generalization, calibration, and robustness to noise, making an ideal proposition from a deployment perspective. The improved generalization, calibration, and robustness to noise are also observed in convolutional architectures while retaining the accuracy. In future work, we plan to investigate the use of IOC-NNs for recurrent architectures. Furthermore, we plan to explore the interpretability aspects of IOC-NNs and study the effect of convexity constraints on the generalization bounds.

References

1. Amos, B., Xu, L., Kolter, J.Z.: Input convex neural networks. In: Proceedings of the 34th International Conference on Machine Learning, vol. 70, pp. 146–155. JMLR.org (2017)
2. Arpit, D., et al.: A closer look at memorization in deep networks. In: International Conference on Machine Learning, pp. 233–242. PMLR (2017)
3. Bach, F.: Breaking the curse of dimensionality with convex neural networks. J. Mach. Learn. Res. **18**(1), 629–681 (2017)
4. Barz, B., Denzler, J.: Do we train on test data? Purging cifar of near-duplicates. J. Imaging **6**, 41 (2020)
5. Boyd, S., Boyd, S.P., Vandenberghe, L.: Convex Optimization. Cambridge University Press, Cambridge (2004)
6. Burges, C.J.C.: A tutorial on support vector machines for pattern recognition. Data Mining Knowl. Discov. **2**, 121–167 (1998)
7. Chen, Y., Shi, Y., Zhang, B.: Optimal control via neural networks: a convex approach. arXiv preprint arXiv:1805.11835 (2018)
8. Chen, Y., Shi, Y., Zhang, B.: Input convex neural networks for optimal voltage regulation. arXiv preprint arXiv:2002.08684 (2020)
9. DeGroot, M.H., Fienberg, S.E.: The comparison and evaluation of forecasters. Stat. **32**, 12–22 (1983)
10. Friedman, J., Hastie, T., Tibshirani, R., et al.: Additive logistic regression: a statistical view of boosting (with discussion and a rejoinder by the authors). Ann. Stat. **28**(2), 337–407 (2000)

11. Guo, C., Pleiss, G., Sun, Y., Weinberger, K.Q.: On calibration of modern neural networks. In: International Conference on Machine Learning, pp. 1321–1330 (2017). PMLR
12. Huang, G., Liu, Z., Van Der Maaten, L., Weinberger, K.Q.: Densely connected convolutional networks. In: Proceedings of the IEEE Conference on Computer Vision and Pattern Recognition (2017)
13. Jacobs, R.A., Jordan, M.I., Nowlan, S.J., Hinton, G.E.: Adaptive mixtures of local experts. Neural Comput. 3(1), 79–87 (1991)
14. Jiang, W.: The VC dimension for mixtures of binary classifiers. Neural Comput. 12(6), 1293–1301 (2000)
15. Kantchelian, A., Tschantz, M.C., Huang, L., Bartlett, P.L., Joseph, A.D., Tygar, J.D.: Large-margin convex polytope machine. In Advances in Neural Information Processing Systems, vol. 27, pp. 3248–3256. Curran Associates Inc. (2014)
16. Kent, S., Mazumdar, E., Nagabandi, A., Rakelly, K.: Input-convex neural networks and posynomial optimization (2016)
17. Krueger, D., et al.: Deep nets don't learn via memorization (2017)
18. Legenstein, R., Maass, W.: On the classification capability of sign-constrained perceptrons. Neural Comput. 20(1), 288–309 (2008)
19. Magnani, A., Boyd, S.P.: Convex piecewise-linear fitting. Optim. Eng. 10(1), 1–17 (2009)
20. Manwani, N., Sastry, P.S.: Learning polyhedral classifiers using logistic function. In: Proceedings of 2nd Asian Conference on Machine Learning, pp. 17–30 (2010)
21. Mukhoti, J., Kulharia, V., Sanyal, A., Golodetz, S., Torr, P.H.S., Dokania, P.K.: Calibrating deep neural networks using focal loss. In: Advances in Neural Information Processing Systems (2020)
22. Nagarajan, V., Kolter, J.Z.: Uniform convergence may be unable to explain generalization in deep learning. In: Proceedings of the 33rd International Conference on Neural Information Processing Systems, 1042, p.12. Curran Associates Inc., Red Hook (2019)
23. Neyshabur, B., Bhojanapalli, S., McAllester, D., Srebro, N.: Exploring generalization in deep learning. In: Advances in Neural Information Processing Systems, pp. 5947–5956 (2017)
24. Neyshabur, B., Li, Z., Bhojanapalli, S., LeCun, Y., Srebro, N.: Towards understanding the role of over-parametrization in generalization of neural networks. arXiv preprint arXiv:1805.12076 (2018)
25. Neyshabur, B., Tomioka, R., Srebro, N.: In search of the real inductive bias: On the role of implicit regularization in deep learning. arXiv preprint arXiv:1412.6614 (2014)
26. Shah, K., Sastry, P.S., Manwani, N.: Plume: polyhedral learning using mixture of experts. arXiv preprint arXiv:1904.09948 (2019)
27. Sjöberg, J., Ljung, L.: Overtraining, regularization and searching for a minimum, with application to neural networks. Int. J. Control 62(6), 1391–1407 (1995)
28. Springenberg, J.T., Dosovitskiy, A., Brox, T., Riedmiller, M.: Striving for simplicity: the all convolutional net. arXiv preprint arXiv:1412.6806 (2014)
29. Takács, G., Pataki, B.: Lower bounds on the Vapnik-Chervonenkis dimension of convex polytope classifiers. In: 2007 11th International Conference on Intelligent Engineering Systems, pp. 145–148 (2007)
30. Zhang, C., Bengio, S., Hardt, M., Recht, B., Vinyals, O.: Understanding deep learning (still) requires rethinking generalization. Commun. ACM 64(3), 107–115 (2021)

UCSL : A Machine Learning Expectation-Maximization Framework for Unsupervised Clustering Driven by Supervised Learning

Robin Louiset[1,2]([✉]), Pietro Gori[2]([✉]), Benoit Dufumier[1,2]([✉]),
Josselin Houenou[1]([✉]), Antoine Grigis[1]([✉]), and Edouard Duchesnay[1]([✉])

[1] Université Paris-Saclay, CEA, Neurospin, 91191 Gif-sur-Yvette, France
`antoine.grigis@cea.fr`
[2] LTCI, Télécom Paris, Institut Polytechnique de Paris, Palaiseau, France
`pietro.gori@telecom-paris.fr`

Abstract. Subtype Discovery consists in finding interpretable and consistent sub-parts of a dataset, which are also relevant to a certain supervised task. From a mathematical point of view, this can be defined as a clustering task driven by supervised learning in order to uncover subgroups in line with the supervised prediction. In this paper, we propose a general Expectation-Maximization ensemble framework entitled UCSL (Unsupervised Clustering driven by Supervised Learning). Our method is generic, it can integrate any clustering method and can be driven by both binary classification and regression. We propose to construct a non-linear model by merging multiple linear estimators, one per cluster. Each hyperplane is estimated so that it correctly discriminates - or predict - only one cluster. We use SVC or Logistic Regression for classification and SVR for regression. Furthermore, to perform cluster analysis within a more suitable space, we also propose a dimension-reduction algorithm that projects the data onto an orthonormal space relevant to the supervised task. We analyze the robustness and generalization capability of our algorithm using synthetic and experimental datasets. In particular, we validate its ability to identify suitable consistent sub-types by conducting a psychiatric-diseases cluster analysis with known ground-truth labels. The gain of the proposed method over previous state-of-the-art techniques is about $+1.9$ points in terms of balanced accuracy. Finally, we make codes and examples available in a scikit-learn-compatible Python package. https://github.com/neurospin-projects/2021_rlouiset_ucsl/.

Keywords: Clustering · Subtype discovery ·
Expectation-maximization · Machine learning · Neuroimaging

Electronic supplementary material The online version of this chapter (https://doi.org/10.1007/978-3-030-86486-6_46) contains supplementary material, which is available to authorized users.

© Springer Nature Switzerland AG 2021
N. Oliver et al. (Eds.): ECML PKDD 2021, LNAI 12975, pp. 755–771, 2021.
https://doi.org/10.1007/978-3-030-86486-6_46

1   Introduction

Subtype discovery is the task of finding consistent subgroups within a population or a class of objects which are also relevant to a certain supervised upstream task. This means that the definition of homogeneity of subtypes should not be fully unsupervised, as in standard clustering, but it should also be driven by a supervised task. For instance, when identifying flowers, one may want to find different varieties or subtypes within each species. Standard clustering algorithms are driven by features that explain most of the general variability, such as the height or the thickness. Subtype identification aims at discovering subgroups describing the specific heterogeneity within each flower species and not the general variability of flowers. To disentangle these sources of variability, a supervised task can identify a more relevant feature space to drive the intra-species clustering problem. Depending on the domain, finding relevant subgroups may turn out to be a relatively hard task. Indeed, most of the time, boundaries between different patterns are fuzzy and may covariate with other factors. Hence, ensuring that resulting predictions are not collapsed clusters or biased by an irrelevant confound factor is a key step in the development of such analysis. For example, in clinical research, it is essential to identify subtypes of patients with a given disorder (red dots in Fig. 1). The problem is that the general variability (that stems from age or sex) is observed in both healthy controls (grey dots in Fig. 1) and disease patients, therefore it will probably drive the clustering of patients toward a non-specific solution (second plot in Fig. 1). Adding a supervised task (healthy controls vs patients) can be used to find direction(s) (horizontal arrow Fig. 1) that discards non-specific variability to emphasize more disease-related differences (subtype discovery in Fig. 1). This is a fundamental difference between unsupervised clustering analysis and subtype identification.

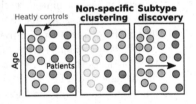

Fig. 1. Subtype discovery in clinical research.

Subgroups identification is highly relevant in various fields such as in clinical research where disease subtypes discovery can lead to better personalized drug-treatment and prognosis [28] or to better anticipate at-risk profiles [26]. Particularly, given the extreme variability of cancer, identifying subtypes enable to develop precision medicine [2,14,16,18,28,28]. In psychiatry and neurology, different behaviour, anatomical and physiological patterns point out variants of mental disorders [13] such as for bipolar disorder [31], schizophrenia [5,8], autism, [24,33], attention-deficit hyperactivity disorder [29], Alzheimer's disease [7,25,27,32] or Parkinson's disease [6]. In bio-informatics, DNA subfolds analysis is a key field for the understanding of gene functions and regulations, cellular processes and cells subtyping [23]. In the field of data mining, crawling different consistent subgroups of written data enables enhanced applications [20].

2   Related Works

Early works [2,6] proposed traditional clustering methods to find relevant subgroups for clinical research in cancer and neurology. However, they were very sensitive to high-dimensional data and noise, making them hardly reproducible [17,18].

 To overcome these limits, [23] and [18] evaluated custom consensus methods to fuse multiple clustering estimates in order to obtain more robust and reproducible results. Additionally, [23] also proposed to select the most important features in order to overcome the curse of dimensionality. Even if all these methods provide relevant strategies to identify stable clusters in high-dimensional space, they do not allow the identification of disease-specific subtypes when the dominant variability in patients corresponds to the variability in the general population. To select disease-specific variability, recent contributions propose hybrid approaches integrating a supervised task (patient vs. controls) to the clustering problem. In [22], authors propose a hybrid method for disease-subtyping in precision medicine. Their implementation consists of training a Random Forest supervised classifier (healthy vs. diseased) and then apply SHAP Algorithm [10,11] to get explanation values from Random Forest classifiers. This yields promising results even though it is computationally expensive, especially when the dataset size increases.

 Differently, a wide range of Deep Learning methods propose to learn better representations via deep encoders and adapt clustering method on compressed latent space or directly within the minimizing loss. In this case, encoders have to be trained with at least one non-clustering loss, to enhance the representations [21] and avoid collapsing clusters [30]. [3] proposes a Deep Clustering framework that alternates between latent clusters estimation and likelihood maximization through pseudo-label classification. Yet, its training remains unstable and designed for large-scale dataset only. Prototypical Contrastive Learning [9], SeLA [1], SwAV [4] propose contrastive learning frameworks that alternatively maximize 1- the mutual information between the input samples and their latent representations and 2- the clustering estimation. These works have proven to be very efficient and stable on large-scale datasets. They compress inputs into denser and richer representations, and successfully get rid of unnecessary noisy dimensions. Nevertheless, they still do not propose a representation aligned with the supervised task at-hand. To ensure that resulting clusters identify relevant subgroups for the supervised task, one could first train for the supervised task and then run clustering on the latent space. This would emphasize important features for the supervised task but it may also regress out intra-class specific heterogeneity, hence the need of an iterative process where clustering and classification tasks influence each other.

 CHIMERA [8], proposes an Alzheimer's subtype discovery algorithm driven by supervised classification between healthy and pathological samples. It assumes that the pathological heterogeneity can be modeled as a set of linear transformations from the reference set of healthy subjects to the patient distribution, where each transformation corresponds to one pathological subtype. This is a strong a priori that limits its application to (healthy reference)/(pathological case) only. [25,27] propose an alternate algorithm between supervised learning and unsupervised cluster analysis where each step influences the other until it reaches a stable configuration. The algorithm simultaneously solves binary classification and intra-class clustering in a hybrid fashion thanks to a maximum margin framework. The method discriminates healthy controls from pathological patients by optimizing the best convex polytope that is formed by combining several linear hyperplanes. The clustering ability is drawn by assigning patients to their best discriminating hyperplane. Each cluster corresponds to one face of the piece-wise linear polytope and heterogeneity is implicitly captured by harnessing the classification boundary non-linearity. The efficiency of this method heavily relies on the prior hypothesis that negative samples (not being clustered) lie inside the convex discrimination polyhedron. This may be a limitation when it does not hold for a

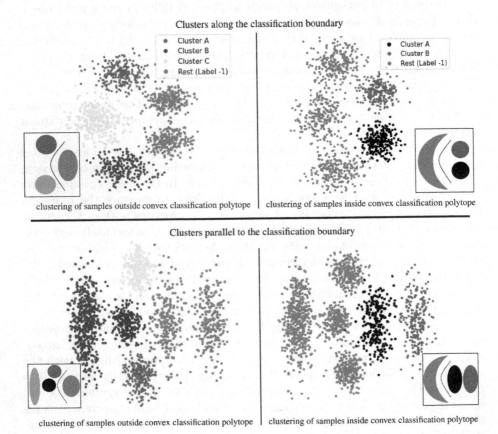

Fig. 2. Toy Datasets - Different configurations we want to address. Grey points represent negative samples. The upstream task is to classify negative (grey) samples from all positive (colored) samples while the final goal is to cluster positive samples. The upper plots show 3 and 2 clusters respectively along the classification boundary. The lower plot show 4 and 2 clusters respectively parallel (and also along on the left) to the classification boundary. Furthermore, plots on the left and right show clusters outside and inside the convex classification polytope respectively.

given data-set (left examples of Fig. 2). Another hypothesis is that relevant psychiatric subtypes should not be based on the disease severity. This a priori implies that clusters should be along the classification boundary (upper examples of Fig. 2). Even though it may help circumvent general variability issues, this strongly limits the applicability of the method to a specific variety of subgroups.

Contributions. Here, we propose a general framework for Unsupervised Clustering driven by Supervised Learning (UCSL) for relevant subtypes discovery. The estimate of the latent subtypes is tied with the supervision task (regression or classification). Furthermore, we also propose to use an ensembling method in order to avoid trivial local minima or collapsed clusters.

We demonstrate the relevance of the UCSL framework on several data-sets. The quality of the obtained results, the high versatility, and the computational efficiency of the proposed framework make it a good choice for many subtype discovery applications in various domains. Additionally, the proposed method needs very few parameters compared to other state-of-the-art (SOTA) techniques, making it more relevant for a large number of medical applications where the number of training samples is usually limited. Our three main contributions are :

1. A generic mathematical formulation for subtype discovery which is robust to samples inside and outside the classification polytope (see Fig. 2).
2. An Expectation-Maximization (EM) algorithm with an efficient dimensionality reduction technique during the E step for estimating latent subtypes more relevant to the supervised task.
3. A thoughtful evaluation of our subtype discovery method and a fair comparison with several other SOTA techniques on both synthetic and real data-sets. In particular, a neuroimaging data-set for psychiatric subtype discovery.

3 UCSL: An Unsupervised Clustering Driven by Supervised Learning Framework

3.1 Mathematical Formulation

Let $(X, Y) = \{(x_i, y_i)\}_{i=1}^n$ be a labeled data-set composed of n samples. Here, we will restrict to regression, $y_i \in \mathcal{R}$, or binary classification, $y_i \in \{-1, +1\}$. We assume that all samples, or only positive samples ($y_i = +1$), can be subdivided into latent subgroups for regression and binary classification respectively.

The membership of each sample i to latent clusters is modeled via a latent variable $c_i \in C = \{C_1, ..., C_K\}$, where K is the number of assumed subgroups. We look for a discriminative model that maximizes the joint conditional likelihood:

$$\sum_{i=1}^n \log \sum_{c \in C} p(y_i, c_i | x_i) \tag{1}$$

Directly maximizing this equation is hard and it would not explicitly make the supervised task and the clustering depend on each other, namely we would like to optimize both $p(c_i | x_i, y_i)$ (the clustering task) and $p(y_i | x_i, c_i)$ (the upstream supervised task) and not only one of them. To this end, we introduce Q, a probability distribution over C, so that $\sum_{c_i \in C} Q(c_i) = 1$.

$$\sum_{i=1}^{n} \log \sum_{c \in C} p(y_i, c_i | x_i) = \sum_{i=1}^{n} \log \left(\sum_{c \in C} Q(c_i) \frac{p(y_i, c_i | x_i)}{Q(c_i)} \right). \quad (2)$$

By applying the Jensen inequality, we then obtain the following lower-bound:

$$\sum_{i=1}^{n} \log \left(\sum_{c \in C} Q(c_i) \frac{p(y_i, c_i | x_i)}{Q(c_i)} \right) \geq \sum_{i=1}^{n} \sum_{c \in C} Q(c_i) \log \left(\frac{p(y_i, c_i | x_i)}{Q(c_i)} \right), \quad (3)$$

It can be shown that equality holds when:

$$Q(c_i) = \frac{p(y_i, c_i | x_i)}{\sum_{c \in C} p(y_i, c_i | x_i)} = \frac{p(y_i, c_i | x_i)}{p(y_i | x_i)} = p(c_i | y_i, x_i). \quad (4)$$

The right term of Eq. 3 can be re-written as:

$$\sum_{i=1}^{n} \sum_{c \in C} \left(Q(c_i) \log \left(p(y_i | c_i, x_i) p(c_i | x_i) \right) - Q(c_i) \log Q(c_i) \right). \quad (5)$$

We address the maximization of Eq. 5 with an EM optimization scheme (Algorithm 2) that exploits linear models to drive the clustering until we obtain a stable solution. First, during the Expectation step, we tighten the lower bound in Eq. 3 by estimating Q as the latent clusters conditional probability distribution $p(c_i | y_i, x_i)$ as in Eq. 4. Then, we fix Q, and maximize the supervised conditional probability distribution $p(y_i | c_i, x_i)$ weighted by the conditional cluster distribution $p(c_i | x_i)$ as in Eq. 5.

3.2   Expectation Step

In this step, we want to estimate Q as $p(c_i | y_i, x_i), \forall i \in [\![1, n]\!], \forall c \in C$ in order to tighten the lower bound in Eq. 3. We remind here that latent clusters c are defined only for the positive samples ($y = +1$), when dealing with a binary classification, and for all samples in case of regression. Let us focus here on the binary classification task. Depending on the problem one wants to solve, different solutions are possible. On the one hand, if ground truth labels for classification are *not* available at inference time, Q should be computed using the classification prediction. For example, one could use a clustering algorithm only on the samples predicted as positive. However, this would bring a new source of uncertainty and error in the subgroups discovery due to possible classification errors. On the other hand, if ground truth labels for classification are available at inference time, one would compute the clustering using only the samples associated to ground-truth positive labels $\tilde{y}_i = +1$, and use the classification directions to

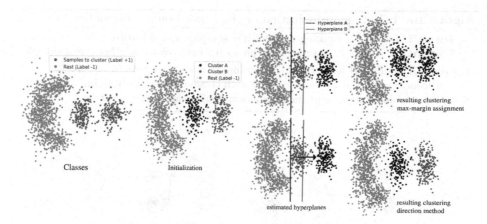

Fig. 3. Limit of maximum-margin based clustering starting from an optimal cluster initialization. When the separation of clusters to discover is co-linear to the supervised classification boundary, the maximum margin cluster assignment (as in [25]) converges towards a degenerate solution (upper figures). Instead, with our direction method (lower figures), the Graam-Schmidt algorithm returns one direction where input points are projected to and perfectly clustered.

guide the clustering. Here, we will focus on the latter situation, since it's of interest for many medical applications.

Now, different choices are again possible. In order to influence the resulting clustering with the label prediction estimation, HYDRA [25] proposes to assign each positive sample to the hyperplane that best separates it from negative samples (i.e. the furthest one). This is a simple way to align resulting clustering with estimated classification while implicitly leveraging classification boundary non-linearity. Yet, we argue that this formulation does not work in the case where clusters are disposed parallel to the piece-wise boundary as described in Fig. 3. To overcome this limit, we propose to project input samples onto a supervision-relevant subspace before applying a general clustering algorithm.

Dimension Reduction Method Based on Discriminative Directions.
Our goal is a clustering that best aligns with the upstream-task. In other words, in a classification example, the discovery of subtypes should focus on the same features that best discriminate classes, and not on the ones characterizing the general variability. In regression, subgroups should be found by exploiting features that are relevant for the prediction task. In order to do that, we rely on the linear models estimated from the maximization step. More specifically, we propose to first create a relevant orthonormal sub-space by applying the Graam-Schmidt algorithm onto all discriminant directions, namely the normal directions of estimated hyperplanes. Then, we project input features onto this new linear subspace to reduce the dimension and perform cluster analysis on a more suitable space. Clustering can be conducted with any algorithm such as Gaussian Mixture Models (GMM), K-Means (KM) or DBSCAN for example.

Algorithm 1. Dimension reduction method based on discriminative directions

> **Input :** $X \in \mathbf{R}^{n \times d}$, training data with n samples and d features.
> **Output :** $X' \in \mathbf{R}^{n \times K}$, training data projected onto relevant orthonormal subspace.
> 1: Given K estimated hyperplanes, concatenate normal vectors in $D \in \mathbf{R}^{K \times d}$.
> 2: Ortho-normalize the direction basis D with Graam-Schmidt obtaining $D^{\perp} \in \mathbf{R}^{K \times d}$.
> 3: Project training data onto the orthonormal subspace, $X' = X(D^{\perp})^T$.

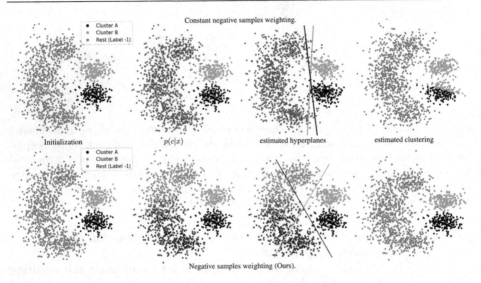

Fig. 4. Starting with an optimal initialization of clusters to discover, constant negative samples weighting (top row) may lead to co-linear discriminative hyperplanes and thus errors in clustering. Conversely, our negative samples weighting enforces non-colinearity between discriminative hyperplanes resulting in higher quality clustering.

3.3 Maximization Step

After the expectation step, we fix Q and then maximize the conditional likelihood. The lower bound in Eq. 5 thus becomes:

$$\sum_{i=1}^{n} \sum_{c \in C} Q(c_i) \log p(y_i|c_i, x_i) + \sum_{i=1}^{n} \sum_{c \in C} Q(c_i) \log p(c_i|x_i) \qquad (6)$$

Here, we need to estimate $p(c_i|x_i)$. A possible solution, inspired by HYDRA [25], would be to use the previously estimated distribution $p(c_i|y_i, x_i)$ for the positive samples and a fixed weight for the negative samples, namely:

$$p(c_i|x_i) = \begin{cases} p(c_i|x_i, y_i) & \text{if } \tilde{y}_i = +1 \\ \frac{1}{K} & \text{if } \tilde{y}_i = -1 \end{cases} \qquad (7)$$

However, as illustrated in Fig. 4, this approach does not work well when negative samples lie outside of the convex classification polytope since discriminative directions (or hyperplanes) may become collinear. This collinearity hinders the retrieving of informative directions and consequently degrades the resulting clustering.

To overcome such a shortcoming, we propose to approximate $p(c_i|x_i)$ using $p(c_i|x_i, y_i)$ for both negative and positive samples or, in other words, to extend the estimated clustering distribution to all samples, regardless their label y. In this way, samples from the negative class ($y_i = -1$), that are closer to a certain positive cluster, will have a higher weight during classification. As shown in Fig. 4, this results in classifications hyperplanes that correctly separate each cluster from the closer samples of the negative class, entailing better clustering results. From a practical point of view, since we estimate $Q(c_i)$ as $p(c_i|y_i, x_i)$, it means that $p(c_i|x_i)$ can be approximated by $Q(c_i)$. $Q(c_i)$ being fixed during the M step, only the left term in Eq. 6 is maximized.

3.4 Supervised Predictions

Once trained the proposed model, we compute the label y_j for each test sample x_j using the estimated conditional distributions $p(y_j|c_j, x_j)$ and $p(c_j|x_j)$ as:

$$p(y_j|x_j) = \sum_{c_j \in C} p(y_j, c_j|x_j) = \sum_{c_j \in C} p(y_j|x_j, c_j)p(c_j|x_j) \qquad (8)$$

In this way, we obtain a non-linear estimator based on linear hyperplanes, one for each cluster.

3.5 Application

Multiclass Case. In the case of classification, we handle the binary case in the same way as [25] does. We consider one label as positive $\tilde{y}_i = 1$ and cluster it with respect to the other one $\tilde{y}_i = -1$. In the multi-class case, we can cast it as several binary problems using the one-vs-rest strategy.

Ensembling: Spectral Clustering. The consensus step enables the merging of several different clustering propositions to obtain an aggregate clustering. After having run the EM iterations N times, the consensus clustering is computed by grouping together samples that were assigned to the same cluster across different runs. In practice, we compute a co-occurrence matrix between all samples. And then we use co-occurence values as a similarity measure to perform spectral clustering. Hence, for example, given two samples i and j and 10 different runs, if samples i and j ended up 4 times in the same cluster, the similarity measure between those 2 samples will be $\frac{4}{10}$. Given an affinity matrix between all samples, we can then use the spectral clustering algorithm to obtain a consensus clustering.

3.6 Pseudo-code

The pseudo-code of the proposed method UCSL (Algorithm 2) can be subdivided into several distinct steps:

Algorithm 2. UCSL general framework pseudo-code

Input : $X \in \mathbf{R}^{n \times d}$, $y \in \{-1, 1\}^n$, K number of clusters.
Output : $p(c|x, y) = Q(c)$, $p(y|x, c)$ (linear sub-classifiers).

1: **for** ensemble in n_ensembles **do**
2:     Initialization: Estimate $Q^{(0)}$ for all samples ($y = \pm 1$) with a clustering algorithm (e.g. GMM) trained with positive samples only ($y = +1$).
3:     **while** not converged **do**
4:         M **step** (supervised step) :
5:             Freeze $Q^{(t)}$
6:             **for** k in $[1, K]$:
7:                 Fit linear sub-classifier k weighted by $Q^{(t)}[:, k]$ (Eq. 6).
8:             **end for**
9:         E **step** (unsupervised step) :
10:             Use Algorithm 1 to obtain $X' \in \mathbf{R}^{n \times K}$ from sub-classifiers normal vectors $D \in \mathbf{R}^{K \times d}$.
11:             Estimate $Q^{(t+1)} = p(c|x, y)$ (Eq. 4) for all samples with a clustering algorithm trained on X' with positive samples only.
12:     **end while**
13: **end for**
14: **Ensembling**: Compute average clustering with the ensembling method (Sec. 3.5).
15: **Last EM** : Perform EM iterations from ensembled latent clusters until convergence.

---

1. **Initialization:** First, we have to initialize the clustering. There are several possibilities here, we can make use of traditional ML methods such as KM or GMM. For most of our experiments we used GMM.

2. **Maximization:** The Maximization step consists in training several linear models to solve the supervised upstream problem. It can be either a classification or a regression. We opted for well-known ML linear methods such as logistic regression or max-margin linear classification method as in [25].

3. **Expectation:** The Expectation step makes use of the supervised learning estimates to produce a relevant clustering. In our case, we exploit the directions exhibited by the linear supervised models. We project samples onto a subspace spanned by those directions to perform the unsupervised clustering with positive samples.

4. **Convergence:** In order to check the convergence, we compute successive clustering Adjusted Rand Score (ARI), the closer this metric is to 1, the more similar both clustering assignments are.

5. **Ensembling:** Initialization and EM iterations are performed until convergence N times and an average clustering is computed with a Spectral Clustering algorithm [25], [18] that proposes the best consensus. This part enables us to have more robust and stable solutions avoiding trivial or degenerate clusters.

4   Results

We validated our framework on four synthetic data-sets and two experimental ones both qualitatively and quantitatively.

Implementation Details. The stopping criteria in Algorithm 2 is defined using the ARI index between two successive clusterings (at iteration t and iteration $t+1$). The algorithm stops when it reaches the value of 0.85. In the MNIST experiment, convolutional generator and encoder networks have a similar structure to the generator and discriminator in DCGAN [19]. We trained it during 20 epochs, with a batch size of 128, a learning rate of 0.001 and with no data augmentation and a SmoothL1 loss. More information can be found in the Supplementary material. Standard deviations are obtained by running 5 times the experiments with different initializations (synthetic and MNIST examples) or using a 5-fold cross-validation (psychiatric dataset experiment). MNIST and synthetic examples were run on Google Colaboratory Pro, whose hardware equipments are PNY Tesla P100 with 28Gb of RAM.

Synthetic Dataset. First, we generated a set of synthetic examples that sum up the different configurations on which we wish our method to be robust: subtypes along the supervised boundary or parallel to it. We designed configurations with various number of clusters, outside or inside the convex classification polytope. UCSL was run with Logistic Regression and GMM. In order to make our problem more difficult we decided to add noisy unnecessary features to the original 2-D toy examples. For each example and algorithm, we performed 10 runs with a different initialization each time (GMM with only one initialization) and we did not perform the ensembling step for fair comparison with the other methods. We compared with other traditional ML methods such as KM GMM, DBSCAN and Agglomerative Clustering. Results are displayed in Fig. 5. For readability, we divided the standard deviation hull by 2. Compared with the other methods, UCSL appears to be robust to unnecessary noisy features. Furthermore, it performs well in all configurations we addressed.

MNIST Dataset. To further demonstrate what an intra-class clustering could be used for, let us make an example from MNIST. We decided to analyse the digit 7 looking for subtypes. To perform this experiment, we trained on 20 000 MNIST digits and considered the digit 7 as positive class. We use a one-vs-rest strategy for classification where input samples are the flattened images.

Visually, digit 7 examples have two different subtypes: with or without the middle-cross bar. In order to quantitatively evaluate our method, we labeled 400 test images in two classes, 7 with a middle-cross bar, and those with none. We ran UCSL with GMM as a clustering method, logistic regression as classification method and compared with clustering methods coupled with deep learning models or dimension reduction algorithms. We use the metrics V-Measure, Adjusted Rand Index (ARI) and balanced accuracy (B-ACC), since we know the expected clustering result.

As it is possible to notice from Table 1, UCSL outperforms other clustering and subtypes ML methods. We also compared our algorithm with DL methods, a pre-trained convolutional network and a simple convolutional encoder-decoder. Only the convolutional autoencoder network along with a GMM on its latent space of dimension 32 slightly outperforms UCSL. However, it uses a definitely higher number of parameters (7500 times more!) and takes twice the time for training. Our model is thus more relevant to smaller data-sets, which are common in medical applications. Please note that UCSL could also be adapted in order to use convolutional auto-encoders or contrastive methods such as in [9] and [3], when dealing with large data-sets. This is left as future work.

Psychiatric Dataset. The ultimate goal of the development of subtype discovery methods is to identify homogeneous subgroups of patients that are associated

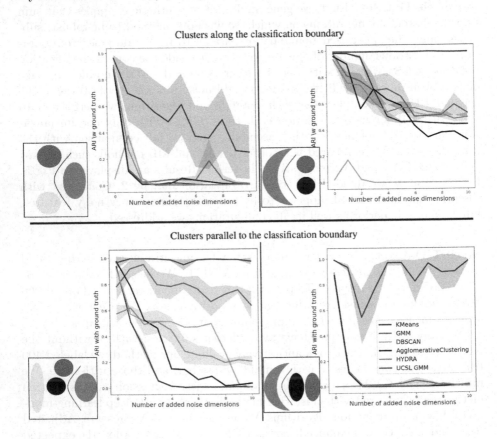

Fig. 5. Comparison of performances of different algorithms on the four configurations presented in Fig. 2. Noisy features are added to the original 2D data. For each example, all algorithms are run 10 times with different initialization.

Table 1. MNIST dataset, comparison of performances of different algorithms for the discovery of digit 7 subgroups. AE : convolutional AutoEncoder; PT VGG11: VGG11 model pre-trained on imagenet; GMM: Gaussian Mixture Model; KM: K-Means. Latent size: dimension of space where clustering is computed. * : to limit confusion, we assign no parameters for t-sne, umap and SHAP. We use default values (15,30,100) for perplexity, neighbours and n estimators in t-sne, umap and SHAP respectively.

Methods	Latent Size	Nb params	Avg Exec Time	V-measure	ARI	B-ACC
AE + GMM	32	3 M	21 m 40 s	0.323 ± 0.013	0.217 ± 0.025	0.823 ± 0.009
UCSL (our)	2	406	12 m31 s	0.239 ± 0.001	0.330 ± 0.001	**0.808 ± 0.001**
PT VGG11 + KM	1000	143 M	32 m 44 s	0.036 ± 0.001	0.087 ± 0.001	0.616 ± 0.001
AE + GMM	2	3 M	13 m 34 s	0.031 ± 0.015	0.033 ± 0.021	0.607 ± 0.025
t-sne* [12] + KM	2	4	2 m 04 s	0.029 ± 0.020	0.049 ± 0.056	0.568 ± 0.033
t-sne* [12] + GMM	2	14	2 m 04 s	0.023 ± 0.021	0.020 ± 0.048	0.566 ± 0.033
umap* [15] + KM	2	4*	24 s	0.050 ± 0.015	0.078 ± 0.015	0.555 ± 0.022
umap* [15] + GMM	2	14*	24 s	0.025 ± 0.006	0.080 ± 0.010	0.547 ± 0.005
SHAP [10]* + KM	196	392*	1 h 02	0.012 ± 0.007	−0.014 ± 0.035	0.540 ± 0.016
KM	196	392	0.32 ms	0.006 ± 0.000	0.010 ± 0.000	0.552 ± 0.000
HYDRA	196	394	9 m 45 s	0.005 ± 0.006	0.024 ± 0.031	0.520 ± 0.018
GMM	196	77 K	0.32 ms	0.0002 ± 0.000	−0.001 ± 0.000	0.510 ± 0.000

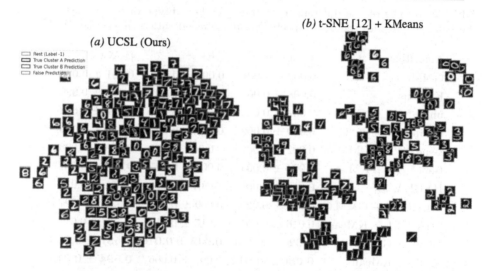

(a) UCSL (Ours)   *(b)* t-SNE [12] + KMeans

Fig. 6. Comparison of latent space visualization in the context of MNIST digit "7" subtype discovery. Differently from t-SNE, our method does not focus on the general digits variability but only on the variability of the "7". For this reason, subtypes of "7" are better highlighted with our method.

with different disease mechanisms and lead to patient-specific treatments. With brain imaging data, the variability specific to the disorder is mixed up or hidden to non-specific variability. Classical clustering algorithms produce clusters that correspond to subgroups of the general population: old participants with brain atrophy versus young participants without atrophy, for instance.

To validate the proposed method we pooled neuroimaging data from patients with two psychiatric disorders, (Bipolar Disorder (BD) and Schizophrenia (SZ)), with data from healthy controls (HC). The supervised upstream task aims at classifying HC from patients (of both disorders) using neuroimaging features related to the local volumes of brain grey matter measured in 142 regions of interest (identified using cat12 software). Here, we used a linear SVM for classification. The clustering task is expected to retrieve the known clinical disorder (BD or SZ). Training set was composed of 686 HC and 275 SZ, 307 BP patients.

We measured the correspondence (Table 2) between the clusters found by the unsupervised methods with the true clinical labels on an independant TEST set (199 HC, 190 SZ, 116 BP) coming from a different acquisition site. As before, we used the metrics V-Measure, Adjusted Rand Index (ARI) and balanced accuracy (B-ACC). Please note that the classification of SZ vs BD is a very difficult problem due to the continuum between BP and SZ. Therefore, performances should be compared with the best expected result provided by a purely supervised model (here a SVM) that produces only 61% of accuracy (last row of Table 2).

Table 2. Results of the different algorithms on the subtype discovery task BP/SZ. The last row provides the best expected result obtained with a supervised SVM.

Algorithm	V-measure	ARI	B-ACC
GMM	0.002 ± 0.001	0.003 ± 0.008	0.491 ± 0.024
KMeans	0.008 ± 0.001	−0.01 ± 0.001	0.499 ± 0.029
umap* [15] + GMM	0.001 ± 0.002	0.000 ± 0.007	0.497 ± 0.013
umap* [15] + KM	0.000 ± 0.002	0.001 ± 0.005	0.502 ± 0.006
t-sne* [12] + GMM	0.002 ± 0.0024	−0.00 ± 0.005	0.498 ± 0.028
t-sne* [12] + KM	0.004 ± 0.004	0.003 ± 0.005	0.505 ± 0.041
HYDRA [25]	0.018 ± 0.009	−0.01 ± 0.004	0.556 ± 0.019
SHAP [22] + GMM	0.004 ± 0.005	0.000 ± 0.006	0.527 ± 0.027
SHAP [22] + KMeans	0.016 ± 0.005	0.017 ± 0.012	0.575 ± 0.011
UCSL + GMM	**0.024 ± 0.006**	**0.042 ± 0.016**	**0.587 ± 0.009**
UCSL + KMeans	**0.030 ± 0.012**	0.004 ± 0.006	**0.594 ± 0.015**
Supervised SVM	*0.041 ± 0.007*	*0.030 ± 0.008*	*0.617 ± 0.010*

As expected, mere clustering methods (KMeans, GMM) provide clustering at the chance level. Detailed inspection showed that they retreived old patients with brain atrophy vs younger patients without atrophy. Only clustering driven by supervised upstream task (HYDRA, SHAP+KMeans and all UCSL) can disentangle the variability related to the disorders to provide results that are significantly better than chance (59% of B-ACC). Models based on USCL significantly outperformed all other models approaching the best expected result that would provide a purely supervised model.

5   Conclusion

We proposed in this article a Machine Learning (ML) Subtype Discovery (SD) method that aims at finding relevant homogeneous subgroups with significant statistical differences in a given class or cohort. To address this problem, we introduce a general Subtype Discovery (SD) Expectation-Maximization (EM) ensembled framework. We call it UCSL : Unsupervised Clustering driven by Supervised Learning. Within the proposed framework, we also propose a dimension reduction method based on discriminative directions to project the input data onto an upstream-task relevant linear subspace. UCSL is adaptable to both classification and regression tasks and can be used with any clustering method. Finally, we validated our method on synthetic toy examples, MNIST and a neuro-psychiatric data-set on which we outperformed previous state-of-the-art methods by about +1.9 points in terms of balanced accuracy.

References

1. Asano, Y.M., Rupprecht, C., Vedaldi, A.: Self-labelling via simultaneous clustering and representation learning. In: ICLR (2020)
2. Carey, L.A., Perou, C.M., Livasy, C.A., Dressler, L.G., Cowan, D., et al.: Race, breast cancer subtypes, and survival in the Carolina breast cancer study. JAMA **295**(21), 2492–2502 (2006)
3. Caron, M., Bojanowski, P., Joulin, A., Douze, M.: Deep clustering for unsupervised learning of visual features. In: ECCV, pp. 139–156 (2018)
4. Caron, M., Misra, I., Mairal, J., Goyal, P., Bojanowski, P., Joulin, A.: Unsupervised learning of visual features by contrasting cluster assignments. NeurIPS **33**, 9912–9924 (2020)
5. Chand, G.B., Dwyer, D.B., Erus, G., Sotiras, A., Varol, E., et al.: Two distinct neuroanatomical subtypes of schizophrenia revealed using machine learning. Brain **143**(3), 1027–1038 (2020)
6. Erro, R., Vitale, C., Amboni, M., Picillo, M., et al.: The heterogeneity of early Parkinson's disease: a cluster analysis on newly diagnosed untreated patients. PLoS One **8**(8), e70244 (2013)
7. Ferreira, D., Verhagen, C., Hernández-Cabrera, J.A., Cavallin, L., et al.: Distinct subtypes of Alzheimer's disease based on patterns of brain atrophy: longitudinal trajectories and clinical applications. Sci Rep **7**, 1–13 (2017)
8. Honnorat, N., Dong, A., Meisenzahl-Lechner, E., Koutsouleris, N., Davatzikos, C.: Neuroanatomical heterogeneity of schizophrenia revealed by semi-supervised machine learning methods. Schizophr. Res. **214**, 43–50 (2019)
9. Li, J., Zhou, P., Xiong, C., Hoi, S.C.H.: Prototypical contrastive learning of unsupervised representations. In: ICLR (2021)
10. Lundberg, S.M., Erion, G.G., Lee, S.I.: Consistent individualized feature attribution for tree ensembles. In: ICML workshop (2017)
11. Lundberg, S.M., Lee, S.I.: A unified approach to interpreting model predictions. In: NeurIps, pp. 4768–4777 (2017)
12. Maaten, L., Hinton, G.: Visualizing data using t-SNE. J. Mach. Learn. Res. **9**(86), 2579–2605 (2008)

13. Marquand, A.F., Wolfers, T., Mennes, M., Buitelaar, J., Beckmann, C.F.: Beyond lumping and splitting: a review of computational approaches for stratifying psychiatric disorders. Biol. Psychiatry: Cogn. Neurosci. Neuroimaging **1**(5), 433–447 (2016)
14. Marusyk, A., Polyak, K.: Tumor heterogeneity: causes and consequences. Biochim. Biophys. Acta **1805**(1), 105–117 (2010)
15. McInnes, L., Healy, J., Melville, J.: UMAP: Uniform Manifold Approximation and Projection for Dimension Reduction. arXiv:1802.03426 [cs, stat] (2020)
16. Menyhárt, O., Győrffy, B.: Multi-omics approaches in cancer research with applications in tumor subtyping, prognosis, and diagnosis. Comput. Struct. Biotechnol. J. **19**, 949–960 (2021)
17. Oyelade, J., Isewon, I., Oladipupo, F., Aromolaran, O., Uwoghiren, E., Ameh, F., Achas, M., Adebiyi, E.: Clustering algorithms: their application to gene expression data. Bioinform. Biol. Insights **10**, 237–253 (2016)
18. Planey, C.R., Gevaert, O.: CoINcIDE: a framework for discovery of patient subtypes across multiple datasets. Genome Med. **8**(1), 27 (2016)
19. Radford, A., Metz, L., Chintala, S.: Unsupervised Representation Learning with Deep Convolutional Generative Adversarial Networks. arXiv:1511.06434 [cs] (2016). arXiv: 1511.06434
20. Rawat, K.S., Malhan, I.V.: A hybrid classification method based on machine learning classifiers to predict performance in educational data mining. In: ICCCN, pp. 677–684 (2019)
21. Saito, S., Tan, R.T.: Neural clustering: concatenating layers for better projections. In: ICLR - workshop (2017)
22. Schulz, M.A., Chapman-Rounds, M., Verma, M., Bzdok, D., Georgatzis, K.: Inferring disease subtypes from clusters in explanation space. Sci. R. **10**(1), 1–6 (2020)
23. Sonpatki, P., Shah, N.: Recursive consensus clustering for novel subtype discovery from transcriptome data. Sci. R. **10**(1), 1–6 (2020)
24. Tager-Flusberg, H., Joseph, R.M.: Identifying neurocognitive phenotypes in autism. Philos. Trans. R. Soc. Lond. B Biol. Sci. **358**(1430), 303–314 (2003)
25. Varol, E., Sotiras, A., Davatzikos, C.: HYDRA: revealing heterogeneity of imaging and genetic patterns through a multiple max-margin discriminative analysis framework. Neuroimage **145**, 346–364 (2017)
26. Wang, Y., et al.: Unsupervised machine learning for the discovery of latent disease clusters and patient subgroups using electronic health records. J. Biomed. Inf. **102**, 103364 (2020)
27. Wen, J., Varol, E., Chand, G., Sotiras, A., Davatzikos, C.: MAGIC: multi-scale heterogeneity analysis and clustering for brain diseases. In: MICCAI. LNCS (2020)
28. Wu, M.Y., Dai, D.Q., Zhang, X.F., Zhu, Y.: Cancer subtype discovery and biomarker identification via a new robust network clustering algorithm. PLOS ONE **8**(6), e66256 (2013)
29. Wåhlstedt, C., Thorell, L.B., Bohlin, G.: Heterogeneity in ADHD: neuropsychological pathways, comorbidity and symptom domains. J. Abnorm. Child Psychol. **37**(4), 551–564 (2009)
30. Yang, B., Fu, X., Sidiropoulos, N.D., Hong, M.: Towards K-means-friendly spaces: simultaneous deep learning and clustering. In: International Conference on Machine Learning, pp. 3861–3870. PMLR (2017)
31. Yang, T., et al.: Probing the clinical and brain structural boundaries of bipolar and major depressive disorder. Transl. Psychiatry **11**(1), 1–8 (2021)

32. Yang, Z., Wen, J., Davatzikos, C.: Smile-GANs: Semi-supervised clustering via GANs for dissecting brain disease heterogeneity from medical images. arXiv:2006.15255 (2020)
33. Zabihi, M., Oldehinkel, M., Wolfers, T., Frouin, V., Goyard, D., et al.: Dissecting the heterogeneous cortical anatomy of autism spectrum disorder using normative models. Biol. Psychiatry: Cogn. Neurosci. Neuroimaging **4**(6), 567–578 (2019)

Efficient and Less Centralized Federated Learning

Li Chou, Zichang Liu$^{(\boxtimes)}$, Zhuang Wang, and Anshumali Shrivastava

Department of Computer Science, Rice University, Houston, TX, USA
{lchou,zl71,zw50,anshumali}@rice.edu

Abstract. With the rapid growth in mobile computing, massive amounts of data and computing resources are now located at the edge. To this end, Federated learning (FL) is becoming a widely adopted distributed machine learning (ML) paradigm, which aims to harness this expanding skewed data locally in order to develop rich and informative models. In centralized FL, a collection of devices collaboratively solve a ML task under the coordination of a central server. However, existing FL frameworks make an over-simplistic assumption about network connectivity and ignore the communication bandwidth of the different links in the network. In this paper, we present and study a novel FL algorithm, in which devices mostly collaborate with other devices in a pairwise manner. Our nonparametric approach is able to exploit network topology to reduce communication bottlenecks. We evaluate our approach on various FL benchmarks and demonstrate that our method achieves 10× better communication efficiency and around 8% increase in accuracy compared to the centralized approach.

Keywords: Machine learning · Federated learning · Distributed systems

1 Introduction

The rapid growth in mobile computing on edge devices, such as smartphones and tablets, has led to a significant increase in the availability of distributed computing resources and data sources. These devices are equipped with ever more powerful sensors, higher computing power, and storage capability, which is contributing to the next wave of massive data in a decentralized manner. To this end, federated learning (FL) has emerged as a promising distributed machine learning (ML) paradigm to leverage this expanding computing and data regime in order to develop information-rich models for various tasks. At a high-level, in FL, a collection or federation of devices collaboratively solve a ML problem (i.e., learn a global model) under the coordination of a centralized server. The crucial aspect is to accomplish the task while maintaining the data locally on

L. Chou and Z. Liu—Equal Contribution.

© Springer Nature Switzerland AG 2021
N. Oliver et al. (Eds.): ECML PKDD 2021, LNAI 12975, pp. 772–787, 2021.
https://doi.org/10.1007/978-3-030-86486-6_47

the device. With powerful computing resources (e.g., cloud servers), FL can scale to millions of mobile devices [13]. However, with the continual increase in edge devices, one of the vital challenges for FL is communication efficiency [3,13,17].

Unlike classical distributed ML, where a known architecture is assumed, the structure for FL is highly heterogeneous in terms of computing resources, data, and network connections. Devices are equipped with different hardware, and are located in dynamic and diverse environments. In these environments, network connections can have a higher failure rate on top of varying communication and connection patterns. For example, 5% and more of the devices participating in the single round of training may fail to complete training or completely drop out of communication [13]. The device network topology may evolve, which can be useful information for communication efficiency. In FL, the central server has the critical role of aggregating and distributing model parameters in a back and forth manner (i.e., rounds of communication) with the devices to build and maintain a global model. One natural solution is to have powerful central servers. However, this setup comes with high costs and are only affordable for large corporations [13]. Moreover, overly relying on a central server can suffer from single point failure [32] and communication bottleneck.

An alternative to the client-server architecture is peer-to-peer (P2P) networking. P2P dramatically reduces the communication bottleneck by allowing devices to communicate with one another. Given this insight, we propose a FL framework, which we dub as `FedP2P`, that leverages and incorporates the attributes of a P2P setting. `FedP2P` significantly reduces the role and communication requirements of the central server. Additionally, P2P naturally utilizes network topology to better structure device connections. It is widely accepted that communication delays increase with the node degrees and spectral gap of the network graph [13]. Explicit consideration of network topology increases communication efficiency such as wall-clock time per iteration. Empirically, we show `FedP2P` outperforms the established FL framework, `FedAvg` [21] on a suite of FL benchmark datasets on both computer vision and language tasks, in addition to two synthetic data, involving three different model architectures. With the same number of communication rounds with a central server, `FedP2P` achieves an 8% increase in accuracy. With the same number of devices participating in the training, `FedP2P` can achieve 10× speed up in communication time.

2   Preliminary and Related Work

FL is a distributed ML paradigm, where data reside on multiple devices, and under the coordination of a centralized server, the devices collaboratively solve an ML problem [13]. Moreover, and importantly, data are not shared among devices, but instead, model aggregation updates, via communication between server and devices, are used to optimize the ML objective. Naturally, what we can expect as corollaries, and as key challenges, from this setting are: a massive number of devices, highly unbalanced and skewed data distribution (i.e., not identically and independently distributed, and limited network connectivity

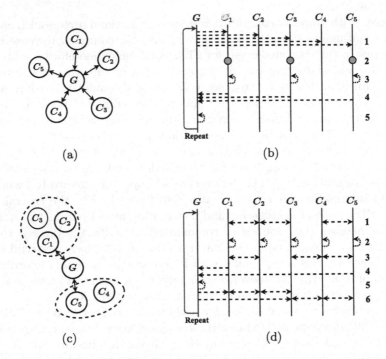

Fig. 1. (a) Structure for centralized FL framework, in which central server directly communicates with all devices for model distribution and aggregation (b) Communication flow for centralized FL (FedAvg). The steps are 1: server G sends global model to clients, 2: G designates client devices for training (C_1, C_3, C_5), 3: selected client devices train on local data, 4: devices send parameters of the trained model back to G, 5: G aggregates the trained parameters, and the process repeats. (b) Structure for our proposal (FedP2P). Dash circle represents local P2P network, in which devices perform pairwise communication. The central server only communicates with a small number of devices, one device within each local P2P network here. (d) Communication flow for FedP2P. The steps are: 1: form P2P network, 2: all devices train in parallel on local data, 3: devices can aggregate/synchronize models via Allreduce, 4: G receives aggregated parameter from each partition (not from all devices), 5: G aggregates the parameters from the partitions to get global parameter, and 6: G sends global parameter to the P2P networks (not to all devices).

(e.g., slow connections and frequent disconnections). In FL, the optimization problem is defined as

$$\min_{\theta} f(\theta) \qquad \text{where} \qquad f(\theta) \triangleq \sum_{i=1}^{N} p_i F_i(\theta). \qquad (1)$$

N is the number of devices, $p_i \in [0, 1]$, and $\sum_{i=1}^{N} p_i = 1$. For supervised classification, we typically have $F_i(\theta) = \ell(X_i, Y_i; \theta)$, where (X_i, Y_i) are the data and

θ are the model parameters. ℓ is some chosen loss function for client devices C_i. We assume data is partitioned across N client devices, which give rise to potentially different data distributions. Namely, given data $\mathcal{D}$ partitioned into samples $D_1, \ldots, D_N$ such that D_i corresponds to C_i, the local expected empirical loss is $F_i(\theta) = \mathbb{E}_{X_i \sim D_i}[f_i(X_i; \theta)]$ where $p_i = |D_i|/|\mathcal{D}|$. D_i varies from device to device and is assumed to be non-IID. The federated averaging (FedAvg) algorithm [21] was first introduced to optimize the aforementioned problem, which we describe next.

2.1 Centralized Federated Learning: Federated Averaging

FedAvg is a straightforward algorithm to optimize Eq. (1). The details are shown in Algorithm 1 and Fig. 1(b). The global objective, orchestrated by a central server G, is optimized locally via the F_i functions applied to the local data on the devices. Let $\mathcal{C} = \{C_1, \ldots, C_N\}$ be the collection of client devices that are willing and available to participate in training a global model for a specific machine learning task. At each training round t, the central server randomly selects a subset $Z \subset \mathcal{C}$ devices to participate in round t of training. G sends to each selected device a complete copy of the global model parameters θ_G^t. Each device trains F_i, via stochastic gradient descent (SGD), initialized with θ_G^t using its own training data D_i for a fixed number of epochs E, batch size O, and according to a fixed learning rate η. Subsequently, the client devices sends the updated parameters $\theta_{C_i}^{t+1}$ back to G, where G performs a model synchronization operation. This process is repeated for T rounds, or until a designated stopping criterion (e.g., model convergence). Unlike traditional distributed machine learning, FedAvg propose to average model parameters (i.e., Aggregate($\cdot$) operation in Algorithm 1) instead of aggregating gradients to dramatically reduce communication cost. FedAvg started an active research field on centralized federated learning, which assumes that all communications occur directly with the central server.

2.2 Multi-model Centralized Federated Learning

Here we focus on related FL settings, where the goal is to learn multiple models using the idea of clustering. These methods mostly utilize the information from model parameters for device clustering. All these methods concentrate on optimization while ignoring practical communication constraints. For example, [5] uses a hierarchical k-means clustering technique based on similarities between the local and global updates. At every round, all devices are clustered according to the input parameter k, which makes this technique not practical in a real-world FL setting. [10] focuses on device clustering. They assume k different data distributions, and the server maintains k different global models. A subset of devices is sampled at each round, and all k models are sent to each of the

---

Algorithm 1. Federated Averaging (FedAvg)

Input: $T, Z, \eta, O, E, \theta_G^0$
for $t = 0$ to $T-1$ **do**
    Server G samples subset of client devices $C_i \in Z$
    and sends θ_G^t to sampled devices: $\theta_{C_i}^t \leftarrow \theta_G^t$
    for each $C_i \in Z$ **in parallel do**
        //Device trains on local data with
        //step size η, batch size O, epochs E
        $\theta_{C_i}^{t+1} \leftarrow \min F_i(\theta_{C_i}^t)$
    end for
    $\theta_G^{t+1} \leftarrow$ Aggregate $\left(\theta_{C_i}^{t+1}, \forall C_i \in Z \right)$
end for

---

devices. The sampled devices then optimize the local objective, using each of the k models, and selects the model with the lowest loss. The updated model with the lowest loss, which also corresponds to the device's cluster identity, is sent back to the server for aggregation based on the clusters. The process is repeated for a given number of rounds. This method can be interpreted as a case of k-means with subsampling. The need to communicate k models to each device creates an information communication bottleneck. We emphasize that our goal is not to solve a clustering problem.

2.3 Decentralized Federated Learning

Decentralized FL is also an active area of research, and we focus on the corresponding FL setting that involves P2P communication in this section. FL utilizing P2P communication has been previous proposed by [15]. However, a graph structure is imposed, and communication is based on the graph structure and limited to one-hop neighbors. Also, graph topology can change over time. The algorithm requires potentially complex Bayesian optimization. In addition, the experiments are limited to 2 nodes.

P2P ML, where the goal is to learn personalized models, as opposed to a single global model, has been proposed by in [32,34] and under strong privacy requirements by [2]. The gossip protocol is a P2P communication procedure based on how epidemics spread [9]. Gossip algorithms are distributed asynchronous algorithms with applications to sensors, P2P, and ad-hoc networks. It has been used to study averaging as an instance of the distributed problem [4], and successfully applied in the area of decentralized optimization [8,14]. Also, several works focus on decentralized FL from the perspective of optimization, decentralized SGD [19,20,28,29] with topology consideration [23]. Note that these works do not consider the non-IID case.

2.4   Efficient Pairwise Communication

Allreduce [26,31] is a collective operation that reduces the target tensors in all processes to a single tensor with a specified operator (e.g., sum or average) and broadcasts the result back to all processes. It is a decentralized operation and only involves P2P communication for the reduction (e.g., model synchronization). Because Allreduce is a bandwidth-optimal communication primitive and is well-scalable for distributed training, it is widely adopted in distributed ML frameworks [25,27].

3   Less Centralized Federated Learning

In this section, we outline our federated learning framework, FedP2P, and provide a detailed discussion of the critical aspects of our design. With FedP2P, we exploit efficient P2P communication in conjunction with a coordinating center that does high-level model aggregation.

3.1   Proposed Framework: FedP2P

The structure of centralized FL follows a star graph, in which the central server directly communicates with all client devices (see Fig. 1(a)). On the contrary, we aim to reorganize the connectivity structure in order to distribute both the training and communication on the edge devices by leveraging P2P communication. To this end, we form L local P2P networks in which client devices perform pairwise communication within each P2P network, which we refer to as FedP2P. With FedP2P, the central server only communicates with a small number of devices, one from each local P2P network (see Fig. 1(c)).

Here, we describe the training process. The details are shown in Algorithm 2 and Fig. 1(d). To better understand our framework, we follow the centralized setup by describing the whole training process in T rounds. We describe the process as three phases for each round as follows.

1. **Form Local P2P Network:** At the start of each round t, the central server randomly partitions N devices into L local P2P networks $Z_1, \ldots, Z_L$, each of size Q. The central servers distribute the global model θ^{t-1} from the previous round to each P2P network. Note that the central server only communicates with one or a few devices from each P2P network for the global model distribution. In practice, L is not a tuning parameter, but can be precisely calculated to minimize communication cost given devices bandwidth and the desired number of total participating devices in each round. We provide more information on the choice of L in Sect. 3.2.
2. **P2P Synchronization:** Within a local P2P network Z_l, part of or all devices train the local P2P network model $\theta^t_{Z_l}$ using its own data in parallel. Once the devices finish training, the model is locally synchronized within Z_l. This is done via an Aggregate($\cdot$) operation, which we define as $\theta_{Z_l^{t+1}} \leftarrow \sum_{C_i \in Z} \gamma_i \theta_{C_i}$, where $\gamma_i = |D_i| / \sum_i |D_i|$. This process can be conducted one or more times,

---

Algorithm 2. Federated Peer-to-Peer (`FedP2P`)

Input: $T, L, Q, \eta, O, E, \theta_G^0$
for $t = 0$ to $T-1$ **in parallel do**
   $Z_1, Z_2, \ldots, Z_L$ `//Form P2P networks`
   for $l = 1$ to L **in parallel do**
      for $C_i \in Z \subseteq Z_l$ s.t. $|Z| = Q$ **in parallel do**
         `//Device trains on local data with`
         `//step size` η`, batch size` O`, epochs` E
         $\theta_{C_i}^{t+1} \leftarrow \min F_i(\theta_{C_i}^t)$
      end for
      $\theta_{Z_l}^{t+1} \leftarrow$ Aggregate $\left(\theta_{C_i}^{t+1}, \forall C_i \in Z\right)$`//Allreduce`
   end for
   $\theta_G^{t+1} \leftarrow$ Aggregate $\left(\theta_{Z_1}^{t+1}, \ldots, \theta_{Z_L}^{t+1}\right)$
end for

---

and we can efficiently accomplish a single P2P network model synchronization using the Allreduce approach. Note that all training and communication inside each local P2P network are conducted independently and in parallel.

3. **Global Synchronization:** The central server G globally aggregates the updated models, $\theta_{Z_l}^{t+1}$ for $l = 1, \ldots, L$, from every local P2P network, and perform model averaging over L models. Namely, $\theta_G^{t+1} \leftarrow L^{-1} \sum_{l=1}^{L} \theta_{Z_l}^{t+1}$. Since each local P2P network is already locally synchronized, G gathers models from one device for each P2P network.

In order to obtain the global model among all P2P networks, our proposed framework also utilizes a central coordinator for global model synchronization. However, note that the communication and workload for the central coordinator in `FedP2P` are significantly reduced. Assume both `FedP2P` and `FedAvg` utilizes P participating devices in a single round of training. The `FedAvg` central server communicates with P devices and perform model synchronization among P models. In contrast, the `FedP2P` central server only communicates with K devices and perform model synchronization among K cluster models where $P \gg K$. From another perspective, we assume the central server has a fixed bandwidth that can be devoted to coordinating federating training. `FedP2P` allows for many more devices to participate within each global round, enabling the global model to train on more data in a single global communication round. `FedP2P` distributes both computation workloads and communication burden to the edge, and we will provide a detailed comparison of communication cost in Sect. 3.2.

Potential privacy and trust problems of `FedP2P` arise from two levels of cooperation: 1) the cooperation between the central server and the device within each local P2P network; and 2) the cooperation within each local P2P network. The first level is the same as standard centralized FL and we expect established solutions such as cryptographic protocols proposed by [3] to work well for `FedP2P`.

The second level follows decentralized FL settings. Existing secure aggregation protocols such as confidential smart contract [13] can be adopted.

3.2 Communication Efficiency

The central server bandwidth becomes the performance bottleneck in FedAvg when there are a large number of sampled devices. FedP2P alleviates this server-device bottleneck through decentralization. Namely, the server only needs to communicate with a subset of the sampled devices, which we refer to as agents. However, it incurs additional communication overhead inside P2P networks because the agents have to communicate with other devices. FedP2P needs to balance the trade-off between the server-agent and agent-device communication overhead.

We model and analyze the communication cost of FedAvg and FedP2P in this section. To simplify the analysis, we define B_d as the bandwidth between each device-device pair for the training, B_s as the total bandwidth capacity from the server to the devices (i.e., uplink bandwidth of server), and M as the model size. We assume all P2P networks have the same P number of devices participating in one single round of model training.

Communication Efficiency of FedAvg: There are two main steps in the data transmission: 1) the model distribution from the server to the sampled devices has the communication time of MP/B_s; and 2) the model aggregation from the devices to the server has the communication time of $\alpha MP/B_s$, where $1/\alpha B_s$ is the total bandwidth capacity from the devices to the server (i.e., downlink bandwidth of server). Note that $\alpha \geq 1$ since the upload bandwidth of devices is typically lower than their download bandwidth. Let H_{avg}, denoting the communication time in FedAvg, be $H_{avg} = (1 + \alpha)MP/B_s$.

Communication Efficiency of FedP2P: There are four main steps in the data transmission: 1) the model distribution from the server to the agent in each cluster has the communication time of LM/B_s; 2) the model distribution from the agent to the sampled devices in each P2P network has communication time of PM/LB_d, where P/L is the number of sampled devices in each P2P network; 3) local model synchronization at the end of each local training round has the communication time of $2M/B_d$ (the exact communication time of Allreduce is $\frac{2(n-1)M}{nB_d}$, where n is the number of workers); and 4) global model synchronization at the end of each global training round has the communication time of $\alpha LM/B_s$. Let H_{p2p} denote the communication time in FedP2P be defined as

$$H_{p2p} = \frac{(1+\alpha)LM}{B_s} + \frac{PM}{LB_d} + \frac{2M}{B_d}.$$

The choice of L is a way to balance the trade-off between the server-agent and agent-device communication overhead. A small L results in the low server-agent communication overhead because the server just communicates with a small number of agents, while it increases the agent-device communication time

780   L. Chou et al.

because each agent needs to broadcast the model to more devices; and vice versa. H_{p2p} reaches its minimum: $\min H_{p2p} = \frac{2M}{B_d}(\frac{P}{L} + 1)$ when $L = A\sqrt{P}$, where $A = \sqrt{\frac{B_s}{(1+\alpha)B_d}}$ is a constant in a federated learning system. Define $R = \frac{H_{avg}}{\min H_{p2p}}$ and $\gamma = \frac{B_s}{B_d}$. Then

$$R = \frac{(1+\alpha)P}{2\sqrt{\gamma(1+\alpha)P} + 2\gamma}. \tag{2}$$

$R > 1$ indicates that FedP2P has lower communication overhead than FedAvg. Equation (2) shows that the value of R increases with 1) the increasing number of sampled devices; 2) the decreasing gap between the bandwidth capacities of the server and devices; and 3) the increasing gap between the server's uplink and downlink bandwidth capacities.

3.3   Theoretical Insight

In this section, we provide some theoretical insight into the performance of FedP2P. We instantiate the framework of [18] (i.e., Theorems 1, 2, and 3; for spacing, we omit the details) to simplify the analysis. Similarly, let F^* and F_i^* be the minimal values of F and F_i. Let $\Gamma = F^* - \sum_{i=1}^{N} p_i F_i^*$. The magnitude of Γ quantifies the degree of non-IID (i.e., heterogenity) such that Γ goes to zero as the number of non-IID samples decrease. We also apply the same assumptions on smoothness, convexity, and bounds on variance and norm of the stochastic gradient. Notation-wise, we substitute $\theta_G = \mathbf{w}$, $J = C$, $U = L$ as the Lipschitz constant, and $V^2 = G^2$ to uniformly bound the expected squared norm of stochastic gradients from the top. First, FedP2P satisfies the following base case.

Corollary 1. *Let $\sum_{l=1}^{L} |Z_l| = N$ and the aggregation at the server G be defined as $\theta_G^t \leftarrow \sum_{l=1}^{L} \psi_l \theta_{Z_l}^t$ such that $\psi_l = |\mathcal{D}|^{-1} \sum_{C_i \in Z_l} |D_i|$. Then the same bound in Theorem 1 [18] holds for FedP2P.*

Corollary 1 is restrictive since we cannot hope to compute all ψ_l, and the total number of devices participating varies due to the stragglers effect in practice. However, we next analyze Theorems 2 and 3 under the FedP2P scheme at the level of server G, which highlights an advantage of FedP2P. To ground the analysis, under the special case of one device P2P network, we have the following.

Corollary 2. *Let $\sum_{l=1}^{L} |Z_l| < N$ and $|Z_l| = 1$, $\forall l$. Then the same bound in Theorem 2 and 3 [18] holds for FedP2P.*

Here we focus on the J term in Theorems 2 and 3. From Theorem 2, we have $J = 4K^{-1}E^2V^2$, where K is number of devices for aggregation. However, note that with FedP2P, effectively, we have $J = 4(K\sum_l |Z_l|)^{-1}E^2V^2$ at G. Therefore, J is reduced by a factor of $\sum_l |Z_l|$ and thus, improves the bound. Similarly, from Theorem 3, we have $J = 4(N-1)^{-1}K^{-1}(N-K)E^2V^2$. However, at G, the terms $(N - K)$ shrinks, and K^{-1} grows significantly under FedP2P since K increases by a factor of $\sum_l |Z_l|$, thereby also improves the bound.

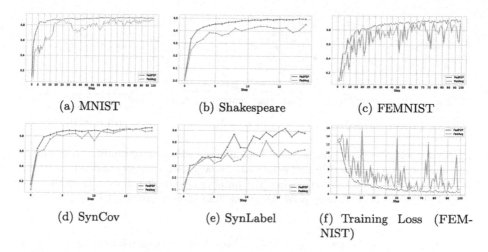

(a) MNIST   (b) Shakespeare   (c) FEMNIST

(d) SynCov   (e) SynLabel   (f) Training Loss (FEM-
                                                            NIST)

Fig. 2. (a), (b), (c), (d), and (e) are average testing accuracy across devices as a function of global communication rounds. FedP2P shows higher and much smoother accuracy curves on all real datasets and two synthetic datasets. (f) plots the loss during training on FEMNIST. FedP2P gives a smooth convergence curve.

4   Experiments

In this section, we present the experiment results that demonstrate the performance between FedP2P and FedAvg from various perspectives. Specifically, we focus on answering the following questions: 1) how does FedP2P perform compared to centralized methods on model accuracy? 2) does FedP2P outperform centralized federated learning in terms of communication efficiency? 3) how robust is FedP2P handling network instability such as stragglers? 4) how does FedP2P perform with different parameters?

4.1   Datasets

Synthetic Datasets: In FL, the data on devices are non-IID. Following the taxonomy described in [13], non-IID is decomposed into five regimes: 1) covariance shift, 2) label probability shift, 3) same label with different features, 4) same features with different labels, and 5) quantity skew. To effectively explore and understand the various non-IID situations, we create two synthetic datasets: SynCov and SynLabel. The setup is similar to [17]. SynCov simulates covariance shift with quantity skew, and SynLabel simulates label probability shift with quantity skew. The high-level data generation process is as follows. First, we designate $N(=100)$ number of client devices C_i. Next, we sample from a lognormal distribution to determine the number of data points for each C_i. Let X be

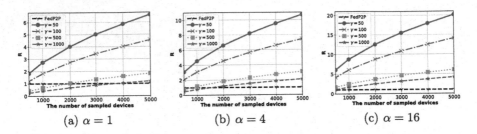

(a) $\alpha = 1$ (b) $\alpha = 4$ (c) $\alpha = 16$

Fig. 3. Numerical comparison between `FedP2P` and `FedAvg` on normalized communication time. Black line is `FedP2P` and others are `FedAvg` with various bandwidth settings. `FedAvg` outperforms `FedP2P` only when the number of sampled devices is small or the device bandwidth is extremely poor. Detailed discussion is in Sect. 4.4.

the features and Y be the class labels, where feature dimension is 60 and number of classes is 10. Then, we sample (X,Y) from the data distribution $P_i(X,Y)$ for C_i. The non-IID data refers to differences between $P_i(X,Y)$ and $P_j(X,Y)$ for two different client devices C_i and C_j. Note that we can factor $P_i(X,Y)$ as $P_i(Y|X)P_i(X)$ or $P_i(X|Y)P_i(Y)$. Using this factorization, the details of the data generation process is as follows.

- **SynCov**: $P_i(X)$ varies and $P(Y|X)$ shared among client devices C_i. We parameterize $P_i(X)$ with a Gaussian distribution $\mathcal{N}(\mu_i,\sigma_i)$ and $P(Y|X)$ with a softmax function, which has weight and bias parameters W and b respectively. First, we sample $W,b \sim \mathcal{N}(0,1)$. Then, for each C_i, we sample $\mu_i,\sigma_i \sim \mathcal{N}(0,1)$ and sample features $x_i \sim P_i(X)$. We obtain y_i from $\arg\max_{y\in Y} P(Y|X=x_i)$.
- **SynLabel**: $P_i(Y)$ varies and $P(Y|X)$ shared among client devices i. Given the number classes $|Y|$, we create a discrete multinomial distribution sampled from a Dirichlet distribution $\mathrm{Dir}(\beta_1,\ldots,\beta_{|Y|})$, where $\beta_k > 0$. We repeat this for each C_i. Then, for each $y \in Y$, we sample $\mu_y,\sigma_y \sim \mathcal{N}(0,1)$ to parameterize $P(Y|X)$ for all i. For each client device we apply logical sampling [11] to obtain $y_i \sim P_i(Y)$ and $x_i \sim P(X|Y=y_i)$ accordingly.

Real-world Datasets: We evaluate on three standard FL benchmark datasets: MNIST [16], Federated Extended MNIST (FEMNIST) [7], and *The Complete Works of William Shakespeare* (Shakespeare) [21]. These datasets are curated from recently proposed FL benchmarks [6,30]. We use the data partition provided by [17]. For MNIST, the data is distributed, via power law, across 1,000 devices and each device has samples of 2 classes. FEMNIST is a 62 image classification dataset, including both handwritten digits and letters. There are 200 devices and 10 lowercase letters ('a' to 'j') is subsampled such that each device has samples for 5 classes. Shakespeare is used for the next character prediction task, consisting of lines spoken by different characters (80 classes) in the plays.

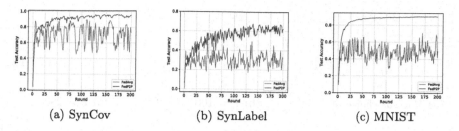

(a) SynCov   (b) SynLabel   (c) MNIST

Fig. 4. Comparison on average test accuracy across clients between FedAvg and FedP2P with 50% Stragglers. FedP2P achieves similar accuracy as when there are no stragglers. However, FedAvg suffers from straggler. FedAvg achieves much lower accuracy. Besides, FedP2P shows a much smoother curve. We can observe accuracy jump up to 20% between two rounds for FedAvg.

Table 1. Best test accuracy on various datasets for FedP2P and FedAvg.

Dataset	FedP2P	FedAvg	Dataset	FedP2P	FedAvg
MNIST	**0.9092**	0.8763	SynCov	**0.9252**	0.9032
FEMNIST	**0.9168**	0.8929	SynLabel	**0.6199**	0.5149
Shakespeare	**0.5019**	0.4563			

4.2 Implementation

We consider centralized FL as our comparison and implemented FedAvg and FedP2P in PyTorch [25]. We evaluate using various model architectures, both convex and non-convex. We use logistic regression for synthetic and MNIST, Convolution Neural Network for FEMNIST, and LSTM classifier for Shakespeare. In order to draw a fair comparison, we use the same set of models and parameters for FedAvg and FedP2P. Specifically, we use 2-layer CNN with a hidden size of 64 and 1-layer LSTM with a hidden size of 256. Models are trained using SGD. ReLU is used as the activation function. Data is split 80% train and 20% test. We use batch size of 10. We use learning rates of .01 for synthetic, MNIST, FEMNIST, and .5 for Shakespeare based on grid search on FedAvg and 20 epochs. Number of devices selected to train per round is fixed to 10. Code located at github.com/lchou/fedp2p.

4.3 Model Accuracy

In this section, we compare FedP2P and FedAvg in terms of test accuracy. We hold test data for each device and follow the standard evaluation metric of classification accuracy. As shown in Table 1, FedP2P outperforms FedAvg on all datasets. Figure 2 shows the test accuracy as a function of the total round of communication with the central server. Here, we control the number of global communication to be the same for both methods. For a fair comparison, we only let one round of training within each local P2P network.

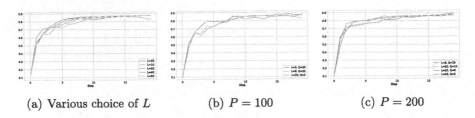

<table>
<tr><td>(a) Various choice of L</td><td>(b) $P = 100$</td><td>(c) $P = 200$</td></tr>
</table>

Fig. 5. Test accuracy on MNIST for `FedP2P` on various parameter settings. (a) `FedP2P` with various numbers of local P2P network. Different choices of L do not affect model performance with `FedP2P`, and thus we can set L to optimize for communication efficiency. (b) and (c) `FedP2P` with various number of total participating devices. `FedP2P` is also robust with diverse choices of L and Q.

On FEMNIST, `FedP2P` achieved 2.6% increase in test accuracy. On MNIST, `FedP2P` achieved 3.7% increase in test accuracy. On Shakespeare, `FedP2P` achieved 9% increase in test accuracy. We see that `FedP2P` achieves higher testing accuracy on every global communication from Fig. 2. Compared to `FedAvg`, `FedP2P` gives a rather smooth accuracy curve. In Fig. 2(f), we provide a plot for training loss on FEMNIST. It worth noting that `FedP2P` also gives a smooth convergence curve. These results demonstrate that `FedP2P` achieves higher model accuracy compared to the centralized approach.

4.4   Communication Efficiency

We numerically compare `FedP2P` and `FedAvg` in terms of the communication efficiency based upon the analysis in Sect. 3.2. The ratio of the download bandwidth to the upload bandwidth of devices varies with different Internet providers, and it covers a wide range. We set the ratio α to $\{1, 4, 16\}$ as in [1]. The number of sampled devices for one round of training is within $[500, 5000]$ [13]. Another parameter that determines the value of R in Eq. (2) is γ, which is the ratio of the server bandwidth to the device bandwidth. Current edge devices have the bandwidth of more than 20 Mbps (e.g., 4K streaming [24]). Also, the advent of 5G enables much higher bandwidth for edge devices [33]. The server bandwidth typically ranges from 10 Gbps to less than 1 Gbps [12]. Therefore, we set γ in the range of $[50, 1000]$ in our simulation.

Figure 3 shows the comparison of the communication time. With $\gamma = 100$, `FedP2P` always outperforms `FedAvg` with the number of sampled devices larger than 500. `FedP2P` can achieve better performance than `FedAvg` with more sampled devices, smaller γ, or larger α. However, if the number of sampled devices is small (e.g., $P < 100$), or if the device bandwidth is extremely poor (e.g., $\gamma > 1000$), then `FedAvg` can potentially outperform `FedP2P` in terms of the communication time because the performance bottleneck is not the server in these cases. In practice, thousands of devices are sampled for the training in each round, and the sampled devices usually have high bandwidth due to the sampling mechanism [13]. For example, the server tends to select devices with powerful computing capacity and high bandwidth to avoid stragglers. As a result, we argue that FedP2P is more scalable and communication-efficient than FedAvg for large-scale FL.

4.5 Stragglers Effect and Choice of L and Q

Stragglers: It is known that FL suffers from stragglers. Specifically, devices fail in on-device training or drop connection due to hardware issues or communication instability. Stragglers cause convergence problems and lead to models with poor performance. We empirically test how robust FedP2P is in the straggler situation in terms of model performance, and we present our result in Fig. 4. We drop 50% of the selected devices to simulate stragglers. FedP2P performs exceptionally well with stragglers. FedP2P archives similar accuracy while the performance of FedAvg dramatically drops. Besides, comparing to FedAvg, accuracy curves are still relatively smooth for FedP2P. For example, the most significant jump on MNIST for FedAvg can exceed over 20%, while we do not observe a noticeable increase with FedP2P.

L and Q: We conduct two sets of experiments on MNIST to illustrate how FedP2P performs under various parameters. FedP2P introduces L local P2P networks, and within each P2P network, Q devices participated in the training. We conduct the first set of experiments with varying L and the same Q and present the result in Fig. 5(a). We do not observe significant differences among various L. Compared with FedAvg from Fig. 2(a), we notice that all FedP2P plots lies above FedAvg. Thus, we conclude that FedP2P is robust to various L setting from the perspective of convergence and accuracy. As a result, in practice, FedP2P allows us to choose L to optimize for communication efficiency. We conduct the second set of experiment with various combination of L and Q, where $P = L \times Q$ and $P = 100$ and 200. We present the result the in Fig. 5(b) and (c). We observe that different combinations of L and Q has negligible effect on classification performance.

5 Conclusion

In this work, we focus on decentralizing FL and propose FedP2P, an approach that utilizes peer communication to train one global model collectively. A unique possibility due to our random partition process is that it allows us to exploit device network topology. If we assume that the data distribution is independent of the network evolving topology, then a random selection of devices to form local P2P networks is identical with any deterministic selection (i.e., Principle of deferred decisions [22]). Effectively, we can partition devices into local P2P networks based on favorable network topology. For example, it is widely accepted that long communication hops create potential problems such as low throughput and high latency due to network congestion. Grouping devices based on communication hops would greatly benefit communication efficiency. For example, in Fig. 1(c), devices within fewer communication hops are grouped into the same local P2P network.

Acknowledgements. This work was supported by National Science Foundation IIS-1652131, BIGDATA-1838177, AFOSR-YIP FA9550-18-1-0152, ONR DURIP Grant, and the ONR BRC grant on Randomized Numerical Linear Algebra.

References

1. Allconnect (2020). https://bit.ly/2SFCrOH
2. Bellet, A., Guerraoui, R., Taziki, M., Tommasi, M.: Personalized and private peer-to-peer machine learning. In: AISTATS (2018)
3. Bonawitz, K., et al.: Towards federated learning at scale: System design. SysML (2019)
4. Boyd, S., Ghosh, A., Prabhakar, B., Shah, D.: Gossip algorithms: design, analysis and applications. In: IEEE Infocomm (2005)
5. Briggs, C., Fan, Z., Andras, P.: Federated learning with hierarchical clustering of local updates to improve training on non-IID data. In: IEEE WCCI 2020 (2020)
6. Caldas, S., et al.: Leaf: a benchmark for federated settings (2019)
7. Cohen, G., Afshar, S., Tapson, J., van Schaik, A.: EMNIST: an extension of MNIST to handwritten letters (2017)
8. Colin, I., Bellet, A., Salmon, J., Clémençon, S.: Gossip dual averaging for decentralized optimization of pairwise functions. In: ICML (2016)
9. Demers, A., et al.: Epidemic algorithms for replicated database maintenance. In: Proceedings of the Sixth Annual ACM Symposium on Principles of Distributed Computing (1987)
10. Ghosh, A., Chung, J., Yin, D., Ramchandran, K.: An efficient framework for clustered federated learning. In: NeurIPS (2020)
11. Henrion, M.: Propagation of uncertainty in Bayesian networks by probabilistic logic sampling. In: UAI (1986)
12. Hsieh, K., et al.: Gaia: geo-distributed machine learning approaching LAN speeds. In: 14th USENIX Symposium on Networked Systems Design and Implementation (NSDI 17), pp. 629–647 (2017)
13. Kairouz, P., et al.: Advances and open problems in federated learning. arXiv:1912.04977 (2019)
14. Koloskova, A., Stich, S.U., Jaggi, M.: Decentralized stochastic optimization and gossip algorithms with compressed communication. In: ICML (2019)
15. Lalitha, A., Kilinc, O.C., Javidi, T., Koushanfar, F.: Peer-to-peer federated learning on graphs. arXiv:1901.11173 (2019)
16. Lecun, Y., Bottou, L., Bengio, Y., Haffner, P.: Gradient-based learning applied to document recognition. Proc. IEEE **86**(11), 2278–2324 (1998)
17. Li, T., Sahu, A.K., Zaheer, M., Sanjabi, M., Talwalkar, A., Smith, V.: Federated optimization in heterogeneous networks. MLSys (2020)
18. Li, X., Huang, K., Yang, W., Wang, S., Zhang, Z.: On the convergence of FedAvg on non-IID data. In: ICLR (2020)
19. Lian, X., Zhang, C., Zhang, H., Hsieh, C.J., Zhang, W., Liu, J.: Can decentralized algorithms outperform centralized algorithms? A case study for decentralized parallel stochastic gradient descent. In: NeurIPS (2017)
20. Lian, X., Zhang, W., Zhang, C., Liu, J.: Asynchronous decentralized parallel stochastic gradient descent. In: ICML (2018)
21. McMahan, H.B., Moore, E., Ramage, D., Hampson, S., Arcas, B.A.: Communication-efficient learning of deep networks from decentralized data. AISTATS (2017). ([v1] arXiv:1602.05629, 2/2016)

22. Mitzenmacher, M., Upfal, E.: Probability and Computing: Randomized Algorithms and Probabilistic Analysis. Cambridge University Press, Cambridge (2005)
23. Neglia, G., Xu, C., Towsley, D., Calbi, G.: Decentralized gradient methods: does topology matter? In: AISTATS (2020)
24. Netflix (2020). https://help.netflix.com/en/node/306?rel=related
25. Paszke, A., et al.: PyTorch: an imperative style, high-performance deep learning library. In: NeurIPS (2019)
26. Patarasuk, P., Yuan, X.: Bandwidth optimal all-reduce algorithms for clusters of workstations. J. Parallel Distrib. Comput. **69**(2), 117–124 (2009)
27. Sergeev, A., Del Balso, M.: Horovod: fast and easy distributed deep learning in TensorFlow. arXiv preprint arXiv:1802.05799 (2018)
28. Stich, S.U.: Local SGD converges fast and communicates little. arXiv:1805.09767 (2018)
29. Tang, H., Lian, X., Yan, M., Zhang, C., Liu, J.: D^2: decentralized training over decentralized data. In: ICML (2018)
30. TensorFlow-Federated (2020). https://www.tensorflow.org/federated
31. Thakur, R., Rabenseifner, R., Gropp, W.: Optimization of collective communication operations in MPICH. Int. J. High Perform. Comput. Appl. **19**(1), 49–66 (2005)
32. Vanhaesebrouck, P., Bellet, A., Tommasi, M.: Decentralized collaborative learning of personalized models over networks. In: AISTATS (2017)
33. Xu, D., et al.: Understanding operational 5G: a first measurement study on its coverage, performance and energy consumption. In: SIGCOMM (2020)
34. Zantedeschi, V., Bellet, A., Tommasi, M.: Fully decentralized joint learning of personalized models and collaboration graphs. In: AISTATS (2020)

Topological Anomaly Detection in Dynamic Multilayer Blockchain Networks

D. Ofori-Boateng[1]([✉]), I. Segovia Dominguez[2], C. Akcora[3], M. Kantarcioglu[2], and Y. R. Gel[2]

[1] Portland State University, Portland, USA
dorcas.oforiboateng@pdx.edu
[2] University of Texas at Dallas, Richardson, USA
[3] University of Manitoba, Winnipeg, Canada

Abstract. Motivated by the recent surge of criminal activities with cross-cryptocurrency trades, we introduce a new topological perspective to structural anomaly detection in dynamic multilayer networks. We postulate that anomalies in the underlying blockchain transaction graph that are composed of multiple layers are likely to also be manifested in anomalous patterns of the network shape properties. As such, we invoke the machinery of clique persistent homology on graphs to systematically and efficiently track evolution of the network shape and, as a result, to detect changes in the underlying network topology and geometry. We develop a new persistence summary for multilayer networks, called stacked persistence diagram, and prove its stability under input data perturbations. We validate our new topological anomaly detection framework in application to dynamic multilayer networks from the Ethereum Blockchain and the Ripple Credit Network, and demonstrate that our stacked PD approach substantially outperforms state-of-art techniques.

Keywords: Anomaly detection · Dynamic multilayer network · Blockchain transaction · Topological data analysis · Clique persistent homology

1 Introduction

Due to the recent spike in popularity of crypto assets, detecting anomalies in time evolving blockchain transaction networks has gained a new momentum. Here anomaly detection in dynamic graphs can be broadly defined as the problem of identifying instances within a sequence of graph observations where changes occur in the underlying structure of the graph. Indeed, these anomalies have significant implications, ranging from emergence of new ransomware (e.g., collecting ransom via cryptocurrencies) to financial manipulation. For example, in

Electronic supplementary material The online version of this chapter (https://doi.org/10.1007/978-3-030-86486-6_48) contains supplementary material, which is available to authorized users.

© Springer Nature Switzerland AG 2021
N. Oliver et al. (Eds.): ECML PKDD 2021, LNAI 12975, pp. 788–804, 2021.
https://doi.org/10.1007/978-3-030-86486-6_48

blockchain transaction networks, e.g., Ethereum, more frequent than expected appearance of particular subgraphs may indicate newly emerging malware or price pump-and-dump trading [55]. Similarly, as recently shown by [53], the flow of coins on the Bitcoin graph provides important insights into money laundering schemes. As criminal, fraudulent, and illicit activities on blockchains continue to rise, with already stolen $1.4B only in 2020, cryptocurrency criminals increasingly employ cross-cryptocurrency trades to hide their identity [41]. As such, [57] have recently shown that the analysis of links across multiple blockchain transaction graphs is critical for identifying emerging criminal and illicit activities on blockchain. However, while there exists a plethora of methods for network anomaly detection in single layer networks [21,44,45], there is yet *no single* method designed to detect anomalies in dynamic multilayer networks.

Why TDA? Motivated by the problem of tracking financial crime on blockchains, we develop a state-of-the-art methodology for anomaly detection on multilayer networks using Topological Data Analysis (TDA). Since crime on blockchains such as money laundering tends to involve multiple parties who possibly move funds across multiple cryptocurrency ledgers, one of our primary goals is to identify anomalous patterns in higher order graph connectivity. We postulate that anomalous higher order patterns can be detected using geometric and topological inference on graphs, that is, via a systematic analysis of the graph shape. To explore latent graph shape, we invoke the TDA machinery of the clique persistent homology (PH). PH allows to systematically infer qualitative and quantitative multi-lens geometric and topological structures from data directly and, hence, to enhance our understanding on the hidden role of geometry and topology in the system organization [9,12,52]. As a result, it may be intuitive to hypothesize that there shall be an intrinsic linkage between changes in the underlying graph structure and changes in the network shape which are then reflected in the extracted network topological characteristics. However, to the best of our knowledge, this paper is the first attempt to introduce TDA to anomaly detection in dynamic multilayer networks.

Why Ethereum and Ripple? Using the Blockchain global events timeline [54], we validate our methodology in application to anomaly detection in two multilayer blockchain network types, Ethereum and Ripple. While cryptocurrencies have already been adopted in payments, the recent surge in financial blockchain activity is largely due to platforms, such as Ethereum, which have brought algorithmic trading of digital assets by using Smart Contracts (i.e. short software code on the blockchain) in what is called Decentralized Finance [15]. Assets include cryptocurrencies and crypto tokens as well. Hence, a given address (i.e. a node) may participate in transactions of multiple digital assets. Looking at an individual asset transaction network alone (i.e. a single layer of the transaction graph) may provide a limited view. As a result, we need to consider multiple layers (e.g., a layer for each crypto token) and their interactions to detect anomalies. Resulting multilayer networks and participant activities are temporal, nuanced in the traded assets (e.g., coins, or fiat currencies), rich in network patterns and encode a new wave of financial heart-beat. The Ripple Credit Network transactions also comprise cross-border remittance transfers and even fiat currency trades, allowing trading Ether, Bitcoin and other currencies on its system.

Our contributions, both in application and theory, are as follows:

1 To the best of our knowledge, this is the first paper on anomaly detection in dynamic multilayer networks.
2 Our new methodology is based on the notion of clique persistent homology. To quantify topology of multilayer graphs, we introduce a multidimensional multi-set object, called the *stacked persistence diagram* (SPD). We prove that SPD is robust against minor input data perturbations w.r.t. bottleneck distance.
3 In the absence of the state-of-the-art anomaly detection methods for dynamic multilayer networks, we benchmark our topological anomaly detection (TAD) tool against a multiple testing framework, based on the strongest state-of-the-art (SOTA) methods for anomaly detection in single layer networks. To control for family wise error rate (FWER) in the multiple testing framework, we use Bonferroni correction. We show that TAD substantially outperforms all competitors based on SOTA single layer solutions and the additional technique based on graph embedding.
4 We demonstrate utility of TAD on Ethereum and Ripple blockchains, where digital assets worth billions of US Dollars are traded daily. We provide Blockchain benchmark data for anomaly detection on multilayer networks which is the first benchmark multilayer network dataset with ground-truth events, thereby further bridging AI with crypto-finance.

2   Related Work

Graph-Based Anomaly Detection: Over recent years, there has been an increase in application of anomaly detection techniques for single layer graphs in interdisciplinary studies [20,58]. For example, [31] employed a graph-based measure (DELTACON) to assess connectivity between two graph structures with homogeneous node/edge attribution, and identified anomalous nodes/edges in the sequence of dynamic networks based on similarity deviations. With DELTACON, an event is flagged as anomalous if its similarity score lies below a threshold. In turn, [51] devised a likelihood maximization tool that extracts a "feature" vector from individual networks, and uses dissimilarity between successive networks snapshots to classify anomalous or normal/regular events. Procedure of [62] segments network snapshots into separate clusters, infers local and global structure from individual nodes and their distribution via community detection and chronological ordering of the results in an effort to single-out potential anomalies. An online algorithm for detecting abrupt edge weight and structural changes in dynamic graphs has been recently introduced by [56], but the method requires a pre-training data set to identify tuning parameters. In turn, [6,36,47] discuss detection of malicious nodes in multiplex/multilayer networks. Finally, [18] proposed a score test for change point detection in multilayer networks that follow a multilayer weighted stochastic block model (SBM). However, the SBM assumption is infeasible for financial networks. To our knowledge (see also the reviews by [21,44]), *there is no existing anomaly detection method designed for dynamic multilayer networks.*

Blockchain: Blockchain graphs have been extracted and analyzed for price prediction [1,25,32], measurement studies [33,50] and e-crime detection [3,14]. Graph anomalies have been tracked to locate coins used in illegal activities, such as money laundering and blackmailing [43]. These findings are known as taint analysis [17]. Typically, a set of features are extracted from the blockchain graph and used in Machine Learning (ML) tasks. Here we bypass such a feature engineering step in learning on Blockchain networks. Ethereum structure has been analyzed by [22,33], while anomalies in Ethereum token prices have been evaluated using TDA tools [35]. In turn, Ripple has been assessed for its privacy aspects [39] and for health of the credit network [38]. However, multilayer analysis of blockchains have not been studied before.

TDA: Multiple recent papers show utility of TDA for developing early warning signals for crashes in the cryptocurrency market [24], cryptocurrency price analytics [35], and ransomware detection on blockchain transaction graphs [3]. While TDA (as any other tool) cannot be viewed as a universal solution, TDA allows us to assess graph properties which are invariant under continuous deformations; hence it is likely to be one of the most robust tools for blockchain data analytics [60]. TDA has been employed for visual detection of change points in single layer graphs [27]. In the multilayer network context, TDA has been used primarily for centrality ranking [48], including analysis of connectivity in the multiplex banking networks [16], and clustering [59]. Application of TDA to anomaly detection in multilayer networks is yet an unexplored area.

Multilayer Network Benchmark Data: Multilayer networks receive an increasing attention in the last few years, due to their flexibility of modeling interconnected systems [4]. There also exist several data repositories with multilayer graphs, e.g. [5,19], but neither of them have publicly available benchmark data on multilayer graphs with ground truth for anomaly detection.

3   The Mechanism of Persistent Homology

Topology is the study of shapes. TDA and, in particular, *persistent homology* (PH) provides systematic mathematical means to extract the intrinsic shape properties of the observed data $\mathcal{X}$ (in our case $\mathcal{X}$ is a multilayer graph but $\mathcal{X}$ can be a point cloud in Euclidean or any finite metric space) that are invariant under continuous transformations. The key postulate is that $\mathcal{X}$ are sampled from some metric space $\mathcal{M}$ whose properties are lost due to sampling. The goal of PH is then to reconstruct the unknown topological and geometric structure of $\mathcal{M}$, based on systematic shape analysis of $\mathcal{X}$. In this paper, we introduce the PH concepts to analysis of dynamic multilayer networks, starting by providing background on PH on graphs.

Definition 1. *Let* $\mathcal{G} = (V, E, \omega)$ *be a (weighted) graph, with vertex set* V, *edge set* $E = \{e_1, e_2, \ldots\} \subseteq V \times V$, *edge weights* $\omega = \omega(e) : E \to \mathbb{Z}^+$ *for all* $e \in E$.

At the initial stages of PH, we select a certain threshold $\nu_* > 0$, and then we generate a subgraph $\mathcal{G}_* = (V, E_*, \omega_*)$, such that $E_* = \{e \mid \omega(e) \leq \nu_*\}$, and $\omega_*(e) = \omega(e)$, for all $e \in E_*$. Then the observed graph $\mathcal{G}_*$ is equipped with a basic combinatorial object known as an *abstract simplicial complex*. Formally, a simplicial complex is defined as a collection $\mathcal{C}$ of finite subsets of $V(\mathcal{G})$ such that if $\sigma \in \mathcal{C}$ then $\tau \in \mathcal{C}$ for all $\tau \subseteq \sigma$. The basic unit of simplicial complexes is called the *simplex*, and if $|\sigma| = m + 1$ then σ is called an m-simplex. Specific to our analysis, we use a simplicial complex type called the *clique complex* to systematically and efficiently extract topological features from the observed $\mathcal{G}$. A clique complex $\mathcal{C}(\mathcal{G}_*)$ is a simplicial complex with a simplex for every clique (i.e., a set of vertices of $\mathcal{G}_*$ such that any two points in the clique are adjacent) in $\mathcal{G}_*$. Furthermore, a k-clique community is formed whenever two k-cliques share $k - 1$ vertices ($k \in \mathbb{Z}^+$). With a range of thresholds $\nu_1 < \ldots < \nu_n$, we can obtain a hierarchically nested sequence of graphs $\mathcal{G}_1 \subseteq \ldots \subseteq \mathcal{G}_n$ for any graph $\mathcal{G}$, where each individual subgraph will generate its own clique complex. Subsequently, the procedure which generates complexes from the nested sequence $\mathcal{G}_1 \subseteq \ldots \subseteq \mathcal{G}_n$ is known as the *network filtration*, and the resultant complex generated by $\mathcal{G}$ is called a *filtered complex* [63]. Particular to cliques, we construct clique complexes and then obtain the *clique filtration* $\mathcal{C}(\mathcal{G}_1) \subseteq \ldots \subseteq \mathcal{C}(\mathcal{G}_n)$.

The mechanism of *clique persistent homology* involves tracking clique complexes over the filtration and quantifying lifespan of topological features/shapes such as loops, holes, and voids that appear and disappear at various thresholds ν_* [46,64]. We say that a topological feature is born at the i-th filtration step if it appears in $\mathcal{C}(\mathcal{G}_i)$, and the topological feature dies at the j-th filtration step, if it disappears at $\mathcal{C}(\mathcal{G}_j)$. Hence, the lifespan of a topological feature is $\nu_j - \nu_i$. The primary objective of TDA will then be to assess which topological features/shapes persist (i.e. have longer lifespan) over the clique filtration and, hence, are likelier to contain important structural information on the graph, and which topological features have shorter lifespan. The latter features are typically referred to as topological noise.

One of the most widely used topological summaries is the *persistence diagram* (PD) [9,63]. The PD is a collection of points $(v_i, v_j) \in \mathbb{R}^2$ with each point corresponding to a topological feature, and the x- and y-coordinates representing birth and death times for the topological feature. Similarity between any two PDs, D_a and D_b, can be computed using the Wasserstein (W_r) or the Bottleneck distances (W_∞):

$$W_r(D_a, D_b) = \left(\inf_{\eta} \sum_{x \in D_a} \|x - \eta(x)\|_\infty^r\right)^{1/r}, \quad W_\infty(D_a, D_b) = \inf_{\eta} \sup_{x \in D_a} \|x - \eta(x)\|_\infty.$$

Here $r \geq 1$, η ranges over all bijections from $D_a \cup \Delta$ to $D_b \cup \Delta$, counting multiplicities, with $\Delta = \{(x, x) | x \in \mathbb{R}\}$ and $\|z\|_\infty = \max_i |z_i|$ [30,52]. We evaluate both distances in the methodological development of the TAD.

4   Persistence Methodology for Network Anomaly Detection

We now introduce the new topological method (TAD) for anomaly detection on multilayer graphs and support its design with relevant theoretical guarantees. Table 1 in Appendix A details all notations we introduced, and we use the terms graph and network interchangeably.

Definition 2 (Multilayer network). *A multilayer network,* $\mathbb{G} = (\mathcal{G}^1, \ldots, \mathcal{G}^L)$, *is a graph structure that consists of L non-overlapping graph layers, where each layer is modeled with a (weighted) graph* $\mathcal{G}_i = (V_i, E_i, \omega_i)$, *with* $i = 1, \ldots, L$.

Problem Statement: Let $\{\mathbb{G}_t\}_{t=1}^T = \{(\mathcal{G}_t^1, \ldots, \mathcal{G}_t^L)\}_{t=1}^T$ be a T sequence of multilayer networks observed over time t, with $1 \leq t \leq T < \infty$. The objective is to locate a time point $t^* < T$, such that an event within the time range $[t^* - m, t^* + m]$, for $0 \leq m < t^*$ causes the structure and shape of $\mathbb{G}_{t^*}$ to differ from the structural properties of the earlier observed networks $\mathbb{G}_1, \ldots, \mathbb{G}_{t^*-1}$. With this search, we include anomalies which cause: ❶ the network system to experience a brief shock at t^*, and ❷ a permanent change in the network system until the next $t^* + m$.

Main Idea: Conceptually, TAD method is designed to associate anomalies in the sequence of multilayer networks to anomalies identified from the time series of their topological summaries. In addition, we introduce our new idea of a specialized persistence diagram for multilayer networks known as the *stacked persistence diagram* (SPD).

Definition 3 (Stacked Persistence Diagram (SPD)). *For a multilayer network* $\mathbb{G} = (\mathcal{G}^1, \ldots, \mathcal{G}^L)$, *we define the associated PD of $\mathbb{G}$ as* $D_{\mathbb{G}}$ $= (D_{\mathcal{G}^1} \oplus \ldots \oplus D_{\mathcal{G}^L})$, *i.e.* $D_{\mathbb{G}}$ *is created as a direct sum of all PDs $D_{\mathcal{G}^l}$ associated with each single intra-/inter-layer network* $\mathcal{G}^l \subseteq \mathbb{G}$, *for* $1 \leq l \leq L$.

Why Do We Stack PDs and Why Not to Average PDs? As our primary focus here is on anomaly detection in multilayer graphs, our goal is to *simultaneously capture joint dynamics* of topological properties exhibited by each graph layer within the interconnected system. As such, currently existing methods based on averaging PDs and their vectorizations [7,40] which are developed for analysis of a single, possibly time-varying object, are not feasible in our context. That is, averaging PDs of the two distinct layers may be viewed as averaging PDs, extracted from apples and oranges. In turn, our idea is to jointly track dynamic topological properties which are demonstrated by apple and orange trees over the same time period, and the SPD structure is motivated by the notion of direct sums of multiple vector spaces which serve as mathematical formalization of very different objects.

Geodesic Densification of Blockchain Graphs: Dynamic networks such as Blockchain transaction graphs tend to be sparse, because a node (i.e. an address) can be inexpensively created without proving identity, which allows

Dimension of Features

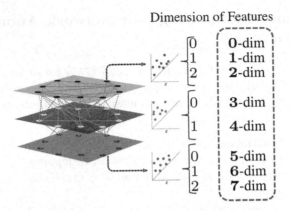

Fig. 1. An example of the formation of the SPD for a multilayer network. The multilayer network has 3 layers with PDs that have unequal topological features (3 in the first, 2 in the second, and 3 in the third). Although the first and third layer PDs contain information about 3-dimensional topological features, they have unequally-positioned points. Essentially, the SPD for the multilayer network will contain information about 7 classes of topological features.

users to hide their transactions behind new addresses for privacy and security purposes. Furthermore, blockchain communities (e.g. Bitcoin) encourage one-time-use addresses (i.e. creating a new address every time a transaction is created). As a result, a sparse and constantly evolving network structure emerges, making it difficult to rely on conventional network connectivity (i.e. adjacency matrix). To address this limitation, we replace the (weighted) adjacency matrix of the single layer graph $\mathcal{G}^l$ of $\mathbb{G}$ with the (weighted) geodesic distance (GD) matrix [8] which redefines the edge weights ω^l as $\omega^{l^+} = \sum_{e \in E(P_{uv})} \omega(e)$, where P_{uv} is the shortest path length between vertex pair u, v. This densification reconnects node pairs that have a common path. Paths encode useful information because nodes (i.e. addresses) may merge their coins into a single address to sell them to leave the Blockchain (and thus pay less transaction fees).

The proposed TAD framework operates according to the following order:

$$\{\mathbb{G}_t\}_{t=1}^T \xrightarrow{T-step} \{\mathbb{D}(D_{\mathbb{G}_{t-1}}, D_{\mathbb{G}_t})\}_{t=2}^T \xrightarrow{AD-step} \{t_1^*, \ldots\},$$

where $\mathbb{D}(D_{\mathbb{G}_{t-1}}, D_{\mathbb{G}_t})$ is any suitable distance metric between two persistence diagrams $D_{\mathbb{G}_{t-1}}$ and $D_{\mathbb{G}_t}$. Note that this distance can either be the Bottleneck or the r-th Wasserstein distance.

T − Step: At this step, we implement the clique PH to convert the sequence of multilayer networks $\{\mathbb{G}_t\}_{t=1}^T$ into a sequence of SPDs. This involves the transformation of all the (weighted) adjacency matrices of $\mathcal{G}_t^l$ into $\mathcal{G}_t^{l^+}$, followed by the filtration of persistent topological features by using clique PH.

AD − Step: While TAD method can be integrated with any user-preferred outlier or change point detection algorithm for univariate time series, we adopt the

---

Algorithm 1: Topological anomaly detection in multilayer networks (TAD)

Input : Sequence of L-multilayer graphs $\{\mathbb{G}_t\}_{t=1}^T = \{\mathcal{G}_t{}^1, \ldots, \mathcal{G}_t{}^L\}_{t=1}^T$.
Output: Anomalous events $\{t_1{}^*, \ldots\}$.

1 **for** $t \leftarrow 1 : T$ **do**
2   **for** $l \leftarrow 1 : L$ **do**
3   Compute GD matrix $\mathcal{G}_t{}^{l+}$ for $\mathcal{G}_t{}^l$
4   Generate the PD $\mathcal{D}_{\mathcal{G}_t^{l+}}$ for $\mathcal{G}_t{}^{l+}$
5   **end**
6   Obtain SPD $\mathcal{D}_{\mathbb{G}_t}$ by chronologically stacking PDs from $\mathcal{D}_{\mathcal{G}_t^{1+}}$ to $\mathcal{D}_{\mathcal{G}_t^{L+}}$
7 **end**
8 **for** $t \leftarrow 2 : T$ **do**
9   With suitable distance metric ($\mathbb{D}$), obtain similarity between $\mathcal{D}_{\mathbb{G}_{t-1}}$ and $\mathcal{D}_{\mathbb{G}_t}$
10 **end**
11 With S-ESD, detect anomalies $(t_1{}^*, \ldots)$ from the series $\{\mathbb{D}(\mathcal{D}_{\mathbb{G}_1}, \mathcal{D}_{\mathbb{G}_2}), \ldots,$
12 $\mathbb{D}(\mathcal{D}_{\mathbb{G}_{T-1}}, \mathcal{D}_{\mathbb{G}_T})\}$

---

recently proposed seasonal extreme studentized deviate test S-ESD [28,49]. For an observed time series, S-ESD filters out the seasonal component, piecewise approximates the long-term trend component (in order to decrease the instances of false positives) and then incorporates robust statistical learning to identify the location of anomalies. S-ESD is our choice due to its sensitivity to both global anomalies irrespective of seasonal trends and intra-seasonal local anomalies. We provide pseudocode for TAD below, and discuss its computational complexity in Appendix C.

4.1 Theoretical Properties of the Stacked Persistence Diagram

As shown by [10], the conventional PD of an object (i.e. a single layer graph or point cloud) is stable under minor data perturbations. Noting that SPD is derived from the direct sum of the persistence modules corresponding to each layer in $\mathbb{G}$ and using the Isometry theorem for individual persistence modules [11], we derive similar theoretical guarantees for SPD.

Theorem 4 (Stability of SPD). *Let* $\mathbb{G}_X = \{\mathcal{G}_X^1, \ldots, \mathcal{G}_X^L\}$ *and* $\mathbb{G}_Y = \{\mathcal{G}_Y^1, \ldots, \mathcal{G}_Y^L\}$ *be two multilayer networks generated from the same space of L-multilayer networks. Then*

$$W_\infty(D_{\mathbb{G}_X}, D_{\mathbb{G}_Y}) \leq \max_{1 \leq l \leq L} \left(d_{GH}(\{\mathcal{G}_X^l, \omega_{\mathcal{G}_X^l}\}, \{\mathcal{G}_Y^l, \omega_{\mathcal{G}_Y^l}\}) \right)$$

where W_∞ is the Bottleneck distance and d_{GH} is the Gromov-Hausdorff distance.

Proof for Theorem 4 is in Appendix B of the Supplementary material. Theorem 4 implies that the proposed new SPD $D_{\mathbb{G}}$ (see Definition 3) for any multilayer network $\mathbb{G}$ is robust with respect to W_∞ under minor input data perturbations. As a result, Theorem 4 provides theoretical foundations to our TAD idea. Hence, under the null hypothesis of no anomaly, we expect to observe similar SPDs over dynamic multilayer networks $D_{\mathbb{G}_t}$, while a noticeable difference between two adjacent SPDs is likely to be a sign of anomaly. Note that stability of SPD in terms of W_1 requires vectorization of SPD and Lipschitz continuity of the associated vectorization. While such vectorization approaches are highly successful for image and graph learning (see, e.g., [2,29,61]), our preliminary studies show lack of sensitivity of such vectorization techniques in conjunction with network anomaly detection.

5   Experiments on Blockchain Networks

5.1   Experimental Setup

Baseline Algorithms: We compare performance of TAD method against the following strong state-of-the-art (SOTA) algorithms for anomaly detection on single layer networks: ❶ DeltaCon by [31] (which we label DC) for weighted/unweighted networks, ❷ Scan Statistics algorithm by [13] (which we label gSeg) for unweighted networks, ❸ Edge monitoring method with Euclidean distance by [51] (which we label EMEu) for weighted networks, and ❹ Edge monitoring method with Kullback-Leibler divergence by [51] (which we label EMKL) for weighted networks. Finally, we also considered an embedding-based algorithm for anomaly detection. That is, we tracked Frobenius norms among embeddings of multilayer blockchain graphs at each time snapshot, delivered by the one of the most widely used algorithms for multilayer graph embedding, MANE of [34]. This ❺-th approach is denoted by Graph-Em. We provide a brief description of the mechanism for each method in Appendix C in the Supplementary material. For all competing methods, we use the default parameters reported in the literature. Wherever applicable, we set a standard level of significance α of 0.05.

Since all competing methods are designed for single layer networks, we implement them (individually) w.r.t. each l layer in all the multilayer graphs $\{\mathbb{G}_t\}_{t=1}^T$ and then combine the detected results, while correcting for the multiple hypothesis testing framework. In Appendix C, we provide two types of multiple hypotheses that specifies how we retain anomalies for the sequence of multilayer graphs and these include: ❶ keep all anomalies identified from at least one $\{\mathcal{G}_t^l\}_{t=1}^T$, ❷ keep all anomalies that are commonly identified from all $\{\mathcal{G}_t^l\}_{t=1}^T$. We provide results for choice (1), and defer the results for (2) to Appendix C in the Supplementary material. Additionally, we construct a single layer version of TAD (which we call S-TAD) and apply this to the same single layers. To be precise, our improvised S-TAD will extract PDs from each l-layer, and without creating SPDs, apply the chosen distance metric to consecutive PDs to obtain a time series of topological summaries for the sequence $\{\mathbb{G}_t^l\}_{t=1}^T$. Therefore, our evaluation will investigate the performance of TAD method against the performance of the chosen techniques (DC, gSeg EMEu, EMKL, Graph-Em), and S-TAD when the (un)weighted multilayer networks are viewed as a multiple hypothesis.

Topological Distances in TAD. We have experimented with various topological metrics, particularly, W_∞ bottleneck and W_1 Wasserstein distances. While our

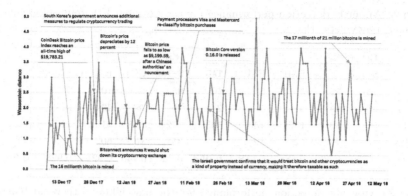

Fig. 2. Anomalous events detected by TAD for the multilayer Ethereum network.

preliminary results do not indicate that W_1 yields substantial gains over W_∞ (i.e. 70% of the true anomalous events are detected regardless of the distance choice), W_1 tends to be slightly more sensitive than W_∞. As such, we proceed with W_1 as the primary choice and consider $W_1(D_{\mathbb{G}_{t-1}}, D_{\mathbb{G}_t})$, between consecutive SPDs $D_{\mathbb{G}_{t-1}}$ and $D_{\mathbb{G}_t}$ for $2 \le t \le T$. We apply the TAD technique to two input data types: weighted and unweighted multilayer networks. Edge weight is defined as a number of transactions between nodes.

Reproducibility and Replicability. The anonymized codes and data sets for this project are available at https://github.com/tdagraphs.

5.2   Ethereum Token Networks

Data Set: The Ethereum blockchain was created in 2015 to implement Smart Contracts, which are Turing complete software codes that execute user defined tasks. Among many possible tasks, contracts are used to create and sell digital assets on the blockchain. The assets can be categorized into two categories: ❶ Tokens whose prices can fluctuate; ERC20 or ERC721 [50], ❷ Stablecoins whose prices are pegged to an asset such as USD [37] (these are also ERC20 tokens). Token networks are particularly valuable because each token naturally represents a network layer with the same nodes (addresses of investors) appearing in the networks (layers) of multiple tokens. For our experiments, we extract token networks from the publicly available Ethereum blockchain, and use the normalized number of transactions between nodes as the edge weights. By principle, a token network is a directed, weighted multigraph where an edge denotes the transferred token value. Although address creation is cheap and easy, most blockchain users use the same address over a long period. Furthermore, the same address may trade multiple tokens. As a result, the address appears in networks of all the tokens it has traded. From our data set timeline, we only include tokens reported by the EtherScan.io online explorer to have more than $100M in market value. Eventually, the data set contains 6 tokens, and on average, each token has a history of 297 days (minimum and maximum of 151 and 576 days,

Table 1. Anomaly detection performance for the weighted Ethereum blockchain and Ripple currency networks.

	Ethereum					Ripple				
	S-TAD	DC	EMEu	EMKL	TAD	S-TAD	DC	EMEu	EMKL	TAD
TP	15	52	3	5	10	95	105	10	10	16
FP	28	69	3	5	2	260	283	40	32	9
TN	99	30	132	130	135	872	837	1152	1161	1187
FN	10	1	14	12	5	35	37	60	59	50
Acc.	0.750	0.539	0.888	0.888	**0.954**	0.766	0.746	0.921	0.928	**0.953**

Table 2. Anomaly detection performance for the unweighted Ethereum blockchain networks.

	Ethereum					Ripple				
	S-TAD	DC	gSeg	Graph-Em	TAD	S-TAD	DC	gSeg	Graph-Em	TAD
TP	17	52	14	3	10	80	105	17	0	11
FP	28	69	21	11	2	241	283	56	1	23
TN	97	30	106	126	135	900	837	1130	1195	1173
FN	10	1	11	12	5	41	37	59	66	55
Acc.	0.750	0.539	0.789	0.849	**0.954**	0.777	0.746	0.909	**0.947**	0.938

respectively). Note that each token has a different creation date, hence token networks have non-identical lifetime intervals.

Ground Truth: As ground truth, we adopt and curate Blockchain events from Wikipedia [54], which lists and explains major events since 2008. In total, there are 72 events that have shaped blockchain networks—some of them in adverse (see the supplementary material for the complete list). However, token networks cannot detect events before 2015 because Ethereum and its tokens did exist before then. Hence, our experiments focused on at most 32 (out of the 72) the token transaction events.

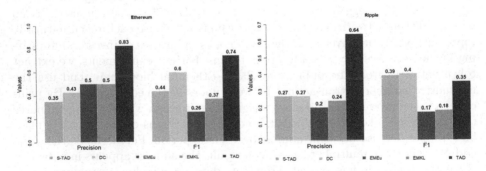

Fig. 3. Precision and F1 scores for the weighted Ethereum blockchain and Ripple currency networks.

Results: Table 1 presents summary statistics for the weighted token multilayer network analysis against the three single-layer SOTA solutions (i.e. DC, EMEu, EMKL) and our topological S-TAD method. We find that TAD delivers lower FP values. In addition, we notice that TAD achieves a significantly higher accuracy (>7% of what EMEu/EMKL report). This is evidenced by the detected points in Fig. 2. From Fig. 3 we notice again that the TAD yields substantially higher Precision (66% more than what DC gets) and F1 (>23% of what DC gets) values, implying that TAD tends to be substantially more efficient in locating relevant anomalies within the multilayer graph sequence than its competitors. In addition, we find that the performance results of the Graph-Em method in Table 2 and Fig. 4 are substantially worse than the ones delivered by our proposed TAD. This phenomenon can be explained by higher data aggregation typically performed by graph embedding tools which results in lower sensitivity to anomalous changes in the graph structure. Altogether, these results indicate that TAD tends to be the most preferred tool for identifying anomalies in the multilayer network setting. Table 2 presents experimental results for the anomalous event detection in the unweighted multilayer Ethereum blockchain networks. We find that TAD delivers the highest detection accuracy (0.954, which is about 20% greater than what gSeg yields). In addition, we notice that TAD attains the lowest FP value (about 10% of what gSeg obtains) and the highest TN value (about 27% more than what gSeg gets). In turn, Fig. 4 suggests that TAD yields the highest precision (93% greater than DC) and the highest F1 score (23% more than DC). These findings suggest that the new TAD method tends to be the most accurate approach for flagging relevant anomalous events.

5.3   Ripple Currency Networks

Data Set: The Ripple Credit Network was created to facilitate remittance across countries, but the network has transitioned to a blockchain-like structure where network approved entities (e.g., banks) issue currencies in I-Owe-You notes, and addresses can trade these currencies in blocks (which are called ledgers). On the Ripple network any real life asset, such as Chinese Renminbi or US$, can be issued by certain participants only but traded by all addresses (nodes). In

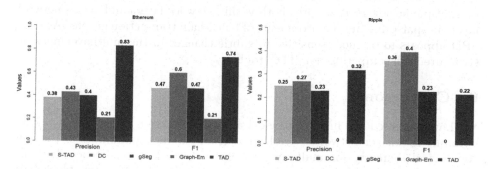

Fig. 4. Precision and F1 scores for the unweighted Ethereum blockchain and Ripple currency networks.

terms of regulatory issues by governments and price movements, Ripple is a part of the Blockchain ecology and the networks are impacted by the global events such as government regulations and trade volume increases [38]. We use the official Data API (https://xrpl.org/data-api.html) and extract the five most issued fiat currencies on the Ripple network: JPY, USD, EUR, CCK, CNY. We construct a multilayer network from the payment transactions of the five currencies that covers a timeline of Oct-2016 to Mar-2020. Similar to the Ethereum token analysis, we use the normalized number of transactions between nodes as the edge weights.

Ground Truth: As ground-truth, we use the same events described in the Ethereum token network experiments. However, since the Ripple data set has a longer temporal span of observations than the Ethereum token networks, there are a total of 66 Blockchain events.

Results: Summaries from Table 1 indicate that TAD attains the highest event detection accuracy (0.953). Furthermore, we find that TAD yields the lowest FP value, which is actually 22.5% of the value by EMEu and about 28% of the value by EMKL. Figure 3 displays detection results for the anomalous events in the multilayer Ripple payment networks. We find that TAD yields the highest precision (more than double what DC/S-TAD get) and is close to the top F1 score. Differing from Ethereum experiment, the best F1 performance is delivered by DC, closely followed by S-TAD and then TAD. Table 2 suggests that TAD delivers the highest detection accuracy (0.938, which is about 3% greater than what gSeg yields) for the unweighted Ripple currency network. In addition, we notice that TAD attains the lowest FP value (about 41% of what gSeg obtains) and the highest TN value (about 3% more than what gSeg gets). In turn, Fig. 4 shows that TAD yields the highest precision (about 18% greater than DC) but the lowest F1 score (55% of what DC gets).

Finally, note that in Ethereum we use 6 tokens, whereas Ripple experiments are performed on 5 currencies. As Tables 2 and 1 suggest, the Ethereum results appear to be better than those of Ripple. That is, detection accuracy substantially improves with a higher number of layers. However, for both cases TAD either outperforms or on par with baseline techniques. The key intuition behind these results is that TAD allows for simultaneous evaluation of subtle changes in multiple homological features both within network layers and across network layers in sparse dynamic environments of blockchain transaction graphs. As such, SPD appears to be more sensitive to subtle changes in the multilayer network structure than competing non-TDA tools.

6   Conclusion

We have proposed the first topological anomaly detection (TAD) framework for dynamic multilayer networks. We have derived stability guarantees of the new topological summary for multilayer graphs, i.e., stacked persistence diagram, which is the key tool behind TAD and validated utility of TAD on two blockchain transaction graphs. Our studies have indicated that TAD yields a highly competitive performance in detecting anomalous events on Ethereum and Ripple blockchains. In the future we plan to advance TAD to anomaly detection in attributed dynamic networks and analysis of evolving communities.

Acknowledgements. This work is supported in part by NSF Grants No. ECCS 2039701, DMS 1925346, CNS 1837627, OAC 1828467, IIS 1939728, CNS 2029661 and Canadian NSERC Discovery Grant RGPIN-2020-05665. The authors would like to thank Baris Coskunuzer for insightful discussions.

References

1. Akcora, C.G., Dixon, M.F., Gel, Y.R., Kantarcioglu, M.: Bitcoin risk modeling with blockchain graphs. Econ. Lett. **173**, 138–142 (2018)
2. Adams, H., et al.: Persistence images: a stable vector representation of persistent homology. JMLR **18**(1), 218–252 (2017)
3. Akcora, C.G., Li, Y., Gel, Y.R., Kantarcioglu, M.: BitcoinHcist: topological data analysis for ransomware detection on the bitcoin blockchain. In: IJCAI, pp. 1–9 (2020)
4. Aleta, A., Moreno, Y.: Multilayer networks in a nutshell. Annu. Rev. Condens. Matter Phys. **10**, 45–62 (2019)
5. Alves, G.A.L., Mangioni, G., Cingolani, I., Rodrigues, A.F., Panzarasa, P., Moreno, Y.: The nested structural organization of the worldwide trade multi-layer network. Sci. Rep. **9**, 1–14 (2019)
6. Bansal, M., Sharma, D.: Ranking and discovering anomalous neighborhoods in attributed multiplex networks. In: ACM IKDD CoDS COMAD, pp. 46–54 (2020)
7. Berry, E., Chen, Y.-C., Cisewski-Kehe, J., Fasy, B.T.: Functional summaries of persistence diagrams. J. Appl. Comput. Topol. **4**(2), 211–262 (2020). https://doi. org/10.1007/s41468-020-00048-w
8. Biasotti, S., Falcidieno, B., Giorgi, D., Spagnuolo, M.: Mathematical Tools for Shape Analysis and Description. Morgan & Claypool (2014)
9. Carlsson, G.: Topology and data. BAMS **46**(2), 255–308 (2009)
10. Chazal, F., Cohen-Steiner, D., Guibas, L.J., Oudot, S.: The stability of persistence diagrams revisited. Technical report, CRISAM - Inria Sophia Antipolis, June 2008
11. Chazal, F., De Silva, V., Glisse, M., Oudot, S.: The Structure and Stability of Persistence Modules. Springer, Heidelberg (2016). https://doi.org/10.1007/978-3-319-42545-0
12. Chazal, F., Michel, B.: An introduction to topological data analysis: fundamental and practical aspects for data scientists. arXiv:1710.04019 (2017)
13. Chen, H., Zhang, N.: Graph-based change point detection. Ann. Stat. **43**(1), 139–176 (2015)
14. Chen, W., Zheng, Z., Cui, J., Ngai, E., Zheng, P., Zhou, Y.: Detecting Ponzi schemes on ethereum: towards healthier blockchain technology. In: WWW, pp. 1409–1418 (2018)
15. Chen, Y., Bellavitis, C.: Blockchain disruption and decentralized finance: the rise of decentralized business models. J. Bus. Ventur. Insights **13**, e00151 (2020)
16. de la Concha, A., Martinez-Jaramillo, S., Carmona, C.: Multiplex financial networks: revealing the level of interconnectedness in the banking system. In: Cherifi, C., Cherifi, H., Karsai, M., Musolesi, M. (eds.) COMPLEX NETWORKS 2017 2017. SCI, vol. 689, pp. 1135–1148. Springer, Cham (2018). https://doi.org/10. 1007/978-3-319-72150-7_92

17. Di Battista, G., Di Donato, V., Patrignani, M., Pizzonia, M., Roselli, V., Tamassia, R.: BitconeView: visualization of flows in the bitcoin transaction graph. In: IEEE VizSec, pp. 1–8 (2015)

18. Dong, H., Chen, N., Wang, K.: Modeling and change detection for count-weighted multilayer networks. Technometrics **62**(2), 184–195 (2020)

19. FBK: Multilayer Network Datasets Released for Reproducibility, June 2020. https://comunelab.fbk.eu/data.php

20. Eswaran, D., Faloutsos, C., Guha, S., Mishra, N.: SpotLight: detecting anomalies in streaming graphs. In: ACM SIGKDD, pp. 1378–1386 (2018)

21. Fernandes, G., Rodrigues, J.J.P.C., Carvalho, L.F., Al-Muhtadi, J.F., Proença, M.L.: A comprehensive survey on network anomaly detection. Telecommun. Syst. **70**(3), 447–489 (2018). https://doi.org/10.1007/s11235-018-0475-8

22. Ferretti, S., D'Angelo, G.: On the ethereum blockchain structure: a complex networks theory perspective. Concurr. Comput. Pract. Exp. **32**, e5493 (2019)

23. Fortunato, S.: Community detection in graphs. Phys. Rep. **486**(3–5), 75–174 (2010)

24. Gidea, M., Goldsmith, D., Katz, Y.A., Roldan, P., Shmalo, Y.: Topological recognition of critical transitions in time series of cryptocurrencies. Phys. A: Stat. Mech. Apps **548**, 123843 (2020)

25. Greaves, A., Au, B.: Using the bitcoin transaction graph to predict the price of bitcoin. No Data (2015)

26. Grossman, J.W., Zeitman, R.: An inherently iterative computation of Ackermann's function. Theoret. Comput. Sci. **57**(2), 327–330 (1988)

27. Hajij, M., Wang, B., Scheidegger, C., Rosen, P.: Visual detection of structural changes in time-varying graphs using persistent homology. In: IEEE PacificVis, pp. 125–134 (2018)

28. Hochenbaum, J., Vallis, O.S., Kejariwal, A.: Automatic anomaly detection in the cloud via statistical learning. arXiv: 1704.07706 (2017)

29. Hofer, C.D., Kwitt, R., Niethammer, M.: Learning representations of persistence barcodes. JMLR **20**(126), 1–45 (2019)

30. Kerber, M., Morozov, D., Nigmetov, A.: Geometry helps to compare persistence diagrams. In: ALENEX, pp. 103–112 (2016)

31. Koutra, D., Shah, N., Vogelstein, J.T., Gallagher, B., Faloutsos, C.: DeltaCon: principled massive-graph similarity function with attribution. ACM TKDD **10**(3), 1–43 (2016)

32. Kurbucz, M.T.: Predicting the price of bitcoin by the most frequent edges of its transaction network. Econ. Lett. **184**, 108655 (2019)

33. Lee, X.T., Khan, A., Sen Gupta, S., Ong, Y.H., Liu, X.: Measurements, analyses, and insights on the entire ethereum blockchain network. In: WWW, pp. 155–166 (2020)

34. Li, J., Chen, C., Tong, H., Liu, H.: Multi-layered network embedding. In: SIAM SDM, pp. 684–692 (2018)

35. Li, Y., Islambekov, U., Akcora, C., Smirnova, E., Gel, Y.R., Kantarcioglu, M.: Dissecting ethereum blockchain analytics: what we learn from topology and geometry of the ethereum graph? In: SIAM SDM, pp. 523–531 (2020)

36. Mittal, R., Bhatia, M.: Anomaly detection in multiplex networks. Proc. Comput. Sci. **125**, 609–616 (2018)

37. Moin, A., Sirer, E.G., Sekniqi, K.: A classification framework for stablecoin designs. arXiv:1910.10098 (2019)

38. Moreno-Sanchez, P., Modi, N., Songhela, R., Kate, A., Fahmy, S.: Mind your credit: assessing the health of the ripple credit network. In: WWW, pp. 329–338 (2018)

39. Moreno-Sanchez, P., Zafar, M.B., Kate, A.: Listening to whispers of ripple: linking wallets and deanonymizing transactions in the ripple network. PoPETs **2016**(4), 436–453 (2016)

40. Munch, E., et al.: Probabilistic fréchet means for time varying persistence diagrams. Electron. J. Stat. **9**(1), 1173–1204 (2015)

41. Nelson, D.: Crypto criminals have already stolen $1.4b in 2020, says ciphertrace, June 2020. https://www.coindesk.com/author/danielnelsoncoindesk-com

42. Nocedal, J., Wright, S.J.: Numerical Optimization. Springer, Heidelberg (2006). https://doi.org/10.1007/978-0-387-40065-5

43. Phetsouvanh, S., Oggier, F., Datta, A.: EGRET: extortion graph exploration techniques in the bitcoin network. In: IEEE ICDMW, pp. 244–251 (2018)

44. Pourhabibi, T., Ong, K.L., Kam, B.H., Boo, Y.L.: Fraud detection: a systematic literature review of graph-based anomaly detection approaches. Decis. Support Syst. **133**, 113303 (2020)

45. Ranshous, S., Shen, S., Koutra, D., Harenberg, S., Faloutsos, C., Samatova, N.F.: Anomaly detection in dynamic networks: a survey. Wiley Interdisc. Rev.: Comput. Stat. **7**(3), 223–247 (2015)

46. Rieck, B., Fugacci, U., Lukasczyk, J., Leitte, H.: Clique community persistence: a topological visual analysis approach for complex networks. IEEE Trans. Vis. Comput. Graph. **24**(1), 822–831 (2017)

47. Suárez, G., Gallos, L., Fefferman, N.: A case study in tailoring a bio-inspired cybersecurity algorithm: designing anomaly detection for multilayer networks. In: SPW, pp. 281–286 (2018)

48. Taylor, D., Porter, M.A., Mucha, P.J.: Tunable eigenvector-based centralities for multiplex and temporal networks. arXiv:1904.02059 (2019)

49. Vallis, O., Hochenbaum, J., Kejariwal, A.: A novel technique for long-term anomaly detection in the cloud. In: USENIX HotCloud (2014)

50. Victor, F., Lüders, B.K.: Measuring ethereum-based ERC20 token networks. In: Goldberg, I., Moore, T. (eds.) FC 2019. LNCS, vol. 11598, pp. 113–129. Springer, Cham (2019). https://doi.org/10.1007/978-3-030-32101-7_8

51. Wang, Y., Chakrabarti, A., Sivakoff, D., Parthasarathy, S.: Fast change point detection on dynamic social networks. arXiv:1705.07325 (2017)

52. Wasserman, L.: Topological data analysis. Ann. Rev. Stat. Appl. **5**, 501–532 (2018)

53. Weber, M., et al.: Anti-money laundering in bitcoin: experimenting with graph convolutional networks for financial forensics. Preprint arXiv:1908.02591 (2019)

54. Wikipedia contribs: History of bitcoin, November 2013. https://en.wikipedia.org/wiki/History_of_bitcoin. Accessed 01 Mar 2020

55. Xu, J., Livshits, B.: The anatomy of a cryptocurrency pump-and-dump scheme. In: USENIX Security, pp. 1609–1625 (2019)

56. Yoon, M., Hooi, B., Shin, K., Faloutsos, C.: Fast and accurate anomaly detection in dynamic graphs with a two-pronged approach. In: ACM SIGKDD, pp. 647–657 (2019)

57. Yousaf, H., Kappos, G., Meiklejohn, S.: Tracing transactions across cryptocurrency ledgers. In: USENIX Security, pp. 837–850 (2019)

58. Yu, W., Cheng, W., Aggarwal, C.C., Zhang, K., Chen, H., Wang, W.: NetWalk: a flexible deep embedding approach for anomaly detection in dynamic networks. In: ACM SIGKDD, pp. 2672–2681 (2018)

59. Yuvaraj, M., Dey, A.K., Lyubchich, V., Gel, Y.R., Poor, H.V.: Topological clustering of multilayer networks. PNAS (2021). https://doi.org/10.1073/pnas.2019994118

60. Zhao, D.: An algebraic-topological approach to processing cross-blockchain transactions. arXiv preprint arXiv:2008.08208 (2020)
61. Zhao, Q., Wang, Y.: Learning metrics for persistence-based summaries and applications for graph classification. In: NeurIPS, pp. 9855–9866 (2019)
62. Zhu, T., Li, P., Chen, K., Chen, Y., Yu, L.: Hyper-network based change point detection in dynamic networks. Semant. Scholar (2018)
63. Zomorodian, A.: Fast construction of the Vietoris-Rips complex. Comput. Graph. **34**(3), 263–271 (2010)
64. Zomorodian, A.: The tidy set: a minimal simplicial set for computing homology of clique complexes. In: SOCG, pp. 257–266 (2010)

Author Index

Printed in the United States
by Baker & Taylor Publisher Services